Physiotherapy for Adult Neurological Conditions

Abraham M. Joshua
Editor

Physiotherapy for Adult Neurological Conditions

Springer

Editor
Abraham M. Joshua
Department of Physiotherapy
Kasturba Medical College of Mangalore
Manipal Academy of Higher Education
Manipal, India

ISBN 978-981-19-0208-6 ISBN 978-981-19-0209-3 (eBook)
https://doi.org/10.1007/978-981-19-0209-3

This Springer imprint is published by the registered company Springer Nature Singapore Pte Ltd.
The registered company address is: 152 Beach Road, #21-01/04 Gateway East, Singapore 189721, Singapore

To my Teachers

Who encouraged me to develop a scientific approach in exercise science and modern medical science and instilled in me the zest for perennial learning

&

To my Students

Who challenge and stimulate me to stay at the cutting edge and are our future hope!

To my Beloved Parents

Who instilled in me the virtues of perseverance and commitment and provided endless encouragement to strive for excellence.

Consider the possibility that any man could, if he were so inclined, be the sculptor of his own brain, and that even the least gifted may, like the poorest land that has been well cultivated and fertilized, produce an abundant harvest.

Santiago Ramón y Cajal

Foreword

I consider it a delight and a privilege to write a foreword for the first edition of this textbook, which I know will stand the test of time as an outstanding publication in its field of expertise. Writing a foreword for a new textbook is an act of faith and I am happy to vouch for it as the editor and the chapter contributors are not only well qualified but have been meticulous and diligent in discharging their onerous responsibility of compiling the extensive chapters of this huge compendium of knowledge. In preparing this first edition, the editor has assembled a distinguished team of chapter contributors, mostly from India and a few from other countries.

The textbook comprises sixteen chapters with one of the chapters having seven subsections. The chapters are stunningly structured and are successful in broadly covering the entire target area and encompass a brief introduction, historical background, neuroanatomical and neurophysiological background, etiopathology, clinical manifestations, medical and surgical management, and an exhaustive and detailed description of physiotherapy management of common adult neurological conditions with sensorimotor dysfunctions.

The purpose of this textbook is to build a strong theoretical foundation on which the principles of management are laid. The textbook provides a "jump start" for all the readers so that they can begin their studies from a strong and knowledgeable vantage point. The textbook is generously illustrated with a suffusion of diagrams, figures, and pictorial representations which are instrumental in enhancing comprehension.

The contents will also enhance the rationale for treatment, thus widening and improvising the scope for overall care and management of patients suffering from adult neurological conditions. This text is so full of comprehensive information and useful material that it is a must-read for all therapists involved in the care and treatment of patients with sensorimotor deficits which specifically affect the functional abilities, posture, balance, and gait. The editor believes that the contents of this textbook will serve as a guide and source of knowledge of both contemporary and advanced treatment techniques for undergraduate and post-graduate students as well as for those therapists practicing worldwide in the field of adult neurological physiotherapy.

I warmly commend this volume to the large gamut of physiotherapy professionals, who will, I believe, find it an invaluable guide and indispensable source of reference!

Department of Physical Therapy,
College of Rehabilitation Sciences,
University of Manitoba,
Winnipeg, Manitoba, Canada

Tony Szturm

Preface

Treating debilitated adult neurological patients might seem daunting at first glance, but for many students and clinicians, it enthralls and inspires them to pursue the knowledge and skills required to handle such patients. The quest for the knowledge and skills required for adult neurological physiotherapy generally makes one realize that only a few good, elaborate reference materials which provide adequate useful information are available in the broad field of rehabilitation.

Reading is an art and in the field of physiotherapy, there are countless materials available, of which many pieces of literature are for quick reference and focus on university or board examinations. Here comes the distinction between a "genuine reader" and a "regular student." A genuine reader has no specific target, except to acquire and improve knowledge and skills. On the contrary, a regular student's mindset is to clear the end-semester or annual examination. When the aforementioned is the final objective, comprehensive and elaborative textbooks are often kept aside and quick reference and class-note books become the primary focus. Though such shortcuts can provide immediate gratification, they often will be at the cost of knowledge, skills, or techniques, which in the final analysis can prove costly!

Overlooking the basics and foundations of neurological physiotherapy also paves the way for students or clinicians to get fascinated with "alternate therapies." A fair good percentage of them even think that such therapies can patch or remove the lacunae in the management of neurological conditions. Interestingly, many have no scientific base and unlike our field, are not evolved from modern medical science or exercise science. As a reader, we should appreciate that exercise therapy, electrotherapy, and the concept and pathophysiology of the disease, are all built on foundations like human anatomy, physiology, and pathology which are modern medical science-based. Ignoring the basic treatment techniques and concepts and relying strongly and vehemently on many "alternate therapies" can often increase the economic burden on the patient and above all may retard the recovery process compared to the judicious way of using proven and effective therapeutic exercises.

With over two decades of a career as an academician and clinician, I have watched several students go through the travails of not having a comprehensive textbook that slices through the minutiae and offers critical information about the adult neurological condition and its management. This compelled me to initiate the writing of a comprehensive textbook on adult neurological physiotherapy. Sadly, due to many compelling reasons, other than completing a few chapters, the contentment of accomplishing the task was left unfulfilled for several years. About two decades later, during the Covid19 outbreak across the globe, when the educational system moved from offline to online teaching and the academicians' and clinicians' focus moved from platform and seminar presentations to webinars, journal publications, and telerehabilitation, my priority was to revive and complete the textbook that I had dreamt of. In March 2020, during the country-wide lockdown phase, the work was resumed. Within a span of twenty-one months, with constant effort and commitment, the task was accomplished with the sole goal of providing a rich, but comprehensive textbook, strengthening the basic concepts and the rationale of treatment. Even the few chapters that were completed earlier underwent extensive overhauling due to the considerable advancements in their respective areas.

This textbook ideally begins with a chapter titled "Neuroplasticity" that forms the foundation for neurological physiotherapy. The next is on "Therapeutic approaches," a chapter discussing the principles and treatment of various approaches developed and in use for adult neurological conditions for at least the last few decades. The majority of the remaining chapters have a general framework, consisting of a brief introduction, historical background for the small fraction of readers who like it, and the remaining section focusing on the clinical features, pathophysiology, diagnosis, physiotherapy assessment, and management. The original idea was to include photographs of patients to demonstrate the exercises or components of exercise. But to restrict the possible spread of infection to patients during the pandemic phase, healthy normal subjects were recruited instead.

By focusing on the integration of clinical and basic scientific knowledge, this book will help to advance the quality and scientific rigor of physiotherapy for adult neurological conditions. I sincerely believe that this book will accomplish our cherished common goal and hope that you will also feel the same.

Manipal, India Abraham M. Joshua

Acknowledgments

I wish to thank the many people without whom this project would not have been possible. First, I want to thank each of the chapter contributors who willingly shared their knowledge and expertise. Thank you for sharing your knowledge to improve therapy and to touch the lives of patients beyond those you treat yourselves.

To add apologetic content in this section may look a bit offbeat, but the question is where else. Hard work, dedication, discipline, and productivity are a must for preparing an excellent chapter. When asked to contribute a chapter, the response from many was "yes" but meeting deadlines and preparing contents with adequate details were difficult for many and that made me ask myself a question, had I made more foes or to put it subtly, had I inadvertently made my friends/colleagues unhappy. My sincere apologies for any bad experiences they may have endured while attempting or preparing the contents of this book.

Since the focus was on preparing a textbook of adequate standards, I had to forgo the great times that I would have spent with my friends and hope that this did not breach the bonds of friendship. I am indebted to my close friends, Anand Venugopal, MD, Medical Superintendent, Vishak Acharya K., DNB, Professor, Pulmonary Medicine, and Raviraj, Assistant General Manager Operations, Manipal Health Enterprises, Kasturba Medical College, Mangalore. I also extend my special thanks to Srikanth Rai, DNB, Medical Superintendent, Tejasvini Hospital, Mangalore.

I express my profound gratitude to P. Venugopal, MCh, Professor of Urology, Vivian D'Souza, MD, Professor of Biochemistry, and Ramdas Naik, MD, Professor of Pathology, for their moral support and encouragement.

A big thank you to the Springer Nature team and explicitly to Naren Aggarwal, Editorial Director, Raman Shukla, Associate Editor, and Kripa Guruprasad, Project Coordinator, Springer Nature, who had faith in me, supported me, led me through the publication process, and reassured me that this onerous task would really come together.

Thanks, Ashish Prabhakar, MPT, and Sampath Kumar A., MPT, Ph.D., Department of Physiotherapy, Kasturba Medical College, Mangalore, for encouraging me through the thick and thin of this venture. A special thanks to Pradeepa Nayak, MPT, Ph.D., Department of Physiotherapy, Kasturba Medical College, Mangalore, who gave his critical comments and helped me with the proofreading.

I would like to place on record my special appreciation of all my students who not only helped me with the proofreading but also provided valuable inputs, encouragement, and motivation to complete the task. Without naming anyone, I also take the opportunity to acknowledge all the post-graduate students who participated in the photo-shoot sessions.

I extend my gratitude to Kasaragod Ashok Kumar, Preetham M. N., Roopa D. Shetty, Sushma Kishan, and Rathna who gave the consent and contributed as models for the textbook photographs. I thank Chandrashekhar Santhosh Kumar of Gold's Photography, Mangalore, for capturing and editing the still images.

Thank you, Dr. Charles V. Furtado, Associate Professor of English Department, St. Aloysius Evening College, Mangalore, for meticulously ironing out the typographical and grammatical errors which crept in while authoring the chapters. I also thank Dayavanthi H., Senior Clerk, Department of Physiotherapy, Kasturba Medical College, Mangalore, for transcribing handwritten chapter notes to word documents.

I remain grateful to my family who supported me through the ups and downs of this seemingly "endless" project.

I would like to close this by praising and thanking the Almighty, who has showered His countless blessings—courage, strength, knowledge, and opportunity, to make this dream project a reality!

—Abraham M. Joshua

Contents

About the Editor

Abraham M. Joshua (MPT Neurosciences, Ph.D.) graduated from the Manipal Academy of Higher Education (MAHE), Manipal, is a consultant adult neuro-physiotherapist in Kasturba Medical College Ambedkar Circle Hospital, and an academician in Kasturba Medical College, Mangalore. Dr. Joshua has over 20 peer-reviewed national and international publications that are Scopus indexed. He serves as a subject expert in the field of stroke rehabilitation and Parkinson's disease. Dr. Joshua is a board of studies member and a member of the Doctoral Advisory Committee for MAHE and other universities. He has conducted several workshops and delivered several oral presentations in national and international conferences and meetings.

Contributors

R. Chandrasekhar Naidu Krishna Institute of Medical Science Hospital, Secunderabad, India

V. Harigovindan Nampoothiri Allied Rehabilitation Services Inc., Brownsville, TX, USA

Abraham M. Joshua Department of Physiotherapy, Kasturba Medical College of Mangalore, Manipal Academy of Higher Education, Manipal, India

Suruliraj Karthikbabu KMCH College of Physiotherapy, Institution of Kovai Medical Center Research & Educational Trust, Coimbatore, India

Karishma H. S. Keswani The Rise Network, Mumbai, India

Neha Kumari SCS Hospital, Upper Bendoor, Mangalore, India

Gargi Kundaliya Aster RV Hospital, Bangalore, India

Amreen Mahmood Department of Physiotherapy, Kasturba Medical College of Mangalore, Manipal Academy of Higher Education, Manipal, India

S. Malarmathi Department of Psychiatry, Father Muller Medical College, Mangalore, India

Zulkifli Misri Department of Neurology, Kasturba Medical College of Mangalore, Manipal Academy of Higher Education, Manipal, India

Rohit Pai Department of Neurology, Kasturba Medical College, Mangalore, Manipal Academy of Higher Education, Manipal, India

Shivananda Pai Department of Neurology, Kasturba Medical College of Mangalore, Manipal Academy of Higher Education, Manipal, India

Sanjay Parmar SDM College of Physiotherapy, SDM University, Dharwad, India

Santosh Rai Department of Radiodiagnosis, Kasturba Medical College, Mangalore, Manipal Academy of Higher Education, Manipal, India

A. T. Safeekh Department of Psychiatry, Father Muller Medical College, Mangalore, India

Anjali Tiwari Department of Physiotherapy, Health City Cayman Island, Grand Cayman, Cayman Islands

B. Unnikrishnan Department of Community Medicine, Kasturba Medical College of Mangalore, Manipal Academy of Higher Education, Mangalore, India

K. Vijaya Kumar Department of Physiotherapy, Kasturba Medical College of Mangalore, Manipal Academy of Higher Education, Manipal, India

Chapter 1
Neuroplasticity

Abraham M. Joshua

1.1 Introduction

Functional recovery after nervous system injury remains a paradox. Following non-fatal brain damage, neural function takes days, weeks, or months to improve, many a time dramatically, and sometimes the process can continue for years. The extent to which the nervous system recovery occurs depends on many factors, including the age, location, amount of tissue damage, rapidity of damage, rehabilitation program, and environmental and psychosocial factors. The ability of the nervous system to adapt or modify functions and compensate for damage has played an imperative role in recovery; however, the importance of this ability has been appreciated only recently.

The term "plasticity," used in neuroscience for well over a century, is a loosely defined concept that stands in contrast to elasticity and brittleness. Following an impact, an elastic object returns to its original form, whereas a brittle object shatters. A plastic object may survive and continue to function but is changed by the impact. Injury to the nervous system causes similar "plastic" changes, which form the physical basis of rehabilitation. Neural plasticity "or" neuroplasticity can be defined broadly or narrowly. In the broad sense, all learning processes can be included in the concept, whereas in the latter, evidence of morphologic changes such as sprouting and synaptogenesis is essential. Neuroplasticity can also be defined as the ability to adapt or modify the neural structural organization and function to the imposed change. The mechanisms underlying neuroplasticity can include

A. M. Joshua (✉)
Department of Physiotherapy, Kasturba Medical College of Mangalore, Manipal Academy of Higher Education, Manipal, India
e-mail: abraham.joshua@manipal.edu

A. M. Joshua (ed.), *Physiotherapy for Adult Neurological Conditions*, https://doi.org/10.1007/978-981-19-0209-3_1

neurochemical, receptor, end plate, and neuronal structural changes. The concept of plasticity is less often stretched to those adaptations to external circumstances that fall outside the normal range. Even such plastic changes can be physiological or structural, which usually require time to form. For instance, regular marathon running can increase the oxidative metabolism in largely anaerobic muscles of a healthy subject who is not used to such physical activity.

It is essential to know the concept of plasticity, which is central to the work of physiotherapists. In clinical practice, many of the problems a physiotherapist faces result from too little or excess amount of plasticity. For instance, damaged motor axons within the spinal cord will not regenerate, causing permanent motor impairment, whereas injured axons in a peripheral nerve may grow easily but indiscriminately, limiting the functional recovery of reinnervated muscles. One must realize that many of the procedures used in physiotherapy exploit the plasticity inherent in the neuromuscular system to maximize rehabilitation.

1.2 Historical Background

In 1890, William James, an American philosopher and psychologist, was the first to suggest the theory of neuroplasticity in his work "Principles of Psychology." He suggested that the human brain is capable of continuous functional changes. In subsequent years, Santiago Ramón y Cajal, a Spanish neuroanatomist who first defined the neuron as the anatomical, physiological, genetic, and metabolic unit of the nervous system, suggested that an increase in the number of connections could augment the capacity of the brain. In 1893, Eugenio Tanzi, an Italian neuropsychiatrist, proposed that during the process of learning or practice, repetitive activity in a neuronal pathway produces hypertrophy, thus reinforcing the already existing connections. Michael Foster and Charles Scott Sherrington, in the year 1897, coined the term "synapse" (known as "junction" by Cajal) and identified it as the device that ensures unidirectional transmission of signals along neural pathways. Interestingly, Sherrington did not elaborate on any possible relations between learning and synaptic plasticity. Later on, in 1906, Ernesto Lugaro suggested the chemical nature of synaptic transmission and formulated the link between Tanzi's theories and Cajal's ideas of neurotropism. Though Tanzi and Lugaro were considered to be the supporters of Cajal's ideas, there remains a mystery about who first coined the term plasticity.

For a very long time, it was believed that the improvement does not occur in man beyond a period of 2 years and that paralysis is permanent following a cerebral accident. In 1915, Shepherd Ivory Franz and coworkers in an article titled "The Possibility of Recovery of Motor Function in Long-standing Hemiplegia," stated that the return of function in a paralyzed segment exists much beyond the time limit set by some neurologists. In the same article, they explained the assumption of function by other portions of the brain as the possible reason for the recoveries. Robert Ogden and Shepherd Ivory Franz, in 1917, provided the initial evidence for functional recovery in the affected upper limb through forced use, following immobilization of the unaffected upper limb of rhesus monkeys. They hypothesized that the return of function was due to behavioral recovery and possible plasticity of the cortex and suggested that the "vicarious" functions of other cerebral parts had to be investigated. In 1923, during the experimental research on monkeys, Karl Lashley demonstrated specific neuronal pathway changes as evidence of plasticity.

The twentieth century witnessed an enormous body of work focused on the properties of synaptic transmission. The Canadian psychologist Donald Olding Hebb, in his publication "The Organization of Behaviour," in 1949, articulated a theory regarding the possible neural mechanisms of learning and memory, which subsequently came to be known as "Hebb's postulates" and had a vast influence on neurophysiological studies. Though "Hebb's postulates" and "Hebbian plasticity" are two terminologies widely used in the literature, Hebb himself expressed a mixture of amusement and irritation and admitted that he did not propose anything new. In 1948, Jerzy Konorski, a Polish neurophysiologist, was the first to define the term "neuroplasticity" and postulated that morphological changes in neural connections could be the substrate of learning. Konorski suggested a theory by which neurons that have been activated by the closeness of an active neural circuit can change and incorporate themselves into that circuit. Stanley Cohen and Rita Levi-Montalcini in the 1950s first discovered the nerve growth-stimulating factor from mice sarcomas. Eric Richard Kandel and Ladislav Tauc, in 1965, published the first evidence linking short-term plasticity to behavioral modifications in *Aplysia*, a large, shell-less marine snail or sea hare.

Several behavioral and neurophysiological experiments suggested that the sensory pathways are plastic. Paul Bach-y-Rita, an American neuroscientist, proposed the concept of sensory substitution and, in 1969, demonstrated sensory plasticity by delivering cutaneous vibratory stimuli to the back of the blind patients. The optical images from a TV camera were real-time converted to cutaneous vibratory stimuli, which helped the blind patients to see.

About 20 years after the introduction of Hebb's theory, experimental evidence supporting Hebb's long-term potentiation was discovered in the dentate gyrus of the rabbit hippocampus. In 1986, Caroline Herron and coworkers demonstrated the involvement of N-methyl-D-aspartate (NMDA) receptors in synaptic plasticity. Wickliffe Abraham and Mark Bear, in the year 1996, introduced the term metaplasticity, a phenomenon that involves the activity-dependent changes in neuronal function that modulate synaptic plasticity, also known as "the plasticity of synaptic plasticity." The role of metaplasticity is unclear and might be serving to maintain synapses within a dynamic range of activity, allowing synapses and networks to respond to a changing environment.

During the late nineteenth and early twentieth century, several scientists had to fight against an academic dogma, which disapproved the existence of neuroplasticity among the adults, except during the developmental phase and younger ages of life. Michael Merzenich identified two distinct periods of brain plasticity: the critical period, a period when the child's brain establishes neural processes for the stimuli to which it is presented, and the period of adult plasticity, a period when the adult brain refines its neural processes as it masters a variety of tasks.

Though a considerable body of work has focused on evaluating dynamic changes in neural circuitry, there is mounting evidence that motor training can induce structural changes, which include changes in gray and white matter density. In 2008, Lynne Gauthier and co-researchers, in a study on constraint-induced movement therapy, a treatment proposed to improve motor function after stroke, demonstrated increased matter density in the affected cerebral hemisphere. An important factor that contributed to substantial advances in the understanding of neuroplasticity has been the development of noninvasive techniques for measurement of plastic changes. Noninvasive techniques like positron emission tomography (PET), magnetic resonance imaging (MRI) and functional magnetic resonance imaging (fMRI), real-time fMRI, magnetoencephalography (MEG), and transcranial magnetic stimulation (TMS) have all played crucial roles in the evaluation of plastic processes associated with functional recovery following central nervous system lesions.

1.3 Plasticity Within the Developing Brain

Though the plasticity in development and young age group is not a topic for elaboration in this textbook, the author believes that a short description of the same is appropriate. The brain is most sensitive to experience during development and childhood, particularly when the changes are most dramatic. For instance, the language acquisition capacity of a toddler is remarkable as compared to an adult. Even the cells within a developing neuromuscular system have a higher capacity for adaptability than mature ones. Such capacity is essential to facilitate interaction between different types of cells within the neuromuscular system and is particularly crucial for matching the function of the different components and promoting specificity of motor control. Fortunately or unfortunately, this has both beneficial and harmful consequences. For example, the programmed cell death associated with neuromuscular system development reduces the number of motor neurons by about 50% to match the number of neurons to the muscle and is partly regulated by a retrograde signal from the muscle. It is assumed that these neurons appear to compete for this signal, and an increase in the number of muscle fibers available for the motor neuron increases the survival possibility, while reduction increases motor neuron death.

Developmental plasticity has four special features. The first feature is found in the cells within the subventricular zone, a region situated on the wall of each lateral ventricle and the dentate gyrus of the hippocampus (Fig. 1.1). Both the abovementioned regions contain stem cells that remain active throughout life. The subventricular zone cells produce both glial and neural progenitor cells, which have the potential to migrate into cerebral gray or white matter, even in adulthood. In humans, these cells appear mostly quiescent but can become activated, largely in response to stress to the health of the brain. The stem cells in the hilus of the dentate gyrus

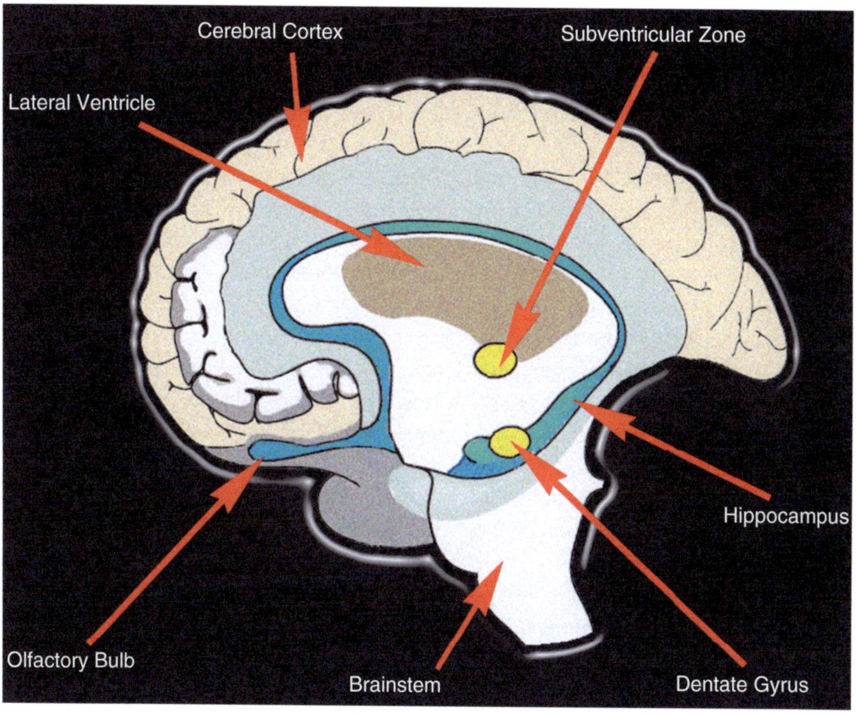

Fig. 1.1 The subventricular zone and the dentate gyrus of the hippocampus

produce new neurons at a slow and steady pace throughout life; however, the potential for generating new neurons (neurogenesis) declines with aging. The functional role of these new neuronal cells is not entirely understood, but these cells do integrate with the existing neurons and may play a role in the formation of new memories.

The second special feature consists of the three types of plasticity distinguished in the developing brain: experience-independent, experience-expectant, and experience-dependent. Experience-independent plasticity involves nervous system changes regardless of the environment and unfolds over time through a tightly regulated series of molecular events. It occurs mostly during the prenatal developmental phase when the genome generates a rough approximation of connectivity that is modified by certain specific events. Neuronal cell connectivity increases when they are active together and weakens their connections when not coincidentally active. Experience-expectant plasticity occurs mostly during early postnatal development and describes the normal, generalized development of neuronal connections and will not unfold until they are triggered by specific environmental cues that the brain expects to encounter. For instance, the visual cortex will not be fully functional until the neonates open their eyes for the first time, and the young children will not learn the language until they hear others' speech. Experience-dependent plasticity refers to ways neural pathways are strengthened through repeated engagement, which begins in early postnatal life and continues for a lifetime. Experience-dependent plastic changes are unique to each person and reflect the range of social and cultural environments and the specific activities in which the person engages. Enriched experience (discussed later in this chapter) is an example of experience-dependent plasticity.

The speed at which dendrites modify their structure to form or delete synapses in response to experience is the third special feature of plasticity. The period of heightened sensitivity to environment and experience is called a critical period for that circuit, which is the fourth special feature. The experiences during these critical periods play a crucial role in shaping the brain and are important for experience-expectant plasticity. Once the critical period ceases, substantial alteration of neural circuits will be difficult. For instance, closure of one eye of a kitten soon after birth, for a period of few months, expands the territory of the open eye at the expense of the closed eye and may result in permanent loss of spatial vision. In the past 5–6 decades, considerable effort has been made to elucidate molecular and cellular mechanisms underlying the activation and regulation of critical periods in the brain. Unraveling these mechanisms may potentially assist scientists to augment the plasticity in the adult brain.

Plasticity in the young has both positive and negative consequences. Clinical and experimental studies had indicated that young animals and children have a remarkable degree of functional sparing after central nervous system (CNS) injuries, which were devastating when they occurred in adulthood. However, on the contrary, the recovery in children can be minimal after certain types of injury, and in others, it may be at a high cost. For instance, sparing of language after early left hemisphere damage is accompanied by substantial intellectual and perceptual deficits, which are not seen after equivalent lesions in adults. In the same way, some extreme plastic responses may exacerbate a developmental abnormality.

1.4 Plasticity: An Ongoing Process

Scientists once believed that the brain's capacity to develop ceases after the first few years of life. They believed that connections between neurons develop only until the first few years of life. This implies that only young brains would be "plastic" and be able to form new connections. Because of this belief, they also thought that if a particular area of the adult brain is damaged, the nerve cells could not form new connections or regenerate, and the functions controlled by that area of the brain would be permanently lost. However, new research on animals and humans has overturned such a mistaken old view. It is recognized that the brain continues to reorganize itself by forming new neural connections throughout life. This phenomenon, called neuroplasticity, allows the neuronal cells within the brain to compensate for injury and adjust their activity in response to new situations or changes in their environment and also plays an essential role in CNS functions such as learning and memory, both throughout the lifetime and post-brain injury.

Human brains have the extraordinary capacity for structural and functional change. The ability of the neuronal cells to change their physical structure as a result of learning is known as structural plasticity. Structural plasticity could be defined in terms of dendritic and axonal arborization, spine density, number and size of synapses, and receptor density. Neurogenesis, the formation of new neurons in certain regions of the brain, and neuronal migration, a process in which neurons travel from their "place of birth" toward their final position within the brain, are instances of structural plasticity during development and growth. The changes of gray matter volume or proportion in the brain following learning can also be considered as instances of structural neuroplasticity. On the contrary, functional plasticity refers to the brain's ability to adapt and/or modify the physiological properties of neurons. Activity-dependent plasticity (changes in response to previous activity) and reactive plasticity (changes in response to damage or malfunction of neuronal cells) are forms of functional and structural neuroplasticity. Activity-dependent plasticity involving the synapses, known as synaptic plasticity, includes long-term potentiation, long-term depression, and intrinsic and extrinsic plasticity. In reactive plasticity, the neuronal functions of the injured brain area are transferred to another region for recovery. Structures like the hippocampus display both structural and functional plasticity into adulthood.

1.4.1 Learning and Memory

For the past many years, research workers had spent enormous time in understanding the biological basis of learning and memory among vertebrates, especially in mammals, including humans. Our memory is very crucial to our sense of self, and diseases like brain injuries and Korsakoff's syndrome (caused by long-term alcohol abuse) prevent the patients from acquiring new memories, and they live in a world where every experience is new and do not benefit at all from their past. Neuroplasticity is a necessity for learning and memory. Advancing knowledge about synaptic transmission has contributed to a better understanding of the cellular and molecular changes associated with learning and memory responsible for changes in behavior in organisms, including humans. About eight decades ago, Hebb proposed that learning was mediated by changes in synaptic strength or "efficacy." According to Hebb, when an animal learned a new act, some synapses became stronger; that is, those particular synapses in the neuronal pathways that were responsible for the learned behavior gave a large postsynaptic response to the stimulation of the presynaptic neuron. However, he did not specify whether this increased efficacy was caused presynaptically (like an increased release of neurotransmitters) or postsynaptically (like an increased number of receptors). Even though this basic hypothesis was very appealing, it took many years to conclude that Hebb was essentially correct.

Evidence is now accumulating that other parts of the brain, mainly those concerned with motor function, which includes the cerebellum and basal ganglia, mediate the learning and memory of motor skills. Most of this evidence comes from studies of brain lesions in humans and animals upon the acquisition of motor tasks, and its validity depends upon distinguishing learning from performance deficits. It is believed that the hippocampus is concerned with the formation of declarative memory, such as remembering words, faces, or facts, and structures like the cerebellum and basal ganglia are required for forming procedural memories, which involve learning tasks like driving a car or playing a musical instrument. Patients with hippocampal damage can often learn new procedures but not the new declarative information. For instance, the patient may learn to drive a car but may fail to remember the rules pertaining to it. The cerebellum and basal ganglia, structures important for procedural memories, show long-term potentiation like the cerebral cortex, where the higher cognitive functions such as comprehension, planning, reasoning, and decision-making are located.

1.4.1.1 Long-Term Potentiation and Long-Term Depression

Decades ago, a form of synaptic plasticity was discovered that was called long-term potentiation (LTP). Many neuroscientists believed LTP as an essential part of the process of learning, and as a result, an enormous amount of time and effort was allotted to understand the cellular changes underlying LTP, especially in a region of the mammalian brain called the hippocampus (Box 1.1).

Box 1.1 Features of Hippocampus

Hippocampus

Phylogenetically, hippocampus is considered as one of the oldest parts of the cerebrum, and the damage to the same will produce striking amnesia. In mammals, it is known to play an important role in memory acquisition. Following damage to the hippocampus, mammals, including humans, have great difficulty learning new things but not recalling things that they already learned. Though it is involved in an early stage of memory formation, it is not involved in the long-term storage or recall of memories.

In addition to LTP, a contrary phenomenon called long-term depression (LTD) occurs in various regions of the brain, including the hippocampus. LTD may be a process involved in selective forgetting or ignoring irrelevant information, which is another kind of learning that is important in an animal's function. Scientific evidence is available to prove that the NMDA and α-amino-3-hydroxy-5-methyl-4-isoxazole propionic acid (AMPA) receptors located within the synapses are involved in both learning new information and actively unlearning old ones. In 1949, Hebb first used the phrase "neurons that fire together, wire together," to describe how pathways in the brain are formed and reinforced through synaptic strengthening. Contrary to that, the "neurons that fire apart, wire apart" can explain synaptic weakening, which is a fundamental feature of LTD.

Experimental studies have found that the neurotransmitter at the LTP synapses is almost always glutamate, and at least two or three types of glutamate receptors, including the NMDA and the AMPA receptors on the postsynaptic neurons, have been identified. The molecular mechanism subserving LTP or LTD is unsettled, but it is known to be mediated by calcium and probably involves changes in both postsynaptic glutamate receptors and presynaptic transmitter release. Studies have also suggested that structural changes in dendritic spines may also be involved in learning. Further details regarding LTP and LTD are given in Boxes 1.2 and 1.3, respectively.

Box 1.2 Details Regarding Long-Term Potentiation

Long-Term Potentiation

The synaptic strength is defined as the average amount of voltage excursion produced in the postsynaptic neuron by an action potential in the presynaptic neuron. LTP is an example of synaptic plasticity, a process by which the synaptic strength between neuronal cells becomes stronger with repeated stimulation. It is believed to be a way in which the brain changes its response to experience and maybe a mechanism underlying learning and memory. There are many ways in which LTP is produced. Glutamate released from the presynaptic neuron first activates a subtype of glutamate receptor known as the AMPA receptor on the postsynaptic membrane. Low-level release of glutamate will not activate the NMDA receptors found nearby the AMPA receptors due to the presence of magnesium ion blocking the ion channel of the NMDA receptor. Repeated and stronger stimulation of AMPA receptors will cause the depolarization of the postsynaptic neuron, eventually causing the voltage-dependent magnesium blockage of the NMDA receptor to be removed, allowing calcium ions to flow through the NMDA receptor (Fig. 1.2). The calcium influx initiates cellular mechanisms that cause more AMPA receptors to be inserted into the postsynaptic neuronal membrane. The new AMPA receptors being more responsive to glutamate will allow more positively charged ions to enter the cell when activated, thus making the postsynaptic cell more sensitive to glutamate, making the synapse stronger and more likely to be activated in the future. In addition to the above, the signals that travel back across the synapse are believed to stimulate greater levels of glutamate release. The process is also associated with changes in gene transcription in the neuron, which can lead to the production of new receptors or modifications of cell structure augmenting the LTP.

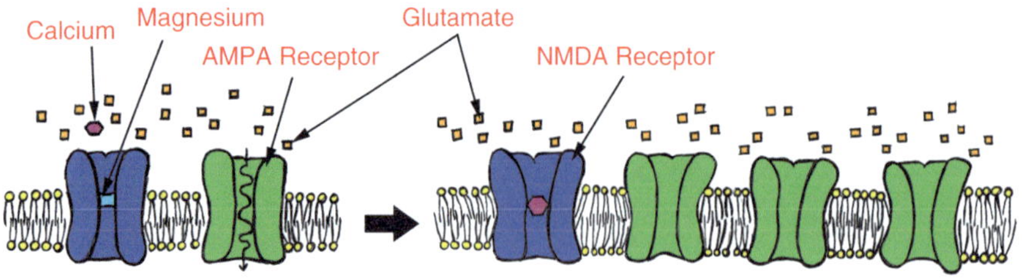

Fig. 1.2 An illustration of long-term potentiation

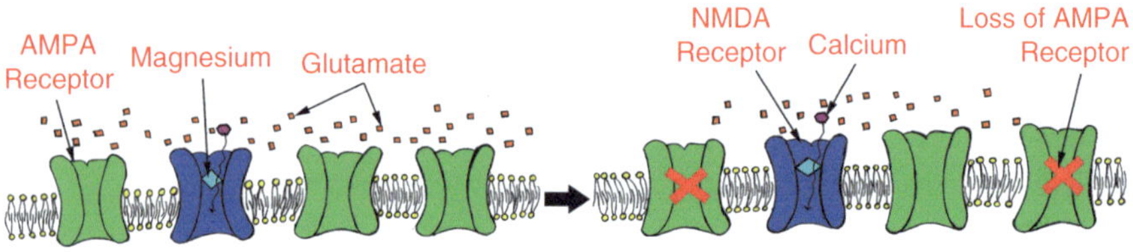

Fig. 1.3 An illustration of long-term depression

Box 1.3 Details Regarding Long-Term Depression

Long-Term Depression

Long-term depression is another example of synaptic plasticity, a process by which the synaptic strength between neuronal cells becomes weaker. It is a process that is opposite to that of LTP and believed to be important for learning and memory, perhaps by resetting previous synaptic changes to allow for new memories. The best-understood mechanism for the production of LTD involves the same glutamate receptors (NMDA and AMPA receptors) involved in LTP. The LTP typically occurs after brief high-intensity stimulation of a postsynaptic neuron, whereas the LTD typically occurs after prolonged low-intensity stimulation. Inadequate depolarization fails to remove the magnesium ions blocking many of the NMDA receptors on the postsynaptic membrane, allowing only a few of the NMDA receptors to pass calcium ions through them. This low-level calcium is insufficient to activate the enzymes that facilitate LTP and, on the contrary, activates a cellular process that causes removal of AMPA receptors, thus reducing the number of glutamate receptors on the postsynaptic membrane leading to a weaker synapse (Fig. 1.3). In addition to the above, the LTD may also decrease the amount of glutamate released from the presynaptic neuron by involving other glutamate receptors.

1.4.2 Muscle Activity and Plasticity

Muscle plasticity is defined as the ability of a skeletal muscle to alter its structural and functional properties secondary to certain conditions imposed on it. It is well known that intense physical training can cause strengthening and hypertrophy of the muscles, and the effects can be further augmented by neuromuscular electrical stimulation. By endurance training, the oxidative metabolism of muscles will increase and tends to convert fast-contracting motor units into slow ones, indicating that the individual muscles can be enlarged and strengthened by selective patterns of intensive training. On the contrary, a reduction in muscle activity can lead to muscle wasting.

Functional adaptations to reduced muscle activity depend upon the nature of the reduction. The absence of weight-bearing tends to cause muscle weakness and atrophy, whereas immobilization can lead to resorption of sarcomeres as an adaptation to being held in a shortened position. In animal model studies, unloading muscle revealed preferential atrophy of Type 1 muscle fibers and increased proportions of Type 2 fibers. Experimental studies on humans revealed significant atrophy of Type 1 and 2 muscle fibers as a major finding. With regard to the type of muscle fibers involved, the results among humans were mixed. The variability seen across human studies could be due to the differences in the choice of muscle studied, the type and intensity of physical activity before the period of inactivity, or the muscle contractions performed during the period of immobility. Studies during space flight have found differences in the pattern of fiber atrophy between rodents and humans. Rodents were found to show more significant atrophy in Type 1 than Type 2A fibers, whereas humans showed more atrophy in Type 2A than Type 1 fibers, and the difference in the pattern may be related to initial fiber size.

Muscle inactivity causes major changes, and such changes are initiated within a matter of few hours following inactivity or immobilization. Reduction in the rate of protein synthesis is one of the dominant characteristics of muscle atrophy. In rodents, a 35% reduction in protein synthesis was recorded within hours after the onset of inactivity. A decline in contractile protein concentration can lead to a reduction in the number of active cross-bridges per muscle volume reducing the electromechanical efficiency. Even prolonged periods of reduced muscle activity can lead to significant reductions in maximal voluntary contraction. It is believed that reduced neural input is also partly responsible for the reduction in maximal voluntary force during periods of muscle inactivity.

Increased muscle activity or training results in a change in expression of myosin heavy chain isoforms characterized by greater expression of slow myosin heavy chain and hence more Type 1 fibers; however, the nature of the activity also determines their expression. For instance, heavy weight-lifting induces a greater expression of fast myosin heavy chain. Even moderate loads are capable of inducing a greater expression, but the minimum threshold required to induce the transformation is yet to be clear. Animal model studies on training have revealed hyperplasia that contributes further to muscle hypertrophy.

Appropriate strategies must be used to counteract the ill effects of immobilization or muscle inactivity. For instance, strengthening exercise programs to hasten the recovery of muscle functions after periods of immobilization or use of electrical stimulation in addition to passive limb exercises in paralytic or paretic neurological conditions can counteract the ill effects of such changes. Similarly, the four main principles of training skeletal muscles, namely, overload, specificity, reversibility, and individuality, have to be used to improve muscle strength and endurance to minimize the deleterious effects of muscle inactivity.

1.4.3 Peripheral Nerve Injury and Plasticity

The response of the neuromuscular system to interruption of nerve axon (excluding neurapraxia or demyelination) can be considered in two different ways. The first one consists of the new growth of neural processes by protoplasmic extension from the severed axon end, and the second one requires synaptic plasticity, which operates locally. Both these mechanisms are necessary for full functional recovery of the neuromuscular system.

1.4.3.1 Nerve Regeneration

Damage to the nerve fiber disrupts nutritive contact between the proximal and distal parts of the axon. As a result, the distal part degenerates along with its myelin sheath. Many mature neurons that survive an injury to their axons respond by attempting to regenerate a new axon. The regeneration process begins with axonal elongation after a latent period of a day or two, the time required for the modification of biosynthetic activity in the cell body. The severed axon stump or the zone just proximal to it may emit 1–20 sprouts and can advance at the rate of approximately 2 mm per day. To guide the regenerating sprouts, the surviving Schwann cells present in the distal nerve segment proliferate and become aligned in strands to form the bands of Büngner (Fig. 1.4). Ultimately, the fate of these sprouts depends upon their success at entering the distal nerve stump and reaching an appropriate peripheral target.

The axonal sprouts are capable of growing long distances along empty nerve sheaths and will even grow through artificial conduits to reach their synaptic targets. In the process of crossing the zone of injury, fibers are strongly influenced by local mechanical factors. Many grow in a retrograde direction back along the proximal nerve, often forming spirals around parent axons or blood vessels, whereas others are arrested or escape from the nerve bundle. However, sprouts that reach the distal stump track along the bands of Büngner toward the peripheral target. The trophic substance picked up by the successful sprout in the periphery may be responsible for shunting the flow of structural metabolites selectively toward the connected sprout to foster its maturation. The mechanism for "culling" the unsuccessful

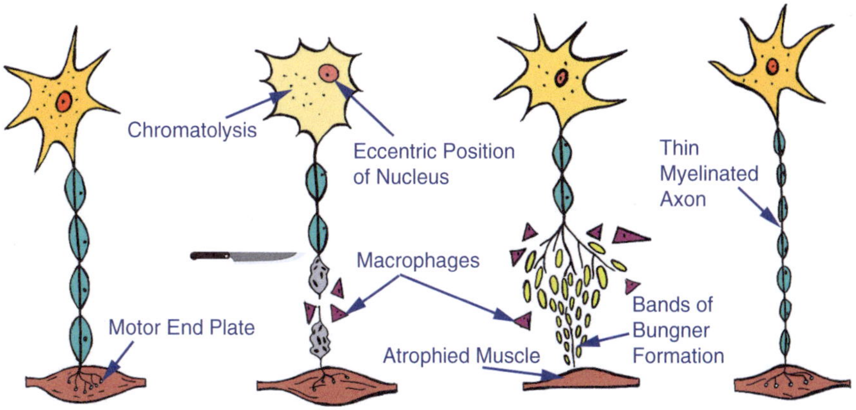

Fig. 1.4 An illustration of the axonal degeneration and regeneration post-trauma

sprouts during regeneration may be similar to those involved in the natural death of excess sensory, motor, and autonomic neurons in developing the nervous system or the elimination of polyneuronal innervation in maturing muscle fibers. Like the normal unmyelinated axons, the young regenerating sprouts conduct impulses in the continuous (non-saltatory) mode. Shortly after reconnection, a wave of diameter increases, and myelination sweeps over the axon from the point of injury distal ward, and conduction becomes saltatory, and, gradually, the velocity of conduction accelerates.

The death of axon till the nearest node of Ranvier, widening of nodes, fragmentation and/or blistering of the myelin sheath, chromatolytic changes in the cell body and dendritic retractions, and stripping off some afferent synapse of the dendrites are certain proximal nerve segment changes seen soon after nerve fiber severance. These changes are considered to be a process of metabolic mobilization in preparation for axon repair. The process is reversed upon successful regeneration, and the absence or delay in successful regeneration can lead to the progression of these changes, and, eventually, the cell may die. The incidence of retrograde cell death depends on several factors, including the particular neural system involved, time after axotomy, age, location of the injury, and the existence of axon collateral branches.

Even if the injured nerve regenerates, it is only half the battle for recovery. Reinnervation of nerve fibers must be specific if it has to be functionally beneficial. Experimental and clinical studies suggest that the mechanisms, which guide the innervation of appropriate targets during embryonic development, do not operate during reinnervation. For instance, following the transection of a large mixed nerve, some sensory axons will innervate muscles, and some motor axons will innervate cutaneous receptors making it functionless. Apart from the abovementioned, individual muscles may be reinnervated by motor neurons from other motor pools and will, therefore, be activated inappropriately. Microsurgical repair of severed nerves by closing or bridging the gap while minimizing tissue obstructions and maintaining as good a fascicular alignment may optimize the specific reinnervation. Studies on reinnervation in animals suggest that the precision of motor reinnervation may improve with time even if the reinnervation may be nonspecific initially, due to the selective withdrawal of misdirected collateral axonal branches over many weeks.

The term "sprouting," a necessary neuronal process that follows nerve injury, can be reactive, regenerative, or compensatory type. The reactive (collateral) sprouting refers to the new growth of intact nerve fibers in response to denervation in the adjacent areas. Regenerative sprouting refers to new growth from near the severed end of the nerve fiber, and compensatory sprouting refers to the new growth of one branch of an axonal tree when a distant branch is cut. It is unclear what induces axons to sprout in the event of injury and to what extent the mechanism underlying sprouting in these three situations is different. Some of the factors believed to regulate the sprouting are given in Box 1.4.

Box 1.4 Factors Believed to Regulate the Neuronal Sprouting

Possible Factors Regulating Sprouting

- Sprouts tend to move in mechanical adherence to preexisting axonal or glial processes.
- The environment through which the sprouts pass can alter neural processes, and abnormal tissue configurations can fail the growth of sprouts.
- Nerve sprouts and cells may recognize their proper target utilizing specific "cytochemical position markers" made up of specific tissue-target recognition molecules.
- Sprouts may be partially directed by preexisting chemical "fields" or gradients.
- By attractive and/or repulsive "forces," individual axons within an outgrowing population may interact with one another to constrain the growth of its neighbor.
- Temporal factors like the time of arrival of earlier axons for connection exclude the connection of later arriving ones.
- In certain systems, neurons and neuronal processes are produced in excess, and those that fail to form functional connections are eliminated.
- Mutual dependence may be seen in certain neural elements, i.e., in the absence of one, the second may fail to differentiate or may atrophy.
- The axonal system competes for the exclusive occupation of available terminal space that is left vacant by the degeneration of neighboring axons.
- Certain axonal populations of the neural system tend to conserve the total amount of their axonal arborization, i.e., if growth in one part is limited, branches in another part will sprout extra collaterals.

At least, weeks or months are taken for the regenerating sprouts to return to the denervated tissue. During this period, there are certain changes in intact fibers that normally share the distribution field of the injured ones or that innervate adjacent zones. It is believed that the motor unit can expand even up to five times its original size, compensating for the considerable deficiencies in innervation. This explains why in conditions like motor neuron disease, muscle weakness does not become apparent until about 50% of motor neurons have degenerated. Even in the early stages of muscular dystrophy, sprouting occurs when newly developing muscle fibers replace degenerating muscle fibers. The trigger for motor neuron sprouting in muscle is unclear but probably involves a local signal from inactive denervated muscle fibers. Such return or improvement of function also occurs in the sensory system, and sprouting of neighboring sensory nerves accounts for the same.

1.4.4 Denervation Hypersensitivity

Another important category of plastic change secondary to axonal damage is the increase in chemosensitivity of postsynaptic elements known as denervation hypersensitivity or denervation supersensitivity. Denervation hypersensitivity results in a permanent increase in neuronal responsivity to diminished input and maybe even a factor in CNS reorganization. An increase in the number of receptors or the receptor site becoming more sensitive can be considered as the reason for the same. For instance, following denervation of muscles, acetylcholine sensitivity spreads over the whole membrane surface (postsynaptic junctional membrane), and the changes produced within the muscle membrane, concomitant cellular changes, and the loss of muscle activity generally tend to persist until innervation is restored.

1.4.5 Synaptic Plasticity

Many researchers believe that synaptic plasticity is central to understanding the mechanisms of learning and memory. Decades ago, the synapse was simply considered a junction for the transfer of information between one neuron and another neuron or between a neuron and a muscle cell. Once established during development, these synapses were believed to be relatively permanent in their strength. Nevertheless, the developments in science during the past five to six decades have realized that most synapses are incredibly plastic, and they are capable of changing their strength as a result of their own activity (intrinsic or homosynaptic plasticity) or through activity in another pathway (extrinsic or heterosynaptic plasticity).

Short-term synaptic depression and short-term synaptic facilitation are two types of homosynaptic plasticity, and, typically, both of them are not found at the same synapse. Such plastic changes are thought to play crucial roles in short-term adaptations to sensory inputs, transient changes in behavioral states, and short-term forms of memory. Most forms of short-term synaptic plasticity are triggered by short bursts of activity, causing a transient accumulation of calcium in presynaptic nerve terminals. The mechanism of synaptic depression varies and is attributed to the depletion of the readily releasable vesicles, and the efficacy of synaptic transmission depends upon the frequency of stimulation. Paired-pulse facilitation and post-tetanic potentiation are a few types of synaptic facilitation. The plausible mechanism contributing to paired-pulse facilitation is the presence of residual calcium ions. Longer-lasting forms of plasticity are observed following repetitive or tetanic stimulation of synapses with prolonged trains of stimulation. Post-tetanic potentiation describes an enhancement of transmitter release lasting for several minutes and may lead to biochemical modifications of proteins in the presynaptic terminal.

Even heterosynaptic plasticity is either of presynaptic inhibition or presynaptic facilitation type. The axoaxonic synapses (synapse made between the axon of one neuron and the axon of another neuron) mediate presynaptic inhibition and presynaptic facilitation. In presynaptic inhibition, a third neuron, which makes an axoaxonic synapse with the presynaptic neuron near its terminal button, reduces the effect of the presynaptic neuron on the postsynaptic neuron. Presynaptic inhibition is quite prominent in the spinal cord and regulates the propagation of potentials to higher brain centers. For instance, the pain gate theory proposed by Ronald Melzack and Patrick Wall in 1965 is based on presynaptic inhibition of pain produced by mechanical stimulation, the basic rationale for transcutaneous electrical nerve stimulation. In presynaptic facilitation, a third neuron, which makes an axoaxonic synapse with the presynaptic neuron near its terminal button, increases the effect of the presynaptic neuron on the postsynaptic neuron.

1.5 Plasticity Within the Adult Central Nervous System

One of the most striking differences between the peripheral nervous system and CNS is that the axonal injury in the CNS does not lead to regeneration, and the consequence is more or less permanent impairment. In the brain of some lower vertebrates, including organisms like a lamprey, even in their adult forms, severed fiber trunks such as optic and spinal tracts may regenerate much as in the peripheral nervous system. The processes of axon outgrowth after peripheral nerve injury in mammalian adults are reminiscent of initial axon outgrowth in the developing nervous system, and for evolutionary reasons that are entirely unclear, the mammalian CNS has lost a great deal of this lability when compared to the peripheral nervous system.

The major difference between a developing and a developed system is that in a developed system, the surrounding milices are mature and do not unfold themselves. Following CNS damage, neuroplasticity in the form of sprouting, denervation supersensitivity, unmasking of latent synapses, or behavioral compensation may occur in the absence of unfolding of surrounding milices. However, the presence of inhibitory environment can make successful regeneration practically impossible for functional restoration. Early pathophysiological processes such as the decline of edema and recovery from ion-imbalances are certain reasons for functional restitution following nervous system injury and will not be dealt with in this chapter.

Unmasking Quiescent neuronal connections that are inhibited in the normal state may be unmasked following brain damage. Unmasking may be another important mechanism of recovery of function and may also produce negative effects. The appearance of "pathologic" reflexes such as a tonic labyrinthine reflex and tonic neck reflex following a brain injury can be considered as instances for the unmasking of reflexes that were normal in infancy but became inhibited during development.

Both in animal and human stroke studies, ipsilesional peri-infarct activation was observed after partial damage of the primary motor cortex. Such a finding suggests the unmasking of preexisting, yet functionally silent, areas in the vicinity of the lesion or the progressive activation of neural networks normally not devoted to the lost function. Changes in glutamatergic transmission and LTP have been reported in the peri-infarct cortex and beyond in the first week after stroke. Evidence suggests that the brain-derived neurotrophic factor (BDNF) appears directly linked to the activity-dependent early unmasking of existing connections. Studies on mammalian CNS have proved the presence of silent synapses that lack functional AMPA-type glutamate receptors but possess NMDA-type glutamate receptors. Even evidence from electrophysiological measurements has established the existence of silent synapses and their emergence as active synapses with appropriate stimulation.

Behavioral Compensation A considerable extent of recovery of function after brain damage can be due to the development of compensatory behavioral strategies, a new combination of behaviors that circumvent impairments. For instance, the patient may use different groups of muscles or cognitive strategies for fulfilling the functional activities or spontaneously develop over-reliance on the uninvolved limb for normal postural support behaviors and transitions. Animal model studies have revealed that unilateral lesions can result in remarkable neuroanatomical changes in the motor cortex opposite and homotopic to the lesions associated with a sequence of changes in the neuronal and glial cells in this region. Soon after the lesion, there will be loss of axonal processes presumably arising from the damaged cortex and reactive changes in glial cells, and there will be an increased presence of neurotrophic factors, followed by major axonal, dendritic, and synaptic connections growth within this region. Synaptic connections also show structural changes that are characteristic of increased potency.

Larger lesions of the sensorimotor cortex can cause the contralesional motor cortex to contribute new axonal projections to subcortical regions underlying the lesion. After unilateral lesions, animals progressively rely on the uninvolved limb.

Animal model studies have also revealed that restriction of uninvolved limb soon after brain lesion for 15 days prevented the dendritic growth usually seen in the contralesional cortex. Though the compensatory behaviors have shown certain improvement in the functional level in animal studies, such behaviors can be stubborn to correction in humans, especially when there is a potential for further improvement.

1.5.1 Brain Plasticity

Even though it is customary to explain brain functions in terms of "centers," it is not known to what extent brain functions are localized to "centers." The functional loss following the damage of a particular group of cells means only that the rest of the brain cannot perform the function without the contribution normally made by the damaged cells. This contribution might be a minor one, say, setting the level of excitability of some other group of cells, or be a major one such as organizing the neural impulse sequence required. There are many ways by which the excitability of the target cells could be restored, but if a specific, localized group of cells that organizes the neural impulse sequences is destroyed, the recovery might require massive and perhaps impossible reorganization of remaining neural aggregates. This implies that when a vital circuit element is destroyed, plastic changes such as the substitution of parallel channels or the mobilization of redundant capacity could still support the need for therapeutic exercise.

Synaptic plastic changes occur in the brain in response to both local and peripheral injuries. It was presumed that such adaptation was possible only when the injury happens in early development or when young. However, more recent works suggest that substantial reorganizations also occur after injury to the adult brains. Recently, positron emission tomography of adult patients who have suffered striatocapsular stroke has demonstrated bilateral activation of motor pathways and the recruitment of additional sensorimotor cortical areas associated with the recovery of motor function.

There is growing evidence that the glial cells are involved in short- and long-term plasticity. The belief that they are passive bystanders in neuronal brain circuits and required for housekeeping and brain metabolism is incorrect, and they play an active role in regulating physiological function and synaptic plasticity. With their intimate association with synapses, astrocytes and perisynaptic Schwann cells are well-positioned to regulate synapses. They have an established role in the clearance of neurotransmitters and control of extracellular ionic gradients and excitability.

Functional radio-imaging studies on post-stroke patients have identified the role of the healthy hemisphere in recovery. Earlier studies on rodents have demonstrated enhanced activity of the contralesional hemisphere in the very acute stage after stroke, followed by perilesional activation at later stages during the recovery phase. Experimental silencing of the healthy hemisphere (using muscimol, an ionotropic gamma-aminobutyric acid [GABA] receptor agonist) within hours after a stroke has demonstrated improvement in functional recovery, and the duration of the inactivation was directly correlated with the improvement. There is scientific evidence that the activity of the healthy hemisphere can worsen motor recovery. For instance, a recent quantitative noninvasive study among acute stroke survivors could relate an increased contralesional hemispheric activity to the negative final outcome. Conversely, when the sensorimotor cortex lesion is considerably extensive, the healthy hemisphere could be important to vicariate lost functions. Studies have shown that stroke patients can demonstrate a significant increase in contralesional motor cortex activity during movement of the affected extremities, and TMS-induced disruption of the function within this area has shown impairments in the recovered movement of the affected extremity. However, controversy still exists whether the healthy hemisphere has a positive or negative impact on recovery.

1.5.2 Experience-Dependent Neuroplasticity

Experience can change the synaptic efficacy, neuronal structure, and rate of neurogenesis and remodel vasculature and glial processes. Experience is also crucial for function following brain lesions, such as stroke and traumatic brain injury. Manipulations of behavioral experience, including physical activities and physical exercises, are the primary tools currently available for the treatment of functional deficits following brain injury. Behavioral experiences, including rehabilitation strategies and interventions and self-taught compensatory strategies, have a powerful impact on post-injury brain remodeling and functional outcomes.

The type and site of brain remodeling vary with different types of experience. For instance, aerobic exercise drives hippocampal neurogenesis and angiogenesis in the cerebellum and hippocampus. Learning a motor skill is associated with changes in neuronal architecture, synaptogenesis, and dendritic spine plasticity in the cerebral motor cortex and cerebellum, which is in addition to the motor map reorganization. Animal model studies have shown that rodents socially housed in a large cage with many objects to explore and manipulate, an enriched environment (Fig. 1.5), resulted in dendritic growth, synaptogenesis, and angiogenesis in cerebral cortices and other regions. Following are the principles of the experience-dependent plasticity which are likely to impact the neurological physiotherapy treatment efficacy for post-brain injury:

Use It or Lose It Neural circuits not actively engaged while performing the task for an extended period of time tend to degrade. Behavioral experiences post-injury can protect neurons and enhance performance and restore neuroplasticity that would otherwise be lost after the injury. The loss can be prevented, and functional reorganization can be promoted by appropriate use of rehabilitation strategies, including the task-oriented approach and the constraint-induced movement therapy.

Fig. 1.5 An illustration of an animal model enriched environment

Use It and Improve It Training in an enriched environment can enhance the function and attribute to synaptogenesis and increased synaptic responses. Primates trained to perform fine movements had an increase in digit representation area within the primary motor cortex. It is hypothesized that similar neural changes occur in response to exercise training and mediate functional improvement. Several research works indicate that behavioral experience can enhance performance and optimize restorative brain plasticity post-injury.

Specificity The nature of the plasticity depends on the nature of the training experience. For instance, individuals trained to perform skilled movements exhibited enhanced corticospinal excitability, compared to those trained to repeat unskilled movements. Learning-induced neuroplasticity also shows regional specificity. For instance, unilateral task training in rodents caused dendritic growth in the contralateral motor cortex with a modest effect on the ipsilateral motor cortex. Skill acquisition is associated with changes in gene expression, dendritic growth, synapse addition, and neuronal activity in the cerebral cortex and cerebellum.

Repetition Matters An adequate amount of repetition is required to induce brain plasticity. Merely engaging a neural circuit in task performance is not adequate to drive brain plasticity. Animal models have revealed that the skilled reaching task performed several days increased the synaptic strength, synapse number, or map reorganization, emphasizing the importance of repetition in driving plasticity and concomitant learning, which is critical for rehabilitation.

Intensity Matters The intensity of stimulation or training is another essential factor for the induction of brain plasticity as indicated by the fact that low-intensity stimulation can weaken the synaptic response (LTD), and high-intensity stimulation can strengthen the synaptic response (LTP). However, care has to be taken to avoid overuse to prevent the possible worsening of function, especially during a vulnerable early period, due to exaggeration of excitotoxicity in the susceptible tissue surrounding the primary injury.

Time Matters The plasticity of the brain, underlying learning can be considered as a process rather than a single measurable event. Neuroplasticity is a complex cascade of molecular, cellular, structural, and physiological events. For instance, during motor skill learning, gene expression precedes synapse formation, which in turn precedes motor map reorganization. Theoretically, if training promotes neural restructuring, then it should work anytime. On the contrary, there appears to be a time frame when it is particularly useful in inducing plasticity. For instance, a rehabilitative regimen initiated 1 month after cerebral infarct tends to be far less effective in improving functional outcomes and promoting cortical dendritic growth compared to the same regimen initiated 5 days post-infarct. The time delay may also encourage establishing self-taught compensatory behaviors that may interfere with rehabilitation.

Salience Matters The training experience must be sufficiently salient to induce plasticity. For an organism to function effectively, there must be a system in place to weigh the importance of any given experience such that it can be encoded. As in classical conditioning, merely playing a tone without the reward does not alter the topography of the auditory maps. Sufficient motivation and attention are essential to promote engagement in the task.

Age Matters Plasticity occurs more readily in younger brains as compared to older brains. Healthy aging is associated with widespread neuronal and synaptic atrophy and physiological degradation and is a known fact that neuroplastic responses such as experience-dependent synaptic plasticity, synaptogenesis, and cortical map reorganization reduce with aging. However, the effects of aging vary with lifespan experiences and are generally better in individuals with greater physical and mental activity. In animal models, ischemic injury triggers an increased production of new neurons (neurogenesis), and such a response is significantly reduced in aged rodents.

Transference Plasticity induced by one training experience can augment the acquisition of similar behaviors. It is the ability of plasticity within one set of neural circuits to promote concurrent or subsequent plasticity. Learning a particular skill can generalize or transfer to similar real-world skills to improve independence in the home environment.

Interference Plasticity induced by one experience can interfere with the acquisition of another one. It refers to the plastic ability within a given neural circuitry to impede the induction of a new one within the same circuitry. For instance, the compensatory strategies, including circumduction gait, developed in the post-brain injury patients are stubborn to correction when training for the appropriate or more effective strategies.

1.5.3 Enriched Environment and Plasticity

Several environmental factors may influence the normal physiological functions of the brain and its ability to counteract pathological changes. Experimental studies have demonstrated that experience shapes the neural circuits, making them more functional, keeping them "young." Experience is a factor that makes our brain more plastic. The brain plasticity represents the biological basis of "cerebral reserves," an apparent protection from the onset of cerebral disease and/or cognitive decline during the process of aging.

There is clinical evidence that individuals with a higher level of education show a slower level of cognitive decline. The two types of cerebral reserves recognized are the brain reserve and the cognitive reserve. The former one is based on the protective potential of anatomical features like brain size, neuronal density, and synaptic connectivity. The brain reserve is passive and is defined as the amount of brain damage that can be sustained before reaching a threshold for clinical expression. On the contrary, the cognitive reserve represents a functional reserve and is the brain's ability to improvise its function or change the way it functions to cope with challenges. It is developed by a lifetime of education and experience and is an active process of coping with brain damage or functional decline by using preexisting cognitive processes or compensatory mechanisms.

In humans, the development of cognitive reserves can be influenced by several factors, such as educational level, physical activity, social integration, and emotional involvement. Cognitively stimulating experiences and regular physical activity are associated with neurogenesis, increased levels of neurotrophic factors, and diminution of neuronal apoptosis. Evidence provides that individuals with a high level of education who maintain regular physical activity and healthy food habits have more cognitive reserve as compared to the rest. Exposing the animals to an enriched environment (Fig. 1.5) is similar to that which occurs in the human lifestyle. In animal models, all these factors are provided by the environmental complexity and novelty to which the animals are exposed. When they are reared in an enriched environment, the animals show significant functional and structural effects consisting of cellular level changes such as neurogenesis, synaptogenesis, dendritic arborization, dendritic spine density, gliogenesis and angiogenesis, and molecular level alterations such as changes in neurotransmitter and neurotrophin expression. In addition to the above, enriched environment and physical activity are proven to have a neuroprotective role against neurodegenerative diseases. Studies suggest that enriched environment exposure up to middle age can provide a reserve-like advantage for specific spatial capabilities in old age.

1.5.4 Physical Activity, Exercise, and Plasticity

The effects of physical activity on the brain are relatively widespread; however, there is also some specificity, such that prefrontal and hippocampal areas appear to be more influenced than other areas of the brain. Six months of light aerobic activity among elderly adults exhibited a 2% increase in the hippocampal volume when compared to the elderly who did not participate in the aerobic activity. Physical fitness appears to prevent cortical decay and improve cognitive performance. In humans, exercise has shown enhancement of spatial learning, pattern separation, working memory, processing speed, and executive function. Physical activity triggers a variety of neurobiological mechanisms to produce both acute and chronic neurological effects on many brain regions.

Exercise modulates neurotransmitters that communicate information in the brain and may play a role in exerting its neurological effects. Physical activity directly influences the central dopaminergic, noradrenergic, and serotonergic systems. Alterations in such systems may cause disorders, including depression. Exercise increases the levels of such neurochemicals and may restore normal brain function, as shown by the transient elevations in plasma tryptophan (a serotonin precursor) following 16 weeks of aerobic training, performed 45 min per session, 3 days per week. Even animal models also demonstrate a similar link between exercise, neurotransmitters, and improvement in brain function.

Exercise has shown that alteration in the levels of monoamines alleviates the symptoms of Parkinson's disease and Huntington's disease in rodents. Dopamine appears to play a crucial role in the acquisition of skilled motor tasks, and studies on rodents have demonstrated the impaired function of learning new tasks after depletion of dopamine. Similarly, an interaction exists between exercise and hormones and may be vital for proper brain function. Physical activity levels are strongly associated with salivary levels of estradiol among regularly menstruating women. Cessation of estradiol and progesterone production by the ovaries, as seen in postmenopausal women, is associated with a variety of symptoms that include short-term memory loss and difficulty concentrating. Hormone therapy is commonly utilized for the treatment of postmenopausal cognitive decline; however, the efficiency is considerable when exercise is combined with hormone therapy.

Neurotrophic factors are peptides or small proteins that support the neuronal growth, survival, and differentiation of both developing and mature brains. Exercise increases the concentration of several neurotrophic factors such as fibroblast growth factor-2, insulin-like growth factor-1, vascular endothelial growth factor, and BDNF, which are likely to support morphological changes in the brain. Evidence suggests that an increase in the levels of BDNF in the spinal cord, cerebellum, cerebral cortex, and hippocampus post-physical activity promotes the differentiation, neurite extension, and survival of neuronal populations. In addition to the above, it potentiates synaptic transmission, participates in gene transcription, modifies synaptic morphology, and enhances neuronal resilience, implicating the importance of BDNF as a prime candidate behind exercise-induced neuronal plasticity and learning enhancement. Forced treadmill training exercises and intense rowing exercises performed by adults have exhibited increased levels of plasma BDNF. Physical activity can also produce neurological changes by altering cerebral blood flow, microvasculature, and vascular endothelial growth factor expression. The presence of angiogenesis in animal studies is considered to be pre-requisition for many forms of neural and behavioral plasticity. Animal models have shown that continuous exercise exposure is essential to produce substantial changes in the microvasculature and blood flow and its neurological effects diminish when physical inactivity returns.

Physical activity or exercise may reduce oxidative stress levels in the brain. Oxidative stress occurs when an organism cannot eliminate the oxygen-free radicals, the chemically reactive molecules that attack and degrade important molecules for biological functions. Animal studies have revealed that exercise increases the levels of antioxidants and reduces the levels of reactive oxygen-free radicals and oxidative protein damage. Studies on rodents have also revealed the significance of exercise in extending the lifespan and improving the behavioral performance on spatial and nonspatial tasks. Exercise is also known to alter the levels of apoptosis in the brain and may cause a transient increase in apoptosis that triggers neuroprotective mechanisms. Animal model studies have shown transient apoptosis at the onset of exercise in the hippocampus, followed by neurogenesis and enhanced function.

1.5.5 Spinal Cord Plasticity

Given its clinical importance, a good deal of research has been devoted to plastic changes related to spinal cord transection. Recovery from a spinal cord injury in mammals is minimal, and the results have been disappointing for those seeking a cure for complete spinal cord injuries (quadriplegia or paraplegia), but they have produced valuable data on fundamental aspects of sprouting. Several researchers have documented the beginning of sprout outgrowth at the severed ends of spinal tracts. These sprouts are arrested at the spinal scar, proliferate within the proximal gray matter and form numerous synaptic complexes. Usually, the sprout outgrowth is soon aborted, and, at best, only a handful of fibers cross the gap at the severed end. The poor response of central neurons to injury is not due to the inability to grow their axons but due to the presence of an inhibitory environment. However, the CNS does undergo synaptic plasticity, and hence some recovery of motor function can be possible. Many authors have found a rapid replacement of degenerating synapses near the transected spinal cord with new synapses emerging from local neuron populations, which is in addition to those emerging from dorsal roots and long descending tracts.

During the past 5–6 decades, research has shown a remarkable degree of plasticity within the spinal cord. Many of the characteristics of hippocampal LTP have been identified in the spinal cord, providing a potential cellular mechanism for central sensitization. Even the glial cells located within the spinal cord have been implicated in spinal LTP. The capacity for glial cells to affect glutamatergic signaling through the release of a host of neuromodulators has led researchers to assess the importance of glia in CNS plasticity. The spinal cord may also mediate plastic changes relevant to the acquisition of motor skills. Alteration of the spinal stretch reflex in response to reward-driven motor training, which is retained independently of supraspinal influences, is an example of such plastic changes in the spinal cord. It may be that all structures in the CNS concerned with motor function undergo plastic changes during the learning of motor skills.

Most studies on cortical plasticity after spinal cord injury have shown that the topographical organization of the sensorimotor cortex is not static but undergoes considerable reorganization after cord injury. Both animal and human studies have shown that the adult body and cerebral cortex are in constant and intimate interaction. Injuries to the spinal cord that convey ascending or descending information disrupt this interaction and cause plasticity changes at the cortical level. Stimulation of a body part above the cord injury level causes activation of the cortex that normally represents a body part below the level. Spinal cord injuries cause structural changes in subcortical substrates, including the thalamus, and the changes generally resemble the cortical changes. At the cellular level, the primary injury causes cell death, and the secondary injury is likely to induce a spread of damage. In the dorsal column and other spinal cord regions, spinal cord injury often leads to degeneration, apoptosis, atrophy, and transneuronal changes. Regardless of the above changes, spinal regeneration and sprouting are often observed simultaneously after chronic spinal cord injury.

Spinal plasticity is not limited to maladaptive plasticity. It also demonstrates several forms of adaptive motor plasticity. It can be considered as a two-edged sword: an adaptive process that can foster sensorimotor recovery and reduce neuropathic pain and a maladaptive mechanism that can lead to pain, spasticity, and lack of functional recovery. Several preclinical studies have claimed functional improvements following spinal cord injury using regenerative approaches. However, most of the studies have shown a weak correlation between regeneration and recovery and "failed to stand the test of time and scrutiny." Even several clinical trials using stem cells, the relatively recent approach which claimed to offer a potential treatment for complete spinal cord injury, have been terminated due to the lack of any noticeable functional improvement. On the contrary, considerable functional recovery can be anticipated for incomplete spinal cord lesions, even in the absence of any regenerative approaches.

The current data suggest that in addition to the regenerated inputs (motor and sensory) and locomotor networks within the spinal cord, which are essential for recovery, the recovery requires appropriate changes and interactions within and between different components of the spinal cord motor system. Though spontaneous recovery from complete spinal cord injury is unheard of in mammals, including humans, experimental studies on lamprey (primitive fishlike jawless vertebrates) have shown remarkable recovery of locomotor function within 12 weeks after a complete spinal cord transection, which is supported by repair of the spinal lesion, axon regeneration, and synapse formation. However, studies on lamprey reveal that several aspects complicate the link between regeneration and recovery of the spinal cord. The regeneration is never complete, axons grow short distances and project to ectopic locations, and regenerated synapses are sparse and small. The robust functional recovery from cord injury in such organisms, when the regenerated synapses are sparse and small, suggests that functional recovery is due to a complex set of compensatory changes throughout the spinal network.

1.6 Neuroplasticity and Neurological Physiotherapy

For a long time, the hardware of the brain was considered "hard," and the structure and the function of the brain were believed to be never regainable or restorable following neurological conditions like stroke or traumatic brain injury. The existing data suggest that neurons possess a remarkable ability to alter their structure and function in response to a variety of internal and external pressures, including rehabilitative training. The gamut of clinical and research evidence strongly suggests that rehabilitative training is the most successful means to enhance functional recovery following such incidences.

Neuroplasticity is considered to be the physical basis of rehabilitation. The key principle of neuroplasticity is that brain activity promotes brain reorganization. In other words, "brain workouts" help the brain to reorganize connections more quickly and stimulate reorganization when the brain is not capable of reorganizing on its own. Presenting oneself to challenging intellectual environments, interacting in social situations, and getting involved in physical activities are some examples of brain workouts. However, generalized stimulation may not be beneficial for rebuilding a specific damaged area of the brain.

Currently, the approaches to improve brain function post-injury focus on limiting the severity of the initial injury to minimize functional loss or reorganizing the brain to restore and compensate for those functions already impaired or lost. Even in the absence of overt rehabilitation efforts after damage, the brain has the potential to adapt. Those patients developing compensatory movement strategies to perform daily activities using the less-involved limb are associated with plastic changes in the contralesional hemisphere. Such behavioral changes can be adaptive and can contribute to functional outcomes. However, they can also be maladaptive and interfere and limit the propensity of individuals to engage in behaviors that improve the function of the involved side using rehabilitation strategies or training.

Brain plasticity can produce a considerable degree of spontaneous recovery, and therapeutic training may modify and boost the neuronal plasticity processes. Experimental studies on animals have extended these findings and provided insight into a broad range of underlying molecular and physiological events. Typically, the best recoveries are associated with the greatest return toward the normal state of functional organization. Reorganization of surviving CNS neuronal cells supports functional recovery through changes in the interhemispheric lateralization, the activity of cortices adjoining the injured zones, and the organization of cortical representational maps. Studies involving the extremity of the rodents have described a shift in laterality of activation after stroke. The rodents, soon after a stroke, exhibited contralesional cortical activation during affected paw stimulation, and later, the activation shifted back to the normal pattern in the ipsilesional cortex. However, larger ischemic insult produced stronger activity in the contralesional primary motor cortex with no functional shift back to the ipsilesional site.

Similarly, the destruction of the forelimb primary motor cortex causes the neurons in the hind limb area to take over the functional recovery of the forelimb. The significant gains in the recovery of forelimb were obtained when exercise training was initiated within 5 days post-stroke as compared to 2–4 weeks post-stroke. Recovery was also associated with increased dendritic branching of layer V neurons of the primary motor cortex in the ipsilesional hemisphere.

Activity-dependent modification of synaptic connections and reorganization within adult brain areas are thought to involve LTP and LTD mechanisms by which information is stored in the mammalian CNS. Synaptic plasticity in cortical horizontal connections has been proposed to underlie cortical map reorganization. Topographical maps are shaped during the early part of life and remain quite stable in adulthood. However, they can change even in the adult by experience-dependent plasticity or post-brain damage. Remapping of the motor cortical areas has been observed in stroke patients using fMRI or TMS.

Animal data show that skilled learning leads to a profound rewiring of the motor cortex, observable both anatomically and physiologically. Such findings are not limited to laboratory animals but can be demonstrated in the human motor cortex. Noninvasive techniques like TMS have demonstrated similar learning-dependent neuroplasticity in the human motor cortex. Individuals trained on a one-handed, five-finger piano playing task demonstrated increased motor cortex area representation for the hand muscles trained during the task. Similarly, studies have shown that highly skilled racket players have a larger representation of muscles of the trained hand in comparison to less proficient players and nonplaying controls.

Building on the principle that neuronal activity promotes reorganization of its complex mass of synapses and neurotransmitters, the therapist should attempt to stimulate those neurons that have not been active for some time. Here, the goal is to promote selective self-repair and reorganization through specific motor activity. Brain reorganization generally becomes more difficult as we age (for reasons not yet fully understood), and, in such a situation, the damaged brain needs a specific "neuroplasticity jump-start" to rebuild. For instance, practicing a particular movement over and over again, as seen in "forced" rehabilitation or constraint-induced movement therapy, helps the brain to form and strengthen the connections necessary for that movement. The use of treadmills and partial body weight support systems for patients who have lost the ability to walk are believed to enable neural reorganization evident clinically by improvement in gait parameters. Like "forcing" the subject to use the affected extremity, "timing" of physical therapy is another important aspect that can influence brain reorganization. If a subject who has suffered from brain damage does not practice a lost movement, the damaged neurons and surrounding neurons starved of stimulation will be unable to reconnect.

Research on nonhuman animals indicates that the use of involved limbs immediately after the injury to the brain can further worsen the area of damage. A higher rate of mortality was observed among acute stroke patients who underwent high-intensity very early mobilization compared to those who received the usual care. Even studies on very early mobilization among stroke patients were not associated with beneficial effects when carried out during the initial 24–48 h after the onset of a stroke.

For rehabilitation to be successful, it is mandatory to commence rehabilitation only when the patient's medical and hemodynamic conditions are stable. Following the stabilization of the condition, the use of the injured limb stimulates damaged connections that would otherwise atrophy without input. Even excessive practice of certain movements can have untoward effects, i.e., if practiced too many times per day for months and years, the pattern of connections can grow so much that it inhibits or "squeezes out" other patterns of connection, resulting in the inability to perform other movements. For instance, some of the compensatory techniques like circumduction gait, if habituated, can impede the possibilities of normal or near-normal gait.

Appropriate therapeutic exercises, right timing and intensity of exercise to facilitate normal movement and purposeful activity, optimal environmental setting, correct sensory and proprioceptive feedback, and strategies to minimize abnormal movements and spasticity are often used in rehabilitation to optimize sensorimotor recovery. Such methods and strategies might strengthen synaptic chains, guide axonal sprouting, facilitate function by unmasking latent synapses, or compensate by behavioral changes to promote optimal or near-optimal motor recovery in patients with motor deficits. Therefore, rehabilitation therapy should take advantage of the brain's natural flexibility for forming new neural connections; however, this is a delicate process that must be done carefully and under professional guidance.

Further Reading

Atwood HL, Wojtowicz JM. Silent synapse in neural plasticity: current evidence. *Learn Mem.* 1999;6:542–71.

AVERT Trial Collaboration group. Efficacy and safety of very early mobilisation within 24 h of stroke onset (AVERT): a randomised controlled trial. *Lancet.* 2015 Jul 4;386(9988):46–55.

Bach-y-Rita P. Brain plasticity as a basis for recovery of functions in humans. *Neuropsychologia.* 1990;28:547–54.

Bach-y-Rita P. Sensory plasticity. Applications to a vision substitution system. *Acta Neurol Scand.* 1967;43(4):417–26.

Baroncelli L, Braschi C, Spolidoro M, Begenisic T, Maffei L, Sale A. Brain plasticity and disease: a matter of inhibition. *Neural Plast.* 2011;2011:286073. Available from: https://doi.org/10.1155/2011/286073.

Belzung C, Wigmore P, editors. Neurogenesis and neural plasticity. New York: Springer-Verlag; 2013.

Berlucchi G, Buchtel HA. Neuronal plasticity: historical roots and evolution of meaning. *Exp Brain Res.* 2009;192(3):307–19.

Bruton A. Muscle plasticity: response to training and detraining. *Physiotherapy.* 2002;88(7):398–408.

Cramer SC, Sur M, Dobkin BH, O'Brien C, Sanger TD, Trojanowski TQ, et al. Harnessing neuroplasticity for clinical applications. *Brain.* 2011;134(6):1591–609.

Citri A, Malenka RC. Synaptic plasticity: multiple forms, functions, and mechanisms. *Neuropsychopharmacology.* 2008;33(1):18–41.

Dimyan MA, Cohen LG. Neuroplasticity in the context of motor rehabilitation after stroke. *Nat Rev Neurol.* 2011;7(2):76–85.

Ding Y, Kastin AJ, Pan W. Neural plasticity after spinal cord injury. *Curr Pharm Des.* 2005;11(11):1441–50.

Erickson KI, Gildengers AG, Butters MA. Physical activity and brain plasticity in late adulthood. *Dialogues Clin Neurosci.* 2013;15(1), 99–108.

Ferguson AR, Huie JR, Crown ED, Baumbauer KM, Hook MA, Garraway SM, et al. Maladaptive spinal plasticity opposes spinal learning and recovery in spinal cord injury. *Front Physiol.* 2012;3:399. Available from: https://doi.org/10.3389/fphys.2012.00399.

Fuchs E, Flügge G. Adult neuroplasticity: more than 40 years of research. *Neural Plast.* 2014;2014:541870. Available from: https://doi.org/10.1155/2014/541870.

Gauthier LV, Taub E, Perkins C, Ortmann M, Mark VW, Uswatte G. Remodeling the brain: plastic structural brain changes produced by different motor therapies after stroke. *Stroke.* 2008;39(5):1520–5.

Hamilton GF, Rhodes JS. Exercise regulation of cognitive function and neuroplasticity in the healthy and diseased brain. *Prog Mol Biol Transl Sci.* 2015;135:381–406.

Hara Y. Brain plasticity and rehabilitation in stroke patients. *J Nippon Med Scheme* 2015;82(1):4–13.

Johansson BB. Brain plasticity and stroke rehabilitation: the Willis lecture. *Stroke.* 2000;31:223–30.

Kania BF, Wrońska D, Zięba D. Introduction to neural plasticity mechanism. *J Behav Brain Sci.* 2017;7(2):41–9.

Kleim JA, Jones TA. Principles of experience-dependent neural plasticity: implications for rehabilitation after brain damage. *J Speech Lang Hear Res.* 2008;51(1):225–39.

Kleim JA. Neural plasticity and neurorehabilitation: teaching the new brain old tricks. *J Commun Disord.* 2011;44(5):521–8.

Langhorne P, Wu O, Rodgers H, Ashburn A, Bernhardt J. A Very Early Rehabilitation Trial after stroke (AVERT): a Phase III, multicentre, randomised controlled trial. *Health Technol Assess.* 2017 Sep;21(54):1–120.

Lawes N. Neuroplasticity. In: Stokes M, editor. Physical management in neurological rehabilitation. second ed. London: Elsevier Mosby;2004.

Mandolesi L, Gelfo F, Serra L, Montuori S, Polverino A, Curcio G, et al. Environmental factors promoting neural plasticity: insights from animal and human studies. *Neural Plast.* 2017;2017:7219461. Available from: https://doi.org/10.1155/2017/7219461.

Mateos-Aparicio P., Rodríguez-Moreno A. The impact of studying brain plasticity. *Front Cell Neurosci.* 2019;13:66. Available from: https://doi.org/10.3389/fncel.2019.00066.

Møller AR. Neural plasticity and disorders of the nervous system. New York: Cambridge University Press; 2006.

Nudo RJ. Recovery after brain injury: mechanisms and principles. *Front Hum Neurosci.* 2013;7:887. Available from: https://doi.org/10.3389/fnhum.2013.00887.

Parker D. The lesioned spinal cord is a "new" spinal cord: evidence from functional changes after spinal injury in lamprey. *Front Neural Circuits.* 2017;11:84. Available from: https://doi.org/10.3389/fncir.2017.00084.

Rossini PM, Dal Forno G. Neuronal post-stroke plasticity in the adult. *Restor Neurol Neurosci.* 2004;22(3–5):193–206.

Sharma N, Classen J, Cohen LG. Neural plasticity and its contribution to functional recovery. *Handb Clin Neurol.* 2013;110:3–12.

Ward NS. Neural plasticity and recovery of function. *Prog Brain Res.* 2005;150:527–35.

Xu T, Yu X, Ou S, Liu X, Yuan J, Chen Y. Efficacy and Safety of Very Early Mobilization in Patients with Acute Stroke: A Systematic Review and Meta-analysis. *Sci Rep.* 2017 Jul 26;7(1):6550. https://doi.org/10.1038/s41598-017-06871-z.

Chapter 2
Therapeutic Approaches

Abraham M. Joshua and Suruliraj Karthikbabu

This chapter has six subsections covering the important neurodevelopmental and non-neurodevelopmental approaches used by the clinicians for handling patients with neurological deficits. Margaret Rood's therapeutic approach (Sect. 2.1) was based on the reflex model of the central nervous system. The concepts and techniques were developed based on neurophysiological theories and experiments. Normalization of tone, use of ontogenic developmental sequence, promotion of purposeful movements, and practice are the basic principles of Rood's approach. Although Rood's rationale for treatment has been criticized, it does not mean that the techniques have no value. This section comprises a brief explanation of the Rood's approach, including the facilitatory and inhibitory techniques, sequences of motor control, treatment program, and techniques for autonomic nervous system stimulation.

The corresponding author of Sect. 2.1 is Abraham M. Joshua, E-mail: abraham.joshua@manipal.edu
The corresponding author of Sect. 2.2 is Abraham M. Joshua, E-mail: abraham.joshua@manipal.edu
The corresponding author of Sect. 2.3 is Abraham M. Joshua, E-mail: abraham.joshua@manipal.edu
The corresponding author of Sect. 2.4 is Suruliraj Karthikbabu, E-mail: karthikbabu78@gmail.com
The corresponding author of Sect. 2.5 is Abraham M. Joshua, E-mail: abraham.joshua@manipal.edu
The corresponding author of Sect. 2.6 is Abraham M. Joshua, E-mail: abraham.joshua@manipal.edu

A. M. Joshua (✉)
Department of Physiotherapy, Kasturba Medical College of Mangalore, Manipal Academy of Higher Education, Manipal, India
e-mail: abraham.joshua@manipal.edu

S. Karthikbabu
KMCH College of Physiotherapy, Institution of Kovai Medical Center Research & Educational Trust, Coimbatore, India

A. M. Joshua (ed.), *Physiotherapy for Adult Neurological Conditions*,
https://doi.org/10.1007/978-981-19-0209-3_2

Section 2.2 deals with the Bobath approach, a concept that revolutionized neurological rehabilitation. It was based on the hierarchical theory and Karl Bobath and Berta Bobath postulated that posture and movement were non-separate entities, sensory input influenced motor output, and muscle strength did not necessarily equal function. Their concept was a clear deviation from the conventional compensatory approach that focused on regaining movement of the more involved side and not just strengthening the less involved side and usage of braces, slings, and canes. Despite over 70 years of clinical use and its questionable effectiveness, the Bobath approach is still widely used in the field of neurophysiotherapy, and the concerned subsection discusses the general neurophysiologic and treatment principles, evaluation concepts, and treatment techniques for adult hemiplegic patients.

Brunnstrom's approach (Sect. 2.3) was primarily developed to address specific concerns of hemiplegic patients. Signe Brunnström's clinical observation and judgment and the use of available motor control literature led to the development of the abovementioned approach. She observed that the movement recovery following stroke is stereotypical in nature, with altered synergistic control of the affected limbs. Brunnstrom enumerated the stages of recovery for the upper and lower extremities and hand. Concepts like basic limb synergies, released postural reactions, use of proprioceptive and exteroceptive stimulations to provoke desired motion or tonal changes during the early stages of recovery, relevance of repetition to promote learned movements, and practice in the context of activities of daily living to enhance learning are the highlights of this section.

Section 2.4, the proprioceptive neuromuscular facilitation approach, covers its philosophy in five subdomains, namely positive approach, functional approach, mobilizes reserves, treating the whole person, and using motor learning and motor control principles. The proprioceptive neuromuscular facilitation philosophy utilizes motor control concepts, and its stages of mobility, stability, controlled mobility, and motor skills. The philosophy addresses not only the level of body structure and function but also activity limitation and participation restriction. The most common basic principles are manual contact, body position, body mechanics, auditory commands, verbal stimulus, traction or approximation, resistance, irradiation and reinforcement, stretch, timing, and patterns. The proprioceptive neuromuscular facilitation techniques are either agonist or antagonist techniques based on the movement patterns aimed at improving the motor control and skills of the patient following neurological dysfunctions.

Janet Carr and Roberta Shepherd developed the motor relearning program (Sect. 2.5). Based on their clinical experience and the principles of motor control and learning, exercise physiology, and biomechanics, the guidelines for the post-stroke training program were developed. It is a task-oriented approach and is firmly based on kinesiological theories that emphasize a distributed motor control model than a hierarchal model of control. According to them, neither exercise therapy nor stimulus-response-based techniques are adequate for the treatment of post-stroke patients. The general principles of the motor relearning program, including the steps for relearning the motor task and an overview of the seven components representing the essential functions of daily life, are discussed.

For several years therapists have been using many therapeutic interventions to remediate the motor deficits and restore motor functions. Neurofacilitatory methods, progressive strength training exercises and strategies or techniques to normalize the tone have been advocated in the past to improve motor control and muscle strength. Though several therapeutic techniques are available to facilitate motor recovery, no single therapeutic technique has shown an overwhelming beneficial effect when compared to the rest. Constraint-induced movement therapy (Sect. 2.6), developed by Edward Taub, is a behavioral approach to tackle the learned non-use post-stroke. Restraining the less involved limb, encouraging intensive therapy characterized by repetition and structured practice, and application of a package of behavioral techniques that transfers gains from the clinical setting to the real-world setting are the three essential components of this approach. In addition to the concept and principles, training protocol, and clinical applications, a brief discussion is also provided regarding the modified constraint-induced movement therapy and training for the more involved lower limb.

2.1 Rood's Sensorimotor Approach

Abraham M. Joshua

2.1.1 Introduction

This approach, based on the reflex model of the central nervous system organization, was developed by Margaret Rood, an American physiotherapist and occupational therapist in the 1940s. The treatment methods used in Rood's approach were based on neurophysiological theories and experiments. Her treatment methods emphasized controlled sensory stimulation, the use of ontogenetic sequence, and the need to demand a purposeful response through the use of activity. Although the treatment techniques of this approach were originally designed for neurological conditions such as cerebral palsy, stroke, and traumatic brain injury, Rood believed that it could be effectively used for musculoskeletal conditions to reduce protective muscle spasms, increase the soft tissue range and elicit normal postural reactions. Some of the neuro-facilitation techniques developed and described by Rood have been used by other rehabilitation specialists, who proposed other neurodevelopmental therapeutic approaches like proprioceptive neuromuscular facilitation technique, Brunnstrom's approach, and Bobath approach.

Though the concept and the theory were originated in the 1940s, based on the ongoing neuroscientific evidence, the approach underwent several revisions, until the death of Margret Rood. According to her, motor functions and sensory mechanisms are interconnected. The approach deals with the activation or deactivation of sensory receptors, which interact with somatic, autonomic, and psychic factors regulating motor behavior. Rood believes that proper sensory stimulation to appropriate sensory receptors can encourage advanced motor patterns from primitive reflexes. She also believed that four components are essential for improvising motor control and they are as follows:

1. Normalization of tone and elicitation of desired muscular responses by appropriate sensory stimuli. Afferent sensory input can influence the movement and postural controls located within the central nervous system and selection of an appropriate stimulus/stimuli is essential for producing the desired response. According to Rood, appropriate sensory stimulation can facilitate and inhibit muscle activity and help achieve normalization of muscular tone.
2. Sensorimotor control is developmentally based and treatment must start at the patient's level of development and progress him sequentially towards higher and higher levels of sensorimotor control. She recommends the use of the ontogenic developmental sequence. The muscular responses reflexively obtained are used in the developmental sequence patterns in an effort to develop supraspinal control over those responses.
3. To gain better motor control, purposeful movements have to be encouraged. The cerebral cortex does not direct each muscle separately or in a sequence, to accomplish the task. While training, the emphasis is given on drawing the patient's attention towards the end goal or purpose rather than the movement sensation accompanying the joint movement, which is fundamental to motor learning. To enhance motor learning, Rood recommends incorporating purposeful movements wherever possible.
4. Repetitive movements encourage better motor learning and the muscular responses need to be repeated to make the movement more permanent.

2.1.2 Controlled Sensory Stimulation

According to Rood, the sensory stimuli and their association with motor functions play a significant role in the analysis of motor dysfunction and the application of the treatment. The relearning of muscular activity can be based on the phenomena of summation of stimuli (both temporal and spatial), which activates or deactivates the sensory receptors, affecting the anterior horn cell of the spinal cord. Based on the clinical experience, Rood theorized that the therapeutic application of sensory stimulation could 'wake up' motor responses from the cortex. According to Rood, four types of receptors can be stimulated to get the desired muscular response, namely exteroceptive, proprioceptive, vestibular, and special sensory receptors.

A variety of afferent stimuli such as cutaneous, thermal, positional, and mechanical can be utilized to achieve proximal control and distal selective movement and these stimuli can be categorized either as facilitatory or inhibitory. Rood suggests that selection of appropriate stimuli is based on the desired motor response or anticipated response to facilitation or inhibition techniques. In any therapeutic procedure, the patient's response is carefully monitored and in order to elicit the desired response, the stimuli can be changed when required. The response to stimuli can be influenced by the internal and external ambiance of the patient. Even the position of the patient during stimulation can affect the patient's response. For instance, tonic labyrinthine reflex and tonic neck reflex may assist or retard the effects of the applied stimulus. Researchers have found that the stretch and kinesthetic stimulations can excite the motor area of the cerebral cortices, which is principally supraspinal. Sensory stimulation of the anterior horn cells through the short and the long latency reflex loop pathways affects the spinal and supraspinal levels. Studies have demonstrated that such sensory stimulations effectively enhance the excitability of the anterior horn cells necessary for generating the appropriate motor responses in patients with motor dysfunctions. Current evidence also suggests that sensory stimulation is effective for the development of movements and skills and researchers have found that kinesthetic stimulation plays a critical role in learning, acquisition, and performance of an extremity skill.

2.1.2.1 Facilitation Techniques

Facilitation techniques are meant to increase or facilitate the muscle output, whereas inhibitory techniques are to reduce or inhibit the same. Once the patient is able to produce the desired voluntarily response, facilitatory techniques like stroking and other forms of sensory stimulation are no longer considered effective or appropriate.

Fast Brushing According to Rood, tactile stimulation can be offered in two different ways. They are fast brushing and light stroking. Fast brushing using a battery-operated brush on the skin overlying a muscle could facilitate muscle contraction. Fast brushing of the skin over the posterior primary rami distribution, which innervates the deep musculature of the back, produces a tonic response of the same. If applied over the rest of the body (supplied by anterior primary rami), it produces a tonic response of superficial muscles. For each area, the brushing should be done for about "5" s, and if no response, brush after "30" s and repeat the procedure three–five times until a response is elicited or else move over to another muscle. Fast brushing over the pinna of the ear can stimulate the vagus nerve, and brushing over the skin supplied by posterior primary rami of L1–L2 can cause voiding and S2–S4 can predispose to bladder retention. Due to the aforementioned reasons, fast brushing is contraindicated for those areas.

Fast brushing is believed to stimulate the "C"-size sensory fibers that have polysynaptic pathways, which might influence the background gamma efferent activity of the muscles. Fast brushing is considered to be a high-intensity stimulus and is believed to sensitize the muscle spindle. Rood stated that the maximum effect of brushing would be achieved 30 min after stimulation. However, evidence suggests that the post-application effect lasts less than a minute in contradiction with what she had stated.

Light Touch Rood advocated the use of light touch over the skin to activate the phasic muscles. Light touch or stroking is believed to activate the low threshold A-beta and A-alpha sensory fibers, which produce superficial phasic muscle activity. The effects are fast and short-lived. It can be applied over specific areas, including the dorsum of the webs of fingers or toes, the palm of the hand, or the sole of the feet, and may elicit a phasic withdrawal response of the stimulated limb and the repetitive use of this stimulus to these areas may result in a crossed extensor reflex pattern. It is hypothesized that the stimulation of low-threshold mechanoreceptors through light touch preferentially activates the phasic fibers.

Thermal Facilitation or Icing Icing is expected to have the same effect as brushing or stroking. She believes that the same neural mechanisms underlie the production of motor response. "C"-icing has the same effect as fast brushing and it should be done by holding the ice cube pressed up to 3–8 s on the skin over the muscle. Researchers have compared C-icing and fast brushing and found that the latter is more effective than the former. They also found that C-icing and fast brushing were most effective during the time of stimulus application.

To produce a phasic motor response, an ice cube can be quickly swept longitudinally over the chosen muscle belly (A-icing). Following each application of ice, a towel can be used to wipe the area to remove the cold water present over the application area. In case if no response is seen even after 5–10 attempts, Rood advices to move forward and continue the same procedure for the next muscle. Quick icing facilitates muscle activity by stimulating A-delta fibers and the effect of A-icing is the same as that of stroking and tends to evoke a reflex withdrawal response.

When using ice for facilitation, the precautions required are the same as for fast brushing and stroking. Icing can be a potent stimulus and results can be unpredictable. For instance, the application of ice over the skin above the sympathetic chain may cause a protective response and therefore it is contraindicated. Placing ice over the lips is not advisable as it may produce undesirable behavioral and autonomic responses. Similarly, the use of ice in the region of the left shoulder can be risky if the subject has a known ischemic heart condition.

Heavy Joint Compression It is hypothesized to facilitate the co-contraction of muscles around a joint. The weight applied must be greater than the weight usually passed through the body part, and the force should be through the longitudinal axes of the bones to approximate the articular surfaces. Care must be taken to keep the joints in correct alignment. This stimulation technique is thought to activate high-threshold joint receptors. Prior to heavy joint compression, the stabilizing muscles need to go through a stage of holding against resistance in a shortened position (isometric contraction). Weight-bearing positions of prone on elbows, prone on hands, quadruped, and standing can offer heavy joint compression, provided patient lifts one or two extremities and weight bear through the rest of the body parts or limbs (Figs. 2.1, 2.2, and 2.3). The use of weighted jackets can also be an alternate to serve the same purpose.

Quick Stretching Quick stretch can activate immediate phasic stretch reflex and facilitate phasic muscle contraction. Even tapping the tendon or belly of the muscle produces the same effect and she recommends 3–5 taps over the muscle's belly.

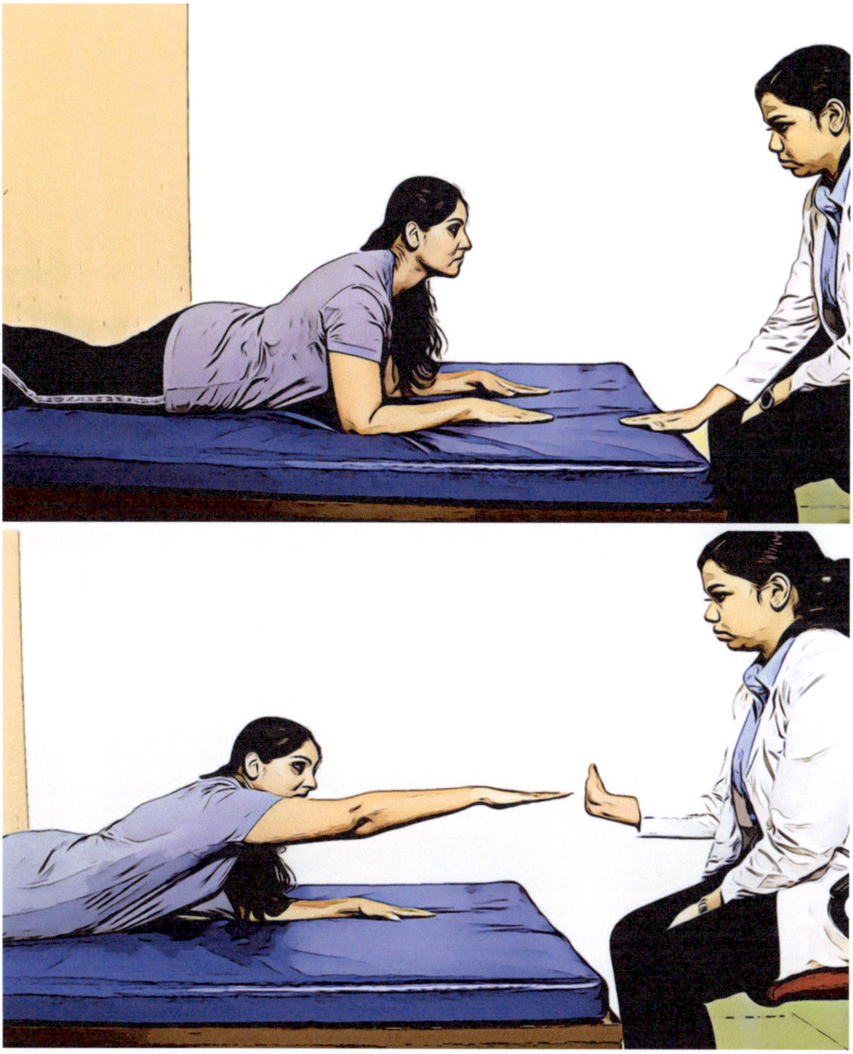

Animated photograph of model and therapist with permission

Fig. 2.1 An illustration of prone on elbows and reach-out

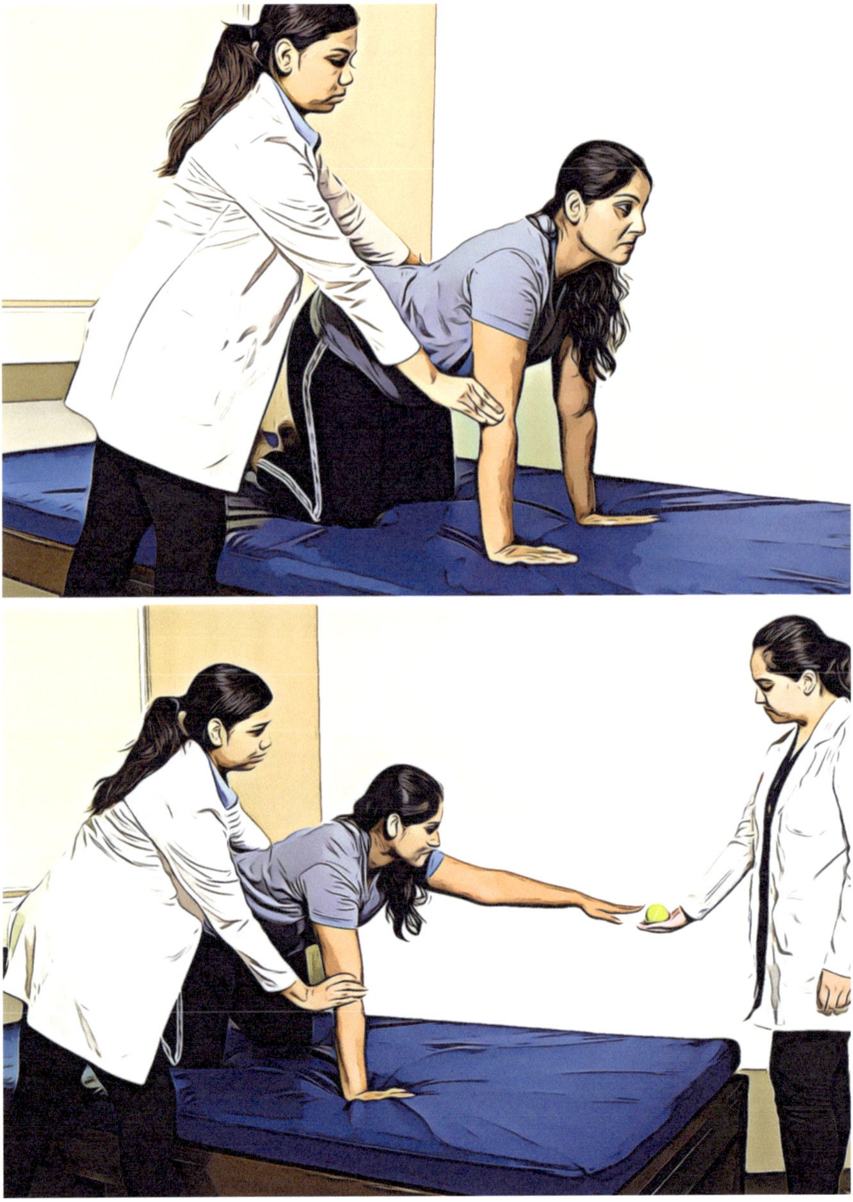

Animated photograph of model and therapist with permission

Fig. 2.2 Illustrates quadruped positioning and reach-out

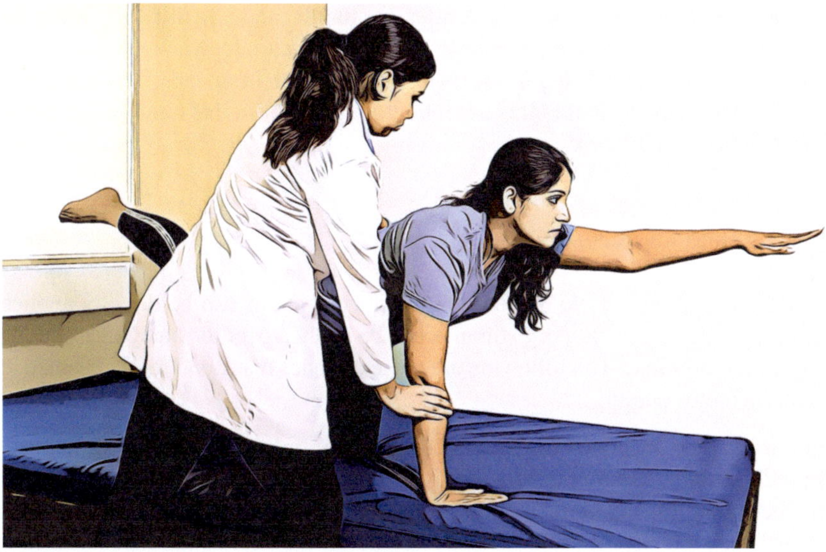

Animated photograph of model and therapist with permission

Fig. 2.3 An illustration of prone positioning with the opposite arm and leg lifting

According to Rood, quick stretch is most effective on the phasic muscles as they respond in a phasic manner. Quick stretch can be used as a technique to facilitate muscle contraction or further augment an already contracting muscle through autogenic facilitation.

Generally, the effect of quick stretch is immediate and short-lived, and to encourage better reactions, repeated stimulation and application of resistance while the muscles are transiently contracting may be required. Resistance is believed to further stretch the muscle spindle receptors, thereby increasing the drive to the extrafusal fibers, thus enhancing the response. The quick stretch works by stimulating the primary muscle spindle receptors and the 1a sensory fibers, and contrary to Rood's belief, the tonic muscles have a greater proportion of these receptors as compared to phasic muscles, which logically could have excelled the latter. She believed that eccentric contraction against resistance could help in recruiting more motor units. Quite the contrary, she also hypothesized that quick stretch to intrinsic muscles of the hand or foot could facilitate the co-contraction of proximal stabilizing muscles.

Pressure on Bony Prominence Receptors found in ligaments and capsules of the joints are known to play a vital role in the control of posture and movement. Pressure on bony prominence has both facilitatory and inhibitory effects. For instance, during normal purposeful movement, pressure over the lateral aspect of the calcaneus facilitates the medial dorsiflexors and inhibits the calf muscles. On the other hand, pressure over the medial aspect of the calcaneus facilitates the lateral dorsiflexors.

Auditory and Visual Stimuli These afferent stimuli can be facilitatory or inhibitory. For example, music with a definite beat would facilitate, while soft music has an inhibitory effect. The therapist's voice and manner of speech also affect the patient's performance. Dull colors are inhibitory and well-lighted and bright colorful areas can be facilitatory.

2.1.2.2 Inhibition Techniques

To reduce or inhibit muscle output, an inhibitory stimulus can be directly applied over the muscle or a facilitatory stimulus can be applied over the antagonist muscle (reciprocal inhibition). The following are the techniques that Rood recommended to inhibit the motor output.

Light Joint Compression Light joint compression or light joint approximation is believed to inhibit spastic muscles and is particularly helpful in reducing shoulder pain secondary to spasticity in hemiplegics. In case of spasticity of the upper extremity or protective muscle spasm, pressure on the heel of the hand with firm rubbing along the posterior border of ulna and compression through the longitudinal axis of the whole arm with the scapula and the glenohumeral joint maintained in normal alignment relieves spasticity or protective muscle spasm and enhances the deep postural muscle tone. Pressure from the top of the skull to the ischial tuberosity helps to activate the deep postural muscles, enhancing head and trunk stabilization. Spasticity of long flexors with ulnar flexion deviation of the wrist can be reduced by the application of pressure through the pisiform bone and gradually, the wrist can be taken into radial extension.

Slow Stroking Slow rhythmic stroking using alternating hands over the paravertebral region, the distribution of the posterior primary rami, applied for 3–5 min or until the patient relaxes, is hypothesized to reduce the sympathetic chain output, which affects the gamma motor neuron firing. Even slow rhythmic movements of the joints are considered to be inhibitory.

Pressure on Tendons Pressure on the tendinous insertion of a muscle is inhibitory. The exact mechanism behind it is unclear. According to some researchers, the tendon pressure stimulates the Pacinian corpuscles and the Golgi tendon organs (GTOs) and by the autoinhibition mechanism relaxes the same muscle. The effect is brief and short-lived and the results last only until the stimulus is present.

Prolonged Passive Stretching Prolonged stretching or maintenance of a lengthened position for a period ranging from minutes to several weeks is inhibitory in terms of the neural response. The effect of prolonged, slow stretching on the muscle is not entirely clear but appears to influence both the neural components of the muscle, via the GTO and muscle spindle (re-biasing of the spindle to a longer position) and the structural components in the long term, via the number and length of the sarcomeres. Following this procedure, the lengthened muscle spindle will not readily react to stretch at shorter ranges. On the contrary, the spindles of a muscle placed in the shortened position may react briskly when stretched beyond the shortened range.

Golgi Tendon Organ Stimulation Intense contraction prior to moving a limb into a lengthened position is believed to produce autoinhibition. Such a technique is believed to activate a large number of GTOs at once to inhibit the same muscle. The effect is more pronounced on the flexor and adductor group of muscles in which the GTOs are abundant and less likely in the extensor or abductor muscles in which there are few or no GTOs.

If the subject has some motor control, spasticity in flexor or adductor muscles can be reduced by stimulating the GTOs in the abovementioned manner. The subject should be taught to repeatedly produce a very small range of contractions of the spastic muscle and its antagonists. Resistance like gravity must be avoided, and the part should be supported with no excessive effort and facilitation. After many repeated contractions in a very small range, stimuli can be given to elicit strong isotonic contraction of the antagonists (extensor and abductor muscles), which will lengthen the spastic group. The process can be repeated using a new starting position, a position in which the spastic muscles are more extended than previously.

Prolonged Icing The prolonged use of ice reduces the transmission of nerve impulses and makes the spindle less sensitive to stretch. The same beneficial effect of this technique can be utilized to reduce spasticity and for that, the muscle spindles must be cooled adequately. Application of ice packs, ice cubes wrapped in a towel, or dipping the extremity in ice-cold water are some of the methods. Yet care has to be taken not to end up with frostbites or cold burns. The gamma efferents are more susceptible to cooling than alpha efferents and cooling is known to cause an overall reduction in the metabolic rate of the tissues and slowing down of impulses in the muscles and motor nerves.

Neutral Warmth Wrapping the body or body parts with a thick towel is believed to be inhibitory. The required range of temperatures that should be utilized for this technique is 35–37 °C. Evidence supports Rood's concept of neutral warmth and has demonstrated the activation of the parasympathetic nervous system by the same. Some researchers believe that the thermal changes, so close to body temperature, created by the wrapping material will not be sufficient to stimulate the thermoreceptors and the stimulation of the low threshold mechanoreceptors by the light pressure exerted by the same as the plausible mechanism of inhibition.

2.1.3 Sequences of Motor Control

Rood's concept is to first reflexively evoke muscle response using sensory stimulation. Such responses are used in developmental patterns and to ultimately gain control over it, the patient should use these responses purposefully. According to her, normal or near-normal neuromuscular integration can occur only if the muscle learns to contract first as it would normally be used, i.e., if the first normal response of a muscle is stabilizing contraction, it should be facilitated to contract in that pattern than in a mobilizing pattern. Similarly, as in ontogenetic development, the movement control should follow through a sequence of flexion, extension, adduction, and abduction, followed by ulnar deviation, radial deviation, and rotation. She stated that the "muscles have two distinct duties or actions," and classified it into light work or heavy work muscle actions. Table 2.1 provides detailed descriptions of light and heavy work muscles. Though some muscles can perform both light and heavy work functions, Rood categorized these muscles as either phasic or tonic muscles with distinct responses to specific stimuli.

Rood introduced two categories of ontogenic sequences (a) The motor development sequence and (b) The vital functions sequence. The motor development sequence finally leads to fine coordinated skilled movements and the vital functions sequence leads to well-articulated speech. The ontogenic motor development sequence consists of supine withdrawal, roll-over, pivot prone, co-contraction of the neck, prone on elbows, all four limbs, erect standing, and walking. Inspiration, expiration, sucking, swallowing, phonation, chewing and swallowing, and speech are the sequences of components for vital functions.

For improvising the muscular response, she recommended the positioning of phasic flexors in a lengthened range and extensors in the mid to shortened range and that the process of strengthening a muscle's response must be continued until the muscle is able to work strongly in all positions. According to her, for attaining

Table 2.1 Descriptions about light and heavy work muscles

Type of muscle work	Descriptions
Light work muscles	Fast glycolytic, rapidly fatigable, fusiform muscles with small area of attachments and with high metabolic cost; typically located on the superficial, lateral and distal aspect of the body and usually multi-arthrodial; subjugated to more voluntary control and render phasic work; activated by light stretch or low threshold exteroceptor stimulation and inhibited by unresisted contraction. According to Rood, these muscles are primarily flexors and adductors and even include the wrist and finger extensors.
Heavy work muscles	Tonic, stability muscles capable of prolonged sustained contraction; belong to slow oxidative motor units, pennate with a large area of attachment, slow to fatigue with a low metabolic cost; deep muscles, located proximally and medially. They are under greater reflex control and are activated by heavy resistance or maintained stretch and high threshold receptor stimulation. Primarily, trunk and proximal limb extensors and abductors and even include the interossei of the hands and feet.

normal motor control, the subject must go through the four phases, which are the mobility phase, the stability phase, the mobility superimposed on stability phase, and the skill phase. Table 2.2 lists the features of all the four motor control phases. She also identified eight phases in the ontogenetic developmental sequence (Table 2.3), which is strictly followed in the application of afferent stimuli. In the ontogenetic developmental sequence, head control is obtained before that of arms and upper trunk and lastly, control of lower trunk and legs; thus, the principles of cephalocaudal development are observed.

Though Rood believes in the ontogenic developmental sequence, developmental studies do not support this sequence concept. According to developmental studies, the adult patterns of movement behavior emerge from a sequence of interactions between inherited tendencies and experience-dependent learning. In addition to the above, developmental studies strongly believe that the relearning of movement among adults neither occurs in cephalocaudal nor returns in style corresponding to normal development seen among children. However, readers should understand that these concepts were based on the neurophysiological theories in the past and Rood's was not the only proponent for neurodevelopmental postural sequences. In the present time, most of these neurophysiological approaches are a topic of criticism, and Rood's approach is not an exception.

Table 2.2 Descriptions of the features of all the four motor control phases

Motor control phases	Features
Mobility phase	It is the first phase to develop and appear as phasic muscle action; such reciprocal shortening and lengthening contraction of muscles sub-serve a protective function. Primarily reflex governed by spinal and supraspinal centers; patterns include supine withdrawal and roll-over; stimulus for this response is quick light stretch or stroking of distal parts or other low threshold A-fiber type of stimulation.
Stability phase	Characterized by the tonic holding contraction; defined as simultaneous contractions of antagonists and agonists, necessary to stabilize and maintain the posture to allow exploration of the environment by the distal segments of the body. Activities include pivot prone, neck co-contraction, prone on elbow, quadruped, and standing. High threshold stimuli, heavy joint compression, stretching of intrinsic muscles of hands or feet help to produce such tonic response.
Mobility superimposed on stability	Characterized by the movement of proximal segments when distal segments are fixed; necessary to develop controlled mobility of proximal joints. Activities include weight shifting and rocking movements in prone on elbows and quadruped.
Skill	The last phase to acquire; examples of skills include reaching, prehensile activities, crawling, and walking; such finely coordinated patterns require control from the highest cortical level.

Table 2.3 Eight phases of ontogenetic developmental sequence and their descriptions

Name	Descriptions of the features
Withdrawal or total flexion pattern	A supine posture characterized by total flexion towards the T10 vertebral level with upper extremities across the chest, the dorsum of the extended hands touching the face, and the lower extremities flexed and abducted. This position is a protective posture and mobility posture, which requires reciprocal innervation. This posture demands heavy work of the trunk and the proximal parts and light work for the distal parts of the extremity. The dominance of extensor response or lack of reciprocal flexion movement can be overcome by promoting flexor response in this posture. It is used to integrate the tonic labyrinthine reflex. To elicit this pattern, fast brushing is carried over the skin supplied by posterior primary rami followed by static stretching of short extensors and then placing the neck flexors and abdominals in a shortened position (to re-bias the spindles). Once the heavy work response of the trunk is obtained, stroking or A-icing of distal parts can elicit a light work response distally. If the tonic labyrinthine reflex is strong, the same technique can be accomplished in a side-lying position.
Roll-over	Characterized by flexion of arm and leg on the same side as the trunk rotates. Roll-over posture is a mobility pattern for upper and lower extremities and it facilitates the activity of the lateral trunk musculature. Attempts at roll-over integrate the tonic neck reflex. Activities that can promote roll-over are like a roll down on incline or roll-over to reach an object. Roll-over promotes segmental mobility of the spine and the extremities.
Pivot-prone	This posture is considered as the first postural or stability pattern, featured by hyperextension of the head, trunk, and legs in prone position. Like the total flexion pattern, this pattern also is centered at the tenth thoracic vertebrae. To assume this posture, a phasic muscle work is required but to maintain that posture, shortened held resisted contraction of extensor muscles is required. This posture is considered both as a mobility pattern and a stability pattern posture. The pivot prone position indicates the integration of symmetric tonic neck reflex and tonic labyrinthine reflex and the development of righting reactions. If the extensor spasticity predominates, this pattern must be avoided. This position plays a key role in preparation for stability in upright postures.
Co-contraction of neck	A pattern to develop head control, which is activated in the prone position. Prior to achieving this posture, it is necessary to activate the flexors if they are inactive. The co-contraction of the neck is essential for lifting the head against gravity.
On elbows	Prone on elbows inhibits symmetric tonic neck reflex and facilitates contraction of the neck and upper trunk muscles and muscles around the shoulder. First, the back and neck extensors and extensors and abductors of the shoulder are C-brushed or iced and then he is asked to assume the pivot-prone position followed by placing the patient prone on elbows. This position gives a better view of the environment and an opportunity to perform weight shifts sideways and activities like watching television while on elbows can be promoted.
All four limbs or Quadruped	The development of this pattern will occur once the neck and upper extremities have attained sufficient stability. This position aids in developing co-contraction of the muscles of the trunk and lower extremities. Once the patient can maintain a static bilateral weight-bearing posture, perturbations or weight shifts should be promoted. Reaching out activities by lifting an arm or leg off the ground is also recommended for better motor control. Performing weight shifts in this position is a preparation for a better balance response for more advanced erect postures.

Table 2.3 (continued)

Name	Descriptions of the features
Erect standing	A sufficient amount of motor control of the head, trunk, and extremities is required for erect standing. Once the subject develops the same, erect standing has to be promoted. Attaining static bilateral weight-bearing posture is a pre-requisite for weight shifts.
Walking	It is the last pattern and is considered as the skill level of standing. Prior to walking, perturbations, and weight shifts, single-limb standing and reach-out has to be promoted.

2.1.4 Treatment Program

Assessment of sensation, perception, postural reaction, quality of movement, distribution of muscle tone, and level of motor control, and noting any local circulatory defects must precede treatment planning. According to Rood, no treatment should follow a set pattern but should be planned to meet individual needs and be adjusted as the evaluation of its effectiveness indicates. Therapy should begin by facilitating the patient's muscles needed to promote the sequence patterns described above. Different types of stimuli may be required to produce the appropriate muscular response. If necessary, the patient should be assisted in the desired pattern, and later a purposeful activity that demands movement could be encouraged. Once the patient can easily manage the task, move to a higher level of motor control. For example, if a head injury patient can roll over in the bed and rotate the trunk while sitting, promote prone with the extension of the head, trunk, and legs, which is the next ontogenetic developmental sequence pattern.

The treatment should start at that point where the patient has difficulty in doing the pattern. For instance, if a stroke patient is having some voluntary flexion and has difficulty to flex the elbow without abduction of the shoulder, then we have to assume that shoulder abductors are contracting in the "phasic" manner, which is improper. In order to overcome this issue, it is essential to promote tonic patterns preventing phasic abduction while flexing the elbow. The selection of the stimuli and activities varies with the clinical presentations. Table 2.4 provides details about some of the stimuli and activities that can be tried for the common neuromotor and neuromuscular conditions.

2.1.5 Autonomic Nervous System Stimulation Techniques

Rood's concept even includes manipulation of the autonomic nervous system as a part of the treatment approach. According to Rood, motivation is crucial to successfully regain movement and she recognized several factors that could aid in enhancing the motivation. When the patient understands the activity as a meaningful one, he or she will participate with maximum interest and enthusiasm in the therapeutic

Table 2.4 Details about the stimulation techniques and activities suggested for the common neuromotor and neuromuscular conditions

Tonal or movement dysfunction	Technique and/or activity suggested
Hypotonicity of the muscle	Overall general stimulation; especially swinging, rolling, spinning in all planes; use of specific exteroceptive and proprioceptive stimulation for the affected muscles
Spasticity of the muscle	Use of inhibitory methods directly over the spastic muscles; use of facilitatory stimulus over antagonist muscle (reciprocal inhibition)
Rigidity of the muscle	Neutral warmth; reciprocal patterns may also help to relieve rigidity
Hyperkinetic movement disorders	Slow stroking; promoting stability patterns like co-contraction of neck and mobility over stability are emphasized
Poor head and neck control and presence of nystagmus	Encourage co-contraction of the neck
Intention tremors	Controlled mobility may reduce the intention tremors in cerebellar dysfunction patients
Weak extensor muscles	Promote holding response; if stability is poor, facilitate the pivot prone, the key pattern which becomes the first treatment goal
Weak flexor muscles	Supine withdrawal response or roll-over pattern to be taught
Facilitation of swallowing and speech:	Developing neck co-contraction to encourage a stable head position is the pre-requisite; other techniques include a light brushing to the upper lip, face, and throat (avoid the undersurface of the floor of the mouth), application of ice to the lips and tongue, resisted sucking, and application of a wipe of ice anteriorly to the lower neck.

activities. The patient's participation can often get hindered by excessive anxiety levels, exaggerated emotional responses, raised blood pressure, heart rate or respiration, and increased muscle tone. All the abovementioned responses are affected by the autonomic nervous system. Rood believes that the dominance of sympathetic or parasympathetic activity will determine how the individual will react to particular sensory stimuli, and for the same reason, she advised manipulation of these stimuli as a strategy for treatment.

Activation of the sympathetic or parasympathetic component of the autonomic nervous system is determined by the intensity and frequency of the stimulus. For hypotonic, drowsy, and lethargic conditions, Rood recommended the activation of the sympathetic nervous system. Whereas for hypertonic, hyperkinetic, and hyperexcitable motor dysfunction conditions, she recommended activation of the parasympathetic nervous system. She believed that the manipulation of these stimuli could be a method of treatment of motor dysfunction patients. Sensory stimulation for activating or manipulating the sympathetic nervous system includes icing, pungent odors, unpleasant tastes, sharp and short vocal commands, bright flashing lights, and fast tempo music. Slow, rhythmic and repetitive rocking movements, gentle rolling or shaking, pleasant odors, stroking the skin over the paravertebral muscles, soft and low voice, neutral warmth, contact on palms of hands, soles of feet, upper lip or abdomen, dim light, and soft music are some of the methods of sensory stimulation for activating or manipulating the parasympathetic nervous system.

2.2 Bobath Approach

Abraham M. Joshua

2.2.1 *Introduction*

Born in Berlin, Germany, in 1907, Berta Bobath got her initial training as a remedial gymnast, during which her keen interest was in understanding normal movement, exercise, and relaxation. Berta Bobath completed her physiotherapy graduation from the Chartered Society of Physiotherapy in 1950. Even before attaining her graduation degree, Bobath had early success in restoring the abilities of Simon Elwes, a popular portrait painter who suffered a large stroke. Karel Bobath, a medical doctor by profession and spouse of Berta Bobath, completed his medical graduation in 1936 and started his career working in the pediatric unit and later with cerebral palsy children. During the late 1940s, Karl Bobath and Berta Bobath together developed the popular approach known as Bobath and Bobath (Bobath) approach. The Bobath approach was empirically developed to evaluate and treat cerebral palsy and was considered neurodevelopmental in nature. The treatment techniques were developed specifically from observations of abnormal tone in children with cerebral palsy. According to the Bobaths, for accurate diagnosis and treatment of these children, comparison needed to be made with normal developmental milestones. They also assumed that the development follows a hierarchical sequence. The treatment of children with motor delay was advocated based on reducing abnormal tone and posture, thus preparing and enabling them to reach the normal development reactions expected based on age and stage of development.

For providing a scientific explanation for treating children with cerebral palsy, Bobath and Bobath had referred to neurodevelopmental theories and neurophysiological principles. Subsequently, the treatment techniques were expanded for utilizing it for the management of adult hemiplegia. The treatment techniques for stroke patients were designed to normalize the affected side's tone through inhibitory techniques and postures and facilitate the correct movement patterns through handling by the physiotherapist. Though the therapeutic concepts were similar for adults and children, the specific guidelines on the actual treatment method of children with motor developmental delay were not well documented by the Bobaths during the earlier years of this approach.

The Bobaths disagreed with the traditional therapy concept of compensatory training that neglected the hemiplegic side's potential for normal function. Till the early 1950s, conventional neurological physiotherapy had a strong orthopedic influence. Use of therapeutic massage, thermal agents, passive and active Range of Motion (ROM) exercises, pulleys, suspensions, weights, prescriptions of splints, and walking aids like calipers and tripods were typical to enable the patient to function. The traditional method emphasized passive stretching, strengthening, and compensation and did not advocate any techniques for normalizing abnormal tone or abnormal coordination. They also disagreed with the concept and techniques of Margaret Knott and Dorothy Voss (proprioceptive neuromuscular facilitation technique) and Signe Brunnstrom and believed that patients have the potential for more normal movement patterns.

Bobath and Bobath described the concept as hypothetical, based on clinical observations, confirmed and strengthened by prevailing research. According to them, this treatment approach underwent many changes during its evolution but the underlying concept has not changed. Discarding the static ways of treatment such as "reflex inhibiting postures" and the addition of new techniques like 'reflex inhibiting movement patterns' and functional activities are some of the major changes introduced. Both believed that the approach is a holistic one, as it dealt with the patient as a "whole" (addressing not only the motor problems but also the sensory, perceptual, and adaptive behaviors).

2.2.2 General Neurophysiological Principles

Charles Scott Sherrington, an English neurophysiologist, had stated that "normal movement needs a background of normal tonus" and the tonus should be of moderate intensity, sufficient to move against gravity but not too high to interfere with the movement. He also stated that tonus and coordination are indivisible and both depend on each other. The abnormal postural tone and the stereotyped mass motor patterns, common in patients with upper motor neuron lesions, can be considered the result of disinhibition (i.e., lack of inhibition—a release of lower patterns of activity from the higher inhibitory control). Such disinhibition not only produces abnormal tone, exaggerated stretch, and tendon reflexes but also abnormal patterns of coordination. Inhibition is considered to play an important factor in the control of posture and movement and lack of inhibition can affect the patient both physiologically and psychologically. During the development of the nervous system, the inhibitory control of the maturing brain enables the organism to gain selective control of the posture against gravity with fractionation of the total movement pattern. Thus the development of more mature postural reactions depends on the increase of inhibitory control on those reflex patterns of posture and movement such as tonic labyrinthine reflex, tonic neck reflexes, and various other primitive reflexes.

The Bobaths believed that the lack of inhibition of reflex patterns of posture and movement as the fundamental difficulty in cerebral palsy and associated them with the abnormal tone. According to them, to retrain normal functional movement patterns, normalization of the tone and activation of postural responses functioning at an automatic level and re-education of muscles for weight-bearing functions are essential. Even for upper motor neuron lesions like stroke, therapy should concentrate on 'normalization' of tone and elimination of abnormal patterns of posture and movement that prevent the patient from regaining normal function on his involved side. They also believed that a permanent reduction of abnormal tone could be possible only if the involved side moves in normal patterns of coordination.

2.2.2.1 Abnormal Tone

As discussed earlier, "normal tone is essential for normal movement," and it must be high enough to allow movement against the pull of gravity but low enough to allow normal speed and timing of movement. Flaccid limbs tend to be more heavy or floppy, have a low tone when moved, and lack placing responses (A momentary holding response seen when a limb is suddenly and unexpectedly relieved of the passive support). In flaccidity, no active adjustment of muscles occurs to the change of posture. In comparison, spastic limbs tend to be stiff and resist movements against the pattern of spasticity. The degree of resistance encountered by the examiner indicates not only the degree of spasticity but also the degree to which it interferes with the possibility of the movement performed by the patient unaided. Spasticity is characterized by the clasp-knife phenomenon, exaggerated stretch response, and lengthening and shortening reactions. Voluntary movements can become labored, with reduced speed. Fear, frustration, unfamiliar environment, and meeting strangers can increase the stiffness due to spasticity. In certain positions, the spastic muscle may exhibit a placing response and demonstrate uncontrolled assistance for those passive movements in the direction towards which it shortens.

The absence of a normal tone is a common feature following central nervous system dysfunction. For instance, stroke patients frequently demonstrate hypertonus of the involved extremities with hypotonus of trunk musculature, and the movements are limited to mass patterns of flexion or extension. The spastic contraction of the flexors and depressors of the shoulder girdle and the leg extensors suppresses the normal postural activity of the antagonists and prevents any reversal of movement. The spastic limb also lacks the normal adaptation of muscles against gravity during movements of the limb. Stress, excessive effort, and certain activities and postures, including associated reactions are some of the common factors which can increase the abnormal tone in these patients.

2.2.2.2 Presence of Associated Reactions

In hemiplegic patients, associated reactions may manifest as an accentuation of the hemiplegic attitude. These reactions can produce a widespread increase in abnormal tone throughout the affected side. In those patients with significant spasticity with associated co-contraction of opposing muscles, these reactions may not produce visible posturing or noticeable movements of the involved limb(s), except for the accentuation of tone detected on palpation. Associated reactions are different from associated movements as the latter is a synonym for synkinetic movements. Associated reactions are not only seen in hemiplegia but also in spastic diplegic and quadriplegic patients. The Bobaths believed that the reinforcement and strengthening of the spastic patterns in such upper motor neuron disorders through associated reactions can lead to contractures and deformities later. These reactions not only emerge from the activity of the sound side to the affected side but also from the affected arm to the affected leg and vice versa. The use of excessive effort, fear of falling, excitation, and anxiety can worsen these reactions.

2.2.2.3 Poor Postural Reactions

Normal postural adjustments form the necessary background for normal movement and functional skills. These normal reflex mechanisms consist of normal postural reactions that work together, reinforce each other and interact for the purpose of protection against falls and injuries. Normal automatic postural reactions give us the ability to counteract gravity without fatigue. These reactions are active movements but automatic and subcortically controlled. Before the execution of voluntary movements, these reactions change our posture automatically and those postural adjustments are known as "postural sets." Normal postural reactions, such as righting reactions (Box 2.1) and equilibrium reactions (Box 2.2), controlled by central neuronal mechanisms, help control the position of the body against gravity. These postural reactions are associated with changes in tone. Furthermore, functional movements, including weight shifts, are associated with automatic postural changes and these changes precede and or accompany them.

In children with cerebral palsy or stroke patients, poor motor control with abnormal tone and postural asymmetry prevents postural reactions from functioning on the affected side, compromising functional movements and safety and facilitating compensatory strategies. The Bobaths assume that the main factor responsible for abnormal postural reflex activity interfering with movements is the released effects of specific postural reactions, such as asymmetrical tonic neck reflex, positive supporting reactions, and associated reactions, as seen in patients with upper motor neuron lesions. Certain additional factors responsible for abnormal postural activity are the presence of associated sensory and perceptual disturbances like homonymous hemianopia, hemianesthesia, and impaired proprioceptive and tactile sensations.

Box 2.1 Descriptions About the Righting Reactions

Righting Reactions

Righting reactions are automatic postural reactions that serve to maintain and restore the normal position of the head in space and its normal alignment and relationship with the trunk and limbs. These normal reactions that develop in the growing infant undergo gradual modification and become integrated into more complex activities, such as equilibrium reactions and voluntary movements. These mature reactions once appeared will persist throughout life and are necessary for transition movements like turning over from supine to prone and back to supine, getting up from the chair or floor, getting out of the bed, and sitting up.

Box 2.2 Descriptions About the Equilibrium Reactions

Equilibrium Reactions

Like the righting reactions, the equilibrium reactions are normal automatic postural reactions and they serve to maintain and restore balance when the center of gravity (COG) is disturbed. It is characterized by tonus change and/or movements that are well-coordinated, swift, well-timed, and adequate in range and their development gradually overlaps with those of the righting reactions. When the subject is on the verge of falling or losing balance, the equilibrium reactions are inadequate and inappropriate to regain balance and therefore to avoid unsafe landing on the floor, protective reactions such as the protective extension of arm and parachute reactions serve as the second line of defense. Such protective reactions are considered automatic in nature and are closely associated with the development of equilibrium reactions. In hemiplegic patients, abnormal tone and coordination prevent both equilibrium and protective reactions from functioning on the involved side.

2.2.2.4 Abnormal Coordination

During the phase of acquisition of motor control, the development of coordination goes hand in hand with the development of postural reactions until more complex and more voluntary skilled activities are acquired. In addition to the above, during the motor control acquisition, one should realize that there is no separating line between posture and movement but only a fluid transition from one to the other. Functional inefficiency of the affected extremity can be the result of abnormal motor coordination. Abnormal coordination means a lack of coordination between agonists, antagonists, synergists, and fixators. Following Central Nervous System (CNS) damage, those normal sequences of muscular activity necessary for skilled movements, stored as motor programs within the CNS, will no longer be available, explaining the poor timing, sequence, and coordination of muscular activity. Those patients with abnormal patterns of movement and incoordination may exhibit abnormal synergies, co-contraction, abnormal timing, and need conscious attention and excessive effort to produce any movement on the affected side.

The absence of reciprocal inhibition, which smoothens the movement, can be another cause for poor motor coordination. Reciprocal inhibition is a central phenomenon that inhibits the antagonist muscle when an adequate stimulus produces excitation of the agonist. In 1913, Sherrington stressed the importance of reciprocal innervation for normal activity. According to him, in intact organisms, spinal reciprocal inhibition becomes modified by higher central nervous influences to allow reciprocal innervation. Reciprocal innervation is required to pit the agonists, antagonists, and synergists against each other, in a finely graded way to promote normal postural fixation, graded movement, and maintenance of equilibrium. For postural fixation, graded movements, and maintenance of equilibrium, normal degrees of reciprocal interaction between muscles of various parts of the body and limbs are necessary. Damage to certain parts of the brain, such as the cerebellar system, can cause deviation of reciprocal innervation towards complete reciprocal inhibition making the movement uncontrolled and hypermetric with the inability to control in intermediate positions. According to the Bobaths, spasticity, on the one hand, can cause deviation of reciprocal innervation towards excessive muscle co-contraction and, on the other hand, can cause deviation towards excessive reciprocal inhibition.

2.2.2.5 Associated Sensory and Perceptual Disturbances

Sensory and perceptual deficits add considerably to a patient's difficulties and can be a serious handicap to effective treatment and adversely influence the recovery chances from functional disability. For generating normal movements, a close and intimate relationship is necessary between the CNS's motor and sensory centers. The sensory messages from the somatosensory, visual, and vestibular organs are integrated at various levels of the CNS to produce a well-coordinated response to meet the demands of the environment. Such motor responses are guided throughout their course by constant feedback from these sensory receptors. According to Thomas Evans Twitchell, an American neurologist, the motor deficit secondary to complete deafferentation has a more disturbing effect than those resulting from Rolandic area ablation, stressing the importance of the sensory system integrity for movement control. Homonymous hemianopia, a common neurological deficit seen among hemiplegic patients affecting the vision associated with hemianesthesia or a state of unawareness of the whole of the hemiplegic side, can cause inattention about the paretic side and ignorance about the objects kept near or around the affected side. Such patients may show little concern about his or her inability or clumsiness to move the affected side. Lesions involving the thalamic or peri-thalamic area can cause loss of appreciation of the joint position sense of the limbs in space and their relationship with the rest of the body leading to sensory ataxia. Cortical lesions, especially, damage to the parietal lobe, can produce cortical inattention and extinction (visual and sensory).

2.2.2.6 Poor Functional Performance

Sherrington once said, "posture follows movement like a shadow" and the normal automatic postural reactions underlie all functional behavior. Unless and until these reactions are fully developed, normal functional activities are impossible. For instance, while standing, to free our hands for manipulative activities, an effective stability of the trunk upon the lower extremities is required. For the normal performance of the functional tasks, integration, interaction, and coordination between the two sides of the body are required. Nevertheless, the patients with CNS dysfunction lack the ability to integrate both sides of the body and have poor coordination between the two sides to perform bilateral functional tasks such as gait, lifting, and carrying objects. In stroke patients, lack of interaction and coordination between the two sides makes even the unaffected side unable to perform the normal movement patterns. Such problems can lead to the development of compensatory techniques, which tend to increase their orientation towards the sound side, increasing the postural asymmetry and neglect of the involved side. Though such compensatory techniques can fulfill certain functional tasks, the same will not be satisfactory for the majority of the bilateral functional activities of daily living.

2.2.3 Treatment Principles

The treatment concept developed by the Bobaths is to promote normal movement patterns in children with cerebral palsy and adults with hemiplegia. The treatment aims to inhibit the abnormal released patterns of coordination, facilitate higher integrated automatic normal postural reactions, and encourage more voluntary activity. By using the patterns that inhibit spasticity, the patient will be able to develop and control the disinhibited action of tonic reflex activity and channelize into more normal patterns of function. The treatment principles are as follows:

- Movements and activities that increase muscle tone or produce abnormal responses on the paretic side should be avoided.
- Therapy should be directed towards the development of normal patterns of posture and movement. There is no need to follow patterns based on the developmental sequences and instead, functional postural patterns can be given more importance.
- The incorporation of the hemiplegic side is essential in treatment activities to encourage and re-establish symmetry and functional use.
- The treatment should produce a change in the quality of movement and functional performance of the paretic side.

2.2.4 General Evaluation and Treatment

Both assessment and treatment should be regarded as related entities and to obtain the best results from treatment, a thorough assessment is a basic requirement. The treatment should be planned based on regular and careful assessment. Unless a measurable change occurs during every treatment session, the treatment must be modified. It is essential to ensure that the patient is free from emotional or physical tension during the evaluation or treatment, as it can promote abnormal tone. The Bobaths believed that the traditional examinations for muscle tone, joint range of motion, and muscle strength will not give sufficient information about the patient's functional status and hardly help to make the treatment plan. For these reasons, the assessment of the patient's motor patterns on the affected side, such as voluntary movement, postural patterns (automatic postural reactions), and abnormal coordination, should be given utmost importance.

Evaluation of children is designed based on the developmental sequence. Once the child's highest level of consistent performance is ascertained, the therapist should begin appropriate treatment to develop the next level of control. Such an evaluation helps to determine the extent and distribution of abnormal tone. It also provides information about the effect of hypertonus on each body part. It can be estimated while placing the patient into developmental sequenced postural patterns. Undue ease of placement and hyper-extensibility suggest hypotonus and undue resistance suggests hypertonus. Earlier, there were several test postures used for evaluating the level of postural control. The therapist placed the child for each test posture. Reflex inhibiting patterns were advised if spasticity was severe and interfered with the placement. The presence of contractures and structural deviations were the other causes that prevented the therapist from placing the child in the test posture. Once placed in the test posture, these children were asked to stay in the test posture and then encouraged to move into the test position independently. A '0–5' grading system was used to grade the patient's ability to assume and sustain each test posture (Table 2.5). Such a system of examination helps to answer the following questions:

(a) Whether the patient's motor abilities are arrested in one or more developmental sequence levels?
(b) What are the abnormal patterns exhibited by the patient?
(c) What is the distribution of the abnormal tone?
(d) How are the abnormal patterns interfering with the activity?
(e) Is there any sign of persistent postural asymmetry?
(f) Are there any contractures or deformities?

Supine, prone, sitting, kneeling, squatting, and standing were the test postures recommended by Bobaths. Further details regarding the evaluation and treatment of cerebral palsy children are not within the scope of this chapter; however, certain common principles concerning treatment are discussed. Treatment, irrespective of

Table 2.5 Feature or description of grades used for the test items

Grades	Feature/Description
Grade 0	Unable to place the child in the test posture.
Grade 1	Able to place in the test posture but cannot hold or assume the test position independently.
Grade 2	Able to hold the test posture momentarily after being placed.
Grade 3	Somehow able to assume an appropriate test posture unaided.
Grade 4	Able to assume and sustain test posture in a near-normal manner.
Grade 5	Normal

being a child or adult, should concentrate on handling the patient in such a way as to inhibit the abnormal distribution of tone and abnormal postures while stimulating or encouraging active motion in the next level of motor control. They believed that once the patient can move easily in and out of normal basic patterns of posture and movement, he will automatically be able to elaborate on these patterns to learn more of the skilled activities required to perform functional tasks. According to them, the treatment should begin at the earliest, as there is a necessity to avoid fixed abnormal patterns or contractures. The main features of their treatment are:

- Reflex inhibiting patterns: Specifically used to inhibit abnormal tone associated with abnormal movement patterns and abnormal postures.
- Sensorimotor experience: To stimulate or encourage active motion in the level from where the patient's motor control is poor. It provides the patient a sense of normal or near-normal tone and movements, away from the existing abnormal tone associated with abnormal movement patterns and abnormal posture.
- Facilitation techniques: To promote mature postural reactions.
- Key points of control: Used by the therapist to inhibit or facilitate an appropriate response.

2.2.4.1 Reflex Inhibiting Patterns

Reflex Inhibiting Patterns (RIPs) are used to inhibit abnormal muscle tone patterns, which are caused by the influence of predominating primitive reflexes such as tonic neck and tonic labyrinthine reflexes. RIPs are partial patterns or postures opposite to the typical abnormal pattern of postural tone that dominates the patient. RIPs are believed to prevent the shunting of sensory inflow into an abnormal pattern and redirect it into normal ones. These patterns have to be adapted and altered to each patient's abnormal postural reactions for which careful analysis of the patient's motor problems is essential. RIPs not only inhibit abnormal activity but, at the same time, give the patient normal "postural sets" to initiate movements, including righting and equilibrium reactions. The developers of this approach also believe that the full-body RIPs have to be avoided, as they may shunt the tone into a reverse pattern. Older cerebral palsy children and severely affected adult hemiplegic patients may require RIPs for a long time before attempting to move them actively. Given below are some instances for RIPs.

- Horizontal abduction or diagonal extension of the humerus to inhibit flexion in the neck, arms, and hands.
- Flexion of the head to encourage flexion of the rest of the body and hyperextension of the head to facilitate extensor tone in the rest of the body.
- External rotation of limb to inhibit flexion and internal rotation to inhibit extension.
- Elevation of arms to inhibit flexor hypertonus and facilitate the extension of the hips and trunk.

- Flexion and abduction of the hip joint with flexion of the knee to inhibit extensor hypertonus.
- Rotation of trunk between shoulder and pelvis to reduce both flexor and extensor hypertonus.
- Symmetrical extension of limbs to inhibit flexor hypertonus.

2.2.4.2 Key Points of Control

The therapist can change parts of the abnormal pattern to reduce the spasticity throughout the affected body parts. Placement of the therapist's hand over the parts or locations to influence the quality of posture and movement of the patient are known as key points of control. According to this concept, holding a key part of the patient in an opposite pattern might redistribute the tone of the whole body more normally. This concept of modifying the tone in distal muscles by controlling the tone of the proximal muscles was interpreted based on Rudolf Magnus's research on reflexes involved in body posture. These points of contact will allow the patient to develop flexible control over his movement and postures. Such points are generally located proximally from which abnormal reflexes seem to originate. The head, neck, shoulder girdle, trunk, and pelvic girdles are the proximal key points, while the thumb, fingers, and toes are the distal key points.

Key points of control allow the RIPs to be maintained and at the same time promote the use of the limb for more normal movement patterns. The Bobaths assumed that the use of proximal key points facilitates movements of the limbs while distal key points facilitate the movements of the trunk. The therapist must carefully select the key points of control and may need to change them constantly during the treatment due to the possibility of adaptation. During a long movement sequence, the therapist's hand can be shifted from one key point to other several times, to assist the patient's posture and movement.

2.2.4.3 Handling

The manner of controlling the patient through RIPs to elicit righting and equilibrium responses can be called handling. It helps to influence the postural tone, regulate coordination, inhibit abnormal patterns, and facilitate normal automatic responses. The therapist should introduce specific handling strategies to encourage the desired aspects of the movement needed and minimize the negative ones. Handling strategies also help to minimize the use of excessive muscle force employed to stabilize body segments and compensations used by the patient to work against gravity.

Handling has to be done slowly, to give the patient time to understand what movements are being performed and organize his response. At first, handle the patient passively in the correct patterns of posture or movement while he is encouraged to co-operate and help as much as he can. Once the patient is in a position to take over the responsibility to move in the correct patterns, the guidance and support by the therapist should be gradually withdrawn. Therefore, handling strategies are always temporary and need to be given with the intent to give less with each movement repetition during the treatment sessions. The patient is not allowed to exert excessive effort to ensure that tone does not increase with the effort, which may shunt into abnormal patterns. Once the normal movement pattern or reaction appears, it is repeated to establish the new sensorimotor patterns.

2.2.4.4 Righting and Equilibrium Reactions

According to the Bobaths, elicitation of righting and equilibrium reactions is essential to gain true inhibition of primitive postural patterns. They are crucial for attaining dynamic control over the normal developmental patterns of movements and postures. These developmental patterns are nothing but the fundamental motor patterns those normal children develop during the first 2 years of life. Righting reactions help the patient to move from supine to prone, then to prone on elbows, quadruped, kneeling, kneel standing, and finally to standing. Using the righting reactions, the patient is moved in and out of the normal developmental postures, which are within his developmental capabilities. Key points are used while eliciting righting reactions. For instance, the head is used as the key point when eliciting neck or head righting and scapula or shoulder for labyrinthine righting. While facilitating movements from the head and neck, the therapist should place one hand lightly under the patient's chin and the other against the back of his head. For facilitating movements from the shoulder girdle, the therapist should place the hand under the patient's axillae with fingers spreading over the scapula to control the girdle.

Once the patient can maintain a posture against gravity, the equilibrium and protective reactions can be elicited. The equilibrium reactions can be elicited by displacing the patient's COG while maintaining a posture. The displacement of the COG can be backward or forward, side to side, and oblique. The displacement must not generate any fear of fall. At first, the reactions need to be elicited slowly and gently. The speed and range of displacement have to be increased with gradual withdrawal of support. The therapist can use a therapeutic mat or gym ball for eliciting equilibrium reactions. Even protective reactions, which are the next level to guard and protect the individual from falling, can be elicited at the same time as equilibrium reactions.

2.2.4.5 Sensorimotor Stimulation

Sensory stimulation is done for those patients who have flaccidity or hypotonic muscles or sensory disturbance. It must be executed in RIPs, to shunt the inflow into the desired channels. Sensory stimulation aims to produce a local response, and these techniques have to be used with great care, as they may result in an abnormal tonic reflex activity instead of producing normal postural tone and normal coordination. Carefully grading the stimulation and simultaneous use of RIPs can avoid muscle hyperactivity and other generalized associated reactions. The common types of sensory stimulation advocated by the Bobaths are mentioned in Table 2.6.

Inhibition techniques are not performed on those patients who lack spasticity or associated reactions. Facilitation techniques are designed to teach the sensation of normal movement by moving the limbs with proper patterns of initiation and sequencing and by stimulating the muscles directly to contract isometrically, eccentrically, and concentrically.

2.2.5 Assessment and Treatment Post-Stroke

Assessment and treatment post-stroke should not be regarded as unrelated entities. A thorough assessment of each hemiplegic patient's problems is a basic necessity if the best results are to be obtained from the treatment. The Bobaths believe that the traditional method of assessing the ROM of the joints may not give any information about abnormal tone or functional abilities. Even traditional muscle strength evaluation can be unreliable as spasticity and certain postures influence strength. Mass patterns, sensory deficit, poor postural and movement control, and abnormal coordination are other reasons for its unreliability. They believe that the traditional rehabilitative concept is concerned with a quantitative rather than qualitative assessment of the patient's functional abilities. In addition to the above, it does not assess the patient's functional abilities, the compensations, and abnormal movements of the affected side.

The Bobaths' evaluation of the patient's motor patterns is qualitative and is based on the observation of the patient's motor functions on the affected side. The limitation of ROM of joints and weakness are considered secondary problems. Assessment of abnormal coordination and postural reactions are considered to be of utmost importance. They assume that the evaluation of the patient's postural patterns not only provides information about the patient's functional abilities but also about the

Table 2.6 Common types of sensorimotor stimulation used in the Bobaths approach

Type of stimulation		Effect/Response
Weight-bearing with pressure and resistance		To enhance postural tone and reduce abnormal movements like ataxia
Placing and holding (to hold the position without assistance once the limb is placed in various positions by the therapist)		To gain movement control and muscle strength
Tapping	Pressure tapping (joint compression)	To increase the postural tone for maintenance of a posture
	Inhibitory tapping	To reduce spasticity using reciprocal inhibition
	Alternate tapping	To stimulate balance reactions

postural tone. Since spasticity is closely associated with typical abnormal postural patterns, a separate assessment of spasticity can be avoided as postural patterns and postural tone are assessed simultaneously. Joint position and movement senses, tactile localization, and stereognosis need to be tested as these patients are likely to have sensory deficits, which are in addition to the inability to move the affected limb. According to the Bobaths, patients generally tend to appreciate the pressure with one finger on the affected limbs better than the light touch, and therefore they recommend pressure examination prior to light touch examination.

A hemiplegic patient's reactions to being moved can be compared to those reactions by a normal person. On the one hand, there will be considerable resistance to being moved against the patterns of spasticity, and on the other hand, there will be undue assistance when moved into the patterns of spasticity. The patient's abnormal tonus and coordination generally will not allow for a normal postural adjustment when they are moved. While sitting, the patient's trunk flexors tend to contract instead of erector spinae muscle and as a result, he or she may fall forwards. When attempting to lean towards the involved side, the lateral flexors of the neck and spine of the same side contract without any contraction of the sound side muscles predisposing to falls towards the affected side.

Clinical examination will reveal the loss of normal adaptation of muscles against gravity during movements of the patient's limbs. He will be unable to arrest the movement of the involved arm or leg when the support for the limb is withdrawn. The Bobaths motor assessment focuses on selective movement control of the involved arm, hand, leg, and foot. Some of the important components covered include the patient's potential to hold the extended arm in elevation after having it placed, lifting the arm to touch the opposite shoulder, supinating the forearm and wrist, opening the hand following grasp, ability to move individual fingers, moving the ankle towards dorsiflexion, and the ability to adduct and abduct the affected leg with the foot on the ground. Inspecting and assessing the control of the pectoral girdle and arm, painful and fixed shoulder, associated reactions, use of walking aids, gait and balance, and other automatic postural reactions are all part of the assessment. A detailed assessment of components such as movement patterns, balance, and postural reactions is not within the scope of this chapter.

According to the Bobaths, routine or conventional rehabilitative treatment during the acute and later stages typically aims to mobilize the patient out of bed and make him or her as much independent as possible for activities of daily living. Such treatment generally tends to emphasize compensation from the sound side for the loss of function of the affected side. Though such treatment can save time and has an economical advantage, it neglects the affected side's potentialities right from the beginning. The Bobaths believe that it is possible to generate a considerable deal of normal activity out of the affected side provided the treatment is designed to systematically prepare the affected side for functional use. Therefore, treatment should emphasize developing the affected side's functional potentialities instead of neglecting them.

The frequent neurological dysfunctions seen among stroke patients include abnormal tone, abnormal patterns of movement, abnormal postural control, poor balance, and protective responses with loss of specific motor abilities such as normal hand functions and gait. The main task of treatment is to improve the tonus and coordination by encouraging normal active reactions of the affected side while being moved. Improving the quality of the affected side ultimately helps both sides to work harmoniously as much as possible within the scope of the cerebral injury.

The Bobaths approach follows a broadly hierarchical treatment regimen, depending on the stage of recovery (i.e., flaccid, spastic, and relative recovery). The patient's recovery can get arrested at any of the abovementioned stages and the therapist should keep in mind that these three stages do overlap without any clear separation. Subsumed within each of these stages are patterns of activities that are intended to prepare the patients for subsequent movement control. For instance, sit-to-stand is achieved through preparation in supine and side-lying, and in such positions, the aim is to facilitate control of the leg, arm, and trunk using various techniques. Theoretically, the patient should not be allowed to progress until normal postural activity can be demonstrated and maintained for that activity. Similarly, volitional movement should be permitted only on the basis of normal automatic postural control. The therapist will determine the patient's successful accomplishment through the use of skillful handling of "key points." Use of appropriate handling techniques and facilitation of normal movement patterns along with sufficient practice of the activity until the patient can perform independently forms the substance of treatment.

2.2.5.1 Acute (Initial Flaccid) Stage

During the initial stage post-stroke, the patient usually will be confused and poorly oriented. The stroke patient typically feels that his or her body is divided into two halves with reduced or absence of sensation from the affected side, different postural tone for either side and reduced or lack of coordination and harmony between the involved and the uninvolved side. All the abovementioned will cause the patient to orient completely towards the sound side and neglect the affected side. This should be counteracted in treatment and not reinforced. The above factors also can create fear of falls and increase the possibility of spasticity. Considering all the above, the common therapeutic goals during this stage should be:

- Regaining balance in trunk patterns, which is essential for function in sitting.
- Incorporating the hemiplegic arm for bed mobility and transfers.
- Developing strategies for self-care that involve the use of the affected arm.
- Improving or restoring and maintaining the alignment and mobility of the affected limbs.
- Retraining normal movement patterns for affected limbs.

The patient should be encouraged to carry weight on the affected side and learn to balance on that side while sitting and standing. Retraining patterns of trunk movement during the initial stage helps the patient regain control of automatic postural patterns essential for the maintenance of normal patterns of posture and balance for a functional task. Similarly, incorporating the hemiplegic arm for bed mobility, transfers, and movements and promoting bilateral activities improvises the sound side's interplay with the affected side. Incorporating the affected side also reduces the neglect of the affected limb, protects the limb from injury, maintains the good alignment of trunk and girdles to prepare the affected limb for normal participation in the tasks he is being trained for.

During this stage, care is taken to position and handle the patient appropriately to prevent any undue increase of spasticity, the chance of shoulder pain and shoulder hand syndrome, the possibility of contractures, retraction of the shoulder and pelvic girdles, and the neglect of the affected side. Since nursing care for the patient is an all-day-long process and the physiotherapist is with the patient for only a short time each day, a good inter-professional cooperation and understanding of the therapist and the nurse is vital for the effective management of the patient. From time to time, the therapist can inform the nurse about the progress the patient has made and learned to perform, with or without minimum help. The patient, the caregivers, and the nursing staff can be shown how to position and handle the patient towards the ultimate goals of maximum independence.

Bed Positioning

The developers of this approach have recommended certain postures for patients in this stage. In supine, to prevent retraction of the shoulder, the affected shoulder should be placed on a pillow with the arm outstretched alongside the body on a pillow somewhat higher than the trunk (Fig. 2.4). The hand should be resting on a pillow with the forearm supinated. As there is a tendency for the head to laterally bend to the affected side, it should be positioned laterally to the unaffected side. If the patient has a flexor tendency of the leg and lack of extensor tone, a pillow can be placed under the pelvis on the affected side (to lift the pelvis and to prevent pelvic retraction), all the way down towards the knee to provide support to the lateral side of the thigh to prevent external rotation of the leg (Fig. 2.5). For those stroke patients with considerable plantar flexion or supination of the ankle, a board can be placed against the foot to keep the ankle in dorsiflexion and pronation. This way of positioning will prevent the possibilities of flexor contractures of the hip, knee, and ankle and pressure sores developing in the lower limbs.

Disinhibited tonic labyrinthine reflex in many stroke patients can maximize the spasticity, especially within the retractor muscles of the shoulder and extensors of the leg, in the supine position. For those patients with a tendency for developing extensor spasticity, positions other than supine, including side-lying positions for both sound and affected side, should be promoted. A mini-pillow or small rubber cushion can be placed under the knee to keep the knee slightly bent (Fig. 2.6). The use of a footboard under the foot is not recommended when patients tend to push the toes against the board. While in the side-lying position, the affected shoulder and

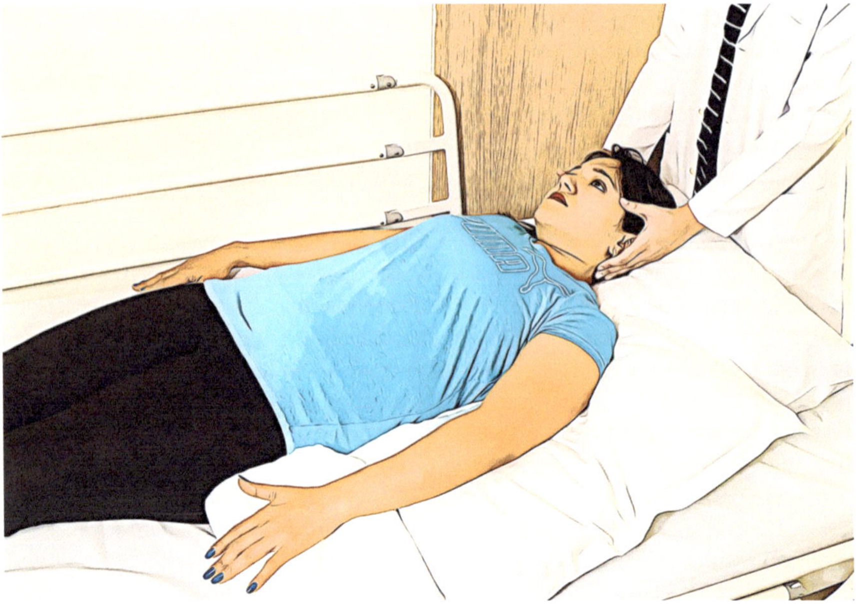

Animated photograph of model and therapist with permission

Fig. 2.4 Recommended posture in supine. Note the position of the arm outstretched alongside the body on a pillow and the hand resting on a pillow with the forearm pronated (for those with increased tone within the pronators, the forearm can be kept in supination)

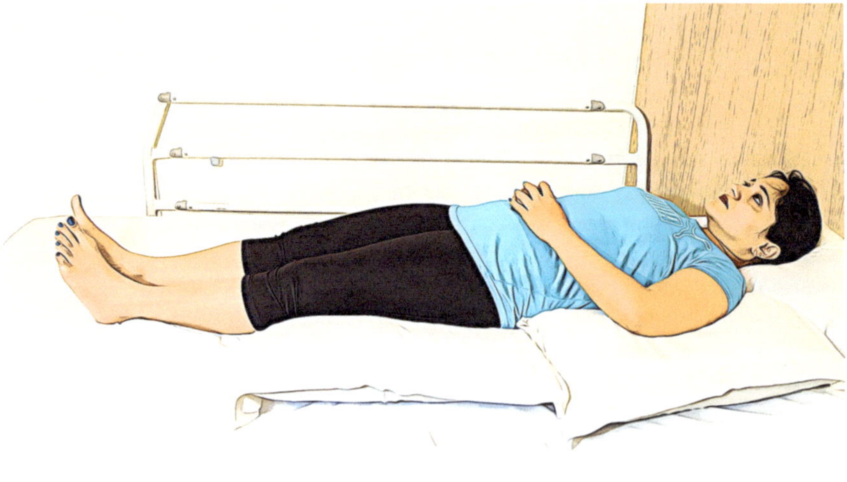

Animated photograph of model with permission

Fig. 2.5 Recommended supine position when there is flexor tendency of the leg associated with lack of extensor tone. A pillow is placed under the affected side pelvis (to lift the pelvis and to prevent pelvic retraction), all the way down towards the knee to prevent external rotation of the leg

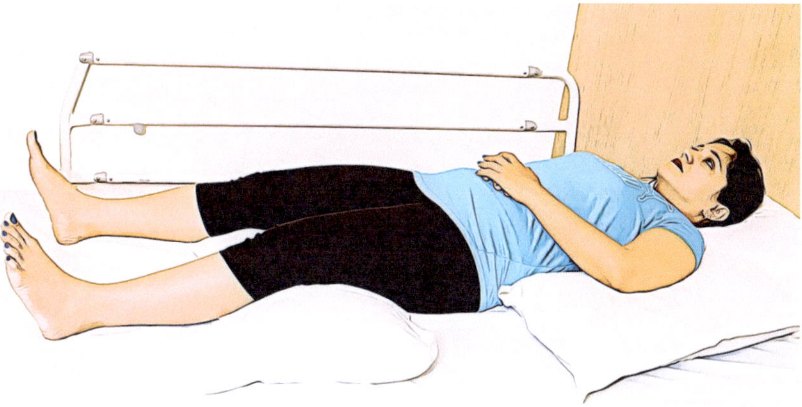

Animated photograph of model with permission

Fig. 2.6 Recommended posture in supine for those with a tendency for developing extensor spasticity. Note the position of a soft pillow under the knee to keep the knee slightly bent

arm should be placed well forward with the elbow extended and forearm supinated, and the leg in the natural semi-flexed position.

Bed Mobility

Turning within the bed encourages the patient to relearn segmental rotation of the trunk. Frequent turning helps to minimize the potential for the development of pressure sores and deformity and promotes the normal sensory feel of the movement. Preferably in the initial stages, the movements should be performed passively. Encourage commencement of turning using the upper part of the body (initiate with shoulder girdle and arm movements). To achieve the above mentioned, the patient should be instructed to locate his or her involved arm in the bed and should lift the involved arm using the good arm with hands clasped or fingers interlocked to one another. This way of clasping the hands will consistently help the patient to hold and move his arms when volitional control of the involved upper limb is poor. Interlocking and clasping the hands keep the palms together with thumbs facing up, forearm in mid-position, and wrist in extension, and enables the comfortable shoulder overhead movement (Fig. 2.7). Patients may have difficulty with the task of interlocking their fingers or maintaining the palm contact necessary for wrist extension. In such situations, a modified way of placing the ulnar border of the affected hand and wrist in the palm of the sound hand with the sound thumb in the affected side palmar arch and the sound fingers clasping the dorsum of the affected hand and wrist can be recommended.

The patient should be promoted to turn in either direction and it should start from a supine position with both knees bent and feet on the bed. Using the sound arm, the patient should extend the elbows and lift his or her hands toward the ceiling until the shoulders are flexed to approximately 90°, and then the therapist should provide

Animated photograph of model with permission

Fig. 2.7 Illustrates the interlocking and clasping of the hands to keep the palms together with the thumbs facing up, the forearm in mid-position, and the wrist in extension to enable overhead movements of the shoulder

Animated photograph of model and therapist with permission

Fig. 2.8 An illustration of therapist-assisted turning towards the affected side. Note the position of the arm (the shoulders flexed to 90°, the elbows extended and the hands lifted towards the ceiling)

assistance to turn the patient's head and shoulders to the side of turning (Fig. 2.8). Rolling is completed by turning the knees and pelvis to the same side until the side-lying position is achieved. Rolling back to supine from side-lying is easy if initiated by pelvis rotation and knees followed by arms and upper body. Figure 2.9 illustrates the therapist-assisted turning towards the sound side. In supine, encouraging bridging exercise (Fig. 2.10) facilitates the patient to move up or sideways on the bed. It

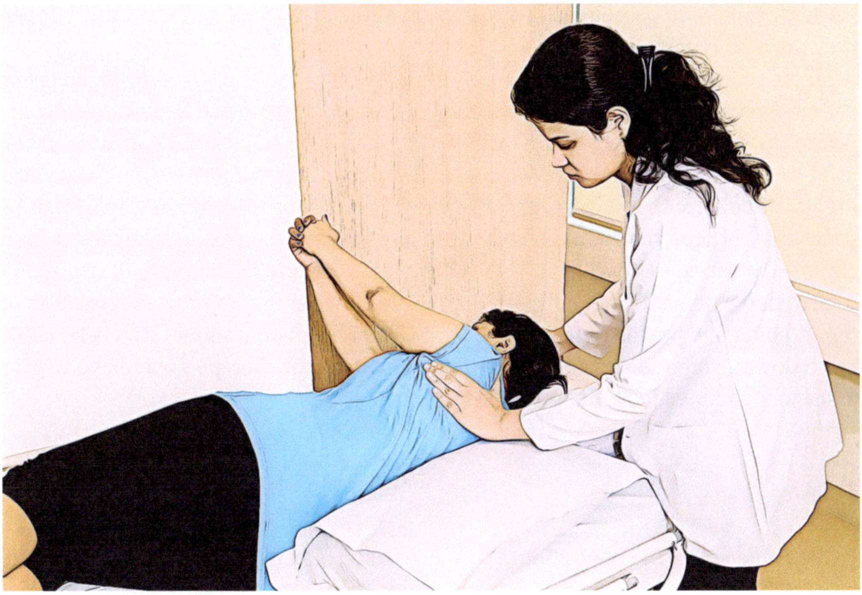

Animated photograph of model and therapist with permission

Fig. 2.9 An illustration of therapist-assisted turning towards the sound side

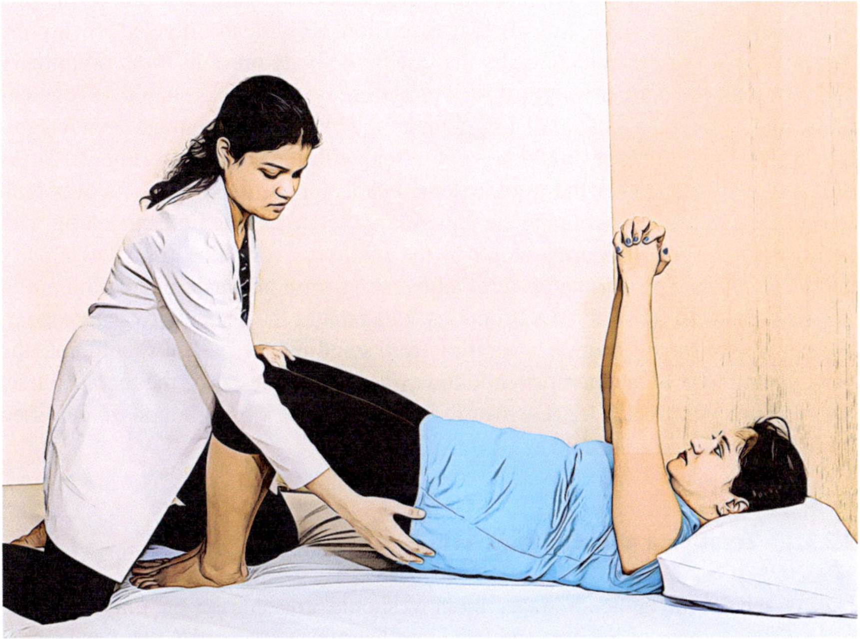

Animated photograph of model and therapist with permission

Fig. 2.10 Encouraging bridging exercise in supine to facilitate bed mobility (to move up or sideways on the bed)

helps the patient to participate or assist in changing bed sheets, skincare, and bedpan use, easing the nurse's job.

To encourage sit up from the side-lying position, the patient should move his or her sound leg over the edge of the bed and the therapist with one hand can assist the patient to bring the affected leg over the edge and the other hand can support the head to move it towards the affected side. Though compared to the sound side turning the affected side turning is easier, sitting up from the affected side-lying can be somewhat difficult but should be practiced and encouraged. While lying down from sitting, the therapist may have to hold the affected arm of the patient externally rotated and extended diagonally forward at shoulder height to prevent retraction of the shoulder and flexion of the affected arm. Then the patient should slowly be made to lie down using the support of the sound arm, and the therapist may have to give assistance to move the affected leg into the bed.

Transfers

In the acute stage, the poor control of the affected side of the patient can make the task of transfer difficult and often frightening. Proper transfer techniques with adequate support and guidance will minimize such problems. To gain balance and reduce the fear of falling, the therapist should encourage the patient to use the affected arm for weight-bearing even while sitting (Fig. 2.11). For preparing the patient to stand up from sitting, the therapist should stand in front, facing the patient, support the hemiplegic arm against the therapist's body and assist him in moving the trunk forwards with the trunk held in extension. Holding the affected arm around the therapist's waist than hanging by the side of the body prevents trunk asymmetry and promotes the correct weight shift. For getting up, the knees should be together in the midline and the feet parallel right under the knees. To encourage even weight-bearing, the affected foot should be kept in line with the sound foot (Fig. 2.12). As the patient's control over the trunk extension and forward weight shifts over both lower limbs improve, encourage the patient to practice getting up from sitting with both hands clasped with arms closer to the midline. Preventing trunk asymmetry and teaching the appropriate postural adjustments with adequate instruction not to use or push with uninvolved extremities will enable the patients to gain a near-normal way of standing up. While sitting from standing, the patient should hold the head up and look at the therapist, not downwards. In some cases, the therapist may need to push the affected leg downwards from above the knee if it tends to get lifted up from the ground (Fig. 2.13).

2.2.5.2 Treatment of Hemiplegic Arm

During the initial or acute stage, therapy should concentrate on improving the patient's awareness of the arm and hand and learning to recognize them as being a part of him or her. Complete orientation towards the sound side should be

Fig. 2.11 Weight-bearing through the affected upper limb with the elbow maintained in extension

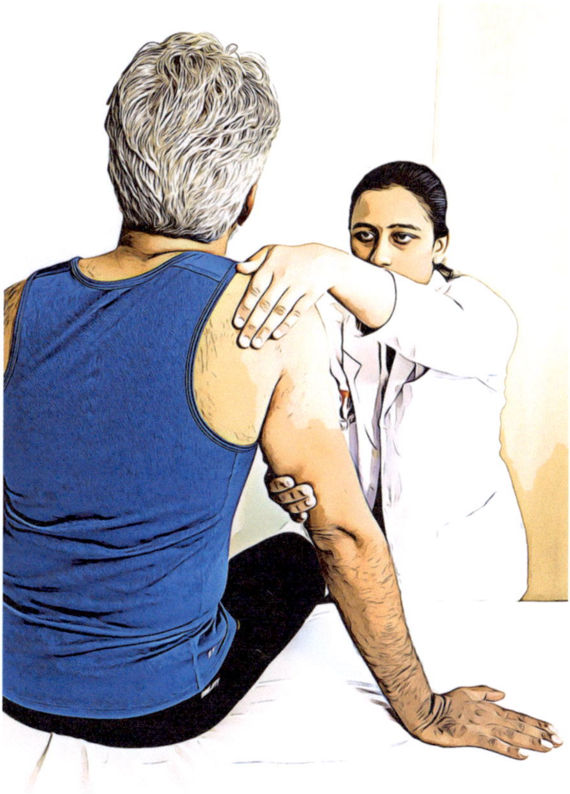

Animated photograph of model and therapist with permission

discouraged as the affected limbs will lose both sensory and motor potential for recovery. Throughout the day, the affected arm and hand should be in front of the patient's body where he or she can see them, rather than hanging helplessly by the side. In the initial stages, to gain control over the shoulder girdle and arm, the supine position is preferable as hip flexion in sitting can promote flexor spasticity in the affected upper extremity. Extensor spasticity or associated reaction of the lower limb while attempting to lift or extend the arm can be minimized by positioning the affected leg in some degree of flexion, with the sole fully supported on the ground with the foot in pronation. The affected leg should be adducted with the pelvis rotated forward towards the sound side to prevent abduction and retraction of the pelvis.

In this stage, scapular mobilization is of utmost importance, as it is required to gain normal alignment and mobility of the scapula, prevent shoulder pain, maintain muscle length around the shoulder and elbow and minimize the development of spasticity in the arm. Mobilization of the shoulder girdle helps the patient to gain better control of his shoulder girdle and arm. Mobilization of the scapula in sitting with the hips flexed may increase the tendency of flexor spasticity in the arm. Hence the mobilization of the shoulder girdle is done primarily in supine (Fig. 2.14) or side-lying on the uninvolved side. Scapular mobilization aims to make painless

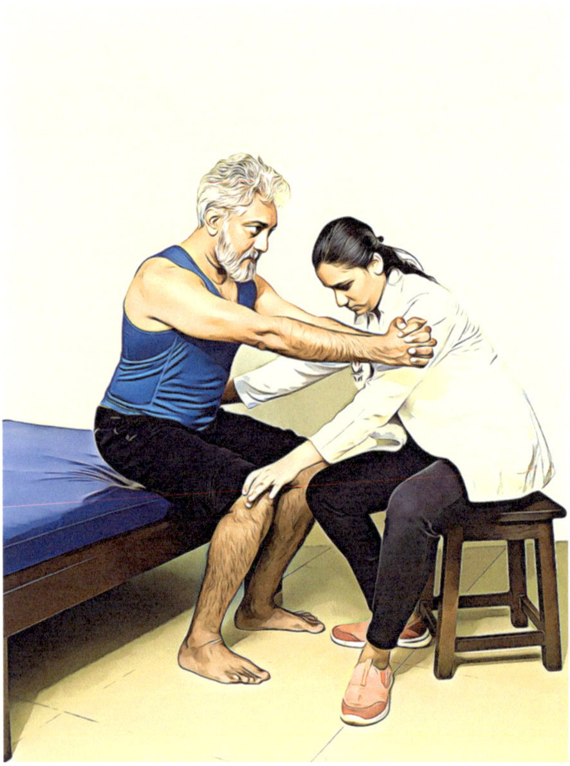

Fig. 2.12 Encouraging getting up from sitting with both hands clasped with the arms closer to the midline and the affected foot placed parallel to the sound foot

Animated photograph of model and therapist with permission

elevation of the affected arm possible. In the supine position, the therapist can support the affected arm, and by using both hands, the shoulder girdle can be mobilized forward, upward, and downward. Care has to be taken to avoid any retraction of the scapula and throughout the procedure, the affected arm needs to be maintained in extension and the forearm in supination. While performing the forward elevation of the arm, the thumb can be positioned in abduction to minimize the flexor hypertonus. Therapist supporting the affecting limb to avoid the flexion of the elbow and the retraction of the shoulder while walking is depicted in Fig. 2.15.

If shoulder retraction is strong, it is preferable to perform the scapular mobilization in side-lying (Fig. 2.16). For mobilization of the scapula in side-lying, the patient should be lying on the sound side and the therapist should stand by the side of the bed facing the patient. The therapist needs to place one hand on the shoulder and scapula and the other hand should support the affected arm maintained in external rotation. Even, the patient's arm can be cradled against the therapist's body and by using both the therapist's hands the scapula can be moved forward, upward, and downward. While mobilizing the shoulder girdle care has to be taken to avoid moving it backwards as this reinforces retraction of the scapula. If the arm can be moved freely, progressively keep the patient's arm in more flexion and the same mobilization technique can be carried out until full elevation is

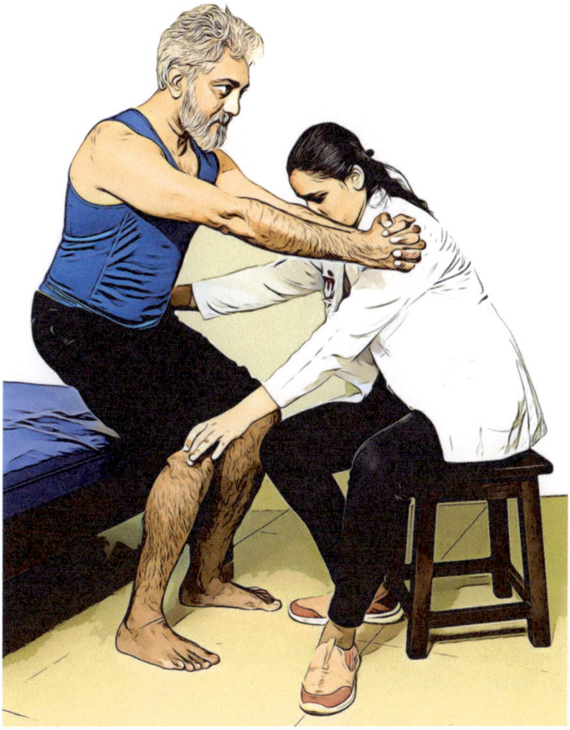

Animated photograph of model and therapist with permission

Fig. 2.13 Encouraging sitting from standing with the head held up and looking forwards. Note the therapist pushing the affected leg downwards from above the knee to prevent the foot from lifting off the ground

possible, the patient should be encouraged to extend the elbow actively, while the therapist still supports his hand. Placing and holding (isometric holding techniques) and eccentric muscle contraction followed by concentric exercises should be introduced to regain more control of the affected arm. Later, the patient should learn to arrest and reverse the arm movement during the intermediate stages of shoulder upward and downward movement. If the patient cannot control the movement eccentrically or concentrically, then the arm has to be placed in the last controlled position and an isometric response is requested. An ample amount of practice is required to regain volitional control of the arm through a large range of movement.

Subluxation of Humeral Head

Subluxation of the humeral head is a common problem for many stroke patients while in upright postures. Separation of the humeral head from the glenoid fossa occurs when the scapula rotates downwardly on the thoracic cage. Flaccidity of the affected side musculature during the acute stage and abnormal posturing of the trunk and lateral flexion of the spine are some reasons for the downward rotation of the scapula. The gravity that pulls the affected arm downwards and spasticity of the depressors, adductors, and medial rotators of the humerus are additional factors that

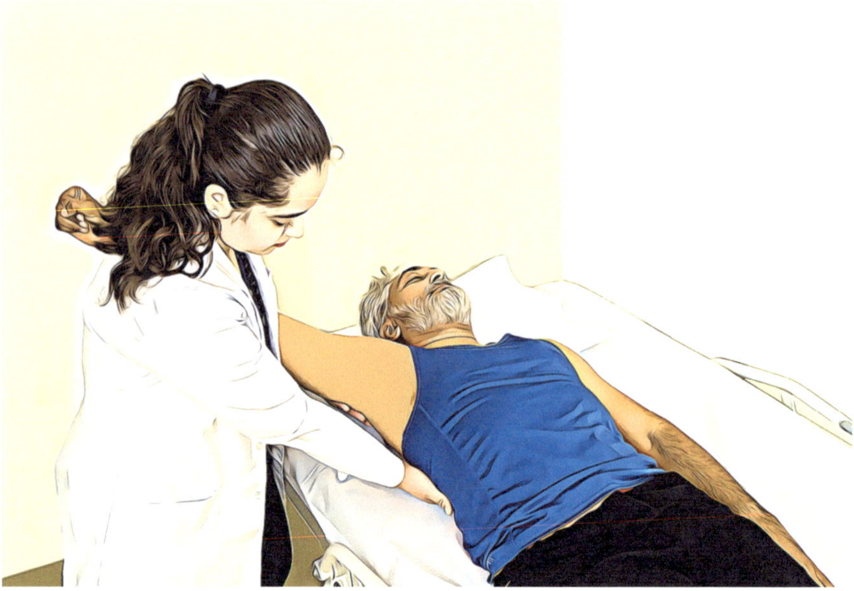

Animated photograph of model and therapist with permission

Fig. 2.14 An illustration of the mobilization of the shoulder girdle performed in the supine position

reinforce the pattern of flexion and depression of the shoulder girdle predisposing subluxation. As long as the scapula is mobile, subluxation per se may not cause shoulder pain for the passive elevation of the arm. If rotation and abduction of the scapula are prevented by spastic scapular retractors and the rest of the muscles, the glenoid fossa will face downwards and passive lifting of the arm above 90° will produce shoulder pain due to impingement of capsule and supraspinatus against the acromion, especially when the arm is internally rotated.

Until the stroke patient can use the supraspinatus and the deltoid to hold the head of the humerus within the glenoid fossa, the patient should be advised to support the arm in the upright postures to prevent stretching of the capsule and the supraspinatus. The use of a "cuff" applied to the upper arm and held by a figure-of-eight bandage can minimize the subluxation. Though the Bobaths earlier recommended the use of a small and soft foam rubber cushion under the axilla, which was believed to abduct the arm slightly, they later found that such placement of foam had a tendency to displace the head of the humerus laterally, affecting the joint integrity. Once the glenohumeral joint has been repositioned, the therapist should encourage normal arm movements by guiding the arm into flexion, abduction, and extension. As the patient's motor control improves, guidance has to be slowly excluded. Even correction of the truncal posture and promoting weight-bearing and weight shifts through the forearm in sitting may improve the alignment, mobility, and activity of the arm and trunk muscles to regain the shoulder girdle musculature integrity.

Fig. 2.15 Therapist supporting the affecting limb to avoid the flexion of the elbow and the retraction of the shoulder, while standing or walking

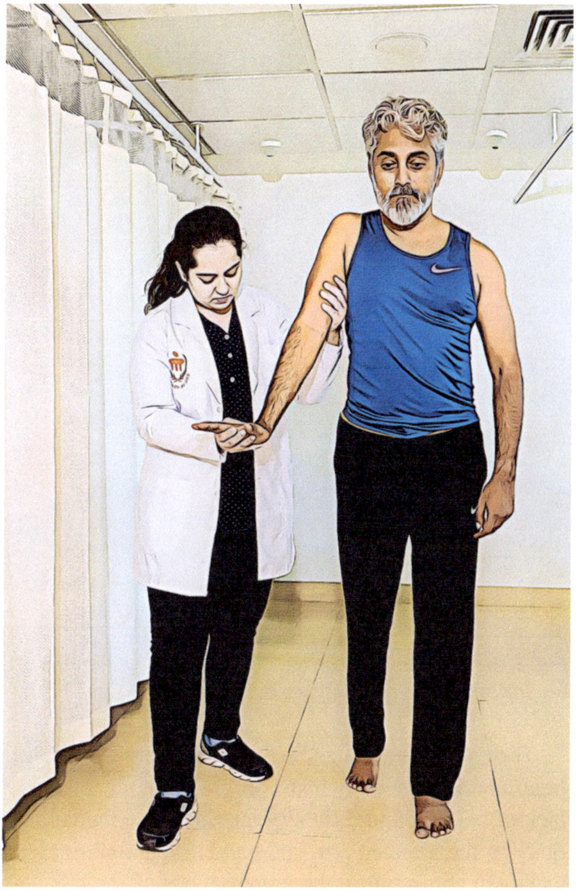

Animated photograph of model and therapist with permission

2.2.5.3 Later Stage

Though this stage is characterized by better control of the trunk and limbs, spasticity can still be a frequent problem compared to the initial stage. During this stage, the patient's balance, posture, and movements can be defective and abnormal. Techniques to inhibit the flexor posturing of arms and extensor posturing of legs and the use of skillful handling of 'key points' to facilitate or re-educate normal patterns of upper extremity movement and lower extremity movements form the essential components of management in the later stage. Therapeutic goals during the relative recovery stage are as follows:

- Reducing abnormal tone
- Promoting normal posture
- Promoting normal righting and balance reactions
- Encouraging or re-educating normal patterns of extremity movements

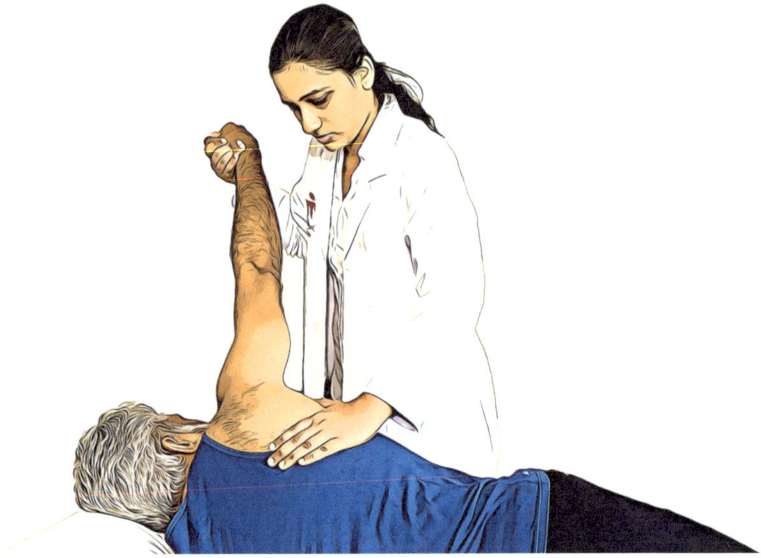

Animated photograph of model and therapist with permission

Fig. 2.16 Scapular mobilization in side-lying with the affected arm maintained in extension and external rotation and the forearm in supination

Though most therapeutic exercises are given in sitting and standing postures, the treatment is essentially just a progression of exercises given in the earlier stage. The lack of balance and difficulty moving the affected leg will increase the flexor spasticity in the affected arm and hand. Even abnormal gait and the effort to stand and walk can compromise the potential use of the affected arm. Reducing spasticity in the trunk and arm will decrease the extensor spasticity of the leg, which facilitates more normal movements of the leg necessary for standing and walking. Throughout this stage, the therapist must hold the alignment and provide postural stability while the patient learns and practices normal movements. It is also essential to teach the patient ways to incorporate the learned normal movement patterns into functional activities and transitional movements. Throughout the stages, care has to be taken not to teach patterns of compensation that can encourage the development of spasticity and associated reactions. For older patients, treatment in prone and kneeling can often be difficult. Much of the benefits offered by such postures, including elongation of musculature, development of balance, and gaining control of affected extremities, can be attained from functional postures like forward lean sitting with forearm supported on a table.

Treatment for Arm and Hand: During this stage, a close association of physiotherapists with occupational therapists is crucial to ensure that there is a carry-over of what the patient learns in physiotherapy into daily life. Both the health professionals preferably should do the first assessment together to ensure that their ideas and ways of treatment are the same in order to attain common goals. In the recovery stage, the emphasis is given on promoting the bilateral use of upper limbs and the use of the affected hand for independent grasp and release, regardless of the position or movement of the arm at the shoulder girdle and the elbow. During therapy, care should be taken not to encourage excessive effort from the patient to avoid spasticity and associated reactions. If the affected arm and hand are extremely spastic, with a poor scope of functional recovery for the affected limb, the occupational therapist may encourage the patient to use self-help techniques using the sound hand for a certain level of functional independence. Even when the hand has no potential functional use, the trunk and arm must be trained for bilateral activity.

A great variety and many different combinations of selective movements are available for normal subjects to perform the manipulation of objects. Such variety and combinations are possible only if the hand's movement becomes independent of the position of the arm at the shoulder and elbow. Even during the later stages, the movement patterns of the patient's arm and hand may show a predominance of flexion of the arm with the pronation of the forearm. Movements requiring flexion of the arm and pronation of the forearm are fairly easy for many patients but may show difficulty in releasing the object due to the inability to extend or move the thumb and the index finger. Those patients, who can hold or manipulate an object kept on a table, may find it difficult to do the same when the arm is lifted up. Such difficulties are due to the inability to dissociate the total patterns of flexion or extension, combine various fragments of movement patterns and inhibit movements that do not belong to the intended activity. Therefore, to regain the skilled movements of the hand, easier and less selective movements have to be mastered during the earlier treatment sessions before more combinations that constitute a skill are relearned.

In stroke patients, many motor problems are associated with sensory deficits and the same applies even to the return of hand functions. Therefore, the evaluation should consist of tests for proprioception, tactile, and special sensations. The focus of treatment should be to improve the discrimination of various sensory modalities such as heat, cold, textures, and shapes along with perceptual and visuomotor training and discrimination of the right from the left side.

Exercises to Improve Motor Control of Leg: The techniques to improve the motor control of the leg essential for standing and walking should begin from the acute stage. To improve the motor control of the leg, the therapy should focus on working with the pelvis and its connections to the spine and shoulder girdle. While undergoing treatment for leg control, care has to be taken to avoid associated flexion of the arm and retraction of the shoulder. Promoting flexion of the leg at the hip and knee without abduction at the hip in supine or flexion of the knee with hip in extension is necessary for walking without circumduction. The patients may find difficulty performing such activities, due to excessive or uncontrolled extension and co-contraction of the leg muscles. To overcome this problem, the Bobaths have recommended controlled extension without encouraging extensor spasticity in supine and later in sitting. Once the patient learns this, reversal of movement (back and forth flexion and extension movements) could be introduced to actively inhibit spasticity.

According to the Bobaths, straight leg raising has no functional significance and may increase extensor spasticity in the leg. When the patient can control his leg in some degree of flexion, with the heel firmly on the support, emphasize and practice active dorsiflexion of the ankle with the foot in pronation. The dorsiflexion of the toes can reinforce the dorsiflexion of the ankle with the foot in eversion. Dorsiflexion of the ankle and the toes can be achieved by sensory stimulation with quick stroking along the plantar aspects of the toes (Fig. 2.17), excluding the big toe. To gain control over the hips with the knee in flexion, a preparatory exercise for weight-bearing, the therapist can support the foot and encourage the patient to bend the leg and move the foot down over the side of the bed to aid hip extension with the knee in flexion (Fig. 2.18). Promoting the forward rotation of the affected side pelvis with leg crossing over the sound leg, which is in extension (Fig. 2.19), gaining isolated flexion and extension of knee in supine and developing control of hip abduction and adduction in supine are other preparatory exercises for weight-bearing.

Control of hip abduction and adduction movements can be regained by making the patient lie supine with both legs bent and feet flat on the support. The affected foot should remain parallel with and near the sound foot and may require support to prevent it from slipping into extension. The sound leg should be kept steady in the midline, and the patient should be helped to move the affected hip towards adduction and abduction in small increments (Fig. 2.20). The same exercise can be practiced in sitting, and the patient should also be promoted to sit with the affected leg

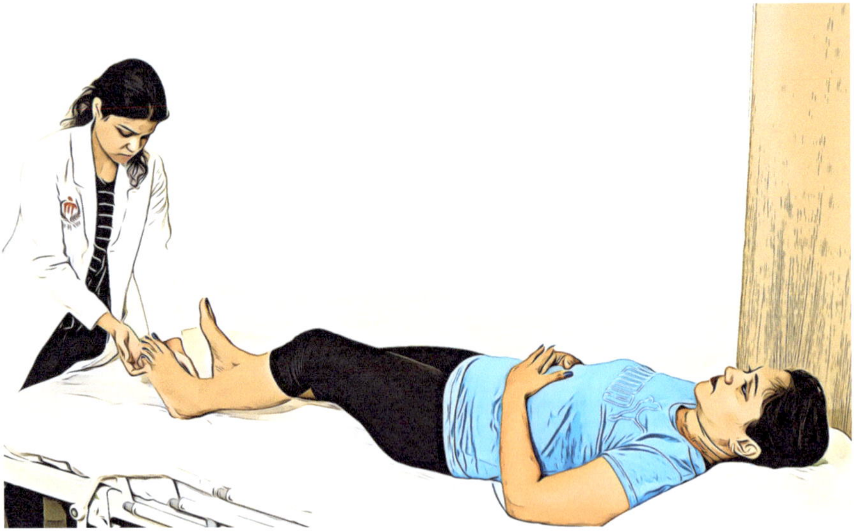

Animated photograph of model and therapist with permission

Fig. 2.17 Quick stroking (sensory stimulation) along the plantar aspects of the toe to encourage ankle and toes dorsiflexion

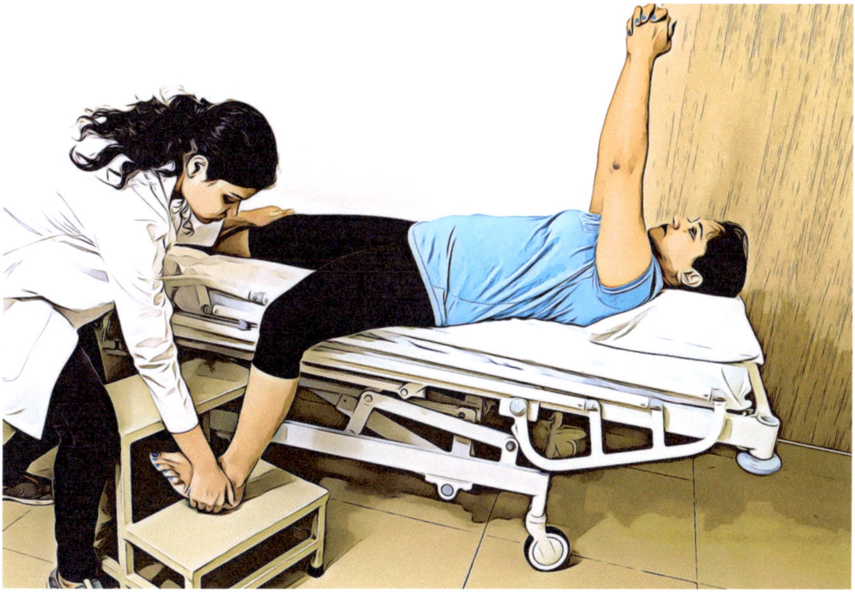

Animated photograph of model and therapist with permission

Fig. 2.18 An illustration that shows how to gain control over the hip in extension with the knee in flexion; a preparatory exercise for weight-bearing with the foot placed on a foot stool or stepper

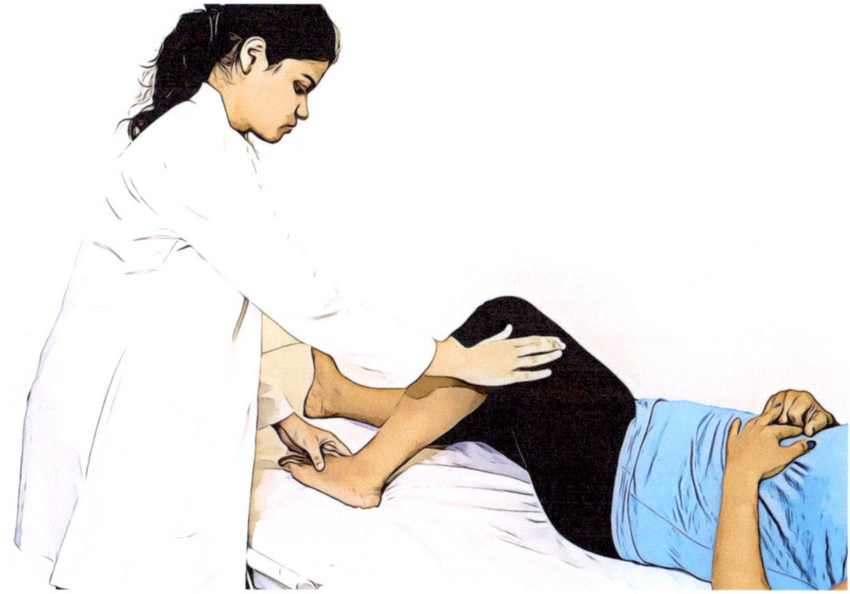

Animated photograph of model and therapist with permission

Fig. 2.19 An illustration of how to promote the forward rotation of the affected side pelvis with the leg crossing over the sound leg held in extension

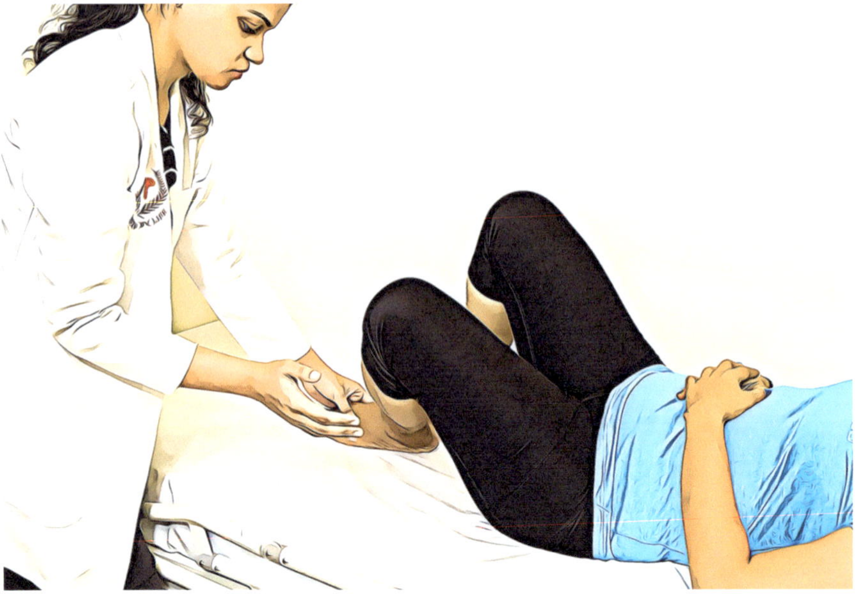

Animated photograph of model and therapist with permission

Fig. 2.20 Controlled hip abduction and adduction movements of the affected lower limb in supine with both legs bent and the sound leg kept steady in the midline

crossed over the sound leg with his hands clasped around the knee (Fig. 2.21). Bending the knee and placing the foot back under the chair with the heel in contact with the floor is essential in preparation for standing up with the weight on the affected leg. This movement pattern also benefits walking when the patient needs to independently flex the knee before attempting a step forward.

The patient should be encouraged to bear more weight on the affected leg and lean well forward at the hips so that he or she starts weight-bearing on both legs before the actual stands. The patient should keep his arms well forward with hands clasped and should not look down. He can also practice standing up from a high plinth by placing the affected leg first for weight-bearing and then the sound leg (Fig. 2.22). Encourage equal weight-bearing on both lower limbs while practicing sit-to-stand. The therapist should stand next to the affected side of the patient, with one hand supporting the axilla to keep the patient's shoulder girdle raised and the other supporting the patient's hand with wrist and elbow in extension to inhibit the flexor spasticity and associated reactions. While in standing, weight transfers need to be practiced, emphasizing the affected side. Later, the patient should practice bending and extending both knees, followed by bending and extending of first one leg and then the other. Reciprocal actions such as straightening one knee while bending the other can be considered as a preparation for walking. However, to avoid circumduction during the swing phase of the gait, the preparation of flexion of the

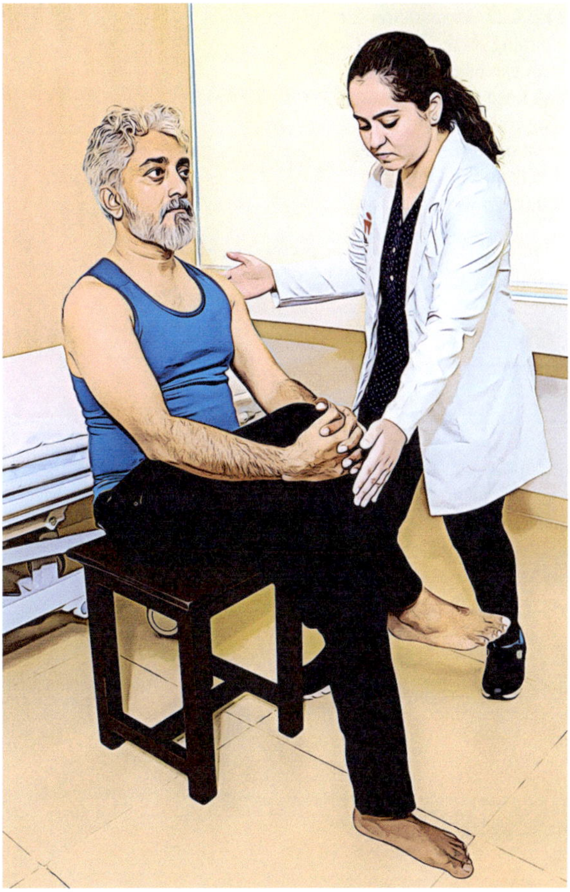

Animated photograph of model and therapist with permission

Fig. 2.21 Promoting sitting with the affected leg crossed over the sound leg with the hands clasped around the knee

knee with hip in extension and ankle in dorsiflexion has to be trained in prone (Fig. 2.23) and then progressed in sitting and standing.

While in standing, with minimal external support by the therapist, the patient should attempt to lift the sound leg's heel and make small steps forward and backward with the sound side. Encourage the patient to perform such steps slowly to provide effective weight-bearing on the paretic lower limb (Fig. 2.24). For those patients with extensor spasticity interfering with dorsiflexion of the ankle and toes, the Bobaths advised the use of a toe-spreader, a foam rubber, to abduct the toes, to inhibit the plantar flexion of the ankle and toes, and extensor spasticity of the affected lower limb. Absence or inadequate dorsiflexion of ankle and toes is a great obstacle for the swing phase of the gait and interferes with weight-bearing and weight transfer from heel to toes. While the patient stands on the heel, placement of

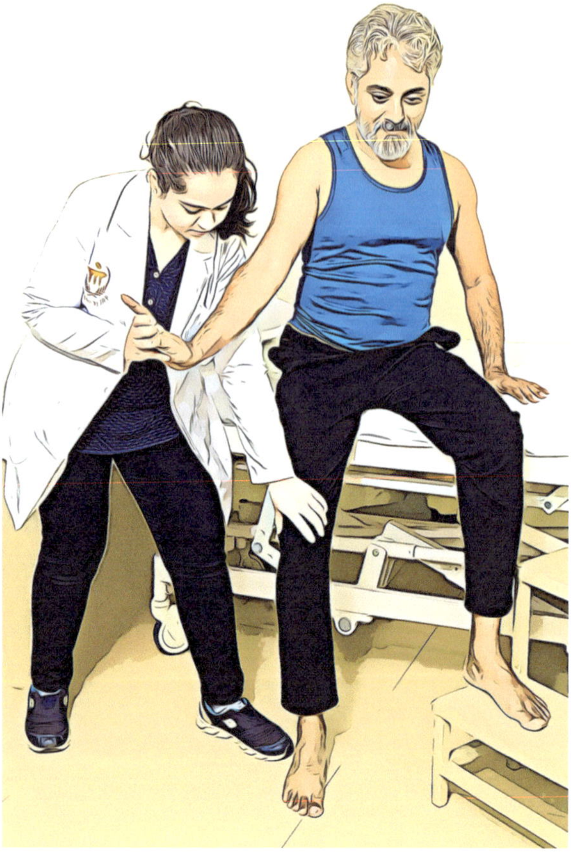

Fig. 2.22 Preparation for standing on the affected leg with extension of the hip and knee and foot firm on the ground. Note the position of the sound leg that is on the stepper to minimize weight-bearing

Animated photograph of model and therapist with permission

the therapist's hand under the ball of the patient's foot can inhibit plantar flexion and encourage the lift of ankle and toes into dorsiflexion, and then, to gain better control over dorsiflexion, the patient can be instructed to lower the foot slowly down to the ground.

To develop symmetry in gait patterns, the use of a stick or cane during gait training must be avoided and till then, the therapist or the caregiver should assist the patient in walking until he or she develops sufficient balance. During the gait training, the therapist should stay on the affected side of the patient. For the swing phase, encourage the patient to lift the leg and take a step. The patient should also learn to perform a controlled extension of the leg while he places his foot down on the ground. Throughout the treatment, proper alignment of the head and neck, trunk,

Animated photograph of model and therapist with permission

Fig. 2.23 Preparation of knee flexion with hip in extension and ankle in dorsiflexion in prone lying

and extremities is crucial. To gain a better walking pattern, all the components mentioned above, including weight shifts and weight transfers, have to be practiced while standing.

For further improvement in gait, selective movements of the knee, ankle, and toes, independent of the position and movement of the hip need to be achieved. Achieving protective postural reactions against falls while standing on the affected leg improves the balance and even the gait. Crossed standing and walking, a preparation for rotation of the pelvis while walking and walking backward, are a few methods that could be used to improve balance and are believed to improve walking forwards. To improve the coordination in walking, the therapist can augment and encourage the pelvis and shoulder girdle rotation by standing behind the patient. Rotation of the shoulder girdle makes arm swing possible and has to be practiced in standing and while walking (Fig. 2.25). This way of rhythmical swinging of the arms and the rotation of the trunk assisted by the therapist can encourage the patient to develop a near-normal gait pattern.

Fig. 2.24 Encouraging effective weight-bearing on the paretic lower limb and promoting small steps slowly forward and backward with the sound side

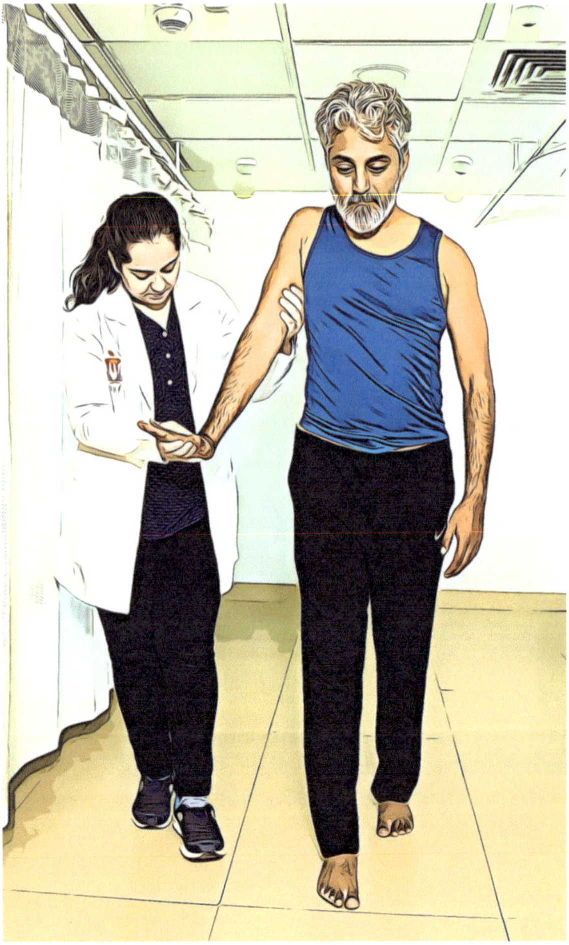

Animated photograph of model and therapist with permission

Fig. 2.25 Therapist assisted pelvic and shoulder girdle rotation while standing and walking

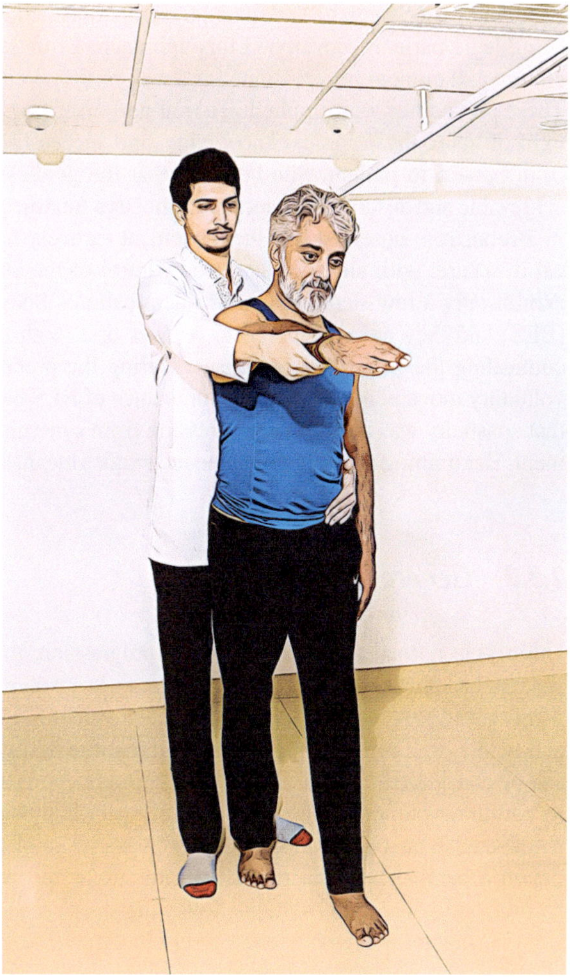

Animated photograph of model and therapist with permission

2.3 Brunnstrom's Approach

Abraham M. Joshua

2.3.1 Introduction

Cerebrovascular accident patients usually exhibit motor disturbance in one-half of the body and the disability is referred to as hemiplegia. The motor disturbance is often accompanied by sensory disturbance. The degree of neurological deficits varies greatly among patients with hemiplegia. In the early 1950s, Anna Signe Sofia

Brunnström, a Swedish-American physiotherapist, concerned with the problems of hemiplegic patients, developed this approach. Brunnstrom used motor control literature and clinical observations for developing exercise techniques or procedures. These procedures were applied in a trial-and-error fashion and were modified in the light of neurophysiological knowledge, and successful procedures were replicated from patient to patient. She believed that the goals set for the patient should be achievable and described the technique not as a treatment but as a training procedure.

Brunnstrom observed that the movement recovery following stroke is stereotypical in nature, with altered synergistic control of the affected limbs. These patients exhibit only a few stereotypic movement patterns known as Basic Limb Synergies (BLS) and they are considered to recover first, with the dominant muscle groups controlling the pattern of responses. During the process of recovery, independent voluntary movements outside the dominance of BLS become possible. She believed that spasticity was the key to progression from synergistic to non-synergistic movement. Her training procedures had no techniques meant to normalize the tone directly.

2.3.2 General Principles

During the normal motor development progression, the spinal cord and brainstem reflexes become modified and their components are rearranged into purposeful movements through the influence of higher centers. According to Brunnstrom, these reflexes and whole-limb movement patterns represent the normal stages of development and can be considered "normal" when the Central Nervous System (CNS) reverts back to an earlier development stage post-stroke. According to her, the hierarchy in the CNS is reflected in normal development as well as in hemiplegia. In this approach, she had drawn a parallel between recovery from stroke and normal development, i.e., the recovery of voluntary movements following stroke proceeds in sequence from reflexive to voluntary movement, mass to discrete movement, and proximal to distal control, which is similar to normal motor development. From this premise, reflex activity is used as the basis for voluntary movement, and training strategies should be based on the stage of recovery, i.e., reflexes and primitive movement patterns should be used in the early stages to facilitate the recovery of voluntary movements post-stroke. She believes that a subcortical motion synergy elicited reflexively can serve as a wedge by which a limited amount of willed movement can be learned.

During the recovery process, these stages need to be achieved but not perfected before the strategies to achieve the next stage begin. She believes that the patient has to be encouraged to undertake activities in the sitting position as soon as possible. Newly generated correct movement patterns must be practiced to be learned, and practicing within the context of daily activities is believed to enhance the learning process. According to Brunnstrom, both proprioceptive and exteroceptive stimuli can be used therapeutically to evoke the desired motion or tonal changes. The stretching of muscles, use of traction, rhythmic repetitions, verbal commands, and cutaneous stimulation are some of the important therapeutic stimulations used in this approach, which is in addition to the use of associated reactions and tonic reflexes to facilitate movement. However, the use of such procedures is considered temporary, and such sensory stimulations, postural reactions, and verbal and manual support need to be withdrawn once the voluntary control of the movement develops.

2.3.2.1 Basic Limb Synergies

The movement produced by a normal subject is characterized by synergistic motor behavior and such normal synergistic movement patterns are almost limitless consisting of coupling together of muscles in an orderly fashion for producing purposeful movements with maximum precision and minimum waste of energy. Unlike normal subjects, stroke patients exhibit only a few stereotypic movement patterns, which do not permit different combinations of muscles, and the group of muscles constituting a BLS act together as a bound unit of motion. Consequently, the patients are unable to recruit the same muscles for different movement combinations and cannot master individual joint movements. BLS consists of either a gross flexor movement (flexor synergy) or a gross extensor movement (extensor synergy). There are four BLS, i.e., a flexor synergy and an extensor synergy for the upper limb and a flexor synergy and an extensor synergy for the lower limb.

These synergies are considered to be primitive, reflexive, automatic, and stereotypic among all hemiplegic patients and reflect the loss of inhibitory control normally exerted by higher centers. The BLS make their appearance during the early spastic period and can be evoked volitionally or reflexively. During the early stages of recovery, movements within the BLS are easier to achieve than non-synergistic movements. Those muscles that form the strong components of flexor or extensor synergy (Table 2.7) are considered to be dominant components, and almost all patients tend to initiate movement patterns using the dominant muscles.

The direction of joint movements of one specific BLS is always opposite to the direction of the antagonistic synergy. However, inversion is a common component for both flexor and extensor BLS of the lower limb. It is observed that the antagonist of the dominant component of one synergy is the weakest of the other synergy. The reciprocal inhibition is the fundamental neurophysiological mechanism underlying the weakness of certain components of BLS, i.e., when movements are attempted, stretching of the spastic strong antagonist causes reciprocal inhibition of the agonist.

When the BLS dominates in the hemiplegic patient, the quantity and the quality of movements depend on the relative strength of individual components of the synergy and how they interact. Even the typical arm posture in hemiplegia, i.e., the attitude of the upper limb at rest, results from the combination of the strongest components of flexor and extensor synergies, i.e., adduction and internal rotation of the shoulder with elbow flexion and forearm pronation. In hemiplegic patients, the flexor synergy is typically weaker among the two lower limb synergies, and the typical attitude of the lower limb is extensor in nature due to the domination of extensor synergy. In addition to the above, a strong extensor synergy makes the patient fail to achieve the hip abduction necessary for standing balance and ambulation.

Table 2.7 Features of upper and lower limb basic limb synergies

Extremity	BLS type	Components of BLS	Strong and weak components
Upper limb	Flexor synergy	• Retraction and/or elevation of the pectoral girdle. • External rotation of the shoulder. • Abduction of the shoulder to around 90°. • Flexion of the elbow to an acute angle. • Complete range supination of the forearm. • Wrist and hand position varies (Flexion of the wrist and the digits are more frequent).	Elbow flexors and shoulder elevators are strong components. Shoulder abductors and external rotators and supinators are the weak components.
Upper limb	Extensor synergy	• Protraction of the shoulder. • Internal rotation of the shoulder. • Adduction of the arm in front of the body. • Elbow extension to full range. • Complete range pronation of the forearm. • Wrist and hand position varies (Extension of the wrist with fist closure are more frequent).	Shoulder adductors and internal rotators are strong components. Elbow extensors are the weak component.
Lower limb	Flexor synergy	• Flexion of the hip. • Abduction and external rotation of the hip. • Flexion of the knee to about 90°. • Dorsiflexion and inversion of the foot. • Dorsiflexion of the toes.	Components of flexor synergy tend to be weaker than extensor synergy. Typically, flexor synergy is initiated by hip flexion.
Lower limb	Extensor synergy	• Extension of the hip. • Adduction and internal rotation of the hip. • Extension of the knee to complete range. • Plantar flexion and inversion of the foot. • Plantar flexion of the toes is frequent but is by no means universal.	Knee extensors are the strongest and usually, the patients tend to use them to initiate the extensor synergy. Even hip adductors and plantar flexors are strong components.

2.3.2.2 Evolution and Dissolution of the Nervous System

According to Hughling Jackson, a British neurologist, the phylogenetic organization of the nervous centers occurs at three levels and these three levels or groups of nervous centers are an integral part of the fully developed CNS of a normal subject. The three levels are labeled as lowest, middle, and higher motor centers. Here, the lowest motor center represents all muscles of the body, which produces few movement combinations that are most automatic in nature. The middle motor center re-represents all muscles of the body in more combinations that are more voluntary and less automatic. The higher motor center "re-re-represents" all the muscles of the body with unlimited combinations which are mostly voluntary.

Following certain pathologies like cerebrovascular accidents, the nervous system reverts to a lower level of evolution, which Jackson termed as "dissolution" of the nervous system or "evolution in reverse". Patients with severe CNS involvement must rely on the lowest motor centers, which provide few movement combinations that are automatic in nature. Less severely involved patients may recover sufficiently to utilize the middle motor centers so that more movement combinations become available. However, full motor recovery needs the normal functioning of the middle motor centers with the least involvement of the highest center.

2.3.2.3 Postural Reflexes

Postural or attitudinal reflexes such as the tonic neck, tonic labyrinthine, and tonic lumbar reflexes get integrated as the nervous system matures. Neck movements evoke the tonic neck reflexes and the reflexes are either symmetrical or asymmetrical. The tonal variation produced by the symmetrical neck reflex is identical on both the left and right sides of the body, i.e., forward flexion of the neck results in flexion of both upper limbs and extension of both lower limbs. Whereas for asymmetrical neck reflex, the tonal variations are opposite on the left and the right limbs, i.e., when the head is rotated to one side, increased extensor tone or extensor movement pattern is produced on the face side limbs, and less extensor tone or flexor movement pattern is elicited on the skull side limbs.

A change in the position of the head in space acts as the stimulus for eliciting tonic labyrinthine reflex. In normal animal models, quadruped posture is associated with minimal extensor tone, and while lying supine, the limbs have maximal extensor tone. Whereas tonic lumbar reflexes are elicited by changes in the position of the upper part of the body with respect to the pelvis. Forward and backward bending, rotation, and side flexion of the upper body with respect to the pelvis are associated with tonal variations in the limbs. For example, rotation of the chest to the right is found to facilitate the flexion of the right upper limb and extension of the right lower limb. The effects seen on the left side are opposite to those on the right side. These attitudinal reflexes become easily demonstrable in the presence of certain CNS pathologies, including stroke, and such released reflexes in a patient may make the voluntary movements difficult or impossible. For instance, in the prone position, the inhibitory effect of the tonic labyrinthine reflex can make the stroke patient's effort to carry out elbow extension impossible. Whereas, in the supine, the facilitating effect of the tonic labyrinthine reflex may abet the stroke patient to perform complete range of elbow extension.

Associated Reactions

Reflex tensing of muscles or involuntary movements known as associated reactions is a frequent observatory finding in hemiparetic patients. In many of these patients, forceful voluntary movements of some other body parts readily elicit such reactions on the affected side. The reflex tensing or movement usually appears within a few seconds and tends to occur in several of the joints of the affected limb or limbs. The muscular tension associated with these reactions may continue even after the stimulus that evoked the reaction ceases. George Riddoch, a Scottish neurologist, and Edward Farquhar Buzzard, a British physician, in 1921, defined it as "automatic activities which fix or alter the posture of a part or parts when some other part of the body is brought into action by either voluntary effort or reflex stimulation." These reactions are more readily elicitable when the limb is essentially spastic but can be occasionally seen when the limbs are flaccid. Francis Walshe, an English neurologist, considers these associated reactions as released postural reactions deprived of voluntary control. These associated reactions are usually flexor or extensor in nature. Even yawning, sneezing, and coughing can make these reactions apparent. The attitudinal reflexes are believed to influence the outcome of associated reactions.

According to Brunnstrom, in the upper limb, the reactions tend to be of the same type as the movement which evoked the responses, i.e., flexion tends to evoke flexion and extension tends to evoke extension. Whereas in the lower limb, flexion tends to evoke extension and extension tends to evoke flexion. In addition to the above, a mutual dependency appears to exist between the synergies of the affected side limbs, which she referred to as 'homolateral limb synkinesis'. In homolateral limb synkinesis, the voluntary flexion of the affected limb tends to evoke or facilitate flexion of the other affected limb and extension tends to evoke or facilitate extension of the other.

Raimiste J., a French neurologist, had reported many associated reactions that were different from those reported by others. He found that strong resistance to the non-involved side hip adductors moved the affected side hip into adduction in the supine position and resistance to non-involved hip abductors elicited abduction movement of the affected hip, which was later termed as Raimiste's hip adduction and abduction phenomenon, respectively. Raimiste also found that adduction appears readily as compared to abduction, which could be due to the dominance of the extensor synergy. A reaction alike Raimiste's phenomenon was observed by Brunnstrom in the upper extremity of the patients with hemiplegia (shoulder adduction and abduction phenomenon). According to Raimiste, such a tendency to move both legs simultaneously is present even in normal subjects. However, the normal subject can easily prevent the movement response if asked to control it, which is difficult or impossible for hemiplegic patients.

2.3.2.4 Hand Reactions and Restoration of Hand Functions

In humans, the evolution of the grasping function of the hand may be traced back to the simple stretch reflex, which is a spinal cord reaction. According to Denny-Brown B. (1956) and Thomas Evans Twitchell (1958), normal maturation and activities of the higher centers on the lower centers of the CNS transform such reactions into more and more elaborate mechanisms until adult human hand functions evolve. Neurophysiologically, the restoration of hand function in stroke patients follows a sequence. In hemiplegic patients, Twitchell identified seven steps in the restoration of the true grasp reflex, which is essential for normal hand functions and proceeds in a manner that closely resembles the evolutionary one. Table 2.8 gives details about the seven steps of attaining the true grasp reflex.

According to Twitchell, the recovery can cease before the last stage is reached and those patients who exhibit the true grasp reflex could progress to full recovery. First to recover is the return of stretch reflex, a monosegmental reflex followed by the appearance of proprioceptive traction response, a plurisegmental response involving the higher centers. Later, the true grasp reflex, a subcortical response appears, followed by the instinctive grasp reaction (Box 2.3) and the instinctive avoidance reaction (Box 2.4) before the total restoration of the hand function.

2.3.3 *Brunnstrom's Recovery Stages*

Following the observation of many hemiplegic patients, Brunnstrom noticed an almost stereotyped sequence of events during recovery post-stroke. Based on this, she enumerated the seven stages of recovery, which are presented in Table 2.9. Depending on the severity of the insult and the degree of sensory involvement, recovery can arrest at any stage of recovery. She believes that the stages in the recovery cannot be skipped but, in some cases, especially when the insult is minor, the recovery may proceed rapidly with no distinct observable stages. She also believes that the stages of recovery post-stroke resemble normal infantile motor development.

2.3.4 *Principles of Evaluation*

Brunnstrom advocates that the purpose of the evaluation is to note the degree of recovery following the CNS insult and that the evaluation procedure should be based on the recovery stages of the hemiplegic patients. The evaluation procedure should be brief and easy to administer, thus minimizing the possibility of overly fatiguing the patient and wasting the time available for treatment. Traditional neurological examination procedures like manual muscle testing for muscle strength evaluation do not give any information about the stage of recovery and real muscle strength. The use of traditional manual muscle testing will create errors in grading

Table 2.8 Details of the seven steps in the development of true grasp reflex

Stage	Features
Stage I	Tendon reflexes return and become hyperactive
Stage II	Characterized by the development of spasticity in the long flexors; hypertonicity felt as resistance felt on passive motions
Stage III	Proprioceptive stimuli facilitate voluntary finger flexion of the affected hand
Stage IV	Characterized by elicitation of proprioceptive traction response, i.e., a stretch of the flexor muscles of any one joint of the limb evokes the contraction of the flexor muscles of all the joints; righting reactions have either facilitating or inhibiting effect on this response
Stage V	Control of hand movements possible without proprioceptive stimuli
Stage VI	In this stage, grasp is greatly reinforced by tactile stimuli in the palm of the hand; also characterized by decline of spasticity.
Stage VII	The true grasp reflex is elicited by a distally moving deep pressure over the reflexogenic areas of the palm and digits (see fig). This reflex has two phases, a catching phase and a holding phase. The catching phase consists of weak contraction of flexors and adductors, and the holding phase appears only if traction is applied on the tendons; the holding phase tends to continue until the traction is maintained; this stage is characterized by a further decline of spasticity.

Box 2.3 Features of Instinctive Grasp Reaction

Instinctive Grasp Reaction

The stimuli needed for this reaction is a stationary contact of any solid object with the palm of the hand. The response is an involuntary fist closure with the patient unable to release the object. The patient may not have any difficulty opening or closing the hand in the absence of the object in hand. This reaction is seen at birth but subsequently disappears as the brain matures. Certain brain pathologies affecting the frontal cerebral lobe like stroke cause the reappearance of the same.

Box 2.4 Features of Instinctive Avoidance Reaction

Instinctive Avoiding Reaction

This reaction is seen in patients with parietal lobe lesions. It can be illustrated in such patients when the arm is held elevated in a forward-upward direction during which fingers and thumb tend to be held in extension. The stimuli required to produce this response is the stroking of a solid object on the palmar surface towards the distal direction. The response is an exaggeration of the hand posture. Neurophysiologically, both instinctive grasp and instinctive avoiding reactions are considered as transcortical release, i.e., pathology can disturb the balance existing between antagonistic reflex systems within CNS and result in overactivity of the intact mechanism.

Table 2.9 Motor features during the seven stages of recovery following stroke

Recovery stage	Features
Stage I	Seen soon after the acute episode; affected limbs are essentially flaccid; no voluntary or reflexive movement possible
Stage II	As recovery begins, the BLS or some of their components appear as associated reaction or with considerable voluntary effort; spasticity begins to develop particularly in the dominant components
Stage III	The patient gains voluntary control on BLS; a full range of all synergy components not mandatory; spasticity reaches its peak; considered as the semi-voluntary stage
Stage IV	Some movement combinations that do not follow the paths of BLS are mastered, first with difficulty and then with ease; spasticity begins to decline, but its influence on non-synergistic movements readily observable
Stage V	More difficult movement combinations away from the paths of BLS are mastered; the dominance of BLS over motor control reducing; spasticity continues to decline
Stage VI	Individual joint movements possible; coordination reaching normalcy; absence of spasticity
Stage VII	Normal motor functions restored

muscle strength, especially when there is inability or difficulty to perform individual joint movements as a result of spasticity or due to the influence of released attitudinal reflexes on tone and movement.

Brunnstrom had developed and tried several evaluation forms and found many of them unsatisfactory or cumbersome. In many of these evaluation forms, a large number of movement combinations and functional activities were tested. Later, she revised the form and produced a simplified one that could be administrated at ease and had only minimal test items which helped to identify the level of motor control and the stage of recovery. In many of the patients, the recovery may progress gradually, and there may be instances when the patient is in the transition phase between stages of recovery. For instance, if the patient begins to perform motions of the next stage, the staging will be like 2 going to 3 or 3 going to 4. Similarly, in many instances, the stage of recovery for the upper limb and lower extremities may not be alike.

Brunnstrom believes that the patient should be physically and mentally comfortable prior to evaluation. When the patient is evaluated, no facilitation techniques are used, and no movements beyond his capabilities are demanded. The evaluation of the upper limb should be performed in sitting. It consists of tests for passive motion sense of the joints and the level of motor control, i.e., whether the limb is flaccid or exhibiting synergy or movement deviating from BLS. Table 2.10 gives the details regarding the test items (movement deviations possible) for stage 4 to stage 6 of the upper limb and the lower limb. According to her, the sensory evaluation should precede the motor evaluation. Clear and simple instructions should be given in functional terms. Each motion has to be demonstrated to the patient before the motor evaluation.

She advised speed test for movement if the patient is in stage 4 and beyond, i.e., instructing the patient to move the hand from lap to chin or lap to opposite knee in

Table 2.10 Test items for stage 4 to stage 6 of upper and lower extremities

Extremity	Stage	Test items
Upper extremity	Stage IV	1. Hand to lower back region or placing the hand behind the body. 2. Elevation of the arm to a forward-horizontal position. 3. Pronation and supination of the forearm with the elbow at 90° flexion.
	Stage V	1. Abduction of the arm to 90°. 2. Elevation of the arm to a forward and overhead position. 3. Pronation and supination of the forearm with the elbow extended with arm forward or side-horizontal position.
	Stage VI	1. Able to perform individual joint movements with the absence of spasticity and presence of near-normal coordination.
Lower extremity	Stage IV	1. Ability to flex the knee beyond 90° with the foot sliding backward on the floor while the patient is sitting. 2. Voluntary dorsiflexion of the ankle without lifting the foot off the floor in sitting.
	Stage V	1. Isolated non-weight-bearing knee flexion with the hip in extension or near extension in standing. 2. Isolated dorsiflexion of the ankle with the knee extended and the heel kept forward as if for heel strike.
	Stage VI	1. Able to perform hip abduction (excluding the elevation of the pelvis) in standing. 2. The ability to perform ankle inversion and eversion while sitting.

5 seconds. Speed test provides information about the extent to which spasticity is interfering with the reversal of direction of movement. Upper limb evaluation also consists of tests for fingertip recognition, wrist stabilization for grasp with the elbow flexed and extended, and wrist flexion, extension, and circumduction in elbow flexed and extended attitude. For digits, hand functions such as mass grasp, mass extension, hook grasp, cylindrical grasp, prehension, and individual finger movements, and skilled activities like the ability to button and unbutton the shirt need to be tested. Table 2.11 gives details about the hand control during various stages of recovery.

Evaluation of the trunk and lower limb is performed with the patient sitting on a chair and in standing. Even for the trunk and lower limb, passive motion sense, sole sensation, trunk balance, and the level of motor control have to be tested. Trunk balance is tested both in sitting with no back support and in standing. Type of ambulation, i.e., whether walks with or without a brace or cane or support or escort, needs to be assessed. Ankle, knee, and hip joints attitude in stance phase and swing phase, arm swing, and cadence are the components analyzed under gait assessment.

2.3.5 Training Procedures

The therapist must have adequate confidence in his or her knowledge, skills, and judgment. For effective treatment, it is essential to have a constructive patient-therapist relationship. Both for evaluation and treatment, if the patient lacks

Table 2.11 Details regarding the hand control during various stages of motor recovery

Stage	Hand functions
Stage I	Forearm and hand muscles are essentially flaccidity
Stage II	Little or no active finger flexion possible; muscle tone developing within the long flexors
Stage III	Mass grasp and or hook grasp possible; but no release; no voluntary finger extension possible
Stage IV	Lateral prehension possible; release by thumb movement; able to perform small range of semi-voluntary finger extension
Stage V	Palmar prehension possible and may awkwardly perform cylindrical and spherical grasp with limited functional use; variable ranges of voluntary mass extension of digits possible
Stage VI	All prehensile movements under control; skills improving; full-range voluntary extension of digits possible; individual finger movements present; accuracy less than the unaffected side
Stage VII	Normal motor functions restored

confidence in the therapist, he may feel ill at ease and will be reluctant to make an effort required. During training sessions, only those tasks that the patient can master or almost master should be demanded, as repeated failures may lead to frustration and disappointment, which may abate his motivation.

2.3.5.1 Recommended Bed Posture and Bed Mobility Exercises

When the flaccid condition prevails, the limbs should be positioned in the most favorable way without any interference from the spastic muscles. In supine, if the affected lower limb tends to extend with the hip in abduction and external rotation, placement of a small pillow under the knee with lateral support extending from the hip to the knee can prevent hip abduction and external rotation. No clothing should be in contact with the foot, as it can further accentuate the extensor synergy predominating in the lower extremity. The affected upper extremity should be supported on a pillow in a position that is comfortable for the patient. At the same time, to avoid inferior subluxation of the glenohumeral joint, attention has to be given not to keep the humerus in undue abduction with respect to the scapula. While handling the patient, traction on the affected arm has to be avoided and the patient should be instructed to support his affected arm with the normal hand while moving around the bed.

When the flaccidity predominates, passive motions of the extremity are first carried out and then developed into active assisted movements. Progressively the program should expand to include head, neck, and trunk movements. The patient should be taught to move around in bed, sit up and turn to side-lying while protecting the affected arm. Turning towards the affected side needs little or no activity of the paretic side. Turning can be promoted with the affected arm kept close to the body, thus rolling over his affected arm. Nevertheless, if the patient complains of pain in the shoulder, then the patient can be taught to roll to the affected side with the affected arm held at the wrist by the sound hand. For many patients, turning to the sound side is more difficult as it requires the active participation of the paretic side. Turning to the sound side should be performed by bringing the affected lower extremity into partial flexion with momentary stabilization of the position by the therapist. Then the patient should attempt to turn over by swinging his arms and the affected knee across the body (Fig. 2.26). Later, as the control improves, it should be performed as a continuous movement from the supine to the side-lying with no assistance.

2.3.5.2 Exercises for Gaining Trunk and Neck Control

As early as possible, the treatment should incorporate trunk and neck exercises in the sitting position. Exercises in sitting aim to improve trunk balance and trunk control, promote voluntary control of affected upper extremity for functional activities, and provide valuable face-to-face interaction, especially when the patient is aphasic. Adequate head, neck, and trunk flexibility is a prerequisite for any trunk balance activity. In sitting, trunk rotations can be performed by placing arms close to the body with the affected limb's elbow supported using the sound hand. Figure 2.27 illustrates head mobility in sitting with affected limb supporting using the sound hand. Initially, trunk rotation should be performed gently within a small range and then the range should be gradually increased. Throughout the movement, the patient should look straight ahead in order to provide a certain amount of neck mobilization, which accompanies the trunk rotation.

Many of the patients may tend to list or lean towards the affected side and may even fall, especially when the assistance in sitting is withdrawn. The listing phenomenon can be as a result of perceptual deficits or due to a lack of activity of the affected side trunk muscles to check the movement when the center of gravity (COG) shifts to the sound side. It is essential to overcome listing, as it promotes uneven weight-bearing in sitting and decreases the likelihood of appropriate standing balance and weight-bearing in the upright position. Evoking balance responses by deliberately disturbing the COG by "pushing the patient" in different directions, particularly in

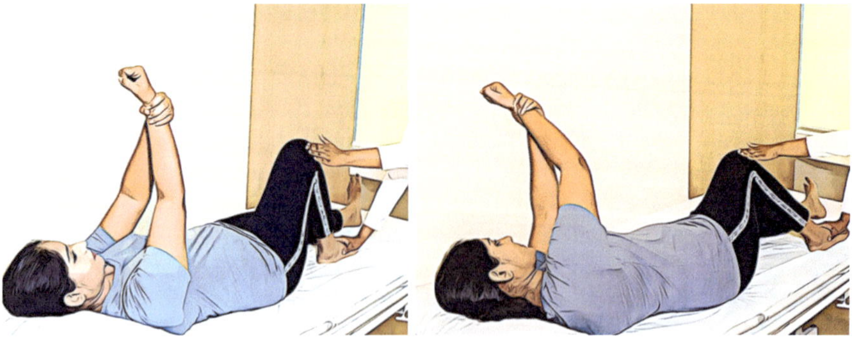

Animated photograph of model and therapist with permission

Fig. 2.26 The therapist momentarily stabilizes the affected lower extremity in partial flexion to encourage rolling towards the sound side

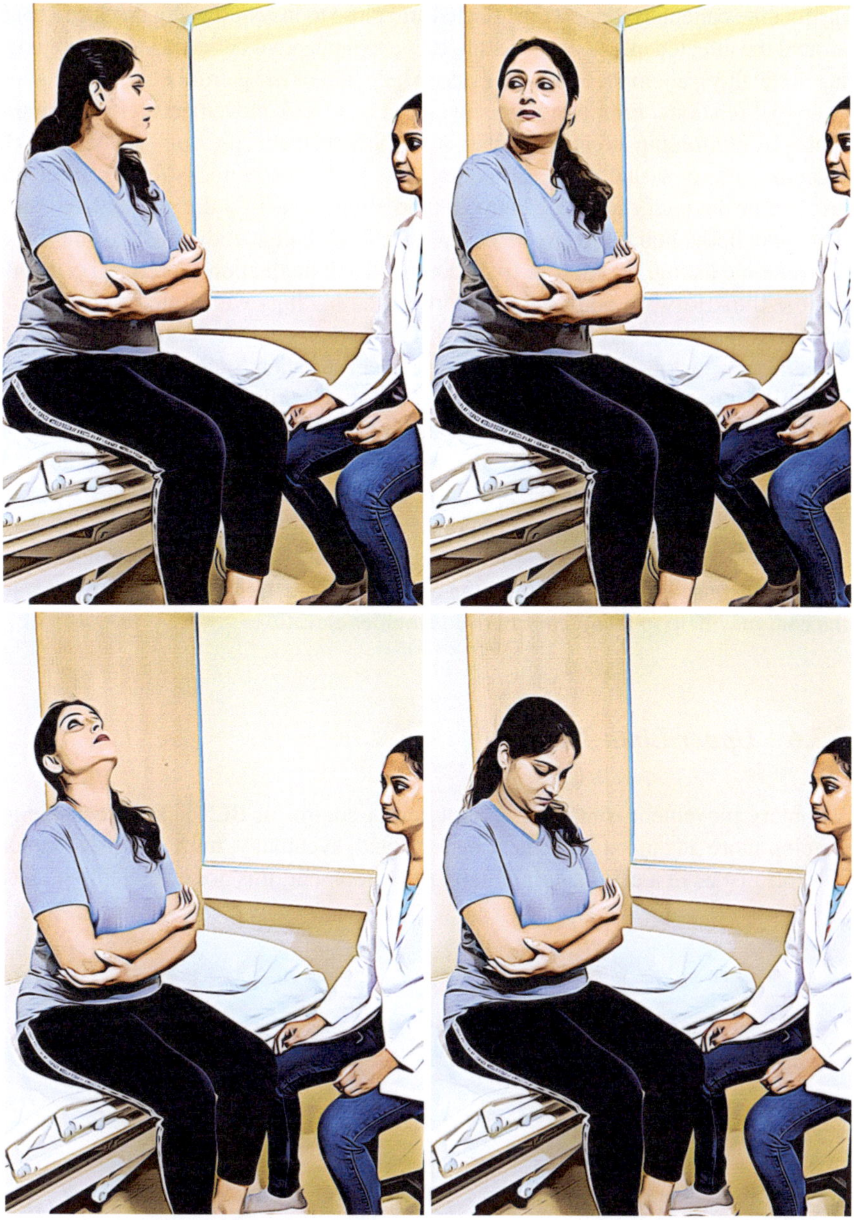

Animated photograph of model and therapist with permission

Fig. 2.27 Encouraging head mobility in sitting with the sound hand supporting the affected limb

the direction towards the side of the listing, helps to minimize the listing tendency and promotes trunk symmetry. While promoting balance responses, the patient should avoid grabbing any chairs or similar objects using the sound hand and instead instruct the patient to support the affected arm close to his chest using the sound hand around the affected elbow. If the listing is due to perceptual deficiency (error in judging verticality), attempts should be made to help him or her gain a better appreciation of spatial relations. Promoting repetitive head and trunk movements, enhancing the senses by reinforcing afferent impulses from the kinesthetic, pressure, and light touch receptors, use of visual clues, temporary raise under the buttocks (either under the affected or the unaffected side buttock), and a light touch by the normal hand on a vertical or horizontal stationary object are specific techniques advised by Brunnstrom to overcome listing. To improve trunk control, trunk inclination exercises, i.e., bending the trunk forward and obliquely forward on his hips with little or no forward flexion of the spine, can be incorporated. The therapist may have to guide the trunk and arm movements by holding her hands under the patient's elbow (Fig. 2.28).

While sitting, head and neck movements may be utilized to facilitate shoulder girdle movements over which patients may have little or no control. For instance, if the patient lacks elevation of the affected side scapula, resistance to lateral head motion towards the affected side may promote the elevation of the scapula. Emphasizing isometric or eccentric muscle contraction by giving commands like "hold" or "do not let me pull your head away from your shoulder," followed by concentric muscle contraction by using commands like "get your shoulder closer to the ear" may help to attain control over shoulder elevation.

2.3.6 *Upper Limb Training*

Voluntary movement control or at least partial control of BLS is a prerequisite to develop more advanced movement patterns and, eventually, more extensive motor control. Just as in a child's normal motor development, it is not essential to perfect

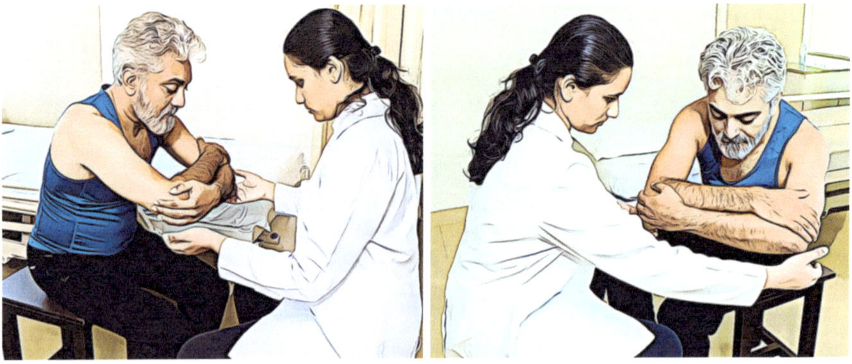

Animated photograph of model and therapist with permission

Fig. 2.28 An illustration of the trunk bending forward and obliquely forward on hips with little or no forward flexion of the spine. Note that the therapist is guiding movements by holding her hands under the subject's elbow the activities before attempts are made for more advanced stages of motor developmental sequences. For instance, in hemiplegic patients, it is not necessary to wait for the completion of the entire range of motion of all the components of the BLS before variations are attempted. For patients in the flaccid stage, to initiate BLS, associated reactions or attitudinal reflexes can be utilized. Such reactions or reflexes enable background tension in the affected limbs and then to gain voluntary control over the BLS, volitional effort should be superimposed on the reflexly produced background tension. Once the patient can initiate movement on a voluntary basis, reflex assistance has to be withdrawn, and movement patterns that deviate from synergy have to be introduced.

According to Brunnstrom, therapy becomes more meaningful only if the patient utilizes whatever control he has gained over BLS or its components for functional activities. The BLS or its components can be used for stabilizing objects like an envelope or a worksheet or carrying some objects like a bag or coat or holding small objects like a toothbrush, while the sound hand is used for manipulating the object. Though the BLS can be utilized for many household activities, there are controversies regarding its long-term usage. For instance, the possibility of one-armedness or neglect of the affected arm can be minimized by early usage of BLS of the affected extremity. On the contrary, the long-term use of BLS or its components for functional activities may promote an excessive reliance on the BLS, which may reinforce the components of BLS and hinder the potential for further recovery. Considering the abovementioned merits and demerits and based on the therapist's knowledge of the individual patient, a careful and judicious manner of selection and usage of BLS for functional activities is advisable.

If a rapid passive range of motion exercise evokes considerable spasticity in the dominant components and hinders further recovery process, efforts must be taken to modulate and use very gentle passive movements to facilitate a normal tone. Initially, shoulder girdle movements such as elevation, depression, retraction, and protraction can be stressed. Later, shoulder flexion movements have to be performed bilaterally and then unilaterally. If the patient is unable to perform shoulder movements, then the therapist will have to assist or guide the movement and simultaneously might use percussion or cutaneous stimulation over the muscles to activate them.

2.3.6.1 Shoulder Pain and Subluxation

For normal glenohumeral rhythm, the therapist may have to guide the patient's scapula towards upward rotation when shoulder movements are attempted. When passively mobilizing the shoulder, if the patient complains of pain secondary to stretching of spastic muscles, then mobilization of the shoulder joint without forceful stretching of tense muscles should be attempted. The sheer anticipation of pain can accentuate the tension within the spastic muscles and aggravate the pain on passive movement. For such patients, attempts have to be made to guide the movement of the arm with respect to the trunk without the patient sensing the movement of the arms (Fig. 2.29). Forward inclination of the trunk and trunk rotations with the

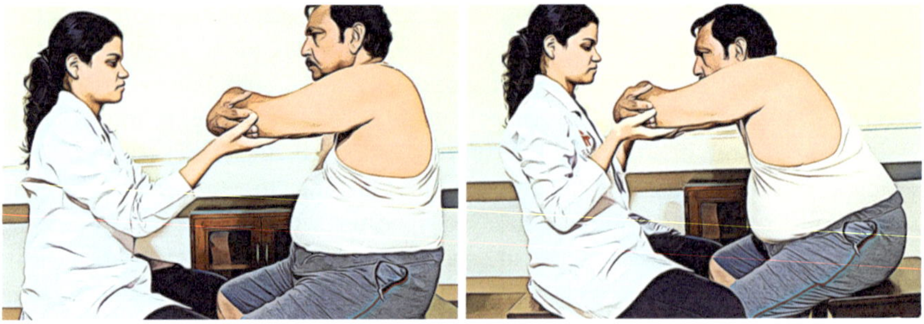

Animated photograph of model and therapist with permission

Fig. 2.29 Forward inclination of the trunk upon the hips, with the therapist supporting the elbows of the patient to promote shoulder flexion

therapist supporting the elbows of the patient promote shoulder flexion and abduction and adduction, respectively. Such trunk rotation exercises also help to utilize the tonic reflexes to alternatively increase and reduce the tension in the pectoralis major muscle enabling a larger range of shoulder abduction without pain. Once there is no pain on passive movement, standard active assisted movements of the arm can be started.

Promoting activities for the components of flexor synergy, especially when resistance is applied, has an inhibitory effect on the pectoralis major muscle, which is essential for a pain-free shoulder movement. As the training progresses, the emphasis is also given in developing movement control over extensor synergy, and in due course, both the synergies have to be sequenced in a "roundabout" fashion for producing painless shoulder movements.

Activation of muscles around the glenohumeral joint is not only needed for the normal functioning of the upper limb but also to guard against joint subluxation. Normal tone and normal activity of the rotator cuff muscles, especially supraspinatus, play an important role in preventing glenohumeral separation. Flexor synergy can be used to encourage background tension in the supraspinatus and other external rotators (Fig. 2.30). Later, a more isolated contraction of the external rotators can be emphasized.

2.3.6.2 Techniques for Improving Motor Control

- Use the dominant component of synergy to strengthen the weaker ones. For example, pronation of forearm or "waist squeezing" (Fig. 2.31) can precede the patient's effort for elbow extension.
- Movements are guided in the beginning before active-assisted, active, and resisted exercises are given. For example, while performing bilateral "classic rowing exercise" (Fig. 2.32), the therapist can guide the movement and later add resistance to the affected side. In bilateral rowing exercise, the pull phase is initiated and executed using the flexor synergy and the push phase using the extensor synergy of the affected extremity.

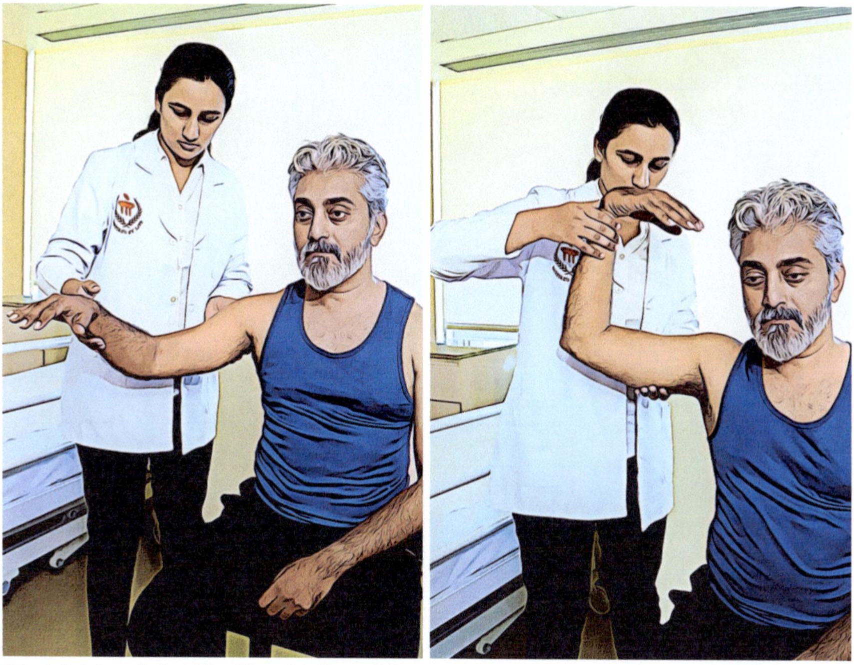

Animated photograph of model and therapist with permission

Fig. 2.30 Use of flexor synergy to encourage background tension in the supraspinatus and other external rotators

Fig. 2.31 An illustration of the "waist squeezing" exercise

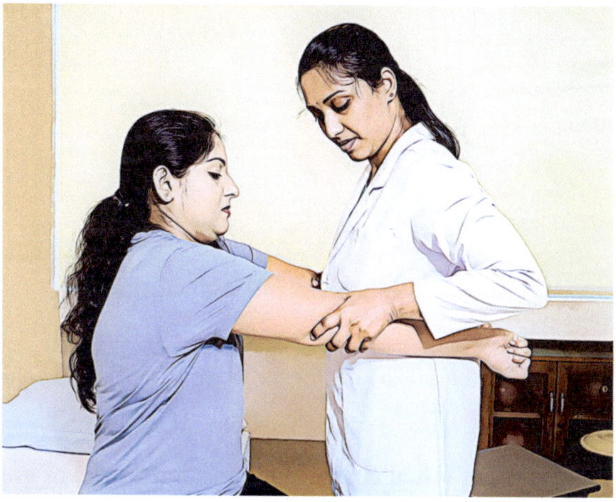

Animated photograph of model and therapist with permission

- Local stimulation (proprioceptive and exteroceptive) can also facilitate muscle contraction. For example, vigorous to and fro stroking of the skin over the weak muscles and use of proprioceptive or proximal traction response may facilitate muscle contraction. These facilitatory techniques may have a cumulative effect if they are repeatedly applied.

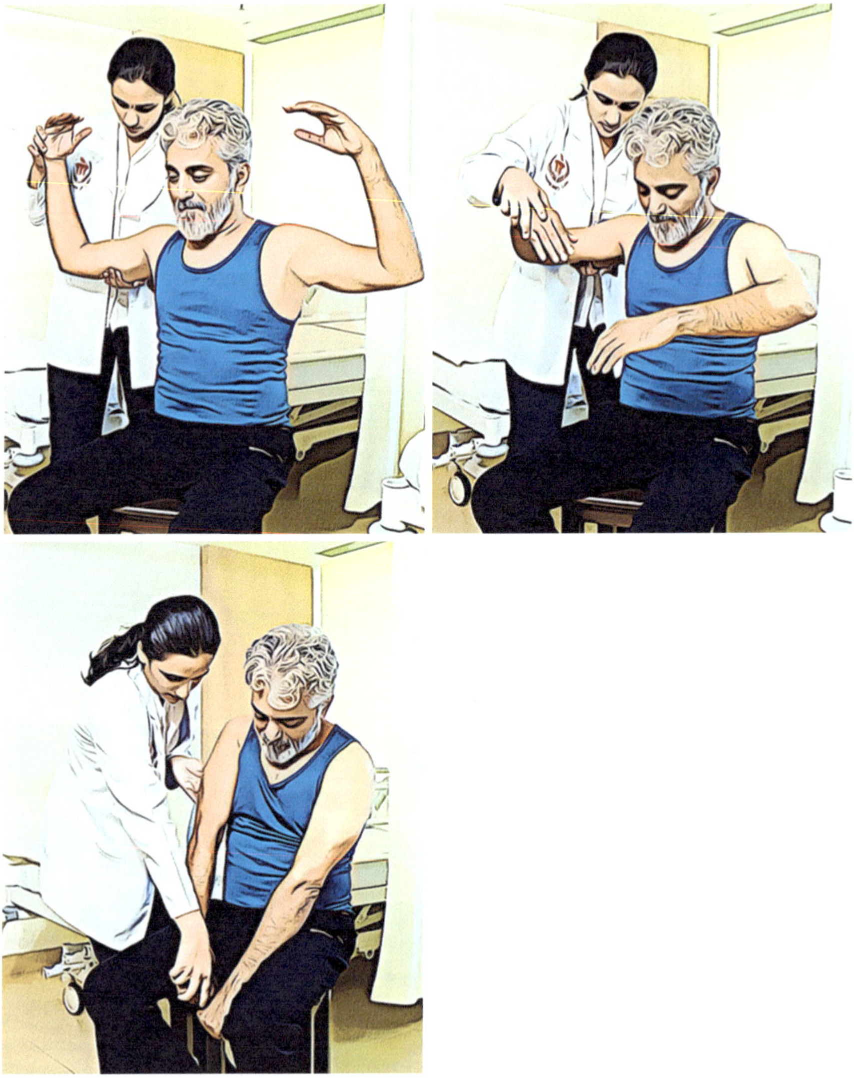

Animated photograph of model and therapist with permission

Fig. 2.32 An illustration of how the therapist should assist the subject to perform the bilateral "classic rowing exercise"

- The use of visual guidance is considered to be of greater benefit than the facilitatory effect of tonic neck reflex.
- In the early stage, in supine and seated positions, the use of associated reactions can initiate components of the synergy.
- The use of postural reactions such as the tonic neck, tonic labyrinthine, and tonic lumbar reflexes can promote background tension in the weak muscles.

- Isometric and lengthening contractions can be attempted before shortening contractions. "Hold after positioning" the affected limb or part of the limb promotes isometric contractions (Fig. 2.33).
- To gain better motor control, the reversal of the direction of the movement can be emphasized. For instance, the "push and pull exercise" in sitting can strengthen weak extensor synergy and promote better control of flexor and extensor synergy.
- Weight-bearing through the affected arm, in sitting or other functional positions promotes background activity in the weak synergic muscles.
- Once the patient gains certain control over the movement, the therapist's assistance and use of facilitatory and postural reflexes should be withdrawn.

2.3.6.3 Serratus Anterior Training

In hemiplegic patients, the flexor synergy may enable them to bring their shoulder joint to 90° abduction. Many times, on visual inspection and palpation, the examiner may identify the strong contraction of trapezius and deltoid muscles with simultaneous winging of scapula, suggesting the idle nature of the serratus anterior. Excessive spasticity in the pectoralis major muscle and the painful range of motion at the shoulder for raising the affected arm overhead are a few of the major factors contributing to the unsatisfactory functioning of the serratus anterior. To encourage the serratus anterior activity, a certain amount of control of muscles belonging to both synergies of the upper extremity is a pre-request. However, prior to facilitating

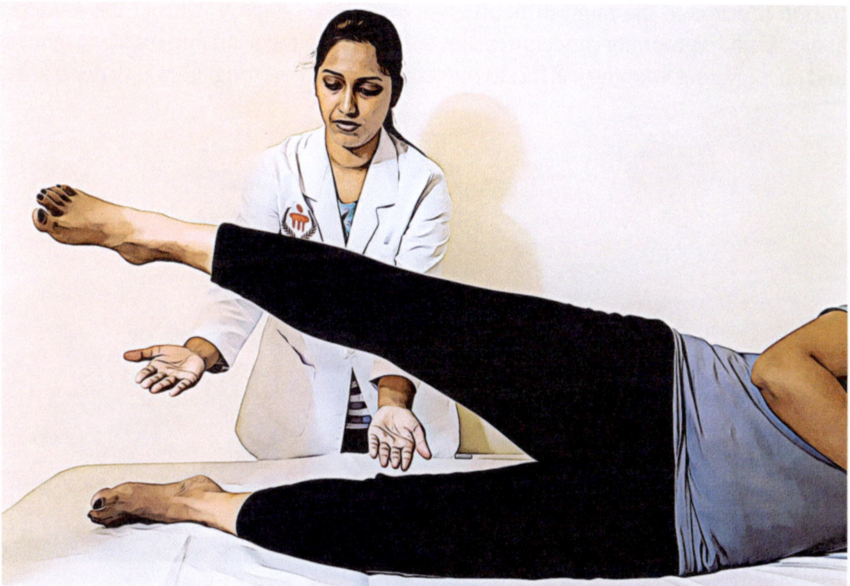

Animated photograph of model and therapist with permission

Fig. 2.33 An illustration of "hold after positioning" to promote isometric contractions contraction of the serratus anterior, mobilization of the scapula to obtain a painless range of motion at the shoulder is imperative.

Facilitation of the serratus may begin with the patient in the sitting position with the arm supported in a forward-horizontal direction. Then, the patient should be requested to reach forward, which may be assisted, and the therapist should briskly push the affected arm in the backward direction to evoke a stretch reflex in the serratus muscle (Fig. 2.34). Commands like "don't let me push your arm back" should be simultaneously given to enable the isometric and later eccentric activity of the muscle. If the pectoralis major muscle overacts, the affected arm should be brought into a more abducted position and the facilitation technique should be repeated. Once the patient can perform isometric and eccentric muscle contractions, concentric exercises as "arm to ear" (Fig. 2.35), "arm to the opposite shoulder," and eventually "arm above the head" should be taught. As the patient gains more voluntary control, "reversal of movement" should be introduced and the guidance, assistance, and resistance offered by the therapist for the movement synergies and their components should be gradually withdrawn.

2.3.6.4 Training Program for IV and V Recovery Stages

A certain level of voluntary movement control of both basic synergies is a prerequisite for training the upper limb for recovery stage IV and V. In recovery stage IV, the effectiveness of the voluntary effort is on the increase. It enables the patient to learn comparatively easy movement combinations that deviate from BLS, thus helping him to proceed to the more difficult combinations of stage V. During the recovery stage IV and V, training procedures aim at modifying the available motor responses and encouraging voluntary effort to overcome the still existing, although decreasing

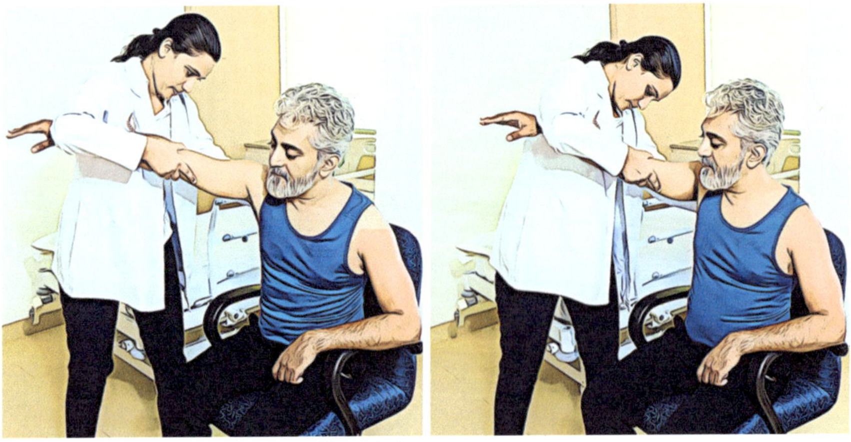

Animated photograph of model and therapist with permission

Fig. 2.34 The therapist briskly pushing the affected arm in the backward direction to evoke a stretch reflex in the serratus muscle

Fig. 2.35 An illustration of performing "arm to ear" isometric muscle contraction to strengthen the serratus anterior

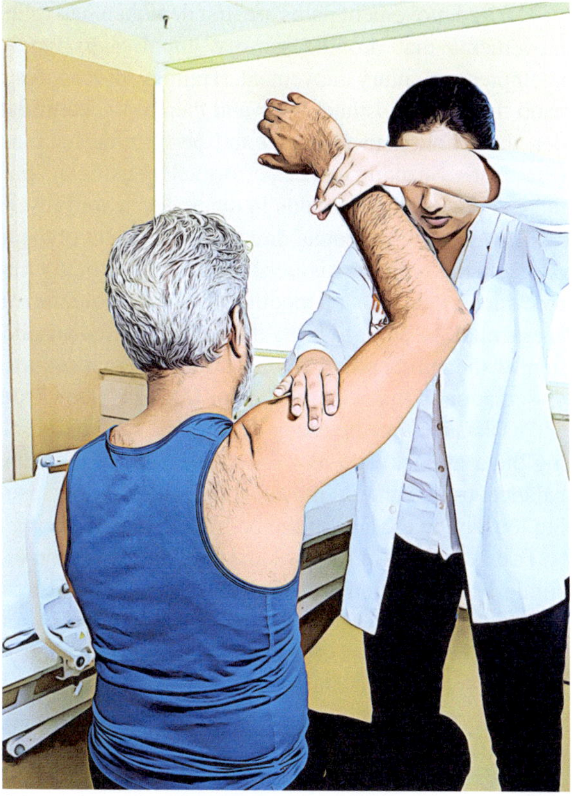

Animated photograph of model and therapist with permission

linkage between the components of BLS. Therefore, the training program should introduce movement combinations that deviate from the BLS paths to break up the strong linkage existing between certain components of BLS. During this phase of training, it is also essential to activate those muscles which do not belong to either of the BLS.

If patients have difficulty initiating movements away from the path of either synergy, functional activities that resemble the BLS can be used as variations in the movement direction. For instance, variation of the "classic rowing" pattern (explained earlier in this chapter) incorporating shoulder adduction during the pull phase and shoulder abduction during the push phase can be used as variations in the movement pattern. Similarly, modification of either synergy is essential to bring the hand to mouth (Usually, the patient brings his hand to the mouth with simultaneous abduction of the arm and the head turns to meet the hand), opposite shoulder, forehead, and top and back part of the head. This can be achieved by activating the pectoralis major muscle during elbow flexion by pressing the elbow firmly against the side of the body, inhibiting the abduction of the shoulder that appears with flexor synergy. If the elbow flexors are weak to bring the hand to the mouth, utilize facilitatory techniques such as proprioceptive and exteroceptive stimulation and hold after positioning technique.

Desired movement paths are first demonstrated to the patient by repeated passive movements that provide sensory impulses to facilitate better control over the attempted voluntary movement. Training procedures should include bringing the hand first to the ipsilateral ear and then to the contralateral ear, the opposite shoulder, the elbow, and the hand and performing back and forth stroking movement (Fig. 2.36). Stroking can be performed by using the palm of the hand, provided that there is no excessive tension in the finger flexor muscles or else with a closed fist.

The forward-horizontal direction movement of the arm with the elbow in flexion or extension requires dissociation of the pectoralis major from the triceps muscle, for which a step-by-step modification of the voluntarily performed extensor synergy is essential. In order to bring the entire affected arm behind the body, the participation of the subscapularis, which is an internal rotator of the shoulder, along with dissociation of the pectoralis major from the elbow extensors, is required. Guidance and assistance may be required for some time before the patient succeeds in bringing the arm backward as a free movement. If the patient has satisfactory standing balance, trunk rotation movements, gradually increasing in range, with arms hanging loosely by the side of the body, also helps to bring the arm behind the body.

Bringing the arm forward to a horizontal position or side-horizontal position and similarly, performing pronation and supination of forearm with the elbow in flexion or extension requires adequate movement control over the upper extremity muscles. Neither the flexor nor the extensor BLS provides the exact movement combination required for the abovementioned. Guidance and assistance, promoting isometric or eccentric muscle activity prior to concentric activity and reversal of movement directions, play a vital role in gaining control over movement combinations away from the path of either synergy. Once the patient moves to recovery stage IV or V, functional activities that are possible with those new muscle combinations away from BLS have to be stressed. Successful performance of functional activities like bringing a piece of bread kept in the affected hand to the mouth or trying to comb the hair with the affected hand gives a boost to the patient's motivation and improves the coordination and control.

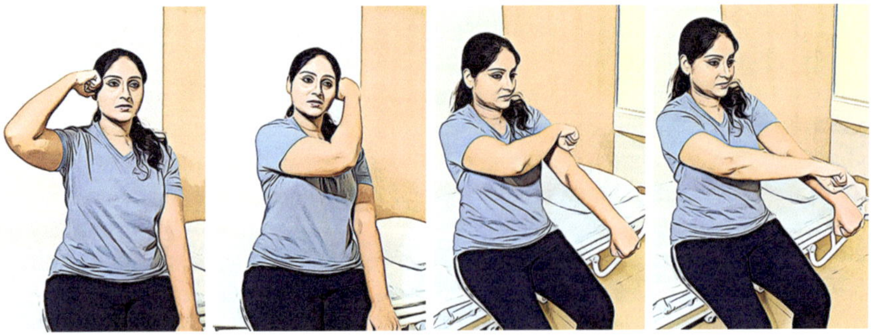

Animated photograph of model with permission

Fig. 2.36 Encouraging the subject to bring the hand first to the ipsilateral ear and then to the contralateral ear, the opposite shoulder, the elbow, and the hand

2.3.6.5 Training Program for Stage VI

Those patients who have rapid recovery following stroke may be expected to achieve this stage of recovery. At this stage, the influence of synergies is least on the movement combinations. Lack of resistance to passive movements can be felt to the movements indicating the disappearance of spasticity. Individual joint movements are performed in the normal or near-normal fashion. However, there is a general tendency for the recovery of the hand to lag behind the rest of the affected upper limb. If the hand functions do not recover completely, the patient may feel those skilled movements performed by the affected hand as awkward and he may continue to use it as a second hand. In this stage, the therapist should motivate the patient to utilize his affected hand for most functional activities so that the coordination and control improves and becomes comparable to the sound hand.

2.3.6.6 Training Program for Hand Functions

The effectiveness of the skilled activities performed by hand is directly related to the function of the entire upper extremity. The restoration of hand function is predominantly a flexor phenomenon and progresses in a somewhat same way as the normal restoration of hand function (discussed earlier). Therefore, the therapist's first aim should be to promote mass grasp and mass release of objects with the hand. Once he can perform the abovementioned movements, more refined prehension activities have to be introduced. The use of symmetrical associated reactions and imitation synkinesis (a tendency to flex or extend the sound side digits when the patient attempts to flex or extend involved side digits) may facilitate functional movements of the hand. If the patient is unable to initiate finger flexion on a voluntary basis, proximal traction response may help to achieve a mass grasp provided that he has some control of the proximal components of the flexor synergy. When utilizing such a response, the therapist should maintain the patient's wrist in extension, and a command like "squeeze the fingers" can be given (Fig. 2.37).

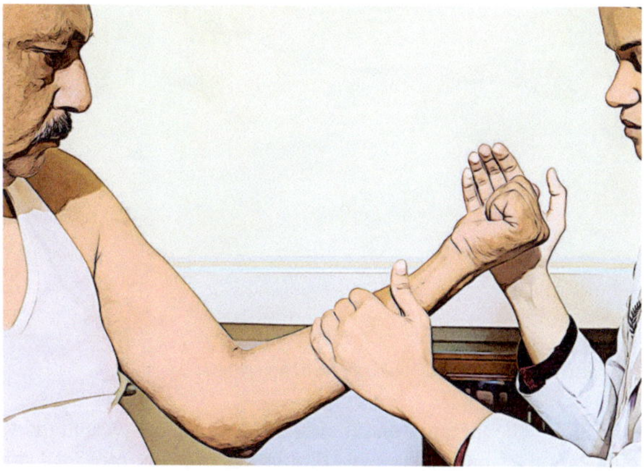

Fig. 2.37 Proximal traction response to achieve mass grasp with simultaneous commands like "squeeze the fingers"

Animated photograph of model and therapist with permission

For grasp to be effective, a strong linkage between the muscles stabilizing the wrist in extension and fingers into flexion is essential. Following the stroke, this linkage is usually disturbed. To reestablish the same, during the early stage of training, the therapist may have to stabilize the wrist in extension with the arm and the elbow in extension. Simultaneous application of proprioceptive or exteroceptive stimulation over the proximal portion of the wrist extensors along with commands like "squeeze your fingers" may promote finger flexion with the wrist in extension (Fig. 2.38). Gradually, the support given can be withdrawn and commands like "hold your wrist" and "do not allow your wrist to drop" to be incorporated to synchronize the contraction of the wrist extensors and the finger flexors. Once the wrist stabilization for grasp is well established, train the patient to perform the same grasp with the elbow in flexion.

Just like facilitating the voluntary mass grasp, the voluntary release of grasp is equally important. If the spasticity of the finger flexors is extensive, the thumb may be held flexed with the rest of the fingers closed tightly over the thumb and such posture may make the release of the fingers impossible. Brunnstrom has advised a series of manipulations, cutaneous stimulation, and favorable body and arm postures to release the tension in the finger flexors and evoke background tension in the extensor muscles to prepare voluntary finger extension. For instance, the therapist should pull out the patient's thumb by gripping its base (around the muscles of the thenar eminence) with simultaneous passive supination of the forearm (Fig. 2.39). If the tension in the flexors is marked, the wrist can be held in some degree of flexion. Next, the therapist should supinate and pronate the forearm slowly and repeatedly with cutaneous stimulation over the dorsum of the wrist and hand when the forearm is in supination. Throughout the manipulations, the grasp around the thumb is maintained but the pressure applied is lessened during forearm pronation and

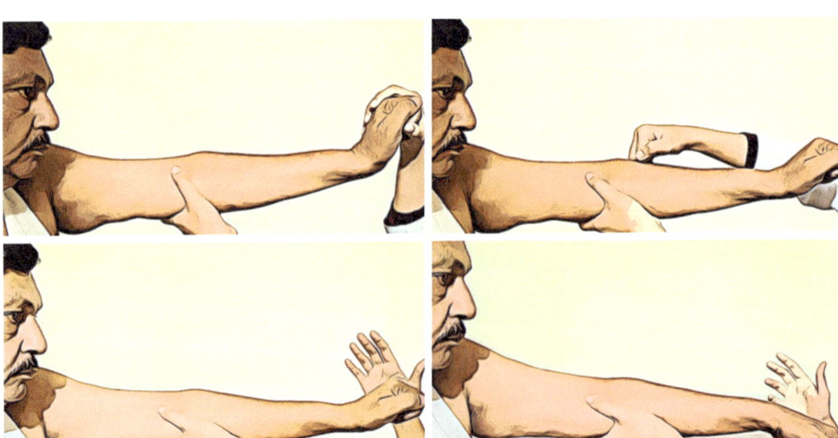

Animated photograph of model and therapist with permission

Fig. 2.38 The therapist stabilizing the arm, the elbow, and the wrist in extension, followed by the application of proprioceptive stimulation over the proximal portion of the wrist extensors with simultaneous commands like "squeeze your fingers"

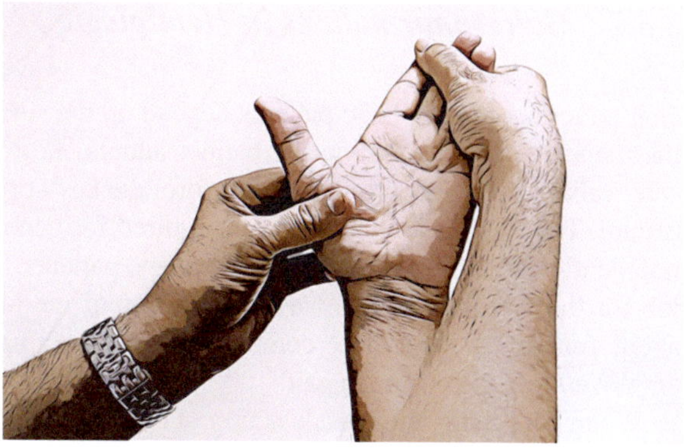

Fig. 2.39 An illustration of how to grip the base of the subject's thumb with the simultaneous passive supination of the forearm

increased during supination. Once background tension is evoked in the extensor muscles, the patient can actively participate in opening the hand partly or completely in the supinated position.

Even rolling of the fingers into flexion may elicit stretch reflex, thus aiding in the transfer of tension from flexors to extensors. Souques's position (the arm forward elevated to forehead level), a position used for eliciting Souques' phenomenon (In stroke patients, the forward elevation of the affected arm above the horizontal tends to evoke involuntary extension of the fingers or reduction of tone within the long finger flexors), with supination of the forearm and friction over the ulnar aspect of the dorsum of the forearm with the arm supported may relieve the tension within the finger flexors. Voluntary effort should be summated with the abovementioned reflex stimulation to promote active finger extension. Once the extensor response of the digits is well established, the patient has to be requested to alternatively open and close the fist with the arm gradually kept in a lowered position. Care has to be taken to avoid contact of the reflexogenic zone of the hand, which may accentuate hypertonicity within the finger flexors. According to Brunnstrom, the indiscriminate use of a sponge ball given to strengthen the handgrip must be discouraged as it may serve only to increase the tension in the wrist and finger flexors and prevent relaxation of the same muscles and volitional finger extension.

Generally, voluntary thumb movements begin to appear when voluntary mass fingers extension becomes possible. Enabling the movement of the thumb away from the contact with the index finger is of utmost importance for learning lateral prehension and advanced prehension for functional activities. While moving the thumb away from the index finger, the patient should be requested not to generate excessive effort as it can flex the thumb and the other fingers. Application of friction or gentle local percussion over the tendons of abductor pollicis longus and the extensor pollicis brevis muscles near the wrist may aid thumb movements.

Depending on the severity of the cerebrovascular accident, the restoration of hand function can halt at any stage and seldom recover completely. Hemiplegic patients with sensory deficiency may find it difficult to maintain grip without conscious effort. Similarly, the presence of instinctive avoidance reaction or instinctive grasping reaction can make prehensile movements difficult or impossible and may continue until the influence of these reactions diminishes.

2.3.7 Gait Abnormalities in Hemiplegic

Gait patterns of hemiplegic patients depend on the severity of the involvement and the manner of compensation the patient adopts. Restoration of safe standing and safe walking in a near-normal pattern form the key components of the training procedure. The combination of muscles required for generating normal locomotion is not identical to those BLS seen in hemiplegic patients. Therefore, it is a challenging job for the therapist to find ways and means of modifying the synergies to bring about functional movement combinations resembling, if not identical, to those employed in normal human gait.

When comparing the muscle action of hemiplegic patients with normal subjects, there are considerable changes in the phasic action of muscles between them. The stance phase on the affected side for these patients is considerably shorter, and the swing phase is correspondingly longer than the normal side. For the involved side, the knee extensors and the plantar flexors are active throughout the stance phase. The energy consumed for locomotion is considerably high, with reduced speed and cadence. The poor performance of the affected extremity necessitates the normal lower extremity to make a considerable number of adjustments causing alteration and intensification of muscle activity of the latter.

The two main factors that cause locomotion difficulties in hemiplegia are the firm linkage of muscle groups within the movement synergies and the slowness of reactions of the muscle groups. Even the rapid rise and fall in muscle tension required for normal locomotion will not materialize in hemiplegic patients as it is slow to build up and slow to fade out. While analyzing the gait, the behavior of the affected ankle, knee, and hip during the stance and swing phase has to be carefully studied. The comparison of the muscle activities in the normal gait and hemiplegic gait is given in Table 2.12.

2.3.8 Gait Training

Stroke patients who cannot access the motor patterns necessary for normal locomotion may utilize components of movement synergies still available to them to overcome the problem of ambulation. During the flaccid and later stages, these patients may use one or other forms of compensation and may become good or fair walkers. Such compensations are characterized by poor gait patterns, which are stubbornly resistant to correction. To avoid such compensatory strategies, right from the early days, the therapist should concentrate on preparation for walking while postponing actual gait, keeping in mind that the aim is to restore safe standing and safe walking as normal as possible. However, in most severely involved patients, when chances to progress beyond the synergy stages appear grim, attempts to modify the motor patterns necessary for normal gait may never succeed, and in such cases, a compromise has to be reached and our objective should be to make them modestly ambulant.

Table 2.12 Comparison of muscle activities between normal gait and hemiplegic gait

Phase of gait	Joint	Motor behavior in normal subjects	Motor behavior in hemiplegic patients
Early stance phase	Ankle	During the heel strike, the angle between the foot and the leg is ≈ 90° and the eccentric contraction of the pretibial muscles helps the sole to gradually and smoothly lower the forward foot to the ground. The activity of these muscles needed to stabilize the ankle prevents sudden plantar flexion.	The dominance of BLS with anticipation of weight-bearing frequently activates all the extensor synergy components. This prevents the normal association of the ankle dorsiflexors and the knee extensors essential for heel strike, causing forefoot or simultaneous contact of heel and forefoot to the ground (based on the severity of spasticity of plantar flexors).
Between early stance and midstance phase	Knee	During the heel strike, the knee is nearly extended. Immediately after the heel strike, a cushioning response consisting of initial flexion of short duration and small range precedes the knee extension. This response prevents the center of gravity of the body from rising and falling abruptly, aiding a smooth forward translatory motion of the body. During the flexion movement of the knee, there is a maximum activity of the quadriceps muscle, which prevents buckling of the knee and it ceases when the limb moves to midstance.	If the limb is flaccid, attempts to support the body weight through the affected limb cause buckling of the knee. On the contrary, if the limb is spastic, mild to severe hyperextension of the knee may occur, making the normal flexion-extension moment impossible. Hyperextension of the knee can also result from excessive tension in the plantar flexors.
Between early stance and midstance phase	Hip	In these phases, the activity of the hip abductors provides lateral stabilization of the pelvis, which prevents the opposite side of the pelvis from dropping. Immediately after heel strike, the hip extensors are found to show a brief burst of activity which ceases soon before midstance.	During weight-bearing, the dominance of extensor BLS (including the hip adductors) causes failure of hip abductor muscles to respond, thus compromising the lateral stability of the pelvis leading to a Trendelenburg limp (pelvis of the normal side lowers when the affected side bears weight). The limp will not be obvious if the patient uses a cane or external support.

(continued)

Table 2.12 (continued)

Phase of gait	Joint	Motor behavior in normal subjects	Motor behavior in hemiplegic patients
Midstance phase	Ankle	Once the metatarsal heads contact the ground, the activity of dorsiflexors reduces. During this phase, the leg pivots forward about the ankle joint, causing a reduction of the angle between the foot and the leg. A gradual and eccentric contraction of the calf muscles begins when the bodyweight advances in front of the ankle joint preventing excessive pivoting of the leg at the ankle.	Attempt to advance the body weight in front of the ankle joint increases the tension within spastic calf muscles preventing forward pivoting motion of the leg at the ankle joint. The patient may find difficulty in advancing the normal foot to proper distance, making the stride length unequal and the swing phase too rapid. If the tension within the calf muscles is marked, the involved limb may lead throughout and the normal foot may advance up to but not beyond the affected one.
Late stance phase	Ankle	This phase is characterized by the rising of the heel from the ground with simultaneous flexion of the knee to prepare the limb for forward swing. In this phase, the rapid concentric activity of the plantar flexors generates an effective "push-off." The activity of the plantar flexors brings the foot to 20° plantar flexion. During "push-off," the activity of the calf muscles, inactivity of the knee extensors, and the gravitational forces are responsible for initiating the forward swing.	Tension within the quadriceps persists in this phase and prevents the knee from flexing. As a result, the combination necessary for push-off and initiation of the swing phase will not materialize. The patient has to rely on compensatory methods such as external rotation and adduction, circumduction, or excessive hip hiking to bring the limb forward.
Late stance phase	Knee	The inhibition of the quadriceps muscle activity enables the knee to flex in order to initiate the forward swing of the limb.	The strong bond existing between the calf muscles and the quadriceps muscles does not allow the muscle combination required for the initiation of the forward swing of the limb.
Late stance phase	Hip	To prepare the limb for forward swing, there will be flexion of the hip. The flexion of the hip accompanies the flexion of the knee and plantar flexion of the ankle. The activity of the superficial hip flexors is responsible for initiating it.	The normal combination required for preparation for forward swing will be difficult or impossible due to the dominance of the BLS. Lack of hip flexion and knee flexion, along with the persistence of calf muscle activity, prevents effective push-off.

Table 2.12 (continued)

Phase of gait	Joint	Motor behavior in normal subjects	Motor behavior in hemiplegic patients
Swing phase	Ankle	Minimal activity of pretibial muscles throughout this phase prevents the foot from dropping down.	The dominance of the extensor BLS prevents the activity of dorsiflexors, causing foot drop and insufficient ground clearance. The ankle may also invert markedly, making the gait unsafe for the patient. The patient may exaggeratedly utilize the flexor BLS to dorsiflex the ankle for clearing the ground.
Swing phase	Knee	Muscular and gravitational forces bring about the forward acceleration of the limb. However, during most of the swing phase, the muscular activity is minimal and inertia is responsible for forward motion.	Persistence of extensor synergy until the late stance phase makes the knee stiff. In such situations, the patient may often use the total flexor synergy for flexing the knee joint and frequently, the onset of knee flexion is delayed due to poor push-off. Therefore, the involved side knee will be lifted forward and upward with the leg hanging more or less vertical and foot, well above the ground, as if the patient is stepping over an obstacle.
Swing Phase	Hip		If extensor synergy dominates, the normal shortening of the limb essential for swing through will not materialize. In such a situation, the patient may circumduct or hike the pelvis to clear the ground. If the hip is held stiff, he may tilt the pelvis backward and use the abdominal musculature, and the gait may resemble that of an ankylosed hip. The patient may also use the total flexor synergy for flexing the hip joint, as mentioned earlier. If the motor control is poor, the patient may use the hip adductors to drag the affected limb forwards. In such situations, the patient will never advance his affected limb beyond the normal foot, and the pelvis will stay behind the affected side with the hip held in external rotation and foot in eversion.

Preparation for gait should commence from the early period by promoting weight-bearing exercises and trunk balance in sitting and standing. Modification of motor response to obtain muscle associations resembling normal gait, the reversal of movement direction to promote a rapid release in the tension of muscle groups following their activation, and strengthening the weak muscles are the essential keys for near-normal gait. To attain adequate control over the movement patterns for normal gait, the treatment should follow a stepwise series of activities with increasing complexity, progressing from the least demanding position (supine) to the most demanding position (standing).

2.3.8.1 Training Procedures to Improve Gait

The dominance of synergies may make the severely involved hemiplegic patients unable to restore their normal gait. Whereas for those patients with minimal involvement, modification of motor responses may not be necessary as the dominance of synergies may wane, leading to the spontaneous return of normal or near-normal gait pattern. Many of the patients who fall between these extremes require modification of motor responses to attain a normal or near-normal gait.

Training procedures to modify the muscular responses should consist of techniques to strengthen the weak components of flexor synergy, namely the hip flexors and abductors, knee flexors, and ankle dorsiflexors. In order to facilitate hip flexion control, bilateral contraction followed by unilateral contraction of affected side hip flexors can be attempted. For instance, inclining the trunk back and forth when the patient sits on a chair with no back support produces a lengthening contraction and shortening contraction of hip flexors, respectively. Trunk inclination movements also activate the abdominal muscles. Commands like 'hold' when the patient inclines forwards will promote isometric contraction of hip flexors (Fig. 2.40). For patients with the absence of hip flexion control, the use of the Bechterev-Marie-Foix reflex (quick passive plantar flexion of the toes in supine position produces flexion of all joints of the leg) can be attempted to elicit a total flexion response including ankle dorsiflexion (Fig. 2.41). Later, voluntary effort should be superimposed on such reflexive stimulation. This should be followed by training activation of dorsiflexors in combination with hip and knee extension, which is essential for promoting the near-normal association required for early stance.

The predominance of extensor synergy will not enable normal association of hip abduction with the hip and knee extension required for preparation of the limb for early stance and gait. Failure of hip abductor response causes the Trendelenburg limp. Therefore, attempts should be made to activate the hip abductors and associate them with hip and knee extension. This can be achieved by reflexive elicitation of hip abductor activity that includes the use of Raimiste's hip abduction phenomenon in supine (Fig. 2.42). Use of local facilitatory techniques, superimposition of voluntary effort on the reflex contraction by commands like "spread your legs," and promotion of isometric contraction of hip abductors in side-lying can be attempted to gain better control over the hip abductors. Later attempts should be made to abduct the affected limb and then the unaffected limb by promoting side kicking movement in standing with support progressing to without support (Fig. 2.43). Hiking or elevating the pelvis of the sound side also necessitates a strong contraction of the hip abductor muscles of the affected side. If the methods of activation of hip abductors are successful, then commands like grow tall on each step can be incorporated to generate a smooth walking pattern with a suitable cadence dictated by the therapist. If there is no potential for return of hip abductor activity and the elimination of the

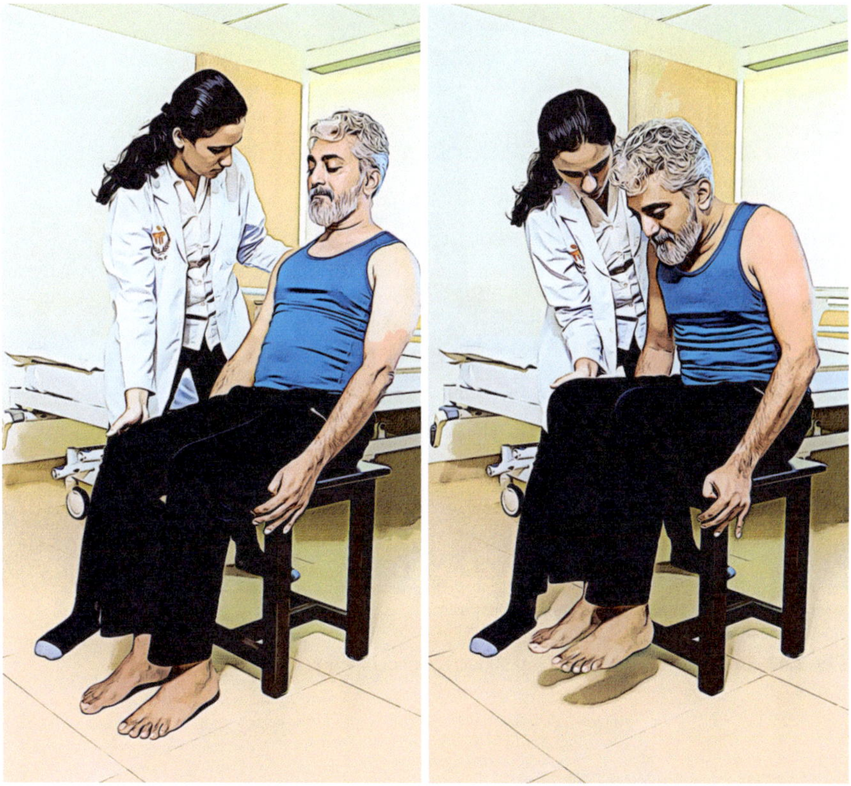

Animated photograph of model and therapist with permission

Fig. 2.40 Back and forth inclination of the trunk while sitting produces lengthening and shortening contractions of hip flexors and abdominals. Commands like "hold" when the subject inclines forwards may promote isometric contraction of the hip flexors

Fig. 2.41 Illustrates the elicitation of the Bechterev-Marie-Foix reflex

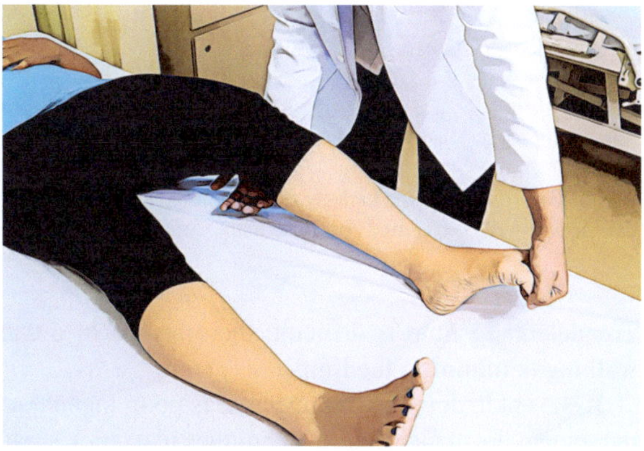

Animated photograph of model and therapist with permission

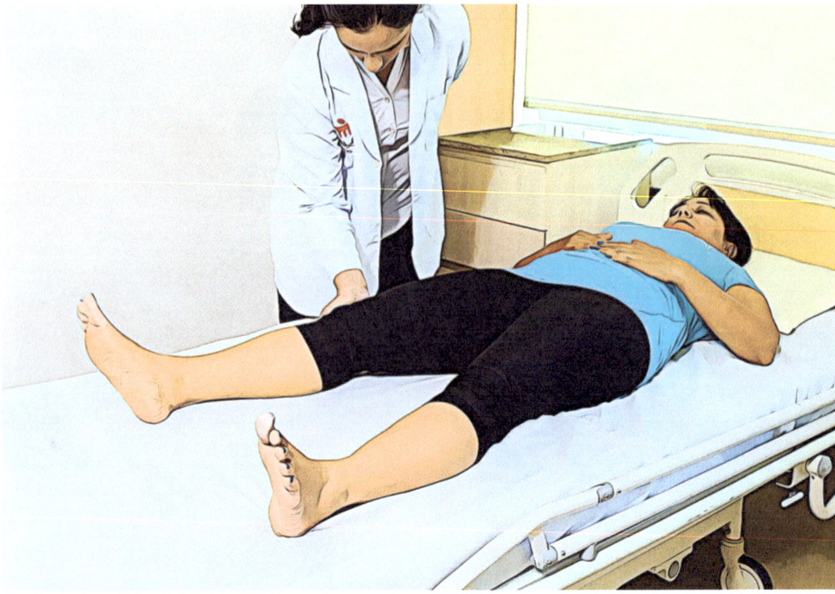

Animated photograph of model and therapist with permission

Fig. 2.42 Eliciting the Raimiste's hip abduction phenomenon in supine

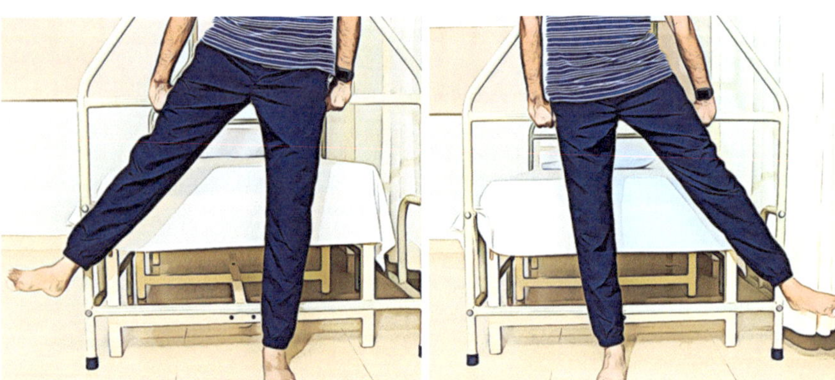

Animated photograph of model and therapist with permission

Fig. 2.43 Promoting side kicking movement in standing with self-support. Note: Additional manual support will be required for those who cannot stand self-supported

Trendelenburg limp is difficult, then the use of a cane can be encouraged while walking to minimize the limp.

If the ankle dorsiflexion response is poor, manual resistance given to hip flexion movement, local facilitatory techniques like quick stretching, exteroceptive stimulation over the belly of the pretibial muscles, or percussion of their tendons as they pass the ankle joint may prove effective. As a progression of the training procedure, attempts should be made to promote the voluntary dorsiflexion of the ankle, while sitting on a higher chair (Fig. 2.44). Later in standing posture without support, voluntary attempts should be made to dorsiflex the ankle to activate and control ankle dorsiflexion with hip and knee extension. Activation of evertors (In most of the patients' peroneal muscles lie idle on the affected limb) prevents the abnormal association of inversion along with dorsiflexion of the ankle, which is commonly seen in the ambulant hemiplegic patients. Use of stretch reflex, percussion, or exteroceptive stimulation over the evertor muscles and efforts to isometrically contract the muscle may promote better motor control of the ankle and subtalar joint.

Fig. 2.44 Promoting voluntary ankle dorsiflexion while sitting on a higher plinth

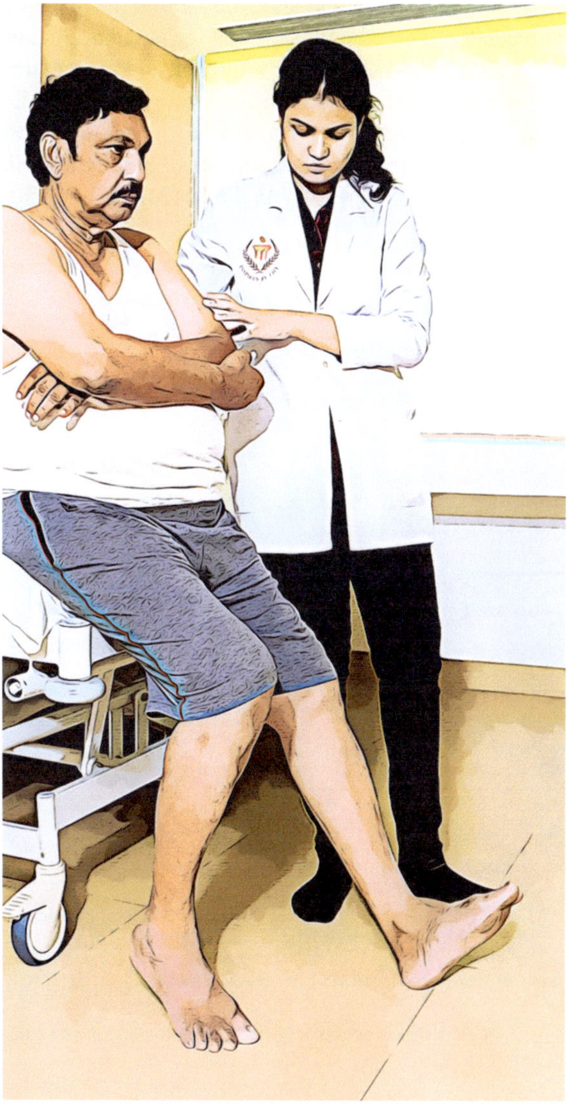

Animated photograph of model and therapist with permission

Techniques to promote alternating knee flexion and extension movement in order to decrease the duration of contraction of the opposing muscle groups and reduce tension within the knee extensors should be attempted for smooth locomotion. To achieve the abovementioned, training of the knee flexor muscles is indicated in supine, side-lying, sitting, and standing positions. If the labyrinthine reflex creates more tension within the knee extensors, the patient may find difficulty initiating knee flexion in the supine position. Associated reactions and the use of exteroceptive stimulation over the hamstring muscles may induce knee flexion; if not, the use of side-lying position may eliminate the influence of gravitational force and minimize the effect of tonic labyrinthine reflex on the affected limb.

Training the patient to slide the foot on the floor while sitting on a chair helps to guide and provide the sensation of isolated knee flexion motion (Fig. 2.45). Sitting on tall chairs and sliding the foot back and forth underneath the chair increases the hip extension and enables further development of reciprocal knee flexion and extension. As preparation for standing and walking, the reciprocal knee flexion and extension exercise should be performed in standing with the patient leaning over the table with a forearm resting on the table (Fig. 2.46). To promote effective heel strike

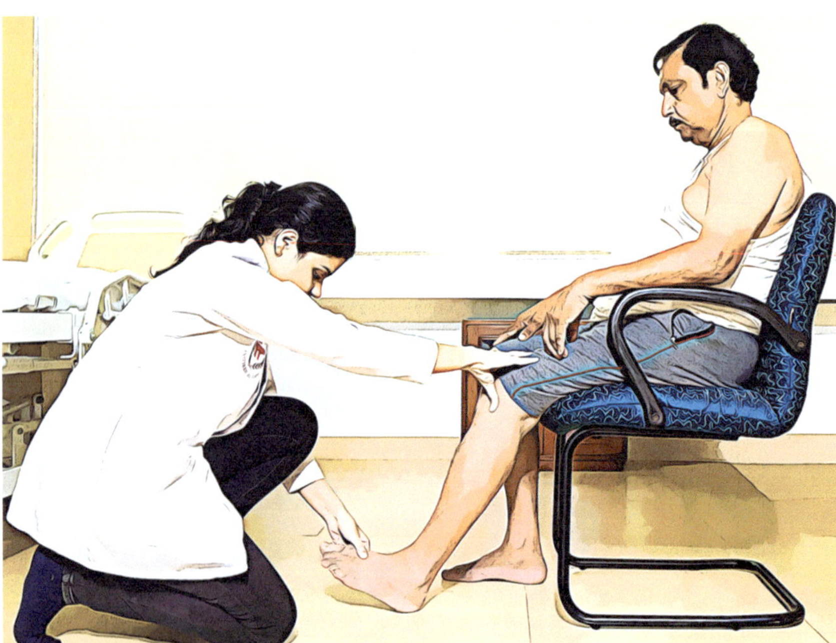

Animated photograph of model and therapist with permission

Fig. 2.45 The therapist assisting the subject to slide the foot on the floor while sitting on a chair

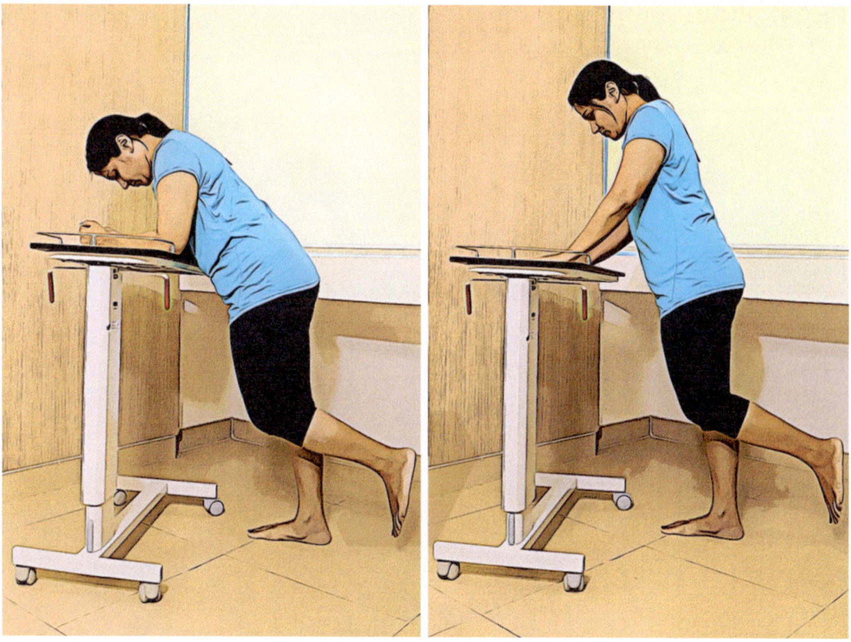

Animated photograph of model and therapist with permission

Fig. 2.46 Reciprocal knee flexion and extension exercise performed in standing while leaning over the table with the forearm or hand resting on the table

and push-off components of gait, reciprocal ankle dorsiflexion with knee extension and ankle plantarflexion with knee flexion can be attempted with the patient in the same position. The position should be gradually modified to a standing position with the patient facing and leaning against higher support.

Even in the presence of increased tension within the quadriceps, while standing, the patient's affected knee may buckle and cause a fall or hyperextension of the knee. Partial squat with even weight-bearing (Fig. 2.47), weight shifts, and lifting of the sound leg with the involved leg on the ground may improve knee stability and minimize buckling or hyperextension of the same.

The degree of spasticity in the affected extremities and dominance of BLS determines the full return of normal arm swing, rotation, and counter-rotation of the pelvis and trunk. Appropriate training may benefit the patient to walk with a certain amount of arm swing along with trunk rotation. For instance, while standing, rotation of the trunk slowly to either side with the patient asked to wrap his arms around

Fig. 2.47 Partial squat with even weight-bearing

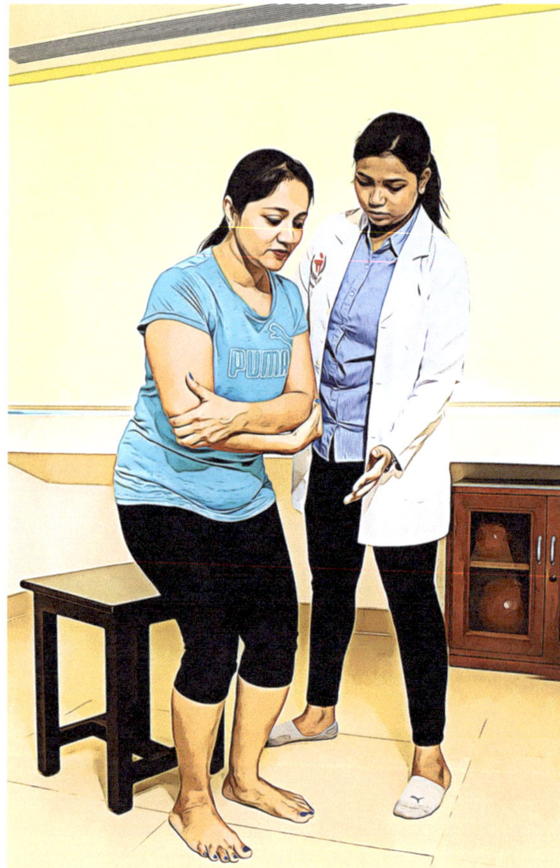

Animated photograph of model and therapist with permission

the body (Fig. 2.48), with or without the therapist's external assistance and guidance to the movement, may promote arm swing and trunk movements. Later, when they learn to voluntarily wrap their arms without any assistance, attempts to walk with exaggerated trunk rotation can be incorporated. However, care should be taken to maintain the rhythm, even if the trunk and arm movements are not perfect. Brunnstrom also advised the use of suitable music at an appropriate speed to encourage the hemiplegic patient to gain a rhythm in walking.

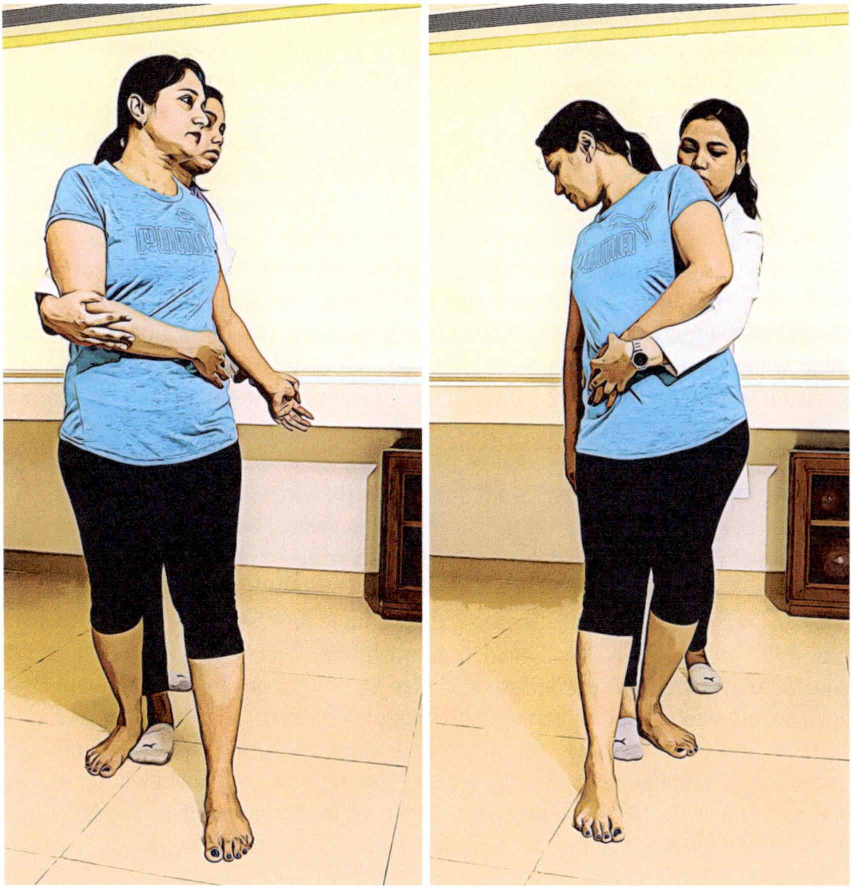

Animated photograph of model and therapist with permission

Fig. 2.48 Therapist assisted promotion of the arm swing and trunk movements

2.4 Proprioceptive Neuromuscular Facilitation in Neurorehabilitation

Suruliraj Karthikbabu and Abraham M. Joshua

2.4.1 History

Proprioceptive Neuromuscular Facilitation (PNF) is one of the traditional neurophysiological approaches widely practiced by physiotherapists in neurological rehabilitation. Proprioception provides sensory information about the movement and position of the body. Neuromuscular means involving the nerves and the muscles.

Facilitation refers to making a motion more comfortable. Herman Kabat, an American neurologist, originally developed this concept. Kabat, along with Elizabeth Kenny, an Australian nurse, and Margaret Knott, an American physiotherapist, conceptualized the treatment ideas of PNF. They developed the PNF concepts based on the understanding of the neuroscience of their times. Kenney introduced specific stretching and strengthening activities among patients with poliomyelitis.

The principles of PNF were developed based on the neurophysiological theories available during the late nineteenth and early twentieth centuries. Charles Edward Beevor, an English neurologist and anatomist, proposed that the brain knows the movement but not the isolated muscle action. Guillaume-Benjamin-Amand Duchenne, a French neurologist, felt that an isolated muscle contraction was not possible while doing daily functional tasks. Ernst Gellhorn, a neurophysiologist, supported it by observing a set of muscle activities following stimulation of the cerebral cortex. It was Charles Sherrington, an English neurophysiologist, who postulated the phenomenon of successive induction, irradiation, reciprocal innervation, and reciprocal inhibition. Sherrington's work was primarily adopted while developing PNF procedures and techniques. Kabat assimilated the manual techniques of Kenney and the neurophysiological rationale proposed by Sherrington in the PNF concept.

Henry Kaiser, a parent of a young person with multiple sclerosis, funded the financial recourses to start the Kaiser-Kabat institute in Washington in 1946. In 1948 and 1950, two more Kaiser Foundation rehabilitation centers were introduced in Vallejo and Santa Monica, California, United States. Kabat employed Knott in 1948 for conceptualizing the patterns and techniques of PNF. Knott started a clinical training program to teach other physiotherapists. Dorothy Voss, an eminent American physiotherapist, joined Kabat and Knott in 1952, and together they refined the foundation of PNF concepts. The first PNF textbook was published by Knott and Voss in 1956. After Kabat had left Vallejo in 1954, Sedgwick Mead, an American physician, joined Knott and Voss. Since 1954, the Kaiser Foundation rehabilitation center had been offering short-term courses of PNF training for clinical therapists. Following the demise of Knott in 1978, Hink Mangold and Sue Adler, two American physiotherapists, continued the work of Knott. In the year 1990, the International PNF Association (IPNFA) was founded by the European instructors, and since then, the objectives of the IPNFA are to foster and continue the clinical practice of the PNF philosophy and look after its education and research-related activities.

2.4.2 Philosophy of PNF Concept

The IPNFA explained the PNF philosophy in five subdomains: namely positive approach, functional approach, mobilization of reserves, treating the whole person, and use of motor learning and motor control principles. As a positive approach, the assessment and treatment aimed at encouraging the patients to perform exercises or activities to quickly achieve the desired goal without any pain and discomfort. Following the successful mental engagement, the patient should experience a feeling of achievement in some part of the activities, and the focus of training should be to work on the more substantial segments of extremities and trunk. The stronger part of the body irradiates the neural drive and muscular forces into the most impaired or weaker sections of the body. The active participation during the intense rehabilitation of a patient with neurological disease mobilizes the reserve or unmasks the potential. The repetition and variations of meaningful PNF movement patterns increase the demand for the selected functional activity. The developers of this approach believe that by incorporating blocked or random practice under supervision, the self-directed functional skill can be achieved both in indoor and outdoor environmental situations.

The PNF philosophy addresses not only the level of body structure and function but also activity limitation and participation restriction. The PNF therapist mostly identifies and treats the involvement of body structure and physiological function related to the limitations in daily activities and participation of a patient with neurological dysfunction. The functionally oriented assessment and treatment enhances motor learning by optimizing the task performance of the patient. Based on the attention ability and the feedback resulting from the knowledge of results and performances, the patient acquires motor skills in different stages of cognitive, associative, and autonomous learning.

The primary objective of PNF treatment is to improve the motor control and skills of the patient following neurological dysfunctions. The PNF philosophy utilizes motor control concepts and its stages of mobility, stability, controlled mobility, and motor skills. The motor learning effects addressing the response of the neuromuscular mechanism and real motor recovery are the critical components of the PNF concept. Besides philosophical rationale, the PNF concept in neurological rehabilitation is well explained by understanding the basic principles and procedures and its techniques. The PNF techniques should be adopted within the context of a functional therapeutic goal and never practiced separately.

2.4.3 Basic Principles and Procedures of the PNF Concept

For an effective PNF treatment, the therapists working with neurological patients should have a sound knowledge of the basic principles, skills, and techniques of the PNF. The most common ten basic procedures are manual contact, body position and body mechanics, auditory commands, verbal stimulus, traction or approximation, resistance, irradiation and reinforcement, stretch, timing, and patterns. The features of the same are as follows:

Manual Contact Manual contact increases the muscle power and controls the movement with proper grip and pressure. The therapist uses the lumbrical grip for guiding the patient's motion. The handgrip stimulates the tactile cutaneous receptors and other pressure receptors. The therapist places the hand against the movement direction in a way that it should facilitate the agonist muscular contraction. When manual contact is on the trunk, it promotes postural stability.

Body Position and Body Mechanics The therapist should consider the proper position and body mechanics that enable providing specific and well-targeted guidance for movement control and stability. The therapist should position himself/herself in the direction of the diagonal movement pattern so that the patient achieves better movement facilitation and neuromuscular response.

Auditory Stimulation (Commands) Verbal commands with optimal vocal volume shall direct the patient performing the desired muscular recruitment. The acoustic stimulation can be either a preparatory or corrective command to prepare muscle activation or correct movement clumsiness, respectively.

Visual Stimulation The patient uses his vision to guide the motion, and the visual feedback modulates the muscle force, as well.

Traction or Approximation The traction facilitates the motor recruitment of the limbs. Traction aids in the elongation of the soft tissues and also relieves joint pain. The joint approximation facilitates the contraction of antigravity muscles, thus promoting better weight-bearing capacity, upright balance reactions, and postural stabilization of the patient.

Resistance The resistance should match the patient's muscle force, which aids muscle contraction, motor control, and learning. Additionally, it increases muscle strength and hypertrophy. The spasticity, cortical drive, firing rate, motor unit synchronization, and agonist-antagonist co-contraction might affect muscular strength. So, the therapist's resistance should match with the patient's efforts while controlling the movement or stabilizing the body segment. The movement situation and the purpose of the action shall determine the optimal resistance to isometric, concentric contraction, and eccentric lengthening of the muscles.

Irradiation and Reinforcement Irradiation and reinforcement spread the neuromotor response upon stimulation. Irradiation can be either excitatory or inhibitory in nature and sprinkles the nerve impulses to the neighboring muscles. When the reinforcement is precisely applied, the simultaneous weaker stimuli to different body parts shall reinforce the intra-muscular excitation, i.e., spatial summation. For instance, while lifting a cup of water, the force from the scapular and shoulder stabilizers in a stroke person may irradiate to reinforce the weak wrist extensors. The effects remain ongoing, and the patient experiences a sense of movement lightness after the end of the irradiation (after-discharge). The after-discharge remains high when there is increase in the muscular effort, demand and time of the stimulus. In addition to the above, the patient should have control over the different muscle groups so that he can reach out his arm and place the cup on the table. For which, an increased agonist excitation followed by the antagonist contraction is critical (successive induction).

Stretch The pre-stretch (initial stretch) elongates the muscles from proximal to distal components of the specific pattern. The therapist applies a stretch stimulus at the beginning of the range of motion. It facilitates muscle contraction and recruits the motor units and reduces muscle fatigue. If the stretch stimulus is in succession (re-stretch), the subliminal weaker stimuli could summate the motor unit excitation, i.e., temporal summation.

Timing The sequence of muscle contractions is called normal timing. The movement usually starts from the distal part of body segments and ends in the proximal, whereas the muscle activation begins proximal and ends distal. In addition to initiating muscle activation, normal timing also increases the muscle contraction. Reciprocal innervation is essential for the coordinated timing for the contraction of the agonist muscle synergy. This innervation becomes reciprocal inhibition when there is a relaxation of the antagonist and activation of the agonist. Conscious efforts can change the normal timing, i.e., the sequence of the muscle contraction and recruitment, known as "timing for emphasis," targeting the desired muscle activity by overflowing the nerve impulses from the healthier to the weaker muscle groups.

Patterns Patterns are the three-dimensional synergistic mass movements of the extremities or torso that are the essential components of daily functional tasks. The PNF pattern involves the rotation component along with the angular displacement of the limbs or trunk. The specific PNF pattern is named after the isotonic activity of agonist muscle groups, considering the end position of the movement. The rotation component of the PNF pattern guides the direction of the diagonal movement. Surface electromyography analysis of the upper limb post-stroke showed that the PNF pattern and limb position had an impact on the initiation of voluntary action and also evoked the motor potential of weak muscles.

2.4.4 PNF Patterns

The segmental PNF patterns available in the human body are scapular, pelvic, arm, leg, head, and trunk patterns. The scapular and pelvic patterns are of anterior-elevation, posterior-depression, posterior-elevation, and anterior-depression.

2.4.4.1 Scapular Patterns

The patient performs the basic scapular pattern in a side-lying position with neutral cervical and lumbar lordosis. To facilitate the anterior-elevation and posterior-depression of the scapula, the pinna of the ear and the acromion process should be in the same line depicting the imaginary 12 o'clock and 6 o'clock positions. The therapist stands behind the patient's scapula and faces the diagonal of 7 o'clock and 1 o'clock positions. The therapist cups his hands with the lumbrical grip around the acromion and anterior shoulder line. Following which, the therapist brings the scapula passively into adduction-depression towards the thoracic spine. After a brief stretch stimulus to the scapula in an inferior-medial direction by the therapist, the patient is advised to pull the scapula into anterior-elevation (Fig. 2.49),

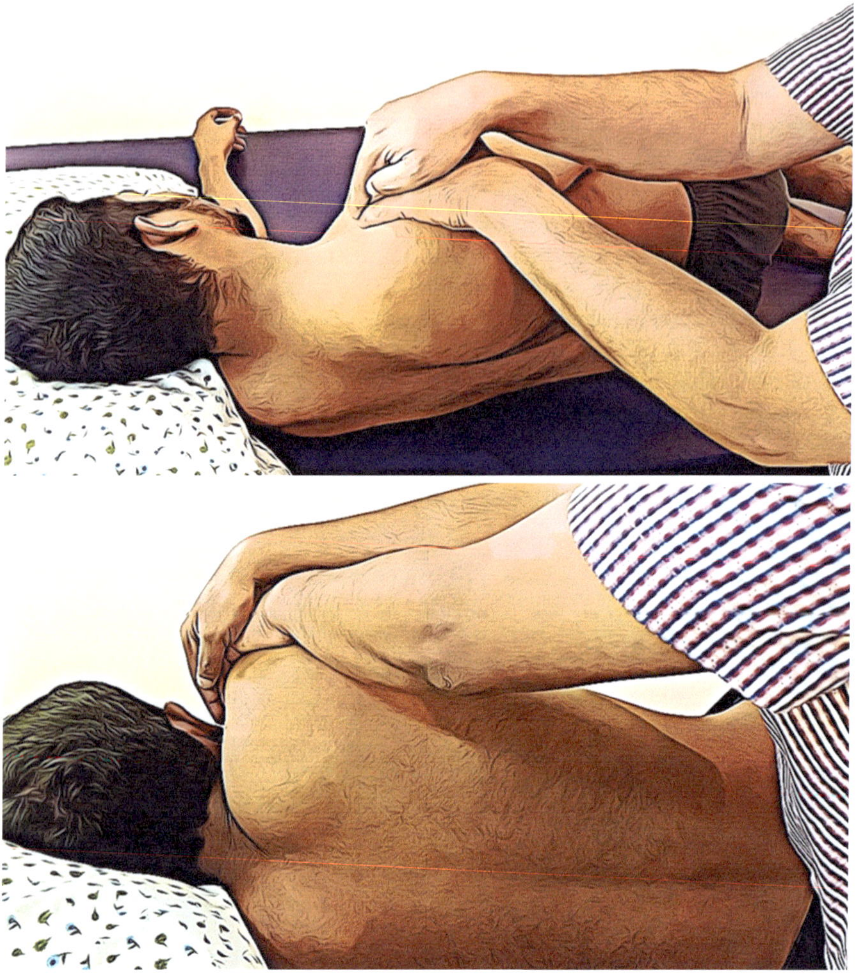

Animated photograph of model and therapist with permission

Fig. 2.49 Anterior-elevation of the scapula

assuming the ideal 1 o'clock position. Verbal commands like "shrug your shoulder in front of your ear" may encourage the patient to initiate and direct the movement. From this position to achieve a posterior-depression of the scapula (Fig. 2.50), the therapist places his hand over the shoulder blade (the base of the palm at the inferior angle and fingers across the spine of the scapula). Then, the patient is instructed to push the shoulder blade downwards and inwards towards the thoracic spine.

For posterior-elevation (Fig. 2.51) and anterior-depression (Fig. 2.52) of the scapula, the therapist stands beside the patient's head, facing the diagonal of 11

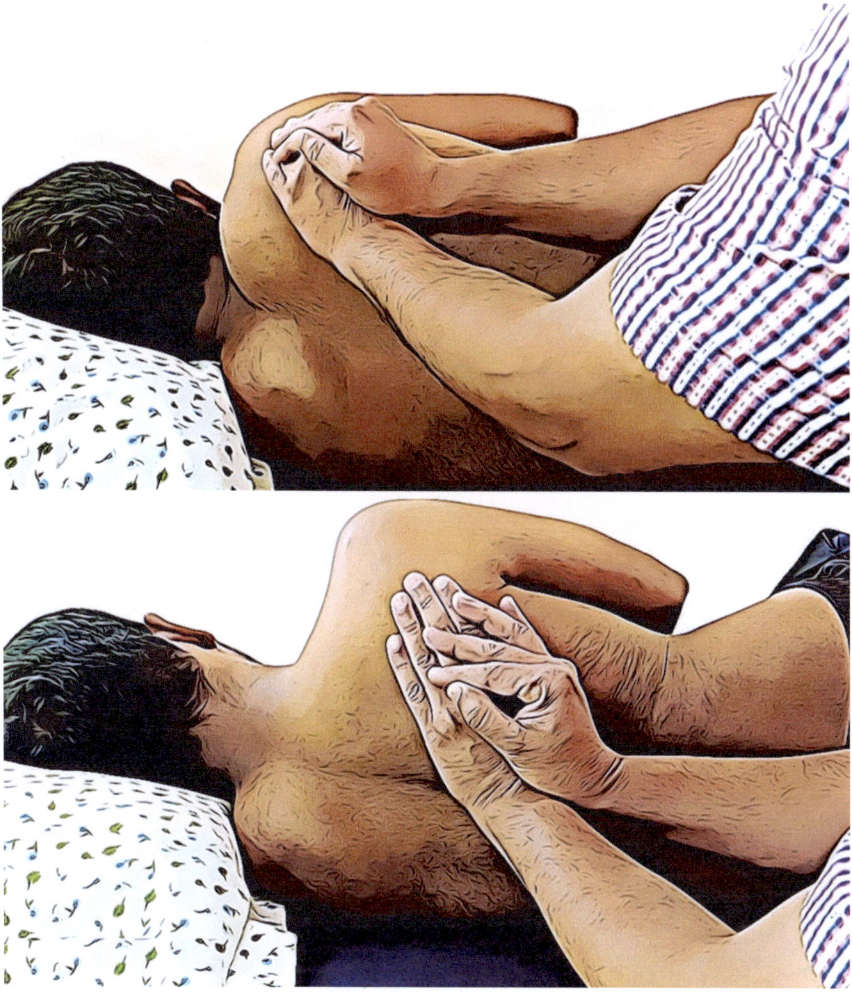

Animated photograph of model and therapist with permission

Fig. 2.50 Posterior-depression of the scapula

o'clock and 5 o'clock positions. To encourage the scapular posterior-elevation, the therapist places his base of hand at the spine of scapula and palm and fingers across the superior shoulder line and lateral part of the clavicle, respectively. The patient shrugs the shoulder and pulls it behind the ear. For anterior-depression of the scapula, the therapist keeps one hand just below the collarbone and another hand lateral to the scapula, fingers pointing towards the chest wall. The patient is asked to bring the top elbow downwards towards the navel, matching the therapist's manual resistance.

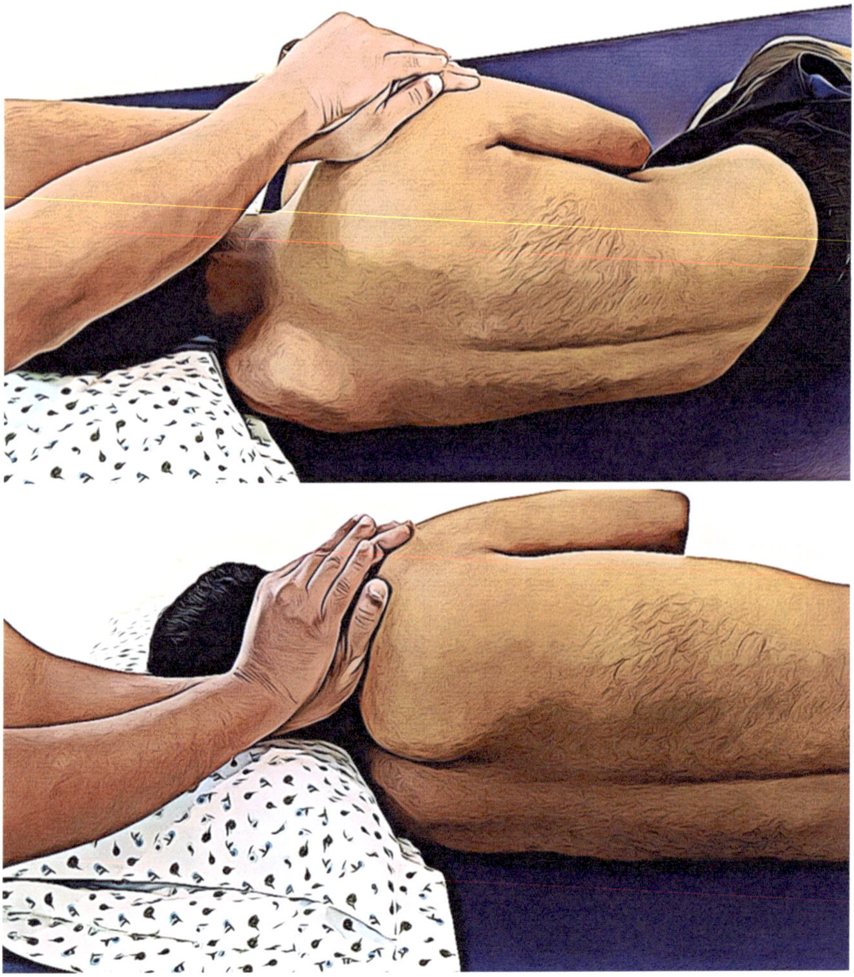

Animated photograph of model and therapist with permission

Fig. 2.51 Posterior-elevation of the scapula

2.4.4.2 Pelvic Patterns

The patient's starting position and the therapist's stance and handgrips for the basic pelvic patterns are similar to the scapular patterns. For anterior-elevation of the pelvis (Fig. 2.53), the therapist uses his lumbrical grip around the anterior-superior iliac spine and the patient is instructed to "pull the hip towards navel." For posterior-depression of the pelvis (Fig. 2.54), the patient pushes the buttock against the therapist's hand kept at the ischial tuberosity. The patient shortens the bottom side trunk concentrically and lengthens the top side trunk eccentrically while performing the

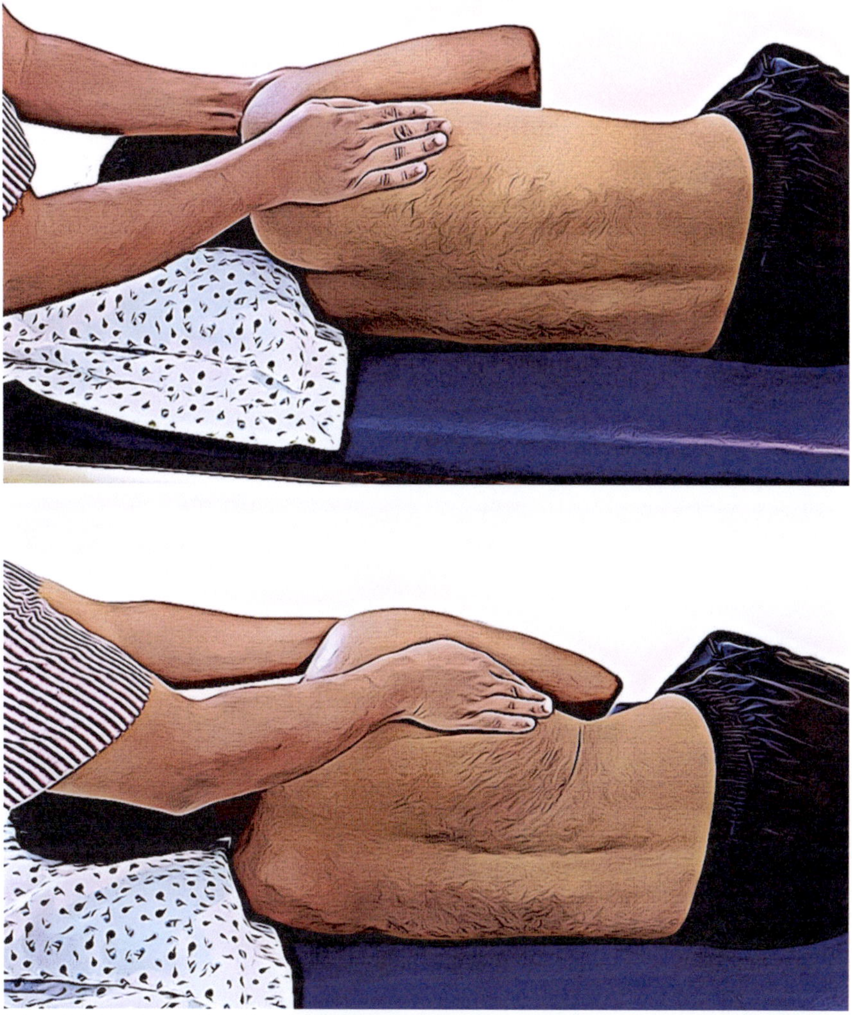

Animated photograph of model and therapist with permission

Fig. 2.52 Anterior-depression of the scapula

posterior-depression of the pelvis. Figure 2.55 depicts the posterior-elevation of the pelvis. For pelvic anterior-depression (Fig. 2.56), the therapist places his hands at anterior-superior iliac spine and ischial tuberosity. The patient then elongates the top thigh bone by pushing the knee ahead of the bottom one. On the other hand, the patient shortens the top leg by hiking the hip backward to achieve the pelvic posterior-elevation. The therapist optimally resists the movement with the hand placed at the posterior-superior iliac spine.

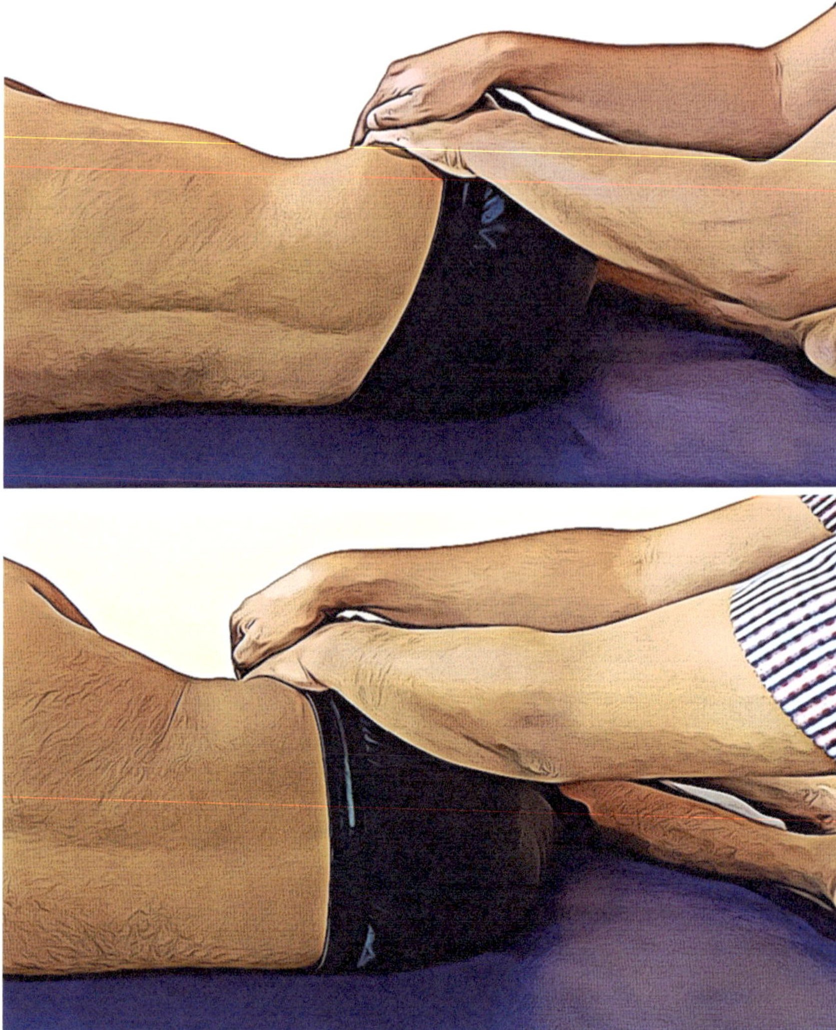

Animated photograph of model and therapist with permission

Fig. 2.53 Anterior-elevation of the pelvis

2.4.4.3 Diagonal Patterns of Extremities

There are two patterns possible in each diagonal movement of the extremity. The patterns of upper and lower extremities are described based on the motion occurring at the proximal shoulder and hip joints, respectively. These patterns are either in the

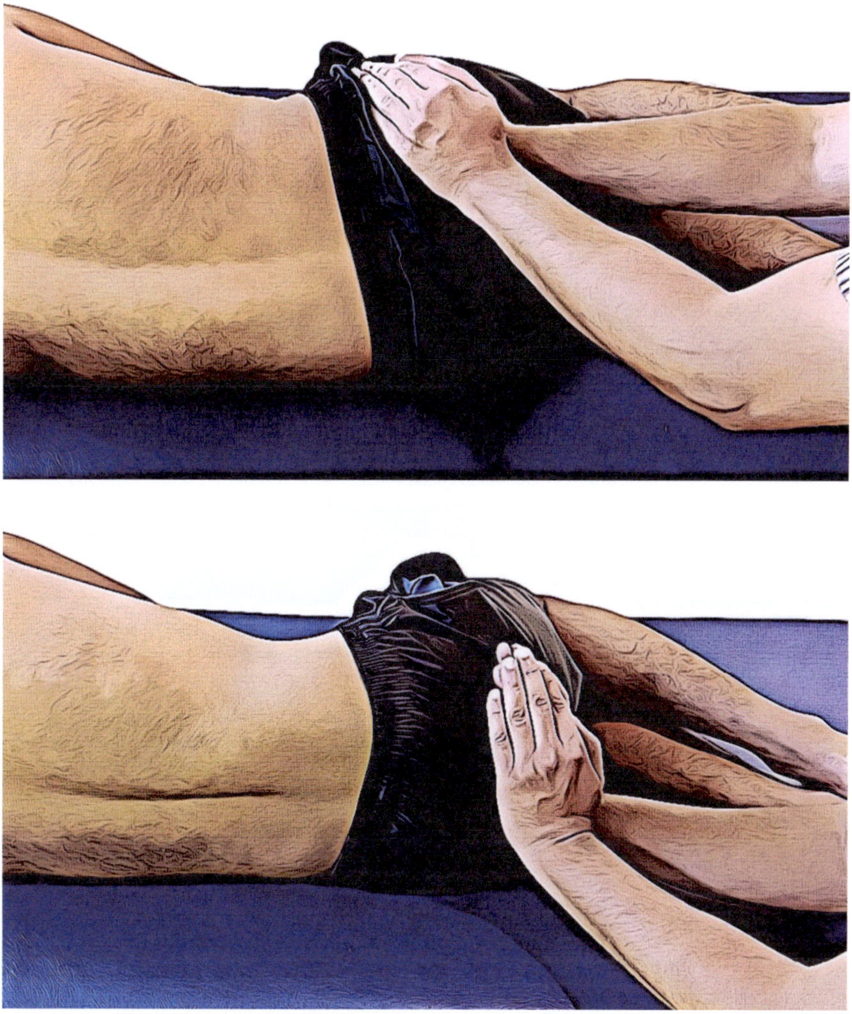

Animated photograph of model and therapist with permission

Fig. 2.54 Posterior-depression of the pelvis

flexion or extension pattern, given the activity of the agonist muscle group. Also, these movement patterns may incorporate the intermediate joints (i.e., elbow and knee) either into flexion or extension. From the context of meeting daily upper limb activities, balancing, walking, and functional mobility tasks of the people with neurological conditions, an optimal motor control over these patterns is essential. While doing most of the daily functioning, the diagonal pattern of extremities harmonizes with scapular and pelvic movements.

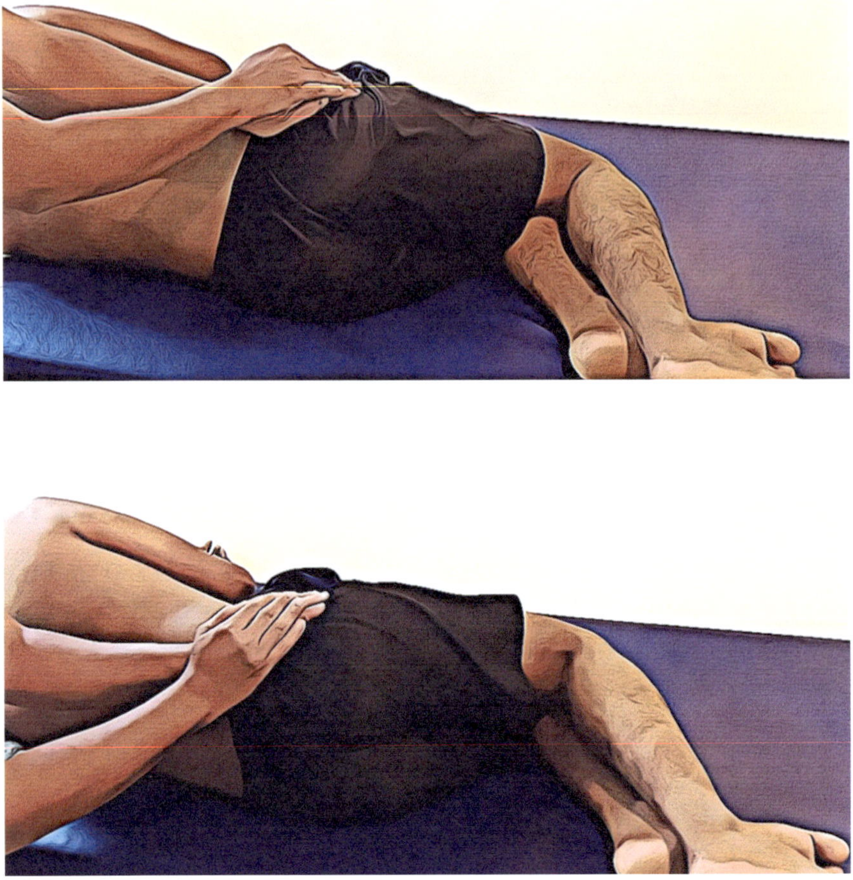

Animated photograph of model and therapist with permission

Fig. 2.55 Posterior-elevation of the pelvis

Arm Diagonal Patterns

The arm patterns of flexion-abduction-external rotation and extension-adduction-internal rotation take place in the same diagonal plane. Similarly, the flexion-adduction-external rotation and extension-abduction-internal rotation of upper limb diagonal patterns occur in another plane. The patient performs all these basic patterns in supine lying. For facilitating the flexion-abduction-external rotation (Fig. 2.57), the therapist uses his distal and proximal hand grips at the dorsum of the hand and forearm (the common belly of long extensors). And then, he passively elongates the arm into extension-adduction-internal rotation with forearm pronated,

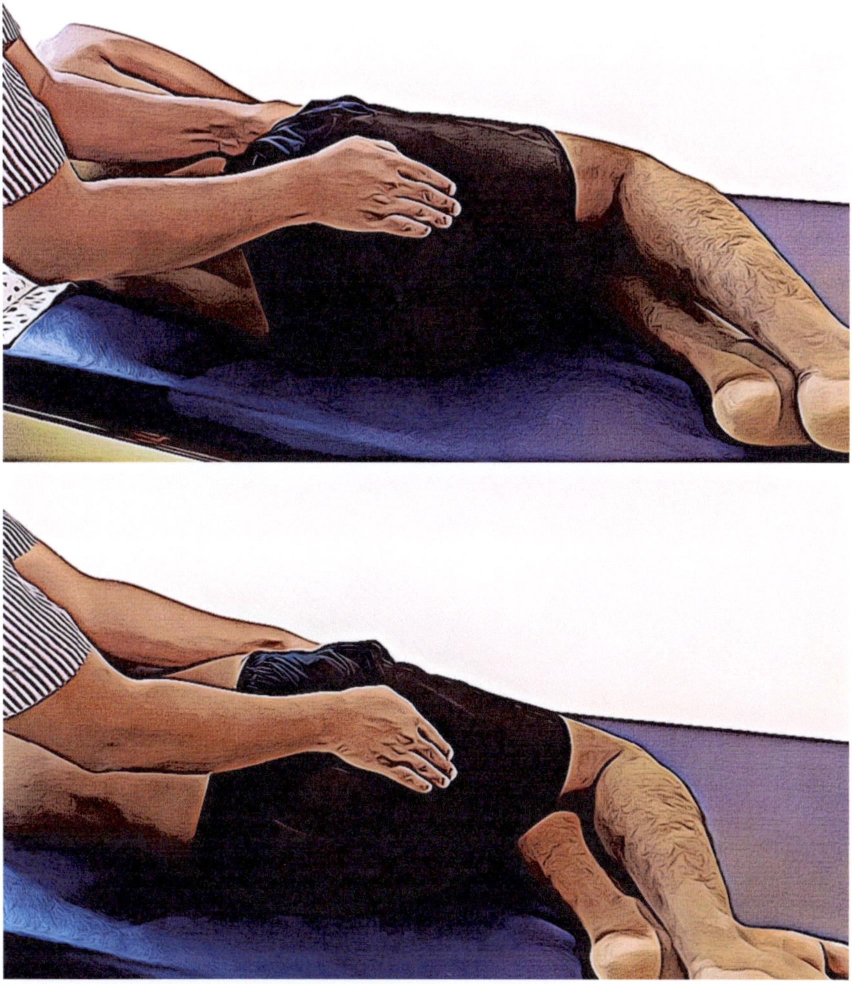

Animated photograph of model and therapist with permission

Fig. 2.56 Anterior-depression of the pelvis

wrist and fingers flexed (5 o'clock position). Following a brief pre-stretch, the therapist asks the patient to open the hand, and while maintaining the muscle tension, the patient is instructed to continue to pull the arm next to the ear and the thumb facing the floor (11 o'clock position). The reversal of movement pattern is extension-adduction-internal rotation (Fig. 2.58). The therapist alters his lumbrical grips at the palm and proximal forearm. The patient makes a hand fist and pushes his arm towards the opposite hip against the therapist's optimal manual resistance (towards the 5 o'clock position).

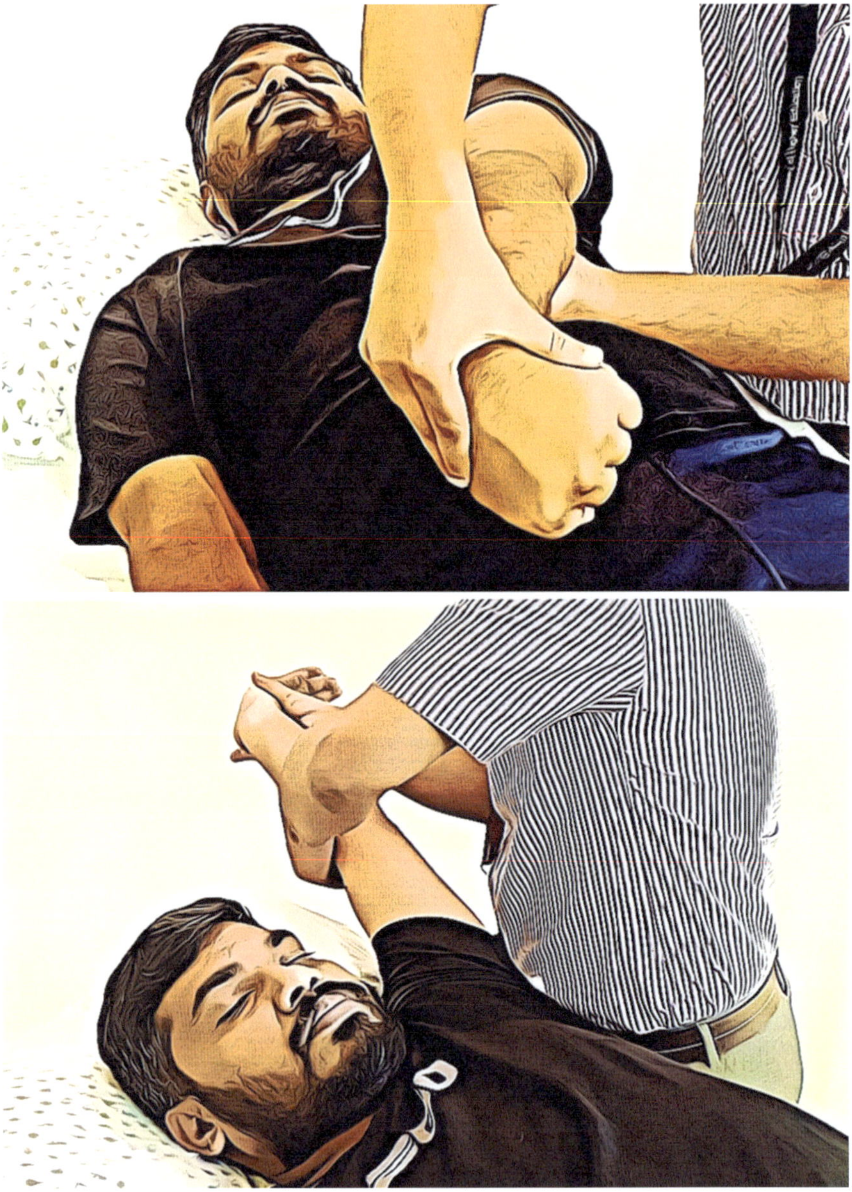

Animated photograph of model and therapist with permission

Fig. 2.57 Flexion-abduction-external rotation of the arm

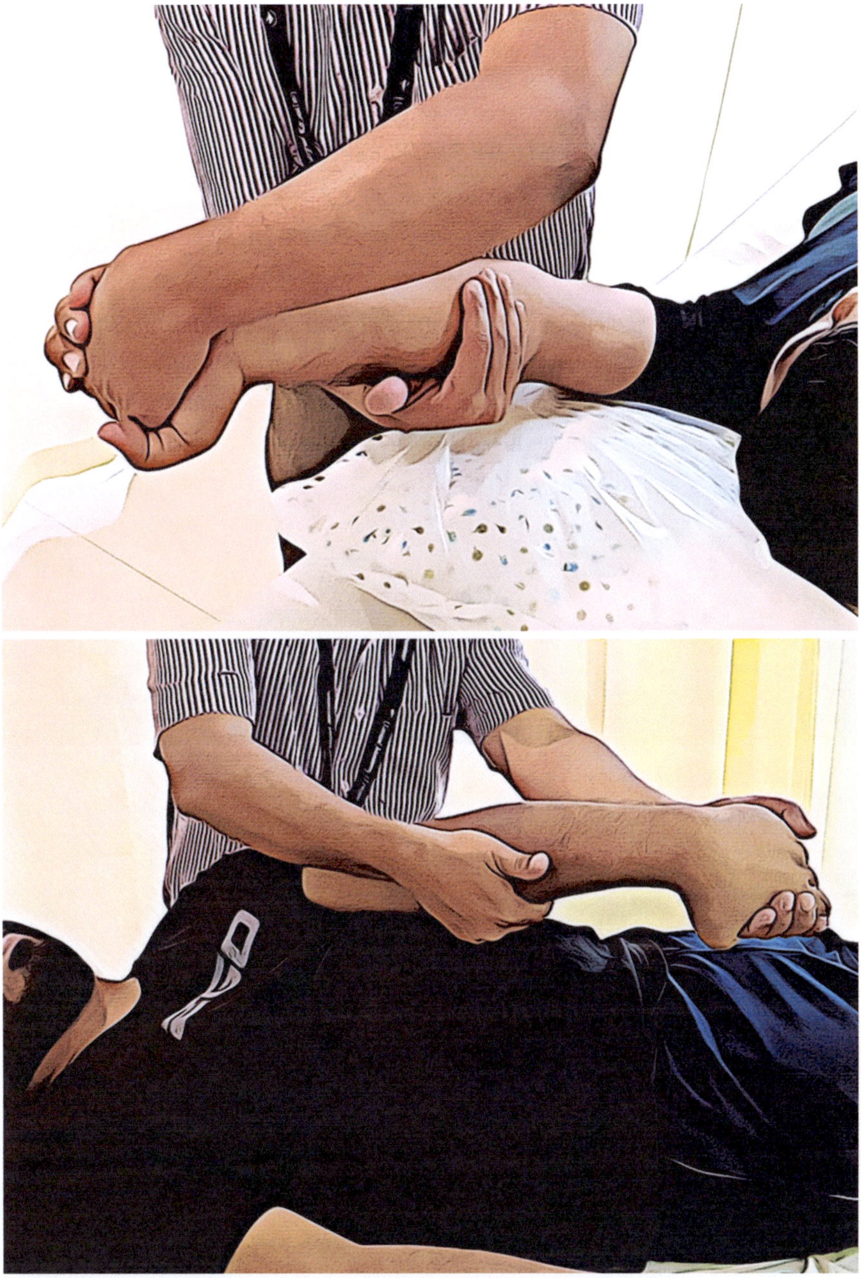

Animated photograph of model and therapist with permission

Fig. 2.58 Extension-adduction-internal rotation of the arm

For flexion-adduction-external rotation (Fig. 2.59), the therapist initially guides the patient to keep the arm at the side of the body with hand opened and then instructs him to make a fist and move as if to punch across the nose (from 7 o'clock to 1 o'clock position). The reversal of movement is the extension-abduction-internal rotation (Fig. 2.60) of the arm. The therapist must manually resist or guide the

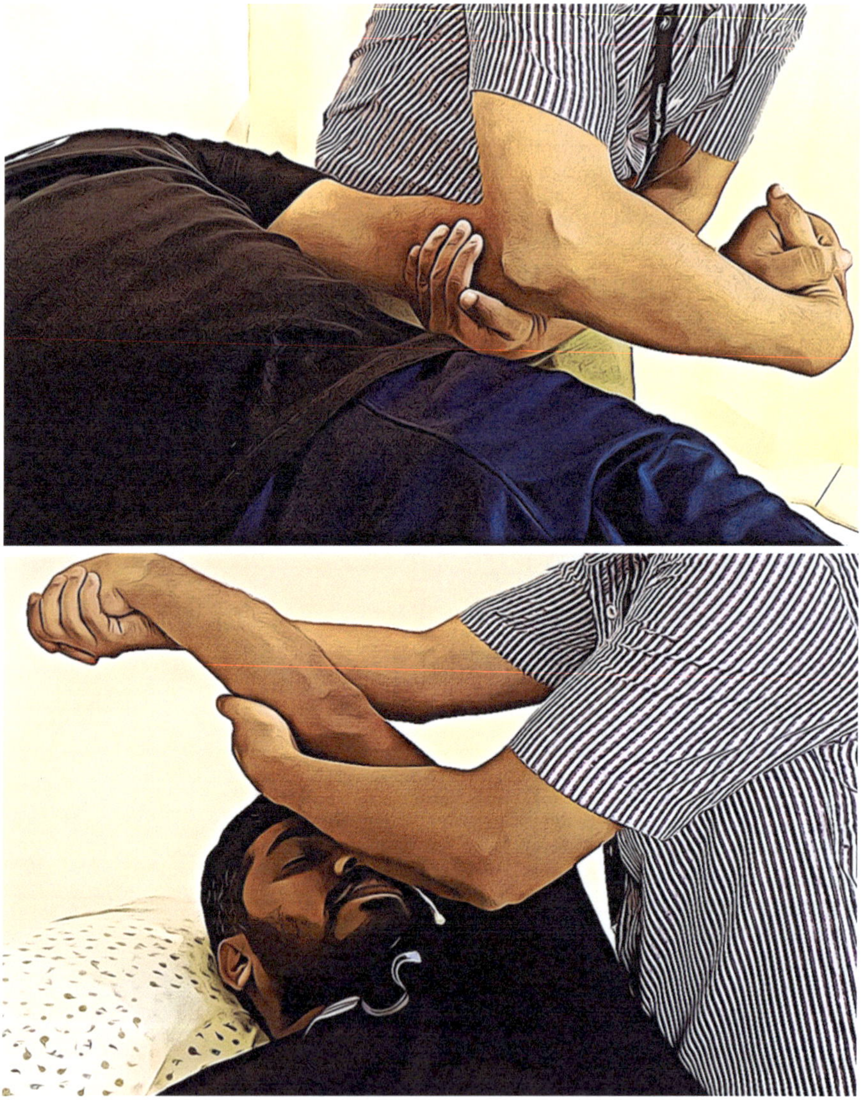

Animated photograph of model and therapist with permission

Fig. 2.59 Flexion-adduction-external rotation of the arm

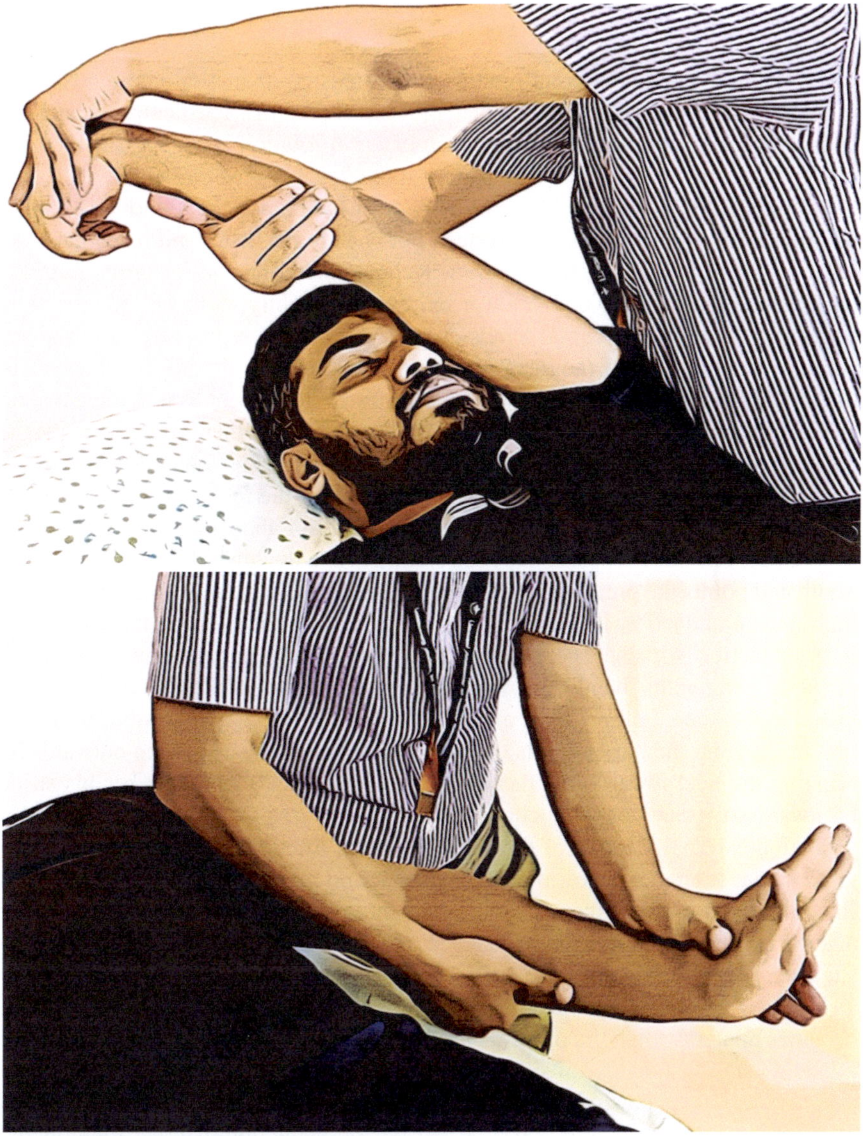

Animated photograph of model and therapist with permission

Fig. 2.60 Extension-abduction-internal rotation of the arm

diagonal movement pattern according to the patient's efforts. The therapist shall wisely use his body position and body mechanics so that he controls the patient's movement pattern within the groove of the diagonal plane.

Leg Diagonal Patterns

The flexion-adduction-external rotation and extension-abduction-internal rotation of the lower limb are co-planar diagonal movement patterns. The flexion-abduction-internal rotation and extension-adduction-external rotation are other diagonal co-planar patterns. Unlike the arm patterns, the leg patterns are relatively more comfortable to guide as the degree of movement freedom at the hip joint is less. In addition to the above, the therapist need not change his distal and proximal hand grips constantly while applying various techniques in diagonal patterns. With the patient in supine lying and the lower extremities slightly abducted, the therapist keeps his distal and proximal hands at the dorsum of the pronated foot and the anterio-medial aspect of the distal thigh, respectively. The therapist applies slight traction to the whole lower limb and then instructs the patient to pull the ankle up and bring the heel above the opposite knee. During this, the therapist must guide the patient's hip into the flexion-adduction-external rotation (Fig. 2.61). The therapist's hand placements are at the plantar aspect of the sole of the foot and posterior-lateral aspect of the distal thigh to encourage the extension-abduction-internal rotation (Fig. 2.62). The patient pushes the limb sideways and downwards to the edge of the plinth and points the great toe towards the floor. The starting position for the lower limb flexion-abduction-internal rotation (Fig. 2.63) is that the patient's lower limb is in adduction with heels touching each other. The therapist stands facing the patient's feet, and his hand placements are on the dorsum of the foot and superior-lateral to the knee. Following an elongation of the leg and a quick stretch to pronation of the foot, the patient pulls the ankle up and then flexes the hip outwards by bringing the heel towards the therapist. Meanwhile, the therapist takes his stride backward to guide the abduction-internal rotation of the hip. For extension-adduction-external rotation of the leg (Fig. 2.64), the therapist changes his hand-grips with a supinated forearm into the plantar and posterior-medial aspects of the foot and distal thigh, respectively. The patient needs to push the lower extremity against the therapist's manual resistance and approximates the heel to the opposite heel.

The PNF pattern is unilateral when performed in a single extremity. The pattern becomes bilateral when the movements are performed by both the extremities. The bilateral patterns could be of symmetrical and asymmetrical combinations. If both arms are performing a similar pattern, i.e., flexion-abduction-external rotation, it is bilaterally symmetrical (Fig. 2.65). When flexion-abduction-external rotation and flexion-adduction-external rotation are performed simultaneously with both arms, the pattern becomes bilateral asymmetrical (Fig. 2.66). In case of both arms moving into the same diagonal pattern but in the opposite direction, for example, the left-sided upper limb performing the flexion-abduction-external rotation and the right-sided arm in the extension-adduction-internal rotation, the pattern becomes symmetrical reciprocal (Fig. 2.67). If the arms show two different diagonal patterns in the opposite direction (flexion-abduction-external rotation and extension-abduction-internal rotation), it is an asymmetrical reciprocal pattern (Fig. 2.68).

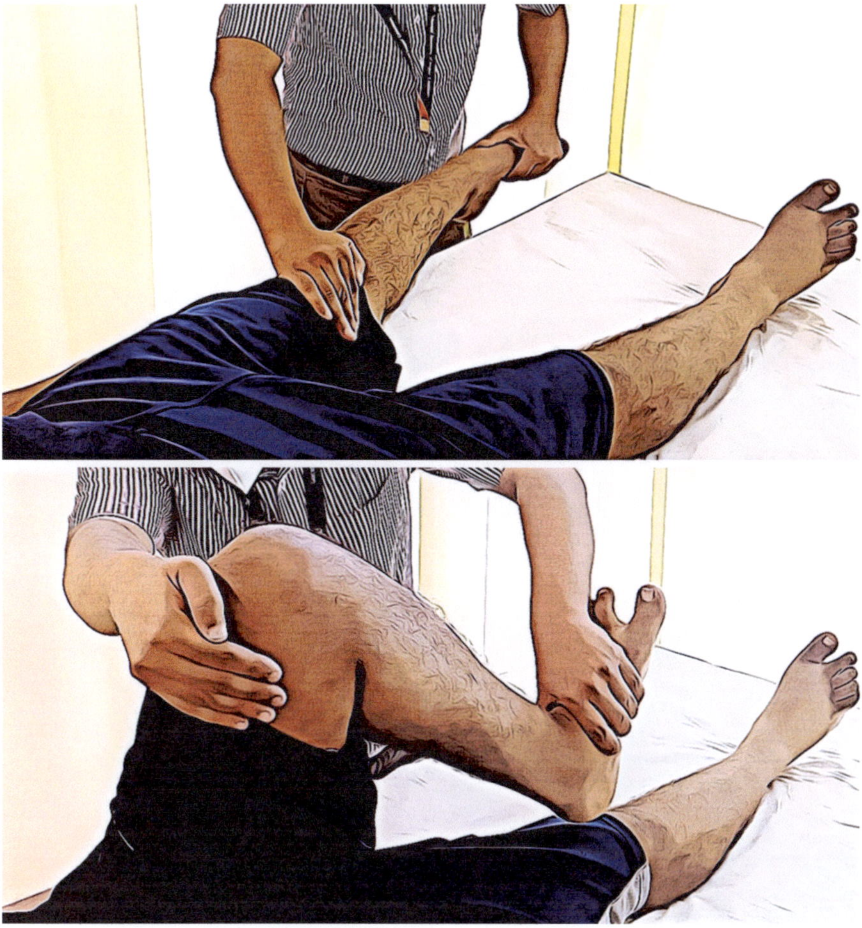

Animated photograph of model and therapist with permission

Fig. 2.61 Flexion-adduction-external rotation of lower limb

2.4.4.4 Neck Patterns

The neck patterns are of flexion or extension, combined with lateral flexion and rotation to the same side. These combined movements take place below the second cervical spine at the zygapophyseal joint. The direction of the lateral flexion-rotation of the neck movement to the right or left side specifies the neck pattern. For an example of left-sided neck flexion-rotation, the therapist stands behind the patient who is in a seated position. The therapist keeps his hand on the patient's chin (thumb in the middle and fingers below the chin) and other hand on the head (ventral-anterior-lateral to the left). From the starting position of neck extension-rotation to the right, the patient tugs the chin and brings it to the left-sided collar bone against the therapist's manual resistance (Fig. 2.69). The therapist slides his top hand on the

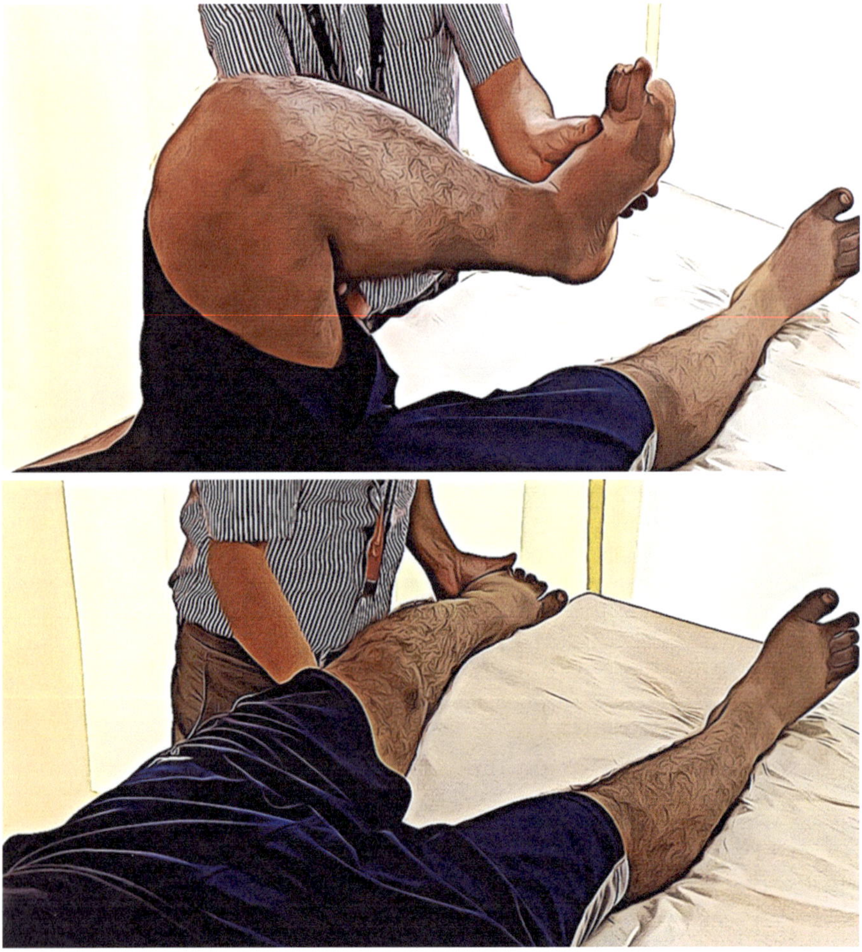

Animated photograph of model and therapist with permission

Fig. 2.62 Extension-abduction-internal rotation of the lower limb

ventral-posterior-lateral aspect of the patient's head to encourage the extension-rotation of the neck. The therapist resists the lateral flexion-rotation of the neck by placing the fingers lateral to the jaw or mandible (Fig. 2.70).

2.4.4.5 Trunk Patterns

The trunk patterns are similar to the neck pattern. The flexion (chopping) or extension (lifting) patterns are the coupled lateral flexion and rotation of the trunk to the same side. The patient shall begin trunk patterns either from the upper or lower

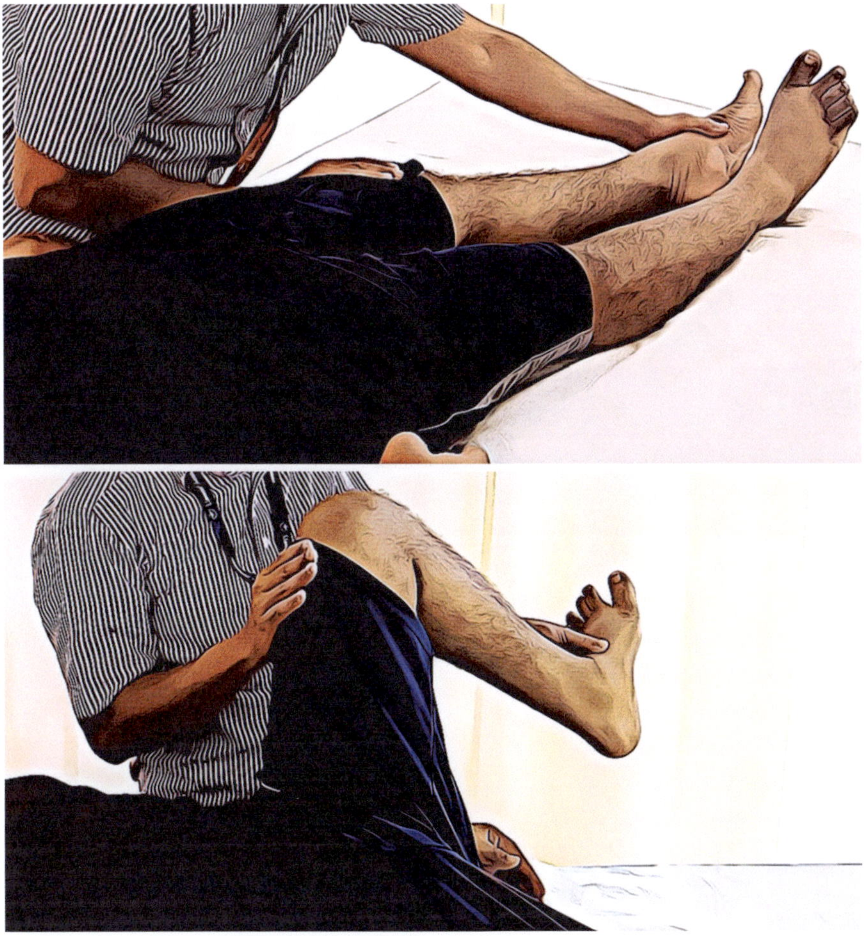

Animated photograph of model and therapist with permission

Fig. 2.63 Flexion, abduction and internal rotation of the leg

torso. The healthy bilateral arm patterns shall initiate the trunk pattern and irradiate the neuromuscular force into a weaker trunk. The therapist approximates the arms in chopping so that it recruits a re-stretch response to the abdominal and chest wall muscles. The chopping with trunk flexion is done with extension patterns of arms while exhaling the air. The patient practices the lifting with trunk extension while inhaling the air and reinforces it further by the bilateral asymmetrical arm flexion pattern. The therapist initially applies traction to the upper torso, followed by an approximation, the moment when the patient starts extending the spine. Figure 2.71 depicts the chop and lift movements performed by the patient and the therapist's stance and manual contact.

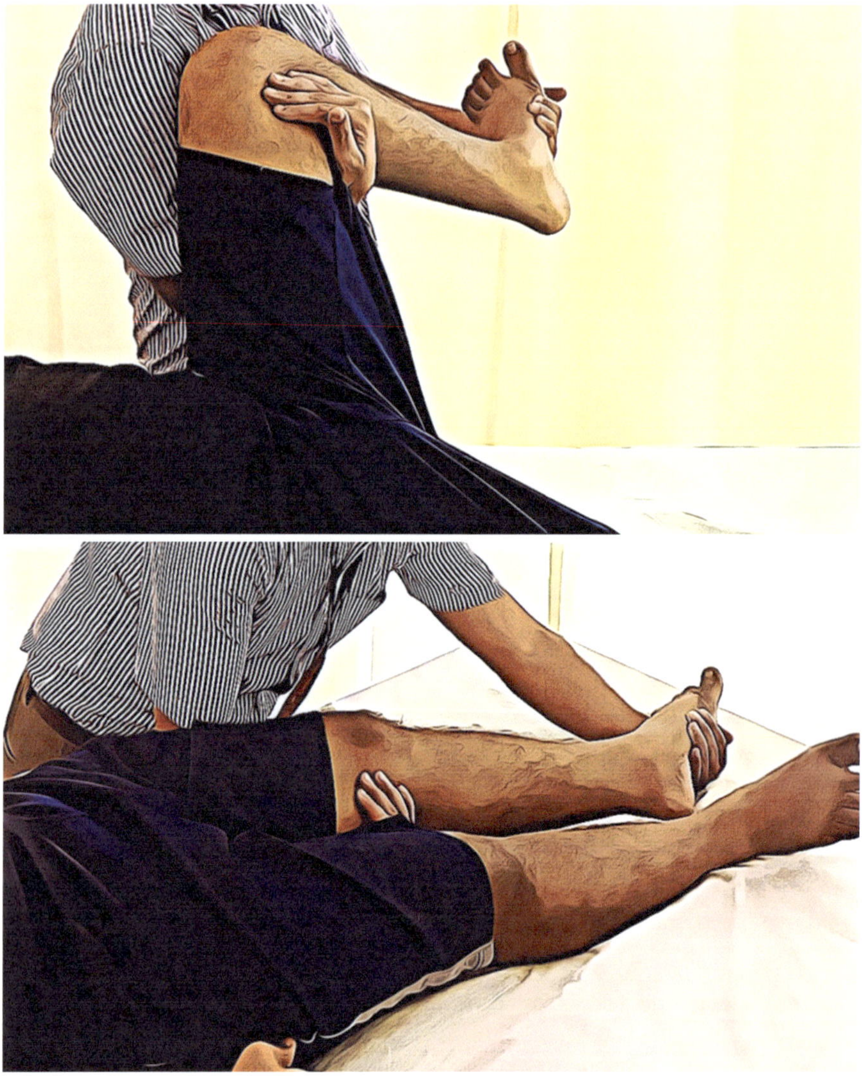

Animated photograph of model and therapist with permission

Fig. 2.64 Extension-adduction-external rotation of the lower limb

2.4.5 *PNF Techniques*

It is common for the patients to exhibit multiple issues of motor control, bed mobility, hand dexterity, balance, and functional mobility after neurological disease. Therefore, the physiotherapist shall choose appropriate treatment techniques in the preferred PNF patterns and help the patient practice them in different starting positions in the real-life functional context. The PNF techniques are either agonist or

Fig. 2.65 Bilateral symmetrical pattern of the arms

Animated photograph of model with permission

Fig. 2.66 Bilateral asymmetrical pattern of the arms

Animated photograph of model with permission

antagonist techniques based on the movement patterns, i.e., the activity of agonist group alone and both agonist and antagonist muscle groups, respectively. The rhythmic initiation, replication, repeated contractions, and combination of isotonic contraction techniques are the techniques used to activate and strengthen only the paralyzed agonist muscle groups. The dynamic reversal, stabilizing reversal, and rhythmical stabilization are the techniques applied to both the agonist and

Fig. 2.67 Symmetrical reciprocal pattern of the arms

Animated photograph of model with permission

Fig. 2.68 Asymmetrical reciprocal pattern of the arms

Animated photograph of model with permission

antagonist muscle groups. The former employs the isotonic muscle contraction, whereas the latter two methods utilize the isometric contraction. In many chronic neurological conditions, the long-lasting clinical problems are pain, maladaptation

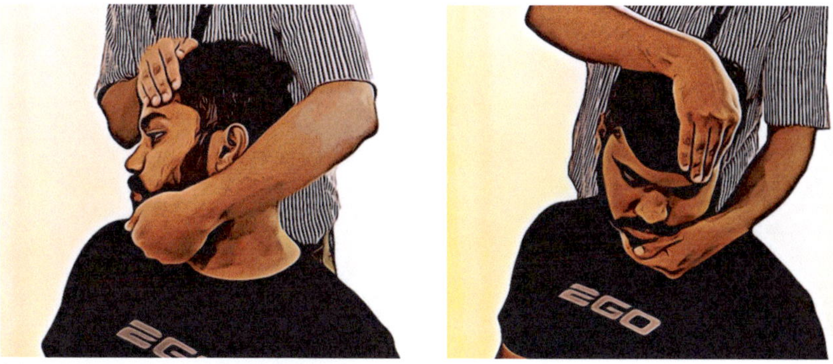

Animated photograph of model and therapist with permission

Fig. 2.69 Performing neck flexion rotation to left

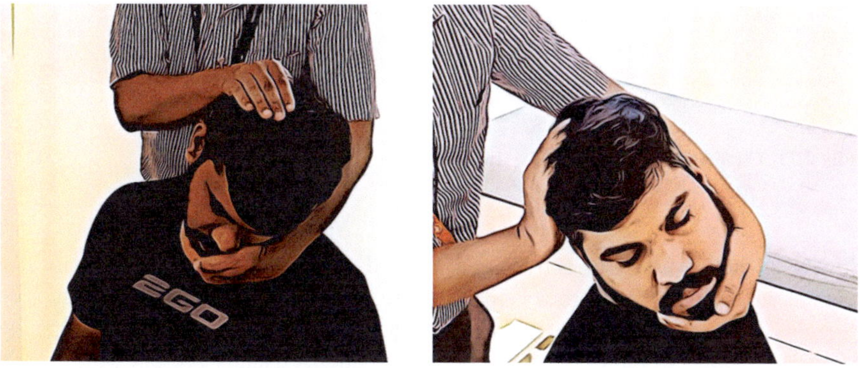

Animated photograph of model and therapist with permission

Fig. 2.70 Performing neck extension rotation to the right

of muscles, and soft-tissue stiffness that restrict the mobility of joint segments and limit the daily functional tasks. Hold–relax and contract–relax are useful soft-tissue elongation techniques to alleviate pain and increase joint mobility.

2.4.5.1 PNF Agonist Techniques

The PNF agonist techniques are aimed to activate and strengthen specific group of muscle synergies. The rhythmical initiation, replication, repeated contraction, and combination of isotonic contraction techniques involve the dynamic contraction of muscles in a particular PNF pattern.

Rhythmic Initiation In the rhythmic initiation technique, the patient performs the rhythmic movement of muscular agonist activity throughout the joint range of motion. Initially, the physiotherapist shows the movement pattern passively fol-

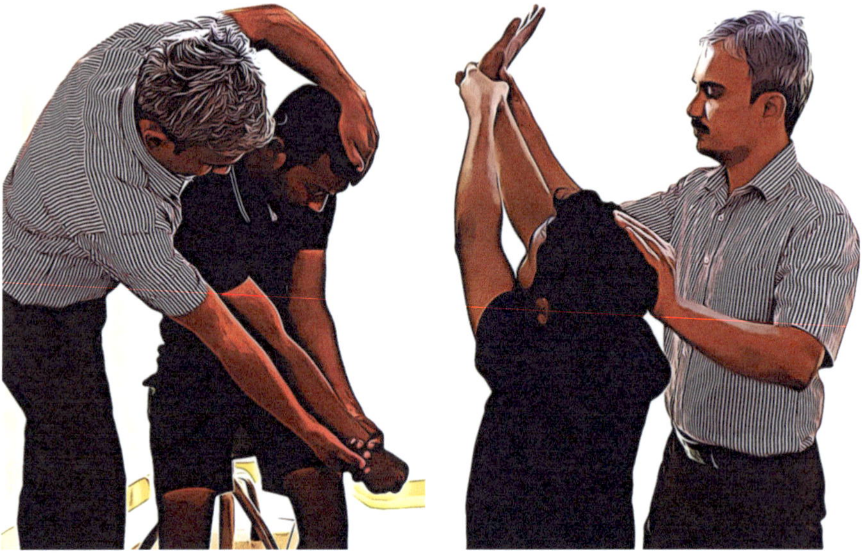

Animated photograph of model and therapist with permission

Fig. 2.71 Chop and lift patterns of the trunk

lowed by active-assisted, active, and resisted movement from the patient. The patient with neuromotor issues is made aware of the purpose of the technique. The therapist must use appropriate manual contact, auditory, and visuals cues. Following a vascular injury to the brain, the majority of patients usually present with muscle flaccidity, weakness, loss of motor control, and praxis issues. During this period, the rhythmic initiation is beneficial in improving the low muscle tone, learning a new movement, and coordination of muscle synergies. The rhythmic initiation can also modulate the hypertonicity in late-stage stroke recovery if performed opposite to the spastic muscle group. The rhythmic extension-rotation pattern of an upper trunk in a seated position is useful for managing the slouched sitting posture in patients with stroke. This is also helpful for treating the axial rigidity, bradykinesia, and stooped posture in patients with Parkinson's disease.

Replication (Imitation) Replication is an agonist technique that creates the kinesthetic awareness of movement patterns and joint position sense. Replication is a useful technique for patients suffering from sensory impairments following stroke, head injury and spinal cord injury involving the dorsal white column, Gillian-Barrie syndrome, and sensory polyneuropathy. The therapist brings the patient's limb passively into a chosen new position. The patient remembers that position by the isometric contraction of muscles. Afterwards the therapist brings the extremity into the starting position and then the patient actively brings the limb to the desired position.

Repeated Contractions (Re-stretch) Repeated contractions or re-stretch is an agonistic technique in which the therapist applies a brief stretch stimulus against the muscle contraction throughout the joint range of movement. Repeated contraction increases the spatial and temporal summation of the motor unit recruitment in patients having poor motor control and hypotonicity. It also improves muscle strength and increases the range of motion. Besides enhancing the resilience of the agonist muscle groups with a re-stretch stimulus, it might also modulate the hypertonicity of antagonist muscle groups by the reciprocal inhibition phenomenon.

Combinations of Isotonic Contraction This is another agonist technique in which the patient performs the pattern in a combination of concentric, static, and eccentric contractions without losing the muscle tension. The patient initially contracts the muscular agonist pattern concentrically against the therapist's resistance. Following a brief static contraction without losing muscle tension, the patient elongates the same agonist muscle group eccentrically. The therapist does not change the handgrip while administering the combination of isotonic technique. The combination of isotonic contraction is useful in gaining the motor control and strength of proximal or distal muscle groups in neurological diseases like myopathies and neuropathies as well.

2.4.5.2 PNF Antagonist Techniques

The PNF antagonist techniques aimed to activate and strengthen the two or more groups of muscle synergies in opposite directions. The dynamic reversal involves the isotonic contraction of muscles in both the flexion and extension PNF patterns. On the other hand, there is a static contraction of muscle groups in stabilizing reversal and rhythmical stabilization techniques.

Dynamic Reversal In a dynamic reversal technique, the patient performs the concentric contractions of the agonist muscle group followed by the antagonist in a reversal way, not having relaxation between those two patterns. The patient with neuromotor dysfunction performs the dynamic reversal beginning with a strong muscle group and ending in weaker ones. The therapist never changes the distal hand grips simultaneously. The therapist alters his distal handgrip after the moment the patient activates the reversal of the antagonist pattern. Dynamic reversal improves the endurance, strength, and inter-muscular coordination of the agonist and antagonist muscle groups.

Stabilizing Reversal Stabilizing reversal is a technique in which isometric contraction of agonist and antagonist occurs without relaxation of both muscle groups; thus, the steadiness of desired positions is encouraged. Resistance to isometric contraction is initially given to the more substantial muscle group and then to the antagonist muscle group. The therapist applies sufficient approximation and optimal resistance to all three components of the diagonal pattern. The patient is required to build static muscle contraction using central processing in response to the therapist's handgrips and resistance. Dysfunction of balance and mobility is common not only in a supratentorial lesion but also in cerebellar ataxia. Hypotonicity and impaired muscular co-activity of trunk muscles and proximal girdle muscles are the possible reasons for truncal instability and incoordination of the extremities. The stabilizing reversal of the scapular and pelvic patterns in a side-lying position and the neck and trunk patterns in the seated position are beneficial in facilitating the proprioceptors and muscular co-activity, therefore, overcoming truncal ataxia and postural instability.

Rhythmic Stabilization In rhythmic stabilization, the isometric (static) contraction of the agonist and antagonist is encouraged in a reversal without losing muscle tension. Contrary to stabilizing reversal, the therapist does not change his hand grips in the rhythmical stabilization while controlling the static contraction of the patient. Rhythmical stabilization can be beneficial for those who are having hypotonicity, poor balance control, and truncal ataxia. The patient uses feed-forward postural control strategies to build up static muscular contraction against the destabilizing force from the therapist.

2.4.5.3 PNF Elongation Techniques

Hold-relax and contract-relax are elongation techniques, also called PNF relaxation techniques. Although both these techniques have similar therapeutic benefits of soft-tissue tightness and pain relief, the former employs isometric contraction at the end feel range, wherein the joint motion is not permissible. Contrary to that, the patient actively contracts an isotonic muscle activity at the end-feel range in contract-relax. There is a voluntary relaxation of muscles in both the techniques post-contraction.

Hold–Relax The hold–relax is an elongation technique in which static contraction is build-up followed by voluntary relaxation. At the end-feel range of movement or painful range, the patient builds up a static contraction in the opposite direction against the therapist's resistance, followed by a voluntary relaxation. After relaxation, the patient actively brings the extremity or neck into a new range in a chosen pattern. Else, the therapist brings it passively into a preferred position. The primary goals are to reduce spasticity, increase joint range of movement, and pain relief. The autogenic inhibition, reciprocal inhibition, and gate control theory are the possible explanations for the pain relief using hold relax.

Contract–Relax The contract–relax is an elongation technique in which isotonic contraction is built-up followed by voluntary relaxation. In contract–relax as opposed to hold-relax, the therapist gives a dynamic command at the end of the movement. The patient performs the isotonic contraction of muscle at the end of the chosen pattern against the therapist's resistance. After voluntary relaxation, the desired action in reverse is achieved actively by the patient or passively by the therapist. The maximum and sub-maximum voluntary contraction have similar effects in increasing the range of motion. The primary goals are to reduce stiffness, increase flexibility and improve joint mobility, and relieve pain. PNF stretching is much better than a static stretch of lower limb muscles in modulating the gait parameters.

2.4.6 PNF in Stroke Rehabilitation

The discharge plan following hospital admission for acute stroke depends on the independent sitting balance and also on the absence of secondary medical complications. Immediately post-stroke, the patients experience flaccid weakness of the extremities and truncal hypotonicity. Due to poor trunk control in the initial stages of stroke recovery, the patients find difficulty even in seemingly simple bed mobility tasks of rolling over within a bed and getting up to sitting over side of the bed and unsupported sitting postures. The patients need to get out of the bed at the earliest to achieve basic self-care needs. Additionally, the muscular inactivity and weakness of lower limb muscles contribute to poor sit-to-stand transitions, impaired standing balance and walking difficulty. The proximal scapular instability, intra-limb muscular incoordination, and clumsiness of hand dexterity could affect the overall arm functioning.

As the patients start regaining muscle tone and some voluntary control in the extremities during the subacute stroke recovery phase, they show compensatory movement behavioral strategies for meeting their daily functional needs. Besides learned non-use of extremities, the muscular disuse and soft-tissue adaptations that the stroke survivors develop in late subacute and chronic phases will potentially affect their true motor recovery, thereby restricting the functional activities. The PNF techniques utilizing the basic principles and procedures are helpful for the motor recovery of post-stroke patients in various stages of disease chronicity. The subsequent paragraphs explain the PNF treatment strategies that aid the motor recovery of stroke patients.

2.4.6.1 Bed Mobility After Stroke

There are many PNF treatment strategies available for bed mobility and an optimal trunk performance is conditional for achieving the same. The patients may prefer rolling over either initiated from the upper trunk or the lower trunk. In a quarter-turned side-lying position with a neutral spine and flexed lower limbs, the recruitment of abdominal muscles is more feasible using scapular and pelvic patterns. The therapist asks the patient post-stroke to bring the elbow towards the umbilicus, a scapular anterior-depression pattern, reinforced by extension-adduction-internal rotation of the arm to overflow the nerve signals to the abdominals. The rhythmic initiation, replication, repeated contractions, and combination of isotonic contraction are the possible agonist techniques to activate and strengthen the abdominal muscles. Alternatively, the patient brings the anterior superior iliac spine by pulling the top knee towards the umbilicus using the flexion-adduction-external rotation pattern of the lower limb. As the patient gains strength in the abdominal muscles, the therapist resists the isometric contraction of the scapular anterior-depression pattern and encourages the patient to dissociate the lower trunk and pelvis into flexion-rotation isotonically and vice versa.

From the elongated trunk (posterior-elevation of scapula and posterior-depression of the pelvis) in quarter-turned supine lying, the patient can roll over into side-lying position by performing the anterior-depression of the scapula and anterior elevation of the pelvis together. The opposite pattern would facilitate the rolling-over back into a supine position. The combination of scapula and pelvis movement patterns in different directions and diagonal planes shall further improve the dissociated movement between the upper and the lower trunk, i.e., the ability of the trunk to coordinate. The dynamic reversal and stabilizing reversal techniques of the scapular and pelvic patterns are beneficial for trunk proprioception and later transferring the benefits on the sitting postural stability. The rhythmical stabilization exercise in the prone sequence positions (prone-on elbow, quadruped, and kneeling) might be useful for the dynamic stability of neck and trunk. Plus, the flexion-rotation of the head-neck pattern can spatially and temporally summate the abdominal muscles' recruitment in the crook-lying position.

2.4.6.2 Upper Limb Stroke Recovery

The primary goals of upper limb rehabilitation early after stroke are to initiate the voluntary contraction of paralyzed muscles and improve scapular stability. The proximal control of the shoulder girdle muscles is conditional for the coordinated synergetic movement patterns of the arm, required for many daily hand–arm functioning. If the stroke patient lacks voluntary initiation of the affected upper extremity, the indirect approach of resisted scapular and arm patterns on the sound side is useful. Based on the concepts from mirror neurons in the cerebral cortices and imitation synkinesis of the upper extremities, the irradiation of the strong muscles overflows the nerve impulses to the weaker ones. The patient will be able to initiate and reinforce the scapular movements in a better way in the side-lying position. The therapist initially administers the rhythmical initiation, imitation, and repeated contraction techniques on the patient's paralyzed scapula, followed by the combination of isotonic contractions and dynamic reversal. In this way, after gaining motor control in scapular patterns, the desired muscle groups are strengthened for better scapular stability. As mentioned in the previous section, supine lying is the best suitable position for practicing the arm patterns since the patient needs to work less against gravity and can keep his vision on the movement.

During the spontaneous and chronic stroke recovery stages, the predominant atypical upper limb flexion is the most deliberating synergy pattern. A stroke patient may develop disuse or learned non-use of muscle weakness in the opposite muscle group, thus limiting the abduction and external rotation of the shoulder joint and over-head arm movement. Both flexion patterns of the upper limb have the external rotation component; therefore, the patient can begin practicing the flexion-adduction-external rotation and flexion-abduction-external rotation for achieving overhead arm and grooming activities. The therapist should consider facilitating the anterior elevation of the scapula while training the arm flexion-adduction-external rotation. Similarly, the posterior-elevation combines with a flexion-abduction-external rotation of the arm. Once the patient gains motor control in the desired movements, these PNF techniques are then practiced in a seated position while doing the daily arm tasks such as reaching and holding a cup, toothbrush, and a book. In case of weak shoulder muscles with difficulty to perform lifting of the arm, the timing for emphasis is beneficial to increase the muscular recruitment and contraction of external rotators or abductors, utilizing the stronger components of arm patterns. The therapist can thus help patients to perform scapula and arm techniques in functional postures other than in lying or seated position.

Another essential aspect of arm recovery is regaining the wrist extension and handgrip for the many daily hand tasks and arm-reaching activities. The therapist can select either flexion-abduction-external rotation or extension-abduction-internal rotation as these two arm patterns have the component of wrist extension. Adding the elbow flexion component into the flexion-adduction-external rotation will improve the grip strength. The motor learning effect on the wrist and fingers extension and grip strength can further be promoted by timing for emphasis and irradiation from the shoulder complex. Many stroke patients can actively flex their wrist and fingers but cannot open the fingers owing to the tightness of long flexors of the forearm. Based on the understanding from the system's model of motor control, the hand dysfunction in chronic stroke recovery is not only affected by the neuromotor performance but also the soft tissue restrictions and joint stiffness of the wrist. For instance, in case of a spastic hand, the therapist moves the patient's wrist into an end-feel range. And then asks the patient to contract the long flexors of the forearm isometrically (hold-relax) or isotonically (contract-relax). Following a brief contraction of the long flexors of the forearm, the patient relaxes it by reciprocal inhibition. And then, the therapist either passively takes the wrist into a new extension range, or the patient actively extends the wrist (successive induction). In a typical presentation of the upper limb in chronic stage stroke recovery, besides spastic hand, the forearm pronators, elbow flexors, shoulder adductors, and internal rotators also develop soft tissue tightness. The patients shall voluntarily contact the agonist muscle group (extension-adduction-internal rotation pattern of the arm with elbow flexion) at the end-feel, followed by active shoulder flexion-abduction-external rotation with elbow extension.

The soft tissue stiffness not only restricts the joint mobility but also results in muscle aches and cramps in those who regain sensation post-stroke. Manual therapists use passive mobilization techniques for lengthening the soft tissue tightness and to overcome joint stiffness and pain. Unlike manual techniques, there is always voluntary participation of the patient while performing the lengthening PNF techniques as opposed to manual strategies. The PNF movement techniques of hold-relax and contract-relax are more applicable in the aforementioned condition, thus promoting the motor control in a newly attained range of movement which is in addition to improvement in joint mobility and muscle flexibility and alleviation of pain due to the possible stimulation of mechano-receptors and increased blood supply.

2.4.6.3 Sitting Balance After Stroke

The PNF treatment concepts in the sitting position not only help the stroke patient to regain postural stability but provide an opportunity for upper and lower extremities recovery. While a stroke patient is seated with 90° hips and knees flexed and feet on the floor, the therapist guides the trunk extension-rotation with rhythmical initiation. The therapist applies strong approximation against the erect torso using stabilizing reversal and rhythmical stabilization techniques further to increase the coactivity of trunk muscles and postural stability.

In a seated position, the indirect approach of a combination of isotonic contraction (flexion-adduction-external rotation of arm with elbow flexion) performed on the strong arm overflows the nerve impulses to paralyzed upper limb (extension-abduction-external rotation of arm with elbow extension), therefore, promoting the weight-bearing capacity towards the more affected side. The dynamic reversal technique of bilateral arms (asymmetrical reciprocal pattern) improves the intra-limb and inter-limb muscular coordination and also postural stability. The bilateral symmetrical extension pattern of upper limbs encourages the stroke patient for forward trunk inclination with better weight-bearing between the feet, thus enabling the sit-to-stand performance.

2.4.6.4 Standing Balance and Walking

Weight acceptance towards the affected lower limb is a precursor for functional mobility since 60% of the gait cycle involves the stance phase. A majority of mobility tasks require a phasic burst of flexor and extensor chain of muscular activity. For example, when stair climbing or stepping over obstacles while walking outdoors, a post-stroke patient finds poor intra-muscular coordination between flexion-adduction-external rotation (with knee flexion facilitating the swing phase) and extension-abduction-internal rotation (with knee extension facilitating the stance phase) muscles synergies. Therefore, the dynamic reversal technique of these leg patterns might be useful for functional walking. In addition to the above, the bilateral symmetrical and asymmetrical reciprocal patterns of lower limbs in a high-sitting position improve the coordination between the legs. The dynamic reversal technique of bilateral arms (asymmetrical reciprocal pattern) in standing facilitates the weight-bearing and stepping capabilities of the affected lower limb. The application of the basic principles and procedures of PNF with manual contact and joint approximation at the iliac crest facilitates the sit-to-stand, forward, and lateral weight transfer in standing, walking, and stair climbing. The author recommends the reader to refer to the standard PNF textbooks for learning the treatment techniques further.

2.4.6.5 Chest PNF After Stroke

The PNF techniques, when applied with proper grips and optimal positions (supine, side-lying, and sitting), are clinically useful for those suffering from breathing dysfunction, particularly following a brainstem stroke. For diaphragmatic breathing, the patient is positioned supine with knees slightly flexed. The therapist places his hands a few inches above the patient's umbilicus area and applies a quick stretch stimulus at the end of expiration to initiate inspiration. The therapist needs to use mild manual pressure or resistance to the abdominal motion for the facilitation of the diaphragm muscle. To facilitate the lower costal breathing pattern, the physiotherapist stands at the head end of the patient and places his hands at the lower part of thoracic-cage fingers pointing along the ribs. The therapist applies pressure towards the caudal and centripetal directions while the patient is inhaling a deep breath. The normal diaphragmatic breathing with optimal chest-wall compliance might probably minimize the use of the accessory muscles of respiration and further prevent from developing the hypoxia cascade. The chop and lift trunk patterns might also drive neural inputs to the intercostal and diaphragmatic muscles. The greater force from the diaphragm apposition zone accounts for better thoracic wall awareness and compliance.

2.4.7 PNF in Other Neurological Conditions

Besides stroke, the PNF treatment concepts and techniques are beneficial in many neurological diseases. The neurological level of spinal cord injury predicts the functional potential of individuals. Incomplete thoracic spinal cord injury patients with functional strength (scoring at least three on manual muscle testing) in 50% of major lower limb muscles can independently walk using assistive aids. Even though many thoracic spinal injury patients can walk with orthosis and crutches, they prefer to use a wheelchair for their functional mobility. The primary therapy goal in these individuals is that of strengthening and conditioning the intact muscles for meeting their daily needs. The description of the PNF treatment ideas in spinal injury is beyond the scope of this chapter. However, a few of the PNF strategies listed below might be useful for understanding the underlying concepts in managing a person with lower cervical spinal cord injury.

A young person with C6-complete spinal cord injury and absence of co-morbidity can perform tenodesis hand function, bed mobility, and wheelchair skills, except for standing. The latissimus dorsi, pectoralis major, serratus anterior, and wrist extensors are the intact muscles in C6-complete spinal injury. The therapist may strengthen these muscles using preferred PNF techniques and patterns. For instance, the latissimus dorsi plays a crucial role in balancing and lifting the body into a seated position. The latissimus dorsi depresses the scapula in a closed-chain activity of lifting and wheelchair transfer, along with an active contraction of wrist extensors and anterior deltoid that shifts the ground reaction force posterior to elbow joint, overcoming the paralyzed triceps. The PNF antagonist techniques are helpful to improve the scapular girdle stability. The combination of isotonic contraction into scapula posterior-depression and shoulder extension-abduction patterns could strengthen the latissimus dorsi, deltoid, and wrist extensors muscles, thereby relieving the pressure on the ischial tuberosity in sitting, as well as enhancing wheelchair transfer skills. The patient needs strong wrist extensors and biceps to maneuver the legs into the support surface during transfers. The patient hooks the extended wrist below the knee and then either lifts the leg or extends it into long-sitting. The combination of isotonic contraction of arm flexion-abduction-external rotation pattern with elbow flexion is useful for achieving the same. This technique is also helpful to hold an object with tenodesis action of the hand and bring it to the face for grooming activities.

With deinnervated triceps and abdominal muscles, the C6 spinal cord injury patient finds it difficult to perform bed mobility. Besides the aforementioned classical PNF patterns, there are two thrust and withdrawal combinations, i.e., radial and ulnar thrusts, which are useful for rolling over in bed and getting up to a seated position. In the ulnar thrust combinations, the patient pushes the shoulder into the flexion-adduction-external rotation along with anterior-elevation of scapula and elbow extension. In contrast, the forearm, wrist, and fingers would pronate and extend with ulnar deviation, respectively, i.e., the opposite to basic pattern (Fig. 2.72). The patient moves the shoulder into extension-adduction-internal rotation with anterior-depression of the scapula during the radial thrust while supinating the forearm and extending the wrist and fingers (Fig. 2.73). The withdrawals are

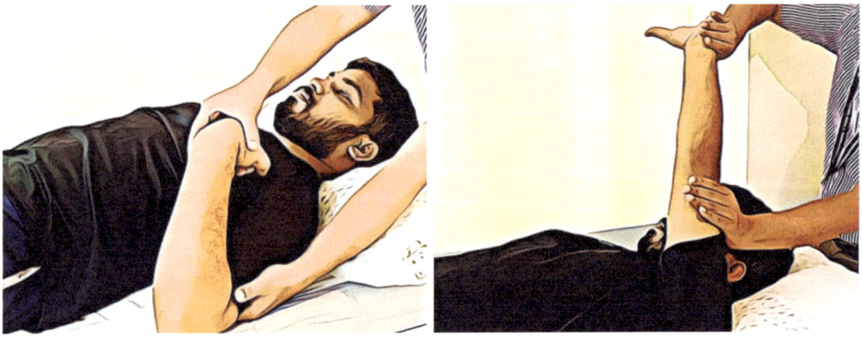

Animated photograph of model and therapist with permission

Fig. 2.72 Ulnar thrust

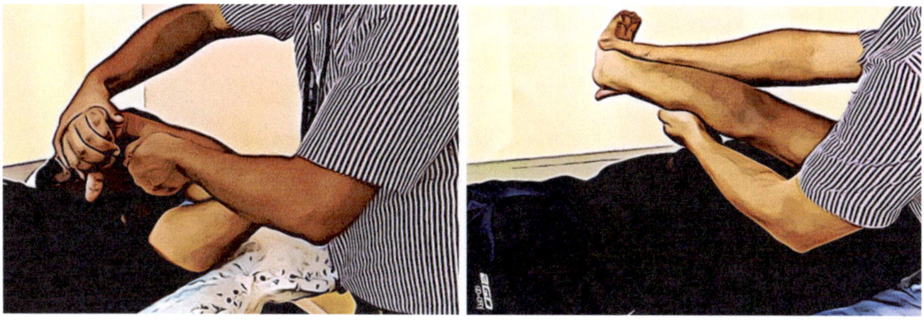

Animated photograph of model and therapist with permission

Fig. 2.73 Radial thrust

pulling movements of the arm and reversals of thrust pattern. For radial withdrawal pattern, the patient brings the shoulder into flexion-abduction-external rotation with posterior-elevation of the scapula with forearm supination and wrist and fingers flexion.

The PNF techniques can also positively influence the bradykinesia, axial rigidity, and balance issues of people with Parkinson's disease. The classical clinical observation in Parkinson's disease is that of limited motion of the torso and extremities. Besides the slowness of movement initiation, they show less arc of joint mobility while performing the majority of daily needs. As the disease status progresses, they experience balance dysfunction and fall risks. The axial rigidity and stooped tendency of the trunk would further limit their sit to stand and turning capacity during functional waking. The neutral extension of the spine is conditional for the axial rotation of the thoracic spine, i.e., the trunk dissociation and thoracic wall compliance. The rhythmical initiation technique, otherwise the rhythmical extension-rotation of the thoracic spine in prone-sequence positions (prone on elbow, quadruped, modified quadruped, kneeling) and sitting could be of interest to improve the postural control and breathing in Parkinson's disease. Since the PNF classical patterns are large amplitude movements of the extremities, these big motions shall address the bradykinesia and rigidity in patients with Parkinson's disease.

To summarize, a sound knowledge of neurophysiology and understanding the philosophical concepts of PNF are a must for physiotherapists. Besides that, they also need to develop hands-on skills for better delivery of PNF techniques in their clinical practice. Contrary to the common belief that PNF is a sensory-motor approach, it addresses the activity limitation and participation restriction of the international classification of functioning, disability, and health frame of reference. The focus of task practice is to enhance the ability of functional performance. The PNF, when combined with neurotherapeutic strategies may show a better quality of motor recovery and daily functioning among people with neurological dysfunction.

2.5 Motor Relearning Program

Abraham M. Joshua

2.5.1 Introduction

Janet Carr and Roberta Shepherd, the two Australian physiotherapists, known for their work in the field of physiotherapy education and research, developed this approach. Carr and Shepherd recognized the importance of motor learning and skill acquisition and motor control mechanisms relevant to rehabilitation practice. With her colleague Shepherd, Carr also wrote academic texts and their main aim in writing such scholarly materials was to reconcile the gap between research and practice in physiotherapy. Several neuro-physiotherapists consider them as the pioneers in guiding academicians and clinicians in the evidence-based practice concepts that are widely used during the present day. Based on their clinical experience and the principles of neuroscience, motor control and learning, exercise physiology, biomechanics, cognitive psychology, and human ecology, Carr and Shepherd (1987, 1989, and 2003) developed this approach to provide guidelines for post-stroke training program.

2.5.2 General Principles

The Motor Relearning Program (MRP) is a task-oriented approach to improve motor control and relearn activities of daily living. The approach is firmly based on kinesiological theories that emphasize a distributed motor control model than a hierarchal model of control. It is an approach based on a model of normal motor learning, which involves the elimination of unnecessary muscle activity, feedback to modify the motor output, practice of the motor activity, and the interrelationship between posture and movement.

Both the developers of this approach had questioned the quality of rehabilitation of their times, which was based on outdated ideas of exercise therapy, the need to wait for recovery to occur, negative expectations, and lack of therapist's enthusiasm to pursue the goal of ensuring the best possible functional recovery. Carr and Shepherd believed that many of the traditional concepts were developed based on the interest in treatment techniques than analyzing the motor problems.

They believed that the result of rehabilitation could be different only if the physiotherapists recognize the real potential of physiotherapy, discard the ineffective techniques, understand the need to develop a new method of treatment, and research the effectiveness of new treatment methods, thus improvising the general framework of stroke care. The MRP promotes an active role for the patient and differs from the Bobaths approach, in which patients often are passive recipients of the treatment. According to Carr and Shepherd, regaining activities of daily living post-stroke requires a relearning process that is similar to the learning process for normal subjects. They claim that neither exercise therapy nor stimulus-response-based neurorehabilitation treatment is adequate for the treatment of post-stroke patients. They believed that practicing needs to be task and context-specific and meaningful for the patient and not a sheer practice of exercises.

Motor control is an essential component of every aspect of life. The program aims to stimulate and/or make the best possible use of the brain's capacity for reorganization and adaptation. It is theorized that if the stroke patient learns to perform the components in a controlled manner and correct sequence, he will be able to perform the whole activity. The MRP is structured to enable the therapist to train the patient to perform the missing key components and then to practice the entire activity.

This program aims at increasing neither the muscle strength nor the activation of a maximum number of motor units but at helping the patient learn to control the muscle activation needed for a particular function. For training muscular activity, the therapist must take into account the complexity of muscle function. The motor activity is influenced by peripheral factors like the initial position of the body part, muscle length, velocity, temperature, joint angle, length of limb segments, and external forces. Throughout the practice of all activities, the therapist should monitor and train the missing components of body alignment as the patient is moved about.

During the initial stages of learning, the MRP also emphasizes using the patient's cognitive functions to progress practice to a more automatic level for skill acquisition. The MRP aims to trigger previously known motor programs by involving the patient cognitively in remembering the movements or activities he could do prior to the brain lesion. The treatment emphasizes the practice of specific everyday activities through cognitive control over the muscles and movement components and conscious elimination of unnecessary muscle activities. The developers of this concept believe that the program stimulates thinking, thereby enhancing alertness and motivation, which itself is an important stimulation for brain recovery. Studies have already proved that such experiences among stroke patients increase the cerebral blood flow by approximately 20%. For patients having problems with memory and mental functions, early experience of erect positions and demanding active participation in the activities encourages mental alertness. Even the environment in which the rehabilitation is performed encourages the learning process. Providing an enriched physical and emotional environment motivates the patient towards a better recovery. Considering the abovementioned, for optimal recovery, therapists should take the responsibility to treat the patients in the most appropriate environment.

Therapists should provide a clear and concise education about the pathophysiology of stroke, meaning of symptoms, brain's potential for adaptation, physical and emotional effects of stroke, methods of communicating with the dysphasic person, planning for discharge, and the role of participation in rehabilitation to the relatives and patients. The patients should be encouraged to practice activities beyond the therapy sessions. The aforementioned should be reinforced by the relatives and staff. Such routine "drills" carried out will essentially strengthen better learning. For patients with bladder and bowel incontinence, the level of anxiety and depression can be reduced by providing reassurance that they will soon regain control of their bladder and bowel.

A confused state of mind due to somatosensory and perceptual-motor problems, common among stroke patients during the early stages, should not be considered a poor prognostic sign or a barrier to rehabilitation. Emphasizing the active use and awareness of the involved side and the space it occupies, early weight-bearing through the involved side, training in upright positions, elimination of unnecessary muscle activity and sensory input to the intact side, and the use of visual and auditory feedback usually corrects such problems and prevents them from becoming established. The patients with perceptual problems are typically easily distractable and often have difficulty ignoring extraneous stimuli, noise, and bright colors. In such situations, the patients may concentrate better in a quiet room without distractions. Carr and Shepherd believe that sensory stimulation techniques like brushing, icing, vibration, or toweling used to improve the patient's awareness of the involved side may worsen the confused state level of the patient. Stereognosis and poor tactile discrimination may interfere in particular with hand function and need to be improved by direct practice. For those patients with tactile or visual inattention (extinction), sensory stimulation to the affected side could be provided with cognitive awareness of the stimulus along with restriction of stimuli to the uninvolved side. For visual field defects like homonymous hemianopia, the patients could be advised to turn the head to compensate for lack of vision.

Spasticity seen in many post-stroke patients is nothing but unnecessary muscular activity, which has become habitual. Even in those patients with the presence of spasticity, the treatment will focus on the elicitation of appropriate muscle activity and the elimination of unnecessary muscle activity. For instance, the knee flexed to 90° will make it easier for the patient to eliminate unnecessary knee extensor activity as the quadriceps muscle is less likely to contract excessively in this position. Controlled hip extension in standing and practicing forward stepping with the intact leg to ensure weight-bearing through the involved leg with the hip joint maintained in correct alignment can help to minimize the unnecessary muscular activity of the quadriceps and plantar flexors.

Unilateral spatial neglect can be persistent for some stroke patients, and the therapy should include strategies to make the patient aware of the problem and provide a solution to overcome it. A constant reminder to regain the body alignment by consciously shifting the weight onto the involved limb, use of a limb load monitor to provide auditory feedback, encouraging visual scanning towards the involved side, addressing or speaking to the patient from his involved side, and verbal feedback and manual or visual guidance reminders are some appropriate strategies to overcome the neglect.

Very little equipment is necessary for this approach. A low bed of convenient height for the patient to practice standing up and sitting down, common small objects for retraining hand function, and a few splints are some of the things required for this training program. Many physiotherapy equipment traditionally used, including parallel bars and canes, may interfere with the relearning of normal body alignment and relearning of normal motor tasks. The use of devices to hold the foot in dorsiflexion like a splint or short leg brace should be avoided as they may prevent plantar flexion, which is necessary at certain stages of the gait cycle.

According to Carr and Shepherd, most approaches delay active treatment in a very early stage of stroke, concentrate on positioning and emphasize concepts that do not apply to the adult stroke patient. They presume that the treatment should begin as soon as the patient's medical condition is stable, and the movement of the involved extremities, including the movement of the hand, must be stimulated following early stroke. The key features that stand prominent for MRP in comparison with traditional approaches are:

- A non-developmental approach and non-hierarchical; does not believe that recovery takes place from proximal to distal.
- Principles and concepts of psychomotor learning are used in this approach.
- Spasticity is not considered a major problem.
- Believes that cognitive processes are much more involved in the acquisition of complex motor behaviors.
- Treatment emphasizes on conscious involvement of the patient in movement control than on a sub-cortical elicitation of movement.

The MRP is designed to discourage compensatory behavior and facilitate learning through the active involvement of the individual. The reduction of spasticity is not seen as central in this approach. It is assumed that the patient's early involvement in MRP will lessen the chance of these habits getting permanent. The goal is to train patients to activate muscles in exactly the same way the muscles perform normal everyday activities. The therapist monitors the patient's performance to correct errors as they occur to ensure that only correct motor responses are practiced.

For optimizing the treatment program, the therapist must develop the five stages of problem-solving skills: recognition, analysis, decision-making, action-taking, and reevaluation. According to Carr and Shepherd, the analysis and decision-making stages are probably the most crucial and effective analysis depends on the therapist's knowledge of movement. For those therapists with an inadequate understanding of muscle action and movement, correct analysis of movement can be difficult. For instance, trouble stepping forward with the involved leg may be incorrectly analyzed as a result of foot drop when it can be due to lack of knee flexion at toe-off. Therefore, the program's effectiveness depends to a large extent on the therapist's ability to analyze and identify the key missing components and provide appropriate strategies to retrain the missing component by clear and concise explanation and demonstration. The patient's performance needs to be monitored with adequate verbal feedback and should be reevaluated each session to know the effectiveness of the treatment. When the patient's performance is unsuccessful, the therapist needs to verify the original analysis of the patient's problems and the technique which was carried out. Many a time, the errors in analysis can be a frequent cause for ineffective treatment.

According to Carr and Shepherd, it is the quality of rehabilitation that matters more than the location or settings. A rehabilitation plan, an early start, consistency of goals, motivation, mental stimulation, and educational programs are the factors that can influence the quality and therefore, the outcome of rehabilitation. The rehabilitation plan should consist of a general program plus a specific program. The rehabilitation plan should be achievable for the patient. The general program is concerned with the way the patient spends the day in his or her surroundings. Specific programs are planned for the individual patient's needs and may include a motor program with the therapist, a self-care program with the occupational therapist, and a communication encouraging program with the speech pathologist. Restricting the activities is known to cause impaired intellectual functioning, and it is believed that the time spent on therapy sessions without reinforcement during the rest of the day to be considered a waste of time. They also believe that the correct method of practicing makes learning more meaningful.

The patient needs reassurance, encouragement, and proof of capacity to overcome the barriers around. Success in treatment sessions counteracts the tendency towards anxiety, apathy, and depression post-stroke, leading to more rapid skill acquisition, particularly in the early stages of relearning a motor activity. Therapists should provide positive reinforcement through praise, affection, and acceptance. Positive reinforcement words like "good" should be offered as a reward only when it is relevant and if unsuccessful, the therapist should use words or gestures to encourage him and to acknowledge his efforts. In case the patient appears unmotivated, the cause for it should be investigated.

2.5.2.1 Steps for Relearning Motor Activities

For learning a complex motor skill, one should identify what is to be learned and should organize the information into correct sequences to carry out the task. Each patient has a limited information-processing capacity, and if this capacity is exceeded, performance will breakdown. Therefore, the therapist must keep the demands within the capacity of the patient. In the later stages of skill acquisition, less concentration and cognition are needed. Feedback given during early skill acquisition must be immediate and specific. Only a successful performance is rewarded by praise so that the patient knows exactly what is correct. Unsuccessful performance needs suggestions to correct it on the next attempt. The program trains the patient to perform key components of movement that are missing and then practice the entire activity. The essential features of an activity must be taught before the final polishing or refinements are attended. The environment is organized to include motivation, positive attitudes, reinforcement from relatives and friends, consistency of feedback, and consistency of practice. Four steps are underlying relearning of motor programs. They are:

1. Analysis of function
2. The practice of missing components
3. The practice of the whole activity
4. Transference of learning

Analysis of Function

Right from the early post-stroke stage, it is essential to search for even the slightest amount of muscle activity. Gross patterns of movement should be avoided during evaluation to prevent masking of the minimal muscle activity present in under-active muscles. A thorough analysis of each motor problem should include anatomical, biomechanical, physiological, and behavioral factors based on which the therapist should make the correct treatment decisions. The therapist should analyze each functional task to determine which component of the task cannot be performed by the patient (missing components in the activity). While analyzing the task, the therapist should ask himself the possible reasons for the patient's altered motor behavior. For instance, if the patient exhibits knee hyperextension while standing, the therapist should ask himself whether the hyperextension is due to poor quadriceps control near the terminal extension or hyperactivity of the plantar flexors.

The Practice of Missing Components

Following identification of the missing component of the movement, the therapist, using the description of the normal movement as a guide and biomechanical necessities, should select those essential movements to perform the activity. Explanation and demonstration are used as key techniques, and the activities are progressed until the patient gains certain control. The patient is encouraged to use his or her observations and experiences during this relearning process.

If the patient has inability or difficulty to activate muscle movement(s), he should be encouraged to initiate muscle movements and if not, should be elicited reflexively. Passive movements should be avoided except to help the patient understand the movement being sought. Support by the therapist may be needed for insufficient muscle activity. Similarly, bilateral movements should be avoided until the patient regains control over the affected limb. Above all, the approach comprises techniques found in other treatment approaches, such as verbal commands, joint approximation, and cutaneous stimulation. However, the therapist is instructed to apply techniques only based on assessment and theoretical applicability.

The Practice of the Whole Activity

As soon as isolated muscle action is elicited; it must be incorporated into meaningful activities. Attempts to perform activities of daily living using those movements possible should begin from the early stage. Motor activities can be either practiced in their entirety or broken-down components. During the early stages, many of the patients may find it easy to practice each component followed by practice of the whole activity. The patient should understand what he or she is preparing, and the preparation must immediately follow the performance of at least part of the activity for which he is preparing. By practicing in this manner, the patient will understand which muscle activity or component fits into the sequence of the whole activity. In some situations, the practice of the whole activity with manual guidance from the therapist with normal speed, rhythm, timing, and sequencing can trigger off the memory of the movements required, even if the patients make a few errors.

Transference of Learning

It refers to a person's ability to carry out the same task in a different environment. Training subjects on a task repeatedly in the clinical setting may improve performance in that particular task but not transfer to any activities of daily living when the patient is back at home. Practicing strategies in multiple environments with varied tasks and demands will encourage the transfer of learning. Training must be specific to the patient's deficits and goals for learning to be effective.

Carr and Shepherd advise that the patient should not waste time by practicing what he or she already knows. As soon as the patient gains certain control over movement, the activities should be further progressed. Brief instructions and appropriate verbal cues can trigger appropriate movements. Verbal feedback should be brief, relevant, and concise, given both during the performance and at its end or continuous throughout the treatment. Manual guidance can ensure correct alignment and guide the movement. Movements can be made more complex by reducing manual guidance and feedback. During the phase of skill acquisition, the patient will transit from the cognitive phase of learning to the automatic phase of learning. The developers of this approach do not support inappropriate methods of progression, which include progressing from passive Range of Motion (ROM) exercises to resisted exercises, from parallel bars to walking aids like quadripod, from a wide base to a narrow base, roll over before practicing sitting balance, crawling before walking, and gaining control over shoulder movements before the patient can move the hand.

2.5.3 Essential Functions of Daily Life

According to Carr and Shepherd, the program is made up of seven components representing the essential functions of daily life. They are:

1. Upper limb function
2. Oro-facial function
3. Motor tasks in sitting
4. Motor tasks in standing
5. Standing up
6. Sitting down
7. Walking

The sequence in which the components appear is insignificant as there is no intent to progress from one component to the other and similarly, there is no necessity for a patient to perfect one part before going on to the next section. Training of each of these functions comprises of the earlier mentioned four steps, namely task analysis, the practice of missing components, the practice of the tasks, and transference of training. The developers of this approach encourage the use of techniques or methods like biofeedback that promote specific muscle action or movements. According to them, the use of auditory or visual evidence of muscle contraction does not conflict with the basic principles underlying the MRP. The patient's daily therapy sessions should range from at least half an hour twice daily in the first few days to daily 1-h sessions or preferably more.

2.5.3.1 Upper Limb Function

Most daily activities involve complex movements of the upper limb. Such activities involve the ability to grasp and release different objects of various shapes, sizes, weights, and textures, grasp and release different objects with the arm in different relationships to the body, manipulate tools for specific purposes, reach in all directions, and use two hands together (one hand-holding and the other manipulating the object). For effective use of the upper limb, the pre-requisites include making postural adjustments while the arm moves, freeing the hands for manipulation and sensory discrimination. Sensory discrimination involves the ability of the person to feel an object in hand, to recognize its size, shape, dimensions, composition, and texture, and to appreciate the compressibility of the object and position of the object in hand. It also involves awareness of the relationship of parts of the hand to each other, their position in space, and appreciation of counter-pressure of the fingertips.

Though upper limb functions are complex, Carr and Shepherd believe that identifying and mastering the essential movement components will allow the performance of many different activities. Shoulder abduction and forward flexion and elbow flexion and extension, accompanied by appropriate pectoral girdle movements and glenohumeral rotations, are a few of the major functions of the arm to enable the hand to be positioned for manipulation. For hand function, the essential components required are wrist extension with some degree of radial deviation, palmar abduction and opposition of the thumb, flexion and opposition of individual fingers towards the thumb, extension of the metacarpophalangeal joints of the fingers with some degree of interphalangeal joints flexion, and supination of the forearm.

In the early stage, the therapist who understands the muscle function should search actively for any small amount of muscle activity. If the conditions are favorable, a muscle that appears non-functional may contract. If the control is poor, muscle activity around the shoulder should be analyzed with the patient in the supine position. According to the developers, with regard to arm and hand function, the common sequelae preventable to a certain extent are the habitual posturing of the limb, neglect, compensation with the uninvolved arm, tendency to move the involved arm with the uninvolved arm, and contracture of the soft tissues of the shoulder and/ or wrist. Common abnormal muscular activities seen in the involved arm are a poor scapular movement with persistent depression of the shoulder girdle, internal rotation of the shoulder, lack of shoulder abduction and forward flexion and excessive elbow flexion, pronation of forearm, and wrist flexion. The typical abnormalities seen in hand include:

- Difficulty to grasp objects with the wrist in extension and radial deviation and tendency for ulnar deviation while using the hand.
- Difficulty to extend the metacarpophalangeal joints with the interphalangeal joints in some degree of flexion.
- Difficulty to abduct and rotate the thumb.
- Inability to release an object without flexing the wrist.
- Tendency to pronate the forearm while holding or picking objects.
- Inability to hold objects while moving the arm.

Reduced motor activity or inactivity of the muscles surrounding the shoulder leads to shoulder pain, stiffness, and subluxation. Three mechanical factors causing the abovementioned are the impingement of soft tissues against the acromion, the friction of soft tissues against bone, and the traction to soft tissues. According to Carr and Shepherd, these mechanical factors are precipitated by the passive ROM exercises, active abduction movement with internal rotation or without external rotation, flexion and elevation with internal rotation, gravity, traction on the involved arm while positioning or transfers and rolling onto the paretic shoulder. Passive ROM exercises, when performed with inactive muscles surrounding the shoulder, can predispose to an abnormal relationship between humerus and scapula (movement of the glenohumeral joint without corresponding movement of the shoulder girdle) damaging the soft tissues around the glenohumeral joint. In case of any shoulder pain during movement analysis, treatment including peripheral joint mobilization, interferential or transcutaneous nerve stimulation is recommended. Hot fomentation and ultrasound can be the options to alleviate the symptoms and encourage normal repair process if chronic inflammation exists and should wait until signs of recovery appear to begin with active therapy.

During the early-stage post-stroke, the motor activity can be elicited in supine position with the involved arm in elevation and should encourage eccentric contraction rather than concentric, particularly within the inner half of range. Arm movements, including movements of the hand, should be stimulated right from the early stage. The upper limb function consists of the complex combinations of meaningful muscle action and as soon as isolated muscle actions are possible, it is essential to practice and extend these isolated muscle actions into meaningful actions. The patient should gain control over various speeds of movement, different ranges of movement, and muscle actions shifting from concentric to eccentric and prime mover, synergist, and fixator. Elicit activity at first in the position of the greatest advantage to the muscle and support provided by the therapist's hand is to be withdrawn once sufficient muscle activity has been retrained. The therapist should avoid encouraging movements that are a part of abnormal synergy, which do not have any functional significance. The objective is to stimulate muscle activity and to train control over the functional activity and not in terms of strengthening muscles. Until the patient regains control over the involved extremity, bilateral movements could be avoided as this may reinforce movements of the intact limb at the expense of the involved extremity.

To encourage movement control for manipulating objects, practice picking up small objects and objects of various dimensions and textures between the thumb and each finger. Encourage patients to pick up objects with a certain amount of wrist extension and also train them to supinate the hand while holding the object. To use the hand effectively, a fine degree of control over the shoulder is essential and for which activities like reaching forward to pick up an object, reaching sideways to pick up an object, grasping and releasing objects with the arm outstretched, and use of both hands for manipulation objects (one for grasping and the other for manipulating) are useful.

2.5.3.2 Oro-Facial Function

Oro-facial functional issues can be disturbing and frustrating for many stroke patients. Activities like facial expression, speech, swallowing, and breathing are certain oro-facial functions that may get involved post-stroke and interfere with feeding, communication, and socialization. Remediation of these issues should begin during the early phase of rehabilitation.

Of the above oro-facial activities, swallowing is a complex and integrated neuromuscular function. The initial stage, which is under voluntary control, comprises of preparation of food bolus. Once the bolus passes into the oral pharynx, the involuntary stage of swallowing begins. The food bolus formed on the tongue is squeezed backward towards the posterior oral cavity by the tongue movement against the hard palate. Typically, contact of the bolus or fluid with the back of the tongue and pharynx will serve as the stimulus for the initiation of the swallowing reflex. During the involuntary stage, under the influence of gravity and partly due to the successive contraction of the constrictor muscles, the food bolus passes through the pharynx to the esophagus. During this stage, the lips and jaw are closed, and the soft palate and uvula are tensed to seal off the nasopharynx. The larynx is pulled upwards behind the hyoid bone and towards the back of the tongue to narrow the lumen of the larynx to protect the bolus entering into the respiratory tract. As the bolus reaches the epiglottis, the pharynx is pulled upwards to propel the food into the esophagus, and it continues downward by peristaltic action. Throughout the brief process of swallowing, there will be a momentary inhibition of breathing.

The jaw and lip closure, the elevation of posterior one-third of the tongue to close off the posterior oral cavity, and the elevation of the lateral borders of the tongue are a few essential components during the initial stage of swallowing. The ability to move the head independent of the body, effective sitting balance, and breath control in relation to swallowing are few of the prerequisites for swallowing. Prior to analyzing the swallowing, observe the sitting posture, lips, jaw, and cheek movements, observe the eating and drinking and perform an intra-oral digital examination of the tongue and cheeks. While analyzing the oro-facial function, the therapist may observe open jaw, poor lip closure, drooling, immobile and/or asymmetrical tongue, hypotonic tongue, and food collected between the cheek and the gums. While examining the oro-facial function, the use of a spatula is not advisable as it may create unpleasant texture compared to the therapist's digits.

The therapist should encourage the patient to sit upright with hips well back in the chair and head and trunk straight. Carr and Shepherd believe that the use of ice for oro-motor stimulation may cause numbness and increase the patient's difficulty to move the tongue within the oral cavity. For training lip closure, the therapist should not use continuous stimulation but only brief use of digits to encourage the patient to concentrate on the movement. Intra-oral techniques should be frequently interrupted, and the jaw is to be held close to allow the patient to move the tongue inside the mouth. The therapist should ensure that he does not push the atlanto-occipital joint into extension during the abovementioned procedure. Commands like "close your mouth" and "keep your lips gently together" can be used to encourage jaw closure. To train tongue movement, horizontal vibration can be given using the therapist's digits (one or two) over the anterior one-third of the tongue with firm downward pressure. The amplitude of the vibratory should be small and the stimulation should not last longer than 5 s. While stimulating, ensure that the posterior aspect of the tongue is not stimulated.

For encouraging the elevation of the posterior one-third of the tongue, which is necessary to close off the posterior oral cavity, the therapist can use the index finger to apply firm pressure to the anterior one-third of the tongue in a downwards direction. Commands like "open your mouth" and "let me help you to encourage swallowing" may help in tongue movements. A cotton bud can be briefly applied to the soft palate to stimulate a hypo-active gag reflex. For the hypersensitive mouth, firm digital pressure starting in the least sensitive area and working towards the most sensitive area, i.e., from lips to gums to anterior one-third of the tongue, can be attempted. To encourage facial movement, the jaw should be held closed and the patient should attempt to smile a little while the therapist instructs the patient to close the lips gently. Liquid is more easily aspirated than solid food. Therefore, to gain the patient's confidence, practice swallowing solids before liquids are attempted. Palatable thickened food is usually handled with relative safety.

2.5.3.3 Motor Tasks in Sitting

Carr and Shepherd believe that early practice of balance in sitting with emphasis on eye and head control fosters the ability to pay attention and to concentrate. Sitting at a table, the patient should practice lowering and lifting the arm repeatedly both in flexion and abduction. The patient should work within a range he or she can control and then gradually increase it. The therapist can give commands like "now reach up to touch my hand" and "don't let it drop" while encouraging the patient to gain involved upper limb control while performing forward flexion of the shoulder. Once the control improves, activities that suit those active movements can be introduced. When required, the therapist can perform brief passive stretches immediately before the exercise session to decrease muscle stiffness. If the muscle stiffness is considerable, frequent brief stretches may be required throughout the exercise session. Encouraging the patient to perform active exercises helps in the active stretching of the musculature. Holding objects of different sizes can stretch the thumb webspace, and typically, the larger the objects provided, the greater the stretch.

2.5.3.4 Motor Tasks in Standing

To perform tasks in standing, standing balance is a prerequisite. Balance in standing involves the ability to stand without undue muscular activity, move about in standing, and move in and out of the standing position without using the arms for support. Normal alignment of body segments ensures less energy requirement compared to a poorly aligned position. The normal alignment of body segments in the standing position consists of the feet a few inches apart, the hips and knees extended, the hips over the feet, the shoulders over the hips, the head balanced on level shoulders, and the head and trunk erect.

The normal balance reaction during the lateral shift in the center of gravity consists of lateral flexion of the neck and trunk, i.e., elevation of pelvis and depression of shoulder, in the opposite direction. On the contrary, the normal reaction during backward shift in the center of gravity consists of extension of neck, forward inclination of trunk on hips, and ankle dorsiflexion. During the analysis of standing alignment and balance reaction, the patient can be asked to look behind, reach forward, sideways, and backward, stand on one leg, pick up an object from the floor and the therapist should note any missing components, if any. Wide base of support, voluntary restriction of movement, shuffling of feet instead of making appropriate adjustments with body segments, sideways or backward stepping whenever the center of gravity moves, and use of arms for grabbing support are some of the common abnormalities in the standing alignment and balance reaction. Standing with weight-bearing through the affected leg in correct alignment may be one of the most important factors in preventing the development of poor standing alignment.

To encourage early weight-bearing through the affected lower limb and to allow the patient to practice hip control, a knee splint can be provided to prevent buckling, especially when the patient has difficulty controlling the knee. The patient should be encouraged to practice within the limits of his ability and always keep trying to extend these limits. The therapist must constantly and actively discourage the patient from holding on or reaching out for support. The patient should be advised to use the pelvis and legs to balance instead. For promoting balance reaction while maintaining the postural alignment, the therapist can move the waist of the patient sideways to encourage lateral flexion of the neck and/or trunk for returning the center of gravity back to its base.

2.5.3.5 Standing Up

Standing up involves placing the feet backward and shifting the body forward and upward with minimal energy expenditure. Placement of the feet backward brings the base under the center of gravity as the body moves forward. Leaning of the extended trunk forward at the hips and movement of the body forward by dorsiflexion at the ankles brings the center of gravity over the feet and enables the body weight to be shifted forward and upward. Foot placement, leaning of the trunk forward at the hips with extended spine, and extension of hips for erect standing alignment are the essential components of standing up.

The therapist should observe the patient's body alignment throughout the sit-to-stand transition movement. Poor weight borne on the involved lower limb, inability to shift the center of gravity sufficiently forward, and an undue flexion of the trunk and head instead of leaning the erect trunk forward at the hip are a few of the common problems seen during standing up. The therapist may need to assist the patient if he is weak, overweight, or unable to initiate the transition movement. In the initial phase of sit-to-stand training, the therapist may use a higher stool to gain control. During the initial training phase, it is advisable to rest the patient's arms on the therapist's shoulders or around the therapist's waist. The therapist guides the movement to ensure that weight is taken through both feet. Ensure that the trunk is inclined far enough forward and that the knees move forwards. Encourage the patient to practice this activity correctly with other members of the staff and relatives. The patient should not be allowed to pivot on his intact leg while transferring himself from bed to chair, from chair to lavatory, and so on as it may encourage non-use of the affected side and prevent him from learning the task.

2.5.3.6 Sitting Down

Sitting down involves the placement of feet backward and shifting the body downward with minimal energy expenditure. Placement of feet backward brings the base under the center of gravity as the body lowers downward. The patient flexes hips, knees, and ankles to lower body mass toward the seat. During sitting down, the trunk flexes forward at the hips as body mass is lowered and the weight remains supported over the feet until the body mass is moved back to enable seat contact. The therapist may need to assist the initiation of knee flexion by moving the knee forward and stabilizing the shank and foot to assist weight-bearing on the involved lower limb. To improve performance, the practice of sitting down with weight through the affected leg can be beneficial.

As discussed earlier, for sit-to-stand training, raising the seat height will reduce the lower limb muscle force demands and may enable the patient to practice sitting down with a certain amount of weight through the involved lower limb or more evenly distributed between the two lower limbs. Extensibility of the soleus muscle is crucial to place the foot backward and weight-bear through the involved leg for sit-to-stand and stand-to-sit activities. Both passive and active stretching can be used to preserve the functional muscle length. A brief passive stretch given just before exercises helps to reduce muscle stiffness. Active stretching occurs during the practice of sit-to-stand and stand-to-sit transitions when the paretic limb is loaded with the ankle in dorsiflexion. Approximately 20 s of stretch hold, repeated 4–5 times with adequate rest periods between the stretch holds are sufficient for the same.

2.5.3.7 Walking

Normal walking involves the movement of the center of gravity through space with the least possible energy expenditure and minimal muscular activity that is rhythmical and symmetrical in nature. Electromyographical studies indicate that during the walking cycle, the muscles act only over brief periods, the limbs being carried forward to a large extent by their momentum. Muscular contraction is more during the deceleration than in the actual progression. The precise relationship between muscle contraction and displacement of the body makes energy storage and recovery efficient during normal walking. The essential components of the stance phase of gait include the extension of the hip throughout, the lateral horizontal shift of the pelvis and trunk, and approximately 15° knee flexion initiated on heel strike, followed by extension and then flexion prior to toe-off. The essential components of the swing phase are the flexion of the hip and knee, lateral pelvic lilt downward in the horizontal plane at toe-off, rotation of the pelvis forward on the side of the swinging leg, and extension of the knee plus dorsiflexion of the ankle just before the heel strike.

The major problems found during analysis of the stance phase of hemiplegic gait include a lack of extension of the hip and poor control of the knee flexion-extension activity during the heel strike and the excessive lateral horizontal shift of pelvis and excessive downwards pelvic tilt on the intact side. Analysis of the swing phase will reveal a lack of knee flexion at toe-off and lack of hip flexion, knee extension, and ankle dorsiflexion on heel strike.

Since the pelvis provides a link between the supporting leg and the torso, control of the pelvis on the leg is essential to assume and maintain a normal aligned position. While weight-bearing through the involved limb, many of the stroke patients may have a tendency to shift their pelvis too far laterally, which is accompanied by a compensatory tilt of the pelvis downwards on the sound side. Difficulty to contract the ipsilateral hip abductors and contralateral trunk side flexors at the appropriate moment, poor awareness of the normal extent of shift, and a lack of hip extension while standing on the involved limb are the possible reasons for the abovementioned deviation. Instead of weight being shifted forward onto the involved limb with hips in extension, the patient flexes the trunk forward on the affected hip and takes a short step forward with the sound limb.

Patients with poor knee control while attempting to weight-bear through the involved lower limb may buckle at the knee and as compensation may learn to passively lever the knee into hyperextension until the end of the stance phase. Encouraging the patient to practice controlled knee flexion and extension between 0 and 15° while weight-bearing through the involved side can improve control over the quadriceps. While practicing the controlled knee flexion and extension in standing, ensure not to perform the knee movements in a jerky manner.

To retrain controlled lateral weight shift, while standing with hips feet apart, the patient should practice weight shifts from one side to the other in small increments. Avoid excessive lateral weight shifts, which may trigger hip flexion and knee hyperextension. Walking sideways is another alternative to stimulate controlled lateral weight shift. With the affected foot on a step, the patient should shift the weight forward and step up onto the step and down again with the sound leg.

The hip flexors will be in a better position to initiate a pendular movement at the beginning of the swing phase, provided the involved leg is positioned behind the sound leg. Lack of knee flexion during the swing phase is another major problem and any compensatory attempts to overcome tend to distort the entire sequence of the swing phase. To train knee flexion, the therapist needs to assist the patient to eccentrically and concentrically control the hamstrings through small ranges of movement in prone lying position. Once sufficient control develops, the same exercise can be performed while standing on the sound leg. The therapist may guide the patient to control the knee flexion while the patient initiates the step. Commands like "bend your knee," "step forward," and "heel down first" may be used.

The practice of walking and components of walking, with the therapist's guidance, usually improves alignment and standing balance. To improve walking skills, the complexity of the training can be increased by performing quick steps in different directions, varying the speed of walking and the spatial confines, practicing stepping over objects of different heights, and walking combined with other activities such as conversation and carrying objects. To improve the rhythm and timing of gait, Carr and Shepherd also recommend treadmill walking.

Climbing upstairs involves similar movement components of walking; however, the ranges of movement at the joints involved and muscle activity required are somewhat different. A larger range of hip and knee flexion is required for climbing up. Forward inclination of the body at the supporting ankle while placing the foot on the step and a forward and upwards shift in the center of gravity over the forward leg are few instances that show the differences between climbing upstairs and walking. Safety can be a major concern while climbing downstairs. Unlike walking or climbing upstairs, for climbing downstairs, the center of gravity needs to be kept back over the supporting leg, and the forward movement is performed by a controlled eccentric contraction of the hip and knee extensors of the supporting leg.

2.6 Constraint-Induced Movement Therapy

Abraham M. Joshua

2.6.1 Introduction

Stroke, the leading cause of disability across the world, presents with a variety of impairments, of which the motor control deficit on the hemiparetic side is a common one that can compromise the quality of life. The motor control deficit may include weakness of the extremities, impaired reaction time, reduced force, and abnormal tone and movement. The typical tendency of hemiparetic patients with the abovementioned motor deficits is to utilize compensatory strategies to achieve a certain degree of immediate independence, and the strong reliance on compensatory strategies may slow or even inhibit functional recovery of the hemiparetic side.

For several decades therapists have been using many therapeutic interventions to remediate the deficits and restore motor function. Neurofacilitatory methods have used sensory stimulations, including quick stretch, muscle tapping, and brushing, to facilitate movement after stroke. Exercises progressing from assistive to active to resistive modes have improved muscle strength and control. Several therapeutic strategies and techniques have been recommended to normalize the tone for improved motor control. Though many therapeutic techniques are available to facilitate motor recovery, no single therapeutic technique has shown an overwhelming effect when compared to the others.

Constraint-Induced Movement Therapy (CIMT), a rehabilitative strategy developed by Edward Taub, an American behavioral neuroscientist, is primarily used to treat motor dysfunction among stroke patients. The therapy was originally introduced to increase the functional use of the neurologically weaker upper limb through massed practice while restraining the uninvolved upper limb. According to Taub, many rehabilitative strategies and treatments focused on compensation rather than restoration of the paretic upper limb function. The use of compensatory strategies, including over dependency and use of the unaffected upper limb for activities of daily living, may slow or even inhibit the potential for recovery of the affected side. CIMT, in contrast, discourages the use of the unaffected extremity and provides a strategy that encourages active use of the hemiparetic arm. The signature form of CIMT consists of concentrated, repetitive training of the affected upper limb for ten consecutive working days, spanning 2 weeks, 6 h daily, interspersed with 1 h of rest, while constraining the use of uninvolved upper limb for the majority of waking hours (90%) to induce an increased use of the more affected limb.

2.6.2 Historical Background

The theoretical foundation of CIMT can be traced back to the behavioral studies performed during the late 1800s and early 1900s. In 1895, Josef Starlinger, an Austrian psychiatrist, in a chapter titled "Die durchschneidung beider pyramiden beim hunde" meaning, the intersection of both pyramids in dogs, trained a dog to give his paw after dissection of both pyramidal tracts. Hermann Munk, a German physiologist, in the year 1909, claimed that it was possible to induce a hungry monkey to reach and grasp food to its mouth using its deafferented upper limb, a clear purposeful movement, by rewarding the attempts and restraining the normal upper limb. Munk concluded that the primary deficit was in sequential functional movements of the limbs and not in the independent, voluntary use of the limb. Based on a case series, in 1915, Shepherd Ivory Franz and coworkers stated that the return of function in hemiplegic patients' paralyzed segments exists much beyond 2 years, the maximum time limit set by some of the neurologists. They commented that the assumption of function by other regions of the brain as a possible mechanism for the functional return.

In 1916, Robert Ogden and Shepherd Ivory Franz conducted a series of animal experiments on rhesus monkeys, which were rendered hemiplegic by lesions in the motor cortex. Perhaps, Ogden and Franz's work can be considered as the first to conceptualize the treatment strategy known as "forced use." In their study, the normal upper limb of the monkey was strapped to the trunk by means of a jacket constructed in a manner not to allow the animal to use the arm for any movements, including feeding and climbing. The purpose of the restriction was to compel the animal to use the corresponding paralyzed upper limb. In addition to the above, passive treatment efforts were made to encourage the animal to move the paralyzed segments, which included the application of noxious and irritating stimuli and friction over to the nerves and paralyzed muscles. The "anger" created by such stimuli endeavored the monkey to escape or prevent the irritating stimuli either by moving away or grasping with the paralyzed arm. A similar strategy was used for the paralyzed lower limb. The animal was encouraged to use the paralyzed arm for support and the paralyzed leg while walking. Though little or no reactions were observed during the initial sessions of treatment, soon, the application of the stimuli produced small but appropriate responses, and days later, the responses to the stimuli were almost equal to those of a normal animal. Ogden and Franz noticed that the monkeys were able to use their arms for picking and grasping food and bringing it to the mouth and their arms and legs for walking and climbing, at first awkwardly and later more appropriately. Following 14 days of intervention, the monkeys were using the arms and legs for all activities with no obvious difference between the normal and paralyzed segments.

Ogden and Franz provided detailed evidence for functional recovery through "forced use" when they immobilized the normal limbs of the monkeys. They believed that motor disabilities, to a large extent, were due to disuse and hypothesized that the return of function of the paralyzed limb was due to behavioral recovery and the plausible plasticity of the cortex. They even stated that the rapid recovery of motor function could be due to the "vicarious" functions of other cerebral parts that need to be investigated. Following the original work by Ogden and Franz, almost 50 years later, in 1963, H. D. Knapp, Edward Taub, and A. J. Berman observed that monkeys do not use their affected upper limb in "free situations" following unilateral upper limb somatosensory deafferentation. They noticed that the monkeys in the conditioned situation (a buzzer lasting for 7 s as a conditioning stimulus and an electric shock of 2–6ma as unconditioned stimulus) could initiate purposive movements with the affected (deafferentiated) limb when the normal limb was restrained; a "forced use" for the deafferented limb to accomplish the function. According to them, the grave motor deficit demonstrated by the animal in the free situation after deafferentation indicated the monkey's heavy dependency upon the cutaneous and proprioceptive sensations. Whereas the independent purposive movements seen in the deafferented limb of those monkeys in the conditioned situation, with the normal limb restrained, indicated the compulsion or "force" to use some other sources of afferent input to substitute the absent somatic sensation from the affected limb.

Until the 1970s, several researchers, including Max Rothmann, Karl Spencer Lashley, Sarah Tower, and Jerzy Konorski, have employed behavioral techniques on animals to study the motor deficits following neurological damage. However, none of the abovementioned investigators in scientific research were directed towards formally explaining the plausible mechanism or prediction of recovery when behavioral training was given after neurological injury. Consequently, these findings remained a set of disconnected observations that received little attention, and no attempts were made to extrapolate and apply the restraining and behavioral strategies on humans in a methodical fashion.

2.6.3 Concept and Principles of CIMT

The general impression among the neurologists and physiotherapists in the field of neurorehabilitation is that once a patient reaches a plateau in motor recovery, typically 6–12 months post-stroke, further administration of therapeutic strategies or interventions will not produce any significant beneficial changes. Taub believed that the use of movement restriction of the uninvolved upper limb and formal training of the involved limb of those patients beyond one-year post-stroke (chronic stroke) would improve the upper limb function further. In 1980, Steven L. Wolf, an American physiotherapist, examined the effect of forced use on the affected upper limb of a chronic hemiplegic patient. The experiment was based on the belief that directing the patient's attention and effort towards the paretic limb without any formal training and excluding the use of the unaffected limb using a restraint would enhance the functional opportunities. Though the efficiency and the quality of movement scores following 1 week of restraint did not reflect significant changes, Wolf noticed an increased frequency of purposeful behaviors and functional use of the affected limb during the period of restraint. Almost a decade later, to know the effect of learned nonuse among chronic stroke and traumatic brain injury patients, Wolf experimented with forced use therapy for 2 weeks and found a remarkable improvement in the functional measures; however, the study lacked any information about the transfer of learning to the real-world environment.

The inclusion of repetitive and adaptive task practice under clinical supervision is the additional structured element that evolved the forced use therapy to the current concept of CIMT. The operant conditioning delivered through therapist feedback and adaptive task practice or "shaping" promotes self-motivation to use the affected limb and improve patient's problem-solving ability. In addition, intensive practice fosters motor relearning. Successful application of CIMT is thought to induce a use-dependent increase in the cortical reorganization of the areas of the brain controlling the most affected limb. Several studies, primarily in mild to moderately impaired survivors of stroke, have demonstrated clinically relevant results.

2.6.3.1 Learned Nonuse

The CIMT approach, developed by Edward Taub, is fundamentally a behavioral science approach, and the original observations were made in the context of motor control literature and the role of sensory feedback in motor learning. This approach is based on the theory of "learned nonuse" and to study the same, during the 1960s to 1980s, Taub did extensive work on non-human primates. For these experimental works, the somatosensory deafferentation of the upper limb of primates was attained by dorsal rhizotomy. Post deafferentation, the animals stopped using the affected limbs and never noticed any spontaneous restoration of the function. According to Taub, repeated attempts to use the deafferented limb often lead to aversive consequences, including loss of food objects, falls, injuries, and incoordination. He stated that punishment is a constitution of such aversive consequences, leading to the suppression of behavior and masks the ability even when the animal has the potential to restore the function. He postulated that the primate did not use the limb due to learned behavior suppression or learned nonuse (Fig. 2.74).

The situation mentioned above changes dramatically when a movement restriction device is placed on the unaffected limb. The restrain imposed on the unaffected limb, and the compulsion to use the deafferentiated limb for useful activities motivated the animal to overcome the learned nonuse or learned behavior suppression imposed on it. After learning effective ways of using the deafferentiated limb, doffing of the movement restriction device within a few days after the commencement of the experiment, made the primate revert to nonuse of the deafferentiated limb and use of the unaffected limb. However, when the restriction of the unaffected limb was extended for several days (few weeks), the use of the deafferented limb persisted, and there was no ostensible diminishment of range or quality of movement after the device removal. Restraint to the unaffected limb over several days while training the animal to use the affected limb is fundamental for the formulation of CIMT. The inability to use the deafferentiated upper limb would be due to cortical mechanisms

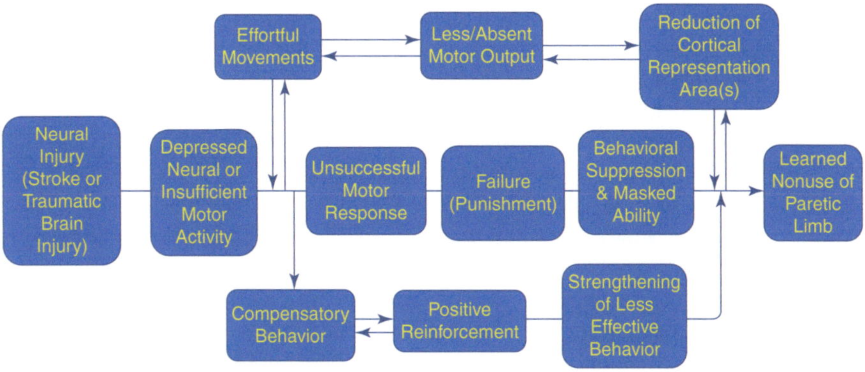

Fig. 2.74 Schematic diagram depicting the development of learned nonuse of the paretic limb

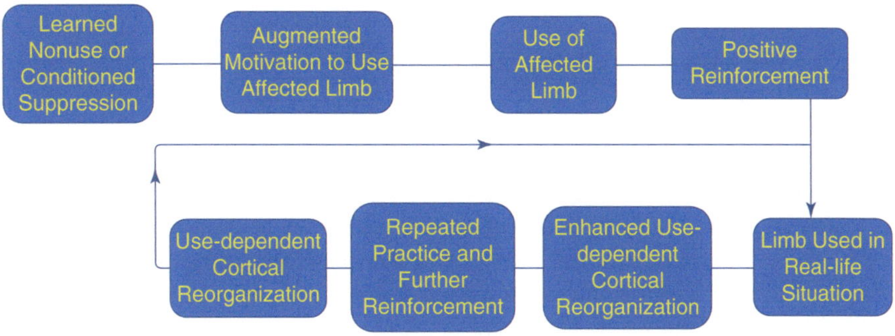

Fig. 2.75 Schematic diagram depicting the method of overcoming learned nonuse of the paretic limb

rather than processes associated with deafferentation at the level of the spinal cord. Taub hypothesized that the deafferentation of the limb led to inactivity, and learning on the part of the animal not to use the deafferentiated limb led to "learned nonuse." He stated that appropriate behavioral training and subsequent improvement in the deafferentiated limb could cause a conditioned suppression of the "learned nonuse" and promote functional recovery (Fig. 2.75).

According to Taub, the learned nonuse could also be relevant for neurological conditions like stroke or brain injury. Though the somatosensory deafferentation and stroke are two different types of lesions, the learned-nonuse mechanism in either situation will come to play soon after the large insult and remains thereafter. Unilateral sensorimotor dysfunction encourages the stroke patients to direct their attention towards the unaffected, contralateral extremity inducing "learned nonuse" of the affected limbs. The learned nonuse further accentuates when the patients are allowed and even trained, to function in activities of daily living using the intact limb. Such compensations and increased reliance on the intact limb during the early stage hinder the recovery of function in the affected limb.

Interestingly, the learned nonuse mechanism is observable even in stroke patients with mild to moderate motor deficits. Many of the mild to moderate hemiparetic patients, capable of using the paretic limb with reasonably good quality of movement when asked to carry out tasks, may demonstrate relative or near-complete nonuse of the paretic limb. Reduction in motor performance outside the rehabilitation or therapy environment is often reported as a source of frustration for many stroke survivors. In an article titled "Stroke Recovery: He can but does he?" Keith Andrews and Jean Stewart reported that 52% of the family members or caregivers felt that the patient was unable to perform two or more activities at home, which were possible in the stroke rehabilitation unit, and attributed it to a lack of motivation. Taub felt that the strategies to overcome learned nonuse in primates following somatosensory deafferentation could motivate and uncover latent motor abilities of many hemiparetic patients and thereby constitute a possible treatment to improve upper limb function.

2.6.3.2 Use-Dependent Cortical Reorganization

In recent years, there has been an explosion of knowledge concerning the brain's ability to reorganize and adapt to neural damage following conditions like a vascular insult or degenerative disease. Scientific evidence of intensive physical retraining after stroke has recently been validated by the work of Randolph J. Nudo and co-workers. To examine the intensive physical retraining plasticity, these investigators used the intracortical microstimulation technique to map out the motor representation area of the hand, wrist, and proximal upper limb of the adult squirrel monkeys before and after focal ischemic infarcts. These primates were given intensive behavioral training procedures that required skilled use of the hand to retrieve food pellets from small wells before and after inducing lesion by bipolar electrocoagulation of small vascular bed over the hand area of the motor cortex. Nudo and co-workers were able to clearly demonstrate the enlargement of the hand and wrist areas in the motor cortex of these animals, indicating that the area surrounding the infarct, which would not normally involve in the control of the hand, came to participate in that function. Nudo and co-workers also demonstrated loss of cortical representation following sensory or motor deprivation and brain injury and the reversal of changes following intense physical training. Following physical training, functional imaging studies in recovering stroke patients have demonstrated associated shifts of activation involving the affected hand to ipsilateral secondary and tertiary motor areas and to contralateral homologous motor areas, which were similar to the findings observed in the animal model experiments.

Several neuroimaging studies have demonstrated use-dependent cortical reorganization in humans. The large cortical somatosensory representation of the digits in string players, as compared to non-musicians, and the enlarged and disordered representation of the digits of blind Braille readers, as compared to blind non-Braille readers are certain instances for use-dependent cortical reorganization. The results of studies relating to use-dependent cortical reorganization suggest that the size of the cortical representation of a body part depends on the amount of use of that part. Massive cortical reorganization following CIMT for sensory deafferentiated primates and brain-injured humans is a piece of strong evidence in the recent past linking the association of use-dependent cortical reorganization with the therapeutic effect of CIMT. The functional changes observed after CIMT are accompanied and correlated with structural changes within the brain, which includes an increase in the grey matter of bilateral cortical motor areas and bilateral hippocampi. However, a fair amount of variability among use-dependent cortical reorganization patterns has been reported in studies, possibly due to factors like the location of the lesion and involvement of the pyramidal tract.

Studies employing brain mapping and functional neuroimaging techniques have demonstrated the alteration of the function of specific regions of the brain following CIMT. A functional magnetic resonance imaging study on chronic stroke patients, following 2 weeks of home-based CIMT, with associated improved hand motor function, revealed motor physiological changes, including an enhanced activation of the premotor and secondary somatosensory cortex of the ipsilesional cerebral cortex. Studies using Transcranial Magnetic Stimulation (TMS) have demonstrated the expansion of motor maps among chronic subjects undergoing CIMT. Even among sub-acute stroke subjects, CIMT has revealed an associated increase in the TMS motor map area in the ipsilesional cortex with constant findings in the contralesional cortex. Research works have also reported expansion of ipsilesional and reduction of contralesional TMS motor maps following CIMT among chronic stroke patients. Research also has proven that ipsilesional enlargement of motor maps following CIMT tends to persist for certain months among 3–9 months post-stroke patients. Though the exact mechanism underlying the cortical reorganization remains unclear, evidence of functional reorganization of areas near the infarcted region, increased activation of the undamaged ipsilesional hemisphere, and activation of collateral pathways in the same hemisphere have been suggested as possible mechanisms underlying functional recovery.

Following CIMT, motor map expansion of the hand area is a consistent phenomenon in both early (3–9 months old cases) and late (1 year and above cases) stroke patients. However, a few studies have revealed that the CIMT-induced cortical reorganization (larger posterior shift in motor maps and a more dramatic increase in map size) was larger among late stroke patients, as compared to the early stroke patients. On the contrary, functional improvement immediately after CIMT was more marked among early stroke patients, as compared to late stroke patients. Such a trend may imply that the role of motor map expansion as an index of recovery may differ according to the time elapsed post-stroke and not be tightly linked to function.

2.6.4 Training Protocol and Clinical Applications

Not all hemiparetic patients are eligible to participate in CIMT. A certain amount of voluntary movement control of the paretic wrist and hand is a prerequisite for the therapy. According to Taub and Wolf, the higher-functioning stroke patients demonstrate a minimum active extension of 20° at the wrist and 10° at the metacarpophalangeal and interphalangeal joints of all digits and the lower-functioning patients demonstrate a minimum active extension of 10° at the wrist and 10° for any two digits, along with 10° abduction/extension of the thumb. Taub considers the latter as the minimum eligibility criteria for including stroke patients for upper limb CIMT. Patients should demonstrate adequate balance and adequate walking ability with the restraint. He or she should be able to perform safe transitions, including sit-to-stand and stand-to-sit, while wearing the restraint. The other eligibility criteria to participate in CIMT include the presence of considerable nonuse of the affected limb, absence of excessive spasticity for the affected limb musculature (a score of 1+ or below on the modified Ashworth scale), minimal cognitive deficit (a score of 24 or above for mini-mental state examination) and the absence of severe pain in any joint of the paretic limb. Unfortunately, only less than one-fourth of the total stroke population may meet the above criteria to benefit from this therapy.

2.6.4.1 Training Protocol

Typically, a splint, mitten, glove, or sling is worn for 2 weeks on the unaffected upper limb to restrain its activity throughout the waking hours. Concentrated and repetitive training of the paretic upper limb must occur during this period for 6 h a day. Though the signature CIMT (details mentioned in the earlier section of this chapter) has changed during the past few decades, the therapy has preserved a large part of the original characteristics, including the three main elements: massed practice, a focus on functional activities, and restricted use of the unaffected limb. Taub also recommends the application of a "transfer package" to guarantee adherence to the training protocol and transfer those gains to the patient's real environment. He also urged the patients to use their paretic limb during the training period, primarily by constraining the unaffected limb. The following are the essential components of the CIMT protocol for the upper limb:

1. Intensive supervised training.
2. Training based on behavior shaping principles.
3. A transfer package.
4. Use of a movement restraint device on the unaffected limb.

Intensive Training

Several research works have validated the scientific basis of intensive motor training following stroke. Intensive training is found to be a crucial tool for motor improvement. Such training provides a sufficient number of repetitions of a specific movement to induce long-lasting neuroplasticity. Evidence exists regarding a positive dose-response relationship between intensive motor training and motor outcomes. Intensive training refers to the frequency and amount of training, the duration of the training session (minutes or hours), and the duration of the training period (days to weeks). In the signature CIMT, each stroke patient is given intensive physical therapy for 10 consecutive weekdays, 6 h of one-to-one contact therapy. The 6 h therapy includes two 45 min sessions of conventional physical therapy, two 45 min sessions of task practice, a lunch break, and adequate rest intervals. Taub believes that intensive practice of correct use of the paretic limb is more crucial for effective learning than the restraint for the unaffected limb.

Behavior Shaping

The concept of shaping was first developed and used by Burrhus Frederic Skinner, an American psychologist, who is known for his theories on learning behaviors through reinforcement. It is defined as the procedure that reinforces behaviors that are successively closer to the target behavior (also known as successive approximations). Behavior shaping is considered to be a variant of operant conditioning. Instead of waiting for an organism to exhibit a desired behavior spontaneously, any behavior leading to the target behavior is rewarded. For instance, to train a rodent to push a lever, any movement in the direction of the lever can be rewarded until it learns to push the lever.

According to Taub, shaping has several advantages over the conditioned response training employed in his earlier work and they are as follows:

- By a slow, step-wise training, the learner is encouraged to gradually lead from a rudimentary initial response to more complex responses.
- Responses being shaped resemble those carried out in daily life, both in complexity and functional importance.
- Shaping takes a longer time and involves considerably more training than the conditioned response situations, encouraging a long-lasting permanent behavior.

Shaping permits an almost complete reversal of the motor disability that may progress from the total absence of the target behavior to a near-normal performance. In non-human primates, restraining the unaffected limb and shaping had improved motor function after unilateral forelimb deafferentation. In chronic stroke patients, Taub felt that shaping may encourage explicit learning and if combined with limb restriction, may complement and synergistically work to overcome learned nonuse.

Performance on tasks is measured quantitatively, and even for the smallest performance improvements detected, verbal reinforcement has to be given. For every occasion, the maintenance of previous gains has to be acknowledged. Any regression in performance is never punished, usually ignored, and encouraged to improve further. The selection of tasks for behavior shaping has to be tailored to address the motor deficits of the individual patient. During the early phase of shaping series, patients may be given physical assistance, the same as "assisted movement," to carry out parts of a movement sequence that are difficult for themselves.

Preferably, the training has to be carried out with the movement of the unaffected upper limb constraint by a restraining device like a mitten or splint. Frequent rest intervals should be introduced during the shaping session. Shaping is a time-intensive procedure that involves one-on-one interaction with the physiotherapist and should be carried out for a period of 3–6 h a day. Several tasks including the use of shuffleboard to cast a shuffleboard disk as far as possible, grasping an inflatable ball and moving over a table surface, tapping the telegraph key, placing ring on a prong, simulating brushing teeth, dot-to-dot drawing, patting powder puff on the face of a dummy, placing graduated weights on different height boxes, building blocks to create a tower, and simulating the use of spoon or fork for handling food pieces can be used for behavior shaping. If the aim is to teach the opening of the hand to release a ball, each attempt has to be timed, frequent feedback and cueing have to be provided regarding the movement with adequate positive reinforcement, and the difficulty of the task to be increased in step-by-step increments. For instance, learn to release a larger ball in the early phase of shaping sessions, before attempts are made for a smaller one.

Transfer Package

More than the ability to use the affected limb, the CIMT emphasizes on the amount of use the affected limb puts up in the real-life situation. Though massed practice appears to be a critical factor for the transfer of learning, behavioral strategies are required to improve adherence to the training program and increase the transfer of functional gains into real-world activities. The term "transfer package" is a collective expression meant for a behavioral contract signed by the patient, therapist, and caregiver, to make the patient accountable for adherence to the requirements of the therapy. It includes daily reports regarding the real-world use of the affected limb, details about home exercises, problem-solving ability to improve performance and overcome barriers, and use of home practice diary. The package is designed to increase the real-world relevance of the therapy program for the patient and urges the patients to use their affected limb during their waking hours, while the unaffected limb is restrained. Few studies on the role of transfer package in CIMT suggest a profuse increase in grey matter in bilateral sensorimotor cortices and bilateral hippocampi, as compared to CIMT without transfer package.

2.6.4.2 Modified CIMT

Though the CIMT is efficacious, a few surveys in the past have reported that stroke patients had a frequent temptation of using the unaffected limb during the training program and often complained about tiredness while wearing the mitt. About one-third of the patients reported compliance with the signature CIMT protocol. Many patients preferred a therapy protocol lasting for more weeks than 2–3 weeks and favored a shorter activity session than 6 h per day and fewer hours of wearing the movement restrictive devices than 90% of the waking hours. Approximately two-thirds of the therapists speculated poor adherence of patients to CIMT and lack of adequate resources to execute CIMT in many facilities. Both therapists and stroke patients also expressed concerns over compromises in independent activities. Longer hours of signature CIMT treatment may cause frustration among patients and suffer discomfort, muscle soreness, and skin lesions.

Due to the shortcomings mentioned earlier, many of the researchers and clinicians felt the need for developing a less intense and more convenient protocol, based on which the current modified CIMT protocols have been developed. Though the modified CIMT protocol developed by researchers have certain differences, the common findings are that the training protocol lasts for several weeks than a few weeks, 30 min to a few hours of intensive therapy instead of 6 h of treatment, and the movement restriction reduced to less than 50% of waking hours instead of 90% of the day.

The signature CIMT protocol uses massed practice as a learning strategy, whereas modified CIMT is based on distributed practice (spaced practice), a learning strategy, where the practice is broken up into many short sessions, spanning over a longer period of time. Learning is more effective when practiced for several sessions, spanning over a long period, than practiced for longer sessions in a short period (massed practice). Distributed practice is the most efficient method of procedural learning, and it increases the priming effect for subsequent practice sessions and directly influences the efficacy of memory recall.

2.6.4.3 CIMT for Acute Stroke

Experimental studies in animal models suggest that very early CIMT can be harmful than helpful and termed it as "forced overuse." Animal studies have proved that the immediate use of forced use following a brain injury can enlarge the lesion, presumably due to excitotoxicity, and may impede the motor recovery of the affected limb. When the restraint and behavioral strategies were imposed beyond 1 week, no expansion of the lesion was noticed; however, the negative impact on behavioral recovery stayed unaltered. Based on these findings, there was skepticism about the role of CIMT in medically stable acute stroke patients. Few researchers stated that high-intensity CIMT started in the first few days after stroke among humans may aggravate limb function deterioration. Whereas, some human studies, during the acute phase, indicated that CIMT is feasible and safe and showed a positive effect on upper limb motor function. According to a few meta-analytical studies in the recent past, both CIMT and modified CIMT may be more beneficial in acute or sub-acute stroke than conventional rehabilitation therapy, provided the intensity is low.

2.6.4.4 Training for Lower Limb

Though the CIMT protocol for the lower limb is based on the upper limb strategies, certain modifications are visible in the lower limb protocol. The bilateral nature of the activities that humans perform with the lower limbs and the fact that the paretic lower limb must be used for gait and balance during the early phase of recovery (even though they are performed incorrectly and ineffectively) leads to a situation similar to learned nonuse known as "learned misuse." Learned misuse can give rise to bad habits of abnormal coordination of lower limbs and abnormal gait patterns, including circumduction during the swing phase and ineffective gait.

Theoretically, it is possible to restrain the unaffected lower limb to induce movement of the paretic limb; however, the predominant bilateral nature of lower limb activities and the need for safety and balance make it impractical to apply restraint to the unaffected lower limb. Moreover, current evidence related to CIMT of the upper limb suggests that non-paretic limb restriction is relatively less relevant with respect to the concentrated or intensive practice and functional activities. It is also observed that there are no widely used CIMT protocols for the lower limb and many of the researchers have used the CIMT principles for upper limb training to study the efficacy of the treatment. Regarding eligibility criteria for the lower limb training, there are no standard protocols available. Few studies have stated some degree of lower limb movement or Brunnstrom recovery stage of 2 or above as the minimum criteria for lower limb CIMT, which is in addition to the absence of excessive spasticity for the paretic lower limb musculature, minimal cognitive deficit, and absence of severe pain in any paretic lower limb joints. Currently, there is a growing body of evidence available, exploring the benefits of CIMT for the upper limb compared to the lower limb CIMT. Investigators are also exploring the possibilities of CIMT in other conditions, including incomplete spinal cord injury, hip fracture, dystonia, cerebral palsy, and multiple sclerosis.

Further Reading

Adlers S, Buck M, Beckers D. PNF in practice. 4th edition. Springer publishers: Berlin, Heidelberg; 2014.

Andrews K, Stewart J. Stroke recovery: he can but does he?. *Rheumatol Rehabil.* 1979;18(1):43-8.

Bassett SW, Lake BM. Use of cold applications in the management of spasticity: report of three cases. *Phys Ther.* 1958 May;38(5):333-4.

Bieber I, Fulton JF. Relation of the cerebral cortex to the grasp reflex and to postural and righting reflexes. *Arch Neurol Psychiatry.* 1938;39:435-54.

Blanton S, Wolf SL. An application of upper-extremity constraint-induced movement therapy in a patient with subacute stroke. *Phys Ther.* 1999;79(9):847-53.

Bobath B. Adult hemiplegia evaluation and treatment. 3rd ed. Oxford: Butterworth-Heinemann; 1990.
Bobath K, Bobath B. The facilitation of normal postural reactions and movements in the treatment of cerebral palsy. *Physiotherapy*. 1964;50:246-62.
Brocklehurst J, Andrews K, Richards B, Laycock PJ. How much physical therapy for patients with stroke? *BMJ*. 1978;1:45–9.
Bruell JH, Peszczynski M, Volk D. Disturbance of perception of verticality in patients with hemiplegia: second report. *Arch Phys Med Rehabil.* 1957;38(12):776-80.
Brunnstrom S. Associated reactions of the upper extremity in adult patients with hemiplegia; an approach to training. *Phys Ther Rev.* 1956;36(4):225-36.
Brunnstrom S. Recording gait patterns of adult hemiplegic patients. *Phys Ther.* 1964;44:11-8.
Brunnstrom S. Walking preparation for adult patients with hemiplegia. *Phys Ther.* 1965;45:17-29.
Bütefisch C, Hummelsheim H, Denzler P, Mauritz KH. Repetitive training of isolated movements improves the outcome of motor rehabilitation of the centrally paretic hand. *J Neurol Sci.* 1995;130(1):59-68.
Canning CG, Ada L, O'Dwyer NJ. Abnormal muscle activation characteristics associated with loss of dexterity after stroke. *J Neurol Sci.* 2000;176(1):45-56.
Carr JH, Ow JEG, Shepherd RB. Some biomechanical characteristics of standing up at three different speeds: implications for functional training. *Physiother Theory Pract.* 2002;18(2):47-53.
Carr JH, Shepherd RB, editors. Early care of the stroke patient-a positive approach. London:Heinemann Medical; 1979.
Carr JH, Shepherd RB, editors. A motor relearning programme for stroke. 2nd ed. London: Aspen Publishers Inc; 1987a.
Carr JH, Shepherd RB, editors. Movement science: foundations for physical therapy in rehabilitation. Rockville: Publishers Inc; 1987b.
Car JH, Shepherd RB, editors. Stroke rehabilitation-guidelines for exercise and training to optimize motor skill. London: Butterworth Heinemann; 2003.
Chan DY, Chan CC, Au DK. Motor relearning programme for stroke patients: a randomized controlled trial. *Clin Rehabil.* 2006;20(3):191-200.
Cirstea MC, Levin MF. Compensatory strategies for reaching in stroke. *Brain.* 2000;123:940-53.
Cromwell FS. Margaret S. Rood: 1909–1984. Am J Occup Ther. 1985 Jan;39:54-5. Available from: https://doi.org/10.5014/ajot.39.1.54.
Crosbie J, Shepherd RB, Squire TJ. Postural and voluntary movement during reaching in sitting: the role of the lower limbs. *J Hum Mov Stud.* 1995;28:103-26.
Dean CM, Richards CL, Malouin F. Task-related training improves performance of locomotor tasks in chronic stroke. A randomized controlled pilot trial. *Arch Phys Med Rehabil.* 2000;81: 409-17.
Denny-Brown D. Disintegration of motor function resulting from cerebral lesions. *J Nerv Ment Dis.* 1950;112:1-45.
Denny-Brown D. Positive and negative aspects of cerebral cortical functions. *N C Med J.* 1956;17(7):295-303.
Dromerick AW, Edwards DF, Hahn M. Does the application of constraint-induced movement therapy during acute rehabilitation reduce arm impairment after ischemic stroke?. *Stroke.* 2000;31(12):2984-8.
Dubey L, Karthikbabu S. Trunk proprioceptive neuromuscular facilitation influences pulmonary function and respiratory muscle strength in a patient with pontine bleed. *Neurol India* 2017;65(1):183-4.
e Silva EMGS, Ribeiro TS, da Silva TCC, Costa MFP, Cavalcanti FADC, Lindquist ARR. Effects of constraint-induced movement therapy for lower limbs on measurements of functional mobility and postural balance in subjects with stroke: a randomized controlled trial. *Top Stroke Rehabil.* 2017;24(8):555-61.
el-Bahrawy MN, Elwishy AAB. Efficacy of motor relearning approach on hand function in chronic stroke patients. A controlled randomized study. *Italian J Physiotherapy*. 2012;2:121-7.

Fleet A, Page SJ, MacKay-Lyons M, Boe SG. Modified constraint-induced movement therapy for upper extremity recovery post stroke: what is the evidence?. *Top Stroke Rehabil.* 2014;21(4):319-31.
Franz SI, Sheetz MK, Wilson WW. The possibility of recovery of motor function in long standing hemiplegia. *JAMA*. 1921;65:2150-4.
Friel KM, Nudo RJ. Recovery of motor function after focal cortical injury in primates: compensatory movement patterns used during rehabilitative training. *Somatosens Mot Res.* 1998;15(3):173-89.
Fritz SL, Butts RJ, Wolf SL. Constraint-induced movement therapy: from history to plasticity. *Expert Rev Neurother.* 2012;12(2):191-8.
Fuzaro AC, Guerreiro CT, Galetti FC, Jucá RB, Araujo JE. Modified constraint-induced movement therapy and modified forced-use therapy for stroke patients are both effective to promote balance and gait improvements. *Rev Bras Fisioter.* 2012;16(2):157-65.
Gelber DA, Josefczyk PB, Herrman D, Good DC, Verhulst SJ. Comparison of 2 therapy approaches in the rehabilitation of pure motor hemiparetic stroke patients. *J NeuroRehabil.* 1995;9:191-6.
Gjelsvik BEB, Syre L. The Bobath concept in adult neurology. 2nd ed. Stuttgart: Thieme Publishers; 2016.
Goff B. The application of recent advances in neurophysiology to Miss M. Rood's concept of neuromuscular facilitation. *Physiotherapy.* 1972;58(12):409-15.
Guiu-Tula FX, Cabanas-Valdés R, Sitjà-Rabert M, Urrútia G, Gómara-Toldrà N. The Efficacy of the proprioceptive neuromuscular facilitation (PNF) approach in stroke rehabilitation to improve basic activities of daily living and quality of life: a systematic review and meta-analysis protocol. *BMJ Open.* 2017;7(12):e016739.
Guldenpfennig WM. Cerebral palsy and the Bobath treatment. *S Afr J Commun Disord.*1965;11(1): 11-3.
Gustavsen M, Jansen R, Kjendahl A, Lorentzen A. Motor relearning program approach improves short-term motor outcomes and reduces hospital stay after stroke. *Aust J Physiother.* 2002;48(1):59. Available from: https://doi.org/10.1016/s0004-9514(14)60285-3.
Hindle KB, Whitcomb TJ, Briggs WO, Hong J. Proprioceptive Neuromuscular Facilitation (PNF): Its Mechanisms and Effects on Range of Motion and Muscular Function. *J Hum Kinet.* 2012 Mar;31:105-13.
Ince LP. Escape and avoidance conditioning of responses in the plegic arm of stroke patients: a preliminary study. *Psychon Sci.* 1969;16:49-50.
Jackson JH. Remarks on evolution and dissolution of the nervous system. *J Ment Sci.* 1887;33:25-48.
Jarus T, Loiter Y. The effect of kinesthetic stimulation on acquisition and retention of a gross motor skill. *Can J Occup Ther.*1995;62(1):23-9.
Johnstone M. Restoration of Motor Function in the Stroke Patient. London: Churchill Livingstone; 1978.
Kleim JA, Barbay S, Nudo RJ. Functional reorganization of the rat motor cortex following motor skill learning. *J Neurophysiol.* 1998;80(6):3321-5.
Knapp HD, Taub E, Berman AJ. Effect of deafferentation on a conditioned avoidance response. *Science.* 1958;128(3328):842-3.
Knapp HD, Taub E, Berman AJ. Movements in monkeys with deafferented forelimbs. *Exp Neurol.* 1963;7:305-15.
Krakauer J. Motor learning: its relevance to stroke recovery and neuro-rehabilitation. *Curr Opin Neurol. 2006a*;19:84-90.
Krakauer JW. Motor learning: its relevance to stroke recovery and neurorehabilitation. *Curr Opin Neurol.* 2006b;19(1):84-90.
Krutulyte G, Kimtys A, Krisciūnas A. Kineziterapijos metodu Bobath ir judesiu mokymo programos efektyvumas reabilituojant ligonius, sirgusius galvos smegenu insultu [The effectiveness of physical therapy methods (Bobath and motor relearning program) in rehabilitation of stroke patients]. *Medicina (Kaunas).* 2003;39(9):889-95.

Kwakkel G, Wagenaar RC, Koelman TW, Lankhorst GJ, Koetsier JC. Effects of intensity of rehabilitation after stroke. *Stroke.* 1997;28:1550–6.

Langhammer B, Stanghelle JK. Can physiotherapy after stroke based on the Bobath concept result in improved quality of movement compared to the motor relearning programme. *Physiother Res Int.* 2011;16(2):69-80.

Langhorne P, Wagenaar RC, Partridge C. Physiotherapy after stroke: more is better? *Physiother Res Int.* 1996;1:75-88.

Larsen GJ. Rehabilitation for dysphagia paralytica. *J. Speech Dis.* 1972;37:187-94.

Leone JA, Kukulka CG. Effects of tendon pressure on alpha motoneuron excitability in patients with stroke. *Phys Ther.* 1988;68(4):475-80.

Levin MF, Panturin E. Sensorimotor integration for functional recovery and the Bobath approach. *Motor Control.* 2011;15(2):285-301.

Linkous LW, Stutts RM. Passive tactile stimulation effects on the muscle tone of hypotonic, developmentally delayed young children. *Percept Mot Skills.* 1990;71:951-4. Available from: https://doi.org/10.2466/pms.1990.71.3.951.

Loiselle C. Rood based program for decreasing pre-feeding behaviours. *Can J Occup Ther.* 1979;46:93-8.

Manning J. Facilitation of movement- the Bobath approach. *Physiotherapy.* 1972;58(12):403-8.

Margaret S. Rood Sixth Mary McMillan Lecturer. *Phys Ther.* 1969;49(6):632-3. Available from: https://doi.org/10.1093/ptj/49.6.632.

Mark VW, Taub E, Uswatte G, Bashir K, Cutter GR, Bryson CC, et al. Constraint-induced movement therapy for the lower extremities in multiple sclerosis: case series with 4-year follow-up. *Arch Phys Med Rehabil.* 2013;94(4):753-60.

McDonough SM. The neurophysiological basis of reciprocal inhibition in man. *PhysTher Rev.* 1997;2:19-28.

Metcalfe AB, Lawes N. A modern interpretation of the Rood Approach, Phys Ther Rev. 1998;3(4):195-212. Available from: https://doi.org/10.1179/ptr.1998.3.4.195.

Morreale M, Marchione P, Pili A, Lauta A, Castiglia SF, Spallone A, et al. Early versus delayed rehabilitation treatment in hemiplegic patients with ischemic stroke: proprioceptive or cognitive approach? *Eur J Phys Rehabil Med.* 2016 Feb;52(1):81-9.

Mudie MH, Matyas TA. Upper extremity retraining following stroke: effects of practice. *J Neurol Rehabil.* 1996;10:167-84.

Nudo RJ, Milliken GW. Reorganization of movement representations in primary motor cortex following focal ischemic infarcts in adult squirrel monkeys. *J Neurophysiol.* 1996;75(5):2144-9.

Odeen I. Reduction of muscular hypertonus by long-term muscle stretch. *Scand J Rehab Med.* 1981;13:93-9.Oxendine A. Psychology of Motor Learning. 2nd ed. New Jersey: Prentice-Hall; 1984.

Ogden R, Franz S. On cerebral motor control: the recovery from experimentally produced hemiplegia. *Psychobiol.* 1917;1(1):33-50.

Ostendorf CG, Wolf SL. Effect of forced use of the upper extremity of a hemiplegic patient on changes in function. A single-case design. *Phys Ther.* 1981;61(7):1022-8.

Paci M. Physiotherapy based on the Bobath concept for adults with post-stroke hemiplegia: a review of effectiveness studies. *J Relhatbil Med.* 2003;35:2-7.

Pandian S, Arya KN, Davidson EWR. Comparison of Brunnstrom movement therapy and motor relearning program in rehabilitation of post-stroke hemiparetic hand: a randomized trial. *J Bodyw Mov Ther.* 2012;16(3):330-7.

Perry CE. Principles and techniques of the Brunnstrom approach to the treatment of hemiplegia. *Am J Phys Med.* 1967;46(1):789-815.

Raine S, Meadows L, Lynch-Ellerington M, editors. Bobath concept theory and clinical practice in neurological rehabilitation. Oxford:Wiley-Blackwell;2009.

Reiss AP, Wolf SL, Hammel EA, McLeod EL, Williams EA. Constraint-induced movement therapy (CIMT): current perspectives and future directions. *Stroke Res Treat.* 2012;2012:159391. Available from: https://doi.org/10.1155/2012/159391.

Reynolds G, Archibald KC, Brunnstrom S, Thompson N. Preliminary report on neuromuscular function testing of the upper extremity in adult hemiplegic patients. *Arch Phys Med.* 1958;39:303.

Richards CL, Malouin F, Wood-Dauphinee S. Task-specific physical therapy for optimization of gait recovery in acute stroke patients. *Arch Phys Med Rehabil.* 1993;74:612-20.

Rood MS: Eleanor Clarke Slagle lecture: everyone counts. *Am J Occup Ther.* 1958;12(6):326-9.

Rood MS. Neurophysiological mechanisms utilized in the treatment of neuromuscular dysfunction. *Am J Occup Ther.* 1956;10:220-5.

Rood MS. Neurophysiological reactions as a basis for physical therapy. *Phys Ther Rev.* 1954;34:444-9.

Rood M. The use of sensory receptors to activate, facilitate and inhibit motor response, autonomic and somatic in developmental sequence. In: Scattely C, editor. Approaches to treatment of patients with neuromuscular dysfunction, third international congress, world federation of occupational therapists. Dubuque: William Brown Group; 1962.

Sawaki L, Butler AJ, Leng X, Wassenaar PA, Mohammad YM, Blanton S, et al. Differential patterns of cortical reorganization following constraint-induced movement therapy during early and late period after stroke: A preliminary study. *NeuroRehabilitation.* 2014;35(3):415-26.

Sawner KA, LaVigne JM, editors. Brunnstrom's Movement therapy in hemiplegia- a neurophysiological approach. 2nd ed. Philadelphia: JB Lippincott Company; 1992.

Seligman ME, Maier SF. Failure to escape traumatic shock. *J Exp Psychol.* 1967;74(1):1-9.

Seyffarth H, Denny-Brown D. The grasp reflex and the instinctive grasp reaction. *Brain.* 1948;71(2):109-83.

Sherrington CS. Flexion reflex of the limb, crossed extension reflex and reflex stepping in standing. *J Physiol.* 1910;40:28-121.

Sherrington CS. The integrative action of the nervous system. 2nd ed. New Haven: Yale University Press; 1947a.

Sherrington CS. the integrative action of the nervous system. 2nd ed. New Haven: Yale University Press; 1947b.

Smedes F, Giacometti da Silva L. Motor learning with the PNF-concept, an alternative to constrained induced movement therapy in a patient after a stroke; a case report. *J Bodyw Mov Ther.* 2019 Jul;23(3):622-7.

Smedes F, Heidmann M, Schäfer C, Fischer N, Stępień A. The proprioceptive neuromuscular facilitation-concept; the state of the evidence, a narrative review. *Phys Ther Rev.* 2016;21(1):17-31.

Smith RH, Sharpe M. Brunnstrom therapy: Is it still relevant to stroke rehabilitation? *Physiother Theory Pract.* 1994;10(2):87-94.

Sneha BS, Karthikbabu S, Syed N, Prasanth LK. Effects of trunk proprioceptive neuromuscular facilitation on dynamic balance, mobility and quality of life in Parkinson's disease. *Parkinsonism & Related Disorders.* 2016;22(2):e68-9.

Stern PH, McDowell F, Miller SM, Robinson M. Factors influencing stroke rehabilitation. *Stroke.* 1971;3:213-8.

Stockmeyer SA. An interpretation of the approach of Rood to the treatment of neuromuscular dysfunction. *Am J Phys Med.* 1967;46:900-56.

Taub E, Crago JE, Burgio LD, Groomes TE, Cook EW, DeLuca SC, et al. An operant approach to rehabilitation medicine: overcoming learned nonuse by shaping. *J Exp Anal Behav.* 1994;61(2):281-93.

Taub E, Miller NE, Novack TA, Cook EW, Fleming WC, Nepomuceno CS, et al. Technique to improve chronic motor deficit after stroke. *Arch Phys Med Rehabil.* 1993;74(4):347-54.

Taub E, Uswatte G. Constraint-induced movement therapy: bridging from the primate laboratory to the stroke rehabilitation laboratory. *J Rehabil Med.* 2003;41(S):34-40.

Taub E, Uswatte G, King DK, Morris D, Crago JE, Chatterjee A. A placebo-controlled trial of constraint-induced movement therapy for upper extremity after stroke. *Stroke.* 2006;37(4):1045-9.

Taub E, Uswatte G, Pidikiti R. Constraint-induced movement therapy: a new family of techniques with broad application to physical rehabilitationDOUBLEHYPHENa clinical review. *J Rehabil Res Dev.* 1999;36(3):237-51.

Taub E, Wolf SL. Constraint induced movement techniques to facilitate upper extremity use in stroke patients. *Top Stroke Rehabil.* 1997;3(4):38-61.

Twitchell TE. The restoration of motor function following hemiplegia in man. *Brain.* 1951;74(4):443-80.

Tyson SF, Selley AB. The effect of perceived adherence to the Bobath concept on physiotherapists' choice of intervention used to treat postural control after stroke. *Disabil Rehabil.* 2007;29(5):395-401.

van Vliet PM, Lincoln NB, Foxall A. Comparison of Bobath based and movement science based treatment for stroke: a randomised controlled trial. *J Neurol Neurosurg Psychiatry.* 2005;76:503-8.

Wade DT, Wood VA, Langton-Hewer R. Recovery after stroke-the first 3 months. *J Neurol Neurosurg Psychiatry.* 1985;48:7-13.

Wagenaar RC, Meijer OG, van Wieringen PC, Kuik DJ, Hazenberg GJ, Lindeboom J, et al. The functional recovery of stroke: a comparison between neuro-developmental treatment and the Brunnstrom method. *Scand J Rehabil Med.* 1990;22(1):1-8. PMID: 2326602.

Walshe FMR. On certain tonic or postural reflexes in hemiplegia, with special reference to the so-called "associated movements." *Brain.* 1923; 46:1-37.

Wanderley D, Lemos A, Moretti E, Barros MMMB, Valença MM, de Oliveira DA. Efficacy of proprioceptive neuromuscular facilitation compared to other stretching modalities in range of motion gain in young healthy adults: A systematic review. *Physiother Theory Pract.* 2019 Feb;35(2):109-29.

Wang RY, Chen HI, Chen CY, Yang YR. Efficacy of Bobath versus orthopaedic approach on impairment and function at different motor recovery stages after stroke: a randomized controlled study. *Clin Rehabil.* 2005;19(2):155-64.

Wartenberg R. Babinski reflex and Marie-Foix flexor withdrawal reflex; historical notes. *AMA Arch Neurol Psychiatry.* 1951;65(6):713-6.

Wolf SL, Lecraw DE, Barton LA, Jann BB. Forced use of hemiplegic upper extremities to reverse the effect of learned nonuse among chronic stroke and head-injured patients. *Exp Neurol.* 1989;104(2):125-32.

Wolf SL, Winstein CJ, Miller JP, Taub E, Uswatte G, Morris D, et al. Effect of constraint-induced movement therapy on upper extremity function 3 to 9 months after stroke: the EXCITE randomized clinical trial. *JAMA.* 2006;296(17):2095-104.

Wu CY, Lin KC, Chen HC, Chen IH, Hong WH. Effects of modified constraint-induced movement therapy on movement kinematics and daily function in patients with stroke: a kinematic study of motor control mechanisms. *Neurorehabil Neural Repair.* 2007;21(5):460-6.

Yu J. Functional recovery with and without training following brain damage in experimental animals: a review. *Arch Phys Med Rehabil.* 1976;57(1):38-41.

Chapter 3
Stroke

Abraham M. Joshua, Zulkifli Misri, Santosh Rai, and V. Harigovindan Nampoothiri

3.1 Introduction and Historical Background

Cerebrovascular disease or stroke refers to a group of disorders of the brain vasculature that can affect the vascular supply of the underlying tissues. Stroke, an important cause of prolonged disability, often makes the survivors unable to return to work or continue their duties as family members or citizens. Stroke has a considerable psychosocial and economic impact worldwide. The term stroke was often known as "apoplexy," a Greek word meaning "struck suddenly with violence." In the vast majority of cases, "stroke undoubtedly alters the history of the world for the survivor" as the loss of function is often instantaneous, totally unanticipated and impairments more or less permanent and devastating. The loss of function may include one or more features like sudden inability to move a limb, stand or walk, speak or understand spoken language, see, read, write, think, and feel.

A. M. Joshua (✉)
Department of Physiotherapy, Kasturba Medical College of Mangalore, Manipal Academy of Higher Education, Manipal, India
e-mail: abraham.joshua@manipal.edu

Z. Misri
Department of Neurology, Kasturba Medical College of Mangalore, Manipal Academy of Higher Education, Manipal, India
e-mail: misri.zk@manipal.edu

S. Rai
Department of Radiodiagnosis, Kasturba Medical College of Mangalore, Manipal Academy of Higher Education, Manipal, India
e-mail: santosh.rai@manipal.edu

V. Harigovindan Nampoothiri
Allied Rehabilitation Services Inc., Brownsville, TX, USA

A. M. Joshua (ed.), *Physiotherapy for Adult Neurological Conditions*,
https://doi.org/10.1007/978-981-19-0209-3_3

For many of the eminent leaders in medicine, science, and politics, stroke had prematurely cut short their productivity. Louis Pasteur, a biologist, microbiologist, and chemist; Russell DeJong, the first editor of the journal *Neurology*; Raymond Escourolle, the neuropathologist; and H. Houston Merritt, the writer of *Merritt's Neurology* were a few who got severely disabled by multiple strokes. Prominent political leaders, including Vladimir Lenin, Franklin D. Roosevelt, Winston Churchill, and Joseph Stalin, had severe cerebrovascular disease during the helm of their careers.

In 400 BC, perhaps Hippocrates, the Greek physician, was the first to write about the medical aspects of the stroke. He was mostly interested in the prognosis of the illness and was a keen observer. In his aphorisms on apoplexy, he stated that most subjects were between the ages of forty and sixty and mentioned that "attacks of numbness" and "pains in the head" were reflections of the "impending apoplexy." The Greeks recognized that interruption of the blood vessels to the brain was the cause for loss of consciousness, and they named those arteries carotid, meaning "deep sleep," from the Greek word *karos*. Roman physician and philosopher Galen, a few hundred years after Hippocrates, described the anatomy of the brain and its blood vessels following dissections of animals. Galen believed the disease to be a disequilibrium between the four body humors: the black bile, yellow bile, phlegm, and blood. Until the fourteenth century, his voluminous writings were blindly followed by physicians, and dissection, experimentation, and observations were discouraged and considered unscholarly.

Andreas Vesalius, a physician and an anatomist in the sixteenth century, challenged the Galenic tradition by dissecting humans and, based on his personal observations, wrote books on human anatomy known as *De humani corporis fabrica libri septem*, meaning on the fabric of the human body. Vesalius is considered the founder of modern human anatomy. In the latter half of the seventeenth century, two eminent physicians, Johann Jakob Wepfer and Thomas Willis, made further anatomical and clinical observations on apoplexy. Wepfer performed meticulous examinations of the brains of patients dying of apoplexy. It was Wepfer who described the carotid siphon and the course of the middle cerebral artery in the Sylvian fissure. He mentioned that apoplexy could be either due to obstruction of the carotid and vertebral arteries or due to a bleed into the brain. Thomas Willis, a physician and a neuroanatomist, best known for his "circulus arteriosus cerebri" or cerebral arterial circle, also recognized transient ischemic attacks and the consequence of embolism. In 1689, William Cole, an English physician, in a medical essay concerning apoplexy, first introduced the word "stroke" in the field of medicine.

In the eighteenth century, Giovanni Battista Morgagni, the father of modern anatomical pathology, focused his attention on the pathology and cause of apoplexy. He recognized that paralysis was on the side of the body opposite to the brain lesion and gave an elaborate description of the intracerebral hemorrhage. His work helped in shifting the emphasis from anatomy to etiology, pathology, and clinical manifestations of the diseases. In 1812, John Cheyne, a British physician and a surgeon, in his book, made a strong attempt to separate apoplexy from the phenomena of lethargy and coma. In the nineteenth century, John Abercrombie, a physician and a philosopher, made a detailed clinical classification of apoplexy and published the book *Pathological and Practical Researches on Diseases of the Brain and the Spinal Cord*, a book regarded as the first textbook in neuropathology. Abercrombie also speculated the etiological mechanisms underlying the spasm of vessels, interruption of the circulation, and rupture of vessels causing hemorrhage. In the later part of the nineteenth century, Rudolf Virchow, a German physician and a pathologist, mentioned certain important experimental and pathological information about vascular disease, particularly relating to the thrombus and infarction. A French neurosurgeon, Henri Düret, made a detailed observation about the distribution of the arteries and veins in the cranium, including the arteries supplying the cranial nerve nuclei. During the same period, several clinicians gathered information on the clinical findings of stroke that involved various regions of the brain. However, minimal concern was given to the pathogenesis, laboratory confirmation, or management for the same. The medical textbooks and scholarly articles written by William Osler, William Richard Gowers, Samuel Alexander Kinnier Wilson, and Charles Foix made detailed clinical descriptions about many stroke syndromes. Charles Miller Fisher, a Canadian neurologist and a neuropathologist, mentioned "...fleeting attacks of paralysis, numbness, tingling, speechlessness, unilateral blindness or dizziness..." as the warning symptoms of carotid artery disease. During the mid- and later years of the twentieth century, Fisher made major pathological and clinical observations of carotid artery disease, intracerebral hemorrhage, lacunar stroke, and vascular lesions of the posterior circulation.

In the last few decades, there has been an explosive growth of interest in and knowledge about stroke. Recent advances in science and technology allowed better visualization of the normal, pathological anatomy, and functional aspects of the brain. Technology has also helped to pool up the databases and registries of large numbers of stroke studies to identify and quantify the most common clinical and laboratory findings in various stroke syndromes. Researches in the field of epidemiology have helped in accurately identifying the risk factors for stroke and plausible prevention strategies for the same. Even significant advances in surgical, medical, and radiological management have shown success in reducing the morbidity and mortality of stroke.

3.2 Normal Arterial Blood Supply of the Brain

The human brain, though weighing over a kilogram, consumes approximately 20% of the body's oxygen supply at rest and must continuously receive a voluminous amount of blood, about one liter per minute. Two pairs of arterial trunks, the right and the left internal carotid arteries, and the right and the left vertebral arteries supply blood to the brain. These four arteries lie within the subarachnoid space, and their branches anastomose at the base of the brain around the optic chiasma, the stalk of the pituitary gland, and the hypothalamus. The following section of this chapter focuses on arteries and their branches supplying the brain parenchyma.

3.2.1 Internal Carotid Artery

Both the right and the left internal carotid arteries are the branches of common carotid arteries. Typically, the right common carotid artery arises from the brachiocephalic artery and the left common carotid artery arises directly from the aortic arch (Fig. 3.1). Each common carotid artery, roughly near the C4 level or the upper level of the thyroid cartilage, bifurcates to form the respective internal and external carotid arteries. Each internal carotid artery consists of cervical, petrous, cavernous, and cerebral segments (Fig. 3.2). The cervical segment traversing through the neck enters the cranial cavity through the carotid canal of the petrous part of the temporal bone (petrous segment) and reaches the cavernous sinus from below (cavernous segment). Subsequently, the internal carotid artery pierces the roof of the cavernous sinus and enters the cranial cavity, flanked by the oculomotor and optic nerves. The cerebral segment of the internal carotid artery is quite short and extends upward and backward to give rise to all the major branches of the internal carotid artery. From the neck to the termination, the internal carotid artery makes several, almost 90° turns, and these turns reduce the pressure and the velocity of the blood it brings to the thin-walled arteries of the brain.

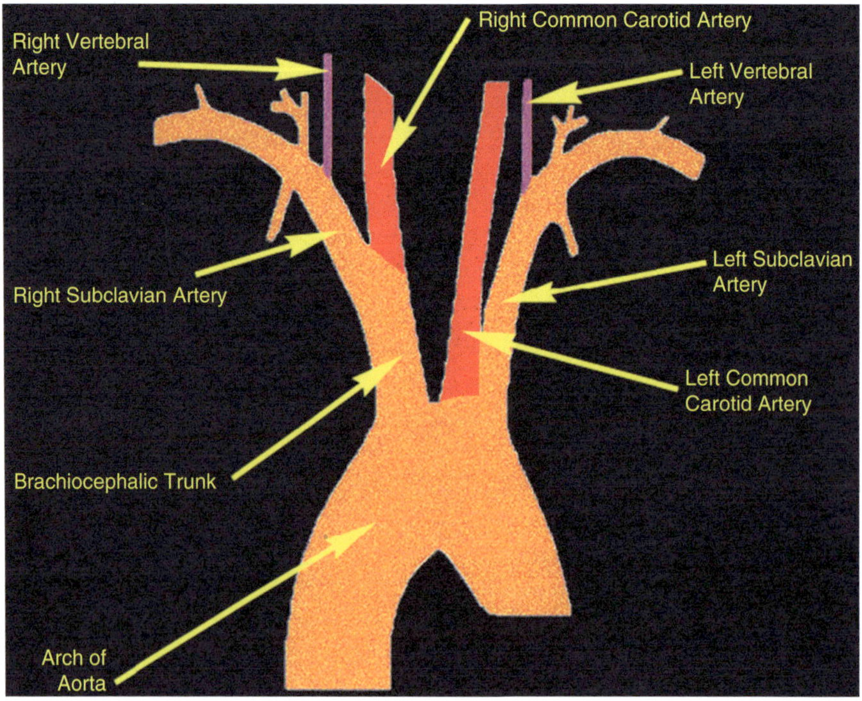

Fig. 3.1 An illustration of the right common carotid artery arising from the brachiocephalic artery and the left common carotid artery arising from the arch of the aorta

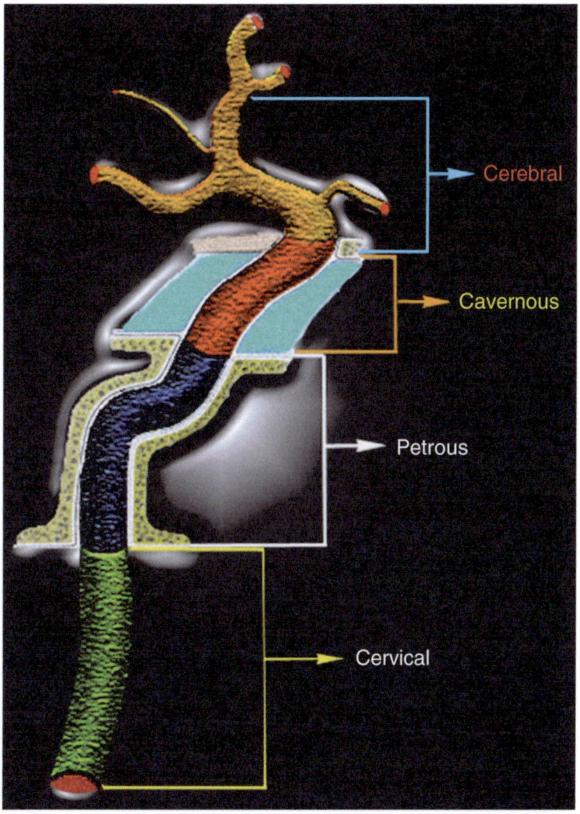

Fig. 3.2 The cervical, petrous, cavernous, and cerebral segments of the internal carotid artery

The cerebral segment of the internal carotid artery branches into ophthalmic, anterior choroidal, posterior communicating, anterior cerebral, and middle cerebral arteries (Fig. 3.3). The ophthalmic artery supplies all the structures in the orbit. The anterior choroidal artery, a small narrow artery arising from the posterior aspect of the internal carotid artery, while traversing along the path of the optic tract toward the lateral geniculate body, supplies the optic tract, lateral geniculate body, optic radiation, hippocampus, posterior limb of the internal capsule, tail of the caudate nucleus, and choroid plexus of the inferior horn of the lateral ventricle. The posterior communicating artery, a small vessel that originates from the cerebral segment of the internal carotid artery, close to its terminal bifurcation, runs posteriorly above the oculomotor nerve to join the posterior cerebral artery forming part of the cerebral arterial circle. Usually, the size of the right and left posterior communicating arteries is not identical; one can be frequently smaller than the other and at times entirely absent or doubled. The main function of these arteries is to ensure sustainable blood supply to the brain in case if the internal carotid or vertebral artery occludes.

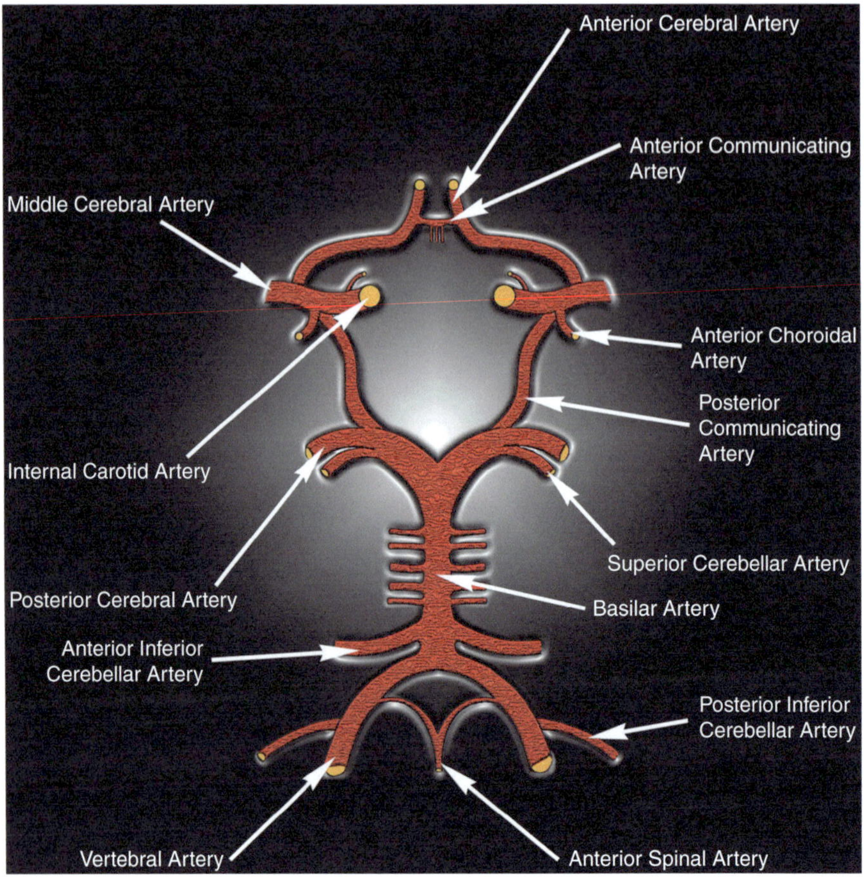

Fig. 3.3 An illustration of the formation of the "circle of Willis" by the anterior and posterior circulations

3.2.2 Anterior Cerebral Artery

The anterior cerebral artery, one of the two terminal branches of the internal carotid artery, runs forward and medially, superior to the optic nerve, to enter the great longitudinal fissure. Here, it connects with its counterpart of the contralateral side by a short branch named the anterior communicating artery. In the longitudinal fissure, the right and left anterior cerebral arteries lie in a close approximation of about 4–5 mm, and these arteries follow the genu and then the superior border of the corpus callosum until they anastomose with their corresponding posterior cerebral arteries. The anterior communicating, the central, and the cortical are three branches of each anterior cerebral artery. The anterior communicating artery is approximately 4 mm long and gives off numerous branches that supply the structures, including the

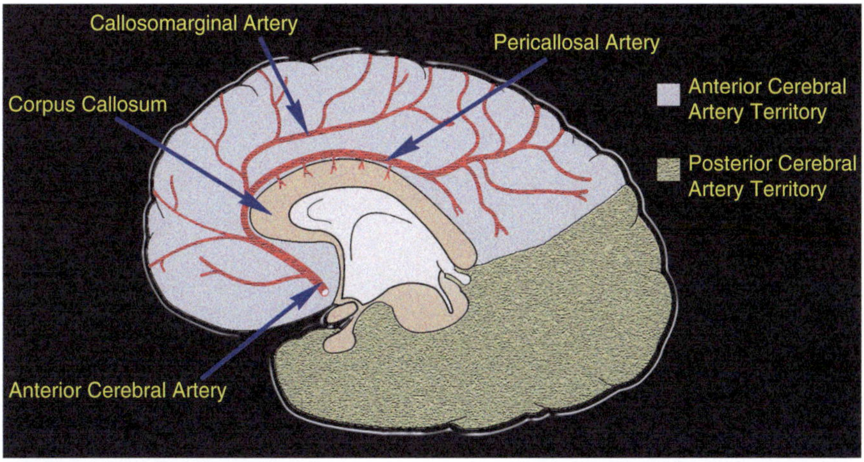

Fig. 3.4 The vascular territory of the anterior cerebral artery and its branches

optic chiasma, hypothalamus, para-olfactory areas, fornix, and cingulate gyrus. The central branches arising from the initial part of the anterior cerebral artery, just before or after the origin of the anterior communicating artery, supply the anterior part of the corpus callosum, head of the caudate nucleus, part of the putamen, globus pallidus, and anterior limb of the internal capsule. Among the central branches, the recurrent artery of Heubner (medial striate artery) is the largest perforating branch from the central branches and is the only one routinely seen on angiography.

The names of the cortical branches of the anterior cerebral artery according to their distribution are orbital, frontal, and parietal. The orbital branch supplies the orbital surface of the frontal lobe, olfactory lobe, medial orbital gyrus, and gyrus rectus. The frontal branch supplies the medial frontal gyrus, cingulate gyrus, paracentral lobule, superior frontal gyrus, middle frontal gyrus, and precentral gyrus. Whereas, the parietal branch supplies the superior parietal lobule (precuneus) and the adjoining lateral surface. To sum up, the cortical branches of the anterior cerebral artery supply the motor and somatosensory cortices representing the lower limb, i.e., the entire medial surface of the cerebral cortex up to the parieto-occipital sulcus (Fig. 3.4) including a small strip of cortex on the adjoining lateral surface (Fig. 3.5).

3.2.3 *Middle Cerebral Artery*

The middle cerebral artery, the larger terminal branch of the internal carotid artery, runs first in the lateral sulcus (also known as lateral or Sylvian fissure) then postero-superiorly on the insula and divides into central and cortical branches. Like the anterior cerebral artery, the central branches supply blood to the deeper structures,

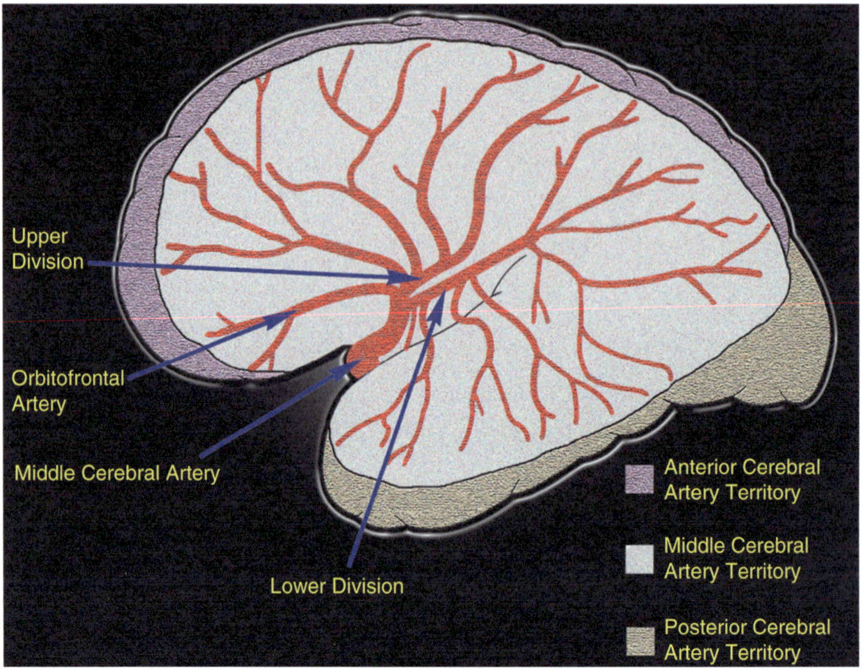

Fig. 3.5 The vascular territory of the middle cerebral artery and its branches

and cortical branches supply blood to the cortical regions. The central branches have 10–15 slender striate arteries (lenticulostriate arteries) that pierce the floor of the lateral sulcus to supply the deeper cerebral structures, including the corpus striatum, most of the caudate nucleus and the lenticular nucleus, and the external and the internal capsules.

The cortical branches, during their course along the lateral sulcus, are named according to their location and distribution as the orbital, frontal, parietal, and temporal branches. The orbital branches serve the inferior frontal gyrus and the lateral orbital surface of the frontal lobe. The frontal branches serve the middle frontal, the precentral, and a part of the inferior frontal gyri, and the parietal branches serve the postcentral gyrus, the whole inferior parietal lobule, the lower part of the superior parietal lobule, and the angular and supramarginal gyri. The temporal branches serve the lateral surface of the entire temporal lobe up to the occipital gyri. Therefore, the cortical branches of the middle cerebral artery supply the whole lateral surface of the cerebral cortex (Fig. 3.5) except the narrow strip supplied by the anterior cerebral artery and the occipital pole and the inferolateral surface of the cerebral cortex supplied by the posterior cerebral artery, i.e., the cortical branches supply the motor and somatosensory cortices representing the entire body, except the lower limb (Fig. 3.4).

3.2.4 Vertebral Artery

Each of the vertebral arteries, arising from the respective first part of the subclavian artery, ascends cephalically through the transverse processes of the upper six cervical vertebrae and pierces the atlanto-occipital membrane and the dura mater to enter the posterior cranial fossa through the foramen magnum. Close to the anterolateral aspect of the medulla, the right and the left vertebral arteries converge medially as they ascend, and approximately at the pontomedullary junction level, the right and the left vertebral arteries unite to form the basilar artery (Fig. 3.6). Together the intracranial branches of the vertebral and basilar arteries supply the spinal cord, brainstem, cerebellum, posterior parts of the diencephalon, and parts of occipital and temporal lobes of the cerebral cortex.

The anterior and posterior spinal arteries, the posterior inferior cerebellar artery, and the medullary divisions are the main branches of the vertebral artery supplying the brainstem and the spinal cord. The anterior spinal artery emerges near the end of the vertebral artery and descends anteriorly to anastomose with its counterpart from the opposite side at the mid-medullary level. The single trunk then descends on the anterior median fissure of the spinal cord. In the majority of the subjects, the posterior spinal arteries arise from the posterior inferior cerebellar artery or may emerge directly from the vertebral artery near the medulla. The right and left posterior

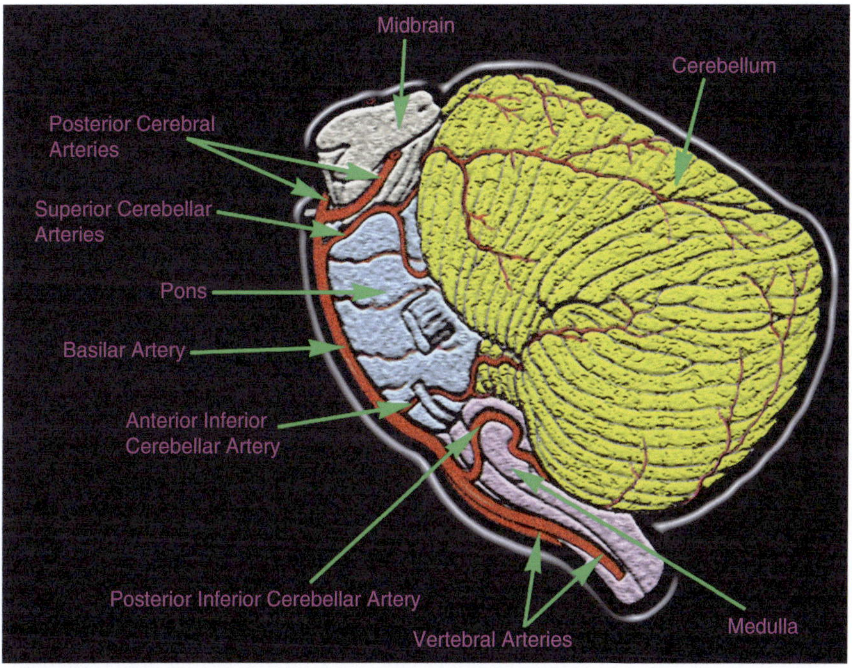

Fig. 3.6 Illustration of the posterior circulation of the brain

spinal arteries descend along the respective posterolateral surface of the spinal cord and supply the ipsilateral grey and white dorsal columns of the spinal cord. The anterior spinal artery and the right and left posterior spinal arteries are reinforced by radicular arteries that enter the vertebral canal through the intervertebral foramina, sequentially from the vertebral, ascending cervical, posterior intercostal, and first lumbar arteries.

The posterior inferior cerebellar artery, the largest branch of the vertebral artery, arises near the lower end of the olivary nucleus, curves posteriorly, and ascends behind the roots of the ninth and tenth cranial nerves. Near the inferior border of the pons, the artery divides, and the medial branch supplies the cerebellar hemisphere and inferior vermis. The lateral branch supplies the inferior cerebellar surface as far as its lateral border and anastomoses with the anterior inferior and superior cerebellar arteries of the basilar artery. The trunk of the posterior inferior cerebellar artery supplies the medulla oblongata dorsal to the olive, lateral aspect of the hypoglossal nucleus near its emerging roots, the dentate nucleus, and the choroid plexus of the fourth ventricle. The medullary divisions, small branches, emerging from the cranial portion of the vertebral artery, supply the medulla oblongata (also served by the posterior inferior cerebellar artery and/or the anterior and posterior spinal arteries).

3.2.5 Basilar Artery

The basilar artery is a large median vessel formed by the union of the vertebral arteries at the mid-medullary level. It lies in the shallow median groove on the ventral pontine surface and extends from the rostral end of the medullary pyramids to the rostral end of the pons. Near the interpeduncular cistern, the artery terminates by dividing into two posterior cerebral arteries (Fig. 3.6). The branches of the basilar artery are the pontine divisions, labyrinthine artery, anterior inferior cerebellar artery, and superior cerebellar artery. The small pontine divisions or branches arising at 90° to the basilar artery penetrate and serve the substance of the pons and the midbrain. The labyrinthine artery, usually a branch of the anterior inferior cerebellar artery and less frequently a branch of the basilar artery, is larger than the pontine branches, accompanies the facial and the vestibulocochlear nerves, and supplies the internal ear.

The anterior inferior cerebellar artery, just rostral to the medullary pyramids, emerges from the basilar artery and follows the sixth cranial nerve and loops around the seventh and eighth cranial nerves and reaches the cerebellopontine angle. The small pontine branches and the medial and lateral branches of the anterior inferior cerebellar artery serve the pons and midbrain and the anteromedial and anterolateral aspects of the cerebellum, respectively. The branches of the artery mentioned above anastomose with branches of the posterior inferior cerebellar and superior cerebellar arteries. The superior cerebellar artery emerges from the basilar artery, caudal to the third cranial nerve, and curves around the cerebral peduncle to supply the superior surface of the cerebellum. The branches also vascularize the pineal body, the midbrain, and the choroid plexus of the third ventricle. The branches of the superior cerebellar artery anastomose with branches of the anterior inferior cerebellar artery.

3.2.6 Posterior Cerebral Artery

The posterior cerebral artery, larger than the superior cerebellar artery, a terminal branch of the basilar artery originates near the third cranial nerve, lateral to the midbrain. The right and left posterior cerebral arteries run laterally, parallel with the superior cerebellar arteries, and join with the corresponding posterior communicating arteries. Each artery then winds around the cerebral peduncle and reaches the superior surface of the tentorium cerebelli to serve the temporal and occipital lobes. In addition to the choroidal branch, like the anterior and middle cerebral arteries, the posterior cerebral artery also has cortical and central branches. Figure 3.7 illustrates the cortical vascular area supplied by the posterior cerebral artery. The temporal, the lateral and medial occipital, and the splenial are the cortical branches of the posterior cerebral artery. The temporal branch supplies the uncus, the parahippocampal, and the medial and lateral occipitotemporal gyri. The occipital branches supply the cuneus, precuneus, lingual gyrus, and posterolateral surface of the occipital lobe.

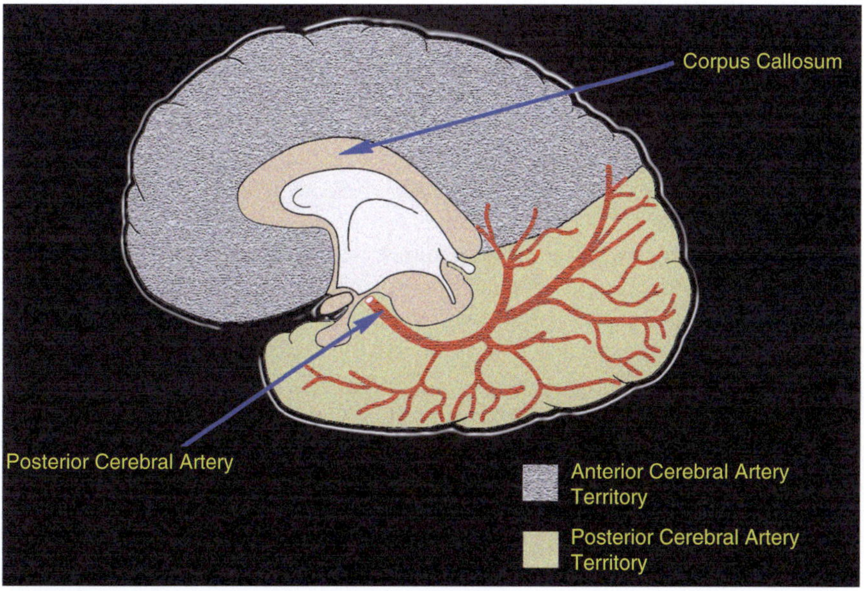

Fig. 3.7 The vascular territory of the posterior cerebral artery

The splenial artery supplies the splenium and posterior portion of the corpus callosum.

The central branches of the posterior cerebral artery vascularize the subcortical structures. Several central branches of the posterior cerebral artery, along branches from the posterior communicating artery, supply the optic tract, the mammillary bodies, and the thalamus. In addition to that, they supply the geniculate bodies, the cerebral peduncles and the interpeduncular area, the corticospinal tracts, the mesencephalic reticular formation, the substantia nigra, and the tegmentum of the midbrain. The choroidal branches supply the choroid plexus of the lateral and third ventricle, the medial and lateral geniculate bodies, the peduncle, the tegmentum, the colliculi, the pulvinar, the posterior and medial portions of the thalamus, and the pineal body.

3.2.7 Cerebral Arterial Circle

The cerebral arterial circle, also known as the circle of Willis or "circulus arteriosus," is a large arterial anastomosis at the base of the brain around the optic chiasma, infundibulum, and other structures of the interpeduncular fossa. This circle forms an anastomosis between proximal portions of the anterior and posterior cerebral arteries, the distal-most portion of the internal carotid artery, and the anterior and posterior communicating arteries (Fig. 3.8). Anteriorly, the anterior cerebral arteries are connected by the small anterior communicating artery, and, posteriorly, the two posterior cerebral arteries are merged to the corresponding side internal carotid artery by a posterior communicating artery. This arterial wreath equalizes the blood flow to various parts of the brain. In normal situations, due to equal blood pressure, only a minimal exchange of blood occurs between the right and left halves of the circle. Following the occlusion of one or more of the arteries contributing to the circle, alterations of blood flow in the arterial circle are bound to occur.

Variations in the caliber of the arteries forming the cerebral arterial circle are common. The hemodynamics of the arterial circle is influenced by variations in the caliber of communicating arteries and the segments of the anterior and posterior cerebral arteries that form a part of the circle. Among them, the greatest individual divergence seen is the presence or absence and the caliber of the posterior communicating arteries.

The circle is also a common site for saccular aneurysms. Congenital defects in the blood vessel wall and high turbulence of blood are a few key factors involved in the development of aneurysms. The most common sites for the development of saccular aneurysms are the origins of the anterior and posterior communicating artery, the bifurcation of the middle cerebral artery, the cavernous segment of the internal carotid artery, the bifurcation of the internal carotid artery, and various locations on the vertebrobasilar arteries. Rupture of these saccular cerebral aneurysms leads to subarachnoid hemorrhages and often results in serious neurological impairments.

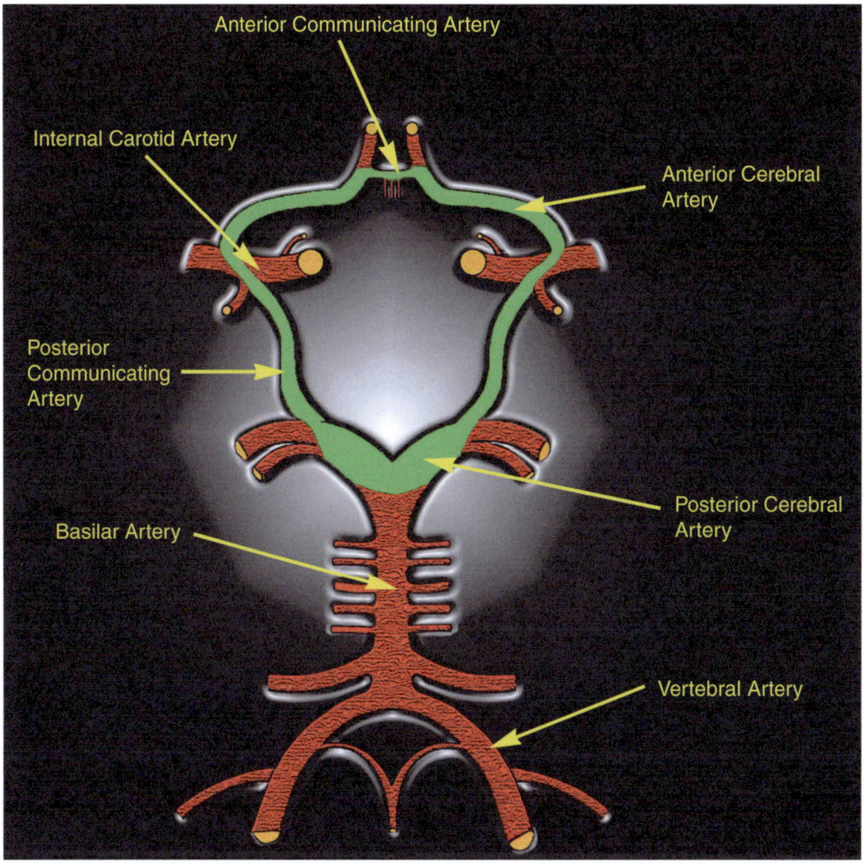

Fig. 3.8 The anastomosis of cerebral vessels forming the "circle of Willis"

3.3 Functional Anatomy of Cerebral Cortex and Internal Capsule

A longitudinal fissure separates the two cerebral hemispheres of the human brain. The outer layer, made up of gray matter, called the cerebral cortex (cerebral mantle), and the inner layer, called the white matter, constitute each cerebral hemisphere. The two cerebral hemispheres are joined beneath the cerebral cortex by the corpus callosum. The cerebral cortex consists of six lobes: the frontal, parietal, temporal, occipital, insular, and limbic lobes. The following section covers only the first four lobes of the cortex as the insular cortex and the limbic lobe connections with the rest of the brain and functions are complex and beyond the scope of this chapter.

3.3.1 Frontal Lobe

Regarding the lobes, the frontal lobe is considered the largest and accounts for approximately 40% of the cerebral cortex. The frontal lobe, the rostral region of the cerebral hemisphere and the youngest part of the brain based on the phylogeny, is located above the lateral fissure or Sylvian fissure and is limited posteriorly by the central sulcus. The primary motor cortex, premotor cortex, supplementary motor cortex, frontal eye field, and Broca's area of speech are the main regions of the motor cortex of the frontal lobe (Fig. 3.9). The dorsolateral, medial, and orbitofrontal regions constitute the prefrontal cortex region of the frontal lobe.

The primary motor cortex (Brodmann area 4), located within the precentral gyrus and anterior to the central sulcus, is involved in the control of voluntary movements through its projection of the cortical neurons to the brainstem and spinal cord. The primary motor cortex contains a topographically organized map known as the motor homunculus of the opposite half of the body, representing the head most laterally and the legs and feet medially on the hemisphere in the paracentral lobule. The representation of body parts is disproportionate to their physical size but proportionate to the ability to produce fine-controlled or fractionated movements. In addition to receiving input from the neighboring primary somatosensory area, premotor cortex, and ventral lateral nucleus of thalamus, the cortex area 4 has major thalamic connections and loops projected from the deep cerebellar nuclei and the basal ganglia. Inputs from these centers modulate the output of the primary motor cortex by promptly providing information about the position, timing, and coordination of voluntary movements.

The premotor cortex (Brodmann area 6), located immediately rostral to the primary motor cortex, assists in the integration of sensory and motor information for

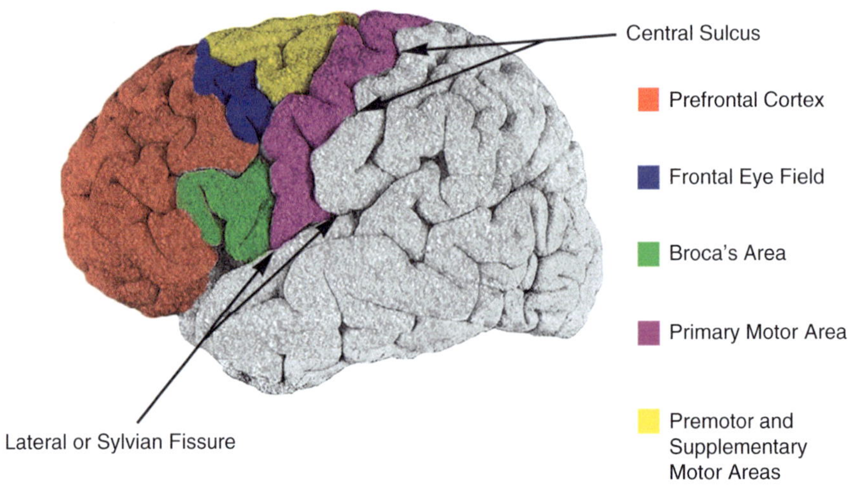

Fig. 3.9 An illustration of the frontal lobe

the performance of an action. The premotor area has neuronal projections from the secondary somatosensory area, the ventral anterior thalamic nucleus, and the premotor area of the contralateral side. The neurons in this cortex have extensive reciprocal connections with the primary motor cortex, which is in addition to those influencing motor behavior directly through the axons projecting to the lower motor neurons (corticobulbar and corticospinal pathways) of the brainstem and the spinal cord. Damage to the premotor cortex may result in apraxia (inability to perform purposeful skilled movements in the absence of paralysis, sensory loss, abnormal posture and tone, involuntary movement, incoordination, or inattentiveness), deficits in performing complex patterns of movements, and initiation and selection of movements.

The supplementary motor cortex lies medial to area 6, and this area has extensive reciprocal connections with the thalamus and the ipsilateral frontal lobe, including the primary motor cortex, the premotor area, the prefrontal area, and the frontal eye field. This cortex even has connections to the ipsilateral superior parietal area and the contralateral supplementary motor area and the motor cortices of the contralateral frontal lobe. Primarily, the role of the supplementary motor area is to control the movement involving complex tasks that require the temporal organization of sequential movements and retrieval of motor memory. The consequences of supplementary motor area damage strikingly resemble the effects of basal ganglia dysfunction, namely, akinesia and impairments in the performance of sequential and complex movements.

The pyramidal tract, the most important output from the motor cortex, consists of approximately 30% fibers from the primary motor cortex, 30% from the premotor and supplementary motor cortices, and the remaining from the somatosensory cortex of the parietal lobe and the cingulate gyrus. The projections from the primary motor cortex to the brainstem and spinal cord are facilitatory. The destruction of the pure pyramidal fibers arising from the primary motor cortex or the interruption of its pyramidal projections to the brain stem causes flaccid weakness. However, in most scenarios, cortical lesions are more extensive and involve premotor and supplementary motor areas. The latter two areas are normally inhibitory to the ventromedial bulbar reticular formation, and the inhibitory influences are conducted down to the spinal cord by the dorsal reticulospinal tract. Lesions involving the premotor and supplementary motor cortices or the projections to the brainstem and spinal cord will cause spastic weakness and hyperactive stretch reflex in upper motor neuron lesion conditions, including stroke.

The frontal eye field (Brodmann area 8) located rostral to the premotor cortex has extensive ipsilateral connections with several visual areas in the occipital, parietal, and temporal lobes. The eye field receives fibers from the prefrontal cortices and projects to the motor and premotor cortices. In addition to the above, there are prominent neuronal projections from the eye field to the superior colliculus, the pontine gaze center within the pontine reticular formation, and the cranial nuclei for extraocular muscles. The primary function of the frontal eye field is to control the voluntary movements of the eyes toward the contralateral visual field, and the eye field damage will cause the inability to move the eyes toward the contralesional side voluntarily.

Broca's area (Brodmann areas 44 and 45), an essential area for language, which helps to put thoughts into words, lies in the inferior frontal gyrus of the dominant cerebral hemisphere. Broca's area is structurally and functionally heterogeneous, and this area has been implicated in diverse cognitive functions, extending well beyond articulation and comprehension. The area is connected to many regions of the brain, including the Wernicke's area (through a neuronal tract known as the arcuate fasciculus), prefrontal cortex, primary motor cortex, premotor cortex, supplementary motor area, parietal lobe (specifically the inferior parietal lobule), and visual cortices of the occipital lobe. The supplementary motor cortex of the frontal lobe is believed to participate in spontaneous and automatic speech and may play a part in the programming of speech by its connections with Broca's area. In addition to the above, the Broca's area also has projections to the basal ganglia. Broca's area functions in the assembly of phonemes into words and words into sentences and for forming the grammatical structure for sentences. The area is also involved in motor-related activities and sensorimotor learning and integration. The lesion in this area mainly leads to reduced fluency, improper or distorted articulation of phonemes, agrammatism, and impaired retrieval of words, resulting in non-fluent and effortful speech, with impaired repetition, word-finding difficulties with relatively preserved comprehension.

The prefrontal cortex (Brodmann areas 9 and 46) located on the lateral surface of the cerebral hemisphere receives major thalamic afferents and corticocortical connections within the frontal lobe, including the supplementary motor area, the premotor cortex, and the frontal eye field. All these thalamic and corticocortical connections are reciprocal. In addition to the above, the prefrontal area receives association fibers from the posterior and middle superior temporal gyrus, including the auditory association areas, the parietal lobe, and the limbic lobe. The prefrontal cortex has commissural connections with the homologous parts of the contralateral hemisphere and with the contralateral inferior parietal cortex. The prefrontal cortex is important for the spatial processing of afferent information, organization of working memory, the mnemonic processing of objects, understanding the value of time, normal expression of emotions, and the ability to predict the consequences of actions. The medial and orbitofrontal regions of the prefrontal lobes are closely linked with the limbic system, and damage to the same regions can cause disinhibition of affective behavior due to which the patient can be unduly jocular and unmindful about his or her intellectual disability. Impaired motivation has an association with lesions of the dorsolateral prefrontal cortex and anterior cingulate. Patients with prefrontal dysfunction show deficits in planning and decision-making, poor judgment, easy distractibility, inability to look ahead in time or generate hypotheses for future events, reduction in total verbal output, restricted range of sentence structures, echolalia (automatic imitative repetition of spoken words made by another person), and inability to appreciate verbal and nonverbal humor. Patients with prefrontal dysfunction will have emotional dysregulation and emotional blunting, behavioral rigidity, poor frustration tolerance, defective social and moral reasoning, lack of self-awareness and thoughts and feelings for others, and increased susceptibility to psychiatric syndromes such as depression, mania, apathy, and obsessive-compulsive disorder. Uninhibited bladder may occur with bilateral brain lesions above the pontine micturition center, especially when the lesion is in the medial prefrontal area and anterior cingulate gyrus of the frontal lobe.

3.3.2 *Parietal Lobe*

The parietal lobe is located posterior to the central sulcus and superior to the lateral fissure (Fig. 3.10). The lobe is clearly demarcated from the occipital lobe by the parieto-occipital sulcus near the medial aspect of the cerebral hemisphere. However, on the posterolateral aspect of the cerebral hemisphere, the lobe has less distinct boundaries with the occipital lobe and the temporal lobe. The primary somatosensory cortex, the secondary somatosensory cortex, and the superior and inferior parietal lobules are the main regions of the parietal lobe. The primary somatosensory cortex of the postcentral gyrus is located between the central and postcentral sulci. Like the primary motor cortex, the primary somatosensory cortex (Brodmann areas 3, 1, and 2) also has its topographical mapping of the contralateral half of the body, with the face, tongue, and lips represented inferiorly, the trunk and upper limbs represented superolaterally, and the lower limbs medially. This area has complex internal connectivity and apparently has a stepwise hierarchical progression of information processing occurring from area 3 through areas 1 and 2. The primary sensory area has ipsilateral association connections with the motor cortices of the frontal lobe and superior parietal lobe. The area has reciprocal corticocortical

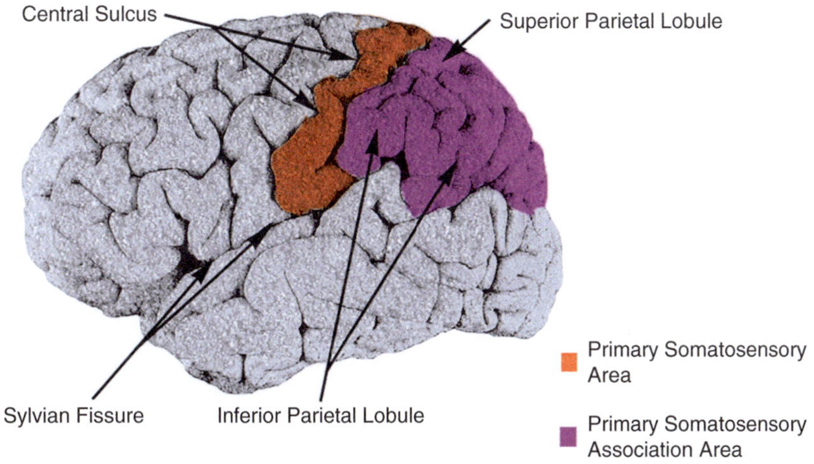

Fig. 3.10 An illustration of the parietal lobe

connections with the secondary somatosensory area, commissural connections with its contralateral homologue, and subcortical connections with the thalamus and basal ganglia. It also has projections to the brainstem via the corticopontine and corticotectal fibers.

The secondary somatosensory area located posterior to the central sulcus, along the upper bank of the lateral fissure, has reciprocal connections with the thalamus. The area has neuronal projections to the dorsal horn of the cervical and thoracic spinal cord, the posterior cingulate gyrus, the primary motor cortex, the superior parietal lobe, the principal trigeminal nucleus, and the periaqueductal gray matter of the midbrain. Within the parietal cortex, the secondary sensory area has reciprocal connections with the primary sensory area and commissural connections with the secondary sensory area of the contralateral hemisphere. Unilateral lesions of the secondary somatosensory area may degrade tactile discrimination abilities, impair some elements of sensory discrimination, and elevate pain thresholds without elimination of pain sensibility.

Posterior to the postcentral sulcus, a large area of the parietal lobe is divided by the intraparietal sulcus into the superior and inferior parietal lobules. The somatosensory association cortex (Brodmann areas 5 and 7) is located posterior to the primary sensory cortex in the parietal lobe. Similar to the primary somatosensory area, the superior parietal lobule has complex and extensive corticocortical, commissural, association fibers, and thalamic connections. The area also contributes to the corticospinal tract. Superior parietal lobule injury can lead to astereognosis (inability to recognize familiar solid objects by touch), a variety of body scheme or body image dysfunctions, hemineglect (neglect of the contralateral side of the world, usually seen in nondominant hemisphere lesion), asomatognosia (inability to recognize the affected limb as one's own), and a variety of syndromes, including dressing apraxia. The inferior parietal lobe (Brodmann areas 39 and 40) of the dominant hemisphere includes parts of Wernicke's speech area, and damage of the dominant hemisphere inferior parietal lobule can present with a tetrad of symptoms known as Gerstmann's syndrome. In adults, this syndrome can arise as a result of impaired blood flow to the inferior parietal lobe of the dominant hemisphere and is characterized by the inability to perform mathematical calculations (acalculia or dyscalculia), inability to write (agraphia or dysgraphia), inability to identify one's own finger (finger agnosia), and inability to make the distinction between the right and the left side of the body. In addition to the above, the patients may have difficulty expressing themselves through speech and difficulty in reading and spelling.

3.3.3 Temporal Lobe

The temporal lobe, the auditory language and speech comprehension area, is located inferior to the lateral fissure. The medial aspect of the temporal lobe consisting of the hippocampus and entorhinal cortex (an area in the medial temporal lobe essential for memory, navigation, and perception of time) and the areas of the neocortex adjacent to these limbic regions are grouped as medial temporal association cortex.

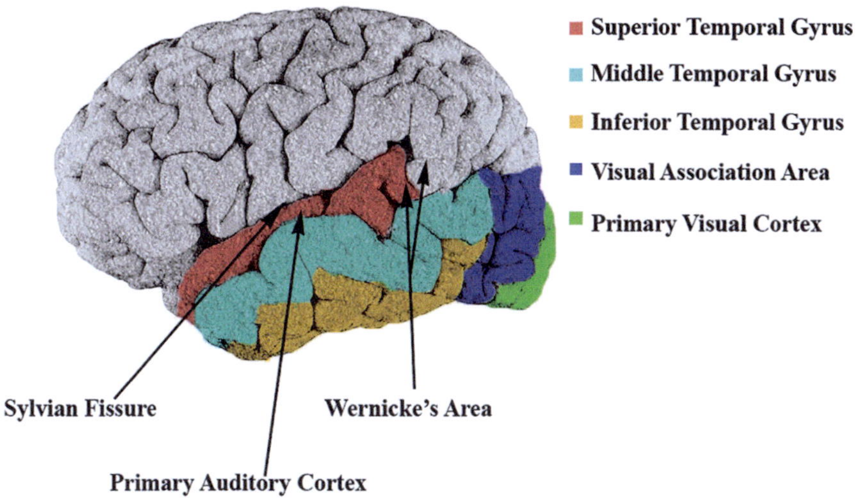

Fig. 3.11 An illustration of the temporal and occipital lobes

Their connections and functions are beyond the scope of this chapter. The superior and inferior temporal sulci located on the lateral surface of the temporal lobe divide the area into the superior, middle, and inferior temporal gyri (Fig. 3.11). The primary auditory cortex (Brodmann area 41 and 42) is located in the superior temporal gyrus, and the posterior area constitutes the Wernicke's area (auditory association area, Brodmann area 22). The auditory association cortex (Brodmann areas 22 and 42), located inferior and posterior to the primary auditory cortex (along the superior bank of the middle temporal gyrus), is the area concerned with memory and classification of sounds. The primary auditory cortex is reciprocally connected with the medial geniculate body and receives additional thalamocortical projections. Injury of the primary auditory cortex produces cortical deafness, verbal and non-verbal auditory agnosias. Both the primary and association areas, located in the auditory cortex of the superior temporal gyrus, have interconnections with the prefrontal cortex and the frontal eye field. It has contralateral connections with the adjacent and same regions of the other hemisphere. The connections of other sensory association pathways converge toward the superior temporal sulcus. Damage to the auditory association area in the superior temporal gyrus of the dominant hemisphere causes receptive fluent aphasia.

The middle temporal gyrus (Brodmann area 21) also is polysensory and has its connections with the auditory, somatosensory, and visual cortical association pathways. It has complex connections with the auditory association area of the superior temporal gyrus and has connections with the prefrontal cortex, the eye field, and the visual association area. In addition to the above, the middle temporal gyrus also has thalamic connections. The inferior temporal gyrus (Brodmann area 20), the higher visual association area, receives major ipsilateral corticocortical fibers from the visual association areas and has neuronal projections to the prefrontal cortex, the limbic areas, and the frontal eye field and reciprocal connections with the thalamus. Commissural connections are between the corresponding areas and the adjacent visual association areas of the contralateral hemisphere. Due to its complexity, damage to the temporal lobe can lead to considerable disturbance of intellectual function, including visuospatial difficulties, prosopagnosia (also called face blindness, an inability to recognize familiar faces), and sensory aphasia, particularly when the dominant hemisphere is involved. Vascular lesions of frontal and/or temporal lobes can cause an acquired communication disorder named aphasia, characterized by impaired language comprehension, oral expression, and use of gestures or symbols to communicate ideas. Table 3.1 explains various types of aphasia and their associated features.

3.3.4 Occipital Lobe

The occipital lobe, located in the rearmost portion of the cerebral hemisphere, being the smallest of the four lobes, lies posterior to the parietal and temporal lobes (Fig. 3.11) and rests on the tentorium cerebelli. The lobe is primarily responsible for processing visual information and is comprised of Brodmann areas 17, 18, and 19. The primary visual cortex (Brodmann area 17) is located mostly on the medial aspect of the lobe and occupies the upper and lower lips and depths of the posterior part of the calcarine sulcus. The visual cortex of each cerebral hemisphere receives impulses from the two retinal halves (temporal half of one side and nasal half of the other side) that represent the contralateral half of the binocular visual field. The inferior half of the visual field represented by the superior retinal quadrants project to the visual cortex above the calcarine sulcus and the upper half of the visual field represented by the inferior retinal quadrants project to the visual cortex below the calcarine sulcus.

The main purpose of the primary visual cortex is to receive, segment, and integrate visual information. The processed information from the cortex is sent to the other regions of the brain to be analyzed and utilized to recognize objects and patterns quickly without a significant conscious effort. The corticocortical fibers from the primary visual cortex project to functional areas (Brodmann areas 18 and 19), the adjacent regions of the parietal and temporal cortices, the middle temporal area, and the medial superior temporal region. The efferent projections connect to the superior colliculus, pretectum, thalamus, and reticular formation of the pons. The commissural fibers reciprocally connect the primary visual cortex to the contralateral side and the geniculo-cortical projections are reciprocal. Injury to the primary visual cortex, being the initial processing center of all visual information, causes loss of vision in the contralesional visual field including homonymous hemianopia and upper (superior) or lower (inferior) quadrantanopia. Figure 3.12 depicts the visual field deficits due to lesions of the visual pathway.

The secondary visual cortex (Brodmann area 18) and the visual association area (Brodmann area 19), located near the primary visual cortex, project to several visual areas in the temporal and parietal association cortices and the frontal eye fields. The thalamocortical connections to these visual cortices arise from the lateral geniculate body. The commissural fibers through the corpus callosum connect both the visual

Table 3.1 Details regarding the types and features of aphasia

Features	Global	Broca's	Wernicke's	Transcortical motor	Transcortical sensory	Mixed transcortical	Anomic	Conduction
Comprehension	No	Yes	No	Yes	No	No	Yes	Yes
Fluency	No	No	Yes	No	Yes	No	Yes	Yes
Repetition	No	No	No	Yes	Yes	Yes	Yes	No
Spontaneous speech	Scant, reduced to few words, mutism possible	Short, slow, effortful, agrammatical, can pass ideas, words often mispronounced	Meaningless but fluent	Non-fluent	Meaningless but fluent	Telegraphic in nature with short, missing function words, brief noun-verb combinations	Grammar, fluency, and phrase length normal	Fluent, expression better than repetition
Accompanying neurological features	Right-side hemiplegia, hemisensory loss, and homonymous hemianopia	Hemiparesis with more involved of the right arm and face; dysarthria	Alexia, apraxia, Hemianopia or quadrantanopia	Leg weakness	Features are variable	Right side weakness or sensory loss	Features are variable	Features are variable
Lesion site	Large perisylvian infarcts involving receptive and expressive speech areas, primary sensorimotor cortices and primary visual and auditory cortices	Broca's area and regions inferior and posterior to Broca's	Wernicke's area, superior temporal gyrus	Area surrounding Broca's: Anterior association cortices	Area surrounding Wernicke's: posterior association cortices	Isolated damage to anterior and posterior eloquent cortex	Variable, possible lesions in the dominant hemisphere	Arcuate fasciculus that connects the Broca's and the Wernicke's areas
Prognosis (for vascular origin)	Poor	Generally good and typically resolves into anomic aphasia	Generally worse than Broca's but may resolve to transcortical sensory aphasia	Generally good and typically resolves into anomic aphasia	May evolve to anomic aphasia	Generally poor but variable	Good, often resolves to word-finding difficulty	May evolve to anomic aphasia

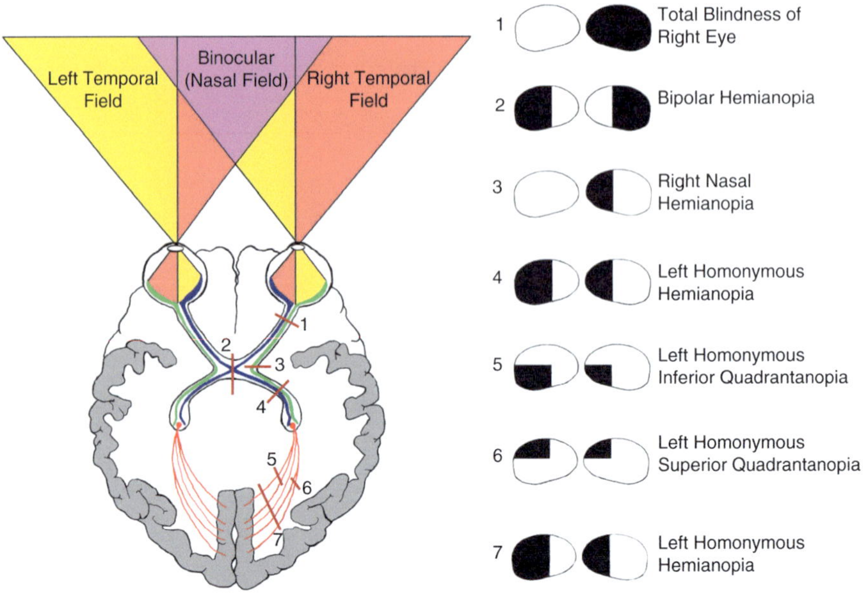

Fig. 3.12 An illustration of visual field deficits due to lesions of the visual pathway

cortices. Once the primary visual area discerns edges, shapes, and movements of the objects, the secondary visual cortex and the visual association areas synthesize these into a recognizable structure or pattern. These areas are also essential for visual depth perception. Damage to these areas disrupts motion perception and causes optic ataxia and may interfere with the learning of visuospatial tasks. In addition to the above, the patients with visual association cortex involvement may develop visual hallucinations (characterized by stereotyped images which seldom resemble fragments of memory), distortion of visual images (metamorphopsia), and failure to recognize the nature of objects seen (visual agnosia).

3.3.5 Internal Capsule

The internal capsule, a broad white matter band, composed of bundles of myelinated fibers located in the inferomedial portion of each cerebral hemisphere, separates the head of the caudate nucleus and thalamus from the lentiform nucleus (Fig. 3.13). The capsule consists of ascending and descending fibers that connect the cerebral hemispheres with the subcortical structures, the brainstem, and the spinal cord, i.e., it is a two-way track for information transmission to and from the cerebral cortex. Fibers projecting cephalically toward the cerebral cortex in a radiant pattern from the internal capsule, known as the corona radiate, are located above the superior border of the lentiform nucleus. The internal capsule has an anterior limb, genu, posterior limb, and retrolenticular and sublenticular segments (Fig. 3.14).

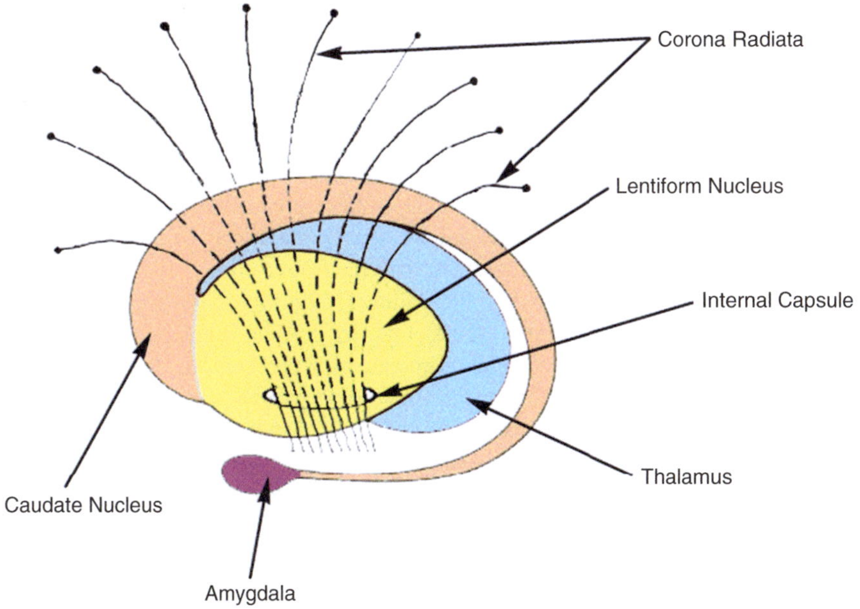

Fig. 3.13 The internal capsule separating the head of the caudate nucleus and thalamus from the lentiform nucleus

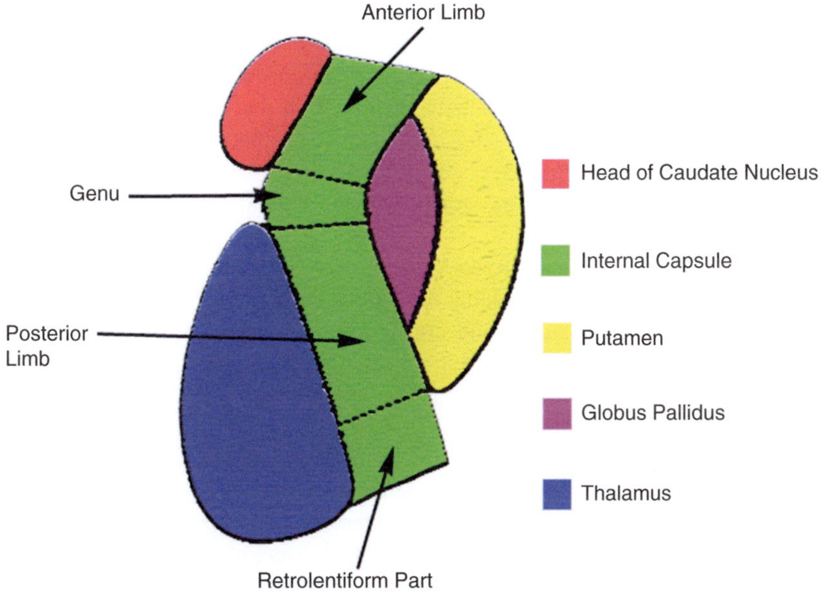

Fig. 3.14 The anterior limb, genu, posterior limb, and retrolenticular part of the internal capsule

Each of the regions mentioned above carries distinct ascending and descending tracts with their discrete functions.

The anterior limb of the internal capsule is bounded by the head of the caudate nucleus medially and the lentiform nucleus laterally. The anterior limb of the internal capsule contains frontopontine fibers arising from the frontal lobe. The fibers synapse with cells in the pontine nuclei, and through the middle cerebellar peduncle, the axons of these cells project to the opposite cerebellar hemisphere. The anterior limb also contains thalamic radiations which interconnect the thalamic and hypothalamic nuclei and limbic structures with the frontal cortex. The anterior limb fibers are known to be associated with emotional, cognitive, decision-making, and motivational processes. The genu of the internal capsule contains corticobulbar fibers derived mainly from the primary motor cortex and some from the premotor and supplementary motor areas to terminate at the appropriate cranial nerve nuclei located on the contralateral side of the brainstem. These fibers are concerned with the voluntary movements of facial muscles, mastication, and deglutition.

The posterior limb of the internal capsule is surrounded by the thalamus medially and the lentiform nucleus laterally. The posterior limb of the internal capsule contains the corticospinal tract. The corticospinal fibers representing the upper limb are located anteriorly. The fibers concerned with voluntary movements of the trunk and lower limbs are more posteriorly located. The descending axons of frontopontine fibers, specifically from the primary and supplementary motor areas of the frontal lobe and the corticorubral fibers that connect the frontal lobe to the red nucleus, are located within the posterior limb of the internal capsule. The somatosensory relays from the thalamus to the primary somatosensory cortex are located in the posterior one-third of the posterior limb. The retrolenticular segment of the internal capsule contains occipitopontine, occipitotectal, and parietopontine fibers. In addition to the above, this segment also contains thalamic radiation and optic radiation arising from the lateral geniculate body and interconnections between the occipital and parietal lobes and the thalamus. The sublenticular segment predominantly contains temporopontine fibers and auditory radiation from the medial geniculate body to the primary auditory cortex.

The vascular supply for the internal capsule arises from the perforating branches of the main cerebral arteries, including the middle cerebral artery, the anterior choroidal artery, and the internal carotid artery. The superior regions of the anterior limb, genu, and posterior limb get their vascular supply from perforating branches of the middle cerebral artery. Whereas, the inferior regions of the anterior limb receive supply from the recurrent artery of Heubner, the largest of the perforating medial lenticulostriate arteries arising from the anterior cerebral artery, and the inferior regions of the genu receive from the perforating arteries of the internal carotid and anterior choroidal arteries. In addition to the above, the perforating arteries of the anterior choroidal artery also supply the inferior regions of the posterior limb, the retrolenticular and sublenticular segments of the internal capsule. Occlusion of any of the perforating arteries supplied by the major vessels can predispose to cerebrovascular accidents known as lacunar strokes. Absence of cortical deficits including aphasia, agnosia, dysgraphia, apraxia, alexia, and amnesia are typical for such deep strokes.

The neurological deficits due to infarctions of the internal capsule correlate with the fiber located within each limb. The anterior limb infarction presents with confusion, impaired attention, agitation, and dysarthria. The genu infarction presents with face and tongue weakness and dysarthria. The posterior limb infarction can cause pure motor hemiparesis contralateral to the lesion site, and the infarction of the posterior one-third of the posterior limb can cause contralateral hemisensory deficits with ataxic hemiparesis. Since the retrolenticular segment largely contains optic radiation fibers, infarctions of this region can lead to visual field deficits like homonymous hemianopia and superior or inferior quadrantanopia. The involvement of auditory radiations creates auditory deficits in lesions of the sublenticular segment.

3.4 Epidemiology

The World Health Organization (WHO) defines stroke as "rapidly developing clinical signs of focal or global disturbance of cerebral function, with symptoms lasting 24 hours or longer or leading to death, with no apparent cause other than of vascular origin." Stroke is one of the leading causes of long-term disability in adults worldwide and is the second leading cause of death after ischemic heart disease. Earlier studies estimated that stroke accounted for 9.6% of all deaths, and 40% of the patients were aged less than 70 years. Existing epidemiological data suggest that one out of four people are at risk of stroke in their lifetime. Beyond 65 years of age, the incidence of stroke increases dramatically with age. Around 30% of strokes occur in people below 65 years of age. Until the age of 85 years, women have a lower predisposition for stroke as compared to men. Approximately 70% of all strokes occur in low- and middle-income countries. The incidence of stroke in high-income countries has declined by around 40%. However, it has doubled in low- and middle-income countries. Demographic data of the developed countries reveals that the average age at which stroke occurs is around 73 years, and the probability of a first stroke is approximately 2 per 1000 population per annum.

Stroke patients are at the highest risk of death during the first week, and the risk of death can be 20% to 50% during the first month depending on the type, location, and size of the lesion, age, comorbidities, level of consciousness, complications, and effectiveness of treatment. Stroke survivors may be left with apparently no disability or with a mild, moderate, or severe disability. Survivors have a 10% risk of a subsequent stroke during the first year, and thereafter the risk reduces to 5% every year. In many of the survivors, a considerable extent of spontaneous recovery is anticipated during the first 6 months, and around 60 to 80% may achieve independence in self-care within a year after stroke. Some of the common disabilities noted among elderly stroke survivors after 6 months include hemiparesis, inability to walk unassisted, dependency in activities of daily living (ADL), aphasia, and depression.

The type of stroke is important in determining survival. Hemorrhagic stroke accounts for around 35% and ischemic stroke accounts for around 10% of deaths in one month. Analysis of the worldwide data indicates that Caucasians have approximately 80% ischemic strokes, 10 to 15% hemorrhagic strokes, and the remaining due to other causes of strokes. The hospital-based studies in the eastern Mediterranean region indicated that the stroke pattern is similar to that of western countries. Studies conducted in Asian and African countries indicate a higher proportion of hemorrhagic stroke (20 to 30%) among them as compared to Caucasians.

Regardless of the type of stroke, it is an accepted fact that stroke is caused by many different disease processes and not a single homogenous disease. Epidemiological studies have identified many risk factors for stroke including medical factors like prior history of stroke, ischemic heart disease, atrial fibrillation, and glucose intolerance; inherent biological traits like age and sex; physiological traits like serum lipoprotein and fibrinogen; behavioral traits like smoking, alcohol consumption, diet, and physical inactivity and social characteristics such as education, social class, and ethnicity; and environmental or geographic factors like altitude and temperature. Table 3.2 depicts the association of common conditions predisposing to stroke among adolescence to elderly population. Certain medical, behavioral, and environmental factors are classified as modifiable risk factors, whereas age, sex, and genetics are non-modifiable factors. Epidemiological studies have revealed that raised blood pressure is the single most important risk factor for ischemic stroke. In addition to the above, the risk of hemorrhagic stroke is directly proportional to the blood pressure and level of low-density lipoprotein. Regular use of anti-hypertensive medications has shown around a 40% reduction in the risk of stroke. Tobacco use, a

Table 3.2 Listing the common conditions causing stroke among adolescence to elderly population

Age category	Common conditions causing stroke
Adolescence and early adult life	Vascular malformations Pregnancy and puerperium Valvular heart disease Premature atherosclerosis Sickle cell anemia Migraine Arteritis Coagulopathies Arterial dissections Cerebral amyloid angiopathy
Middle age	Atherosclerotic thrombosis and embolism Cardiogenic embolism Hypertensive cerebral hemorrhage Rupture of saccular aneurysm Dissecting aneurysm
Late adult life	Atherosclerotic thrombotic occlusive disease Embolic occlusive disease Lacunar stroke Cerebral amyloid angiopathy

modifiable risk factor, increases the risk of both ischemic and hemorrhagic strokes by about twofold, indicating that heavy smokers are at a higher risk of stroke than light smokers. Epidemiological evidence reveals that many of the risk factors, including blood pressure and tobacco use, are somewhat similar in developed and developing countries.

3.5 Etiology

Among the vital organs, the brain depends continuously on an adequate volume of oxygenated blood. The total cessation of blood flow to the brain for more than 5 minutes can cause permanent brain damage. Under the control of brainstem centers, baroreceptors and vasomotor reflexes assure constancy in the cerebral circulation. The majority of strokes have a rapid or sudden evolving onset. Occlusion of an artery by thrombus or embolus deprives the brain of oxygenated blood, due to which the brain tissues may undergo ischemic necrosis or infarction. Occlusion or stenosis of the vessels to the brain can be the result of disease of the arterial wall, embolism from the heart, and hematological disorders. Leak or rupture of an artery can release blood into the brain, causing a hematoma within the brain substance and/or ventricles or subarachnoid spaces within the cranium. As the blood leaks or spills into the brain, the area the artery supplies will be deprived of oxygenated blood, causing a hemorrhagic stroke. The pressure created by the enlarging hematoma or the blood within the intracranial space and the biochemical substances released by the hemorrhage may adversely affect the adjacent vascular and brain tissues creating further localized or generalized injury to the brain tissue. Though the classification of stroke mechanism depends on the presence of risk factors and etiologies for stroke, the presence of more than one risk factor and etiology are frequently seen among stroke patients. Thrombosis, embolism, and decreased perfusion are considered the three commonly recognized mechanisms of ischemic stroke.

Thrombosis refers to an obstruction of blood flow due to a localized occlusive process formed by a thrombus or an atherosclerotic plaque (with or without a superimposed thrombotic occlusion). Atherosclerosis is the most common vascular pathology in which fibrous and muscular tissues overgrow in the endothelial lining layer of tunica intima of the arterial walls. The platelets which adhere to the atherosclerotic plaque can form clumps that serve as "nidi" for the deposition of fibrin, thrombin, and formation of the clot. The gradually growing atherosclerotic plaque can progressively occlude the lumen of the artery. Though atherosclerosis tends to affect the larger extracranial and intracranial arteries, the plaques are typically seen near the vessel-branch points where the turbulence of blood flow is maximum. The most common sites for atherosclerotic plaque formation are the carotid bifurcation, the origin of the middle or anterior cerebral arteries, vertebral artery origin, or vertebrobasilar junction (Fig. 3.15). Fibromuscular dysplasia, arteritis, dissection of the vessel wall, and hemorrhage into a plaque are the other less plausible causes for vascular pathological obstruction.

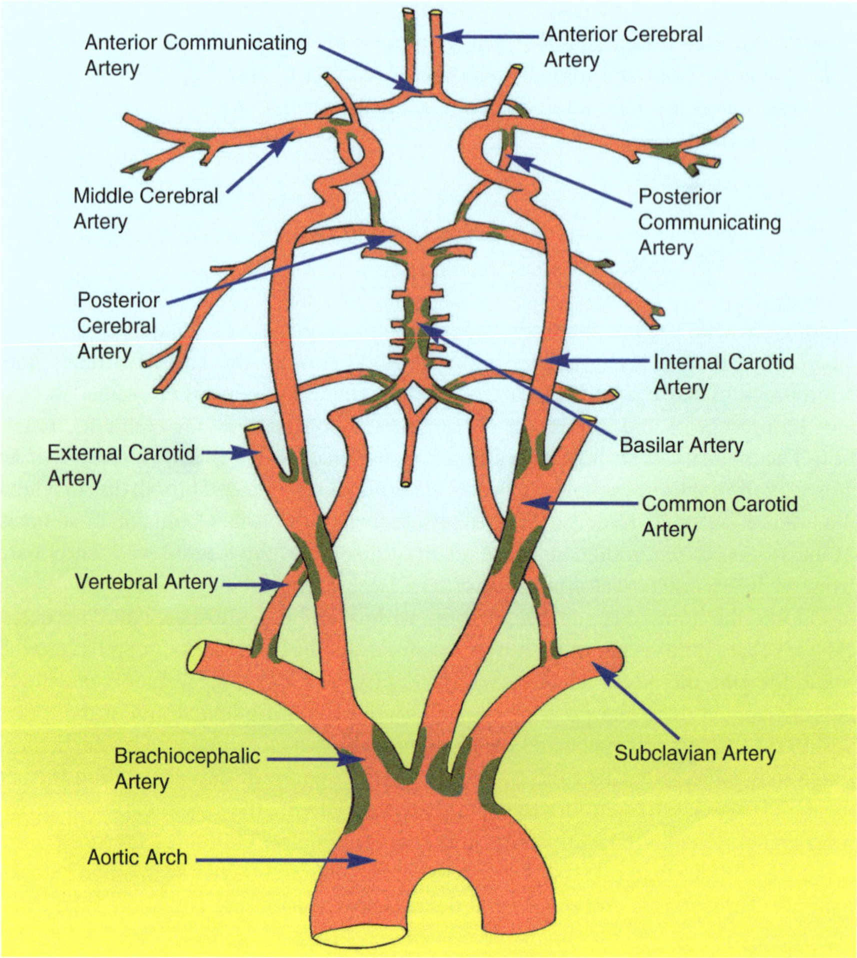

Fig. 3.15 Common sites of atherosclerotic plaque formation (grayish-green area) within the extracranial and intracranial arteries

In embolism, material or the embolus is formed elsewhere within the vascular system and dislodged to block the vasculature supplying the brain. The embolus typically originates from the heart, major arteries, and systemic veins. The frequently recognized cardiac sources for embolism include atrial fibrillation, sinoatrial disorder, acute myocardial infarction, bacterial endocarditis, and valvular disorders. For embolic strokes after acute myocardial infarction, the emboli are likely to arise from mural thrombi on the left ventricular wall, specifically in the anterior wall myocardial infarction. Large myocardial infarcts, left ventricular dilation, and congestive heart failure are risk factors for left ventricular mural thrombosis. The majority of the strokes tend to occur in the first weeks after the acute myocardial infarction. Atrial fibrillation can occur after acute myocardial infarction, and fibrillation can be an independent risk factor for embolic stroke. The common source for the embolus is the left atrial appendage. In a patient with atrial fibrillation who has an acute stroke, recurrent stroke risk may be particularly high during the first week or two after the acute infarction. Cardioembolic stroke is a well-known complication of valvular disorders. The embolus can arise from the diseased native valve or the prosthetic valve. Rheumatic mitral stenosis with or without mitral regurgitation is commonly associated with thromboembolism. The risk of thromboembolism in rheumatic valve stenosis is related to age and low cardiac output and not to the severity of mitral stenosis, mitral calcification, or left atrial size. Atherosclerotic plaques of the aortic arch constitute another source of atherothrombotic embolism. Highly mobile thrombi in the lumen of the aortic arch and the presence of plaques of more than 4 mm size in the proximal arch are independent risk factors for atherothrombotic embolism.

Diminished blood flow to the brain tissue caused by systemic hypoperfusion either due to cardiac pump failure or systemic hypotension can critically affect the border zone or watershed regions at the periphery of the major vascular territories. In such cases, the effect of decreased perfusion can be more generalized, affecting the brain diffusely and bilaterally. Severe stenosis of the carotid and basilar artery and microstenosis of the small deep arteries are specific causes of stroke due to perfusion failure. More than 70% stenosis of the main artery by atherosclerotic plaque increases the risk of brain infarct in the distal zones. This arteriolar abnormality remains the most common defect in lacunar infarcts. Though the nature of the intrinsic arteriolar abnormality remains unclear, microatheroma, inadequate cerebral blood flow, or vasospasm may be the factors attributing or leading to lacunar infarcts.

The most common causes for hemorrhagic strokes are hypertension, arteriovenous malformation, aneurysm, bleeding disorders, use of cocaine, and anticoagulation therapy. For the prevention of recurrent thrombosis or thromboembolism, anticoagulation therapy is a crucial component of management of several medical conditions, including ischemic stroke, cardiac valvular disorders, and deep vein thrombosis. For those undergoing anticoagulant therapy, the International Normalized Ratio (INR) value of 4.9 and above increases the risk of hemorrhagic stroke.

3.6 Pathophysiology of Stroke

Brain tissue is easily vulnerable to the effects of ischemia due to its complete dependence on aerobic metabolism and low respiratory reserve. Tissue damage tends to occur more rapidly during hyperthermia compared to hypothermia. The spectrum of severity of tissue damage in the affected region depends mainly on the presence of patent collateral circulation, which in turn is determined by the normal variations in the circle of Willis, caliber of various collateral vessels, atherosclerosis, and other acquired arterial lesions. Ischemia of the brain tissue activates an ischemic cascade progressing from local depletion of oxygen or glucose to failure of production of adenosine triphosphate (ATP), in turn adversely affecting the energy-dependent processes necessary for brain tissue survival and setting off a series of interrelated events ending in tissue injury and death. The extent of tissue injury generally depends on the severity, duration, and location of ischemia.

Thrombotic or embolic occlusion to the brain tissue creates an inner core of severe ischemia with blood flow below 10–25% and an outer layer of less severe ischemia known as the penumbra supplied by collaterals. The inner core area (zone of infarction) will display necrosis of both neuronal as well as supporting glial elements, whereas the neuronal cells in the penumbra may recover if reperfused in time. The extent of angiogenesis within the penumbral zone is mediated by the vascular endothelial growth factor and angiogenic growth factor secreted by the inflammatory infiltrates like leucocytes, blood platelets, and macrophages. Typically, the zone between the inner core area and the penumbral zone is critically hypoperfused, and appropriate treatment given within a few hours may salvage the same.

The various mechanisms involved in tissue damage are the following:

(i) Failure of mitochondria to replenish the depleting cellular energy store may trigger cell death.
(ii) Loss of ion pump functioning of the membrane causing loss of potassium in exchange of sodium, chloride, and calcium ions and the deleterious effects accompanied by the inflow of water and rapid swelling of neurons and glia (cytotoxic edema).
(iii) Release of excitatory neurotransmitters, mainly glutamate and aspartate. Uncontrolled release of glutamate into the ischemic areas mediates excitotoxic synaptic transmission via activation of N-methyl-D-aspartate (NMDA), which allows calcium and sodium influx causing disordered activation of a wide range of enzyme systems including proteases and lipases. These enzymes and their metabolic products including free radicals, damage cell membranes, genetic material, and structural proteins in the neurons, eventually leading to cell death.
(iv) Oxygen-free radicals and other reactive oxygen species produced in the ischemic zone will damage and destroy the cellular and extracellular elements, including the vascular endothelium.
(v) Apoptosis or programmed cell death may occur in the peripheral zone (penumbral zone).

In contrast to necrosis occurring in the ischemic core, apoptosis-inducing factors from mitochondria and proapoptotic molecules may trigger apoptosis in the peripheral zone. The ischemic cascade also activates various neuroprotective mechanisms against necrotic and apoptotic cell death such as the release of antiapoptotic and proapoptotic counter molecules, prion protein, and neurotrophin-3.

Cytotoxic or cellular edema tends to advance with time (minutes to hours) and can be potentially reversible. Failure of ATP-dependent calcium and sodium ion transport and release of oxygen-derived free radicals cause cellular edema of the brain cells, including neurons, glial cells, and endothelial cells. With the progression of infarction, cytotoxic edema caused by the initial acute hypoxia can give way to vasogenic edema. Vasogenic edema tends to occur over a few hours and days and is generally irreversible. The extravasation of macromolecular serum proteins caused by increased permeability of brain capillary endothelial cells results in increased extracellular fluid volume along with increased intracranial pressure. The raised intracranial pressure may displace the brain hemisphere, shift the compartment of the brain, and compress the tracts, cranial nerves, and cerebral arteries. A sustained increase in pressure may cause irreversible damage to the brain and may lead to cerebral herniation and potential death.

Compared to ischemic strokes, hemorrhagic strokes are more dangerous. Intracerebral hemorrhage and subarachnoid hemorrhage are the two types of hemorrhagic stroke. The former generally occurs in small arteries or arterioles and is frequently due to hypertension, bleeding disorders, trauma, amyloid angiopathy, use of illicit drugs like amphetamines or cocaine, and arteriovenous vascular malformations. Typically, subarachnoid hemorrhage occurs due to rupture of an aneurysm near the base of the brain or a bleed from the vascular malformations near the pial surface.

3.7 Clinical Features of Supratentorial Stroke

Cerebrovascular accidents present clinically as neurologic deficits of sudden onset. The clinical features depend upon the area or the region of the brain affected, which in turn is defined by the arterial anatomy involved. Anatomically, the word supratentorial means the region of the brain located above the tentorium cerebelli, and almost 90% of strokes are supratentorial. The supratentorial area contains the cerebral hemispheres, subcortical structures including diencephalon and basal ganglia, optic nerve, limbic structures, and the lateral and third ventricles. The common symptoms of the left hemisphere stroke include right hemiparesis, aphasia, and right hemianopia. On the other hand, the symptoms of right hemisphere stroke include left hemiparesis, left hemispatial neglect, and left hemianopia. Infratentorial stroke due to posterior circulation impairment causes additional symptoms, including altered or reduced levels of consciousness, diplopia, bulbar palsies, dysmetria, and incoordination.

Though additional symptoms can be present in both supra- and infratentorial strokes, the most important historical feature is the "suddenness" of its onset. Sudden weakness or numbness in the face, arm, or leg (especially on one side of the body), confusion or trouble speaking or understanding speech, vision problems in one or both eyes, difficulty to walk, dizziness, loss of balance, or problems with coordination and severe headache with no known cause are the common five sudden warning signs of a stroke. Knowledge about the warning signs of a stroke can make a difference between recovery and disability. The BE-FAST algorithm (balance, eyes, face, arm, speech, and time) helps the public for the identification of the warning signs of persons having an acute stroke. Early recognition and timely medical or surgical management may save lives and certainly reduce the morbidity associated with the stroke.

Two terminologies that require special attention under this section are the transient ischemic attack (TIA) and the stroke in evolution or evolving stroke. The signs and symptoms of TIA usually resemble those of acute ischemic stroke. The presenting symptoms of TIA depend on the arterial territory involved and usually last for a few minutes to a maximum of 24 hours. Ischemic stroke and TIA share the same causative factors, including blockage caused by atherosclerotic plaques in the arteries or embolus from the heart. The TIA is often called a "mini-stroke," and several studies in diverse populations have shown that TIAs carry a very substantial risk of imminent brain infarction. Since TIA is a warning sign of impending stroke, seeking immediate medical attention following it can essentially prevent the imminent stroke.

Stroke in evolution is characterized by the deterioration of the neurologic deficits hours to few days after the initial clinical presentation of ischemic stroke. This concept of stroke in evolution or evolving stroke initially appeared in the 1950s. Several mechanisms might be potentially contributing to this phenomenon, including thrombus propagation, development of brain edema and its associated mass effect, metabolic disturbances such as hyperglycemia or hyponatremia, hemorrhagic transformation of the ischemic lesion, and pyrexia. Progressive damage of the ischemic penumbral zone due to the abovementioned factors could be the reason for the deterioration in evolving stroke.

The following section will cover the clinical features of common stroke syndromes occurring supratentorially, and the latter part of this chapter will cover the clinical presentation and management of infratentorial stroke (specifically, brainstem strokes).

The anterior cerebral artery stroke occurs due to the occlusion of the anterior cerebral artery, the smaller and the first terminal branch of the internal carotid artery. The artery supplies the medial aspect of the cerebral hemisphere, subcortical structures including the basal ganglia and anterior limb of the internal capsule, the anterior fornix, and the ventral four-fifths of the corpus callosum. The occlusions that occur in the first or A1 segment of the anterior cerebral artery (part of the artery between the internal carotid artery and the anterior communicating artery branch) are less likely to produce any symptoms because of patent collateral circulation, and occlusions in the distal segments of the artery are more likely to produce significant deficits. Table 3.3 lists the segments of the anterior cerebral artery. Table 3.4 gives the clinical features of the anterior cerebral artery syndrome and the structures involved.

The middle cerebral artery stroke occurs due to occlusion of the middle cerebral artery, the largest and the second terminal branch of the internal carotid artery. The artery supplies the whole of the lateral surface of the cerebral hemisphere and subcortical structures, including the basal ganglia, corona radiata, and posterior limb of the internal capsule. Table 3.5 lists the segments of the middle cerebral artery. Occlusion of the proximal segments of the middle cerebral artery can produce extensive brain tissue damage and significant cerebral edema. A large lesion with considerable cerebral edema can produce brain herniation and possible death. Table 3.6 depicts the clinical features of middle cerebral artery syndrome and the structures involved.

Table 3.3 The segments of the anterior cerebral artery

Name of the segment	Origin and termination
A1: Horizontal or Pre-communicating segment	• Originates from the terminal bifurcation of the internal carotid artery and extends approximately 15 millimeters in length • Terminates at the anterior communicating artery
A2: Vertical, post-communicating, or infracallosal segment	• Originates at the anterior communicating artery, extending anterior to the lamina terminalis; along the rostrum of the corpus callosum • Terminates either at the genu of the corpus callosum or the origin of the callosomarginal artery
A3: Precallosal segment	• Originates around the genu of the corpus callosum or the origin of the callosomarginal artery • Terminates where the artery turns directly posterior above the corpus callosum
A4: Supracallosal segment	• Originates above the body of the corpus callosum • Terminates anterior to the plane of the coronal suture
A5: Postcallosal segment	• Originates above the body of the corpus callosum posterior to the plane of the coronal suture

Table 3.4 The clinical features and the structures involved in the anterior cerebral artery syndrome

Clinical features	Location/structure involved
Contralateral hemiparesis; lower extremity more involved than upper extremity	Predominant involvement of motor area of lower extremity (paracentral lobule)
Paresis or lesser involvement of opposite arm	Partial involvement of motor area of the arm or corona radiata through which those fibers are descending
Contralateral hemisensory loss mainly involving lower extremity	Primary sensory area for lower extremity located on the medial aspect of cortex (paracentral lobule)
Frontal "gait apraxia" (broad-based gait, short stride, "magnetic" or shuffling type, freezing, falls or tendency to fall, en bloc turns, and an inability to walk and talk)	Probable location: inferomedial frontal-striatal projections
Urinary incontinence	Superior frontal gyrus: posteromedial region
Contralateral grasp reflex, sucking reflex, palmomental reflex, paratonic rigidity (subjects involuntarily resist passive movements)	Medial surface of the posterior frontal lobe
Akinetic mutism (abulia), lack of spontaneity, motor inaction, slowness, delay, whispering, Reflex distraction to sights and sounds	Probable location: superomedial lesion near subcallosum
Dyspraxia of left limbs; difficulties with imitation and bimanual tasks; tactile aphasia in left limbs	Corpus callosum
Perseveration and amnesia	Uncertain localization
Cerebral paraplegia	Bilateral occlusion of anterior Cerebral arteries and involvement of bilaterally motor area of the lower extremities

Table 3.5 The segments of the middle cerebral artery

Name of the segment	Origin and termination
M1: Horizontal or sphenoidal segment	• Originates at the terminal bifurcation of the internal carotid artery • Terminates either at the genu adjacent to the limen insulae (anteroinferior aspect of the insular cortical surface) or at the main bifurcation
M2: Insular segment	• Originates at the genu near the limen insulae or the main bifurcation • Terminates at the circular sulcus of insula
M3: Opercular segment	• Originates at the circular sulcus of the insula • Terminates at the external or superior surface of the Sylvian fissure
M4: Cortical segment	• Originates at the external or superior surface of the Sylvian fissure

Table 3.6 The clinical features and the structures involved in the middle cerebral artery syndrome

Clinical features	Location/structure involved
Contralateral hemiparesis; upper limb and face more involved than lower limb	Prominent involvement of motor areas of face and upper limb and partial involvement of the fibers descending from the leg area entering the corona radiata
Contralateral hemisensory loss mainly involving upper limb and face	Somatosensory area for the face and arm and thalamoparietal projections within the internal capsule
Contralateral homonymous hemianopia or homonymous quadrantanopia	Optic radiation in the internal capsule
Contralateral paralysis of conjugate gaze	Frontal eye field or projecting fibers
Sensory ataxia of contralateral limb or limbs	Parietal lobe
Motor speech disorder: Broca's aphasia	Broca and adjacent motor area of the dominant hemisphere
Sensory or receptive speech disorder: Wernicke's aphasia	Wernicke's area located posterior third of the superior temporal gyrus of the dominant hemisphere
Global aphasia	Large perisylvian lesions involving receptive and expressive speech areas
Perceptual deficits: Unilateral neglect, depth perception, inaccurate localization in the half field, impaired ability to judge distance, agnosia, amorphosynthesis, dressing apraxia, constructional apraxia	Usually parietal lobe sensory association cortex of the nondominant hemisphere.
Word deafness, anomia, jargon speech, alexia	Central language area and parieto-occipital cortex of the dominant hemisphere
Gerstmann's syndrome	Inferior parietal lobe of the dominant hemisphere
Pure motor hemiplegia	Lacunar stroke in the upper portion of the posterior limb of the internal capsule and the adjacent corona radiata
Gait apraxia or Bruns ataxia	Frontal lobes (bilateral)
Limb-kinetic apraxia	Premotor or parietal cortical damage
Loss or impairment of optokinetic nystagmus	Supramarginal gyrus or inferior parietal lobe

Table 3.7 The segments of the posterior cerebral artery

Name of the segment	Origin and termination
P1: Pre-communicating segment	• Originates at the termination of the basilar artery • Terminates at the posterior communicating artery
P2: Post-communicating segment	• Originates from the posterior communicating artery around the midbrain • Terminates as it enters the quadrigeminal cistern
P3: Quadrigeminal segment	• Travels posteromedially through the quadrigeminal cistern • Terminates as it enters sulci of the occipital lobe
P4: Cortical segment	• Originates within the sulci of the occipital lobe

The internal carotid artery syndrome occurs due to occlusion of the internal carotid artery. It frequently produces a massive zone of infarction in the region of the brain parenchyma supplied by the middle and anterior cerebral arteries, especially when the collateral circulations are weak. Extensive cerebral infarction in these areas can cause significant cerebral edema and mass effect, including uncal herniation, coma, and death.

The posterior cerebral artery syndrome occurs due to occlusion of the posterior cerebral artery, the terminal branch of the basilar artery. The artery supplies the respective occipital lobe and medial and inferior temporal lobe, the rostral part of the brainstem, and the posterior diencephalon, including most of the thalamus. Table 3.7 lists the segments of the posterior cerebral artery. Blockages of the proximal segment of the posterior cerebral artery (from the bifurcation site to the origin of the posterior communicating artery) can produce minor neurological deficits owing to the collateral blood supply from the posterior communicating artery. Blockages of the posterior cerebral artery beyond the proximal segment can cause a variety of deficits depending on the cortical and central branches occluded. The clinical features of posterior cerebral artery syndrome and the structures involved are mentioned in Table 3.8.

3.8 Complications of Stroke

Medical and neurological complications are frequent among post-stroke patients. Complications tend to increase the length of hospitalization, hinder functional recovery, interfere with rehabilitative therapies, and increase the costs of care. In addition to the above, the complications are a major cause of death in the early phase post-stroke. Those patients with severe disabling strokes are more vulnerable to develop complications. Pyrexia, systemic inflammatory response, hypoxia, hyperglycemia, and certain medications used for the management of the complications may have a damaging physiological effect on an injured brain or may compromise its capability for plastic change. Most of the complications develop during the acute phase (first few weeks) of stroke. Cardiac abnormalities, dysphagia, and pneumonia are a few of the early complications, whereas others, such as decubitus ulcers,

Table 3.8 The clinical features and the structures involved in the posterior cerebral artery syndrome

Territory	Location/structure involved	Clinical features
Cortical	Contralateral homonymous hemianopia	Calcarine cortex or optic radiation
	Bilateral homonymous hemianopia, cortical blindness, denial of blindness, inability to perceive objects not centrally located, apraxia of ocular movements, inability to count objects	Bilateral occipital lobe, possibly with involvement of parieto-occipital region
	Dyslexia without agraphia, color anomia	Calcarine lesion on the dominant hemisphere and posterior part of corpus callosum
	Memory defect	Lesion of inferomedial portions of temporal lobe bilaterally or less frequently the dominant side
	Visual agnosia	Dominant occipital lobe
	Prosopagnosia or face blindness (inability to recognize familiar faces)	Nondominant hemisphere fusiform or lateral occipitotemporal gyrus; visual association area
	Simultagnosia (inability to perceive more than a single object at a time)	Dominant visual cortex
	Topographic disorientation	Nondominant calcarine and lingual gyri
	Unformed visual hallucinations, metamorphopsia, photophobia	Calcarine cortex
Central	Thalamic syndrome: spontaneous pain and dysesthesias and loss of all sensory modalities	Ventral posterolateral nucleus of thalamus
	Choreoathetosis, hemiballismus	Subthalamic nucleus or its pallidal connections
	Intention tremor, contralateral ataxic, postural tremor	Dentatothalamic tract
	Claude's syndrome: ipsilateral third cranial nerve palsy, contralateral hemiparesis, and ataxia	Dentatorubral fibers, corticospinal and bulbar fibers, and oculomotor nerve fibers in midbrain
	Weber's syndrome: contralateral hemiplegia with third cranial nerve palsy	Midbrain infarction with involvement of third nerve and cerebral peduncle
	Contralateral hemiplegia	Cerebral peduncle
	Vertical eye movement paralysis, skew deviation, sluggish pupillary reaction to light, miosis, and ptosis	Supranuclear fibers to third cranial nerve; high midbrain tegmentum ventral to superior colliculus

deep vein thrombosis, and falls, can occur after several days. Many of the complications are preventable and early recognition and treatment can effectively ameliorate the deleterious effects of the same. Only the major systemic medical and neurological complications are discussed here.

Cerebrovascular accidents and cardiovascular diseases share several risk factors. As discussed earlier in the chapter, cardiac diseases such as atrial fibrillation, valvular disorders, or congestive heart failure can increase the risk of stroke, and this can predispose patients to develop cardiac complications. Autonomic dysregulation and physiological stress induced by cerebrovascular accidents can be plausible causes for the development of cardiac complications. Timely and accurate recognition of patients who are at high risk for cardiac complications might help to prevent such complications. Stroke patients with established cardiovascular diseases, severe stroke, diabetes mellitus, and peripheral vascular disease are at high risk of myocardial infarction during the early recovery phase of stroke than those stroke patients without any of these comorbidities. Cardiac arrhythmias, including atrial fibrillation, ventricular tachycardia, supraventricular tachycardia, and ventricular ectopic beats have been reported post-stroke. Hemodynamic instability due to cardiac arrhythmias and thromboembolism due to atrial fibrillation can predispose to second or multiple strokes.

Bronchopulmonary pneumonia is a frequent medical complication of stroke and the most common cause of pyrexia within the first few days after stroke. According to the current evidence, pneumonia has a threefold increased risk of death. Most of the stroke-related cases of pneumonia develop as a result of aspiration. Empirical use of antimicrobial treatment for gram-positive and negative bacilli can effectively manage aspiration pneumonia. Severe stroke, old age (>65 years), speech impairment, dysphagia, cognitive dysfunction, reduced or altered level of consciousness, severe facial palsy, mechanical ventilation, brainstem strokes, and multiple strokes are certain independent risk factors of pneumonia. The presence of these conditions can alert the clinician and anticipate the risk of pneumonia in stroke patients to decide about further surveillance, diagnostics, and feeding. Weak expiratory muscles and poor or ineffective cough can also predispose to pneumonia. Paying close attention to oral care and dental hygiene can minimize the possibility of the development of pneumonia. The use of oral antiseptics and frequent dental hygiene regimen for mechanically ventilated stroke patients may lower the ventilator-associated risk of pneumonia.

Desaturation of oxygen and apnea within the first few days post-stroke can aggravate brain injury by compromising oxygen delivery to the penumbral zone of the brain tissues. The reasons for stroke patients developing hypoxemia can include alteration in the central regulation of respiration, weakness of the respiratory muscles, and added complications such as pneumonia, aspiration, atelectasis, and pulmonary emboli, which can impair air exchange at the pulmonary segments level. Stroke patients with larger brain lesions, old age, preexisting cardiac and pulmonary disease, and swallowing impairments are more likely to develop oxygen desaturation. Maintaining the normal oxygen saturation is the logical goal to minimize the desaturation effects, and acute stroke patients should be monitored by pulse oximetry maintained at a target oxygen saturation level of $\geq$92%.

Pyrexia is a complication seen among post-stroke patients, particularly during the first 72 hours, and is more common among hemorrhagic strokes than ischemic strokes. Fever can indicate systemic stress, underlying infection, or presence of intraventricular blood in a hemorrhagic stroke. Central thermoregulatory impairment can also be a reason for the fever. The central hyperthermia presents with rapid onset of high fever and notable temperature variation. It is commonly seen in brainstem strokes associated with high mortality. Fever can increase the metabolic demands of the injured brain tissues and exacerbate the neuronal injury. Prompt treatment of hyperthermia with antipyretic medications can effectively improve neurological outcomes in many stroke patients. Though inducing hypothermia can be an alternative, it can lead to cardiac arrhythmias, infections, hemorrhagic transformation of infarcts, and venous thrombosis.

Hyponatremia is a common electrolyte disorder encountered in acute stroke patients. It develops either due to the syndrome of inappropriate secretion of antidiuretic hormone or cerebral salt wasting syndrome. Antidiuretic hormone is a posterior pituitary gland hormone that primarily regulates body water by acting on the kidneys. Normally, a drop in plasma volume or an increase in serum osmolality causes the release of the antidiuretic hormone to increase total body water. In the former syndrome, failure of the negative feedback mechanism that usually controls the release of the antidiuretic hormone causes persistent production of the same, despite body fluid hypotonicity and expanded circulatory volume. Cerebral salt wasting syndrome, described by Peter J.P. and co-workers in 1950, is characterized by the development of excessive natriuresis (excretion of an excessively large amount of sodium in the urine) and subsequent hyponatremia, particularly in patients with subarachnoid hemorrhage. Though theories for the pathophysiology of cerebral salt wasting include the release of brain natriuretic peptide or damage to the hypothalamus with subsequent sympathetic system dysfunction, the exact mechanism is largely unknown. The treatment strategies for either syndrome are not the same. For cerebral salt wasting syndrome, the patient is given fluids and sodium supplementation, whereas for the syndrome of inappropriate secretion of antidiuretic hormone, the patient is fluid restricted. Typically, with appropriate management, the hyponatremia tends to resolve within days to weeks but can remain a chronic issue.

Approximately 40% to 75% of stroke patients develop dysphagia. Reduced oral intake due to dysphagia increases the risk of poor nutrition and dehydration. As discussed earlier, dysphagia is a major risk factor for pneumonia. Usually, the nutritional requirements of stroke patients with dysphagia are met either by insertion and placement of the nasogastric tube (Ryle's tube) or surgical placement of percutaneous endoscopic gastrostomy (PEG). Evidence states that during the first few weeks post-stroke, nasogastric tube feeding is preferable as it can enable some patients to recuperate their swallowing functions. Ryle's tube is easy to insert but can be uncomfortable and get easily dislodged. On the contrary, PEG is an invasive procedure, and peritonitis and bowel perforation are a few complications of the same. However, both of these managements don't offer any protection against aspiration pneumonia.

Gastrointestinal hemorrhage has been reported in some prospective studies as a less common complication after stroke. The risk of bleeding is higher in stroke patients of Asian descent. The severity of the stroke, prior history of peptic ulcer disease, cancer, renal failure, sepsis, and presence of abnormal liver function are independent predictors of gastrointestinal bleeding among acute stroke patients. Late gastric emptying, stress ulcers, and mucosal irritation secondary to gastric feeding are certain mechanisms speculated to cause gastrointestinal hemorrhage. Judicious prophylactic use of antacids, H2 receptor antagonists, and sucralfate is effective in reducing the risks of bleeding.

Approximately 30% to 50% of acute stroke patients develop bowel incontinence. Bowel and bladder incontinence can coexist in many of the patients. Old age, large and disabling strokes, and reduced or altered level of consciousness are certain important predictors of bowel incontinence. Stroke patients having reduced functional mobility, loss of manual dexterity, visual impairment, cognitive dysfunction, and communication difficulties are likely to have persistent bowel incontinence even at 3 months post-stroke. Certain medications with anticholinergic properties such as antipsychotics or antiemetics can considerably increase the risk of fecal incontinence. Encouraging early mobilization, improving toilet access, dietary modification, avoiding dehydration, and minimizing polypharmacy, may facilitate bowel control after a stroke.

Urinary tract infection is another important and frequent complication seen among acute stroke patients. Older age, use of urinary catheters, severity of the stroke, and female gender are independent predictors of urinary tract infections after stroke. The majority of the patients may develop uncomplicated urinary tract infections. Escherichia coli frequently cause urinary catheter-related urinary tract infections. Meticulous catheter care and avoidance of unnecessary catheterization can help prevent urinary tract infections. For suspected infection, urine cultures can confirm the diagnosis and guide appropriate antibiotic treatment. Urinary incontinence and retention occur in 30% to 60% of stroke patients. Urodynamic studies have revealed high rates of detrusor hyperreflexia post-stroke, and high detrusor hyperreflexia suggests damage to corticospinal pathways. Large infarcts, altered sensorium, cognitive impairment, aphasia, and severe functional impairments are independently associated with bladder dysfunction. The incontinence can adversely affect the self-esteem and confidence of stroke patients, impose burden on family members or caregivers, delay hospital discharge, and lead to institutionalized care. Methodical assessment and management of urinary symptoms by specialized professionals may tackle the issues of incontinence during the acute phase of stroke.

Deep vein thrombosis (DVT) is a major concern during the post-stroke stage, especially when the lower limb is paralyzed. The prevalence is as high as 50% during the first two weeks after stroke, in the absence of DVT prophylaxis. In a prospective study, magnetic resonance imaging (MRI) for the pelvis and lower extremities detected venous thromboembolism in 40% of acute stroke patients and pulmonary emboli in 12%, despite aspirin and graded compression stockings usage. In general, DVT develops early and possibly within the first few weeks post-stroke. The severity of paralysis, advanced age, and dehydration are important risk factors for DVT development. The most typical symptoms of DVT include edema of the paretic or paralyzed limb, painful and tender area, and a certain amount of redness and localized rise in the skin temperature. In approximately 50% of cases, such clinical symptoms are undetectable and can be identified only by Doppler ultrasonography, contrast venography, or impedance plethysmography. In undiagnosed and/or untreated proximal DVT, the most feared complication is the fatal pulmonary embolism, often occurring between the second and fourth weeks post-stroke, which accounts for 15% of early death. Low molecular weight heparin or low dose unfractionated heparin used prophylactically can be effective in preventing DVT and pulmonary embolism and the risk of death; however, it can increase the risk of intracranial bleeding, especially in hemorrhagic stroke or hemorrhagic transformation of infarcts. Placement of an inferior vena cava filter can be an option for preventing pulmonary embolism in such high-risk patients, though this approach can encourage further DVT formation. The use of graded compression stockings and intermittent pneumatic compression can be effective for preventing DVT, particularly when the risks of bleeding associated with anticoagulants in stroke patients are high. However, in some patients, especially with a history of peripheral vascular disease, the use of compression stocking poses serious concerns including skin ulcers and necrosis, as it might further reduce the blood flow to the lower extremities.

Survivors are generally anxious about the possibility of recurrence of stroke, as it is an important cause of morbidity and mortality. Patients with a prior history of stroke are four times more likely to have another stroke compared to matched controls. Stroke patients are most at risk of recurrence during the first 6 months. Patients below 65 years of age are at lower risk as compared to older age groups. The absolute risk of recurrence is approximately 30% for all types of strokes during the first 5 years, and the risk of recurrence following primary hemorrhagic stroke is somewhat similar to that of ischemic stroke. Stroke recurrence can have a devastating effect on morale and can prevent patients from regaining independence. The underlying mechanisms of recurrent stroke are not very clear, and the etiology can be multifactorial. Patients with large vessel atherothrombosis and cardiovascular conditions are at higher risk of recurrent stroke. Stroke survivors need to seek information and guidance from the neurologist or physician about the risks of a recurrent stroke and measures required to prevent them. Based on the causative and risk factors for recurrent stroke, the preventative measures may include the use of antithrombotic agents, antihypertensive medication, hypoglycemic agents, statins for reduction of elevated low-density lipoprotein and triglycerides, anticoagulation for atrial fibrillation and cardiac conditions, and cessation of smoking.

Stroke patients are at increased risk for fractures, mostly involving the hip joint. In general, hip fractures are associated with high morbidity and mortality, especially among the elderly. There is a sevenfold higher risk of fractures for stroke patients during the first year. Elderly stroke patients are also likely to have age-associated osteopenia. Reduced lower limb muscle strength, inadequate weight-bearing, and use of anticoagulants accelerate bone loss after stroke, especially in the paretic limbs. Osteopenic bones are brittle and are more susceptible to fractures from trivial trauma or fall, and such fractures are frequent on the paretic side. Associated arm weakness, loss of protective and balance reactions, neglect, seizures, cognitive and motor impairments, sedative medications, or drugs affecting sensorium are the independent factors associated with falls and fractures. The use of hip protectors as shock absorbers, assistive and orthotic devices for ambulation, gait training for safe ambulation, and promoting balance reactions may prevent fractures in stroke patients.

Pain is another frequent complication seen among stroke patients. Among stroke patients, pain mainly involves the paretic limbs, especially the affected side shoulder. It can interfere with physiotherapy, interrupt sleep, and contribute to fatigue and depression. Almost one-third have moderate to severe pain in the first few months post-stroke. However, in most cases, the pain improves spontaneously, although in some it may worsen with time. Preexisting painful disorders like arthritis and decreased mobility, changes in gait, advanced age, and postural changes can contribute to pain. Complex regional pain syndrome, previously known as reflex sympathetic dystrophy, is a chronic painful condition that may affect the paretic extremity in some stroke survivors. The patient may complain about the severe and constant pain in the affected area and often describes it as burning or pins and needles type and may experience changes in skin temperature, skin color, and swelling of the joints. The affected limb may feel warmer or cooler compared to the unaffected limb. Changes in skin texture (skin may appear shiny and thin), abnormal sweating patterns, changes in nail and hair growth patterns, and stiffness in affected joints are certain common features of complex regional pain syndrome. Early identification, physiotherapy, and medical management using chemical sympathetic blocks and oral or intramuscular corticosteroids may help to minimize the consequence of this debilitating condition.

Approximately 10% of stroke patients can develop central post-stroke pain as a direct consequence of a lesion affecting the central somatosensory system or its pathway. The thalamus is believed to play a key role in the underlying pathophysiology of central pain. The central pain can be focal, segmental, or affecting half the body and can present as severe and persistent burning or aching pain, intermittent and spontaneous lacerating or shooting pain, evoked by mechanical or thermal stimuli. Typically, the central pain tends to develop during the first few months post-stroke, and medications like amitriptyline and lamotrigine may show certain efficacy in ameliorating it. Regarding pain involving the shoulder joint, muscle paralysis, spasticity, joint inflammation, soft tissue contracture, and nerve injury can be the possible causes. The hemiplegic shoulder pain syndrome can be due to the weakness of the vertical stabilizers of the shoulder joint needed for glenohumeral stabilization. In such patients, biceps and supraspinatus are frequently tender, and the "Neer Impingement Sign" is positive. Use of analgesics, pain management using electrotherapeutic modalities, foam supports for shoulder or shoulder strapping, and daily passive range of motion (ROM) exercises can help prevent shoulder pain.

Stroke patients may report fatigue as a frequent complaint. Fatigue can cause functional limitations and can contribute to depression. The cause of fatigue among stroke patients is multifactorial. General physical deconditioning, associated medical ailments, side effects of medications, and certain central mechanisms are the possible factors attributing to fatigue. Management of fatigue among stroke patients needs to be individualized with assessment and treatment of concomitant disorders, including infection, hypothyroidism, anemia, depression, and adrenal insufficiency.

Emotional incontinence can be a feature of lesions of the brain affecting the frontal lobe, hypothalamus, and limbic system. Stroke patients with pseudobulbar affect (pathological laughing and crying) are known to develop emotional lability or emotional incontinence. Such patients demonstrate emotional outbursts of uncontrolled or exaggerated laughing or crying that are inappropriate with the mood. They may quickly change from laughing to crying with minimal provocation. Often the patients may express that they are unable to control or inhibit such episodes. Frequent crying may also accompany depression. Antidepressants and coping and support may help to control symptoms of pseudobulbar affect.

Based on the available literature, about 33% of stroke patients have depression as another complaint. Young stroke patients, women, aphasia, and those with severe stroke and greater disabilities are at a higher risk of developing depression. Post-stroke depression can contribute to mortality, including suicide. Depressed patients are less likely to participate in rehabilitative therapies, less compliant with medications, and tend to have poorer recovery following stroke. The use of antidepressants, counseling, and cognitive behavioral therapy has been attempted for treating post-stroke depression with variable results.

Stroke patients are susceptible to develop bedsores, particularly when bedridden for a prolonged time. Immobility or reduced mobility, age, bowel incontinence, bladder incontinence, poor nutrition, poor cognition, inadequate perfusion, presence of chronic diseases, and reduced skin sensation increase the risk of skin breakdowns. In stroke patients, pressure ulcers are frequently developed over prominent bony areas such as sacral, ischial, trochanteric, malleolar, and heel. Early mobilization, regular two-hour change of posture within the bed, use of padded heel boots, and alternating pressure air mattresses can prevent or minimize their development.

Stroke is the most common cause of seizures among the elderly population. Post-stroke seizure and epilepsy (multiple episodes of seizures) are common causes of hospital admissions or readmissions. Approximately 10% of the post-stroke population is at risk of developing seizures within the first 5 years, and around 5% of patients will have a seizure within the first few weeks post-stroke. Hemorrhagic stroke patients are more likely to have seizures after a stroke than those with ischemic stroke. One-third of cases can present with generalized tonic-clonic seizures and the remaining can present with partial seizures. Early-onset seizures usually present with a focal onset, while generalized tonic-clonic seizures are more common with late-onset seizures. Around 10% of post-stroke epilepsy patients can develop status epilepticus. Though the prognosis of status epilepticus is poor, the use of antiepileptic medications, including phenytoin, carbamazepine, and phenobarbital, remains the mainstay of management of seizures post-stroke.

3.9 Clinical Diagnosis

If approached systematically, the clinical diagnosis can be easier and more logical. The neurologist or the physician must decide on certain key questions to diagnose and plan the medical treatment: (1) What is the mechanism (hemorrhage or ischemia, including their subtypes) of stroke? (2) Where is the anatomical location of the lesion? While gathering the information from history and medical records, the clinician should also identify whether the signs and symptoms are due to any non-vascular processes such as a brain tumor, infection, trauma, intoxication, metabolic abnormality, or seizure disorder, which mimics a stroke. The information (data collected from the patient, family members or bystanders, and medical records) regarding the past and present personal and family illnesses, presence and nature of past strokes or TIA, activity at the onset of the stroke, temporal course and progression of the findings and presence of any accompanying symptoms like headache, altered level of consciousness, vomiting, and seizures can aid in identifying the mechanism of stroke.

Analysis of the neurological symptoms and their distribution, neurological examination findings, and brain and vascular imaging findings provides information about the diagnosis of stroke location. Mechanisms of stroke and anatomical diagnoses need not be always absolute. In a stroke patient, hemorrhagic stroke may be by far the most possible diagnosis, but embolism and thrombosis are also likely and should not be eliminated from consideration. Whether it is the mechanism or the anatomical location, the process of diagnosis should include hypothesis generation and testing. The clinician must decide on ischemia versus hemorrhage before hypothesizing the subtypes and similarly thrombosis versus embolism versus reduced perfusion before distinguishing subtypes of thrombosis and thus proceeding systematically from the general to the specific. Raised blood pressure, murmurs or arrhythmia, vascular bruits, and cardiac enlargement are certain examples of physical findings that can influence the identification of the stroke mechanism.

A raised blood pressure of 250/130 mmHg would increase the likelihood of hemorrhagic stroke than an ischemic stroke. A normotensive patient with prior history of several intracerebral hemorrhages in different locations has a high probability of having cerebral amyloid angiopathy or bleeding diathesis as the plausible cause. The presence of fever and cardiac murmur suggests the likelihood of a systemic condition such as infectious endocarditis. The presence of diabetes mellitus and coronary artery disease strongly favors a diagnosis of associated atherosclerosis of the larger vessels and a thrombotic mechanism of stroke. A prior history of heart disease raises the possibility of arrhythmia, mural thrombosis, and valvular heart disease, which are potential sources of brain embolism. Repeated history of two or three strokes during the past one year, in the absence of heart disease, can suggest brain embolism probably from a hypercoagulable state or an aortic source. A young stroke (ischemic or hemorrhagic stroke occurring in adult subjects aged less than 65 years) is more likely to be a hemorrhagic mechanism than a thrombotic stroke. However, a middle-aged person with a history of tobacco smoking, a family history of cardiac disease, and a high level of low-density lipoproteins have the possibility of premature atherosclerosis and occlusion of the large artery as the mechanism of the stroke.

The nature of the symptoms and their suggested localization are important in making a diagnosis. With left hemiparesis and the presence of a left visual field deficit or left hemineglect, the anatomical location of the lesion can be to the right cerebral hemisphere. The presence of gaze palsy to the right, nystagmus, and internuclear ophthalmoplegia would favor a brainstem site lesion in the pons or midbrain. Repeated TIAs in the same vascular territory are virtually diagnostic of thrombotic stroke. For instance, if a patient who presents with aphasia and right limb weakness had an attack of right face and numbness of right hand one week earlier and a TIA characterized by right-hand weakness three weeks prior, then it could be relatively sure that the stroke is a result of thrombotic occlusion of the left anterior circulation. If the same patient also gives a history of a black shade descending over the left eye, causing temporary blindness, then it would be certain that the left internal carotid artery before its ophthalmic artery branch is the anatomical location of the occlusive lesion.

During the night or a nap, the circulation is least active and most sluggish, and many of the thrombotic stroke patients may notice the symptoms soon after the sleep or nap. On the contrary, patients with embolic or hemorrhagic strokes tend to develop the symptoms when the circulation is more active or when the blood pressure is high. However, the current data state that ischemic and hemorrhagic strokes tend to occur during the morning hours till noon, especially when the patient has begun the daily activities. Though hemorrhagic stroke can occur at night and thrombotic stroke can occur during the active time of the day, chances of developing a thrombotic or a lacunar stroke during vigorous physical activity are less likely. The Valsalva maneuver or vigorous sneeze or coughing can loosen an embolic section, resulting in brain embolism. Physical trauma to the neck and sudden neck movements including neck manipulations, following weight lifting, and during the postpartum and after labor should raise the suspicion of arterial dissection.

Clinical improvement shortly after the onset of the neurological deficit argues strongly against an intracerebral hemorrhage. The gradual and progressive development of focal neurological deficit accompanied by the gradual and progressive development of symptoms of raised intracranial pressure suggests intracerebral hemorrhage. The maximal neurological deficit at the onset unassociated with headache is most likely an embolic mechanism. Sudden and severe headache, accompanied by vomiting and an altered level of consciousness right at the onset, suggests a subarachnoid hemorrhage. Unlike intracerebral hemorrhage that tends to have a progressive focal neurological deficit at the onset, subarachnoid hemorrhage causes sudden raised intracranial pressure responsible for the severe headache and the decreased level of consciousness. Altered sensorium is unlikely in lacunar stroke, a subtype of thrombotic stroke. Loss of consciousness is not uncommon in large ischemic stroke and bilateral brainstem stroke. The presence of seizures in the early period after stroke onset argues for intracerebral hemorrhage or an embolic stroke. Vomiting is rare in anterior circulation ischemic strokes. However, it is common in posterior circulation strokes, presumably because of the involvement of the vomiting center located on the floor of the fourth ventricle.

To diagnose and treat cardiogenic stroke, which is different from the intrinsic disease of the intracranial and extracranial arteries, a carefully detailed history of cardiac symptoms, myocardial infarction, rheumatic heart disease, palpitations or arrhythmia, and congestive heart failure, and a detailed examination of the cardiac system is essential. Clinical examination of the systemic and extracranial arteries may provide clues regarding the presence of atherosclerosis or diminished blood flow, possibly supported by the eye examination findings, as the eyes provide further clues regarding the body's vascular system and the possible stroke mechanism.

The localization of the brain lesion is mainly from the patient's historical description of the neurological symptoms and the neurological examination findings. The most important and frequent brain dysfunctions that need special emphasis are the higher cortical function, level of alertness, visual and oculomotor systems, and gait. If the patient has symptoms or signs present in the right limbs or the right visual field, the bedside tests of the higher cortical function should include the examination of language functions like asking the patient to write a few lines about the town where he or she lives in, to read a paragraph from a newspaper, to name objects in the environment, and to repeat the spoken language. When the symptoms and signs of dysfunction are present in the left limbs or the left visual field, it is specifically important to check the visuospatial functions and look for neglect of the left side of space. Bedside tests like drawing a clock or a house, copying a single two-dimensional figure, reading a brief paragraph or headline, and identifying objects in a picture may reveal findings like omitting words, phrases, or people on the left side of the picture suggesting left neglect.

Memory can be affected by a focal brain lesion involving the posterior cerebral artery territories. Large lesions in the posterior aspect of the cerebral hemispheres may produce only visual impairments and may leave speech, motor, and somatosensory systems unscathed. The confrontation test may provide further clues regarding the visual field deficit. The presence of conjugate-gaze paralysis during clinical examination indicates a frontal or deep hemispheric lesion opposite the side of gaze palsy. The presence of dysconjugate palsies or paralysis of eye movement muscle(s) and the presence of horizontal or vertical nystagmus during gazes are usually diagnostic of vertebrobasilar stroke. If an apparently normal clinical examination finding is revealed in a recumbent or seated position, with the absence of any bulbar findings, and the same patient demonstrates incoordination and imbalance while walking, such a finding may suggest cerebellar hemorrhage or infarction. Table 3.9 depicts the general categories of neurological signs and symptoms and the vascular territories or areas involved.

3.9.1 Neuroimaging and Laboratory Diagnosis

Neuroimaging is an integral part of the evaluation for all stroke patients. For such a potentially devastating condition, clinicians need all of the objective data available to diagnose, prognosticate, and treat individual stroke patients. Rapid identification of the occluding artery or hemorrhage, including its size, shape, and extent of the brain lesion, and estimation of the central necrotic core size and the penumbra can guide the clinician in the appropriate management of stroke. Various neuroimaging modalities, including computed tomography (CT) and MRI, are available for the evaluation of patients specifically presenting with acute ischemic stroke,

Table 3.9 Listing the general categories of neurological signs and symptoms and the vascular territories or areas involved

Location	Vascular territory/area involved	Features
Left hemisphere (anterior hemisphere) lesion	In the territory of the internal carotid artery or its middle and anterior cerebral arteries tributaries	Right hemiplegia, right limb hemisensory loss, aphasia, visual field defect of right, right conjugate gaze palsy, difficulty reading, writing, and calculating
Right hemisphere (anterior hemisphere) lesion	In the territory of the internal carotid artery or its middle and anterior cerebral arteries tributaries	Left hemiplegia, left limb hemisensory loss, left visual space neglect, difficulty drawing and copying, visual field defect of left, left conjugate gaze palsy, left tactile extinction, left visual extinction
Left posterior cerebral artery lesion	Posterior cerebral artery territory	Visual field defect of right, difficulty reading with retained writing ability, difficulty naming colors and objects presented visually, normal repetition of spoken language, numbness, and sensory loss in the right limbs
Right posterior cerebral artery lesion	Posterior cerebral artery territory	Visual field defect of left, often with neglect, left limb numbness, and sensory loss
Vertebrobasilar infarction	Vertebrobasilar territory	Vertigo, diplopia, vomiting, headache in the occiput, mastoid or neck region, weakness or numbness of all four limbs or bilateral regions, crossed motor or sensory findings (i.e., weakness or numbness of one side of the face and the opposite side of the body), ataxia of limbs, bilateral blindness or reduced vision, nystagmus, dysconjugate gaze, gait or lower limb ataxia out of proportion to weakness, crossed signs, bilateral visual-field defects, amnesia
Pure motor stroke or ataxic hemiparesis	Lacunar etiology; vascular supply for internal capsule or basis pontis	Weakness of the face, arm, and leg of one side of the body, without higher cortical dysfunction, sensory or visual dysfunction, or reduced alertness, may present with mixed weakness and incoordination or ataxia on the same side of the body
Pure sensory stroke	Lacunar etiology; usually thalamic area; non-thalamic areas include brainstem, internal capsule, and somatosensory cortex	Numbness or reduced sensibility of the face, arm, and leg on one side of the body, sensory ataxia, absence of motor weakness, visual or higher cortical function abnormalities

intracerebral hemorrhage, and subarachnoid hemorrhage. Both CT and MRI are noninvasive and safe new generation scanners of which the former one is readily available in many of the hospitals. Both for CT and MRI, the image findings depend on the time of the scan and the clinical event.

3.9.1.1 CT and MRI in Stroke

CT scan will show well-circumscribed areas of high density (hyperdense) with smooth borders immediately after the onset of intracerebral bleeding. In case of continued bleed, the sequential scans taken hours to days later may demonstrate enlarging hematoma. Edema developed during the first few days may present as a dark rim around the hyperdense (white in appearance) hematoma. For minor subarachnoid hemorrhagic bleed, CT cannot be a reliable diagnostic tool. When subarachnoid hemorrhage is suspected, lumbar puncture spectrophotometry may reveal xanthochromia (a yellowish tinge color of the Cerebrospinal Fluid [CSF] due to degeneration of red blood cells) in the CSF which is generally confirmatory for the same. The absence of blood in the CSF on lumbar puncture excludes the diagnosis of subarachnoid hemorrhage.

For large ischemic lesions, due to large-artery occlusive disease or cardiogenic brain embolism, a CT scan may show infarction as a low density (hypodense) lesion. Often during the first few hours (up to 6 hours) of the onset of symptoms, the scan may appear normal. CT scans taken at a later time for such large hemispheric ischemic lesions will show a clearly demarcated hypodensity and surrounding edema with mass effect. A CT axial image showing diffuse hypodensity in the cerebral cortex suggestive of ischemic stroke is depicted in Fig. 3.16. Following 2 to 3 weeks after the stroke onset, some infarcts that had been hypodense may become isodense and may obscure the lesion for some time. This phenomenon is called the "fogging effect," and within weeks, a repeat scan will turn the image back to hypodense. CT is generally not useful in detecting and delineating brainstem, cerebellar, and spinal cord infarcts and is not accurate in delineating lesions adjacent to bony surfaces.

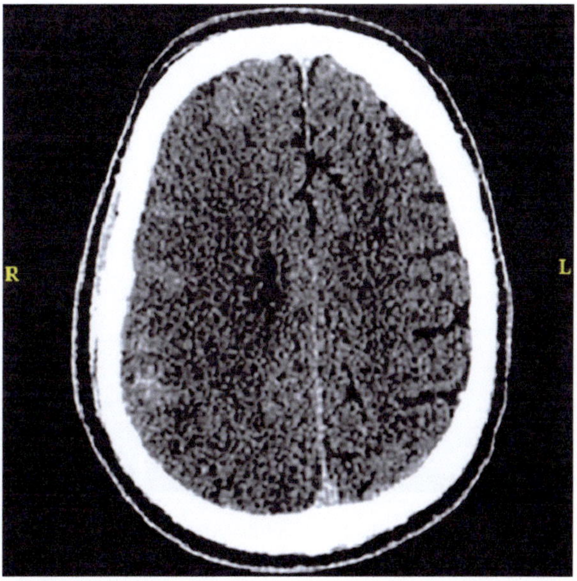

Fig. 3.16 CT axial image showing diffuse hypodensity in the right cerebral cortex with loss of gray white matter interface and cortical effacement suggesting ischemic stroke

Well-defined borders, considerable hypodensity, and shrinkage of the infarcted brain region all suggest months' old chronic infarct, whereas poorly defined hypodensity with surrounding edema and mass effect usually suggests an acute process. For those patients admitted with pure motor hemiparesis due to a lacunar infarct, the CT and MRI may not show any lesions involving the descending corticospinal system.

MRI is the investigation of choice for acute onset stroke and patients with TIAs. MRI is more sensitive than CT in detecting early ischemic changes. Cerebral ischemia alters water content in the affected region, prolonging the T1 and T2 relaxation constants and due to which the lesions will appear as a dark region (hypointense) in T1-weighted image sequences and as bright region (hyperintense) in T2-weighted images. With time (a few days later), lesions will become more hypointense on T1-weighted and more hyperintense on T2-weighted images. Diffusion-weighted MRI or diffusion-weighted imaging (DWI) is specifically sensitive for the detection of acute brain infarcts as it can detect the water shift (extracellular to intracellular) occurring in the cytotoxic edema phase within minutes of stroke onset. Infarct areas appear bright on DWI (same as hyperintense on T2-weighted images) and dark on apparent diffusion coefficient (ADC) images. The DWI and the ADC images of an early stage infarct (hyperacute stage) are depicted in Fig. 3.17. DWI can detect the earliest ischemia, and some hours later, T2 fluid-attenuated inversion recovery (FLAIR) can also show hyperintensity. Figure 3.18 shows patchy hyperintensity on the T2 and FLAIR images suggesting early (acute infarct) stroke. For those TIAs with no residual symptoms, the DWI scans may show brain infarcts, and such lesions are regarded as "small strokes." Both for the anterior and posterior circulations, the location, pattern, and multiplicity of DWI sequences of the lesions can help in suggesting the causative stroke mechanism, including the cardiogenic multiple embolic strokes or multiple strokes due to hypercoagulable state.

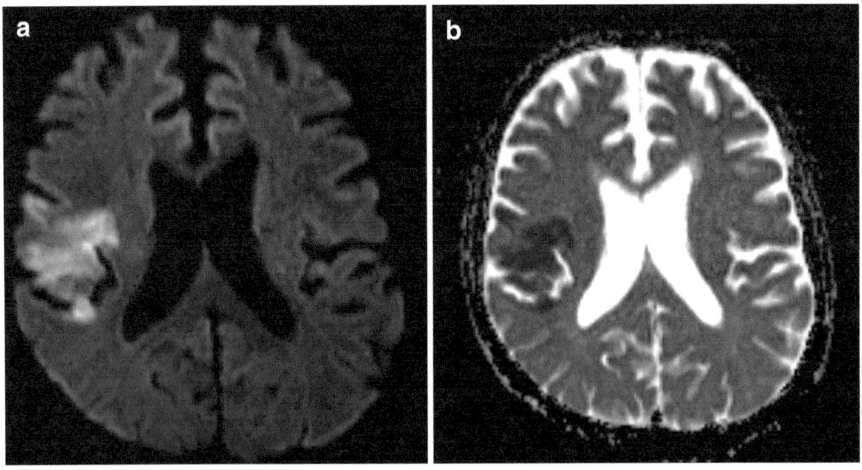

Fig. 3.17 Depicting the DWI and the ADC images of an early stage infarct. (**a**) Magnetic resonance (MR) axial images showing patchy zones bright on DWI suggesting early stage infarct. (**b**) MR axial images showing patchy dark zones on ADC suggesting early stage (hyperacute) infarct

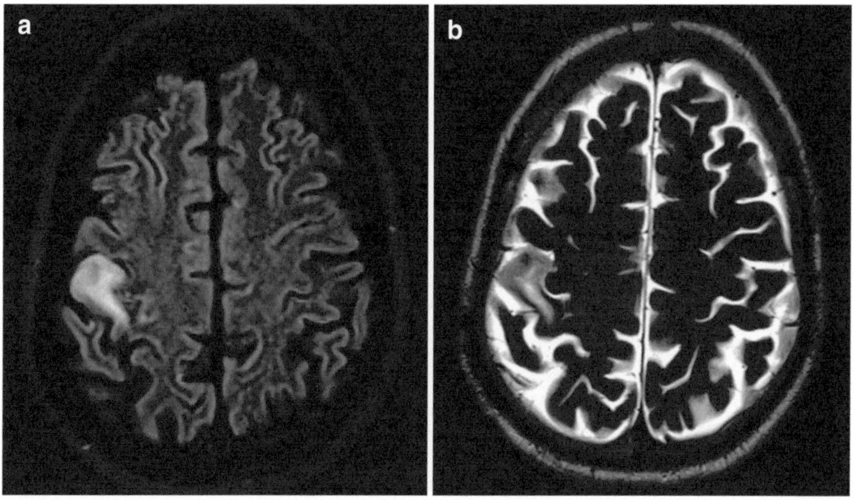

Fig. 3.18 MR axial images showing patchy hyperintensity on the T2 (**a**) and FLAIR (**b**) images, respectively, suggesting early (acute infarct) stroke

Fig. 3.19 CT axial image showing ill-defined hypodensity in the left posterior cerebral cortex with hyperdensities within suggesting hemorrhagic infarct

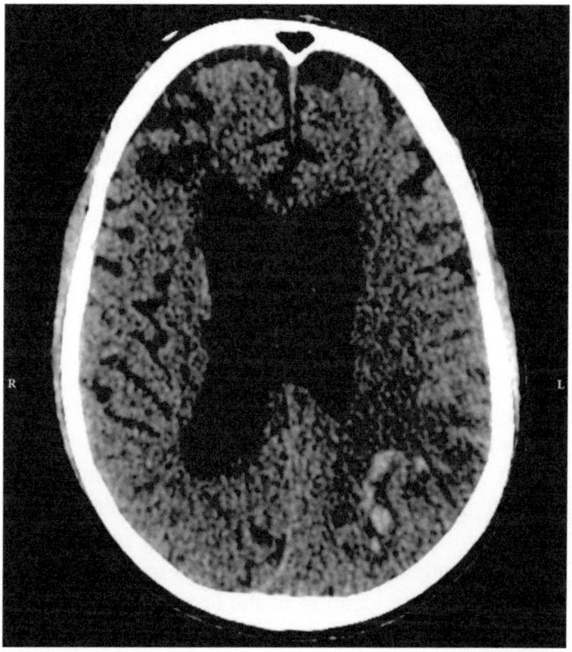

Figure 3.19 demonstrates a CT image of a hemorrhagic infarct. A CT axial image showing a large hemorrhage with ventricular extension is depicted in Fig. 3.20. Though CT scans of intracerebral hemorrhage are easier to interpret than MRI scans, susceptibility-weighted imaging (SWI), a sensitive technique under MRI,

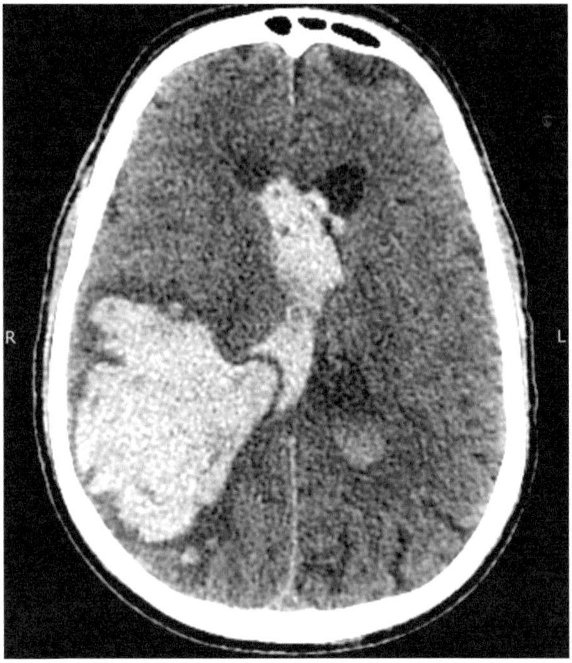

Fig. 3.20 CT axial image showing large hemorrhage in the right cerebral cortex and ventricle extension

can accurately demonstrate the bleed as an area of blooming lesion. Hemoglobin derivatives (oxyhemoglobin, methemoglobin, hemosiderin, or ferritin) at the site of bleed have a paramagnetic effect, and the appearance of the image (hypointense or hyperintense) depends on the nature of the compound, which is time-bound.

Though the size of the lesion has considerable prognostic value, as large lesions in the anatomical area are more likely to cause severe deficits than small lesions in the same area, the severity of the clinical deficit need not always be directly proportional to the infarct size as shown on CT or MRI. While comparing the CT or MRI findings with the clinical deficits, it would not be surprising to see a discrepancy between them. A discrepancy might suggest that certain tissues, although not morphologically damaged enough to show on scans, are not functioning normally. MRI and perfusion CT studies can give detailed information about the neuronal tissue at risk of death by a given vascular lesion. While comparing the DWI and the perfusion CT images, if the perfusion defect appears larger than the diffusion-weighted zone of infarction, the tissue displaying perfusion deficit is at imminent risk of infarction unless blood flow improves to that zone. The clinician should also confirm whether the location of the lesion(s) based on CT and MRI and the location based on the clinical findings are proportionate and correlating.

CT and MRI images can provide details about the mass effect, including the displacement of midline structures, effacement of gyri, cisternal spaces encroachment, brain tissue herniation, and brainstem displacement. Cerebral edema can develop around infarcts and can even be secondary to the reperfusion of a blocked

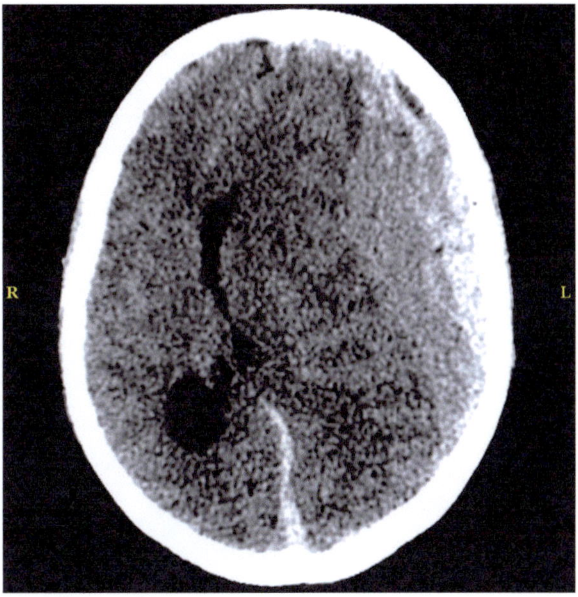

Fig. 3.21 CT axial image showing large subdural hemorrhage with the displacement of midline structures and effacement of gyri

artery. In young patients, edema can be more threatening than in aged patients as the atrophied brain in the latter can provide extra space for brain expansion.

The mass effect created by enlarging hematomas (Fig. 3.21) can be a more serious concern than the edema and mass effect following brain infarction. CT angiography can reveal a "spot sign" a feature indicating the dynamic nature of intracerebral bleed. The spot sign is a unifocal or multifocal contrast enhancement(s) within an acute primary bleed, disjointed from adjacent normal or abnormal blood vessels, and it corresponds to the site(s) of active hemorrhage and is a signature of active intracerebral bleed. The presence of a "spot sign" indicates a greater risk of hematoma expansion, especially if it drains into the CSF space and is an adverse prognostic factor. Infratentorial bleeds have a worse prognosis than supratentorial bleeds. The mass effect produced by the hematoma within the brain and blood within the ventricular system can obstruct the flow of CSF and cause hydrocephalus, which requires surgical decompression by temporary drainage or permanent shunting of CSF.

Blood may accumulate around a bleeding aneurysm or the subarachnoid spaces and cisterns and can provide clues about the site of bleeding. Bleeding from the anterior communicating artery aneurysm causes the presence of blood in the suprasellar cisterns and frontal interhemispheric fissure. Bleeding from the right or left middle cerebral artery bifurcation aneurysm leads to blood in the respective Sylvian fissure. A posterior fossa aneurysm bleeds into the pontine and cerebellopontine angle cisterns (Fig. 3.22). The thickness of blood seen on CT or MRI sequences will provide information about the degree of bleeding. Large subarachnoid bleeds can cause hydrocephalus, and the vasoconstriction of the blood vessels in the vicinity

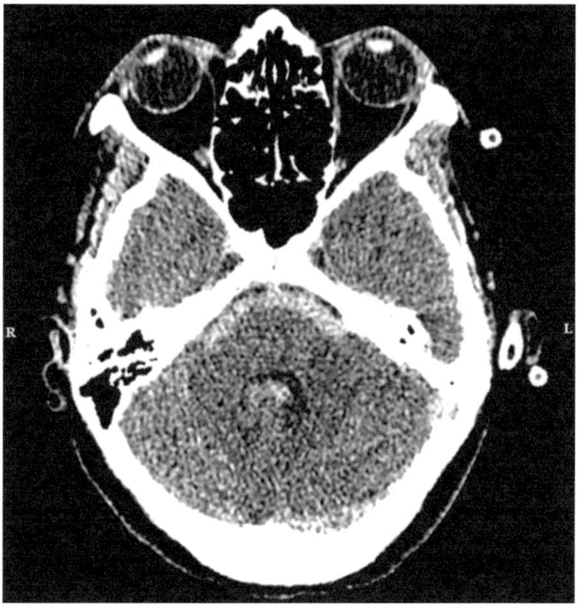

Fig. 3.22 CT axial image showing hyperdensities into the pontine and cerebellopontine angle cisterns

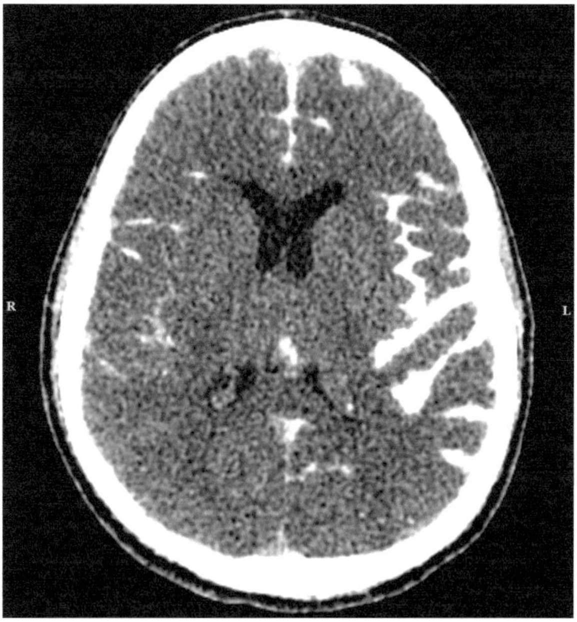

Fig. 3.23 CT axial imaging showing hyperdensities suggestive of large subarachnoid bleeds (left more than the right)

can cause delayed cerebral infarction. Figure 3.23 depicts a CT axial image showing hyperdensities suggestive of large subarachnoid bleeds. Both CT and MRI are not particularly sensitive for the detection of subarachnoid hemorrhage, especially when bleeding is minor or in the recent past. Restless, claustrophobic, or critically ill patients often have difficulty holding still, the time required to produce high-quality MRI images, and for the same reasons, such patients will not be suitable candidates for MRI.

3.9.1.2 Ultrasound and Vascular Imaging in Stroke

Ultrasound and vascular imaging techniques can reveal the vascular structures involved. Transcranial Doppler ultrasonography, a noninvasive technique, is used to check the patency of extracranial components of the carotid arteries and the posterior circulation of the brain. MR angiography offers many advantages over other noninvasive methods of vascular imaging. Angiography in patients with hypertensive bleeds will show the predominance of microbleeds in the deep structures, including the subcortex, basal ganglia, lateral ganglionic region, pons, and cerebellum. In such a case, the likelihood of an etiology other than hypertension is relatively low. Angiography in amyloid angiopathy will demonstrate the predominance of microbleeds in the lobar regions of the cerebral cortex. Hematomas resulting from aneurysms are invariably contiguous to the aneurysms at the base of the brain or surface of the brain. Intracerebral hematomas secondary to anticoagulants are mostly located to the lobar or cerebellar and tend to evolve and enlarge gradually. Arteriovenous malformations can be located anywhere within the brain, particularly in subependymal regions. Figure 3.24 depicts the MR and T2 axial angiography showing a cluster of vessels suggesting arteriovenous malformation.

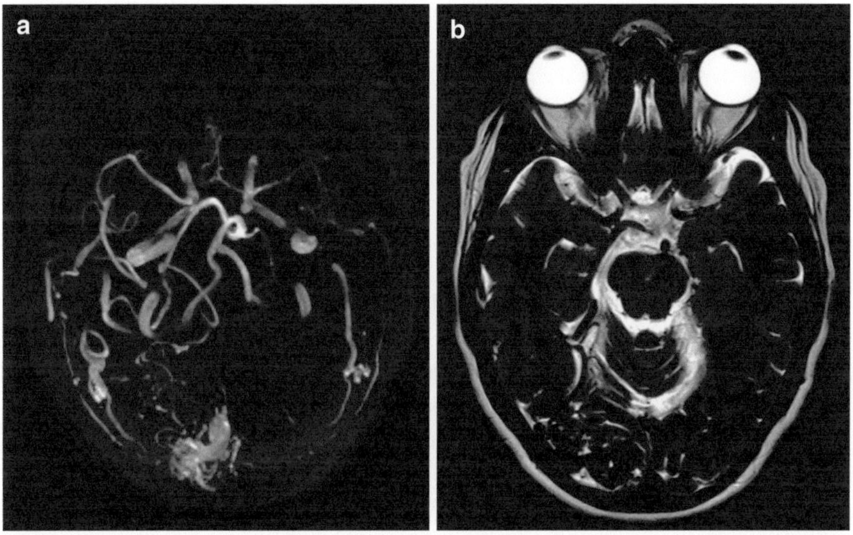

Fig. 3.24 MR and T2 axial angiography showing a cluster of vessels suggesting arteriovenous malformation. (**a**): MR axial angiography showing a cluster of vessels with arterial feeders suggesting arteriovenous malformation in the posterior occipital cortex (**b**): MR axial T2 angiography showing a cluster of vessels as signal voids suggesting arteriovenous malformation

3.9.1.3 Blood Investigations

Based on the nature, location, and severity of the vascular lesion, it is crucial to find out whether the abnormalities of blood constituents are causing or contributing to ischemic or hemorrhagic strokes. Abnormalities in the clotting system can predispose to hypercoagulability and thrombosis. Infections, cancer, and other diseases can release acute phase reactants that may alter coagulability sufficient to promote thromboembolism. Increased viscosity of blood can alter blood flow, especially in small arterioles and capillaries of the brain, and can cause or contribute to regional decreases in blood flow and potentiate cerebral ischemia. Box 3.1 displays the standard set of blood analyses consisting of hematological studies, serum electrolyte levels, and renal and hepatic tests essential to identify the hematological, metabolic, renal, and hepatic abnormalities underlying the stroke or TIA.

Box 3.1 The Standard set of Blood Tests for Stroke and TIA

Standard set of blood tests for stroke and TIA
• Blood sugar • Hemoglobin • Hematocrit • Homocysteine • Total and differential WBC count • Platelet count • Erythrocyte sedimentation rate • Activated partial thromboplastin time • Prothrombin time-international normalized ratio (INR) • Serum fibrinogen level • Total cholesterol and lipid profile • Serum calcium • Blood urea nitrogen • Blood electrolytes • C-reactive protein

3.10 Medical and Surgical Management of Stroke

Technological advancements in diagnostics have made it possible to quickly and safely determine the cause of most strokes. Even though socioeconomic and psychological factors can influence stroke treatment, the presence of comorbid medical conditions can often limit the choice of available treatments. For instance, the clinician may not recommend surgery for stroke patients with advanced cancer. Abnormalities of coagulation functions and blood constituents can indicate certain therapeutic strategies and contraindicate other treatments. However, age is never an absolute contraindication for stroke management provided that the elderly are cautiously handled.

The nature, location, and severity of the vascular lesions are important factors determining the selection of possible and optimal therapeutic strategies. In addition to the above, the choice of treatment also critically depends on the nature of the causative cerebrovascular process, i.e., management of internal carotid artery occlusion will undoubtedly be different from the management of severe internal carotid artery stenosis or carotid atherosclerotic plaque without stenosis. Reperfusion can be potentially hazardous when the vascular territory is necrotized. On the contrary, if the vascular territory is ischemic and the neurons are stunned and dysfunctional, reperfusion or augmenting the blood flow can substantially save the neuronal tissue from the imminent infarction.

The pace of progression also dictates the urgency and speed of stroke management. For instance, a patient with an acute episode of TIA has a greater probability of stroke tomorrow as compared to a patient with three-month-old case of a single TIA. A stroke patient worsening under medical supervision requires more urgent management than a stable patient with a week old stroke. Box 3.2 displays the reasons for the worsening of acute stroke symptoms. The therapeutic strategies are either general or specific. The general strategies (Box 3.3) apply to all stroke patients, whereas specific therapeutic strategies focus on the specific concerns of the patient, which include management of raised intracranial pressure, potential threat for a second embolism, and recurrent threat of subarachnoid hemorrhage. The general care goals for the clinicians are to limit suffering, give comfort, and prevent complications. Even when no specific therapy seems warranted, stroke patients deserve excellent nursing and general care. Since stroke patients are at risk for future stroke, prevention strategies should begin as early as possible.

Box 3.2 Reasons for Worsening of Acute Stroke Symptoms

Reasons for worsening of symptoms
• Hemorrhagic transformation of an ischemic infarct • Failure of collateral circulation • Progressive occlusion of the arterial lumen • Expansion of hemorrhage • Reduced perfusion due to hypovolemia or systemic hypotension • Cerebral edema • Embolization or propagation of thrombus • Seizures • Infections, particularly pneumonia and urinary tract infection • Cardiac arrhythmias • Pulmonary embolism • Depression

Box 3.3 General Therapeutic Strategies for Stroke Management

General therapeutic strategies
1. Prevent stroke complications such as urinary infections, DVT and pulmonary embolism, phlebothrombosis, bedsores, and contractures and joint stiffness.
2. Prevent further strokes and vascular disease by controlling the stroke risk factors.
3. Maintain adequate blood pressure to minimize the risk of hypotension and hypertension.
4. Restore/maintain fluid and electrolyte balance.
5. Maintain nutrition.
6. Maintain blood glucose levels within the normal range.
7. Maintain bowel and bladder function, which may include a urinary catheter.
8. Improve neurological function and facilitate recovery.

Different medical and surgical strategies are available to improve the circulation to the ischemic region distal to the occluding lesion. During the early stage of acute care, a reduction in blood flow caused by postural changes can reduce the cephalad flow through the stenotic vessel or collateral channels and may worsen the ischemia. For minimizing the possibility of inadequate blood perfusion, ischemic stroke patients are generally advised to be treated in the supine position as sitting, standing, and propped up position of head may increase ischemic symptoms. The therapeutic strategies used for reperfusion include angioplasty, stenting, thrombolysis, mechanical clot retrieval or clot aspiration, surgical endarterectomy, and vasodilator treatment.

Neurointerventional techniques have become an important therapeutic strategy for managing many stroke conditions. Since the 1980s, interventional procedures have used devices including coils, catheters, balloons, and glues, for the treatment of patients with intracranial aneurysms and arteriovenous malformations. Percutaneous transluminal angioplasty, with or without vascular stents, first developed for coronary artery, has become an important option for cervical and cerebral atherosclerotic arteries and fibromuscular dysplasia. Vascular stenting, in conjunction with percutaneous balloon angioplasty, has shown significant improvement in outcomes including lowering the rate of recurrent stenosis and event-free survival post 1 year, compared to angioplasty alone. In addition to the above, stents generally reduce the risk of plaque dislodgement and significant intimal dissection and improve the elastic recoil of the vessel wall.

Increasing evidence is available regarding the benefits and effectiveness of angioplasty, with or without stenting, to treat carotid atherosclerotic disease occluding the vessel or generating embolic debris. An important technical advancement was the development of embolus protection devices (distal protection devices) that consist of baskets or umbrellas placed distal to the target atherosclerotic lesion to trap debris dislodged while performing the angioplasty. Angioplasty and vascular stenting can manage occlusive lesions and symptomatic lesions in neck arteries and intracranial arteries. Patients with long and smooth lesions in the large arteries, especially with coronary artery disease, tend to do better with neurointerventional techniques. On the contrary, patients with focal irregular ulcerated atherosclerotic lesions, tortuous arteries, and arteries with catheter access difficulty tend to do better with surgical endarterectomy.

Thrombolysis means the chemical lysing of the clots within the vessel. Following thrombus formation, the vascular system stimulates the formation of endogenous fibrinolytic mechanisms for thrombolysis. At the sites of fibrin deposition, the release of tissue plasminogen activator (Factor XII) and other chemical substances converts plasminogen to plasmin, the active fibrinolytic enzyme. The endogenous formation of plasmin is probably responsible for the spontaneous recanalization of thrombosed arteries. Recombinant tissue plasminogen activator (r-TPA), streptokinase, and urokinase are certain thrombolytic drugs used in the past. These drugs were administered either intra-arterially into the clots or intravenously. The presence and extent of reperfusion depended mostly on the location of the clot and the stroke mechanism. The main stem and divisional middle cerebral artery occlusions respond better to intra-arterial thrombolysis, compared to internal carotid artery occlusions. The proximal segment of the middle cerebral artery responds better than distal branch occlusions. Embolic occlusions were more successfully recanalized compared to thrombosis engrafted upon atherosclerotic narrowing. For those thrombolysis cases with re-occlusion, transluminal angioplasty might help to maintain the patency of the lumen. Recanalization is generally better when angiography suggests good collateral circulation prior to administration of r-TPA. Though the recommended therapeutic window ranges from 3 to 4.5 hours, patients treated with r-TPA) within 3 hours have shown better response compared to those treated between 3 and 6 hours and controls. Significant infarction, mass effect, edema, hemorrhage, and the onset of stroke beyond 4.5 hours are specific contraindications for r-TPA. Overall, the thrombolytic drug will not facilitate recovery if it fails to recanalize the vessel. Intracranial hemorrhage and death are a few of the complications associated with thrombolysis.

Catheter-based mechanical thrombectomy techniques have made rapid advances in the recanalization of the occluded arteries during the past few decades. Compared to thrombolytic drugs, mechanical thrombectomy is more effective and substantial in reperfusion, more effective in the removal of large clots in the proximal vessels, and less risky for intracerebral and systemic hemorrhages. During mechanical thrombectomy, to minimize the possibility of vasoconstriction, intra-arterial infusion of papaverine, an alkaloid antispasmodic drug, may be used. The endovascular mechanical intervention devices can be categorized as angioplasty and/or stent devices, suction thrombectomy devices, and thrombus retrieval devices. Angioplasty and/or stenting are highly effective in the recanalization of local atherosclerotic occluded brain arteries. Suction thrombectomy devices use vacuum aspiration to remove the occlusive clot out of the artery. Stent retrievers are the most recent and most successful among the family of thrombectomy devices. The mesh columns of the stent retrievers expand and entangle inside the target clot, and while withdrawing the stent, the mesh will bring out the thrombus. The two major advantages of stent retrievers over the rest are their substantially higher recanalization rate and instant restoration of blood flow immediately upon deployment within the target artery.

The local reconstruction endarterectomy or surgical endarterectomy is the common direct surgical method of unblocking an occluded vessel. Surgical endarterectomy not only augments the blood flow but also removes the source of intra-arterial emboli. Endarterectomy is considered when arterial stenosis is between 60 and 99% as it can reduce the risk of imminent stroke by as much as 55%. Revascularization following endarterectomy of symptomatic carotid stenosis has a much higher therapeutic index as compared to asymptomatic occlusive lesions. However, symptomatic patients with arterial stenosis between 50 and 69% tend to have modest benefits. A careful selection of patients can significantly lower the risk of neurological and cardiac morbidity and mortality associated with this procedure. The complications associated with endarterectomy include stroke, cerebral hyperperfusion syndrome, bleeding, and infection.

The search for effective neuroprotective agents is mostly unsuccessful. Neuroprotective agents are likely to stabilize the threatened tissues and salvage vulnerable brain tissues before the thrombolytic agents are administered. Magnesium sulfate and nitroglycerin are a few of the neuroprotective agents studied. Researchers have been exploring the potential beneficial effects of hypothermia in combination with thrombolysis. In addition, they are also exploring the beneficial effects of standard anticoagulants and antiplatelets as adjuncts to thrombolytic drugs. The use of ultrasound to increase the transport of r-TPA) into the thrombus and binding of r-TPA to fibrin has also been explored.

Unlike brain infarction, a large hemorrhagic stroke or subarachnoid hemorrhage always adds extra volume into the rigid and closed cranial cavity. Such extra volume often increases the pressure inside the cranium. Herniations into other dural compartments, midline shifts in the intracranial contents, and the generalized rise in the intracranial pressure are the common causes of death in such strokes. Treatment should include strategies to control and prevent elevation of intracranial pressure. Subarachnoid hemorrhage patients almost always have increased intracranial pressure due to the expanding intracranial volume. Increased volume, either in the form of blood or edema surrounding the hematoma, can obstruct the CSF flow and drainage or reduce the cerebral perfusion pressure, further escalating the brain parenchymal damage. Serial neuroimaging studies have reported that hematomas tend to expand during the first few hours (3–6 h) after the onset of the symptoms, whereas edema around the hematoma tends to develop around the first 48 hours. Usually, larger hematomas have more surrounding edema. Hemoglobin products and thrombin are a few factors that may promote edema formation. The aim of therapy should be to limit the size of the hemorrhage by limiting the bleeding or draining the hematoma and treating the accompanying edema. In case of hemorrhage caused by arteriovenous malformation or aneurysm, removing the offending vascular lesion prevents recurrent hemorrhage. Another important strategy to stop the bleeding is to reduce arterial tension.

Endovascular coiling, surgical clipping, or bypass or flow diversion are the frequent procedures performed for intracranial aneurysms. Endovascular coiling is a minimally invasive fluoroscopy-aided procedure, where the platinum coils are released into the aneurysm through a catheter. These coils induce embolization of the aneurysm and seal off the opening of the aneurysm, thus preventing blood from getting into the aneurysmal sac. The number of coils required depends on the size of the aneurysm and are left permanently inside the aneurysm. Endovascular coiling procedures are generally indicated for unruptured brain aneurysms than ruptured aneurysms, as well as for older patients and for those patients for whom surgery is contraindicated.

Clipping is a surgical procedure to treat an aneurysm. Following a craniotomy, the goal is to isolate and clip the aneurysm from the rest of the vasculature without blocking any of the adjoining large or small arteries nearby. The clip, made of titanium, is placed permanently across the base or neck of the aneurysm to prevent the blood from filling the aneurysm. Clips are made in a variety of shapes, sizes, and lengths due to variations in size and shape of the neck of aneurysm (part of aneurysm near the origin on the main artery) among patients. A ruptured aneurysm is life-threatening, and the risk of repeated bleeding is high during the next few weeks and is an indication of surgical clipping within 72 hours of the first bleed. Vasospasm, stroke, seizure, bleeding, and an imperfectly placed clip are a few of the complications which could arise from aneurysm clipping. Size, location, and neck geometry of the aneurysm, age of the patient, general health, and comorbidities are certain essential factors that decide the various treatments like surgical clipping or bypass or endovascular coiling or flow diversion.

3.11 Physiotherapy Management of Supratentorial Stroke

A coordinated interdisciplinary team consisting of neurologist, nurse, physiotherapist, speech therapist, occupational therapist, dietician, clinical psychologist, and medical social worker should oversee the comprehensive plan of care to address various issues of stroke patients and the concerned family members. To provide a supportive environment and assist patients and their family members, communication between the interdisciplinary team members is critical. Among the health professions, physiotherapy plays an important role in promoting functional abilities, reducing disabilities, preventing or minimizing the complications of stroke, promoting independence, and achieving a better quality of life.

Patients who sustained cerebrovascular accidents may have several impairments. These impairments can interfere with their functional capabilities, and the extent and severity of impairments will depend on the anatomical location of the lesion and the mechanism of stroke. A detailed assessment of the patient covering the medical history, course of the disease, impairments, activity limitations, and participation restrictions helps to identify the priorities and concerns of the patient, his or her abilities, and the resources available. The intervention should include restorative strategies to reduce impairments, activity limitations and participation restrictions. Preventive strategies will minimize the complications and the secondary sequels of the disease, and compensatory strategies will improve function by modification of the task or environment. For patients with more disabling sensorimotor deficits, which is common among moderate-to-severe strokes, long-term planning is essential as they are less likely to be resolved during the acute or subacute phase management in a hospital-based setup (acute care or inpatient rehabilitation).

The acute phase therapeutic intervention should begin as soon as the patient is medically and hemodynamically stable, which is generally 48–72 hours after the onset of symptoms. During this phase, rehabilitation consists of low-intensity exercises. By reviewing the medical record and communicating with the medical team, the therapist should regularly update himself about the neurological and medical status of the patient. Early mobilization during this phase prevents or reduces the ill effects of bed rest and deconditioning, may improve the level of consciousness of the patient, and speed up the return to independence. Encouraging the use of the paretic side and facilitation or stimulation of the same side are likely to promote functional reorganization, reduce the possibility of learned non-use of the paretic extremities, and minimize the potential for maladaptive patterns of movement. Interventions should not only include early mobilization, positioning, bed mobility, transfer techniques, ROM exercises, splinting, and encouraging the ADL but also instructing, educating, and training the patient and family members.

The acute phase is a quite stressful time for the patient and the family members, and therefore the information and instructions given to the patient and family members need to be sorted. Appropriate, but effective, communication is crucial during this stage. Providing a less distractible environment may encourage the patient to draw his or her attention for effective communication. The location from which the therapist interacts with the patient should also consider visual field defects like homonymous hemianopia and quadrantanopia or perceptual disorders like spatial neglect if any. In this phase, the physiotherapist should be vigilant about the potential risk of medical emergencies and complications, including uncontrolled or accelerated hypertension, cardiac arrhythmias, DVT, and aspiration.

During the subacute phase of rehabilitation, patients usually require more intense intervention, particularly when they have considerable residual impairments. Generally, an inpatient rehabilitation facility will be more suitable for such patients. The intensity and timing of rehabilitation efforts, medical stability, severity of cognitive and perceptual deficits, and patient's motivation and endurance are certain factors influencing the functional outcomes. Many of the interventions begun during inpatient rehabilitation tend to continue and progress beyond the inpatient rehabilitation facility. Typically, post-discharge, the rehabilitation services are continued and progressed either on an outpatient basis or for subjects with limitations in accessing hospital facilities; the same services are rendered through a community outreach program. The intervention programs focusing on flexibility, strength, balance, locomotion, endurance, and upper limb function have shown meaningful outcomes for those treated in the outpatient facilities.

Historically, motor functional recovery was thought to be complete within 3 to 6 months post-stroke. However, recent evidence proves that functional recovery after a stroke can continue for months or years. The acute phase functional gains following stroke is generally attributed to improved perfusion or recanalization of the ischemic or penumbral zone, reduction of cerebral edema, and absorption of damaged tissue. However, improvement during the long-term functional recovery is ascribed to neuroplastic mechanisms like collateral sprouting, synaptogenesis, and unmasking of neural pathways, believed to circumvent rather than repair the damaged brain tissue.

Regarding training programs for stroke patients, the specificity of training and increased intensity of training are important factors in the process of learning. Practice sessions beyond the therapy session are recommended for better learning. It is also important to create an enriched environment that supports learning and provides the typical challenges of everyday life. The enriched environment tends to improve the activity level and enhance recovery. The strategies include computers with Internet access, reading material, puzzles, board games, tablets, access to music and books, encouraging family members to bring in hobbies and activities, and access to communal areas. In cases of severe sensorimotor deficits with a limited scope of recovery or stroke with multiple comorbidities, compensatory training strategies should be encouraged to resume functions using the less involved extremities and alternate movement patterns.

During the chronic phase, 6 months following the onset of symptoms, the patient and family are instructed about the home exercise program and are educated about the importance of maintaining physical fitness, health promotion, lifestyle modification, modification of risk factors, fall prevention, and safety. For those patients who cannot avail outpatient rehabilitation facilities due to severe disability or immobility or being bed-confined, home care rehabilitation services can provide the intervention strategies. Finally, the patient's recovery and eventual outcome not only depends on the location of the vascular lesion, mechanism of stroke, and time and type of medical or surgical intervention but also on the presence of preexisting medical conditions, appropriate rehabilitation strategies, and amount of family support and financial resources available.

3.11.1 Clinical Examination

A detailed clinical examination helps to identify the impairments pertinent to the stroke syndrome, identify the stroke patients that require extensive rehabilitation intervention programs, and develop an appropriate plan of care, which includes the anticipated goals, interventions, expected outcomes, and prognosis. In addition to the above, the clinical examination also helps to monitor the recovery and treatment progress toward projected goals and expected outcomes, prescription of appropriate walking aids, splints, and adaptive equipment and develop a home-based exercise program at the time of discharge.

Lesions in the frontal and temporal lobes of the dominant hemisphere can lead to specific communication deficits. Around 30% of stroke patients will have a certain degree of language dysfunction. Different types of aphasia and its features have already been addressed earlier in this chapter. Evaluating and treating patients with receptive and global aphasia can often be challenging as they may fail to comprehend spoken words and gestures. Appropriate speech therapy, along with time and patience, may help to develop some basic communication methods to interact with these patients. Other communication deficits like dysarthria and emotional lability may also affect the patient's ability to interact with individuals.

The clinical examination should include tests and measures to quantitatively or qualitatively assess the impairments, functional performance, and activity limitations. Subjective information obtained by interviewing the patient or family members and from the medical records should cover details such as general demographic data, present and past medical and surgical history, socioeconomic status, risk factors, and social and health habits. The presence of blurring of vision, visual field deficits, facial muscle weakness, hemisensory loss of facial sensation, hearing difficulties, swallowing difficulties, and slurring of speech necessitates the examination of cranial nerves. Since, perceptual dysfunctions like visual neglect or inattention can mimic visual field defects, it is essential to rule them out before the test for visual field is executed using the confrontation method. Examination of the third, fourth, and sixth cranial nerves may disclose ocular motility disturbances, including diplopia, strabismus, oscillopsia, and conjugate gaze paralysis.

Many of the stroke patients may have partial impairments, as opposed to a total loss of sensory perception. These sensory impairments may affect the patient's ability to control and coordinate movement. Such patients may also lose the ability to perceive upright postures and may face difficulty in performing normal weight shifts and sequenced motor activities. Deficits in type and extent of somatic sensation impairments among stroke patients depend on the location and size of the vascular lesion. Usually, for cortical lesions, sensory impairments will be specific to a local area, unlike the diffuse involvement of one side of the body which suggests deeper lesions involving the thalamus, internal capsule, or nearby structures. In addition to the above, cortical lesions are characterized by a loss of combined and cortical sensations and relative preservation of the primary modalities of sensations. Crossed sensory loss, i.e., anesthesia over ipsilateral face and contralateral trunk and limb, is characteristic of a brainstem stroke. Profound involvement of the somatosensory system will adversely affect the motor performance, motor learning, and therapeutic outcomes and contribute to hemineglect and learned nonuse of limbs. During the evaluation, it is also essential to look for central post-stroke pain as it can debilitate and limit the participation of the patients during the rehabilitation programs. Lesions affecting the somatosensory cortex or its pathways can result in sensory ataxia with the muscle strength somewhat preserved. Vascular lesions involving the cerebellum typically produce cerebellar ataxia and hypotonicity. The presence of Romberg's sign, absence of incoordination with eyes open, and the presence of incoordination with eyes closed (finger-to-finger or finger-pointing test with eyes closed) distinguish sensory ataxia from cerebellar ataxia. A detailed sensory evaluation may not be practical or possible in many patients when cognition and communication skills are affected.

The degree and the extent of motor recovery depend on several factors, including the anatomical location of the lesion, severity of the lesion, mechanism of stroke, rehabilitation strategies used, and motivation of the patient. The muscles of the involved side, soon after the onset of stroke, due to cerebral shock, will be typically flaccid (flaccid stage-stage 1 of Brunnstrom recovery stages). This early stage is short-lived, and during stage 2 of recovery, the flaccidity will be replaced by progressive development of spasticity, hyperreflexia, and the appearance of basic limb synergies (mass movement patterns). These synergistic movements are generally stereotypical and primitive and can either be elicited reflexively or volitionally. Muscles involved in the basic limb synergies are strongly linked and are relatively easier to recruit during the early and middle stages (stage 2 and stage 3) of recovery. The mass movements produced by the synergistic action of these muscles are highly stereotypic among stroke patients. The spasticity will accentuate, and movements outside the basic synergies will not be possible (stage 3) until the recovery progresses further. In the later stages of recovery (stage 4 and stage 5), the spasticity progressively wanes, and the strong linkage between the muscles of synergies reduces characterized by movement patterns away from the basic synergies. Isolated joint movements (fractionation of movements) and normal synergies with near-normal coordination and speed are the features of stage 6, and the last stage (stage 7) is characterized by the return of normal motor function, including fine motor skills, speed, and coordination. The extent of recovery need not be the same for the upper and lower extremities as the vascular territories for the concerned areas are not the same. Though the pattern of motor recovery is somewhat similar among most of the supratentorial strokes, individual variations are common that include the speed of recovery (from few days to several months) and extent of recovery (full recovery to incomplete recovery). The presence of basic limb synergy, increased tone, hyperreflexia, and Babinski sign warrants the need for noting the Brunnstrom recovery stage for the upper and lower extremities of the involved side. For further details about the basic limb synergies and stages of recovery, refer to Brunnstrom's approach under the chapter titled "Therapeutic Approaches." In addition to identifying the motor recovery stage of the stroke patient, the ability of the patient to voluntarily move those muscles which are not associated with basic limb synergy needs to be evaluated, including latissimus dorsi, serratus anterior, finger extensors, and ankle evertors. Evaluation of fine motor and dexterity skills (writing, dressing, and feeding) is a part of the examination when the stage of motor recovery is above 5.

Evaluation of the patient's deep tendon reflexes will provide valuable information about the presence of abnormal muscle tone. The reflexes are hypoactive in hypotonia and absent in flaccidity. The presence of hyperreflexia, one of the features of upper motor neuron lesion, generally tends to emerge when the cerebral shock dissipates soon after the onset of stroke. A percussion or knee hammer can help the therapist to grade the hyperactivity of the deep tendon reflexes, and while examining the tendon reflex, it is wise to know whether there is any presence of sustained or non-sustained clonus. The most common sites for eliciting clonus are the ankle, patellar, and wrist. The grades and descriptions of deep tendon reflexes are mentioned in Box 3.4. The clinical examination may reveal the presence of released postural reflexes like tonic neck reflex and tonic labyrinthine reflex and associated reactions. Associated reactions are readily elicitable when hyperreflexia, spasticity, and mass synergies are present.

Box 3.4 The Grades and Descriptions of Deep Tendon Reflex

Deep tendon reflex grades
0: No response or not elicitable 1+: A diminished or slight response or a response brought out only with reinforcement 2+: A normal response; same as the contralateral unaffected side 3+: A brisk response; more than seen on the contralateral unaffected side 4+: A hyperactive, exaggerated or very brisk response; presence of clonus
It is preferable to document the clonus site (ankle, patellar, or wrist) and the type of clonus (sustained and non-sustained) separately. More than 5 beats is considered as a sustained clonus.

Spasticity, a feature of upper motor neuron lesion, tends to occur predominantly in the antigravity muscles like scapular retractors and depressors, shoulder adductors and internal rotators, elbow flexors, forearm pronators, wrist and finger flexors, pelvic retractors, hip adductors, extensors, and internal rotators, knee extensors, plantar flexors, and invertors. Abnormal processing of the sensory (afferent) input reaching the spinal cord level and defect in inhibitory modulation from higher cortical centers and spinal interneuron pathways are the possible mechanisms for spasticity in many of these patients. In addition to the hindrance created to voluntary movements, the spastic muscles can cause typical posturing of the affected limb(s), painful spasms, degenerative changes, and fixed contractures. Lack of phasic activity of these spastic muscles also causes impairment in the automatic postural adjustments required for normal transitions and mobility, translating to balance impairments and increased risk for falls. Therefore, bedside examination of tone is essential to determine the presence of flaccidity, hypotonicity, or spasticity. If clinical examination reveals the presence of spasticity, the severity of spasticity can be graded based on the resistance offered to the passive stretch. In the acute stage, the severity of the spasticity is graded by an ordinal scale named the modified Ashworth scale (Table 3.10). In subacute and chronic stages, to differentiate spasticity from soft tissue changes (possible tightness and contracture,) the modified Tardieu scale is the most appropriate.

The use of a standard or electronic goniometer can provide the passive ROM reading of joints as flexibility can be affected by tonal abnormalities of the muscles and soft tissue changes. Testing the active ROM may be practically less important, especially when the spasticity is profound, and the influence of basic limb synergies on isolated voluntary movements is considerable. Regarding the starting position opted, the released postural reflexes like tonic neck and tonic labyrinthine reflexes may cause inconsistent active ROM measurement findings. Contractures can develop as a result of poorly managed or unmanaged spasticity, and further when it

Table 3.10 Grades of modified Ashworth scale

Grade	Features
0	No increase in muscle tone
1	Slight increase in muscle tone, manifested by a catch and release or by minimal resistance at the end of the range of motion when the affected part(s) is moved in flexion or extension
1+	Slight increase in muscle tone, manifested by a catch, followed by minimal resistance throughout the remainder (less than half) of the ROM
2	More marked increase in muscle tone through most of the ROM but affected part(s) easily moved
3	Considerable increase in muscle tone, passive movement difficult
4	Affected part(s) rigid in flexion or extension

Courtesy Professor Richard Bohannon

progresses, the edema and pain can worsen the mobility of the joint(s). For subacute and chronic stroke patients, the evaluation should include examination for tightness and contractures, especially the shoulder adductors, elbow flexors, wrist and finger flexors, forearm pronators, hip adductors, knee flexors, and plantar flexors.

Patients may exhibit difficulty in performing purposeful movements in the absence of sensory impairments or motor weakness. Motor praxis, the ability to plan and execute coordinated movement, is often impaired or absent in premotor frontal cortex lesions of right or left hemispheres, corpus callosal lesions, and injuries of the left inferior parietal lobe. Generally, apraxia is more prevalent in the left hemisphere lesion as compared to the right hemisphere lesion. Apraxia can be evident when the stroke patient attempts to perform self-care activities, and these patients may not remember how to hold and use a comb or a toothbrush or don a piece of cloth. Some of these patients may perform the functional movements spontaneously but may fail to perform on command or remember the steps or sequence of movements necessary to achieve the goal. Types, locations, and features of apraxia are listed in Table 3.11.

A considerable number of stroke patients will have paresis of the affected side, which is a major factor for motor function impairment, activity limitation, and participation restriction. The force necessary for initiating and controlling movement is affected, and increased effort and fatigability are the frequent complaints of those patients demonstrating weakness. The degree of muscular weakness is related to the anatomical location of the lesion and the severity of the lesion. It can range from total inability to produce any contraction, as seen in hemiplegia, to a reduction in-force production, as seen in hemiparesis. Most stroke patients will not be able to generate normal levels of muscular force, tension, or torque necessary to initiate and control functional movements and posture. Maintenance of a constant level of force production to control extremity movements can also be difficult. Generally, weakness is more profound in the distal aspect compared to proximal. The normal phasic activity of the proximal and distal muscles is replaced by tonic activity. Changes can also be noticed in the muscle fibers characterized by selective loss of Type II muscle fibers associated with a percentage-wise increase in the number of Type I muscle

Table 3.11 Details regarding types, locations, and features of apraxia

Type	Location	Features
Ideomotor apraxia	Damage to parietal association areas; lesions of intrahemispheric white matter fibers interconnecting parietal and frontal areas; less frequently, lesions of the premotor and prefrontal cortices and supplementary motor area	On verbal instruction, unable to perform complex commands like a salute, snap the fingers, and wave goodbye usually with the involved extremity; unable to pantomime how to use common objects like hammer, toothbrush, and comb or how to kick or throw a ball; may substitute a hand or finger for the imagined object; unable to carry out the act on command but may imitate it
Ideational apraxia	Damage to the left posterior temporoparietal junction	Able to carry out individual components of a complex motor task but unable to perform the entire sequence properly; while attempting to sequence, the patient omits steps or gets the steps out of order; inability to plan the series of steps; for instance, trying to put the car in drive before starting the engine or sealing an envelope before inserting the letter
Buccofacial apraxia	Usually, affecting the infero-anterior portion of the left supramarginal gyrus or the infero-posterior portion of the left posterior central gyrus	On request, unable to execute complex motor acts involving the lips, mouth, and face; unable to execute activities like whistling, pursing the lips, sticking out the tongue, pretending to blow out a match stick, or demonstrate a flying kiss
Constructional apraxia	Most commonly associated with lesions in the parietal-occipital lobes	Impaired visuospatial skills causing inability to copy geometric forms of any complexity; able to draw individual shapes with inability to draw more complex geometric figure: for instance, able to draw a square but not a cube; inability to comprehend spatial relationships; often associated with hemineglect
Dressing apraxia	Usually right parietal lobe lesions	Inability to don or doff clothing correctly; inability to manipulate the clothing in space and understand its 3D relationships; often associated with hemineglect
Limb kinetic apraxia	Mild lesions involving the corticospinal tract that are not severe enough to cause detectable weakness	Difficulty with fine motor control in the absence of obvious weakness; characterized by difficulty to type on the computer or tying shoelaces; not a true apraxia due to damage of the primary motor pathway
Gait or Bruns apraxia	Bilateral frontal lobe disorders	Inability to initiate the process of walking, despite the power and coordination of the legs being normal; characterized by broad-based gait, short stride, "magnetic" or shuffling type gait, freezing, falls or tendency to fall and en bloc turns

Table 3.11 (continued)

Type	Location	Features
Apraxia of speech	Left premotor and supplementary motor areas	Patients have forgotten how to make the sounds of speech; no weakness of the vocal tract; prosody may be impaired and speech may be stuttering; speech pattern may change and sounds like a foreign accent; may be able to repeat short common words but fails to repeat polysyllabic words: for instance, if asked to repeat "potatoes," he or she may repeat it as "tapotos or posatos"; grammar and syntax intact

fibers. In addition to the above, a reduced number of functional motor units, abnormal recruitment of motor units, and decreased firing rates account for lesser force production and difficulty in initiating rapid voluntary movements. Though muscle strength evaluation is an integral component of neurological examination, the presence of spasticity, hyperreflexia, released postural reflexes, and dominant basic limb synergies, and an inability to produce isolated joint movement will pose problems in executing the traditional manual muscle test (MMT) for many of the stroke patients. Till the motor recovery stage 6, when the isolation of joints is not possible, it is preferable to use functional strength testing instead of MMT. Activities like the 30 sec sit-to-stand test; Timed Up and Go (TUG) test; observing the contribution of paretic upper extremity while bridging, hitching, or hiking the pelvis; and swinging of the paretic upper limb while walking are ways of assessing the functional strength of the paretic upper limb and lower limb muscles. Beyond the motor recovery stage 5, when isolations are possible, traditional MMT and objective tools like the hand-held dynamometer and the computerized isokinetic dynamometer are more appropriate for grading or quantitatively measuring the muscle strength.

Postural impairments in alignment, symmetry, and stability and impairments in dynamic balance control are common issues seen in stroke patients. Poor reactive postural control to destabilizing external forces and reduced proactive or anticipatory postural control during self-initiated movements predispose these patients to lose balance. During changing tasks and environmental demands, disruptions in the central sensorimotor processing further contribute to ineffective recruitment of postural strategies and inappropriate postural movements. Errors in timing and sequencing of muscle activity, abnormal coactivation of muscles, and delays in initiating motor activity will result in disorganized postural reactions. Stroke patients will typically demonstrate uneven weight-bearing, increased postural sway, and a tendency to fall toward the affected side. The routine examination should cover both static and dynamic balance control in sitting and standing. The usual postural deviations seen among stroke patients are presented in Table 3.12. Asking the patient to reach while sitting or standing will provide information about the limits of stability.

Table 3.12 Presenting the usual postural deviations among stroke patients

Part of the body	Postural deviations
Head and neck	Forward head posturing; lateral flexion of the head with rotation away from the affected side
Trunk	Flattened lumbar curve with an exaggerated thoracic curve (round back posture); lateral flexion with trunk shortening on the affected side
Pectoral girdle	Shoulders are of unequal height and affected shoulder depressed; if the upper extremity is more involved, presence of humeral subluxation with scapula facing downwards and adducted; presence of scapular winging
Upper limb	Affected side shoulder held in flexion, adduction, and internal rotation, elbow in flexion, forearm in pronation, and wrist and fingers in flexion (typical attitude of the hemiplegic upper limb)
Pelvic girdle	Asymmetrical weight-bearing; majority of the weight borne by the unaffected side; fear or reluctance to shift body weight toward the affected side; sacral sitting with posterior pelvic tilt; affected side pelvis retracted and elevated during standing
Lower limb	While sitting, the affected side hip has a tendency to be held in abduction and external rotation with knee in flexion; while standing, the affected side hip is held in extension, adduction, and internal rotation, with knee in extension, and ankle and toes in plantar flexion; unequal weight-bearing on feet and more weight borne by the unaffected side

The therapist should also note the alignment of body segments and trunk control while performing the weight shifts. Berg Balance Scale (BBS), Brunel Balance Assessment (BBA), Tinetti Performance-Oriented Mobility Assessment (Tinetti-POMA), Function in Sitting Test (FIST), Functional Reach Test (FRT), Lateral Reach Test (LRT), and TUG test are some of the qualitative or quantitative scales available to measure the static and dynamic balance of stroke patients. Trunk Control Test (TCT), Trunk Impairment Scale (TIS), and Postural Assessment Scale for Stroke (PASS) are tests or scales developed to examine the postural abilities of stroke patients. The components, grading or scoring methods, procedure, and psychometric properties for these tools are not within the scope of this book. Tables 3.13 and 3.14 present the details of the BBS and Tinetti-POMA scale, respectively.

Pusher syndrome, also known as lateropulsion or contraversive pushing, common after right hemisphere damage, due to altered perception of the body's orientation with respect to gravity, is not uncommon among stroke patients. Typically, the patients with pusher behavior use the stronger extremities of the non-paretic side to push themselves toward their hemiparetic side. These patients have a postural preference toward the affected side, and any attempt to transfer weight over to the non-paretic side or return the body to a neutral or midline position will be strongly resisted. Depending on the severity, the syndrome can substantially hamper maintenance of lying, sitting, or standing postures and transitions. While standing, the strong pushing behavior creates significant postural instability, and these patients demonstrate no fear when active pushing is inducing instability. During ambulation, in addition to the lateropulsion, the affected side lower extremity may scissor, and the cane or walker may pose added trouble than support. Functional skills are

Table 3.13 Berg Balance Scale

Name: ________________________________ Date: __________________
Location: ______________________________ Rater: __________________
Item Description Score (0–4)
1. Sitting to standing ________
2. Standing unsupported ________
3. Sitting unsupported ________
4. Standing to sitting ________
5. Transfers ________
6. Standing with eyes closed ________
7. Standing with feet together ________
8. Reaching forward with outstretched arm ________
9. Retrieving object from floor ________
10.Turning to look behind ________
11.Turning 360 degrees ________
12.Placing alternate foot on stool ________
13.Standing with one foot in front ________
14.Standing on one foot ________
Total ________

General Instructions

Please document each task and/or give instructions as written. When scoring, please record the lowest response category that applies for each item. In most items, the subject is asked to maintain a given position for a specific time. Progressively, more points are deducted if:

- The time or distance requirements are not met.
- The subject's performance warrants supervision.
- The subject touches an external support or receives assistance from the examiner.

Subject should understand that they must maintain their balance while attempting the tasks. The choices of which leg to stand on or how far to reach are left to the subject. Poor judgment will adversely influence the performance and the scoring. Equipment required for testing is a stopwatch or watch with a second hand, and a ruler or other indicator of 2, 5, and 10 inches. Chairs used during testing should be a reasonable height. Either a step or a stool of average step height may be used for item # 12.

(continued)

Table 3.13 (continued)

1. Sitting to standing Instructions: Please stand up. Try not to use your hand for support. ( ) 4 Able to stand without using hands and stabilize independently ( ) 3 Able to stand independently using hands ( ) 2 Able to stand using hands after several tries ( ) 1 Needs minimal aid to stand or stabilize ( ) 0 Needs moderate or maximal assist to stand **2. Standing unsupported** Instructions: Please stand for two minutes without holding on. ( ) 4 Able to stand safely for 2 minutes ( ) 3 Able to stand 2 minutes with supervision ( ) 2 Able to stand 30 seconds unsupported ( ) 1 Needs several tries to stand 30 seconds unsupported ( ) 0 Unable to stand 30 seconds unsupported If a subject is able to stand 2 minutes unsupported, score full points for sitting unsupported. Proceed to item #4 **3. Sitting with back unsupported but feet supported on floor or on a stool** Instructions: Please sit with arms folded for 2 minutes. ( ) 4 Able to sit safely and securely for 2 minutes ( ) 3 Able to sit 2 minutes under supervision ( ) 2 Able to able to sit 30 seconds ( ) 1 Able to sit 10 seconds ( ) 0 Unable to sit without support 10 seconds **4. Standing to sitting** Instructions: Please sit down. ( ) 4 Sits safely with minimal use of hands ( ) 3 Controls descent by using hands ( ) 2 Uses back of legs against chair to control descent ( ) 1 Sits independently but has uncontrolled descent ( ) 0 Needs assistance to sit	( ) 4 Can reach forward confidently 25 cm (10 inches) ( ) 3 can reach forward 12 cm (5 inches) ( ) 2 can reach forward 5 cm (2 inches) ( ) 1 reaches forward but needs supervision ( ) 0 loses balance while trying/requires external support **9. Pick Up Object from the Floor from a standing position** Instructions: Pick up the shoe/slipper, which is placed in front of your feet. ( ) 4 Able to pick up slipper safely and easily ( ) 3 Able to pick up slipper but needs supervision ( ) 2 Unable to pick up but reaches 2–5 cm(1–2 inches) From slipper and keeps balance independently ( ) 1 Unable to pick up and needs supervision while trying ( ) 0 Unable to try/needs assist to keep from losing Balance or falling **10. Turning to look behind over left and right shoulders while standing** Instructions: Turn to look directly behind you over toward the left shoulder. Repeat to the right. Examiner may pick an object to look at directly behind the subject to encourage a better twist turn. ( ) 4 Looks behind from both sides and weight shifts well ( ) 3 Looks behind one side only other side shows less weight shift ( ) 2 Turns sideways only but maintains balance ( ) 1 Needs supervision when turning ( ) 0 Needs assist to keep from losing balance or falling **11. Turn 360 degrees** Instructions: Turn completely around in a full circle. Pause. Then turn a full circle in the other direction. ( ) 4 Able to turn 360 degrees safely in 4 seconds or less ( ) 3 Able to turn 360 degrees safely one side only 4 seconds or less ( ) 2 Able to turn 360 degrees safely but slowly ( ) 1 Needs close supervision or verbal cuing ( ) 0 Needs assistance while turning

5. Transfers Instructions: Arrange chair(s) for pivot transfer. Ask subject to transfer one way toward a seat with armrests and one way toward a seat without armrests. You may use two chairs (one with and one without armrests) or a bed and a chair. ( ) 4 Able to transfer safely with minor use of hands ( ) 3 Able to transfer safely definite need of hands ( ) 2 Able to transfer with verbal cuing and/or supervision ( ) 1 Needs one person to assist ( ) 0 Needs two people to assist or supervise to be safe **6. Standing unsupported with eyes closed** Instructions: Please close your eyes and stand still for 10 seconds. ( ) 4 Able to stand 10 seconds safely ( ) 3 Able to stand 10 seconds with supervision ( ) 2 Able to stand 3 seconds ( ) 1 Unable to keep eyes closed 3 seconds but stays safely ( ) 0 Needs help to keep from falling **7. Standing unsupported with feet together** Instructions: Place your feet together and stand without holding on. ( ) 4 able to place feet together independently and stand 1 minute safely ( ) 3 Able to place feet together independently and stand 1 minute with supervision ( ) 2 Able to place feet together independently but unable to hold for 30 seconds ( ) 1 Needs help to attain position but able to stand 15 seconds feet together ( ) 0 Needs help to attain position and unable to hold for 15 seconds **8. Reaching forward with outstretched arm while standing** Instructions: Lift arm to 90 degrees. Stretch out your fingers and reach forward as far as you can. (Examiner places a ruler at the end of fingertips when arm is at 90 degrees. Fingers should not touch the ruler while reaching forward. The recorded measure is the distance forward that the fingers reach, while the subject is in the most forward lean position. When possible, ask subject to use both arms when reaching to avoid rotation of the trunk.)	**12. Place alternate foot on step or stool while standing unsupported** Instructions: Place each foot alternately on the step/stool. Continue until each foot has touched the step/stool four times. ( ) 4 Able to stand independently and safely and complete 8 steps in 20 seconds ( ) 3 Able to stand independently and complete 8 steps in > 20 seconds ( ) 2 Able to complete 4 steps without aid with supervision ( ) 1 Able to complete >2 steps needs minimal assistance ( ) 0 Needs assistance to keep from falling/unable to try **13. Standing unsupported one foot in front** Instructions: (Demonstrate to subject.) Place one foot directly in front of the other. If you feel that you cannot place your foot directly in front, try to step far enough ahead that the heel of your forward foot is ahead of the toes of the other foot. (To score 3 points, the length of the step should exceed the length of the other foot and the width of the stance should approximate the subject's normal stride width.) ( ) 4 Able to place foot tandem independently and hold 30 seconds ( ) 3 Able to place foot ahead independently and hold 30 seconds ( ) 2 Able to take small step independently and hold 30 seconds ( ) 1 Needs help to step but can hold 15 seconds ( ) 0 Loses balance while stepping or standing **14. Standing on one leg** Instructions: Stand on one leg as long as you can without holding on. ( ) 4 Able to lift leg independently and hold >10 seconds ( ) 3 Able to lift leg independently and hold 5–10 seconds ( ) 2 Able to lift leg independently and hold ≥3 seconds ( ) 1 Tries to lift leg unable to hold 3 seconds but remains standing independently ( ) 0 Unable to try of needs assist to prevent fall
( ) **Total Score** (maximum = 56)	

With permission from Professor Katherine Berg.

Table 3.14 Tinetti performance-oriented mobility assessment scale

Tinetti Performance-Oriented Mobility Assessment (POMA) – Balance Tests Initial instructions: Subject is seated in hard, armless chair. The following maneuvers are tested		
1.	**Sitting balance**	Leans or slides in chair =0 Steady, safe =1 _____
2.	**Arises**	Unable without help =0 Able, uses arms to help =1 Able without using arms =2 _____
3.	**Attempts to arise**	Unable without help =0 Able, requires >1 attempt =1 Able to rise, 1 attempt =2 _____
4.	**Immediate standing balance** (first 5 seconds)	Unsteady (swaggers, moves feet, trunk sway) =0 Steady but uses walker or other support =1 Steady without walker or other support =2 _____
5.	**Standing balance**	Unsteady =0 Steady but wide stance(medial heals >4 inches apart) and uses cane or other support =1 Narrow stance without support =2 _____
6.	**Nudged** (subject at maximum position with feet as close together as possible, examiner pushes lightly on subject's sternum with palm of hand 3 times)	Begins to fall =0 Staggers, grabs, catches self =1 Steady =2 _____
7.	**Eyes closed** (at maximum position of item 6)	Unsteady =0 Steady =1 _____
8.	**Turing 360 degrees**	Discontinuous steps =0 Continuous steps =1 _____ Unsteady (grabs, staggers) =0 Steady =1 _____
9.	**Sitting down**	Unsafe (misjudged distance, falls into chair) =0 Uses arms or not a smooth motion =1 Safe, smooth motion =2 _____
Balance score:		**_____/16**
Tinetti performance-oriented mobility assessment (POMA) – gait tests Initial instructions: Subject stands with examiner, walks down hallway or across room, first at "usual" pace, then back at "rapid, but safe" pace (using usual walking aids)		
10.	**Initiation of gait** (immediately after told to "go")	Any hesitancy or multiple attempts to start =0 No hesitancy =1 _____
11.	**Step length and height**	Right swing foot Does not pass left stance foot with step =0 Passes left stance foot =1 _____ Right foot does not clear floor completely with step =0 Right foot completely clears floor =1 _____ Left swing foot Does not pass right stance foot with step =0 Passes right stance foot =1 _____ Left foot does not clear floor completely with step =0 Left foot completely clears floor =1 _____

Table 3.14 (continued)

Tinetti Performance-Oriented Mobility Assessment (POMA) – Balance Tests Initial instructions: Subject is seated in hard, armless chair. The following maneuvers are tested		
12.	**Step symmetry**	Right and left step length not equal (estimate) =0 Right and left step length appear equal =1 _____
13.	**Step continuity**	Stopping or discontinuity between steps =0 Steps appear continuous =1 _____
14.	**Path** (estimated in relation to floor tiles, 12-inch diameter; observe excursion of 1 foot over about 10 ft. of the course)	Marked deviation =0 Mild/moderate deviation or uses walking aid =1 Straight without walking aid =2 _____
15.	**Trunk**	Marked sway or uses walking aid =0 No sway but flexion of knees or back or spreads arms out while walking =1 No sway, no flexion, no use of arms, and no use of walking aid =2 _____
16.	**Walking stance**	Heels apart =0 Heels almost touching while walking =1 _____
Gait score = **_____/12**		
Total score (gait + balance) =		**_____/28**

significantly impaired, and patients are vulnerable to repeated falls toward the affected side. The disorder typically is associated with unilateral lesions of the left or the right posterolateral thalamus. Early identification of lateropulsion helps the physiotherapists to plan and guide the therapeutic strategies for managing this unique motor dysfunction. Though the clinical assessment scales, Burke Lateropulsion Scale (BLS) and Scale for Contraversive Pushing (SCP), are few of the outcome measures available for the Pusher syndrome, currently there are no gold standard, reliable, and valid outcome measures for clinicians and researchers to consistently quantify the degree and presence of contraversive pushing.

Though the majority of stroke patients learn to walk independently by 6 months post-stroke, gait abnormalities tend to persist throughout the chronic stages of the disease. Primary muscle weakness, disinhibited and hyperexcitable brainstem descending pathways (reticulospinal and vestibulospinal tracts), and influence of spastic synergistic activation organized into fewer motor synergies are the plausible reasons for the impaired postural stability and mobility. Gait impairments seen among stroke survivors are the mechanical consequences of interactions of weakness, spasticity, and spastic synergistic activations of the trunk, pelvis, and lower limb muscles. During the stance phase of gait, the hip and knee extension is generated by the synergistic activation of the spastic muscles, which prevents the flexion of the hip and knee for foot clearance. To overcome these impairments, during the swing phase, the patients will typically hike the pelvis and circumduct the affected lower limb (circumduction gait) for foot clearance. Depending on the severity of weakness and spasticity and the extent of spastic synergistic activations, a wide spectrum of gait impairments can be clinically observed.

In routine clinical practice, to identify the gait deviations, the observational analysis should cover the movements occurring at the trunk, pelvis, hip, knee, ankle, and foot during the gait in different planes of motion. Gait assessment should address the stance phase control of the involved lower limb during the initial weight acceptance, midstance, and weight advancement on the involved limb. Similarly, the assessment should address knee and foot control during the swing phase for the toe clearance and position of the foot. The common gait deviations with respect to body part or location and the possible reasons are summarized in Table 3.15. Quantitative analysis of the distance, time, cadence, velocity, and step and stride length and

Table 3.15 Lists of common gait deviations with respect to body parts and the possible reasons

Location	Deviation	Possible reasons
Trunk	Forward trunk leaning during stance	Weak hip extensors/flexion contracture
Pelvis	Retraction	Weak abdominals and increased tone in extensor synergy muscles of lower extremity
	Hiking	Inadequate hip and knee flexion; increased tone in the trunk and lower extremity extensor muscles
Hip	Circumduction	Increased extensor tone, weakness of the hip and knee flexors, increased plantar flexion in the ankle or foot drop
	Inadequate hip flexion during swing phase	Increased tone in hip and knee extensors, weakness or flaccidity of lower limb flexor muscles, poor proprioception
	Trendelenburg limp	Weakness of hip abductors
	Tendency for affected leg scissoring	Spasticity of hip adductors
	Exaggerated hip-knee-ankle flexion during swing phase seen in some stroke patients	Use of strong flexor synergy, spastic hamstrings
Knee	Decreased knee flexion during swing	Increased lower extremity extensor (mainly quadriceps) tone, weak hip flexion
	Excessive flexion during stance	Weak hip and knee extensors, poor proprioception, flexion contracture
	Hyperextension during stance	Severe spasticity in quadriceps, hip retraction, plantar flexion contracture, impaired proprioception; knee extensors weakness and compensatory locking of knee in hyperextension
Ankle	Equinus gait and presence of foot drop	Spasticity or contractures of gastrocnemius and soleus; weakness of dorsiflexors
Foot and toes	Inversion of the subtalar joint	Hyperactive or spastic tibialis anterior and posterior muscles, weak peroneal muscles
	Clawed toes and hammer toes	Toe flexors spasticity; weak toe extensors

kinematic and kinetic analysis of various joints in different planes of motion using various instruments including GAITRite®, 2D or 3D video motion analyzers, optical motion analyzers, electrogoniometers, EMG sensors, and force plates have helped researchers to analyze gait deviations among stroke patients. In addition to the above, several mobility and gait assessment tools are in use for both clinical and research fields including Dynamic Gait Index (DGI), Functional Ambulation Profile (FAP) Community Balance and Mobility Scale (CB & M), Functional Gait Assessment (FGA), Wisconsin Gait Scale (WGS), and Modified Gait Abnormality Rating Scale (GARS-M).

Stroke patients spending more time in bed, diminished consciousness levels, severe neurological deficits, and very elderly patients with delicate and thinner skin are more susceptible to develop pressure ulcers. Spasticity and contractures contributing to increased friction, faulty transfer techniques contributing to sheer force, and bowel or bladder incontinence causing maceration can be causative factors for pressure sores. Studies have shown that poor nutrition and pressure sores can increase the morbidity and mortality of stroke patients. Poor nutrition can retard wound healing due to the reduction of fibroblasts, angiogenesis, collagen synthesis, granulation tissue formation, and tissue remodeling. Periodic use of the Braden scale, a scale to predict bed sores, can prevent or minimize the possibilities of hospitalized stroke patients developing pressure ulcers. Inculcate routine inspection of pressure-prone areas among nurses and caregivers for those patients who are vulnerable to developing pressure sores.

For those stroke survivors with associated cardiovascular disease, a supervised exercise tolerance test is advisable during the early phase of rehabilitation. Standard methods of monitoring the vital signs, electrocardiogram (ECG) monitoring for any ECG changes, use of a pulse oximeter for oxygen saturation, and use of Borg's Rating of Perceived Exertion (RPE) can be the measures to assess the exercise tolerance. The Borg's RPE is depicted in Table 3.16. Standard cycle ergometry, semirecumbent cycle ergometry, and treadmill walking with or without harnesses are certain modes used for testing. The test protocols should be individualized, submaximal, with intermittent rest if required, and gradually progressed in intensity. The presence of dysrhythmias, systolic blood pressure more than 200 mm Hg, diastolic blood pressure more than 120 mm Hg, ST-segment elevation or depression of ≥2 mm, and severe fatigue are considered to be the clinical endpoints of testing. For independent ambulatory stroke patients with or without walking aid, 2- or 6-minute walk test can measure the walking endurance. The therapist should note the total distance covered, the number of rest periods taken, and any symptoms reported while walking or at rest.

Most stroke survivors experience difficulties with self-care activities and mobility tasks, and some may experience the inability to perform activities like feeding, bathing, bed mobility, sitting up, and walking. These functional limitations are the result of sensorimotor dysfunctions caused by the stroke. Many functional outcome measurement tools are used to objectively measure the functional gains and abilities of the patient during the phases of rehabilitation and post-discharge from the hospital. The outcome measures used for functional abilities help to determine the impact

Table 3.16 Borg's rating of perceived exertion scale

Borg RPE Scale®		
Use this scale to tell how strenuous and tiring the work feels to you. The exertion is mainly felt as fatigue in your muscles and as breathlessness or possibly aches. When the exercise is hard it also becomes difficult to talk. It is your own feeling of exertion that is important. Don't underestimate it, but don't overestimate it either. For common exercise, such as cycling, running or walking, 11-15 is a good level. For strength and high-intensity interval training (HIIT), 15-19 is good. If you are sick follow your doctor's advice. Look at the scale and the descriptions and then choose a number. Use whatever numbers you want, even numbers between the descriptions.		
6	**No exertion at all**	No muscle fatigue, breathlessness or difficulty in breathing.
7	**Extremely light**	Very, very light.
8		
9	**Very light**	Like walking slowly for a short while. Very easy to talk.
10		
11	**Light**	Like a light exercise at your own pace.
12	**Moderate**	
13	**Somewhat hard**	Fairly strenuous and breathless. Not so easy to talk.
14		
15	**Hard**	Heavy and strenuous. An upper limit for fitness training, as when running or walking fast.
16		
17	**Very hard**	Very strenuous. You are very tired and breathless. Very difficult to talk.
18		
19	**Extremely hard**	The most strenuous effort you have ever experienced.
20	**Maximal exertion**	Maximal heaviness.
		Borg RPE Scale® Ratings (R) of Perceived (P) Exertion (E). © Gunnar Borg, 1970, 1998, 2017 English

of impairments and activity limitations, decide the short- and long-term goals and objectives of treatment, plan and monitor the effectiveness of treatment, and predict community reintegration and return to work. Functional mobility skills including bed mobility, transfers, locomotion, basic and instrumental ADL, personal care and hygiene, and communications are some of the components that are included in most of the functional scales. The Barthel Index (BI), Functional Independence Measure (FIM), Fugl-Meyer Assessment (FMA), Wolf Motor Function Test (WMFT), Action Research Arm Test (ARAT), Stroke Rehabilitation Assessment of Movement (STREAM), and Stroke Impact Scale (SIS) are some of the common functional assessment tools used in clinical and research areas. The outcome scales are either impairment-based or are activity limitation or participation restriction based. Most of the tools mentioned above have demonstrated excellent reliability, validity, and sensitivity. Readers should note that there is sufficient literature available dealing exclusively with the components, standard procedures, and psychometric properties of such scales which are not within the scope of this book.

3.11.2 Physiotherapy Treatment

Before going into the details of physiotherapy management, the author would like to specify some valid points. A shotgun approach is not an appropriate method to tackle the sensorimotor dysfunctions of the stroke patient. Instead, an eclectic program that is tailor-made to the individual patient's problems is more meaningful. The therapist must have adequate knowledge, both theoretical and practical, about various therapeutic approaches to provide an eclectic treatment program for stroke patients. Just the way each motor control theory explains a certain number of observations and fails to explain other motor behaviors, neurodevelopment-based and non-neurodevelopment-based approaches (discussed in chap. 2) when delivered as a standalone has provided therapeutic benefits but not for all the motor dysfunctions. The fact as mentioned above has been substantiated by current evidence, which states that none of the therapeutic approaches, as a standalone, is superior or more effective in promoting recovery of extremity function or postural control post-stroke, as compared to the other. Since all these approaches have strengths and weaknesses of their own and collectively have the edge over a standalone, an eclectic approach makes it all the more meaningful to tackle most of the stroke patient's sensorimotor dysfunction issues.

The existing gamut of research with regard to the physiotherapeutic management of stroke patients, if carefully inspected, will reveal several shortcomings including selection criteria, inappropriate outcome measures, and biasing at various levels of the study to prove the superiority of the experimental group over the conventional or control group. Deliberately underperforming conventional or routine therapy group in certain instances and claiming benefits produced by an existing clinically or statistically proven intervention while piggyback riding on it are common findings in many works. All the above aspects have to be kept back of one's mind before instinctively selecting or following any technique.

Unlike patients with other neurological disorders, stroke patients belong to a diverse group with variable levels of function. Based on the patient's abilities and requirements, the therapists need to carefully select the strategies that have the greatest chance of successfully remediating existing impairments and promoting functional recovery. Pitting a few of the effective techniques of specific approaches will not convert a standalone approach into an eclectic approach. As discussed earlier, a true eclectic model needs sound knowledge, both theoretical and practical, about both the traditional and recent approaches. The author believes that the knowledge earned from each scientific therapeutic approach must serve as a reservoir from which a clinician can wisely choose the necessary tools to customize the eclectic treatment program to suit the specific needs of a patient. Choosing the best treatment methods to address the patient's sensorimotor issues must be the most rational approach when substantial evidence for the effectiveness of any single approach over the others is unavailable. The choice of interventions should also consider other factors, including the stage or phase of stroke, age of the patient, existing comorbidities, potential discharge plans, and social and financial status.

The therapist, based on an accurate examination, should interpret the clinical findings and utilize educational skills and manual handling techniques to retrain movement and solve the problems. The therapist needs to develop the expertise of recognizing correct and incorrect responses to therapeutic strategies so that positive outcomes are encouraged and unwanted results are avoided. In many stroke patients, with a potential for recovery, functional task-specific training should be the mainstay of treatment to regain control of the functional movements. In the early phase of management, the focus should be to improve the motor control and strength of the trunk and proximal girdle muscles, emphasizing more on the affected side, through specific treatment strategies. With respect to gaining trunk control, the author would like to mention the relevance of a textbook titled *Right in the Middle*, a comprehensive work by an eminent physiotherapist, Patricia M. Davies. According to Davies, gaining control over the hemiplegic arm and leg is dependent upon the patient's ability to control the trunk. The trunk and the girdles act as stable anchorages for the extremity muscles, and the lack of trunk control predisposes to the development of primitive and stereotypic mass patterns. She advised a series of exercises explicitly emphasizing the trunk like "bridging and tentacle" exercises on stable and unstable surfaces to activate the core muscles, specifically the abdominal obliques. For a better understanding of the selective trunk and extremity control, the author advises the readers to refer to the textbook mentioned above.

The aim of physiotherapy management following a stroke is to maximize the return of functional movement and independence and to minimize the possibilities of secondary complications. Emphasizing early functional independence improvement provides an important source of motivation for both the patient and the family members. Though the intervention commences in the acute stage following admission to the hospital, active participation in the relearning of mobility and independence generally takes place during the subacute and the chronic stages post-stroke.

Despite the common patterns, the acute, the subacute, and the chronic stages are rarely distinct, frequently tend to overlap, and do not follow the same time frame for every patient. In the acute stage, the stroke patients' levels of consciousness and neurological deficits can vary, ranging from unconscious to fully conscious, intubated or on oxygen support to room air-breathing, and complete communication breakdown to sound communication and from total paralysis to normal muscle strength and control. Depending on the presentation, during the acute stage, the treatment strategies should ensure normal respiratory function, general mobility, and facilitation of movements of the paretic side, encourage motor relearning, and prevent sequels including DVT, pressure sores, and painful shoulder.

3.11.2.1 Pulmonary Care

Respiratory muscle weakness among acute stroke patients may increase the frequency of chest infections and contribute to weak cough and poor airway clearance. Strategies to improve the strength and endurance of the diaphragm include the application of manual resistance by placing the therapist's hand on the patient's upper abdomen in the semi-recumbent position and then, while inspiring, instructing the patient to lift the weight of the therapist's hand. The therapist can apply a quick stretch to the diaphragm before an active inspiration to facilitate a stronger contraction of the inspiratory muscle. By increasing the manual resistance or changing the patient's starting position, the exercise should be made more challenging to improve the patient's performance. Segmental breathing exercises can also be encouraged to minimize the possibility of atelectasis of the lower lobes of the lungs. Expansion exercises for the lateral lobes should be encouraged by placing the therapist's hands on the patient's lateral lower rib cage, and the patient should be instructed to "breathe in" against the manual pressure. Activities to improve pulmonary function should include the use of incentive spirometers, paper blowing exercise, blow bottle exercise, and inspiratory or expiratory muscle trainers. In addition to the above, general trunk flexibility or mobility exercise may also improve breathing efficiency, especially when there is a presence of lateral chest wall tightness. For those stroke patients who are unarousable following cerebrovascular accidents, routine care should include chest manipulations and physiotherapy to maintain pulmonary hygiene.

3.11.2.2 Bed Positioning

Proper positioning is an important component of physiotherapy during acute stage management, and it is the responsibility of all the members of the interdisciplinary team to ensure that the appropriate position is maintained throughout. Proper positioning minimizes the possibilities of abnormal tonal development, discourages mass movements, stimulates normal motor functioning, improves sensory awareness, enhances oromotor and respiratory functions, and prevents the possibilities of

ROM restrictions, musculoskeletal deformities, and pressure ulcers. The position of the patient should be changed every two hours between lying on the back, the affected side, and the unaffected side. Both the girdles require extra attention and need to be placed in mild protraction, as the rhomboids and gluteus maximus tightness secondary to immobility and tonal abnormality can contribute to retraction of the shoulder and pelvic girdles.

A 1.5-inch thick towel roll spanning from the affected scapula to the pelvis can promote protraction in the supine position. If the thickness of the towel roll is considerably more than the recommended, it can be counterproductive and may encourage postural asymmetry. In supine, keep the shoulder in 10–20° abduction with external rotation, elbow in extension and forearm supinated, and the wrist and fingers in a functional position. Protracted pelvis with mild hip knee flexion (a pillow kept under the knee joint) minimizes the likelihood of extensor synergy dominating the lower limb. The use of an additional pillow for the head and neck in the supine position can promote cervical flexion and worsen forward head posturing and therefore is not recommended.

While lying on the unaffected side, the stroke patient's trunk should be in neutral (neither bent forward nor unduly extended), and with the support of a pillow or two, the affected upper extremity should be in protraction, elbow extended, and forearm in neutral with the wrist and fingers in a neutral or extended position. With the support of a pillow, position the affected lower extremity with the pelvis in protraction, hip and knee in flexion, and ankle in neutral (Fig. 3.25). A similar kind of position is recommended for the affected upper extremity, without pillows, while lying on the affected side. Ensure that the affected shoulder is protracted sufficiently to

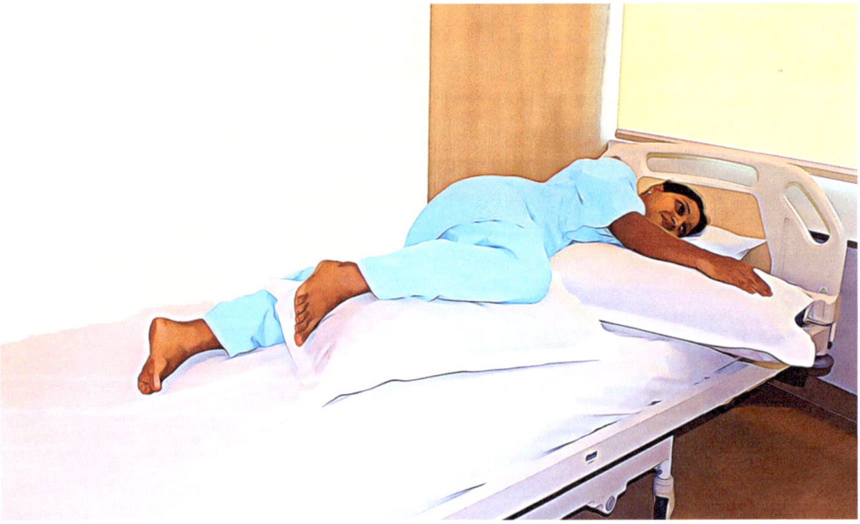

Animated photograph of model with permission

Fig. 3.25 Recommended bed posture in side-lying on the unaffected side. Note: The affected upper extremity is kept in protraction, the elbow in extension, and the forearm in neutral with the wrist and fingers in a neutral or extended position minimize the possibility of impingement and excessive weight falling over the shoulder if directly under. The affected side pelvis should be protracted, with the hip in extension and knee in slight flexion. Since neglect is often a feature of cortical stroke, increased sensory input provided by side-lying position on the affected side reduces the effects of neglect in these stroke patients.

3.11.2.3 Management of Sensory Impairments

Patients with significant sensory impairments may demonstrate impaired or absent spontaneous voluntary movements of the hemiplegic side limbs. Encouraging the patient to use the affected side provides a greater chance of improving the awareness and function of the side limbs. Refusal by the patient to use the affected side accentuates the problems caused by the absence of normal sensorimotor experience and contributes further to the learned nonuse phenomenon. Sensory retraining or sensory stimulation approaches are certain interventions used to address these sensory impairments. Mirror therapy, repetitive functional activity practice augmented by sensory cues, and bimanual movements are examples of sensory retraining approaches. Weight-bearing techniques, manual compression, use of inflatable pressure splints, mobilizations, electrical stimulation, and use of thermal stimulation are examples of sensory stimulation approaches. With the current evidence neither proving nor refuting these techniques, a careful and judicious way of choosing one or other techniques based on the therapist's knowledge and experience may help to improve the sensory impairments in such patients.

The physiotherapist should encourage the stroke patients to weight bear and undergo compression (joint approximation) of the sensory-deficient limbs during the functional training in sitting (Fig. 3.26), standing, or modified plantigrade position. Care has to be taken to keep the joints in proper alignment while performing weight-bearing or joint approximation in various positions, to minimize the possibility of injuries to the affected sensory-deficient limb. Besides, a safety education program has to be instituted for the patient and family members to improve sensory awareness and protection of the affected limbs, mainly while performing transfers and wheelchair mobility.

3.11.2.4 Management of Visual Field Deficit and Unilateral Neglect

Visual field deficits and perceptual impairments can often worsen the poor awareness of the paretic side. Though more prevalent among patients with right hemisphere stroke, hemispatial neglect and anosognosia, two distinguishable and dissociable syndromes which often coexist, can cause total unawareness of the disability or the extent of the problems. Such patients may benefit from strategies that encourage awareness of paretic limbs, use of the environment on the paretic side, and use of the paretic limbs. The use of visual, verbal, and motor cues, promoting active visual scanning movements, encouraging the patient to look at his or her

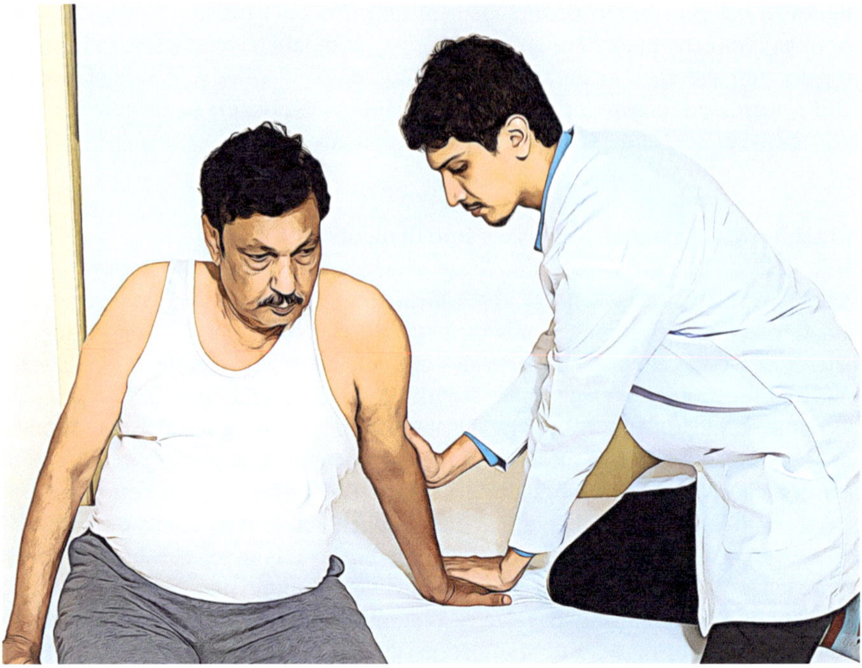

Animated photograph of model and therapist with permission

Fig. 3.26 Weight-bearing through the affected upper limb in sitting

paretic limbs, and exteroceptive and proprioceptive stimulation of the affected limbs by brushing, stroking, muscle tapping, and vibration are strategies to overcome poor awareness of the paretic limb. Reach out activities, Proprioceptive Neuromuscular Facilitation (PNF) diagonal patterns and chopping movements crossing the midline, and functional activities that encourage bilateral interaction of the paretic and non-paretic limbs can also be helpful to overcome the poor awareness of the paretic side. Even encouraging family members to address or interact from the paretic side and placing the commonly used objects like mobile phones, spectacles, napkins, and tissue wipes on the paretic side tends to increase the awareness and attention given to that side of the body.

3.11.2.5 Management of Joint Flexibility and Integrity

About two-thirds of stroke patients develop spasticity and may result in muscle tightness and joint stiffness, eventually affecting the mobility and functional abilities. Spasticity, in addition to muscle tightness and joint stiffness, can cause soft tissue pain and antagonist muscle weakness. Considerable evidence is now available, suggesting the beneficial effects of flexibility exercises among stroke survivors. The flexibility exercise program aims to relieve spasticity, improve ROM, prevent tightness and contractures, and improve motor functions. The flexibility exercise program consists of stretching exercises and joint ROM exercises performed by an external force or self. Often the flexibility exercises are used for warm-up and may be combined with aerobic training and strength training that can help paretic patients improve the ROM. The flexibility exercises are preferably initiated in the early phase of management to maintain joint mobility and integrity. Providing terminal stretch for passive, active, or active-assisted ROMs, without inducing unbearable pain (except the bearable stretch pain), positional strategies to maintain the soft tissue length, and flexibility exercise program performed two to three times daily are a few strategies that can prevent or minimize the potential development of tightness and contracture. The proper positioning of the paretic extremities encourages normal or near-normal joint alignment and minimizes the potential development of typical abnormal attitudes of the limb seen among stroke patients.

Most stroke patients will have impaired scapulohumeral rhythm due to tonal abnormalities of the scapular muscles and inactivity or poor strength of the muscles in the scapular region. Before shoulder ROM exercises, the passive mobilization of the scapula on the thoracic wall emphasizing upward rotation and protraction will help minimize the potential soft tissue impingement in the subacromial space while performing glenohumeral movements. Gentle traction to the shoulder with external rotation further minimizes the possibility of impingement. The use of overhead pulley, a self-ROM exercise, traditionally practiced in many centers, is unwarranted due to the possibility of impingement during overhead activity. Weight-bearing exercise given in sitting, with paretic shoulder kept in partial abduction and externally rotation, elbow in extension, forearm in supination, and wrist and fingers in extension can relieve the spasticity and improve the soft tissue flexibility, thus minimizing the potential for elbow flexors, forearm pronators, and long flexors developing tightness.

Self-performed ROM exercises of the trunk and the girdles, especially with cradling of the affected arm, well elaborated in chapter "Therapeutic Approaches" under Brunnstrom's approach and Bobath approach, further encourages scapular mobility over the thoracic wall, a prerequisite for activities like bed mobility and functional activities in sitting. In case of inadequate scapulohumeral rhythm, overhead shoulder movements with hands clasped with the shoulder in flexion, elbow in extension, and forearm in supination is contraindicated. Promote normal scapulohumeral rhythm (scapular mobilization techniques) before passive ROM can be commenced. Despite the regular flexibility exercises, a certain number of patients may find it difficult to maintain the joints in proper alignment, particularly when spasticity is considerable. For such patients, the use of a resting splint, which maintains the position of the forearm, wrist, and fingers in the functional position, is recommended.

In the case of paretic lower extremity, many stroke patients may find it challenging to dorsiflex their foot due to spasticity of plantar flexors and/or weakness of the dorsiflexors. The traditional stretching technique for the tight or spastic plantar flexors, prolonged positional stretches using toe or foot wedges, weight-bearing activities with affected lower extremity maintained in proper alignment, and weight-shifting activities in standing and modified plantigrade position can be beneficial to regain the ankle ROM. Following stretching of the plantar flexors, procedures like passive positioning of the hip and knee in near-complete flexion in the supine lying may encourage patients to voluntarily activate the dorsiflexors and reciprocally inhibit the activity of the spastic plantar flexors (Fig. 3.27).

Similarly, overactivity and/or tightness of the gluteus maximus with spasticity of quadriceps may make it difficult for many stroke patients to flex the hip and knee in the supine position. For such patients, in the crook lying or supine lying position, passive stretching (15–20 sec) of the hip extensors with the knee held in flexion, followed by commands to activate the hip flexors in the inner range, may help to initiate the hip flexors voluntarily. Commands like "do not allow your leg to drop" or "'move your thigh toward the chest" with or without manual assistance or support may encourage the patient to gain control over the hip flexors (Fig. 3.28). The

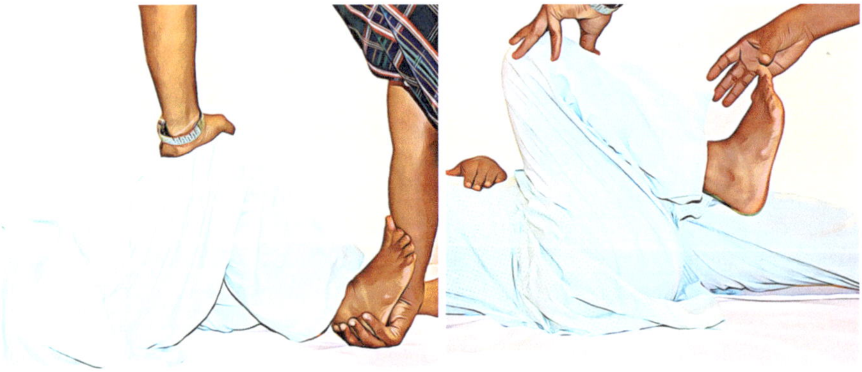

Animated photographs of model with permission

Fig. 3.27 Passive stretching of the plantar flexors in supine followed by the placement of hip and knee in near-complete flexion to facilitate the voluntary activation of the dorsiflexors

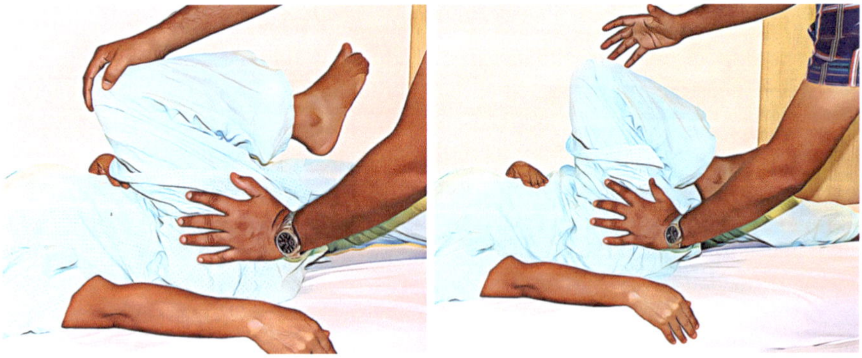

Animated photographs of model with permission

Fig. 3.28 Passive stretching of hip extensors with the knee held in flexion, followed by commands to encourage activation of hip flexors in the inner range

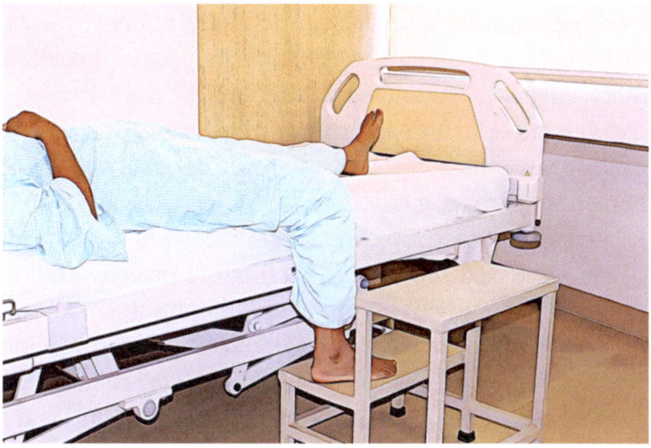

Animated photograph of model with permission

Fig. 3.29 Placement of affected side hip in abduction, knee in flexion, and leg off the side of the bed with foot flat on the stool/stepper

technique mentioned above may be repeated certain times to encourage the patient to eccentrically or isometrically contract the hip flexors before he or she can advance to concentric contraction. Following the initial phase of gaining control over the inner range, progress toward gaining control over the middle and outer ranges of motion. Eventually, when the patient gains sufficient control and strength, he or she can be advised to perform hip and knee flexion for tasks like crossing the affected leg over the flexed unaffected leg in a crook lying position.

For those patients who are wheelchair-bound, regular stretching of the hip flexors is advisable. If the extensor synergy dominates, position the affected lower limb of the patient with the hip in abduction, knee bent, leg off the side of the bed, and foot flat on the floor or stool (Fig. 3.29). Such a placement of the paretic lower limb may help to break the strong dominance of extensor synergy, which is typical in many patients, and prevent the possibility of limb scissoring while attempting to stand or walk.

3.11.2.6 Spasticity Management

Since normal joint flexibility and integrity is an amalgamation of independent factors like the normalization of tone, soft tissue extensibility, and the volitional activation of antagonistic muscle groups, the readers can expect a certain degree of overlap in the strategies or techniques utilized for encouraging joint flexibility and integrity and those aimed at mitigating spasticity. Familiarizing themselves with the contents in the preceding sections of the text will ensure a complete understanding of the nuances in the management of tone.

Early mobilization and daily stretching exercises help to maintain the length of spastic muscles and soft tissues. Once the flexibility exercise program has achieved full ROM, positioning the limb in the lengthened position reduces the hyperactivity of the spastic muscles. The use of weight-bearing exercises in kneeling or quadruped positions on a therapeutic mat helps to normalize the muscle tone of the paretic limb. The PNF techniques like rhythmic initiation or rhythmic rotation incorporating axial rotation and the various truncal mobility exercises mentioned in literature like "right in the middle" can reduce truncal stiffness and improve trunk control. Sustained or prolonged stretching helps to relax the spastic muscles through the mechanisms of autogenic inhibition. Slow vestibular stimulation induced by slow rocking movements can provide relaxation effects and may relieve the hyperactivity of the spastic muscles. It is advisable to teach and encourage the family members to perform the right method of flexibility and stretching exercises to ensure that the flexibility exercise program extends beyond the training given by the therapist.

As discussed previously, facilitating or activating the weak antagonist will also reduce the hyperactivity of the spastic muscles. The use of exteroceptive or proprioceptive techniques like quick stretching, muscle tapping, and brushing over the weak antagonist muscles can facilitate or enhance the action of the same. Such facilitation techniques need to be given only for those muscles which are weak and antagonist to the spastic group and should be discarded once the patient gets voluntary control over those muscles. While attempting to facilitate or voluntarily initiate weak muscles, avoid excessive effort from the patient's side which can often worsen the tone of the spastic muscles.

Modalities including the application of cold and use of neuromuscular electrical stimulation (NMES), can also help temporarily relieve the spasticity of the muscles. A cold or ice pack applied over the spastic muscle for 10–15 minutes duration tends to slow the nerve conduction and reduce the hyperactivity of the muscle spindle. Electrical stimulation of the weak antagonist muscles may help to reduce the tone of spastic muscles through reciprocal inhibition. However, it is essential to ensure that electrical stimulation does not activate the spastic muscles. In addition to the above, certain monophasic currents like interrupted galvanic currents are not advisable for the stimulation of weak antagonists as they may pose safety concerns due to possibilities of inducing chemical burns. The use of orthotic devices, including resting splints, serial casts, and inflatable pressure splints to maintain the spastic muscle in its lengthened position, can reduce the hypertonia and improve or preserve the passive ROM.

The therapist should discourage the placement of soft or squeeze ball and cloth roll in the patient's palm, especially when there is a likely chance of spasticity and tightness developing within the wrist and finger flexors. Such strategies may facilitate palmar grasp reflex, encourage imbalance between the flexors and extensors muscle strength, and predispose to a poor functional return of hand, which is typically characterized by difficulty to extend the wrist and open the hand, simultaneously.

3.11.2.7 Shoulder Dysfunction and Pain Management

Subluxation of the glenohumeral joint, spasticity, soft tissue impingement in the subacromial space, rotator cuff tears, adhesive capsulitis, and complex regional pain syndrome are specific reasons for hemiplegic shoulder pain. Flaccidity of the rotator cuff muscles, lack of normal tone and absence of voluntary movements, impaired proprioception, and inactivity of certain muscles like serratus anterior alter the normal position of the scapula and shoulder biomechanics. In the absence of supporting musculature, glenohumeral movements like forward flexion or abduction with scapula rotated downward and depressed can predispose the humerus to sublux from the glenoid fossa. Mechanical stresses created by the repeated traction and gravitational forces eventually can change a painless subluxation to a painful one.

Even spasticity of certain muscles, like the scapular depressors and pectoralis major along with inactivity of muscles like serratus anterior, can contribute to depression, retraction, and downward rotation of the scapula and predispose to subluxation and restricted ROM. The friction and compression stresses occurring between the humeral head and superior soft tissues while performing the passive or active-assisted ROM in the absence of normal scapulohumeral rhythm can also predispose to shoulder pain. In stroke patients, undue tightness and thickening of the soft tissues and the joint capsule can lead to conditions like adhesive capsulitis. Improper handling techniques, poor positioning of the shoulder joint, improper alignment of upper limb joints during functional weight-bearing exercises, passive ROM with inadequate scapular mobilization, and undue traction of affected upper extremity during transfers can cause microtrauma to the soft tissue structures and lead to shoulder pain. Use of scapular mobilization prior to glenohumeral passive ROM, preventing undue weight falling on the shoulder joint while lying on the involved side, minimizing traction on the shoulder during transfer techniques, taping techniques, NMES of the weak shoulder muscles, and use of slings are some strategies recommended to minimize or prevent the shoulder pain.

Impaired glenohumeral joint biomechanics and trauma to the affected shoulder have been implicated in the development of complex regional pain syndrome post-stroke. Glenohumeral subluxation is a common finding in stroke patients with complex regional pain syndrome. Complex regional pain syndrome, also known as Sudeck's atrophy or shoulder hand syndrome in stroke patients, is characterized by pain; edema; stiffness and limitation in ROM; vasomotor and sudomotor changes; trophic changes in the hair, nails, and skin; and patchy bone demineralization of the affected extremity. Muscle weakness, spasticity, sensorimotor deficits, and coma during the initial days post-stroke are a few clinical factors associated with its development. In the early stage (first stage), the pain is typically intermittent and limited to the shoulder, and later, the pain intensifies and extends to the whole extremity. Due to joint stiffness and limitation, the wrist tends to assume a flexed posture, and while attempting to extend the wrist, the patient may report intense pain and guard against any movement attempts. Pale-pink discoloration of the skin and cold skin are vasomotor changes. The skin can be hypersensitive to touch, pressure, or temperature variations. Pain subsides during the second stage. Dystrophic changes of the muscle, atrophy of the skin, vasospasm, hyperhidrosis, coarse hair and nails, and radiographic evidence of patchy osteoporosis are the characteristics of the second stage. In the third stage, pain and vasomotor changes are rare, and pericapsular fibrosis and articular changes and progressive atrophy of the skin, muscles, and bones are its features. The hand is typically held in a clawed position due to the contracture of muscles, with marked atrophy of the thenar and hypothenar muscles. If identified during the early phase, the signs and symptoms of this disabling condition are quite reversible. On the contrary, the beneficial effect of interventions is least likely once the condition advances to the third stage.

The therapeutic interventions are based on the clinical examination findings of complex regional pain syndrome. The goals of treatment are to reduce pain, maintain joint mobility, and restore function. Gentle mobilization of the joints, passive ROM exercises, strengthening of the affected extremity muscles, management of edema, and desensitization techniques are the mainstay of care for these patients. Preventative strategies to minimize early joint injury can reduce the likelihood of complex regional pain syndrome. The strategies should include providing support for the affected arm during the flaccid stage, proper positioning and handling of the extremity, preventing traction injury, promoting passive ROM following scapular mobilization, and advising the concerned nursing staff to minimize venous infusions into the hemiplegic hand. A gradual process of desensitization (by applying systematically different sensory stimuli on the skin over the affected area) may help normalize the sensation by resetting the altered central processing in the nervous system. For the restoration of the upper extremity function, promoting active movements of the affected shoulder, strengthening the weak muscles, and edema relieving techniques need to be considered.

Flaccidity or hypotonicity of the shoulder muscles can increase the risk of traction injury to the soft tissues in and around the glenohumeral region. Slings can minimize or prevent soft tissue injuries, relieve pressure on the neurovascular bundle, improve the anatomical alignment of the shoulder, and ease the stress and the gravitational pull on the glenohumeral joint. Conventional triangular sling, Bobath sling, Rolyan humeral cuff sling, and arm pouch sling are some of the shoulder slings used for hemiplegic patients with glenohumeral subluxation. During the initial phase transfers and gait training, the therapist can focus his attention on the patient's posture or trunk control, if the sling is worn to support and safeguard the affected shoulder. However, many of the slings have a modest effect in reducing the subluxation of the glenohumeral joint or improving the shoulder function. Most of the slings can encourage typical upper limb posturing characterized by shoulder adduction and internal rotation, elbow flexion, forearm pronation, and wrist flexion, promote tightness and contracture, discourage active voluntary contraction of shoulder muscles, encourage learned nonuse, increase flexor hypertonicity, and contribute to body schema dysfunction and neglect. The Rolyan humeral cuff sling has an arm cuff on the humerus supported by a "figure of eight" harness. It encourages better symmetry, maintains elbow in extension, provides a certain reduction of subluxation, does not restrict the distal limb functions, and can be worn for a longer period. Such humeral slings are preferred over those conventional slings encouraging shoulder adduction and internal rotation and elbow flexion. Taping techniques for minimizing shoulder subluxation, neuromuscular electrical stimulation of weak rotator cuff muscles (especially supraspinatus) and the middle fibers of the deltoid, use of proper support to the shoulder and elbow while sitting or standing, and use of lateral elbow guard and/or straps while the patient is on a wheelchair are the other alternatives to the use of a sling.

3.11.2.8 Strategies to Improve Voluntary Movement Control

In the initial phase post-stroke, the affected limbs are flaccid or hypotonic. Following the cerebral shock, when the deep tendon reflexes return and patients progressively develop control over the basic limb synergies, activities that promote voluntary movement control, postural control, and functional use of the extremities should be the primary focus of training. The presence of obligatory basic synergy patterns can hinder the patient in re-attaining the fractionated movements of the extremities. Released postural reactions, including associated reactions, can affect the inter-and intra-limb movement control. During the training phase, if the stroke patient can initiate voluntary movements, the focus of therapy should be toward encouraging movement patterns away from the basic synergy and fractionation or dissociation of body segments. Use of reflex inhibiting pattern and key point of control, selection of appropriate postures, adequate stabilization of the body part, and guidance can help the patient to perform movements without inducing excessive tone or abnormal synergy pattern. Advice the patient not to perform movements too quickly or with excessive effort as it may accentuate the hypertonia of the muscles.

For those patients with the inability to initiate movement, techniques or strategies discussed in chapter "Therapeutic Approaches" (such as the facilitatory techniques of Rood's approach and exteroceptive and proprioceptive stimulation and use of postural reflexes as stated in Brunnstrom approach can be an option. In addition to those techniques, stretching of the spastic antagonists prior to the facilitation of agonists or NMES of the weak agonists can be tried. The aim of the treatment should be to gain adequate control over the movement initiated by the patient so that it can be further fractionated or dissociated for functional movements.

The functional movements, including reaching, walking, and climbing, require phasic activity of the muscles, smooth and coordinated actions of muscles in different roles (agonist, antagonist, synergist, and fixator), and variation in the type of contractions (eccentric, isometric, and concentric). Reversal of direction of movement encourages the phasic switching over action of muscles, and practice of tasks utilizing variations of contractions helps in improvising the interaction between agonists and antagonists crucial for normal coordination and function. For instance, practicing partial squats or modified wall-squats (Fig. 3.30) within a limited range with or without the support or assistance of the therapist helps to sequentially contract the hip and knee extensors between eccentric to isometric to concentric. For those stroke patients with considerably weak muscles, to encourage better recruitment of motor units, eccentric and isometric contractions are emphasized before concentric contractions.

3.11.2.9 Strategies to Improve Muscle Strength

Hemiparesis, the most common impairment seen in stroke patients, can persist for months to years, and the weakness can significantly contribute to activity limitations and participation restriction. The inability of muscles to generate adequate

Fig. 3.30 An illustration of the modified wall-squat

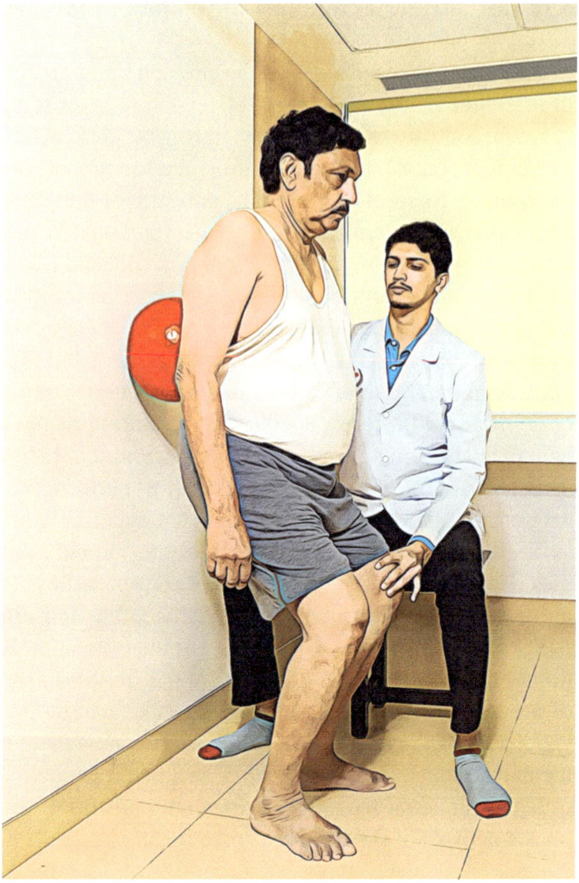

Animated photograph of model and therapist with permission

force post-stroke can be due to increased stretch reflex excitability, increased antagonist muscle coactivation, reduced rate of motor unit firing, and alterations in the muscle property (decreased muscle mass and muscle fiber length and changes in pennation angle and tendon length). Recent evidence on progressive resistance strength training of the hemiparetic side muscles has shown improvement in muscle strength and functional performance with no apparent worsening of spasticity or reduction of ROM.

The modalities used for strength training include manual resistance, free weights, elastic bands, and machines. Use of reeducation board, sling suspension therapy, or aquatic exercises is recommended for those patients who have muscle strength less than that required to move against gravity. The structured strength training exercise is advisable for those who demonstrate independent movement in the gravity-resisted plane. Resistance training consisting of 8–12 repetitions, 2–3 sets per treatment session, a minimum of 2–3 days per week performed with free weights, bands, or machines is adequate to improve the strength of the weak muscles of the paretic limbs. Task-oriented progressive resistance strength training has shown a carryover effect in functional activities in addition to improvement in lower limb strength.

Performing task-oriented functional activities or circuit training using body weight or wearing weighted cuffs with a progressive increase in the number of repetitions, complexity, or duration of exercise can improve muscle strength and endurance.

For recent stroke and stroke patients with comorbidities, including ischemic heart disease, valvular disorders, and poorly controlled blood pressure, high-intensity strengthening exercises are contraindicated. Strength training activities, especially isometric exercise, can induce breath-holding that can elevate the blood pressure to dangerous levels and impose excessive cardiac load. Low-intensity exercises (30–50% of 1 repetition maximum [RM]), dynamic exercises performed in sitting position, exhaling during the lifting phase and inhaling during the lowering phase of the strength training, adequate rest periods between the sets, and repeated monitoring of the blood pressure are certain strategies to avoid the abovementioned.

3.11.2.10 Improving Upper Limb Functions

Though sensorimotor impairments and functional issues vary from patient to patient, it tends to be more severe in the middle cerebral artery syndrome and the dense lesions affecting the internal capsule. In such conditions, since the sensorimotor recovery is often limited, compensatory training strategies and environmental modifications in addition to the standard treatment strategies, like early mobilization, flexibility exercise program, and positioning strategies, are preferable to maximize the functions. For those stroke patients with a better scope of motor recovery, the training strategies should focus on promoting fractionation of the movements, reversal of direction of movements, and repetitive task-specific practice.

To enhance proximal stabilization and to counter the effects of excessive flexor hypertonicity and dominance of flexion basic limb synergy, weight shift with weight-bearing on the affected side extended arm while stabilizing the hand on a supporting surface should be encouraged as an early activity. Such joint approximation activities performed while sitting, standing, or modified plantigrade position can help to increase the activity of pectoral girdle stabilizers, and, meanwhile, muscle tapping or exteroceptive stimulation over the triceps muscle belly can facilitate the elbow extensors to gain better control over the elbow joint.

The focus should be on encouraging the activity of the serratus anterior and upper fibers of the trapezius muscles to gain control over the scapular upward rotators. In the supine position, encouraging the patient to push the arm up toward the ceiling with shoulder flexed to 90°and elbow extended can initiate the serratus anterior muscle. The therapist will often have to support the arm in the abovementioned position (Fig. 3.31) to encourage him/her to initiate the movement. Commands like "do not allow me to push your arm downward" or "push your arm upward" just after a quick stretch to the serratus anterior in supine may activate the muscle. Once the patient can initiate the movement, manual resistance can be offered to further build the strength of the muscle. Overactivity of the pectoralis major muscle (sternal fibers) can hinder the voluntary initiation of serratus anterior in many stroke patients. Positional stretch, myofascial release, or traditional stretching techniques can be

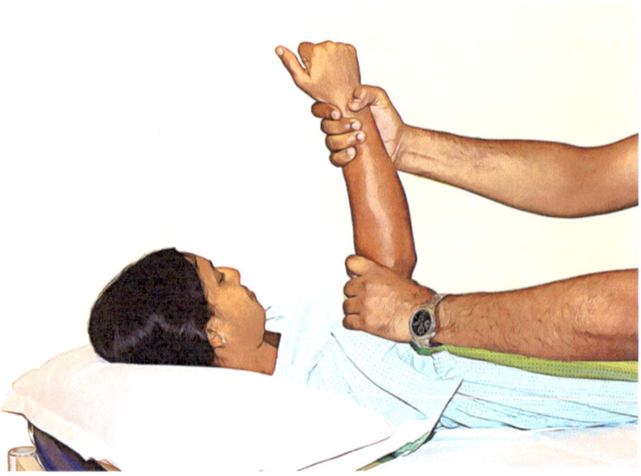

Animated photograph of model with permission

Fig. 3.31 Proprioceptive stimulation and commands like "do not allow me to push your arm downward" to activate serratus anterior in supine position

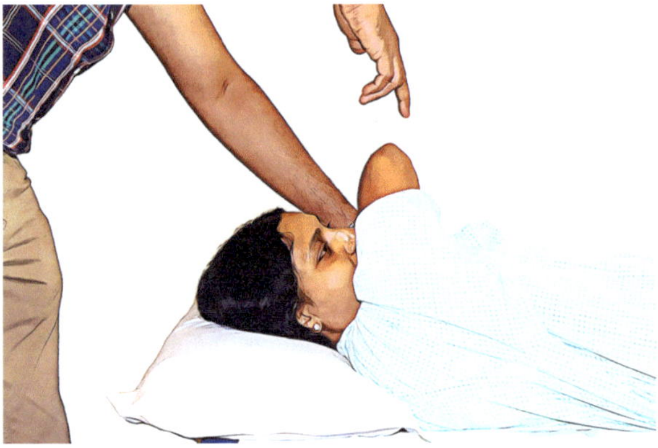

Animated photograph of model with permission

Fig. 3.32 Short lever exercises for the shoulder flexors in the supine position

attempted in such a situation to minimize the overactivity of the pectorals and encourage the initiation of serratus anterior muscle. The activity of the trapezius muscle can be initiated in the sitting position. Support and guidance to encourage shrugging of the shoulders (initially bilaterally and then advancing to affected side shoulder alone) and commands like "do not drop your shoulder" can help in initiating the same muscle.

To further build up the control over the affected side shoulder, short lever exercises for shoulder flexors (Fig. 3.32) and horizontal adductors (Fig. 3.33) should/can be encouraged in the supine position. Reaching out activities with the affected upper limb requires synergistic activity of the serratus anterior, shoulder flexors, and elbow extensors. To prepare for the same, facilitation techniques can be attempted to activate the elbow extensors with the patient's arm held in 90° flexion (Fig. 3.34). Once the patient can initiate the contraction, encourage the patient to eccentrically, isometrically, and concentrically control the elbow extensors. Following which attempts should be made to perform reach-out activities in supine with the elbow held in extension or near extension (Fig. 3.35). In the supine position, attempts to

Fig. 3.33 Short lever exercises for the shoulder horizontal adductors in the supine position

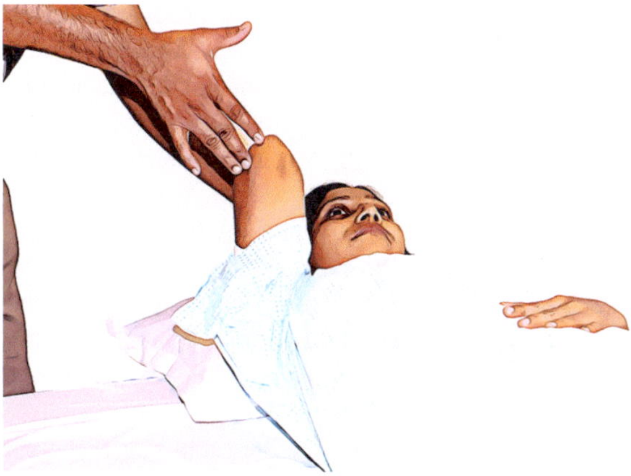

Animated photograph of model with permission

Fig. 3.34 An illustration of the facilitation technique (muscle tapping) to activate the elbow extensors with the subject's arm held in 90° flexion

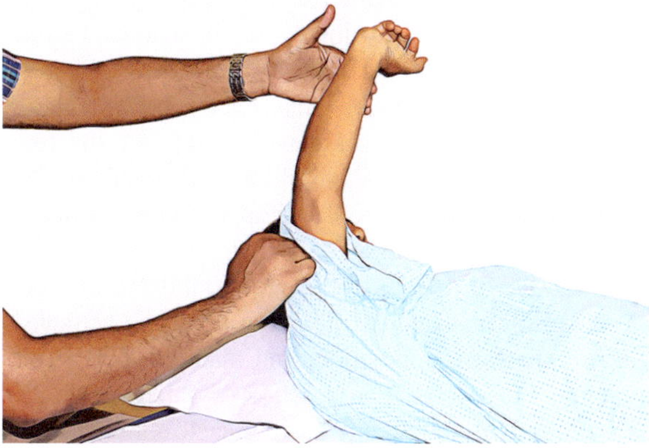

Animated photograph of model with permission

Fig. 3.35 Reach-out activity performed in supine with the elbow held in near extension

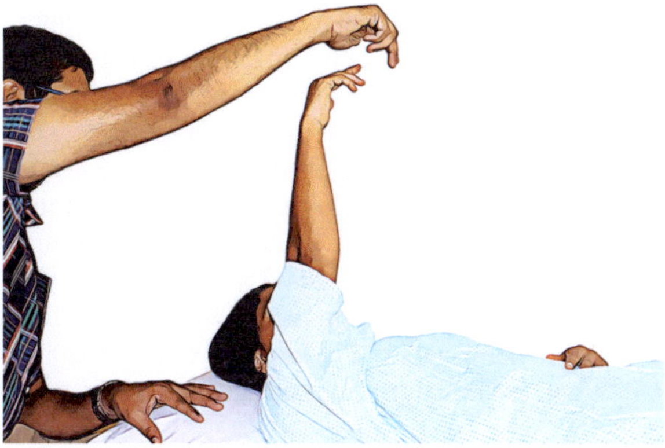

Animated photograph of model with permission

perform circumduction movement in a short circle and later advance toward a larger circle with adequate support and guidance may further improve the control of the shoulder. Patients with limited control over the shoulder and elbow muscles, while sitting, should be encouraged to slide the hand forward over a tabletop or re-education board. Practicing wiping or polishing activities and pushing or sliding a ball forward-backward help to recruit the scapular protractors, shoulder flexors, and elbow extensors. Visual guidance will help in improving eye-hand coordination during the reach-out activities. The patient should practice independent lifting and multiplanar reaching with increasing resistance (weight of the object held by the hand or secured by the Velcro) and varying height and distance in sitting, modified plantigrade, and standing positions based on the motor recovery potential. Trick movements or substitution movements and encouraging patterns similar to basic limb synergies need to be avoided as they can hinder the potential for further recovery.

For hand function, grasp, manipulation, and release are the movements required for the finer motor activities of the hand. During the initial phase of hand function recovery, voluntary release can be difficult as compared to the grasp. Weight-bearing activities through the hand may normalize the tone of the overactive long flexors of the wrist and hand. Positional stretch, sustained stretch, and other inhibitory techniques for the spastic wrist and finger flexors along with facilitatory techniques for wrist and finger extensors may encourage extension movements. NMES for the wrist and finger extensors can also improve the function of the wrist and finger extensors, similar to the deltoid and supraspinatus muscle functions for shoulder subluxation. Task-oriented activities can be practiced with the weaker hand, which will encourage the patient to use the affected limb for ADL. For those patients with potential for further recovery, wrist stabilization exercises (initially with the elbow in extension and later in flexion) should be achieved prior to hand activities like finger extension, opposition, and manipulation of objects. Careful observation of functional tasks like manipulation of objects and use of utensils or tools may provide clues to eliminate unwanted muscle activities or movement patterns interfering with the functions. For those patients in stage 3 and above of Brunnstrom motor recovery stage, therapeutic approaches like constraint-induced movement therapy (CIMT) and motor relearning program can promote or enhance further hand function recovery. CIMT consists of a multifaceted intervention designed to promote upper extremity function by engaging patients in intense task-oriented practice wearing a mitt on the uninvolved upper extremity for $\leq$6 hours per day for 10 consecutive days. The physiotherapist uses feedback, shaping techniques, and encourages the patients to modify and progress performance during practice. The patients undergoing CIMT are encouraged to self-monitor the target behaviors and use problem-solving abilities to identify the obstacles and generate potential solutions. The less intense modified CIMT (mCIMT) and CIMT have shown neuroplastic changes, including a shift in motor cortical activation toward the areas in the contralesional hemisphere. Evidence does exist regarding the beneficial effects of CIMT and mCIMT for acute, subacute, and chronic stroke patients.

Certain literature reports that bimanual training is as effective as but not superior to unimanual training of the paretic upper extremity post-stroke. It is hypothesized that similar movement in the less involved upper limb facilitates movement in the involved upper limb. Functional task training (FTT), Bilateral arm training with rhythmic auditory cues (BATRAC), and robot-assisted training (RAT) are the three main types of bimanual training interventions. In terms of outcome measures, compared to standard or routine therapy, bimanual training was not shown to be significantly better than the former.

In the past, electromyographic (EMG) biofeedback has been extensively used to improve motor function in stroke patients. The EMG biofeedback helps to recruit motor units in weak or hypoactive muscles, train voluntary inhibition of spastic muscles, and increase the kinesthetic awareness of voluntary movements. Existing evidence reports that the benefit of biofeedback is considerable when it is used as an adjunct with task-specific training. For moderate to severe motor impairments of the upper extremity, task-oriented training with robotic devices has been attempted to reach and grasp and/or release movements. Rehabilitation using robotic devices enables high levels of intensive practice and is generally well-tolerated by stroke patients. Though it is a useful adjunct to therapy, the high cost is a major limitation for widespread use of these units worldwide.

For the past two decades, virtual reality has been used in the field of stroke rehabilitation. Virtual reality is the simulation of a real environment that enables the user to interact with certain elements in the virtual environment using a man-machine interface. It provides the user visual, auditory, tactile, and motion information and a variety of interfaces for interacting with the simulation setting, ranging from common devices like mouse or joystick to more complex devices with motion capture systems or haptic devices, providing users sensory feedback and experience close to the real task. The degree of immersion varies depending upon the virtual environment setting. The term "immersion" refers to the extent to which the user perceives the virtual environment as the real world. The virtual environment can be immersive, non-immersive, or semi-immersive. Semi-immersive and non-immersive systems are the most widely used systems for the management of motor symptoms among stroke patients. Various virtual reality programs have been developed and used for upper and lower limb training, gait training, and posture and balance training for subacute and chronic stroke patients. Current evidence suggests that virtual reality training for stroke patients is safe and cost-effective for improving upper and lower limb functions, dynamic balance, prevention of falls, stair climbing, ROM, muscle strength, and gait speed. Virtual reality is hypothesized to motivate and promote practice-dependent reorganization. Transcranial magnetic stimulation and functional MRI studies have demonstrated reduced ipsilateral cortical activation and increased contralateral cortical activation as a result of intensive practice of the affected limb using virtual reality. Further analysis has also revealed that the activation was more in the contralateral primary sensorimotor cortices than in the ipsilateral premotor cortex or contralateral supplementary motor areas.

3.11.2.11 Improving Lower Limb Functions

Activities specifically emphasizing the breaking up of the obligatory basic synergy patterns of the affected lower extremities are the key strategy for preparing the patient for the gait. PNF diagonal pattern (D1 extension pattern), which associates hip extensors and abductors with knee extensors, can encourage effective midstance, which otherwise is less likely to occur due to the strong linkage of the dominant extensor synergy. The use of elastic bands around the upper thigh to encourage hip abduction with the knee in extension in supine or standing positions and encouraging lateral side-steps in standing are a few other alternatives. Hip adduction should be stressed during hip and knee flexion to facilitate the normal swing phase, which otherwise is characterized by abduction and flexion of the hip with knee flexion, a feature of flexor basic limb synergy. Activities like PNF D1 flexion pattern in supine and standing and crossing and uncrossing the affected lower limb over the unaffected limb in crook-lying position or sitting can associate hip adduction with hip and knee flexion and break the linkage between hip abductors and the flexors. Bridging and encouraging knee flexion in standing or modified plantigrade standing postures are activities that can promote hip extension along with knee flexion essential for the swing phase of the gait.

Several therapeutic exercises are available to improve the trunk and pelvic control of stroke patients. These exercises can be practiced on firm ground, therapeutic mat, or Swiss ball. Swiss ball exercises can be used for encouraging pelvic rolling in supine and pelvic shifting in sitting. Dissociation of the upper trunk to the lower trunk can be practiced in supine, crook-lying, and kneeling positions on a therapeutic mat or standing positions on firm ground. The smooth reversal of direction of movement produced by reciprocal muscle action initiated in supine (like foot sliding up and down in crook-lying position or leg sliding in and out in supine), then in sitting (like foot slides under the chair), and finally in standing (partial squats or partial wall squats) and the use of proprioceptive training (including dynamic squats and single-limb squats) for the affected lower limb initially on a firm surface and later on foam surface can improve eccentric and concentric muscle control of the knee musculature.

As a preparation for gait, it is also essential to initiate and strengthen the activity of the dorsiflexors and evertors of the foot. Stretching of the overactive plantar flexors is a prerequisite for initiating the activity of dorsiflexors and evertors. The use of exteroceptive or proprioceptive techniques can often help in initiating the evertors, which are neither a component of flexor nor an extensor basic limb synergy. Care has to be taken while selecting the technique or strategy to initiate and strengthen the muscles, as wrong selections can further create overactivity of the plantar flexors and invertors accentuating the muscular imbalance at the level of ankle and foot.

3.11.2.12 Strategies to Improve Postural Control and Balance Reactions

Sensorimotor dysfunctions can significantly change the postural control and balance reactions of stroke patients. Absence, delay, or variability in the balance reactions and impairments in latency, amplitude, and timing of muscle activity are certain typical features seen in these patients. Poor postural control and balance reactions can predispose these patients to falls and fractures and can further reduce their level of confidence in balance and locomotor skills. Balance training programs progressively proceeding from less challenging to more challenging postures performed overground and on unstable surfaces like a gym ball or a wobble board will boost the patient's level of control and confidence. Gradually increasing the level of difficulty like the range of displacement and speed of displacement, encouraging self-initiated movements and emphasizing consistency, postural symmetry, and maximizing the use of the affected side should be the goals of training. To assist and instill confidence and prevent falls, during the early phase of balance training in standing, the therapist may encourage the patient to use appropriate supportive devices like a harness or gait belt. Indiscriminate and prolonged use of such supportive devices can often be counterproductive and needs to be discarded, especially when patients seem likely to improve their postural and balance control.

Following effective and consistent maintenance of symmetrical aligned static upright postures, the stroke patients should be encouraged to self-displace their center of mass to explore the limits of stability. The use of verbal and tactile cues and verbal instructions can progressively encourage the patient to achieve larger ranges of displacement to further improve the limits of stability. Provide an ample amount of practice for the patient to master the skill of safely moving in any direction while aligning the center of mass over the base of support. Encouraging weight shift activities in sitting and standing with more weight-bearing on the affected side pelvis and lower limb, respectively, and discouraging the overuse of unaffected limbs will provide the opportunity for the affected side musculatures to develop appropriate postural and balance responses.

To regain balance following the unexpected displacement of the center of mass, ankle, hip, and stepping strategies are crucial. Single limb standing (affected side) on a foam surface with minimal upper limb support, preferably with the affected side knee maintained in minimal flexion, can encourage the ankle and foot muscles to contract actively. Standing on a wobble board or a half-foam roller can be the alternatives to activate these muscles to promote ankle strategies required for regaining balance; however, such exercises may be too advanced for stroke patients during early rehabilitation. Perturbations in the anteroposterior direction or weight shifts in the same direction can promote anteroposterior hip strategies. Tandem stance or near tandem stance on floor or foam can promote mediolateral hip strategies, whereas displacement of the center of mass beyond the base of support can promote stepping strategies.

Advancing the exercises from stable to an unstable surface, increasing the amplitude of displacement of the center of mass within and beyond the base of support, reducing the base of support, progressing from uniplanar to multiplanar reaching outs, incorporating head and trunk rotation for the task, encouraging dual-task activities, walking sideways, backward and braiding, and proceeding from a closed to an open environment are certain strategies to improve the level of difficulty for the balance training program. To maintain and regain balance during the balance training program, the stroke patient should be allowed to identify the potential problems and encouraged to actively solve the problem by recruiting the appropriate safety strategies.

3.11.2.13 Treatment Strategies for Pusher Syndrome

The postural and balance issues among the Pusher's syndrome patients are distinct from those seen among the rest of the stroke patients. Asymmetry in sitting and standing, excessive weight-bearing through the affected side, tendency or strong urge to push consistently toward the affected side using the unaffected limbs, apprehension about falls toward the unaffected side with absolute "no botheration" about the repeated instability, and falls toward the affected side are certain hallmarks of "Pusher's behavior" or ipsilateral pushing. Typically, the therapist's efforts to passively correct the tilted posture will result in a stronger resistance from the patient and often can be counterproductive. Visual stimuli, environmental prompts and boundaries, self-initiated corrections, active efforts to achieve vertical (midline/neutral) position, use of mirrors, verbal, and tactile cues for postural orientation, training activities on a Swiss ball to promote symmetry, and the use of air splints and walking aids are the strategies available to correct the pushing behavior. However, the author (Abraham M. Joshua) believes that strategy emphasizing overcorrection along with carefully guided and instructed activities can be a better practical solution than the abovementioned strategies. The strategy developed and encouraged by the author is distinct from those training programs which emphasize upright positions with active movement shifting toward the stronger side or for vertical orientation.

During the early phase of management of Pusher's behavior, a clear and concise explanation about the pushing behavior and its consequences needs to be addressed to the stroke patient and the family members. The information should even emphasize the safety issues and the roadblocks of rehabilitation. Advising and encouraging the patient to lie on the unaffected side for as much time as possible should be the first component of the strategy. Meanwhile, emphasis should be given to discourage the patient from lying on the affected side. Once the patient becomes comfortable with side-lying on the uninvolved side, he or she should be encouraged to lie on the uninvolved side propped up on the elbow. The posture mentioned above can be actively maintained, if not by pillows. Once the patient can actively maintain the posture, encourage him or her in a controlled fashion to perform repeated short-range propping up from the side-lying position. Reach-out activities for the affected upper limb can be introduced in the side-lying with propped up elbow position (Fig. 3.36). The direction of reach-outs should be toward the uninvolved side. If the patient tends to flex and rotate the neck toward the affected side, the therapist should dissuade those attempts and should encourage rotation toward the unaffected side. Once the pushing tendency subsides, the patient should be advised and/or assisted in bringing both legs off the side of the bed. The reach-out activities should continue for this posture (Fig. 3.37) before advancing to sitting up over the side of the bed. While sitting, a partial amount of lean toward the unaffected side should be encouraged. Instructions like "shrug your shoulder" for the affected side and "drop your shoulder" for the uninvolved side can further encourage the partial tilt toward the uninvolved side. An alternate way to encourage the leaning toward the uninvolved side is to instruct or guide the patient to bring the shoulder or forehead closer toward the therapist or family member sitting by the side of the uninvolved side.

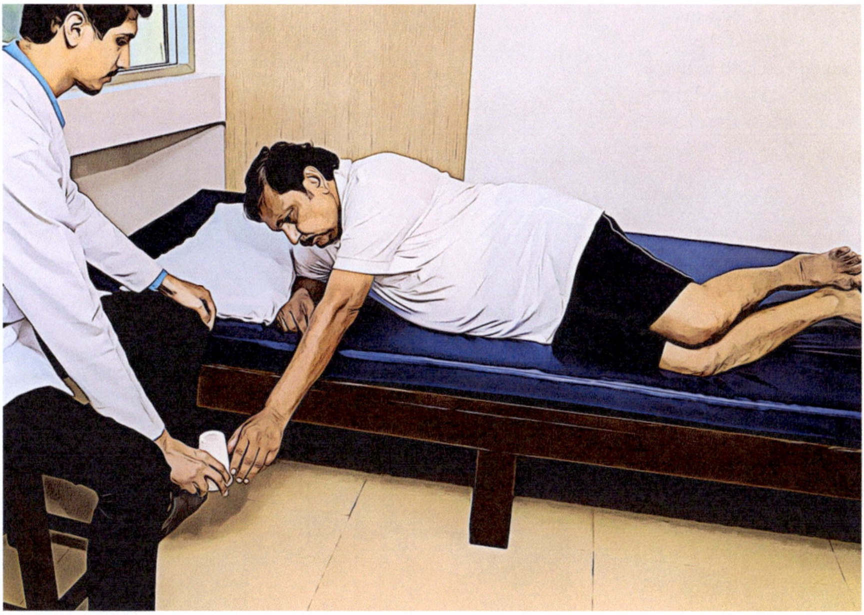

Animated photograph of model and therapist with permission

Fig. 3.36 Reach-out activity performed in side-lying with propped up elbow of the uninvolved side

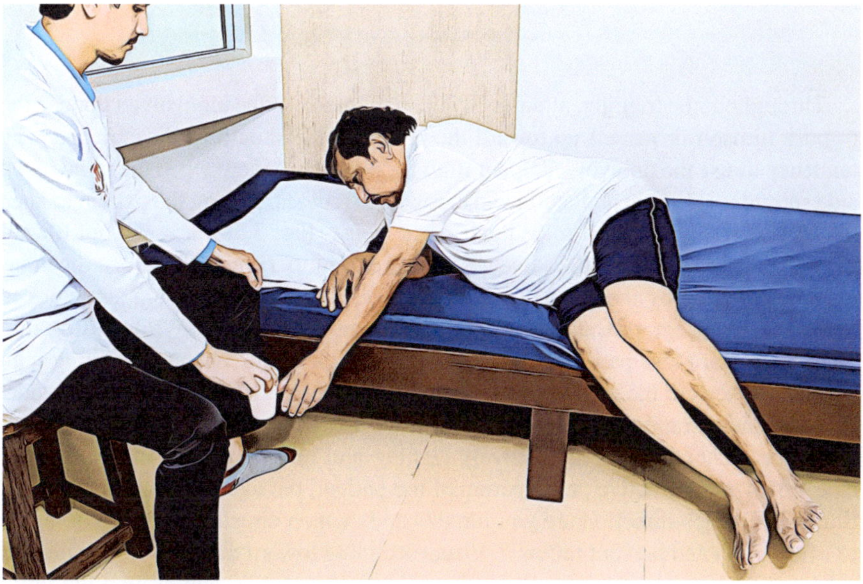

Animated photograph of model and therapist with permission

Fig. 3.37 Reach-out activity in side-lying with the legs off the bed

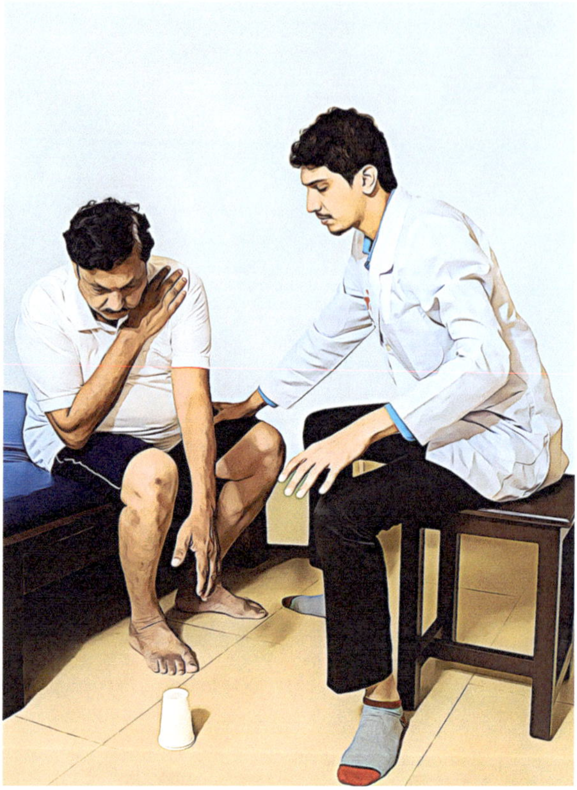

Fig. 3.38 Reach-out activity toward the uninvolved side with the uninvolved upper limb folded across the chest

Animated photograph of model and therapist with permission

Throughout the training, attempts by the patient to use the uninvolved upper limb to push himself or herself up toward the affected side should be discouraged. The tendency to use the uninvolved upper limb for pushing is more when the uninvolved side shoulder is kept in abduction and elbow in extension. Asking the patient to keep his uninvolved upper limb folded across the chest (Fig. 3.38) or reach-out toward the uninvolved side using the same side upper extremity (Fig. 3.39) or rest the uninvolved arm hang by the side (Fig. 3.40) or over the thigh are certain strategies to minimize the usage of the uninvolved upper limb for pushing. While sitting in a wheelchair, for those patients with severe pushing behavior, encourage sitting with partial lean toward the uninvolved side with head and neck turned to the same side.

Often the therapist needs to repeatedly educate and encourage the patient on the relevance of not pushing and actively leaning and turning toward the uninvolved side. To encourage active participation of the patient, the therapist should ask questions like "which direction are you tilted?" and "which direction you need to move to overcorrect and prevent falling?" Visual scanning toward the uninvolved side and asking the patient to give details of the objects placed near and around the uninvolved side of the patient can further encourage the overcorrection. Once able to sit unsupported with partial leaning, the patient should be encouraged to perform

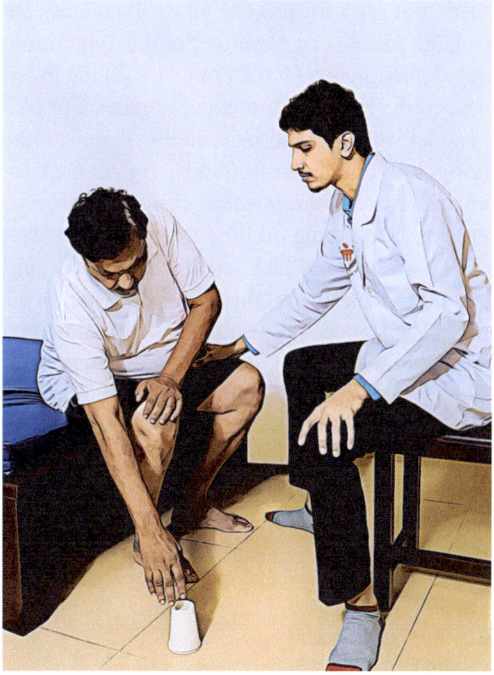

Animated photograph of model and therapist with permission

Fig. 3.39 Reach-out activity in sitting using the upper extremity of the uninvolved side

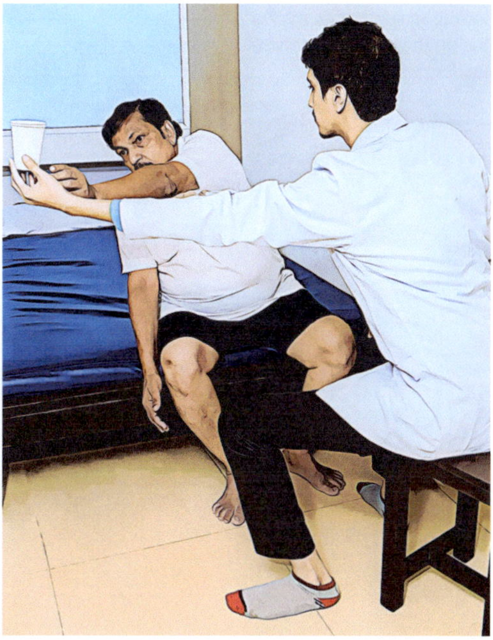

Animated photograph of model and therapist with permission

Fig. 3.40 Reach-out activity in sitting with the uninvolved arm hanging by the side reach-out only toward the uninvolved side. Depending upon the strength and control, the patient can use either of the upper limbs for reach-outs. The therapist should avoid making any tactile contact with the involved side trunk as the patient can lean toward the therapist's manual contact. For those patients who have achieved independent sitting, small increments of self-initiated movements can be encouraged to bring the trunk to the midline and then back to the leaning posture toward the uninvolved side. As the postural control improves, self-initiated to and fro movements passing beyond the midline toward the involved side can be encouraged.

For sit-to-stand task, cues and manual guidance can be given to encourage the patient to lean toward the uninvolved side while transferring (Fig. 3.41). The therapist's hand can act as a cue and instruction, including "touch your head on my hand" and "follow my hand," can help the patient to lean and get up toward the uninvolved side. While standing, the same strategy used in sitting, including overcorrection, reach-out activities, and weight shifts toward the uninvolved side, needs to be encouraged (Fig. 3.42). Once the subject can stand with partial lean to the uninvolved side, self-initiated to and fro movements initially up to the midline and later beyond midline toward the involved side need to be incorporated. In order to make the task more interesting, the therapist can ask the patient to do "head butting" on

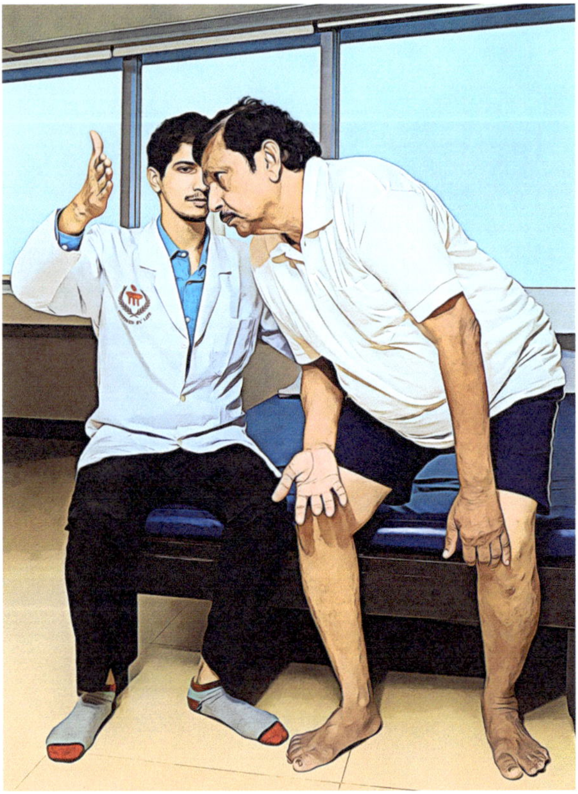

Animated photograph of model and therapist with permission

Fig. 3.41 Guidance to encourage the subject to lean toward the uninvolved side during the sit-stand transition. Note: The upper limb of the uninvolved side is kept in supination over the thigh to minimize the tendency of pushing

Fig. 3.42 Providing visual guidance to maintain leaning toward the uninvolved side while standing

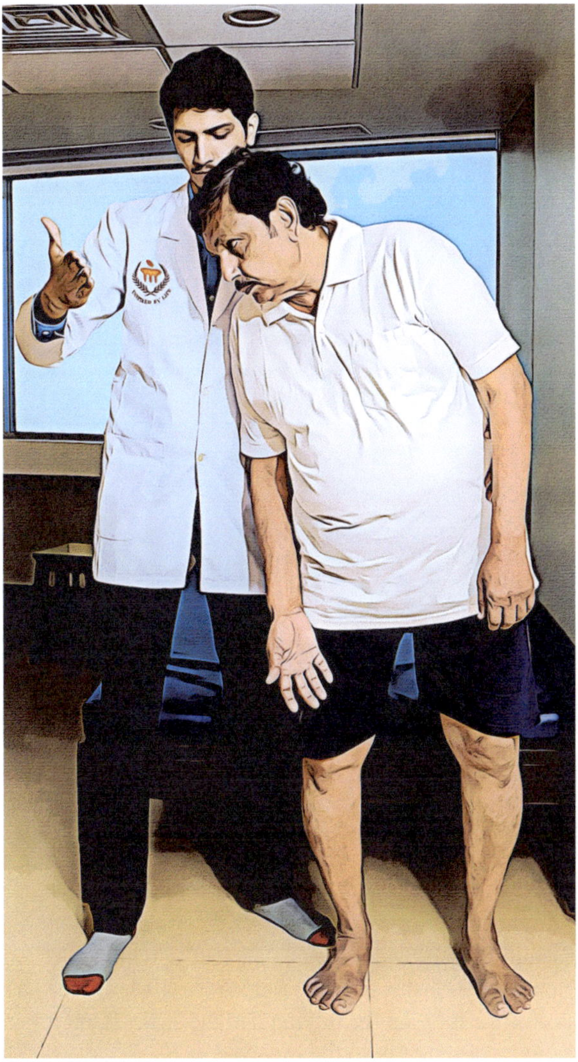

Animated photograph of model and therapist with permission

the therapist's hand to encourage a faster correction. The position of the patient's head, neck, trunk, and extremities, appropriate instructions, the therapist's hand that serves as a cue, and manual guidance and support provided will ensure the correct way of maintaining and regaining the erect posture without pushing. Once the patient can maintain an erect posture, weight shifts and marching in place (Fig. 3.43) and stepping forward, backward, and sideways can be attempted as preparation for walking. In case if patient tends to revert to pushing behavior in standing, overcorrection strategies can be reattempted to minimize the behavior. For those patients with a tendency to push while attempting to walk, instruct and guide them to walk with mild-to-moderate lean toward the uninvolved side. Figure 3.44 illustrates how

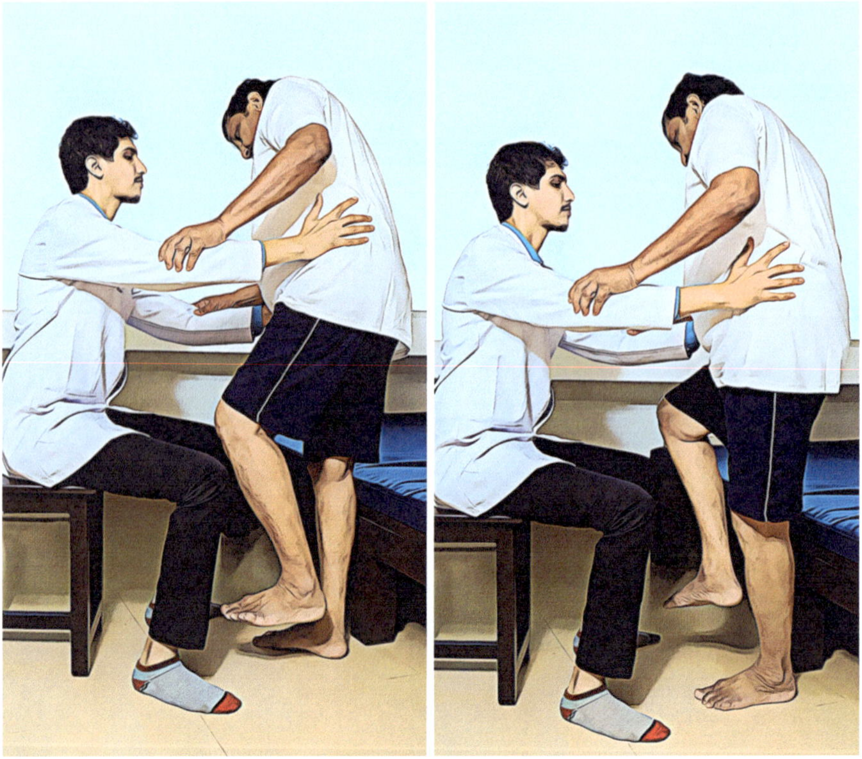

Animated photograph of model and therapist with permission

Fig. 3.43 Marching in place with the trunk mild to moderately tilted toward the uninvolved side

the patient can be guided to walk while maintaining the mild-to-moderate lean toward the uninvolved side. Typically, within 2–3 weeks, with the strategy mentioned above, the pushing behavior will subside, and the patient will be able to perform sitting up from lying and the sit-to-stand transfer without losing balance or falling consistently to the involved side.

The overall strategy is to overcorrect the pushing behavior until he or she can maintain an erect posture without the tendency to fall toward the affected side. For instance, if the patient tends to push toward the right side and backward, the overcorrection should be toward the left side and forward. In the initial phase of training, the use of mirrors may not provide much benefit. For patients with minimal pushing behavior, cognitive strategies and mirrors can be helpful in self-initiating weight-bearing toward the uninvolved side. Even motor learning strategies can be of help in reducing ill effects. Demonstrating correct vertical orientation, providing consistent feedback about the body's orientation, practicing correct weight shifts, and involving the patient for problem-solving may help in tackling the pushing behavior.

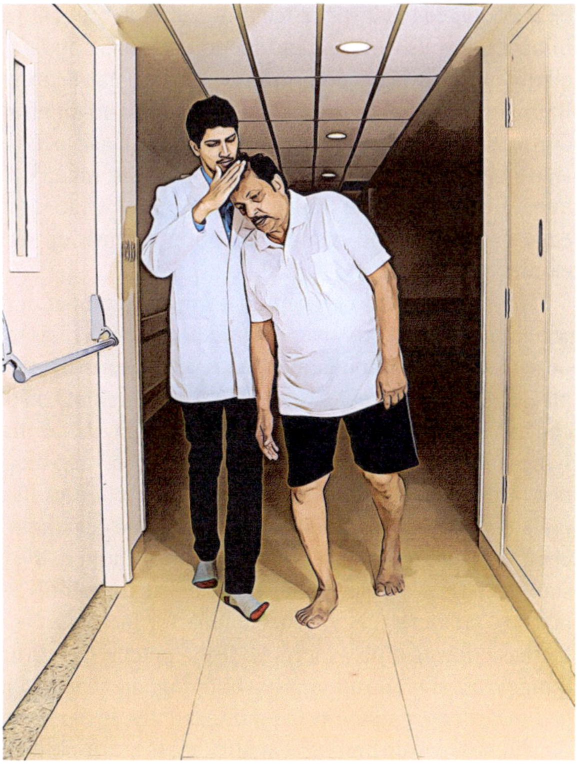
Animated photograph of model and therapist with permission

Fig. 3.44 An illustration depicting the strategy to walk with a mild-moderate lean toward the uninvolved side

3.11.2.14 Promoting Functional Independence

Restoring or improving functional independence in ADL is an important factor for the quality of life post-stroke. Stroke survivors who need assistance for ADL always feel themselves as a burden for their family members and often are socially isolated, overwhelmed, and abandoned. In addition to restoring the sensorimotor function, the therapy should include effective strategies aimed at improving functional independence in basic, instrumental, and everyday activities. The sensorimotor dysfunction of the paretic side can impose a tremendous challenge for relearning functional mobility and independence. During the early phase of therapy, strategies focusing on improving truncal symmetry and usage of both sides of the body should be the priority. For independent control, exercise should progress from assisted to guided movements and then eventually active movements.

Concerning bed mobility, rolling to either side should be encouraged right from the early stage of rehabilitation. Attempts to roll to the unaffected side can be difficult as compared to rolling to the affected side. While practicing to roll, care must be taken not to leave the affected upper extremity behind; instead, the patient must be encouraged to clasp the hands together to bring the affected extremity forward to initiate the roll. Rolling onto the affected side and encouraging the side-lying on-elbow position helps to promote early weight-bearing through the affected upper limb. Stroke patients should practice supine-to-sit after rolling from both sides, but during the bed mobility training, more emphasis needs to be given to the supine-to-sit using the involved side. Verbal commands, manual guidance, and assistance need to be given during the transition, especially while moving the affected leg over the edge of the bed and while attempting to push up to sit using the affected upper extremity.

The joints need to be kept in normal alignment during the transition and undue weight falling directly over the affected shoulder has to be avoided as it may predispose to shoulder impingement and pain. Rather than practicing the task as a whole, block practicing should be encouraged. Attempts also should be made to perform controlled lowering as it may help in gaining better motor control during the side-lying-to-sit transition. The rhythmic initiation and diagonal patterns that encourage bed transitions can also be incorporated. Activities like bridging and tentacle exercises need to be promoted as those exercises can help in developing trunk and hip control necessary for early weight-bearing through the foot. In addition to the above, such activities also promote scooting, bed mobility, sit-to-stand transfers, use of bedpan, pressure-relieving technique for the buttocks, and movements away from the basic lower limb synergies. If the patient has difficulty maintaining the crook-lying position while attempting bridging, the therapist may need to stabilize the foot and the affected side lower limb manually. If the hip extensors are weak, bridging the affected lower limb and unaffected lower limb in a tentacle position can be difficult. In such situations to encourage the contribution of affected side hip extensors, placing a small ball or a soft roll under the unaffected foot or crossing the unaffected limb and placing it over the affected limb can be tried for plain bridging before bridging along with tentacle exercise.

Achieving a symmetrical posture with the spine and pelvis in proper alignment should be the focus during the early phase of training in sitting. The therapist should encourage the patient to hold his or her spine erect, pelvis in neutral, and feet flat on the floor. Most of the stroke patients will sit with asymmetric posture bearing more weight on the unaffected side, pelvis posteriorly tilted, and kyphosis of the upper trunk with or without lateral flexion of the trunk to the affected side. Verbal or tactile cues, use of visual feedback, and guidance can encourage these patients to correct such posturing. Encouraging early sitting with bilateral support of upper extremities on the tabletop, large gym ball or the therapist's shoulder is meaningful. Exercise for encouraging the sitting control should progress from static to dynamic, stable to the unstable surface (gym ball or wobble board), dissociations of upper trunk from the lower trunk movements, and eventually dynamic challenges like reach-out activities in multiplanar direction.

Bridging and tentacle exercises and gym ball exercises encourage the activity of the abdominal obliques, generally improve trunk and pelvic control, and translate to better weight shifts while sitting. Practicing scooting in sitting or "butt walking" at the side of the bed and encouraging push-ups in sitting with both upper extremities (provided the affected upper limb has a certain amount of strength and control) can be attempted before sit-to-stand transitions are introduced for functional training.

Sit-to-stand transitions should be practiced with a focus on symmetrical weight-bearing, coordinated muscular responses, and appropriate timing. Placement of feet well behind, with the heel still in contact with the floor or the supporting surface, allows the ankle dorsiflexors to assist the transition. Improper placement of feet, poor alignment of the lower limbs, reduced forward momentum, premature activation of spinal extensors, asymmetrical weight-bearing through the feet, and the improper or poor sequencing of body components required for the sit-to-stand transition can make transfers unsafe or difficult. During the initial days of training for sit-to-stand transfers, to minimize the demand and extensor force required for the transition, a higher bed or seat is preferred to lower platforms. During the training, stroke patients will be instructed and/or assisted to place both feet on the floor right under the knees at hip-width apart. Verbal instructions like "sit tall" will encourage the patient to sit erect. Verbal and tactile cues are required to encourage the patient to flex his or her erect spine over the hips before initiating standing up by extending the lower limbs and then by extending the spine over the hips. Tactile and proprioceptive cues can also assist knee extension during the transition. The patient can be either advised to keep both upper extremities forward with hands clasped together or placed over the therapist's shoulders with the therapist sitting directly in front. Premature extension of the spine during this training process will make standing up unsafe or difficult. Similarly, uneven or excessive pushing off with both hands on the supporting surface should be discouraged as it may lead to uneven weight-bearing or excessive forward momentum. For smooth execution of sit-to-stand transitions, proprioceptive training (including static and dynamic partial squats) on a firm surface (Fig. 3.45) advancing toward a foam surface, weight shifts, single-limb stance using stall bars or manual assistance, and strengthening exercises for the affected lower extremity muscles are recommended.

During the sit-to-stand training, it is not uncommon to see stroke patients relying heavily on the unaffected lower limb. The tendency to lean to the unaffected side, more weight-bearing through the unaffected lower limb, and placement of the normal foot behind the weak foot indicate the patient's overdependence on the unaffected side for such tasks. Encouraging the patient to weight-bear more through the affected limb can only be possible if the affected foot is positioned behind the normal foot. Strengthening of affected side lower limb muscles and proprioceptive training for affected lower limb, including single limb partial squats with adequate support, can further enhance weight-bearing and confidence of the patient. Repeated practice, encouraging the patient to execute the movement with increasing speed, paying attention to the sequence of the components for sit-to-stand, and avoiding pauses between the components, need to be encouraged to improve the sit-to-stand transition. Proper positioning of the upper extremity with the affected elbow in extension and hands clasped together and learning to control the lower extremities eccentrically for a gradual descend to sit-down can further improve the overall control of sit-to-stand transition.

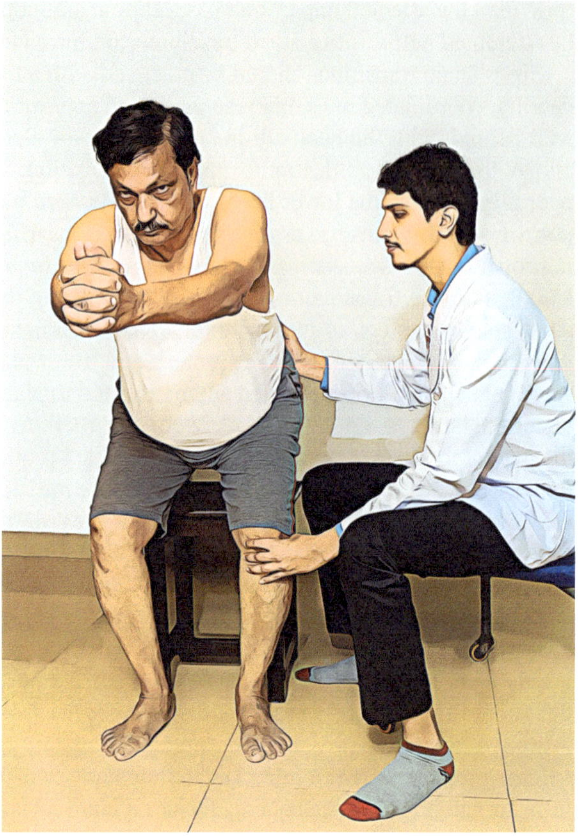

Fig. 3.45 Performing dynamic partial squats on a firm surface to improve knee control and strength

Animated photograph of model and therapist with permission

In the early phase of rehabilitation, while training transfer techniques (transfers like the bed to a wheelchair and wheelchair to a tub seat), the stroke patients may require maximal assistance. Adjusting the height of the hospital bed or wheelchair can reduce the difficulty level for transfer. Placing the wheelchair next to the unaffected side, instructing the patient to scoot the buttock forward to the edge of the bed, and assisting the patient to stand and pivot a quarter turn on the unaffected lower limb before sitting down is the sequence of components required for safe transfer from one surface to other. Though this strategy promotes early transfers, it neglects and minimizes the contribution of the affected side and encourages learned non-use. To avoid such compensatory strategies, the therapist should encourage the patient to transfer to both sides, with the emphasis more toward the affected side. If the upper extremity control is poor, with or without shoulder subluxation or pain, stabilization of affected arm in extension and external rotation against the therapist's body, cradling the hand and holding in forward flexion, and placement of the affected upper extremity over the therapist's shoulder are the appropriate strategies for the same. The therapist's knee can be used to exert a counterforce on the patient's knee during the transfer if the knee control is inadequate.

Both standing and modified plantigrade standing postures can facilitate functional activities. Modified plantigrade can be a safe posture to teach weight-bearing activities for both upper and lower limbs, especially when prone or prone progression postures are not advisable. If the potential for motor recovery is good and symmetrical weight-bearing over the base of support is possible, the patients should be encouraged to practice standing with one hand support and eventually with no upper extremity support. Activities that promote weight shifts, multiplanar reach outs, and activities that encourage the dissociation of the upper trunk to the lower trunk are strategies to improve dynamic stability. PNF techniques like rhythmic stabilization and slow reversal hold technique may help to improve poor postural stability in some stroke patients. Use of additional postures such as prone on elbows, quadruped, kneeling, and half kneeling on the therapeutic mat can increase the level of difficulty and improvise the postural control further. However, such postures can be inappropriate or unsafe for those patients with comorbidities such as ischemic heart disease, chronic obstructive pulmonary disease, and severe osteoarthritis of the hip or knee joint.

3.11.2.15 Strategies to Improve Gait

To develop effective gait training strategies, the therapist should understand the impairments primarily determining the walking ability of stroke patients. Muscle strength, motor control, and balance are the common impairments strongly related to gait. Of all the treatment strategies, exercise is the most common therapeutic intervention used to improve gait in hemiplegic patients. Graded muscle strengthening can improve the ability to generate force but generally does not transfer to improved walking ability. Providing task-specific practice in addition to resistance training to those stroke patients with impaired sensorimotor coordination may help to extract the benefits of strength gains. Early task-specific gait retraining with more repetition may facilitate the development of new motor programs or refinement of existing programs necessary to accommodate these deficits. Treadmill training with or without a harness system and task-specific overground locomotor training (practicing a wide variety of functional mobility tasks such as walking, sit-to-stand transfer, turning, obstacle training) are two main approaches for task-specific practice.

Treadmill practice can be considered as a "forced use" that maximizes the use of the hemiparetic limb by encouraging more number of steps, a greater amount of paretic limb loading, and better activation of the paretic muscles at different speeds. The treadmill with a harness or bodyweight support system enables lower functioning stroke patients to practice early walking when supervised traditional techniques are unsafe to practice. Concerning improvements in gait performance, the evidence is conflicting between treadmill training with or without bodyweight system and standard treatments.

Overground gait training provides a more natural stimulus to challenge the different components necessary for walking. Overground gait training promotes anticipatory postural control like avoiding an obstacle or changing the course to avoid bumping and reactive control, like responding to a slip or nudge. Treadmill training may not allow participants to experience the normal postural demands or visual sampling that occur during walking. In addition to the above, treadmill training will not offer many of the functional variations in gait, such as turning, rising from a chair to walk, and starting or stopping of gait. The successful overground practice of a variety of tasks that represent community walking, including stepping up a curb, walking in a crowded hallway, and walking on even and uneven surfaces may enhance self-efficacy and consequently walking ability. A combination of treadmill and overground task-specific locomotor training should be encouraged to gain the benefits of both approaches if centers have such facilities or infrastructure.

In addition to strengthening exercises for the lower limb muscles, stretching of the appropriate muscles, particularly, the calf muscles should be considered. Functional task-specific skills should include walking forward, backward, and sideways, braiding, and side-stepping. Throughout the training program, care has to be taken not to encourage those moves that can reinforce the basic limb synergies. As the patient gains more locomotion control, he or she should be encouraged to improve the rhythm and speed of walking. Verbal and auditory cues can facilitate rhythm and speed. For those patients with a scope of better motor recovery, prolonged and indiscriminate use of parallel bars and ambulation aids like hemi walkers and quadripod sticks are not advisable as those strategies can hinder the patient's potential to walk without the device, develop appropriate balance mechanisms, and encourage postural symmetry. Patients who are entirely dependent on such aids tend to walk slow with impaired locomotor rhythm.

The therapeutic interventions for gait training should incorporate the repetitive practice of a wide variety of mobility tasks. Such intensive mobility training should contain the components of graded strengthening using functional tasks including repetitive rise from a chair, stepping up and down a stepper, aerobic component including graded walking activity, cycle ergometer, and performing continuously moderate intensity functional tasks and a variety of challenging walking activities with substantial postural control demands. The use of a circuit of workstations with the components mentioned above and adequate rest will generally improve the performance of gait. In the back of the mind, the therapist also should realize that the optimal gait training program should include balance and agility training, which is in addition to the practice of upper extremity and trunk mobility tasks.

During the stance phase, knee hyperextension, an abnormal movement away from the anatomical neutral position into extension, is reported in approximately 65% of stroke subjects. Decreased ability to activate the knee extensor muscles, reduced eccentric control of the knee extensors, spasticity of quadriceps and plantar flexors, reduced strength of hamstrings, and proprioceptive deficits are the possible causes for the same. Longstanding knee hyperextension can lead to the posterior capsule and anterior cruciate ligament laxity predisposing to early degenerative changes of the knee joint, poor proprioceptive control during terminal knee extension, joint deformity, chronic knee pain, and reduced independence in daily activities. During the swing phase, knee hyperextension makes knee flexion difficult for effective ground clearance, in turn promoting circumduction and excessive energy consumption while walking. A novel technique "prowling along with proprioceptive training" developed by the author had shown significant improvement in reducing knee hyperextension, improving the ankle dorsiflexion range, and augmenting spatiotemporal parameters of gait. The word "prowling" means walking in a predatory manner, characterized by walking with bilateral knee bent attitude, which provides the dual advantage of activating the quadriceps muscles and changing the direction of moment-arm of quadriceps and increasing mechanical advantage. The proprioceptive training given for the involved lower limb consists of partial squats, single limb stance, and single limb dynamic partial squats while standing on firm ground as well as on a foam mat. The technique can be beneficial for reducing knee hyperextension while standing and walking for those stroke patients who are cognitively sound with the Brunnstrom recovery stage of ≥3, with no severe plantar flexor tightness.

The author believes that the optimal program to improve gait should involve the repetitive and intensive practice of tasks that loads more weight on the paretic limb, functional strengthening, and balance training, which is progressively incremented in difficulty according to the tolerance of the stroke participant. For community ambulation, tasks like walking on different terrains, negotiating curbs, climbing up and down the stairs, stepping over obstacles, turning, and quick stops and starts while walking should be incorporated. If space or resources are the constraints, the use of virtual reality simulating realistic visual stimuli can be an effective alternative.

3.12 Brainstem Stroke Syndromes

Brainstem stroke accounts for approximately 10% of all strokes. Brainstem stroke syndromes are a group of syndromes that are classically caused by the occlusion of small perforating arteries of the posterior circulation supplying the brainstem. Typically, these syndromes occur in the territories of the basilar or vertebral arteries. Posterior cerebral artery, superior, anterior inferior, and posterior inferior cerebellar arteries, basilar artery, and vertebral artery are main vessels that provide arterial supply to the brainstem. Except for basilar artery occlusion, all the remaining posterior circulation strokes have a low mortality of 5%. The outcome for bilateral vascular lesions of the brainstem due to occlusion of the basilar artery is generally bleak.

The brainstem plays a crucial role in controlling balance, coordination, eye movements, hearing, speech, and swallowing, and strokes within the regions of the brainstem will have a different set of clinical symptoms and challenges as compared to the hemispheric strokes. Ipsilateral cranial nerve palsy(ies) and contralateral weakness and/or hemisensory loss are the typical characteristic clinical pictures of all brainstem strokes. Dysarthria and dysphagia, headache, vomiting and nausea, ocular eye movement abnormalities and diplopia, vertigo, nystagmus, hemiplegia or quadriplegia, ataxia, and change in the level of consciousness are some of the specific impairments resulting from brainstem strokes. Hemineglect, perceptual dysfunction, aphasia, and apraxias are absent as the hemispheric cortical and subcortical areas are spared in brainstem stroke syndromes. In the following section, the author would like to brief some of the common brainstem stroke syndromes occurring at the midbrain, pons, and medullary levels.

3.12.1 Weber's Syndrome (Medial Midbrain Syndrome)

Medial midbrain syndrome (Fig. 3.46) occurs due to occlusion of the penetrating branches of the posterior cerebral artery that supplies the medial aspect of the midbrain. Contralateral paralysis of the lower part of the face, tongue, arm, and leg are its clinical manifestations. The third cranial nerve involvement results in ipsilateral gaze palsy or ophthalmoplegia. It is characterized by the inability to voluntarily move the ipsilateral eye upward and inward with an abnormal downward and outward resting position of the eye. In addition to the above, the involvement of the third cranial nerve also results in unresponsive dilated pupil, diplopia, and ptosis.

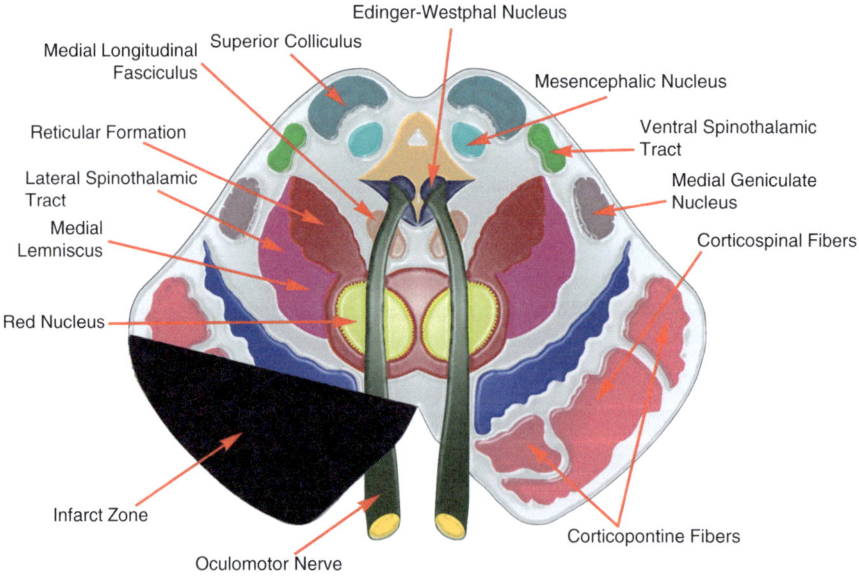

Fig. 3.46 An illustration of midbrain lesion in Weber's syndrome (medial midbrain syndrome)

3.12.2 *Benedikt Syndrome*

Lateral midbrain syndrome, also known as Benedikt syndrome (Fig. 3.47), is another brainstem stroke caused by the occlusion of the penetrating branches of the posterior cerebral artery. The location of the vascular lesion is somewhat lateral to the location for Weber's syndrome and involves the medial lemniscus and red nucleus. Occlusion of those penetrating branches causing Benedikt syndrome produces contralateral hemianesthesia, involuntary movements of the opposite side limbs (involvement of red nucleus), and tremor (involvement of dentatorubrothalamic tract) which is in addition to third cranial nerve palsy. Radiologically, it can often be difficult to distinguish Benedikt from Weber's syndrome.

3.12.3 *Locked-In Syndrome*

Also known as cerebromedullospinal disconnection or pseudo-coma or de-efferented state is caused by the occlusion of the basilar artery, which supplies the pons. The syndrome is easy to identify as a collection of bilateral long tract signs (motor and sensory) supplemented by fifth to eighth cranial nerve dysfunctions, which result in quadriplegia and inability to speak. Locked-in syndrome patients are aware and awake but cannot move or communicate verbally due to

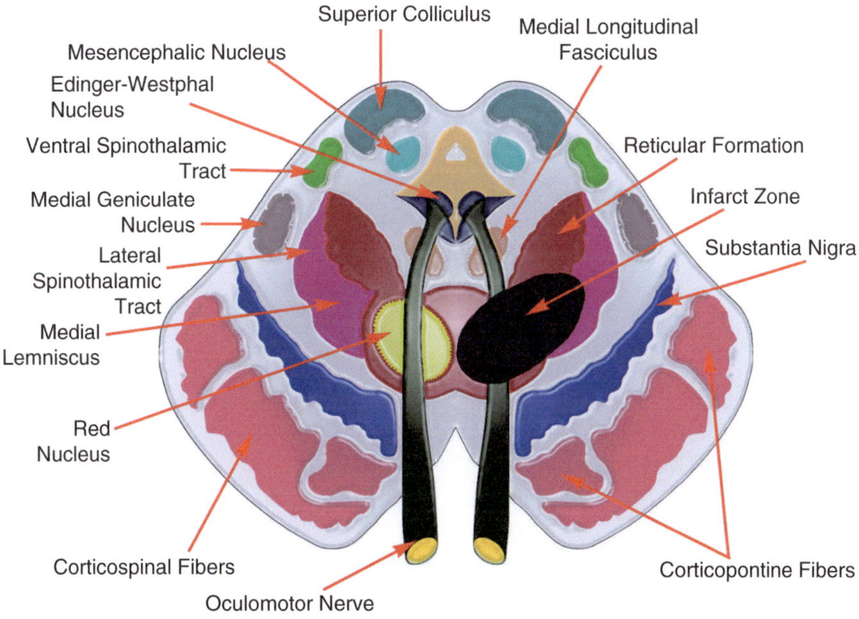

Fig. 3.47 An illustration of midbrain lesion in Benedikt syndrome

complete paralysis of nearly all voluntary muscles in the body except for the eyes. These patients may use their spared movements of the eyes and blinking to communicate with others. Bilateral corticospinal tracts involvement causes weakness or paralysis of both upper and lower extremities, and bilateral corticobulbar tracts involvement causes facial muscle weakness, dysarthria, and dysphagia. Extensive lesions involving bilateral sixth cranial nerves will cause horizontal gaze weakness, and in such cases, the vertical gaze will be the only eye movement possible. Respiratory failure is the common cause of death for many patients with locked-in syndrome.

3.12.4 Medial Pontine Syndrome (Foville's Syndrome)

The medial pontine syndrome (Fig. 3.48) most commonly results from the occlusion of the paramedian branches of the basilar artery, which supplies the medial territory of the pons. Except for the different distinguishing cranial nerve features, the clinical features of patients presenting with the medial pontine syndrome is similar to those presenting with the medial medullary syndrome. The spastic weakness of the contralateral arm and leg (involvement of corticospinal tract), loss of vibration, kinesthetic and position sense of the contralateral arm and leg (medial lemniscus involvement), and strabismus and lateral gaze palsy (involvement of ipsilateral sixth cranial nerve) are the common features of this syndrome.

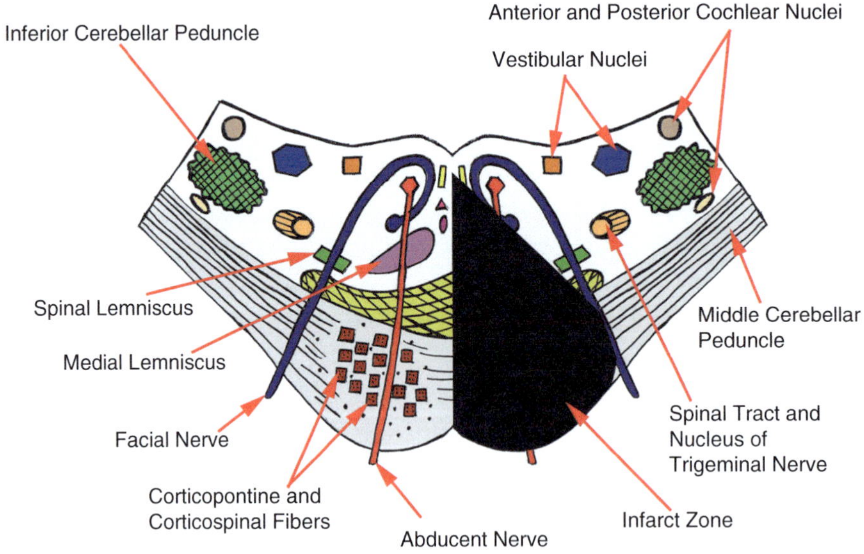

Fig. 3.48 Illustrating the pontine lesion in medial pontine syndrome

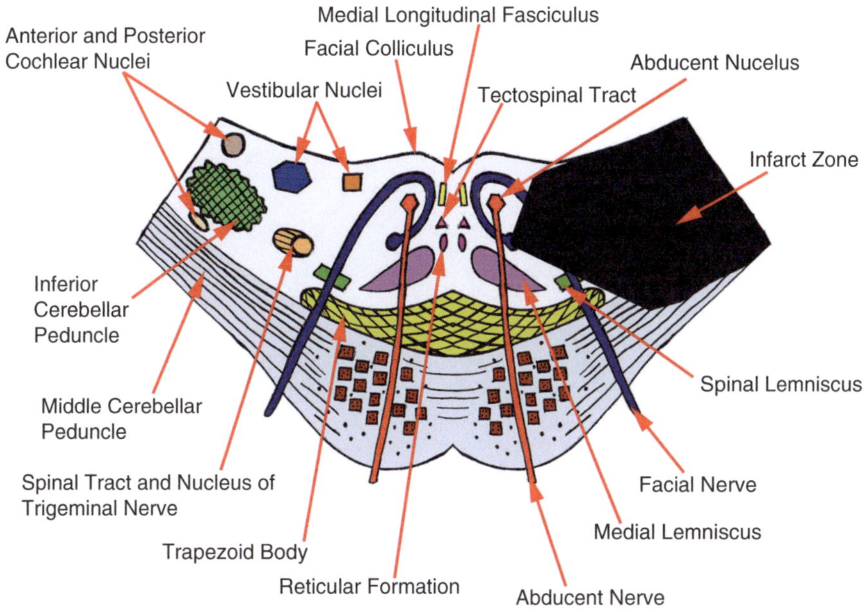

Fig. 3.49 Illustrating the pontine lesion in lateral pontine syndrome

3.12.5 Lateral Pontine Syndrome (Marie-Foix Syndrome)

The occlusion of the anterior inferior cerebellar artery that supplies the lateral territory of the pons causes the lateral pontine syndrome (Fig. 3.49). The clinical manifestations of this syndrome are very similar to those of lateral medullary syndrome, except for different distinguishing cranial nerve features. Lateral pontine syndrome patients typically present with pain and temperature loss of the contralateral trunk and extremities, ataxia of the ipsilateral extremities, gait ataxia, ipsilateral pain and temperature loss over the face, ipsilateral lower motor neuron facial palsy, reduction of ipsilateral lacrimation and salivation, loss of taste from the anterior two-thirds of the tongue, loss of corneal reflex, ipsilateral hearing loss, nystagmus, nausea, vertigo and vomiting, and Horner's syndrome.

3.12.6 Millard-Gubler Syndrome

Millard-Gubler syndrome, also known as ventral pontine syndrome (Fig. 3.50), is caused by the unilateral vascular lesion of the ventrocaudal pons involving the basis pontis and the fascicles of sixth and seventh cranial nerves. The occlusion of the short circumferential or paramedian branches of the basilar artery results in this

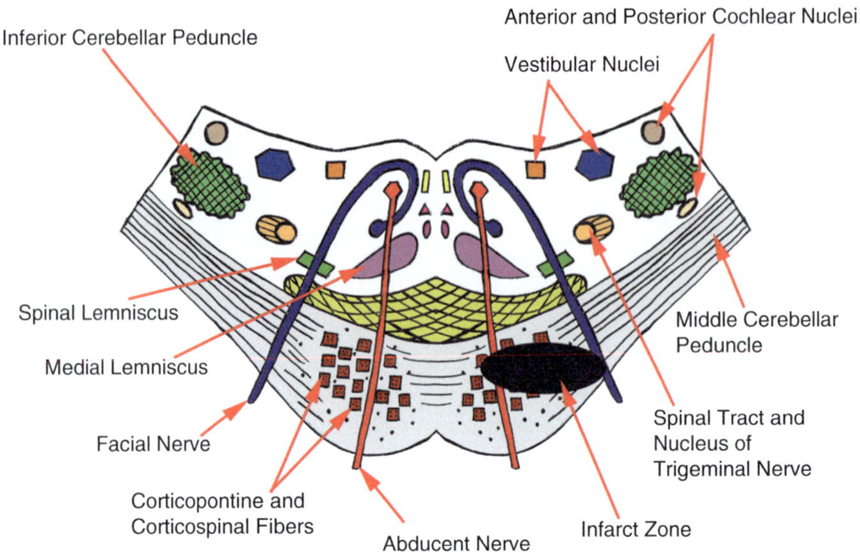

Fig. 3.50 An illustration of pontine lesion in Millard-Gubler syndrome

syndrome. Paralysis of the sixth cranial nerve leads to diplopia, internal strabismus, and inability to rotate the affected eye outward. The involvement of the seventh cranial nerve causes ipsilateral lower motor neuron facial paralysis with loss of corneal reflex and contralateral hemiplegia (sparing the face) due to pyramidal tract involvement. The unique feature of the Millard-Gubler syndrome is the "crossed hemiplegia," i.e., ipsilesional lower motor neuron facial paralysis and contralesional hemiplegia.

3.12.7 Lateral Medullary Syndrome of Wallenberg

Lateral medullary syndrome (Fig. 3.51), also known as Wallenberg syndrome, usually occurs due to the occlusion of the posterior inferior cerebellar artery, if not the vertebral artery branch supplying the vascular territory. It is the commonest of brainstem strokes and involves the dorsolateral medulla and the cerebellum. The clinical features of the lateral medullary syndrome consist of ipsilateral hemisensory loss of pain and temperature sensation of the face due to involvement of the fifth cranial sensory nucleus (the spinal nucleus of the trigeminal nerve) and contralateral hemisensory loss of pain and temperature due to involvement of lateral spinothalamic tract. The involvement of the inferior cerebellar peduncle leads to ipsilateral ataxia of the extremities, gait ataxia, and postural asymmetry. In addition to nausea, vomiting, and vertigo, the involvement of the vestibular nucleus causes ipsilateral nystagmus. Paralysis of the ipsilateral palatal and laryngeal muscles

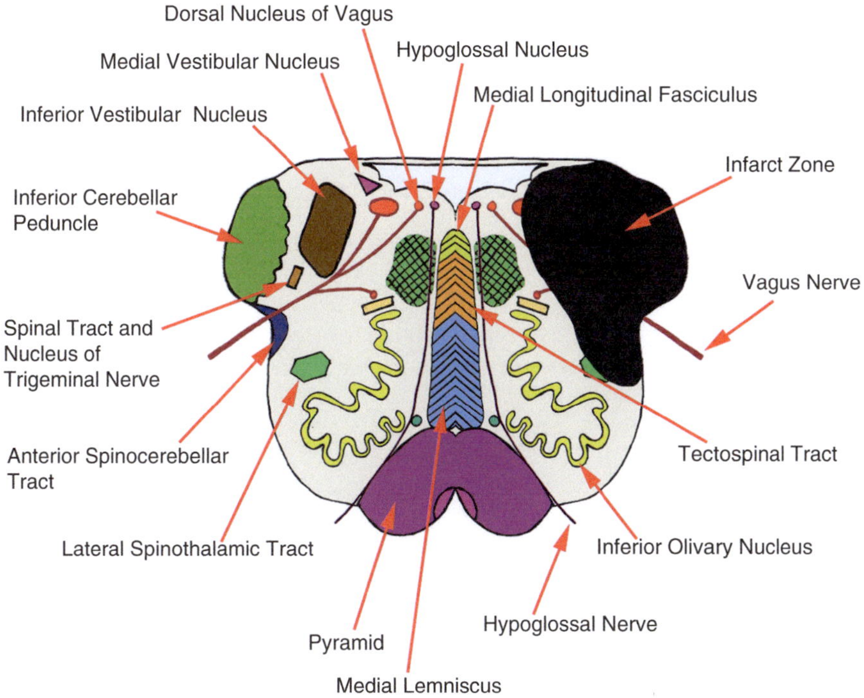

Fig. 3.51 Illustrating the medullary lesion in lateral medullary syndrome of Wallenberg

causes dysphagia and dysarthria, which is in addition to hoarseness and uncontrollable hiccups caused by the involvement of the nucleus ambiguus. The lateral medullary syndrome patients will also have ipsilateral Horner syndrome due to the involvement of the descending sympathetic fibers.

3.12.8 Medial Medullary Syndrome of Wallenberg

Inferior alternating syndrome, hypoglossal alternating hemiplegia, lower alternating hemiplegia, and Dejerine syndrome are the synonyms of the medial medullary syndrome. The medial medullary syndrome (Fig. 3.52) syndrome commonly occurs due to the occlusion of the vertebral artery, which supplies the medial aspect of the medulla. The syndrome presents with contralateral hemiparesis due to the involvement of the pyramidal tract. Medial lemniscus involvement causes contralateral impairment of position and kinesthetic sensations and discriminative sensations. The 12th cranial nerve involvement results in ipsilateral paralysis of tongue muscles and deviation of the tongue to the ipsilateral side.

Table 3.17 provides information about some of the selective brainstem stroke syndromes. Many times the infarctions within the brainstem are patchy and neither

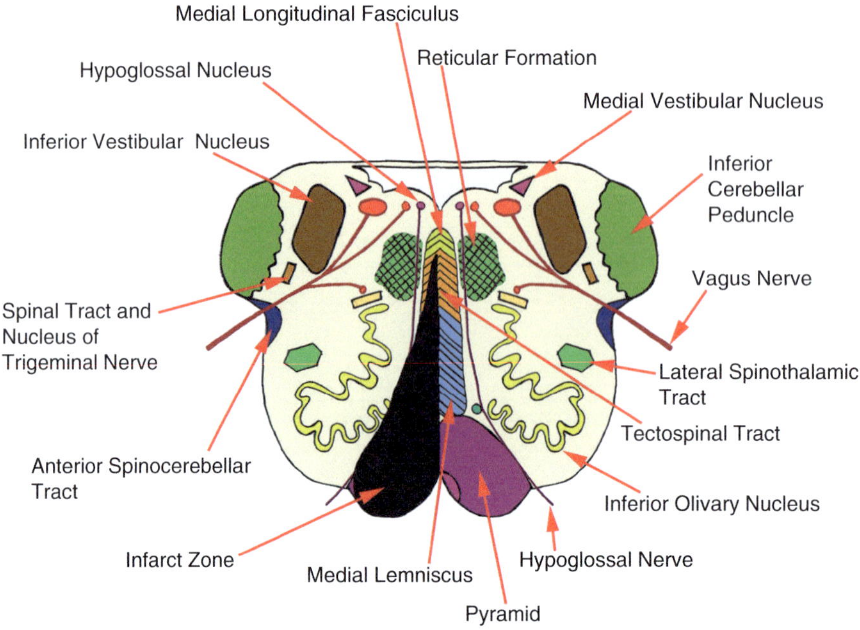

Fig. 3.52 An illustration of medullary lesion in medial medullary syndrome

Table 3.17 Details of certain selective brainstem stroke syndromes

Syndrome name	Lesion location	Artery involved	Cranial nerves involved	Tracts involved	Signs/features
Weber	Base of midbrain	Paramedian branches of posterior cerebral artery or perforating branches from basilar bifurcation	3rd cranial nerve	Corticospinal	Oculomotor palsy, crossed hemiplegia
Claude	Midbrain tegmentum	Perforating branches of posterior cerebral artery	3rd cranial nerve	Red nucleus and brachium conjunctivum	Oculomotor palsy, contralateral cerebellar ataxia and tremor
Benedikt	Midbrain tegmentum	Paramedian penetrating branches of posterior cerebral artery or basilar artery	3rd cranial nerve	Red nucleus	Oculomotor palsy, contralateral abnormal movements

Table 3.17 (continued)

Syndrome name	Lesion location	Artery involved	Cranial nerves involved	Tracts involved	Signs/features
Nothnagel	Midbrain tectum	Penetrating branches of basilar artery	Unilateral or bilateral third cranial nerve	Superior cerebellar Peduncles	Oculomotor palsy, ipsilateral cerebellar ataxia
Locked-in	Bilateral pons	Basilar artery	5th to 12th cranial nerves	Corticobulbar and corticospinal tract	Quadriplegia and fifth to 12th cranial nerve palsies
Millard-Gubler	Caudal ventral medial pons	Short circumferential or paramedian branches of basilar artery	Fascicles of the 6th and seventh cranial nerves	Corticospinal tract	Abduction and peripheral facial palsy, contralateral hemiplegia
Foville	Caudal tegmental medial pons	Perforating branches of basilar artery	6th and 7th cranial nerves	Corticospinal tract, medial lemniscus, medial longitudinal fascicle	Gaze and peripheral facial palsy, contralateral hemiparesis (and hypesthesia, internuclear ophthalmoplegia)
Marie-Foix	Lateral caudal pons	Anterior inferior cerebellar artery	Nil	Middle cerebellar peduncle, corticospinal and spinothalamic tracts	Ipsilateral ataxia, contralateral hemiparesis and spinothalamic sensory loss
Wallenberg	Medulla, lateral tegmentum	Posterior inferior cerebellar artery or rarely vertebral artery	Spinal nucleus of 5th nerve, 9th to 11th cranial nerves	Lateral spinothalamic tract, descending sympathetic fibers, spino- & olivo-cerebellar tracts	Ipsilateral 5th and 9th to 11th cranial nerve palsies, Horner's syndrome, cerebellar ataxia; contralateral pain and temperature deficit
Déjerine	Medial medullary region	Vertebral artery and/or paramedian branches of the anterior spinal artery	12th cranial nerve	Corticospinal, medial lemniscus	Ipsilateral tongue palsy, contralateral hemiplegia, and dorsal column sensory loss

follow the theoretic distribution of a major artery nor present the classical features of complete clinical syndromes. Individual variations in collateral circulation and overlap between the arteries that supply the brainstem are the possible reasons for such diversities.

3.13 Physiotherapy for Brainstem Stroke Syndromes

Though there is extensive literature on physiotherapy for strokes affecting the cerebral hemispheres, there is a paucity of literature on the physiotherapy of brainstem strokes. Overall the brainstem strokes are less frequent than hemispheric strokes and many of the patients with brainstem strokes may experience minimal neurological or functional deficits after stroke. Alternatively, a certain percentage of patients may experience catastrophic brainstem stroke, typically affecting bilateral pontine vascular territory, with low survival rates or profound neurological or functional deficits that preclude their participation in rehabilitation programs. Though conditions like locked-in syndrome have a bleak prognosis, the current literature indicates that within the first year post-stroke, the brainstem stroke survivors have a 35% probability of returning to independent living compared to a 22% probability of hemispheric stroke survivors.

In addition to gaze palsies and cranial nerve deficits, brainstem strokes can lead to several physical impairments, including hemiplegia or quadriplegia, somatosensory deficits, and ataxia leading to gait deviations, reduced balance, and safety. Sensorimotor deficits can also reduce the strength of muscles of the affected side extremities. The brainstem stroke patients can have dissociated sensory loss like pain and temperature loss in lateral medullary syndrome or loss of vibration and kinesthetic sensations in medial medullary syndrome. The impairments mentioned above can lead to functional limitations in bed mobility, transfers, self-care, and gait.

The therapist should perform a detailed evaluation of the patient to derive the treatment objectives and goals and means of treatment. Since the sensorimotor impairments and functional deficits are diverse in various brainstem stroke syndromes, the strategies selected for treatment should be based on the clinical presentation, the problem list, and the functional limitations of the patient. To a certain extent, the standard physiotherapy exercises designed to improve sensation, ROM, muscle strength, postural control, balance, transfers, locomotion, and stair climbing can address most of the issues seen in these patients. The treatment strategies should incorporate activities that encourage postural correction, trunk and pelvis activation, sitting and standing balance, and shoulder and scapular mobility. The overground exercises should advance from the stable to the unstable surfaces, and if the facility for aquatic therapy is available, the latter can be used for improving the posture, balance, and strength of the patients. Task-related motor training can improve the patients' balance capability while sitting, standing, and performing reaching activities. For those patients with mild to moderate paresis, a restorative approach focusing on improvement of the motor function should incorporate strengthening exercises, overground exercises, treadmill training with partial body weight support, and functional electrical stimulation. The use of a biofeedback unit, training using a mirror and balance training with a balance board can be attempted for those patients with a scope of further balance improvement. Facial muscle strengthening exercises and exercises to improve the tongue muscle strength may reduce the dysarthria of the patients.

Approximately 70% to 80% of brainstem stroke patients are likely to be ataxic and these patients suffer from movement inaccuracies including errors in timing, irregular trajectories, delayed initiation of movements, movement decomposition, and imprecision during reaching activities. The therapeutic exercises specific to cerebellar ataxia can include Frenkel's exercise, other exercise strategies for dysmetria, postural and balance control training, and strengthening exercises. For further understanding of cerebellar ataxia and the strategies related to it, the author advises the readers to read the chapter "Cerebellar Dysfunction".

Locked-in syndrome patients will be near completely dependent with considerably reduced mobility and are most likely to develop serious complications due to immobilization. Though the benefits of exercise after stroke have been widely reported, the present evidence concludes that exercises have no adverse effects and may benefit locked-in syndrome patients. The quality of the existing evidence for the benefits of exercise for the locked-in syndrome is relatively low, as the study designs and interventions were heterogeneous; therefore the outcomes cannot be generalized. Typically, these patients require the same care and management strategies that are given for those patients who have quadriplegia following higher cervical spinal cord injuries.

Further Reading

Aroor S, Singh R, Goldstein LB. BE-FAST (balance, eyes, face, arm, speech, time): reducing the proportion of strokes missed using the FAST Mnemonic. *Stroke.* 2017;48(2):479–81.

Balaban B, Tok F. Gait disturbances in patients with stroke. *PMR.* 2014;6(7):635–42.

Bjorklunds A, Fecht A. The effectiveness of constraint-induced therapy as a stroke intervention: a meta-analysis. *Occup Ther Health Care.* 2006;20(2):31–49.

Bohannon RW. Muscle strength and muscle training after stroke. *J Rehabil Med.* 2007;39(1):14–20.

Bohannon RW, Smith MB. Interrater reliability of a modified Ashworth scale of muscle spasticity. *Phys Ther.* 1987 Feb;67(2):206–7.

Bovend'Eerdt TJ, Newman M, Barker K, Dawes H, Minelli C, Wade DT. The effects of stretching in spasticity: a systematic review. *Arch Phys Med Rehabil.* 2008;89(7):1395–406.

Broetz D, Karnath H. New aspects for the physiotherapy of pushing behavior. *Neurorehabil.* 2005;20:133–8.

Burn J, Dennis M, Bamford J, Sandercock P, Wade D, Warlow C. Long-term risk of recurrent stroke after a first-ever stroke. The Oxfordshire community stroke project [published correction appears in stroke 1994 Sep;25(9):1887]. *Stroke.* 1994;25(2):333–7.

Caplan LR. Caplan's stroke: a clinical approach. fifth Cambridge: Cambridge University Press; 2016.

Chayer C, Freedman M. Frontal lobe functions. *Curr Neurol Neurosci Rep.* 2001;1:547–52.

Chen MD, Rimmer JH. Effects of exercise on quality of life in stroke survivors: a meta-analysis. *Stroke.* 2011;42(3):832–37.

Cuoco JA, Hitscherich K, Hoehmann CL. Brainstem vascular syndromes: A practical guide for medical students. *Edorium J Neurol.* 2016;3:4–16.

Dalal KK, Joshua AM, Nayak A, Mithra P, Misri Z, Unnikrishnan B. Effectiveness of prowling with proprioceptive training on knee hyperextension among stroke subjects using videographic observation- a randomised controlled trial. *Gait Posture.* 2018;61:232–7.

Davies PM. Right in the middle: selective trunk activity in the treatment of adult hemiplegia. Berlin: Springer-Verlag New York Inc.; 1990.

Deb P, Sharma S, Hassan KM. Pathophysiologic mechanisms of acute ischemic stroke: An overview with emphasis on therapeutic significance beyond thrombolysis. *Pathophysiology.* 2010;17(3):197–218.

Donaghy M, editor. Brain's Diseases of the nervous system. 12th New York: Oxford University Press; 2009.

Dromerick AW, Edwards DF, Hahn M. Does the application of constraint-induced movement therapy during acute rehabilitation reduce arm impairment after ischemic stroke?. *Stroke.* 2000;31(12):2984–8.

Eng JJ, Tang PF. Gait training strategies to optimize walking ability in people with stroke: a synthesis of the evidence. *Expert Rev Neurother.* 2007;7(10):1417–36.

Fisher M, Garcia JH. Evolving stroke and the ischemic penumbra. *Neurology.* 1996;47(4):884–8.

Flansbjer UB, Miller M, Downham D, Lexell J. Progressive resistance training after stroke: effects on muscle strength, muscle tone, gait performance and perceived participation. *J Rehabil Med.* 2008;40(1):42–8.

Ford A, McGregor KM, Case K, Crosson B, White KD. Structural connectivity of Broca's area and medial frontal cortex. *Neuroimage.* 2010;52(4):1230–7.

Gray V, Rice CL, Garland SJ. Factors that influence muscle weakness following stroke and their clinical implications: a critical review. *Physiother Can.* 2012;64(4):415–26.

Hammer A, Nilsagård Y, Wallquis M: Balance training in stroke patients–a systematic review of randomized, controlled trials. *Adv Physiother.* 2008;10:163–72.

Han P, Zhang W, Kang L, Ma Y, Fu L, Jia L, et al. Clinical evidence of exercise benefits for stroke. *Adv Exp Med Biol.* 2017;1000:131–51.

Harrison RA, Field TS. Post stroke pain: identification, assessment, and therapy. *Cerebrovasc Dis.* 2015;39(3–4):190–201.

Hesse S. Treadmill training with partial body weight support after stroke: a review. *Neurorehabilitation.* 2008;23(1):55–65.

Jovin TG, Demchuk A, Gupta R. Pathophysiology of acute ischemic stroke. *CONTINUUM Lifelong Learning in Neurology.* 2008;14:28–45.

Karnath H, Ferber S, Dichgans J. The origin of contraversive pushing: evidence for a second graviceptive system in humans. *Neurology.* 2000; 55(9):1298–304.

Kruger E, Teasell R, Salter K, Foley N, Hellings C. The rehabilitation of patients recovering from brainstem strokes: case studies and clinical considerations. *Top Stroke Rehabil.* 2007;14(5):56–64.

Kumar S, Selim MH, Caplan LR. Medical complications after stroke. *Lancet Neurol.* 2010;9(1):105–18.

Law YM, Feng LF, Liang Q, Meng LJ, Shen P, Yu SJ, et al. Effect of exercise on physical recovery of people with locked-in syndrome after stroke: what do we know from the current evidence? A systematic review *Cerebrovasc Dis Extra.* 2018;8(2):90–5.

Li S, Francisco GE. New insights into the pathophysiology of post-stroke spasticity. *Front Hum Neurosci.* 2015;9:192.

Li S, Francisco GE, Zhou P. Post-stroke Hemiplegic Gait: New Perspective and Insights. *Front Physiol.* 2018;9:1021. https://doi.org/10.3389/fphys.2018.01021.

Lubetzky-Vilnai A, Kartin D. The effect of balance training on balance performance in individuals poststroke: a systematic review. *J Neurol Phys Ther.* 2010;34:127–37.

Macko RF, Ivey FM, Forrester LW, Hanley D, Sorkin JD, Katzel LI, et al. Treadmill exercise rehabilitation improves ambulatory function and cardiovascular fitness in patients with chronic stroke: a randomized, controlled trial. *Stroke*. 2005;36(10):2206–11.

Mancall EL, Brock DG, editors. Gray's clinical neuroanatomy: the anatomic basis for clinical neuroscience. Philadelphia: Elsevier Saunders; 2011.

Mohr JP, Albers GW, Amarenco P, Babikian VL, Biller J, Brey RL, et al. Etiology of stroke. *Stroke*. 1997;28:1501–6.

Musuka TD, Wilton SB, Traboulsi M, Hill MD. Diagnosis and management of acute ischemic stroke: speed is critical. *CMAJ*. 2015;187(12):887–93.

Na EH, Yoon TS, Han SJ. Improvement of quiet standing balance in patients with wallenberg syndrome after rehabilitation. *Ann Rehabil Med*. 2011;35(6):791–7.

Odier C, Michel P. Common stroke syndromes. In: Brainin M, Heiss WD, editors. Textbook of stroke medicine. second Cambridge: Cambridge University Press; 2014 155–68.

Ojaghihaghighi S, Vahdati SS, Mikaeilpour A, Ramouz A. Comparison of neurological clinical manifestation in patients with hemorrhagic and ischemic stroke. *World J Emerg Med*. 2017;8(1):34–8.

O'Sullivan SB, Schmitz TJ, Fulk GD, sixth ed. Physical rehabilitation. Philadelphia: FA Davis Co; 2014.

Parent A, editor. Carpenter's human neuroanatomy. ninth Baltimore: Williams & Wilkins; 1996.

Patestas MA, Gartner LP. A textbook of neuroanatomy. Oxford: Blackwell Publishing; 2006.

Pedersen PM, Wandel A, Jørgensen HS, Nakayama H, Raaschou HO, Olsen TS. Ipsilateral pushing in stroke: incidence, relation to neuropsychological symptoms, and impact on rehabilitation. The copenhagen stroke study. *Arch Phys Med Rehabil*. 1996;77(1):25–8.

Ropper AH, Samuels MA, Klein JP, Prasad S, editors. Adams and Victor's Principles of neurology. tenth New York: McGraw Hill Education; 2014.

Saleem S, Yousuf I, Gul A, Gupta S, Verma S. Hyponatremia in stroke. *Ann Indian Acad Neurol*. 2014;17(1):55–7. Available from: https://doi.org/10.4103/0972-2327.128554.

Schott M, Dachi L, Cassol MC, Brondani JE, Colpo E. Risk of pressure ulcer in hospitalized patients after stroke:relation of nutritional factors and of morbidity. *Clin Pract*. 2018;15(1):424–32.

States RA, Salem Y, Pappas E. Overground gait training for individuals with chronic stroke: a Cochrane systematic review. *J Neurol Phys Ther*. 2009;33(4):179–86.

Stewart KC, Cauraugh JH, Summers JJ. Bilateral movement training and stroke rehabilitation: a systematic review and meta-analysis. *J Neurol Sci*. 2006;244:89–95.

Teasell R, Foley N, Doherty T, Finestone H. Clinical characteristics of patients with brainstem strokes admitted to a rehabilitation unit. *Arch Phys Med Rehabil*. 2002;83(7):1013–6.

Tenny S, Thorell W. Cerebral salt wasting syndrome. Treasure Island (FL): StatPearls Publishing; 2019.

Tinetti ME. Performance-oriented assessment of mobility problems in elderly patients. *J Am Geriatr Soc*. 1986 Feb;34(2):119–26.

Truelsen T, Begg S, Mathers C. The global burden of cerebrovascular disease. WHO Discussion Paper. Geneva, Switzerland: World Health Organization; 2006.

Wasner G, Schattschneider J, Binder A, Baron R. Complex regional pain syndrome-diagnostic, mechanisms, CNS involvement and therapy. *Spinal Cord*. 2003;41(2):61–75.

Yang YR, Wang RY, Lin KH, Chu MY, Chan RC. Task-oriented progressive resistance strength training improves muscle strength and functional performance in individuals with stroke. *Clin Rehabil*. 2006;20(10):860–70.

Chapter 4
Extrapyramidal Disorders

Abraham M. Joshua and Zulkifli Misri

4.1 Neuroanatomy and Neurophysiology

The portions of the brain and brainstem that are not a part of the direct cortico-pyramidal system contributing to motor control are collectively known as the extra-pyramidal system. The system includes neural pathways through basal ganglia, reticular formation of the brainstem, vestibular nuclei, and red nuclei. Though the basal ganglia and their associated diencephalic and mesencephalic structures have been traditionally referred to as the extrapyramidal system, it is difficult to ascribe specific neurophysiological functions to these diverse groups of motor control areas and for the same reason, the use of "extrapyramidal" is beginning to fade, both clinically as well as neurophysiologically.

The basal ganglia and its related nuclei primarily engage in motor control and have certain roles in motor learning, behavior, emotions, and executive functions. Anatomical and functional components of the basal ganglia differ in each literature. The corpus striatum, amygdala, and claustrum are the three anatomical components of the basal ganglia. The corpus striatum is the largest structure present in the basal ganglia and because of its functional association, literature considers substantia nigra, subthalamic nucleus, red nucleus, and reticular formation as the physiological parts of the basal ganglia.

A. M. Joshua (✉)
Department of Physiotherapy, Kasturba Medical College of Mangalore, Manipal Academy of Higher Education, Manipal, India
e-mail: abraham.joshua@manipal.edu

Z. Misri
Department of Neurology, Kasturba Medical College of Mangalore, Manipal Academy of Higher Education, Manipal, India
e-mail: misri.zk@manipal.edu

A. M. Joshua (ed.), *Physiotherapy for Adult Neurological Conditions*,
https://doi.org/10.1007/978-981-19-0209-3_4

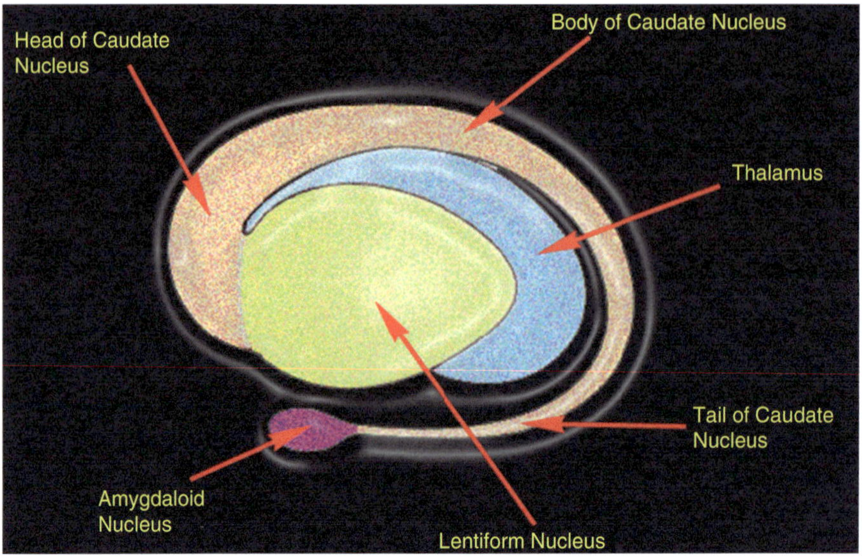

Fig. 4.1 An illustration of the anatomical structure, the corpus striatum

Anatomically, the corpus striatum is made up of two structures: caudate nucleus and lentiform nucleus (Fig. 4.1). The caudate nucleus, the "C"-shaped part of basal ganglia that lies lateral to the thalamus, closely related to the lateral ventricle, has a wide head (caput) anteriorly, a tapered body (corpus) posteriorly, and a tail curled forwardly. Located lateral to the caudate nucleus and thalamus and deeply buried within the white matter of each cerebral hemisphere is a lens or wedge-shaped subcortical structure called the lentiform nucleus. The lentiform nucleus is comprised of the globus pallidus (pallidum) medially and the putamen laterally. The anterior limb of the internal capsule partially separates the caudate nucleus from the putamen of the lentiform nucleus (Fig. 4.2). The globus pallidus, phylogenetically an older structure, is somewhat pale in appearance owing to the presence of myelinated fibers. The pallidum is further divided into globus pallidus interna medially and globus pallidus externa laterally, by a medial medullary lamina.

Situated in the temporal lobe, close to the uncus, a small oval or almond-shaped structure, closely related to the hypothalamus, the hippocampus, and the cingulate gyrus is the amygdaloid nucleus. Being a part of the olfactory and limbic systems, the nucleus plays an essential role in the sense of smell, motivation, and emotional behavior. The claustrum, the third anatomical component of basal ganglia, is located between the insular cortex and the putamen. Though the claustrum has extensive connections with many regions of the cerebral cortex, its functions are largely mysterious and may play a certain role in synchronizing different perceptual, cognitive, and motor functions.

Functionally, the basal ganglia have an afferent and an efferent region. The caudate nucleus and the putamen, together known as the "striatum," considered to be

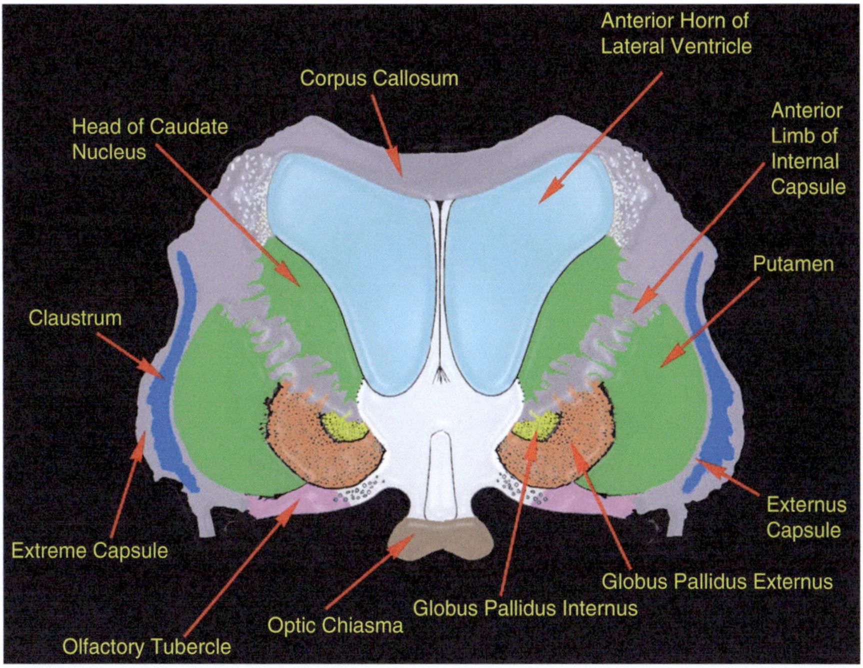

Fig. 4.2 The coronal section of the basal ganglia, demonstrating the partial separation of the caudate nucleus from the putamen by the anterior limb of the internal capsule

the input center or region of basal ganglia, receive input from most parts of the cerebral cortex (corticostriate fibers), the intralaminar thalamic nuclei, the substantia nigra, and the dorsal raphe nucleus (Fig. 4.3). The corticostriate fibers from each region of the cortex project to a specific part of the striatum. Most of the striatal inputs for the caudate region originate from the association cortex and most of the striatal inputs for the putamen region originate from the sensorimotor cortex. The extensive somatotopically arranged projections from the cerebral cortex, predominantly from the ipsilateral sensorimotor cortex, suggest a close inter-functional relationship between the striatum and the cortex. The projections from the thalamus are also somatotopic and predominantly originate from the centromedian nucleus, a part of the intralaminar nucleus of the thalamus.

Information received from the different regions of the brain is processed in the striatum and then sent to the globus pallidus and the substantia nigra, the functional output or efferent regions of the basal ganglia (Fig. 4.4). Information from these structures are transmitted to the ventroanterior and ventrolateral nuclei of the thalamus via the pallidothalamic tracts and then to the cortex. In addition to the above, the pallidal and nigral outputs also project to the superior colliculus and the reticular formation of the brainstem. Like the cerebellum, the basal ganglia have no direct descending pathway to the alpha or gamma motor neurons of the spinal cord.

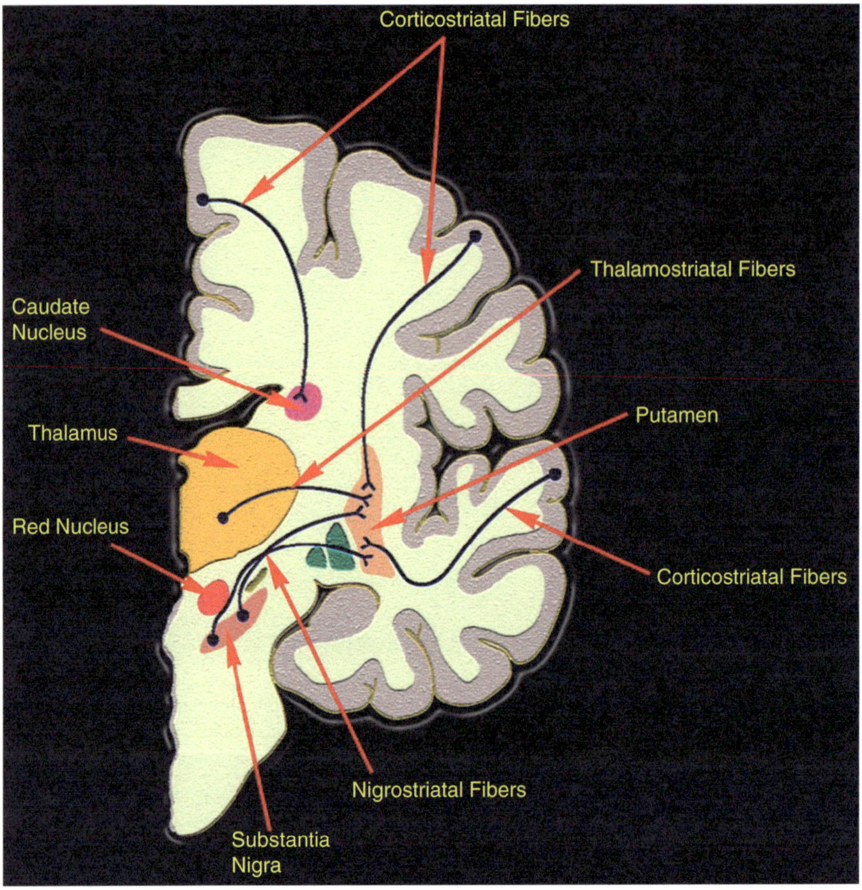

Fig. 4.3 A pictorial illustration of the afferent connections of the basal ganglia

However, the ganglia can influence the motor system through its thalamic and subthalamic projections, and its projections to the brainstem networks, together with those via the thalamocortical loops, contribute to the control of postural muscle tone and locomotion.

Except for the corticospinal pathway (pyramidal tract) arising from the cerebral cortex, four out of five crucial descending tracts from the brain arise from the brainstem and constitute the extrapyramidal tracts. The medial and lateral reticulospinal tracts are largely uncrossed and terminate on the interneurons located at the spinal cord and mainly influence the axial and proximal muscles of the extremities and are primarily responsible for locomotion and postural control. The vestibulospinal tract fibers that primarily arise from the medial and lateral vestibular nuclei are largely uncrossed and terminate on the interneurons and are responsible for extensor muscle activity and postural control. The rubrospinal tract fibers arising from the red

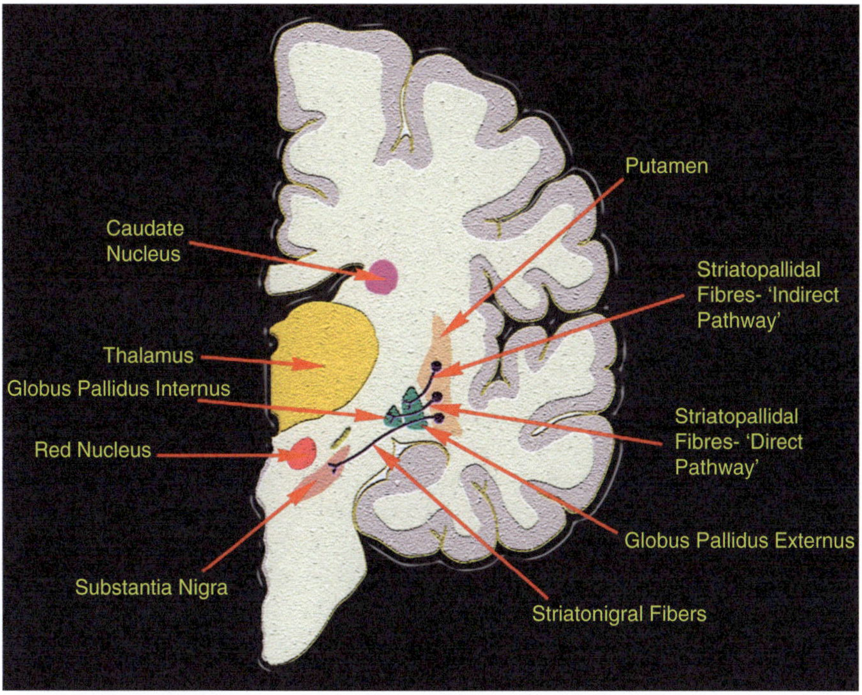

Fig. 4.4 A pictorial illustration of the efferent connections of the basal ganglia

nucleus are crossed and mostly terminate on interneurons in the cervical cord. These fibers are believed to control muscle tone and have an excitatory effect on the flexor muscles and an inhibitory effect on the extensor muscles. Projections from the globus pallidus to the red nucleus along with inputs from the motor cortex and the deep cerebellar nuclei are believed to modulate the tone of flexor muscles. The tectospinal tract that arises from the midbrain tectum, consists of crossed fibers, ends on the interneurons, and influences the movements of the head and body with respect to visual stimulus. In general, the extrapyramidal tracts that influence the axial muscles are largely uncrossed while those influencing the muscles of the limbs are mostly crossed. Such an arrangement permits the independent control of the limbs and axial muscles so that the manipulations can proceed while the posture is maintained.

Basal ganglia were believed to exert an inhibitory influence on the motor system. However, recent researches suggest the presence of both excitatory and inhibitory influences. In addition to the above, the input from the cortex also seems to have priority over the input from the thalamus and substantia nigra and provides certain evidence that the cortex is involved in regulating the responsiveness of the caudate neurons. Basal ganglia stimulation may prepare the cortex for subsequent inputs and activate only the most necessary pathways and inhibit all unnecessary pathways.

4.1.1 Direct and Indirect Pathways

The direct and indirect pathways are the two distinct pathways that process signals through the basal ganglia. Both pathways work in conjunction with each other. Excitation of the direct pathway has the net effect of exciting the thalamic neurons and the excitation of the indirect pathway has the net effect of inhibiting the thalamic neurons. Though the pathways have opposite net effects, a fine balance exists between them for the normal functioning of the basal ganglia. Current literature believes that an imbalance between these two pathways is responsible for the hypokinetic or hyperkinetic movement disorder seen in basal ganglia disorders. For instance, a less active direct pathway or an overactive indirect pathway can reduce cortical activation, causing bradykinesia and akinesia (hypokinetic movement disorders). Alternatively, an overactive direct pathway or less active indirect pathway can facilitate cortical activation, causing hyperkinetic movement disorders.

The direct pathway is a cortico-basal ganglia-thalamo-cortical loop that passes through the striatum, pallidum, substantia nigra, and thalamus to the cortex (Fig. 4.5). By default, the thalamic nuclei (Ventral Anterior nucleus [VA] and Ventral Lateral nucleus [VL]) are inhibitory to the cerebral motor cortices. In a normal

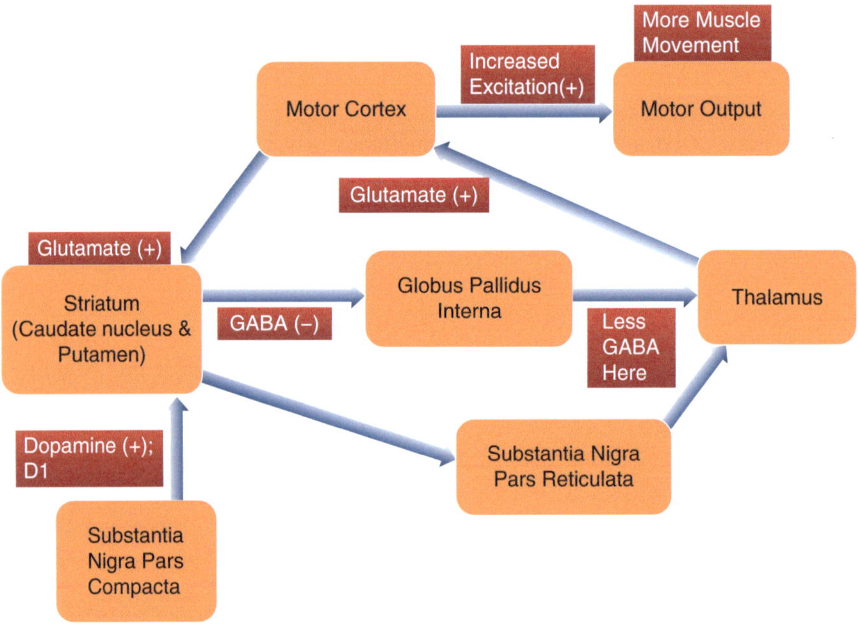

Fig. 4.5 Flowchart depicting the direct pathway

situation, during the activation of the direct pathway, the cerebral cortex excites the striatum through glutamate release. Inhibitory neurotransmitters (Gamma-aminobutyric acid [GABA]) released from the excited striatum inhibit the globus pallidus interna, a part of pallidum, and the pars reticulata, a part of substantia nigra. Inhibition of the globus pallidus interna and the pars reticulata reduces the inhibitory signals to the thalamus nuclei. Inhibition of the inhibitory neurons that project from the thalamic nuclei to the cerebral cortices thereby facilitates and promotes muscular action.

The indirect pathway passes through the striatum, globus pallidus externa, traverses the subthalamic nucleus, enters the globus pallidus interna and then the thalamus, and ends in the cerebral motor cortex (Fig. 4.6). In a normal situation, when the indirect pathway is activated, the release of glutamate from the cerebral cortex will excite the striatum. The excited striatum will release GABA, which in turn will inhibit the globus pallidus externa. The inhibition of the externa reduces the release of GABA, which is inhibitory to the subthalamic nucleus. Less inhibition of the subthalamic nucleus will cause more glutamate release, which is excitatory for the globus pallidus interna. Excitation of the globus pallidus interna will inhibit the thalamic nuclei, in turn inhibiting the cerebral cortices leading to the reduction of muscular activity.

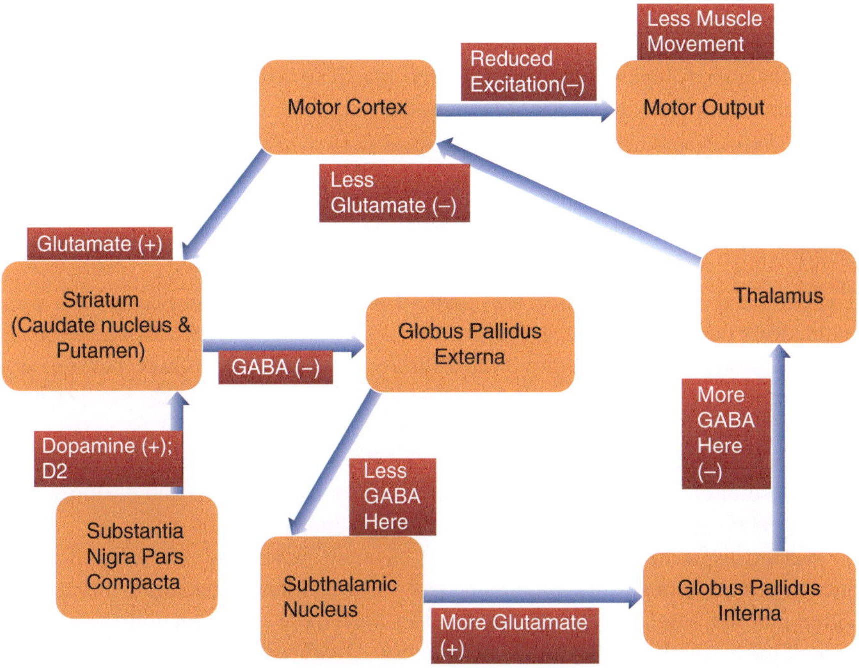

Fig. 4.6 Flowchart depicting the indirect pathway

4.1.2 Neurotransmitters Within Basal Ganglia

Dopamine, acetylcholine, GABA, substance P, enkephalins, and endorphins are certain common neurotransmitters seen in high concentrations within the basal ganglia. Dopamine, a major neurotransmitter of the nigrostriatal pathway, is produced by the pars compacta of the substantia nigra. The axon terminals of these dopaminergic neurons are located in the caudate nucleus. Dopamine is excitatory to the neurons in the direct pathway and inhibitory to the neurons in the indirect pathway. The loss of dopamine in the direct pathway leads to loss of excitation of the direct pathway and excess of excitation of the indirect pathway and reduction in activation of the thalamocortical pathway. Basal ganglia consist of several dopamine receptors and dopamine is also known to modulate the effects of other neurotransmitters, including glutamate.

Acetylcholine (ACh), another neurotransmitter found in the small interneurons of the striatum, is presumed to inhibit the action of dopamine in this region and is classically in tune or balance with the latter. An imbalance between ACh and dopamine is believed to be a plausible cause for the development of movement disorders like Parkinson's disease (PD). GABA, an inhibitory neurotransmitter found throughout the brain, is synthesized by the caudate nucleus and is transmitted to the globus pallidus and substantia nigra. The reduction of GABA in the indirect pathway and imbalanced dopamine activity are possible reasons for the choreiform movements in the early stage of Huntington's disease. Decreased GABA in the direct pathway and imbalanced dopamine activity are the plausible reasons for the rigidity and bradykinesia in the later stages of Huntington's disease.

4.1.3 Basal Ganglia: Posture and Movements

Experimental ablations of different parts of basal ganglia in animal models were extensively performed in the eighteenth and nineteenth centuries. Such ablations in animal's motor models have produced a variety of motor behavioral responses varying from no visible response to abnormal twisting and posturing of the body or body parts. By the nineteenth century, the use of electrical stimulation provided more insight into the functioning of basal ganglia. With sufficient intensity, the caudate nucleus stimulation could produce total body movement patterns and postures, usually a generalized flexion response of the head, trunk, and limbs.

Basal ganglia are implicated in aspects of the initiation and execution of movements, as well as in postural processing and postural adjustments. Many researchers agree that the basal ganglia are involved in movement initiation and preparation. Few of the researchers even recorded a possible "readiness potential" from the scalp of normal subjects that were absent in PD patients. They hypothesized that these potentials were generated from the basal ganglia and not from the motor cortex. More recent studies implicate that lack of these readiness potentials could be the possible cause of initiation difficulty in PD. People with basal ganglia disorders typically assume flexed or stooped postures and have decreased postural stability and a higher risk for falls. These patients often have reduced static postural adjustments. Research studies have also pointed out abnormalities in the reflexes involved in postural adjustments and spotted deficits in the longer loop reflexes but not in the short-latency reflex associated with the stretch reflex. Such deficits may reduce the person's ability to precisely modify the postural responses to the environmental demands and may predispose to abnormal postural reactions and even falls.

The integration of sensory information is essential to perform cognitive activities. Lesions of basal ganglia can produce poor sensory integration and cognitive functions, predisposing to abnormal movements and postures. Studies had proven impaired kinesthesia in basal ganglia disorder patients and also found that the impairment increased when they moved their limb further away from the body's center. Studies also found that learned movements were more affected by ganglia lesions than reflexes, particularly the procedural learning that is often needed to develop habits. Habits are easy to perform and such activities can proceed without thought or a conscious effort, allowing the person to think and react freely to new situations, enabling the person to select the appropriate movement in the proper environmental context.

The bilateral integration of the body is important for movements to be controlled and appropriately sequenced. Anatomical evidence suggests that the basal ganglia may have certain means of bilateral control, for the same reason, persons with basal ganglia disease may reveal function deficits even on the unaffected side and may exhibit difficulty in performing bilateral asymmetric movements simultaneously.

The primary function of the basal ganglia is to control and regulate the activities of the motor and premotor cortical areas and help to choose the right motor behavior from several possibilities at any given time. The functions may also include preparing the cortex for approximate time activation, setting appropriate postural reactions, sensory input organizing to produce an appropriate motor response that is context-specific, and inhibiting unnecessary motor activity. Earlier, the functional organization of the basal ganglia was believed to be a loop mechanism where afferent activity from the cortex is processed and modulated by the basal ganglia and subsequently sent back to the cortex to either facilitate or inhibit motor activity. However, the current concepts believe that the basal ganglia have several loops, where the cortical and the subcortical projections interact with the internal loops, forming a complex network, designed for selecting and inhibiting simultaneously occurring events and signals (Fig. 4.7).

Pathology affecting any of the motor projections or loops encompassing the basal ganglia can result in a spectrum of movement abnormalities ranging from excessive involuntary movements that may interfere with normal functioning (hyperkinetic or dyskinetic) to slowness or total poverty of voluntary movements (hypokinetic or akinetic). The details regarding the common movement disorders seen in basal ganglia pathologies are listed in Table 4.1.

The extrapyramidal disorders are either primary or secondary in origin. Progressive idiopathic neurodegenerative disorders like idiopathic Parkinson's

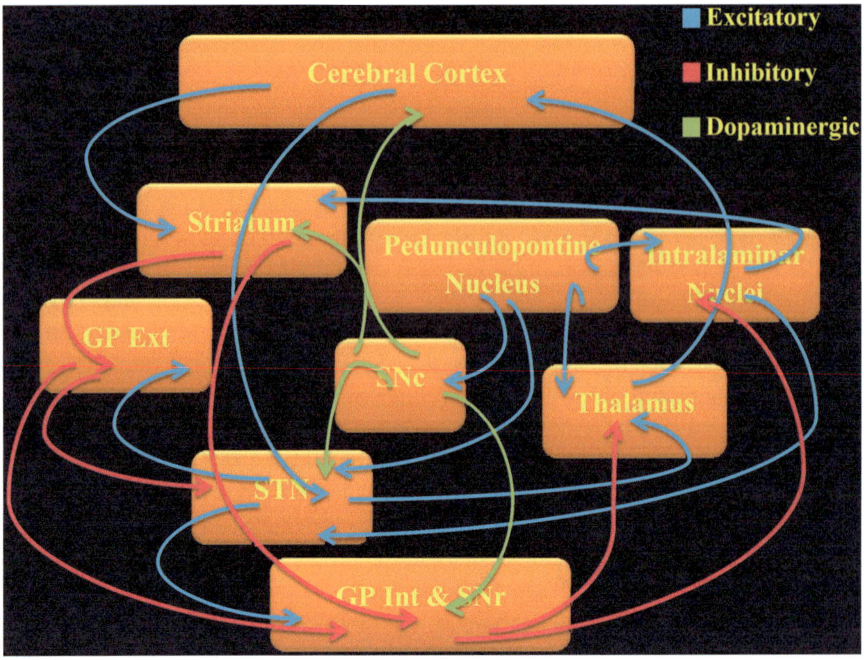

Fig. 4.7 Facilitatory, inhibitory, and dopaminergic loops operating at the basal ganglia

disease, Shy–Drager syndrome, progressive supranuclear palsy, and progressive genetic neurodegenerative disorders like Wilson's disease and Huntington's chorea are primary extrapyramidal disorders. The secondary causes of extrapyramidal disorders can be due to toxins or drugs, metabolic, vascular, space-occupying, trauma, and infections. Table 4.2 shows the clinical differences between extrapyramidal and pyramidal (corticospinal) disorders. This chapter will cover all the common and some unique disorders of the basal ganglia which affect the adult population.

4.2 Idiopathic Parkinson's Disease

In 1817, James Parkinson, a general practitioner and a surgeon in the countryside of London, published a monograph called "the shaking palsy." In his original description about shaking palsy, also known as paralysis agitans, features like involuntary tremors, reduced muscle power, bend or stooped trunk, tendency to pass from a walking to a running pace, and intact sense and intellect were mentioned. He also considered that the disease was of brainstem origin. Several decades later, Jean-Martin Charcot, also known as the father of modern neurology, renamed this condition as Parkinson's disease.

Table 4.1 Listing the common movement disorders of the basal ganglia, their features, and the location of the lesion

Movement disorder	Definition/features	Location/possible location/ cause
Ballismus	Involuntary, purposeless, jerky, wild, flinging movements of an entire limb; ballistic movements are usually unilateral (hemiballismus); flinging movements can be continuous or intermittent and can be unsafe or often cause falls.	Lesion in and around the surrounding areas of subthalamic nucleus
Tics	Stereotyped, repetitive irresistible movements that tend to change in type and anatomical location over long periods of time; typically seen as actions like repeated clearing the throat, protrusion of chin, sniffing, and blinking; often compelled to make these movements to relieve perceived tension; can be suppressed for a short time by will and may reappear when the subject's attention is diverted.	No known pathologic location in the basal ganglia or other locations of the brain
Chorea	Involuntary, purposeless, frequent, brief, sudden, twitch-like movements that flow from body part to body part; seen proximally or distally; large or small in amplitude; intermittent or nearly continuous.	Usually in the caudate nucleus or the striatum
Athetosis	Involuntary, purposeless, slow, often complex, writhing movements of the extremities; distal more than proximal; particularly of the hand, forearm, and arm and in some cases, lower extremity, neck, and tongue.	Common in lesions of striatum; pathological changes are seen more in the putamen than in the caudate nucleus
Myoclonus	Involuntary, quick, irregular twitching of a muscle or a group of muscles; intensity and frequency can be variable from case to case; may affect one or more body parts.	Usually diffuse or predominantly in cortical, subcortical, or cerebellum
Resting tremor	Involuntary repetitive, rhythmic, oscillatory movements; observed when the patient is at rest; usually seen in distal parts of extremities, jaws, lips, and tongue; frequency in the 4–6 Hz range; episodic in time; suppressed or enhanced by motor or cognitive activity	Localization of the central oscillator is still debatable; either basal ganglia or thalamic origin; may extend beyond basal ganglia circuitry
Bradykinesia	Slowness in both initiation and execution of movement	Dopamine (produced by nigral cells) is essential for modulation and facilitation the striatum; Reduced GABAergic transmission through the direct pathway and increased GABAergic transmission through the indirect pathway

(continued)

Table 4.1 (continued)

Movement disorder	Definition/features	Location/possible location/ cause
Dystonia	Involuntary muscle spasms, producing repetitive or twisted postures of different parts of the body; spasms can range from mild to severe; can affect one body part focal dystonia), two or more adjacent parts (segmental dystonia), or all parts of the body (general dystonia); precipitated by certain specific actions like handwriting; worsened with stress, fatigue or anxiety.	Focal lesions in the putamen, globus pallidus, or thalamus; different nuclei of the basal ganglia are emphasized
Rigidity	A form of muscle hypertonia, characterized by stiffness of muscles and difficulty to move the joints passively; resistance is felt in both the agonists and antagonist, however, can be more profound in one group-usually flexors; unlike spasticity, the resistance is not velocity dependent; the presence of normal tendon jerks; may reduce the range of motion of the joints; either cogwheel or lead pipe in nature; stress and anxiety may worsen rigidity	Hyperexcitability in long loop reflex pathways
Akinesia	An inability to initiate with absolute poverty of movement, characterized by a complete loss of voluntary movement in the absence of paralysis; considerable effort and mental concentration required to produce even the simplest motor activity.	

Table 4.2 Listing the clinical differences between extrapyramidal and pyramidal disorders

Clinical feature	Extrapyramidal	Pyramidal
Muscle tone	Plastic, lead pipe rigidity (resistance equal throughout the passive movement), or cogwheel rigidity (resistance with waxing and waning)	Clasp-knife type and velocity-dependent hypertonicity
Distribution of tone	Generalized hypertonicity, however, predominates in flexors of limbs and trunk	Flexors of arms, extensors of legs
Abnormal or involuntary movements	Presence of resting tremor, ballismus, chorea, tics, athetosis, and dystonia	Absent
Deep tendon reflex	Normal or occasionally brisk	Brisk or exaggerated; Sustained or transient clonus may accompany
Babinski sign	Absent	Present
Voluntary movements	Present, but can be hyperkinetic or hypokinetic	Absent/reduced; based on the extend of corticospinal tract lesion.

Idiopathic Parkinson's Disease (IPD) is the most common progressive degenerative neurological condition after Alzheimer's disease. Typically, the natural history of the disease begins between the fifth and seventh decades of life. The disease is more frequent among men than women and the ratio is approximately 3:2. Though the cause of IPD is unknown, physical trauma, exposure to certain unidentified environmental toxins, rigid personality, overwork, exposure to cold, emotional troubles, and stress can be some of the important factors predisposing to the disease. However, there is no conclusive evidence to support such assertions. The IPD is seen in all parts of the world and all races. At any given time, IPD affects 1–2 per 1000 of the population and affects approximately 1% of the population over 65 years of age. In most cases, the disease has no genetic or familial predisposition. At the same time, the role of genetic factors cannot be ruled out as dopamine metabolism utilizing Positron Emission Tomography (PET) scanning on asymptomatic twins of diagnosed IPD cases revealed evidence of striatal dysfunction.

4.2.1 Clinical Features of Idiopathic Parkinson's Disease

Typically, the symptoms of IPD emerge slowly. Resting tremor, bradykinesia, rigidity, and postural reflex disturbance are the major motor symptoms. The core symptoms consist of resting tremor, poverty and slowness of voluntary movements (bradykinesia), stooped posture, mask-like expressionless face (hypomimia), rigidity, axial instability, and Parkinsonian or festinating gait. During the very early stage of the disease, the symptoms are often difficult to perceive and both the patient and the family members may attribute those to the effects of aging. Often during the early stage, the patient may have vague complaints of fatigue and weakness, cervical and low back pain, shoulder and hip pain, and a general slowing up of physical activity. Gradually typical features like resting tremor and voice change will emerge. The patient's voice may progressively become soft and monotonous. Mild stiffness and slowness of movement, reduction in the natural swing of arms while walking are usually ignored or not bothered with until one day it draws the attention of the physician or family members as the possible inroads of the disease.

As the disease progresses, the progressive changes in gait may become noticeable, including short steps, hurrying, and shuffling. Facial expression reduces and often, the mouth is left slightly open. The blink rate will be reduced to 5–10 instead of a normal blink rate of 12–20 per minute. In addition to the hypomimia, a reduced blink rate with widened palpebral fissures creates a staring gaze known as the Stellwag sign. The weight shifts and adjustment movements, which are seen in normal subjects while seated, are progressively and markedly reduced in IPD patients. In most patients, olfactory dysfunction is one of the earliest non-motor features of IPD; however, total anosmia is rare.

The patients or family members may report the tremors involving the distal aspect of the upper extremity, mainly the hand. Such tremors can be an early sign and in the early stage of its development, be mild and intermittent, often seen only in a few fingers or one hand. Over a period, the tremor can develop into a full form, including the classical pill-rolling tremor of the thumb and fingers. Typically, these tremors, known as "resting tremors," are present when the body part (usually distal: hand or foot) is at rest and motionless. Sleep can abolish, and relaxation and voluntary movements can considerably reduce or eliminate these tremors. Distraction, simple mental arithmetic, or counting with the eyes closed can bring out these tremors when they are less observable. Though the resting tremors are common in the hand and foot, careful observation may reveal such tremors even in the arm, jaw, tongue, and eyelids. The frequency of the resting tremor is usually between 4 and 6 Hz. The resting tremor may worsen with gait or emotional upset; however, the frequency will remain largely unchanged.

The IPD patients may present with another tremor known as action tremor. Action tremors are finer and have a frequency of 7–8 Hz. They tend to persist throughout the voluntary movement and are easily suppressible by relaxation and unnoticeable when the limb is essentially at rest. IPD patients may have resting and/or action tremors. Electromyographic studies of both the tremors mentioned above have proven that the muscle activities recorded were not exactly the same. However, studies have shown certain similarities between the action tremor and the essential tremor.

Unilateral onset is typical of IPD, i.e., the motor impairments and resting tremors tend to develop in one limb or one side of the body and may spread to other regions as the disease progresses. Both rigidity and muscle hypertonicity are not early findings and tend to develop as the disease advances. Once they appear, they are constantly present. Hypertonicity tends to predominate in the flexor muscles of the trunk and limbs and is possibly responsible for the flexed or stooped posture, typically seen in IPD. Even when the patient is well relaxed, on palpation, the muscles may appear unusually firm and hard. Irrespective of the direction of the movement, when the limbs are passively moved, a mild resistance will be felt from the start and continues evenly throughout the movement. The resistance felt during the passive movement can be lead pipe (resistance is equal throughout the passive movement) or ratchet-like cogwheel (resistance with waxing and waning) sensation.

Difficulty or inability to perform ballistic movements is another volitional movement abnormality seen in IPD patients. Typically, the movements are initiated slowly and may demonstrate a series of attempts to initiate, when quick or fast movements are tried. While performing alternating movements, observation of the ongoing movement may reveal a progressive failure in continuing alternate movement or adaptation of a rhythm similar to the patient's own tremor. Further, during the advancement of the disease, the patient will develop difficulties in executing two or more motor tasks simultaneously.

As the disorder progresses, the movement abnormalities, specifically bradykinesia, worsen and affect a variety of functional activities. The handwriting progressively becomes small (micrographia), tremulous, cramped, and illegible. The voice softens further and speech becomes hurried, more monotonous, less audible, indistinct, and mumbling ("hypokinetic dysarthria") and eventually, the patient can only whisper. Bradykinetic deficits, along with rigidity, can cause swallowing difficulties and slowness of chewing and as a result, the patients may take considerable time to finish feeding.

Camptocormia, a severe flexion attitude of the thoracolumbar spine in the sagittal plane, typically more than 45°, while standing and walking and almost completely resolving in the recumbent position, can be a feature of postural abnormality in the advancing stage. The patient may develop other postural abnormalities like scoliosis and antecollis. Antecollis consists of severe forward flexion of the head in the sagittal plane, with the inability to voluntarily fully extend the neck against gravity. It is more frequent in multiple system atrophy than in PD. Another frequent and disabling postural abnormality seen among PD patients is the Pisa syndrome. It is characterized by a marked lateral flexion of the trunk (a Cobb angle of more than 10°), which is completely reversible by passive mobilization or supine position.

In addition to reducing postural adjustment capacities, as the disease advances, righting reactions and the body's defense mechanisms like equilibrium reactions and protective reactions become progressively faulty. Postural instability and gait failure are the most disabling features of IPD. Progressive changes in gait such as a tendency to pass from a walking pace to a running pace with forward bent trunk posture, increased cadence, reduced step and stride length, narrow base of support while walking, lack of arm swings, absence of rotation and counter-rotation of the upper trunk with respect to the lower trunk, reduced velocity, poor ground clearance and tendency to shuffle are certain important features of the festinating gait. The patients may tend to fall forward or backward if pushed or pulled (propulsion or retropulsion). While walking, they may topple forward with faster steps and have difficulty stopping (festination). A history of falls, either forward or sideways, is not uncommon. Walking forward or backward is difficult and may be accomplished with a series of short steps and as the disease advances, the gait develops into the typical slow shuffling walk with small steps on a narrow base.

In the advanced stage, negotiating obstacles, turns, and cluttered spaces, and walking in crowded places or narrow passages causes freezing of gait. Even basic activities like turning over in bed will be difficult, effortful, and time-consuming owing to the bradykinesia, rigidity, and impaired righting reactions. Persistent clawing of the toes or jaw clenching and similar features of dystonia, which are often painful, may be evident in the advanced stages. Even eye movements are bradykinetic, characterized by a delay in the initiation of gaze, slowness in producing conjugate gaze movements, and hypometric in saccadic movements with a breakdown of pursuit movements into small saccades.

The advent of dementia during the advancing stage further complicates the disease. It is estimated that 20–30% of patients will become demented. The incidence of dementia tends to increase with the advancement of disease and age. Approximately 65% of PD patients above the age of 80 years have a certain amount of dementia. Often, depression and neuropsychiatric side effects of medications can make the clinical detection of dementia difficult. Development of dementia, along with cognitive slowing, depression, and memory impairment in the absence of aphasia and agnosia, is typical in the advancing stage.

In the later stage, the drooling of saliva (sialorrhea) can be troublesome. Reduced swallowing due to bradykinesia and rigidity of the muscles of deglutition and excess flow of saliva are the possible reasons for this. These patients often have seborrhea (excessive production of sebum from the sebaceous glands) that makes the skin more greasy or oily. Hyperactivity of the parasympathetic system and/or overactivity of androgens are possible factors attributing to seborrhea. In addition to the above, these patients also have excessive sweating due to lack or failure to cleanse the face regularly and sufficiently, and owing to constant motor activity. Orthostatic hypotension and syncope can be clinical features due to neuronal cell loss in the sympathetic ganglia. Constipation, another common autonomic dysfunction, if severe, may require frequent hospital admission. However, it is not a prominent feature, as seen in the Shy–Drager syndrome.

Clinical examination of the IPD patient will reveal the presence of Myerson sign or glabellar tap sign (inability to resist blinking of the eyes when tapped repetitively over the bridge of the nose or glabella), absence of palmomental, grasp, and suck reflexes and absence of exaggerated jaw jerks. The deep tendon reflexes vary and range from barely elicitable to brisk. Whether symptoms are confined to one side or equal on either side of the body, the plantar responses are flexor in nature.

The disease course is quite variable. The mean time from the commencement of symptoms of the disease to a wheelchair-bound state is approximately 7.5 years in the majority of patients. Whereas, as many as one-third of the cases with relatively mild features may remain stable for 10 years or more, indicating a vast variation in the course duration of the disease. The cognitive decline associated with dementia and the older age of onset were recognized as the predictors of reduced survival time.

4.2.2 Clinical Diagnosis

If one strictly adheres to the definition of PD, in the majority of cases, the clinical diagnosis is quite straightforward. The presence of two out of the three cardinal features—bradykinesia, rigidity and tremor, presence of postural instability, a good clinical response to levodopa therapy and absence of any "atypical" features suggestive of another Parkinsonian syndrome makes errors in diagnosis less likely. The presence of small signs such as reduced blink rate, the Myerson sign and the Stellwag sign and absence of the Babinski sign, hyperactive deep tendon reflexes in the affected limbs, and suck, grasp, and palmomental reflexes may further help in the diagnosis.

Occasionally, the clinicians may face difficulty in distinguishing a typical IPD from the many Parkinsonian syndromes, particularly when all the signs and symptoms are not evident at the outset. In such situations, re-examining such cases after several months may bring better clarity about the impending medical condition. The presence of symmetrical features and rapid onset and progression of the clinical features right from the inception are unlikely to be IPD. Reduction of clinical features such as gait and postural instability and bradykinesia post lumbar puncture and absence of features like forward stooped posture, resting tremor, and a positive response to levodopa therapy helps to distinguish normal-pressure hydrocephalus from IPD. Similarly, some of the features of progressive supranuclear palsy can often be mistaken for IPD. The presence of early falls, with a predilection to fall backward, paralysis of upward and downward gaze, and eventually the lateral gaze with retention of reflex movements of the eye are definite features of progressive supranuclear palsy, unlike IPD.

4.2.3 Etiology, Neuropathology, and Neuropathogenesis

The etiology of PD is most likely multifactorial. Mitochondrial dysfunction, along with free-radical damage and the presence of neuroinflammatory changes, appears to play an important role in the pathogenesis of PD. A presynaptic neuronal protein named α-synuclein, genetically and neuropathologically linked to PD, may also be playing a central role. Misfolding and aggregation of the α-synuclein is a hallmark of PD. The questions of whether neuroinflammation triggers α-synuclein misfolding and aggregation or does misfolding and aggregation of α-synuclein cause microglia activation and neuroinflammation are largely unclear. Similarly, the exact mechanism by which α-synuclein causes cellular toxicity and contributes to neuronal death remains unclear.

Typically, the naked eye examination of the brain will not show any remarkable findings. However, IPD patients with associated dementia may have a mild to moderate degree of cerebral atrophy. The most persistent and relevant neuropathological findings in IPD are the loss of pigmented dopaminergic neuronal cells in the substantia nigra pars compacta associated with the presence of intraneuronal inclusions called Lewy bodies.

Though aging contributes to nigral cell loss, the cell depletion is considerably marked in IPD, indicating that some factors other than aging must be operative. It is estimated that at least 50–70% of nigral cell degeneration is required to produce symptoms of the disease. The nigral cell loss is initially rapid and later slows down with time. In addition to the nigral cell loss, there is a significant neuronal loss within the locus ceruleus, the dorsal motor nucleus of the vagus, the raphe nuclei, and the nucleus basalis. The presence of reactive gliosis accompanying neurodegeneration, Lewy bodies in other locations, including subcortical structures and neuromelanin released from dying neurons of pigmented nuclei, are some additional microscopic observations. The presence of Lewy bodies is not absolutely specific to IPD as these bodies are also seen in other neurological disorders such as the Parkinsonian variant of multiple system atrophy and Lewy body dementia.

4.2.4 Medical and Surgical Management

Immediate drug treatment is often not required or postponed for elderly patients who present late, with no marked disability at the time of diagnosis. To date, no definitive treatment, medical or surgical, that can reverse or halt the progressive neuronal cell loss underlying IPD has been neither discovered nor invented. A careful explanation of the diagnosis, treatment options, and prognosis by the neurologist and reassurance that early medications would not have influenced the situation and no treatment will completely suppress all symptoms, helps to avoid premature drug therapy and pointless overmedication.

Neuroprotective therapy for those patients not requiring symptomatic medications in the early stage, symptomatic therapy when symptoms begin to disable or disturb, and management of treatment-related complications and drug-resistant features in the advanced stage of the disease are the preferred three stages of treatment. Selegiline, a neuroprotective drug, widely used in early Parkinson's, is believed to prevent intracerebral metabolic degradation of dopamine. Clinical trials conducted by the Parkinson Study Group revealed a slow progression of the disease; however, subsequent observations of other trials could not corroborate with the earlier findings. Even the use of vitamin antioxidants like Vitamin E to reduce nigral oxidative damage has not revealed any beneficial effect on the disease progression. Coenzyme Q10, known as a mitochondrial nutrient with free radicals scavenging capacity, is another neuroprotective drug. Administration of coenzyme Q10 is found to benefit several mitochondrial-defective diseases, including PD, particularly in the early stages.

When the symptoms are disabling, the use of Parkinson's medication to control the symptoms is preferred. Levodopa, non-dopaminergic drugs such as anticholinergics and amantadine, and dopamine agonists are the classes of Parkinson's medications. Non-dopaminergic drugs are preferred for mildly affected patients and help to delay the early prescription of levodopa or dopamine agonist. Anticholinergics help prevent the imbalance between ACh and dopamine in the striatum and are effective in suppressing resting tremors when tremors are a major concern than bradykinesia. The mode of action of amantadine is more complex and is believed to increase the presynaptic dopamine re-uptake and release. Insomnia, agitation, confusion, and pedal edema are some of its side effects.

For patients with significant disabilities, dopaminergic treatment is the choice. The theoretical basis for the use of levodopa, the most effective Parkinson's medication, rests on the observation that the ability of the remaining nigral cells to convert levodopa to dopamine is not significantly reduced by the progressive loss of nigrostriatal cells. To increase the bioavailability of orally administered levodopa, reduce peripheral conversion of dopamine, and minimize the side effects like nausea and vomiting, a peripherally acting decarboxylase inhibitor (carbidopa or benserazide) is given as a combination. Levodopa is highly effective against tremor, rigidity, and akinesia and is undoubtedly the most effective treatment for IPD. However, with the progression of the disease, the drug response reduces, and higher dosages when required are associated with serious long-term issues. When the number of remaining nigral cells becomes inadequate and the receptivity to dopamine becomes excessive due to denervation hypersensitivity, the beneficial effects of levodopa will reduce and adverse effects like dyskinesia will begin. Within 5 years post-diagnosis, 50% of IPD patients will experience adverse effects and almost all by the next 5 years. Progressive shortening of the response to each dose leads to the "wearing off effect" or "end of dose deterioration" and is characterized by the reappearance of Parkinson's features before the next dose is due. Sooner or later, the patient frequently switches between drug "on" state (associated with dyskinetic movements) and drug "off" stage (severe rigidity and akinesia). Increasing dopaminergic nigrostriatal cell destruction and reducing presynaptic storage of dopamine are the possible reasons for the drug on-off phenomenon.

To prevent the early onset of long-term levodopa-related adverse effects and as an alternative to levodopa, dopamine agonist monotherapy is employed. Bromocriptine, pergolide, and lisuride are certain dopamine agonists. Overall, dopamine agonist monotherapy is associated with a much lower incidence of dyskinesia and motor fluctuation; however, it has its own side effects like nausea, hypotension, and confusion. Dyskinesias are often difficult to treat and are characterized by restlessness, head wagging, grimacing, lingual-labial dyskinesia, and dystonic posturing of extremities and spine. Only about 50% of the patients, especially patients with early Parkinson's, can tolerate dopamine agonists.

Severe off periods can be extremely unpleasant with rigidity and akinesia, restless limbs, sweating, pain, and autonomic abnormalities. In advanced PD, increasing the frequency of levodopa, finding a levodopa dose that provides the best balance between Parkinsonism and dyskinesia or the introduction of a dopamine agonist and reduction in levodopa dosage may reduce the levodopa-related complications and emerging drug-resistant features. Parkinson's medication can improve gait abnormalities, freezing of gait, and falls to some extent, but the improvement is usually modest at best.

Early neurosurgical treatment, even before the advent of levodopa therapy, revealed only limited improvement and had considerable morbidity. Thalamotomy was effective for tremor but had little impact on bradykinesia and rigidity. Later, after the advent of stereotactic surgeries, stereotactic thalamotomy of the nucleus ventralis intermedius continued to show a certain role in early Parkinson's with severe tremor. Stereotactic pallidotomy is increasingly used in suppressing dyskinesia; however, studies have shown no significant improvement for postural, gait, and speech abnormalities.

Deep brain stimulation was a recent addition and alternative management to stereotactic brain surgeries. The advantage of deep brain stimulation over stereotactic lesions is that the former is reversible and is safer. Deep brain stimulation of the thalamic nucleus ventralis intermedius has demonstrated success in reducing tremor and dyskinesia. Stimulation of the globus pallidus interna or subthalamic nucleus has shown a reduction in all symptoms. Some of the complications that have been reported post-stimulation include weight gain, mild dystonia, eyelid apraxia, hemorrhage, infection, and cognitive impairment. Another neurosurgical development has been the use of transplanted fetal mesencephalic neurons into the striatum of patients with advanced PD. Post-transplantation, levodopa requirements were reported less in some patients, but few studies have shown only a modest improvement.

4.2.5 *Physiotherapy for Parkinson's Disease*

Physiotherapy has a central role in reducing activity limitations, promoting participation and functional independence, and enhancing safety and well-being. To deliver optimal care to PD patients and address all the issues, the physiotherapist needs adequate knowledge about the disease, the clinical manifestations, and specific physiotherapy skills. The common principle for physiotherapy is that the therapy should be patient-centered and tailored to the patients' specific needs and preferences. It is also essential to empower the patients by education and enable them to continue the training program even when the therapist-supervised session is completed for the day.

The exercise training program should consider the history, disease course, severity of impairments, activity limitations, and participation restrictions. Depending on the stage of the disease, the physiotherapy interventions can be restorative (for reducing impairments, activity limitations, and participation restrictions), preventive (for minimizing potential complications and indirect impairments), and compensatory (for functional improvement by modifying the task, activity, or environment). During the early phase (1–2.5 stages of Hoehn and Yahr scale) of PD, the majority of the patients are functionally independent with minimal impairments. The goals of the therapeutic intervention for the early phase are to prevent inactivity and preserve or improve physical capacities such as aerobic capacity, muscle strength, and joint mobility. A referral for physiotherapy at this stage will certainly benefit the patient to improvise his or her physical fitness levels and delay or prevent the early onset of indirect impairments. At this stage, such patients are mostly seen on an outpatient basis.

In the middle stage or phase (2.5–4 of Hoehn and Yahr scale) of the disease, symptoms are more readily noticeable and the emergence of activity limitations also characterizes it. The patient is typically still independent in gait and Activities of Daily Living (ADL), although slow and less efficient in performance. The goal of therapeutic intervention in this phase is to preserve or encourage activities. The therapy should focus on specific issues like transitions within the bed and off the bed, static and dynamic postures, balance, and gait. Cognitive movement strategies and cueing strategies can be effectively utilized during this stage and even the family members can be actively involved during this stage of treatment. Usually, patients in the middle stage of PD are treated on an outpatient basis or following a brief inpatient admission or as a part of home care. During the late phase (fifth of Hoehn and Yahr scale), the disease progression leads to more severe complications and impairments. For many of the functional mobility skills and ADL, the patients are more or less completely dependent and are either wheelchair-bound or bed-confined. For such patients, the physiotherapy goals need to be restructured and should focus on preventative care to avoid secondary complications, including life-threatening complications like bronchopulmonary pneumonia or decubitus ulcers.

Resting tremor, bradykinesia, and rigidity are considered to be the direct or primary effects caused by the progressive degeneration of the nigrostriatal dopaminergic cells of the central nervous system. The involvement of a system other than CNS, like musculoskeletal or cardiopulmonary, gives rise to symptoms like kyphosis, contracture, reduced mobility, urogenital issues, and cardiopulmonary deconditioning, known as the indirect or secondary effects of the disease. Symptoms like abnormal posture, faulty balance, swallowing difficulty and fatigue, arise due to composite involvement of CNS with one or more other systems, known as the composite effects of the actual disease. The therapist should have a comprehensive understanding of the direct, indirect, and composite effects of the disease. Concerning the therapy, in the early and middle stages of the PD, symptoms arising due to secondary and composite effects of the disease are often amenable to corrective or preventive strategies. However, in the later stages (fourth and fifth of Hoehn and Yahr scale), "little can be done for symptoms due to actual disease" and the benefits of corrective or preventive strategies for secondary and composite effects are rather short-lived.

4.2.5.1 Assessment

The information obtained while taking the patient's history should help to formulate the objectives of treatment and be the basis for focusing the physical examination on specific areas of functioning, namely physical capacity, transfers, body posture, balance, and gait. A comprehensive clinical examination is essential to determine the level of impairments and activity limitations and participation restrictions (Table 4.3). Periodic re-examinations at specific intervals (approximately 3–6 months for early and middle stage PD) are worthwhile to distinguish the changes in physical and functional status and for restructuring the goals and means of therapy.

Considering the attention and emotional deficits, the clinical examination of the PD patients should be preferably executed in a quiet room. The fear of falls and the associated anxiety can be alleviated by performing the examination on a low height, broad examination table. During the physical examination, the therapist should note whether the patient is in a drug "on" or "off" phase. Also, it is essential to note down the details regarding the medications, specifically emphasizing the dosage, timings of medication, and the time of the last administration. The selection of procedures and instruments for examination is determined by the patient's age, status, severity of complaints, stage of the disease, phase of rehabilitation, and location of rehabilitation. For both the early and middle stages of PD, physical performance and impairment measurements are relatively stable. However, in the late stage, fluctuating symptoms and adverse effects of pharmacological intervention make measurements less reliable.

Table 4.3 Details regarding the level of impairments and activity limitations and participation restrictions in Parkinson's disease

Impairments in functions			Activity Limitations and restrictions in participation
Neuromusculoskeletal and movement-related functions	Reduced joint mobility, muscle endurance, muscle power, postural responses and control of voluntary movements; impaired muscle tone functions and motor reflex functions; involuntary movement functions and gait patterns; on/off periods		• Mobility: Changing and maintaining body positions; carrying, moving, and handling objects; walking and moving; using transportation • General tasks and demands: Undertaking multiple tasks; handling stress and other psychological demands • Self-care: Washing oneself; toileting; eating; drinking • Learning and applying knowledge: Acquiring skills; making decisions; writing solving problem • Interpersonal interactions and relationships • Communication: Speaking; non-verbal and writing messages • Domestic life: Preparing meals; doing homework • Major life areas: Education; work; employment; economic life • Community, social and civic life: Community life; recreation and leisure; religion; political life
Mental functions	Delirium; Dementia; Impairments in temperament and personality, energy and drive functions, sleep, emotion, perceptual functions higher-level cognitive functions and mental functions of language; reduced attention and memory		
Voice and speech functions	Reduced pitch, loudness of voice and fluency of speech; impaired articulation		
Cardiovascular and respiratory systems functions	Impairments in blood pressure; reduced exercise tolerance		
Sensory functions and pain	Impairments in seeing and smell; dizziness; proprioceptive function; tingling; central pain		
Digestive system functions	Impaired ingestion; constipation; reduced weight maintenance		
Genitourinary and reproductive functions	Impaired urination and sexual functions		
Skin and related structures functions	Impairments in sweating and sebum production and sensations related to the skin		

Physical capacity assessment should focus on the mobility of cervical, thoracic and proximal joints, length of muscles, and strength of major muscle groups. For the assessment of mobility, the examination of joint ROM and general flexibility is important. Musculoskeletal impairments tend to begin proximally, affecting the contractile and non-contractile tissue length and flexibility first of the spine and girdle and then of the more distal musculature. As the disease progresses, PD patients are likely to develop a reduction in joint ranges, specifically for hip and knee extension, dorsiflexion, spine and neck extension, axial rotation and lateral flexion of the spine, shoulder flexion, and elbow extension. Using a digital or standard goniometer, the therapist can document the passive range of motion of the larger joints and spinal inclinometers, cervical range of motion instruments, and cervical or head-mounted laser for the spine. In clinical settings, spinal mobility can also be examined using a series of functional movements, in seated or standing positions, like looking behind and turning the neck sideways or upwards.

Generally, the muscle groups of the spine and limbs become shortened and limit extension throughout these structures. Passive straight leg raise with the pelvis and opposite thigh stabilized can be used as an easier method to check the hamstring length. Testing the tightness of hip flexors, specifically the iliopsoas and hip adductors, and plantar flexors, can be incorporated into the routine assessment. However, care has to be taken not to create undue stretch and pain while examining the muscle tightness.

The deconditioning effect, the reduced physical activity, and the hypertonicity predominantly in the flexors of the spine and limbs cause weakness of muscles, specifically the anti-gravity muscles, contributing to the flexed posture in the middle and advanced stages of PD. Traditional manual muscle testing can be used for the assessment of muscle strength and can be performed using the Medical Research Council (MRC) grading system (Table 4.4). Alternatively, handheld and isokinetic dynamometry can be used for the quantitative measurement of muscle strength, including peak force, torque output, power, and angle of maximal force.

Table 4.4 Medical Research Council Grades for Muscle Power

MRC grade	Description
0	No contraction
1	Flicker or trace of contraction
2	Active movement, with gravity eliminated
3	Active movement against gravity
4	Active movement against gravity and resistance
5	Normal power

Grades 4−, 4, and 4+, may be used to indicate movement against slight, moderate, and strong resistance, respectively.

Used with the permission of the Medical Research Council

As the disease progresses, the movements are slowed (bradykinesia), then movements decrease in range and amplitude (hypokinesia), and in the later stage, movements are characterized by start hesitations and are eventually akinetic. Stopwatch and use of grooved pegboards can help quantify the bradykinesia due to slowness and hesitancy of movement.

The clinical assessment of rigidity in PD is largely qualitative. In the early stage of the disease, the distribution of rigidity is often asymmetrical and can vary particularly during the course of the day, with the medication timing and stress. The level of resistance offered during the passive movement and the Range of Motion (ROM) available can provide information about the severity of rigidity. The grades of rigidity (Box 4.1) can also be documented in the motor examination component of the Unified Parkinson's Disease Rating Scale (UPDRS). In addition to the grading of rigidity mentioned in the UPDRS, Columbia University Rating Scale (CURS) can be used to grade the rigidity. The grade of rigidity is judged by passively moving the major joints with PD patients, preferably in sitting position for upper and lower limbs and neck. The Webster rating scale is another clinical assessment tool to grade the degree of rigidity in PD patient's limbs and neck.

Box 4.1 Grades of Rigidity

Grades	Features
0	Rigidity absent
1	Rigidity slight or detectable only when activated by mirror or other movements
2	Mild to moderate rigidity
3	Marked rigidity, but full range of motion easily achieved
4	Severe rigidity, range of motion achieved with difficulty

Camptocormia, scoliosis, and antecollis are a few of the spinal deformities seen in PD. Assessment should focus on posture, both static and dynamic, in lying, sitting, standing, and while walking. It should also cover the ability to actively correct posture and manipulate objects while sitting and standing. Posture grids, plumb lines, and software analysis of the 2D photographic images or captured videos can be used for the same purpose.

Patients with Pisa syndrome are more likely to have considerable asymmetry in symptoms, increased muscular activity, back pain, postural and balance dysfunction, and a poorer Quality of Life (QOL) when compared with PD patients without Pisa syndrome. The prevalence of Pisa syndrome among PD is approximately 9%. The tendency to lean worsens while walking. The onset of Pisa syndrome symptoms can be acute (within a few days or weeks), subacute (over a few months), and chronic (insidious). In the early stage, the majority of PD patients are less likely to be aware of the lateral trunk deviation. In the advanced stage, the patients often experience debilitating pain, dyspnea, unsteadiness, and falls while walking. It is important to emphasize that the terms "scoliosis" and "Pisa syndrome" are not synonyms. The former cannot be improved by passive movement or supine positioning and is characterized by radiological evidence of a structural curve with axial vertebral rotation that persists when the effect of gravity has been eliminated. The pathomechanism(s) underlying Pisa syndrome is not fully understood. Central and peripheral are the two hypotheses proposed for the same. The central hypothesis considers Pisa syndrome as a consequence of the basal ganglia dysfunction along with altered sensorimotor integration and possibly exacerbation by dopaminergic treatment. Whereas, the peripheral hypothesis considers alteration of the musculoskeletal system as the possible mechanism for the development of the same. The perception of verticality is affected in PD patients and abnormal perception, together with dysfunction of central graviceptive pathways can alter the sensorimotor integration essential for postural stability. Research has also suggested that the asymmetric functioning of the basal ganglia circuit could be the primary cause of the development of Pisa syndrome.

The syndrome is also reported in other conditions including dementia, Parkinsonism, normal-pressure hydrocephalus, Alzheimer's disease, Lewy body disease, and subdural hematoma. It can be a result of exposure to antipsychotic, dopaminergic, and antidepressant drugs. Though there is no consensus on the diagnostic criteria for the Pisa syndrome, the diagnosis is based on clinical examination and accurate measurement of lateral deviation of trunk, using a standard goniometer. Early recognition and either withdrawal of offending medication or revision of drug regimen can prevent veering of this syndrome toward the chronic irreversible variant.

Though not always as troublesome as bradykinesia and rigidity, the tremors may often interfere with routine functional activities. The location and the severity of resting and/or action tremor and to what extent the functional skills like drinking, feeding, dressing, and writing are affected, need to be noted. Due to impairments in motor planning and programming, PD patients may have difficulty to perform multitask (cognitive and motor) and dissociated movements in the middle and advanced stages. The assessment should include observation of the patient's ability to perform dissociation movements of trunk and limbs, dual and multitasks (a combination of cognitive and/or motor), sequential activity, continuous movements without visual guidance, and spontaneous switching from one to another activity.

Though PD is considered a motor disorder, during the progression of the disease, non-motor issues like cognitive deterioration and digestive and genitourinary dysfunctions also emerge. Memory impairments, attention deficits, language issues, and abstract reasoning, problem-solving, and judgment issues are some of the important cognitive deficits that need evaluation. Speed of information processing, attention, and concentration are particularly important when bradyphrenia (inattentiveness and delayed response) is suspected. For the abovementioned, cognitive function scales like Mini-mental State Examination (MMSE) scale or Montreal Cognitive Assessment (MoCA) can be used as a screening tool. In addition to cognitive dysfunction, the presence of depressive symptoms like sadness, apathy, insomnia, anorexia, inactivity, dependency, difficulty concentrating, and impaired memory can be worrisome. The geriatric depression scale or the Beck depression inventory can be used as an instrument to self-report it.

The presence of sway during quiet stance, both medial-lateral and anterior-posterior planes, indicates a poor postural control. Normal subjects typically respond to small shifts of the center of mass using ankle strategy and for larger shifts, by hip and stepping strategies. PD patients, when destabilized, typically respond with postural strategies involving the hip joints or stepping, more than ankle joints. PD patients, especially in the early stage, may not demonstrate balance impairments in a steady stance with a normal base of support or self-initiated movements. However, when competing attentional demands are instituted like talking while balancing or mental calculations while balancing, may demonstrate instability. The common and often used balance measurement tools like the Berg Balance Scale (BBS), the Functional Reach Test (FRT), the Timed Up and Go (TUG) test and the Dynamic Gait Index (DGI) have been reliable and sensitive in the examination of functional balance in PD patients. The retropulsion test or the pull test will evaluate the patient's response to an unexpected, quick, and firm backward pull on the shoulder. Taking multiple steps (three or more) to regain balance as a response to the pull test clinically suggests impairment of balance. Usually, the measurement values of balance using the abovementioned scales or tests are quite stable during the "on" phase of the medication and are less reliable and more fluctuating during the "off" period.

Gait examination should cover unobstructed walking on a level surface, start time or gait initiation, speed of walking, stride length, cadence, stability, variability, and safety. Increased difficulty in walking is also experienced in response to varying attentional demands and dual-task interference. Changes in gait parameters like speed, stride length, and cadence can be observed while simultaneously performing a secondary cognitive task or motor task. A 10-m walk test can help to determine the speed, average stride, and cadence of the patient. Force plates, body markers, 3D videographic and computerized equipment, and software analysis can be used for sophisticated kinetic analysis of the gait. Gait should be examined for kinematic or qualitative changes, including reductions in the hip, knee, and ankle motions that result in a short-stepped, shuffling (festinating) gait pattern. Postural abnormalities that contribute to the development of a festinating gait pattern should also be documented (i.e., flexed, stooped posture).

Several gait impairments, including rhythm control, gait symmetry, bilateral coordination of gait, dynamic postural control, and step scaling, have been associated with Freezing of Gait (FOG) with PD. FOG is a stronger predictor of activity limitation than gait hypokinesia. In the early and middle stages of the disease, the freezing phenomenon is often difficult to assess due to its less predictable nature. In these stages, freezing episodes are less likely to occur during the "on" time and are usually levodopa sensitive; however, during the advanced stage, the episodes are often seen during "on" and "off" times. The Freezing of Gait Questionnaire (FOG-Q), a self-administered measurement tool, is appropriate for assessing the FOG in clinical practice. There is a strong association between the FOG and the risk of falls. The patient or the family members/caregivers may use a fall risk diary to record the fall events. The events recorded should cover information such as activity at the time of the fall, direction and method of landing, number and frequency of near-misses, type of footwear used, fatigue, the timing of medication, food intake, and other possible intrinsic or extrinsic risk factors for fall. To gauge the fear of falling, a short version of the full 16 item Falls Efficacy Scale-International (FES-I) is recommended in clinical practice.

Clinical examination also should address the problems with autonomic dysfunction such as drooling (salivation) or sweating, greasy or oily skin, and abnormalities in thermoregulation. Excessive sweating and flushing during the "on" state may have an association with dyskinesia. Signs and symptoms of orthostatic hypotension like lightheadedness or dizziness, confusion, blurring of vision, fatigue, nausea, and syncope need to be documented, if any, after standing up. A drop of 20 mm Hg in the systolic blood pressure or a drop of 10 mm Hg in the diastolic blood pressure within 2–5 min of standing up is diagnostic of the condition.

Sedentary nature with decreased physical activity and impaired cardiorespiratory function may reduce the endurance of PD patients. In the advanced stage of the PD, respiratory dysfunction can be a major concern as the presence of upper airway obstruction and co-existing chronic obstructive pulmonary disease can predispose to pulmonary complications and even death. Careful observation of rib cage compliance, mobility of the chest wall, breathing patterns, and monitoring of ventilator-derived parameters like Forced Vital Capacity (FVC), Forced Expiratory Volume in 1 s (FEV1), maximal expiratory flow, total lung capacity, and residual volume and objective measures including measurement of chest cage mobility and respiratory rate, are to be incorporated when required. In addition to the above, PD patients are known as "silent aspirators" with weak cough reflex. The presence of tongue tremor, hesitancy in initiating swallowing, long time to finish a meal, difficulty in bolus formation, coughing when eating or drinking, change of voice while feeding, abnormal gag reflex, and reduced pharyngeal mobility are certain clinical observatory findings of swallowing dysfunction which may predispose to aspirations.

Individuals with minimal symptoms of PD (1 and 2 of the Hoehn and Yahr scale) may demonstrate aerobic exercise capacities similar to healthy adults. Individuals with more advanced disease (2.5 and above of Hoehn and Yahr scale) may demonstrate a greater variability and lower aerobic capacities compared to healthy adults. To determine the endurance capacity and the walking velocity, 6 or 12-min walk test can be used. While determining the endurance capacity, the vital parameters like heart rate, respiratory rate, and blood pressure, and exertional symptoms like pallor, dyspnea, dizziness, and fatigue need to be documented. A 2 or 3-min walk test is a sensitive and feasible tool to evaluate the walking endurance of advanced PD patients, while Borg's rating of perceived exertion (RPE) is a feasible scale for perceived exertion. The Borg's scale allows participants to subjectively rate their level of exertion during exercise or exercise testing, starting with a minimal score of 6 which means "no feeling of exertion" and ending with a maximum score of 20 which means "very, very hard." Table 4.5 provides instances of tasks performed by a normal healthy adult below 65 years of age for different Borg's RPE scores.

Table 4.5 Examples of activities or tasks with respect to Borg's rating of perceived exertion scores

Borg's RPE score	Description about the exertion	Examples of tasks performed by a normal healthy adult below 65 years age
6	No exertion	Listening to the radio, reading a book, or watching a movie
7–8	Very, very light exertion	Rolling the sleeves of a shirt or tying the shoe laces
9–10	Very light exertion	Daily chores like laying dinnerware on a dining table or folding clothes
11–12	Fairly light exertion	Walking through the grocery store or similar activity that requires some effort but not enough to speed up the breathing
13–14	Somewhat hard	Brisk walking or other activities that require moderate effort and speed your heart rate and breathing but do not make you out of breath
15–16	Hard	Running, swimming, bicycling, or other activities that take vigorous effort and get the heart pounding and make breathing very fast
17–18	Very hard	Highest level of activity one can sustain
19–20	Extremely hard, maximum exertion	A finishing kick in a race or other burst of activity that you cannot maintain for long

4.2.5.2 Staging of Parkinson's Disease

Not all PD patients experience all the symptoms of the disease. Even in typically progressing PD, the patients may experience the symptoms in quite the same order but not necessarily at the same intensity. The condition does not progress in a straight line and it is often difficult to pin down the exact progression, pace, and severity of the disease. PD rating scales are a means of assessing the symptoms and providing information about the course of the condition and/or QOL. Such scales may also help to evaluate the management strategies and can be useful for clinicians, research scholars, patients themselves, and caregivers. In the past, based on the degree of disability or progression of the disease or the person's ability to perform daily activities in terms of speed and independence, several PD rating scales have been introduced. The Webster rating scale, the Hoehn and Yahr scale, and the Schwab and England ADL scale are some of them.

The Hoehn and Yahr scale, widely used in clinical settings and research, originally described in 1967 by Margaret Hoehn and Melvin Yahr, describes the symptoms progression and the level of disability. The original scale had stages 1 through 5 and has been modified (Box 4.2) with the addition of 1.5 and 2.5 to the former stages. In the recent past, this rating scale has been complemented by the UPDRS which covers both motor and non-motor symptoms. The UPDRS scale was introduced in 1987 and it combines the elements of other Parkinson's rating scales to produce a comprehensive and flexible tool to monitor the disease course and the degree of disability. Box 4.3 provides parts and constituents of UPDRS.

Box 4.2 Modified Hoehn and Yahr Scale

Stage 0: No signs of disease
Stage 1.0: Symptoms are very mild; unilateral involvement only
Stage 1.5: Unilateral and axial involvement
Stage 2: Bilateral involvement without impairment of balance
Stage 2.5: Mild bilateral disease with recovery on pull test
Stage 3: Mild to moderate bilateral disease; some postural instability; physically independent
Stage 4: Severe disability; still able to walk or stand unassisted
Stage 5: Wheelchair-bound or bedridden unless aided

Box 4.3 Unified Parkinson's Disease Rating Scale: Parts and Constituents

Part I: Evaluation of Intellectual Function, Mood, and Behavior:	**Part II: Self-evaluation of Activities of Daily Living:**
Forgetfulness, disorientation in time and space; vivid dreaming; hallucinations; delusions and paranoia; depressed mood; anxious mood; apathy; features of dopamine dysregulation syndrome; nighttime sleep problems; daytime sleepiness; pain and other sensations; urinary problems; constipation problems; lightheadedness on standing; fatigue	Speech; salivation and drooling; chewing and swallowing; handwriting; cutting food and handling utensils; dressing; hygiene; trouble doing hobbies and other activities; tremor impact on activities; getting in and out of bed; walking, balance, and falling; freezing
Part III: Motor Examination:	**Part IV: Evaluation of Motor Complications:**
Speech–volume, diction; reduced facial expression; rigidity; finger tapping; slow hand movements; rapid alternating movements of hands; toe tapping; leg agility; rising from chair; posture-stooped; gait; postural stability; body bradykinesia; tremor at rest	Dyskinesia including time spent with dyskinesia, functional impact of dyskinesia, and painful off-state dystonia; motor fluctuations, including time spent in the off state, functional impact of fluctuations, and complexity of motor fluctuations

4.2.5.3 Exercises and Training Interventions for Parkinson's Disease

The gradual deterioration of functional capacity is a well-known feature of PD. Advancement in recent drug therapy has considerably improved functional expectancy. Current evidence suggests that medical management, in conjunction with physiotherapy, can potentially delay the onset of physical disabilities. Despite the direct effects of the disease, the combination of pharmacological intervention with physiotherapy may provide some psychological well-being for the PD patient.

To maximize functional ability and minimize the complications, a variety of interventions including pharmacological intervention, physiotherapy, semi- and/or unsupervised home exercise program taught by the therapist and performed by the patient under the supervision of family members, environmental modification for the home setting, and supportive counseling are required. Early intervention may play a crucial role in averting or minimizing the musculoskeletal impairments, which PD patients are bound to develop. Physiotherapy should focus on the improvement of motor performance, functional capacity, and activity participation.

Application of Principles of Learning in Training Impairments in motor planning and programing and the absence of internal cueing mechanisms cause gradual deterioration and loss of both formal and informal learned skills. It is essential to emphasize to the patient that the conscious incorporation of appropriate movements is required for skill "reacquisition" and permanent gains are possible only if the patient "relearns" the desired maneuver. Learning, "a change in the internal state of the individual causing a relatively permanent improvement in performance which occurs as a result of practice or experience," is a process that is not directly observable, but can be implied from the behavior or performance of the participating subject. The cognitive (associated with intellectual activities), the affective (emotional control and moral judgments) and the psychomotor (deals with the acquisition of motor or perceptual-motor skills) domains are the three domains of learning and all of them are interdependent.

In PD, the principles of psychomotor learning are applicable for re-mastering or relearning the skills which are faulty or missing. The cognitive, the fixation, and the autonomous are the three phases through which learning occurs in the psychomotor learning domain. The cognitive phase is the initial, more theoretical phase of learning, where the learner focuses on understanding the principles, details, and components of the skill. In this phase, the learner tries to process the information in an attempt to cognitively understand the requirements and parameters of the motor skill. Keen observation and attention are required from the patient to process the information and the physiotherapist should pay attention to key parts of the skill and may need to demonstrate the movement for a better mental picture. In the fixation or associative phase, fundamentally a practical phase, the learner will be attempting to translate declarative knowledge into procedural knowledge until skill acquisition is made. This phase is characterized by conscious performance, smaller gains in performance, lesser verbal information, corrections, disjointed movement, and a longer time to complete the task. This is the phase where learners attempt to reduce the number of errors from feedback. Both knowledge of result (externally generated feedback either by a therapist or trainer) and knowledge of performance (ongoing feedback about the performance, usually from a monitoring device or a videotape replay) play a central role in motor skill acquisition. For realistic motor learning, sufficient and frequent practice with appropriate feedback is critical. "Practice does not make the act perfect but makes it permanent," implies the role of the correct method of practicing for optimal learning. Generally, during the early stage of skill acquisition, a considerable amount of feedback in the form of knowledge of result or knowledge of performance is required; however, the frequency of feedback reduces as the learner masters the skill.

In the autonomous phase, the final phase of skill acquisition, the motor performance becomes largely automatic, cognitive processing demands are minimal and the learner is capable of attending other information (information related to other skills or tasks). In this phase, the learner is able to implement the skill at the highest level, with the least conscious effort like a habit. Interestingly, due to progressive impairments in motor planning and programming, even in the autonomous phase, the patients can typically lose the ability to carry out the learned or relearned motor skills. The breakdown of skills can be visible in the form of small errors during motor performance. For instance, several sessions or days after re-mastering the gait, the therapist may notice a slight stooping of the trunk that was not there earlier. Such errors will grow in numbers and size with repetition and time. Therefore, to salvage the deterioration in the acquired skills, the early detection of such errors is crucial. The therapist or the caregiver who is being sensitized about the same should watch for such movement errors and intervene in a timely manner. To eliminate the newly detected errors, the patient should pass through all the phases of psychomotor learning, beginning with the cognitive phase. Mental practice, another psychomotor learning concept, consists of cognitive rehearsal in the absence of overt physical practice. It can be used as an adjunct to augment relearning of motor skills. However, mental practice may not be an appropriate strategy for patients in the advanced stage when cognitive capabilities are diminishing.

In the early and middle stages of PD, externally generated cues and performance through practice can be helpful in improving their capacity to learn complex movements and sequences. The absence of internally generated cues makes the learning of complex movements and sequences difficult. For learning complex movement sequences, the sequence has to be broken into simple component parts and each part has to be practiced separately and in sequential order with the help of appropriate visual, auditory, and somatosensory cues. Lengthy movement sequences and random practice order have to be avoided and block practice of the component parts has to be encouraged to reduce the effects of contextual interference. The home setting should be the preferred treatment setting if performing complex sequential tasks and movements are problematic in other settings. However, in the advanced stage with pronounced cognitive deficits, such training is less likely to be successful. PD patients may often demonstrate motor learning deficits like slower learning rates, increased dependence on context-specific learning, and lack of generalizability. Repetitive practice and use of structured instructional sets while training have shown improvement in movement speed and consistency even in the advanced stage of disease or with cognitive deficits. To develop procedural skills, practicing with a large number of repetitions is crucial. Providing instruction to focus the attention on the desired movement and an environment that minimizes distraction and competing attentional demands may help learn the skill more efficiently before attempts are made for a different environmental context.

Kinesiological and Mechanical Perspectives for Parkinson's Disease Training Posture and movements are constrained by kinesiological imperatives and mechanics laws. From a kinesiologic and mechanical viewpoint, postural changes cause compensatory postural strategies and alterations in muscular activities. In PD patients, the vertebral segment mobility is typically reduced or lost throughout the cervical, thoracic and lumbar spine. Analysis of posture and movements based on the mechanical and kinesiologic perspectives may reveal movement constraints imposed by the disease, mainly, the lack of segmental mobility of the spine (spinal rotation, lateral flexion, and spinal extension). Alterations like the above may force PD patients to depend on compensatory strategies like long sitting from supine, instead of a roll to one side for sitting. The analysis may also reveal the overuse of rectus abdominis, instead of the abdominal obliques for roll to one side to sit. Therefore, therapy should focus on segmental mobility activities, specifically emphasizing segmental rotation patterns (i.e., isolated upper and lower trunk rotations). Rhythmic initiation using diagonal patterns of Proprioceptive Neuromuscular Facilitation (PNF), on a therapeutic mat and encouraging the PD patients to practice rolling on a firm to soft surfaces may benefit the patient to improve bed mobility skills and trunk flexibility.

The developmental sequence of postural control (a natural and predictable movement progression a normal infant follows during early life to develop abilities to roll, crawl, kneel, stand, and walk) progresses from kinesiologic simple to kinesiologic complex movements. Lower developmental postures within the neurodevelopmental sequence provide the necessary strength, stability, and coordination required for more challenging postures. Such postures minimize the risk of falls and maximize movement efficiency. With the perspectives mentioned above, initial treatment including management of rigidity and encouragement of mobility or flexibility should preferably commence in a fully supported supine position, even if the patient is ambulant and somewhat independent.

Strategies to Reduce Rigidity Generally, rigidity tends to be more in the axial and proximal muscle groups and may predispose for postural and musculoskeletal alteration, which may be unilateral in the earlier stage. Slow rocking can be used to promote the generalized relaxation of skeletal muscles secondary to rigidity. Rocking the patients while being comfortably seated on the rocking chair can induce relaxation and sleep. Both vestibular and somatosensory stimulation associated with rocking may exert a synchronizing action in the thalamocortical networks that reinforces endogenous sleep rhythms. Neurophysiological research supports the notion that vestibular stimulation associated with rocking is related to the relaxation of muscles. The known functional interrelationships of vestibular structures with the cerebellum and reticular activating system and vestibular structures to autonomic centers can explain the association between vestibular stimulation and relaxation and sleep. Prior to flexibility exercises, posture correction and functional training, upper and lower trunk rotation and counter-rotation (rotating the upper trunk to one side while the lower trunk is rotated in the other side) movements or diagonal patterns of PNF like slow, rhythmic, chopping movement performed in the supine position can be helpful to promote relaxation. The rhythmic initiation technique of PNF, moving progressively through passive, active-assisted, and active movements, may reduce the rigidity. Relaxed diaphragmatic breathing exercises and incorporation of bilateral symmetrical diagonal PNF patterns into relaxed breathing exercises can promote relaxation and improve chest expansion. To achieve relaxation, PD patients can also try music therapy and meditation techniques either in lying or well-supported seated positions.

Flexibility and Stretching Exercises Flexibility exercises improve ROM and physical function. The flexibility exercise should go hand in hand with the relaxation techniques for rigidity. If the patients have achieved adequate relaxation, the therapist will be able to feel the segmental mobility at the vertebral, girdle, and proximal joints when passive flexibility exercise is attempted. Flexibility exercises should progress through passive to active-assisted to active. A gentle warm-up exercise or use of hot fomentation can be useful to minimize the possibility of capsular or ligament injury while performing stretching or flexibility exercises of those tight joints. To counter the thoracic kyphosis and promote upper trunk extension, bilateral symmetrical diagonal PNF patterns can be utilized.

Positional stretching techniques can be useful to stretch tight muscles and soft tissues. In late-stage PD, patients are likely to develop severe tightness and flexion contractures of the trunk and limb muscles. Prone-lying for 20–30 min on a daily basis can be beneficial to provide positional stretch for the trunk and limbs flexors. For tight hip flexors, if prone lying is uncomfortable, then high-kneeling with hip in extension or hook-lying bridging can be used as an alternative strategy to stretch the same. Unilateral PNF pattern that emphasizes hip and knee extension can counter the typical flexed, adducted position of the lower extremities. In the advanced stage, following the onset of cardiorespiratory impairments and the development of significant postural deformity, the patient may not be able to tolerate the prone position. For such patients, mechanical stretching can be achieved through the use of a tilt table.

Vigorous and quick passive ROM exercises need to be avoided due to the chance of soft tissue injury. Such stretches may create apprehension in PD patients and also stimulate pain receptors to cause rebound muscle contraction, further worsening the muscle hypertonus. If the bones are osteoporotic, vigorous stretches and quick passive movements may predispose to fractures among elderly PD patients.

Strategies to Reduce Tremor In the ambulatory stage (up to the fourth stage of Hoehn and Yahr scale), resting and/or action tremor, bradykinesia and hypokinesia, rigidity, postural alterations, gait abnormalities, and loss of facial expression can be present, but in varying severity. In some PD patients, tremors may predominate over bradykinesia, rigidity, and postural alterations and make gross motor skills like bed mobility, transitions, and gait, and fine motor skills like dressing and handwriting difficult or impossible to perform. Pharmacological and non-pharmacological treatment strategies are available when tremors are more disturbing than bradykinesia and rigidity. Some of the non-pharmacological strategies for tackling resting tremors include maintenance and stabilization of the arms (proximal joints) as close as to the body when carrying out activities and avoiding multitasking like shaving, brushing, or cooking while standing. Managing stress, anxiety, frustration, and fatigue that are known contributors to resting tremors and the use of adaptive equipment like powered toothbrush, adaptive utensils, and weighted cutlery may help to reduce the tremors.

Training to Improve Posture Following flexibility exercises and stretching exercises, the focus of training should be on improving the posture of the patient. Often the correction of posture further improves the mobility of the joints, specifically the rotation of the spine. In addition to altering the balance and gait, the typical stooped posture seen in IPD also causes swallowing difficulty, weak and soft speech, reduced strength of postural muscles, and pain in the joints. Verbal instructions like "chin-tuck" to reduce the forward head posture, "sit tall" to improve the thoracic kyphosis, and "bring the shoulder blades closer" to improve scapular retraction and minimize the rounding of the shoulders, can be incorporated for posture correction. Such exercises need to be performed regularly with a 1–2 min hold. Posture correction exercises should progress from a seated position with back support to without back support. To correct the posture in standing, the therapist can advise the IPD patient to stand against the wall or stand with the back facing the wall. Meanwhile, verbal instructions like "stand as close as possible to the wall," "tuck the chin," "allow the back of the head, the trunk and the buttocks to be in contact with the wall," and "knees straight and feet as close as to the wall," can help to improve the posture. Regular and repeated practice of such posture correction exercises will help the patient to consciously correct the posture until the posture correction becomes a habit. In addition to the above, the use of mirror, cueing, and biofeedback can complement or be an alternative to the standard methods of posture correction. Exercises with an elastic band can also be incorporated as a method of strengthening the muscles and for improving the posture.

For those PD patients with the Pisa syndrome, to minimize the potential for permanent postural deformities, early rehabilitation, emphasizing stretching exercises for the external oblique and paraspinal muscles, is strongly recommended. An individualized program consisting of proprioceptive and tactile stimulation, combined with stretching, flexibility exercise, cueing, and postural re-education, can be a beneficial, safe and feasible strategy to overcome the postural abnormality. Functional strengthening exercises, especially the extensor muscles, bridging exercises, trunk flexion and rotation exercises, in addition to gait training and balance training, can also be attempted. Investigators have reported that the use of Kinesio® taping for the trunk muscles as an adjunct treatment had no long-lasting beneficial effect compared to conventional therapy. The exercise regimen (overcorrection strategy) used for correcting the listing tendency in stroke patients, as seen in Pusher's syndrome (mentioned in Chap. 3—"Stroke") can be attempted as an alternate strategy to minimize the postural abnormality due to Pisa syndrome.

Strength Training Strategies The PD patients who demonstrate poor posture and functional deficits like the inability to get out of a chair and difficulty performing overhead activities using the upper limb, secondary to the trunk and proximal limb muscle weakness may benefit from strength training. Reduced physical activity, a consequence of aging, and reduced basal ganglia output leading to lower levels of cortical and spinal motor activation are the possible reasons for muscle weakness. Studies on strength training among PD patients have shown improvement in muscle force, bradykinesia, and QOL. Strength training, specifically, progressive resistance strength training, based on the progressive overload principle using free weights, elastic bands, manual resistance, or strength-training equipment, can be used in the clinical or home setting for strengthening the muscles. For each muscle group, a set of 10–12 repetitions and 14–15 rate of perceived excursion on the Borg's scale can be recommended for most of the PD patients. Bridging exercise performed on a stable and unstable surface (gym ball), strengthens the hip extensors and the spinal extensors. Functional training exercises can be an alternative to improve muscle strength. Unlike progressive resistance strength training, functional training exercises are multi-joint and multiple planar movements consisting of compound exercises requiring more than one muscle group to work together. Functional training exercises typically mimic everyday movement patterns like squats and pull-ups. Incorporating rotational components into functional training, in addition to developing muscle strength, will improve flexibility. The disease process and aging are known to cause deconditioning effects. General conditioning programs like aerobic exercises, strengthening exercises and functional training may retard the process of deconditioning effects and maintain the optimal level of physical function.

Training Sit to Stand Transitions PD patients typically sit with a stiff trunk with a posterior pelvic tilt. Exercises designed to improve pelvic mobility and tilt (including anterior and posterior tilt movements and side-side pelvic tilt) along with the extension of the spine can be practiced while sitting on a therapeutic gym ball and later on a firm, stable surface like a therapeutic mat. Exercises in sitting should include weight shifts emphasizing trunk rotations and reach outs. For sit-to-stand transition training, the PD patient will be instructed to scoot his or her trunk forward to the edge of the therapeutic mat. Following this, the patient will be instructed to place the feet on the floor right under the knees at a hip-width apart. Verbal instructions like "sit tall" are to be given to encourage the patient to sit with an erect spine. While maintaining the spine in extension, the patient should be asked to flex the trunk over the hips. Guidance and cues can be used for encouraging the flexion of the trunk over the hips. Assistance and support may be given to perform the standing-up by extending the lower limbs, followed by an extension of the spine. Premature extension of the spine during this training process will make the standing-up difficult or faulty. Proprioceptive training on a firm surface, including static and dynamic partial squats, weight shifts and single-limb stance using stall bars or manual assistance of the therapist and strengthening exercises for lower extremity muscles can help the PD patients to smoothly execute sit-to-stand transitions. The sit-to-stand transitions should be progressively practiced from a standard height chair to a lower chair. For those patients where safety is a concern while standing, the use of a harness system or chair with armrests can be an alternative. Once even weight-bearing through lower extremities is possible, activities like weight shifts and rotational movements of the trunk should be introduced in standing.

Role of Dissociation Movements In addition to the bradykinesia and deficits in motor learning, PD patients typically exhibit undue difficulties in the execution of simultaneous or sequential complex movements of two or more joints of the limbs. Many goal-directed motor acts require such a series of movements, sequential and/or simultaneous, at different joints and of different limbs. Even PD patients can have motor issues like the inability to switch from one motor act to another or execute two different motor acts simultaneously with opposite limbs. The motor planning and programming impairments due to the involvement of the basal ganglia circuitry are the plausible cause for this. Training the patients to perform simultaneous symmetrical movements of limbs (Fig. 4.8) progressing toward simultaneous asymmetric movements [alternating or dissociating in nature] (Fig. 4.9) and adding more complexity by incorporating both upper and lower limbs [simultaneous dissociation movements of shoulder girdle with respect to pelvic girdle] (Figs. 4.10 and 4.11) may improve the ability of the PD patient to execute two different motor acts simultaneously with opposite limbs. In the beginning, the therapist may need to demonstrate the movement and may need to guide and assist the patient in performing such simultaneous alternating movements. For the training program mentioned above, the use of the therapeutic mat with the patient in a supine position is desirable before more advanced developmental postures are attempted. Like the training given for transitions and balance and gait, to master dissociations or alternative movements, the use of relevant instructions and adequate repetitive practice is critical.

Balance and Gait Training Strategies In PD, automatic motor behaviors or adjustment movements like arm swing while walking or weight shifts while standing that accompany voluntary movements and righting and balance are either impaired or faulty. Typically, balance issues are likely to develop within 5 years after the onset of initial symptoms and within another 5 years, recurrent falls will be a major concern. Balance training should be an integral part of the exercise training program, especially when the PD patients' assessment reveals balance impairments. PD patients can take advantage of the balance training programs designed to improve balance among the healthy elderly provided their specific problems are taken into consideration, i.e., prior to the commencement of balance training, it is essential to address issues such as rigidity, flexibility, and postural abnormalities. Balance training should be given in both the seated and standing positions. The training should focus on controlling the center of mass and improving the limits of stability. While maintaining a near-normal or normal postural alignment, the patient should be encouraged to expand the limits of stability. The use of verbal, tactile, or proprioceptive cues and appropriate and well-timed feedback can encourage the

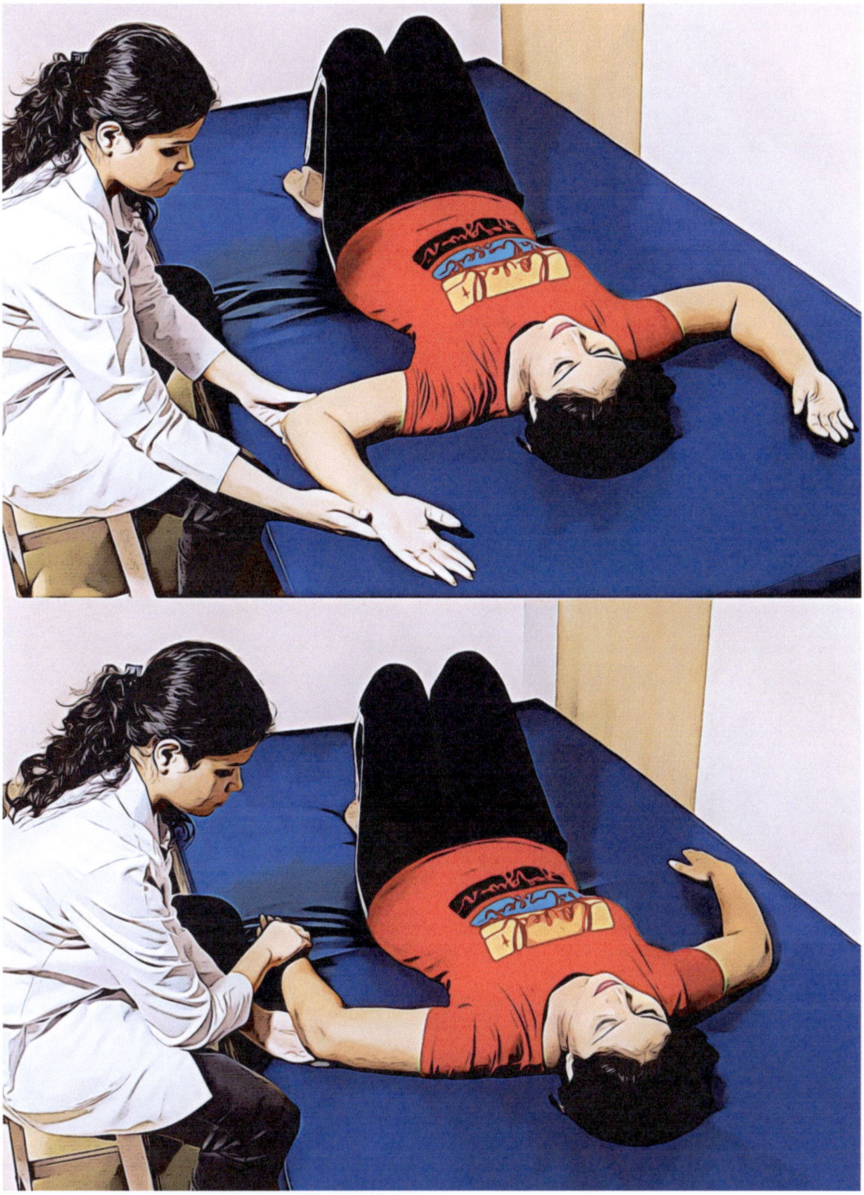

Animated photographs of model and therapist with permission

Fig. 4.8 An illustration of the symmetrical movements of the upper limbs

patient to achieve balance without postural disturbances and falls. Balance training should progress from stable to an unstable surface, static to dynamic positions of limbs or trunk, seated activities to activities in standing, a calm environment to a busy environment, open eyes to closed eyes, and a larger base of support to a small base of support like progression through quadruped, kneeling, half-kneeling and

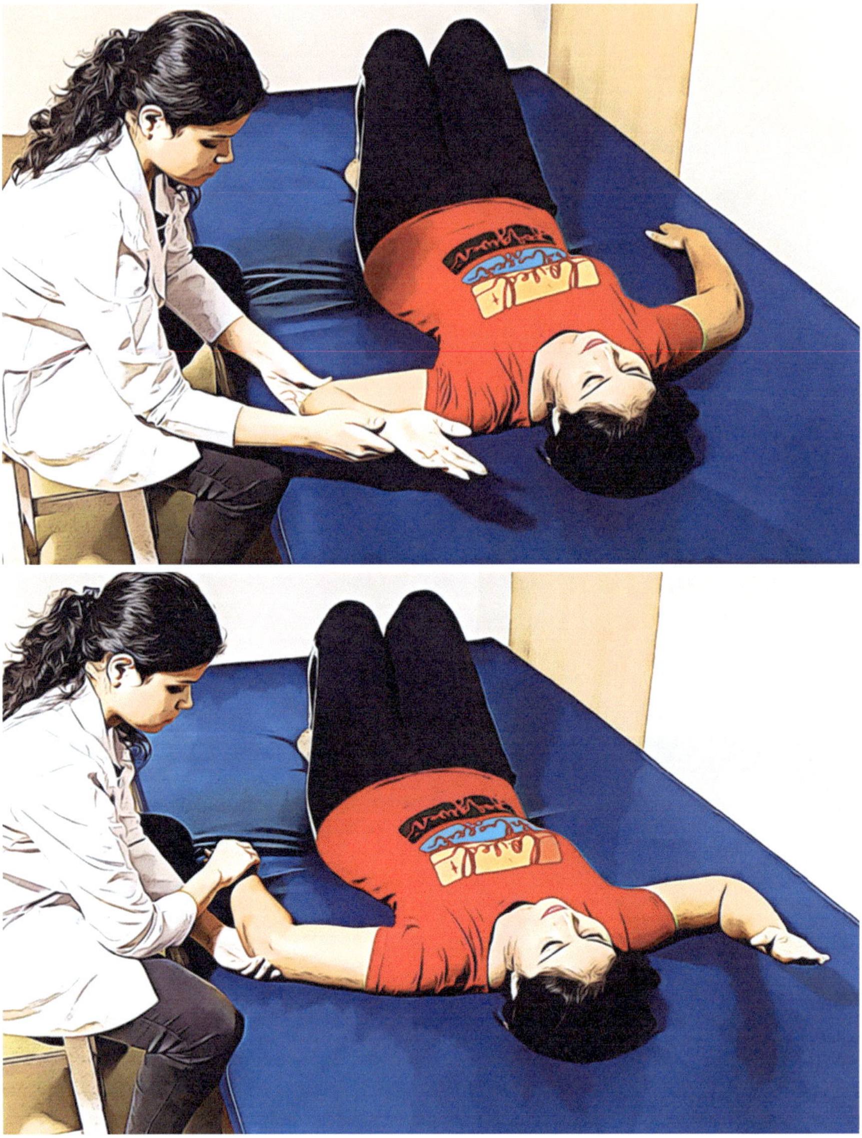

Animated photograph of model and therapist with permission

Fig. 4.9 An illustration of the asymmetrical movements of the upper limbs

standing. Verbal instructions like "stand tall" or "do not bend" and use of a mirror(s) for postural feedback are often encouraging. During balance training, encourage the PD patients to initiate and execute movement as fast as they can. Self-initiated displacements are preferred over perturbations given by the therapist or the caregivers as the latter can create more apprehension and worsening of rigidity and postural

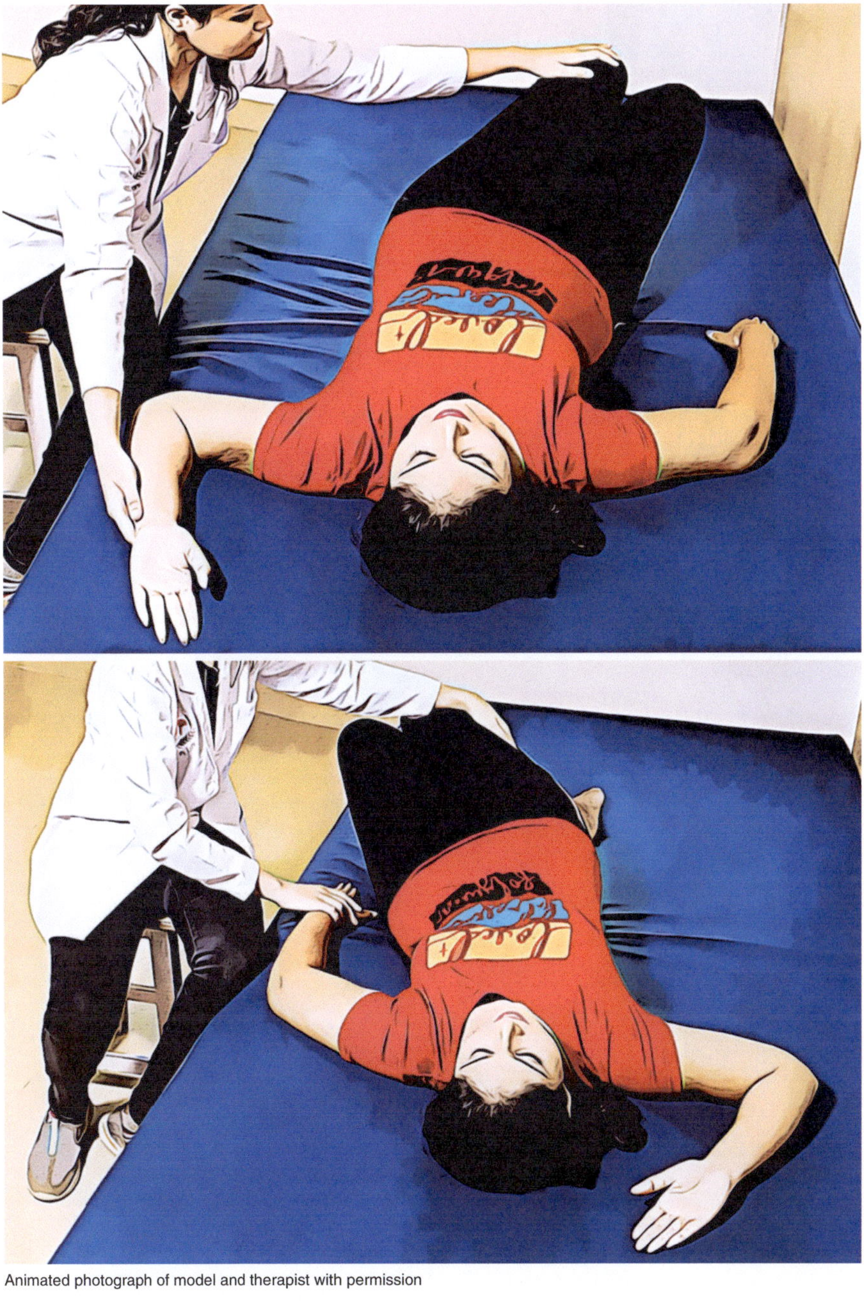

Animated photograph of model and therapist with permission

Fig. 4.10 An illustration of the asymmetrical movements of the upper and lower limbs

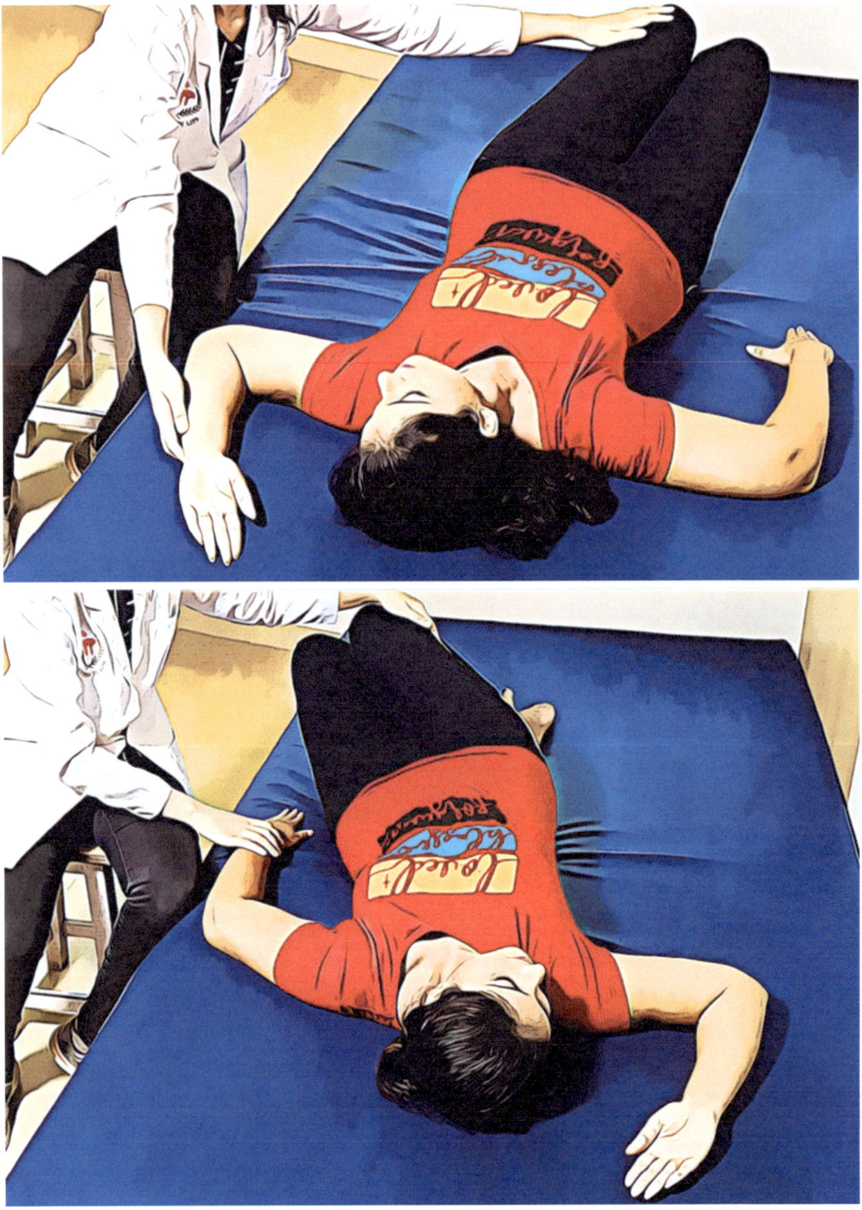

Animated photograph of model and therapist with permission

Fig. 4.11 An illustration of the asymmetrical movements of the upper and lower limbs and the neck

stiffness. For PD patients with recurrent episodes of falls, finding safe ways of standing, identifying key activities causing falls, environmental adaptations, strategies to reduce the fear of falling, and teaching cognitive strategies to avoid or manage situations that can challenge balance are certain recommendations to minimize the episodes of falls.

The basal ganglia circuitries play a key role in running complex motor sequences that make up skilled, largely automatic, movements like gait. For the lower extremity, gait is a complex and sequential automatic activity. Gait training goals focus on reducing primary gait impairments (including reduced velocity, reduced step and stride length, lack of trunk rotation, lack of arm swing, festinating pattern, lack of normal heel-toe pattern, and the tendency to stoop while walking) and improving the PD patient's ability to safely perform functional mobility and avoid falls while walking. Verbal instructions such as "walk fast," "big step," "do not bend," and "swing arms" can enhance velocity, step length, posture, and arm swing. Practicing marching in place and progressing toward an exaggerated high stepping walking may improve the ground clearance if the patient has a tendency to walk with short steps, poor ground clearance, and shuffling.

Role of Cues in Gait Training In basal ganglia disorders like PD, a key motor problem about sequential movements is the deficiency of automatic maintenance of appropriate scale and timing. Festinating or shuffling gait in PD is arrhythmic, small in amplitude, and variable, and often exhibits reduced automaticity of movement and increased attention directed toward the gait. Cueing is a well-established compensation strategy for improving locomotion when internal cueing mechanisms are defective. Unlike clues (hints), cues are prompts and can be auditory, visual, tactile, verbal, or others. Normally, basal ganglia, pre-supplementary motor cortex, and dorsolateral prefrontal cortex use internal cues to initiate self-generated movements.

The external cues can compensate for errors in the internal cueing mechanisms. External cues, either visual or auditory, appear to facilitate movement by utilizing the different regions of the brain, including the premotor cortex and cerebellum. Unlike the internal cueing responsible for learned movements that are more automatic, external cueing bypasses the defective basal ganglia and uses cortical circuitry with heightened attention for generating movements. Therapeutic cueing uses external temporal or spatial parameters to facilitate movement initiation and continuation. Cues given in the early stages of PD tend to improve or maintain quality and avoid deconditioning, while in the later stages help to compensate for the reduction of automaticity. The benefits of cueing are dependent on the type (auditory, visual or other) and the parameters (spatial or temporal) used. For instance, cues that focus on step amplitude primarily affect step size, whereas cues that focus on step rhythm primarily affect the number of steps. Similarly, auditory cues tend to have a greater influence on the temporal components such as cadence and stride synchronization, than on the spatial components of gait.

Since attention plays a critical role in the efficacy of external cueing, the presence of medication instability and disease fluctuations in the advanced stage of PD, and cognitive impairments can often make cueing ineffective. In the early and middle stages of PD, external cues are found to be effective in triggering sequential movements. Visual cues, including floor markings and dynamic stimuli like strobe lights or laser light from the cane or walker, have shown significant improvement both in stride length and velocity and reduced tendency for freezing, but not in cadence. Rhythmic auditory stimulation, including metronome beats or steady beats from a musical device, has shown an improvement in gait speed, cadence, and stride synchronization. Auditory cues should be consistent and rhythmical and the speed of the beat for auditory cueing should preferably be set 25% faster than the patient's preferred pace. With regard to visuospatial cues, transverse visuospatial cues (across the gait path) may be more beneficial than parallel cues (alongside the gait path) in improving gait velocity, stride length, and stance time. The use of floor markers or footprints on the floor can be strategies to improve foot placement. Existing literature suggests that visual cues are relatively better than rhythmic auditory cues, despite the extra attention required for the former. However, literature also reveals that PD patients in the home environment prefer auditory cues over visual cues.

Management during the Advancing Stage In the advancing stage of disease (3–4 of Hoehn and Yahr scale) many of the PD patients will experience freezing episodes. The episodes are sudden, short, and transient and primarily occur while attempting to initiate a walk, navigate through obstacles or negotiate narrow spaces or curbs. The freezing episodes are often triggered by emotional stress, environmental constraints, and dual tasking and may lead to reduced physical activity levels and functional capacity. Defective perceptual judgment, impaired bilateral coordination of gait, and impaired integration of vision with spatial memory can be a few of the contributing factors for the freezing phenomenon. Studies have noted reduced step length, increased cadence, and premature timing of anterior tibialis muscle and gastrocnemius muscle prior to the episodes. Physiotherapy can be helpful to reduce or overcome freezing episodes. The strategies that can prevent or overcome freezing may include the use of metronome, change of direction by stepping sideways or backward, weight shifting before attempting a step, trying to march in place, verbalizing "1-2-3-go" and then stepping forward, initiating a movement of any body part like swinging an arm and stepping over an imaginary line in front. Often, the strobe lights from the cane or the walker can work as a visual cue, and verbal cues like "heel down" or "stand tall" may help the patient overcome the freezing episodes.

Patient and Family Members Education and Advanced Stage Management Regarding symptom progression and advancement of disability, PD is more obscure compared to other progressive neurological conditions like motor neuron disease and Alzheimer's disease. This often necessitates the need to educate the patient and family members about the disease, the medications, dosage, and their side effects, and the preventive strategies to avoid or minimize the complications. The therapist should teach the family members or caregivers the basic and safe techniques of lifting, maneuvering, and transferring the patient and exercises like bridging and pelvic rolling that can assist in the day-to-day care of the patients. Both the patient and the family members should be taught the strategies that can circumvent the movement abnormalities to solve functional problems. The family members should also be encouraged to augment the patient's motivation for performing functional activities.

In the advanced phase (fifth of Hoehn Yahr stage) of the disease, pharmacotherapy becomes less effective and the complications are likely to increase both in magnitude and frequency. Symptoms, including dyskinesia, are more pronounced during the end of dose failure and drug on-off phases making patients functionally dependent. The therapist should keep track of the on-off phases and should provide therapy before the beneficial effects of medications wear off. Preferably the therapy should begin 45 min to 1-h post-medication when the patient is at his or her best. Intensive therapy to improve the facial and oromotor functions, breathing exercises and chest manipulations for lung dysfunction, frequent changes of bed position to prevent decubitus ulcers are certain treatment objectives during the late stage.

4.2.5.4 Alternate Therapies

Aquatic Physiotherapy or Hydrotherapy Aquatic physiotherapy or hydrotherapy can be an alternative to land-based exercise training protocols. The aquatic environment can promote significant therapeutic results such as a reduction in muscle tone, improvement in postural stability, and increment in functional mobility. The warm water used in the aquatic pool may have a potential therapeutic effect on rigidity. Current literature also states that both land and water-based protocols are useful for improving balance. Supervision during aquatic exercise is a must as there is a potential hazard for slips, falls, and drowning. Close monitoring of hygiene and infection control is required to minimize the possibility of infection. Dehydration can be yet another important concern for older adults. The patients should be encouraged to drink 250 ml of water 1 h prior to the pool therapy.

Treadmill Training A considerable amount of literature is available with regard to the use of treadmill in neurorehabilitation. Current literature shows a greater improvement in motor performance compared with conventional therapy for stride length, gait speed, and symptoms. The locomotion training on the treadmill also has a positive effect on balance and QOL. "With the treadmill, there is no getting around it; the patient must match his/her pace to the treadmill." The treadmill acts as an external cue by setting the walking pattern and reinforcing the neuronal circuits (provides an external rhythm that compensates for the defective internal rhythm). Increasing the walking demand and gait speed is believed to cause adaptive neural plasticity which improves gait parameters. Training on the treadmill can be progressed by reducing the body weight support given through the harness, increasing the treadmill speed, reducing the physical assistance, and increasing the duration of time on the treadmill. The use of a safety harness is mandatory for PD patients with balance deficits or FOG during treadmill training. For such patients, if the purpose of treadmill training is to improve endurance but not the gait, arm or leg cycle ergometer is a better and safer alternative.

Lee Silverman Voice Treatment (LSVT) 'Big' Program Also known as "training big" program, the LSVT consists of a therapist-guided high intensity, large amplitude, multiple repetitions, and whole-body exercises, based on the concept that repetitive high-amplitude movements yield greater improvements in motor performance. This exercise is believed to have a neuroprotective effect and is performed 1 h per session, four sessions per week, and studies have shown improvement in the UPDRS motor scores and the gait parameters of 1–3 Hoehn and Yahr scale stage PD patients.

In addition to the above, several alternate therapies, including high-intensity indoor cycling, dance therapy, music therapy, tandem cycling, and Nordic walking, have been studied and have shown promising results for PD patients in the alleviation of symptoms and gait parameters improvement. However, further details regarding those alternative therapies are beyond the scope of this chapter.

4.3 Other Extrapyramidal Disorders

4.3.1 Wilson's Disease (Hepatolenticular Degeneration or Westphal–Strümpell Pseudosclerosis)

Wilson's disease is an inherited metabolic progressive lenticular degenerative disorder associated with liver cirrhosis. Wilson's disease is transmitted as an autosomal recessive trait and the prevalence is about 1–2 per 100,000 of the general population. Siblings of the affected one have a 25% chance of developing the disease. The classical description of the disease was published by Samuel Alexander Kinnier Wilson, a British neurologist, in the year 1912. Wilson entitled his article "Progressive lenticular degeneration: a familial nervous disease associated with cirrhosis of the liver," in which he gave a detailed description of four cases. Prior to the detailed description by Wilson, in 1883, Adolph Strümpell, a German neurologist, and in 1898, Karl Friedrich Otto Westphal, a German psychiatrist under the title named "pseudosclerosis" had given details about the same; however, none of the descriptions, including Wilson's, addressed the association of liver cirrhosis with lenticular degeneration. About a decade before Wilson presented his monograph, in 1902, Bernhard Kayser, a German ophthalmologist, first described a rusty-brown ring, a corneal abnormality and within a year, Bruno Fleischer, another German ophthalmologist, related it to hepatic degeneration. In 1913, A. Rumpel demonstrated the increased copper content within the liver of Wilson's disease patient. The clinical and histopathologic studies by H. C. Hall and W. Spielmeyer (1920–1921) clearly established the association of lenticular degeneration with liver cirrhosis. The same year, H. C. Hall proposed genetic inheritance as an autosomal recessive pattern and in 1953, A. G. Bearn confirmed it by genetic ratio analysis calculation. It was A.J. Glazebrook in 1945 and John Cumings in 1948, in their independent studies, linked copper accumulation with the basal ganglia and hepatic pathology. In 1952, Herbert Scheinberg and David Gitlin demonstrated a low level of serum ceruloplasmin in this disorder.

4.3.1.1 Etiology, Neuropathology, and Pathogenesis

Mutation of a gene on chromosome 13q14.3, which encodes a copper transporting ATPase, is the genetic cause of the disease. A reduced rate of incorporation of copper into ceruloplasmin and a reduction in biliary excretion of copper are the two copper metabolism dysfunctions caused by the genetic defect. Abnormal accumulation of copper in the liver and the brain is caused by defective cellular transport and failure of the liver to excrete copper into bile. The hepatic cellular damage and neuronal cell damage secondary to the abnormal accumulation of copper cause the hepatic and neurological manifestations of the disease.

Pathological findings include enlargement of the ventricles, flattening of the convexity of the head of the caudate nucleus, and lesions in the middle zone of the putamen. The caudate nucleus and putamen appear atrophic, shrunken, and crumbly and in untreated patients, cavitation can also be seen. Rarely the cavitations can be seen in the thalamus, subthalamic nuclei, dentate nuclei, and cerebral cortex. In some cases, multifocal demyelination affecting the cerebral white matter with no primary evidence of inflammatory pathology can also be seen. The myelin staining reveals central pontine myelinosis. Pontine myelinolysis can be the most striking pathology in the brainstem. In general, the pathological involvement of the lentiform nucleus is more prominent than that of the caudate nucleus and pallidum.

4.3.1.2 Neurological Manifestations and Investigation Findings

Generally, the onset of the disease is in the second or third decade of life. Acute or chronic hepatitis and splenomegaly are the first expressions of the disease due to the deposition of copper (Cu) in the liver. The patient can have a history of several attacks of symptomatic jaundice or may present with neurological signs and symptoms in the absence of jaundice. Usually, the neurological manifestations are gradual and insidious, but sometimes with intermittent acute deteriorations or rapid progression. Less frequently, the early manifestations of this disorder can be psychiatric, renal, hematological, or musculoskeletal. The early neurologic manifestations are most often extrapyramidal, consisting of tremors in the limbs and head and bradykinesia. Tremors can be resting and/or postural in nature and when the limbs are outstretched, they may display a coarse, "wing-beating" tremor. Bradykinesia of tongue, lips, pharynx, larynx, and jaw results in dysarthria, dysphagia, and hoarseness. Saccadic movements of the eyes are often slow and have the limitation of upward gaze. Often the patients may display choreiform movements or dystonic posturing of the limbs with flexed or stooped posture, dystonic gait, and sometimes with stridor. During the progression of the disorder, elements of cerebellar ataxia and intention tremor of variable degree may get added to the existing symptoms.

Cerebellar features may include cerebellar ataxic gait, scanned and/or staccato speech, limb ataxia, intention tremor, and titubation of the head. Kayser–Fleischer (KF) rings (rusty brown or golden brown ring visible around the corneo-scleral junction/limbus) in the eyes become more evident as the neurologic manifestation of the disease evolves. In addition to the development and advancement of motor impairments, in the later stage, intellectual functions will also progressively decline. Timely diagnosis and treatment are of utmost importance and if delayed, the disease may progress and the patient can become extremely rigid, dystonic, mute, immobile, and mentally slowed.

The presence of reduced serum ceruloplasmin level within the blood, low serum copper, increased excretion of copper through urine, high copper content in hepatic cells after liver biopsy are certain laboratory findings. The liver function test generally is abnormal. The Computed tomography (CT) scan may reveal atrophic changes in the basal ganglia, cortical and cerebellar regions, associated with dilatation of lateral ventricles and third ventricle and widening of cerebral and cerebellar sulci. T1 weighted Magnetic Resonance Imaging (MRI) scans can demonstrate hyperintensity in lentiform nuclei and mesencephalic regions (Fig. 4.12). T2 weighted scan may demonstrate hyperintensity which typically involves the putamen, globus pallidus, caudate nucleus, and the ventrolateral part of the thalamus. The axial T2 weighted scan at the level of the brainstem may reveal the "double panda sign" (a face of a panda at the midbrain level and a cub of a panda at the pons level).

4.3.1.3 Medical and Surgical Management

If the condition is diagnosed before the onset of the neurologic manifestations, the use of copper (Cu) chelating agents and reduction of dietary intake of Cu can reverse the changes caused by the disease. D-penicillamine remains the principal Cu chelating agent and to prevent anemia, pyridoxine should be given daily. Trientine or ammonium tetrathiomolybdate can be a substitute for D-penicillamine. The use of medication has to be a lifetime to avoid disastrous clinical deterioration as there are no pharmacological interventions that can reverse or stop the defective copper metabolism. The patient should abstain from copper-rich foods, including liver, cocoa, chocolate, mushrooms, and shellfish. With the decoppering therapy, the KF rings usually disappear and the liver function test may return to normal. Early diagnosis and prompt use decoppering agents may prevent or reverse otherwise permanent and fatal clinical manifestations of this disease. Even for those patients with neurological signs and symptoms, the commencement of decoppering therapy may reduce the neurologic manifestations. For those patients with severe liver damage, liver transplantation will cure the underlying metabolic defect and may reverse some of the clinical manifestations.

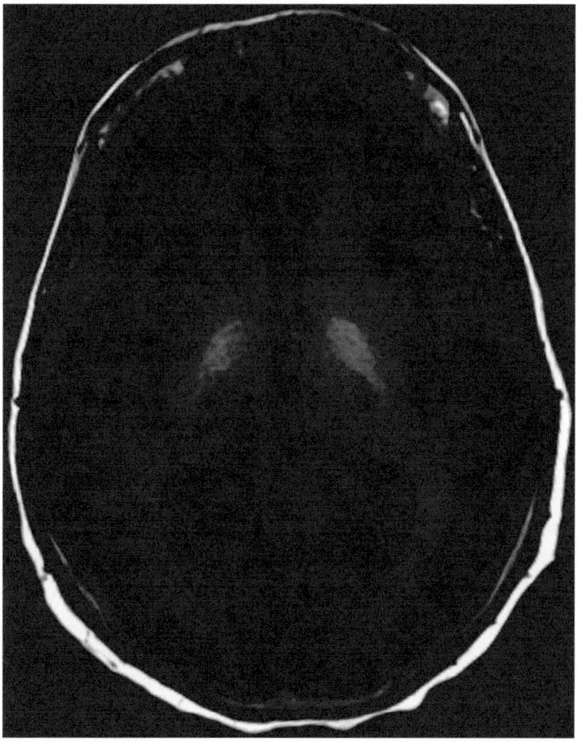

Fig. 4.12 MR axial T1 weighted scan demonstrating hyperintensity typically involves putamen, globus pallidus, caudate nucleus, ventrolateral part of the thalamus consistent with Wilson's disease or metabolic disease

4.3.1.4 Physiotherapy Management

Dystonia, incoordination, tremors, balance issues, and gait abnormalities are the common manifestations of this disorder and the therapy can play an effective role in the treatment of Wilson's disease. Positional stretch and gentle passive stretching or flexibility exercises can be helpful to retard or prevent the progression of contractures secondary to dystonia. For acute or dynamic contractures, serial cast application can be helpful to regain range and flexibility. In cases where cerebellar movement abnormalities are more worrisome for the patient, exercises to minimize dysmetria of limbs or trunk can be given. Isometric exercises, along with slow reversal hold technique progressing toward rhythmic stabilization exercises given for the neck and trunk musculature may reduce the titubation movements and truncal ataxia.

The balance training used for PD or cerebellar (discussed in Chap. 5—Cerebellar Dysfunction) patients can be of benefit to improve balance reactions and limits of stability of patients with Wilson's disease. For patients with mobility and safety issues, walking aids and environmental modifications of the home settings can be helpful. If general weakness and fatigue and/or aches and pains of muscles are the concerns, either due to the extrapyramidal or liver dysfunction, exercises to improve the patient's functional capacities, energy conservation techniques, and pacing strategies to manage fatigue and pain management using electrotherapeutic modalities or thermal agents can be the preferences.

4.3.2 *Multiple System Atrophy*

Multiple System Atrophy (MSA) is an adult-onset, often fatal neurodegenerative disorder. It was originally described under different diagnostic terms, namely Olivopontocerebellar atrophy, Striatonigral degeneration, and Shy–Drager syndrome. In 1969, J.G. Graham and D.R. Oppenheimer, both British neurologists, first introduced the term MSA to combine these clinicopathological disorders. Based on the predominant clinical features, the MSA is categorized into the Parkinsonian (MSA-P), cerebellar (MSA-C), and autonomic (MSA-A) subtypes. It is estimated that the prevalence is approximately 3–5 cases per 100,000 general population. MSA is a disease of middle-age life and usually, the onset is between 40 and 70 years. With regard to gender preference, males have a higher predisposition over females and the survival is approximately 6–10 years from the initial onset of symptoms.

4.3.2.1 Etiology, Neuropathology, and Pathogenesis

Though the etiology of MSA is largely unknown, genetic causes including COQ2 mutation, SHC2 copy number loss, and unidentified autosomal dominant or recessive gene mutations and environmental causes including organic solvents, pesticides, plastic monomers, and metal dust may act as causative factors for the neuropathogenesis of this disorder. Pathologically, along with the characteristic glial cytoplasmic inclusions, the disease is characterized by a variable extent of neuronal loss and gliosis in the striatum, cerebellum and middle cerebellar peduncles, substantia nigra, locus ceruleus, pontine nuclei, inferior olivary nuclei, and intermediolateral columns and Onuf's nucleus of the spinal cord. In addition to the above, the disorder is also characterized by the presence of α-synuclein within the glial cytoplasmic inclusions. In MSA-P, the neuronal loss and gliosis are more prominent in the substantia nigra and striatum, whereas in MSA-C, the loss is more in the cerebellar cortex, middle cerebellar peduncle, and brainstem.

4.3.2.2 Neurological Manifestation and Investigation Findings

This disorder is known for its marked clinical variability. The presence of one or more features of autonomic symptoms, including orthostatic hypertension, bladder and bowel incontinence, and sexual dysfunctions, dysphonia and/or stridor, slowness of movement, rigidity, Babinski sign, and cerebellar ataxia, and less possibility of resting tremor are some of the common features of MSA. Usually, the disease begins with asymmetric striatonigral-parkinsonian features or with prominent autonomic manifestations often associated with disabling orthostatic hypotension. The cerebellar features may rarely dominate the initial stages of the disease. In MSA-P, the extrapyramidal features can be more severe than PD and within 5 years, the patient can be wheelchair-bound or bed-confined. The relative symmetry of the signs and symptoms, rapid course, poor response to levodopa, absence or minimal amount of tremor, presence of prominent autonomic dysfunction, and the less obvious eye movement abnormalities make it easier to distinguish MSA from PD. Imaging techniques may frequently reveal cerebellar and pontine atrophy, aiding the diagnosis of the condition. The T2-weighted MRI will show putaminal hypodensity and increased deposition of iron. The PET scan may disclose impaired glucose metabolism in the striatum and the frontal cortex.

4.3.2.3 Medical Management

No definite management exists as there are no pharmaceutical interventions available to halt or reverse the progression of MSA. For MSA-P, symptoms may alleviate with levodopa, dopamine agonists, or anticholinergics medications. Administration of botulinum toxin can be helpful to reduce dystonia of the hand, foot, or trunk. If autonomic dysfunctions are worrisome, fludrocortisone or desmopressin for nocturnal polyuria, antimuscarinic agents or anticholinergic medication for bladder dysfunction, fluids and laxatives for constipation, tracheostomy for stridor and intracavernosal papaverine for impotency can be administered. Clonazepam, GABApentin, or buspirone can be used for myoclonus or action tremor.

4.3.2.4 Physiotherapy Management

Due to the complex nature of MSA, the best treatment is gained from a multidisciplinary team approach. The team, a group of health and social care professionals, may include a neurologist, family physician, nurse, physiotherapist, speech therapist, clinical psychologist, and dietician. Physiotherapy treatment objectives and means vary for each subtype and stage of the disease. Adequate knowledge about the disease helps the therapists choose the appropriate clinical skills to manage this condition. The goals of therapy should be realistic and appropriate and the therapy must be integrated into daily living.

The motor symptoms of MSA-P (rigidity, bradykinesia, tremor, and poor balance) are similar to those observed in typical PD and for the same reason, the therapist can use the same strategies and treatment for handling the abovementioned issues. Physiotherapy can often be challenging when postural hypotension is accompanied by fatigue. Graduated programs of activity, avoiding prolonged stance and activity, and pacing can minimize fatigue over time. Use of 30° elevation of the head-end of the bed while sleeping, progressive acclimatization using a tilt table, use of abdominal binders, and compressive stockings for the lower limbs can be strategies to manage orthostatic hypotension. Recommending a sufficient amount of fluid (up to 2–3 l/day) intake, especially before exercise, avoiding warm environments and activities that may elicit the Valsalva maneuver will possibly minimize the postural hypotension. If festination and FOG are issues for MSA-P patients, the use of cues (discussed earlier) can be beneficial. Cognitive strategies can be encouraged right from the early stage of the disease as they will act as adjuncts for safe ambulation and for overcoming freezing. Since gait ataxia, truncal ataxia, and dysmetria of the limbs are certain common features of cerebellar disorders, strategies and training programs for cerebellar dysfunction can be extrapolated for the management of MSA-C.

4.3.3 Progressive Supranuclear Palsy

In 1964, John Steele, Clifford Richardson, and Jerzy Olszewski, in their seminal report of nine cases with autopsy confirmation, first described Progressive Supranuclear Palsy (PSP). Following their pioneer work, this disease was known as the Steele–Richardson–Olszewski syndrome. As originally described, PSP is characterized by progressive supranuclear ophthalmoplegia, gait disturbances, postural instability, dysarthria, dysphagia, rigidity, and frontal cognitive disturbance. Under atypical Parkinsonism, PSP is considered to be the most common degenerative form. The prevalence of PSP is approximately 6 per 100,000 population. The average age of onset of this disease is around 60 years. Other than age, there are no proven risk factors for the development of PSP. Males have a higher predisposition for this disease (8:1). In most cases, the disease progression is relatively rapid and relentless and most PSP patients will become dependent on care within 3–4 years after the onset of symptoms and the patient usually succumbs to death within 6–12 years after the onset of symptoms.

4.3.3.1 Etiology, Neuropathology, and Pathogenesis

Most cases of PSP appear to be sporadic. Similarly, there is no conclusive evidence that exposure to industrial wastes, specifically phosphate and chromate ore, has any association with PSP. The PSP belongs to the family of disorders known as "tauopathies" (a group of neurodegenerative diseases having abnormal aggregation of tau protein in neurofibrillary or gliofibrillary tangles within the brain). It is the abnormalities in the protein tau (a microtubule-associated protein, the main component of the intracellular filamentous inclusions) that lead to damage in the cortical, subcortical, cerebellar, and brainstem areas of the brain. Histologic examination reveals neuronal loss, gliosis, and the presence of tau-positive filamentous inclusions in specific anatomic areas involving astrocytes, oligodendrocytes, and neurons. The tau cytoplasmic inclusions in surviving neurons, known as globose neurofibrillary tangles, are classical findings in PSP. Autopsy examination has disclosed a bilateral loss of neurons and gliosis in the superior colliculus, periaqueductal gray matter, tegmentum of the brainstem, substantia nigra, oculomotor nucleus, red nucleus, globus pallidus, and subthalamic nucleus and to some extent in the dentate nucleus and vestibular nuclei. Hypopigmentation of the substantia nigra and locus ceruleus and enlargement of the third ventricle are the additional findings. Neurochemical studies have shown the degeneration of dopaminergic neurons in the striatum, cholinergic and GABA interneurons, and efferent neurons in the basal ganglia and the brainstem.

4.3.3.2 Neurological Manifestation and Investigation Findings

Richardson phenotype or syndrome is the most common and "classic" phenotype of PSP. Broad-based gait with short step length, shuffling steps, gait freezing and lurching with postural instability, and frequent falls are the early and usual presentations of this phenotype. Supranuclear ophthalmoparesis (largely downward gaze palsy) is the hallmark of PSP. Initially, the patients may show slowing of vertical saccades, but with time (3–7 years), all voluntary eye movements will be completely lost, beginning with the vertical gaze and then the horizontal ones (ophthalmoplegia). However, if the patient is instructed to fix the eyes on a target and the head is turned, the clinician can observe the full range of eye movements, proving the supranuclear, nonparalytic character of the gaze palsy.

Dysarthria, dysphagia, pseudobulbar palsy, rigidity, frontal cognitive abnormalities, and sleep disturbances are additional common clinical features. Retracted upper eyelids, wide-opened eyes, and reduced blink with a stare impart an expression of "perpetual surprise" for PSP patients. In some PSP cases, blepharospasm and involuntary eye closure can be prominent issues. Rigidity in PSP patients will be more apparent in the axial musculature than in the limb musculature, specifically the neck and upper trunk. PSP patients, while walking, tend to hold their trunk and lower limbs extended, arms slightly abducted and on pull test for postural instability, classically stagger backward uncontrollably. The rigidity of the trunk musculature and postural instability may cause the PSP patient to consistently lean and fall backward (retropulsion). Resting tremor is an unusual finding in PSP and the patients will not have limb ataxia, Romberg sign, or postural tremor. In a few cases, Babinski signs can be present. Impaired abstract thinking, reduced verbal fluency, and motor perseveration (continuation or recurrence of a motor activity with difficulty to switch between actions; a sign of frontal lobe lesion) are the cognitive dysfunctions seen in PSP. Apathy, disinhibition, irritability, anxiety, impulsivity, and dysphoria are certain common behavioral symptoms of PSP. In the later stage, some degree of dementia can be present in PSP patients and may complain about sleep disturbances and increased urinary frequency and urgency. In the advanced stage, the patients become anarthric, immobile, fully dependent, and wheelchair-bound or bedridden.

Currently, the diagnosis of PSP is based upon clinical features. A gradually progressive disorder, with the age of onset 40 years or above, progressive upward or downward gaze supranuclear palsy and prominent postural instability with falls, with no evidence of symptoms or features suggestive of other diseases like hallucinations, history of encephalitis, cortical sensory deficits, cortical dementia and dysautonomia help in the clinical diagnosis of the condition. Neuroimaging studies will demonstrate generalized atrophy of the brainstem, particularly involving the midbrain. MRI of the brain certainly will reveal the classical "penguin silhouette" or "hummingbird" sign (Fig. 4.13), a sign resulting from the prominent midbrain atrophy with the relative preservation of pons. On axial imaging, MRI can reveal the "Mickey Mouse" sign (Fig. 4.14) due to the atrophy of the dorsal mesencephalon (superior colliculi, red nuclei). As the earliest sign of PSP, the PET scan may reveal reduced glucose metabolism in the midbrain.

4.3.3.3 Medical Management

Like MSA or IPD, no therapeutic intervention can alter the natural course of this disease. Medications that provide significant symptomatic benefits, as seen in IPD, have less beneficial effects in PSP. Poor or short-lived response to levodopa therapy is the usual finding in PSP and can help clinicians to distinguish PSP from IPD. Zolpidem, a GABAergic agonist of benzodiazepine receptors, has been reported to ameliorate rigidity and akinesia. Focal dystonia and drooling issues can be managed by the local administration of botulinum toxin. In some cases of PSP, amantadine can provide a transient therapeutic benefit for dyskinesia and drooling issues. Davunetide, a neuroprotective agent, is assumed to maintain microtubule function, reduce tau phosphorylation, and inhibit apoptosis in PSP. The concerns like dysphagia and dysarthria, to a certain extent, can be managed by health professionals like dietitians and speech therapists.

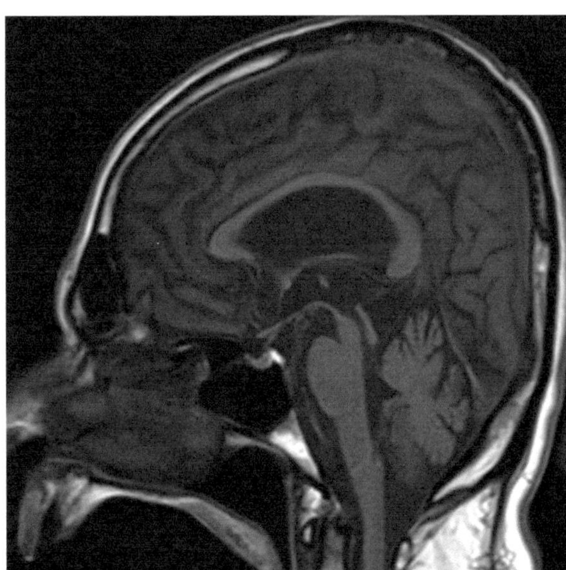

Fig. 4.13 MRI T1 sagittal of the brain revealing the "penguin silhouette" or "hummingbird" sign, a sign resulting from the prominent midbrain atrophy with relative preservation of pons consistent with the diagnosis of PSP

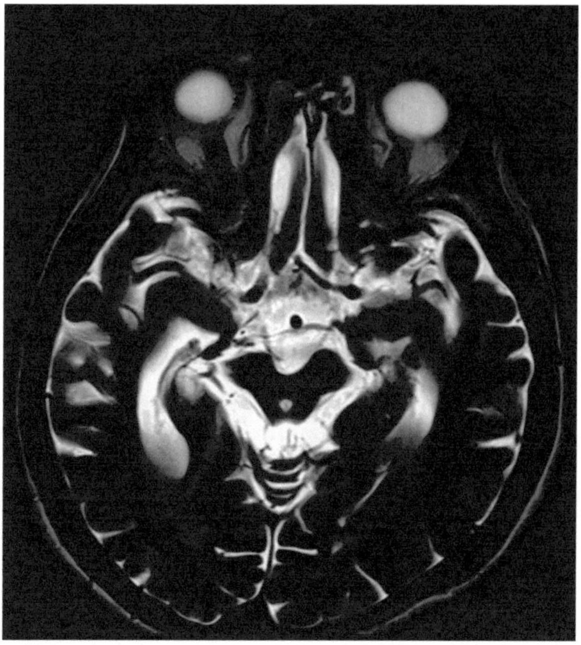

Fig. 4.14 MRI T2 axial imaging revealing the "Mickey Mouse" sign due to the atrophy of the dorsal mesencephalon (superior colliculi, red nuclei) consistent with the diagnosis of PSP

4.3.3.4 Physiotherapy Management

Though the neuropathological findings are somewhat different between IPD and PSP, evidence is available, citing the commonality of certain clinical features among them. From a logical point of view, the treatment modalities proven to be successful in improving motor deficits in PD can be relevant and useful for PSP patients also. Studies have shown that exercise programs focusing on gait, balance, and prevention of falls are feasible and safe for PSP patients. Certain studies reported that aerobic, intensive, motor-cognitive, goal-directed exercises along with treadmill training with or without robotic assistance had improved the spatio-temporal parameters of gait (including cadence, step/stride length, and gait velocity) in addition to the reduction of fall risk and improvement of balance. Due to the high cost and the low benefit-cost ratio, for many of the clinical settings and community-dwelling PSP or PD patients, the robotic-assisted treadmill or treadmill with the bodyweight supporting system may not be a viable option. In addition to treadmill training with or without robotic assistance, several other approaches, including balance training with a tilt board, resistive or isokinetic strength training exercises, and coordination exercises, have been used for gait training.

The treatment strategies like exercises to improve aerobic capacity, balance training for improving balance, sit-to-stand transitions for safe and effective transitions, flexibility exercises to improve mobility of the joints and encourage segmental rotations and movements, motor-cognitive exercises, and use of cueing for improving gait and functional capacities may have beneficial effects as seen in

IPD. Strategies to overcome freezing and prevent falls can also be trained to minimize freezing episodes and the potential for injury, respectively. For patients with ophthalmoplegia, head and neck movements, specifically emphasizing downward and sideways movements, have to be encouraged to see the surroundings and the floor while walking to avoid obstacles.

If retropulsion and tendency to keep the spine in extension are quite prominent in PSP patients, activities like bridging and spinal extension exercises are not advisable. In such cases, strengthening the abdominals (rectus abdominis and external and internal oblique muscles) in supine and encouraging forward reaches and oblique forward reaches while seated can reduce the tendency of retropulsion. Most of the patients have a tendency to prematurely straighten the spine during the sit-to-stand transition, predisposing then to fall backward. Practicing sit-to-stand transition, specifically with trunk bend forward over the hips may reduce the tendency to fall backward until the postural controls are severely affected. Even activities like reaching and picking up objects from the floor while seated on a therapeutic mat (low height) may encourage the patient to bend the trunk adequately for sit-to-stand transitions. During gait training, unlike for IPD patients, instructions to bend the trunk upon the hips and to keep the knees slightly bend may minimize the possibility of leaning backward or falling backward.

4.3.4 Huntington's Disease

A concise description of a syndrome, likely to be Huntington's disease, was first given by Charles Oscar Waters, an American physician, in 1841. In 1872, George Huntington, another American physician, first gave a complete description of Huntington's disease (HD), including the salient clinical features of the disease, the pattern of transmission, and the dismal prognosis. "Huntington's chorea" is a cardinal sign of the disease. Though chorea is an important clinical sign, not all HD patients may present with chorea. Movement abnormalities, affective disturbance, and cognitive impairment are the clinical triad of this disorder. The disorder is autosomal dominant inherited, with a 50% chance of the affected individual's offspring developing the disease during their adult stage. Though genetic studies can identify the pre-symptomatic subjects, no treatment can prevent or ameliorate progressive neurodegeneration. In most countries, the prevalence rate is approximately 4–10 per 100,000 population. The HD is reported from all parts of the world, but it is more common in certain regions like Scotland and Venezuela. Though the disease is rare, the burden the disease can put on the patients, the family members and their offspring is indescribable.

4.3.4.1 Etiology, Neuropathology, and Pathogenesis

Mutation of the gene known as "huntingtin gene" located on the short arm of chromosome "4" is the cause for HD. The gene product of the huntingtin gene is a protein called "huntingtin." Although the exact function of the huntingtin protein is unclear, this protein may be involved in chemical signaling, transporting materials, binding to other proteins, and protecting the cell from apoptosis. The mutation involves a DNA segment known as a CAG (cytosine, adenine, and guanine) trinucleotide repeat in the huntingtin gene. Abnormal expansion of a repeating CAG triplet series leads to an increase in the size of the CAG segment, causing the production of an abnormally long version of the huntingtin protein. The elongated huntingtin protein disintegrates into smaller toxic fragments that bind together and accumulate in neurons (specifically in GABAergic medium spiny neurons) within the striatum and certain parts of the cortex. The early disruption of GABAergic medium spiny neurons in the indirect pathway of basal ganglia, normally presumed to inhibit movements, results in choreiform movements. The disruption of medium spiny neurons in the direct pathway is a plausible explanation of the rigidity and bradykinesias in the later stage of the disease.

Pathologically, HD is characterized by marked atrophy and prominent neuronal cell loss in the bilateral basal ganglia, especially the striatum. The atrophy is associated with a widening of the anterior horns of the lateral ventricles. The anterior parts of the putamen and caudate are more affected than the posterior parts. In addition to the above, neuronal cell loss also occurs in other regions of the brain, including the third, fifth, and sixth layers of the cerebral cortex, especially the frontal and temporal regions, with a certain amount of gliosis. Studies revealed that the neuronal cell loss in HD patients made their brains smaller and lighter as compared to the age-matched controls. The degree of neuropathological changes is generally milder in older onset cases compared to younger onset cases; however, the relationship between the neuropathological changes and the clinical features is unreliable. Histological examination of the striatum reveals the involvement of the smaller neurons before the larger ones, early loss of dendrites of the small spiny neurons, and replacement of the lost cells by fibrous astrocytes. PET studies have revealed impaired glucose metabolism of the striatum prior to the visible atrophy of the same.

4.3.4.2 Neurological Manifestation and Investigation Findings

The disorder usually starts between 35 and 55 years of age. If the age of onset is below 20 years, then those cases are of "juvenile" form or phenotype, which accounts for about 5% of the total cases. Unlike the adult form of HD, during the initial onset of symptoms, the juvenile form tends to present with bradykinesia and rigidity over chorea. The initial manifestation of symptoms can be either neurological or psychiatric or both. The neurological manifestations of the disease are usually insidious in nature. Rarely, HD patients may present with frank neurological and/or psychiatric features. Often, the initial manifestations can be non-specific, including complaints like feeling depressed, forgetful, or clumsy. In general, chorea is the early movement abnormality manifestation seen among HD patients. In the early stage, chorea may be barely perceptible and first noticeable in the hands and face. As it evolves, the severity of chorea may vary from mild restlessness to an intermittent exaggeration of gestures, fidgeting movements, unstable dance-like gait, to a continuous flow of disabling violent movements. These choreiform movements cannot be voluntarily suppressed and stress and anxiety can worsen them. The chorea of HD causes a lurching, stumbling, stuttering, dipping, and bobbing gait with steps forward, backward, or to one side. The gait of HD is complex and dance-like, with the velocity of gait slow, stance time varied from step to step, the base of support wide, and excessive trunk sway while stepping. Dystonic posturing of the hip or knee, either in flexion or extension, can punctuate the ongoing stepping motion. Despite the apparent gait deviation, balance and equilibrium are usually preserved till late-stage and surprisingly the frequency of falls is also less.

In HD, the eye movements are usually slow from the early stage, characterized by difficulty in initiating saccades and broken pursuits. Unlike PD, the frequency of blinking of the eyes is increased in HD. The unwanted darting movements of the tongue may constantly interrupt when the voluntary protrusion of the tongue is attempted. Speech may reveal a marked variation in rate, loudness, and timing, distortion of vowels, harshness of voice, sudden stoppages of speech, and poor coordination with breathing (hyperkinetic dysarthria). The speech abnormalities are indeed the result of choreiform movements affecting the coordination of phonatory and respiratory muscles. Typically, as time passes, the chorea will cease to progress and get overshadowed by dystonia, rigidity, and bradykinesia. Eventually, a progressive increase in rigidity and bradykinesia causes immobility, postural instability, and inability to walk.

Spasticity, brisk deep tendon reflexes, and extensor plantar responses are unlikely in HD. Impairment of movement can affect the laryngeal and respiratory muscles leading to dysphonia. Dysphagia and cachexia are common issues during the middle and late stages of the disease and often become the usual cause of choking and aspiration of oral contents leading to death. Patients may develop insomnia at night and somnolence during the day. The bowel and bladder incontinence common during the late stage may be a consequence of dementia than a specific neurological cause.

In HD patients, managing the cognitive and affective impairments can be more challenging than managing the movement disorder. Patients themselves may be unaware or unconcerned about the cognitive, affective, and motor changes and often it is the family members that bring them to medical attention. Approximately one-third of HD patients report psychiatric and behavioral symptoms as the early manifestation of the disease. Depression is a very common problem and may precede the onset of a variety of neurological dysfunctions. A low threshold for anger, a minimal provocation to react, fear, temper tantrums, irritability, apathy, and sleep disturbances are some of the psychological and behavioral problems seen. Due to the high rates of depression and fear of the disease, the risk of suicide is more common in HD than in other neurological disorders. Mania and hypomania are less commonly seen when compared to depression. Though psychosis is rare, it can be a common feature among early-onset HD, often difficult to treat. Some HD patients may display obsessive-compulsive features or altered sexual behavior. Behavioral and psychiatric manifestations may predate the onset of overt HD by as long as a decade, reflecting early pathological changes in the non-motor areas of the striatum. The general decline in intellectual functions during the course of the disease is a cognitive impairment manifestation of the disease. The dementia of HD (subcortical dementia) presents with cognitive slowing, memory retrieval deficits, attentional difficulties, and executive dysfunction in the absence of aphasia and apraxia. Right from the early stage, the cognitive impairments, specifically the executive dysfunction, may render HD patients unable to work, drive and manage family finances. Even though there is dysarthria and slowness of thinking, many patients may retain the ability to comprehend until the late stage of the disorder.

In the adult phenotype with early-onset, affective disturbances are more often the initial complaints and precede the motor and cognitive impairments by years. In the middle years, the motor and cognitive impairments tend to appear nearly at the same age, whereas, in older age onset, choreiform features are more often the initial problems. The mean survival is 17 years from the onset of the symptoms. Due to the diverse and complex problems associated with HD, a multidisciplinary team (including neurologist, nurse, psychiatrist, physiotherapist, occupational therapist, dietician, speech therapist, and social worker) approach is essential to manage such patients.

In HD, the radiological investigation may reveal atrophy of the caudate nucleus (mainly the head), putamen, and frontal lobes. As the disease progresses, diffuse cortical, thalamic and limbic atrophy may also appear. In addition to the above, T2 weighted MRI images may reveal hyperintense or hypointense signal changes typically in the putamen. The T2 hyperintense signal change is usually attributed to marked neuronal loss and gliosis. In the early stage of disease, when the chorea is more prominent, the T2 weighted MRI images may not reveal any change in signal intensity.

4.3.4.3 Medical Management

With a typical clinical picture and a confirmed family history, the diagnosis of HD is often straightforward. Rarely, vague family history, inaccurate parental information, denial, and obfuscation can make the diagnosis difficult. No pharmacological treatment can reverse, undo or halt the neurodegenerative process. Since the condition is characterized by changes in motor, affective and cognitive symptoms, the medications are likely to evolve over the course of the disease. Currently, the treatment for HD is supportive and/or symptomatic in nature. Tetrabenazine and deutetrabenazine can be used to suppress the chorea associated with HD. Ironically, medications to alleviate the chorea may make bradykinesia, dystonia, and apathy worse. Haloperidol, fluphenazine, risperidone, olanzapine, and quetiapine can be used as antipsychotic drugs. However, these antipsychotic drugs may worsen chorea, dystonia, and drowsiness. Citalopram, escitalopram, fluoxetine, and sertraline are antidepressants, and divalproex, carbamazepine, and lamotrigine can be the mood-stabilizing drugs for HD.

4.3.4.4 Physiotherapy Management

Developing a standard or conventional treatment program for HD patients is difficult due to the heterogeneity of clinical signs and symptoms among patients. As seen in IPD or PSP management, the staging of the disease process into early, middle, and late may provide a general framework for intervention to tackle the possible impairments, activity limitations, and participation restrictions of HD patients. However, the therapist should also understand that such staging is not watertight due to the progressive nature of the disease. In early-stage HD, patients may have issues such as poor endurance, limited physical activity, lack of motivation, anxiety and depression, apathy and sleep disturbance in the absence of obvious motor impairment, or specific limitations in functional activities. In the early stage, the therapy should aim at health education, advice for general promotion of strategies including education on safety, fatigue, and the timing of exercises, referral for exercise prescription, and evaluation of the baseline fitness level. With safety taken into account, gym or home-based aerobic exercises and strength training and task-specific functional activities can be encouraged. It is advisable to have a warm-up and a cool-down for the exercise program. Aerobic exercises may include walking, jogging, swimming, and biking, and strengthening exercises may include the use of progressive resisted strength training using weights or elastic bands or isokinetic devices. Frequency, duration, and intensity should be based on the baseline fitness level of the patient. Activities like balance training, core stability training, video-game and virtual reality-based exercises, and relaxation techniques can be recommended.

During the early toward the middle stage, impaired motor planning and programming may result in increased dependency on routine activities, difficulty in initiation and slowness in performing functional activities, including dressing and hygiene activities, and difficulty managing automatic tasks, such as sit-to-stand or walking. Therapy should aim to improve the ability to perform functional tasks and the speed of the tasks, maximizing safety. Visual, verbal, or other forms of cueing may help bypass the internal cueing mechanism to improve functional abilities. Cognitive movement strategies can also be advised during this phase to speed up the movements. When the mobility, balance, and risk of fall gradually become a concern due to hyperkinetic or hypokinetic movement abnormities, impaired balance reactions, reduced muscle strength, fatigue, and poor cognition, the physiotherapy should aim toward the maintenance of independent mobility, safe transfers, and reduction of fear of fall. Treatment should focus on strengthening exercises, general conditioning exercises, endurance training, balance training, and practicing transfers like sit-to-stand several times. Activities like throwing and catching a ball, self-initiated perturbations, and reach-out activities can be encouraged to improve the postural and balance reactions. The need for repeated practice of tasks in a specific environment, including gait training outdoors and obstacle training, may improve the functional mobility of the patient. Cues, including a metronome and lines on the floor, may promote the rhythm, step initiation, step length, gait velocity, and gait symmetry. It is also advisable to train strategies on how to get up from the floor in case they fall. Periodical assessment is required to re-evaluate the balance and mobility of HD patient. At the appropriate time, the therapist should prescribe assistive devices and protective gear like helmets or elbow/knee pads to maximize safety during mobility.

In the middle stage, when secondary and adaptive changes and deconditioning develop, the patient's physical fitness level decreases, and participation in ADL reduces. Physical deconditioning is due to musculoskeletal and/or respiratory changes secondary to rigidity, dystonia, and chorea. Loss of ROM and loss of strength due to physical inactivity are certain musculoskeletal changes. During this stage, cognitive issues like memory deficits and lack of insight, psychological issues like anxiety and depression, and apathy may become additional challenges for rehabilitation. During this stage, promoting functional activities is the most effective way of maximizing the remaining ability. The frequency of falls also may increase due to the further progression of balance and gait dysfunctions. Stretching and flexibility exercises, encouraging patients to continue aerobic exercises and balance training, and teaching breathing exercises to maintain full respiratory function are certain treatment strategies that are advisable during this stage. Due to the involuntary movements, rigidity, and bradykinesia, the patient often tends to develop abnormal posturing even while sitting. Asymmetrical posturing may further reduce the joint ROM, flexibility and may induce further adaptive changes within the soft tissues. Improving postural control, specifically head control while sitting, is vital for communication, feeding, visual fixation, and balance activities. The treatment should also focus on preventing or limiting soft-tissue adaptation and skin breakdown. Positional stretching and use of custom-made chairs or wheelchairs with lateral support and straps can encourage a better alignment of the spine while sitting and also minimize the possibility of falls from the chair. In the later phase of the middle phase, active, assisted and passive range of motion exercises should be performed frequently to maintain the mobility of the joints. The use of splints is also recommended.

In the late stage, impaired respiratory function and capacity and poor airway clearance further restrict the functional abilities of the patient and predispose to chest infections. During this phase, optimizing the respiratory function for functional activities, improving the cardiorespiratory function, and augmenting secretion clearance are the general aims. Promoting functional exercise, positioning to manage breathlessness, promoting breathing exercises and use of airway clearance techniques, and encouraging relaxation techniques are some of the strategies to be adopted. In the latter part of the advanced stage, the patient will be completely dependent on all the functional skills with limited volitional control of limbs and trunk. The use of pressure-relieving mattresses and cushions for optimal positioning and regular and frequent change of position can reduce the possibilities of pressure sores and may provide a certain amount of comfort for the patient.

Further Reading

Allen NE, Canning CG, Sherrington C, Lord SR, Latt MD, Close JCT, et al. The effects of an exercise program on fall risk factors in people with Parkinson's disease: a randomized controlled trial. *Mov Disord.* 2010;25(9):1217-25.

Barone P, Santangelo G, Amboni M, Pellecchia MT, Vitale C. Pisa syndrome in Parkinson's disease and parkinsonism: clinical features, pathophysiology, and treatment. *Lancet Neurol.* 2016 Sep;15(10):1063-74.

Barry G, Galna B, Rochester L. The role of exergaming in Parkinson's disease rehabilitation: a systematic review of the evidence. *J Neuroeng Rehabil.* 2014;11:33.

Bello O, Sanchez JA, Lopez-Alonso V, Márquezet G, Morenilla L, Castro X, et al. The effects of treadmill or overground walking training program on gait in Parkinson's disease. *Gait Posture.* 2013;38(4):590-5.

Ben-Shlomo Y, Sieradzan K. Idiopathic Parkinson's disease: epidemiology, diagnosis and management. *Br J Gen Pract.* 1995;45:261-8.

Bunner KD, Rebec GV. Corticostriatal dysfunction in Huntington's disease: the basics. *Front Hum Neurosci.* 2016;10:317. Available from: https://doi.org/10.3389/fnhum.2016.00317.

Busse ME, Khalil H, Quinn L, Rosser AE. Physical therapy intervention for people with Huntington disease. *Phys Ther.* 2008;88:20-31.

Daroff RB, Jankovic J, Mazziotta JC, Pomeroy SL, editors. Bradley's Neurology in clinical practice. 7th ed. London: Elsevier; 2016.

Doherty KM, van de Warrenburg BP, Peralta MC, Silveira-Moriyama L, Azulay JP, Gershanik OS, et al. Postural deformities in Parkinson's disease. Lancet Neurol. 2011;10(6):538-549.

Donaghy M, editor. Brain's Diseases of the nervous system. 12th ed. New York: Oxford University Press; 2009.

Cano-de-la-Cuerda R, Pérez-de-Heredia M, Miangolarra-Page JC, Muñoz-Hellín E, Fernández-de-las-Peñas C. Is there muscular weakness in Parkinson's disease? *Am J Phys Med Rehabil.* 2010;89:70-6.

Clerici I, Ferrazzoli D, Maestri R, Bossio F, Zivi I, Canesi M, et al. Rehabilitation in progressive supranuclear palsy: effectiveness of two multidisciplinary treatments. *PLoS One.* 2017;12:e0170927.

Delwaide PJ. Parkinsonian rigidity. *Funct Neurol.* 2001;16:147-56.

Fok P, Farrell M, McMeeken J, Kuo YL. The effects of verbal instructions on gait in people with Parkinson's disease: a systematic review of randomized and non-randomized trials. *Clin Rehabil.* 2011;25:396-407.

Frenkel-Toledo S, Giladi N, Peretz C, Herman T, Gruendlinger L, Hausdorff JM. Treadmill walking as an external pacemaker to improve gait rhythm and stability in Parkinson's disease. *Mov Disord.* 2005;20(9):1109-14.

Fritz NE, Rao AK, Kegelmeyer D, Kloos A, Busse M, Hartel L, et al. Physical therapy and exercise interventions in Huntington's disease: A mixed methods systematic review. *J Huntingt Dis.*2017;6:217-35.

Galletly R, Brauer SG. Does the type of concurrent task affect preferred and cued gait in people with Parkinson's disease?. *Aust J Physiother.* 2005;51(3):175-80.

Gökçal E, Gür VE, Selvitop R, Yildiz GB, Asil T. Motor and non-motor symptoms in Parkinson's disease: effects on quality of life. *Arch Neuropsychiatry.* 2017;54(2):143.

Hass CJ, Buckley TA, Pitsikoulis C, Barthelemy EJ. Progressive resistance training improves gait initiation in individuals with Parkinson's disease. *Gait Posture.* 2012;35(4):669-73.

Huh YE, Kim K, Chung WH, Youn J, Kim S, Cho JW. Pisa syndrome in Parkinson's disease: pathogenic roles of verticality perception deficits. *Sci Rep.* 2018;8:1804. Available from: https://doi.org/10.1038/s41598-018-20129-2.

Keus SH, Munneke M, Nijkrake MJ, Kwakkel G, Bloem BR. Physical therapy in Parkinson's disease: evolution and future challenges. *Move Disord.* 2009;24:1-14.

Keus S, Munneke M, Graziano M, Paltamaa J, Pelosin E, Domingos J, et al. Physiotherapy guideline for Parkinson's disease. Netherlands: KNGF/ParkinsonNet (2014).

Kim HF, Hikosaka O. Parallel basal ganglia circuits for voluntary and automatic behaviour to reach rewards. *Brain.* 2015;138:1776-1800.

Lanciego JL, Luquin N, Obeso JA. Functional neuroanatomy of the basal ganglia. *Cold Spring Harb Perspect Med.* 2012;2:a009621. Available from: https://doi.org/10.1101/cshperspect.a009621.

Lee RG. Physiology of the basal ganglia: an overview. *Can J Neurol Sci.* 1984;11:124-8.

Lena F, Iezzi E, Etoom M, Santilli M, Centonze D, Foti C, et al. Effects of postural exercises in patients with Parkinson's disease and Pisa syndrome: a pilot study. *NeuroRehabilitation.* 2017;41:423-8.

Maiarú M, Garcete A, Drault ME, Mendelevich A, Módica M, Peralta F. Physical therapy in Wilson's disease: case report. *Phys Med Rehabil Int.* 2016;3(2):1082.

Marsden CD. Slowness of movement in Parkinson's disease. *Mov Disord.* 1989;4(1):S26-37.

McAuley JH, Daly PM, Curtis CR. A preliminary investigation of a novel design of visual cue glasses that aid gait in Parkinson's disease. *Clin Rehabil.* 2009;23(8):687-95.

Mirelman A, Herman T, Nicolai S, Zijlstra A, Zijlstra W, Becker C, et al. Audio-biofeedback training for posture and balance in patients with Parkinson's disease. J Neuroeng Rehabil. 2011;8:35.

Nieuwboer A, Rochester L, Muncks L, Swinnen SP. Motor learning in Parkinson's disease: limitations and potential for rehabilitation. *Parkinsonism Relat. Disord.* 2009;15:53-8.

O'Shea S, Morris ME, Iansek R. Dual task interference during gait in people with Parkinson disease: effects of motor versus cognitive secondary tasks. *Phys Ther.* 2002;82(9):888-97.

O'Sullivan S, Bezkor EW. Parkinson's disease. In: O'Sullivan SB, Schmitz TJ, Fulk GD, editors. 6th ed. Physical rehabilitation. Philadelphia: FA Davis Co; 2014. p.807-58.

Petzinger GM, Fisher BE, Van Leeuwen JE, Vukovic M, Akopian G, Meshul CK, et al. Enhancing neuroplasticity in the basal ganglia: the role of exercise in Parkinson's disease. *Mov Disord.* 2010;25(Suppl 1):141-5.

Pohl M, Rockstroh G, Rückriem S, Mrass G, Mehrholz J. Immediate effects of speed-dependent treadmill training on gait parameters in early Parkinson's disease. *Arch Phys Med Rehabil.* 2003;84(12):1760-6.

Poujois A, Mikol J, Woimant F. Wilson disease: brain pathology. *Handb Clin Neurol.* 2017;142:77-89.

Quinn L, Kegelmeyer D, Kloos A, Rao AK, Busse M, Fritz NE. Clinical recommendations to guide physical therapy practice for Huntington disease. *Neurology.* 2020;94:217-28.

Rochester L, Burn DJ, Woods G, Godwin J, Nieuwboer A. Does auditory rhythmical cueing improve gait in people with Parkinson's disease and cognitive impairment? A feasibility study. *Mov Disord.* 2009;24(6):839-45.

Ropper AH, Samuels MA, Klein JP, Prasad S, editors. Adams and Victor's Principles of neurology. 10th ed. New York: McGraw Hill Education; 2014.

Schenkman M, Butler RB. A model for multisystem evaluation treatment of individuals with Parkinson's disease. *Phys Ther.*1989;69:932-43.

Slade SC, Finkelstein D, McGinley J, Morris ME. Exercise and physical activity for people with progressive supranuclear palsy: A systematic review. *Clinical Rehabil.* 2020;34(1):23-33.

Steiner H, Tseng K-Y. Handbook of basal ganglia structure and function. London: Academic; 2010.

Strouwen C, Molenaar EA, Münks L, Keus SHJ, Bloem BR, Rochester L, et al. Dual tasking in Parkinson's disease: should we train hazardous behavior? *Expert Rev Neurother.* 2015;15:1031-9.

Takakusaki K, Saitoh K, Harada H, and Kashiwayanagi M. Role of basal ganglia-brainstem pathways in the control of motor behaviors. *Neurosci Res.* 2004;50:137-51.

Turnbull GI, editor. Physical therapy management of Parkinson's disease. New York: Churchill Livingston;1992.

Turnbull GI, Millar J. A proactive physical management model of Parkinson's disease. *Top Geriatr Rehabil.* 2006;22(2):162-71.

van der Kolk NM, King LA. Effects of exercise on mobility in people with Parkinson's disease. *Mov Disord.* 2013;28(11):1587-96.

Vojciechowski AS, Zotz TGG, Loureiro APC, Israel VL. The international classification of functioning, disability and health as applied to Parkinson's disease: a literature review. *Advances in Parkinson's Disease.* 2016;5:29-40. Available from: https://doi.org/10.4236/apd.2016.52005.

Walshe JM. History of Wilson's disease: 1912 to 2000. *Mov Disord.* 2006;21:142-7.

Willems AM, Nieuwboer A, Chavret F, Desloovere K, Dom R, Rochester L, et al. The use of rhythmic auditory cues to influence gait in patients with Parkinson's disease, the differential effect for freezers and non-freezers, an explorative study. *Disabil Rehabil.* 2006;28(11):721-728.

Yen CY, Lin KH, Hu MH, Wu RM, Lu TW, Lin CH. Effects of virtual reality-augmented balance training on sensory organization and attentional demand for postural control in people with Parkinson disease: a randomized controlled trial. *Phys Ther.* 2011;91(6):862-74.

Zampieri C, Di Fabio RP. Balance and eye movement training to improve gait in people with progressive supranuclear palsy: a quasi-randomized clinical trial. *Phys Ther.* 2008;88:1460-73.

Chapter 5
Cerebellar Dysfunction

Abraham M. Joshua, Karishma H. S. Keswani, and Rohit Pai

5.1 Introduction and Historical Background

The cerebellum, also known as the "little brain," is a part of the central nervous system that functions largely outside the realm of conscious awareness and plays an important role in motor control. The mature cerebellum represents only 10% of the total brain volume; however, it contains the majority of human neurons. It receives input from various parts of the brain, brainstem, and spinal cord and integrates these inputs to fine-tune motor activities. Though it may also be involved in certain cognitive functions such as attention, language, learning, and regulation of many behaviors, its movement-related functions, like its role in coordination, balance, and muscle tone, are well established. Pathologies, including injuries to cerebellar structures, can produce disorders in fine movement, equilibrium, posture, and motor learning. This chapter gives the readers a concise explanation about the anatomy and physiology of the cerebellum and the pathophysiology, clinical features, and management of cerebellar disorders.

A. M. Joshua (✉)
Department of Physiotherapy, Kasturba Medical College of Mangalore, Manipal Academy of Higher Education, Manipal, India
e-mail: abraham.joshua@manipal.edu

K. H. S. Keswani
The Rise Network, Mumbai, India
e-mail: karishma@therisenetwork.in

R. Pai
Department of Neurology, Kasturba Medical College, Mangalore, Manipal Academy of Higher Education, Manipal, India
e-mail: pai.rohit@manipal.edu

A. M. Joshua (ed.), *Physiotherapy for Adult Neurological Conditions*,
https://doi.org/10.1007/978-981-19-0209-3_5

In the sixteenth century, Andreas Vesalius, a Belgian anatomist and a physician, wrote an essay on the cerebellum, a mass that was considered as a structure for power, memory, and talent. During the eighteenth century, Vincenzo Malacarne, an Italian researcher, in his scientific knowledge on cognitive functions of the cerebellum, stated that the intelligence of human beings depends on the cerebellar structures. Only by the beginning of the nineteenth century, based on clinical observation and experimental ablation studies, investigators realized that the cerebellum was involved in motor control and coordination.

The description of motor dysfunctions in cerebellar disorders will be complete only if the pioneering works of the earlier researchers are cited. The clinical manifestations of several cerebellar disorders were described at the turn of the nineteenth and twentieth centuries by the pioneers, including Luigi Luciani, Gordon Holmes, and Joseph Babinski. The Czech anatomist Jan Evangelista Purkinje in 1837 discovered and provided the distinctive and specific morphology of the Purkinje cells. In 1874, the granule cells were first described by Camillo Golgi, an Italian biologist and pathologist. In 1883, Golgi, with his silver impregnation method, revealed the extent and spatial orientation of the dendritic arbor, and Santiago Ramón y Cajal completed the description of the same in 1888. In 1891, Luigi Luciani, an Italian neuroscientist, published the results of his experimental studies on the cerebellum, based on his observational studies on dogs and primates. Luciani stated the three elemental deficits ipsilateral to the cerebellar lesion, formulating his triad of cerebellar symptoms: atonia, asthenia, and astasia. Later, he added a fourth sign, namely, dysmetria.

The concept of asynergia, characterized by complete decomposition of movement, was coined by Joseph Babinski, a French-Polish neurologist. Babinski also gave the first description of adiadochokinesia. In 1917, Gordon Morgan Holmes, an Anglo-Irish neurologist, provided details about the clinical deficits of cerebellar lesions in First World War soldiers who survived gunshot injuries to the occipital region of the brain. He stated that these survivors had severe problems with balance and voluntary movement execution. Holmes grouped the signs and symptoms into five categories: hypotonia, static tremor, asthenia, fatigability, and astasia, which included dysmetria and intention tremor.

Almost a century after the discovery of Purkinje cells, it was Jan Jansen and Alf Brodal who first described the topographical organization of the Purkinje cells and their projections to the cerebellar nuclei. By the 1950s, the emotional and behavioral function of the cerebellum became evident. R. S. Snider and colleagues worked out the connection of the cerebellar nuclei to the limbic system and deduced the role of the cerebellum in psychiatric problems. In the past four decades, advancements in neuroanatomical, neurophysiological, and clinical neuropsychological research have provided insights into the non-motor role of the cerebellum, thus changing the traditional view of the cerebellum as the sole coordinator of somatic motor function.

5.2 Basic Neuroanatomy and Neurophysiology of the Cerebellum

The cerebellum is situated in the posterior cranial fossa and forms the largest part of the hindbrain. It lies posterior to the pons, the medulla, and the fourth ventricle and is separated from the occipital lobes superiorly by the tentorium cerebelli. The cerebellum is ovoid in shape with a median constriction called the vermis that joins the two cerebellar hemispheres. The cerebellum is divided into three lobes: the anterior, the middle, and the flocculonodular lobes (Fig. 5.1).

The anterior lobe occupies the superior surface of the cerebellum and is separated from the middle lobe by a V-shaped fissure called the primary fissure. The middle or the posterior lobe forms the largest part of the cerebellum and lies between the primary and the uvulonodular fissures. The flocculonodular lobe is situated posterior to the uvulonodular fissure. Another fissure called the horizontal fissure separates the superior and inferior surfaces but does not have any functional significance. Other superficial fissures further divide these three lobes into ten lobules, each of which is subdivided into narrow folia. The vermis, the unpaired, median portion of the cerebellum, is composed of nine lobules (Fig. 5.1). The cerebellar tonsils are situated in the inferior aspect of the hemispheres, above the foramen magnum.

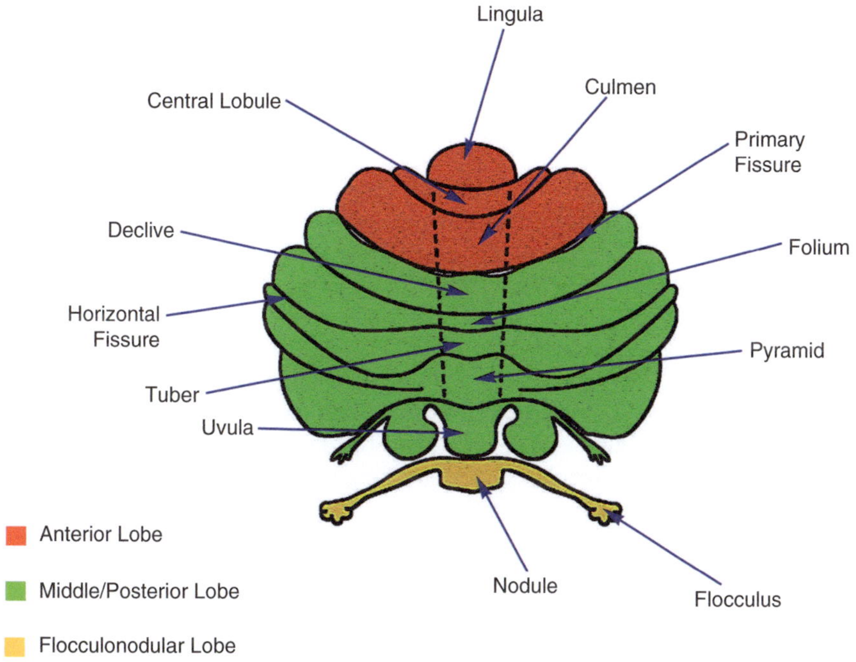

Fig. 5.1 The anterior, middle, and flocculonodular lobes of the cerebellum

5.2.1 Functional Anatomy

The human cerebellum may be considered rationally in terms of transverse or longitudinal oriented compartments. Based on the developmental history and seniority, the transverse compartments of the cerebellum are known as the archicerebellum, paleocerebellum, and neocerebellum.

Archicerebellum The flocculonodular lobe, which is phylogenetically the oldest portion of the cerebellum, constitutes the archicerebellum. It receives input from the vestibular apparatus and the vestibular nuclei located in the brainstem. The output is back to the vestibular nuclei and is responsible for hand-eye coordination and maintenance of balance and equilibrium.

Paleocerebellum Phylogenetically, the paleocerebellum is developed after the archicerebellum and includes the anterior lobe, uvula, and pyramid of the vermis and the paraflocculus. It receives inputs from the spinocerebellar, cuneocerebellar, and trigeminocerebellar pathways and visual and auditory systems and projects into the deep cerebellar nuclei. It is concerned with the gross movements of the head and body and muscle tone.

Neocerebellum It is the largest and the newest component of the cerebellum and comprises the middle lobe (except the uvula and pyramid). It receives input from the sensory, motor, and premotor cortices and projects to the thalamus and the cortex via the dentate nucleus and, in turn, influences and modulates the descending impulses passing through the corticospinal and reticulospinal tracts. The neocerebellum controls the planning and timing of voluntary movements.

Based on functional significance, the cerebellum is also longitudinally compartmentalized into vestibulocerebellum, spinocerebellum, and corticocerebellum (Fig. 5.2). The vestibulocerebellum consists of the flocculonodular lobe and corresponds to the archicerebellum. The vestibulocerebellum is responsible for controlling the balance during gait and stance and coordinating the eye and body movements. The spinocerebellum consists of the vermis (median zone) and the intermediate cortical (paravermal or paramedian zone) region spanning the anterior and posterior lobes and controls the coordination of ongoing limb movements. The cerebrocerebellum (corticocerebellum) is composed of the middle portion of the vermis and the lateral portions of the cerebellar cortex spanning the anterior and posterior lobes. The cerebrocerebellum helps in the planning and preparation of the intended movements.

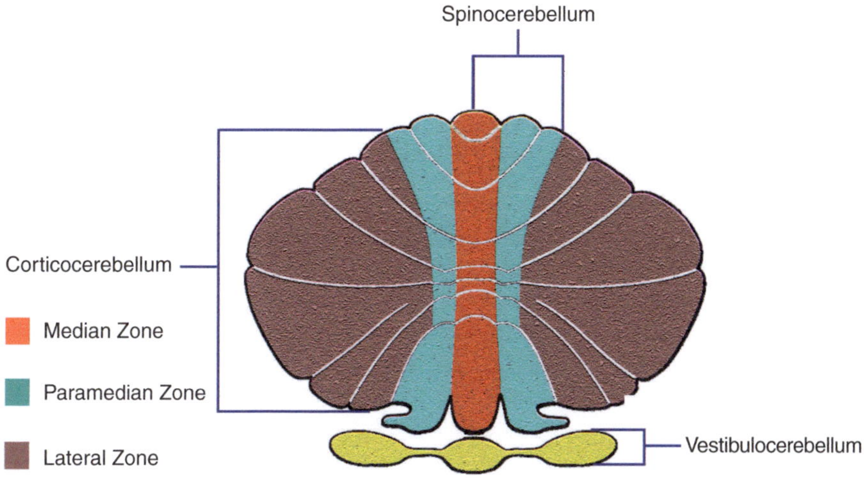

Fig. 5.2 The longitudinal compartments of the cerebellum

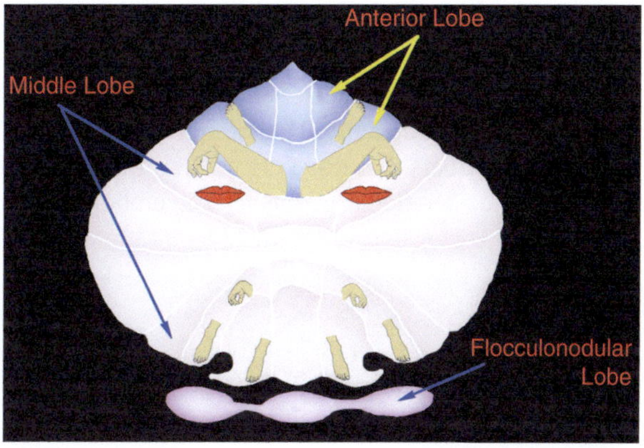

Fig. 5.3 The somatotopic representation of the foot, hand, and lip movements in the cerebellum

5.2.2 Internal Structures and Anatomical Connections of the Cerebellum

Like the cerebrum, the cerebellum is also made up of gray matter and white matter (medullary substance). The gray matter forms the superficial surface of the cerebellar cortex. Small aggregations of gray matter, called the intracerebellar nuclei, are found deep within each cerebellar hemisphere. The cerebellar cortex is folded into many cerebellar folia. Each one of these contains a core of white matter surrounded by gray matter. Functional neuroimaging techniques have discovered the somatotopic representation of the foot, hand, and tongue/lip movements in the cerebellum (Fig. 5.3).

The cortex has a uniform structure throughout its extent and is divided into three layers:

1. An external layer called the molecular layer
2. A middle layer called the Purkinje cell layer
3. An internal layer called the granular layer

The white matter of the cerebellum is made up of three different groups of fibers called the intrinsic, afferent, and efferent fibers. The intrinsic fibers connect the different regions of the cerebellum but do not leave it. The afferent fibers form a greater part of the white matter and proceed toward the cerebellar cortex. The efferent fibers constitute the output of the cerebellum and consist of the axons of the Purkinje cells from the cerebellar cortex and the axons arising from the intracerebellar nuclei.

5.2.2.1 Intracerebellar Nuclei

On either side of the midline, embedded in the white matter of the cerebellum, reside four masses of gray matter called the intracerebellar nuclei (Fig. 5.4). From the lateral to the medial is the dentate, the interposed (consisting of the nucleus emboliform and the globose), and the fastigial or the roof nucleus. The dentate nucleus is the largest intracerebellar nucleus, and it resembles a crumpled bag with its opening facing medially. The interior of the bag is filled with white matter made up of efferent fibers that leave the nucleus through the opening to form a large part of the superior cerebellar peduncle. The emboliform nucleus is wedge-shaped, the

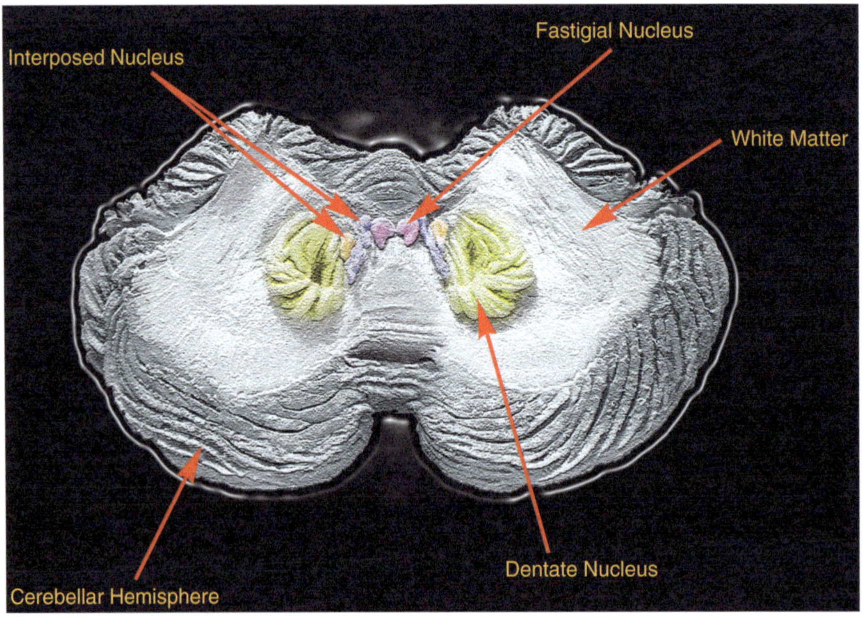

Fig. 5.4 The intracerebellar nuclei of the cerebellum

Table 5.1 Specific functions of the intracerebellar nuclei

Nucleus	Input	Output	Function	Lesion
Fastigial	Vestibular system	Vestibular system	Assists stance and gait; controls muscles in sitting, standing, and walking	Abasia (inability to either stand or walk, caused by poor coordination of muscles)
Interposed	Spinocerebellar tract (muscle spindle and Golgi tendon organ information)	Red nucleus	Assists segmental reflexes; speeds initiation of movements	Truncal titubation; abnormal rapid alternating movements; action tremor; limb ataxia
Dentate	Sensorimotor cerebral cortex	Ventrolateral nucleus of the thalamus to the premotor cortex	Fine dexterity	Delay in initiating and terminating movements; temporal incoordination of multi-joint movements; spatial incoordination of hands and fingers; intention and terminal tremors

globose nucleus is round, and the fastigial nucleus is larger than the globose. The majority of the efferent fibers (Purkinje cell axons) pass to and synapse with the neurons of the intracerebellar nuclei, which then leave the cerebellum. Some Purkinje cell axons in the flocculonodular lobe and parts of the vermis bypass the intracerebellar nuclei and leave the cerebellum without synapsing. Fibers from the dentate, emboliform, and globose nuclei leave the cerebellum through the superior cerebellar peduncle. Fibers from the fastigial nucleus leave through the inferior cerebellar peduncle.

The specific functions of each nucleus are depicted in Table 5.1. Based on the cerebellar cortical connections with the intracerebellar nuclei, the cerebellum can be divided into three zones. They are:

(i) A midline vermal zone projecting to the fastigial nucleus
(ii) An intermediate paravermal zone projecting to the interposed nucleus
(iii) A lateral zone projecting to the dentate nucleus

5.2.2.2 Cerebellar Peduncles

The cerebellum is linked to the different parts of the central nervous system by numerous efferent and afferent fibers that are grouped together on each side into three bundles called the cerebellar peduncles. They are:

1. The superior cerebellar peduncle (brachium conjunctivum)
2. The middle cerebellar peduncle (brachium pontis)
3. The inferior cerebellar peduncle (restiform body)

The superior cerebellar peduncle connects the cerebellum to the midbrain and consists of efferent fibers arising from the intracerebellar nuclei. In addition to the above, the superior cerebellar peduncle consists of afferent fibers, namely, the anterior spinocerebellar tract, the rubrocerebellar fibers, and the tectocerebellar fibers. The middle cerebellar peduncle, which is the largest peduncle, connects the cerebellum to the pons and is made up of extensive corticopontocerebellar projecting fibers. The inferior cerebellar peduncle connects the cerebellum to the medulla oblongata and is made up largely of afferent fibers, namely, the posterior spinocerebellar tract, the cuneocerebellar tract, the olivocerebellar tract, the reticulocerebellar tract, and the vestibulocerebellar tract. The efferent fibers of the inferior cerebellar peduncle are essentially cerebello-vestibular and cerebello-reticular. The following tables summarize the afferent (Table 5.2) and efferent (Table 5.3) pathways of the cerebellum and their function.

Table 5.2 Cerebellar afferent pathways

Pathway	Peduncle	Function
Corticopontocerebellar	Middle	Input from the frontal, parietal, temporal, and occipital lobes
Cerebro-olivocerebellar	Inferior	Somatosensory information from the contralateral inferior olivary nuclei
Cerebroreticulocerebellar	Inferior	Input from the lateral reticular and paramedian nuclei of the medulla
Anterior spinocerebellar	Superior	Input from muscle spindles, tendon organs, and joint receptors from levels below the midthoracic cord
Posterior spinocerebellar	Inferior	Proprioceptive input from the trunk and ipsilateral lower extremity
Cuneocerebellar	Inferior	Proprioceptive input from the upper extremity and neck
Vestibulocerebellar tract	Inferior	Input from the utricle, saccule, and semicircular canals on both sides about the position of the head and movement
Arcuatocerebellar tract	Inferior	Arises from the arcuate nuclei of the medulla oblongata
Trigeminocerebellar tract	Superior and inferior	Arises from the spinal and main sensory nuclei of the trigeminal nerve
Tectocerebellar tract	Superior	Arises in the superior and inferior colliculi and carries auditory and visual information
Cerulocerebellar tract	Superior	Carries fibers from the nucleus ceruleus

Table 5.3 Cerebellar efferent pathways

Pathway	Peduncle	Function
Globose-emboliform-rubral and dentatorubral	Superior	Ipsilateral motor activity
Dentatothalamic	Superior	Ipsilateral motor activity
Fastigial vestibular	Inferior (uncinate bundle of Russell)	Ipsilateral extensor muscle tone
Fastigial reticular	Inferior (uncinate bundle of Russell)	Ipsilateral muscle tone

5.2.3 Microscopic Structure of the Cortex

The molecular layer, the superficial layer, consisting of outer stellate cells and inner basket cells contacts the dendrites and the cell body of the Purkinje cells, respectively. The Purkinje cells (intermediate layer) are the only route of exit from the cerebellum. The granular layer, the lower layer, is composed of three main types of neurons: the granule cells, the Golgi cells, and the Lugaro cells. The granule cells form most of the granular layer and send excitatory signals to the Purkinje cells. The Golgi cells found in this layer are inhibitory to the granule cells. The Lugaro cells are the primary sensory interneurons, and the axons of the Lugaro cells only contact the inhibitory interneurons such as the basket, stellate, and Golgi cells. These cells sample information from the Purkinje cell axon collaterals and forward the information to the molecular and granular layers.

5.2.3.1 Integrated Action of the Cerebellar Cortex

The climbing and mossy fibers (Fig. 5.5) constitute the two main input lines to the cerebellar cortex and are excitatory to the Purkinje cells. Afferent fibers from the vestibular, spinal, and cortical sources terminate in the cerebellar cortex as mossy

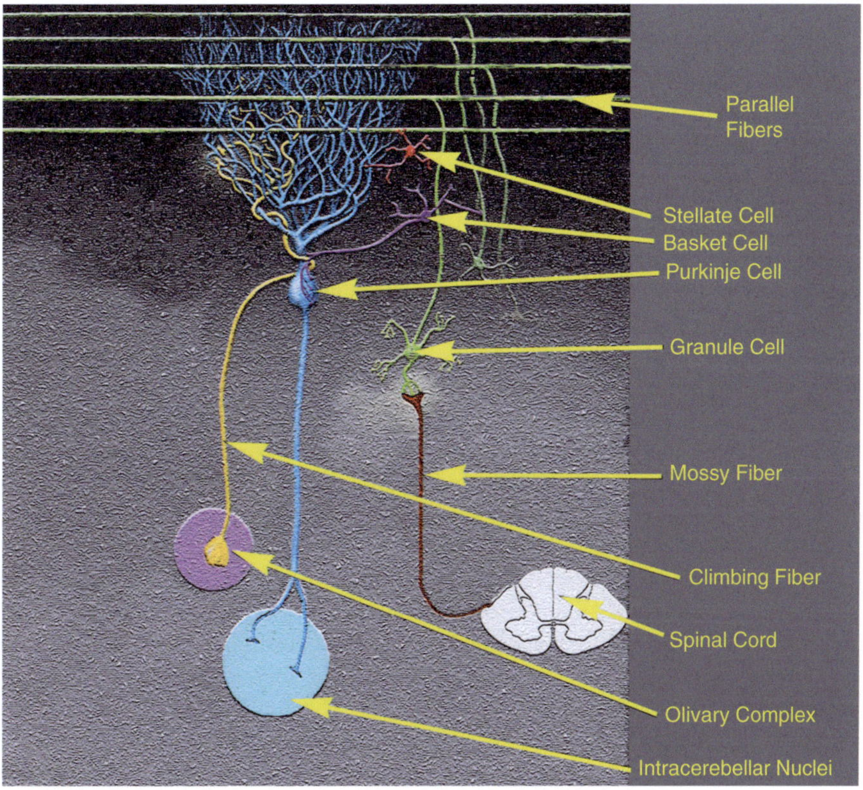

Fig. 5.5 An illustration of the cellular connections at the cerebellar cortical level

fibers. These mossy fibers make excitatory glutamatergic synapses with all the cell types of the granular layer. The granule cell axons spread longitudinally as a dense beam of parallel fibers that activate the Purkinje, stellate, basket, and Golgi cells. The granule cells are the major neurons sending excitatory outputs to the Purkinje cells and interneurons located in the molecular layer. The Golgi cells located in the granular layer receive excitatory synaptic inputs from mossy fibers and parallel fibers, and their axons provide inhibitory input to the granule cells generating a complex combination of feed-forward, feedback, and lateral inhibition effects.

The inhibitory effect of the Golgi cell axons extends far beyond their input field, causing a massive lateral inhibition. By making use of the specific cellular mechanisms controlling excitation, inhibition, oscillation, and plasticity, the granular layer is proven to transform incoming signals. As a result of the Golgi cell lateral inhibition, the granular layer's response to mossy fiber bursts is spatially organized with excitation prevailing in the center and inhibition in the surrounding areas. Because of the Golgi cell feed-forward inhibition, the granular layer generates a time-window effect limiting the duration and intensity of the output. Extended feedback inhibition through the Golgi cells causes the granular layer to have sustained coherent synchronous oscillations of large granule cell fields.

Thus, functionally, the circuit is designed to allow a massive spatiotemporal reconfiguration of the input, with several control mechanisms provided by local connectivity, long-term synaptic plasticity, intrinsic excitability, and neuromodulatory systems. However, there are several clarifications prevailing the mechanisms that contribute to signal decorrelation and understanding of the cerebellar functioning. Scientists are using a realistic computational model and other models to answer the role of the granular layer in the cerebellar functional mechanism. An emerging novel hypothesis is that the granular layer network behaves like a complex set of filters operating in the space, time, and frequency domains.

The signals generated by the granule cells are conveyed along the parallel fibers activating beams of the Purkinje cells, known as the "beam theory." The transverse running axons of the basket and the stellate cells are believed to inhibit rows of Purkinje cells on either side of an excited row of Purkinje cells, thereby giving rise to a band of excited Purkinje cells flanked on each side by zones of inhibition. While molecular layer mechanisms could enhance the vertical transmission of high-frequency bursts, the inhibitory systems in the molecular layer would prevent diffusion of such amplified responses along the parallel fibers. This result indicates that both spots and beams of excitation can coexist, although with different dynamic properties.

The climbing fibers arise from the contralateral inferior olivary nucleus and make an intimate connection between the Purkinje cells and the inferior olivary nucleus neurons. The olivary nucleus receives inputs from the spinal cord and the small pyramid cells of the motor cortex. Unlike the mossy fibers, each climbing fiber makes contact with only one Purkinje cell. The conduction time through the climbing fiber pathway is about 10 milliseconds longer than the mossy fiber pathway. The inferior olivary nucleus functions as an "error detector," and its error detection results in the powerful activation of the target Purkinje cells through the climbing fibers, thus inhibiting the deep cerebellar nucleus neurons, possibly terminating the errors in the movement component.

The output from the cerebellar cortex is by way of the Purkinje cells, which are inhibitory in action. The intracerebellar nuclei of the cerebellum have both excitatory and inhibitory actions on the reticular formation and vestibular nuclei, which are thus modulated by the inhibitory control of the Purkinje cell output. As mentioned earlier, the stellate, the basket, and the Golgi cells serve as inhibitory interneurons. They not only limit the area of the cortex excited but also influence the degree of Purkinje cell excitation produced by the climbing and the mossy fiber input. Thus, the fluctuating inhibitory impulses are transmitted by the Purkinje cells to the intracerebellar nuclei, which in turn modify muscular activity through the motor control areas of the brainstem and cerebral cortex.

5.2.4 Vascular Supply

The blood supply of the cerebellum is derived from the three arteries, namely, the posterior inferior cerebellar artery, the anterior inferior cerebellar artery, and the superior cerebellar artery (Fig. 5.6).

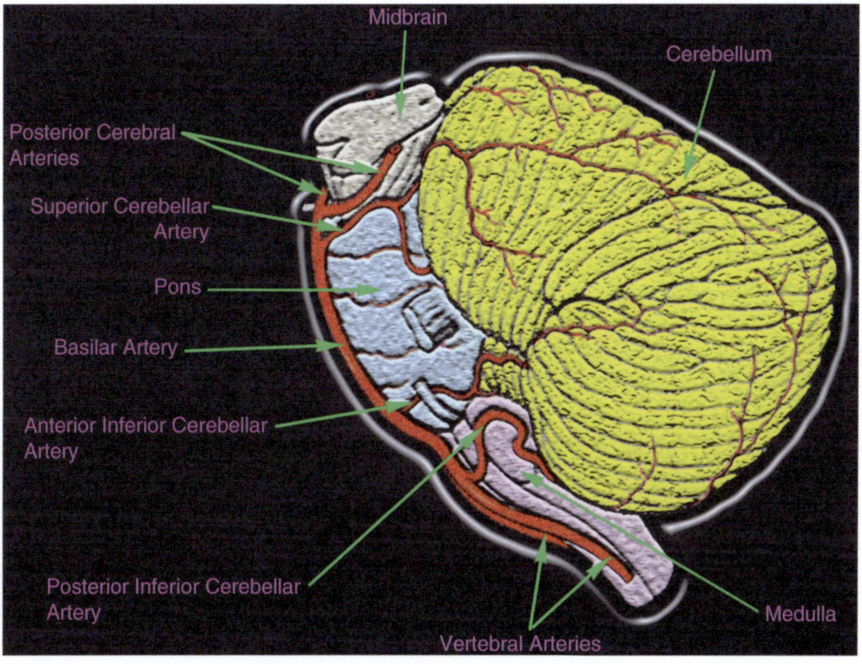

Fig. 5.6 An illustration of the vascular supply of the cerebellum

(i) The posterior inferior cerebellar artery (PICA), arising from the intracranial vertebral artery, supplies blood to the posterior inferior portion of the cerebellum and the inferior cerebellar peduncle.
(ii) The anterior inferior cerebellar artery (AICA), arising from the basilar artery, supplies blood to the anterior portion of the inferior cerebellum and the middle cerebellar peduncle.
(iii) The superior cerebellar artery (SCA), arising from the basilar artery, supplies blood to most of the cerebellar cortex, the cerebellar nuclei, and the superior cerebellar peduncles.

5.3 Theories of Cerebellar Function

The cerebellum is one of the most intriguing parts of the brain, and the concepts related to the cerebellar mechanisms of function are largely related to clinical observations and experimental ablation experiments in the past. It was clear that the cerebellum is fundamentally involved in timing and controlling the ordered and precise execution of motor sequences. In the past, knowledge about the cerebellum was exclusively related to movement. However, during the past few decades, the results of research works have supported the role of the cerebellum in coordinating emotional and visceral functions in making sensory predictions and in elaborating certain aspects of cognition.

5.3.1 Skilled Movement

The cerebellum controls the onset, level, and rate of force production by muscles. There are at least three theories pertaining to the function of the cerebellum, and of which, one theory suggests that the cerebellum is essentially a comparator between sensory input and motor output. The cerebellum has a characteristic double connectivity with the central and peripheral structures and allows it to compare the cortical signals with sensory patterns from the periphery. The intermediate zone of the cerebellum compares the voluntary command from the motor cortex for movement (often referred to as "efference copy" or "corollary discharge") with sensory signals produced by the evolving movement (re-afference). If there is any mismatch between the intended movement and the ongoing movement, the cerebellum will provide corrective feedback to the motor pathways capable of influencing the movement.

Another theory suggests that the cerebellum is a compensator. Rather than providing corrections to ongoing voluntary movement, the cerebellum is assumed to perform the predictive compensatory modification of reflexes in preparation for movement. For instance, if the stretch reflexes of a limb are too sensitive, a high-speed movement may be impossible because of the evoked stretch reflexes.

Therefore, the muscle spindle activity will have to be reduced before and during such movement. This reduction can be produced by an inhibition of gamma motor neurons or interneurons in the stretch reflex pathway. The cerebellum is in a position to make these compensatory changes given the connections of the lateral zones with the cerebral cortex. The first step in initiating a voluntary movement is the intent of performing it, which is generated in the association areas of the cortex, which is linked to the lateral cerebellum. This idea is consistent with the fact that the neurons in the dentate nucleus change their firing rates well before any activity is seen at the cortical level. This may also explain why damage to the dentate nucleus causes delays in initiating movements.

It also plays an important role in the proper timing of initiation and termination of the sequential motor steps required in purposeful movements. The timing and smooth progression of movement, as well as an orderly transition from one movement to the next, are disrupted in cerebellar lesions. The cerebellum may learn small and simple programs that are brought into action in the required order whenever a complex movement is initiated. The lateral and intermediate cerebellum sequences these programs in order during the complex learned motor acts.

5.3.2 Learned Motor Behavior

The cerebellum has also been theorized to learn effective motor behavior. The cerebellum acts as an adaptive feed-forward control system (an open-loop control system), which programs or models the voluntary movement skills based on a memory of the previous sensory input and motor output. Following cerebellar lesions, these motor programs will no longer be available, and the movement has to be guided by long delay sensory feedback loops through the cerebrum, just as in learning a new skill, resulting in incoordination. However, the mechanisms and areas for learning within the cerebellum are not very clear.

Considering the theories mentioned above, if the cerebellum no longer functions as a comparator, the movement will obviously be dysmetric. In such a situation, the patients may consciously be able to correct movement with practice, or the remaining central nervous system (CNS) may be able to assume the role of automatic correction. If the cerebellum cannot function as a compensator, the patients may display abnormal muscle tone, inappropriate postural adjustment, dysmetria, and loss of associated limb movements. In that situation, the therapist needs to assist the patients to develop postural stability and progression of movement. Whereas if the patients have lost learned motor programs controlled by the cerebellum, they will obviously be dysmetric, and due to the inability to learn, they may not benefit from therapy. Currently, uncertainty exists as to whether the cerebellum functions in all three of these capacities or predominantly in one. Apart from the theories mentioned above, the cerebellum is also believed to function in the cognitive, visceral, and emotional aspects.

5.3.3 Visceral Function

The cerebellum is shown to have connections with various visceral control centers in the brainstem as well as reciprocal connections with the hypothalamus. Experimental manipulation of the cerebellum has shown responses like alterations in blood pressure, heart rate, regional blood flow and electrocardiogram, pupillary size and reactivity to light, accommodation of lens, rate and depth of respiration, urinary and gastrointestinal motility, and piloerection. However, the actual physiological role of the cerebellum in the normal regulation of these systems is yet unknown.

5.3.4 Cognitive Function

Interestingly, full growth of the cerebellum occurs at 15–20 years of age, the same time taken for the maturation of a person's mental abilities. Studies in normal subjects have shown changes in the vascular and metabolic activities of the cerebellum during tasks like thinking of words, silent speech, thinking of movements, or doing mathematical calculations. Sophisticated neuropsychological assessments have shown defects in nonverbal intelligence, memory, and higher functions in individuals with cerebellar lesions. It may also play a role in associative learning, especially classical conditioning. Research has also shown that the neocerebellum plays an important role in effortlessly shifting attention from one domain of thought to another, so also shifting motor coordination to changing environmental demands. Reciprocal connections existing between the cerebral cortex and the cerebellum and maintenance of a similar anatomical organization in all the cerebellar regions may be responsible for the organization and modulation of cognition and emotion, similar to its organization and modulation of motor coordination and control. However, further research is needed to ascertain the role of the cerebellum in cognitive functions.

5.3.5 Emotion

The cerebellum has reciprocal anatomical connections with the reticular formation, hypothalamus, and limbic system, which are involved in reactions to emotion. Although stimulation and ablation studies have elicited rage, pleasure-seeking, fear, and aggressive behaviors, much needs to be discovered about the precise role of the cerebellum in regulating these emotional responses. A less known sign called "dysmetria of thought," a consequence of the damage of the corticocerebellar connections relating the cerebellum to the prefrontal and limbic areas, leads to the cerebellar cognitive-affective syndrome. The syndrome is characterized by disturbances of executive functions, like difficulty in planning, abstract reasoning, working memory, set-shifting, and decreased verbal fluency. The impaired spatial cognition consists of visuospatial disorganization and impaired visuospatial memory and personality change, including the flattening or blunting of affect and inappropriate behavior, are features of the same. In addition to the above, the patient may present with linguistic difficulties, like disprosodia, agrammatism, and mild anomia.

5.4 Movement Disorders Associated with Cerebellar Dysfunction

Knowledge of the anatomical and physiological role of cerebellar structures is essential to understand the clinical deficits. The movement disorders associated with cerebellar dysfunction result from a large number of etiologies, including congenital malformations, genetically transmitted ataxias, infections, and immune-mediated, vascular, and traumatic conditions. Table 5.4 shows the classification of

Table 5.4 Classifications of cerebellar disorders with examples

Types	Name of the disorder(s)
Developmental anomalies	• Cerebellar agenesis and hypoplasia (unilateral and bilateral congenital defects) • Arnold-Chiari malformation (tongue of cerebellar tissue extends through the foramen into the spinal canal) • Platybasia (a congenital malformation of bone around the foramen magnum, with the atlas and axis protruding into the posterior fossa; the condition is characterized by cerebellar disturbance and hydrocephalus)
Metabolic disorders	• Myxedema causing cerebellar ataxia • Wilson's disease may present with cerebellar signs and symptoms
Nutritional deficiency	• Deficiency of vitamin B_{12} (cobalamin), thiamine, and riboflavin may produce cerebellar dysfunction
Tumors	• Angioma • Hemangioblastoma • Astrocytoma • Medulloblastoma • Metastatic lesion • Paraneoplastic degeneration
Trauma	• Traumatic brain injury • Repeated head injury (punch-drunkenness)
Hypoxia	• Asphyxia at the time of birth leading to "floppy baby syndrome"
Vascular disease	• Cerebrovascular accident or transient ischemic attack of the cerebellum • Arteriovenous malformation
Demyelinating disease	• Multiple sclerosis • Neuromyelitis optica (Devic's disease) • Acute disseminated encephalomyelitis

(continued)

Table 5.4 (continued)

Types	Name of the disorder(s)
Infections	• Cerebellar abscess • Encephalitis
Immunological disorders	• Rheumatic chorea • Hashimoto's disease
Toxins	• Drugs • Heavy metals • Severe chronic alcohol consumption
Hereditary cerebellar degeneration/atrophy	• Spinocerebellar ataxia • Friedreich's ataxia • Ataxia telangiectasia • Episodic ataxia
Sporadic cerebellar degeneration/atrophy	• Multiple system atrophy • Late-onset cortical cerebellar atrophy

cerebellar disorders based on etiologies. Cerebellar disorders result in tonal and postural abnormalities, incoordination, and disturbances of the accuracy of movements, causing a constellation of symptoms and motor signs, which can be grouped into three categories: hypotonicity, asthenia, and cerebellar ataxia.

5.4.1 Hypotonicity

Muscle tone, defined in terms of the ease with which a muscle may be passively stretched, is determined by the level of gamma input to the muscle spindle. The cerebellum is thought to have an excitatory influence on the gamma support of the stretch reflex. It is then understandable how damage to the cerebellum might result in decreased excitation, leading to decreased sensitivity of muscle spindles and hence decreased muscle tone.

Hypotonicity, a typical symptom, usually occurs ipsilateral to the site of the lesion but may occur bilaterally or contralaterally, if the intermediate area is involved. The intensity of hypotonia is more bilateral than in unilateral lesions. Even though all the muscles of the extremities are involved, the proximal muscles are affected more than the distal ones and are more prominent in the upper limbs. Following a cerebellar lesion, hypotonia tends to dissipate with time, depending on the magnitude of the lesion.

On palpation, the hypotonic muscles appear less firm, and upon passive shaking, the involved limb may move through a greater arc of motion than the normal. The unexpected withdrawal of support given by the examiner will cause the affected extremity to fall heavily. In addition to the features mentioned above, the patient may tend to drop the arm slowly when attempting to maintain the arm against gravity. Similarly, if the patient is distracted while holding an object in his hand, he may drop the object from his grasp. The deep tendon reflexes in the involved limb may be subnormal, and the limb may briefly oscillate back and forth after the tendon tap (knee jerk in sitting and triceps jerk in prone). This phenomenon is called "pendular jerk" and is typically associated with hypotonia. Another important sign of hypotonia is the appearance of a wider base and flatter barefoot print on the ipsilateral side. Hypotonicity can cause postural asymmetries, which can lead to incoordination.

Even though the precise mechanism underlying hypotonicity is unclear, the following are a few important mechanisms suggested for cerebellar hypotonicity:

1. The neocerebellum sends fibers to the dentate nucleus, and the efferent fibers from the dentate traverse through the ventrolateral nucleus of the thalamus to terminate in the motor cortex. This pathway is facilitatory, and damage of the same may cause a lack of facilitation of the motor cortex, ultimately reducing the cortical facilitation of spinal motor neurons causing hypotonia.
2. The medial and the intermediate regions of the cerebellum have a facilitatory effect on the vestibulospinal, rubrospinal, and reticulospinal systems. Depression of the abovementioned facilitatory influence due to cerebellar pathologies may lead to hypotonia.

5.4.2 Asthenia

The word asthenia means generalized weakness. As a result of asthenia, the posture may be poorly maintained, and the patient may complain of fatigue, heaviness, and excessive effort for simple tasks. Asthenia is not a common symptom for all patients, and hypotonia and asthenia may not accompany each other. Holmes in 1922 proposed this term representing a delay in initiating muscle contraction and slowness in attaining exertion of full power. However, the concept of asthenia was originally proposed by Luciani based on animal experiments. In his research work, Luciani described a muscular force production deficit during voluntary movements in hemicerebellectomized animals.

Though the term asthenia has been clinically overlooked in spite of its clinical significance, the exact mechanism underlying asthenia is largely unknown. Suppression of a large proportion of Purkinje cells before the onset of movement, associated with the disinhibition of the dentate nucleus cells, could be a possible mechanism for asthenia. Thus, asthenia is more of fatigability and may affect an individual body part or globally. Neurophysiological studies of fatigue in cerebellar ataxia patients have shown reduced postexercise facilitation of motor potentials evoked from transcranial magnetic stimulation. The normal perception of effort or heaviness is proportional to the intensity of the supraspinal signals required to produce the movement or perform the task. In asthenic patients, any increase in the supraspinal drive to produce voluntary movement may be perceived as an increased effort and may present as fatigue or asthenia.

5.4.3 Cerebellar Ataxia

Ataxia is a general term used to describe the lack of coordination. The word ataxia has a prefix "a," meaning "without," and the Greek word "taxis," meaning "order," i.e., ataxia means "without order." It is an umbrella term encompassing asynergia, dysmetria, and dysdiadochokinesia, described further in the chapter. Causes of ataxia are many, and the principles and management of ataxia due to neuromuscular pathologies other than cerebellar lesions are not within the scope of this chapter. Cerebellar ataxia resulting from cerebellar dysfunction includes many distinct traits. However, all of them need not be present in order for a person to be described as ataxic. The following section discusses the important traits of ataxia caused by a cerebellar lesion.

5.4.3.1 Abnormal Posture and Imbalance

The vestibular connections of the cerebellum enable the coordination of head and eye movements and the maintenance of posture and equilibrium. Maintenance of equilibrium is particularly important during body movements and involves instantaneous correction of postural signals depending on the sensory feedback to the vermis. Lesions in the flocculonodular lobe cause unsteadiness of gait and inability to stand without swaying or falling, with little change in muscle tone, tremor, or dyssynergia.

Impairments in posture and balance are common problems seen in patients with cerebellar dysfunction. Acute and chronic unilateral lesions of the cerebellum can result in the lateral curvature of the spine (postural asymmetry). In contrast, bilateral lesions of the cerebellum may cause the individual to assume extreme slouching and leaning positions when seated without support. While standing, these patients may tend to spread their feet apart and may use their arms for balance. Some patients may consistently fall to one side, and lateral deviation or rotation of the head is fairly common.

Patients with cerebellar lesions may have difficulty maintaining a stationary position (particularly in standing), and problems with balance may become evident whenever they perform any action in which the center of gravity moves vertically or horizontally beyond a certain limit. Such patients have a tendency to use the upper extremities for support in standing and walking. The inability to time and grade muscle force appropriately results in an impaired balance. In addition to the above, vestibular dysfunction and derangement of peripheral retinal information, which may be seen in these patients, may further impair balance control.

In cerebellar dysfunction, opening or closing eyes will not demonstrate any significant effect on patient's ability to maintain balance. Romberg's sign, a typical feature of sensory ataxia, is negative in such patients unless the cerebellar dysfunction is associated with the damage of sensory fibers carrying proprioceptive signals to the cerebral cortex.

The exact pathomechanism behind postural instability is unclear and might be the result of distorted proprioceptive control loops that operate via the cerebellum. Proprioceptive control loops are those pathways that carry proprioceptive information from certain parts of the CNS to the cerebellum and the output back to the brain and the brainstem. The anterior and posterior spinocerebellar tract and the spino-olivary tract carry proprioceptive information from the spinal cord to the midline cerebellar vermis. Like the midline vermis, the intermediate region of the cerebellum also receives proprioceptive input from the spinal cord. The output of the vermis is to the lateral vestibular nucleus or the fastigial nucleus, and both these nuclei can affect the excitability of the alpha and gamma motor neurons of the spinal cord via the reticulospinal and vestibulospinal tracts. Damage to the circuit mentioned above or loop may lead to postural dysfunction.

In addition to the midline and intermediate vermis, the cerebellar cortex also receives proprioceptive information. It has its output to globose and emboliform nuclei. Damage to this pathway, along with their connections to the cerebral motor cortex and red nucleus, may affect posture. Evidence suggests that the ablation of the cerebellar cortex evokes postural impairments only in the ipsilateral limbs and the ablation of the medial zone (vermian and intermediate region) causes postural instability as well as a disruption of the equilibrium of the body as a whole. Evidence also suggests that the cerebellum has the ability to adjust the gain or sensitivity of these proprioceptive reflexes that operate over segmental or suprasegmental paths. If this gain modulation is altered by cerebellar disease, the automatic postural adjustments may become distorted.

Static Tremor Static or postural tremor is considered to be another disturbance of posture. The patient's body may oscillate back and forth while standing, or the limb may oscillate up and down when the person attempts to hold it against the force of gravity. The static tremor may decrease if support is provided. The frequency of oscillation is usually around 3 Hz. Postural tremor is an infrequent symptom of cerebellar dysfunction. The mechanism for postural tremor and intention tremor may not be identical. A rhythmic discharge center in the thalamus may be responsible for postural tremor, suggesting the importance of the cerebello-thalamo-cortical connections for normal posture and movement. The disruption of proprioceptive feedback loops may be another possible mechanism of postural tremor. The normal proprioceptive feedback loops for postural corrections are as shown in Fig. 5.7. A delay in processing sensory input or motor output secondary to a cerebellar lesion may cause oscillation of the abovementioned feedback system, causing the postural tremor. Thus, clinically, we see the limb's static tremor when the patient attempts to hold it steady against gravity.

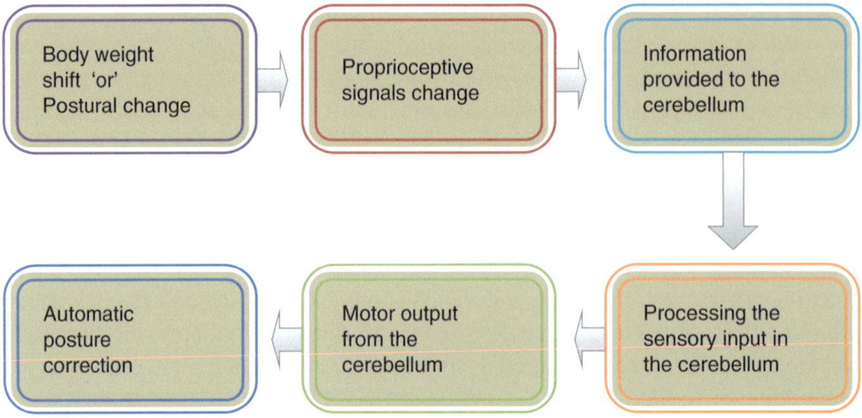

Fig. 5.7 The proprioceptive loop system operating at the cerebellar level

5.4.3.2 Dysmetria

Dysmetria is an important feature commonly seen in patients with cerebellar dysfunction and refers to errors in judging distance, speed, power, and direction of movement. The definition of dysmetria is given in Box 5.1. The presence of dysmetria causes difficulty in placing the affected limbs correctly during voluntary movement. These patients typically overestimate or underestimate the range of motion needed. Such errors committed while executing movements may force the patients to restrict their movements and hold themselves stiffly with a wide base of support and arms outstretched. Patients may also shorten the range through which they move and may minimize the number of joints to be controlled. For instance, while manipulating objects, these patients may confine the use of the hands to a position closer to the body or with forearms supported on a table. For reaching, they may use the supporting surface or object itself to break the limb's movement. Such motor behaviors are compensatory and may reduce the apparent dysmetria. Error in the onset, level, and rate of force production is believed to be the cause of dysmetria.

Box 5.1 Definition of dysmetria

Dysmetria
Dysmetria is the inability of the body part(s) to gauge the distance in the course of a movement. It refers to errors in judging or gauging the distance, speed, power, and direction of movement. Usually, the limb may fall short of the target (hypometria), pass it (hypermetria), or deviate to one or the other side, commonly to the side of the lesion.

Another error in the rate of force production displayed by these patients is the rebound phenomenon (Box 5.2), initially described by Gordon Holmes. Here, the patient loses his ability to check the movement if the resistance applied is unexpectedly removed (see the rebound test given in the box below). Typically, there is a "check factor" (a sudden contraction of the antagonist muscle/muscles), which functions to halt forceful active movements. With cerebellar involvement, the patient is unable to check the motion, and the limb will move suddenly when the resistance is released, and as a result, he may strike or hurt himself.

Box 5.2 The procedure for rebound test

Rebound Test (Stewart-Holmes Sign)

The patient flexes his elbow (with shoulder adducted, forearm supinated, and fingers clenched) against the examiner's manual resistance to produce an isometric contraction of the elbow flexors. Then, the resistance applied by the examiner is unexpectedly removed. Normally, the antagonist muscle (here, the elbow extensors) contracts and the agonist muscle relaxes and checks the movement of the limb, preventing a sudden and violent elbow flexion. While performing the test, care has to be taken to block the violent movement of the arm that may hurt the patient. One may also test elbow extension in a similar manner and likewise flexion and extension movements at the hip, knee, and ankle.

Patients with cerebellar dysfunction may also display intention tremor, in which the distal part of the extremity (hand or foot) oscillates back and forth as they try to move toward the target. The intention tremor is usually 3–5 Hz and is typically worsened during the termination of a goal-directed movement. The absence of neocerebellar servomechanism causes delays in initiating corrective movements leading to intention tremor and damage of the dentate nucleus or the corticopontocerebellar and the cerebello-thalamo-cortical (long-loop feedback) pathways may eventuate in it. Deficits are also seen in coordinating tasks involving reach and grasp. Patients with cerebellar dysfunction have difficulties in adjusting the handgrips with the changing environmental demands.

Impaired Ballistic Movements It is not uncommon to see these patients losing their ability to perform normal ballistic (quick) motions. Normal ballistic motions have a triphasic pattern of muscle activity. First, the agonist undergoes a burst of muscle activity, and the amplitude of the agonist burst depends on the distance moved. Sooner the antagonist undergoes a burst for checking the movement, and it is then followed by a second burst from the agonist for correcting the movement. Ballistic movements are assumed to be centrally programmed, and the brain automatically signals the timing and amplitude of the activity. The cerebellum may be an important part of this neural program, and clinically, a patient with a cerebellar lesion exhibits overshooting. There are delayed reaction time, delayed inhibition of antagonists, and errors in timing the duration of muscle bursts, which cause an inappropriate and inefficient acceleration and deceleration of the limb.

Dysdiadochokinesia The inability to perform rapid alternating movements is called dysdiadochokinesia. When compared to a normal person, movements performed by patients with cerebellar dysfunction appear slow and quickly lose range and rhythm. Dysdiadochokinesia develops as a result of the inability to stop the ongoing movement rather than to a reduced velocity of motion. The antagonist muscle activity typically overlaps the agonist, resulting in a braking action for movement. The patients may also exhibit an abnormal co-activation of the antagonists, making the reversal of direction of movement difficult. Both dysdiadochokinesia and dysmetria occur as a result of the inappropriate timing of muscle activity.

5.4.3.3 Dyssynergia

The term dyssynergia is used to describe the disruption of the normal smooth control of movement provided by the graduated contraction of synergic muscles with the relaxation of their antagonists. In cerebellar dysfunction, the time required to start and stop any voluntary movements is prolonged. In addition to the above, the voluntary movements lose their fluency and break up into simple component parts. Smooth and complex movements disintegrate into a series of slow and jerky movements. The cerebellum functions to sequence and time simple movements into one smooth, complex act. In the absence of this function, the movement becomes separated into individual components (decomposition of movement). For example, when a normal individual is asked to take the heel to the opposite knee and then slide the heel on the shin down to the ankle, the movement will be smooth and continuous. In the case of cerebellar dysfunction, the patient will perform a jerky and noncontinuous movement with a general tendency to flex the hip and then the knee, followed by adduction to bring the heel over the opposite knee for executing the heel-shin movement. Loss of fast internal cerebellar updates forces the system to depend on slow, long-loop feedback from the periphery in response to the movement and results in the disintegration of a normal smooth sequence of movements. The dysfunction of the dentate nucleus has been implicated in movement decomposition.

5.4.3.4 Speech Abnormalities

Scanning speech or slurred speech seen in cerebellar dysfunction is a typical example of cerebellar dysarthria. Cerebellar dysarthria results from dysmetria and dyssynergia as in extremities. Slow pronunciation, misplaced accents, and inappropriately short or long pauses in speech are some of the features of cerebellar dysarthria. This may also result from laryngeal hypotonicity, causing difficulty in varying the pitch and volume of speech. Such patients may also display explosive speech or staccato speech. The damage of the cerebellum can affect the autoregulation of the breath and particularly its integration with speech. The left cerebellar hemisphere regulates melody and continuity of speech, given its connection with the nondominant right cerebral hemisphere as opposed to the right cerebellar hemisphere, which is concerned with language function. The lesion in these specific areas of the cerebellum may also manifest as cerebellar dysarthria.

5.4.3.5 Locomotion Abnormalities

The cerebellum, especially the medial and intermediate regions, controls the timing of locomotive movements, interlimb coordination, and coordination of spinal reflexes with the ongoing locomotion. It provides important feedback to the locomotor pattern generators in the spinal cord regarding the changing circumstances in which locomotion is being performed.

One of the most striking clinical features of cerebellar dysfunction is the staggering gait, which resembles the intoxicated gait (drunkard's gait). The patient cannot walk in a straight line without lurching. The step length is uneven, and the feet may be too close or too far apart, and the legs are lifted without a regular rhythm. The arm swing is typically absent or reduced. The gait pattern is distorted, and patients experience difficulty walking in a small circle or walking backward or sideways.

Gait disturbance is considered to be a general manifestation of ataxia (ataxic gait). Chronic severe alcoholism and late atrophy of the cerebellum are typically characterized by ataxic gait without changes in limb movement, muscle tone, or equilibrium, and both these conditions selectively involve the cortex of the anterior lobe of the cerebellum. In patients with relatively mild cerebellar signs, walking may be faster than appropriate, with large steps and a relatively wide base during double support. It is said that, at a certain level of disability, walking faster may be easier than walking slower. In such patients, ataxia and balance problems may become obvious when they are asked to walk slowly. However, walking may also be slower in some patients, particularly when the balance is poor. While walking, such patients may take small steps to help restore balance and may display similar behavior when attempting to stand still. While turning around, the patients may take small steps instead of turning the body and pivoting on the feet.

Gait disturbance can be a result of errors in the rate and absolute level of force of muscle contraction and can accompany dysmetria of isolated body or body part movement. The cerebellar lesions that affect posture also influence the person's ability to walk. The cerebellum's control of gait extends beyond the integration of sensory input and correction of errors occurring in movement and has been theorized to play a significant role in the generation of the pattern of locomotion. Animal experiments following cerebellectomy have shown the disappearance of normal rhythmic discharge of the rubrospinal pathway during the swing phase and the reticulospinal and vestibulospinal tracts during the stance phase, suggesting the cerebellum's importance in controlling the rhythmic discharge in these descending tracts while walking.

5.4.3.6 Disturbance of Eye Movements

Although the management of deviations of eye movement is beyond the realm of physical therapy intervention, awareness of such problems and their mechanisms enable a better understanding of the cerebellar lesions. Some of the important features of eye movement dysfunctions are the abnormal resting position of the eye, gaze-evoked nystagmus, abnormal vestibulo-ocular reflexes, and abnormal optokinetic nystagmus.

Following cerebellar lesions, just like the posture of the limbs or the body, the resting position of the eyes can be affected. The acute cerebellar hemisphere lesion causes both eyes to deviate toward the contralateral side. Patients with cerebellar dysfunction may exhibit abnormal voluntary movements of the eyes and frequently gaze-evoked nystagmus. Gaze-evoked nystagmus, which is considered as ocular dysmetria, is a rhythmic oscillatory movement of the eye when gazing at an object in the periphery of the vision. Nystagmus is believed to be due to the cerebellum's influence on the synergy and tone of the extraocular muscles. Patients with cerebellar atrophy may display a permanent bilateral gaze-evoked nystagmus. The acute unilateral lesion may display nystagmus temporarily to the ipsilateral side, and rebound nystagmus often appears if the gaze deviation is maintained "20" seconds or longer.

Normally, eyes move in a rapid step called a saccade to accurately place the intended image on the retina's fovea. After cerebellar damage, the saccadic movement of the eyes can become too large or too small, and corrective saccades have to be made, resulting in ocular dysmetria. Ocular dysmetria can also occur with pursuit (slow) movements of the eyes, and the patient's eye may move only in saccades and will continue to move after the object stops. The pontine paramedian reticular formation is the area of the brain primarily responsible for the generation of saccades, and a lesion in this area causes a permanent loss of saccadic eye movements. The feedback of eyeball movements is integrated into the brainstem and cerebellum. The flocculus and the paraflocculus are presumed to be involved in this function, and damage of these areas disrupts smooth pursuit movements.

Similar to the distortion of the voluntary movements of the eyes, the reflex movements (Boxes 5.3 and 5.4 for vestibulo-ocular reflex and optokinetic nystagmus, respectively) may also be distorted in the cerebellar lesions. Patients with cerebellar lesions may complain of visual defects like blurred vision, diplopia, and difficulty to visualize when their body or head is in motion. These patients may also exhibit difficulty in holding the head motionless when they are asked to shift the eye within 30° from neutral. Normally, a healthy subject can shift his eyes 30° from the neutral without any accompanying head motion.

Box 5.3 Details regarding vestibulo-ocular reflex

Vestibulo-ocular Reflex

Vestibulo ocular reflex (VOR) is abnormal in patients with cerebellar lesion. Intact VOR is needed to visualize the image of the object while the head is in motion. The VOR pathway comprises the primary vestibular afferents, vestibulo ocular relay neurons in the medial longitudinal fasciculus, and the neurons innervating the extraocular muscles. The VOR prevents retinal slip during head movements by moving the eyes at the same velocity as the head but in the opposite direction.

Box 5.4 Details regarding the optokinetic nystagmus

Optokinetic Nystagmus

Optokinetic nystagmus is a phenomenon seen when a normal person views black and white stripes on a revolving drum (Fig. 5.8) moving horizontally or vertically. Such nystagmus consists of a sequence of fast phase (a quick eye movement) opposite the direction of the revolving drum and a slow phase (a slow drift backward) in the direction of the revolving drum. These back-and-forth movements of the eyes will continue until the person stops visualizing the revolving drum. Following acute cerebellar lesions, the optokinetic nystagmus reduces, while chronic lesions increase it.

Fig. 5.8 A subject viewing a horizontal revolving drum

5.5 Signs and Symptoms of Cerebellar Disorders

Though the most prominent sign of cerebellar dysfunction is incoordination, the signs and symptoms vary with the cause and often help in the clinical diagnosis of the condition. The following section will discuss the typical features of archicerebellar, paleocerebellar, and neocerebellar syndromes and certain cerebellar vascular syndromes.

5.5.1 Archicerebellar Syndrome

The common signs and symptoms of archicerebellar syndrome are ataxia of gait, vertigo, and nystagmus. The involvement of the flocculonodular lobe by a discrete lesion, such as medulloblastoma in a child or a secondary neoplasm in an adult, produces ataxic gait as its earliest and chief manifestation, often accompanied by a sensation of vertigo and nystagmus. Heel-shin coordination may be normal, and the fast component of nystagmus is typically directed to the side of the lesion. A rotated posture of the head is not uncommon with cerebellar tumors, probably due to damage to the vestibular system and/or its cerebellar projections.

5.5.2 *Paleocerebellar Syndrome (Midline or Vermis Syndrome)*

Ataxia of gait and incoordination of the lower limbs are the common signs and symptoms of paleocerebellar syndrome. The cortex of the anterior lobe of the cerebellum is selectively affected in alcoholic cerebellar atrophy, and the condition is characterized by a broad-based gait with a tendency to lose balance and fall to either side. At first, the coordination of legs may appear normal when the patient is lying down. Sooner or later, difficulty in running the heel down the shin of the opposite leg may become apparent. Muscle tone usually remains normal, and the knee jerks are not pendular. Late atrophy of the cerebellum, a paleocerebellar syndrome, also causes similar features. The patient may be unable to sit erect or hold the neck upright; when severe, standing and sitting balance disturbance leads to to-and-fro nodding movements of the head and trunk known as titubation. These are usually anteroposterior (yes-yes) and have a frequency of 3–4 Hz. The head may be abnormally rotated and tilted, and nystagmus may be present. Dysarthria is often present. Common causes of this syndrome are alcoholic cerebellar degeneration and medulloblastoma.

5.5.3 *Neocerebellar Syndrome (Hemispheric Syndrome)*

The signs and symptoms of neocerebellar syndrome are hypotonia, dyssynergia, and dysmetria affecting the arm more than the leg and foot, speech dysarthria and slurring, gait ataxia, and nystagmus. The distal movements are affected more than the proximal and fine movements are affected more than the gross ones. Usually, patients having neocerebellar syndrome will exhibit rebound phenomenon, dysdiadochokinesia, postural tremor, intention tremor, and a tendency to fall to the side of the lesion. Tumors, multiple sclerosis, vascular disease, and trauma of the neocerebellum can cause characteristic features of the neocerebellar syndrome. Diffuse insult of the bilateral midline and hemispheric structures (diffuse cerebellar dysfunction) may present with nystagmus, gait and truncal ataxia, and appendicular incoordination, features similar to those of the neocerebellar syndrome.

Cerebellar dysfunction should be distinguished from other disorders leading to tonal abnormality and incoordination, as seen in pyramidal or extrapyramidal disorders. Hyperkinetic disorders may disturb the timing and accuracy of movements. Frontal lobe ataxia due to dysfunction of the contralateral frontal lobe may mimic cerebellar ataxia but show signs of upper motor neuron lesion such as hyperreflexia, increased tone, and presence of the Babinski sign. On the contrary, ataxia due to pure cerebellar lesions may cause hypotonia, diminished deep tendon reflexes, pendular jerk, and presence of other cerebellar features. Some of the known cerebellar vascular syndromes and their clinical features are shown in Table 5.5. In addition to the classical cerebellar ataxia features, cerebellar dysfunction may cause cerebellar cognitive and affective syndromes characterized by disturbed executive function, visuospatial disorganization, impaired visuospatial memory, personality change, and linguistic difficulties.

The occlusion of blood supply to various regions of the cerebellum may result in diverse clinical presentations, as depicted in Table 5.6. The cerebellar infarction may result from the thrombotic or embolic occlusion of vessels supplying the cerebellum. Two distinct clinical syndromes recognized as a result of cerebrovascular accidents of the cerebellum are:

Table 5.5 Some cerebellar syndromes and their clinical features

Type	Clinical features
Anterior vermis syndrome	Wide-based and titubating gait Gait ataxia Normal or impaired arm coordination Hypotonia, nystagmus, and dysarthria
Posterior vermis syndrome	Axial disequilibrium and staggering gait Little or no limb ataxia Spontaneous nystagmus
Cerebellar hemispheric syndrome	Ipsilateral limb incoordination
Pancerebellar syndrome (involving midline vermis and lateral cerebellar hemispheres)	Bilateral cerebellar signs affecting the trunk, limbs, and cranial muscles

Table 5.6 Clinical features based on the involvement of arteries supplying the cerebellum and the adjoining brainstem structures

Artery	Clinical features	Areas supplied
Superior cerebellar artery (SCA)	Vertigo (involvement of vestibular nuclei and connections), nystagmus (medial longitudinal fasciculus and cerebellar pathways), ipsilateral Horner's syndrome (descending oculosympathetic pathway involvement), ipsilateral ataxia and dyssynergia (superior cerebellar peduncle and cerebellum), ipsilateral intention tremor (dentate nucleus and superior cerebellar peduncle), contralateral trunk and limb hypoalgesia and thermoanesthesia (lateral spinothalamic tract), contralateral hearing impairment (involvement of crossed fibers of lateral meniscus), and contralateral fourth nerve palsy (involvement of pontine tectum)	Superior surface of cerebellar hemispheres, superior vermis, dentate nucleus, cerebellar white matter, parts of the midbrain, superior cerebellar peduncle, and middle cerebellar peduncle

(continued)

Table 5.6 (continued)

Artery	Clinical features	Areas supplied
Anterior inferior cerebellar artery (AICA)	Vertigo, nausea, vomiting, nystagmus (involvement of vestibular nuclei), ipsilateral facial hypoalgesia, thermoanesthesia, corneal hypesthesia (involvement of trigeminal nucleus and tract), ipsilateral Horner's syndrome, contralateral hypoalgesia and thermoanesthesia (lateral spinothalamic tract), ipsilateral ataxia and dyssynergia (middle cerebellar peduncle), ipsilateral deafness, and facial palsy	Lateral pons, anterior inferior cerebellum, including the flocculus, middle cerebellar peduncle, and 7th and 8th cranial nerves
Posterior inferior cerebellar artery (PICA)	Acute vertigo, truncal ataxia, unsteadiness, limb ataxia, dysmetria, facial hypoalgesia, contralateral trunk and limb hypoalgesia, hoarseness, dysphonia, dysphagia, hiccups, Horner's syndrome	Posteroinferior cerebellar hemispheres including cerebellar tonsils and nucleus gracilis, the inferior portion of the vermis, the lower part of the medulla, and inferior cerebellar peduncles

(i) Space-occupying cerebellar infarcts with the fourth ventricular or brainstem compression: This may present with sudden onset of occipital headache, vertigo, nausea, vomiting, unsteady gait and trunk ataxia, and dysarthria. The cerebellar tissue could herniate downward (tonsillar herniation) or upward (transtentorial herniation). The compression of the fourth ventricle by the edematous cerebellum could lead to obstructive hydrocephalus.
(ii) Cerebellar infarcts without the fourth ventricular or brainstem compression: This includes small (border zone) infarcts not easily localizable and confining to well-defined arterial boundaries. Physical findings may be absent or include wide-based gait, lateropulsion, mild dysmetria, dysarthria, or dysdiadochokinesia.

5.6 Recovery Following Lesions of the Cerebellum

It is a proven fact that a certain level of spontaneous recovery or compensation is expected in most brain lesions. The level of spontaneous recovery or compensation depends upon the severity of the lesion and its location. Recovery can stop at any stage along the continua, which is not predictable; however, the speed of early spontaneous recovery offers a clue to the ultimate level of function that shall be gained.

In the past, to learn the effects of cerebellar lesions, extensive studies were done on animals, including primates. In primates, the most severe problems resulting from total cerebellectomy were truncal ataxia, dysmetria of the limbs, hypotonia, and postural tremor. Within a few months post cerebellectomy, the severity of these problems reduced and the improvement then reached a plateau. If only one cerebellar hemisphere is damaged with no nuclear involvement, the primate displays ipsilateral dysmetria, postural tremor, and ataxic gait, which becomes essentially normal over the next few months. If the vermis or midline structure is involved, the chief problem is truncal ataxia, which improves over the first 3–5 months but never disappears. Such animal studies have provided the following information:

1. Recovery is generally very poor from a total cerebellectomy than a partial cerebellectomy.
2. A bilateral lesion with damage to the deep nuclei is grimmer than that of a cortical lesion.
3. Spontaneous recovery and/or compensation will be completed within 6 months to 1 year.

Response to acute cerebellar injury may not be the same as in primates, but the above information provides a general framework for humans. A degenerative cerebellar disease or a tumor produces milder signs and symptoms in humans than those produced by acute lesions. Milder signs and symptoms could be due to the effect of concurrent compensation with the steadily progressing lesions. In the case of partial destruction of the cerebellum, a certain extent of neural adaptation may occur because of the undamaged adjoining regions of the cerebellum. However, the deficits are more severe if a second lesion occurs in those adjoining areas where prior compensation was adapted. The cerebral motor cortex is also considered to be an essential structure upon which compensation for a cerebellar lesion depends.

Functional recovery following cerebellar damage is mainly ascribed to the neuroplastic properties of the brain, though the close linkage between neuroplasticity and functional recovery is not yet fully clarified. Single photon emission computed tomography and other neuroimaging studies have demonstrated neurological and functional recovery in the form of improved perfusion in other regions of the cerebral cortex, including the frontal and temporal lobes. Positron emission tomography and functional neuroimaging techniques have consistently demonstrated relative overactivations in motor-related brain regions during the movement of the affected limb compared to control subjects. Considering the experimental and clinical evidence, recovery can occur spontaneously, following morphologic reorganization or due to transfer of function by the spared tissue.

5.7 Medical Management of Cerebellar Disorders

Cerebellar disorders mainly result in motor dysfunctions, which can lead to significant and serious restrictions in activities of daily living (ADL). Except for certain specific diseases and symptoms, the possibilities of medical interventions are rare and limited. Since the causes for cerebellar impairments vary, functional recovery heavily depends on the cause and site of the lesion. Among the spectrum of causes that includes stroke, cerebellar tumors, multiple sclerosis, and degenerative disease, the degenerative cerebellar diseases are hard to treat, despite the greatly improved understanding of the genetic underpinnings. For such conditions, no medications to ameliorate ataxia or decelerate disease progression are yet available.

Focal cerebellar disorders, such as cerebellar stroke, cerebellar abscess, multiple sclerosis lesions, or tumors, are treated according to the respective evidence-based clinical practice guidelines. Effective medications are available to treat extracerebellar symptoms in acquired and degenerative cerebellar disease. However, no medication is known to ameliorate the clinical symptoms of cerebellar ataxia. For several years, a number of different drugs have been tested for symptomatic treatment of cerebellar ataxia. Those studies that described positive effects were open-label trials, performed in a small number of patients; however, large study populations failed to reproduce the results. Serotonergic drugs including 5-hydroxytryptophan and buspirone, nutritional supplements like creatine and L-carnitine, antioxidants including vitamin E and coenzyme Q10, and medications such as amantadine and gabapentin are certain drugs routinely used for managing cerebellar dysfunction. Even the use of antiepileptic drugs like clonazepam, topiramate, primidone, and levetiracetam for cerebellar tremors has not shown any clear effects. During recent years, the focus of interest is on coenzyme Q10, riluzole, and aminopyridines as possible treatment options in cerebellar ataxia.

Idebenone, a short-chain derivative of the coenzyme Q10, an antioxidant and electron carrier, is believed to improve the ataxic features in Friedreich's ataxia, a rare hereditary degenerative disorder characterized by a loss of frataxin that leads to respiratory chain defects and accumulation of iron in the mitochondria with increased oxidative stress. Though few recent studies have suggested that a high dosage of idebenone may ameliorate neurological symptoms, the results of two randomized placebo-controlled double-blind drug trials in large study populations failed to support the effectiveness of high-dosage idebenone.

Riluzole, a neuroprotective agent, has been administered in different neurodegenerative disorders, including conditions like autosomal dominant spinocerebellar ataxia, types 1, 2, and 28, recessive conditions like Friedreich's ataxia, X-chromosomal ataxias, ataxias associated with immunological disease, and sporadic diseases like sporadic adult-onset ataxia and multiple system atrophy type C, with varying success. In the recent past, aminopyridines have been used to treat episodic ataxia type 2 and downbeat nystagmus. Aminopyridines are potassium channel blockers believed to increase the inhibitory drive of cerebellar Purkinje cells and have shown both reductions of episodic attack and interictal cerebellar signs. Downbeat nystagmus, with known or unknown etiology, leads to the unsteadiness of gait and vertigo, and for such conditions, provided no structural lesions of the cerebellum and brainstem, aminopyridine reduces nystagmus in approximately 50% of the patients.

5.8 Evaluation and Goals of Treatment

The primary goal of physiotherapy is to make the patient as functional as possible, ensuring maximum safety, reasonable energy cost to the patient, and cosmesis. The evaluation includes the determination of baseline basic functional capabilities (Box 5.5).

Box 5.5 Components of basic functional capabilities

Basic Functional Capabilities
• To move or roll within the bed • To transit from a supine/side-lying/prone position to sitting • To maintain a sitting posture on a stable surface • To transit from sit to stand and vice versa from bed or a chair • To maintain a standing posture • To ambulate in his/her usual and normal environment • To feed, groom and dress, and perform basic personal hygiene activities

The description of the functional performance may include the assistance needed, level of effort involved, time to complete the activity, potential hazards to the patient, and the accompanying abnormal movements. The therapist should determine the causes for the difficulties in carrying out the above basic functions. Similarly, it is also essential to examine both sides of the body even if a unilateral cerebellar lesion has been diagnosed. Given below (Table 5.7) are some of the

Table 5.7 Cerebellar movement abnormalities and assessment findings

Movement abnormalities	Test or examination	Findings
Hypotonia	Deep tendon reflex	Subnormal; pendular (seen in the knee or elbow) in nature
	Muscle palpation	Firmness reduced when compared to the unaffected side or age- or gender-matched population
	Resting posture	Asymmetrical, slack, and may have round back posture in sitting and a tendency to lean
	Footprint	Footprints are flat and tendency to have wide base on standing
	Voluntary flexion and extension of the elbow/knee	Presence of limb ataxia while unsupported
	Passive shaking of the distal extremity (wrist or foot)	Limbs move through a greater arc of motion than the normal limb
Instability and postural dysfunction	Unanticipated perturbations	Unable to resecure balance immediately
	Hold the extremity against gravity	Presence of static tremor
	Posture while standing	Tends to stand with a wide base of support, trunk leaning, or may use hands for support and presence of static tremor
	Single limb stance using the affected lower limb	Unable to secure and maintain balance; stance time quite reduced when compared to age-matched population
	Sideways or backward walk and walk in circles	Unable to walk and/or secure balance

(continued)

Table 5.7 (continued)

Movement abnormalities	Test or examination	Findings
Asthenia	Muscle strength	Affected side muscles unable to work against strong resistance, when compared to the contralateral unaffected side or age- or gender-matched population
	Muscle endurance	Gets tired easily of activities that require repetition
	Active holding of the arm at 90° forward elevation or abduction	Gets tired easily
	Posture while sitting or standing	Postural asymmetry, may slack or have round back posture in sitting and tendency to lean consistently to one or other side
Dysmetria	Finger to nose, finger to finger, placement of peg in a peg hole; trace a circle using a pen or pencil; trace a circle using great toe; sliding the heel down and up the shin unhurriedly; placing the foot on the footprint marks on the floor while walking	Presence of intention tremor and hypometria or hypermetria
	Unexpected release of resistance offered by the examiner	Rebound phenomenon
	Forward elevation of the arm to 90°; then quickly elevate beyond and then return to 90° position without error	Unable to return to 90° position without error and may exhibit overshooting or undershooting
	Ballistic movements	Delay in the initiation and requires a longer time to execute
	Supported or unsupported voluntary flexion or extension of the knee/elbow	Limb ataxic especially in an unsupported position
Dysdiadochokinesia	Rapid alternating supination and pronation of the forearm; while sitting, rapid alternate forefoot and heel tapping on the floor	Unable to perform the movement rapidly and careful observation also reveals loss of movement rhythm and range
	Activities that require rapid alternating movements such as gait, brushing teeth, and combing hair	Unable to perform such activities; may cause injuries and safety concerns
Dyssynergia	Observation of heel-shin test	The smooth and continuous flow of movement is lost and typically broken into separate components
	Complex upper limb activities	May appear slow and robotic in nature

Table 5.7 (continued)

Movement abnormalities	Test or examination	Findings
Gait dysfunction	March in place or to music	Not able to maintain the rhythm and tendency to lose balance quickly
	Uneven ground walking; toe or heel walking	Loses balance and may fall
	A figure-of-eight walking or walking in circles	Loses balance and stumbles while walking

Table 5.8 Nonequilibrium and equilibrium tests for cerebellar dysfunction

Nonequilibrium test or test for coordination	Equilibrium or balance test
1. Finger-nose test 2. Finger-to-finger (fingertips in the midline) test 3. Rapid alternating pronation and supination of the forearm or opening and closing fist 4. Rapid alternate forefoot and heel tapping against the floor 5. Hand tapping on the lap with speed gradually increasing 6. Fingers held in opposition and tapping one after the other, sequentially, the thumb 7. Foot tapping with speed gradually increasing 8. Past pointing to check hypermetria 9. Heel-shin test 10. Drawing circles or tracing a circle on a blackboard using the upper limb 11. The figure-of-eight drawing using the foot on the floor 12. Rebound test	1. Normal comfortable stance 2. Standing with a narrow base of support 3. Standing with feet together with eyes open and then with eyes closed 4. Near tandem and tandem stance 5. Single limb stance 6. Forward and sideways reaching 7. Unanticipated or unexpected perturbations 8. Walking a straight line 9. Walking forward, backward, and sideways 10. March in place 11. Abruptly starting and stopping the gait on command 12. Heel walking and toe walking

specific tests and movement abnormalities in cerebellar dysfunction. Table 5.8 provides details regarding the equilibrium and nonequilibrium tests for cerebellar dysfunction.

5.8.1 Scales for Cerebellar Ataxia

The development, evaluation, and validation of scales for the assessment of severity and progression of disease in neurological conditions have gained increasing impact during the last few decades. Many of the scales have served as outcome measures in clinical trials. When evaluating the quality of clinical scales, several essential aspects have to be considered, namely, reliability, validity, sensitivity, and ease of application. Box 5.6 provides the definitions of the psychometric properties. A good scale is also characterized by the absence of significant floor or ceiling effects. For longitudinal studies, a clinical scale should also retain the sensitivity to detect changes in disease progression over time. Scales should be easy to apply and compact to avoid fatigue in the patient and the examiner. The International Cooperative Ataxia Rating Scale (ICARS), the Brief Ataxia Rating Scale (BARS), the Friedreich Ataxia Rating Scale (FARS), and the Scale for the Assessment and Rating of Ataxia (SARA) are the common rating scales used in cerebellar disorders.

Box 5.6 Definitions of common psychometric property terminologies

Psychometric Properties—Definitions

- Reliability: Consistency of a clinical rating scale
  - Interrater reliability: Refers to the variation in measurement when different investigators are using the same rating scale or tool
  - Test-retest reliability: Refers to the variation in measurement on the same item under the same condition
  - Internal consistency: Assesses the consistency of results across items that are part of the same scale or subscale
- Validity: Describes the extent to which a scale measures what it intends to measure
  - Construct validity: The degree to which a test measures what it claims or means to measure
  - Content validity: A subjective form of measurement and it is the extent to which a measure represents all facets of a given construct
  - Criterion-related validity: It is also known as instrumental validity. It is the extent to which a measure is related to an outcome
  - Internal validity: It is the extent to which a cause-and-effect relationship established in a study cannot be explained by any other factor(s)
  - External validity: It is the extent to which the results of a study can be generalized to and across other settings or situations
- Sensitivity: The ability of a scale or tool to correctly identify those with the disease
- Specificity: The ability of a scale or tool to correctly identify those without the disease

In 1997, the ICARS was developed, and it offers a semiquantitative quantification of cerebellar ataxia. The scale emphasizes the specific components of ataxia, namely, the static functions attributed to the vermis and anterior lobe, limb ataxia resulting from cerebellar hemisphere impairment, and oculomotor deficits reflecting the dysfunction of the vermis and flocculus. The 19 items of the ICARS are grouped into four multi-item subscales related to the functional anatomical compartmentalization of the cerebellum, and they are as follows:

1. Posture and gait disturbances (items 1–7)
2. Limb ataxia (items 8–14)
3. Dysarthria (items 15 and 16)
4. Oculomotor disorders (items 17–19)

A total score of 100, which is the maximum for this scale, indicates the most severe cerebellar dysfunction. The test takes about 20 minutes to complete, and training is required to practice the scale. The interrater reliability, test-retest reliability, internal consistency, criterion-related validity, and internal construct validity were found to be very good. The scale has no significant ceiling effects and no floor effects.

Brief ataxia rating scale is a concise five-item scale, covering gait, kinetic function of the leg, kinetic function of the arm, speech, and eye movements. A total score of 30, which is the maximum for this scale, indicates the most severe ataxia. The BARS has been derived from a modified version of the ICARS. The interrater reliability was found to be excellent. The FARS, originally developed for rating Friedreich's ataxia with special emphasis on the neurodegenerative pattern and non-neurological symptoms, is an ordinal grading scale. It takes approximately 30 minutes to complete the assessment using the FARS. The scale is composed of a physical examination emphasizing the bulbar, upper and lower limbs, peripheral nerve, and upright stability/gait. In addition to these, it also functionally stages the overall mobility and assesses the ADL, speech, coordination of the upper limb, and timed walk of 50 feet. The FARS's interrater variability is found to be high for most of the parameters.

The SARA (Table 5.9) is a semiquantitative tool for the assessment of cerebellar ataxia. The test items of the SARA are specific for ataxia and do not consider extra-cerebellar involvement. The test items are essential components of a standard neurological examination and do not require technical equipment or extensive training for its administration. The test has eight items, with a maximum total score of 40, which indicates the most severe ataxia. Gait, stance, sitting, speech disturbance, finger chase, nose-finger test, fast alternating hand movements, and heel-shin slide are the eight components of the SARA. No more than 40 minutes are required to complete the test. Its interrater variability and test-retest reliability are very high, and the internal consistency is also found to be excellent.

5.8.2 Treatment

The involvement of the cerebellum may lead to damages in several essential functions, such as movement coordination, balance, gait, speech, hearing, and ocular movement. Ataxia, gait abnormalities, and reduced balance ability, associated with decreased joint mobility, muscle tone, and proprioception, lead to unsafe mobility and difficulty performing ADL. With no available disease-modifying pharmacological treatments for cerebellar ataxia, physical therapy intervention is the primary treatment option to restore normal movement functions. Though therapeutic intervention is the mainstay for the management of the cerebellar ataxia, there is limited high-quality research into the effectiveness of treatments. Evidence suggests that a wide range of physiotherapy interventions are used for ataxia management, with many of them reporting positive effects. Patients with cerebellar degenerative disease exhibit progressive worsening of motor coordination. Recent evidence suggests that patients with such degenerative diseases may benefit from long-term

Table 5.9 The scale for the assessment and rating of Ataxia

Rater: ______________________ date: ____________ patient: ____________________

Scale for the Assessment and Rating of Ataxia (SARA)

1) Gait Proband is asked (1) to walk at a safe distance parallel to a wall including a half-turn (turn around to face the opposite direction of gait) and (2) to walk in tandem (heels to toes) without support. **0** Normal, no difficulties in walking, turning and walking tandem (up to one misstep allowed) **1** Slight difficulties, only visible when walking 10 consecutive steps in tandem **2** Clearly abnormal, tandem walking >10 steps not possible **3** Considerable staggering, difficulties in half-turn, but without support **4** Marked staggering, intermittent support of the wall required **5** Severe staggering, permanent support of one stick or light support by one arm required **6** Walking > 10 m only with strong support (two special sticks or stroller or accompanying person) **7** Walking < 10 m only with strong support (two special sticks or stroller or Accompanying person) **8** Unable to walk, even supported	**2) Stance** Proband is asked to stand (1) in natural position, (2) with feet together in parallel (big toes touching each other) and (3) in tandem (both feet on one line, no space between heel and toe). Proband does not wear shoes, eyes are open. For each condition, three trials are allowed. Best trial is rated. **0** Normal, able to stand in tandem for > 10 s **1** Able to stand with feet together without sway, but not in tandem for > 10s **2** Able to stand with feet together for > 10 s, but only with sway **3** Able to stand for > 10 s without support in Natural position, but not with feet together **4** Able to stand for >10 s in natural position only with intermittent support **5** Able to stand >10 s in natural position only with constant support of one arm **6** Unable to stand for >10 s even with constant support of one arm
Score	**Score**
3) Sitting Proband is asked to sit on an examination bed without support of feet, eyes open and arms outstretched to the front. **0** Normal, no difficulties sitting >10 sec **1** Slight difficulties, intermittent sway **2** Constant sway, but able to sit > 10 s without support **3** Able to sit for > 10 s only with intermittent support **4** Unable to sit for >10 s without continuous support	**4) Speech disturbance** Speech is assessed during normal conversation. **0** Normal **1** Suggestion of speech disturbance **2** Impaired speech, but easy to understand **3** Occasional words difficult to understand **4** Many words difficult to understand **5** Only single words understandable **6** Speech unintelligible / anarthria

Table 5.9 (continued)

Score			Score		
5) Finger chase Rated separately for each side Proband sits comfortably. If necessary, support of feet and trunk is allowed. Examiner sits in front of proband and performs 5 consecutive sudden and fast pointing movements in unpredictable directions in a frontal plane, at about 50 % of proband's reach. Movements have an amplitude of 30 cm and a frequency of 1 movement every 2 s. Proband is asked to follow the movements with his index finger, as fast and precisely as possible. Average performance of last 3 movements is rated. **0** No dysmetria **1** Dysmetria, under/ overshooting target < 5 cm **2** Dysmetria, under/ overshooting target < 15 cm **3** Dysmetria, under/ overshooting target > 15 cm **4** Unable to perform 5 pointing movements			**6) Nose-finger test** Rated separately for each side Proband sits comfortably. If necessary, support of feet and trunk is allowed. Proband is asked to point repeatedly with his index finger from his nose to examiner's finger which is in front of the proband at about 90 % of proband's reach. Movements are performed at moderate speed. Average performance of movements is rated according to the amplitude of the kinetic tremor. **0** No tremor **1** Tremor with an amplitude < 2 cm **2** Tremor with an amplitude < 5 cm **3** Tremor with an amplitude > 5 cm **4** Unable to perform 5 pointing movements		
Score	Right	Left	**Score**	Right	Left
mean of both sides (R+L)/2			mean of both sides (R+L)/2		
7) Fast alternating hand movements Rated separately for each side Proband sits comfortably. If necessary, support of feet and trunk is allowed. Proband is asked to perform 10 cycles ofrepetitive alternation of pro-and supinations of the hand on his/her thigh as fast and as precise as possible. Movement is demonstrated by examiner at a speed of approx. 10 cycles within 7 s. Exact times for movement execution have to be taken. **0** Normal, no irregularities (performs <10s) **1** Slightly irregular (performs <10s) **2** Clearly irregular, single movements difficult to distinguish or relevant interruptions, but performs <10s **3** Very irregular, single movements difficult to distinguish or relevant interruptions, performs >10s **4** Unable to complete 10 cycles			**8) Heel-shin slide** Rated separately for each side Proband lies on examination bed, without sight of his legs. Proband is asked to lift one leg, point with the heel to the opposite knee, slide down along the shin to the ankle, and lay the leg back on the examination bed. The task is performed 3 times. Slide-down movements should be performed within 1 s. If proband slides down without contact to shin in all three trials, rate 4. **0** Normal **1** Slightly abnormal, contact to shin maintained **2** Clearly abnormal, goes off shin up to 3 times during 3 cycles **3** Severely abnormal, goes off shin 4 or more times during 3 cycles **4** Unable to perform the task		
Score mean of both sides (R+L)/2	Right	Left	**Score** mean of both sides (R+L)/2	Right	Left

With permission from Professor Thomas Klockgether

motor training. However, patients with cerebellar disease with associated motor learning deficits are less likely to get benefits from rehabilitation that relies on motor learning. The following are the common goals toward which a therapist must work with a patient having a cerebellar lesion:

1. Postural stability
2. Balance and functional gait
3. Accuracy of limb movement

5.8.2.1 Postural Stability

Patients with postural instability need to be assisted sequentially to each level of independently maintained posture, beginning from recumbent postures to the more challenging and upright postures against gravity. For those with poor sitting balance, developing sufficient head and trunk control in the prone position is a prerequisite for gaining stability in sitting. Similarly, developing adequate sitting balance is a prerequisite for standing.

Preparation for Independent Sitting To develop head control, the patient may be positioned in prone while propped on forearms with pillows (Fig. 5.9) or a wedge bolster under his chest. The goal is to have the patient lift his head and hold it steady. Even facilitating pivot prone position by stimulating neck and upper back muscles (using C-brushing or 3–5 seconds of C-icing to the neck extensors; refer to Chap. 2 A) followed by stretching and then resistance to neck extension helps to promote head holding. To progressively promote trunk control, the support provided by pillows or bolster should be weaned and removed. The patient should attempt to prop on both forearms and progress to use of just one forearm for support to promote weight shifts at the shoulders. One may also incorporate dynamic activities of one upper extremity while the body is supported on the other upper extremity. Biofeedback also can be used to promote an upright head position and trunk position in a severely involved patient.

Once the patient gains control for head and trunk in the prone posture, he can be offered progressive-challenging postures like sitting, followed by quadruped, kneeling, and then standing. Rhythmic stabilization for trunk and joint approximation at the patient's hips or shoulders (preferably weight-bearing on these joints rather than manual compression by the therapist) may help to achieve sustained contraction of trunk muscles and, in turn, increase the trunk stability. If the patient cannot sustain a co-contraction of the trunk muscles, a slow-reversal-hold pattern over a steadily decreasing range of trunk rotation might be attempted instead (refer to Chap. 2 D).

If the patient is heavy or weak, promote rolling to one side and push up to sitting. For those patients who have difficulty getting into side-lying, providing resistance in side-lying to flexion, adduction, and external rotation of the shoulder and trunk rotation, the flexion pattern of proprioceptive neuromuscular facilitation (PNF) may initiate the transition (Fig. 5.10). As the strength of the shoulder muscles improves, resistance can be applied to these movements during the transition. In side-lying,

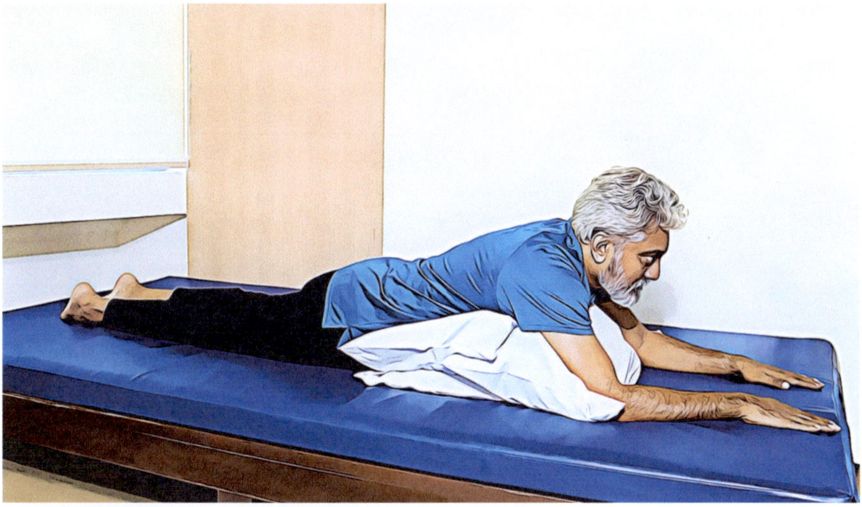

Animated photograph of model with permission

Fig. 5.9 An illustration of the subject in prone while propped on forearms with pillows under his chest

the patient should be instructed to drop his legs off the edge of the table and push up with his arms to a sitting position. The therapist may provide adequate guidance and support during the early phase of training transition. Ideally, the patient should be taught to roll in both directions. Rolling toward the affected side is easy, but the patient may find it difficult to push up to sitting with the involved arm. Straight rising from the supine position using abdominal muscles (rectus abdominis) may also be taught. However, a straight way of rising from the supine can be considered strenuous or demanding for weak or heavy patients. It is also advisable to encourage the patient to assume a sitting position from a prone position, provided he or she is not frail, obese, or with certain underlying comorbidities that are not suitable for prone lying.

The exercises challenging the patient's balance should preferably be done in a safe place such as a therapeutic mat, especially when the involvement is considerable. As the patient's balance and postural control develop, he or she should be progressed from the supine or prone position to sitting. While sitting, the patient should be made to practice weight shifts in all directions, initially with both hands support and then weaning the support off from the hands. The patient should try to maintain balance with arms overhead and while rotating his trunk. The exercise regime depends on the patient's weight, side of involvement, and muscle strength.

The use of trunk stabilization exercises, namely, trunk extension, flexion, and side bridges (Fig. 5.11) on a mat or therapeutic ball, along with the rest of the overground exercises, is advisable for patients with a higher potential or scope for improvement. The function of the truncal musculature is often impaired in patients with cerebellar dysfunction and is frequently overlooked or not addressed in

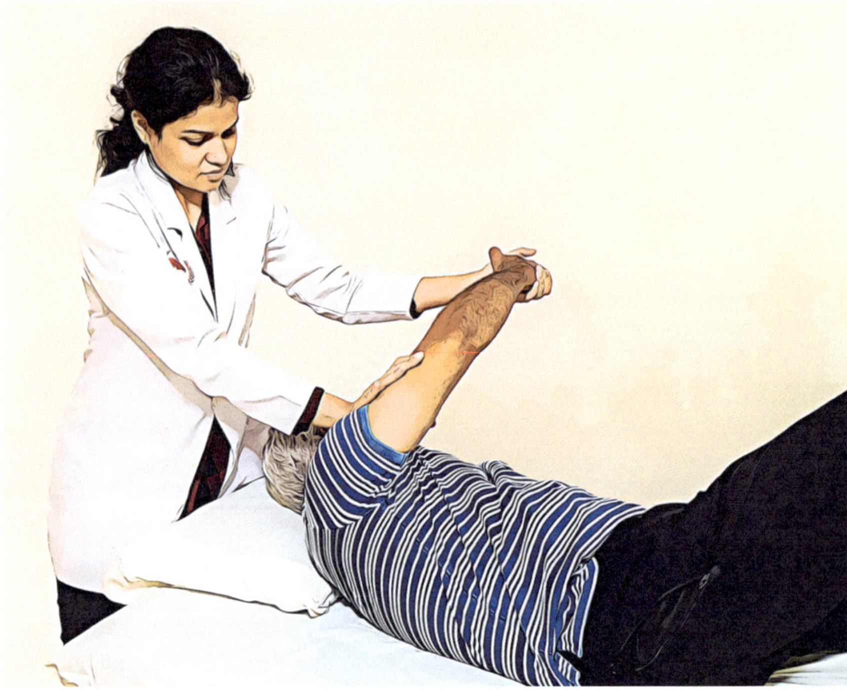

Animated photograph of model and therapist with permission

Fig. 5.10 Providing resistance in side-lying to flexion, adduction, and external rotation of the shoulder and trunk rotation, the flexion diagonal pattern of PNF

rehabilitation. Truncal stability is required for the maintenance of balance and efficient limb control during gait. Muscle strength, endurance, and control are important factors in trunk stability. During kinetic chain activities, rectus abdominis, oblique abdominal muscles, transverse abdominis, erector spinae, and multifidus act as a corset to provide stability and a mobility function. Studies have reported that trunk muscle activity, specifically of transverse abdominis and oblique abdominal muscles, frequently occurs before the activity of the lower extremity musculature, suggesting the role of trunk muscles for a stable foundation for lower extremity movements. Electromyographic studies have specifically demonstrated tonic activity of the transverse abdominis throughout the gait cycle and phasic activity of the remaining muscles during the gait. Cerebellar dysfunction patients may demonstrate truncal ataxia and difficulty in regulating the force and speed of trunk muscle contraction, thus emphasizing the need for trunk stabilization exercise to improve limb coordination, balance, and gait.

If the patient develops adequate sitting balance but is not considered a safe candidate for ambulation, he should be taught as many independent transfers as

Animated photograph of model with permission

Fig. 5.11 An illustration of side bridging

possible. If the patient is safe for ambulation, then improving stability at the hips is a prerequisite for assuming standing posture, and this can be developed by working initially in the crawling or kneeling position (Fig. 5.12) on a mat table. It is also essential to practice weight shifts in the crawling (Fig. 5.13) and kneeling positions.

Preparation for Independent Stance A patient preparing to ambulate should be able to assume a quadruped position from sitting on a mat or prone on elbows position. The therapist can help the patient to achieve this position from sitting or prone position. Stall bars or assistance of a therapist can be used to assume the kneeling position from sitting. Also, encourage patients to shift between kneel sitting and all four limbs. The elderly population may find it difficult to practice such exercises to bring them to crawling, kneeling, and prone positions, and in such cases, the exercise regime may be tailored appropriately to suit the strength and abilities of the patient.

Assuming a standing position from sitting without help is a difficult task for patients with cerebellar dysfunction. The cerebellar ataxic patients may have difficulty in controlling trunk movements while standing. The patient might flex the trunk too forward and lose balance or extend the trunk too early and drop back into the chair or might consistently fall to any side. Stall bars can be used to stand from a kneeling position. Precise instructions, demonstrations, and feedback with plenty of repetitions for sit-to-stand movement may help to minimize the errors while standing.

A tilt table may be more suitable for elderly patients who are wheelchair-bound or bed-confined in order to prepare the person's cardiovascular system for being upright. Once the postural hypotension subsides and the patient can tolerate standing,

Fig. 5.12 Demonstrating an activity to improve the limits of stability in the kneeling position

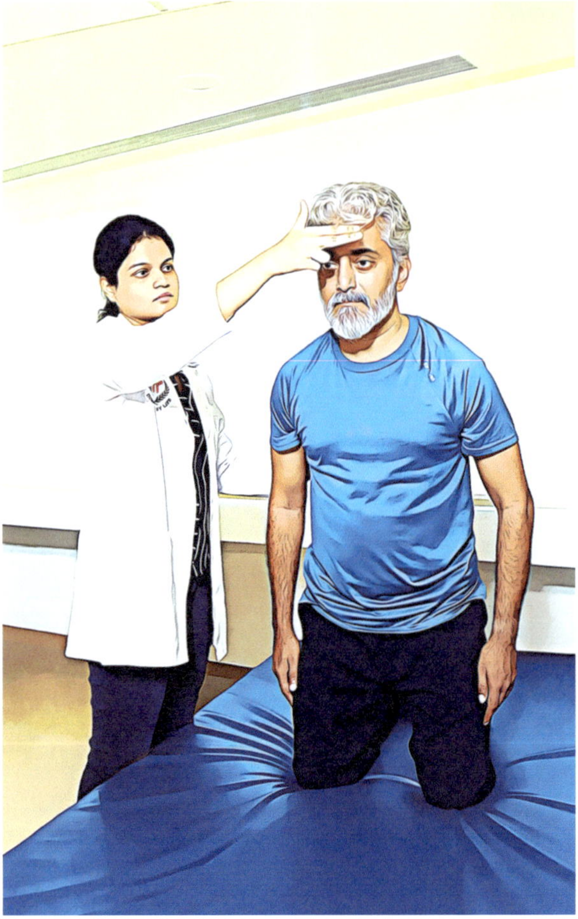

Animated photograph of model and therapist with permission

activities in standing, including weight shifts, reach-outs, and perturbations, should be introduced within the parallel bar. Standing balance can be further improvised by practicing joint approximation through the hips and shoulders and performing rhythmic stabilization for trunk rotators. If rhythmic stabilization is difficult to practice, slow-reversal-hold over a steadily decreasing range of trunk rotation can be used. Vestibular ball exercises in sitting and wobble board exercises in sitting and standing can also be used to improve the balance and stability of the patient. The use of perturbations and activities like reach-out on a firm surface advancing toward an unstable surface like a foam mat or surface can further facilitate balance reactions in standing. For reducing postural tremors, which are often seen in cerebellar patients, the therapist can advise the use of ankle weights or weighted belts. Promoting co-contraction of the axial and proximal girdle muscles using slow-reversal-hold and rhythmic stabilization techniques may help in reducing postural tremors.

Animated photograph of model and therapist with permission

Fig. 5.13 Demonstrating weight shifts in the quadruped position

5.8.2.2 Balance and Gait Retraining

Physiotherapy intervention is the primary treatment for gait ataxia and imbalance in cerebellar dysfunction patients. While walking, the gait ataxia can be evident as uneven step and stride length, the tendency for high step pattern, and decreased push-off and veering. Such gait abnormalities can often be taxing and unsafe for mobility. Few months of intense training have shown improvement in motor skills, balance, and gait, suggesting that patients with cerebellar dysfunction may benefit from long-term motor training.

The exercise program should focus on static and dynamic balance activities for those patients whose balance deficits are the major reason for gait ataxia than the lower limb coordination deficits. Exercises for highly functional cerebellar dysfunction patients should include balance training in sitting and standing. More severely affected patients may require hand support or belts or harness for safety while training in standing. Evidence suggests that the level of challenge to balance is more important than the duration of balance training in producing positive effects. Since balance and limb coordination vary from patient to patient, an individuated tailored training program is important to optimize the challenge and ensure safety. Evidence suggests that caregiver supervised home-based balance exercise programs can improve the locomotor and balance abilities in patients with degenerative cerebellar disease. For those patients with safety issues, the balance exercise can begin in sitting. Initially, exercise should be performed while sitting on a firm surface and then progressively challenge on 6-inch dense foam, then an exercise ball, and finally on a balance disc. Stabilizing exercises should precede the mobility (dynamic) exercises within a challenging position. The use of whole-body controlled video games (exergaming) training for acquired and mild-to-moderate cerebellar degenerative ataxias is a highly motivational and playful strategy to train dynamic balance.

If the patient can stand without losing balance, he should be promoted to practice lifting the feet off alternately. Music can be used for promoting rhythm. A mirror is beneficial in providing visual feedback to the patient regarding the posture. It enables the patient to correct any deviation or swaying in any direction by providing additional sensory input. To begin ambulation in parallel bars, the patient may need precise verbal feedback about the step length, body rotation, accessory movements, and trunk position. It is essential to isolate one problem at a time to correct the errors. Plenty of practice is usually required in these cases.

Once the patient is ready to walk outside the parallel bars, the therapist should decide whether the patient needs any ambulatory aid for walking. An ataxic patient may have difficulty controlling his own body, and the aid may become an added obstacle. Walker is the most stable walking aid, but the patient should have adequate control to place his legs and the walker at the correct distance and with proper timing. Crutches or canes need the reciprocal movement of the arms and legs with appropriate timing and placement, which can be problematic for the patient. The patient may do better with tall poles for support rather than canes or crutches. The selection of walking aid should consider both the safety and the mobility aspects and finalize the appropriate walking aid based on the patient's requirements and the abovementioned aspects.

Finally, the outcome of therapy can be assessed by measuring the progression in ambulation by recording the number of times they lose their balance in a treatment session, the frequency of a specific error, the distance ambulated, or the level of assistance needed. For those patients who have a potential for community reintegration and participation, encourage walking on various surfaces (even and uneven terrain, hard and loose soil and lawn), promote negotiation of curbs and obstacles, and provide opportunities to walk at various speeds, directions, and while carrying objects. In addition to the above, adding lightweights on extremities while walking, walking in a crowded place, and stair climbing can further improvise their independence.

5.8.2.3 Exercises to Minimize Incoordination and Dysmetria

Activities necessary to increase proximal stability and improve balance are essential prerequisites for reducing limb or trunk ataxia. The details regarding the strategies to improve postural stability and balance have already been dealt with. To date, there is no therapeutic regime that can completely overcome or eliminate dysmetria. Frenkel's exercise can be useful as a strategy to minimize dysmetria of the upper and lower extremities. These exercises were originally developed in 1889 to treat tabes dorsalis patients with problems of incoordination due to loss of proprioception (sensory ataxia). Although Frenkel's exercises are classically described for the lower extremities, they can be modified for the upper extremities. Table 5.10 provides details about Frenkel's exercises performed in supine, sitting, and standing positions for the lower extremities.

It is rare that all the available neural pathways required for the generation of coordinated movements are blocked or damaged following cerebellar lesions. Improvement that results from Frenkel's exercises is possibly due to increased use of the pathways which remain or use of other regions of the brain, including the

Table 5.10 Frenkel's exercises for cerebellar dysmetria and movement incoordination

Starting position	Exercises
Supine, on a therapeutic mat or treatment table	• Abduction and adduction of the hip as smooth as possible with the knee and hip held in extension • Flexion and extension of the affected lower limb at the knee, sliding the heel down and up in a straight line on the treatment table or therapeutic mat • Abduction and adduction of the hip as smooth as possible with the knee bent, the heel on the therapeutic mat • Flexion of the hip with the knee held in extension, heels off the therapeutic mat, toward a target nearby (examiner's fingertip or palm) • Flexion and abduction of the hip with the knee held in extension, heels off the therapeutic mat, toward a target nearby like the examiner's fingertip or palm (Fig. 5.14) • Flexion and abduction of the hip along with flexion of the knee, heels off the therapeutic mat, toward a target nearby like the examiner's fingertip or palm (Fig. 5.15) • Alternate hip knee flexion and extension of each leg, with heels on the table • Simultaneous, but alternate hip knee flexion-extension of both legs, with heels on and off the table (like cycling movement) • Flexion and extension of one leg while the other moves toward abduction and adduction
Sitting with back support	• Placing a foot on a mark or target (Fig. 5.16). Distance and the position of the target can be changed as the treatment sessions advance • Sit-to-stand transition with knees approximated, feet with hip width apart • Subcomponents of sit-to-stand transitions: scooting or hitchhiking the buttocks forward, placing the feet with hip width apart, forward-leaning of trunk upon pelvic, the extension of legs followed by extension of the trunk can be separately trained to improve the timing, range, and accuracy before the complete task of sit to stand is taught
Standing	• Weight shifts forward and backward and to either side • Placing the foot on a mark or target or a footprint (Fig. 5.17) • Marching in place with emphasis on timing • Walking within parallel bars with footprints in a straight line • Walk out of the parallel bars with footprints marked on the floor • Walking between two parallel lines (Fig. 5.18) • Walking on a trace of circle • Walking sideways and backward

Fig. 5.14 A targeted activity for the lower extremity. Note the flexion and abduction of the hip with the knee held in extension, the heels off the therapeutic mat, toward the examiner's fingertip

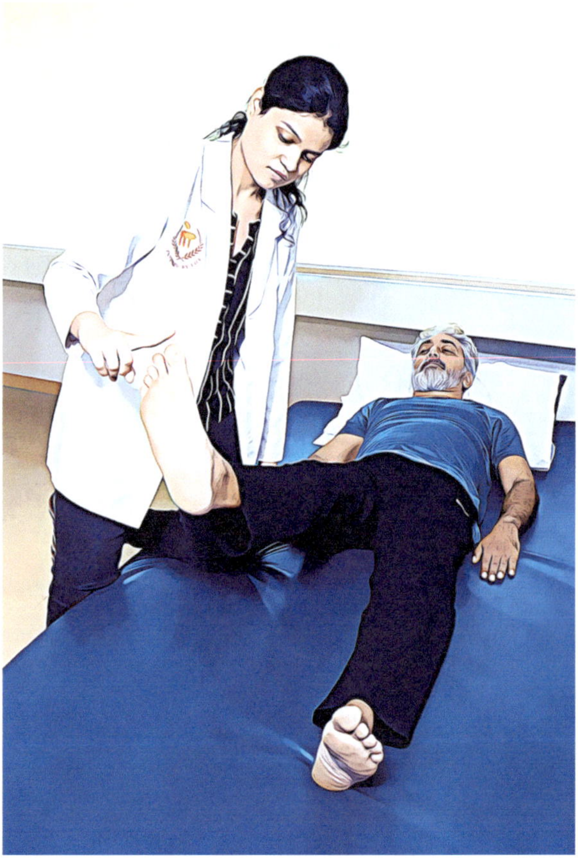

Animated photograph of model and therapist with permission

cerebral cortex, that may compensate to a certain extent the loss of cerebellar function. Concentration or attention, precession, and repetition, being the key elements of Frenkel's exercise, may provide an opportunity to use such available neural circuits or develop an alternative route. Eventually, it is the repetition or practice of the task that helps in establishing the new or alternate pathway for normal or near-normal coordinated movements.

Frenkel's exercises for the upper extremities include tracing circles or squares, coloring pages, arranging cubes one over the other, fixing the pegs in the pegboard, and placing one plastic cup into another cup. It may be worthwhile to practice functional task-specific movements with ample repetitions. The tasks must be progressed by increasing the accuracy demands depending on the performance of the patient. Feedback in the form of knowledge of result or knowledge of performance is invaluable in enabling the patient to relearn motor skills.

For the lower extremities, weight-bearing exercises help to improve coordination. Other interesting, challenging actions like throwing a ball to a target,

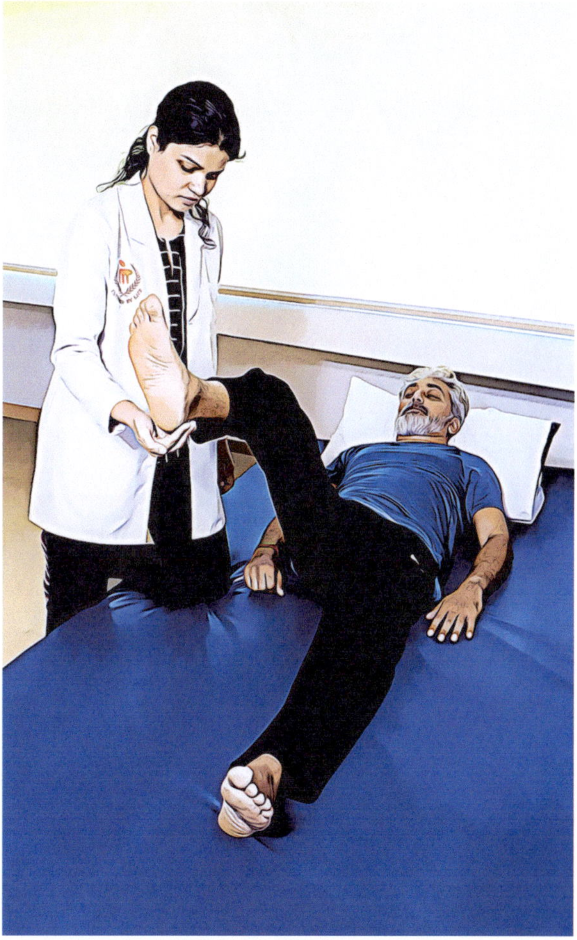

Fig. 5.15 Another targeted activity for the lower extremity. Note the flexion and abduction of the hip with the knee held in flexion, heels off the therapeutic mat, toward the examiner's palm

defending a wicket using a cricket bat (the bat can be used as an intermittent prop for balance), and walking on a treadmill may also be practiced. Treadmill walking imposes a constant and predictable external timing, and the amount of weight-bearing through the legs can be controlled by the use of arms for support or by the suspended harness. Each of these activities makes certain demands on the individual, which help to re-establish control with practice.

Both for acute and chronic ataxia, a training program can only be successful if ample opportunity of supervised and/or unsupervised practice is provided. As the person gains more control of a particular task, it is also essential to increase the complexity of the task. There are several ways in which the therapist can increase the complexity of the exercise or activity to push the individual to the best limits of their performance (Box 5.7).

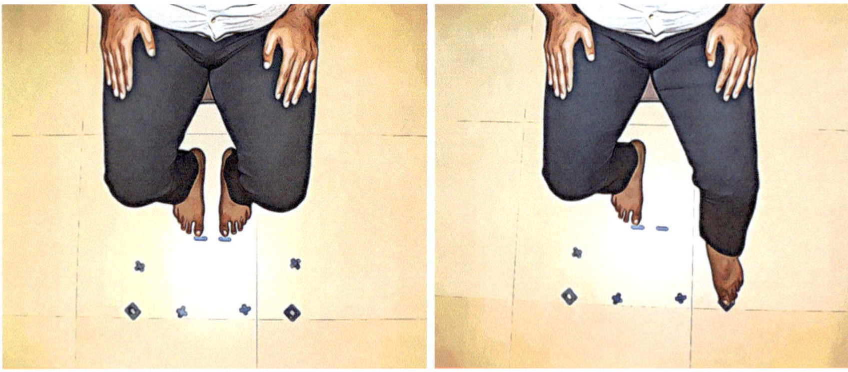

Animated photograph of model with permission

Fig. 5.16 Placing the foot on a mark or target while sitting with back support

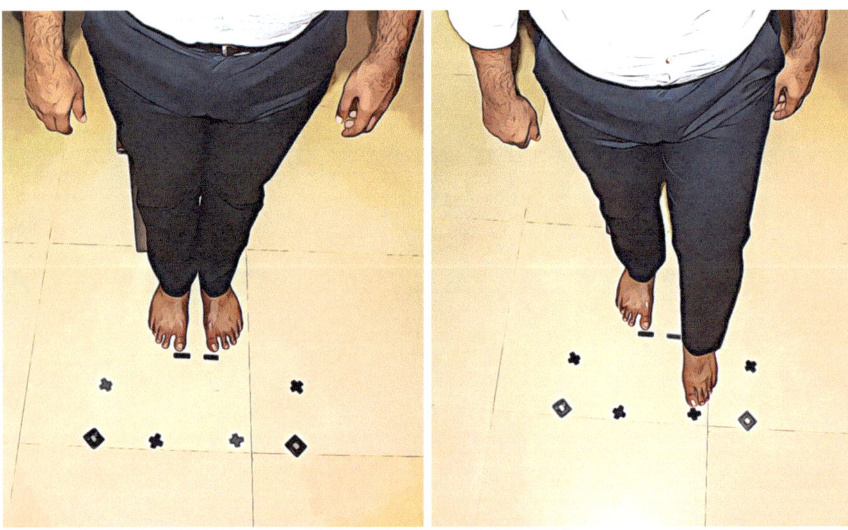

Animated photograph of model with permission

Fig. 5.17 Placing the foot on a mark or target while standing

Box 5.7 Methods to increase the complexity of activities

Methods to Increase the Complexity of the Activities
1. Withdrawal of external control and guidance
2. Reduce and wean off the external support provided
3. Encourage changes in amplitude, direction, and force
4. Add complexity in the task that requires speed alterations
5. Introduce starts and stops at different points in the movement
6. Encourage bilateral involvements and dissociated exercises
7. Increase the balance requirements

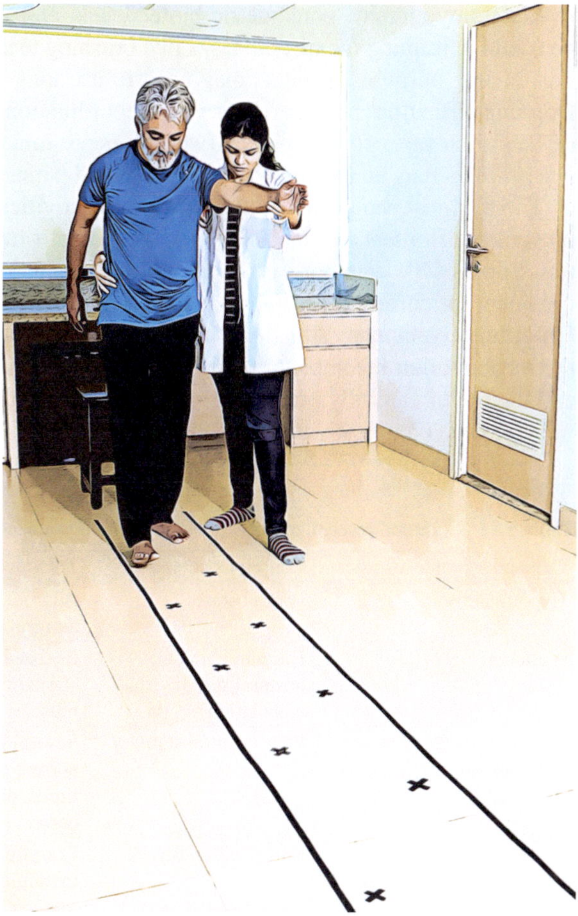

Animated photograph of model and therapist with permission

Fig. 5.18 An illustration of walking between two parallel lines with marks to place the feet at an equal distance

PNF patterns for the arms at varying speeds (repeated contraction) may be useful to improve the control for functional tasks. The use of PNF techniques such as rhythmic stabilization, slow-reversal-hold, joint approximation, and facilitation of isometric contraction in various developmental postures may help to improve the posture and balance, thus reducing the dysmetria secondary to postural instability. Perturbations and wobble board exercises may also help to minimize the dysmetria accompanying balance impairment.

Weights added to the extremities may reduce distal dysmetria. Coordination exercises with intense gait training and perhaps the use of weighted wrist bracelets and ankle bracelets to decrease the excursion of ataxia may be necessary for those patients who have a severe intention tremor. The weights required vary from patient to patient, and the optimal amount of weight has to be determined by trial and error. Usually, heavier weights are needed for greater movement error, but if the weight is beyond the required, the dysmetria may become worse.

Electromyography, goniometric biofeedback, and virtual reality exercises may be useful in training for specific tasks like brushing teeth, combing hair, and eating. For feeding purposes, patients may need to use long handle spoons and utensils. Dressing difficulties produced by severe incoordination may require the substitution of Velcro closures for buttons and shoelaces. Automatic devices such as electric page-turners may be useful for severely affected patients.

The therapist should select the therapeutic activities that would best correct the movement disorders and hence improve the patient's functional mobility and activities. Some of the common cerebellar motor dysfunctions, their consequences, and treatment for correcting or improving the motor behaviors are given in Table 5.11. Unfortunately, in many situations, the task for the therapist is often challenging and the rehabilitation incomplete. However, with patience, persistent training in ADL, and the use of carefully prescribed self-help devices, the patient may be made partially or even completely independent.

Table 5.11 Common cerebellar motor dysfunctions, consequences, and management

Cerebellar motor disorder	Consequence(s)	Treatment
Hypotonia	Poor posture	Activities to enhance the tone of the antigravity muscles
Asthenia	Fatigability, poor postural stability, and ambulation	Resisted exercises to enhance the strength of antigravity muscles and muscles for locomotion
Dysmetria in the absence of hypotonia and asthenia	Poor postural stability	Several sessions of practice consisting of correct posturing of the trunk, upper limbs, or legs in an attempt to make correct posture an automatic function
Dysmetria	Gait abnormality	Coordination exercises for the extremities and gait training
Selective dysfunction of cerebellar motor program for gait in the absence of dysmetria, hypotonia, or asthenia	Gait disturbance	Several sessions of gait training may reduce the gait disturbance
Improper timing	Impaired reaction time, movement time, and termination time	Practicing movements under time constraints; perform movements to music or metronome; timing tasks; external feedback; brisk icing or tapping
Scaling problems	Inability to grade muscle forces to task demands	Picking a paper cup full of water; lifting lightweight objects; attempts to perform short-range movements (difficulty level based on the patient's potential)

Further Reading

Belas dos Santos M, Barros de Oliveira C, dos Santos A, Pires GC, Dylewski V, Arida RM. A comparative study of conventional physiotherapy versus robot-assisted gait training associated to physiotherapy in individuals with ataxia after stroke. *Behav. Neurol.* 2018:2892065. Available from: https://doi.org/10.1155/2018/2892065

Bodranghien F, Bastian A, Casali C, Hallett M, Louis ED, Manto M, et al. Consensus paper: revisiting the symptoms and signs of cerebellar syndrome. *Cerebellum.* 2016 Jun;15(3):369–91. Available from: https://doi.org/10.1007/s12311-015-0687-3.

Brodal P: The Central Nervous System: Structure and Function. New York: Oxford University Press, 1992.

Carr JH, Shepherd RB. Neurological rehabilitation-optimizing motor performance. Oxford: Butterworth-Heinemann; 1998.

D'Angelo E. Neural circuits of the cerebellum: hypothesis for function. *J Integr Neurosci.* 2011;10(3):317–52.

De Jong RN. The neurologic examination. 6th ed. Baltimore: Lippincott Williams & Wilkins; 2005.

Donaghy M, editor. Brain's Diseases of the nervous system. 12th ed. New York: Oxford University Press; 2009.

Freund JE, Stetts DM. Use of trunk stabilization and locomotor training in an adult with cerebellar ataxia: a single system design. *Physiother Theory Pract.* 2010;26:447–58.

Gill-Body KM, Popat RA, Parker SW, Krebs DE. Rehabilitation of balance in two patients with cerebellar dysfunction. *Phys Ther.* 1997;77:534–52.

Gillen G. Improving activities of daily living performance in an adult with ataxia. *Am J Occup Ther.* 2000;54:89–96.

Hausen HS, Lachmann EA, Nagler W. Cerebral diaschisis following cerebellar hemorrhage. *Arch Phys Med Rehabil.* 1997;78:546–9.

Hewer RL, Cooper R, Morgan MH. An investigation into the value of treating intention tremor by weighting the affected limb. *Brain.* 1972;95:579–90.

Ilg W, Synofzik M, Brotz D, Burkard S, Giese MA, Schols L. Intensive coordinative training improves motor performance in degenerative cerebellar disease. *Neurology.* 2009;73:1823–30.

Keller JL, Bastian AJ. A home balance exercise program improves walking in people with cerebellar ataxia. *Neurorehabil Neural Repair.* 2014;28:770–8.

Kelly G, Shanley J. Rehabilitation of ataxic gait following cerebellar lesions: applying theory to practice. *Physiother Theory Pract.* 2016;32:430-7. Available from: https://doi.org/10.1080/09593985.2016.1202364

Knott M, Voss DE. Proprioceptive neuro-muscular facilitation. 3rd ed. Philadelphia: Harper & Row; 1985.

Lance JW, McLeod JG. A physiological approach to clinical neurology. 3rd ed. London: Butterworths; 1981.

Manto M, Schmahmann JD, Rossi F, Gruol DL, Koibuchi N. editors. Handbook of the cerebellum and cerebellar disorders. Dordrecht: Springer; 2013.

Marsden J, Harris C. Cerebellar ataxia: pathophysiology and rehabilitation. *Clin Rehabil.* 2011;25:195–216.

Martin C, Tan D, Bragge P, Bialocerkowski A. Effectiveness of physiotherapy for adults with cerebellar dysfunction: a systematic review. *Clin Rehabil.* 2009;23:15–26.

Martins C, Rodrigues E, Oliveira L. Physical therapy approach to spinocerebellar ataxia: a systematic review. *Fisioter e Pesqui.* 2013;20:293–8.

Ropper AH, Samuels MA, Klein JP, Prasad S, editors. Adams and Victor's Principles of neurology 10th ed. New York: McGraw Hill Education; 2014.

Sartor-Glittenberg C, Brickner L. A multidimensional physical therapy program for individuals with cerebellar ataxia secondary to traumatic brain injury: a case series. *Physiother Theory Pract*. 2014;30(2):138–48.

Schmitz-Hübsch T, du Montcel ST, Baliko L, Berciano J, Boesch S, Depondt C, et al. Scale for the assessment and rating of ataxia: development of a new clinical scale. *Neurology*. 2006 Jun 13;66(11):1717–20.

Snell RS. Clinical neuroanatomy for medical students. 4th ed. Philadelphia: Wolters Kluwer; 2019.

Synofzik M, Ilg W. Motor training in degenerative spinocerebellar disease: ataxia-specific improvements by intensive physiotherapy and exergames. *Biomed Res Int*. 2014;27:889–96.

Trombly CA, Scott AD. Occupational therapy for physical dysfunction. 7th ed. Baltimore: Lippincott Williams & Wilkins; 2014.

Umphred DA, Burton GU, Lazaro RT, Roller ML, editors. Umphred's neurological rehabilitation. 6th ed. St. Louis: Elsevier Mosby; 2013.

Chapter 6
Traumatic Brain Injury

Abraham M. Joshua, K. Vijaya Kumar, and R. Chandrasekhar Naidu

6.1 Introduction

Traumatic Brain Injury (TBI) is defined as an alteration or disruption in brain function caused by an external force, which can be a bump, blow, jolt, or penetrating injury to the head. Damage to brain tissue may be either primary or secondary. Primary damage involves brain tissue distortion at the time of injury and direct trauma to the blood vessels, nerves, axons, and glial cells. Secondary damage is caused by a cascade of biochemical, cellular, and molecular level events that develop over a period due to hypoxia, ischemia, edema, hydrocephalus, and increased intracranial pressure (ICP), which are linked to the initial injury. Both primary and secondary injuries are not distinct and can occur concurrently and result in temporary or permanent neurologic deficits.

6.1.1 *Epidemiology*

Globally, TBI is considered a "silent epidemic" and among various traumatic injuries, TBI is one of the most devastating types, causing varying degrees of paralysis, loss of consciousness, amnesia, and even death. TBI also creates financial burdens

A. M. Joshua (✉) K. Vijaya Kumar
Department of Physiotherapy, Kasturba Medical College of Mangalore, Manipal Academy of Higher Education, Manipal, India
e-mail: abraham.joshua@manipal.edu; vijay.kk@manipal.edu

R. C. Naidu
Krishna Institute of Medical Science Hospital, Secunderabad, India

A. M. Joshua (ed.), *Physiotherapy for Adult Neurological Conditions*,
https://doi.org/10.1007/978-981-19-0209-3_6

for families and societies through lost productivity and high healthcare costs. The global burden of TBI increased significantly between 1990 and 2016. Definitional and diagnostic ambiguity surrounding TBI and underreporting of TBI have resulted in inconsistent reports and difficulty in accurately estimating the prevalence and incidence rate in many parts of the world.

Traumatic injury of the brain affects all communities, all age groups, and all societies worldwide. The burden caused by TBI to the patients, relatives, caregivers, and society at large is increasing exponentially. TBI, by definition a heterogeneous disease, has important regional variations in its epidemiologic characteristics and outcomes. There is a great lack of adequate, comprehensive data about the incidence of TBI in any community, most of all in communities that have the major burden of neurotrauma. TBI is a particular burden on developing countries, which have the least capacity to manage it. Successful prevention and reduction in the incidence of brain trauma can occur only by greater political action and public awareness and engagement of societies.

A 2018 study titled "Estimating the Global Incidence of Traumatic Brain Injury" notified that each year nearly 69 million people endure TBI worldwide. The study reported that road traffic accidents were the principal cause of brain injuries, followed by a fall from heights. Lower- and middle-income countries reported three times more TBI cases as compared to higher-income countries. The proportion of TBIs resulting from road traffic accidents was greatest in Africa and Southeast Asia and lowest in North America. However, the overall incidence of TBI per 100,000 population was greatest in North America (1299 cases) and Europe (1012 cases) and least in Africa (801 cases). Epidemiological studies from India reported that TBI is a leading cause of mortality, morbidity, and socioeconomic loss and occurred primarily in males, with no obvious predilection for the younger and the older age groups. Studies also reported that road traffic accidents were the main cause of TBI among young people, whereas falls were the main reason among the elderly.

6.2 Etiology of TBI

Though various mechanisms cause TBI, the commonest causes include road traffic accidents, falls, sports-related injuries, assaults, and penetrating trauma. Road traffic accidents or motor vehicle accidents and sports and leisure activities related to TBI are frequent among young adults and middle-aged people. Within the above-mentioned age groups, the predilection for sustaining TBI is higher among men than women and the ratio is approximately 2:1. Trivial fall-related TBI is common among the elderly. Brain injuries related to child abuse and bicycle accidents are the commonest among children and adolescents. Among military personnel, penetrating injuries, blast injuries, and assaults are the common causes of TBI.

TBI is a serious and frequent consequence of violence and is more prevalent in the USA. Since the 1980s, studies in developed countries have noticed a steady decline in the mortality rates and hospitalization following TBI. The decline in TBI-related deaths in road traffic fatalities is perhaps due to the presence of safety equipment in vehicles and increased use of seat belts and helmets.

6.2.1 Mechanism of Injury in TBI

The mechanism of TBI refers to the cause of injury and the resultant physiological or structural damage. There are four primary mechanisms of TBI and they are the direct impact of mechanical force (contact forces), injuries related to rapid acceleration and deceleration (inertial forces), blast overpressure injuries, and penetrating injuries.

Direct Impact Injuries Brain injury resulting from the impact of the head on the object or vice versa, i.e., the contact phenomena appear when the head strikes or is struck by an object. Contact phenomena include local deformations of the skull, which result in local or remote compressive, shear, and tensile strains in the underlying skull and brain. It can occur with or without fracture of the skull. In the case of direct impact injuries to the head, two types of injuries are likely to occur and they are the coup and the contrecoup injuries. In a coup injury, the brain damage occurs right under the impact site, whereas in contrecoup, the injury occurs opposite to the impact site (Fig. 6.1).

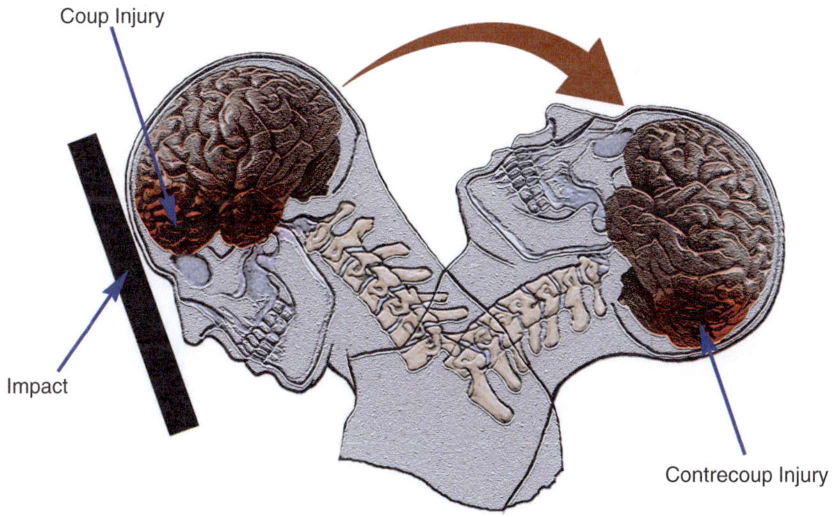

Fig. 6.1 Coup and contrecoup injuries of the brain

Head Injuries Related to Acceleration and Deceleration Forces The severe rotational or angular acceleration-deceleration forces (inertial forces) cause dynamic shearing, tensile, and compression deformation of the brain tissue and result in the diffused axonal injury. Typically, inertial loading occurs in concert with contact phenomena (e.g., the head is struck by or strikes a rigid object). Head accelerations or decelerations of appropriate magnitude, rates of onset, and direction, together contribute to the type and severity of inertial injuries. Though the impact is a common cause for acceleration-deceleration forces, it is not always necessary to cause sufficient head motion to result in mild, concussive-type injuries. Acceleration-based injuries include concussion, diffuse axonal injury, and the most acute subdural hematomas and contrecoup injuries.

Normally, the neuronal tissue is compliant and resilient to mechanical stretch, but the pliable axons will not be able to withstand the sudden uniaxial stretch or elongation, causing damage to the axonal cytoskeleton. The large size of the brain along with variable densities of gray and white matter creates tissue tension in the gray and white matter interface to cause tissue damage. Generally, the damage is more profound in unmyelinated axons, and the events which trigger axonal pathology may eventually lead to delayed axotomy. Though the diffuse axonal injuries are typically deceptive, the lesions can occur in distinct regions, namely the parasagittal white matter near the cerebral cortex, the corpus callosum, and the pontomesencephalic junction adjacent to superior cerebellar peduncles.

Blast Overpressure Injuries The blast injuries are usually found in warfare, where detonation of explosive devices causes an instant rise in pressure and creates a blast wave that starts at the site of the explosion and travels outward. The impact of the blast waves and the extent to which it inflicts brain injury depends on the type and strength of the blast, the distance between the subject and the blast, and whether the exposure is in an open or a closed environment. Blast-induced overpressure injury is the mechanism of injury for which research is emerging, but the exact mechanism(s) that cause brain damage from this loading is largely unclear. The TBI caused by the blast waves can be categorized into primary, secondary, and tertiary.

1. Primary: Blast overpressure has a strong effect on the brain. When the blast wave hits the body, the organs surrounded by fluid, such as the brain and the spinal cord, are susceptible to injury, just like the air-filled organs, such as the ears, lungs, stomach, and intestines.
2. Secondary: Shrapnel or flying debris can cause penetrating or direct impact trauma.
3. Tertiary: When the victim is physically thrown away by the blast and the head strikes the ground or a wall sustaining a direct impact trauma

Penetrating Injuries In a penetrating head injury, the dura mater, the outer layer of the meninges, is breached by high- or low-velocity projectile objects, including shrapnel or knife, or bone fragments driven into the brain following a skull fracture. The penetrating injury produced by high-velocity projectiles causes initial laceration and crushing of brain tissue and cavitation. Damage to blood vessels caused by the high- or low-velocity projectile objects can lead to intracranial hematomas and ischemia.

6.2.2 *Classification*

Based on the pathophysiology, TBI damage can be categorized into primary and secondary injuries. The primary injury occurs during the initial insult and results from the physical displacement of the brain structure(s), whereas the secondary injury occurs gradually and may result from an array of cellular processes.

6.2.2.1 Primary Injury

Direct physical injury to the brain, such as compression, deformation, displacement, shearing, tearing, and crushing, causes primary injury. The contact injuries, the result of brain tissue coming into contact with an object like the inner surface of the bony skull or external object, typically cause contusions, lacerations, and intracerebral hematomas. The primary injury can be either focal or diffuse. The damage due to a contact injury is generally focal and predisposed to certain areas such as anterior temporal poles, frontal poles, lateral and inferior temporal cortices, and orbital frontal cortices. The compression, tension, shearing caused by acceleration, deceleration, or rotational forces within the brain, can cause diffuse axonal injury, tissue tearing, and intracerebral hemorrhages.

Contusions A contusion can occur with or without associated skull fractures. It results from the back and forth movement of the brain within the boundaries of the skull. Contusion results from the mixture of vascular and tissue damage and typically consists of a discrete region of inflamed brain tissue and the presence of ruptured blood vessels with the blood spilled into the inflamed area. Based on the location of the damage with respect to physical impact, the contusions can either be of the coup or the contrecoup type. Most of the energy of insult from a bigger object's impact causes a smaller amount of injury at the collision site and a contrecoup contusion opposite site. In contrast, the insult from a smaller but hard object is inclined to distribute at the insult position to produce coup contusion at the collision site. The contusions occur primarily in the cortical tissue near the sharp ridges on the inside of the skull. The collision of the soft brain surface against the bony protuberances on the inside surface of the skull is the plausible reason for brain contusion. Since the protuberances are more profound under the frontal and temporal lobes and the roof of the ocular orbit, those areas of the brain are the most vulnerable sites for contusions. The contusions are frequently associated with edema and if extensive, are likely to cause raised ICP.

Lacerations Lacerations are tears in the brain parenchyma caused by penetrating sharp foreign objects or pushed-in bone fragments from a skull fracture. Even lacerations can cause blood vessel rupture, bleeding, and swelling in the brain. Lacerations of dura and arachnoid matters can predispose to Cerebrospinal Fluid (CSF) rhinorrhoea, a condition characterized by CSF discharge through the nose worsened by the flexion of the neck or coughing. Both contusion and laceration injuries can damage the cranial nerves. The optic, oculomotor, abducens, facial, and vestibulocochlear nerves are the most common cranial nerves damaged in TBI.

Epidural Hematoma Approximately 10% of the TBI cases are epidural hematomas. It occurs mainly in adults when the meningeal vessel bleeds and causes blood to accumulate between the skull and the dura mater. Most cases of epidural hematomas are accompanied by skull fracture (Fig. 6.2). It typically results from a blow or trauma to the side of the head and is found on the same side of the skull impacted by the blow. The epidural bleeding is often caused by ruptured arteries and tends to progress rapidly. Typically, fracture of the temporal bone can rupture the middle meningeal artery and lead to epidural hematoma (EDH). In such cases, the condition is characterized by a transient phase of loss of consciousness, followed by a brief regaining of consciousness and then slipping to loss of consciousness. A "lucid interval," a feature characterized by a temporary improvement in a patient's condition after a head injury, after which the condition deteriorates, is especially indicative of an EDH. During the brief phase of consciousness, the patient may complain of headache and present with confusion, vomiting, slurred speech, and paresis of the limbs. The condition is a surgical emergency and if not timely diagnosed and managed (evacuation of hematoma through a burr-hole or craniotomy), death is inevitable.

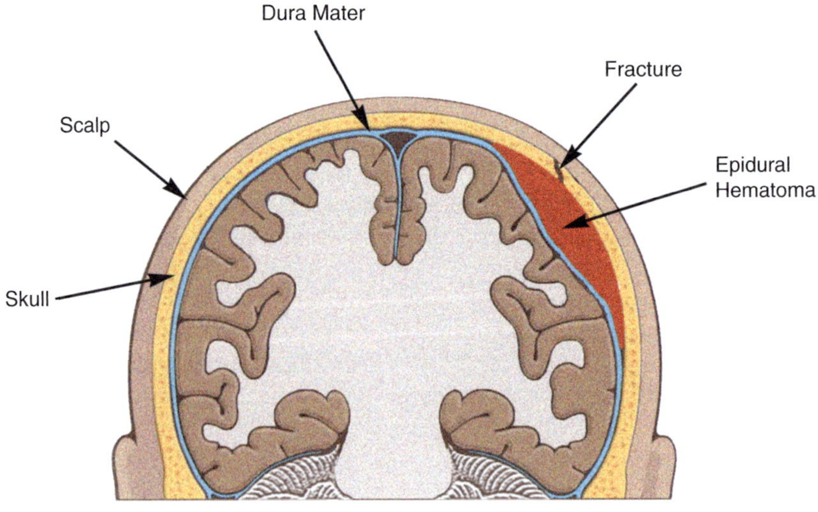

Fig. 6.2 An illustration of an epidural hematoma

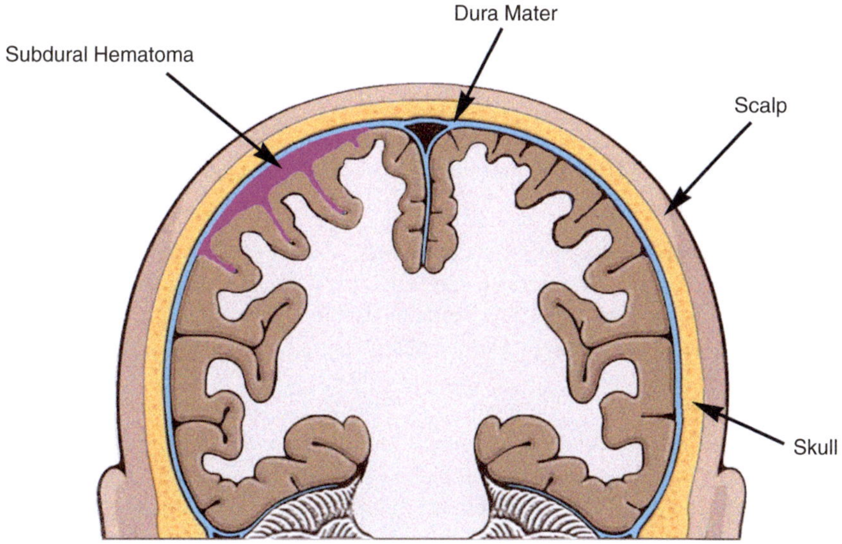

Fig. 6.3 An illustration of subdural hematoma

Subdural Hematomas Blood leaking from a ruptured blood vessel into a space below the dura mater results in a subdural hematoma (Fig. 6.3). The clinical features are similar to those of EDHs. The headache can be dull and unilateral. Acceleration-deceleration injuries typically rupture the bridging veins directed to the superior sagittal sinus leading to a subdural hematoma (SDH). Unlike epidural arterial hemorrhage, which is steadily progressive, venous bleeding is usually arrested by the rising ICP. The blood pooled in the subdural space leads to symptoms of exhaustion and lethargy. The acutely occurring symptoms, such as fatigue and lethargy, are life-threatening. In case of delayed bleeding, it may take several weeks for the clinical symptoms to progress. Compared to EDH, SDHs are not as rapid as the former; however, if not timely diagnosed and managed, the condition can progress to mass lesions resulting in severe morbidity or mortality.

Diffuse Axonal Injuries Diffuse axonal injury (DAI), a severe form of TBI, results from a blunt injury to the brain. The most common etiology of DAI involves high-speed road traffic accidents and primarily affects the white matter tracts in the brain. The shearing forces created by the brain's rapid acceleration and deceleration motion cause microscopic and gross damage to axons near the junction between the gray and white matter. Axonal injury leads to disconnection or malfunction of the neuronal interconnection and mechanical disruption of cytoskeletons, causing proteolysis, swelling, and other microscopic and molecular changes to the neuronal structure. Typically the patients present with bilateral neurological deficits, frequently affecting the frontal and temporal white matter, corpus callosum, brainstem, basal ganglia, and cerebellum.

Approximately 10% of TBI cases admitted to the hospital have a certain degree of DAI and around 25% of diagnosed cases have a fatal outcome. The severity of a DAI determines the clinical presentation. Patients with mild DAI present with signs and symptoms of concussion and the symptoms most commonly include headache, loss of memory, reduced concentration and attention span, and poor sleep. Those patients with a severe DAI may present with loss of consciousness and remain in a persistent vegetative state.

Penetrating Objects Because of shock waves, objects piercing with high speed, such as bullets or explosive shrapnel, can add to the damage distant from the impact areas. Sharp foreign objects at low velocity usually damage the tissues near the vicinity of the penetrating object. Certain details regarding penetrating injuries have already been discussed.

Blast Injuries As previously mentioned, these injuries occur when an explosive object, solid or liquid, explodes and turns into a gas that expands to form a wave of high pressure that moves at an extremely high velocity. This pressure then falls, producing a relative vacuum (blast under pressure wave) resulting in airflow reversal, accompanied by another high-pressure wave in turn. Blast injuries can happen through a variety of mechanisms. The primary blast wave produces severe changes in pressure, which can cause discomfort and damage.

6.2.2.2 Secondary Injury

Though primary injury can be fatal or give rise to severe disability, subsequent deterioration in many cases could be due to cerebral ischemia from brain edema, hematoma formation, hypoxia, and hypotension, which are the leading secondary causes of brain injury. The secondary injury causes are either intracranial or extracranial and can develop several hours to days after the primary injury. The events leading to secondary damage include glutamate neurotoxicity, calcium ion influx into the cells, free radical release, cytokines, and inflammatory responses. Direct trauma, ischemia, or restoration of blood flow following the primary injury can cause the liberation of chemicals, disruption of the blood-brain barrier, edema, and neuronal cell death. Excessive release of glutamate and other excitatory neurotransmitters often worsens the ion channel leakage leading to swelling of the brain and a further increase of ICP.

Depending on the site, the hematoma can be epidural, subdural, or intracranial and is a common factor in secondary brain injury. Hypoxic-ischemic injury can be another factor for secondary brain injury. Systemic hypotension, anoxia, or disruption of the vascular system can contribute to loss of oxygenated blood supply to the brain and result in hypoxic-ischemic injury. Swelling of the brain tissue within the solid skull structure surrounding it, abnormality in brain fluid dynamics, or intracranial hematoma can increase the ICP and predispose to secondary injury.

Rupture of the blood vessels and leak and collection of blood in the ventricular space can lead to acute hydrocephalus. Progressive collection of blood increases the size of the ventricles and pressurizes the cerebral structures between the cranium and the fluid-filled ventricles and worsens the pressure on brain tissue. The high-pressure building up within the cranium will alter the partial pressure of carbon dioxide ($PaCO_2$) level, further damaging the brain tissue.

Increased Intracranial Pressure The brain, CSF, and blood are the three main components within the rigid cranium. According to the Monroe-Kellie doctrine, the contents of the cranium are in a state of constant volume and are fixed. Any increase in the volume of its contents will increase the pressure within the cranial vault and an increase in the volume of one component will be at the expense of one or two of the other components. Normal ICP is between 5 and 15 mm Hg. The dramatic elevation of ICP beyond the normal limits, due to increased volume of the content, can result in decreased cerebral blood flow and/or herniation of the brain tissue, which requires immediate medical and surgical care.

Raised ICP is typically associated with worse outcomes and higher rates of mortality. Raised ICP secondary to cerebral edema or hematoma will distort the brain parenchyma, as the brain gets trapped in the unyielding and rigid bony skull. If the pressure keeps mounting and not relieved, the raised ICP will cause some segments of the brain to herniate. The common sites of herniation include subfalcine, uncal, central, and transtentorial. Even brainstem herniation down the foramen magnum is another usual site. Among the common herniations, subfalcine herniation is the most common and occurs when the brain parenchyma displaces under the falx cerebri.

Cerebral Hypoxia or Ischemia Being a highly metabolically active organ and consuming a significant amount of energy compared to its weight and size, the brain is exquisitely sensitive to hypoxia and hypoperfusion. If prompt and timely intervention is not executed within minutes, the cellular injury can lead to permanent brain injury. Epidemiological studies have revealed that approximately 45% of severe TBI patients experience cerebral hypoxia and have been associated with adverse neurological outcomes. Cerebral hypoperfusion, apnea, and hypoventilation, mostly related to brainstem injury, are the usual causes of TBI-induced cerebral hypoxia. Systemic hypoxia caused by injuries such as obstructed airways, lung puncture, and excessive blood loss can be the extracranial causes for cerebral hypoxia.

Intracranial Hemorrhage Intracerebral hemorrhage is a common complication of TBI and is often difficult to differentiate from hemorrhagic contusion. It results in poor neurologic outcomes and high mortality. The hemorrhage of the intracranial vessels will lead to brain tissue hypoxia supplied by those vessels and besides, will cause strain and distortion to the brain tissue. Within minutes after the injury, the brain bathed by metabolites from impaired cells and blood will trigger tissue edema and ischemia, predisposing to cell death and brain tissue necrosis.

Electrolyte Imbalance and Acid-Base Imbalance Electrolyte imbalances are common after TBI. Potassium derangements after severe TBI affect the normal buffering mechanisms needed to maintain ionic homeostasis and resting membrane potential. Prolonged ionic dysfunction leads to cytotoxic swelling, a major cause of raised ICP and poor outcome after severe TBI. In many TBI patients, the incidence of sodium disorders is quite high during the first few days. Around 15–30% of TBI patients have hypothalamic-pituitary dysfunction, characterized by growth hormone deficiency and adrenocorticotropic hormone, thyroid-stimulating hormone, and gonadotrophin deficiency. Inappropriate secretion of antidiuretic hormone secretion or cerebral salt-wasting syndrome characterized by natriuresis may result in hyponatremia. The activity of brain natriuretic peptide, a potent diuretic, natriuretic, and vasodilator agent, can also cause hyponatremia. Severe TBI patients can also develop hypernatremia over the course of their stay due to impaired sensorium, altered thirst, and central diabetes insipidus.

Reduced cerebral blood flow, cerebral hypoxia, and excessive release of excitatory amino acids, mainly due to the failure of oxidative energy metabolism, may predispose to the development of cerebral acidosis. Acidosis has been associated with the formation of intracellular edema, increased calcium ion concentrations, enzyme inactivation, mitochondrial impairment, and free radical formation. The activation of sodium or hydrogen ion antiporter system together with a passive chloride or bicarbonate ion antiporter enables an increased inflow of sodium and chloride ions and osmotically obligated water into the cells, which is believed to contribute to edema formation during acidosis. In addition to the above, acidosis may interfere with mitochondrial ATP generation. Dislodgment of Ca^{+} from its protein-binding sites increases the free calcium ion levels, which can impair mitochondrial metabolism. Lower pH values can promote the formation of free radicals, leading to delayed cell death after TBI. Cell death associated with electrolyte imbalance and acidosis can occur days, weeks, or months following the primary injury.

Infection Following Head Injury The TBI can be either of the closed or open types. The skull is intact in a closed head injury, whereas open injury (penetrating injury) is characterized by a fracture of the skull. A fractured skull increases the risk of developing an infection as it can enable bacteria to enter the brain and cause infection of the meninges and could spread to the rest of the nervous system if untreated. Though intracranial infection, such as meningitis, encephalitis, and brain abscesses, is a less common aftermath of TBI, it is considered to be a poor prognostic factor for a head injury. In a study by Lin et al., the intracranial infection rate was 6.54%, and the CSF leak and lumbar or ventricular drainage were noted as the risk factors for the infection. The duration of the drainage also increased the risk tendency, particularly when the drainage exceeded 3 days. Invasive ICP monitoring and multiple craniotomies may also pose a threat to intracranial infection.

The occurrence of meningitis is estimated to be around 1.5% in TBI but could be higher in the case of unstable skull fractures. The risk of developing meningitis post-brain injury is the highest during the first few weeks. In response to infection when inflammatory mediators like tumor necrosis factors are released into the CNS, the chemotactic and adhesion molecules along with bacterial components can lead to the influx and/or activation of leukocytes and glial cells. The tissue-damaging substances (proteases and oxidants) produced by the polymorphonucleocytes and macrophages trigger several processes, including vascular changes and cytotoxic or vasogenic edema contributing to brain damage. The CSF block, secondary to inflammation and exudation, can trigger hydrocephalus and raised ICP, predisposing further brain damage. The TBI patients are also considered to be at a higher risk for nosocomial infections compared to other neurosurgical and non-neurosurgical critically ill patients and the infection-related mortality rates can be as high as 28%.

Post-TBI Seizures Seizures are typically an indication of a severe TBI and are mostly seen acutely within 6 months following the TBI. Seizures that occur shortly after a TBI may further damage the vulnerable brain as a result of reduction in the amount of oxygen available to the brain, release of excessive excitatory neurotransmitters, increased metabolic need of the brain, and raised ICP. Approximately 7% of patients suffer at least one episode of seizure post-brain injury. The post-TBI seizures are divided into immediate seizures (occurring within 24 h of brain injury), early seizures (between a day and a week after brain trauma), and late seizures (occurring beyond 1 week after brain trauma). Immediate seizures are believed to be a direct effect of brain injury. Cerebral edema, intracranial hemorrhage, contusion, and laceration are the plausible causative factors for early seizures. Restructuring of the neural networks, presence of blood within the brain, changes in the blood-brain barrier, excessive release of excitatory neurotransmitters, and free radical injuries to the brain tissue are the usual cause for late seizures. TBI patients may develop a chronic condition characterized by repeated episodes of seizures, known as post-traumatic epilepsy, which is a type of epilepsy.

6.2.2.3 Severity of Brain Injury

Classification of the severity of brain injury guides its management and contributes to the determination of prognosis. The Glasgow Coma Scale (GCS) scores, the duration of Loss of Consciousness (LOC), and the duration of Post-traumatic Amnesia (PTA) are the common indicators of TBI severity. The GCS ratings provide guidance for early care and predict early outcomes (mortality and morbidity) as well as late functional outcomes (functional status and return to employment). A study by Sherer et al. found a strong association between the PTA and the PTA-LOC interval (defined as the interval in days from the end of the period of coma and the return of orientation), a strong association between the LOC and the PTA, and a moderate association between the GCS and the LOC. In addition to the above, more severe GCS categories were associated with longer median LOC, PTA, and PTA-LOC intervals. Age of the patient, pupillary reaction to light, neuroimaging findings, and cranial operations are the other factors that determine the outcomes after TBI. Table 6.1 provides the relationship between TBI and the severity indicators.

Table 6.1 Relationship between TBI and the severity indicators

TBI severity	Glasgow coma scale	Duration of loss of consciousness	Post-traumatic amnesia
Severe	GCS score ≤8	>24 h post-TBI	>7 days
Moderate	GCS score between 9 and 12	>30 min but <24 h	>24 h and <7 days
Mild	GCS score between 13 and 15	<30 min	<24 h

6.3 Clinical Manifestations

The type (diffuse or focal) and severity of TBI, the neuroanatomical structures involved, and the neurophysiological, cognitive, and behavioral changes associated with TBI can lead to a spectrum of symptoms. Among the neurological symptoms of TBI, the state of consciousness is the most important.

Coma and Altered Sensorium Bilateral widespread and diffuse cerebral hemispheric depression of function or depression or destruction of the brainstem-activating system or a combination of the both can be the possible factors contributing to coma or changes in consciousness. The duration of unconsciousness is typically more prolonged in those having moderate to serious injury. Konovalov et al., based on the degree of consciousness, have distinguished the clear consciousness, obtundation, stupor, and coma stages (Table 6.2). The anatomical proximity of the structures responsible for the level of wakefulness and innervation of the eye determines the close relationship between pupillary symptoms, mobility of the eyeballs, and changes in the level of wakefulness. Disorders of the innervation of the pupils are of great importance in predicting the treatment outcome of severe TBI patients. The oculocephalic reflex is of great prognostic value in patients with a head injury. According to Plum and Posner, the neural systems that regulate consciousness differ and are anatomically distant from those regulating motor function.

When a concussion occurs after a minimal impact, LOC may or may not occur with a negligible possibility of retrograde amnesia. Concussion at times can cause DAI with associated temporary or permanent damage. The patients who sustain a concussion can be irritable or distractible, with registration and memory issues, and may complain about headache, dizziness, generalized weakness, alterations in behavior, and emotional states.

In case of a coma, a state of unresponsiveness with complete paralysis of cerebral functions, many patients may show signs of arousal within 2–4 weeks post-brain injury. In the coma state, the oculomotor and pupillary signs may aid in localizing the lesion and determining the depth of the coma. Eliciting brain stem responses which may include grimacing to pain, flexor or extensor motor response, observation of palatal and tongue movements, gag reflex, and other cranial nerve reflexes may further help in localizing brain stem damage.

Table 6.2 The features of different states of consciousness and the relationship with Glasgow coma scale scores

State of consciousness	Description	Glasgow coma scale score
Clear	Arbitrary eye opening, targeted reactions, performance of instructions, correct orientation, free speech	15
Moderate obtundation	Drowsiness, slowness when performing instructions, elements of disorientation, difficult speech	13–14
Deep obtundation	Severe drowsiness, execution of only elementary instructions, disorientation, minimal verbal contact	10–12
Stupor	Eye opening to a pain stimulus, location of pain, inability to perform instructions, no verbal contact	8–9
Moderate coma	Unarousable unresponsiveness, uncoordinated motor response to a pain stimulus, possible spontaneous motor activity, impaired sphincters regulation	6–7
Deep coma	Unarousable, unresponsiveness, no motor activity	4–5
Terminal (atonic) coma	Unarousable, unresponsiveness, deep muscle atony, areflexia	3

Delirium, a state when a severe TBI patient recovers from the unconscious level, is characterized by unrealistic fears, disoriented behaviors, misinterpretation of stimuli, and agitated mood. The delirious patients demonstrate confusion, distractibility, memory issues, and delayed responses. Persistent vegetative state, another disorder of consciousness, is a condition where the patient with severe brain injury is in a state of partial arousal (a condition of wakefulness without awareness) but has reduced responsiveness and absence of cortical functions. In the persistent vegetative state, patients cannot speak or respond to verbal stimulation and may stay in that state for weeks, months, or years and are unlikely to achieve an independent outcome following 3 months of vegetative state. Many persistent vegetative state patients are unresponsive to external stimuli, may often open their eyes in response to feeding, and may track with their eyes and show minimal spontaneous motor activities, which may appear purposeful. Though cortical functions like speech and purposeful movements are lost, the brainstem functions such as breathing, cardiovascular, and hemodynamic functions are preserved. Some patients may demonstrate sleep-wake cycles and may reveal behaviors like grinding the teeth, crying, grimacing, moaning, or screaming without any definite association with the external stimulus. The persistent vegetative state might result from diffused cerebral hypoxia or severe and diffused damage to the white matter following a physical impact. One year post-TBI, the irreversible persistent vegetative state is classified as a permanent vegetative state.

The comatose and vegetative state following TBI occurs in one out of eight patients and has a long-lasting impact on the social and economic aspects of their lives. Coma refers to a state of absent perception of oneself and one's environment, from which one cannot be aroused. It is a state of complete unresponsiveness, without spontaneous eye opening and inability to awaken even after vigorous sensory stimulation. The World Federation of Neurosurgical Societies (WFNS) based on the pertinent neurological disturbances, has proposed four grades of coma. Grade 1 coma is characterized by coma without any of the neurological disturbances. Grade 2 coma features coma with lateralizing signs, unilateral fixed and dilated pupil, or hemiparesis. Grade 3 has coma with pathological extensor responses, whereas grade 4 has coma with bilateral fixed and dilated pupils.

In 1974, two neurosurgeons, Graham Teasdale and Bryan Jennett, first published the GCS. The GCS is used to objectively describe the level of impaired consciousness in all types of medical and trauma patients. By the 1980s, the use of the GCS became widespread and the first edition of Advanced Trauma and Life Support recommended its use in all trauma patients. Additionally, in 1988, the WFNS used it in its scale for grading patients with subarachnoid hemorrhage. Since then, the GCS and its aggregate score have been incorporated in numerous clinical guidelines and scoring systems for trauma victims and critically ill patients. Although GCS was proposed in 1974, the scale did not contain a precise definition of coma, however, several years later based on the numerical value, the coma has been defined as a GCS score of eight or below.

During the early phase after TBI, the patient's emergence from a coma is as unpredictable as a sudden worsening of consciousness from full alertness to coma. As a rule, in the first 24 h after brain injury, solely on clinical grounds, GCS is inappropriate to prognosticate. Early clinical classification of TBI as mild, moderate, or severe, based on the GCS score at the time of the injury and at variable times (6, 12, or 24 h) thereafter, is widely used but not adequately correlated with outcomes. Age of the patient and duration of coma are of major prognostic significance. Additional neurological disturbances noted during the first 24 h can be of prognostic relevance. For instance, mortality is 90% or above if both the pupils are nonresponsive and dilated, 50–60% if abnormal extensor responses are present, 30–50% if one of the pupils is fixed and dilated, and 5–10% if none of these are the case. DAI could be the cause of post-traumatic coma if the CT reveals no hematoma exerting a mass effect and the patient remains in a coma for 6 h or more after the brain trauma. Coma could be attributed to damage of neural pathways that ascend from the brainstem to the hemispheric white matter. Current evidence suggests that post-traumatic coma is due to brainstem dysfunction rather than hemispheric axonal disruption.

Cognitive, Personality, and Behavioral Changes Cognitive, affective, and behavioral symptoms are the common sequels of TBI. The cognitive dysfunctions following mild TBI or concussion, irrespective of any age group, include inattention, reduced information processing speed, executive dysfunction, and learning and memory deficits. Evidence supports the understanding that cognitive recovery following mild head injury is good among adolescents. In most cases, cognitive deficits are typically resolved within 10 days. However, the cognitive recovery trajectory is less clear and predictable among moderate-to-severe TBI patients. The cognitive deficits secondary to moderate-to-severe brain injury also include language disabilities, visuoperceptual dysfunctions, reduced attention and concentration, and motor dysfunctions.

To a certain degree, the association between cognitive deficit and brain injury is moderated by the mechanism of injury, extent of cerebral tissue loss, area of the brain affected, and premorbid intellectual functioning. The majority of the cognitive recovery, following moderate-to-severe brain trauma, occurs within the first year, whereas less dramatic recovery and flattening of the recovery curve are likely to occur during the second year post-brain injury. Although older adults demonstrate less tendency to sustain severe brain injuries, their potential for improvement during the rehabilitation course is less compared to younger adults. A study also stated that 60% of army veterans who sustained TBI were more likely to develop dementia compared to veterans without any history of TBI. The potential mechanisms of action accounting for the increased risk for dementia among brain injury include structural changes, accumulation of amyloid precursor protein and β-amyloid plaques, and tauopathy.

The affective and behavioral effects associated with TBI include mood and anxiety disorders, aggressive behavior, and post-traumatic stress and post-traumatic stress disorder. Among the abovementioned, anxiety and depression are more frequently reported than the rest. Few studies do suggest that depressive symptoms are more frequent than those associated with anxiety. Though the definite etiology of these deficits is unclear, poor psychosocial functioning has been hypothesized to be the cause and the consequence of depression. In some TBI patients, depression may reflect an organic etiology. For instance, lesions in the left dorsofrontal cortex, left basal ganglia or right posterior hemisphere have a strong association between depression and TBI. The patients recovering from TBI have three-to-four times increased risk of committing suicide compared to the general population. Severe TBI associated with cerebral contusions or intracranial hemorrhages demonstrated the highest risk of suicide.

Aggressive behavior following a brain injury can be a major concern for caregivers and complicate rehabilitation. Aggression may include agitation and/or violence, especially with the intimate partner, and suicide attempts. Such behaviors in TBI patients are likely to get associated with lower psychosocial functioning. The aggressive episodes are often triggered by staff prompts or erupt with no apparent provocation. Aggressive behavior post-TBI can co-exist with other psychiatric and psychosocial issues, such as anger, hostility, impulsivity, depression, substance abuse, post-traumatic stress, and post-traumatic stress disorder. Though premorbid factors such as alcohol use may influence the presence and etiology, these factors seem to be less predictive of post-injury aggression. Lesions of the orbitofrontal regions have been associated with alterations in behavior, including impulse control. Behavioral changes like anger and general irritability associated with frontal cortex lesions need to be distinguished from pseudobulbar lesions contributing to the emotional imbalance of uncontrollable and inappropriate laughing or crying.

Raised Intracranial Pressure Raised ICP is a common finding in moderate and severe TBI. The pressure exerted by the bleed or swelling of the brain against the solid structure of the cranium following an impact causes raised ICP. By definition, an ICP that exceeds 20 mm of Hg is considered high and necessitates immediate treatment, as the rising pressure can cause secondary damage to the brain, thus worsening the condition of the patient and delaying the recovery further. If the raised ICP goes untreated, two major complications are likely. The first being a temporary or permanent loss of vision and the second being the development of a severe headache that may last 48 h or more, with features like irritability, lethargy, slow cognitive processes, abnormal behavior, near-unconsciousness, coma, or even death. Another concern for patients with raised ICP is the possibility of incurring brain herniation which can be fatal if it compresses the vital structures that connect the brain to the spinal cord. The clinical presentation of patients with ICP resulting from TBI is similar to patients presenting solely with elevated ICP due to other factors. Typical elevated ICP patients present clinically with headache, vomiting, nausea, reduced state of consciousness, and blurred vision. Raised ICP causes certain physiological responses of the body that are displayed in the form of signs and symptoms (Table 6.3).

Autonomic Nervous System Dysfunction The Autonomic Nervous System (ANS) governs homeostatic control over different organs in the body. The regulation of blood pressure, gastrointestinal responses, contraction of the urinary bladder, focusing of the eyes, and thermoregulation are all controlled by the ANS. The ANS, composed of the sympathetic and parasympathetic pathways, works in concert with the endocrine system to regulate cardiac, renal, adrenal, homeothermic, and enteric function. The central ANS network is a complex network involving the cerebral cortex (the insular and medial prefrontal regions), amygdala, stria terminalis, hypothalamus, and brainstem centers (periaqueductal gray, parabrachial pons, nucleus of the tractus solitaries, and intermediate reticular zone of the medulla). Direct or indirect trauma affecting the brainstem, can disrupt the functions of the ANS leading to alterations in the heart rate, respiratory rate, elevation in the basal body temperature,

Table 6.3 Early and late manifestations of raised ICP

Early manifestations	Late manifestations
• Headache • Change in level of consciousness: restless, confusion, agitation, and irritability • Reduced GCS score • Convulsions • Lethargy • Slow/slurred speech • Nausea/vomiting • Focal neurological signs/symptoms • Pupillary changes: unilateral change in size and shape, sluggish reaction to light	• Severe headache • Projectile vomiting; usually occurs after waking, and frequently accompanies morning headache • Progressive decline in the level of consciousness: progressive stupor or coma • Significant speech impairment • Cushing's triad (hypertension, bradycardia, and irregular respiration in response to increased ICP) • Abnormal limb posturing: Decorticate and decerebrate posturing • Unilateral/bilateral pupil: dilated and fixed • Impaired brainstem function: irregular respiration, cardiac arrhythmias

variations in the blood pressure, dilated pupils, vomiting, or nausea and secretory signs like increased perspiration, salivation, tearing, and sebum secretion.

Injury to the brain stimulates a complex systemic response, characterized by profound alterations in the neuroendocrine and immune functions to restore homeostasis. Activation of the hypothalamic-pituitary-adrenal axis and the sympathetic nervous system leads to the secretion of glucocorticoids and catecholamines respectively, along with complex neuroimmune interactions. The term paroxysmal sympathetic hyperactivity (also known as "storming"), a syndrome characterized by episodes of increased activity of the sympathetic nervous system manifested as increased heart rate, increased respiration, increased blood pressure, diaphoresis, and hyperthermia, frequently complicates the clinical picture. The incidence of paroxysmal sympathetic hyperactivity ranges from 8% to 33% in patients with TBI in the intensive care unit.

The hypothalamic-pituitary-adrenal axis is a crucial contributor to autonomic function. In severe TBI, the hyperadrenergic state can be maladaptive and result in myocardial or other critical organ damage. Circulating proinflammatory cytokines can play a pivotal role in autonomic and cardiovascular function. The surge of cytokines after brain injury is part of an adaptive response to deal with damaged tissues and maintain brain function. However, higher brain cytokine levels compared to serum levels can magnify the response and can contribute to secondary injury.

Sodium imbalance and the concept of the cerebral salt-wasting syndrome and syndrome of inappropriate antidiuretic hormone secretion resulting from posterior pituitary damage are well-established in brain injury. The dysregulation of renal function leads to hypernatremia and hyponatremia. Sodium imbalance plays a role in both secondary damage to the brain and acute renal injury. Overall, brain injury may cause various systemic abnormalities from increased sympathetic activity to immune system depression. The presence of autonomic dysfunction has also been shown to correlate with increased morbidity and mortality in moderate and severe TBI.

Sensory, Motor, Perceptual and Functional Impairments Moderate and severe TBI patients commonly exhibit impaired motor function. The unilateral or bilateral weakness of limb(s), abnormal tone, impaired coordination, poor posture, and balance control, and impaired gait are certain impairments that are typically seen. Though static tremors and intentional tremors are common abnormal involuntary movements, choreiform movements and dystonia are less frequent among TBI patients. Brain injury patients, depending on the location of the lesion, may also exhibit impaired somatosensory function(s).

Occasionally the TBI patient may present with hemiparesis on the ipsilesional side, a false-localizing sign known as the Kernohan–Woltman notch phenomenon. A false-localizing sign reflects the dysfunction distant or remote from the expected anatomical locus of pathology and often challenges the traditional clinicoanatomical correlation paradigm. The mechanism of the Kernohan-Woltman notch phenomenon is as follows. An expanding supratentorial mass lesion or hemorrhage can cause uncal or transtentorial herniation. The pressure created by the uncal or transtentorial herniation on the cerebral peduncle of the midbrain can result in impingement and indentation of the contralateral cerebral crus by the tentorium cerebelli (Fig. 6.4). The damage to the crus cerebri and its descending corticospinal tracts leads to the ipsilesional motor deficit with possible deterioration in the level of consciousness. Table 6.4 provides the list of false-localizing signs in TBI.

In addition to paralysis or paresis of limb(s) and abnormal posturing (decerebrat e/decorticate), the involvement of the cranial nerves may result in extraocular muscles weakness, paralysis of the facial muscles, paralysis of the tongue muscles, vestibular and vestibulo-ocular reflex dysfunctions, dysarthria, and dysphagia. The phasic stretch reflex can be hyper- or hypo-reflexic and changes in muscle tone may

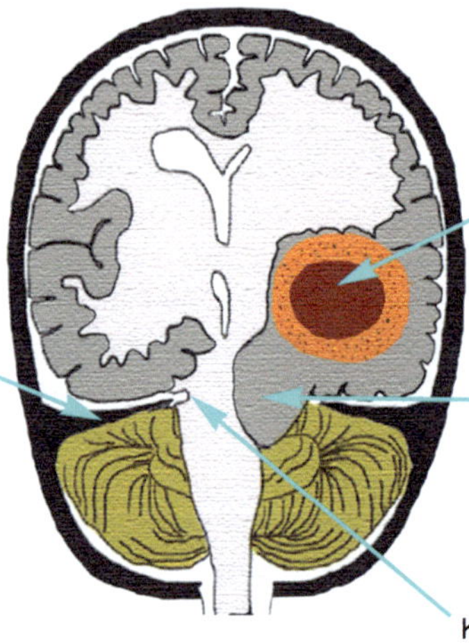

Fig. 6.4 A pictorial depiction of the Kernohan-Woltman Notch phenomenon

Table 6.4 Instances of false-localizing signs in head injuries

Sign	Feature	Reason
Kernohan-Woltman Notch phenomenon	Hemiparesis ipsilateral to the lesion	When the contralateral cerebral peduncle is compressed by the free edge of the tentorium.
Abducens nerve palsy	A common false-localizing sign; Seen in supratentorial and infratentorial lesions.	Compression of the sixth cranial nerve against the petrous ligament or the ridge of the petrous temporal bone
Oculomotor nerve	Unilateral fixed dilated pupil (Hutchinson's pupil)	Transtentorial herniation of the brain compressing the oculomotor nerve against the free edge of the tentorium
Papilledema	A usual feature of elevated intracranial pressure	Can be due to ischemic optic neuropathy
Cerebellar	Features of cerebellar dysfunction	Frontocerebellar pathway damage may result in incoordination of the contralateral limbs, mimicking cerebellar dysfunction.

reflect the anatomical location and the physiological effects of brain tissue compression or irritation. The pyramidal, extrapyramidal, and cerebellar signs can be asymmetric. Focal or diffuse brain lesion(s) in addition to motor dysfunctions can result in bowel or bladder disturbances. Sensory and perceptual dysfunctions might occur when the right hemisphere or the parietal and occipital lobes of the brain are damaged. The somatosensory and special sensory issues may include hypersensitivity towards sound or light, visual field deficits, hearing impairment, vision loss, paraesthesia in the extremities, somatosensory dysfunction, dizziness, and vertigo. The perceptual issues may include body scheme issues, visuospatial alterations, agnosia, apraxia, and visual discrimination skills dysfunction.

The complexity of the visual system makes its anatomical and functional connectivity vulnerable to TBI. Many of the tracts are long and some are quite small and delicate, particularly those in the brain stem. Considering that almost 50% of the brain is involved in visual function, it is not surprising that visual problems are some of the most common complaints after TBI. Vision and visual-perception functions include oculomotor control, visual acuity, and visual field. The convergence insufficiencies, visual field deficits, visual agnosia, accommodation dysfunction, oculomotor deficits, and diplopia are some of the common visual and visuoperceptual deficits seen among TBI patients.

Disorders of Communication Communication comprises verbal as well as nonverbal expressions like gestures and body language, facial expressions, visual or auditory signs, figures, objects, and colors, which ensures an efficient conversation process. For both these components of communication, various parts of the cortex are collectively activated. Communication is defined as "the capacity to exchange or discuss ideas, to dialogue, to converse with the aim of an understanding between the parties." It is of great importance in the realization and development of work, leisure, education, relationships, conversation, and negotiation.

Communication is characterized by the capacity to exchange information, express oneself and interact with others, developing knowledge bases and expanding one's concepts to develop language. Language is conceptualized as a "privileged instrument of inter-human communication and the privileged vehicle of thought." The symbol comes from imitation, and the representation of words is the use of the symbolic function of language. Scientific evidence suggests that the ability to develop language is innate and that the universal aspects of language acquisition are established in the structure of the human brain and are susceptible to sociocultural influences. Among other forms of communication, nonverbal communication (like gestures and codes) is also a means of transmitting information.

In the healthcare field, communication is extremely relevant. TBI patients may have cerebral injuries affecting communication and language skills, which make their social lives difficult. Aphasia, dysarthria and apraxia are among the chief alterations in communication that are caused by TBI-related disorders. Aphasia is a multimodal disorder that affects reading, writing, auditory comprehension, and orally-expressed language. Aphasia is essentially a linguistic processing disorder in which the mechanisms that transform thought into language are blocked. Table 6.5 provides details about the various types of aphasias.

Table 6.5 Types and descriptions of aphasias

Name	Location of lesion	Type	Description
Broca's Aphasia	Frontal language center in the dominant hemisphere (Broca's area)	Emissive (expressive) aphasia	Expressive aphasia associated with an inability to translate spoken concepts into meaningful sounds or in other words, to produce speech. The result is speech that is not fluent, with pauses between words or phrases. Comprehension is preserved and repetition is affected. Spontaneous speech is effortful, short, slow, agrammatical, and can pass ideas but words are often mispronounced
Conduction Aphasia	Supra-marginal gyrus region, arcuate fasciculus	Emissive aphasia	Characterized by phonemic paraphasias, anomies, and semantic paraphasias during the conversation. The speech may appear with hesitation and self-corrections. A striking feature of this type of aphasia corresponds to errors found in the repetition test. Comprehension and fluency preserved.
Transcortical Motor Aphasia	Area surrounding Broca's: anterior association cortices	Emissive aphasia	Main feature is the reduction of speech. Spontaneous language is extremely reduced and its expression is slow and short. Comprehension and repetition preserved but fluency lost.
Wernicke's Aphasia	Posterior language area in the dominant hemisphere.	Receptive aphasias	Sensorial aphasia, the most serious comprehension aphasia, is characterized by problems in the comprehension and formulation of speech. Comprehension and repetition lost but fluency preserved.
Transcortical Sensory Aphasia	Area surrounding Wernicke's: posterior association cortices	Receptive aphasias	Oral expression is fluent; at the same time, severe and moderate comprehension deficits appear; and there are semantic paraphasias, anomies, and circumlocutions. Repetition preserved
Anomic Aphasia	Varies, multiple regions of dominant hemisphere	Receptive aphasias	Comprehension, fluency, and repetition preserved. Characterized by semantic changes, paraphrases, and anomies.
Mixed Transcortical Aphasia	Isolated damage to anterior and posterior eloquent cortex	Mixed aphasias	Comprehension and fluency lost. Repetition preserved. Spontaneous speech is of the telegraphic type characterized by short, missing function words and brief noun-verb combinations
Global Aphasia	Large perisylvian, infarcts involving receptive and expressive speech areas	Mixed aphasias	Comprehension, fluency, and repetition lost. Spontaneous speech is scant, reduced to few words, and mutism possible

Affective processing is an important component of social communication and its impairment can produce problems in understanding and communicating emotions. Diminished social contact, impaired self-regulation, mood disorders, physical impairments, and diminished social interaction opportunities can lead to reduced psychosocial functioning. Verbal apraxia, a motor planning disorder, can also be a consequence of traumatic lesions of the CNS and can be characterized by grabbling articulatory movement, self-correction attempts, aprosody, prolonged irregular intonation, and sometimes trouble to initiate an expression.

Traumatic injury of the brain is one of the most frequent etiologies for dysarthria. Dysarthria, an articulation disorder, is characterized by slowness, weakness, and/or lack of muscle coordination related to speech function. Dysarthria's main consequence is a reduction in speech intelligibility, which limits the speaker's communicative ability and social participation. Impairment in the motor apparatuses necessary for oral production, which includes breathing, phonation, resonance, articulation, and prosody, following a central or peripheral neurological change leads to dysarthria. In accordance with the degree and location of the injury, several types of dysarthria have been identified. They are flaccid dysarthria, spastic dysarthria, hypokinetic or hyperkinetic dysarthria, ataxic dysarthria, or mixed dysarthria. The most common symptoms range from a decreased rate of speech, vagueness, articulation, slow, irregular speech to a lack of change in pitch or intensity. With regard to TBI, flaccid dysarthria is the most common type, and the injury is located in the lower motor neuron, which is peripheral but can also emerge due to some cranial nerve lesions. Spastic dysarthria can also emerge after traumatic injury of the brain and can be either unilateral or bilateral upper motor neuron lesion, causing an increase in muscle tone, spasticity, and weakness. Spastic dysarthria's main characteristics are a rough, stressed voice, tight, choked sounds, monotony, imprecise articulation of consonants, and hypernasality. Damage to the superior cerebellum or the superior cerebellar peduncle is believed to produce ataxic dysarthria. Ataxic dysarthria is characterized by abnormalities in articulation and prosody, and the speech is typically explosive or staccato, scanning or telegraphic or slurred type. The clinical features include abnormalities in speech modulation, consonantal misarticulations, and rate of speech. In severe cases of TBI, it is not uncommon to see the co-existence of dysarthria and dysphagia.

6.4 Investigations

Within the first 48–72 h post-injury, prompt management, including the use of imaging techniques to determine the presence and extent of the injury, can significantly alter the course of TBI sequelae. In addition to the above, the neuroimaging findings also serve as an important prognostic indicator. In the acute setting, early diagnosis and aggressive and appropriate management may minimize or prevent secondary injuries and may significantly reduce mortality and morbidity, while minimizing length of hospital stay and health care costs. By providing anatomic localization and navigation information, neuroimaging essentially helps to determine extracranial landmarks to surgically plan the skin incision and guide the placement of burr-holes.

Though imaging plays an important role, not all head trauma patients require neuroimaging. Worsening level of consciousness, LOC for more than 5 min, focal neurological deficits, seizure, mental status in status quo over time, penetrating skull injuries, signs of basal or depressed skull fracture, and confusion or aggression on examination are certain circumstances that suggest major injury and warrant neuroimaging. Investigators suggest that the imaging should not be performed unless the GCS score is below 13. According to the New Orleans criteria, all TBI patients over 60 years of age should undergo neuroimaging. TBI patients with headache and vomiting or a history of two or more episodes of vomiting post-trauma might indicate an impending surgical emergency and hence should be imaged.

Conventional Computed Tomography (CT) being readily available, more cost-effective and easier to perform, and requiring a shorter imaging time for ventilator supported, in traction, or agitated patients, is the initial imaging modality of choice during the first 24 h post-injury. It is also superior in evaluating skull fractures and detecting acute subarachnoid hemorrhage (SAH) or acute intraparenchymal hemorrhage. Though CT is better at detecting bony pathology and certain types of early bleeds, as the blood composition changes, the ability of Magnetic Resonance Imaging (MRI) to detect hematomas improves over time. For the same reason, MRI is generally considered superior to CT, 48–72 h post-injury.

Typically, minimal brain injury patients may not show any abnormality on MRI, however, the abnormalities such as hemorrhagic cortical contusions, petechial, or foci of altered signal that represent white matter shear injury might be detected. MRI is superior to CT in detecting axonal injury and subtle neuronal damage and is also superior at imaging the brainstem, basal ganglia, and thalami. New MRI technology and acquisition sequences have further improved MRI sensitivity. Fluid-attenuated Inversion Recovery (FLAIR) MRI suppresses the high signal from CSF by using a long inversion time (T1-weighed), to detect traumatic lesions and hematomas. Studies on intraventricular hemorrhages also suggest that FLAIR and fast spin-echo FLAIR MRI may be superior to non-contrast CT.

Neither Positron Emission Tomography (PET) nor SPECT imaging is used routinely in the acute management of head trauma. Both require a fair amount of time to complete and provide only functional information than anatomic information and are unlikely to replace CT or MRI in the acute setting of head trauma. Longer and more severe amnesic episodes imply a greater chance of hemorrhage and for amnesia lasting more than half an hour, Single Photon Emission Computed Tomography (SPECT) studies may disclose bilateral cerebral hypoperfusion.

Repeat imaging is usually indicated if hemorrhage or edema worsens or enlarges to produce a significant mass effect with neurological deterioration, which is a surgical emergency. In the acute stage, CT is more sensitive than MRI, as the clot signal can be indistinguishable from the brain parenchyma on MRI. CT is superior to conventional MRI in detecting acute SAH due to the low hematocrit and low deoxyhemoglobin in the blood, which makes acute SAH appear similar to brain parenchyma on T1- and T2-weighted images. However, FLAIR sequences may find small acute or subacute SAH missed by CT and conventional MRI. On non-contrast CT, cerebral contusions appear as low attenuation, provided hemorrhage is absent and mixed or high attenuation if the hemorrhage is present.

The absence of CT findings does not completely exclude elevated ICP. But the presence of midline shift, SDH, intraparenchymal hematoma, herniation, change in ventricular shape or size, or loss of gray-white junction indicating cerebral edema should raise suspicion of ICP. Both CT and MRI can effectively diagnose cerebral herniation, however, better soft tissue definition of MRI and multiplanar imaging ability are particularly important in identifying descending transtentorial herniation, making MRI superior in certain situations.

Depending on the location, size, and type of fracture, fractures may need to be surgically repaired to relieve or prevent CSF leakage, infection, hemorrhage, or vascular compromise. Though plain X-ray films of the skull may detect fractures, CT is the neuroimaging modality of choice. A follow-up CT may help to exclude pneumocephalus in patients with basilar skull fractures. Radionuclide with 111-Indium, CT cisternography, and MRI using a 3D-constructive interference in steady-state sequence are certain imaging modalities available to identify CSF leaks.

Based on size and velocity, foreign bodies can cause direct laceration, shock-wave transmission, and cavitation. Non-contrast CT remains the imaging modality of choice for identifying foreign objects and determining whether removal is necessary. Imaging can also help track the path and subsequent movement of the foreign body and anticipate the corresponding complications. A few additional diagnostic investigation techniques that could be used depending on the individual circumstances are given below.

Angiography Although the actual incidence of vascular damage in head trauma is unclear, many such lesions are asymptomatic. Angiography should be performed if trauma disrupting the arterial wall (dissections, aneurysms, or fistulae) is suspected. CT angiography of the neck has become an important screening exam for blunt injuries of the carotid and vertebral arteries. CT angiography of the head also helps to distinguish non-traumatic pathology, which could be the cause of injury from vascular injury associated with TBI.

ICP Monitor ICP is a valuable indicator of injury severity after TBI. The relationship between ICP and mortality after severe TBI is well established, and the relationship becomes proportionally greater as the ICP value increases. The risk of death is sixfold when ICP is greater than 40 mm Hg. ICP monitoring is recommended if two or more of the following are present at the time of admission: Above 40 years of age, unilateral or bilateral abnormal posturing of the limb(s), or episodes of low systolic blood pressure (below 90 mm Hg). Though both invasive and noninvasive monitors are available for ICP monitoring, intraparenchymal strain gauge or fiber optic monitor and intraventricular monitor using a ventriculostomy are the preferred and recommended techniques.

Electroencephalograph (EEG) Conventional EEG refers to the standard analog or digital recording of the electrical activity generated by the brain, using scalp electrodes, presented as raw traces of bioelectric waveforms and inspected visually by a qualified electroencephalographer. Long before the advent of neuroimaging techniques, EEG was the first clinical neurodiagnostic assessment that revealed abnormal brain function following TBI. It is a painless bedside procedure and is mostly performed in mild TBI to diagnose post-traumatic epilepsy and in identifying post-concussive syndrome.

6.5 Medical and Surgical Management

Generally, a neurosurgeon or an emergency medicine/trauma care specialist shoulders the main responsibility of the head injury patient arriving to the hospital. Soon after the admission to the trauma center, the first priority is to resuscitate the patient if required. Meanwhile, the history is obtained and the baseline assessments are also made. Depending on the severity of the brain trauma and/or the presence of blood and necrotic tissue in the cranial vault, surgery will be indicated.

6.5.1 Initial Care and Interventions

During the early phase of medical/surgical management, the initial concerns may include the management of respiratory dysfunction, re-establishment of cardiovascular homeostasis, and treatment of raised ICP. In addition to the above, the general treatment may include fluid and electrolyte balance management, nutrition, eye and skin care, contracture prevention, oral and bronchial hygiene and care, and safety considerations. The need for such care gradually lessens as the patient starts responding, whereas if the unconsciousness continues to persist, the treatment will be further extended.

The GCS score is used to initially determine the level of consciousness and to measure the brain stem activity. The details regarding the three components of GCS: eye, motor, and verbal responses, are provided in Table 6.6. The GCS score can range between 3 and 15. The lower scores correlate to the lower levels of brain function, denoting severe brain injury, and higher scores correlate to higher levels of function, denoting minimal brain injury. A GCS score of 13–15 denotes mild TBI, 9–12 suggests a moderate TBI, and a score of 8 or below suggests severe TBI.

Approximately 80% of TBI is comprised of mild TBI. A panel of experts convened by the Centers for Disease Control and Prevention (CDC) has further defined mild TBI as an injury to the head that results in any of the following: Any period of confusion, disorientation, or impaired consciousness; any memory dysfunction after the trauma; LOC lasting less than 30 min; any signs or symptoms of neurological or neuropsychological dysfunction after the trauma. Further, the inclusion and

Table 6.6 The Glasgow Coma Scale score chart

Response	Description	Score	Score obtained
Best eye response (E)	None	1	E=
	To pressure	2	
	To sound	3	
	Spontaneous	4	
Best verbal response (V)	None	1	V=
	Sounds	2	
	Words	3	
	Confused	4	
	Orientated	5	
Best motor response (M)	None	1	M=
	Extension	2	
	Abnormal flexion	3	
	Normal flexion	4	
	Localizing	5	
	Obeys commands	6	
Minimum GCS score is 3 and maximum score is 15. The most up-to-date information about the GCS Aid is available at https://www.glasgowcomascale.org			Total score (E+V+M)

With permission from the University of Glasgow and Sir Graham Teasdale

exclusion criteria for the same were established by the American Congress of Rehabilitation Medicine (ACRM). According to ACRM, the inclusion criteria (more than one should be present) are (1) Duration of LOC >30 min, (2) Any loss of memory of incidents immediately before or after the injury as far back as 24 h, (3) dazed, disoriented, or confused mental state at the time of trauma, and (4) Focal neurological deficit(s) that may or may not be brief. The exclusion criteria (minimum of one manifested) are (1) LOC >30 min, (2) PTA persisting longer than 24 h, and (3) A drop of GCS below 13 after 30 min post-TBI.

According to the Brain Injury Association of America (BIAA), a moderate TBI patient experiences few minutes to a few hours of LOC; days to weeks of confusion; and temporary (for months) or permanent physical, cognitive, or behavioral impairments. A severe TBI patient experiences a prolonged state of unconscious or coma lasting for days, weeks, or months. One-third of the TBI patients are likely to have associated extracranial injuries, for which a thorough physical examination and appropriate special tests and investigations are required.

In addition to the investigation procedures discussed in the prior section, measurement of cerebral blood flow and metabolism, monitoring of cardiorespiratory and cardiovascular function, CSF analysis, and other biochemical studies may provide additional important information about the medical status of the patient. The findings obtained from neurophysiological tests such as visual evoked potential studies, brainstem auditory evoked potential studies and somatosensory evoked potential studies when combined with other clinical and laboratory examinations may provide further insight regarding the presence, evolution, and resolution of the lesion.

With regard to the bedside examination of unconscious patients, the reflex motor responses can be assessed by applying noxious stimulus like pressure on the nail bed or supraorbital area and recording the motor reaction, which can be appropriate (localizing the pain), inappropriate (abnormal flexor or extensor response) or absent. The cognitive and behavioral functions are usually tested by neuropsychological tests, which provide more information about a patient's cognitive/behavioral capabilities than a basic neurological evaluation. These tests evaluate a variety of functions, including attention span, orientation, memory, concentration, language, learning capacity, mathematical reasoning, spatial perception, abstract and organizational thinking, problem-solving, social judgment, motor abilities, sensory awareness, emotional characteristics, and general psychological adjustment. Such tests often help to differentiate the cognitive and behavioral functional changes caused by TBI from post-traumatic stress syndrome, hysterical reactions, malingering, depression, and anxiety.

6.5.2 Basic Principles of Pharmacological Management

Pharmacological management in TBI patients begins with the analysis of the current medications. Since it is not essential that all cognitive deficits are organic and many are related to medications, "minimalization" of pharmacological agents is the primary principle. It involves the stopping of any redundant medications, particularly drugs used for sedation or those affecting neurological recovery. "Substitution" is the second principle and means that the medication(s) with the fewest side effects and least impact on neurological recovery should be the drug of choice if medications are essential. The use of drugs to augment the cognitive and functional processes forms the last principle of medical management.

The pharmacological agents are chosen based on the symptoms. An osmotic agent like mannitol is the medication of choice to control the raised ICP and maintain the $PaCO_2$ at around 30 mm Hg. The mannitol helps to pull the fluid from brain parenchymal cells back to the blood, thus lowering the ICP. It promotes diuresis by increasing the rate of filtration and blocking the reabsorption of water by the renal tubules. The commonly reported side effects during or after mannitol infusion include pulmonary congestion, electrolyte imbalance and loss, acidosis, marked diuresis, and headache. To control ICP, propofol, a barbiturate is recommended especially when other means are not effective. Hyperventilation is a known method to rapidly lower the raised ICP by decreasing the $PaCO_2$, which subsequently leads to arterial vasoconstriction, thus lowering the cerebral blood flow, cerebral blood volume, and ICP. Few studies have pointed out that the dramatic reduction of $PaCO_2$ by hyperventilation had worsened the outcome compared to that of the patients managed with medication, and therefore it is currently recommended only for non-responsive cases and for short durations.

To manage cerebral edema, glucocorticoids like dexamethasone and methylprednisolone have been used. However, most studies have shown no long-term changes in outcome, and the mortality rate was reported to be high for those managed with methylprednisolone. Similarly, therapeutic hypothermia management remains controversial in patients with TBI. Blood pressure control is important in patients with brain injury. Cerebral Perfusion Pressure (CPP) to maintain cerebral blood flow against increased ICP is calculated by subtracting the ICP from the mean arterial pressure. A pressure greater than 60 mm Hg is the recommended level of CPP. If fluid management cannot keep the blood pressures elevated, then vasopressor drugs such as phenylephrine can be used to constrict the peripheral vessels.

6.5.2.1 Potential Harmful Effects of Medications

Harmful side effects of the pharmacological agents used in TBI can be of two categories: (1) The medication-specific side effects of a particular drug, regardless of the disease. For instance, the sedative effect of drugs like clonidine, irrespective of the TBI type. (2) The side effects are explicit to a disease-specific population. For instance, the slow rate of motor recovery following benzodiazepine's use among stroke patients. Emerging human research and animal studies suggest that four classes of medications can potentially impair the neurological recovery process following TBI. These are discussed below:

Neuroleptic Agents It is preferable to avoid traditional antipsychotic medications, like haloperidol, thiothixene, and chlorpromazine, which are pharmacological agents that block dopamine receptors in the brain. The literature on animal studies explicitly suggests that haloperidol has a negative effect on neurological recovery. The milder pharmacological variants, such as risperidone and olanzapine, may be less detrimental and have a reduced risk of causing extrapyramidal symptoms and tardive dyskinesia, as suggested by the animal studies. Studies on atypical agents like quetiapine, ziprasidone, and aripiprazole have shown further milder side effects but need to be further researched as the studies are limited among brain injury patients.

Benzodiazepines Benzodiazepines act on the gamma-aminobutyric acid (GABA) receptor and readily cross the blood-brain barrier. These pharmacological agents cause a sedative effect and affect memory and learning. Clinical studies reveal that these agents can delay motor recovery after a stroke. If indicated to relieve anxiety, then buspirone, a serotonergic agent, can be a preferable alternative instead of benzodiazepines. To facilitate sleep, if benzodiazepines are used, it is advisable to avoid daytime naps and caffeinated drinks. When pharmacological intervention is needed, zolpidem or trazodone can be appropriate alternatives for benzodiazepines.

Anticonvulsants Anticonvulsant agents like phenobarbitone and phenytoin are known to delay or alter the neurological recovery in rodents and may potentially impair human cognitive function. For the same reason, phenobarbitone should be used only as a last recourse for seizure management. Following brain injury, it is advisable to discontinue phenytoin usage for seizure prophylaxis beyond 1 week. Carbamazepine or valproic acid can be the appropriate pharmacological agents for managing partial and generalized seizures, respectively. Growing evidence of recent anti-epileptic drugs like lamotrigine and levetiracetam suggests that sedation is still attainable with lesser cognitive side effects.

Centrally Acting Antihypertensive Agents Rodent studies reveal that centrally acting antihypertensive drugs like methyldopa, clonidine, and prazosin, alter the metabolism of norepinephrine and have been shown to retard neurological recovery. Calcium channel blockers, angiotensin-converting enzyme inhibitors, angiotensin receptor blockers, and hydrophilic β-blockers can be reasonable substitutes to control hypertension. If rate control is desired, calcium channel blockers along with digoxin or hydrophilic β-blockers (atenolol) can be used.

6.5.2.2 Medications for Cognitive and Functional Augmentation

Hypoarousal Arousal is the fundamental cognitive function, without which sensory information and motor responses cannot be processed. Research data on the treatment of hypoarousal in brain injury patients is limited due to a dearth of controlled trials and comprehensive dosage protocols. Amantadine, a dopaminergic drug used for Parkinson's disease patients, is noted to improve arousal among TBI patients. Other dopaminergic medications like bromocriptine, levodopa, and selegiline, may also be beneficial to promote arousal, however, they are primarily used for the management of attention deficits. Modafinil, an exclusive psychostimulant, originally approved for narcolepsy management has also shown beneficial effects in augmenting alertness.

Attention Deficits Methylphenidate, a neurostimulant, is used widely to enhance attentiveness in patients suffering from brain injuries. Controlled trials have reported that this pharmacological agent enhances concentration and improves processing speed. A sustained-release formulation of methylphenidate helps to minimize the peak and trough effects encountered by some patients. Patients with a prior history of drug abuse have a high risk of addiction to this agent. Bromocriptine, protriptyline, dextroamphetamine, and Dexedrine are a few other pharmacological agents that are used to treat attention deficits following TBI.

Initiation Deficits For those TBI patients with initiation deficits, dopaminergic agonists like amantadine, bromocriptine, dextroamphetamine, methylphenidate, and protriptyline can be beneficial.

Memory Deficits Frontal cortical cholinergic system and hippocampus have a vital role in attention, learning, storage, and recovery of novel information.

Cholinergic dysfunction plays a significant role in memory impairment following TBI. Few studies on donepezil an acetylcholinesterase inhibitor, originally introduced for Alzheimer's treatment, have reported improvement in memory deficits following the traumatic injury of the brain. Other medications under acetylcholinesterase inhibitors like rivastigmine and galantamine have been studied, however, the results are largely still inconclusive. Methylphenidate and bromocriptine may help in enhancing memory deficits secondary to poor attention.

6.5.2.3 Spastic Hypertonia

Spasticity need not be detrimental in all situations. Some patients may functionally utilize the spasticity of the musculature for transitions and gait. The physicians must assess the severity of spasticity and functional limitations caused by it prior to the management of tone. Among the oral pharmacological agents, dantrolene sodium acts peripherally by preventing the release of calcium from the sarcoplasmic reticulum and is the preferred choice of therapy for spasticity following head injury. Drugs like baclofen, clonidine, and valium are alternatives but are frequently known to cause drowsiness and cognitive side effects at higher doses, and a lower dose may not be adequate to relieve the spasticity.

When oral medicines are ineffective or poorly tolerated and to restrict spasticity to a few functional muscle groups, phenol injections and muscle botulinum toxin injections may be preferable. Phenol is typically administered near the motor point of the affected muscle or nerve branches. Phenol application causes instant demyelination of the gamma nerve fibers and reduces the stretch sensitivity of the muscles and post-phenol application muscles can be lengthened at ease. Serial casting after injections will improve the effectiveness and increase the length of the shortened muscle. Phenol injections though painful are cost-effective and the beneficial effect lasts for about 6 months. The musculocutaneous, obturator, femoral, and tibial nerves are usually treated with phenol injections.

The injection of Type A or B botulinum toxin into the muscle prevents the release of acetylcholine from the presynaptic area of the neuromuscular junction, thus minimizing the conversion of neural impulses into muscle action potentials. The botulinum injection takes a few days to kick in, and the effect stays for 3–6 months until the collateral sprouting of the axon leads to a gradual reappearance of hypertonia. Compared to phenol, these drugs are quite expensive but are easy to administer and fairly painless with less possibility of dysesthesia. Due to the possibility of antibody formation and to maintain the toxin's beneficial effect, the botulinum injection is recommended once every 3 months.

Implantation of an intrathecal baclofen pump is the treatment of choice when the spasticity is severe and diffusely involves many muscles. Prior to implantation, a small dose of baclofen (test dose) is administered to note the positive reaction and predict the dose required to alleviate the spasticity. A catheter with a pump is subcutaneously tunneled posteriorly around the abdomen and is placed in the intrathecal space. The speed of the baclofen administration can be adjusted by small increments at either a constant or variable rate throughout the 24 h. Since only a small dose is required to produce the desired effect, the extent of sedation seen with oral baclofen is less likely when intrathecally administered. Typically, post-implantation, the baclofen dose is slowly titrated using an external program until the desired effect is achieved. Percutaneously, every 1–6 months, the pump reservoir needs to be refilled and due to limited battery life, every 5–7 years, the pump needs to be replaced.

6.5.3 Surgical Management

The decision for craniotomy or other surgical procedures is determined by the type of injury and the patient's neurological state. Studies in the early 1990s reported that approximately 10% of mild TBI patients had conclusive neuroimaging impressions, and only 1% needed definite surgical treatment. Typically, moderate-severe acute TBI may require neurosurgical intervention. If a blood clot leads to elevated ICP, surgical removal of the clot will be undertaken within hours or days post-injury. When the risk of clot dislodgement is low, surgical removal is deferred.

An elevated ICP, presence of cerebral edema, and non-communication (obstructive) or communicating hydrocephalus, can substantially affect the TBI recovery process. Monitoring the ICP has a critical role in medical and surgical decision-making. In case the ICP is raised considerably, medications are given to transfer fluid from the neural tissues into the blood vessels, decrease the metabolic demand of the brain and improve the circulation to the damaged structures. TBI patients may also require artificial ventilation to warrant adequate oxygenation to facilitate the repair.

Decompressive craniectomy (surgical removal of a portion of the skull) is considered a life-saving procedure to prevent further compounding of secondary injuries and central herniation. If the cerebral edema is extreme and diffuse, the raised ICP can temporarily be alleviated by decompressive craniectomy which provides adequate space or room for the swollen tissues to bulge out, minimizing the possibility of secondary injury caused by stress. In severe TBI, decompressive craniectomy is indicated when the sustained elevated ICP is not responding to medication. Decompressive craniectomy is typically considered for diffuse brain injury compared to focal extra-axial or parenchymal injury. If acute hydrocephalus, with or without obstructive hydrocephalus, is complicating the clinical picture, external ventricular drainage is indicated. Ventricular drainage clears the excess fluid accumulated and helps the pulsatile CSF flow to get the uninvolved brain cells back into function.

Brain parenchymal hemorrhages and SDHs can lead to an elevated ICP and might require emergency surgeries. A CT axial image showing a hyperdense collection suggesting subdural hemorrhage is depicted in Fig. 6.5. Typically, SDHs are treated using burr-hole surgeries. In case of an acute SDH, surgery is indicated if the clot thickness is greater than 10 mm and the midline shift is 5 mm or more on CT

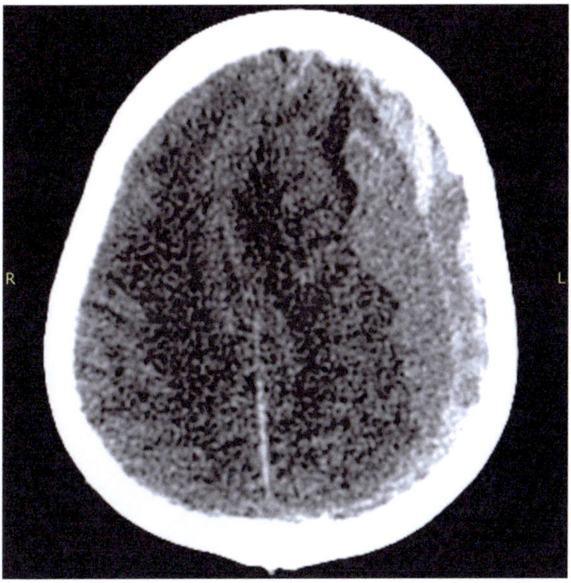

Fig. 6.5 CT axial image showing a semilunar hyperdense collection with a layering of hemorrhage in the left extra-axial region suggesting subdural hemorrhage

scan, irrespective of the GCS score. For those SDH patients with a GCS score lower than 9, monitoring of ICP is mandatory to prevent any worsening of the condition. SDH is conservatively managed if the thickness is less than 10 mm or the midline shift is less than 5 mm, provided repeated imaging reveals no worsening. From the time of brain trauma, if the GCS score drops by 2 or more, ICP shoots to 20 mm Hg or higher, or abnormal pupillary responses are observed, surgery is recommended. A craniotomy, with or without duraplasty, is effective in the evacuation of the hemorrhage. Ambiguïty still exists with regard to performing craniotomy or craniectomy for acute SDH.

The treatment options for EDH can be either immediate surgical intervention or initial, conservative, close clinical observation followed by delayed evacuation. An acute EDH classically presents with LOC followed by a lucid interval and obtundation in sequence. EDH is characterized by a "bi-convex lens" shaped hemorrhage on CT (Fig. 6.6), and the source of hemorrhage is typically the middle meningeal artery. Treatment is generally by urgent surgery in the form of a craniotomy or burr-hole. Surgical management of EDH is based on neuroimaging and patient status. Surgery is indicated if imaging reveals more than 30 cc of a blood clot, irrespective of the GCS score. EDH is conservatively managed if the volume is less than 30 cc, less than 15 mm thickness, midline shift of less than 5 mm, and a GCS score of 9 or above. Table 6.7 provides the management options for intracranial hemorrhage.

Certain types of skull fractures warrant surgical intervention to prevent infection and further damage to the underlying brain parenchyma. In most cases, closed depressed skull fractures are treated non-operatively. Generally, surgery is indicated if the open skull fracture is depressed to a degree more than the skull's thickness.

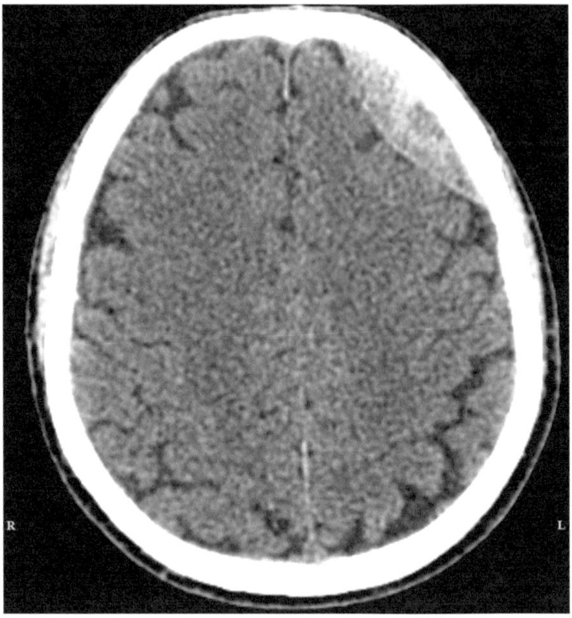

Fig. 6.6 CT axial image showing a bi-convex hyperdense collection in the left extra-axial region suggesting epidural hemorrhage

Table 6.7 Management options for intracranial hemorrhage

Type	Norms or criteria	Management
Subdural hematoma	>10 mm thick hematoma and >5 mm midline shift	Surgical evacuation
	<10 mm thick hematoma and <5 mm midline shift but with a GCS score <9 associated with ≥2 point drop in the score and/or pupillary dysfunction and/or ICP >20 mm Hg	Surgical evacuation
	GCS score <9	ICP monitoring
Epidural hematoma	>30 ml volume, irrespective of the GCS score	Surgical evacuation
	<30 ml volume and <15 mm thickness, with <5 mm midline shift and GCS score of >8 without focal neurological deficit	Intensive monitoring, conservative management, and serial radioimaging
Intraparenchymal hemorrhage (intracerebral hemorrhage)	Progressive neurological deterioration, mass effect on radioimaging, medically refractory intracranial hypertension	Surgical evacuation
	Any lesion >50 ml	Surgical evacuation
	Frontal or temporal contusions >20 ml and GCS score between 6 and 8 and ≥5 mm midline shift and/or cisternal compression evident on radioimaging	Surgical evacuation
	No evidence of neurological compromise; controlled ICP and no significant signs of mass effect radioimaging	Intensive monitoring, conservative management, and serial radioimaging

Surgical management may include elevation of the bone flap and debridement of the involved tissue, removal of foreign objects if any, followed by copious irrigation. Conservative management is the choice of treatment if the open depressed skull fracture is less than 1 cm, provided, there is an absence of dura mater damage, pneumocephalus, significant intracranial hematoma, frontal sinus involvement, gross deformity, and signs of wound infection. Surgical intervention of skull fractures must be accompanied by anti-epileptics and antibiotics treatment. For open, complex, and comminuted fractures, the fractured cranium may be removed, and the defect may be repaired later to prevent subsequent infection. In the case of penetrating injuries, surgery is primarily to manage ICP, remove the foreign objects with care if possible, and prevent secondary injuries. In penetrating injuries management, the primary objective of surgery is not to recover the projectile but to debride the involved tissue, remove the hematoma, repair the vessel due to trauma and prevent the infection.

6.5.4 Complications Following TBI

6.5.4.1 General Medical Complications

TBI-associated hypertension is seen in 10–15% of patients. Excessive release of catecholamine from the adrenal glands, which leads to increased output, and vasoconstriction or injury to the central blood pressure control centers within the brain are possible mechanisms responsible for brain injury-associated hypertension. Considerable release of catecholamines during the adrenergic surge after brain trauma can lead to myocardial injury. Beta-blockers, calcium channel blockers, angiotensin-converting enzyme inhibitors, or angiotensin receptor blockers are a few reasonable choices to control hypertension. Orthostatic hypotension is a frequent complication associated with prolonged bed rest or medications. The presence of anemia or dehydration and the use of antihypertensives should be ruled out before orthostasis is attributed to prolonged bed rest. Support stockings, abdominal binders, and reclining wheelchairs can minimize the possibilities of orthostatic hypotension.

Around 20% of TBI patients demonstrate endocrine complications, which include the syndrome of inappropriate antidiuretic hormone, diabetes insipidus, anterior hypopituitarism, cerebral salt-wasting, and primary adrenal insufficiency. The syndrome of inappropriate antidiuretic hormone, the most common endocrine complication following TBI, causes hypervolemic hyponatremia due to inappropriate renal water conservation. Fluid restriction is the treatment of choice for many if the syndrome of inappropriate antidiuretic hormone is the cause for hyponatremia. The clinical features of diabetes insipidus include polyuria and polydipsia and desmopressin acetate is the drug of choice to treat diabetes insipidus. The cerebral salt-wasting syndrome is caused by impaired renal tubular function, which leads to the inability of the renal system to conserve salt, and intravenous normal saline administration is the treatment for the same.

Anterior hypothalamus damage in severe TBI patients may develop central fever or central hyperthermia. It is a rare entity developed due to the autonomic dysregulation in the absence of infectious etiologies or other causes and manifested as fever, tachycardia, paroxysmal hypertension, dilated pupils, tachypnea, and extensor posturing. Physical cooling and medications like bromocriptine or baclofen can be the management of central hyperthermia. In case of peripheral nerve injuries, such as ulnar nerve, common peroneal nerve, or brachial plexus injuries, associated with TBI, nerve conduction studies and electromyography can be useful in the evaluation. Peripheral nerve injuries can be mild, moderate or severe and based on the severity of injury it will be managed nonsurgically or surgically.

6.5.4.2 Depression and Agitation

Up to 50% of TBI patients are likely to develop depression within the first few years post-trauma. Psychotherapy is the most appropriate compressive treatment program to address depression. The therapy should involve both the patient and the caregivers and should focus on encouraging self-esteem, reducing denial, and promoting emotional support. Selective serotonin reuptake inhibitors and neurostimulants are a few of the medications used to treat depression.

Agitation, a subtype of delirium, indicated by unrestrained behaviors such as hostility, restlessness, lack of inhibition, and emotional instability are likely to occur during the PTA period. In most cases, the agitation is quite short-lived. Damaged cognition is closely associated with agitation. Agitation is a manifestation and not a disorder. Pain, infection, hypoxia, metabolic abnormalities, urinary obstruction, withdrawal of certain medications, or fresh intracranial lesions are some of the causes that need to be explored. Nonpharmacological treatments can be the first line of choice and may include reducing environmental stimulation and redirection. With regard to pharmacological agents, propranolol is probably the best studied, and the possible side effects include hypotension and bradycardia. Other agents like amantadine, carbamazepine, and benzodiazepines are alternatives. In general, first-generation neuroleptics should be the last line of treatment after all the other agents have failed.

6.5.4.3 Heterotopic Ossification

The term ectopic bone formation and myositis ossificans are often used interchangeably with the term heterotopic ossification. It is the formation of normal, mature bone in the soft tissues around large joints of the body like hips, elbows, knees, and shoulders. The incidence of ectopic bone formation is between 10% and 75% among TBI patients. It usually occurs 2 weeks to 4 months post-brain trauma, and the chances are higher with considerable spasticity, prolonged coma, and fractures near larger joints. The involved limb will have joint Range of Motion (ROM) restriction and pain. The area will be inflamed, warm, and erythematous. Around 10–20% of patients may develop permanent functional limitations. Deep vein thrombosis, fracture, infection, and hematoma are the certain differential diagnosis of heterotopic ossification. Radiological (X-ray and CT scan) studies can be helpful in the diagnosis of the condition. Alkaline phosphatase can be elevated but may not be confirmatory.

Symptomatic treatment should be initiated with indomethacin, acetaminophen, and disodium etidronate, which may minimize the progression of the condition. Few pieces of evidence suggest that aspirin and warfarin may also help. Rest to the involved joint(s) is recommended until the acute inflammatory stage settles, after which gentle passive ROM and active ROM exercises in the pain-free range are advised to minimize any functional limitations or ankylosis. For impending ankylosis, the involved joint(s) should also be kept in the most functional position. Surgical treatment and low-dose radiation therapy can be the options for mature heterotopic ossification when alkaline phosphatase levels are back to the baseline level.

6.5.4.4 Hydrocephalus

Though ventriculomegaly is widespread after a severe head injury, the incidence of overt hydrocephalus is around 5%. Sometimes it can be difficult to distinguish between hydrocephalus and hydrocephalus secondary to brain parenchyma atrophy. Acute obstructive hydrocephalus is managed with ventriculostomy. The clinical triad of gait disturbance, urinary incontinence, and cognitive function decline is less beneficial in identifying the condition as these symptoms can be the result of traumatic injury to the brain itself, among these patients.

6.5.4.5 Seizure Disorder

The occurrence of seizures following head injury depends heavily on the severity of trauma. The details regarding the post-traumatic seizures are already dealt with in the earlier section of the chapter. Large bi-parietal contusions and dural penetration with bone fragments or foreign bodies, subcortical contusions, SDH, intracranial procedures, and a midline shift of more than 5 mm have a significant risk of developing seizures. Phenytoin and carbamazepine were effective in reducing early post-traumatic seizures. However, these medications have side effects, including a possible adverse effect on the speed of neurological recovery, hence medications like valproate are desirable for managing existing seizures but not as seizure prophylaxis.

6.5.4.6 Vegetative State and Minimally Conscious State

The brain injury patients in vegetative and minimally conscious states usually need gastrostomy tube feeding and tracheostomy and are more likely to face several medical complications. A thorough assessment of neuro-medical reasons for diminished responsiveness should include examination for covert infections such as retroperitoneal abscess, sinusitis, and osteomyelitis, endocrine disorders, and hydrocephalus. Round-the-clock nursing care, proper positioning, and regular physiotherapy are quite important. In an attempt to boost arousal among severely injured patients, several medications (neurostimulants or dopaminergic agents) can be attempted. The recent multi-centric study has indicated that amantadine could be of certain help for these patients. Uncertainty still exists regarding the beneficial effects of "sensory stimulation" or "coma stimulation" as a clinical intervention.

6.5.5 Prognosis

Due to the influence of multiple factors, predicting the outcome following moderate-to-severe TBI is often challenging for the rehabilitation team. The final outcome of head injury is related to the severity of brain trauma and can be predicted based on the GCS score and the duration of LOC or PTA. Generally, the prediction of outcome for moderate-to-severe TBI is difficult during the initial days, post-injury. Since the meaning of recovery and expectations varies from person to person, prediction can be more difficult. Prognostication is further restricted by poor assessment and selection of inappropriate outcome measures. The clinical usefulness of the Glasgow Outcome Scale (GOS) is limited because of its broad categories of classification. The GOS classification grades are (1) death, (2) persistent vegetative state (unable to interact with environment), (3) Severe disability (unable to live independently but able to follow commands), (4) Moderate disability (able to live independently but unable to return to work), and (5) Mild or no disability (able to return to work). The Glasgow Outcome Scale Extended (GOSE) is an expanded version of the GOS and has eight grades instead. Age, general health, prior psychiatric issues, substance abuse, and premorbid medical conditions can affect the prognosis. Associated complications like hypoxic injury of the brain, considerable mass effects on initial neuroimaging, and the presence of cardiopulmonary dysfunctions are poor prognostic factors.

6.6 Physiotherapy Assessment and Management

Brain injury patients often have complex and diverse symptoms with an impact on many areas of their lives. Since the neurological consequences are numerous and complex, these patients require specific and extensive rehabilitation. The acute phase of rehabilitation often needs to be followed by a very long, psychologically challenging and intense period of rehabilitation and can make the work of the team members, including physiotherapists, difficult. There is strong evidence that early, intensive physiotherapy has many positive effects, including utilization of the brain plasticity for maximum recovery within the constraints imposed by the injury. Rehabilitation several years after the primary brain injury also may have a long-term but less pronounced effect.

Several different assessment and treatment methods have been suggested by physiotherapists working with TBI patients. The treatments differ considerably between the acute phase of rehabilitation and the later stages. The acute phase commonly focuses on respiratory care, passive ROM treatment, and reduction of spasticity, whereas the later phase consists of functional training and compensatory strategy training. With regard to the patient's functional status, current evidence could neither identify any major difference between the various treatment methods nor an optimal treatment method.

6.6.1 History and Review of Acute Care Medical Records

History is an essential facet in the evaluation of a patient having TBI. Details about the mechanism of injury and its complications should be recorded. Initially, the level of consciousness should be determined using the GCS and any medical records for fractures, nerve injuries, or bleeds should be noted. If the patient is on ventilator support or ICP monitor or is advised to restrict ROM or weight-bearing owing to a musculoskeletal injury, a thorough medical record review may provide a comprehensive perspective about the patient's condition. It also helps to understand the precautions and contraindications that must be observed during the physical examination and subsequent treatment.

History of the use of any addictive substance can be relevant while planning post-discharge services with respect to adherence to the rehabilitation program. Chronic and heavy use of addictive substances like alcohol might lead to withdrawal symptoms and are more likely to hinder the rehabilitation process as it might affect the alertness, attention, comprehension, memory, cooperation, and behavior of the patient. Surgical history is also imperative to understand the possible post-surgical complications like brain edema and hemorrhage and to be cautious with respect to the position of the head (immediately after surgery, the head end of the bed is kept elevated at 15–30° with the neck in neutral), to facilitate the CSF drainage and aid the cerebral venous outflow to manage the raised ICP.

6.6.2 Physical Examination

Prior to handling the patient, it is important to collect all critical information, as many patients may not be medically stable and are likely to have various precautions and complications. Vital signs, general appearance, cardiopulmonary, genitourinary, musculoskeletal, and neurological examinations are the components of physical examination. Signs like heart rate, blood pressure, respiratory rate, and oxygen saturation should be noted to know whether the patient is medically stable. Fever with or without associated autonomic dysfunction can be an indication of infection, pulmonary embolism or pain. Prolonged bed rest, autonomic dysfunction, and certain medications can lead to reduced venous return and cause orthostasis.

General observation should include inspection for any devices like tracheostomy, gastrostomy, and chest tube(s), and urinary catheter. It is essential to note if the patient is on any intravenous or central lines. The patient's psychological status with respect to affect, mood, and cooperation must be noted. Any abnormal posturing or movements like dystonia and myoclonus should be documented. In case if the TBI patient is non-cooperative, confused, or agitated, a thorough evaluation may not be possible. Observe the skin for any lesions like lacerations or ecchymoses, which may provide information about the mechanism of trauma or previous undetected trauma. The skin over bony prominent areas must be observed for any discoloration and presence of wound. It is also essential to observe for dry, leathery, and scaly skin which can be due to autonomic dysfunction. Observe for conjunctival hemorrhage and exophthalmos. Medical records may provide information about ear, nose, and throat injuries.

X-rays may provide information about any fractures of the rib(s) or any other thoracic bony structures. Observe the type of breathing pattern and in addition auscultate the chest to note for any atelectasis, reduced air entry, or pulmonary consolidation. It is essential to note chest excursion by palpation. Patients with a tracheostomy must be assessed for the cough reflex and the cough strength must be noted. In cardiac evaluation, attention should be given to arrhythmias if present. Observation of limbs for long-standing skin changes, absence of pulses indicating occlusion of arteries or compartment syndrome, and swelling indicating venous thrombosis, covers assessment of the peripheral vasculature. Any bony or joint deformities, swelling, limb-length discrepancy, or amputation must be observed and documented. Passive and active ROM of joints should be performed to assess the mobility, weakness, pain, and tightness or contracture. Muscle strength evaluation using Kendall's manual muscle testing system or Medical Research Council (MRC) grading system may not be possible in many of the patients due to the altered sensorium or poor cognitive function. The following are the components assessed with regard to the neuromuscular system:

- Assessment of the LOC
- Assessment of cranial nerves
- Sensory system assessment
- Upper motor neuron (UMN) and lower motor neuron (LMN) injury assessment
- Reflex and tone assessment
- Balance and coordination assessment
- Functional assessment

6.6.2.1 Assessment of the Level of Consciousness

Grading the LOC is essential for both assessment and treatment purposes. Neurobehavioral criteria, the guidelines for the level of consciousness to differentiate coma, vegetative, minimally conscious, locked-in syndrome, and akinetic mutism states were recommended by ACRM. Table 6.8 provides the neurobehavioral features of coma, vegetative, and minimal conscious states. The Western Neuro Sensory Stimulation Profile (WNSSP) was established to evaluate the cognition in severely affected head injury patients. The scale has nine subsets ascribed to assess the behavioral aspects namely, arousal attention, auditory response, auditory comprehension, expressive communication, visual comprehension, visual tracking, object manipulation, tactile response, and olfactory response. There are 33 components, and the highest achievable score is 113. The test is reliable (0.95), and the concurrent validity is strong in relation to the Rancho Los Amigos (RLA) levels of cognitive functioning scale.

The Disorders of Consciousness Scale (DOCS), designed to measure arousal and neurobehavioral recovery among patients with altered conscious levels, is a valid and reliable tool. The scale consists of 23 items, and the components include olfactory function, swallowing, tactile sensation, proprioception, auditory function, visual function, and social knowledge. The Coma Recovery Scale-Revised (CRS-R), another tool consisting of six subscales with 23-items, is valid and reliable and assesses the level of consciousness.

The commonest cognitive deficits associated with traumatic injury of the brain are in attention, memory, and executive functioning. Attention is best assessed by asking the patient to repeat digits randomly in ascending and descending order. For

Table 6.8 Neurobehavioral features in coma, vegetative, and minimal conscious states

State	Neurobehavioral features
Coma state	• Does not open eyes spontaneously or to external stimulation • Does not follow any commands • Not able to utter recognizable words • Does not demonstrate purposeful movement but may show reflexive movement like flexor withdrawal or decerebrate or decorticate posturing to noxious stimulus, involuntary smiling • Non-sustained visual pursuit movements of the eyes through a 45° arc in any direction when the eyes are manually held open
Vegetative state	• Same as all the abovementioned points except the ability to open the eyes spontaneously or after stimulation
Minimally conscious state	• Unequivocally meaningful behavioral response after a specific command, question, or environmental prompt like an attempt to wave hand to say goodbye • Evidence of equivocal meaningful response, which occurs significantly less often to specific command, question, or prompt • Few responses observed on at least one occasion during a period of formal assessment

instance, the examiner can ask the patient to repeat seven digits forwards and five digits backward. The bedside test for orientation consists of asking the patient to state his name and personal identity, the city or place where he belongs, and the time of the day, date, and the ongoing season. If oriented with the present situation, the patient should be able to state the reason why he or she is in the healthcare center. The Galveston Orientation and Amnesia Test (GOAT) can be used to measure the orientation to person, place, and time, and memory. Evidence suggests that persisting altered GOAT scores seen in bilateral or diffused injury patients have longer PTA as compared to focal unilateral lesions. Duration of unconsciousness is usually the period from the time of injury till the individual achieves a score of 8 or more on GCS. PTA existing for more than 2 weeks could possibly indicate a severe trauma.

The Rancho Los Amigos (RLA) scale for levels of cognitive functioning is a descriptive scale used to examine cognitive and behavioral functional changes when the TBI patient emerges from a coma. Though the scale does not address specific cognitive deficits, the general cognitive and behavioral statuses are useful for communicating and planning treatment. The cognitive and behavioral recovery among TBI patients may plateau at any level. The RLA scale (Table 6.9) is a reliable and valid measure of cognitive and behavioral function for patients with TBI. The original scale had eight categories, and the revised RLA scale has ten categories that describe typical cognitive and behavioral progression following brain trauma. The revised scale provides a better understanding of the highest levels of recovery, with two additional components (9 and 10) defining superior functioning.

Memory depends on accurate encoding and retrieval of data. The patient is asked to recollect a sequence of 3–4 words of random objects (for instance, orange, cycle, bat, and tea), to evaluate immediate and delayed memory. Ensure that the patient recites all items at least once, and after a short distraction, the patient should be asked to recollect the sequence. Document the number of words he or she can recall spontaneously or after giving clue(s). The ability to recall with clues indicates retrieval issues than encoding issues.

Abstract reasoning can be assessed by asking the patient to interpret common phrases/proverbs or list similarities or differences between two objects. It requires the participant to draw conclusions based on his/her thought process rather than facts. To assess judgment, an imaginary situation can be given, and the patient can be asked how they would react in that particular situation, especially with respect to the safety of self and people around. Communication can be assessed to identify the presence of any disorders like aphasia, apraxia, dysarthria, dysphonia, or cognitive-linguistic dysfunctions.

6.6.2.2 Assessment of Cranial Nerves

With respect to TBI, olfactory, oculomotor, trochlear, abducens, facial, and vestibulocochlear are the frequently injured cranial nerves. Among the abovementioned, the olfactory nerve is the most frequent and is usually associated with trauma to the cribriform plate. Anosmia is noted in about 2–38% of patients with brain injury. It

Table 6.9 The Rancho Los Amigos scale

Level of cognitive functioning	A person at this level will or may	What family and friends can do
Cognitive level I No response	• Be unresponsive to sounds, sights, touch, or movement.	• Keep the room calm and quiet. • Keep comments and questions short and simple. • Explain what is about to be done using a "calm" tone of voice.
Cognitive level II Generalized response	• Begin to respond to sounds, sights, touch, or movement. • Respond slowly, inconsistently, or after a delay. • Respond in the same way to what they hear, see, or feel. Responses may include chewing, sweating, breathing faster, moaning, moving, and increasing blood pressure.	• Same approach as for Level I.
Cognitive level III Localized response	• Be awake on and off during the day. • Make more movements than before; react more specifically to what they see, hear, or feel. For example, they may turn towards a sound, withdraw from pain, and attempt to watch a person move around the room. • React slowly and inconsistently. • Begin to recognize family and friends. • Follow some simple directions such as "look at me" or "squeeze my hand". • Begin to respond inconsistently to simple questions with "yes" and "no" head nods. • Respond more consistently to familiar people.	• Limit the number of visitors to 2–3 people at a time. • Allow the person extra time to respond, but don't expect responses to be correct. • Give the person rest periods. • Tell the person who you are, where they are, why they are in the hospital, and what day it is. • Bring in favorite belongings and pictures of family members. • Engage the person in familiar activities, such as listening to their favorite music, talking about their family and friends, reading out loud to the person, watching TV, combing their hair, putting on lotion, etc.

(continued)

Table 6.9 The Rancho Los Amigos scale

Level of cognitive functioning	A person at this level will or may	What family and friends can do
Cognitive level IV Confused, agitated	• Be very confused and frightened. • Not understand what they feel or what is happening around them. • Overreact to what they see, hear, or feel by hitting, screaming, using abusive language, or thrashing about. In some cases, they may need to be restrained to prevent hurting themselves or others. • Be highly focused on their basic needs, i.e., eating, relieving pain, going back to bed, going to the bathroom, or going home. • May not understand that people are trying to help them. • Not pay attention or be unable to concentrate for more than a few seconds. • Have difficulty following directions. • Recognize family/friends some of the time; with help, be able to do simple routine activities, such as feeding themselves, dressing or talking.	• Allow the person as much movement as is safe. • Allow the person to choose activities, and follow their lead, within safety limits. Do not force the person to do tasks or activities. • Give the person breaks and change activities frequently especially if they are easily distracted, restless, or agitated. • Keep the room quiet and calm. For example, turn off the TV and radio, don't talk too much and use a calm voice. • Limit the number of visitors to 2–3 people at a time. • Experiment to find familiar activities that are calming to the person, such as listening to music, eating, etc. • Bring in family pictures and personal items from home to make the person feel more comfortable. • Tell the person where they are and reassure the person that they are safe. • Take the person for rides, if the person uses a wheelchair. If ambulatory, take the person for short walks in a safe environment.

Cognitive level V Confused, inappropriate, nonagitated	• Be able to pay attention for only a few minutes. • Be confused and have difficulty making sense of things around them. • Not know the date, where they are or why they are in the hospital. • Need step-by-step instructions to start or complete everyday activities, such as brushing their teeth, even when physically able. • Become overwhelmed and restless when tired or when there are too many people around. • Have a poor memory. They will remember past events which happened prior to the accident better than their daily routine or information they have been told since the injury. • Try to fill in gaps in memory by making things up. • May get stuck on an idea or activity and need help switching to the next step. • Focus on basic needs such as eating, relieving pain, going back to bed, going to the bathroom, or going home.	• Repeat questions or comments as needed. Don't assume they will remember what you have told the person previously. • Tell the person the day, date, name, and location of the hospital, and why they are in the hospital when you first arrive and before you leave. • Keep a calendar and list of visitors available. • Keep comments and questions short and simple. • Help the person organize and get started on an activity. • Limit the number of visitors to 2–3 people at a time. • Give the person frequent rest periods when they have problems paying attention. • Limit the number of questions you ask. Try not to "test" the patient by asking a lot of questions. • Help the person connect what they remember with what is currently going on with their family, friends, and favorite activities. • Bring in family pictures and personal items from home. • Reminisce about familiar and fun past activities.

(continued)

Table 6.9 The Rancho Los Amigos scale

Level of cognitive functioning	A person at this level will or may	What family and friends can do
Cognitive level VI Confused, appropriate	• Be somewhat confused because of memory and thinking problems. Will remember main points from a conversation, but forget and confuse the details. For example, they may remember they had visitors in the morning but forget what they talked about. • Follow a schedule with some help, but become confused by changes in the routine. • Know the month and year, unless there is a severe memory problem. • Pay attention for about 30 min, but have trouble concentrating when it is noisy or when the activity involves many steps. For example, at an intersection, they may not be able to step off the curb, watch for cars, watch the traffic light, walk, and talk at the same time. • Brush their teeth, get dressed, feed themselves, etc., with help; know when they need to use the bathroom. • Do or say things too fast, without thinking about potential consequences. • Know that they are hospitalized because of an injury, but will not understand all of the problems they are having. • Be more aware of physical problems than thinking problems. They often associate their problems with being in the hospital and think they will be fine at home.	• Repeat things. Discuss things that have happened during the day to help the person improve their ability to recall what they have been doing and learning. • Encourage the person to repeat information that they need or want to remember. • Provide cues to help the person start and continue activities. • Encourage the person to use familiar visual and written information to help the person with their memory (e.g., calendar). • Encourage the person to participate in all therapies. They will not fully understand the extent of their problems and the benefits of therapy. • Encourage the person to write down something about what they have done each day.

Cognitive level VII Automatic, appropriate	• Follow a set schedule. • Be able to do routine self-care without help, if physically able. For example, they can dress or feed themselves independently. • Have problems in new situations and may become frustrated or act without thinking first. • Have problems planning, starting, and following through with activities. • Have trouble paying attention in distracting or stressful situations. For example, family gatherings, work, school, church, or sports events. • Not realize how their thinking and memory problems may affect future plans and goals. Therefore, they may expect to quickly return to their previous lifestyle or work. • Continue to need supervision because of decreased safety awareness and judgment. They still do not fully understand the impact of their physical or thinking problems. • Think more slowly in stressful situations; be inflexible or rigid, and they may seem stubborn. These behaviors are common after brain injury. • Be able to talk about doing something, but will have problems actually doing it.	*Approach for levels VII and VIII are the same:* • Treat the person as an adult while still providing guidance and assistance in decision-making. Their opinions should be respected and their feelings should be validated. • Talk with the person as an adult. Use a natural and respectful tone of voice and attitude. You may need to limit the amount of information or the complexity of the vocabulary, but do not talk down to the person. • Be careful when joking or using slang, because the person may take things literally and misunderstand the meaning. Also, be careful about teasing the person. • Be sure to check with the physicians on the person's restrictions concerning driving, working, and other activities. Do not rely only on the person with the brain injury for information, since they may feel they are ready to go back to their previous lifestyle right away. • Help the person participate in family activities. As the person begins to see some of the problems they have in thinking, problem-solving, and memory, talk with the person about how to deal with these problems without criticizing the person. Reassure the person that the problems are caused by the brain injury. • Strongly encourage the person to continue with therapy to increase their thinking, memory, and physical abilities. They may feel that they are completely normal. However, they are still making progress and may benefit from continued treatment.

(continued)

Table 6.9 The Rancho Los Amigos scale

Level of cognitive functioning	A person at this level will or may	What family and friends can do
Cognitive level VIII Purposeful, appropriate	• Realize that they have a problem with their thinking and memory skills. • Begin to compensate for their problems; be more flexible and less rigid in their thinking. For example, they may be able to come up with more than one way to solve a problem. • Be ready for driving or job training evaluation. • Be able to learn new things at a slower rate. • Still become overwhelmed in difficult, stressful, rapidly changing or emergency situations. • Show poor judgment in new situations and may require assistance. • Need some guidance to make decisions. • Have thinking problems that may not be noticeable to people who did not know the person before the injury.	• Discourage the person from drinking or using drugs, due to medical complications. If substance abuse is an issue, encourage the person to seek outside help. • Encourage the person to use note taking as a way to help with their remaining learning problems. • Encourage the person to do their self-care and other daily activities as independently as possible. • Discuss what kinds of situations make the person angry and what they can do in these situations. • Talk with the person about their feelings. • Help the person think about what they are going to do before they do it, and practice before they actually do it. Afterward, talk about how it went and what might work better next time. • Consult with social work and psychology. Learning to live with a brain injury is difficult. It may take a long time for the person and family to adjust.

With permission from Rancho Los Amigos National Rehabilitation Center
Hagen C, Malkmus D, Durham P. (1979). Levels of Cognitive Functioning, Rehabilitation of the Head Injured Adult; Comprehensive Physical Management, Downey, CA: Professional Staff Association of Rancho Los Amigos National Rehabilitation Center

Table 6.10 Fundus and pupil examination findings of an unconscious patient

Structure	Features
Fundus	• Papilledema indicates raised ICP but might not be noted in the acute stage • Sub-hyaloid hemorrhages are usually noted with SAH • Presence of microaneurysms, blot hemorrhages, microvascular abnormalities, and "cotton wool" spots suggest diabetic retinopathy • May show signs of hypertensive retinopathy
Pupil size	• Small (pinpoint) pupils can occur following opioids and lesions in the pons • Semi-dilated (unresponsive) pupils indicate lesions in the midbrain • Large (unresponsive) pupil can be an early sign of third cranial nerve compression
Pupil asymmetry	• Large, non-reacting pupil can be due to ipsilateral lesion to the third cranial nerve • Adie's pupil—a pupil with parasympathetic denervation which constricts poorly to light, responds better to accommodation • Small, slow dilating pupil seen in Horner's syndrome
Pupillary light reflex	• May be sluggish in unconscious individuals

is associated with frontal skull fractures and post-traumatic rhinorrhoea. While performing bedside tests, avoid pungent odors like camphor or menthol and separately perform the test for each nostril using familiar odors like lemon oil, eucalyptus, or coffee extracts.

Though the optic nerve is less frequently involved, the injury can be due to a fracture passing through the orbit or secondary to prolonged ICP. It can result in visual loss or field defect, and the chiasmal lesion can present with bitemporal hemianopia. Table 6.10 provides findings regarding the fundus and pupil examination of an unconscious patient. Third cranial nerve palsies are usually associated with direct trauma or severe head injury. Compression of the nerve due to expanding EDH or SDH or herniation can be the other possibility. The injury is associated with ptosis, mydriasis, strabismus, and ophthalmoplegia. The long course of the trochlear nerve (from the midbrain to orbit) makes it particularly vulnerable to injury in association with severe TBI. The contrecoup forces can compress the trochlear nerve against the rigid tentorium over which it lies, and the injury can cause superior oblique palsy and diplopia typically worsened in downward gaze. The abducens nucleus or subarachnoid space lesion and petrous apex orbit and basilar skull fractures may cause traumatic sixth nerve palsy.

Fracture of the base of the skull and transverse temporal bone fracture can result in facial and vestibulocochlear nerve injuries. Cortical or subcortical lesions involving corticonuclear fibers of the seventh cranial nerve result in the UMN type of facial palsy. The UMN facial palsy is characterized by the paralysis of the contralateral lower half of the face with the sparing of the muscles on the upper half of the face. Damage to the facial nucleus or the nerve emerging from the nucleus causes the LMN type of facial palsy characterized by weakness of all muscles on the ipsilateral side of the face. Injury to the eighth cranial nerve leads to vestibulocochlear features like hearing deficits, vertigo, dizziness, and tinnitus. If the bedside "whisper test" suggests deafness, Rinne's test and Weber's test can be carried out to distinguish conductive deafness from sensorineural deafness. In middle ear deafness or conductive hearing loss, for Rinne's test, the bone conduction appears better than air conduction. Whereas for sensorineural deafness, when Weber's test is performed, the sound is heard best in the normal ear when compared to the involved ear.

6.6.2.3 Sensory System Assessment

Materials like a cotton tip applicator, hot and cold water in separate test tubes, a two-point aesthesiometer, and a paper pin can be used for standard bedside testing of the somatosensory system. Careful assessment is essential to identify the radicular and peripheral nerve distribution pattern impairments. For the joint proprioception test, distal joint(s) are tested first, and if any impairment is noted, more proximal joints are further tested. The Romberg's test is a specific test for joint position and kinesthetic sensation and is performed by asking the patient to stand with feet together, with eyes open and then with eyes closed. The patient with considerable proprioceptive loss will be able to stand still with eyes open by effective compensation using vision for the loss of position sense but will sway or fall when the eyes are closed.

6.6.2.4 UMN and LMN Injury Assessment

Following TBI, patients may suffer damage to both the central and peripheral nervous systems. Therefore, identifying and differentiating UMN from LMN injuries is critical during the neurological examination. Post-TBI, weakness involving one-half of the body, both lower extremities, and both upper and lower extremities, including the trunk, are the most likely results of an UMN injury. Whereas segmental, plexus, or radicular presentation can be the result of a LMN injury. Table 6.11 depicts the difference between the UMN and LMN lesions.

6.6.2.5 Reflex and Tone Assessment

Asymmetry in muscle activation or localized muscle atrophy can be an indication of peripheral nerve damage. The presence of pathological reflexes, like Babinski sign and Hoffman's sign(examiner flicks the fingernail of the middle finger downward, and the response is an involuntary flexion and adduction movement of the thumb and/or flexion of the index finger) are indicative of UMN lesions. An increase in resistance to a passive stretch of a muscle, which is velocity-dependent, is an important feature of spastic hypertonia and modified Ashworth Scale (MAS) will be the preferred tool to grade the spasticity. However, in subacute or chronic cases, Tardieu's scale will be most appropriate as it helps to differentiate neural components from the soft tissue changes (tightness and contractures).

Table 6.11 The difference between the UMN and LMN lesions

Component	UMN feature	LMN feature
Site of lesion	Cerebral hemisphere, brainstem, and spinal cord	Anterior horn cells, ventral roots, plexus, and peripheral nerves
Weakness	Weakness present and group muscles are involved, typically seen generalized to one half of the body, both lower extremities or both upper and lower extremities, including the trunk	Weakness present but restricted to individual or selective muscles locally.
Muscle bulk	Bulk is usually normal but may present with disuse atrophy in longstanding cases	Significant atrophy of the muscles
Type of paralysis	Spastic paralysis	Flaccid paralysis
Tone	Hypertonia	Atonia/hypotonia
Abdominal reflex	Absent	Present
Deep tendon reflex	Hyperreflexia	Absent/reduced (hyporeflexia)
Babinski sign (plantar reflex)	Present (upgoing plantar response)	Absent (down going response)
Hoffman's sign	Present	Absent
Clonus	Present	Absent
Fasciculation	Absent	Present
Sensory loss	Cortical sensations lost	Peripheral sensations lost

6.6.2.6 Balance and Coordination Assessment

The lists of equilibrium and non-equilibrium tests mentioned in Chap. 5 "Cerebellar Dysfunction" helps to comprehensively test the balance and coordination of the patient. The coordination test includes the finger to nose test, the heel to shin test, the test for dysdiadochokinesia, and the rebound phenomenon. The extent to which the incoordinated movements are unsafe and hinder the routine functional activities needs to be documented. Other than the observational gait analysis of the ambulatory patient, for natural and tandem gait (if possible), the gait parameters like speed, safety, type of gait deviation, features of gait phases, balance, and coordination between the limbs should be noted. For those patients who are non-ambulatory, transfers, wheelchair mobility, and upper limb strength need to be assessed.

6.6.2.7 Functional Assessment

The ability to perform transfers helps the patient feel a sense of independence. The assessment should include bed mobility, transitions like supine-to-sit, sit-to-stand, weight shifts in sitting and standing, transfer from bed to the wheelchair, and independent transfers. The Functional Independence Measure (FIM), an 18-item assessment scale, is a common tool administered for the level of independence in mobility, self-care, and cognition. FIM may not be sensitive for those patients who are very low or highly functional and hence may not be an appropriate outcome measure for patients with minimal brain injury or severe brain injury.

The Disability Rating Scale (DRS), when used in TBI, helps to track the recovery process from coma towards full recovery. The initial three components (eye opening, ability to communicate, and motor response) of the DRS represent the GCS and measure the impairments. The potential to feed, groom, and perform personal hygiene gives information about the extent of disability. The items of function are the modified version of a tool utilized by Scranton et al. With respect to DRS scoring, the lowest score of 0 denotes no disability, and the highest score of 29 represents a severe vegetative state. The tool can be administered by interviewing the patient or family member, and being concise and easy to administer, DRS is a popular outcome measure for assessing TBI.

6.6.3 Acute Phase Intervention

The physiotherapeutic care of the TBI patient should begin as soon as the patient is clinically stable. Promoting the arousal level, preventing the development of secondary impairments, improving the patient's function, and educating the patient and the caregivers regarding the condition are some of the important early goals of intervention. In minor and moderate TBI, with the absence of any medical complications, the length of stay in the acute care facility is generally short. However, in severe TBI with associated medical complication(s), the length of stay can stretch beyond the average length of hospitalization, which is usually less than 2 weeks.

Moderate-to-severe TBI is likely to exhibit abnormal tone and postures. Hence, early intervention should address the patient's positioning. In the supine position, the impact of tonic labyrinthine reflex and dominance of extensor tone is more profound, and to minimize the influence of the tonic labyrinthine reflex, side-lying and semi-prone positions are desirable. However, for those with respiratory complications, care should be taken while positioning in semi-prone or prone positions. Whether in supine, side-lying or prone position, the upper extremities should be placed in slight abduction and external rotation to exert an inhibitory influence on the abnormal muscle tone. It is desirable to educate the family members and nursing staff regarding how to position the patient out of the decerebrate or decorticate posture. The use of firm towels and pillows bolsters, or half rolls can be used for maintenance of optimal position. In case of significant abnormal muscle tone, contractures can develop quickly, especially in the elbow and ankle. Proper positioning, gentle stretching, ROM exercise, and static splinting, may alleviate or retard such potential complications.

The treatment strategies to minimize or prevent the development of secondary complications include strategies to overcome the effects of prolonged bed rest and respiratory deficits. Prolonged immobilization is a common denominator in all cases of heterotopic ossification. Though the incidence of heterotopic ossification is between 11% and 76%, no effective treatment method is available and difference of opinion does exist regarding the continuation of physiotherapy following the diagnosis. Many experts do agree that the ROM exercises should be continued to prevent possible ankylosis and positioning, splinting, and spasticity management, including the use of reflex inhibiting patterns of movement may be beneficial.

Regarding respiratory care, cardiorespiratory physiotherapy interventions may sometimes have a deleterious consequence on the blood pressure, cerebral blood supply, and oxygenation and may lead to secondary cerebral ischemia. Hence, a quick assessment should be performed to weigh the risks involving ICP and CPP before any physiotherapy interventions are executed. ICP of less than 15 mm Hg carries low risk as the CPP tends to be stable at around 70 mm Hg. The risk becomes moderate when the ICP raises to 15–20 mm Hg and high when it crosses 20 mm Hg. Hence it is advisable to keep the intervention time short to minimize the deleterious effects of raised ICP or unstable CPP. Interacting with the nursing staff can give valuable information regarding the patient's response to handling or interventions. Local policies should be checked regarding monitors, equipment, patient parameters, and post-surgical procedures. The common respiratory issues and treatment options are listed in Table 6.12. Table 6.13 lists the effects of respiratory physiotherapy interventions.

During this acute stage of rehabilitation, even for comatose patients, activities targeted towards improving the level of awareness are important. Though a coma patient may not be able to converse or move limbs on command, it should not be assumed that the patient is unable to hear or understand the information provided. Instead, it should be assumed that the patient can hear and understand information relevant to his or her care. During the therapy sessions, it is vital that the therapists should explain to the patient what they are doing at all times. Communicating with the patient on a personal level also establishes that therapists are caring and respectful and are trying to develop a good rapport with the patient.

Table 6.12 Common respiratory problems and treatment options

Common respiratory issues	Treatment options
Reduced airway protection	• Head and neck maneuvers to optimize ventilation • Removal of obstruction if caused by the tongue, foreign objects, tissues, and bodily fluids like blood and gastric contents • Supraglottic (oropharyngeal and nasopharyngeal airways) infraglottic (tracheal intubation), and surgical methods (cricothyrotomy and tracheotomy).
Retention of secretion	• Oral/tracheal suctioning • Postural drainage • Manual hyperinflation • Chest vibrations and percussion techniques • Neurophysiological facilitation for respiration • Intermittent positive pressure breathing
Hypoventilation/at electasis	• Manual hyperinflation • Intermittent positive pressure breathing
Type II respiratory failure	• Communicate with a critical care specialist or intensivist • May require intubation or NIV

Table 6.13 Respiratory physiotherapy interventions and their effects

Strategy or technique	Possible effects
Manual hyperinflation technique	• Manual hyperinflation tends to increase the intrathoracic pressure and leads to a reduced venous return to the heart, increased cerebral blood volume, and raised ICP • Manual hyperinflation reduces the filling pressure to the right atrium from the inferior vena cava leading to reduced stroke volume and a drop in blood pressure • The rate and intensity of manual hyperventilation will affect the cerebral vasculature due to CO_2 retention or removal • For raised ICP, rapid, small volume breaths will help to reduce ICP by removal of CO_2, along with intermittent larger volume manual hyperinflation breaths.
Intermittent positive pressure breathing	• Tends to increase the intrathoracic pressure, and the effects are similar to manual hyperinflation • May reduce the mean arterial pressure
Positioning	• 15–30° head-up position is advisable to reduce ICP • For patients with raised ICP, the head should be kept in the midline with chin in line with the sternum to reduce pooling of venous blood within the brain due to venous obstruction near the neck region. • For those who cannot tolerate turning, bolus sedation may be required. • Gentle and slow changing of the patient's position is advisable for raised ICP • In case of bulbar weakness or paralysis risk of aspiration is high.
Chest manipulation techniques (vibrations and percussions)	• Ensure adequate padding (use of a towel or folded bed-sheet) over the chest region prior to chest manipulation • If the patient cannot tolerate the manipulations, he or she may require bolus sedation to ensure adequate analgesia prior to manual techniques • May induce bronchospasm • Vigorous manipulations and shaking not advisable if the ICP is high
Suction	• Can cause transient hypoxia • Transient hypercapnia leading to raised ICP • Suction may induce coughing, which can increase the intrathoracic pressure and lead to increased ICP • Can induce the Valsalva maneuver (forced exhalation against a closed glottis.) • May trigger a vasovagal response leading to bradycardia (heart rate below 60 beats per minute)

6.6.3.1 Coma Stimulation

Coma stimulation is an approach wherein a therapist or a caregiver systematically applies stimulation to one or more of the patient's special or somatosensory senses to aid or improve the patient's responsiveness. The rationale for coma stimulation is that repeated exposure to different types of sensory stimulation may facilitate dendritic growth and improve the synaptic connectivity of the damaged nervous systems. Such stimulation can improve the patient's cognitive functioning, environmental awareness, and interaction. It is a low invasive, harmless, inexpensive, and simple-to-apply technique and is considered a potentially attractive rehabilitative method for severe TBI patients with poor arousal levels.

Coma stimulation consists of a structured sensory stimulation presented in a systematic fashion, which may include visual, auditory, olfactory, gustatory, vestibular, or tactile senses. For a coma stimulation program, the application of stimuli can be either unimodal or multimodal. In unimodal stimulation, one of five different senses mentioned here can be used. Many studies have used auditory stimuli as their primary stimuli for the program. In multimodal stimulation, all of the five senses are stimulated repeatedly at a regular interval.

Principles of Coma Stimulation The therapist should carefully apply the stimuli to ensure that the technique is harmless. Before the introduction of stimuli, it is essential to record the baseline values of heart rate, blood pressure, and respiratory rate. The program should be implemented only on those who are medically and hemodynamically stable. Coma stimulation programs for ventriculostomy patients with raised ICP or lower CPP can be minimized or avoided. When indicated for such patients, pre and post-session ICP and CPP monitoring are mandatory.

Meaningful stimuli should be used, such as familiar voices or sounds, music, and scents. The surroundings should be less distracting, simple, and organized, with limited persons near the patient. The stimuli should be organized and presented in a sequential manner, and only one or two types of senses should be involved at a time. To increase the probability of obtaining responses in later sessions, stimuli that are emotionally associated with the patient should be reinforced. The therapist should identify the list of stimuli to which the patient gives the correct response. The purpose of rehabilitation is directed to raise the rate of reaction, the duration of alertness, the variations in the reactions, and the attention quality. When administering sensory stimulation to unresponsive TBI patients, it is advisable to limit the time of the exposure. To prevent habituation for stimuli, it is essential to control the intensity and frequency of stimulation. In consideration of the delayed processing of information, extra time should be provided for the patient to respond appropriately.

The patients' responses to the different sensory modalities administered must be observed. For instance, the responses may include changes in heart rate, blood pressure or respiration rates, diaphoresis; increase or decrease in muscle tone, head-turning, eye movements, grimacing, or vocalizations. Pressure on the patient's nail bed and the use of a pin to elicit noxious stimulus may elicit a pain response, however, such stimuli are only used if the patient is not responding to other forms of stimulation. The therapist's voice can also be used as a tool to influence the patient's response. For those patients with a heightened state of awareness, the use of a soft tone of voice may calm the patient. On the contrary, for those patients who are lethargic, brief, concise, loud voice commands can be utilized to arouse them.

Overstimulation can agitate the patient and may cause increased fatigue and, therefore should be avoided. Increased perspiration, flushing of the skin, signs of distress, tendency to keep the eye closed, unexpected drop in the arousal, increased muscular tone, and sustained increase in respiration rate are indications of overstimulation and if observed, should be rectified by providing an adequate period of rest. Articles on coma stimulation have used stimulation programs varying from 3 min to 2 h duration per day, with a frequency ranging from 2 days to 32 weeks. A study conducted by Megha et al. showed a considerable improvement in the level of consciousness after the application of multimodal stimulation programs over unimodal stimulation. They also concluded that frequent, short-duration stimulation (15–20 min per session, 4–5 times a day) had a positive effect on the recovery of the conscious level of comatose patients.

Techniques of Coma Stimulation On approaching the patient, for the purpose of identification, the therapist must introduce himself. The therapist should talk to the patient softly with an optimal tone of voice, and the words and statements used during the instructions should be short and simple for better comprehension. Extra time should be given to the patient to think about what has been instructed during the treatment session.

The bedside visual stimulation should be highly stimulating and should include colorful familiar objects, family photographs, and music videos for a few minutes at a time. The patient should be positioned upright to have an appropriate orientation to the visual stimuli. Distracting objects have to be reduced or removed to enable the patient to focus on the given visual stimulus. Once the patient focuses on the stimulus, visual tracking needs to be encouraged. With regard to auditory stimulation, only one person should be permitted to converse at a time. Music, television, and voice recordings can be provided at frequent intervals throughout the day. Soft music and instrumental music should be avoided as these are more likely to encourage sleep than to facilitate arousal. When providing bedside auditory stimulation, strong stimulus that evokes startle response should be avoided. To enhance the patient's response, it is advisable to change the direction and location of the stimulation. For instance, calling the patient's name, whistling, clapping, or playing music from different locations in the room, encourages the patient to improve the alertness and attention towards the auditory stimulus.

Tactile stimulation can be facilitatory or inhibitory in nature. For instance, light touch and pain produce an inhibitory response, while sustained touch, pressure to the oral area tends to produce facilitation. Variety of textures and temperatures such as patient's own clothes or blankets, warm or cold objects, and metal dipped for a few seconds in hot or cold water should be used for stimulation. Direct application of ice should be avoided to prevent a sympathetic response of the body. Stimulation should be employed on one of the lower limbs and then advanced to the other limb. ROM exercises can serve as movement stimulation. Activities like rolling, tilt-table to acclimatize a patient to an erect posture, change in the bed posture, rocking in a chair, and therapeutic mat activities can serve as movement stimulation. The therapist must note any early protective responses or late balance responses while doing these tasks. Slow alterations in position or movements tend to be inhibitory and faster movements facilitate arousal. Spinning activities may trigger seizures, and the use of mechanical interference or hydraulic systems to raise and lower the patient's bed has little functional meaning and limited response.

The use of strong odors like perfume, flavored extracts, coffee grinds, and lemon or peppermint oil is recommended for olfactory stimulation. To avoid overstimulation, each stimulus should be presented for not more than 10 s. Avoid contact of extracts or oils with the skin to prevent any habituation to the scent. Garlic and mustard are used as noxious stimuli. Vinegar and ammonia should be avoided because they irritate the trigeminal nerve. For those patients with possible or definite olfactory nerve injury and patients with tracheostomies, olfactory stimulations are meaningless, and the presence of a nasogastric tube blocking a nostril can worsen the sense of smell.

The stimulus for taste is provided, only if the patient is not susceptible to aspiration. A cotton swab dipped in a sweet, salty, or sour solution can be used. For those patients with difficulty to handle oral secretions, the sweet taste is avoided as it may increase salivation. Sponge-tip, glycerin swab, or a soft toothbrush can be used to reduce increased sensitivity, and the unnecessary oral reflexes advocated. A flavored cleansing liquid or mouthwash (mint or lemon) can be used to provide a stimulus while performing everyday mouth care. In case if the patient is defensive to touch and tends to purse the lips or close the mouth with a withdrawal, the stimulation is continued in a gentle manner to prevent any defensive responses and progressed carefully.

6.6.4 Cognitive Rehabilitation Therapy

Cognition refers to conscious mental activities, like thinking, remembering, comprehending, or learning. It is a broad term that refers to the mental processes which include perception, reasoning, judgment, intuition, memory, attention, problem solving, executive functioning, language, visual-spatial skills, awareness, and comprehension. It is a process by which sensory input is transformed, reduced, elaborated, stored, recovered, and used. Attention, recognition, discrimination, identification, maintenance of order of stimuli and responses, and organization are all part of the processes.

Cognitive rehabilitation therapy can sometimes be confused with cognitive behavioral therapy. Though not mutually exclusive and sometimes delivered conjointly, both the therapies are undoubtedly distinct and dissimilar, differing in both treatment goals and techniques. The former consists of rehabilitation of thinking skills like attention and memory following a brain injury, whereas the latter is commonly used to rehabilitate affective and psychiatric issues, which include anxiety, mood, chronic pain, and psychotic disorders. Cognitive behavioral therapy typically centers on modifying maladaptive thoughts and emotional behaviors using psychoeducation. It also helps to recognize and reappraise distorted negative thoughts and anxiety-provoking or distressing stimuli, with the intent of forming new adaptive emotional associations with the feared stimuli.

To benefit from cognitive rehabilitation, several factors should be considered. The CNS should be metabolically optimized to adequately support rehabilitation efforts and learning. During the early stages after TBI, the metabolic function may be compromised due to medications, brain parenchymal edema or neuromodulator, neuroendocrine, biochemical, and metabolic imbalances. Comorbidities like pulmonary, infectious, renal, hepatic, nutritional, or endocrine dysfunctions commonly seen after TBI may further compromise the cerebral function. General anesthesia and iatrogenic complications of certain pharmacological agents can also delay the recovery following brain injury. Even the metabolic status can impact the degenerative process (anterograde and retrograde axonal degeneration and deafferentation) that causes the death of adjacent and distant cells. If a metabolic compromise occurs while the cells are in a state of metabolic paralysis, they are likely to succumb even to the metabolic events that are normally harmless to the functioning cells. The role of neuroendocrine function as a precursor to cognitive function is being increasingly recognized. Around 40% of the TBI patients have hypopituitarism due to neuroendocrine dysfunction following head injury. The cognitive function has already been linked to the anterior pituitary function and should be considered while providing cognitive rehabilitation.

6.6.4.1 Therapeutic Intervention

Unlike unilateral focal brain disorders like stroke, brain abscess, or focal neoplastic lesion, the neuropathology of TBI is complex as it consists of both focal and diffuse cortical and subcortical lesions along with a cascade of neurobiological changes. Therefore, the cognitive disruption observed in TBI can be bilateral and extensive, and the challenge of cognitive rehabilitation training is to implement efficient and effective treatment modalities that will enable the patient to maximize his or her level of functioning in the face of this diffuse systemic disruption.

Cognitive rehabilitation therapy can be categorized into two major components: remedial and compensatory. Remedial rehabilitation is based on neuroanatomical and neurophysiological models of learning and suggests that neuronal growth and synaptogenesis are the direct results of repeated exposure and stimulation. The compensatory retraining model assumes that certain functions cannot be restored completely and hence methods for optimizing functional efficiency need to be used and strategically planned. During cognitive rehabilitation, both remedial and compensatory retraining strategies are used in combination to provide an effective approach.

Cognitive training strategies are designed to follow a hierarchical order that is created in accordance with the linguistic and cognitive components of normal developmental sequences. Attention, memory, perceptual feature processing, categorization, and cognitive distance are the five primary areas of cognition. Attention is presumed to be the most basic among cognitive skills, and it impacts and underlies all aspects of cognitive function. The ability to identify and interpret perceptual features effectively is instrumental in developing categorization skills. Treatment should begin with the performance of physical tasks in a controlled and enclosed environment and then progress towards mental tasks in a stimulus-rich environment. The treatment room should be designed in such a way that the therapy materials, furniture, ambiance and lighting, and temperature can be modified. For instance, if a TBI patient exhibits severe attentional deficit, he may require an environment with little visual and auditory stimuli present, like a room without furniture, low lighting, and minimal therapy materials. As the attention improves, the environment should be modified to implement a step-wise progression through the hierarchy of attention.

Once the patient can perform simple tasks in the controlled environment, complexity in auditory and/or visual stimuli may be gradually introduced. It is advised to move gradually from the least to the most salient and advance in one sensory modality at a time. For instance, therapy may begin with listening to instrumental or soul music that is of the soft and relaxing type and then progress to more energetic music at higher volume levels, ending with the type of music the patient enjoys most and knows well.

"Task complexing" is an important part of the therapeutic process and in consistency with the bottom-up approach framework; it should start with physical activities and then gradually progress to mental activities, utilizing a taxonomy of cognitive distance. Physical activities may include categorizing or sorting by iconic features like color or size, whereas symbolic categorization can be utilized for abstract activities. Complexity in tasks can be increased by adding more objects or allotting or engaging more time. When working with severe TBI patients, it is often necessary to perform a single therapeutic activity repeatedly for a considerable extent of time to build basic attention skills. The therapeutic environment should continuously challenge the patient to perform activities with satisfactory accuracy and sustained attention in a stimulus-rich environment.

Attention Impairment in attention is one of the major cognitive deficits post-TBI. It includes complaints like inability to concentrate and difficulty to alternate between tasks or multitask. Though the manifestation of attentional deficits can be readily noticed in a clinical setting, it may be difficult to identify the underlying cause precisely. For instance, when a TBI patient presents with "distractibility," we may acknowledge the need to simplify the therapeutic environment but may fail to adequately address the complex nature of the attentional disorder. There are several ways to address attention issues but designing a bottom-up therapeutic program based on a developmental approach helps to build attention skills. Such an approach will ensure that all the attention skills are acquired in a developmental and sequential manner, setting the base for higher-level cognitive processes. At the base of the hierarchy lies sustained attention, which is the individual's ability to direct and maintain focus with regard to a task across a period of time. Physical or concrete tasks, such as sorting, scanning, and finding the path or direction, should be introduced in the beginning. For those with very poor attention begin with simple, auditory, sustained, attention or vigilance tasks. For instance, asking a patient to listen to a string of stimuli targeting a specific number, letter, or word for short periods of time. Once the attention and accuracy improve, the duration of time can be extended. For visual attention tasks, the same hierarchy and concept can be used.

Selective attention is the next step in the hierarchy. It is the ability to sustain concentration on a specific task with good precision in a stressful environment. Once the patient exhibits consistency in performing tasks in a quiet and controlled environment, a hierarchy of distractors should be introduced. For instance, the patient can be moved from the controlled environment to a familiar environment with minimal distractors, and to further challenge the attentional system, he can be moved to a highly distractible familiar setting like a lobby or lounge, and finally, to provide maximal distractors, an unfamiliar and high traffic area like a mall or crowded place. Once the patient maintains attention in a maximum distractor-laden environment with good accuracy, abstract or mental tasks can be introduced as a progression of concrete tasks. Working memory tasks like reordering a string of random numbers from the smallest to the largest or the largest to the smallest are more cognitively challenging.

Alternating attention is the next component in the hierarchy and it is the ability to switch focus from one task to another with minimal interference to sensory stores, task sequencing, and task accuracy. Alternating attention is a complex cognitive skill in the hierarchy and is often impaired in TBI. To attain this cognitive shift skill, a relative intactness of the basic level of attention is crucial. Divided attention is the highest level of attention and it refers to the ability to simultaneously attend to two or more different tasks. For instance, normal subjects, divide their attention while driving, performing routine chores, or watching television. Performing these concrete and abstract tasks while simultaneously answering a series of questions differing in levels of complexity can promote alternating attention. For example, the patient may be asked to sort a pack of cards into the four suits (hearts, spades, diamonds, and clubs) while simultaneously responding to open-ended or true or false questions of varying difficulty.

Perseverative behaviors are a type of attentional deficit often seen among severe TBI patients and are characterized by the inability to shift a focus of attention among perceptual features. The activities that reduce perceptual salience and establish the use of iconic and symbolic feature identification skills typically result in a reduction of perseverative responses. Strategies like presenting an object to the patient and then diverting the patient's attention to various perceptual features of the object like color, shape, and size may help to overcome perseverative behaviors.

Memory Memory is a component of cognition that involves the ability to encode, store, retain and recall information. Richard Atkinson and Richard Shiffrin in their model described it as a three-stage sequence from sensory memory to working or short-term memory to long-term memory, as opposed to a single entity. The ability to retain impressions of sensory information once the original stimuli have ended is known as sensory memory. The sensory memory acts like a buffer but is retained accurately only for a brief period. Therapeutic tasks that address auditory or visual sensory memory include the echoic store and the iconic store tasks. For instance, the task may include the verbal presentation of random numbers, and the patient is requested for a forward recall or to organize the numbers from the smallest to the largest (latter assesses both immediate recall and working memory). Such tasks should typically progress from the presentation of three numbers to seven numbers. The goal is to achieve at least 80% accuracy over three consecutive trials for the seven numbers in a quiet environment and then progress to a distracting environment. For the visual memory task, numbers can be displayed on the television screen for 3–5 s, and then the patient can be asked to place those numbers in the same order as they appeared earlier on the screen. For the aforementioned task, the goal is to achieve at least 80% accuracy in recalling.

Working or short-term memory is the ability to retain information for a short period of time and process that information simultaneously. It acts like a "scratch pad" for the temporary recall of information that is being processed. Working memory is important for reasoning and decision-making. It has a limited capacity, and normally most subjects can only hold about seven pieces of information in their working memory. "Chunking," a term often used in working memory, refers to the ability to organize information into shorter meaningful groups to make them more manageable and improve the working memory capacity. For example, chunking a mobile number into groups of three and four digits is easier than remembering a string of ten numbers.

Long-term memory means the ability to store information over a long period of time. By the process of consolidation, which involves rehearsal and meaningful association, the short-term memories are converted into long-term memories. In many TBI patients, memory for "old" information is typically preserved. These patients usually do not forget their place of birth, the school where they studied, or other major events that occurred in their life. Usually, their ability to learn, store, or recall new information is more affected. Prospective memory is a type of memory that needs special emphasis. It refers to the ability to remember information or actions to be completed in the future, meaning "remembering to remember." Providing assignments to the patient to complete a task for a given date or time in the future and strategies to aid the patient recall the information, including the use of day planners, organizers, calendars, and sticky notes for mobiles, may help to improve prospective memory.

Feature Identification During the feature identification, the patient is trained to attend and identify different perceptual features of real objects. The perceptual features can be divided into seven iconic and one symbolic feature. Color, shape, size, construction, weight, and texture are some of the common iconic features used, whereas identifying the function of objects is the symbolic feature. Cognitive distance (an essential entity for reestablishing the subject's ability to not only take in sensory information when it is readily available, but also to call upon information when the available information is reduced or absent), the first level of feature identification consists of the patient describing the iconic and symbolic features of real objects. Progressively the cognitive distance is built by moving the patient through a hierarchy of sublevels consisting of color photographs of objects, black-and-white photographs of objects, line drawings of objects, written words, and ultimately, spoken words. Eventually, when the objects are no longer physically present, the patient has to rely on the mental representation of objects to describe them.

During the next level, the patient is encouraged to expand feature identification skills which consist of identifying eight features one by one and providing an extended feature. For instance, when describing a traffic signal, the patient should verbalize that the signal sign is green and should identify another object of the same color, such as a green apple or avocado. Abstract negation is the next level and during this stage, the patient further expands the feature identification skills through negative categorization. For instance, the signal sign means "go or proceed" and the "go" sign is not red and red is not the color of the sky. The cognitive distance hierarchy ranging from real objects to spoken words should again be followed during this phase. Often, severe TBI patients may find it difficult to provide extended and negative feature identification due to reduced visual imagery skills and mental flexibility.

Categorization In this level of the cognition module, the patient is required to identify the iconic and symbolic features of objects grouped together (the objects are arranged in three rows with three objects in each row). Here again, the concept of cognitive distance hierarchy needs to be maintained, which consists of activities beginning with real objects and then advancing towards spoken words. In the categorization stage, the patient should identify one perceptual feature in common across the three rows. For instance, if rows of red, blue, and green objects are set on a table, the patient should recognize "color" as the common perceptual feature. The subsequent stage of the cognition module includes symbolic categorization and the purpose is to develop the ability to categorize objects based on the function. The use of real objects is preferred when the patients have difficulty in following abstract photographs. The final step of categorization consists of the patient shifting his or her perspective and identifying three functions the object cannot be used. For instance, a pencil cannot be used as a screwdriver, cloth hanger, or comb.

Prioritization of perceptual characteristics is then carried out to tackle more complex language and cognitive concepts while encouraging creativity. This task requires the patient to identify whether each of the perceptual features is important or not important to the function of the object and provide a rationale. For instance, if the object is a pencil, "color or shape" is not important as long as it can be used for writing. Weight and size are important as the pencil cannot be used for writing if it is too heavy or bulky.

Next, a multisensory visualization task is given to further develop cognitive skills. The task requires the patient to describe a given experience using the five senses as well as generating emotionally based responses relevant to the situation. For instance, the patient might be prompted to give details of his or her cricket match watching experience. Here in this scenario, the correct response may include statements about the presence of two teams, the color of their jersey, the details of the cricket pitch, crowd cheering, smell of the popcorn, the rain which delayed the match, and the hot dog eaten by the fellow spectators. As the visual imagery skills improve, the cognitive distance can be increased by having the patient describe situations or experiences he or she is not familiar with. During the training, the processing speed needs to be monitored by timing the patient's responses in the different levels of the cognition module. When progressing to higher levels of complexity, response times may become lengthier; however, the response time tends to improve with repetition.

The cognition module helps with the overall thought organization process and it enables the ability to switch freely between perceptual attributes, eliminates the perceptual fixation, improves the registration of available perceptual information, and improves encoding, retrieval, problem-solving speed, attention, categorization, processing efficiency, and cognitive distance. Therefore, it is important to consistently cue the patient to a specific order allowing for improved organization and efficiency of information processing. The clinical methods discussed earlier are intended to regain cognitive abilities at the basic level. Higher-level thought processes and memory cannot be adequately addressed if the basic level cognitive skills are not first put into place. The cognition module is not meant to be the only treatment activity; rather, it is an essential part of the overall rehabilitation program.

6.6.5 Subacute Phase Intervention

Once the TBI patient is medically stable, for further rigorous intervention, he will be transferred to an inpatient rehabilitation setting. The chief problems during this stage include reduced ROM, potential contractures, abnormal tone and posturing, lower levels of awareness and responsiveness to the environment, presence of released reflexes, reduced functional mobility, poor tolerance to upright posture, decreased endurance, reduced sensory awareness, impaired communication, and lack of knowledge of about one's own condition.

The therapeutic strategies used in the subacute phase of rehabilitation focus on developing postural control and mobility. However, strategies deployed in the acute phase like the use of appropriate position and regular interval change of position for prevention of skin breakdown, improvement in lung function, and reduction of abnormal tone, should be continued during this phase, if necessary. Once the patient becomes medically stable, acclimation to an upright position is important. Sitting on a bed or chair improves the patient's orientation and aids endurance and bronchial hygiene.

Throughout the rehabilitation, the concepts of motor control and motor learning are applied. Neuroplasticity is the basis of the recovery process, and multiple systems of the body along with the environment are collectively focused during the rehabilitation. Both the environment and patient's motivation play a vital role in forming neural circuitries. Complex tasks as opposed to simple tasks are responsible for a greater number of synaptic formations with a significant improvement in function. Motor patterns required for performing bed mobility activities like rolling, supine to sit, sitting, sit to stand, static and dynamic standing, and walking as well as upper limb activities like reach-outs, grasping, and throwing, are also addressed in order to build independence. Along with various approaches, mental practice combined with physical practice can be effective in muscle group activation.

For those with poor head and trunk control with a low level of functioning, a tilt-in-space wheelchair or reclining wheelchair can be used. Unlike a reclining wheelchair, a tilt-in-space wheelchair allows the patient to recline while maintaining 90-degree angles at the hips, knees, and ankles. The tilt-in-space wheelchair simply shifts the patient from upright to reclining position as a single unit on its base. A wheelchair with straps, seat belts, and lateral trunk support can promote an upright posture for low-level functioning patients. For patients with fair trunk and head control, the use of a standard wheelchair can be encouraged. Lap trays securely attached to the wheelchair can support the patient's upper limbs and help maintain better alignment while sitting. Careful monitoring of the patient is required to avoid issues like orthostatic hypotension, pressure sores, and fatigue when activities in sitting are initiated.

Improper positioning in the wheelchair or bed can lead to contracture development and worsen the abnormal muscle tone. Continuation of the routine ROM exercises and stretching exercises can minimize the likelihood of contracture development. It is essential to dedicate adequate time to manually stretch soft tissue structures like the hamstrings and the heel cords. Though stretching of individual joints is essential, it is time-intensive and has only limited short-term benefits. To achieve greater therapeutic benefits and increase patient flexibility, patients have to be encouraged to use functional positions and different developmental postures on a therapeutic mat which include prone on elbow, kneeling, and kneel sitting to stretch the hip and knee musculatures. Even tilt-table standing can be encouraged to promote joint approximation, weight-bearing through the foot, and stretching of the posterior tibial muscles.

For those patients with yieldable contractures, static splinting or serial casting can be a more effective intervention instead of a vigorous stretching program. Typically each cast is applied for a period of 7–10 days. Routine inspection of skin and observation of voluntary movements of joint(s) below the cast is required to ensure that the cast is not applied too tight. Three to four casts may be required to achieve the desired results, and eventually, the final cast should be bivalved so that it can serve as a permanent splint for the patient.

TBI patients can often be dependent on all aspects of mobility. Promoting functional mobility tasks is another important aspect of treatment. When handling an extremely low functioning level patient, there is no doubt that the therapist will require an extra set of hands. For the in-patient setting, the therapist can take the help of fellow staff, physiotherapy assistant, rehabilitation aide, or caregiver. Prior to promoting mobility, the therapist needs to spend adequate time to normalize the tone and improve the posture. Prolonged stretch, weight-bearing, approximation, slow rhythmic rotation, and pressure over the tendon are some of the therapeutic techniques to inhibit abnormal tone and enhance normal tone. Apart from the strategies to elongate the shortened spastic muscles, facilitation and strengthening of the antagonistic muscles are crucial. Post inhibition of the abnormal muscle tone, normal movement patterns must be facilitated to promote motor relearning. However, in those with severe TBI, characterized by poor postural and motor control with the inability to initiate voluntary movements, presence of abnormal muscle tone and reflex activity, and lack of dissociation of extremity movements from the trunk, head and trunk control must be developed before the patient can attain control over the distal extremities. Interventions performed with the patient in prone or prone progression positions on a therapeutic mat with appropriate use of wedge, pillow, bolster, or ball may provide excellent opportunities to address head and trunk control. Such prone progression positions help the patient work the cervical extensors against gravity and also provide inhibition to the supine tonic labyrinthine reflex. Though the use of a prone position on a Swiss ball is challenging, the activity has a profound effect on reducing the abnormal tone. Gentle rocking with the patient positioned on a ball may also reduce the abnormal tone; however, such postures and activities on the Swiss ball are contraindicated in patients with seizure disorders and increased ICP. All the patients need to be carefully monitored to ensure the adequacy of ventilation during the prone activities.

Repeated practice of task enhances motor learning. TBI patients may find it difficult to master new motor tasks but may respond well to those activities that they are thorough with. Therefore, the therapist should select those activities which are meaningful to the patient while executing the task learning program. Therapeutic hand-over-hand guidance and visual guidance are valuable sensory modalities as they provide proprioceptive and kinesthetic feedback as he performs the functional movement pattern. Encouraging the patient to maintain eye contact with the therapist or to look at a specific object can be beneficial to improve head control for those with difficulty to hold the head steady and erect in a sitting position. Vision can also help the patient to guide the movement during activities like rolling or turning. It is also crucial to encourage the patient to optimally utilize the functional improvements achieved.

Feedback has a role to play in the outcome of any task or activity. It can be intrinsic where the internal systems like proprioception and vestibular apparatus are the sources of feedback, whereas extrinsic feedback comprises visual inputs and verbal cues. TBI patients with cognitive deficit may find knowledge of results more effective as compared to knowledge of performance. Such feedbacks prove beneficial in long-term learning. Once learning takes place, practice is essential to obtain the right functional outcome. Different types of practice are used for learning, and they are constant, variable, random, massed, and distributed. The constant practice consists of completing a task in the same manner, under the same conditions each time it is performed. Whereas variable practice consists of completing a task in a variety of ways, under varying conditions. In blocked practice, the participant will work on the same task in one session and will repeat this over and over again until he or she learns a predetermined level of competence. For random practice, discrete or serial skills are practiced in random order and here the participant does not practice the same task on two consecutive attempts. The massed practice consists of training sessions that are long and intense with minimal rest period as opposed to distributed practice which uses shorter and less intense sessions with adequate rest period.

Distributed practice is more commonly used in the initial phases of the rehabilitation and patients with lower endurance and high fatigability.

Early acclimation to sitting can improve arousal and promote postural alignment and righting and equilibrium reactions. Arousal is of utmost importance when it comes to attention in rehabilitation. Motivation also plays a role in attaining optimal arousal. Patient-specific goals, interesting activities, and support from the family can aid in keeping the patient motivated throughout the therapy. For those with a short attention span, the environment can be modified to minimize distractions, and later the same distractions can be introduced to challenge their attention. Memory can be addressed in a similar manner by including puzzles or activities while performing functional tasks.

For those patients with reduced muscle strength, neuromuscular electrical stimulation and functional electrical stimulation can be attempted to augment strength. When muscle activity is limited, EMG biofeedback and eccentric and isometric muscle contractions progressing to concentric contractions can be used in the training. Once muscles can work against gravity, task-specific activities with added resistance can be incorporated. If the patient has the potential for standard strengthening techniques, the use of resistance bands, weight cuffs, and dumbbells, can also be encouraged. Studies have also proven that functional training can be another alternative to strengthen the muscles, which may include activities like sitting-to-standing, stepping, and squatting. Speed of the movements can be trained using isokinetic equipment, manual resistance, PNF techniques, and force platforms. Endurance can be improved by increasing the number of repetitions, duration, and intensity. Cardiovascular endurance can be improved with aerobic exercises. To improve endurance, upper limb ergometers can be an option for those who cannot ambulate or use a cycle ergometer. For patients having difficulty with gait, treadmill training with or without body weight support can be useful.

Sensory systems also need to be addressed in the therapy process. Along with the motor systems, the somatosensory, vestibular, visual, and auditory systems contribute to the movement. Based on the evaluation, different tactile inputs can be given while performing motor tasks, especially for the hand function. Inputs can be altered by using variety of textures on the surfaces and different shapes and sizes for the objects. Objects and textures can be rough initially and progress to soft or smooth based on the patient's adaptability. Objects of different colors along with shapes and sizes can be incorporated for addressing the visual system. Impaired visual perception will need visual rehabilitation where the professional will use objects of varying sizes, shapes, colors, and contrasts to train the perception. Objects with sound and verbal cues can be a good input while dealing with patients having auditory integration issues. Along with conventional methods, EMG biofeedback can be used to train proprioception.

Vestibular system deficits will require the therapist to work in conjunction with the visual and somatosensory systems. Patients may require considerable aid from the visual and somatosensory system initially, which can be gradually reduced as part of therapy. Training where the environment moves while the patient is stable causes the visual system to think that the patient is moving and hence challenges the somatosensory and vestibular systems. Perturbations with eyes open at the beginning and later with eyes closed will challenge the somatosensory and vestibular systems. Equipment like wobble boards and tilt tables can be used to train the vestibular system to improve balance.

For low functional level patients, to assist and guide transfers, the therapist may require assistance from a fellow therapist or physiotherapy assistant. During the transfers, the therapist should keep track of the patient's level of awareness and muscle tone. The main goal for the activity in sitting is to achieve a neutral pelvic position with an erect trunk and head. Activities that promote gentle anterior and posterior weight shifts should be encouraged as soon as the patient gains certain control over static sitting. Such weight shifts help the therapist to access the patient's positional responses and also serve to improve the patients' awareness. The sitting posture also aids weight-bearing through the upper limbs, thus reducing the abnormal tone and promoting proximal joint stability. The patient should be encouraged to perform reaching activities and tasks like throwing and catching and tying the lace of the shoe in a sitting position. To maximize the comprehension for those TBI patients with cognitive issues having difficulty processing multiple verbal and sensory cues, it is advisable to designate only one therapist to interact with the patient. Besides, the instructions provided should be straight, concise, and simple.

For low functioning TBI patients with poor trunk control, a therapist-assisted sit-pivot transfer is recommended. Sit-to-stand transition training can be a difficult process for low-functioning patients. With sufficient guidance and support and adequate training, the therapist can encourage patients to perform sit-to-stand transitions. Proper foot placement, adequate head and trunk control, correct alignment of hip, knee, and foot are crucial for developing independent sit-to-stand transition. Standing is an important posture that provides opportunities for various functional tasks while promoting weight-bearing and sensory input. For low-functioning patients during the initial stage, a tilt table may be required to provide necessary stabilization to maintain a standing posture. Activities that improve awareness, alertness, and cognition can be introduced while the patient is on the tilt table. As the treatment advances, weight shift and reach-out activities, stepping, and gait training activities can be encouraged.

Aggressive behaviors among TBI patients can often be an area of concern. The RLA scale provides details of the possible patient responses at the confused-agitated level. The goal is to assist the patient to deal with stressful and anxiety-producing situations and develop self-controlling behaviors to manage the behavior. Such behavioral changes can occur when the patients become anxious, frightened, threatened, or fatigued. Inability to manage stress and frustration successfully can pave the way to a crisis, characterized by heightened sympathetic activity (increased heart rate, blood pressure, and respiration rates), impaired communication skills, reasoning and judgment, and depressed cognitive skills. Early identification and use of appropriate strategies can assist the patient to deal with stressors and prevent crisis development. Certain strategies to avoid such aggressive behaviors are mentioned below:

- Identify and withdraw the stimulus which might be making the patient anxious or overstimulated.
- Assess the demands of the activity and if they are too demanding, reduce them.
- Watch for any physical changes like pacing, tapping of the feet or wringing of the hands, or changes in the patient's tone of voice.
- If found getting stressed, provide rest and offer emotional support, and redirect the patient to another task.
- Provide a structured and controlled environment if the patient is likely to exhibit aggressive behaviors
- Allow the patient to vent the increased energy to help him to calm down.
- Reorientation may be beneficial as disorientation is often the underlying factor in severe behavior disturbances.

During a crisis, the TBI patient can be dangerous to himself or others. He or she can lose control over verbal and physical responses and may exhibit destructive and attacking behaviors. In such a situation, the therapist needs to remain calm and composed. The therapist's primary emphasis in such a scenario should be to protect the patient from harming himself or others. However, if it becomes excessively stressful to handle, it is advisable to remove ourselves from the situation. Typically the episode will run a short course and during the crisis minimize the audience and once the patient recovers from the event, it is essential to re-establish the rapport and emotional support provided to the patient. It is best to allow the patient to rest following the event. Once the patient has returned to the baseline behavioral level, the therapist can brief the patient about the episode and may enquire about the event, object, or the individual that triggered the episode. Identification of the stressful event, object or trigger helps to minimize future possibilities of such responses. It is also important to reassure the patient that the therapist will continue to provide support and care. In the midst of the crisis, the therapist has no reason to take the event personally. Internalizing the event can affect the patient-therapist relationship and may ultimately affect the care that is provided.

Constraint-Induced Movement Therapy (CIMT) Although it has been extensively studied in the stroke population, it is gaining evidence in individuals with TBI. Any function can be built up by either addressing the impairments in the affected limb or by forming compensatory strategies with the uninvolved limb. Compensation is the quickest response to such impairments, but it leads to more severe impairments that might remain permanent. CIMT works on the basis of using the paralyzed limb by keeping the unaffected limb at a disadvantage, typically by covering the unaffected limb for multiple hours through the day thus reinforcing the use of the weak limb. There are two essential concepts to understand the beneficial effects of CIMT. Firstly, after a neurological insult, patients face difficulty in using and controlling the weak limb. This frustration can lead to a conceptual shutdown of the neuronal circuit, also termed as "learned non-use." Secondly, following a head injury, the cerebral structures undergo a reorganization depending upon the used patterns. Repetition with the task-specific practice of the involved extremity may aid in inducing reorganization of the cortex, subsequently progressing the function. CIMT strives to target both these issues. However, patients usually find CIMT stressful and less compliant, and to address the same, modifications in the intensity and duration of treatment may be required.

Telerehabilitation Advancement of information and communication technologies has been the cornerstone for many telehealth-care systems to carve its path to patients and health care providers. Telerehabilitation services may entail assessment, supervision, education, counseling, skills training, case management, and monitoring. The benefits include access to high-quality care, reduction in time and cost of travel, cost-effectivity, and improved collaboration among health care professionals. Literature suggests that telehealth can be a potentially efficient and successful alternative to hospital-based treatment to provide outcomes fulfilling patients' expectations in health care. It also helps clinicians to offer rehabilitative services to patients who are unable to have access to these professionals due to physical, financial, and logistical obstacles. Telephone calls, text messages, voice chatting, and emails may help motivate and inform patients and caregivers who are not acquainted with the rehabilitation protocol.

Virtual Reality Virtual reality is an advanced form of human and computer interface that helps in creating a simulated environment in its most natural form using computer graphics to mimic the experiences perceived in real-life situations. Virtual reality can be more stimulating than traditional forms of rehabilitation and may provide a more pleasant experience for the patient to maximize the progress. Such protocols have a high potential in enhancing neurocognitive presentation, upper limb function, gait, balance, and routine activities of patients with TBI. A systematic review indicated a substantial body of evidence supporting the scope of virtual reality for the cognitive recovery of TBI patients. Another systematic review reported moderate support for the use of virtual reality-enhanced rehabilitation techniques.

6.6.6 Chronic Phase Intervention

The International Classification of Functioning, Disability and Health (ICF) format guides the intervention and makes it customized as per the patient's requirement. When any given case is analyzed in the ICF format, the management is specific to the impairments and limitations. Once the impairments are identified, therapy is targeted in overcoming them, and thus limitations are addressed. Eventually, task-oriented practices and activities aid in improving participation in the community.

Participation cannot be single-handedly tackled with tasks and activities; hence adaptation and modifications in the environment in solidarity aid in obtaining an improvement in the functional outcome. These adjuncts include the use of functional electrical stimulation, braces, wheelchairs, walking aids, railings, ramps, foam paddings, and modifications in environmental changes.

Disability is the loss of physical and mental functional abilities as a result of neurological impairments. Everyday scenarios include the inability to socially interact appropriately, impaired ambulation, difficulty in manipulating objects, inability to perceive various features of objects, inability to register, process, and integrate information, difficulty in vision, and poor coordination of extremities. Participation restriction is the overall inconvenience a person with multiple impairments may go through, and it includes personal and professional relationship problems, difficulty participating in social events and recreational activities, and hindrance in gaining job benefits.

In many moderate-to-severe TBI patients, the injury-related impairments can be permanent. For them, variations or modifications in the therapy and combined therapy can be the mainstay of treatment several years post-injury. In addition to the above, the near and dear ones along with the caregivers, need to participate in formulating and executing the rehabilitation protocol. The two categories addressed under chronic treatment are community-based rehabilitation to reinstate the patient back to work and social participation and treatment of long-lasting complications.

Community-Based Rehabilitation The aim of community-based rehabilitation should be to get the patient participation in society. While some patients might require individualized programs, others might need a multidisciplinary approach, and the interventions may include providing motivation, promoting support from the family, advocacy, and development of the community outreach program to monitor, assess, treat, refer, and collaborate with other professionals. It requires a team of professionals that comprises therapists, social activists, a case supervisor, and an expert in vocational rehabilitation to execute the program, and the rehabilitation approach may vary with regard to the age and location of the patient.

Treatment of Chronic Consequences of TBI Residual symptoms might not grossly affect the patient but can hinder their daily function hence they require to be addressed by neurologists, physiatrists, neuropsychologists, and other health care professionals. The commonly observed issues include the management of abnormal muscle tone, which may require physiotherapy, medications, or minor surgical intervention. For chronic pain, which might affect the mental status of the patient, medications, physiotherapy, and psychotherapy are the possible line of treatment. Psychological problems like depression, anxiety, or behavioral issues are generally managed using pharmacological agents and counseling, while convulsions and headaches may need medication.

Further Reading

Ashley MJ, editor. 4th ed. Traumatic brain injury: rehabilitation, treatment and case management. Boca Raton: CRC Press, Taylor & Francis Group; 2018.

Aulisio MC, Han DY, Glueck AC. Virtual reality gaming as a neurorehabilitation tool for brain injuries in adults: A systematic review. *Brain Inj.* 2020 Aug 23;34(10):1322-30.

Barthélemy EJ, Melis M, Gordon E, Ullman JS, Germano IM. Decompressive craniectomy for severe traumatic brain injury: A Systematic Review. *World Neurosurg.* 2016 Apr;88:411-420.

Bullock MR, Chesnut R, Ghajar J, Gordon D, Hartl R, Newell DW, et al.; Surgical management of traumatic brain injury author group. Surgical management of acute subdural hematomas. *Neurosurgery.* 2006 Mar;58(3 Suppl):S16-24

Coello AF, Canals AG, Gonzalez JM, Martín JJ. Cranial nerve injury after minor head trauma. *J Neurosurg.* 2010 Sep;113(3):547-55.

Cooper G, editor. Essential physical medicine and rehabilitation. Totowa, New Jersey: Humana Press Inc; 2006.

Cossu G. Therapeutic options to enhance coma arousal after traumatic brain injury: state of the art of current treatments to improve coma recovery. *Br J Neurosurg.* 2014 Apr;28(2):187-98.

Davanzo JR, Sieg EP, Timmons SD. Management of traumatic brain injury. *Surg Clin North Am.* 2017;97:1237–53

Deepika A, Devi BI, Shukla D. Predictive validity of disability rating scale in determining functional outcome in patients with severe traumatic brain injury. *Neurol India.* 2017 Jan-Feb; 65(1):83-6.

Dewan MC, Rattani A, Gupta S, Baticulon RE, Hung YC, Punchak M, et al. Estimating the global incidence of traumatic brain injury. *J Neurosurg.* 2018 Apr;1:1-18. Available from: https://doi.org/10.3171/2017.10.JNS17352.

Elbaum J, Benson DM. Acquired brain injury an integrative neuro-rehabilitation approach. New York: Springer; 2007.

Esterov D, Greenwald BD. Autonomic dysfunction after mild traumatic brain injury. *Brain Sci.* 2017 Aug 11;7(8):100. Available from: https://doi.org/10.3390/brainsci7080100.

Fong R, Konakondla S, Schirmer CM, Lacroix M. Surgical interventions for severe traumatic brain injury. *J Emerg Crit Care Med.* 2017;1:28–28. https://doi.org/10.21037/jeccm.2017.09.03.

Galgano M, Toshkezi G, Qiu X, Russell T, Chin L, Zhao LR. Traumatic brain injury: current treatment strategies and future endeavors. *Cell Transplant.* 2017 Jul;26(7):1118-30.

Hagen C, Malkmus D, Durham P. Rancho Los Amigos level of cognitive functioning scale. Downey, CA: Professional Staff Association, Rancho Los Amigos Hospital; 1972.

Hardman JM, Manoukian A. Pathology of head trauma. *Neuroimaging. Clin N Am.* 2002 May;12(2):175-87.

Hellweg S, Johannes S. Physiotherapy after traumatic brain injury: a systematic review of the literature. *Brain Inj.* 2008 May;22(5):365-73.

Hong S, Kim DK, Jeon SR, editors. Primary management of polytrauma. Singapore: Springer Nature; 2019.

Jain S, Iverson LM. Glasgow Coma Scale. 2020 Jun 23. In: StatPearls [Internet]. Treasure Island (FL): StatPearls Publishing; 2020 Jan. PMID: 30020670.

Kaur P, Sharma S. Recent advances in pathophysiology of traumatic brain injury. *Curr Neuropharmacol.* 2018;16(8):1224-38.

Khalid F, Yang GL, McGuire JL, Robson MJ, Foreman B, Ngwenya LB, et al. Autonomic dysfunction following traumatic brain injury: translational insights. *Neurosurg Focus.* 2019 Nov 1;47(5):E8. Available from: https://doi.org/10.3171/2019.8.FOCUS19517.

Koehler R, Wilhelm E, Shoulson I. Cognitive Rehabilitation Therapy for Traumatic Brain Injury. Cognitive Rehabilitation Therapy for Traumatic Brain Injury: Evaluating the Evidence. Washington, DC: The National Academies Press, 2011.

Lee B, Newberg A. Neuroimaging in traumatic brain imaging. *NeuroRx.* 2005 Apr;2(2):372-83.

Lennon S, Stokes M, editors. Pocketbook of neurological physiotherapy. London: Elsevier Science; 2008.
Li J, Cheng Q, Liu FK, Huang Z, Feng SS. Sensory stimulation to improve arousal in comatose patients after traumatic brain injury: a systematic review of the literature. *Neurol Sci.* 2020 Sep;41(9):2367-76.
Lin C, Zhao X, Sun H. Analysis on the risk factors of intracranial infection secondary to traumatic brain injury. Chin J Traumatol. 2015;18(2):81-3.
Manivannan S, Al-Amri M, Postans M, Westacott LJ, Gray W, Zaben M. The effectiveness of virtual reality interventions for improvement of neurocognitive performance after traumatic brain injury: a systematic review. *J Head Trauma Rehabil.* 2019 Mar/Apr;34(2):E52-65.
Martin S, Kessler M, editors. Neurologic interventions for physical therapy. 2nd ed. St. Louis, Missouri: Saunders Elsevier; 2007.
Massenburg BB, Veetil DK, Raykar NP, Agrawal A, Roy N, Gerdin M. A systematic review of quantitative research on traumatic brain injury in India. *Neurol India.* 2017 Mar-Apr;65(2):305-14.
Mckee AC, Daneshvar DH. The neuropathology of traumatic brain injury. *Handb Clin Neurol.* 2015;127:45-66.
Megha M, Harpreet S, Nayeem Z. Effect of frequency of multimodal coma stimulation on the consciousness levels of traumatic brain injury comatose patients. *Brain Inj.* 2013;27(5):570-7.
Murthy TV, Bhatia P, Sandhu K, Prabhakar T, Gogna RL. Secondary brain injury: Prevention and intensive care management. *Indian J Neurotrauma.* 2005;2:7-12.
Naraghi L, Larentzakis A, Chang Y, Duhaime AC, Kaafarani H, Yeh DD, et al. Is CT angiography of the head useful in the management of traumatic brain injury? *J Am Coll Surg.* 2015 Jun;220(6):1027-31.
O'Sullivan SB, Schmitz TJ, Fulk GD, editors. 6th ed. Physical rehabilitation. Philadelphia: FA Davis Co; 2014.
Ownsworth T, Arnautovska U, Beadle E, Shum DHK, Moyle W. Efficacy of telerehabilitation for adults with traumatic brain injury: a systematic review. *J Head Trauma Rehabil.* 2018 Jul/Aug;33(4):E33-46.
Pavlovic D, Pekic S, Stojanovic M, Popovic V. Traumatic brain injury: neuropathological, neurocognitive and neurobehavioral sequelae. *Pituitary.* 2019 Jun;22(3):270-82.
Pinto VL, Tadi P, Adeyinka A. Increased Intracranial Pressure. 2020 Jul 20. In: StatPearls [Internet]. Treasure Island (FL): StatPearls Publishing; 2020 Jan. PMID: 29489250.
Røe C, Tverdal C, Howe EI, Tenovuo O, Azouvi P, Andelic N. Randomized controlled trials of rehabilitation services in the post-acute phase of moderate and severe traumatic brain injury – a systematic review. *Front Neurol.* 2019 Jun 6;10:557. Available from: https://doi.org/10.3389/fneur.2019.00557.
Schweitzer AD, Niogi SN, Whitlow CT, Tsiouris AJ. Traumatic brain injury: imaging patterns and complications. *Radiographics.* 2019 Oct;39(6):1571-1595.
Shekhar C, Gupta LN, Premsagar IC, Sinha M, Kishore J. An epidemiological study of traumatic brain injury cases in a trauma centre of New Delhi (India). *J Emerg Trauma Shock.* 2015 Jul-Sep;8(3):131-9.
Sherer M, Struchen MA, Yablon SA, Wang Y, Nick TG. Comparison of indices of traumatic brain injury severity: Glasgow Coma Scale, length of coma and post-traumatic amnesia. *J Neurol Neurosurg Psychiatry.* 2008 Jun;79(6):678-85.
Taha MM, Ammar MG. Electrolyte and acid base disturbances in patients with severe closed traumatic brain injury. *IJNS.* 2015;11:1. Available from: https://doi.org/10.5580/IJNS.32774
Tsao JW, editor. Traumatic brain injury: A clinician's guide to diagnosis, management, and rehabilitation. New York: Springer; 2012.
Umphred DA. Neurological rehabilitation. 6th ed. St. Louis: Elsevier Mosby; 2013.
Vasilyeva Y.B., Talypov A.E., Petrikov S.S. Clinical Features of Traumatic Brain Injury in Various Kinds of Brain Damage. *Russian Sklifosovsky Journal "Emergency Medical Care".* 2019;8(3):295-301. https://doi.org/10.23934/2223-9022-2019-8-3-295-301

Walker MC, O'Brien MD. Neurological examination of the unconscious patients. *J R Soc Med.*1999;92:353-5.

Winn HR, editor. Youmans & Winn Neurological Surgery. 7th ed. Philadelphia: Elsevier; 2017.

Zanier ER, Zoerle T, Di Lernia D, Riva G. Virtual reality for traumatic brain injury. *Front Neurol.* 2018 May 16;9:345. Available from: https://doi.org/10.3389/fneur.2018.00345

Zhang D, Xue Q, Chen J, Dong Y, Hou L, Jiang Y, et al. Decompressive craniectomy in the management of intracranial hypertension after traumatic brain injury: a systematic review and meta-analysis. *Sci Rep.* 2017 Aug 18;7(1):8800. Available from: https://doi.org/10.1038/s41598-017-08959-y.

Zhou L, Parmanto B. Reaching people with disabilities in underserved areas through digital interventions: systematic review. *J Med Internet Res.* 2019 Oct 25;21(10):e12981. Available from: https://doi.org/10.2196/12981.

Chapter 7
Vestibular Rehabilitation

Abraham M. Joshua and Shivananda Pai

7.1 Introduction and Historical Background

Orientation in space and being able to walk upright are critical functions of the vestibular system. The use of vestibular exercises to treat vertigo and balance dysfunction is not novel. Due to the self-limiting nature of the disorder and central nervous system compensation, most of the vestibular dysfunctions, especially the peripheral lesions, have a benign etiology and a remarkable spontaneous recovery. Considerable vestibular functional adaptation can even make it difficult for a casual observer to distinguish a patient with a prior history of dysfunction from someone without a vestibular lesion. Neuroplasticity consisting of neural adjustments that restore original function and supplemention of other sensory input or internal estimates are some of the key factors governing vestibular repair mechanisms.

Recovery following vestibular dysfunction has been studied for over 100 years. In the early 1950s, Terence Cawthorne and Frank S. Cooksey, to treat patients with labyrinth lesions post-surgical or post-head injury, had developed a set of eye, head, and trunk movements to decrease vertigo and improve vestibular function. Years later, Norré M. E., Steven A. Telian, and Neil T. Shepard expanded the principles of Cawthorne and Cooksey exercises to develop what is known as vestibular rehabilitation.

A. M. Joshua (✉)
Department of Physiotherapy, Kasturba Medical College of Mangalore, Manipal Academy of Higher Education, Manipal, India
e-mail: abraham.joshua@manipal.edu

S. Pai
Department of Neurology, Kasturba Medical College of Mangalore, Manipal Academy of Higher Education, Manipal, India
e-mail: dshivanand.pai@manipal.edu

A. M. Joshua (ed.), *Physiotherapy for Adult Neurological Conditions*,
https://doi.org/10.1007/978-981-19-0209-3_7

7.2 Vestibular System: Anatomy and Physiology

The vestibular system detects motion and gravity and is responsible for maintaining balance, stability, and spatial orientation. This system is divided into two main subsystems: the peripheral vestibular system and the central vestibular system.

7.2.1 *Peripheral Vestibular System*

The peripheral vestibular system lies within the inner ear, known as the labyrinth. The peripheral vestibular system (Fig. 7.1) has a bony and membranous labyrinth. The bony labyrinth consists of three semicircular canals, a cochlea, and a chamber between the two called the vestibule, which is filled with perilymphatic fluid with a chemical composition similar to cerebrospinal fluid. Suspended within the perilymphatic fluid of the bony labyrinth and supported by connective tissue is the membranous labyrinth. The membranous labyrinth consists of the membranous portions of the three semicircular canals (SCCs) and the two otolith organs, the "utricle and saccule" (Fig. 7.2). The diameter is widened on one end of each SCC, and the widened end is called the ampulla. The chemical composition of endolymphatic fluid that fills the membranous labyrinth is similar to that of intracellular fluid. Normally, the endolymph and perilymph compartments are watertight and have no direct communication existing between them.

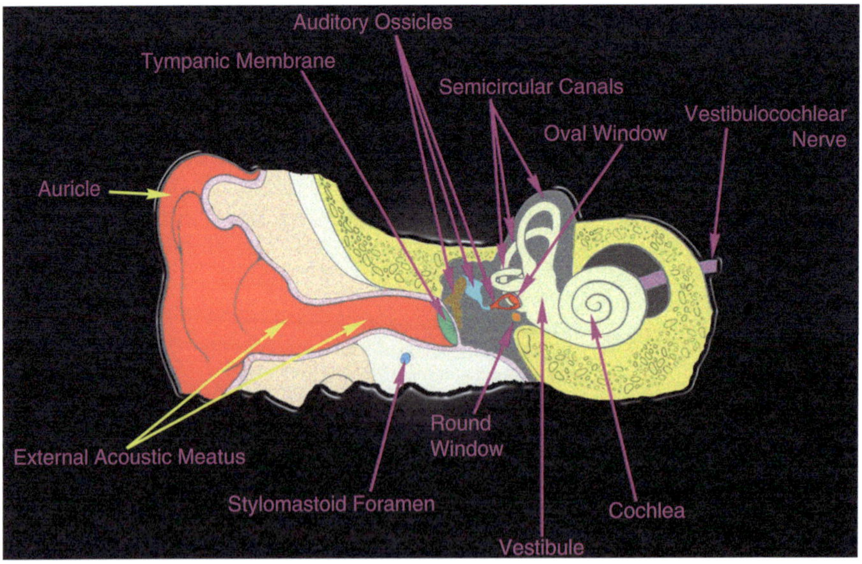

Fig. 7.1 Anatomical structures of the peripheral vestibular system and the adjoining regions

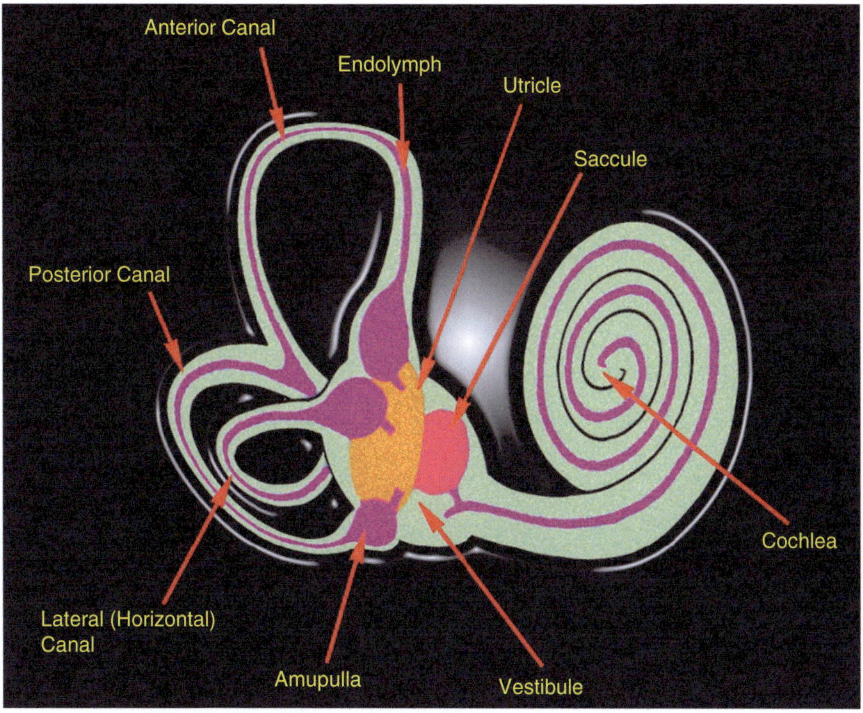

Fig. 7.2 The semicircular canals and the otolith organs

Hair cells located within each labyrinth are the motion receptors of the vestibular system. Hair cells located within the ampulla rest on crista ampullaris, which contains a tuft of blood vessels, nerve fibers, and supporting tissue. Cupula, a flexible, diaphragmatic membrane, lies over each crista (Fig. 7.3) and completely seals the ampulla from the adjacent vestibule. The direction of the bend or deflection of the hair cells, either toward the longest process or away from it, determines the frequency of the neural discharge. The macula (Fig. 7.4), hair cells of the otolith organ, is located on the medial wall of the saccule and the floor of the utricle. Structurally, the otolithic membranes are similar to the cupulae but contain calcium carbonate crystals called otoconia which have markedly more mass than the cupulae.

Hair cells located within the SCCs and otolith organs convert mechanical energy into neural discharges. The orientation of the hair cells and differences in fluid mechanics help the cells to selectively respond to the motion of the head in specific directions enabling the canals to respond to angular velocity and the otoliths to linear acceleration.

Angular velocity sensed by the SCCs enables the vestibulo-ocular reflex (VOR) for gaze stabilization during head movement. Biophysical properties of the SCC loops are responsible for converting head velocity into displacement. It is the viscosity or fluidic friction which creates resistance to endolymph movement so that

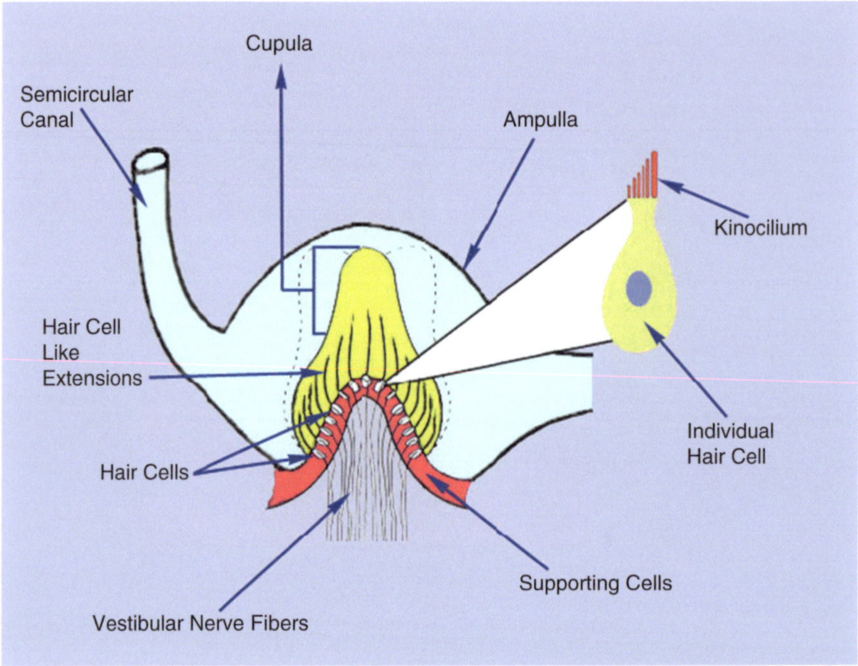

Fig. 7.3 Anatomy of the ampulla with its crista ampullaris, cupula, and hair cells

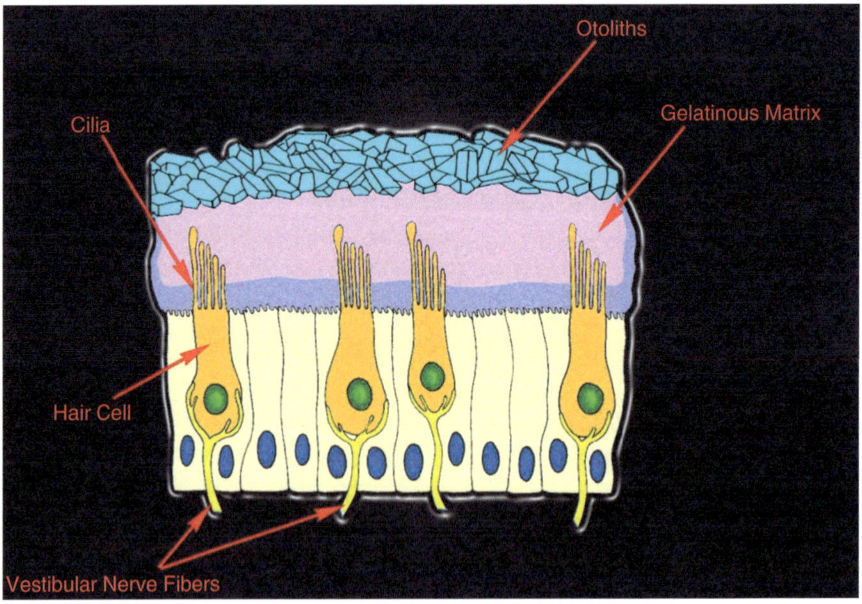

Fig. 7.4 A labeled diagram of the macula

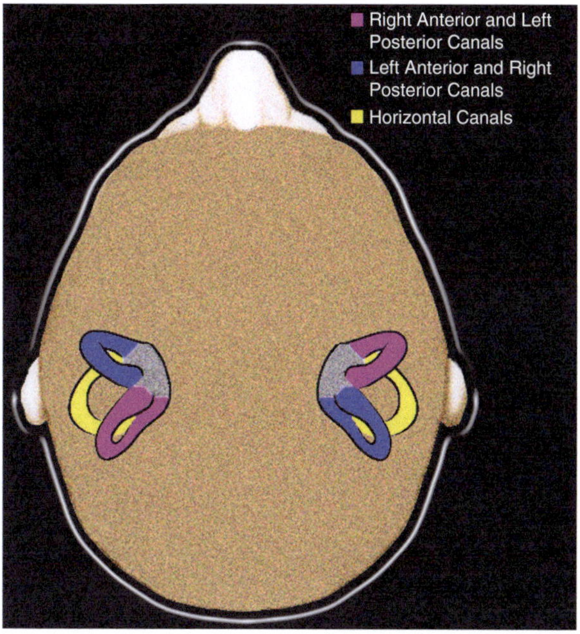

Fig. 7.5 The coplanar orientation of the semicircular canals

trans-cupular pressure and displacement become more closely proportional to the head velocity. Though the canals respond readily to the rotational stimuli, for prolonged rotation, following few seconds, the response decays exponentially due to the spring-like action of the cupula to restore its resting position.

The orientations and the spatial arrangements of the canals need special attention. The planes of each canal located within the labyrinth are perpendicular to the planes of other canals within the labyrinth. Another finding is that the paired planes of the canals between the labyrinths conform very closely to each other, i.e., six individual SCCs form three coplanar pairs, namely, the right and left lateral (horizontal) canals, the left anterior and right posterior canals, and the left posterior and right anterior canals. Figure 7.5 depicts the coplanar orientation of the SCCs. In addition to the above, even the planes of the canals are closely aligned with the axes of the planes of the extraocular muscles.

The coplanar pairing of canals is associated with a "push-pull" (firing increases in one vestibular nerve and decreases on the opposite side) change in the neural output of the canals. Sensory redundancy, common-mode rejection, and overload compensations are a few advantages of the "push-pull" mechanism. Sensory redundancy helps the central nervous system to rely more on the neural impulses arising from the intact canal. Common-mode rejection allows the brain to ignore changes in neural firing that occur on both sides simultaneously, while the overload compensations help the head to rotate swiftly.

The saccule and the utricle, the otolith organs, record linear acceleration, and they respond readily to both the linear head motion and the static tilt (Fig. 7.6). Like

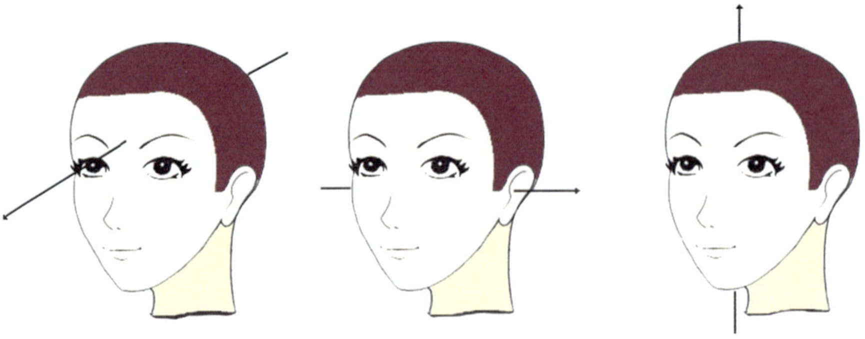

Fig. 7.6 An illustration of the linear acceleration and the static tilt

Fig. 7.7 An illustration of the lateral and forward tilts

the SCCs, the otoliths respond to the motion in all three dimensions. The saccule senses linear acceleration in the sagittal plane, i.e., a sense of forward pitch of the head, and the utricle senses acceleration in the horizontal plane, i.e., a sense of roll (lateral tilt) of the head. Figure 7.7 illustrates the lateral and forward tilts sensed by the otolith organs. Both the saccule and the utricle can encode all possible vectors of linear acceleration. Like the SCCs, the push-pull processing is also integrated into the geometry of each of the otolithic membranes.

7.2.2 Central Components of the Vestibular System

The neural impulses, generated by the hair cells located within the peripheral vestibular system, are carried by the vestibular component of the 8th cranial nerve and enter the brainstem at the pontomedullary junction. Post entry, the two main targets for vestibular inputs are the vestibular nuclear complex and the cerebellum. Inputs are primarily processed by the vestibular nuclear complex, and the necessary motor outputs are delivered to the concerned areas of the central nervous system by faster neural networks. On the contrary, the cerebellum, when required, acts as a main adaptive processor and monitors the vestibular sensory inputs and readjusts the central vestibular processing. Vestibular inputs that reach the nuclear complex and the cerebellum are processed in association with the somatosensory (specifically kinesthetic and proprioceptive) and the visual sensory inputs. Complex and elaborate connections exist between the nuclear complex, cerebellum, motor nuclei innervating extraocular muscles, and the reticular activating system of the brainstem, to communicate appropriate signals to the skeletal and the extraocular muscles.

The vestibular nuclear complex, located within either side of the pontomedullary region, consists of medial, lateral, superior, and inferior nuclei. The superior and medial vestibular nuclei are concerned with VOR, and the lateral nucleus is primarily concerned with the vestibulospinal reflex (VSR). Though the inferior nucleus has no primary outputs of its own, it has connections with the rest of the vestibular complex nuclei and the cerebellum. The vestibular nuclei located on either side of the brainstem are laced together by the commissural fibers, which are mutually inhibitory. These fibers share information between the nuclei on either side of the brainstem and also play their role in the push-pull pairing of the SCCs.

The cerebellum, being a major recipient of discharge from the nuclear complex, is essential for the calibration and efficacy of vestibular reflexes. The cerebellar flocculus adjusts and modulates the gain of the VOR and the lesions of the flocculus either reduce or increase the gain of the VOR as seen in conditions like the Arnold-Chiari malformation. The cerebellar nodulus adjusts the duration of VOR responses, and its lesions lead to gait ataxia and nystagmus. Similarly, the anterior-superior vermian lesions affect the VSR, and typically, patients exhibit gait ataxia with profound truncal instability.

7.2.2.1 Vestibulo-ocular Reflex

The VOR (Fig. 7.8), when activated, produces eye movements in the direction opposite to the head movement and functions in the stabilization of the images on the retinas while the head is in motion. The angular VOR mediated by the SCCs compensates for rotation, and the linear VOR, mediated by the otolith organs, compensates for translation. The angular VOR is primarily responsible for gaze stabilization, whereas the linear VOR helps in viewing near targets when the head is in motion. The output of the VOR goes to the motor neurons supplying the extraocular muscles. These muscles function in pairs and are oriented in planes just like the SCCs. This kind of geometrical arrangement enables a single pair of canals to predominantly connect with a single pair of extraocular muscles resulting in conjugate eye movements in the same plane as head motion. The ascending tract of Deiters and the medial longitudinal fasciculus are the two tracts that carry output from the nuclear complex to the motor nuclei of the extraocular muscles.

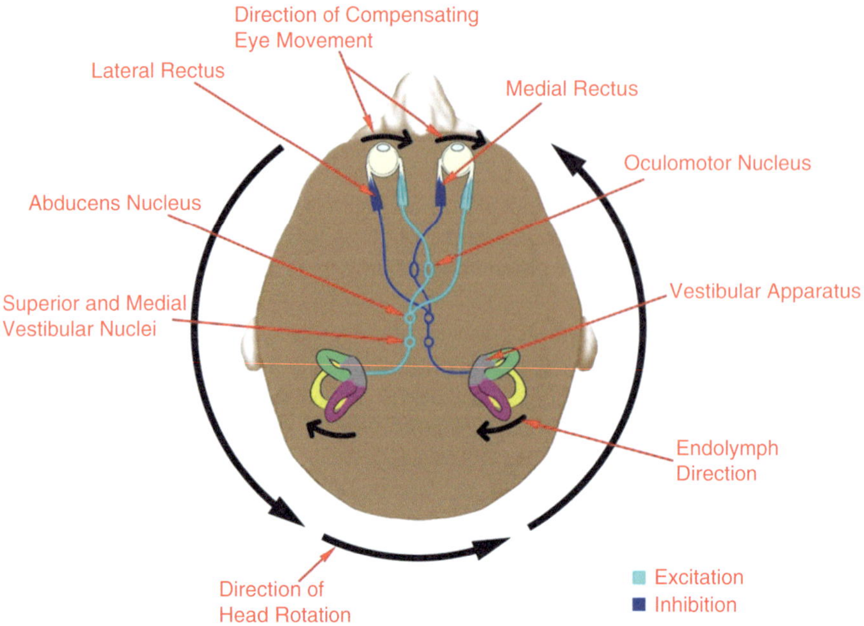

Fig. 7.8 An illustration of the vestibulo-ocular reflex

7.2.2.2 Vestibulospinal Reflex

The output from the VSR pathway (Fig. 7.9) ends in the anterior horn cells (AHC) of the spinal cord, and the connections between the nuclear complex and the AHC are more complex than for VOR. VSR outputs are concerned with strategies to prevent falls such as hip or ankle strategy, a stepping strategy, or a combination of the above.

The lateral and the medial vestibulospinal tracts and the reticulospinal tract are the pathways that connect the nuclear complex to the AHC. The lateral vestibulospinal tract arises from the ipsilateral lateral vestibular nucleus and receives inputs from the otolith organs and the cerebellum. It generates extension or antigravity postural motor activity, primarily in the lower extremities, in response to the positional changes of the head with respect to gravity. The medial vestibulospinal tract arises from the contralateral medial, superior, and inferior vestibular nuclei and produces postural changes of the head or head righting in response to angular motion of the head. The reticulospinal tract has both crossed and uncrossed pathways and is concerned with the maintenance of balance and postural adjustments. It receives input from all the four vestibular nuclei and all the sensorimotor systems of the brain.

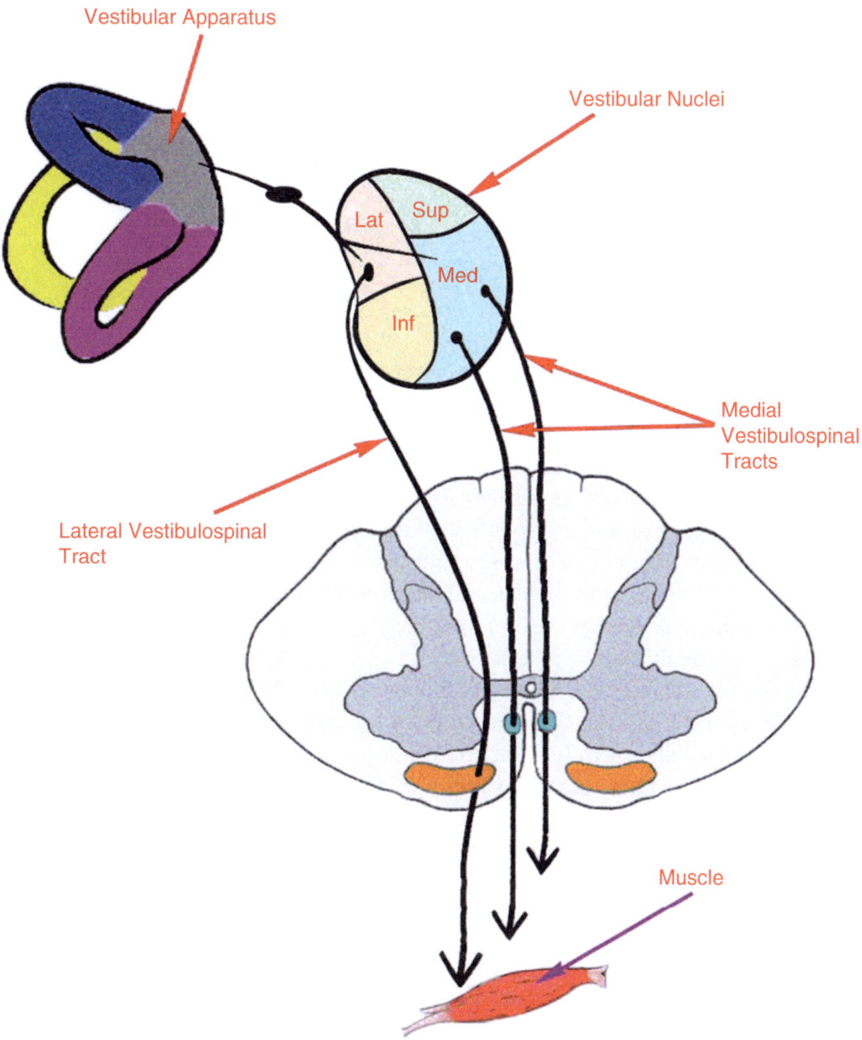

Fig. 7.9 An illustration of the vestibulospinal reflex

7.2.2.3 Cervico-ocular Reflex and Cervico-spinal Reflex

The cervico-ocular reflex (COR) is an ocular stabilization reflex elicited by neck movement. In conjunction with the VOR and the optokinetic reflex, the COR can prevent visual slip of images over the retina during head movements. Though the gain of the COR is low when compared to VOR, under certain circumstances, especially following vestibular apparatus injury, the neck proprioceptors can supplement the VOR to produce reflex movements of the eyes. A neck muscle vibration following unilateral vestibular loss induces very strong nystagmus and vertigo, indicating the compensatory role of COR. The cervico-spinal reflex (CSR), comparable to COR, produces postural changes in the limbs reflexive to cervical or neck proprioceptor activity. Under certain circumstances, the CSR may supplement the VSR for the stabilization of the body.

7.2.3 Vascular Supply

The vertebral-basilar arterial system supplies both the peripheral and the central vestibular systems. The labyrinthine artery, often a branch of the anterior inferior cerebellar artery or seldom a direct branch of the basilar artery, supplies the peripheral vestibular system. The labyrinthine artery divides into the anterior vestibular artery and the common cochlear artery. The former supplies the vestibular nerve, the majority of the utricle, and the ampullae of the lateral and anterior SCCs. The latter divides into the main cochlear artery and the vestibulocochlear artery. The main cochlear artery supplies the cochlea and the ampulla of the posterior SCC, while the vestibulocochlear branch of the common cochlear artery supplies the saccule.

The posterior inferior cerebellar artery, an important artery of the vertebral-basilar system, supplies the central vestibular system. The posterior inferior cerebellar artery supplies the inferior portions of the cerebellar hemispheres and dorsolateral medulla, as well as the inferior aspects of the vestibular nuclear complex. The basilar artery, the main artery of the pons, supplies central vestibular structures via its perforator branches and the anterolateral and dorsolateral pons by its circumferential branches. The anterior inferior cerebellar artery also supplies the ventrolateral cerebellum and the lateral tegmentum of the pons.

7.3 Vestibular System and Postural Control

The sensory component of the vestibular system senses gravity and head acceleration and the motor component, specifically the descending motor pathways such as the vestibulospinal tracts and reticulospinal tracts, play an important role in postural control. As a standalone, the vestibular sensory system can only provide correct information about position and movement of the head in space and vague information about body position or movement. However, the information provided by the vestibular sensory system is closely integrated with the somatosensory and visual information by the central nervous system (CNS) to estimate the position and movement of the whole body and the surrounding environment.

During certain situations, the vestibular system can fail to differentiate a whole-body tilt over the feet from a head tilt on a stationary trunk or a linear acceleration caused by head movement from gravitational acceleration. Such uncertainties can be mitigated only if additional details from the somatosensory and the visual systems are sorted.

The visual, somatosensory, and vestibular systems contribute important information about the body position and motion. The visual system provides information about the position and movement of the head with respect to the surrounding objects, the depth, the direction of verticality, the dimensions and distances between objects and self, and information about slow movements. The somatosensory system provides information about the position and motion of the body or body parts with respect to one another or the supporting surface and sense of fast movements. Even though all three systems provide important information, none can provide all the necessary information required for the position and motion of the body on their own.

7.3.1 Perception of Self-Motion

The vestibular system plays a very important role in maintaining the orientation of the whole body to the vertical. Graviceptors are sensory receptors that sense the direction of gravity. The vestibular system, specifically the otolith organs, provides information about the direction of gravity and is combined with information from the somatosensory system to perceive the gravitational and inertial forces necessary to orient the body. Vertical linear accelerations of the head are sensed by the saccular while the utricles sense horizontal linear accelerations.

The perception of self-motion and orientation depends on sensory cues and the subject's predictions, knowledge about the sensory environment, and past experiences. Given the vestibular system's importance in the perception of self-motion, it is not astounding to hear patients often reporting things like spinning, dropping, or rocking of the room. It is the conflicts in the motion and position sense provided by the defective vestibular system that causes the typical abnormal sensation of self-motion (head spinning/floating of objects around).

7.3.2 Poor Postural Alignment and Balance

The vestibular inputs play an important role in postural alignment, and its importance further increases when the visual information is unavailable. In acute unilateral vestibular lesions, the head and the body may tilt toward the side of the lesion. Within weeks, the extent of asymmetrical posturing may diminish, and the patient may regain a normal postural alignment and control owing to vestibular compensation. Usually, the effect of unilateral vestibular lesions on postural alignment is variable but short-lived.

Bilateral loss of vestibular function may be associated with a forward head position. Subjects with bilateral vestibular loss while attempting to stand on a tilted surface with eyes closed will align their trunk in relation to the tilted surface rather than to gravity. The altered postural alignment seen in patients with vestibular deficits may be the result of the altered internal mapping of body orientation or the limits of stability in space. Often, patients with vestibular deficits may misinterpret the movement of external objects as self-motion and throw themselves into disequilibrium as an attempt to orient themselves to the moving visual object.

Upon a stable surface with adequate room light, the details provided by all vestibular, somatosensory, and visual sensory modalities for postural orientation are consistent and valuable. Theoretically, normal subjects can rely equally on all the abovementioned sensory inputs for postural orientation. However, the literature suggests that normal subjects rely primarily on somatosensory information from the supporting surface in such normal conditions. If a normal person stands on an unstable surface without vision, the CNS will switch its dependency on vestibular information from the unreliable somatosensory inputs and unavailable visual information to maintain the orientation.

Normal subjects will rely about 70% on the somatosensory information from the supporting surface, 20% on the vestibular information, and 10% on the vision for postural orientation while standing on a stable surface with adequate light. It is also essential to know that on a stable surface, a well-compensated vestibular dysfunction patient may use the intact somatosensory information to orient the body, and the same reason might explain the lower sensitivity and specificity values for standard Romberg's test on a stable surface in vestibular dysfunction.

For an unstable surface like foam rubber, sensory weighting changes and subjects will rely around 70% on the vestibular information, 20% on the vision, and the remaining 10% on the somatosensory information. The same reason might explain higher sensitivity and specificity (79% and 80%, respectively) for Romberg's test when performed over a foam rubber for patients with unilateral and bilateral vestibular loss.

The sensory organization test (SOT) is a test developed by Lewis M. Nashner, designed to quantitatively assess a subject's ability to use vestibular, somatosensory, and visual cues to maintain postural stability while standing. The six different sensory environments of this test (Fig. 7.10) reveal that the normal subjects rely heavily on vestibular cues to orient the body when visual cues and somatosensory cues are

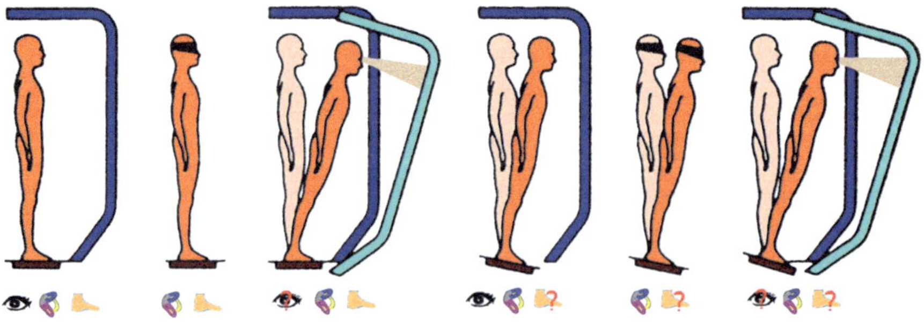

Fig. 7.10 Six different sensory environments of the sensory organization test

unreliable and not corroborating and complementing each other. A minimum of two intact sensory systems is essential for the effective stability of posture while standing. Patients with unilateral vestibular function may use the remaining intact vestibular system for postural orientation or may use the other systems to compensate. Diagnosed cases of bilateral vestibular disorders while undergoing SOT, with both the surface and vision having been altered, will lose the sense of orientation and may consistently fall.

The orientation of the body and equilibrium reactions represents two distinct postural goals. If the balance is disturbed, the muscles of the limbs are activated at short latencies, which are shorter than a voluntary reaction time to restore equilibrium, and these reactions, which are known as automatic postural responses, are primarily triggered by the proprioceptive input than the vestibular inputs. When the balance is disturbed in quiet stance, the vestibulospinal mechanisms may play a certain role in automatic postural responses. However, such mechanisms may not play a large role in the recovery of equilibrium following a slip or trip.

Other than playing a role in aligning the body segments, stabilizing the body in space, and scaling postural responses, the vestibular signals may also play a role in contributing to the selection of appropriate postural strategies for the various environmental conditions. Though the selection of postural strategies for balance is centrally programmed and automatic in nature, some automatic postural responses require more vestibular involvement than the rest. Normally, an ankle strategy is used to recover from smaller perturbations; however, for a larger perturbation, narrow base of support, or quick correction of the center of mass, a hip strategy is required.

When the ankle or hip strategy fails to recover equilibrium, the stepping strategy is opted to increase the base of support over the falling center of mass. The proprioceptive information is sufficient to control the ankle and stepping strategies; however, the vestibular information may play an important role in controlling hip strategy. Vestibular dysfunction may lead to failure to initiate or control hip strategy, and such patients may rely on stepping strategies instead. Patients with bilateral vestibular dysfunction may fail to perform tasks like tandem stance, single-limb stance, and heel-toe walking.

7.4 Peripheral Vestibular Disorders

The involvement of the peripheral components of the vestibular system can produce a variety of signs and symptoms. A detailed evaluation by the clinician is required to identify the specific cause for the key complaints like vertigo and imbalance. A thorough history is the most vital part of the diagnosis, supported by otoneurologic examination findings. Only the common peripheral vestibular disorders are discussed in the following section.

7.4.1 Benign Paroxysmal Positional Vertigo

For patients with benign paroxysmal positional vertigo (BPPV), vertigo is the most frequent symptom. Typically, the episodes of vertigo are brief in nature, triggered by rapid changes or assumptions of specific head posture such as lateral tilting of the head toward the affected ear, rapid turning of the head to the side, and extension of the neck. Most of the patients can easily identify the offending head position and may avoid such positions which precipitate vertigo. They may also complain of mild imbalance between the episodes of a vertigo attack. Typically, each attack will last for half a minute to a few minutes (not more than 2 minutes) and will not reappear even if the triggering position is maintained. Tinnitus, loss of hearing, and feeling of fullness in the ear are not the features of BPPV. The symptoms of BPPV can occur spontaneously, and the exact etiology is largely unknown. However, the onset of the disorder, in less than 50% of cases, can be secondary to head trauma, labyrinthitis, or anterior vestibular artery ischemia.

In 1897, Adler D. first described the clinical presentation of BPPV and recognized it as a possible posterior and anterior SCC structural dysfunction and also noticed that head movements toward the affected side elicited severe vertigo. Later, in 1921, Barany R., an otologist, described in detail the clinical syndrome of recurrent positional nystagmus. Margaret Dix and Charles Hallpike, two British otologists, in the year 1952, were the first to describe the technique to elicit the positional nystagmus, which they termed as "positional vertigo of the benign paroxysmal type." In 1962, Harold Schuknecht introduced the term "cupulolithiasis" to attribute this disorder to the adherence of otoconia crystals to the cupula. In 1979, Hall S. F., Ruby R. R., and McClure J. A. proposed the term "canalolithiasis or canalithiasis" to explain the possible presence of fragments of otoconia freely floating in the SCC, as another mechanism for BPPV.

The disorder is commonly seen among subjects who are elderly and may be precipitated by mild or moderate head trauma. The disorder has a higher incidence of approximately 100 cases per 100,000 population every year. Women have a higher preponderance over men, and the ratio is approximately 2:1. Similarly, the involvement of the right labyrinth is slightly more common when compared to the left. The incidence of bilateral involvement is rare and mostly be secondary to head trauma. Though spontaneous remission is a common feature, the disorder is characterized by recurrence (the rate of recurrence is approximately 50% within the same decade) and may worry the patient for several years.

Normally, otoconia are dislodged, absorbed, and renewed constantly; displacement of the same into the canal (canalithiasis) or adhesion to the cupula (cupulolithiasis) is the possible mechanism of BPPV. Among the three SCCs, the posterior canal is the most commonly involved. Dix-Hallpike tests, observation of nystagmus using Frenzel lenses or video-oculography, and electronystagmography when the measurement of vertical and torsional eye movements is difficult are used for the diagnosis. While inducing vertigo by rapidly assuming a head-hanging position toward the affected ear, the nystagmus typically begins within a latency of few seconds, and the severity of the vertigo increases further in about 10 seconds, associated with a sensation of discomfort and apprehension and even a tendency to move out of the offending position. These symptoms usually diminish within a minute, even if the offending head position is maintained.

Nystagmus, a mixed type, predominantly vertical and torsional in nature, corresponds to the plane of the offending SCC. The direction of the nystagmus changes with the direction of gaze, becoming more torsional when the subject looks toward the dependent ear and more vertical as the patient looks toward the higher ear with the gaze line. Repeating the maneuver several times may cause adaptation or habituation and may decrease the nystagmus. Adaptation of response is of diagnostic value, as central vestibular dysfunctions are not characterized by habituation of the response with repetitive testing. In addition to the above, for central vestibular dysfunctions, the nystagmus fails to subside when the head is maintained in the precipitating position and the direction of the nystagmus may also change with different head positions. Even downbeat nystagmus may be observable in the head-hanging position.

The presence of horizontal nystagmus indicates a lateral canal variant of BPPV, best provoked when positioned in the plane of the lateral SCC. The Pagnini-McClure maneuver is for the abovementioned. In this maneuver, the subject is positioned supine and the head is moved either toward the right or toward the left ear-down position. The horizontal nystagmus induced is of two types. In the canalolithiasis type, geotropic nystagmus, a nystagmus beating toward the lowermost ear and more pronounced when lying on the offending ear, is a common finding, whereas in the less common cupulolithiasis, ageotropic nystagmus, characterized by upbeating toward the uppermost ear, is mostly seen. The geotropic nystagmus seen in canalolithiasis is transient, and the ageotropic nystagmus is long-lasting and tends to be less vigorous while lying on the affected side ear.

Positional nystagmus seen in Ménière's disease, perilymph fistulas, alcohol intoxication, and vestibular migraine are certain differential diagnoses for BPPV. The BPPV is a self-limiting disorder and usually resolves spontaneously within a matter of a few weeks to months. Vestibular exercises or maneuvers that aim to remove the debris from the canal or disperse the otolithic debris from the cupula can speed the process of recovery. Liberatory maneuvers, in conjunction with anti-vertiginous drugs, usually help in resolving the symptoms of the disorder and speed up the recovery. Surgical procedures like singular neurectomy or laser technique may help if the condition is nonresponsive to exercises and anti-vertiginous medications.

7.4.2 Ménière's Disease

Ménière's disease is a disorder of the inner ear and causes considerable hearing and vestibular symptoms. The disease is seen equally in either sex, and the onset of the disease is usually during the fourth to sixth decades of life. The initial sensation of aural fullness, reduced hearing, and tinnitus followed by rotational vertigo, nystagmus, postural imbalance, nausea, and vomiting within a few minutes are the typical features of the attack. Vertigo may typically persist for several minutes and may last for approximately 24 hours. The symptoms will gradually reduce, and the patient will be able to ambulate within a span of 3 days. The sensation of postural imbalance may persist for days to weeks before normal balance returns. During the convalescence stage, the hearing will gradually return and the tinnitus will reduce. Following each attack and as the disease advances, the hearing may fail to return to the baseline level.

The development of endolymphatic hydrops, due to the malabsorption of endolymph in the endolymphatic duct and sac, is the cause of Ménière's disease. The malabsorption of endolymph may be the result of the disturbed functioning or mechanical obstruction of components comprising the endolymphatic duct and sac or altered anatomy in the temporal bone. Ears affected by Ménière's disease may demonstrate hypodevelopment of the endolymphatic duct and sac, periaqueductal cells, and mastoid air cells and may support the hypothesis relating constricted anatomy in the temporal bone and malabsorption of the endolymph.

Electronystagmography, audiometry, MRI, and brainstem auditory evoked potential studies are a few tools used for the diagnosis. Typically, the audiogram may reveal ipsilateral sensorineural hearing loss involving the lower frequencies. Electronystagmography may demonstrate a unilateral vestibular weakness on caloric testing for the affected ear. MRI and brainstem auditory evoked potential studies may help to rule out conditions like central nervous system pathologies and acoustic schwannomas. A histamine derivative called betahistine and the use of vestibular suppressants along with psychological support may help to alleviate or cope up with the symptoms of the disease. For those patients where vertigo is disabling, surgical procedures, including the vestibular nerve section, are proven to be effective.

7.4.3 *Vestibular Neuritis*

Vestibular neuritis, also known as idiopathic acute unilateral vestibulopathy, is the second most common cause of vertigo. A definite etiology for the disease is unknown and probably has an autoimmune origin secondary to a viral infection in the form of an upper respiratory or gastrointestinal tract infection. Prolonged severe rotational vertigo, exacerbated by head movements, which is acute in origin, oscillopsia, nausea, spontaneous horizontal rotatory nystagmus beating toward the good ear, and postural unsteadiness with a tendency to fall toward the affected side are certain features of the disease.

Hearing loss is unusual and the disorder mainly affects subjects aged in the age group between 30 and 60 years. In the acute phase, the caloric test may reveal an ipsilateral hypo-/non-responsiveness suggesting horizontal canal paresis. The clinical features usually subside within 2–3 days, and the balance will return to a near-normal state within 6 weeks. Central compensation of the vestibular tone imbalance aided by physical exercise and peripheral restoration of labyrinthine function are a few of the plausible reasons for recovery. Medical treatment for vestibular neuritis can be accomplished by the use of vestibular suppressants such as the antihistamine dimenhydrinate or the anticholinergic scopolamine. To speed up the process of recuperation, gaze stabilization and postural stabilization exercises may facilitate the compensatory mechanisms of the CNS for adaptation.

7.4.4 Disabling Positional Vertigo

Disabling positional vertigo, also known as vestibular paroxysmia, develops as a result of the neurovascular compression of the vestibular nerve at the root entry. Such compression can produce local demyelination of the root. A diverse collection of signs and symptoms are seen in these patients. The disorder is characterized by short and frequent attacks of to-and-fro or rotational vertigo lasting from seconds to minutes. The episodic attacks frequently develop as a result of assuming or changing head positions. Tinnitus and/or loss of hearing can be a permanent feature or might be present only during the attack. Neurophysiological tests may help in quantitatively measuring the auditory or vestibular dysfunction. The presence of vascular loops around the most proximal part of the vestibular nerve for a T2-weighted MRI scan is a diagnostic finding. Since the disorder usually responds positively to anti-epileptic drugs, microvascular decompression is often not contemplated.

7.4.5 Perilymphatic Fistula

Perilymphatic fistula is a condition characterized by episodic vertigo, loss of hearing, loud tinnitus, aural fullness, disequilibrium, or a combination of the above symptoms due to an abnormal connection between the perilymphatic fluid of the inner ear and the air-filled middle ear. The rupture of the round window or oval window ligaments separating the inner and middle ear causes these fistulas to occur at the round and oval windows of the bony labyrinth. The patient may typically give a history of penetrating injury to the tympanic membrane, barotrauma, mastoid or stapes surgery, minor head trauma, and vigorous straining or Valsalva maneuver before the onset of symptoms.

Patients may often report "something popping in the ear" during the precipitating events like sneezing or straining. They may describe the imbalance as true rotational vertigo, lightheadedness, disequilibrium, intolerance to motion, or any combination of these symptoms. Tullio phenomenon – vertigo, oscillopsia, postural imbalance, and nystagmus induced by an auditory stimulus – can be a less common feature of perilymphatic fistula. Applying manual pressure over the tragus or pressure to the tympanic membrane with a pneumatic otoscope can exacerbate or recreate vertigo (Hennebert's sign) or elicit nystagmus.

These fistulas usually heal spontaneously, and the symptoms may subside in a matter of few weeks, following absolute bed rest, specifically with the head elevated. However, the symptoms can reappear following vigorous blowing of the nose, sneezing, or while straining. Diagnosis can often be difficult due to the variability of clinical signs and symptoms and a lack of a definite pathognomonic test. A positive history, presence of relevant symptoms, physical examination findings including otoscopy, and additional tests, such as audiometry and electronystagmography with the caloric test, may help in the diagnosis of the condition.

7.4.6 Bilateral Vestibular Disorders

The involvement of both right and left peripheral vestibular systems can often be a feature following a variety of conditions like meningitis, type-2 neurofibromatosis, infections of the labyrinth, otosclerosis, Paget's disease, endolymphatic hydrops, vestibular neuritis, use of ototoxic drugs, congenital malformations, and autoimmune disorders affecting the inner ear. In most cases, a clear-cut cause may not be identifiable. The presence of bilateral vestibular dysfunction, often accompanied by progressive bilateral loss of hearing, is a usual clinical feature and helps in the diagnosis of the condition. Medical history clubbed with the blood and other investigative findings will certainly help in narrowing down the condition. Appropriate medications and physical exercises that recruit non-vestibular sensory capacities like the proprioceptive, visual, and cervico-ocular reflex for improving balance and gait may benefit patients with permanent bilateral vestibular function loss.

7.5 Central Vestibular Disorders

Different forms of vertigo of central vestibular dysfunction origin are caused by lesions of the vestibular pathways anywhere from the vestibular nuclei to its projections or networks to the motor nuclei innervating extraocular muscles, the integration centers in the brainstem, the vestibulocerebellar complex, the thalamus, and the vestibular cortex located in the temporoparietal area of the cerebrum. Such forms of vertigo with typical ocular motor, postural, and perceptual manifestations are clearly defined in a variety of clinical syndromes affecting the brainstem, subcortical, and cortical structures. The analysis of nystagmus may help localize the level or site of the lesion.

Vascular and space-occupying lesions and degenerative disorders like multiple sclerosis are a few of the important causes of central vestibular syndromes. Auditory signs, brainstem signs, certain provoking factors like movements of the head, and history may guide in differentiating peripheral from central vestibular disorders. The duration of the symptoms may help to differentiate central forms of vertigo from the rest. A short, rotatory, or postural vertigo lasting for seconds to few hours is usually caused by transient ischemic attacks of vertebrobasilar territory, basilar or vestibular migraine, and multiple sclerosis. A prolonged, persisting, rotatory vertigo lasting for hours to several days can be caused by vascular lesions or multiple sclerosis plaques in the brainstem. A vertigo lasting days to weeks characterized by permanent positional or seldom rotatory vertigo and a tendency to fall is usually caused by vascular or space-occupying lesions affecting the brainstem or the bilateral cerebellum. Such disorders may also exhibit downbeat nystagmus, as seen in Arnold-Chiari malformation, or upbeat nystagmus, as seen in paramedian pontomedullary or pontomesencephalic damage.

7.5.1 Vertigo of Central Vestibular Origin and VOR

One of the most important structures for vertigo of central vestibular origin is the neuronal pathways that mediate the VOR. This reflex makes compensatory eye movements during rapid head and body movements. Damages to the neuronal circuitries concerned with VOR are typically characterized by ocular motor (3rd, 4th, and 6th cranial nerves) deficits, perceptual dysfunction due to impaired functioning of the vestibulocortical projections, and postural abnormalities due to impaired vestibulospinal projections.

A simple clinical classification of the central vestibular brainstem syndromes can be based on the three major planes of action of the VOR: yaw, roll, and pitch. The tendency to past-point, horizontal nystagmus, lateropulsion to the affected side, ipsilateral caloric hyporesponsiveness, and horizontal gaze deviation are certain yaw plane signs. A pure central yaw plane (horizontal plane) syndrome is rare, and the symptoms may mimic the features of the acute peripheral vestibular lesion, and so for the same reason, this syndrome is known as vestibular pseudoneuritis. The lesions are usually seen in the entry zone of the vestibular nerve in the medulla oblongata, the medial and/or superior vestibular nuclei, and the neighboring integration centers for horizontal eye movements. The most common causes of lesions include vascular lesions or multiple sclerosis plaques. Central compensation, along with concurrent management of the underlying illness, may resolve the yaw plane symptoms within days to weeks.

Torsional nystagmus (also known as rotary nystagmus, when the eyeball rotates about the anteroposterior axis or the line of sight), skew deviation (a vertical misalignment of the visual axes) of the eye, ocular torsion, and head and body tilts are the roll signs. The syndromes of roll plane (frontal plane) signs are caused by lesions of the graviceptive vestibular pathways that run via the medial and the superior vestibular nuclei, the contralateral medial longitudinal fasciculus to the ocular motor nuclei, and the integration centers in the rostral midbrain. The causes of such lesions can be of vascular origin, affecting the brainstem, the paramedian thalamus, and the rostral midbrain.

Upbeat or downbeat nystagmus, a tendency for backward tilt and falls, and vertical deviations of perceived straight-ahead are the common pitch plane (sagittal plane) signs. The lesions are usually located in the bilateral paramedian region of the medulla, pontomedullary and pontomesencephalic regions of the brainstem, adjoining areas of the cerebellar peduncle, and the bilateral cerebellar flocculus. Pitch plane syndromes with downbeat nystagmus have fixation nystagmus (peripheral vestibular nystagmus tends to suppress by voluntary fixation of the eye on the object; however, central nystagmus is either unaffected or worsened by fixation) that beats downward in primary gaze position, exacerbated on lateral gaze and in head-hanging position. These subjects may exhibit visual and vestibulocerebellar ataxia, with a tendency to fall backward and past-pointing upward. Downbeat nystagmus is often due to bilateral lesions of the flocculus, paraflocculus, or the caudal part of the fourth ventricle. Craniocervical junction anomalies, multiple sclerosis, and vascular or space-occupying lesions affecting the abovementioned regions may cause the downbeat nystagmus.

Pitch syndromes with upbeat nystagmus are rarer and are characterized by fixation nystagmus that beats upward in the primary gaze position, combined with the vertical smooth pursuit eye movement disorder, visual and vestibulospinal ataxia, and a tendency to fall backward and past-pointing downward. Lesions are usually located near the medial tegmentum of the pontomesencephalic junction, the brachium conjunctivum, and probably the anterior vermis of the cerebellum. Bilateral lesions due to central demyelinating disorders, vascular or space-occupying lesions, cerebellar degeneration, and cerebellar dysfunction secondary to alcoholic intoxication are certain causes for the same. For both syndromes with roll and pitch plane signs, the course and prognosis depend on the etiology of the underlying illness.

7.5.2 Paroxysmal Central Vertigo

The pathological excitation of the various vestibular structures may result in non-epileptic paroxysmal vertigo. Multiple sclerosis, brainstem abscess, brainstem infarction, and arteriovenous malformation are certain pathologies that can cause paroxysmal central vertigo. Dysarthria, ataxia, vertigo, ocular tilt reaction, and room tilt illusion, which are paroxysmal in origin, are the typical manifestations of paroxysmal vestibular syndromes of the brainstem due to multiple sclerosis.

7.5.3 Central Vestibular Falls Without Vertigo

A few of the central vestibular disorders present with a history of falls in the absence of paresis, sensory deficits, or cerebellar deficits. Such patients are unable to maintain an unsupported upright posture and will not complain of vertigo. Conditions like thalamic astasia and lateropulsion seen in the lateral medullary syndrome of Wallenberg's and ocular tilt reaction in pontomedullary or rostral midbrain lesions are a few examples of central vestibular falls without vertigo.

Patients with thalamic astasia, caused by vascular lesions of the thalamus or thalamotomy, have a postural imbalance with a transient tendency to fall. When instructed to sit up, they will typically grasp the rails of the bed using the unaffected hand or both hands instead of the axial musculature to pull themselves up. Astasia is transient and usually subsides in a matter of days or weeks. The disorder primarily involves the superoposterolateral portions of the thalamus.

7.6 Neurological Assessment for Vestibular Disorders

Effective management of subjects with vertigo depends on proper history, accurate bedside clinical examination, and appropriate laboratory testing. History, along with clinical examination, helps in distinguishing signs and symptoms originating from central or peripheral vestibular dysfunction and also may help in finding the extent of dysfunction. Laboratory testing helps to confirm the provisional diagnosis, which is based on history and bedside clinical findings.

7.6.1 History

A proper history is essential to list out the problem(s). A detailed description of the symptoms in chronological order and documentation of how the symptoms interfere with the subject's day-to-day life are essential. Though it is a well-known fact that history is the most vital part of the evaluation, collecting and documenting a good history can be extremely tedious, particularly when the complaints are vague or details are incomplete. The crucial components that can pave the way toward diagnosis are the speed or pace of progression, the symptoms, and the provoking situation or posture.

Whether the vertigo attack is acute, chronic, or in spells may provide hints about the pace of the disorder. An attack within 3 days of duration is generally considered to be acute in onset and beyond 3 days is considered chronic in nature. The clinician should ask the patient to elaborate on the details regarding the first onset of symptoms. History should include details regarding symptoms such as sudden or slow onset, precipitated or provoked by any specific situation, posture or spontaneously, any recent and relevant history of cold, upper respiratory tract infection, or fever. If the history suggests spells of vertigo, it is necessary to note the average duration of the spells and should collect a detailed description regarding the first spell, the most severe spell, and the last spell.

"Dizziness," the first and the most common symptom the patient may express to the clinician, needs clarity. It is of utmost importance to ask the subject to give a clear explanation of the word "dizziness" as this term may describe a variety of symptoms, each having a different pathological mechanism and importance. Despite repeated prompting, if the subject fails to give a clear picture regarding dizziness, the clinician can ask whether the problems caused by dizziness are primarily in the head or with balance. A sense of faintness and being close to passing out, a symptom commonly known as lightheadedness, caused by a momentarily decreased blood flow to the brain, a sudden drop of blood pressure, low blood sugar, or often as a symptom of anxiety or depression can be easily confused with dizziness. Similarly, the sensations of floating or swimming of objects inside the head are frequent symptoms of anxiety disorders, depression, or certain psychiatric disorders; care has to be taken to rule out such symptoms.

Vestibular dysfunction can be one of the factors for imbalance or unsteadiness while standing or walking. However, care has to be taken while documenting and examining, to rule out other causes of imbalance such as proprioceptive impairments, reduced vision, double vision, musculoskeletal conditions causing joint deformities, pain and reduced mobility, defects of the central or peripheral nervous system other than the vestibular origin, and psychological factors.

Nausea with or without vomiting, which may occur during rapid head movements, requires special attention. For peripheral vestibular disorders, nausea and vomiting are usually mild or moderate and often are proportional to the degree of vertigo. The severity of these symptoms varies in central vestibular dysfunction. For anterior inferior cerebellar artery syndrome, mainly involving the pons, the degree of nausea and vomiting is comparable to that seen in peripheral vestibular dysfunction. On the contrary, for posterior inferior cerebellar artery syndromes, predominantly involving the medullary region, nausea and vomiting can be severe and disturbing and out of proportion to the degree of vertigo.

Patients may infrequently interpret an oscillopsia as vertigo. Oscillopsia, an illusion of an unstable vision, occurs when the subject's eyes are open, whereas the symptom of vertigo occurs when the subject's eyes are either open or closed. Vertigo, an illusion of self or surrounding moving, triggered by a sudden imbalance of tonic neural activity in the vestibulocortical pathway, can be normal or pathological in origin. Lesions in the peripheral or central vestibular pathways and mechanical issues, as seen in benign paroxysmal positional vertigo, are the common pathological causes of vertigo. The direction of vertigo may hint about the structures involved or the location of the lesion. Horizontal plane rotational vertigo is typically seen in dysfunction of the horizontal SCC, like vestibular neuritis and Ménière's disease. Torsional plane, clockwise or counterclockwise direction, rotational vertigo is typically seen in anterior and posterior SCC dysfunction or unilateral central lesion in the dorsal medulla. Lateral translation or lateropulsion is seen in utricle dysfunction and central lesions.

Adapting certain postures or certain movements may provoke dizziness, such as standing up from the recumbent posture or movements of the head in the vertical or oblique direction. Dizziness induced by eye movements in the absence of head movement is less likely to be vestibular in origin. However, spontaneous (without provocation) dizziness is vestibular in origin and may worsen with head movements.

In addition to the written or narrative details of the patient's subjective complaints, for specific symptoms, the intensity or severity can be quantified using tools like the visual analog scale (VAS). A disability scale can be used to note the perceived level of disability caused by dizziness. The use of a fall diary may help to collect details such as fall description, frequency of falls, associated injuries due to fall, and near fall instances. The activities-specific balance confidence (ABC) scale, a self-administered or interviewed scale, which assesses the subject's level of confidence and the fear of falls while performing various ambulatory activities, can also be used for this population. The history should also include details regarding the effects of the dizziness on routine daily activities such as dressing, personal hygiene, driving, and light housekeeping.

7.6.2 Clinical Examination

On every subject with a history of vertigo, to facilitate diagnosis, certain examinations such as observation of spontaneous nystagmus, skew deviation of eye, elicitation of nystagmus and vertigo by specific maneuvers, and evaluation of balance and gait are essential. Visual fixation reduces or suppresses horizontal and vertical spontaneous nystagmus generated by unilateral peripheral vestibular dysfunction. Unopposed neural activity in the intact vestibular pathways is the cause of spontaneous nystagmus in unilateral peripheral vestibular dysfunction. Visual fixation does not reduce and may often increase if the nystagmus is of central origin. Lack of asymmetry between the right and the left sides explains the reason for the absence of spontaneous nystagmus in bilateral vestibular lesions.

In peripheral vestibular nystagmus, the direction of jerky nystagmus is usually mixed (both in the horizontal and the torsional planes) and tends to increase with a gaze toward the direction of the fast phase of the nystagmus. Traditionally, the direction of nystagmus is named based on the fast phase. In peripheral vestibular nystagmus, the fast phase of the nystagmus will be toward the uninvolved ear. On the contrary, in the central pathologies, the direction of nystagmus is single-plane horizontal (either torsional or vertical) and stays unaltered or does not change direction with respect to gaze.

Skew deviation of eyes, a vertical misalignment of the visual axes, can often be an initial manifestation of central or peripheral vestibular system disorders. The only bedside clinical finding that may readily be appreciated is the presence of pathological ocular tilt response with skew eye deviation. Three different types of skew deviation occur in ocular tilt reaction. Acute unilateral lesion of the utricle causes a pathological ocular tilt response characterized by an upward deviation of both eyes with different amplitudes. In the lateral medullary syndrome, there will be hypertropia (visual axis of one eye is higher than the fellow fixating eye) of one eye, while the other eye remains in the primary position, whereas in lesions of the midbrain tegmentum, the skew is characterized by a simultaneous hypertropia of one eye and hypotropia (visual axis of one eye is lower than the fellow fixating eye) of the other eye.

Head-impulse test and head-shaking nystagmus test are a few bedside clinical tests that may reveal reduced or abnormal VOR in vestibular dysfunction. The head-impulse test, also known as the head thrust test, is performed with both the examiner and patient seated face-to-face. The patient is instructed to fix his eyes on the examiner's nose. Then, the examiner grasps the patient's head, flexes the neck to 20–30° and gently rotates the head side to side 4–5 times, and quickly and unpredictably rotates the neck to 10–15° toward the right or left side while observing the eyes. For subjects with normal vestibular function, the eyes will stay fixed. The presence of corrective or compensatory saccades indicates a deficient VOR on the same side of the head turn, indicating a peripheral vestibular lesion on the same side. For the head-shaking test, the subject is to be made to sit, preferably wearing a Frenzel or IR goggles. Then, the patient is instructed to close his eyes. With the head flexed to 20–30°, the examiner holds the patient's head and rapidly oscillates at a speed of 1–2 Hz, 20 times horizontally. Once the oscillations are completed, the examiner instructs the patient to open his eyes and should observe for nystagmus. The presence of nystagmus immediately after this procedure suggests a vestibular imbalance either of peripheral or central unilateral vestibular origin. In this test, the fast phase of nystagmus will be directed toward the normal ear.

7.6.2.1 Dix-Hallpike Maneuver

Other than observing for spontaneous nystagmus, certain maneuvers are available to evoke nystagmus. The Dix-Hallpike maneuver is a positional test that evokes transient nystagmus (not more than 1 minute) in the most common vertigo-causing disorder called BPPV. If the nystagmus persists beyond 1 minute, then it is more likely a central positional vertigo. For this test, the patient is made to sit with both knees extended (long sitting) on the examination table, and the examiner holds the head of the patient and rotates the neck horizontally to 45° toward the side to be tested. Maintaining the same neck posture, the examiner quickly brings the head and trunk straight back "en bloc" with the head hanging over the edge of the examination table with 20° neck extension (Fig. 7.11). Instruct the patient to keep his or her eyes open for the examiner to observe any signs of nystagmus. Once the nystagmus subsides or after 60 seconds, the examiner still maintaining the 45° turn can bring the patient back to an upright sitting position and continue observing for nystagmus. The entire maneuver needs to be repeated with the head turned 45° in the opposite direction for testing the other side. In the clinical setting, to minimize apprehension, it is advisable to begin the maneuver testing the unaffected side.

7.6.2.2 Eye Movements and VOR Cancellation

While performing the smooth pursuit movements of the eye, the head is typically kept still; however, for the VOR cancellation, the head moves synchronously with the target. During the VOR cancellation, the VOR has to be suppressed to ensure that the image is maintained on the fovea. The VOR cancellation can be clinically tested by holding the patient's head with both hands and gently oscillating the neck side to side at 1 Hz. The examiner should move his head synchronously with

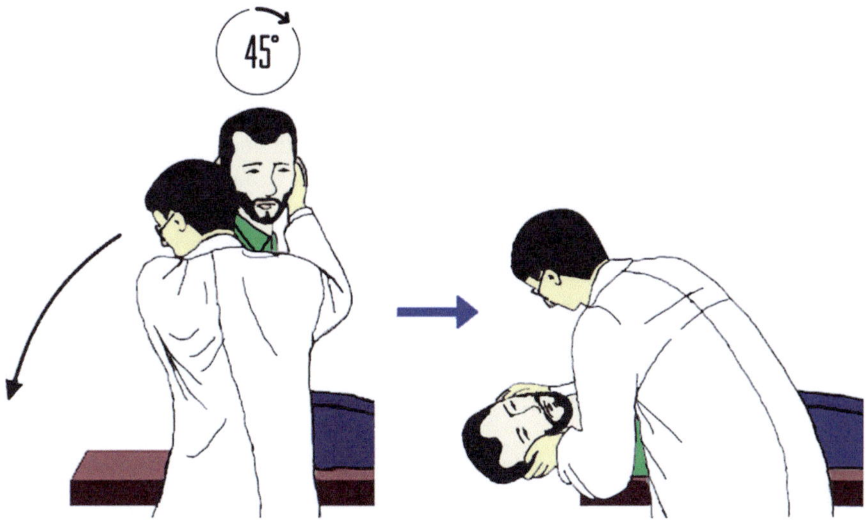

Fig. 7.11 An illustration of the Dix-Hallpike maneuver

the patient's head and should instruct the patient to fix his or her eyes on a target (usually the examiner's nose). Unilateral peripheral vestibular dysfunction generally causes no impairments to the smooth pursuit movements of the eyes or VOR cancellation. Lesions in the parieto-occipital cortex, frontal cortex, cerebellar vermis or flocculus, and pontine nuclei may cause impairments in pursuit and VOR cancellations.

For saccadic movements, the patient is instructed to rapidly change the position of eyes between the examiner's nose and the examiner's finger held approximately 20° away, and the test is repeated several times for both vertical and horizontal directions. During this procedure, the examiner observes the amplitude and velocity of each saccade. Normally, the velocity of saccades will be brisk and the amplitude will be equal; however, in cerebellar disorders, the saccadic movements will exhibit hypermetria or hypometria in one or the other direction. Impairments in saccades are less likely in peripheral vestibular disorders as compared to cerebellar disorders. The midbrain lesion typically causes the slowing of the vertical saccades; on the contrary, the pontine lesion causes the slowing of the horizontal saccades. Slowing of the saccades can also be a feature of internuclear ophthalmoplegia.

7.6.2.3 Balance and Gait

Romberg's test, Fukuda's stepping test, and the test for retropulsion should be performed to evaluate the posture and gait of patients with vertigo. Romberg's test is performed in standing. The patient is asked to stand upright and steady on a stable floor, unsupported with feet together, eyes open and arms by the side or folded across the chest for approximately 15–30 seconds and is then instructed to close the eyes and continue standing for another 15–30 seconds. The examiner, while performing this test, should stand in close proximity. The examiner should have a keen presence of mind and readiness to catch the falling patient if the need arises, especially when the patient attempts to stand erect with eyes closed. A positive Romberg's test result (eyes open, able to stand erect, associated with no or minimal sway; eyes closed, unable to stand, losing balance consistently, associated with worsening of sway) is a classical feature seen in subjects with severe proprioceptive defects. In cerebellar disorders, with eyes open, the subject sways and may lose balance, and with eyes closed, the amplitude of sway and the tendency to fall may increase, indicating that the opening or closing eyes may not demonstrate any significant effect on the patient's ability to maintain balance unlike in proprioceptive loss. In a well-compensated vestibular defect, Romberg's test may not reveal any obvious findings. However, in acute vestibular defects, the test may show findings similar to those seen in proprioceptive loss.

In Fukuda's stepping test, a test that is performed in standing, the subject is instructed to march in place for 50 steps with arms extended and eyes closed. Progressive turning toward one or other side, with a degree of rotation more than 45°, is abnormal and may suggest asymmetric VSR tone due to peripheral or central vestibular pathologies. However, this test may generate false-positive results in non-vestibular disorders, particularly in musculoskeletal conditions such as limb-length discrepancy, sciatica, and hip joint conditions.

For pull test or retropulsion test, the patient is asked to stand erect, unsupported with the feet hip-width apart on a stable floor, and is instructed to just take one step back if pulled backward at the shoulders with a minimal force. The test turns positive when the patient takes three or more steps or falls consistently backward while attempting to regain postural stability. Progressive supranuclear palsy, idiopathic Parkinson's disease, and normal-pressure hydrocephalus are some of the conditions where the pull test is positive. Like in Romberg's test, care has to be taken to prevent any falls or injuries while executing the test.

Patients with unilateral vestibular lesions may present with gait ataxia and a tendency to lurch toward the side of the lesion due to the functional imbalance of the lateral vestibulospinal system. It may also cause reduced head-on-body movements during turns, as seen in vestibular hypofunction. Observation and bedside evaluation of spatiotemporal qualitative parameters of gait may help in distinguishing various neurological gait abnormalities from vestibular ataxic gait. The following are some of the common gait abnormalities seen in neurological conditions:

- Drunkard's gait with dysmetria of limbs in cerebellar disorders
- Festinating gait with a tendency to fall forward in Parkinson's disease
- Shuffling gait with retropulsion and tendency to fall backward in progressive supranuclear palsy
- Presence of washbasin sign (history of falls while washing face next to washbasin/sink or imbalance when splashing water on the face or imbalance while pulling a shirt over the head) and features of sensory ataxia along with a stomping gait as seen in dorsal column disorders

7.7 Physiotherapy for Benign Paroxysmal Positional Vertigo

Benign paroxysmal positional vertigo, a mechanical peripheral vestibular dysfunction, a common cause for recurrent and episodic vertigo, usually results from displacement of otoconia into the SCCs. “Cupulolithiasis” and “canalithiasis” are the two proposed mechanisms of BPPV.

7.7.1 Mechanisms for BPPV

In 1962, Harold Schuknecht, an otologic surgeon, proposed the adhesion of otoconia or calcium carbonate crystals to the cupula (cupulolithiasis) as a possible mechanism of BPPV. The presence of such degenerative debris adhering to the cupula of the SCCs considerably increases the density of the cupula and produces an inappropriate deflection of the cupula resulting in symptoms like vertigo, nystagmus, and nausea. Cupulolithiasis is characterized by the immediate onset of vertigo and nystagmus when moved into the provoking position, and both features will persist until the offending position is maintained.

The presence of free-floating fragments of otoconia (degenerative debris from the utricle) in the long arm of the SCC was proposed as an alternate mechanism (canalithiasis) by Hall for BPPV. The density of otoconia is roughly double that of the endolymph within the SCCs, and any positional change of the canal with respect to gravity may cause movement of the otoconia resulting in deflection of the cupula, causing vertigo and nystagmus. Canalithiasis is characterized by the delayed onset of vertigo and nystagmus (1–40 seconds) after placing in the offending position. The intensity of the symptoms is related to the degree of deflection of the cupula, and the symptoms generally cease within 1 minute. Cessation of endolymph movement and settlement of otoconia crystals in the most dependent portion of the canal attributes to the disappearance of the symptoms even when the provoking position is maintained.

7.7.2 Test Procedures or Maneuvers for BPPV

For clinical diagnosis and to manage BPPV, three different maneuvers are used. To determine the SCC involved and to plan the treatment, a keen observation of the direction and duration of the nystagmus is crucial. Prior to performing the maneuver, the patient should be instructed to keep his eyes open. To prevent the patient from moving away from the offending position, the examiner should convince the patient that the symptom of dizziness is transient and will stop or decrease if the patient remains in the position. For those patients with a history of nausea and emesis associated with vertigo, the diagnostic maneuvers should preferably begin with the unaffected side. The maneuver should be performed with utmost caution for those patients with a prior history of craniovertebral junction anomalies and vertebrobasilar artery insufficiency. Limiting or avoiding neck extension and rotation by using a tilt table is the preferred method rather than executing the standard technique on the treatment table.

The posterior SCC is the most frequent canal involved in BPPV. The Dix-Hallpike maneuver or Barany maneuver is a gold standard test used for the diagnosis of posterior canal BPPV. History suggestive of episodic vertigo with certain provoking positions and a positive Dix-Hallpike test are considered to be the best clinical practice guidelines for the diagnosis of BPPV. The nystagmus will be provoked only when the affected ear is positioned downward and not upward. For the posterior canal, the maneuver consists of placement of the posterior canal of the affected ear in line with the pull of gravity. This provoking posture helps in shifting the debris, adhering to the cupula, or freely floating in the long arm of the canal, away from the cupula, resulting in vertigo and nystagmus. The degenerative debris within the posterior canal induces an upbeating torsional nystagmus toward the tested side.

For those subjects with known cervical or low back injuries, who are unlikely to tolerate the Dix-Hallpike position, the "side-lying test" (Fig. 7.12) is an effective

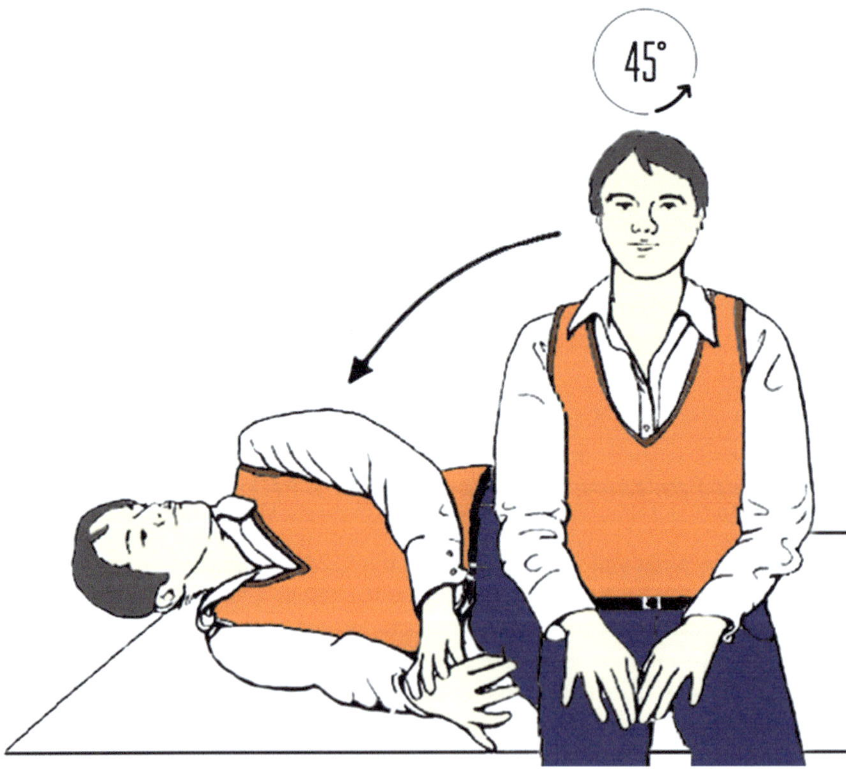

Fig. 7.12 An illustration of the side-lying test.

alternative. For this maneuver, the patient is made to sit up with legs hanging over the side of the treatment table. Then, the examiner instructs the patient to turn his head 45° away from the side to be tested, and while maintaining the same head rotation, he is asked to immediately lie down toward the testing side. Meanwhile, the examiner should observe the direction and type of nystagmus. After about 30–60 seconds, the examiner should instruct the patient to slowly return to the sitting position; after this, the test has to be repeated for the opposite side.

For horizontal canal BPPV, the Dix-Hallpike test or side-lying test may not induce vertigo and nystagmus, and the only appropriate maneuver which may provoke vertigo and nystagmus is the "roll test." The position used in the roll test maneuver (Fig. 7.13) helps to place the horizontal canal in the line of gravity. The nystagmus provoked by this test is horizontal and may have a torsional component. For the canalolithiasis form of horizontal canal BPPV, the nystagmus will be geotropic (the fast phase beating toward the ground or earth) and the duration will be brief in nature. On the contrary, for cupulolithiasis form, the nystagmus is apogeotropic (the fast phase beating away from the ground) and the duration will be prolonged in nature. The procedure is performed by positioning the patient in supine with the head elevated to 15–20° using a pillow, and then the examiner holds the head and quickly rolls the head to the left to see if that provokes geotropic or apogeotropic nystagmus along with vertigo. After the nystagmus ceases or after 30–60 seconds, the head is brought back to the neutral position, and then the test procedure is further continued for the right side. Though roll test can provoke vertigo and geotropic or apogeotropic nystagmus in horizontal canal BPPV, the intensity and degree of symptoms may not help in identifying the side of involvement.

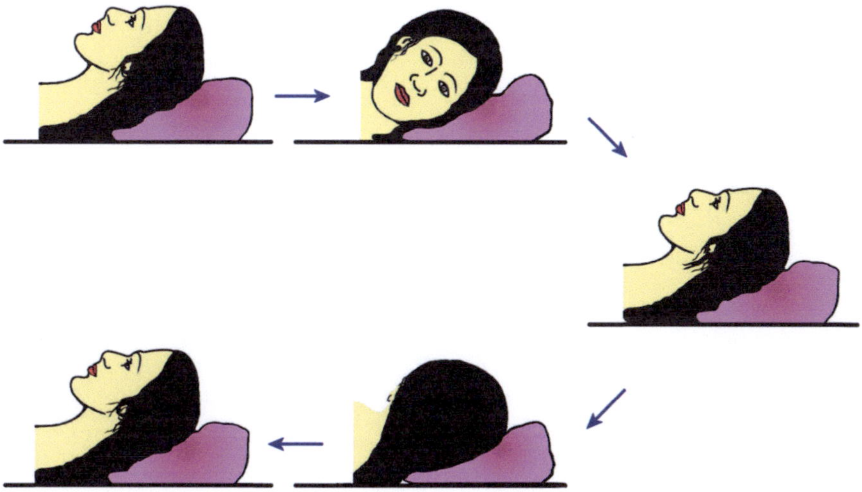

Fig. 7.13 An illustration of the roll test maneuver

When a positive result is obtained for the roll test, the bow and lean test helps determine whether the right or left side horizontal canal is the pathological one. In the "bow" position (Fig. 7.14), for the cupulolithiasis type, the nystagmus beats away from the affected side, and for the canalithiasis, toward the affected side, whereas in the "lean" position (Fig. 7.15), for cupulolithiasis, the nystagmus beats toward the affected side, and for canalithiasis, away from the affected side.

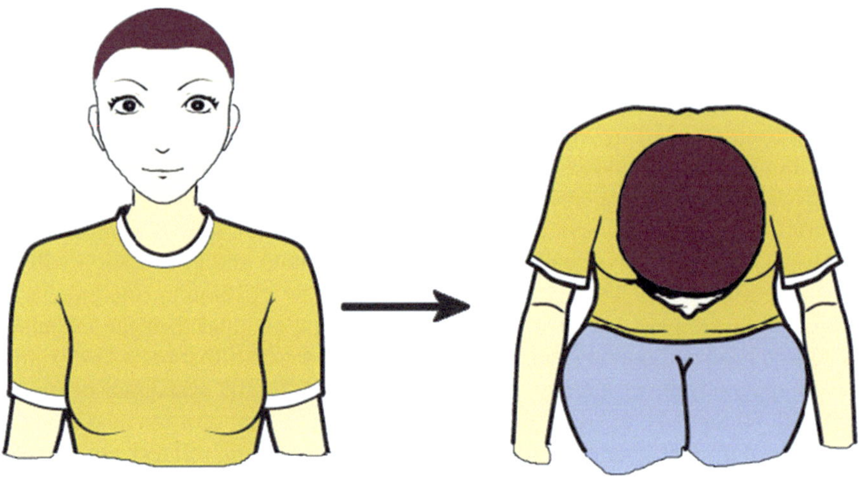

Fig. 7.14 An illustration of the bow position

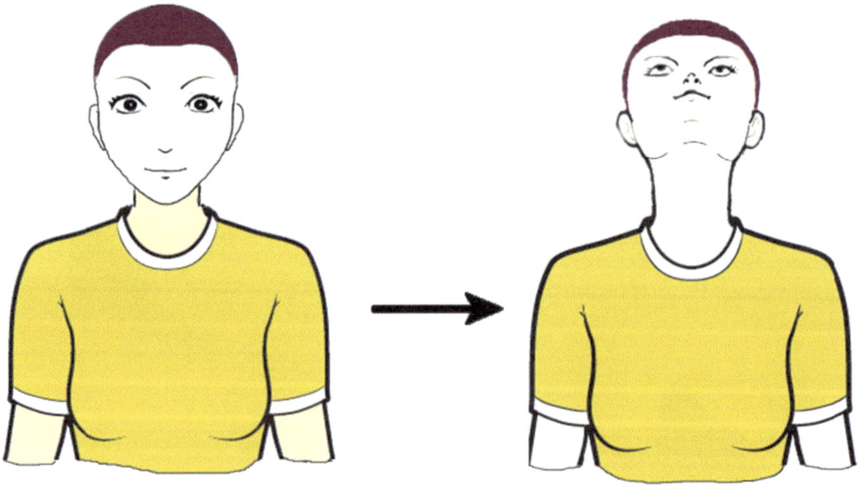

Fig. 7.15 An illustration of the lean position

7.7.3 Treatments for BPPV

Identifying the side and the canal involved and the location of the debris (canalithiasis or cupulolithiasis type) greatly influences the success of the BPPV treatment. The canalith repositioning maneuver, the liberatory maneuver, and the Brandt-Daroff habituation exercises are the three clinical treatment techniques indicated for BPPV.

The canalith repositioning maneuver (Epley's maneuver), a technique proposed and developed by John McNaughton Epley, an American otolaryngologist, theorized that the freely floating degenerative debris or otoconia located within the long arm of the canal could be moved back into the utricle using a specific maneuver. This maneuver is used for the treatment of the canalithiasis form of posterior canal BPPV. Studies on the efficacy of this treatment have shown a strong evidence of remission following this maneuver in the posterior canal BPPV. The specific positions used in the maneuver, the time for moving from one position to another, and the post-maneuver instructions are certain valid points to be considered. The use of anti-emetic medication about an hour prior to the maneuver is advised for those patients with history of severe nausea and vomiting with vertigo. The specific sequence of head positions in 90° increments helps to migrate the otoconia from the long arm of the posterior canal to the utricle. For each position, the patient may experience a brief spell of vertigo due to the migration of degenerative debris. The timing for changing the position during the sequential head position maneuvering is based on the latency of the nystagmus and the time taken for the nystagmus to stop. Usually, each new position is maintained until the nystagmus slows down or stops (30–60 seconds). Post-treatment, patients are advised to keep their heads upright for the next 48 hours and avoid provoking postures and vigorous shaking movements of the head to try and prevent the migration of otoconia back from the utricle to the posterior canal.

Epley's canalith repositioning maneuver (Fig. 7.16) begins with the patient in the Dix-Hallpike position (affected ear in the dependent position and head rotated 45° toward the affected ear and the neck extended to 20–30°), supine on the treatment table. Once the nystagmus stops or slows down (30–60 seconds), maintaining the 20–30° neck extension, the patient's head is slowly rolled by the clinician toward the unaffected side (45° toward the unaffected side from the neutral position of the head) and kept in the new position until the nystagmus stops or slows down. Next, the patient is slowly assisted by the clinician to roll over the patient's shoulder, until the tip of the nose of the patient is pointing toward the floor or ground. This position is also maintained until the nystagmus stops or slows down, and maintaining the same rotated position of the head toward the unaffected side, the patient is slowly assisted to sit up and is then allowed to straighten the head.

The Semont maneuver, also known as a brisk treatment or liberatory maneuver, an alternative for Epley's canalith repositioning maneuver, was proposed by Alain Semont, a French physiotherapist, in 1988, and theorized that a single brisk maneuver using inertia and positional changes against gravity will help to migrate the debris mechanically from the posterior canal back into the utricle. In the Semont liberatory maneuver (Fig. 7.17), the patient is made to sit up with the feet flat on the

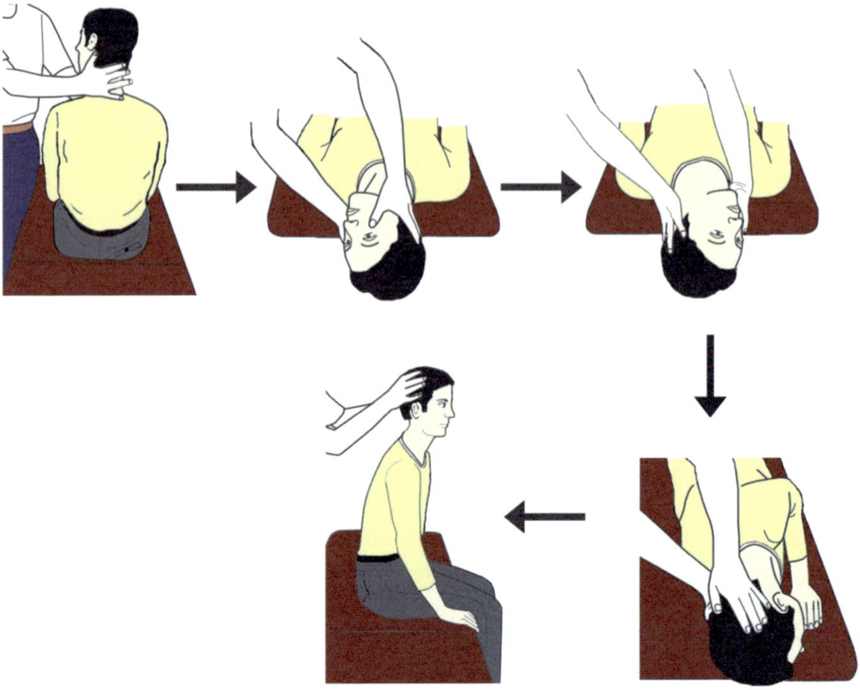

Fig. 7.16 An illustration of Epley's canalith repositioning maneuver

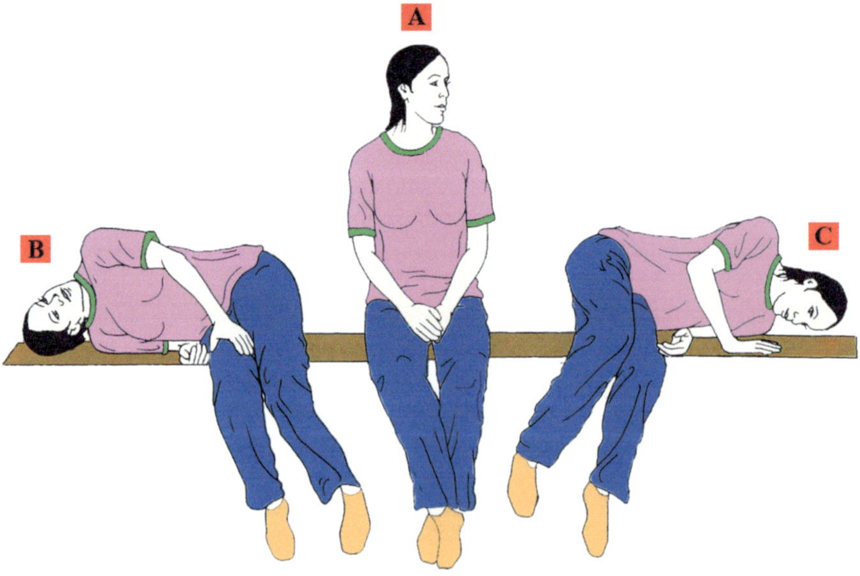

Fig. 7.17 An illustration of the Semont liberatory maneuver

ground (sits up sideways on the examination table), and the head is rotated 45° toward the unaffected side (position **A** of Fig. 7.17). The patient is quickly moved onto the affected side side-lying (**B**), and after 60 seconds, the patient is quickly moved with assistance to the opposite side, side-lying, holding the head still in 45° rotation toward the unaffected side (**C**). This position is maintained for another 60 seconds, and then slowly, maintaining the head turned toward the unaffected side, the patient is assisted to come to the sitting position. Studies have shown that this maneuver also has a remarkable success rate in the treatment of posterior canal BPPV.

The canalith repositioning treatment for the horizontal canal or the 360° barbecue maneuver is a modified version of the original Epley's maneuver used for the posterior canal to treat horizontal canal BPPV. For this maneuver (Fig.7.18), the patient is initially positioned supine with the affected ear down on the examination table. During the progressive sequence of head rotation, each position is maintained for about 15 seconds or until nystagmus or vertigo stops. The head of the patient is assisted slowly to roll 90° towards the unaffected ear until the patient's face is forwards with nose-to-ceiling position. Post maintenance of 15 seconds, the head is slowly rolled another 90° in the same direction until the unaffected ear is down. After 15 seconds of maintenance, roll the head and instruct the patient to roll over the shoulder until the patient is in prone with face downwards (nose-to-ground position). Finally, roll another 90° until the head returns to the same position with the

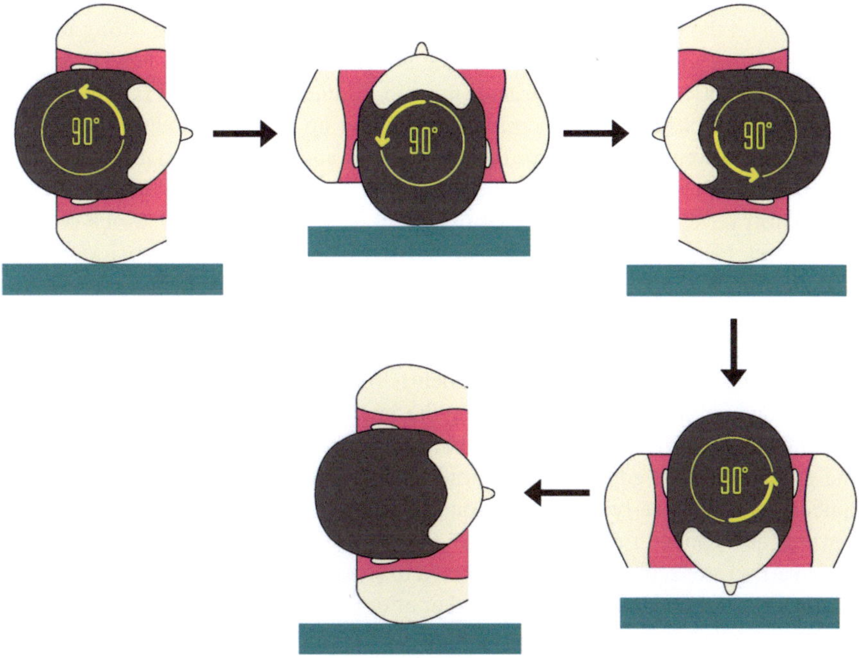

Fig. 7.18 An illustration of the 360° barbecue maneuver

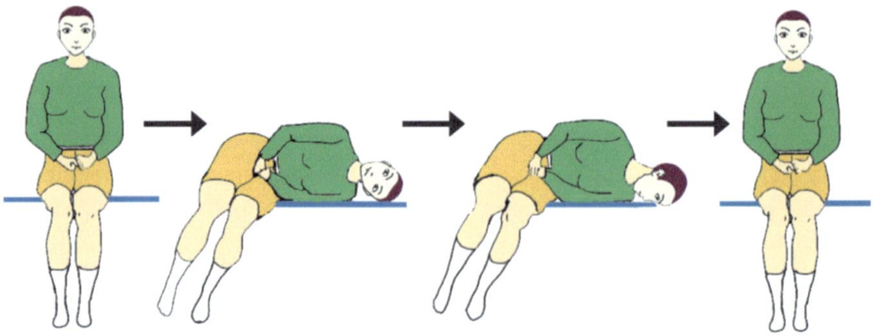

Fig. 7.19 An illustration of the Casani or Gufoni maneuver

affected ear down. If the treatment is effective, the symptoms will subside and then the patient should be assisted to come to side-sitting and to sit-up with legs hanging over the side of the examination table.

The Semont maneuver for cupulolithiasis, also known as the Casani or Gufoni maneuver (Fig. 7.19), is the treatment for horizontal canal cupulolithiasis. Post maneuver, 75% of the patients have reported remissions of symptoms. This maneuver is designed to rapidly move and dislodge the otoconia from the cupula of the horizontal canal. Once this maneuver is completed, the 360° barbeque roll maneuver should follow to move the freely floating otoconia in the horizontal canal back into the utricle. The Casani or Gufoni maneuver is performed with the patient seated sideways on the treatment table. The patient is moved quickly from sitting to side-lying position with the affected ear facing down. Then the patient's head is quickly turned 45°—60° towards the ground (nose-to-ground position) and is instructed to remain in that position for 2–3 min before sitting up again. A modified version of Gufoni maneuver is available for the treatment of apogeotropic variant.

In addition to the abovementioned maneuvers or treatment techniques, Brandt and Daroff developed a home-based treatment technique for BPPV which consists of a series of repeated movements in and out of offending head positions. This maneuver (Fig. 7.20) is especially indicated when the side of BPPV is unclear. The technique may be effective in ceasing or reducing the episodes of vertigo; however, the exact mechanism is unclear. Some evidence suggests that the exercises help in loosening and dispersing the otolithic debris from the cupula and others suggest a possible neural mechanism in reducing the intensity following repeated exposure. For the Brandt-Daroff treatment, the patient is made to sit up with the feet flat on the ground near the edge of the treatment table. The patient is instructed to move quickly into the side-lying position on the affected side with the head turned 45° toward the opposite side and the nose pointed to the ceiling. After 30 seconds or once vertigo stops, the patient is instructed to sit erect with head in neutral position. Following 30 seconds of waiting period or after vertigo stops, he or she is instructed to continue the same series of events but to the opposite side (side-lying on the unaffected side with the head turned 45° toward opposite side and nose-to-ceiling position). The patient is asked to stay for another 30 seconds or wait until vertigo

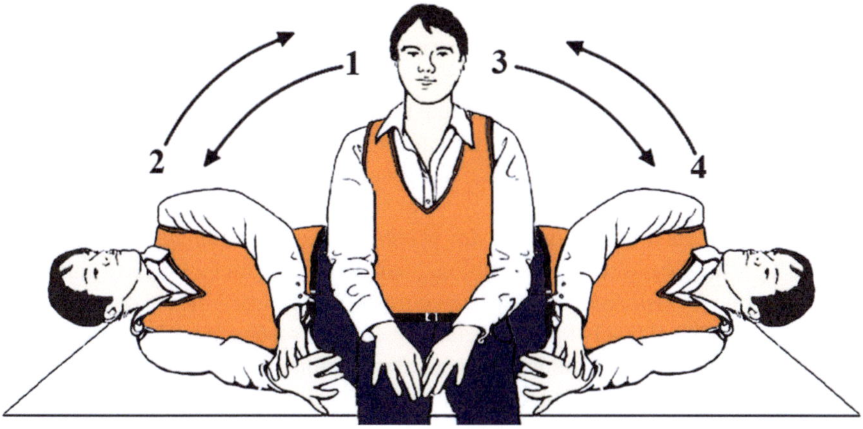

Fig. 7.20 An illustration of the Brandt-Daroff treatment maneuver

stops, before coming back to sitting with the head in a neutral position. The abovementioned maneuver is repeated 10–20 times per session, three times a day, until the patient reports no vertigo for two consecutive days.

7.8 Physiotherapy for Vestibular Hypofunction

Depending on the type and extent of the vestibular deficit, the clinical course and presentation and the level of recovery of peripheral vestibular hypofunction may differ. Many of such patients may have symptoms like vertigo, lightheadedness, nystagmus, blurred vision, unsteadiness, fear of fall, gait abnormality, anxiety, depression, and a reduced physical activity level. Functional deficits in the vestibulo-ocular and vestibulospinal systems, results of sensory mismatch, and physical deconditioning are plausible causes of the abovementioned symptoms. Postural unsteadiness is due to the impairment in the function of the VSR. Impaired or reduced vestibulo-ocular and vestibulospinal reflexes result in imbalance, diminished gait speed, and an increased risk of falling.

Acute asymmetry of the vestibular function, due to vestibular system dysfunction, causes the brain to interpret the abnormal signal as vertigo. For acute pathologies, within days to a few weeks, this asymmetry resolves and the spontaneous dizziness sensation ceases. However, in chronic vestibular dysfunction, vertigo tends to persist with the actual movement of the head.

Assessment should include the subject's history of present illness, fall history, oculomotor and vestibulo-ocular testing, positional testing like Dix-Hallpike, roll test and motion sensitivity, and gait and balance evaluation. The examiner should also find out the type, frequency, duration, and intensity of symptoms and whether the symptoms are of a fluctuating nature. The range of motion evaluation of the neck (rotation, extension, flexion, lateral flexion) needs special emphasis as cervical range limitations are a common clinical finding in vestibular hypofunction. Patients with vestibular dysfunction may self-restrict head movement to minimize the symptoms that may arise as a result of the positional change of head or with head movement. Patients may often restrict head movements to see things clearly or maintain balance. Assessment should include tests for the sensation of the extremities to rule out pathologies affecting kinesthesia and proprioception and coordination to rule out dysmetria of cerebellar origin.

The main goals of physiotherapy for vestibular hypofunction are to improve the patient's postural steadiness and functional balance during transitions and gait, minimize blurring of vision during head movements, and promote and encourage them to return to a normal level of physical activity. Several mechanisms are proposed for the recovery of function following vestibular dysfunction. They include cellular recovery, spontaneous central reestablishment of the tonic firing rate, habituation of unpleasant sensations, substitution of alternative strategies for the lost vestibular function, and adaptation of residual vestibular function.

Gaze stabilization exercises, habituation exercises, and postural stabilization exercises are the three different approaches advocated for the treatment of vestibular hypofunction. In the gaze stabilization exercises, the subject is first instructed to fix his or her eyes on a stationary target kept right in front at an arm distance. The subject is then asked to slowly move his head up and down or right to left to and fro for 10–20 repetitions. The speed of head motion should be increased as long as the target remains in focus. The second component consists of the head and the target moving in opposite directions, while the subject fixes eyes on the target. As a progression, the range and speed of movement can be increased without losing focus on the target. The third exercise involves fixing the eyes on a target 3–6 feet distance in front. Then, closing the eyes and while maintaining the eyes on the remembered target, the head is moved in a to-and-fro manner either up and down or in the right to left direction. The gaze stabilization exercises are believed to induce adaptation of residual vestibular function by creating small retinal slips or by a different mechanism centrally programmed or by facilitating the use of cervical inputs to generate the eye movement to keep the visual target on focus.

Habituation exercises are based on the concept that repeated exposure to a provocative stimulus will result in a reduction in the pathological response to that stimulus. Cawthorne and Cooksey, in the 1940s, were the first to develop a series of habituation exercises that addressed complaints of vertigo and impaired balance. The exercises include pursuit and saccadic eye movements, head movements, tasks demanding coordination of eyes with the head, whole-body movements, and balance. They recommended that the exercises be performed in various positions and at various speeds and emphasized performing them routinely. According to the developers, performing exercises with the eyes closed reduces the patient's reliance on visual information and plausibly forced effective compensation by somatosensory or vestibular mechanisms. Studies have revealed that 84% of the patients responded favorably to Cawthorne-Cooksey exercises. Examples of Cawthorne-Cooksey exercises performed progressively from supine toward standing and walking are provided in Table 7.1.

Table 7.1 Examples of Cawthorne-Cooksey exercises

Posture	Exercises	Features	Supervision
Supine	Eye movements with no head movements	• Vertically scanning the eye upward and downward (10 times) • Side-to-side scanning of eyes (10 times) • Focusing on a finger moved (both up and down and side-to-side movements) at an arm distance away	Usually not required
	Head movements; initially performed slowly and then gradually more quickly	• Eyes closed and performing flexion and extension of the neck (10 times) • Eyes closed and performing side-to-side rotation of the neck (10 times)	
	Head movements with eyes fixed (may ask the patient to fix his or her eyes at a spot right in front)	• Performing flexion and extension of the neck (10 times) • Performing side-to-side rotation of the neck (10 times)	
Sitting	Shoulder exercises in sitting	• Shrugging the shoulders (10 times) • Protracting and retracting the shoulders (10 times)	May be required during the beginning sessions
	Sitting preferably with no back support	• All the exercise components performed in supine to be repeated while sitting • Forward bending of the trunk up on the hips to pick up objects from the ground	
Standing	Transitions	• Sit to stand transitions with eyes open (10 times) • Sit to stand transitions with eyes closed (10 times)	Required during the early sessions
	Standing with feet shoulder-width apart	• All the exercise components performed in sitting to be repeated while standing • Tossing a tennis ball from hand to hand at the eye level • Tossing a tennis ball from hand to hand at the knee level • Turn around to look backward (both sides)	
Walking	Walking straight in a 10 meter walkway	• Walk straight with eyes open (10 times) • Walk with eyes fixed to a point in front and with the head moving upward and downward (10 times) • Walk with eyes fixed to a point in front and with the head moving side to side (10 times) • Walk straight with eyes closed (10 times) • Circle round a person who will throw a large ball and to whom it should be returned • Forward bending of the trunk up on the hips to pick up objects from the ground	Strictly required until the patient performs activities at ease with no loss of balance or fear of falling

- Complexity, including the repetitions, needs to be individualized
- Progression should be based on the extent of habituation
- Few sessions of exercise may not give any noticeable beneficial effects
- Diligence and perseverance are required to return to normal activity

Postural stabilization exercises (Figs. 7.21 to 7.28) utilize the use of visual, somatosensory, and vestibular cues to promote a central preprogramming to improve postural stability. The postural stabilization exercises should emphasize working with and without visual cues or altering the somatosensory cues. Removing or altering cues forces the patient to use the remaining sensory cues. The intensity of stability exercises can be increased by reducing the base of support, adding head movements, unpredictable changes in the direction of gait, and perturbations

Fig. 7.21 Spot marching on a foam surface

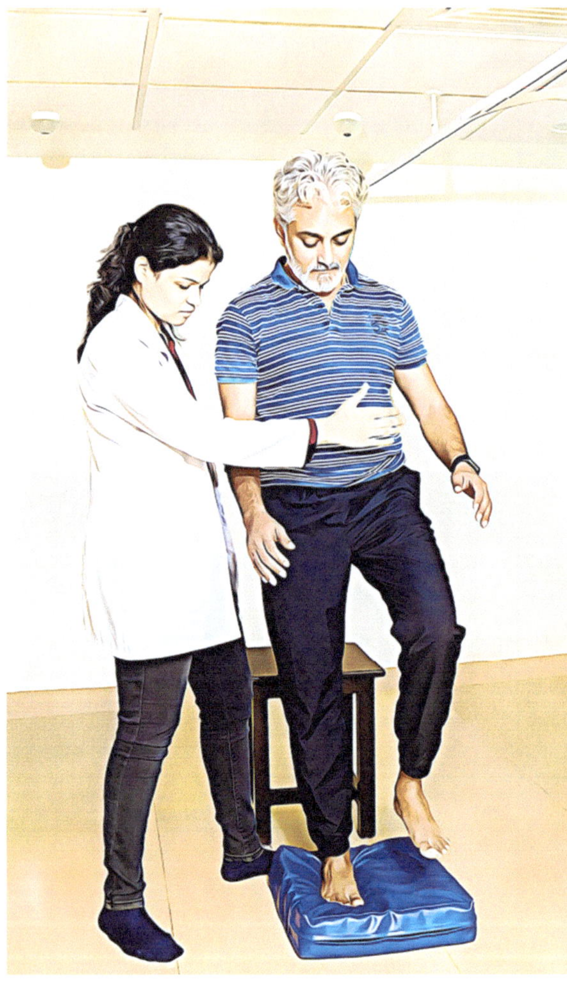

Animated photograph of model and therapist with permission

Fig. 7.22 Clearing an obstacle kept on the ground

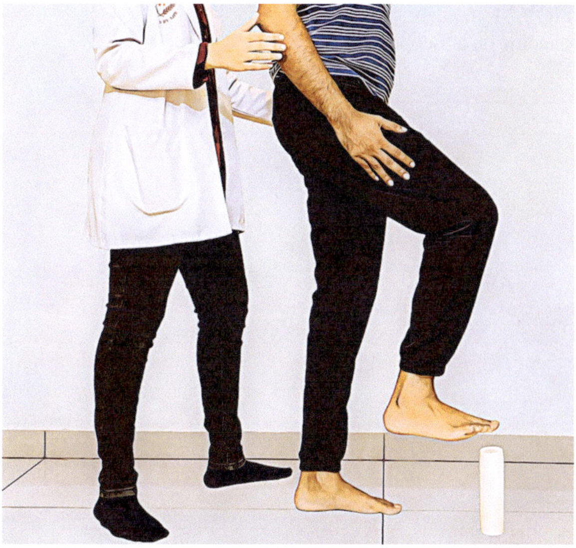

Animated photograph of model and therapist with permission

Fig. 7.23 Standing on a trampoline

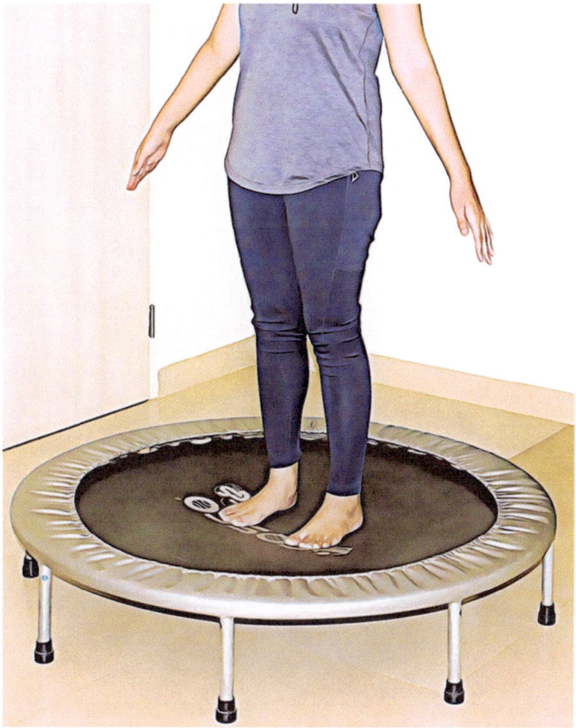

Animated photograph of model with permission

Fig. 7.24 Tandem standing on a trampoline

Animated photograph of model with permission

Fig. 7.25 Single-limb standing on a trampoline

Animated photograph of model with permission

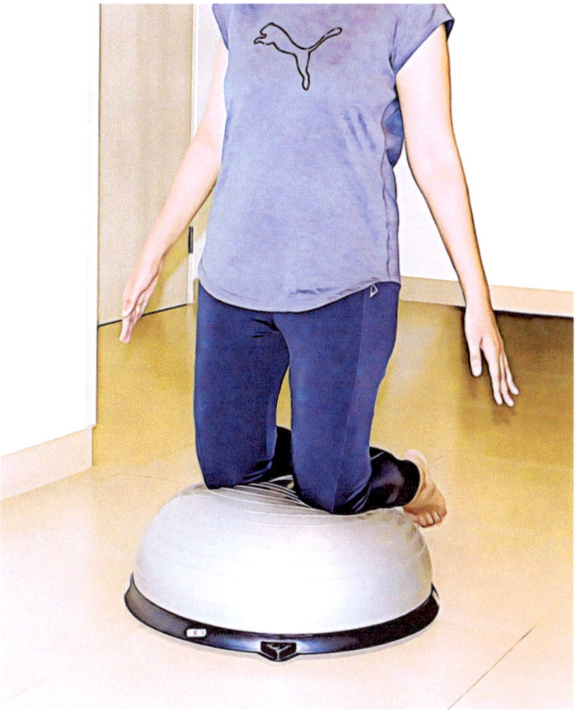

Animated photograph of model with permission

Fig. 7.26 Balancing on a BOSU® ball while kneeling

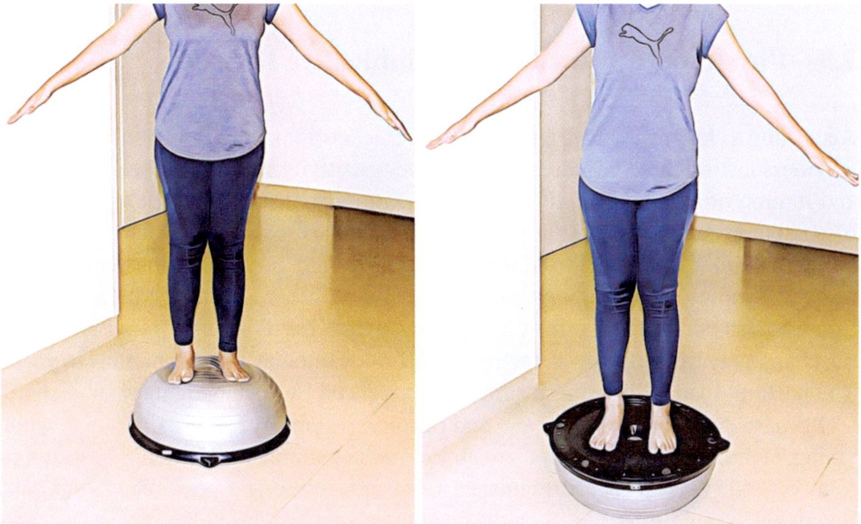

Animated photograph of model with permission

Fig. 7.27 Balancing on a BOSU® ball in standing

Animated photograph of model and therapist with permission

Fig. 7.28 Postural control training with one foot over the stepper while attempting to catch a ball

7.9 Physiotherapy for Chronic Subjective Dizziness

According to Jeffrey P. Staab and Michael J. Ruckenstein, persistent non-vertiginous dizziness lasting three or more months, hypersensitivity to patient's own movement and motion of objects, and difficulty to perform visual precision tasks like the use of computer or reading are the three basic features of chronic subjective dizziness (CSD). Anxious and introverted personality traits may predispose the development of CSD. Studies on rehabilitative treatments have observed that vestibular habituation exercises are effective for CSD management. The CSD may remit spontaneously and may demonstrate asymptomatic periods often lasting for months to years. The postural instability or unsteadiness experienced by such patients is subjective.

Primary symptoms of dizziness and postural instability are usually worse while sitting, standing, or walking and are absent in the recumbent position. These symptoms of CSD need to be distinguished from orthostatic and positional symptoms. Positional symptoms occur when the head moves into or through a specific orientation in space like the provoking head postures of BPPV. The factors that may trigger CSD symptoms are those head or body motions that are not direction specific, visual stimuli such as fluorescent lighting, and computer screens and social stimuli like busy environments and crowds.

Peripheral vestibular disorders such as BPPV and vestibular neuritis are known to precipitate more cases of CSD than central vestibular conditions. In the majority of cases, the onset of CSD is sudden and may start after a single event. Patients may identify the triggering events as a single bout of vestibular neuritis or an episode of vestibular migraine. For patients with most of the neuro-otologic conditions, holding still or removing themselves from motion-rich environments may alleviate their symptoms. However, minimization or removal of the provocative stimuli is likely to prolong and exacerbate the experience of dizziness and postural unsteadiness, lasting hours to days, for CSD patients.

Persistent failure to readapt when the initial postural, balance, and oculomotor impairments resolve and long-standing use of high-risk balance control strategies that are out of proportion to what is necessary to compensate for chronic neuro-otologic deficits are the possible mechanisms of CSD. Anxiety during the early phase of balance disruption may impair the readaptation process and may pave the way to CSD, further worsening anxiety or depressive disorders. Vestibular habituation exercises and balance training exercises in addition to certain specific medications like serotonergic medications and cognitive behavior therapy are some of the treatment options for CSD.

Further Reading

Baloh R, Honrubia V. Clinical neurophysiology of the vestibular system. Philadelphia: FA Davis; 1990.

Brandt TH, Dieterich M. Different types of skew deviation. *J Neurol Neurosurg Psychiatry.* 1991 Jun;54(6):549–50.

Brandt T, Steddin S. Current view of the mechanism of benign paroxysmal positioning vertigo: cupulolithiasis or canalolithiasis?. *J Vestib Res.* 1993;3(4):373–82.

Carpenter M, Allum J, Honegger F. Vestibular influences on human postural control in combinations of pitch and roll planes reveal differences in spatiotemporal processing. *Exp Brain Res.* 2001;140(1):95–111.

Cohen HS, Kimball KT. Effectiveness of treatments for benign paroxysmal positional vertigo of the posterior canal. *Otol Neurotol.* 2005;26(5):1034–40.

Corcoran JP, Jones JD. The role of physiotherapy in the management of the dizzy patient. *The Otorhinolaryngologist.* 2013;6(1):16–24.

Dieterich M. Central vestibular disorders. *J Neurol.* 2007;254(5):559–68.

Eggers SD, De Pennington N, Walker MF, Shelhamer M, Zee DS. Short-term adaptation of the VOR: non-retinal-slip error signals and saccade substitution. *Ann N Y Acad Sci.* 2003 Oct;1004:94–110.

Epley JM. The canalith repositioning procedure: for treatment of benign paroxysmal positional vertigo. *Otolaryngol Head Neck Surg.* 1992;107(3):399–404.

Fukuda T. The stepping tests. *Acta Otolaryngol.* 1959;50:95–108.

Han BI, Song HS, Kim JS. Vestibular rehabilitation therapy: review of indications, mechanisms, and key exercises. *J Clin Neurol.* 2011;7(4):184–96.

Herdman SJ, Clendaniel RA, editors. Vestibular rehabilitation. 4th ed. Philadelphia: FA Davis Company; 2014.

Highstein SM. Role of the flocculus of the cerebellum in motor learning of the vestibulo-ocular reflex. *Otolaryngol Head Neck Surg.* 1998;119(3):212–20.

Horlings CGC, Carpenter MG, Honegger F, Allum JHJ. Vestibular and proprioceptive contributions to human balance corrections. *Ann NY Acad Sci.* 2009;1164(1):1–12.
Hornibrook J. Benign Paroxysmal Positional Vertigo (BPPV): History, Pathophysiology, Office Treatment and Future Directions. *Int J Otolaryngol.* 2011;2011:835671.
Ishikawa K, Edo M, Yokomizo M, Terada N, Okamoto Y, Togawa K. Analysis of gait in patients with peripheral vestibular disorders. *ORL J Otorhinolaryngol Relat Spec.* 1994;56(6):325–30.
Jannetta PJ, Møller MB, Møller AR. Disabling positional vertigo. *N Engl J Med.* 1984;310(26):1700–5.
Lin SI, Woollacott MH. Postural muscle responses following changing balance threats in young, stable older, and unstable older adults. *J Mot Behav.* 2002;34(1):37–44.
Mergner T, Schweigart G, Fennell L, Maurer C. Posture control in vestibular-loss patients. *Ann NY Acad Sci.* 2009;1164(1):206–15.
Minor LB, Goldberg JM. Vestibular-nerve inputs to the vestibulo-ocular reflex: a functional-ablation study in the squirrel monkey. *J Neurosci.* 1991;11(6):1636–48.
Nashner LM, Black FO, Wall III C. Adaptation to altered support and visual conditions during stance: Patients with vestibular deficits. *J Neurosci.* 1982;2:536–44.
Paparella MM. Pathogenesis of Meniere's disease and Meniere's syndrome. *Acta Otolaryngol Suppl.* 1984;406:10–25.
Peterson B, Richmond F, editors. Control of head movement. New York: Oxford University Press; 1988.
Schubert MC, Hall CD, Das V, Tusa RJ, Herdman SJ. Oculomotor strategies and their effect on reducing gaze position error. *Otol Neurotol.* 2010 Feb;31(2):228–31.
Shumway-Cook A, Horak FB, editors. Rehabilitation strategies for patients with vestibular deficits. In: Neurologic Clinics. Vol 8.; 1990:p.441–57.
Shumway-Cook A, Horak FB. Vestibular rehabilitation: an exercise approach to managing symptoms of vestibular dysfunction. *Seminars in Hearing.* 1989;10:196–208.
Semont A, Freyss G, Vitte E. Curing the BPPV with a liberatory maneuver. *Adv Otorhinolaryngol.* 1988;42:290–3.
Viirre E, Draper M, Gailey C, Miller D, Furness T. Adaptation of the VOR in patients with low VOR gains. *J Vestib Res.* 1998;8(4):331–4.
von Brevern M, Radtke A, Lezius F, Feldmann M, Ziese T, Lempert T, et al. Epidemiology of benign paroxysmal positional vertigo: a population based study. *J Neurol Neurosurg Psychiatry.* 2007;78(7):710–5.
Whitney SL, Sparto PJ. Principles of vestibular physical therapy rehabilitation. *NeuroRehabilitation.* 2011;29(2):157–66.
Zacharias GL, Young LR. Influence of combined visual and vestibular cues on human perception and control of horizontal rotation. *Exp Brain Res.* 1981;41:159–71.

Chapter 8
Motor Neuron Disease

Abraham M. Joshua and Zulkifli Misri

8.1 Introduction

Motor neuron diseases are a group of neurological disorders that present in adult life. They are defined as progressive degenerative disorders of the motor neurons in the spinal cord, brainstem, and motor cortex manifested clinically by muscle weakness, wasting, and corticospinal tract signs in varying combinations. The disorder is characterized pathologically by degenerative changes in the anterior horn cells, motor nuclei of the brainstem and corticospinal tract, and clinically by the progressive muscular wasting combined with symptoms of corticospinal tract degeneration.

As the name implies, both the upper motor neurons and the lower motor neurons can be affected. Most patients will show evidence of both upper motor neuron (UMN) and lower motor neuron (LMN) patterns of weakness, although one may predominate. The time course of the illness varies from person to person. This condition is commonly seen in middle life. The majority of patients will deteriorate and succumb to the disease within 2–6 years after the onset of symptoms; however, some may live for several years.

A. M. Joshua (✉)
Department of Physiotherapy, Kasturba Medical College of Mangalore, Manipal Academy of Higher Education, Manipal, India
e-mail: abraham.joshua@manipal.edu

Z. Misri
Department of Neurology, Kasturba Medical College of Mangalore, Manipal Academy of Higher Education, Manipal, India
e-mail: misri.zk@manipal.edu

A. M. Joshua (ed.), *Physiotherapy for Adult Neurological Conditions*,
https://doi.org/10.1007/978-981-19-0209-3_8

8.2 Historical Background

The existence of such a syndrome characterized by progressive muscular weakness and muscular atrophy was first confirmed in the medical community during the mid-nineteenth century. Probably, François-Amilcar Aran, a physician who published his description of the syndrome in 1848, did the pioneer works. Many authors credit Aran and Guillaume-Benjamin-Amand Duchenne, a French neurologist, for the earliest description of progressive spinal muscular atrophy; however, they believed the disorder to be of myogenic origin. Such an interpretation was incorrect, and within years, Jean Cruveilhier, a French surgeon and anatomist, based on his findings, brought this disorder into line with motor neuron disease (MND), which he considered as a myelopathic or spinal origin.

Based on the signs and symptoms, several clinical variants of MND were identified during the later years. In 1858, Duchenne gave the earliest description of labio-glosso-laryngeal paralysis, a term which Adolf Wachsmuth, a German physician, in 1864, changed to progressive bulbar palsy, a variant of MND. The original description of amyotrophic lateral sclerosis, another variant of MND, was given by Jean-Martin Charcot, a French neurologist, in 1874. An earlier work by Jacob Augustus Lockhart Clarke, a British physiologist and neurologist, and Charles Bland Radcliffe, an English physician, in 1862, though failed to name the disease formally, described the pathological changes of a patient affected with amyotrophic lateral sclerosis in their work tilted "An Important Case of Paralysis and Muscular Atrophy, with Disease of the Nervous Centers" Their work is probably the first detailed autopsy study of amyotrophic lateral sclerosis. Jean-Martin Charcot, along with Alix Joffroy, a French neurologist and psychiatrist, and Albert Gombault, a French neurologist, studied the pathological aspect of the disease and between 1872 and 1874, gave clear account of the clinical and pathologic findings of the same. By 1874, the disease was referred to as Charcot's disease, and it was Charcot who coined the term amyotrophic lateral sclerosis, the most common variant of MND. In 1875, after studying several cases with distinctive neuropathologic features, Wilhelm Heinrich Erb, a German neurologist, designated the term primary lateral sclerosis for yet another variant of MND. In 1962, Walter Russell Brain, a British neurologist and the principal author of the standard textbook of neurology, *Diseases of the Nervous System*, used the term motor neuron disease for this group of disorders.

8.3 Clinical Variants or Types

As rightly stated by Walter G. Bradley in his textbook *Neurology in Clinical Practice*, it is important for clinicians to make the distinction between the terms "motor neuron disease" and "motor neuron diseases (MNDs)." The first term refers to a specific disorder of both upper and lower motor neurons, otherwise known as

Table 8.1 Types of motor neuron disease or amyotrophic lateral sclerosis

Types	Occurrence	Variants
Acquired or sporadic ALS	90–95%	Classic ALS (spinal onset) Progressive bulbar palsy Progressive muscular atrophy Primary lateral sclerosis Mills hemiplegic variant Flail-arm variant Flail-leg variant Monomelic amyotrophy (Hirayama disease) UMN onset LMN onset Bulbar onset Dyspnea onset Guam ALS (Western Pacific ALS)
Familial ALS	5–10%	ALS1 to ALS20 (most of them are autosomal dominant and few are autosomal recessive) X-linked dominant ALS ALS-FTD (frontotemporal dementia) ALS due to rare mutations

amyotrophic lateral sclerosis (Table 8.1), and the second term refers to the broader family of disorders that may affect the upper and/or lower motor neuron system as well as non-motor systems. The latter include familial and sporadic disorders; inflammatory, infective, and immune disorders; and others of undetermined cause. Table 8.2 depicts all the primary and secondary MNDs.

Though there are several typical variants or forms, amyotrophic lateral sclerosis is considered the most frequent form of MND. The other variants are progressive muscular atrophy, progressive bulbar palsy, and primary lateral sclerosis. In addition to the above, many atypical variants have been reported from different parts of the world.

Amyotrophic lateral sclerosis (ALS) is characterized by muscle atrophy and hyperreflexia. Amyotrophy refers to the atrophy of muscle fibers, which are denervated as their corresponding anterior horn cells degenerate, and lateral sclerosis refers to the hardening of the anterior and lateral columns of the spinal cord secondary to degeneration of the corticospinal tract. Progressive muscular atrophy, a less frequent clinical variant, has no apparent evidence of corticospinal dysfunction. Here, the pathological changes are confined to the anterior horn cells of the spinal cord with clinical features of the LMN lesion. Whereas, in progressive bulbar palsy, the muscles supplied by the lower brainstem are predominantly affected, and weakness and wasting predominate in muscles of the jaws, face, tongue, pharynx, and larynx. In primary lateral sclerosis, which is the least frequent form, UMN features predominate with no evidence of LMN features.

Spinal muscular atrophy (SMA), another motor system disorder having close clinical resemblance with MND, is an autosomal recessive disorder, frequently seen in infancy, childhood, and adulthood. It is also the leading cause of heritable infant

Table 8.2 Depiction of primary and secondary motor neuron diseases

Primary motor neuron diseases		
Type	Disease/disorder	Inheritance (if any)
Idiopathic	Sporadic ALS and its variants	–
Inherited	Spinal muscular atrophy Neuroaxonal dystrophy Fazio-Londe disease Juvenile onset ALS Brown-Vialetto-Van Laere syndrome Familial ALS (less common) Hereditary spastic paraplegia	Autosomal recessive
	Familial ALS (more common) Juvenile onset ALS Hereditary spastic paraplegia	Autosomal dominant
	Hereditary spastic paraplegia Kennedy's syndrome	X-linked
Multisystem neurodegenerative diseases affecting motor neurons	ALS-FTD Western Pacific ALS-dementia complex Spinocerebellar ataxia including Machado-Joseph disease Neuroacanthocytosis	Autosomal recessive/ autosomal dominant
Secondary motor neuron diseases		
Type	Disease/disorder	
Infective	Acute poliomyelitis Post-polio syndrome Neurosyphilis HIV infection Human T-cell leukemia/lymphoma virus infection Prion disease	
Metabolic	Hyperparathyroidism Hyperthyroidism Hypothyroidism Hexosaminidase A deficiency (HEX A deficiency)	
Immune	Paraproteinemia Disimmune motor system degeneration	
Environmental/radiation	Lead, arsenic, cadmium, or thallium intoxication, postirradiation syndrome, neurolathyrism	
Paraneoplastic	Hodgkin's disease Non-Hodgkin's lymphoma	

mortality. Despite the clinical heterogeneity seen in heritable variants of SMA, they are genetically homogeneous (gene defect for all types of SMA is on chromosome 5). Though familial forms of MND are rare, both familial and sporadic forms are genetically distinct from SMA. Another feature that distinguishes SMA from MND is the presence of "amyotrophic weakness" in the absence of hyperreflexia.

8.4 Natural History

MND is a rare neurological condition with a worldwide distribution and is seen in all races. The estimated incidence rate is approximately 2 per 100,000, and the prevalence is around 6 per 100,000 population. The disorder is generally more common in males than in females. The onset of symptoms of the disease is usually in mid to late adult life, with the average age being between 50 and 70 years. The incidence of MND tends to increase with advancing age. Occasionally, it can present in the third decade or even in the eighth decade of life. If symptoms arise in early adult life, it may resemble the SMA, especially when signs of corticospinal tract dysfunction are least prominent.

MND is a relentlessly progressive chronic disorder. Many times, the onset of symptoms is insidious, but at times, the disorder may run a subacute course. The nature of the earliest symptoms depends upon the region of the nervous system first affected. In many of the patients, the abnormalities are first noticed in the hand, such as weakness, stiffness, and clumsiness of movements of fingers, which may draw the patient's attention, and at times, even the twitch contractions (fasciculations) and wasting may be the first to be spotted. The rate of progression is usually steady for any given patient; therefore, the accrual of disability tends to become more predictable over time and often helps in planning management and discussing prognosis with patients and their families.

Evidence exists for a preclinical phase in MND, and until at least 30% of anterior horn cell motor neurons are lost, the patient will not be aware of the weakness. While collecting the history, the sporadic ALS patients often may hint about the muscle fatigue years before the onset of weakness, wasting, and fasciculations. For most of the variants, the pattern of disease spread is predictable. When onset is in one arm, the spread is often first to the contralateral arm, then the ipsilateral leg, the contralateral leg, and, finally, the bulbar region. If the weakness begins in the leg, the progression of the pattern is similar to that of the upper limb pattern. However, for the bulbar-onset variant (progressive bulbar palsy), the weakness tends to spread to the hands first, then to thoracic myotomes, and eventually to the legs. Overall, the rostral-caudal pattern progresses faster than the caudal-rostral spread. Older age at onset is a poor prognostic factor in MND. Subtypes of ALS, including Mills hemiplegic variant, flail-arm variant, flail-leg variant, primary lateral sclerosis, and progressive muscular atrophy, usually harbor a better prognosis.

8.5 Etiology

The exact cause of the disease is unknown, and many researchers and clinicians agree that various factors, possibly a combination of some or all the below-mentioned processes, may lead to the development of MND, and they are:

- Over-activation of glutamate receptors
- Autoimmunity to calcium ion channels

- Oxidative stressors linked to free radical formation
- Cytoskeletal abnormalities such as intracellular accumulation of neurofilaments
- Mitochondrial dysfunction
- Neurofilament dysfunction
- Pro-inflammatory cytokines

A significant body of research strongly supports and proposes the processes mentioned above as the possible cause for the pathogenesis of MND. Even genetic factors may play a role in the sporadic disease of ALS. Glutamate, the most abundant free amino acid in the central nervous system and a major excitatory neurotransmitter, produces neuronal excitation and participates in many neuronal functions, including neuronal plasticity. Excess amount of glutamate or impaired glutamate transport may leave excessive amounts of free excitatory neurotransmitters to repeatedly stimulate the glutamate receptor allowing the calcium ions to enter the neuron, causing neurotoxicity and neuronal cell death.

Though scientific investigators had debated the role of high-intensity physical training as a possible cause for neuronal cell loss due to excessive glutamate release, the authors observed that no retrievable data has emphasized the role of rest between the high-intensity exercises in minimizing the possible neuronal cell damage. The authors postulate that inadequate rest between sessions of high-intensity training for several days may predispose excessive glutamate release, rendering neuronal cells vulnerable to excitotoxicity, particularly in the presence of defective transport or reuptake of glutamate.

Evidence also implicates an inflammatory process in the pathogenesis of ALS. Even disturbances in mitochondrial structure or function have been implicated as a plausible cause for ALS pathogenesis. Abnormalities in axonal transportation are also considered to play a significant part in the pathogenesis of ALS. Mutations in the genes for neurofilament subunits appear to increase the risk for the development of sporadic ALS.

Convincing evidence for any viral etiology is yet to be found, and attempts to isolate a causal virus have been generally unsuccessful. Attempts to isolate any definite environmental factors triggering or contributing or causing the disease have failed. Several environmental elements have been evaluated as potential causative agents for ALS, including selenium, aluminum, iron, manganese, copper, and zinc, but there is no convincing evidence that any one of these plays a major part in ALS pathogenesis. Reduced amounts of trace elements like calcium and magnesium in soil and water may be the possible environmental factors; however, studies related to them are inconclusive. Some do believe MND is a form of chronic poliomyelitis, but virological studies and histological appearances are again not suggestive. The premature aging process of motor neurons, perhaps precipitated by unknown environmental factors, has received support from many workers, and some of the possible precipitating factors are exposure to cadmium, lead, or mercury, chronic exposure to electromagnetic fields, high levels or excessive physical exertion, high dietary intake of glutamate, physical trauma, and surgery.

8.6 Pathology

Spinal Cord The naked eye examination of the spinal cord may not suggest any abnormality. The cross section of the cord may reveal a reduction in the size of the gray matter of the anterior horn compared to normal. The large anterior horn cells are more involved than the smaller ones. Microscopic studies have shown evidence of severe degeneration of anterior horn cells with a certain degree of gliosis and proliferation of microglial cells. The anterior horn cell degeneration is more marked in cervical enlargement, but it is always widespread. Fortunately, the relative preservation of the small anterior horn cells in the sacral region (Onuf's nucleus) that controls the bowel and bladder function accounts for the preservation of sphincter control until the late stages of the disease.

Corticospinal tract degeneration is another finding that is usually minimal in the upper thoracic region compared to cervical enlargement. Also, the degeneration of the anterolateral column of the spinal cord is accompanied by a variable degree of gliosis. The degeneration of the crossed and the direct corticospinal tract is more severe in the lower thoracic and lumbosacral region, which can be revealed by the myelin staining of the spinal cord.

Ultrastructurally, the disorder is characterized by the presence of many neurofilaments in the anterior horn cells of the spinal cord or motor neurons of the brainstem. Many of the surviving neurons are small, shrunken, and filled with lipofuscin. Another important microscopic finding is the presence of intraneuronal inclusions in the degenerating neurons and glial cells, especially the small eosinophilic cytoplasmic inclusions called Bunina bodies. In the spinal cord, the posterior columns, spinocerebellar tracts, and Clarke's column are generally spared, though Clarke's column can be affected in the familial form of the disease. Evidence of degeneration of the spinocerebellar tract, especially the anterior spinocerebellar tract, is not uncommon; however, the degeneration varies in different segments. A variable degree of extrapyramidal tract degeneration may be seen. Even a mild degree of degeneration of the posterior column of the spinal cord in the absence of sensory symptoms and signs may be present.

Brainstem The degeneration of the motor neuron cells similar to the degeneration of anterior horn cells of the spinal cord is an important pathologic finding. The degeneration is more marked in hypoglossal nuclei, dorsal nucleus of the vagus, nucleus ambiguus, and motor nucleus of the trigeminal nerve. The extent of involvement of the facial motor nuclei is less when compared to the abovementioned nuclei. In typical MND, a certain set of motor neurons are spared, mainly the oculomotor, trochlear, and abducens nerves. A cross section of the brainstem reveals the degeneration of the corticobulbar and corticospinal tracts and is more severe in the medulla and negligible in the pons and cerebellar peduncles.

Cerebral Hemisphere The naked eye examination of the cerebral hemispheres appears normal; however, microscopic examination may reveal degeneration of ganglion cells in the frontal and precentral region of the cerebral cortex, and the affected areas may show glial overgrowth. Degenerative changes in the middle one-third of the corpus callosum and corticospinal tract in the posterior limb of the internal capsule are other findings.

Roots, Peripheral Nerves, and Muscles In the ventral root, large myelinated fibers of anterior horn cells are selectively lost. Wasting of ventral roots emerging from the cord indicates Wallerian degeneration of lower motor neurons. Wallerian degeneration also occurs peripherally, and the collateral branches of the surviving axons in the surrounding area can be seen attempting to reinnervate the denervated muscle fibers. Various stages of atrophy due to denervation and subsequent reinnervation of muscle fibers are revealed in the muscle biopsy; however, in long-standing denervation, the affected muscles may show myopathic changes.

8.7 Clinical Presentation

8.7.1 Amyotrophic Lateral Sclerosis

ALS is also known as Lou Gehrig's disease in America (after the professional baseball player who died of ALS in 1941) and Charcot's disease in Europe. It is the most prevalent form of MND, with an annual incidence rate of 2 per 100,000 general population. Men are affected somewhat more frequently than women, and most patients are 50 years of age or older. The incidence increases with each decade of life. Other clinical variants of MND have more restricted presentations and can evolve into sporadic ALS. ALS may also be seen in combination with Parkinsonism-dementia complex or Alzheimer's disease.

ALS can be divided into familial and sporadic forms. Most are sporadic, accounting for approximately 90–95% of the cases. The remaining cases are familial, and the presentation of symptoms is almost the same as sporadic, except for the tendency to have an earlier onset of symptoms, a slightly shorter survival, and a greater tendency for weakness to begin in the legs.

The symptoms of ALS will primarily depend on the group of nerve cells most affected. In most cases, the symptoms of the LMN dominate over the UMN and are often more severe in the upper than in the lower limbs during the early stage. The onset of weakness is more common in the upper than the lower extremities (typical ALS), but in around 25% of patients, the weakness begins in the bulbar-innervated muscles (bulbar-onset ALS), and rarely (1–2%) the weakness starts in the respiratory muscles (dyspnea or respiratory onset). Interestingly, in some patients, the muscle weakness can be restricted to one side of the body (Mills hemiplegic variant), and in around 10%, weakness can appear with bilateral upper extremity wasting (flail-arm variant) or bilateral lower extremity wasting (flail-leg variant). In typical ALS cases, the earliest indication of the disease will be features like awkwardness in tasks requiring fine finger movements (such as buttoning clothes, picking up small objects, or turning a key), stiffness of the fingers, and weakness and wasting of hand muscles. The weakness of intrinsic muscles occurs before the long flexors' involvement, leading to the hollowing of dorsal interosseous space producing a cadaveric or skeletal hand appearance. Whereas, for patients with initial involvement of the lower limbs, tripping, stumbling, and awkwardness while running are the usual early indications of the disease. Patients may complain of instability of gait and fatigue while walking. Cramps and fasciculations are features during the advancing stage. Muscle cramps are common, and in some patients, persistent cramping of muscles and related joint stresses can cause diffuse and continuous aching of the limbs and back. Usually, physicians are the first to recognize the fasciculations that are common in the muscles of the forearm, upper arm, and shoulder girdle, but it may not be the sole presenting feature.

During the early stage, symptoms can be asymmetrical and, within weeks to months, become symmetrical. Muscle weakness in ALS typically begins in a focal area and then spreads to the contiguous muscles in the same region before the involvement of another region. As the disease advances, there will be involvement of the arm and the shoulder girdle. Progressively, the muscle atrophy becomes more apparent, and spasticity may complicate gait and manual dexterity; however, extreme spasticity of lower extremity muscles is uncommon. The bulk of muscles and strength diminishes in parallel. Those changes occurring in the upper extremities occur, while the thigh and leg muscles seem relatively normal. Patients may walk with useless dangling arms. During the advancement of the disease, the atrophic weakness spreads to the neck, tongue, pharyngeal, and laryngeal muscles, and eventually the trunk and lower extremities. Head drop can be a feature due to the weakness of the cervical and thoracic paraspinal muscles. The involvement of bulbar muscles will lead to symptoms like slurred speech, hoarseness, or an inability to sing or shout, soon followed by progressive dysphagia. The ALS patients may present with pseudobulbar palsy characterized by inappropriate or forced crying or laughter, and the presence of excessive forced yawning may also be a manifestation of pseudobulbar palsy. In the later stage, the triad of disease consisting of the atrophic weakness of hands and forearm muscles, mild spasticity of the arms and legs, and generalized hyperreflexia becomes apparent.

Since the disease primarily involves the motor neurons, the sensory function is typically preserved; however, some patients may have poor position sense. Patients may complain of numbness and paresthesias, and the affected part may ache and may feel cold but most often never have sensory symptoms. Surprisingly, the sphincteric control is well maintained till the late stage, even when both lower limbs are spastic and weak. The possibilities of pressure sores are also rare in many of the patients. Preservation of sensory function, continued control of bowel and bladder function, and complex morphological changes in the skin that are poorly understood may account for the preservation of skin integrity preventing pressure sores.

The abdominal reflexes may be elicitable. The presence of the Babinski sign and brisk or exaggerated deep tendon reflexes are other signs. In the initial stages, deep tendon reflexes may be normal to brisk, but as the disease progresses, reflexes may be diminished or lost. Notable deep tendon reflexes of the upper extremities are another feature seen in ALS, despite amyotrophic changes. Usually, within 2–6 years after the onset of symptoms, patients will succumb to the disease. Dysphagia with possibilities of aspiration pneumonia and respiratory muscle paralysis are the common late-stage complications responsible for the death.

8.7.2 Progressive Muscular Atrophy

In progressive muscular atrophy (PMA), only LMN features are seen. This variant of MND is more common in men than in women, and it comprises approximately 5% of the MND. Compared to ALS, such subjects with purely nuclear amyotrophic presentations have a slower rate of progression of the disease. The average age of onset of PMA is approximately 3 years older than that of ALS. In about half of PMA patients, the symmetrical wasting of intrinsic muscles of the hand is an important finding. The weakness and wasting tend to advance slowly to the more proximal parts of the arms. Less often, the legs and thighs are the sites of the onset of atrophic weakness. Usually, the distal parts of the limbs are affected before the proximal parts. Rarely, patients may survive 15 years or more, and younger patients tend to have a more benign course compared to the older. Most of the very longest-duration cases of MND have the PMA variant, and some of the researchers and subject experts have questioned whether PMA is an independent disease or represents one end of the spectrum of ALS. Over time, the clinical features of UMN disease are likely to emerge in many PMA patients. The familial incidence of PMA is quite rare. The tendon reflexes are diminished or absent, and signs of corticospinal tract disease are typically not detectable. Fasciculations and cramps may or may not be present.

8.7.3 Progressive Bulbar Palsy

The first and dominant symptoms of progressive bulbar palsy (PBP) are the weakness of muscles innervated by the motor nuclei of the lower brainstem. Such weakness leads to difficulty to articulate, causing difficulty to pronounce lingual, labial, palatal, and dental consonants. Slurred speech, hoarseness, and decreased volume of speech are some of the early problems of the patients. In many patients, spasticity of the tongue, pharyngeal, and laryngeal muscles with atrophy cause slurring of speech. As the disease progresses, syllables lose their clarity, and eventually, the speech becomes unintelligible. The voice becomes nasal, and during phonation, the palate and vocal cord move imperfectly or not at all. Once the speech is lost, the patient may need to depend on other forms of communication, such as writing, communication boards, or other assistive devices. Progressive bulbar muscle weakness also leads to dysphagia, and the weakness of masseter muscles prevents chewing of food. Patients usually complain of more difficulty swallowing liquids than solids and often find difficulty forming food bolus and the food may get lodged between the check and teeth. Drooling is common and results from a combination of excessive salivation and poor labial control. The weakness of the lower facial muscles is not uncommon and may cause sagging of the facial muscles.

Initially, the pharyngeal or gag reflex may be notable, but in advanced stages, it will be lost due to the progressive degeneration of dorsal motor nuclei of the vagus that supplies the pharyngeal muscles. The spastic weakness of the tongue makes transport of the food bolus to the back of the mouth difficult. Subsequently, weakness of the pharyngeal muscles may not allow the bolus to enter the esophagus; instead, it may enter the trachea or the food may regurgitate through the nasal passage. Such patients with severe dysphagia are at high risk for aspiration and its sequels.

PBP is the only common clinical situation in which spastic and atrophic bulbar palsy coexists. At times, the signs of spastic weakness may surpass those of atrophic weakness, and the signs of pseudobulbar palsy may reach extreme degrees. Such patients may have exaggerated emotional responses, resulting in frequent and rapid alterations in emotions like episodes of intense laughter followed immediately by tears that may not correspond to the apparent social stimulus or the psychosocial situation.

Fasciculations of the tongue and focal loss of tissue of the tongue are frequent early manifestations of PBP. The facial muscles may also fasciculate. In the later stage, the tongue appears shriveled and lies useless on the floor of the mouth. The presence of jaw jerk or an exaggerated jaw jerk (a slight tap on the chin may evoke clonus or blinking and attempts to open the mouth may elicit a “Bull Dog” reflex characterized by the involuntary snap shutting of the jaw) despite the weakness of the masticatory muscles is an essential sign of PBP. In PBP, the ocular muscles are spared and sensory loss is absent. Like other forms of MND, PBP is progressive but tends to have shorter survival. The involvement of respiratory muscles during the late stage, together with swallowing difficulties that may lead to aspiration pneumonia, causes death and may occur anywhere between 2 and 3 years after the onset of symptoms.

8.7.4 Primary Lateral Sclerosis

Primary lateral sclerosis (PLS) is a rare UMN disease variant that accounts for approximately 2% of ALS cases. This variant begins with pure spastic paraparesis followed by involvement of the arms and pharyngeal muscles and is traditionally distinguished by a lack of LMN involvement. For the abovementioned reason, some do argue that PLS is a separate entity than a distinct form of ALS. The onset of symptoms is insidious, and the average age of onset of symptoms ranges from 35 to 65 years. Usually, patients present with gradual onset, progressive lower extremity stiffness, and pain due to spasticity. Spasticity predominating over weakness is an early manifestation of PLS, and the gait is "robotic" and slow. Due to the above, the affected PLS patient may develop an imbalance with a tendency to fall and may take the cane's support to ambulate until the spasticity becomes severely disabling. Surprisingly, the muscle power is found to be good even when the patient has difficulty walking.

Occasionally, the patient may present with progressive spastic hemiparesis which later extends to the contralateral side. Axial muscle involvement may result in lower back and neck pain, further aggravating back or neck pain from other causes like degenerative disc disease and osteoporosis. As the disease progresses, the upper extremities become involved and patients may have difficulties performing activities of daily living (ADL). Later, the involvement of the organs of speech may result in spastic dysarthria, which initially may be mild. Swallowing and breathing may be compromised late in the disease. Signs of upper motor neuron dysfunction may include limb and trunk spasticity, pathological spread of deep tendon reflexes, clonus, and Babinski sign. No clinically detectable sensory changes are seen in PLS patients. The disease typically progresses for 3 years or more without evidence of LMN dysfunction, and the median duration of PLS can span around 20 years post-diagnosis.

8.8 Lab Investigations and Findings

Laboratory investigations are helpful to confirm the evidence of disease. Lab test results generally are normal and are performed primarily to rule out other disease processes resembling MND. Biochemical markers in blood or cerebrospinal fluid (CSF) are used almost routinely to identify diseases that could mimic MND. Genetic testing may be performed to identify gene defects in some familial types of ALS, as well as other inherited motor system disorders like SMA.

Under electrophysiological tests, electromyographic (EMG) studies reveal widespread fibrillation and positive sharp waves in the weak muscles. Fibrillation potentials and positive sharp waves represent active denervation. Even those muscles with normal strength may show such abnormalities. Abnormality in the recruitment of motor units (an increased recruitment frequency) during minimal to maximal contraction suggests a reduction in the number of motor units due to progressive denervation. The signs of both active and chronic denervation are likely to be observed in EMG studies. Chronic denervation is demonstrated by evidence of large motor unit potentials with polyphasia, increased duration, and amplitude, a reduced interference pattern with firing rates higher than 15 Hz, and unstable motor unit potentials. Evidence of denervation over many segments favors the diagnosis of MND, and it is good practice to insist that denervation be demonstrated in at least three limbs before concluding that the process is ALS. Electromyographic studies may display a widespread denervation of the paraspinal, facial, and genioglossus muscles, which is a strong indicator of the disease. Fasciculation potentials are characteristic of the disease, but their presence is not specific. Similarly, complex repetitive discharges or high-frequency discharges are common in MND of long duration.

Motor and sensory nerve conduction studies (NCS) are performed primarily to rule out other disorders. Motor nerve conduction studies may show a slight slowing in the conduction velocity owing to the depletion of large motor nerve fibers. Sensory nerve conduction studies are usually normal. In the late stages of the disease, LMN involvement may be extensive, and in such cases, the compound muscle action potentials may be reduced. Hallmark findings in the electrodiagnosis of MND are normal sensory NCS and abnormal motor NCS with reduced compound muscle action potentials.

The CSF studies may reveal normal or marginal elevation in the protein level. The serum creatinine phosphokinase will be normal or slightly elevated. Neuroimaging may include computed tomography scan or magnetic resonance imaging (MRI) of the brain and spinal cord. These studies may help rule out structural lesions or neurological conditions that may mimic MND. The results of these studies are generally normal in patients with MND; however, MRI studies may disclose slight atrophy of the motor cortex. Wallerian degeneration of the motor tract in the brainstem and spinal cord is another possible MRI finding.

8.9 Diagnosis and Differential Diagnosis

A thorough history, physical examination, and diagnostic workup by the physician are needed for the diagnosis. The clinical presentation of MND may resemble many neuromuscular disorders, and Table 8.3 shows certain distinctive features of the latter with the former. The presence of fasciculations, especially in the absence of weakness and atrophy, can mislead the diagnosis. The diagnosis should be made according to the established and well-accepted criteria laid by the World Federation of Neurology, which ensures that no treatable or nonfatal disorders accounting for the patient's symptoms are overlooked. The criteria for the diagnosis of ALS are available at http://www.wfnals.org. The highlights of the criteria are as follows:

- Signs of LMN and/or UMN degeneration by clinical examination
- Signs of LMN degeneration by neurophysiologic tests or neuropathologic examination
- Evidence of progressive spread of signs within or other region(s)—bulbar, cervical, thoracic, or lumbosacral
- Absence of neurophysiologic evidence of other disease that might explain the signs of LMN and/or UMN degeneration
- Absence of neuroimaging evidence of other disease that might explain the observed clinical signs and neurophysiologic evidence

Table 8.3 The possible differential diagnosis and the distinctive features of ALS

Differential diagnosis	Distinctive features
Central spondylotic bar or central disc prolapse	Pain in the neck and shoulders Neck movements' limitation Sensory disturbances LMN features limited to one or two spinal segments Absence of active denervation in three limbs, in the paraspinal and bulbar muscles Differences in EMG and NCS studies
Multiple sclerosis (may present with single limb or hemi pattern simulating ALS).	Optic neuritis Sensory disturbances Cerebellar symptoms Remissions and relapses
Peroneal muscular atrophy	Presence of positive family history Sensory changes Distinguishable differences in EMG patterns
Muscular dystrophy	Proximal muscular involvement in muscular dystrophy Presence of familial history in muscular dystrophy Age group EMG patterns suggestive of myopathic disorder Elevated CPK level in blood serum
Polymyositis	Subacute onset Proximal weakness History of infection or vaccination Distinguishable differences in EMG patterns Elevated CPK level in blood serum Definite improvement with corticosteroid therapy
Progressive supranuclear palsy	Vague personality changes Forgetfulness and apathetic appearance Ocular palsies Distinguishable difference in EMG studies Differences in MRI
Diabetic neuropathy	Presence of sensory changes Deep tendon reflex diminished or lost in the lower extremities Other systemic effects of diabetes mellitus may be seen Changes in sensory NCS
Myasthenia gravis	Fluctuating weakness in myasthenia gravis Definite improvement in condition during the neostigmine test Presence of ACh receptor or anti-MuSK antibodies in blood serum Repetitive electrical stimulation and single-fiber EMG revealing neuromuscular junction transmission impairment

8.10 Medical and Surgical Management

Like other progressive neurodegenerative disorders such as Parkinson's disease or Alzheimer's disease, there is no specific treatment for MND. "Riluzole," a small-molecule benzothiazole, an antiglutamate agent, is believed by some to slow the progression of the disease but not markedly. A recent Cochrane review reported a probable prolongation of median survival by about 2–3 months in ALS patients. In 1995, riluzole was approved by the United States Food and Drug Administration (FDA) for the treatment of ALS. It acts as a neuroprotective and glutamate modulator agent.

Oxidative stress induced by free radicals is believed to play an important role in the pathogenesis of ALS. Recently, in the year 2017, edaravone, a low-molecular-weight antioxidant, was approved for the treatment of the same. By donating one electron to a free radical, edaravone, a free radical scavenger, prevents the oxidation of phospholipids, which cause damage to the cell membranes. Edaravone exists both as an anion and a neutral molecule in the plasma and tissues. Edaravone's highly reactive anionic form gives it an advantage over the rest of the antioxidants, as it can react with many different radical species, including peroxynitrite. The drug is delivered by infusion, and several sessions of infusion are required. Certain studies on edaravone have demonstrated efficacy in decreasing the decline of motor function in ALS patients, but its effect on survival, respiratory function, grip strength, and quality of life are still unclear. In addition to the above, the exorbitant cost of the medication makes edaravone less affordable for most patients.

Guanidine hydrochloride, cobra venom injections, gangliosides, interferons, high-dose intravenous cyclophosphamides, and thyrotropin-releasing hormones are some drugs or agents which may arrest the disease process, but none of them are a definite treatment for MND. Since there is no definite treatment, alleviation of the symptoms and supportive measures are imperative for the management of MND and are best achieved through a multidisciplinary team approach.

For improving the patient's quality of life, expertise is required in respiratory function, nutrition, and rehabilitative and occupational measures. Besides, social work counseling is also important. Respiratory devices that actively assist the inspiratory phase of respiration and use of nasogastric tube feeding or surgical procedures like percutaneous endoscopically placed gastrostomy, given at the appropriate time, significantly reduce the complications and increase the survival rate. Excessive salivation and thickened mucous secretion are major problems for patients suffering from MND. Drugs in the form of a transdermal patch or small dose irradiation on the salivary glands may provide a marked relief from excessive salivation. However, when excess salivation becomes a persistent problem, the use of suction apparatus at home becomes mandatory.

8.11 Physiotherapeutic Management

Being a part of the multidisciplinary team, in light of the progressive nature of MND, our aim is to address the progressive loss of function and independence in persons with MND. The goals and means of physiotherapy are based on the physical symptoms and rate of progression of the disease, irrespective of the MND variants. Regular, periodic, and careful monitoring of the subject's existing signs and symptoms and developments of new ones may help the therapist to determine the rate of progression of the disease.

Physical inactivity can cause cardiovascular deconditioning and atrophy from disuse and may further superimpose the weakness caused by the ALS disease itself. In the past, MND patients were frequently advised or suggested to avoid regular physical exercise in order to preserve muscle strength. The inevitable deterioration associated with the disease, heterogeneity of MND presentation within the study population, and high dropout rates led to a dearth of quality physiotherapy-related research in the past. Most of the evidence for physiotherapy was based on the level of clinical opinion, and the role of exercise in the MND population became even more controversial when studies on the transgenic mouse model of ALS revealed deleterious effects following high-intensity exercise. Such studies revealed that high-intensity or vigorous exercise could facilitate some of the pathogenic mechanisms implicated in ALS, including glutamate-mediated excitotoxicity, excessive inflammation and swelling, and disturbances in free radical homeostasis. Currently, evidence suggests that moderate-intensity aerobic exercises are beneficial both in the animal and the human models of ALS. A recent European population-based case-control study concluded that physical activity is not a risk and may eventually be protective against the disease. There is growing evidence that submaximal resistance and endurance exercises in these patients are safe and efficacious in delaying the decline in skeletal muscle strength and the functional level and maintaining respiratory muscle strength should be the key focus for the physiotherapist.

Moderate exercise programs are generally recommended for many neurodegenerative diseases and should include strengthening to address weakness, flexibility exercises to minimize contractures, walking programs, and/or cardiovascular routines to maintain endurance. Moderate exercises are believed to improve neuronal plasticity by mediating trophic factors such as brain-derived neurotrophic factor, insulin-like growth factor-1, and glial cell-derived neurotrophic factor. For conditions like MND, moderate exercises can create dendritic restructuring, increase neurotransmitter release, enhance protein synthesis, strengthen synaptic connections, mediate intracellular calcium balance, and improve axonal transport within the healthy motor neurons. A systematic review on moderate-intensity exercise programs provided moderately strong evidence in support of the same for early stage ALS patients and revealed a marginal betterment of outcomes with respect to the respiratory function, the muscle strength, and the function compared with the early stage ALS patients who did not participate in the exercise. The review also revealed that moderate exercises were not associated with any rapid disease progression or other harmful or deleterious effects.

A set therapeutic protocol makes management easy for the therapist, provided the staging of the disease based on the progression is possible. Though staging MND is not impossible (such as early, middle, and late stages), relative rapid course, variations in the distribution of muscle weakness and clinical course, and a lack of any set pattern of progression can make it difficult. Therefore, in such neurological conditions, an individualized therapy based on the patient's issues is favored that may or may not include strengthening or aerobic exercise prescription. Maintenance of maximal independence in daily activities, as long as possible, within the limits imposed by the disease forms the general goal of physiotherapy. At the same time, the specific goals are:

- To maintain maximal muscle strength within the limits imposed by the disease
- To prevent or minimize secondary complications during the progression of the disease
- Prompt initiation of therapeutic intervention to prolong the functional abilities, once the diagnosis is established

8.11.1 Evaluation

Whether the physiotherapist works as a member of a multidisciplinary team or an independent therapist, the evaluation of referred MND patients is of utmost importance. Close association and interaction between the physiotherapist and other team members, particularly occupational therapists, may help to solve many problems faced by these patients, especially those related to gross motor function or ADL.

Careful, concise, and regular evaluation helps to identify the problems of the patient, determine the rate of progression, and extent and area of involvement, and set the objectives of treatment. The essential components which require periodical evaluation irrespective of the therapist working within a multidisciplinary team or in a community or rural setup are:

- Muscle power: To find the baseline muscle strength
- Range of motion (ROM): To note any tightness or contractures
- Chest function: To spot early respiratory deterioration
- Bulbar function: To prevent complications and for comfort
- Pain evaluation: To promote comfort
- Functional abilities: To maintain or improve the functional level

The insidious onset of spastic weakness can promote joint stiffness compromising the normal ADL. Such stiffness can also give rise to pain and discomfort.

Therefore, the functional assessment should include the level of comfort and specific problems like fatigability or pain during activity. The type and site of pain, the aggravating or relieving factors, and the use of a visual analog scale or body chart index to objectively quantify the intensity of pain are the essential components of pain assessment. The assessment chart also should have a briefing about the patient's medical and activity record, safety issues, psychosocial issues, and family support issues.

For patients with predominant bulbar involvement, it is imperative to assess the following:

1. Tongue—includes the inspection of fasciculations at rest and speech and ROM, i.e., the ability to protrude and perform rapid lateral motion of the tongue
2. Lip—for sucking, smiling or curl lip over the teeth, blowing or whistling
3. Palate—includes inspection of the palate during phonation, pharyngeal reflex, and nasal voice
4. Muscles of mastication—should include observation of muscle wasting and jaw movements to both sides, collecting information regarding undue fatigue while chewing food, and palpation of masseter when the patient clenches the teeth
5. Neck and shoulder—observation of bilateral shoulders and neck posture, along with muscle strength evaluation of the same
6. Respiratory status—measurement using pulmonary function test and handheld respirometer and noting the strength of coughing

8.11.2 Treatment

It is strongly recommended to discuss the goals of the physiotherapy and exercise program with the patient. However, such discussions should not falsely generate any undue expectations about his condition. Meanwhile, when discussing the condition or the exercise program, care has to be taken not to make any devastating statements like the disease is invariably fatal.

In 1987–1988, Sinaki and Mulder developed certain rehabilitation guidelines for ALS persons based on progressive functional limitations, which consists of three stages with substages. In the early stage of MND, the goal is to maintain and optimize mobility and function, and the program should be functional, goal orientated, and performed at intensities without provoking fatigue. In the middle stage, the goal is to continue to maintain functional mobility, manage pain, and carefully monitor and address any symptoms of respiratory compromise. The late-stage goals are designed to maximize the quality of life for the patient. Therapists should realize that not all patients will precisely fit into a specific stage as described in Table 8.4 but will provide a general framework and also suggest the plausible interventions based on the degree of impairment, functional limitations, and level of disability.

Overall, the intervention strategies for MND patients aim to facilitate the healthy innervation of neurons and synaptic networks, slow the degeneration of neurons, and maintain the muscle strength and endurance, and cardiopulmonary efficiency without unduly straining the mechanical, bioelectrical, and metabolic properties of the muscle fibers, particularly the fast-twitch fatigable muscles. Regular exercise maintains muscle elasticity and joint mobility, prevents stiffness, and maximizes the muscle power for as long as possible. Vigorous exercises may damage the muscle, and at the same time, inadequate exercises can lead to disuse atrophy. Carefully designed exercises and functional activity programs can be beneficial and be given to patients irrespective of the type of MND, rate of progression, and pattern of muscle involvement. Such exercise programs help to maximize motor unit

Table 8.4 Treatment strategies for ALS patients based on the stage of the disease

Stage	Substage features	Treatment strategies
Independent/ early stage	Mild weakness Clumsiness of hands Independent ambulation Independent in daily activities	Encourage to continue normal activities If sedentary, promote activities to prevent disuse atrophy Introduce ROM exercise program in the form of stretching and flexibility exercises Begin strengthening exercise program, consisting of low-to-moderate intensity resistance exercises to all musculature with caution not to cause overwork fatigue
	Moderate, selective weakness Independent ambulation Slight dependence in ADL characterized by difficulty to climb stairs, overhead activities, and buttoning clothing	To avoid contractures, continue the stretching exercises For a muscle power grade of 3+ (based on manual muscle testing), continue the strengthening program cautiously without causing overwork fatigue Aerobic exercises are appropriate Advice on energy conservation Appropriate footwear and provide orthotic support if required for the foot/wrist/thumb To facilitate ADL, encourage the use of adaptive equipment
	Severe selective weakness in the ankles, wrists, and hands Independent ambulation possible but long-distance ambulation causing undue fatigability Moderate dependence in ADL Respiratory effort marginally increased	Continue the exercise program as tolerable as possible but be cautious not to fatigue/exhaust to the point of decreasing ADL independence Fatigue and anxiety management, including teaching relaxation techniques Psychological support to the patient and caregiver Introduce pleasurable activities like level ground walking with adequate amount of rest Encourage deep breathing exercises, chest stretching, and postural drainage if required Falls prevention strategies

(continued)

Table 8.4 (continued)

Stage	Substage features	Treatment strategies
Partially independent/ middle stage	Arms dangling by the side of the body with possible shoulder pain and edema in the hand Spasticity leading to deformity and pain Wheelchair dependent Severe lower extremity weakness Able to perform some of the daily activities but fatigues easily	Prescribe wheelchair (standard or motorized), preferably with reclining back support, headrest, and detachable arm and leg support Motorized wheelchair for better independence Hot fomentation and massage to control spasm Gentle stretching program as an adjunct to antispasticity medication to relieve spasticity Electrotherapy modalities for pain relief Measures to prevent edema Active assisted/passive ROM exercises wherever appropriate Encourage isometric contractions of the musculature based on the tolerance Use of arm slings and wheelchair arm supports Home environment modifications to aid mobility and independence
	Significant weakness of the lower limbs Moderate to severe weakness of the upper limbs Increasingly dependent on daily activities Wheelchair dependent Potential skin breakdown due to poor mobility	Teach and encourage caregivers to learn proper positioning, turning, and transfer techniques Encourage the use of hoist and harness to aid transfer, mobility, and independence Electric recliners and tilt table Provide supportive devices like soft/headmaster collar Advice on positioning for comfort and air/ pneumatic mattress to prevent pressure sores Monitor signs and/or symptoms of respiratory impairment Encourage inspiratory muscle training and lung volume recruitment training in addition to the breathing exercise
Dependent/ late stage	Bedridden Completely dependent in ADL	Semisolid, hot palatable food and use of long spoons for dysphagia and once difficult, tube feeding/percutaneous gastrostomy advisable Manually assisted coughing to remove or clear the secretions For ventilated patients, chest manipulations and airway clearances Comfortable positioning of the limbs to minimize pain

recruitment and promote or maintain the ADL and maintain a delicate balance between the extremes of excessive exercise and inadequate exercise, preventing early deterioration in the performance level. Care has to be taken to prevent overuse fatigue when the patient is performing the exercises. During the advancement of the disease, a reduction in muscle strength or deterioration in the performance level during exercises will necessitate the redesigning of the exercises. However, the therapist should not overlook similar deteriorations in strength and performance due to overwork or fatigue.

The selection of exercises or activities should also consider the premorbid functional status. In the presence of assistance, guidance, and environmental modifications, some patients with slow progressive weakness may continue to perform ADL for an extended period. Exercise, when prescribed appropriately, especially in the earlier stages of the disease, can be physically and psychologically beneficial for ALS patients. The use of elastic bands with a sequential system of progressive resistance can be recommended as a resisted exercise. Isometric exercise is preferred over concentric or eccentric exercise. The risk of orthopedic and muscular injuries is unlikely while using elastic bands. The use of functional weight training is another alternative when training with weights or elastic bands is difficult. Activities like swimming, cycling, and walking can be advised to overcome or minimize disuse atrophy and the negative effect on bones. However, activities like swimming should be performed strictly under supervision and with adequate precautions.

In the advancing stage, when the weakened muscles fail to respond to strengthening exercises, free and active assisted exercises are advisable, i.e., when the muscles are still able to contract but are weaker, then free and active assisted exercises are favorable. It is also essential to encourage the patient to move the limbs actively throughout the day in as full a range as possible. Emphasis should also be given to the mobilization of the head, neck, and trunk. If the patient cannot achieve full ROM of one or more joints or if he cannot move the limb at all, then passive ROM exercises and stretching exercises are recommended instead to prevent the reduction in joint mobility and shortening of soft tissues. Similarly, patients with severe respiratory and bulbar complications may not benefit much from active exercise programs and may instead require passive ROM exercises and stretching exercises. It is essential to educate the caregivers on the relevance of routine passive movements and stretching exercises. To minimize shoulder pain, a common problem in MND, early training on correct manual handling, positioning, shoulder care, and ROM exercises are advisable. The use of appropriate electrotherapy modalities like interferential therapy or transcutaneous nerve stimulation, splints, or collars should be considered for relieving pain.

The level of fatigue experienced is significantly higher in MND patients as compared to their healthy age- and gender-matched counterparts. Excessive activity, stress, pain, depression, reduced cardiopulmonary function, medication, and sleep disturbance can further contribute to fatigue, and the therapist should be mindful to ask patients about their level of fatigue. Contractures or imbalance between antagonist groups can also impact functional mobility and fatigue. For those patients suffering from overwhelming fatigue, resisted exercises and exercises with many repetitions may worsen the condition. Careful designing of exercise programs and the use of energy conservation techniques while performing ADL may help such patients to avoid the excessive wasting of energy and early fatigue. Also, it is essential to note that the intensity of resistance or aerobic exercise should not compromise the reserve required to carry out daily activities and such exercises are not advisable during any phase or stage of ALS.

The patient's ability to walk and stand has to be maintained for as long as possible. Often, during the advancing stage, he or she may require assistance to stand when self-standing with support is impossible. Timely prescription and a regular review of walking aids and appliances are essential in maximizing the patient's function and minimizing falls. Lightweight orthosis is recommended to prolong ambulation and to maintain posture. The therapist should also advise the patient to use anti-slip floor mats for safe walking. Once locomotion becomes inevitably tiring, the wheelchair has to be prescribed. Many patients prefer electric or motorized over a self-propelled wheelchair, as muscles of the upper limbs are more often paralyzed during the later stage. However, the prescription of a motorized wheelchair has to be delayed until no option is left for further ambulation. As the disability progresses, the application of proper lifting and handling techniques ensures safety for both the patient and the caregiver. Early prescription of hoists or manual handling belts for transfers can minimize the caregiver's burden and possible musculoskeletal injuries.

During the terminal stage (bedridden stage), the therapist may find physiotherapeutic interventions such as stretching and active exercise programs ineffective. In such a stage, the effort to reduce muscle and joint pain, which is related to immobility, has to be given more priority. Careful positioning and nursing, in conjunction with physiotherapy, helps to prevent deformity and maintain mobility. While positioning the patient, care has to be taken to minimize the effect of gravity on the body. Good alignment in supine or sitting posture can reduce or prevent the abnormal muscle tone, contractures, and pain, prevent the development of pressure sores, and minimize the effort required for respiration. Judicious use of sheepskin with a foam cushion, foam wedges for support, pillows, and small neck supports, in addition to the use of muscle relaxants, may help to alleviate discomfort. Customized seats and wheelchair also provide comfort and positioning. Recliner chairs, sleep systems, profiling beds or mattresses, and high-back wheelchairs may be required as the disease progresses to support and correct body positioning. In the later stage, it is advisable to perform regular standing from sitting position by the support and assistance of the therapist or by the use of standing equipment such as a tilt table or a hoist with harness support to stretch the antigravity and weight-bearing muscle groups. Such sit-to-stand movement transitions itself can be considered as a stretching exercise program for the whole body.

For patients with predominant bulbar involvement, hot plates and heated food dishes make slow eating more palatable. The use of anti-slip tablemats and thickened handles on cutlery, pens, and toothbrushes aids independence. A call system sensitive to light-touch operated by the head, hands, or feet to draw the attention of caregivers when they are not near him can be of a lot of help. Velcro straps and zippers on the dress can aid dressing. Even the use of a communications system such as an emergency messaging system, answering machines, and fax machines can be of great help in case of emergency. A soft neck collar to keep the head from rolling also provides comfort and facilitates eye contact. Eventually, the success of a physiotherapeutic intervention depends on the exercise program, motivation, and persistence of the patient in carrying out the program, support from the family members, and patients' and caregivers' knowledge about the disease process.

During the terminal stage, attempts to maintain the chest expansion by deep breathing exercise may fail. In a study, yoga-style deep breathing exercise has not revealed any significant improvement in forced vital capacity (FVC), standard progressive hypercapnic ventilatory response, respiratory magnetometry, or quality of life. Techniques like the active cycle of breathing and forced expiratory techniques are generally not practical. Even postural drainage can be too traumatizing for some patients, and repeated suctioning has to be minimized. Respiratory adjuncts like acapella® and positive expiratory pressure therapy are generally not indicated to assist the clearance of the secretion. Studies have shown that inspiratory muscle training, lung volume recruitment training, and manually assisted coughing technique have moderate strength effectiveness in improving respiratory outcome measures and participation level and increasing the survival of ALS patients. Manually assisted coughing techniques can be taught to the family members to perform at home. During the advancing or later stages, aspiration of oral fluid or semisolid food may provoke repeated coughs and exhausts patients very often. In such situations, a gentle shaking of the thorax during expiration or pressure over the abdomen using hand or bending forward to aid coughing and clearing of secretions can be attempted. However, such maneuvers are not often required as many of the patients switch to nasogastric tube feeding.

Further Reading

Andersen PM, Borasio GD, Dengler R, Hardiman O, Kollewe K, Leigh PN, et al. Good practice in the management of amyotrophic lateral sclerosis: clinical guidelines. An evidence-based review with good practice points. EALSC Working Group. *Amyotroph Lateral Scler.* 2007 Aug;8(4):195–213.

Brooks BR. El Escorial World Federation of Neurology criteria for the diagnosis of amyotrophic lateral sclerosis. Subcommittee on Motor Neuron Diseases/Amyotrophic Lateral Sclerosis of the World Federation of Neurology Research Group on Neuromuscular Diseases and the El Escorial "Clinical limits of amyotrophic lateral sclerosis" workshop contributors. *J Neurol Sci.* 1994 Jul;124 Suppl:96–107.

Carreras I, Yuruker S, Aytan N, Hossain L, Choi JK, Jenkins BG, et al. Moderate exercise delays the motor performance decline in a transgenic model of ALS. *Brain Res.* 2010;1313:192–201.

Dal Bello-Haas V, Florence JM. Therapeutic exercise for people with amyotrophic lateral sclerosis or motor neuron disease. *Cochr Database Syst Rev.* 2013, Issue 5. Art. No.: CD005229. Available from: https://doi.org/10.1002/14651858.CD005229.pub3.

Dal Bello-Haas V, Kloos AD, Mitsumoto H. Physical therapy for a patient through six stages of amyotrophic lateral sclerosis. *Phys Ther.* 1998;78:1312–24.

Daroff RB, Jankovic J, Mazziotta JC, Pomeroy SL, editors. Bradley's Neurology in clinical practice. seventh ed. London: Elsevier; 2016.

Donaghy M, editor. Brain's Diseases of the nervous system. 12th ed. New York: Oxford University Press; 2009.

Garg N, Park SB, Vucic S, Yiannikas C, Spies J, Howells J et al. Differentiating lower motor neuron syndromes. *J Neurol Neurosurg Psychiatry.* 2017;88:474–83.

Harwood CA, McDermott CJ, Shaw PJ. Physical activity as an exogenous risk factor in motor neuron disease (MND): a review of the evidence. *Amyotroph Lateral Scler.* 2009;10:191–204.

Kato N, Hashida G, Konaka K. Effect of muscle strengthening exercise and time since onset in patients with amyotrophic lateral sclerosis: a 2-patient case series study. *Medicine.* 2018; 97 (25):e11145. Available from: https://doi.org/10.1097/MD.0000000000011145

Lewis MRS. The role of physical therapy and occupational therapy in the treatment of amyotrophic lateral sclerosis. *NeuroRehabilitation*. 2007;22:451–61.

Liebetanz D, Hagemann K, von Lewinski F, Kahler E, Paulus W. Extensive exercise is not harmful in amyotrophic lateral sclerosis. *Eur J Neurosci.* 2004 Dec;20(11):3115–20.

Ludolph A, Drory V, Hardiman O, Nakano I, Ravits J, Robberecht W, Shefner J; WFN Research Group On ALS/MND. A revision of the El Escorial criteria - 2015. *Amyotroph Lateral Scler Frontotemporal Degener.* 2015;16(5–6):291–2.

Lui AJ, Byl NN. A systematic review of the effect of moderate intensity exercise on function and disease progression in amyotrophic lateral sclerosis. J Neurol Phys Ther. 2009;33:68–87.

Macpherson CE, Bassile CC. Pulmonary physical therapy techniques to enhance survival in amyotrophic lateral sclerosis: a systematic review. *J Neurol Phys Ther.* 2016;40:165–75.

McCrate ME, Kaspar BK. Physical activity and neuroprotection in amyotrophic lateral sclerosis. *Neuromolecular Med.* 2008;10:108–17.

Merico A, Cavinato M, Gregorio C, Lacatena A, Gioia E, Piccione F et al. Effects of combined endurance and resistance training in amyotrophic lateral sclerosis: a pilot, randomized, controlled study. *Eur J Transl Myol.* 2018;28:72–78.

Partington T. Motor neurone disease. *Postgrad Med J.* 1962;38:383–95.

Riach CL, Starkes JL. Stability limits of quiet standing postural control in children and adults. *Gait Posture.* 1993;1(2):105–11.

Ropper AH, Samuels MA, Klein JP, Prasad S, editors. Adams and Victor's Principles of neurology. tenth ed. New York: McGraw Hill Education; 2014.

Sinaki M. Physical therapy and rehabilitation techniques for patients with amyotrophic lateral sclerosis. *Adv Exp Med Biol.* 1987;209:239–52.

Trombly CA, Scott AD. Occupational therapy for physical dysfunction. seventh ed. Baltimore: Lippincott Williams & Wilkins; 2014.

Chapter 9
Multiple Sclerosis

Abraham M. Joshua, Anjali Tiwari, Neha Kumari, and Shivananda Pai

9.1 Introduction

Among the demyelinating disorders, multiple sclerosis is the commonest one, and it has also been referred to as "disseminated sclerosis" by the British physicians and "sclérose en plaques" by the French. Chronicity, tendency to recur, and propensity to attack young adults had made this condition one of the most widely recognized neurological disorders. It is clinically characterized by episodes of focal disorders of the optic nerve(s), spinal cord, and brain with the tendency to remit with varying degrees and relapse within several years. The neurological manifestations are numerous, being determined by the varied location and degree of the demyelinating foci. The lesions have an inclination for specific parts of the central nervous system (CNS); hence, the resulting clinical features and radiological appearances can be characteristic of multiple sclerosis.

Multiple sclerosis (MS) is one of the leading causes of nontraumatic neurological disability among young adults. It is a chronic inflammatory demyelinating disease of the CNS and has been considered as an autoimmune T-cell-mediated

A. M. Joshua
Department of Physiotherapy, Kasturba Medical College of Mangalore, Manipal Academy of Higher Education, Manipal, India
e-mail: abraham.joshua@manipal.edu

A. Tiwari (✉)
Department of Physiotherapy, Health City Cayman Island, Grand Cayman, Cayman Islands
e-mail: anjali.tiwari@healthcity.ky

N. Kumari
SCS Hospital, Upper Bendoor, Mangalore, India

S. Pai
Department of Neurology, Kasturba Medical College of Mangalore, Manipal Academy of Higher Education, Manipal, India
e-mail: dshivanand.pai@manipal.edu

A. M. Joshua (ed.), *Physiotherapy for Adult Neurological Conditions*,
https://doi.org/10.1007/978-981-19-0209-3_9

inflammatory process leading to secondary macrophage recruitment and progressive myelin destruction. The clinical course of MS is extremely variable and unprecedented and is usually characterized by episodes of reversible neurological deficit followed by progressive neurological deterioration.

9.2 Historical Background

The historical description of MS can be traced back to the nineteenth century where the disease was entangled in superstition and ancient belief systems. Until the mid-nineteenth century, the disease was not scientifically proven or recognized. The diaries and letters written in the early nineteenth century by Augustus Frederick d'Este, son of Prince Augustus Frederick and a British royal family member, are recognized as the oldest written description of MS. In his diaries (1822–1848), d'Este sequentially explained the symptoms he experienced throughout the phase of illness with evidence of relapse and remission and progressive deterioration of the symptoms. In 1828, Robert Carswell, a British pathologist, investigated the pathology of possible MS but did not attempt to explore any clinicopathological correlation for the disease and described it as "a peculiar diseased state of the cord and pons to the grey substance." Jean Cruveilhier, a French surgeon and anatomist, in 1835, was the first one to give an illustration about the MS lesion and was the first to record the patient's history and correlate it to the focal demyelinating lesion. He observed and identified these lesions as reddish-gray islets and described these lesions as "grey degeneration in the cord, brainstem, and cerebellum." However, Cruveilhier's version was not published until 1842.

In 1849, Friedrich Theodor von Frerichs, a German pathologist, with his colleagues described nystagmus as a symptom of the disease and revealed that remission was an important characteristic of the disease. Subsequent research in the field of MS had given the cardinal features of the disease, and it was Jean-Martin Charcot, a French neurologist, in 1868, who gave a clear understanding about the clinicopathological features of MS. His thesis gave insights into the pathological lesion around the ventricle in the cerebral cortex and the sclerotic patches in the brain and spinal cord to correlate limb paralysis. He had also recognized the course of illness, early remission, and the disseminated sclerotic patches. By the late nineteenth century, most of the physicians had sufficient background knowledge about this disabling disease. James Dawson, a Scottish pathologist, in 1916, examined the brain tissue of an MS patient using Camillo Golgi and Santiago Ramón y Cajal's staining techniques and showed the pathogenesis in the form of inflammatory damage to the myelin and blood vessels. In 1919, Bruce Frederick Cummings, an English author, who recorded his own presentation of symptoms in terms of motor impairment, locomotor ataxia, speech abnormality, and fluctuation symptoms, later identified as possible MS, published his life and struggle with the disease in a book titled *The Journal of a Disappointed Man*. This was considered to be the best known personal account of MS until d'Este's diary was revealed in 1948.

In 1919, abnormality in the cerebrospinal fluid (CSF) was first noted, although its significance was confounded. Following the First World War, research in the field of MS grew enormously. An important milestone was achieved in 1925 when the role of myelin in nerve conduction was clarified, which gave further insights into the mechanism of pathogenesis and its correlation with the clinical presentation. The Austrian neurologist Karl Johannes Thums, in the 1930s, based on studies on identical and fraternal twins, concluded that there was no evidence for any genetic contribution toward MS. The foremost breakthrough occurred just before the Second World War when it was observed that the vaccines given for viral illness have precipitated MS-like symptoms, which further revealed that the partial inactivated viral preparations could affect the myelin, thus further broadening the picture of MS. Based on the viral attacks post administration of vaccines, Thomas Rivers in 1935 explored the relation of modern theories of human immunity, molecular mimicry, and the immunosuppressive treatment of MS.

In 1946, the National Multiple Sclerosis Society (NMSS), a nonprofit organization, was established for advanced research in MS. In 1947, Elvin Abraham Kabat, an American biomedical scientist and immunologist, identified the abnormal immunological patterned protein in the CSF known as the oligoclonal band which majorly demonstrated the relation of MS with the immune system. In 1960, a chapter on MS was first issued by the NMSS for MS patients and their families. Since the 1950s, many attempts were made to increase the sensitivity and specificity of MS diagnostic criteria. It was in 1965 that the first comprehensive clinical criterion was proposed by George Adam Schumacher, an American neurologist. To increase the reliability of the diagnosis, Charles M. Poser and co-researchers included paraclinical tests to these criteria in 1983.

Throughout the history of MS, the disease was untreatable; however, in the last couple of decades, various treatment strategies have been developed to suppress the immune attack and alter the course of the illness. In 2001, the first set of criteria was published as "McDonald's criteria" by William Ian McDonald, a New Zealand neurologist and academician. This was comprehensively revised and the guidelines were regularly updated (revised in 2005, 2010, and 2017) for an accurate diagnosis of MS in the early phase of the disease.

9.3 Etiology and Epidemiology

MS is a chronic demyelinating disease with no definite cure and has been found to cause neurological impairment among adults. Though the exact cause of MS remains undetermined, the epidemiologic facts are well established and will eventually have to be incorporated in any etiologic hypothesis. Approximately 2.3 million people are affected worldwide with MS, with a global prevalence of 33 per 100,000. Nevertheless, the incidence and prevalence depend on the geographical location. The rate of MS is higher among women compared to men, which significantly shows women have a higher predisposition for this disease. Other than gender, many risk factors contribute to MS, including age, area, ethnicity, ecological components, and hereditary qualities. The initial symptoms of MS are less likely to occur before the age of 10 or after the age of 60. The frequency of MS among children is extremely low, and the disorder has a unimodal age-specific curve which shows a high predisposition between 20 and 40 years of age.

The prevalence of MS has shown variation among different populations. In tropical areas, MS is less prevalent with the predominance of 1 per 100,000, while in Northern Europe, the figure ranges from 120 to 200 per 100,000. Within the United Kingdom, the prevalence ranges from 80 to 120 per 100,000. There is a comparable but less clearly characterized relationship of expanding predominance of MS with latitude. For example, the prevalence rate of MS is higher in North America with an estimate of 140 per 100,000 individuals in contrast with the sub-Saharan region in Africa where an estimated prevalence rate is 2.1 per 100,000. In the same reference, Japan encompasses very low incidence and prevalence. Though latitude is considered an important risk factor, ethnicity has shown a relevant contribution to the incidence of MS. This is often illustrated with high prevalence among the white Europeans in the Northern European countries and North America, in comparison to the Inuit populations of Greenland. In India, the current estimated incidence of MS is 8 per 100,000 and the course of the disease in India is broadly similar to that of the west with no well-defined data on its clinical progression among different populations.

Environmental factors have a clear-cut influence on the occurrence of MS. It has been found that migrants moving from a high risk to a lower risk geographical zone had a lower prevalence of MS, especially when the migration occurred before the age of 15 years, whereas migrants moving from a lower to a higher risk area had a tendency to retain the lower MS risk of their country of origin, with no clear age-at-migration effect. Correspondingly, heredity is an important factor for the risk of developing MS; details for all these factors are further explained in the pathogenesis. Though a wide range of plausible potential causes have been acknowledged, no single definite cause has been identified to date. The influence could be due to the combined effect of environmental factors, gene susceptibility, and hereditary risk factors.

9.4 Pathophysiology

The pathological lesion that best correlates with acute clinical exacerbations of disease features foci of inflammation with active myelin degradation, phagocytosis, and partial axonal preservation. Though the precise cause for MS is still under research, the epidemiologic data point to a relationship between MS and some environmental factors. This factor encountered in childhood, with years of latency, might either evoke the disease or contribute to the cause of the disease. During the past few decades, speculation has grown that this factor is most probably a viral infection. A large body of indirect evidence has been gathered in support of this idea, based on the demonstration of alterations in the humoral and cell-mediated immunity to viral agents. Interestingly, to date, no virus has been observed or isolated from the neural tissues of MS patients. Even bacterial agents *Chlamydia pneumoniae* and herpesvirus type 6 have been implicated due to the presence of their genomic material in MS plaques, but the evidence for their direct participation in the disease is less convincing like any other infectious agents. Some of the hypotheses framed to justify the pathogenesis for MS are listed below.

9.4.1 Immune-Initiated Disease Hypothesis

The infiltration of the focal immune cells and the release of cytokines are the cause of the inflammatory changes in the white and gray matter tissue. According to the immune-initiated disease hypothesis, the autoreactive T cells generated in the systemic compartment and then accessing the CNS can induce an inflammatory cascade injuring the previously normal neural tissues. Various animal model studies related to this mechanism have shown a marked heterogeneity with respect to the topography and extensiveness of the demyelination or axonal disruption indicating the variation in the mechanism of inflammatory demyelination and CNS injury.

The apparent increased frequency of myelin-reactive CD4+ T cells in the peripheral blood of MS patients, most likely derived from the memory T cells, implicates a prior sensitization with the disease-relevant antigens. The immune response is of two different types: innate and adaptive immune responses. The innate system plays an important role in the initiation and progression of MS by influencing the effector function of T and B lymphocytes. Many researchers also believe that the adaptive immune responses instigated by the interaction between the antigen-presenting cells (APCs) with the T lymphocytes play a crucial role in the commencement and progression of the disease. The pathogen-associated molecules (antigen) bind to the receptor of APCs and lead to the production of interleukin (IL)-12, IL-23, and IL-4 cytokines. These cytokines induce CD4+ T cells to differentiate into Th1, Th2, or Th17 phenotypes. The gamma interferon and the tumor necrosis factor-alpha, proinflammatory cytokines produced by Th1 cells, are critical for innate and adaptive immunity. Th2 cells secrete the anti-inflammatory cytokines (IL-4 and IL-13), whereas Th17 induces a large number of cytokines (IL-17, IL-21, IL-22, and IL-26) capable of promoting inflammation. Figure 9.1 depicts the pathogenesis of MS.

The regulatory T cells, also known as T reg cells, are another CD4+ T-cell phenotype involved in the pathogenesis of MS. These cells function to regulate effector Th1, Th2, and Th17 cells. Though the number of T reg cells is the same among MS patients and normal subjects, studies have proved a reduced functioning of T reg cells in MS patients.

In addition to the T lymphocytes, the B lymphocytes and their products are involved in the pathogenesis of MS. Certain cytokines (transforming growth factor-beta and tumor necrosis factor-alpha) produced by the B lymphocytes promote inflammation, whereas IL-10, another cytokine produced by B lymphocytes, has an

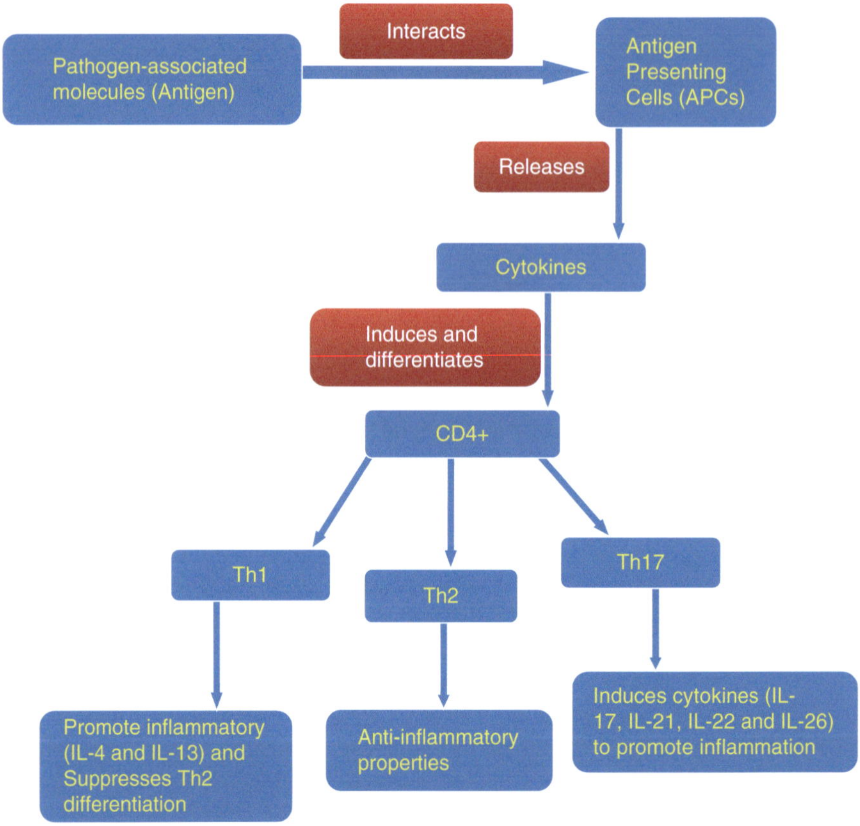

Fig. 9.1 Flowchart depicting the pathogenesis of MS

anti-inflammatory function. Therefore, B lymphocytes have both positive and negative impacts on the progression of MS.

Besides the activation of the CD4+ T cells, the APCs can also activate the CD8+ T cells. Many studies have demonstrated the presence of the CD8+ T cells and stated that these cells by the production of cytolytic proteins like perforin can mediate the suppression and inactivation of the CD4+ T cells. These cells increase vascular permeability, destroy glial cells, leave the axons exposed, and trigger the death of the oligodendrocytes and are believed to play a critical role in the pathogenesis of MS. Fas ligand, a product of the lymphocyte cell, can bind to the Fas receptor located on the oligodendrocyte cell surface and facilitate the apoptosis of the same, thus reducing the number of oligodendrocyte cells and impairing the synthesis of the myelin sheath.

In the experimental model, myelin-directed antibodies could contribute to the tissue injury (demyelination) but not initiate any similar disorder. Depending on the subtype of MS, the disease-specific antibodies are observable in the MS lesions, and these antibodies are known to bind to the neural cell and direct the potential injury mediators. Though the disease-specific antibodies are likely to be seen in the blood and the CSF, their presence does not actually contribute to the course or phenotype of the disease.

9.4.2 Neural-Initiated Disease Hypothesis

This hypothesis suggests that an event within the CNS initiates the MS disease process. It postulates that the persistent and progressive acquired infection in the neural cells may result in the formation of tissue antigens that could initiate the chain of autoimmune responses. Another mechanism whereby the acquired infection could impact includes the molecular mimicry responses, perturbing systemic immune regulatory properties, as might occur with the Epstein-Barr virus infection of the B cells or the MS retrovirus. It has been hypothesized that the release of cytotoxic mediators via the viral attack leads to inflammatory changes in oligodendrocytes which initiate the events in the formation of the MS lesion.

9.4.3 Environmental Factors

The environmental factors including pathological as well as chemical agents, depending on the exposure, have been hypothesized to be a causative agent for MS. The pathological agents include viral and bacterial agents such as the Epstein-Barr virus, human herpesvirus type 6, and *Mycoplasma pneumoniae*. In addition to the above, the chemical agents from smoking (nitric oxide and carbon monoxide) also play an important role in MS. The pathological concentration of the toxic soluble gases causes lipid peroxidation and mitochondrial damage and can lead to the apoptosis of the oligodendrocytes, axonal degeneration, and neuronal demyelination. Exposure to carbon monoxide can cause a blockage of tissue oxygenation, myelin protein degradation, and axonal injury, subsequently triggering an inflammatory response via the activated microglial and CD4+ lymphocyte invasion and further demyelination.

Exposure to infectious agents as well as sunlight (the UV rays essential for vitamin D production) is felt to account for the changing risk of MS when a person migrates from one risk area to another before the age of 15. Pathological agents like the human herpesvirus type 6, Epstein-Barr virus, and *Mycoplasma pneumoniae* are speculated to produce MS by molecular mimicry. These pathogens may have peptides with direct sequence homologies with myelin components. Evidence also suggests that common viral infections like upper respiratory tract infections and bacterial urinary tract infections can trigger MS relapses.

In addition to the above, vitamin D and B_{12} deficiencies are considered to be important risk factors for MS. Vitamin D is responsible for cell proliferation and differentiation. Additionally, it has an important role in gene expression and regulation of immunity along with IL-10 synthesis and suppression of pro-inflammatory cytokines. Vitamin B_{12} is an important factor for the formation of myelin, and deficiency can be a contributing factor for a demyelinating disease like MS.

9.4.4 Genetic Susceptibility

Genetic factors can influence the pathogenesis and susceptibility of the disease. Depending on the amount of genetic information shared among the family members, a 40-fold increased susceptibility among first-degree relatives was reported in certain studies on families and twins of MS patients, suggesting a genetic basis of the disease. The risk rate of developing MS was around 25%, in monozygotic twins with 100% genetic similarity. In subjects with 50% genetic similarities, which included dizygotic twins and first-degree relatives, the risk was approximately 2–5%. The human leukocyte antigen locus on chromosome 6p21 has been linked to MS susceptibility. In addition to these, the IL-7 and IL-2 receptor alphas are the other susceptible genes associated with MS.

9.5 Classification of Multiple Sclerosis

Based on the presentation and progression seen among patients, MS can be classified into four subtypes (Fig. 9.2). The most common form of MS is the relapsing-remitting type which accounts for 85% of the total MS cases. The relapsing-remitting

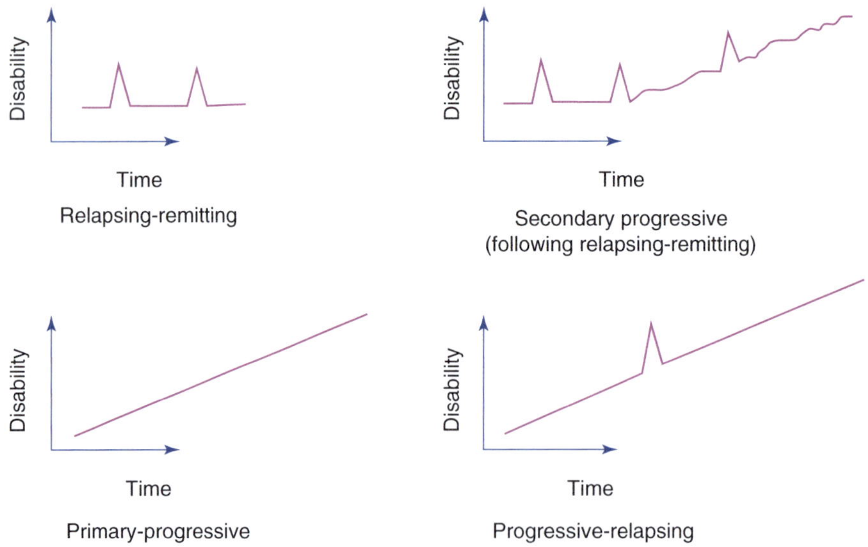

Fig. 9.2 Graphical representation of the multiple sclerosis subtypes

MS (RRMS) is characterized by the relapse and exacerbation of the symptoms followed by a period of remission. In the early stage of RRMS, the patient may present with optic neuritis, transverse myelitis, or an isolated brainstem syndrome. These isolated single attacks may not evolve into MS, and such acute monophasic episode is commonly known as a clinically isolated syndrome (CIS). Those cases where the clinical presentation does not evolve and stays as a single event are classically termed as "true CIS."

The primary progressive MS (PPMS) approximately affects 10% of the remaining MS patients. In this subtype, the symptoms continue to worsen gradually with no relapse or remission and occasionally can reveal plateaus. The secondary progressive MS (SPMS) presents with at least one relapse followed by the progressive clinical deterioration of symptoms. Relapse may develop in some patients where the disease continues to worsen with or without the period of remission or plateau of symptoms. The progressive relapsing MS (PRMS) is a rare form that affects less than 5% of MS patients. It is a progressive form with no period of remission but with intermittent flaring up or worsening of the symptoms.

9.6 Clinical Features

Typically, the focal neurological deficits affecting the CNS may evolve over a period of weeks or months. A study done by McAlpine and coworkers in 1972, while analyzing the mode of onset of MS, reported that approximately 20% of patients fully developed the neurological symptoms in a matter of minutes and another 20% developed them in a matter of hours. For about 30%, the symptoms evolved slowly over a period of a day or several days, and in 20%, the symptoms evolved more slowly over several weeks to months. The remaining 10% had an insidious, slow, and steady or intermittent progression over months and years.

Initially, the patient might present with generalized symptoms of fatigue, weight loss, and vague joint pain which could be present several months before actual neurological deficits evolve. In the early phase, in addition to general weakness, the presentation may include tingling sensation or pain simulating tic douloureux over the facial region, diplopia, vertigo, paresthesia, spasticity, features of optic neuritis, and difficulties with gait and balance.

9.6.1 Optic Neuritis

The early clinical manifestations that are typical of MS can be optic neuritis, transverse myelitis, cerebellar ataxia, or features of brainstem syndromes. In about 25% of MS patients, particularly among children, the initial manifestation can be a single episode of optic neuritis, characterized by a partial or complete loss of vision in one eye. Unilateral optic neuritis is a common initial presentation of RRMS. Typically, before the onset of visual loss, these patients may give a prior history of pain within the orbit, worsened by eye movement or palpation of the globe. In some, the visual loss can be gradual and steadily progressive for several weeks and may resemble a space-occupying lesion like an intrinsic tumor of the optic nerve. Visual field evaluation may demonstrate a wide variety of field defects including a scotoma involving the macular area and blind spot and rarely even a homonymous hemianopia. In some patients, both the optic nerves can be simultaneously or sequentially involved within a few days or weeks or following repeated attacks. In about 50% of the patients with optic neuritis presentation, serial examinations of the fundus may disclose evidence of swelling or edema of the optic nerve head (papillitis). Acute visual loss accompanying papillitis differentiates it from a papilledema secondary to raised intracranial pressure.

For those with considerable visual loss, the pupillary response to light can be diminished, and if the optic neuritis is unilateral, the consensual light reflex from the normal eye can be elicitable. Typically, the majority of patients with optic neuritis recover completely and the remaining improve significantly even if the initial visual loss is profound. The Marcus Gunn pupil can be another clinical manifestation of optic neuritis, and the swinging flashlight test is the test to elicit the same. The test is performed in a dimly lit room, with the patient instructed to fix his gaze on a distant object. The examiner swings a flashlight back and forth over the bridge of the nose, between the eyes. Normally, shining the light in one eye will elicit brisk, direct, and consensual pupillary responses. However, in unilateral optic neuritis, while swinging the light back and forth between the eyes, the examiner will find that the diseased eye reacts much less briskly and even dilates when the light shines on it.

The incidence of uveitis and sheathing of the retinal veins due to T-cell infiltration is higher among MS patients. Dyschromatopsia (the inability to discriminate colors by hue), characterized by a perceived desaturation of colors, may persist and tends to improve with corticosteroid treatment. In the majority of MS patients presenting with optic neuritis, the improvement can commence and continue for several months; however, in due course, many are likely to develop other features of the disorder.

9.6.2 Features of Brainstem Syndromes

Brainstem symptoms like diplopia, vertigo, vomiting, and dysarthria can be early manifestations in some MS patients. Diplopia can be a common presenting complaint and occurs most often due to the involvement of the medial longitudinal fasciculus, producing an internuclear ophthalmoplegia. The internuclear ophthalmoplegia is characterized by paresis of the medial rectus on attempted lateral gaze, with a coarse nystagmus in the abducting eye. Bilateral internuclear ophthalmoplegia in young adults is virtually diagnostic of MS. Paralysis of gaze due to interruption of supranuclear connections to third, fourth, and sixth cranial nerve nuclei can occur but less frequently. Other manifestations of brainstem involvement include myokymia, paralysis of facial muscles, deafness, tinnitus, auditory hallucinations, vertigo, vomiting, and rarely stupor and coma. Involvement of intramedullary fibers of the fifth cranial nerve in young adults can result in transient facial hypesthesia or anesthesia or trigeminal neuralgia. Facial palsy along the lines of Bell's palsy is seldom a sign of MS.

9.6.3 Sensory and Motor Disturbance

Transverse myelitis (complete or partial) can appear as the first symptom of MS. Sensory symptoms can be an initial manifestation of cord involvement. Typically, the cord involvement in MS is patchy than being central, uniform, or symmetric, and the extent and type of sensory loss depend on the magnitude of the various sensory tracts involved. Sensory loss with tingling paresthesia is a common initial presentation of MS. Patients may complain about tingling sensations of the limbs and tight band-like sensations around the trunk or extremities, which are usually associated and are probably due to the involvement of the dorsal columns of the spinal cord. In about 50% of MS patients, weakness and/or numbness in one or more limbs can be the initial symptom(s). As a result, the clinical presentations can vary anywhere from a mere dragging or poor control of one or both lower limbs to a spastic or ataxic paraparesis. Clinically, the condition is characterized by rapidly evolving (several hours or days) symmetrical or asymmetrical paraparesis or paraplegia, ascending paresthesias, loss of deep sensations, a sensory level on the trunk, sphincteric dysfunction, and bilateral Babinski signs.

Involvement of the corticospinal tract leads to signs and symptoms of upper motor neuron lesions. Spasticity in MS presents with increased tone, hyperexcitability of the stretch reflex, stiffness, and contracture. Worsening of the spasticity often signifies the disease progression and also accounts for the morbidity associated with the condition.

Flexion of the neck may induce a tingling or electric-like sensation down the shoulders and back, a phenomenon known as Lhermitte's sign. The sign was originally described by Babinski in the cervical cord trauma condition. Increased sensitivity of demyelinated axons of the spinal cord to stretch or pressure induced by neck flexion attributes to the electric-like sensation. Within a span of 5 years after the initial onset of acute myelitis, about 50% of the patients may develop evidence of an asymptomatic lesion elsewhere in the CNS or the clinical evidence of dissemination.

9.6.4 Cerebellar Ataxic Features

Nystagmus and ataxia can be another frequent and prominent feature of MS, reflecting the involvement of the cerebellar tracts. Cerebellar ataxia can be recognized by the presence of scanning speech, postural or static tremor, dysmetria and intention tremor of the extremities, and balance and gait abnormalities. A combination of nystagmus, scanning speech, and intention tremor, known as Charcot's triad, is often seen in the advanced stages of the disease. Among patients with a long-standing disease with involvement of tegmentum of the midbrain or the dentatorubrothalamic tracts and adjacent structures, the slightest attempt to move the trunk or limbs may precipitate violent and uncontrollable ataxic tremor. The additional involvement of the dorsal columns of the spinal cord or medial lemnisci of the brainstem can lead to a combination of sensory and cerebellar ataxia. In such cases, either the signs of the spinal cord or the cerebellar signs may predominate over the other.

9.6.5 Bowel, Bladder, and Sexual Dysfunctions

Approximately 75% of MS patients present with bladder dysfunction. The extent of bladder, bowel, and sexual dysfunctions often matches the degree of motor impairment in the lower extremities. The type of neurogenic bladder is determined by the location of the lesions within the CNS. Urgency is the most common complaint related to bladder dysfunction, and it usually results from an uninhibited detrusor contraction, reflecting a suprasegmental lesion. As the disease progresses, the urge incontinence becomes more frequent. Involvement of the sacral segments of the spinal cord leads to symptoms of hypoactive bladder which is characterized by reduced urinary flow, interrupted micturition, and incomplete bladder emptying.

Loss of perception of bladder fullness, usually associated with urethral, anal, and genital hypoesthesia, results in atonic dilated bladder with overflow incontinence. A lesion in the spinal cord above the sacral micturition center can present with a combined disorder of detrusor overactivity and detrusor sphincter dyssynergia. A dyssynergic voluntary sphincter that interrupts bladder emptying can lead to frequent but small-volume micturition with large post-void residual volume. The MS patient may present with detrusor hyporeflexia, characterized by incomplete bladder emptying if the focal lesion is located in the pontine region. Cognitive dysfunction, bladder outlet obstruction, stress incontinence, functional incontinence, and side effects of certain medications including opioid analgesics and tricyclic antidepressants are the additional causes for bladder dysfunction. In MS patients, particularly in women, urinary tract infections are common and may present with atypical patterns. Constipation is not uncommon and is usually due to a combination of factors like cord involvement, reduced functional mobility, dietary issues, and restricted fluid intake to minimize the frequency and urge incontinence.

Approximately 40–80% of MS patients have sexual dysfunction. Men may experience various degrees of erectile dysfunction and women may experience trouble having an orgasm and reduced libido. Direct effects of lesions on the motor, sensory, and autonomic pathways within the spinal cord as well as psychological factors like poor self-image, self-esteem, and fear of rejection by the sexual partner are the plausible causes for sexual dysfunction. The mechanical difficulties created by increased tone, weakness of the lower extremities, and urinary incontinence may further aggravate these problems.

9.6.6 Cognitive and Psychological Issues

In about 50% of patients with long-standing MS, some degree of cognitive decline is often present and is characterized by reduced attention, poor memory, and diminished processing speed and executive skills. Typically, the language skills and other intellectual functions are preserved. The cognitive functional decline may correlate with neuroimaging findings like the loss of white matter volume, thinning of the corpus callosum, and brain atrophy. Frequent relapse and remission and subclinical lesions are likely to cause psychosocial issues and sleep disturbances among MS patients. The disabling features of the disease in a considerable number of patients can lead to depression, irritability, and short-temperedness. The percentage of MS patients who are likely to have emotional disturbance and depression during periods of relapse is estimated to be as high as 25–40%. Few of the MS patients may show a euphoric state characterized by pathologic cheerfulness or elation indicating a definite inappropriateness between the abovementioned mental state and the deficit he or she may be exhibiting. Such a status also suggests a cerebral impairment and probably a pseudobulbar palsy manifestation.

9.6.7 Fatigue

Fatigue is seen in 80% of MS patients and is considered to be one of the main causes of impaired quality of life (QOL). It is most commonly seen among the progressive subtypes of MS. For many MS patients, fatigue is considered to be the single most debilitating symptom, surpassing pain and even physical disability. Fatigue is a subjective symptom, and the common definitions include a sense of exhaustion, lack of energy or tiredness, and lack of physical and/or mental energy. Though it can be a persistent complaint and a source of considerable distress, fatigue is often transient and more likely to occur when there is fever or other evidence of infections.

The mechanism of fatigue in MS can be multifactorial (Fig. 9.3) as the disorder is associated with an increased prevalence of other conditions that contribute to fatigue like depression and sleep disorders. The most accepted primary mechanisms of fatigue involve the immune system or sequelae from the CNS damage. Cytokines like interferon-γ and tumor necrosis factor-α are thought to be strong mediators of fatigue. Lower levels of dehydroepiandrosterone and its sulfated compound demonstrated in MS patients suggest a possible endocrine contribution to fatigue. Many patients have reported increased energy while taking corticosteroids further supporting a possible hormonal influence. However, the long-term use of steroids is not encouraged as a treatment for fatigue. Functional MRI has demonstrated compensatory reorganization and increased brain recruitment (activation of cingulate gyri and primary sensory cortex) in fatigued MS patients compared to non-fatigued patients indicating the aforementioned as another possible cause.

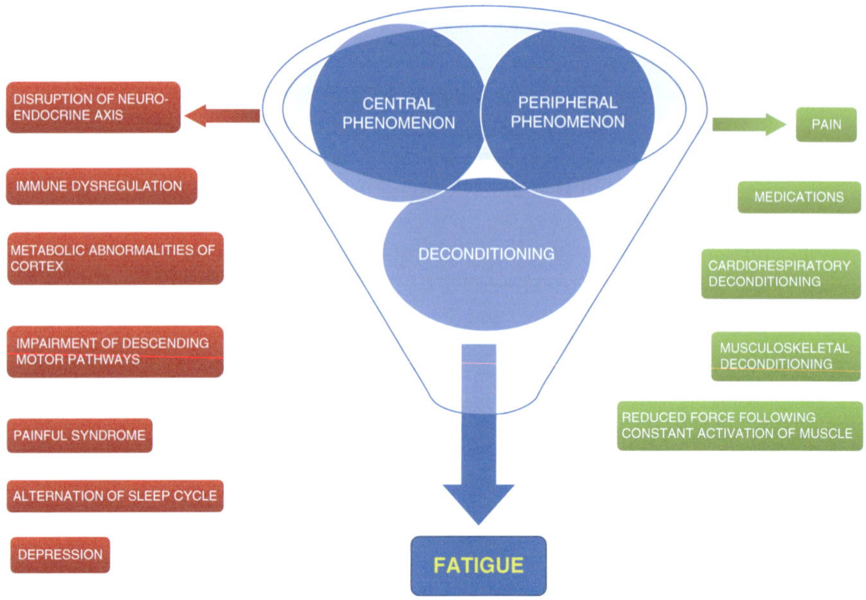

Fig. 9.3 Mechanism of fatigue in patients with multiple sclerosis

Cognitive deficit, neuropsychiatric symptoms, and fatigue can be seen between the episodes of flares due to accumulation of subclinical lesions. The MS patients may provide a history of exacerbation of symptoms with exposure to increased temperature. The Uhthoff phenomenon, another classic feature of MS, is characterized by a temporary induction of symptoms like unilateral visual blurring, tingling, and muscle weakness due to heat or exercise. The phenomenon represents an extreme sensitivity of conduction in the demyelinated nerve fibers to an elevation in temperature. A rise of 0.5°C can block electrical transmission in the thin myelinated or the demyelinated fibers. Similarly, hyperventilation has demonstrated a slowing of the visual evoked potentials. The remarkable sensitivity of the demyelinated and remyelinated regions of the axons to subtle metabolic and environmental changes explains the rapid onset or apparent fluctuation of symptoms in the absence of active inflammatory changes in the CNS. Therefore, smoking, hyperventilation, fatigue, and a rise in room temperature can temporarily worsen the neurological functioning and should not be confused with the relapses of the disease.

9.7 Diagnosis

Though the history and physical examination findings are the keys to diagnosis, the condition has no unique clinical feature which specifies the diagnosis. Hence, additional laboratory tests and neuroimaging findings are essential to support the diagnosis of MS. Electrophysiological tests, oligoclonal bands in CSF, blood investigations, and magnetic resonance imaging (MRI) findings along with physical and neurological examination findings have shown substantial sensitivity toward the diagnosis of MS. Over the last two decades, McDonald's criteria have been used as a guideline for the diagnosis of MS. McDonald's criteria were proposed in 2001 by Ian McDonald, and the updated version of the criteria encourages a more precise and advanced diagnosis of MS.

9.7.1 Laboratory Findings

In approximately 35% of MS patients, particularly in those with acute onset or acute exacerbation, a mild to moderate mononuclear pleocytosis with the cell count below 50 per cubic millimeter is reported. In rapidly progressive MS and neuromyelitis optica, the total cell count may exceed 100 cells per cubic millimeter with a greater proportion of polymorphonuclear leukocytes. The abovementioned pleocytosis may be the only proof of the activity of the disease.

Currently, the presence of oligoclonal bands within the CSF is the most widely accepted confirmatory test for the diagnosis of MS. The oligoclonal bands are abnormal discrete populations of gamma globulin proteins present in the CSF, post-migration by electrophoresis, after production within the CNS. Though demonstration of several oligoclonal bands in the CSF is a crucial finding, the presence of bands is not specific to MS as they can be seen in other conditions like syphilis, Lyme, and subacute sclerosing panencephalitis. The demonstration of bands in the CSF and not in the blood is quite helpful in confirming the diagnosis. In about 40% of the cases, the total protein content of the CSF can be marginally elevated with a proportional elevation of gamma globulin. The IgG index (ratio of CSF gamma globulin or albumin to serum gamma globulin or albumin) of 1.7 or more may suggest MS. The use of a radioimmunoassay can demonstrate high concentrations of the myelin basic protein during the acute exacerbations of MS. The assay though not particularly useful as a diagnostic test simply reflects the destruction of central myelin.

9.7.2 Magnetic Resonance Imaging

Neuroimaging, by virtue of its ability to reveal the asymptomatic lesions in the cerebrum, the brainstem, and the spinal cord, is a widely appreciated and most helpful ancillary examination in the diagnosis of MS. It also plays a wide extending role in the prognosis and treatment of MS. With the advent of newer imaging techniques like ultrahigh field MRI and double inversion recovery MRI, the potential to demonstrate lesions has increased to around 90%. In general, chronic MS plaques are hyperintense (white) on the T2-weighted images and even more strikingly obvious on the fluid-attenuated inversion recovery (FLAIR) images (Fig. 9.4). Those lesions with a certain degree of cavitation appear hypointense on the T1-weighted images. Longitudinal analyses of the hypointense T1 images may reveal a gradual increase of lesion(s) over time. Though MS does not have any signature MRI appearance, the presence of several asymmetric and well-demarcated lesions adjacent to the ventricular surface on the T2-weighted images usually suggests the same. The sagittal view of the images may reveal lesions emanating radially outward from the corpus callosum in a fimbriated pattern, a radiological finding termed as "Dawson fingers" (Fig. 9.5). The presence of such lesions in the callosum is diagnostically useful, as it is usually spared in many other disorders.

In acute MS, the enhancement of lesions following the administration of gadolinium suggests the disruption of the blood-brain barrier, presumably as a consequence of inflammation. The gadolinium enhancement tends to last for 3–4 months following the development of the acute lesion. MRI studies can also help to delineate processes such as plaque remyelination, axonal degeneration, and brain atrophy. The remyelinated foci appear hyperintense on the T2-weighted images.

Volumetric MRI has been a clinical marker for brain atrophy, a feature of longstanding MS. The basis for brain atrophy is not completely understood and probably reflects the loss of glial cells, Wallerian degeneration, and axonal loss triggered by inflammation and neurodegenerative stimuli. Further to support the investigation of brain atrophy, volumetric studies have been supplemented by spectroscopic biochemical analyses. The foci of periventricular hyperintensity observed in MS can

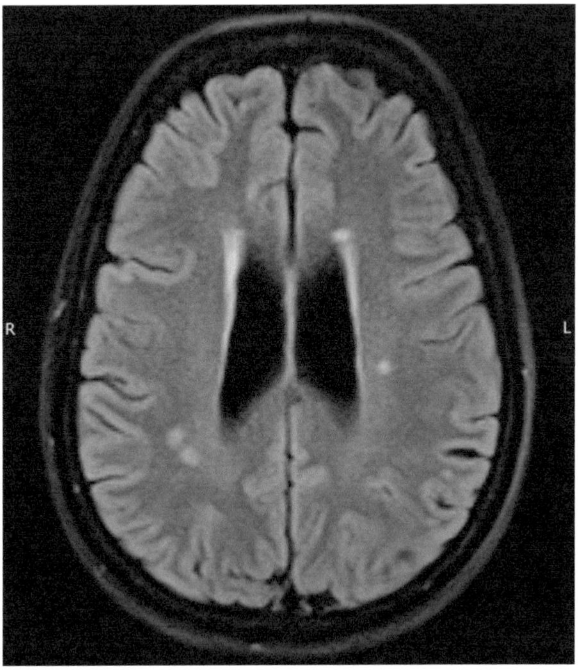

Fig. 9.4 Magnetic resonance (MR) axial FLAIR image showing focal transverse hyperintense lesions in the periventricular region consistent with multiple sclerosis

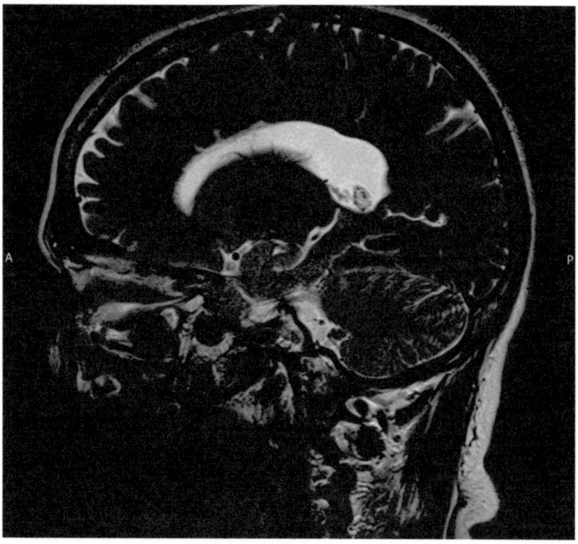

Fig. 9.5 MR sagittal FLAIR image showing focal radially oriented hyperintense lesions in the periventricular region consistent with multiple sclerosis. Note the subtle presence of "Dawson fingers" emanating radially outward from the corpus callosum in a fimbriated pattern

also be an MRI feature in normal subjects, particularly among the elderly; however, the periventricular changes among the elderly are usually milder in degree, less dense, and smoother in outline compared to the MS lesions.

The MRI finding(s) is an essential component of McDonald's diagnostic criteria and is based on two factors: dissemination in space and dissemination in time. The dissemination in space can be demonstrated by lesions in at least two locations (juxtacortical, periventricular, infratentorial, and spinal cord) with a minimum of one symptomatic lesion. The dissemination in time can be demonstrated by the gadolinium-enhancing lesion(s) on follow-up in reference to the baseline scan. Gadolinium-based contrast scans can also help in the diagnosis of CIS, soon after the patient presents with the first clinical attack. In some cases, MRI can be suggestive of MS even before obvious neurological findings and such cases are referred to as a radiologically isolated syndrome. The conversion rate of radiologically isolated syndrome into MS is 20% at 2 years and 30% at 3 years.

9.7.3 Evoked Potential Studies

During the early stages of the disease or in the spinal form of MS, when there is only a discrete lesion in the CNS, a number of sensitive electrophysiological and radiological tests may help to establish the existence of additional asymptomatic lesions. These electrophysiologic tests include visual, auditory, and somatosensory evoked potential studies, electrooculography, and blink response studies. An increase in the visual evoked potential latency can be an evidence for optic neuritis. Plaques located in the auditory pathways can delay the auditory evoked potentials and may suggest the central demyelination of the brainstem. The demyelination involving the dorsal column of the spinal cord can cause a delay in the somatosensory evoked potentials. Besides their clinical significance, the findings are not suggestive of MS, and for the same reason, the evoked potential studies are not included in the recent McDonald's criteria. According to some studies, around 70% of patients with definite MS and 60% of patients with probable or possible MS demonstrate abnormal visual evoked potentials. A similar trend was also reported but with a lower percentage for somatosensory and brainstem auditory evoked responses.

9.7.4 *Differential Diagnosis*

The presentation of MS can be variable, depending on the location of the lesion and the clinical manifestation. The differential diagnosis can be broadly divided into four categories, and they are other idiopathic inflammatory demyelinating diseases, idiopathic inflammatory non-demyelinating diseases, infectious diseases, and metabolic disorders.

9.7.4.1 Other Idiopathic Inflammatory Demyelinating Diseases

Idiopathic inflammatory demyelinating diseases (IIDDs) represent a broad spectrum of CNS disorders, and this spectrum includes monophasic, multiphasic, and progressive disorders ranging from highly localized forms to multifocal or diffuse variants of central demyelination. The four variants of MS discussed earlier form the commonest type of IIDD. Fulminant forms of IIDD include a variety of disorders that have in common the acute clinical course, the severe clinical symptoms, and the atypical findings on neuroimaging. The Marburg disease is the most classic fulminant IIDD, and the other fulminant IIDDs are acute disseminated encephalomyelitis (ADEM), Schilder's disease, and Baló's concentric sclerosis. Monosymptomatic IIDDs like optic neuritis, transverse myelitis, and brainstem demyelinating syndromes are typically the initial manifestation of MS, although a significant percentage of patients never develop the disease. Some IIDDs have a restricted topographical distribution, like the neuromyelitis optica (Devic's disease) and relapsing transverse myelitis (RTM), which can have a monophasic or, more frequently, a relapsing course.

The Marburg disease, believed to be a variant of MS, also known as acute fulminant malignant MS, a potentially fatal IIDD, is characterized by acute extensive fulminant demyelination with rapid progressive deterioration of the sensorium eventually leading to coma, decerebration, and death. The patient succumbs to the disease within a year after the onset of clinical signs, and the condition usually does not respond to corticosteroids which help to differentiate it from MS. MRI shows rapidly progressing supratentorial, infratentorial, and spinal cord lesions.

ADEM is an immune-mediated inflammatory demyelinating disorder of the CNS, which is classically monophasic and rarely recurrent. It is usually associated with vaccination or systemic viral infection and can affect both the adult and the pediatric population. ADEM was originally described in association with rabies and smallpox vaccines. The subsequent modification of such vaccines has reduced the incidence of ADEM. The parainfectious variant of ADEM has been associated with measles, rubella, mumps, and certain viral infections. The bacterial or viral epitopes resembling myelin antigens through molecular mimicry activate the myelin-reactive T cells to elicit the CNS-specific autoimmune response. In the majority of the ADEM cases, a history of preceding infection or vaccination before attacks is reported. The clinical picture is typically an acute or subacute onset encephalopathy that may range from lethargy to coma, accompanied by other focal or multifocal signs. The lesions of ADEM resemble those of MS. On MRI, multifocal white matter hyperintense lesions are seen in the cerebral hemisphere, the cerebellum, and the brainstem. The hemispheric lesions are often subcortical. Generally, with time these lesions tend to resolve. There is no specific biomarker or laboratory test to establish the diagnosis of ADEM. CSF findings with respect to cell count, protein, and glucose are similar to those of MS; however, oligoclonal bands are rarely present in ADEM, and similarly, the involvement of gray matter is more common in ADEM than in MS.

Schilder's disease is a rare, progressive, demyelinating disorder of the CNS. It is characterized by large supratentorial lesions and is more common among children than in adults. The radiological images show large lesions associated with edema and peripheral contrast enhancement. In the acute stage, Schilder's disease can be misdiagnosed as MS, but the rapid progression of the radiological lesions with deterioration in the level of consciousness helps to differentiate the clinical scenario from MS. Baló's concentric sclerosis is another rare, rapid, progressive, and severe monophasic demyelinating disease which presents with headache, cognitive deficit, encephalopathy, and epileptic seizures. Radiologically, Baló's concentric sclerosis presents with a helical intertwined hyper- or hypointense ribbon-like lesion. The histopathology of the lesion shows an onion ring appearance of the demyelinated and normal myelinated layers.

Neuromyelitis optica (NMO), also known as Devic's disease, is a rare autoimmune severe demyelinating inflammatory disorder, which typically presents with clinical episodes of optic neuritis and transverse myelitis and the demonstration of contiguous lesions in the spinal cord. The disease is characterized by recurrent episodes of severe attack eventually causing permanent disability. The disease was originally considered to be a form of MS; however, the current evidence supports a distinct pathogenesis and response to treatment. The presence of serum antibodies against the aquaporin-4 water channel present on the glia limitans at the blood-brain barrier has been a sensitive and specific marker of NMO. Radiologically, NMO presents with unilateral or bilateral optic nerve or chiasma involvement and transverse myelitis of the spinal cord typically involving three or more vertebral segments with longitudinal expanding edema. The brain MRI shows periependymal lesions surrounding the ventricular system. In the initial stage, NMO can mimic the features of CIS; however, the clinical presentation can differentiate it from MS due to its rapid progression of deficits and poor responses or worsening to certain MS medications like beta interferon, fingolimod, and natalizumab. Though there are no Food and Drug Administration (FDA)-approved medications for NMO, intravenous steroids are typically used for acute relapses and are administered daily for 3–5 days. For acute attacks, plasmapheresis may be used as the first or second line of treatment. Post plasmapheresis, the antibody titers generally tend to reduce, indicating the antibody link for relapses. Intravenous immunoglobulin is also used as third-line therapy, and medications like rituximab, azathioprine, and mitoxantrone are used as prophylactic agents for NMO.

9.7.4.2 Idiopathic Inflammatory Non-demyelinating Disease

Neuro-Behçet's disease, systemic lupus erythematosus, Sjögren syndrome, and paraneoplastic neurological syndrome are some of the disorders under idiopathic inflammatory non-demyelinating diseases. Neuro-Behçet's disease is a chronic recurrent multisystemic idiopathic vasculitic disease affecting mainly the veins. Three to nine percent of Neuro-Behçet's disease patients develop neurological involvement which includes ophthalmoparesis, cranial neuropathy, and cerebellar or pyramidal dysfunction. The condition may mimic MS clinically due to its ongoing course with attacks and remissions and focal or multifocal hyperintense CNS lesions for T2-weighed or FLAIR images and a good response to steroid treatment. However, the lesions are typically seen in the mesencephalo-diencephalic region, extending up to the basal ganglia, which is pathognomonic for Neuro-Behçet's disease.

Systemic lupus erythematosus (SLE), an autoimmune disorder with indistinguishable etiology, with possible crucial roles played by genetic, hormonal, immune, and environmental factors may present with neurological involvement in addition to multisystem manifests like arthritis, renal dysfunction, and hematologic and immunological disorders. Patients may present with headache, meningitis, dementia, epilepsy, and cognitive decline. It majorly affects young adults and particularly females. In SLE patients with pure neurological manifestations, MRI may show focal lesions in the periventricular and subcortical white matter and may mimic MS. Cerebral vessel vasculitis can be an important feature of SLE, and angiography may reveal it. In suspected patients, systemic symptoms of SLE must be investigated, and autoantibody tests can help in distinguishing and confirming the SLE diagnosis.

Sjögren syndrome is a chronic, autoimmune, inflammatory, systemic disorder characterized by exocrine gland dysfunction resulting in symptoms of dry eyes and dry mouth. In addition to the above, the condition is characterized by a wide range of multisystem involvement presenting with arthritis, Raynaud's phenomenon, and pulmonary, gastrointestinal, vascular, and CNS manifestations. Though 20–25% of patients may present with neurological features, peripheral neuropathy is more prevalent compared to CNS involvement. Features like myelitis and optic neuropathy might mimic MS, and MRI might reveal hyperintense lesions for focal neurological deficits. As the oligoclonal bands are more prevalent in MS, CSF analysis will help to distinguish Sjögren syndrome from MS. Exocrine dysfunction tests like Schirmer's test (a test to determine the production of an adequate amount of tears to keep the eye moist) for dry eyes, salivary gland biopsy, and positive autoantibodies might further help to distinguish this syndrome.

Paraneoplastic syndromes are immune-mediated disorders associated with metastatic cancer, and the signs and symptoms are not due to the direct invasion of cancer, metastasis, treatment complications, or opportunistic infections. The immune response driven against the cancer cells through molecular mimicry might be the possible reason for CNS involvement. The typical presentation includes features of encephalomyelitis, myelitis, cerebellar degeneration, opsoclonus-myoclonus-ataxia, limbic encephalitis, and cognitive and behavioral changes. Of the aforesaid features, some of them may resemble those of MS; however, neuroimaging can clearly distinguish the paraneoplastic syndromes from MS.

9.7.4.3 Infectious Disease

Lyme disease, a chronic vector conveyed infectious disease, transmitted by a tick bite and caused by *Borrelia burgdorferi*, a gram-negative bacteria, is endemic in North America and Northern Europe. Primarily, the disease presents in three stages, i.e., the first stage characterized by local infection and fever, the second stage by cardiac and neurological manifestations, and the final stage by chronic hematological, cardiac, and cerebral vasculitis with spastic paraparesis. The neurological manifestations include meningitis, meningoencephalitis, and multifocal peripheral nerve involvement, and rarely even features resembling the demyelinating disease. Lyme disease can be easily differentiated from MS in terms of the history of presentation and detection of antibodies against *Borrelia burgdorferi*. The CSF analysis will show lymphocytic pleocytosis along with increased protein level and absence of oligoclonal bands.

The highly active antiretroviral treatments given for human immunodeficiency virus (HIV) syndrome patients can cause HIV-related neurological syndromes, which typically present with meningitis, encephalopathy, myelopathy, cranial neuropathy, or peripheral neuropathy. A slowly progressive spastic paresis and radiological features can confuse the diagnosis, but the presence of HIV antibodies helps to differentiate the symptoms from MS. Untreated spirochetal infection caused by *Treponema pallidum* can cause neurosyphilis. It typically involves multiple sites of the CNS and the peripheral nervous system. Neurological manifestations like cranial nerve involvement, ataxia, gait disturbance, and bowel and bladder dysfunctions can mimic MS. The presence of lymphocytic pleocytosis, elevated protein level, and the absence of oligoclonal bands in CSF and serum confirmation of antibodies using venereal disease research laboratory (VDRL) and rapid plasma reagin (RPR) tests distinguish neurosyphilis from MS.

9.7.4.4 Metabolic Disorders

Inherited or acquired metabolic disturbance can cause damage to the neural structures. Adrenoleukodystrophy and subacute combined degeneration are a few metabolic disorders that can present with the features of MS. Adrenoleukodystrophy is an X-linked recessive neurodegenerative disorder due to the defective ABCD1 gene. The defect in the ABCD1 gene results in the accumulation of very-long-chain fatty acid and can damage the myelin sheath causing myelopathy and demyelination. It is usually seen in the pediatric age group. Radiological findings and absence in elevation of plasma cortisol levels in response to adrenocorticotropic hormone (ACTH) test can diagnose the disorder.

Subacute combined degeneration (SACD), an acquired metabolic disorder due to chronic vitamin B_{12} deficiency, causes the demyelination of the spinal cord and intracranial (frontal) white matter. The condition is manifested by sensory ataxia due to the involvement of posterior columns, spastic paraparesis due to involvement of lateral columns, sexual dysfunction, and Lhermitte's sign. Low serum vitamin B_{12} level, presence of megaloblastic anemia and methylmalonic aciduria, and a higher level of homocysteine confirm the diagnosis of vitamin B_{12} deficiency.

9.8 Medical Management

Several forms of treatment have been proposed during the past decades, and many were thought to be effective because of the remitting nature of the disease. However, based on the controlled trials, only ACTH, prednisone, methylprednisolone, cyclophosphamide, beta interferon, glatiramer acetate, mitoxantrone, and natalizumab have shown beneficial effects both radiologically and clinically. According to some investigators, the selection of therapy should be based on the definition of distinct clinical subtypes of MS. The typical relapsing-remitting type of MS characterized by episodic inflammations probably may respond better to immunomodulatory therapy. On the contrary, these measures may be less effective for the chronic progressive subtypes.

The main aim of medical treatment is to reduce the episodes of relapses and limit the disability due to the ongoing progression of the disease. Corticosteroids are the drug of choice to counteract the inflammatory effect during an acute attack, including an attack of optic neuritis. With regard to the dosage of corticosteroids for an acute attack, an initial high dose is preferred but disputed by others. An intravenous administration of massive doses of methylprednisolone, followed by high oral doses of prednisone, later tapered to a lower dosage is generally effective in aborting or limiting an acute or subacute exacerbation. However, there is limited evidence that corticosteroids have a significant effect on the ultimate course of the disease or prevention of recurrences. For those patients requiring steroid treatment for several weeks, Cushing's syndrome, hypertension, hyperglycemia, osteoporosis, aseptic necrosis of the acetabulum, cataracts, gastrointestinal hemorrhage, and reactivation of tuberculosis are certain side effects of the long-term use of corticosteroids.

Medical treatment for MS can be broadly classified into disease-modifying therapies and symptomatic therapies. Disease-modifying therapy is MS specific which alters the course of the disease and maintains suppression of the inflammation, and it includes immunosuppressants (fingolimod, natalizumab) or immunomodulatory agents (beta interferon, glatiramer acetate, teriflunomide), whereas symptom-based therapy is mainly used to treat the general symptoms secondary to neurological dysfunction.

9.8.1 Disease-Modifying Therapy

The prime focus of the early treatment strategies in MS patients is to prevent long-term disability. "No evidence of disease activity" (NEDA) is a modern concept in the treatment of MS. This concept of the treatment is developed to explore the underlying inflammatory process other than the clinical relapse which is in addition to brain atrophy that may progress in the absence of obvious inflammatory disease activity. NEDA is a new goal that is emerging in MS treatment, and it aims to reach a point where the MS patient is free from silent disease activity as well as the visible changes of the disease. Generally, NEDA is defined by the absence of relapses, disability progression, and MRI activity. In clinical scenarios, the use of NEDA has led to early treatment with highly active therapies. Immunosuppressants are medications to reduce the strength of the body's immune system, and immunomodulatory therapy reduces the relapse rate and the burden of disease. Immune reconstitution therapy (IRT), by definition, is given intermittently and not continuously and can induce long-term remission and in some cases the possibility of a cure. IRT reconstitutes/resets the immune system to regain its ability to respond to infections. Figure 9.6 depicts the chart for disease-modifying therapy. During the initial phase of the disease, immunomodulating drugs are preferred, and during the highly active phase, a combination of immunosuppressant and partial IRT drugs is preferred. For the rapidly evolving phase or if radiological findings are suggestive of brain atrophy, a combination of immunosuppressant and complete IRT is the preferred line of management.

Natalizumab, originally used in rheumatoid arthritis, which blocks lymphocyte and monocyte adhesion to endothelial cells and migration of these cells through the vessel wall, has demonstrated a reduction in the number of relapses and marked effect on MRI lesions. A recent double-blind, placebo-controlled study of RRMS revealed an approximately 70–80% reduction in relapses and new or enlarging lesions. This intravenous treatment has the advantage of single monthly administration with virtually limited side effects. Several clinical trials have shown that the use of beta interferon decreases the frequency and severity of the relapses by about one-third and also reduces the number of fresh or enlarging lesions. An added advantage of this agent is that it can be taken once weekly as an intramuscular injection. The

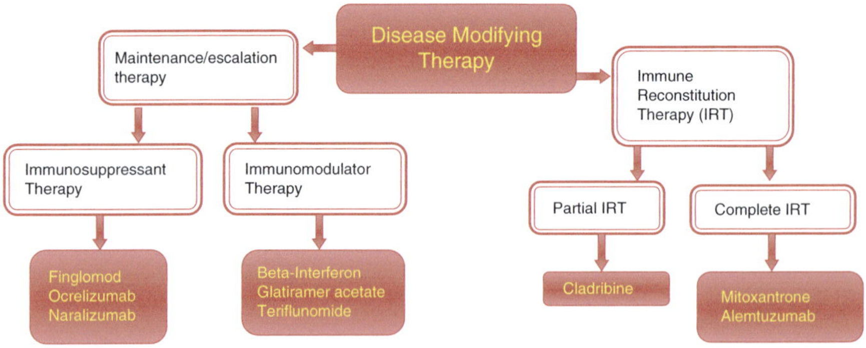

Fig. 9.6 Chart depicting the disease-modifying therapy

side effects of beta interferon are modest and mainly include flu-like symptoms, sweating, malaise, exacerbation of headaches among a few migraineurs, and depression in susceptible patients beginning several hours after the injection.

With regard to immunosuppressive agents like azathioprine and cyclophosphamide which modify immune reactivity, limited success has been reported. Prolonged use of such agents can predispose for neoplastic change, and therefore a widespread use of such agents is discouraged. Mitoxantrone, an agent with broad immunosuppressant and cytotoxic activity used originally to treat various cancers, has shown some beneficial effects on the progressive form of MS. This agent is well tolerated and has the advantage of requiring the administration of one dose every 3 months.

9.8.2 Symptomatic Treatment

Symptomatic treatments basically address the general symptoms due to the CNS involvement and the secondary effects of the disease. General measures include the provision of adequate bed rest during the convalescence stage to ensure maximum recovery, prevention of excessive fatigue, use of rehabilitative measures to postpone the ill effects of the bedridden stage, and prevention and management of complications like bedsores and infections.

Fatigue, a common complaint of MS patients, responds to some extent to amantadine, modafinil, pemoline, and potassium channel blockers. Bladder dysfunction can raise serious problems in the management, and bethanechol chloride can be helpful for urinary retention. Those with severe urinary retention may benefit from intermittent catheterization. In case of urgency and increased frequency, the use of propantheline or oxybutynin may serve to relax the detrusor muscle. If the patient complains of severe spastic paralysis or flexor spasms of the lower limbs, the oral or intrathecal infusion of baclofen can be of a certain value. The selective injection of botulinum toxin into the hypertonic muscles may help some patients. For severe and disabling tremors, ventrolateral thalamotomy or deep brain stimulation technique used for the treatment of Parkinson's disease can be an option.

9.9 Physiotherapy Management

Being a chronic, progressive, variable, and unpredictable disorder with episodic relapses and remissions, the physical and cognitive impairments can potentially lead to inactivity and deconditioning and can hinder and challenge the rehabilitation. Though the direct effects of the disease (direct impairments) are less yieldable to therapeutic strategies, the role of rehabilitation in enhancing significant gains in levels of activity and participation is compelling. A Cochrane review on multidisciplinary rehabilitation for MS concluded that therapeutic strategies produce both short- and long-term gains in activity and participation for inpatient and outpatient programs. The study also stated that low-intensity exercise programs had a stronger evidence for improving the QOL.

According to the NMSS advisory board, the referral and the commencement of rehabilitation should start as soon as there is a sudden or gradual worsening of function or an increase in impairment significantly impacting the patient's independence, ambulation, safety, and QOL. MS patients may benefit from restorative strategies aimed at remediating impairments, activity limitations, and participation restrictions. Restoration strategies are aimed at increasing the capacity of the system, whereas compensatory strategies are targeted to use adaptive behaviors. The compensatory strategies aim at modifying the activity or environment to provide optimal function within the scope of existing impairments and limitations. The focus to restore or compensate for the lost function is case to case dependent. However, in most cases, a combination of either strategy is preferred, and the challenge is to sort out the inactivity and deconditioning that needs a certain level of restoration and dysfunction(s) arising from the neurodegeneration that requires compensation.

Due to the widespread CNS lesions and variable clinical presentations, no single therapeutic approach can serve as a gold standard for rehabilitation management. Whether it is a remedial, compensatory, or mixed approach, the evidence is growing that rehabilitation is beneficial. An intensive therapeutic program can provide long-term benefits in functional skills, participation, and QOL but may fail to improve the fundamental impairments. Early rehabilitation with appropriate compensatory changes and energy conservative techniques has shown brain plasticity with positive changes in the physical abilities of MS patients. The main focus of physiotherapy is to symptomatically manage the effects of spasticity, muscular weakness and tightness, cognitive dysfunction, fatigue, and neurogenic bowel and bladder.

High energy consumption in daily activities, respiratory muscle dysfunction, and deconditioning are the important factors contributing to low tolerance in MS patients. Historically, patients with MS were discouraged from exercise training due to possible overheating which triggers exhaustion and fatigue. Presently, physiotherapy programs are rewarding and popular in reducing disability and improving the condition of MS patients. Depending on the progression of the disease, a tailor-made exercise program will be more beneficial than a structured exercise protocol.

Improving the QOL is the most meaningful outcome for these patients, particularly when the disease is advancing. Preventive strategies are aimed at minimizing complications, impairments, activity limitations, or disabilities. These strategies are geared toward decreasing the duration and severity of the symptoms or delaying the emergence of the disease sequelae through early detection and timely intervention. A coordinated interdisciplinary team consisting of a physician, nurse, physiotherapist, occupational therapist, speech and language pathologist, medical social worker, and dietician is necessary to oversee the comprehensive examination and manage the patient's complex and multifaceted problems. The patient is the central figure and the family and caregivers are the key members of this team. In the advanced stage of the disease, maintenance therapy and a series of occasional clinical, educational, and administrative services designed to maintain the patient's present level of function will be essential.

The primary aim of physiotherapy is to maximize the function in the phase of progressive disability. A clear understanding of the clinical course is essential to maximize functional ability. To plan a rehabilitation protocol, the aim of therapy should be clearly defined and should preferably encourage a tailor-made therapy than a structured program. The primary goals of physiotherapy in MS are to:

- Provide comprehensive evaluation and assessment
- Encourage cardiac and respiratory conditioning
- Promote an optimum level of functional independence
- Conserve energy and manage fatigue
- Promote self-management
- Prevent or manage possible secondary complications
- Educate the patient and family

9.9.1 Evaluations and Assessment

A careful but comprehensive examination is essential to determine the extent of neurological and functional involvement. Frequent reexaminations are often required at regular intervals to distinguish changes in functional or neurological status and monitor the treatment effects. In some patients, due to variability of symptoms, brief observations of performance over a few consecutive days may also be helpful to provide a representative sample of baseline functioning. Even while scheduling the examination, factors like fatigue and causes for exacerbation should be taken into account. The stage and severity of the disease, age of the patient, and rehabilitation setting are some of the additional factors that must be taken into account while structuring the examination.

Preferably the initial interview should be brief and must include a quick screening or questioning about body systems and covering common areas of impairment such as muscle strength, muscle tone, balance, coordination, sensory disturbances, bladder control, cognition, and depression. The results of the interview will help the therapist to develop hypotheses about the potential impairment(s) that might be contributing to the patient's physical or cognitive issues so that the examination can be designed only to observe the problematic tasks and test the hypotheses developed. In case the problem identified by the patient or therapist does not fit within the scope of the physiotherapy practice, the therapist should re-refer the patient to an appropriate health-care professional or the primary consultant. The physical examination should start by testing the patient's functional abilities which are problematic as per the patient or family member. The testing includes activities like transfers, gait, activities of daily living (ADL), or cognitive tasks that can compromise the patient's work, home, and recreational life. Besides, the therapist can use certain QOL measurement tools that will cover participation issues pertinent to MS patients.

9.9.1.1 Subjective Examination

The subjective examination components consist of a data form interview with the patient or caregiver and the information from the medical record like demographic data, medical or surgical history, history related to comorbidities and medications, social and economic status, family history, living environment, general health status, and habits or addictions. The therapist should ascertain the patient's primary or chief complaints and the present functional status and activity level. In addition to the above, a brief observation of position or attitude of the limbs, skin integrity, and asymmetry or inaccuracy in voluntary movements can be noted.

9.9.1.2 Objective Examination

The objective examination can be broadly classified into sensory and motor examination. Due to the potential involvement of the ascending sensory tracts of the spinal cord or the white matter within the cerebrum, a detailed examination of the superficial and deep sensations is required. Such a detailed sensory examination may help to identify the tracts involved and the possible location and extent of the lesion(s). The Nottingham Sensory Assessment (NSA) scale can be used as an additional tool for the sensory deficit and its effect on the QOL.

The motor system examination should include superficial and deep tendon reflex, muscle tone assessment, muscle strength, and endurance testing and balance, coordination, and gait evaluation. The presence of brisk or hyperactive deep tendon reflexes associated with the upgoing plantar response signifies upper motor neuron lesion. In case of MS with transverse myelitis like presentation, abdominal reflexes can be altered with a band-like sensation around the trunk. Central demyelination is likely to produce muscle weakness, and manual muscle testing can help grade muscle strength. Depending on the site and extent of the lesion, the muscle tone can be altered in the upper and lower limbs. If the clinical examination reveals muscle spasticity, the modified Ashworth scale (MAS) can be used for grading the same.

The presence of ataxia and hypotonicity suggests the involvement of the cerebellum. The detailed evaluation of hypotonicity, ataxia and asthenia is covered in Chap. 5 and the author advises the readers to refer to the same. Specific scales are available for evaluating cerebellar ataxia. The Scale for the Assessment and Rating of Ataxia (SARA), developed by Schmitz-Hübsch et al., has good validity and reliability and can be an appropriate tool for rating ataxia. The test consists of eight components: gait, stance, sitting, speech disturbance, finger chase, nose-finger test, fast alternating hand movements, and heel-shin slide.

The therapist should examine the static and dynamic postural control in different positions including sitting and standing. Document the postural abnormalities or postural tremors which are present, if any. Standard methods like the use of posture grids, plumb lines, and photogrammetry can analyze the static posture. The balance which is fundamental to upright posture and movement is produced by a complex interaction of sensory inputs, central processing, and motor outputs. The therapist should examine the static and dynamic balance, reactive and anticipatory control, sensory interaction, and synergistic strategies. Though there are many balance evaluation tools, some of them focus on static tasks (including single-leg stance test, Romberg's test, tandem stance, and computerized platform posturography) and others on dynamic balance (including Berg Balance Scale (BBS), Star Excursion Balance Test (SEBT), and Tinetti Performance-Oriented Mobility Assessment (POMA)). Balance Evaluation Systems Test (BESTest) can be another alternative to evaluate the components of postural stability, orientation, and postural reaction.

Gait can be measured in different ways. The parameters usually examined are the gait speed, distance covered, kinematics, stability, safety, and endurance. For those subjects with considerable ataxia or fatigue, the use of videography will be more appropriate to analyze the gait after capturing the video. Observational gait analysis is the gold standard for qualitative measurement of gait. Quantitative measurement tools like the GAITRite® system and the video-based 3D motion analysis can provide temporal and spatial gait parameters including stride length, step length, step width, cadence, and single- and double-support times. Making a patient walk a certain distance can be an easy way of calculating the gait speed and velocity. Using the abovementioned principle, several short-distance timed tests were developed including a 10 m gait test, 6 min walk test, and 25 feet walk test. The NMSS Task Force on Clinical Outcome Measures has recommended the 6 minute walk test as a sensitive tool for walking endurance. The Timed Up-and-Go (TUG) test and the Dynamic Gait Index (DGI) are two tests that combine walking with functional tasks. The TUG is an easy-to-use tool for both clinical and research settings and is reliable in measuring function in MS patients. According to McConvey and Bennett, DGI is a reliable and valid tool for MS patients. The Multiple Sclerosis Walking Scale-12 (MSWS-12) is a disease-specific 12-item self-rated questionnaire that measures the impact of the disease on walking ability. The MSWS-12 has good reliability and validity.

Approximately 50% of MS patients have cognitive deficits involving problem-solving, short-term memory, visuospatial perception, and conceptual reasoning, of which only 10% have cognitive deficits severe enough to interfere with ADL. Cognitive impairment reduces physical independence and is a barrier to rehabilitation and a predictor of functional status. It can also lead to anxiety and stress. A Brief Repeatable Battery of Neuropsychological (BRB-N) test and the Minimal Assessment of Cognitive Function in MS (MACFIMS) are the two clinical batteries that can assess the cognitive function among MS patients with good validity and reliability.

The therapist should consider how personal and environmental factors can hinder or help the achievement of rehabilitation goals. Heat intolerance, depression, and sleep deprivation secondary to bladder incontinence are a few examples of personal factors. Environmental factors include living in a hot climate or having access to cooling equipment like air conditioners or cooling garments. Disease-specific outcome measures should incorporate such factors to determine the best strategies for effective rehabilitation. For those patients with fatigue as a major complaint, the therapist should interview and collect details related to it. The interview should address the type of fatigue (mental or physical), time of onset, diurnal variations, precipitating factors, and aggravating and relieving factors. In addition to the above, fatigue-related self-report scales or questionnaires like the Modified Fatigue Impact Scale (MFIS) and the Fatigue Severity Scale (FSS) can help understand the perceived impact of fatigue and measures to optimize the rehabilitation.

Several factors such as muscle weakness, spasticity, ataxia, vertigo, balance deficit, cognitive decline, and fatigue can lead to functional limitations and compromise the QOL. Sensorimotor impairments of the upper limbs can result in a reduced ability to perform functional activities. Evaluation of basic functional activities provides a better insight into the functional limitations and strategies to improvise rehabilitation. Standardized tests such as the Box and Block Test (BBT) or the Nine-Hole Peg Test (NHPT) can provide objective data about manual dexterity.

The QOL measures are numerical indices or scores that reflect the patient's perception of the effects of a disease process on his or her life. The QOL questionnaires assess the general QOL as well as the health-related QOL. Among the many different types of QOL questionnaires, Motl and Gosney found that disease-specific scales or questionnaires detected larger changes than generic measures. The Multiple Sclerosis Quality of Life-54 (MSQOL-54) and the Multiple Sclerosis Quality of Life Inventory (MSQLI) are two disease-specific scales or questionnaires for evaluating the QOL of MS patients. MSQOL-54 is based on the Short Form-36 (SF-36) survey questionnaire with an addition of 18 items specific to MS covering fatigue, cognition, and sexual functioning. The MSQOL-54 takes approximately 15 min to complete and requires another 15–20 min to score. The reliability of MSQOL-54 is good to excellent in MS patients. Contrary to MSQOL-54, the MSQLI is a battery consisting of 10 individual scales providing a QOL measure that is both generic and MS specific. Ironically, to administer the complete set of questionnaires approximately 45 min are required, and the scale does not provide a sum score for all tests. The test-retest reliability of MSQLI is good for patients with or without cognitive dysfunction.

There is growing evidence that exercise may modify disease progression in MS patients. Hence, a disease progression measure may be useful to assess the impact of an intervention on the patient's perceived level of disability. The Kurtzke Functional System Score (FSS) and the Kurtzke Expanded Disability Status Scale (EDSS) measure the progression of the disease and also help to assess the impact of the intervention. Though the FSS and the EDSS are scales to measure the severity of the disease, these standard tools have to be administered by a trained person.

9.9.2 Physiotherapy Intervention

Like any chronic and progressive neurological disorder, the therapeutic goals are to maximize or maintain function and minimize or prevent complications. Irrespective of the type of intervention chosen, growing evidence suggests that an increase in physical activity may have a neuroprotective effect. Few studies have reported that a combination of muscle strengthening and balance training decreases the disability in MS patients and reduces the inflammatory markers (levels of pro-inflammatory immune system mediators) in the blood. The reduction of inflammation and thereby alteration in the disease process suggests the role of exercise in neuroprotection beyond symptomatic management. Though there is sufficient evidence to prove the benefits, the exact dosage, intensity, and type of exercise required to produce activity-dependent plasticity are still largely unclear.

While prescribing an exercise program, each patient's level of physical fitness like strength and endurance, muscle tone, sensation, balance, coordination, level of fatigue, heat sensitivity, and cognitive resources like memory and judgment should be considered. A Cochrane review on exercise intervention for MS patients has shown a noticeable improvement in muscle strength, endurance, and mobility-related actions and a minimal betterment in emotional mood when compared to the controls. There was no evidence suggesting the superiority of any specific exercise program over the rest, and the study also reported that the interventions were safe with minimal adverse effects.

The rehabilitation can be successful only if the program fits into the framework of the patient's life. The exercise intervention preferably should begin during the early phase of the disease and not when the disability becomes prominent. Such timely intervention helps to restore function by reducing physical or cognitive decline by disuse or deconditioning. Since depression can affect the adherence to an exercise program, the therapist should provide constant reinforcement and a positive environment. For those RRMS patients experiencing an exacerbation, exercise should be temporarily discontinued until remission manifests, and for those with progressive worsening of weakness, the exercise program should be reinstituted once the deterioration has stabilized and no new symptoms are visible. For PPMS patients, to minimize further deterioration and optimize remaining function, exercise should be prescribed within the limits of their capabilities.

Group exercises can provide motivation and social support. Circuit training can be incorporated to alternatively work the upper extremities and the lower extremities. The environmental temperature should be monitored and adjusted, and the infrastructure should have air-conditioning and climate control units. In the absence of the same, the use of fans, air coolers, water sprays, and aquatic exercises can be encouraged. Free weights and dumbbells are not recommended and are unsafe if the patient has tactile or proprioceptive loss and severe incoordination or tremors. For those patients with cognitive and memory impairments, supervised exercises are more appropriate and may require posters or illustrations for easy understanding of the correct use of equipment, the preferred posture for performing the exercise, and the number of repetitions.

Functional training activities using closed chain exercises can promote strength and endurance. Those patients with ataxia and balance issues may require large stable surfaces like a therapeutic mat or plinth to perform the strengthening exercises.

The problem lists in terms of body function and structure are poor muscle strength and endurance, spasticity, postural instability and balance deficit, ataxia and tremor, cognitive decline, fatigue, and gait abnormality. All the abovementioned problems contribute to functional and physical limitations.

9.9.2.1 Poor Muscle Strength and Endurance

Muscle weakness and reduced endurance are common findings among MS patients. Due to the weakness and reduced endurance, these patients may adopt a sedentary lifestyle and limit physical activity. Maximal muscle force during isometric or isokinetic exercise is typically lower in these patients due to the reduced recruitment of motor units, decreased muscle mass, poor metabolic responses, spasticity, and disuse. Prescribing an appropriate exercise program to improve muscle strength and endurance can be challenging and needs to be individualized for each patient. The exercise program needs to be prepared based on the FITT (frequency of exercise, intensity of exercise, type of exercise, and time or duration) principle. The strengthening exercises and endurance training should preferably be scheduled on alternate days. Training should be performed when the core temperature of the body is low and before the fatigue sets in. Weights, pulley weights, therabands, and machines can be used for resistance training. It is essential to monitor the effects of fatigue, manage the core body temperature, and prevent overheating and overwork. It is also important to incorporate adequate rest periods for each session and progress the exercise gradually with adequate caution.

Resistance training has shown a marked improvement in the QOL, fatigue, and motor output. Studies on the effects of combined resistance and endurance training have shown an improvement in terms of the QOL compared to a single exercise regimen with mild to moderate disability. Supervised progressive resistance training is found to be effective. The resistance program can be done using free weights like sandbags, weight cuffs, or dumbbells progressing to isokinetic movement therapy, focusing on the proximal musculature. As per the American College of Sports Medicine's guideline, two to three sets of 8–12 repetitions for each muscle group performed 3 days a week are adequate to improve the muscle strength without any adverse effects. Even moderate-intensity endurance training, performed using a static cycle, arm ergometer, or treadmill, with 50–60% of VO_2 max is believed to be neuroprotective and has shown neuroplastic changes.

9.9.2.2 Spasticity

In MS patients, the antigravity muscles like the hip adductors, knee extensors, and ankle plantar flexors are often spastic. A similar kind of trend is seen among the shoulder adductors and elbow, wrist, and finger flexors. Generally, MS patients demonstrate more spasticity in the lower limb musculature when compared to the upper limb musculature. If the muscle hypertonicity is considerable, it can cause functional limitation and contribute to many secondary impairments like contractures, deformities, and decubitus ulcers.

The use of cryotherapy, stretching, positioning, certain therapeutic exercise, hydrotherapy, or combinations of the above are the interventions to manage spasticity. The therapeutic interventions should be optimized to the dosing cycle of the antispastic medications. For instance, patients will respond to stretching techniques better when they are applied in the middle of the dosing cycle of baclofen than at the end or beginning. The therapist should also look out for spasticity contributing factors like infection or fever. The topical application of ice packs or cool bath, known to slow the nerve and muscle impulses, can temporarily reduce the spasticity by decreasing tendon reflex excitability and clonus. The effects of cryotherapy are usually relatively short-lived and may last from minutes to a few hours. The proprioceptive neuromuscular facilitation (PNF) stretching techniques like hold-relax and contract-relax are effective in gaining ROM. Patients and caregivers should be taught to perform stretching exercises, and it should be an integral part of the home program. Quick stretching and ballistic stretching movements are contraindicated, as the spastic muscles are velocity sensitive. A sustained stretch from 30 min to 3 h can be helpful to reduce the stretch reflex activity.

Encouraging voluntary contraction or neuromuscular electrical stimulation of the antagonist muscles alleviates the hypertonicity of the spastic muscles through the mechanisms of reciprocal inhibition. Functional activities aimed at reducing the tone should concentrate on the trunk and proximal segments because many patterns of hypertonus seem to be fixed from the action of the stronger proximal muscles. In those patients with extensor hypertonicity, activities that encourage lower limb flexion and trunk rotation may alleviate extensor spasticity. Prolonged supine positioning is not advisable for those patients with a tendency to develop spasticity in the antigravity muscles of the lower limb. Light joint compression may normalize the tone of spastic muscles and the use of night splints can provide prolonged stretch in case the spasticity is considerable and deformities are setting in. Though the abovementioned approaches or strategies are used routinely in clinical practice, strong and conclusive research evidence is still unavailable with regard to their efficacy.

For those patients with mild to moderate spasticity, active stretching has shown better neuroplastic changes, as the brain understands functional movement better than any unidirectional passive stretching movement. Functional movements are multidirectional and also stretch the spastic muscle in a functional pattern. PNF patterns have diagonal movements resembling the functional activities and can improve the flexibility of muscle and coordinated action of the agonist and the antagonist. In long-standing cases of spasticity, sustained stretching post-myofascial release may elongate the soft tissues, re-bias the muscle spindle, and reduce the motor neuron excitability and sensitivity to stretch. A few studies state that functional stretching in weight-bearing positions has better outcomes when compared to conventional stretching. Such weight-bearing techniques also have an additive effect in improving bone density, enhancing vascularity, and preventing muscle atrophy.

9.9.2.3 Ataxia and Tremor

Weighted cuffs, boots, bracelets, or anklets can dampen the intention tremors of the limbs, and a weighted belt or jacket can reduce truncal ataxia. The use of latex resistance bands through proprioceptive loading and resistance may minimize the limb ataxia. When weights are used to dampen the tremor, use the least amount of weight to achieve the desired effect. Loading the axial skeleton is generally more preferred than the extremities. For those patients who require walking aids for ambulation, a weighted assistive device may reduce the dysmetria of the upper limbs. Weighted cutlery sets can minimize the tremors and spilling of food while eating. Wearing a soft cervical collar can reduce the titubation of the head. All the abovementioned strategies are compensatory, and therefore ataxia and tremor will return or in some cases temporarily worsen when these devices are removed. Stress, anxiety, and excitement often worsen these abnormal movements. Since heightened arousal can cause an adrenalin rush to increase the existing tremors, stress management can be another strategy to alleviate the tremors.

Limb(s) and/or trunk ataxia is a common clinical feature for MS patients. Ataxia can be sensory, cerebellar, or of the mixed type, associated with or without muscle weakness. Truncal and lower limb ataxias contribute to postural instability, balance deficit, and gait disturbances and hinder daily activities. The primary objective of therapeutic intervention in such cases should be to improve the joint stability and movement control which in turn will help to improve ataxia. Frenkel's coordination exercises, PNF techniques, prone progression exercises, and balance training exercises have shown a clinical improvement of ataxia and tremor among MS patients. Task- or target-oriented activities, repeated several times with verbal cues and feedbacks, like in Frenkel's exercise, are effective to improve the movement control and errors secondary to the dysmetria of the limbs or the trunk. Frenkel's exercises are progressively performed in four standard positions: lying, sitting, standing, and walking. These coordination exercises are intended to be done slowly with even timing, initially with adequate limb support, and then progressed toward independent unilateral and bilateral movements.

Encouraging isometric contractions, slow reversal hold techniques, and rhythmic stabilization techniques of PNF can improve postural stability and reduce truncal and limb ataxias. The use of chopping and lifting patterns can also help to improve neuromuscular coordination. The use of biofeedback helps to improve the posture and movement control. To encourage co-contraction and coordination of the muscle action, the exercises can be progressed from a stable surface to an unstable surface. The use of foam surface, Swiss ball, or wobble board is recommended to improve the truncal stability and dynamic balance, provided the patients are capable of performing the exercises safely on unstable surfaces.

9.9.2.4 Postural Instability and Balance

The therapeutic intervention programs should be based on a comprehensive understanding of the patient's impairments and personal and environmental factors. Postural stability is a key determinant of functional mobility. Abnormal tonicity, poor coordination between the agonist and antagonist muscles, dysmetria and ataxia of the trunk, and weakness of the postural muscle due to demyelination can predispose to postural instability. Additionally, lack of specific sensory interaction such as somatosensory, visual, or vestibular interrupts the feedback mechanism and predisposes to poor postural stability. Figure 9.7 depicts the postural instability mechanisms. The therapeutic intervention should include strategies to improve the sensory and motor interaction, which would facilitate both the anticipatory and reactive movements of the trunk and the limb musculature.

Trunk being the key predictor of postural stability should be the prime focus during the early phase of rehabilitation. Severe weakness of the trunk musculature leads to difficulty in performing basic ADL and affects overall functional mobility. During the initial phase, the strengthening program for trunk musculature should include activities like bridging and spinal extensor strengthening. The use of a Swiss ball will help to progress the exercise from a stable surface to unstable support. Interventions directed at promoting postural control should focus on static control for antigravity postures: prone progression postures; sitting, modified plantigrade and standing postures. Prone progression is a series of postures that gradually acclimatize the patient to an upright posture by increasing the postural demands (varying the base of support, raising the center of mass, and increasing the number of body segments to be controlled). Heavy joint compression through proximal and axial joints and rhythmic stabilization promotes postural stability. For those patients with severe ataxia and weakness of the proximal musculature, isometric exercises, resistance training, and slow reversal hold in an ever-decreasing range can be encouraged prior to rhythmic stabilization to promote the co-contraction of the muscles. Once the static posture is satisfactory, the dynamic postural control can be challenged by incorporating activities like weight shifts, upper limb reach-outs, and lower limb stepping. Strengthening the core muscles on a therapeutic mat can also improve posture and postural stability. A progressive resistance training for 8 weeks (progressing from 50% of 1 RM to 80% of 1 RM, systematically) has shown a marked improvement in the balance and functional status of the patient.

Throughout the postural stability and balance training programs, the patient's safety is of utmost importance. The training program should incorporate a variety of everyday functional tasks that challenge balance. During the training process, to promote the adaptation of skills, the tasks should be modified with respect to the base of support, supporting surface (stable vs. unstable), and environmental variations. For those patients with vestibular dysfunction secondary to MS, vestibular rehabilitation may improve the impaired balance and the disability due to dizziness or disequilibrium. Water can also serve as an important therapeutic medium to practice static and dynamic postural control. It can provide graded resistance which will slow down the ataxic movements of the trunk or limb(s); meanwhile, the buoyancy of the water will aid the

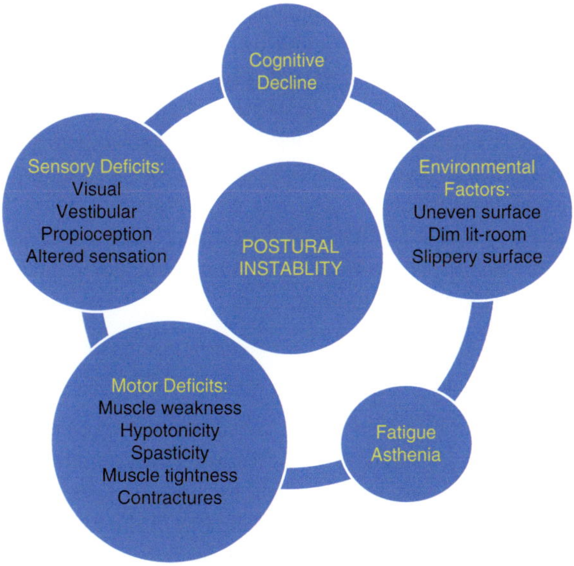

Fig. 9.7 Illustration of postural instability mechanisms

maintenance of an upright posture. Aquatic exercises can be effective in increasing the strength, reducing the muscular fatigue, improving the endurance, and improving the QOL of MS patients. For aquatic exercises, to minimize the heat intolerance issues and fatigue related to it, the pool temperature should be set between 80 and 85 °F.

Balance is an integral component of functional activity, and it synchronizes the coordination between the environment and the task. To maintain the center of mass with the limits of stability, a well-coordinated functioning of the multisensory unit is required. Impaired static posture, poor postural control, loss of anticipatory muscle reaction, and an inability to maintain the center of mass within the base of support will cause imbalance. Balance deficit increases the risk of falls among MS patients. Due to its multifactorial etiology, the strategies have to be diverse in terms of treatment approaches and should include multisensory training, motor strategy training, and resistance and aerobic exercise training.

Functional weight-bearing exercise augments the proximal joints and girdle control, increases the patient's awareness about body orientation, and improves lower limb strength, balance, and mobility. Functional reach-out activities performed in sitting and standing, progressed from a stable to an unstable surface, can improve the joint stability, enhance the dynamic mobility (eccentric and concentric controls of muscles), and augment the balance and anticipatory postural reaction. Overall, such activities improve the kinesthetic sensations of the trunk and lower limbs and are essential prerequisites for the maintenance of balance. Encouraging activities like partial squats and single-limb stance not only improves the strength of the weak lower limb musculature but also augments the proprioceptive input to further enhance the balance and the limits of stability.

The ultimate goal of balance training is to attain an automatic response without any conscious effort. The cognitive deficit has an additive effect on the balance; it could also cause a lack of attention and also affect the ability to comprehend, understand, and execute verbal instructions. Repetition of activity and practicing the components of the task or exercise while using simple commands helps to progress the intervention among the patients. Biofeedback training using augmented feedback can improve the balance reactions. The use of augmented visual and/or auditory feedback displayed by machines like force platforms, video game platforms, or consoles like Nintendo Wii can be useful tools for improving the balance of MS patients with somatosensory deficits and balance issues.

9.9.2.5 Gait Training

Inability to walk safely is a major problem for many MS patients. Gait issues often include poor balance, heaviness and/or stiffness of the lower limb musculature, fatigue, sensory system impairments, and weakness of the musculature. Due to the abovementioned, gait deviations such as circumduction, high stepping, intoxicated, and Trendelenburg gait patterns can be seen. Some patients may demonstrate immediate compensations like a widened base of support while standing and walking, some may hold their knees stiff to increase stability, and others may compensate by bending the knees to lower the body's center of gravity. Such patients may use their arms for support to counteract the increased postural sway.

The primary objective of gait rehabilitation should be to encourage effective weight-bearing through the lower extremities and promote adequate weight shifts with anticipatory postural reactions. Promoting pelvic tilt during weight shifts, adequate knee flexion and ankle dorsiflexion during the gait cycle, and maintaining the COG within the base of support are all important components of the gait training. A well-designed exercise program consisting of tone management, strengthening exercises, and balance and postural control exercises can improve gait and promote safe and energy-efficient walking. Functional activities like walking forward, backward, and sideways, braiding, obstacle gait training, stepping and climbing stairs, and negotiating curbs and ramps further enhance the walking ability for safe community ambulation. For those patients with predominant ataxia, a weighted walker may be the best option as it affords stability and mobility. A wheeled walker with hand brakes and a seat can be provided for those patients who have fatigue and weakness. Studies using treadmill with a body weight support system in MS patients have shown improvements in muscle strength, tone, and endurance, balance, walking speed, and QOL. Studies have shown that functional electrical stimulation (FES) is effective in improving the spatiotemporal parameters of gait, gait speed, QOL, and activity participation.

Since the gait abnormality in MS is multifactorial, a variety of different therapeutic interventions or strategies can be helpful to improve ambulation. Stretching should be an integral part of the preparation for exercise, especially when the muscles are hypertonic. Typically, the plantar flexors, adductors, and quadriceps are involved in the lower extremity. Self-stretching or passive stretching of the tight soft tissues combined with slow rhythmic rotation can be beneficial to gain range. Supervised exercise training for a few months can significantly improve the gait parameters in many MS patients. For those MS patients who do not respond well to the therapeutic interventions, compensatory strategies like the prescription of mobility aids and orthotic devices like canes, crutches, walkers, wheelchairs, lightweight ankle-foot orthosis (AFO), or hip flexion assist orthosis may help to restore the function, energy expenditure, safety, and independence of the patient. Among the lower limb orthoses, AFO is the most common orthosis used by MS patients. In addition to saving energy and providing greater ankle stability, it also improves the foot clearance, the knee hyperextension, and the overall gait pattern.

9.9.2.6 Fatigue Management

Fatigue can be a debilitating symptom for many MS patients, and it is characterized by a sense of weakness, excessive tiredness, and sleepiness. In many of these patients, the fear of bringing on fatigue may cause aversion or reluctance to perform routine activities. Therefore, a balancing act of activities to promote functional independence on one hand and avoidance of overwork and fatigue on the other is essential while prescribing an exercise program. The therapist should know the difference between MS-related fatigue and exercise-related fatigue. Fatigue secondary to heat intolerance is MS related and can be managed with adequate rest and the use of cooling and precooling treatments during exercise. To avoid exercise-related fatigue, a submaximal level of exercise with a discontinuous schedule of training is appropriate. The use of an activity diary may help to identify the rate of perceived level of fatigue for all the routine activities and the time consumed for each activity. The activity diary also helps to differentiate the easier activities from the difficult ones, the important from the less important ones, and the quality of sleep at night. Based on the information obtained from the activity diary, therapists can initiate training sessions emphasizing energy conservation strategies like activity pacing and environmental modifications.

Energy conservation techniques use strategies to reduce the overall energy requirements of the task and the level of fatigue. For further understanding of these techniques, the author recommends the readers to glance through those contents specified under Chap. 13. Activities that are difficult or have high energy demands need to be broken down into components. Activity pacing refers to the balancing of activity with rest periods interspersed throughout the day. The overall levels of energy can be improved if patients learn to set priorities and limit their activities, saving their energy for those activities that are truly important to them. The abovementioned aspects are typically addressed by the occupational therapist who can provide valuable suggestions in terms of planning, work simplification, and developing energy-efficient activities for self-care and home management. Several adaptations may be considered to improve the efficiency and safety, which may include air-conditioning, home or work modifications, or ergonomic equipment. Proper communication between the occupational therapist and the physiotherapist along with, appropriate doses of the pharmacological agents may help to sort out the fatigue-related issues of the MS patient in various circumstances and activities.

In addition to the management of fatigue, a rehabilitation plan should include educating the patient and the caregiver about the preventive strategies to minimize fatigue. Since fatigue is a subjective issue, a tailor-made program consisting of a combination of resistance and aerobic exercises will be more appropriate than a structured protocol. Circuit training with a combination of aerobic and resistance training improves the physical endurance, provided the work stations have an adequate rest period between the activities and the room temperature is regulated to reduce the thermal stress. Evidence related to therapeutic strategies for fatigue management states that a tailor-made exercise protocol with regulated activity not only helps to improve the conditioning of the patient but also improves the fatigue level.

Further Reading

Amato MP, Portaccio E. Management options in multiple sclerosis-associated fatigue. *Expert Opin Pharmacother.* 2012 Feb;13(2):207–16.

Armutlu K, Karabudak R, Narulu G. Physiotherapy approaches in the treatment of ataxic multiple sclerosis: A pilot study. *Neurorehabil Neural Repair.*2001;15:203–11.

Bar-Or A, Oliveira EM, Anderson DE, Hafler DA. Molecular pathogenesis of multiple sclerosis. *J Neuroimmunol.* 1999 Dec;100(1–2):252–9.

Braley TJ, Chervin RD. Fatigue in multiple sclerosis: mechanisms, evaluation, and treatment. *Sleep.* 2010 Aug;33(8):1061–7.

Calabresi PA. Diagnosis and management of multiple sclerosis. *Am Fam Physician.* 2004, Nov 15;70(10):1935–44.

Cañellas AR, Gols AR, Izquierdo JR, Subirana MT, Gairin XM. Idiopathic inflammatory-demyelinating diseases of the central nervous system. *Neuroradiology.* 2007 May;49(5):393–409.

Dalgas U, Stenager E, Hansen TI. Review: Multiple sclerosis and physical exercise: recommendations for the application of resistance-, endurance- and combined training. *Mult Scler.* 2008;14:35–53.

Doson R, Giovannoni G. Multiple sclerosis- a review. *Eur J Neurol.* 2019,Oct 4;26:27–40.

Etoom M, Khraiwesh Y, Lena F, Hawamdeh M, Hawamdeh Z, Centonze D, et al. Effectiveness of Physiotherapy Interventions on spasticity in people with multiple sclerosis: a systematic review and meta-analysis. *Am J Phys Med Rehabil.* 2018 Nov;97(11):793–807.

Ferguson B, Matyszak MK, Esiri MM, Perry VH. Axonal damage in acute multiple sclerosis lesions. *Brain.* 1997; 120, 393–9.

Frohman EM, Racke MK, Raine CS. Multiple sclerosis- the plaque and its pathogenesis. *N Engl J Med.* 2006, Mar 2; 354(9):942–55.

Ganesh A, Stahnisch FW. A history of multiple sclerosis investigations in Canada between 1850 and 1950. *Can J Neurol Sci.* 2014 May;41(3):320–32.

Ghasemi N, Razavi S, Nikzad E. Multiple sclerosis: pathogenesis, symptoms, diagnosis and cell-based therapy. *Cell J.* 2017, Apr;19(1):1–7.

Goldenberg MM. Multiple sclerosis review. *P T.* 2012 Mar;37(3):175–84.

Golzari Z, Shabkhiz F, Soudi S, Kordi MR, Hashemi SM. Combined exercise training reduces IFN-γ and IL-17 levels in the plasma and the supernatant of peripheral blood mononuclear cells in women with multiple sclerosis. *Int Immunopharmacol.* 2010;10:1415–9.

Halabchi F, Alizadeh Z, Sahraian MA, Abolhasani M. Exercise prescription for patients with multiple sclerosis; potential benefits and practical recommendations. *BMC Neurol.* 2017 Sep 16;17(1):185. https://doi.org/10.1186/s12883-017-0960-9.

Hashimoto SA, Paty DW. Multiple sclerosis. *Disease-a-Month.* 1986;32(9):517–89.

Karussis D, Petrou P. Immune reconstitution therapy (IRT) in multiple sclerosis: the rationale. *Immunol Res.* 2018 Dec;66(6):642–8.

Khan F, Turner-Stokes L, Ng L, Kilpatrick T. Multidisciplinary rehabilitation for adults with multiple sclerosis. *Cochrane Database Syst Rev.* 2007 Apr 18;(2):CD006036. Available from: https://doi.org/10.1002/14651858.CD006036.pub2.

Kubsik-Gidlewska AM, Klimkiewicz P, Klimkiewicz R, Janczewska K, Woldańska-Okońska M. Rehabilitation in multiple sclerosis. *Adv Clin Exp Med.* 2017 Jul;26(4):709–715.

Loma I, Heyman R. Multiple sclerosis: pathogenesis and treatment. *Curr Neuropharmacol.* 2011 Sep;9(3):409–16.

MacAllister WS, Krupp LB. Multiple sclerosis-related fatigue. *Phys Med Rehabil Clin N Am.* 2005 May;16(2):483–502.

Martin S, Kessler M, editors. second ed. Neurologic interventions for physical therapy. St. Louis, Missouri: Saunders Elsevier; 2007.

Miller DH, Weinshenker BG, Filippi M, Banwell BL, Cohen JA, Freedman MS, et al. Differential diagnosis of suspected multiple sclerosis: a consensus approach. *Mult Scler.* 2008 Nov;14(9):1157–74.

Milo R, Kahana E. Multiple sclerosis: Geoepidemiology, genetics and the environment. *Autoimmunity Reviews.* 2010;9:387–94.

Ömerhoca S, Akkas SY, Icen NK. Multiple sclerosis: diagnosis and differential diagnosis. *Arch Neuropsychiatry.* 2018;55(1):S1–9.

O'Sullivan SB, Schmitz TJ, Fulk GD, editors. sixth ed. Physical rehabilitation. Philadelphia: FA Davis Co; 2014.

Paltamaa J, Sjögren T, Peurala SH, Heinonen A. Effects of physiotherapy interventions on balance in multiple sclerosis: a systematic review and meta-analysis of randomized controlled trials. *J Rehabil Med.* 2012 Oct;44(10):811–23.

Pearce JMS. Historical description of multiple sclerosis. *Eur Neurol.* 2005 Aug10;54:49-53.

Pandit L. No Evidence of Disease Activity (NEDA) in Multiple Sclerosis - Shifting the Goal Posts. *Ann Indian Acad Neurol.* 2019 Jul-Sep;22(3):261–3.

Prat A, Antel J. Pathogenesis of multiple sclerosis. *Curr Opin Neurol.* 2005;18:225–30.

Presas-Rodríguez S, Grau-López L, Hervás-García JV, Massuet-Vilamajó A, Ramo-Tello C. Myelitis: differences between multiple sclerosis and other aetiologies. *Neurologia.* 2016 Mar;31(2):71–5.

Rietberg MB, Brooks D, Uitdehaag BM, Kwakkel G. Exercise therapy for multiple sclerosis. *Cochrane Database Syst Rev.* 2005 Jan 25;2005(1):CD003980. https://doi.org/10.1002/14651858.CD003980.pub2.

Ropper AH, Samuels MA, Klein JP, Prasad S, editors. Adams and Victor's Principles of neurology tenth ed. New York: McGraw Hill Education; 2014.

Thompson AJ, Banwell BL, Barkhof F, Carroll WM, Coetzee T, Comi G, et al. Diagnosis of multiple sclerosis: 2017 revisions of the McDonald criteria. *Lancet Neurol.* 2018 Feb;17(2):162–73.

Umphred DA, Burton GU, Lazaro RT, Roller ML, editors. Umphred's neurological rehabilitation. sixth ed. St. Louis: Elsevier Mosby; 2013.

Chapter 10
Infections of the Central Nervous System

Abraham M. Joshua and Rohit Pai

10.1 Introduction

Infections of the central nervous system (CNS) are common in both developed and developing nations and account for a significant percentage of morbidity and mortality. The causes of these infections are bacterial, viral, fungal, parasitic, protozoan, and helminthic. Such infections result in CNS illnesses such as meningitis, brain abscess, subdural empyema, and granulomatous infections like tuberculosis and neurosyphilis, causing altered sensorium, coma, and focal neurological disorders.

Septic or pyogenic infections reach the intracranial structures by hematogenous spread or by extension from cranial structures. Such infections can be of iatrogenic origin too. The mechanisms behind the hematogenous spread are largely unknown. Bacteremia secondary to pneumonia or endocarditis can be a forerunner of pyogenic meningitis. Among the adult population, pneumococcus, meningococcus, staphylococcus, and streptococcus are some of the most common pathogenic organisms which cause pyogenic infections of the CNS. Opportunistic infections of CNS related to human immunodeficiency virus (HIV) are not uncommon and are the leading cause of death in adults in several countries.

For brain abscess, infarction of brain tissue by arterial occlusion or venous occlusion may be the first step required for the virulent bacteria to produce the abscess.

A. M. Joshua (✉)
Department of Physiotherapy, Kasturba Medical College of Mangalore, Manipal Academy of Higher Education, Manipal, India
e-mail: abraham.joshua@manipal.edu

R. Pai
Department of Neurology, Kasturba Medical College, Mangalore, Manipal Academy of Higher Education, Manipal, India
e-mail: pai.rohit@manipal.edu

A. M. Joshua (ed.), *Physiotherapy for Adult Neurological Conditions*,
https://doi.org/10.1007/978-981-19-0209-3_10

Infection can be due to a single type of virulent organism or can be a mixed one. Mixed infections pose difficult problems both in management and isolation or demonstration of the causative organisms from the pus collected from the abscess.

10.2 Acute Bacterial Meningitis

Meningitis, which is also known as leptomeningitis, consists essentially of infection of the pia mater, arachnoid mater, and the cerebrospinal fluid (CSF) occupying the subarachnoid space. Infective agents gaining entry to any one of the abovementioned regions will spread rapidly to the rest of it through the CSF space, and considering the above, meningitis is always cerebrospinal. Infection can reach the ventricles either directly from the choroid plexuses or retrograde from the subarachnoid space through the foramen of Magendie and Luschka. Acute meningitis is either aseptic, which is mostly viral in origin, or septic or pyogenic, which is caused by bacteria. Bacterial meningitis can be classified clinically as either acute or chronic. Acute bacterial meningitis (ABM) refers to acute infections caused by pyogenic bacteria and it evolves within hours or days, whereas chronic meningitis evolves over weeks.

Virulent organisms, antibiotic resistance, crowded living conditions, extremes of age, HIV infection, malnutrition, sickle cell disease, splenectomy, recent traumatic brain injury with fracture, post-head injury or post-neurosurgery CSF leakage, middle ear infection, and pneumonia are some of the risk factors for ABM.

10.2.1 Pathology

The lipopolysaccharide capsule that covers the bacterial surface helps them to survive and reach the meninges through the bloodstream or by direct invasion. Though the likelihood of developing a clinical disease is rare, penetration of these organisms across the blood–brain barrier may cause infection of the meninges and CSF space and may occur in association with bacteremia and septicemia. Once the organism reaches the CSF space, the multiplication of bacteria will trigger a massive host immune response and inflammatory reaction (release of inflammatory cytokines, activation of macrophages, invasion of neutrophils, presence of immunoglobulins, and other markers of inflammation) in the pia and arachnoid mater, the CSF space, and all the structures that lie within or adjacent to them. The abovementioned pathophysiological activities may further breakdown the blood–brain barrier and can result in thrombosis, vasculitis, infarction, raised intracranial pressure, focal neurological deficits, and death.

During the first few days of the inflammatory reaction, neutrophils predominate, and later the lymphocytes, plasma cells, and mononuclear cells (macrophages) predominate. A rapid increase in the exudate formation following massive inflammatory reactions can predispose the exudate to get collected over the base of the brain and extend into the sheaths of the cranial and perivascular spaces of the cortex. By around the second week, the exudate may organize into two layers—an outer layer, just beneath the arachnoid membrane, composed of neutrophils and fibrin, and an inner layer consisting of lymphocytes, plasma cells, and macrophages. Such an organization of exudate may result in fibrosis of the arachnoid and loculation of the pockets of exudate.

The exudate encircling the cranial nerves may cause focal cranial neuropathies. The exudate collected around the spinal and cranial nerves may cause the infiltration of inflammatory cells into the perineural sheath and endoneurium and may result in the degeneration of myelinated fibers and proliferation of the Schwann cells and fibroblasts. Usually, little or no damage happens to such nerve fibers. Exudate formed over the base of the brain, which blocks the foramina of Magendie and Luschka, can lead to hydrocephalus. In the later stages, the fibrous subarachnoid adhesions can cause additional interference to CSF circulation and may rarely give rise to chronic adhesive arachnoiditis or chronic meningomyelitis. Cortical neuronal damage secondary to meningitis may arise as a result of diffusion of toxins from the meninges or to a circulatory disturbance or an increase in intracranial pressure.

Swelling of the endothelial cells, infiltration and migration of neutrophils and lymphocytes from the adventitia to the subintimal region, subintimal fibrosis, focal necrosis of the vessel walls, and thrombosis within the small and medium-sized subarachnoid vessels are certain vascular changes during meningitis. Thrombosis of veins is more common than arterial thrombosis due to the thinner walls and the slower current of the blood. By the end of the second week of the infection, even larger veins may undergo thrombophlebitis.

The process of resolution of inflammation is characterized by the disappearance of inflammatory cells, almost in the same manner as they appeared. However, the resolution depends to a large extent on the stage at which the infection is arrested. For instance, if the infection is controlled in the very early stage, there may not be any residual change in the arachnoid. Contrary to that, if the infection continues for several weeks, there may be a permanent fibrous overgrowth of the meninges, resulting in thickened, cloudy, or opaque arachnoid and adhesions between the meningeal layers.

10.2.2 Clinical Features

Fever, severe headache, seizures, altered level of consciousness, and neck stiffness (meningismus or meningism) on forward bending are some of the important early clinical manifestations of meningitis. Photophobia, nausea, vomiting, and backache can be the other symptoms of the disease. Altered level of consciousness can be either irritability or lethargy. The incidence of altered sensorium is approximately 80% and adults with pneumococcal meningitis may present in a coma or semi-comatose state at the time of admission. In the elderly, fever and altered level of consciousness may be the only clinical findings. Raised intracranial pressure is one of the major causes of altered sensorium.

Bacterial meningitis seems to be multifactorial in origin. Increased permeability of the blood–brain barrier leading to interstitial edema, the release of cytotoxic mediators, elevated intracranial blood volume, and disturbance in CSF flow are some of the important causative factors for raised intracranial pressure. A mild to moderate increase in the intracranial pressure can cause headache, irritability, confusion, nausea, and vomiting. A more severe increase can produce coma, papilledema, cranial nerve palsy (usually the sixth cranial nerve), and cerebellar tonsil herniation, which may lead to death.

Focal neurological deficit is another common complication of meningitis and may account for up to 50% of all neurological complications. Weakness following meningitis is usually generalized (quadriparesis) or on one side (hemiparesis). Rarely, the patients may present with monoparesis or paraparesis. Vasculitis, cortical or sagittal vein thrombosis, cerebral artery spasm, cerebral edema, cerebral infarct, hydrocephalus, subdural effusion or empyema, and brain abscess result in such motor deficits. The majority of focal deficits tend to resolve with appropriate treatment; however, long-term disability may persist.

The abducens nerve (sixth cranial nerve) with its longest intracranial route near the brainstem is more vulnerable to increased intracranial pressure and exudates related compression. The third, fourth, and seventh cranial nerves may also be affected. Strabismus, while gazing toward the paralyzed side, can be an early presenting feature. The involvement of the optic nerve can lead to transient or permanent visual loss.

Evidence of infection outside the CNS or presence of underlying conditions predisposing to meningitis such as pneumonia, middle ear infection, HIV, and head injury can be additional findings. Neck stiffness (resistance to passive flexion of the neck while bringing the chin toward the chest) and Kernig's sign are the cardinal signs of meningitis. The presence of Brudzinski's sign, i.e., involuntary flexion of the hip and knee joints in response to passive forward flexion of the neck, and Kernig's sign, i.e., inability to completely extend the legs when the neck is passively forward flexed may be demonstrable among adult patients. Both the signs are resultant of a flexor protective reflex.

Petechial (hemorrhagic) rashes, which are non-blanching in nature, are frequently seen in approximately 50% of the subjects with meningococcal infections, whereas pneumococcal infections are often preceded by the lung, heart valve, and ear or sinus infection and are prone to have cranial nerve abnormalities. Focal cerebral signs are frequent in pneumococcal and Haemophilus influenzae disease. In addition to the above, seizures are frequent in Haemophilus influenzae meningitis. Seizures in bacterial meningitis, which are either partial or generalized, may develop at any time during the course of the disease. Inflammatory exudates, chemical mediators, bacterial toxins, and neurochemical changes within the brain parenchyma are some of the possible causes for seizures among adults. The clinical presentation and area or structures involved in acute, subacute, and chronic bacterial meningitis are depicted in Table 10.1.

Table 10.1 Clinical presentation and area or structures involved in acute, subacute, and chronic bacterial meningitis

Type	Area or structure involved	Clinical presentation
Acute bacterial inflammation	Pure pia-arachnoiditis	Headache, neck stiffness, Brudzinski's sign, and Kernig's sign
	Sub-pial encephalopathy	Confusion or drowsiness, convulsions, stupor, and coma
	Inflammation of cranial nerve roots	Ocular palsies, focal muscle weakness, and deafness
	Thrombosis of meningeal veins	Focal seizures, focal neurological deficits such as hemiparesis
	Cerebellar or cerebral hemisphere herniation due to increasing intracranial pressure (ICP)	3rd cranial nerve injury, hemiparesis or quadriparesis, and death
	Vasculitis of veins or arterioles	Focal neurological deficits
Subacute meningitis	Tension hydrocephalus	Impaired consciousness, decorticated posture, sphincter incontinence, and unsteady gait
	Subdural effusion	Impaired alertness, vomiting, fever, reduced food intake, and immobility
	Venous or arterial infarct	Focal neurologic deficit
Chronic meningitis	Meningeal fibrosis around the second cranial nerve and spinal cord roots	Blindness, optic atrophy, and spastic paraparesis
	Chronic meningoencephalitis with hydrocephalus	Dementia, stupor or coma, and paralysis

10.2.3 Laboratory Findings

The diagnosis of ABM will be based on clinical and laboratory findings. Total blood count, blood glucose, malaria parasite test, blood culture, and an HIV test are certain laboratory tests for ABM. Computed tomography (CT) with contrast may reveal meningeal enhancement. Superior imaging and higher resolution of magnetic resonance imaging (MRI) scans make it a preferred choice over CT scans for the early detection of inflammations. MRI axial post-contrast images may show sulcal enhancement consistent with the clinical diagnosis of meningitis (Fig. 10.1).

10.2.3.1 CSF Examination

Lumbar puncture, a simple and safe test, can be the key investigation, provided there is no evidence of any clear contraindication. It is contraindicated if there is any presence of raised intracranial pressure (clinically characterized by an altered level of consciousness, coma, focal neurological deficit, and papilledema). Pleocytosis (an increase in the cell count) may be an important finding of the CSF examination.

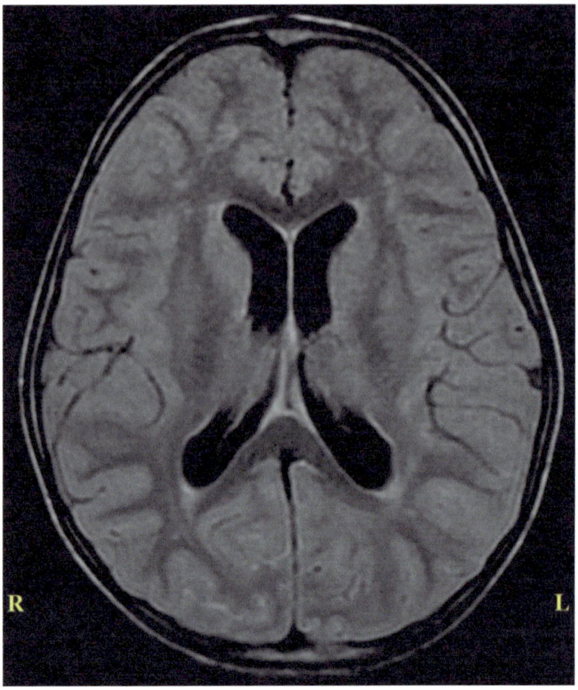

Fig. 10.1 MRI axial post-contrast image showing sulcal enhancement in the right posterior cortex consistent with the clinical diagnosis of meningitis

The number of leukocytes may range anywhere between 250 and 100,000 per cubic millimeter. A considerable increase in CSF cell count may also suggest the possibility of rupture of a brain abscess into a ventricle. Partially treated meningitis may present with the predominance of mononuclear cells. In the majority of ABM, the CSF pressure will be elevated and will be above 15 mm Hg (180 mm H_2O). CSF protein content will be more than 45 mg/dL and the usual range will be between 100 and 500 mg/dL. CSF findings for ABM, tuberculous meningitis, and viral meningitis are shown in Table 10.2. Gram stain and culture of the CSF may help in identifying the aerobic pathogen and may assist in drug sensitivity.

10.3 Complications of Bacterial Meningitis

Hearing loss, motor deficit, cognition defect, and impaired speech are common neurological complications seen among adults. Such complications may occur at any time during the course of the disease and even after the completion of drug therapy. Neurological complications may be sudden or gradual in onset and may present as focal or generalized. In some cases, the complications may manifest or persist as a long-term sequel. Post-meningitis complications may occur in about 50% of all cases, of which 80% may present with the neurological sequel. The list of neurological and non-neurological (systemic) complications of meningitis is depicted in Table 10.3.

Table 10.2 CSF findings for ABM, tuberculous meningitis, and viral meningitis

Parameters	Acute bacterial meningitis	Tuberculous meningitis	Viral meningitis
Opening pressure (*n* = <20 cm in adults)	Elevated	Elevated	Normal or elevated
Appearance (*n* = clear)	Cloudy and purulent	Yellow or cloudy	Clear or cloudy
Cells/mm^3 (*n* = <5/mm^3)	High >2000/mm^3	Increased 50–500	Normal or increased 0–500
Cell type	Neutrophils	Lymphocytes	Lymphocytes
CSF glucose (*n* = >50% plasma)	Very low or absent <1 mmol/L	Low	Normal
CSF protein (*n* = <0.5gm/L)	High 1–2	High or very high 1–5	Normal or moderately high 0.5–1.0
Diagnosis confirmation	Gram stain and culture	Ziehl–Neelsen stain, polymerase chain reaction (PCR), and culture	PCR and/or culture

Table 10.3 Neurological and non-neurological (systemic) complications of meningitis

Meningitis	Neurological complications	Non-neurological (systemic) complications
Acute onset	• Altered sensorium/Coma • Elevated intracranial pressure and cerebral edema • Seizures of acute onset • Cranial nerve palsies • Subdural empyema or effusion • Hydrocephalus • Sensorineural deficit • Motor deficit such as hemiparesis or quadriparesis • Blindness	• Prolonged pyrexia • Sepsis progressing toward septic shock and disseminated intravascular coagulation • Vasomotor depression or collapse • Loss of airway reflexes • Respiratory arrest • Pericardial effusion • Endocrine dysfunction • Hyponatremia
Late onset	• Late-onset epilepsy disorder • Hearing loss • Ataxia • Neurobehavioral, cognitive, and learning disabilities	

10.4 Medical Management

ABM is a medical emergency. Prompt diagnosis and early treatment with antimicrobials is the mainstay of management of ABM. The early treatment is based on a presumed diagnosis of ABM, and within a matter of 20–30 minutes after initial evaluation of the patient, it is important to straight away start treatment with antimicrobials. The entire possible diagnostic test should be carried out at the earliest.

Ongoing investigations, including a lumbar puncture or CT scan, should not be a factor for the delay of treatment using antimicrobials.

Ceftriaxone or another extended-spectrum cephalosporin and cefotaxime are the drugs of the first choice for adult patients. Soluble penicillin, in combination with chloramphenicol, can be an alternative if the abovementioned medications are not available. However, a previous history of anaphylaxis is a contraindication for penicillin but not a history of a rash. Patients who are quite elderly may need additional antibiotic cover, including gentamycin, for some gram-negative strains. Modifications or additions in antimicrobials can be based on the laboratory-based bacteriology stains and cultures. In addition to the abovementioned antimicrobial treatments, supportive measures during the early stage may include oxygen, maintenance of normal blood pressure, careful rehydration (as less than 1–2 liters in the first 24 hours), control of fever, urinary output, electrolyte balance, and control of pain.

The duration of initial medical management can range anywhere between 1 and 2 weeks. In general, the prognosis of this medical condition depends on prompt and early diagnosis and appropriate treatment. However, fulminant meningococcemia and untreated bacterial meningitis have a high mortality rate, and similarly, around 20–30% ABM cases may develop residual neurologic defects. The case-fatality ratio in ABM is highest in elderly adults and neonates. Loss of hearing, motor dysfunctions, impaired cognition, visual disturbance, and seizures are some of the main neurological deficits following ABM.

10.5 Physiotherapy Management

ABM produces a wide range of symptoms, which may range from a conscious and cognitively sound patient with no neurological deficits to a patient who is delirious or deeply comatose with or without decorticated or decerebrated rigidity. Even the motor deficits seen in patients can range from monoplegia to quadriplegia and may present with a certain amount of abnormal muscle tone, i.e., flaccidity, spasticity, or rigidity.

Each patient must be carefully assessed in order to collect a clear and definite clinical picture. Since the symptoms and features are different from one to another patient, the treatment should be symptomatic and individualized. In broad terms, it consists of keeping the airway clear, keeping the joints mobile during the early phase of management, and treating the residual disabilities when the meningitis resolves. For all practical purposes, it is considered that the patients can be treated as having hemiparesis or quadriparesis and that rehabilitation should essentially be along functional lines.

10.5.1 Evaluation

Evaluation forms an integral component of physiotherapy management. For unresponsive and non-mobile patients, the focus during initial sessions of management includes an assessment of the vital functional status, level of consciousness, and responses to sensory inputs. For arousable patients, the initial session should focus on the assessment of cognitive functions, behavior, motor abilities, and identifying the intactness of the sensory channels, with a superficial assessment of vital parameters and level of consciousness. A more detailed evaluation is to be incorporated as the treatment sessions continue. In the later stage, the purpose of evaluation should be focused on observing and analyzing the functional abilities of the patients.

ABM, viral encephalitis, and meningoencephalitis are certain neurological conditions that can change the patient's level of consciousness (LOC). For patients in vegetative or stupor or confused status, the Glasgow coma scale (GCS) can be used to assess the LOC. A constant record of patient's LOC may display a pattern of peak awareness at a particular point in the day and may also provide information about patients' ability to communicate. For the optimal or near-optimal benefit of treatment, the sessions should be delivered when the patient's awareness is at the maximum.

The therapist should monitor the baseline heart rate, respiratory rate, and blood pressure prior to the commencement of therapy. The vital functions value during the session and the time taken for the parameters to return to baseline should also be monitored. The pattern of respiration, variations in the pattern of breathing, and the body temperature are few other parameters to be noted. The autonomic nervous system's reaction to diverse incoming sensory stimuli can be evaluated if the patients show features of autonomic dysfunction. For those with considerable neurological manifestations, a gross assessment of the oromotor function can be included.

With regard to motor abilities, the assessment should focus on both quantitative and qualitative aspects of motor performance. The patient's ability to move within the bed (bed mobility), perform transitions such as lying to sitting and sitting to standing, maintain static and dynamic posture, and ability to perform basic functional activities within each posture should be assessed. If any limitations are observed, then the causative factors need to be explored, which may include abnormal tone, muscle weakness, and reduction of joint range of motion.

The general goals for managing inflammatory or infective conditions of CNS like ABM, tuberculous meningitis, or viral encephalitis are as follows:

1. Facilitating homeostasis of the autonomic nervous system and vital functions
2. Promoting integration of sensory input
3. Promoting normal posture and postural symmetry
4. Promoting normal righting and equilibrium reactions and protective reactions
5. Facilitating transitions from lying to sitting and sitting to standing and ambulation
6. Encouraging normal movement patterns for functional independence
7. Encouraging normal psychosocial and cognitive responses

10.5.2 Acute Stage Management

During the acute stage, if the patient is unresponsive or unconscious, the positioning of the head and neck, trunk, and extremities in the most favorable way without interference by the spastic muscles is essential. Regular evaluation of the joint range of motion followed by a full range of gentle passive movements on an everyday basis, specifically emphasizing certain areas where tightness or contractures for muscles are anticipated, such as the calf and hamstring muscles, should be incorporated. Night splints, positioning of limbs, footboard, prolonged stretching, and serial castings may yield to some extent in correcting severe tightness and mild contractures. Frequent monitoring of the skin where the splints are applied is essential as abnormal muscle spasms or increased tone may cause skin abrasions. However, for those with severe contractures and deformities which are not amenable to therapy, surgical correction may be required. Vigorous passive movements or stretching exercises are not advisable as they may lead to myositis ossificans. In case of suspected myositis ossificans, passive movements or active or active-assisted movements of the affected area need to be discontinued.

Chest physiotherapy is of utmost importance for those patients who have difficulty in clearing secretions. Sputum can be mobilized toward the proximal airways by postural drainage and chest manipulations. Once the secretions are moved more proximally, they should be successfully coughed or suctioned out. Mucolytic agents may be useful if the secretions are thick and viscous. During chest physiotherapy, a pulse oximeter can be used to monitor the peripheral oxygen saturation (SpO2).

Numerous sensory inputs can be used to move the patient from sympathetic to parasympathetic states or vice versa. For facilitating normal movements and postures, an alert and action-oriented state (sympathetic) is essential. To move the patient to a more sympathetic state, brief bursts of sensory input either presented singly or in rapid, repetitive, irregular sequences are required. Exteroceptive inputs such as a light touch, moving touch, or quick icing may facilitate the desired arousal. A movement response or increased attention to the sensory input is a positive response (sympathetic state) that the patient may exhibit to the sensory input. Rapid vestibular input (such as rocking in a rocking chair or certain vestibular ball exercises), gustatory inputs like sour taste, bright lights and colors, clear and loud voice that is not monotonous, and music with rich beats can be tried to promote arousal. Further details regarding enhancement of arousal and alertness of the patient are deliberated under the coma stimulation section of chapter "Traumatic Brain Injury."

For those patients who cannot tolerate vestibular activities, care must be taken not to facilitate the same as it may induce nausea and vomiting. Similarly, care has to be taken not to cross the threshold, as overstimulation may elicit a sympathetic fight or flight response. In such a situation, the patient may perceive the sensory inputs as a threat and may exhibit maladaptive responses. For promoting a calm state, noise-free environment, soft voice communication, gentle manual rocking, slow rhythmic and repetitive movements, and slow and melodious music can be attempted. Thus, a careful manipulation of the autonomic nervous system is essential to promote appropriate adaptive behaviors, i.e., the ability to respond appropriately to a single sensory system input as well as to respond to the input in the presence of multiple systems input. Begin with a controlled introduction of sensory inputs so that the patient progresses toward the ability to deal with multiple inputs. Visual cues and verbal cues, along with proper handling techniques, including proprioceptive and exteroceptive stimulation, will help the patient to produce the correct response.

Augmentation of normal tone is essential for facilitating normal postural reactions. The presence of hypertonicity (spasticity or rigidity), hypotonicity, and dystonia may hinder or prevent the attainment of normal postures and movement patterns. Facilitating tone and movements of the antagonist muscle groups, reflex inhibiting patterns of movements, prolonged icing, prolonged stretching, weight-bearing, and slow and rhythmical vestibular stimulation are treatment strategies or techniques advised for managing spasticity. Relaxation techniques, slow rhythmic passive movements, and gentle rocking movements including rotation and counter-rotation of the upper trunk with respect to the lower trunk are some strategies that can alleviate rigidity. In case if the muscle tone is reduced or absent, facilitatory techniques followed by strengthening exercises can be tried.

10.5.3 Intermediate Stage Management

When the acute stage settles and consciousness returns, the indications on whether the patient requires further therapy become apparent. Some patients show spontaneous improvement with the rapid return of normal functions. Others may show definite physical deficits, and for such patients, daily therapy is warranted. The physiotherapist aims to inhibit abnormal movement patterns and to facilitate the automatic reactions and normal movements essential for functional activities. Treatment should be directed toward postural control, bed mobility, and early sitting. Early sitting encourages the patients to engage in activities that help in improving head and neck control, trunk balance, and weight shifts and even facilitates eye-hand coordination.

Exercises to improve the core strength like bridging and tentacle exercise and pelvic rolling are to be incorporated soon as the patient can actively participate in the therapy sessions. If the patient has hemiparesis, rolling toward the affected side can be at ease, yet the therapist should encourage the patient to roll to either side. Prone progression on a therapeutic mat may encourage postural control and stability. Conversely, such postures may not be advisable for frail or elderly patients. Reach-out activities in each position are also to be encouraged for effective weight shifts.

10.5.4 Later Stage Management

Physiotherapy needs to be continued along the lines described earlier, and in this stage, the aim is to obtain optimum functional ability. In the small proportion of patients with the presence of motor deficits, the treatment can also be directed toward strengthening the muscles of the trunk, shoulder, and pelvic girdle and upper and lower limbs for facilitating functional movements. For those subjects with ataxic features, coordination exercises and postural stability exercises including rhythmic stabilization, slow reversal, and slow reversal hold techniques to be used. While a number of the patients learn to walk satisfactorily without walking aids or assistive devices, a few may need short leg calipers or foot drop splints to hold the foot in a better position for an acceptable gait.

10.6 Encephalitis

Encephalitis is defined as the presence of an inflammatory process in the brain parenchyma associated with the clinical evidence of brain dysfunction. This acute inflammation is caused by viral, bacterial, or fungal infection or due to a defective immune system. Among all, viral infection is the most common cause. Encephalitis should be distinguished from encephalopathy, where the latter is defined as a disruption of brain function that is not due to direct structural or inflammatory process. Encephalopathy is mediated through metabolic processes and can be caused by systemic organ dysfunction, intoxications, or drugs.

10.6.1 Pathology

In viral encephalitis, the entry of viruses to the CNS may occur by either hematogenous or neuronal routes. The hematogenous spread of viruses may result in an altered blood–brain barrier. Viruses such as the Japanese encephalitis virus and the West Nile virus gain access to the nervous system via hematogenous route. In acute encephalitis, inflammation of cortical vessels is a striking pathological finding, occurring primarily in the gray matter or gray-white junction. Passive transfer of virus across the choroid plexus or active replication of the virus in capillary-endothelial cells leads to perivascular lymphocytic infiltration. An abnormal increase in the number of glial cells (gliosis) due to the destruction of nearby neurons is a prominent histopathological finding during the progression of the disease.

Contrary to hematogenous routes taken by abovementioned viruses, the herpes simplex virus and rabies virus enter the nervous system by intraneuronal routes. The olfactory tract is one of the possible routes via which the herpes virus accesses the brain. Once the virus reaches the brain, subsequent replication may occur within the neuronal cells or other cells or in extracellular space.

10.6.2 Clinical Features

The condition generally begins with fever and headache and the symptoms may rapidly worsen. Based on the severity, the patients may have symptoms such as photophobia, nuchal rigidity, nausea, vomiting, stiffness of limbs, clumsiness and slow movements, memory loss, speech abnormalities, hearing issues, and hallucinations. Confusion, drowsiness, convulsion, altered level of consciousness, and coma are other important symptoms. Herpes simplex virus, Epstein-Barr virus, and arboviruses are the common causative viral agents. Among adults, older adults and subjects with weakened immune systems are more likely to get encephalitis. The condition can be life-threatening, and the mortality depends on some factors, including the severity of the disease, specific pathogen, and age. The immunological status of the patient is also another factor that determines the mortality of the disease. Rabies, a zoonotic disease caused by a rhabdovirus, to date remains one of the very few human infections with a near 100% mortality rate. Conversely, the judicious use of passive and active immunization, even after exposure to the virus infection, can prevent the disease.

10.6.3 Laboratory Findings

CSF examination following a lumbar puncture may reveal higher than normal levels of proteins and pleocytosis. It can be inconclusive or might show normal findings too. Polymerase chain reaction (PCR) technology for nucleic acid detection and viral culture from CSF may help in identifying the causative virus. CT scan may detect changes in brain structure. However, MRI being more sensitive and specific than CT detects early changes in the inflammatory processes due to viral encephalitis. The superior anatomical resolution, non-ionizing radiation, multiplanar imaging capability, and the enhanced contrast of soft tissue are certain advantages of MRI over CT as a diagnostic tool. Although the electroencephalograph (EEG) is generally regarded as a nonspecific investigation, in certain situations, it can be a useful diagnostic tool. For instance, the presence of periodic lateralized epileptiform discharge from one or both of the temporal lobes, which is in addition to background slowing, is a common feature in herpes simplex encephalitis. A blood test may be helpful for identifying certain viruses.

10.6.4 Medical Management

Though specific therapy is limited to certain viral agents, immediate and accurate diagnosis and the introduction of appropriate treatment have improved the survival and reduced the extent of permanent brain injury in survivors. With regard to symptomatic treatment, corticosteroids to reduce the cerebral edema, anticonvulsants to control the seizures, sedatives for seizures, restlessness, and irritability, an adequate amount of fluid administration, mechanical ventilator or oxygen support for breathing difficulty, and medications to control pyrexia and headache are the standard means.

10.6.5 Physiotherapy Management

To date, there is a dearth of evidence of literature or research articles discussing the role, guideline, or effectiveness of physiotherapy for encephalitis patients. Considering the similarity in the clinical presentation with meningitis, the therapy can be in line with that of ABM management. Based on the clinical presentations and the problem lists, the short- and long-term realistic goals have to be set before the commencement of the therapy. In the early stage of rehabilitation, preventing chest complications, bedsores, and deep vein thrombosis, correcting the possible deformities secondary to tonal abnormalities and lack of mobility, and normalizing the tone can be the goals of treatment. In the later stages, the focus can be on encouraging normal postural reactions, promoting voluntary movement patterns, and improving the functional capacities.

Chest manipulations and suctioning will be necessary for those patients who are on ventilatory support for maintaining bronchial hygiene. For conscious patients either on room air or on oxygen supplementation, breathing exercise, thoracic mobility exercise, and inspiratory muscle training can be provided to minimize the possibility of pulmonary complications like retention of secretions and atelectasis. For comatose patients, coma stimulation can be given to improve their level of arousal. Early mobilization is advised as soon as the patient is medically and hemodynamically stable. Therapists need to follow the same strategies advocated in other neurological conditions to prevent bedsores, deep vein thrombosis, and deformities. Once the level of cognition improves, it is advisable to encourage sit-to-stand transitions and other functional activities to promote functional independence within the limits imposed by the disease.

10.7 Brain Abscess

Brain abscess is defined as a focal, intracerebral infection that begins as a localized area of cerebritis and develops into a collection of pus surrounded by a well-vascularized capsule. The brain is remarkably resistant to bacterial and fungal infection and brain abscesses are quite uncommon despite the frequency of both overt and occult bacteremia. Rich blood supply and the relatively impermeable blood–brain barrier formed by the capillary-endothelial tight junctions are the possible factors that enhance the resistance of the brain.

10.7.1 Pathology

Certain brain pathologies such as prior stroke, intracerebral hematoma, and neoplasm may serve as a point of origin or focus of infection for abscess formation. However, in most cases, there is no apparent predisposing brain lesion. Based on the likely entry point of infection, the brain abscesses are classified into direct or indirect cranial infection. Direct extensions can be from the adjoining structures such as paranasal sinuses, teeth, and middle ear. Seeding of the brain may occur when the infecting bacteria traverse through the valveless emissary veins that drain these regions into the venous drainage systems of the brain. Direct entry of bacteria can occur subsequent to penetrating brain injury. However, the odds of brain abscess following such injuries are remarkably low. Hematogenous spread from the primary infection in the extracranial source such as the heart, lung, abdomen, or pelvis may cause an indirect or metastatic seeding of the brain. Despite all these potential routes, approximately 30% of cases are identified as "cryptic" brain abscesses for which no obvious source can be identified.

Streptococcus aureus, *Staphylococci aureus*, *Klebsiella pneumoniae*, and *Mycobacterium tuberculosis* are some of the common pathogens responsible for pyogenic brain abscess. Species of Aspergillus or Candida and mucormycosis are the usual pathogens causing fungal brain abscess. Several protozoans and helminths have been reported to produce brain abscesses, including neurocysticercosis, caused by the larval form of Taenia solium.

10.7.2 Clinical Features

The virulence of the organism, the site and size of the brain abscess, and the presence of any underlying systemic conditions are the usual factors influencing the clinical presentation. In pyogenic brain abscess, for almost all patients, headache is the most obvious presenting symptom. In most cases, the headache may not have any distinguishing features and can be dull aching, which is poorly localized. Less than 50% of cases may present with fever during the initial clinical presentation. The absence of fever during the time of admission should not be used as a criterion to exclude the diagnosis. An extremely severe headache of abrupt onset is more likely to be a symptom of ABM or subarachnoid hemorrhage. Sudden worsening of a preexisting headache with a new onset of meningismus can be an indication of the rupture of the brain abscess into the ventricular space, which can often be fatal. Signs of increased intracranial pressure such as nausea, vomiting, drowsiness, lethargy, and stupor may occur and indicate an advanced situation that requires immediate attention. Approximately, in 30–50% of cases, depending on the location of the brain abscess, the patient may present with focal neurological deficits like hemiparesis, hemisensory loss, aphasia, and ataxia. In one-fourth of the cases, papilledema can be present and indicates severe intracranial pressure requiring immediate CT scanning and neurosurgical consultation. The prognosis tends to be poor if presented with significant alterations in the mental status.

10.7.3 Laboratory Findings

Similar to the clinical presentations, there are no laboratory data that are pathognomonic of brain abscess. Patients frequently present with normal leukocyte counts and erythrocyte sedimentation rate. Blood cultures tend to be negative in most situations but rarely may yield in identifying the causative pathogen. A review of the older medical records of the patients may allow the clinicians to uncover evidence of prior bacteremia or infection that led to the current brain abscess. CSF examination usually reveals nonspecific elevation of proteins and cell count, but these parameters may be normal. Rarely, the CSF cultures may identify the causative organism. Due to the potential risk of brainstem herniation, the performance of lumbar puncture can be fatal for patients with brain abscess and may not provide further useful clinical information. The culture of abscess material obtained following surgery provides the best opportunity to make a microbiological diagnosis.

Contrast CT scanning is more helpful in visualizing the lesions than the noncontrast scan. Scans can be difficult in differentiating metastatic brain abscess from metastatic tumor lesions. Soft-tissue resolution and superior imaging details in the MRI scan make it the first imaging procedure of choice for patients suspected of having brain abscess. On T1-weighted sequences of MRI scans, the abscess appears as a hypointense lesion, and following administration of intravenous gadolinium, a ring enhancement will be visible. For T2-weighted sequences, the abscess reveals a hyperintense central area of pus surrounded by a well-defined hypointense capsule and surrounding edema. Magnetic resonance (MR) axial fluid attenuated inversion recovery (FLAIR) and T2 images showing focal ring-like lesions consistent with brain abscesses or tuberculomas are depicted in Fig. 10.2. Figure 10.3 depicts the MR axial post-contrast FLAIR image illustrating tiny focal ring-enhancing lesions consistent with neurocysticercosis lesions (parasitic).

10.7.4 Treatment

Therapeutic drainage of the abscess contents by stereotactic needle aspiration provides diagnostic specimens for culture and studies. Based on the gram stain results and the presumptive source of the abscess, empirical antimicrobial therapy can be started. However, for a better result, attempts have to be made to isolate the causative pathogen from the specimen for a more selective antimicrobial therapy. Surgical treatment (open craniotomy and drainage or excision of the lesion) of brain abscess have largely been replaced by a minimally invasive, closed-drainage procedure performed under mild sedation and local anesthesia. For the deeper and critical

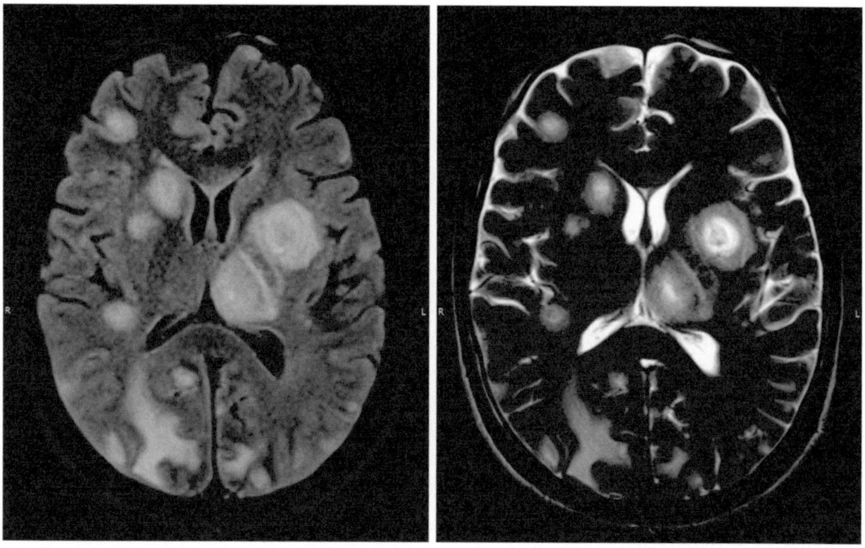

Fig. 10.2 MR axial FLAIR and T2 images showing focal ring-like lesions consistent with brain abscesses or tuberculomas

Fig. 10.3 MR axial post-contrast FLAIR image illustrating tiny focal ring-enhancing lesions consistent with neurocysticercosis lesions

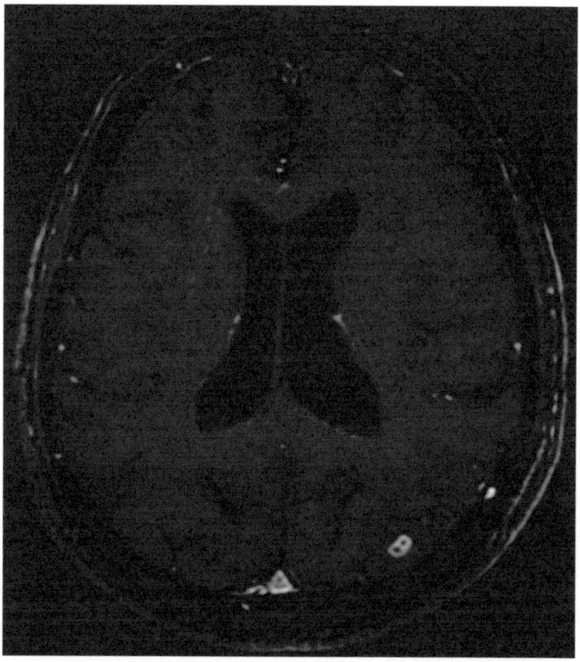

regions, stereotactic biopsy and drainage procedure can be performed. In cases of more resistant pathogens or multiloculated abscesses, a complete excision of the abscess and the surrounding capsule is preferred over the minimally invasive closed-needle aspiration procedures. Following appropriate medical or surgical intervention, many of the brain abscess patients typically have limited deficits and may require less intense physiotherapy. Patients may present with mono/hemiparesis and the strategies used in conditions like stroke can be incorporated to rehabilitate such patients. The author would like to sensitize the readers that overall sparse evidence is available with regard to physiotherapy management of patients following infections of the central nervous system. Until the research findings regarding the relevance and importance of physiotherapy treatment for these patients are more definitive, integrating standard principles of exercise prescription with physiotherapy examination findings, rehabilitation framework, and a thorough understanding of the nature and course of the disease continues to be the best practice.

Further Reading

Bamberger DM. Diagnosis, initial management, and prevention of meningitis. *Amn Fam Physician.* 2010;82:1491–8.

Christie S, Chan V, Mollayeva T, Colantonio A. Systematic review of rehabilitation intervention outcomes of adult and paediatric patients with infectious encephalitis. *BMJ Open.* 2018;8(5):e015928. Available from: https://doi.org/10.1136/bmjopen-2015-010754.

Daroff RB, Jankovic J, Mazziotta JC, Pomeroy SL, editors. Bradley's Neurology in Clinical Practice. 7th ed. London: Elsevier; 2016.

Gray LD, Fedorko DP. Laboratory diagnosis of bacterial meningitis. *Clin. Microbiol. Rev.* 1992;5:130–45.

Howlett WP. Neurological Infections. Neurology in Africa: Clinical Skills and Neurological Disorders. Cambridge: Cambridge University Press; 2015. p. 119–58.

Jelsma J, Lilley H, Smith H, Mielke J, Hakim J, Gangaidzo I, et al. The impact of meningitis on the motor function of adult survivors in Harare, Zimbabwe. *S Afr J Physiother.* 1999 Feb;55(1):6–10.

Liggins CA. The role of physiotherapy in the management of tuberculous meningitis in children. *S Afr J Physiother.* 1977 Mar;33(1):5–8.

Mathisen GE, Johnson JP. Brain abscess. *Clin Infect Dis.* 1997;25:763–9.

Mustafa M, Iftikhar M, Latif MI, Munaidy RK. Brain abscess: pathogenesis, diagnosis and management strategies. *Int J Res Appl Nat Soc Sci.* 2014;2:299–308.

Ropper AH, Samuels MA, Klein JP, Prasad S, editors. Adams and Victor's Principles of neurology 10th ed. New York: McGraw Hill Education; 2014.

Siddiqui EU. Neurologic complications of bacterial meningitis. In: Wireko-Brobby G, editor. Meningitis. Croatia: InTech Publishing; 2012:36–44.

Steiner I, Budka H, Chaudhuri A, Koskiniemi M, Sainio K, Salonen O, et al. Viral encephalitis: a review of diagnostic methods and guidelines for management. *Eur J Neurol.* 2005;12:331–43.

Tunkel AR, Hartman BJ, Kaplan S L, Kaufman B A, Roos KL, Scheld WM, et al. Practice guidelines for the management of bacterial meningitis. *Clin Infect Dis.* 2004;39:1267–84.

Whitley RJ, Gnann JW. Viral encephalitis: familiar infections and emerging pathogens. *Lancet.* 2002;359:507–13.

Chapter 11
Peripheral Nerve Disorders

Abraham M. Joshua and Zulkifli Misri

11.1 Introduction

The peripheral nerves are composed of sensory, motor, and autonomic components. Most peripheral nerves are mixed and contain sensory and motor as well as autonomic fibers. Each peripheral nerve resembles a "cable-like" structure consisting of bundles of myelinated and unmyelinated axon fibers, with their supporting elements. The Schwann cell wraps around the myelinated fibers several times, thereby insulating the axon with multiple layers of lipid-rich cell membrane. Except at the consistent microscopic gaps called the nodes of Ranvier, the myelinated axons are covered entirely by myelin and Schwann cells. The thick myelin sheath with low capacitance and high resistance to bioelectric current and high concentration of voltage-gated sodium channels at the nodes of Ranvier causes saltatory conduction to propagate impulses from one node of Ranvier to the next. The unmyelinated fibers are surrounded only by the plasma membrane of a Schwann cell. The unmyelinated axons have voltage-gated sodium channels along the entire length of the membrane and the nerve impulses travel continuously along the axons.

Peripheral nerve disorders are neurological impairments caused by peripheral motor, sensory, or autonomic nerve dysfunction. The causes of the disorders are dissimilar and their clinical manifestations are highly variable. Entrapments,

A. M. Joshua (✉)
Department of Physiotherapy, Kasturba Medical College of Mangalore, Manipal Academy of Higher Education, Manipal, India
e-mail: abraham.joshua@manipal.edu

Z. Misri
Department of Neurology, Kasturba Medical College of Mangalore, Manipal Academy of Higher Education, Manipal, India
e-mail: misri.zk@manipal.edu

A. M. Joshua (ed.), *Physiotherapy for Adult Neurological Conditions*,
https://doi.org/10.1007/978-981-19-0209-3_11

systemic diseases, autoimmune, inflammatory, infective, or inherited disorders, ischemic situations, nutritional deficiencies, and toxins are some of the common causes of the same. A detailed history, comprehensive neurological examination, and appropriate electrophysiological studies including ancillary testing like skin biopsy, nerve biopsy, and tests for autonomic dysfunction, comprise a sound systematic diagnostic approach to peripheral nerve disorders. Such an approach helps to identify the deficit type, fiber type, pattern, and time course of the peripheral neuropathies and shortens the list of diagnostic and causative possibilities.

11.2 Historical Background

In the second century, Aelius Galenus, a physician, surgeon, and philosopher of the Roman Empire, discovered that nerves conveyed sensation and motility to the extremities. Much later, in the sixteenth century, Andreas Vesalius, the Flemish anatomist and physician, provided the detailed anatomical course of the nerves. It was Antonie van Leeuwenhoek, a Dutch businessman and the pioneer microscopist, who probably first documented the microscopic observations of myelinated nerve fibers. Felice Gaspar Fontana, a physicist and naturalist, in 1781 teased a nerve and using a magnifying lens revealed the elementary cylinders. Fontana's observations were corroborated by Louis-Antoine Ranvier, the French physician, anatomist, histologist, and pathologist, who discovered the nodes of Ranvier.

Till the nineteenth century, the physiological function and the pathological states affecting the nerves were never clearly understood. In 1822, French physiologist François Magendie demonstrated limb paralysis following a section of the anterior spinal roots and loss of sensation following a section of posterior spinal roots. Robert Remak, a German-Polish physiologist was the first to describe nuclei on unmyelinated nerve fibers but Theodore Schwann, the German scientist and a colleague of Remak, got the credit for discovering them. In 1839, Schwann postulated that the long chains of nucleated cells merged to form a syncytium with a continuous band of protoplasm at the center. In 1851, Augustus Volney Waller, a British neurophysiologist, demonstrated the degeneration and regeneration of nerves in frogs.

The symptoms of diabetic neuropathy, an acquired neuropathic condition, were first narrated by Sushruta, an ancient Indian Ayurvedic physician. By the 1700s, Western physicians started studying diabetes mellitus and its related complications. John Rollo, a Scottish military surgeon, best known for his work on the diabetic diet, wrote a book titled "Cases of Diabetes Mellitus" in 1798. He mentioned the nervous system involvement in diabetes mellitus and also stated the presence of pain and paresthesia, predominantly in the legs. In 1864, Charles Jacob Marchal de Calvi, a French pathologist and physiologist, recognized diabetes as the cause than the result of nervous system disturbance. Frederick William Pavy, a British physician and physiologist, in 1885, provided a detailed description of the neuropathic features of diabetes which included numb feet, deep-seated pain in the feet, heavy legs, impaired knee jerk, muscle tenderness, and hyperesthesia. Davies Pryce, a

British surgeon, was the first to report the macro- and microscopic changes in the peripheral nerves of diabetic patients and suggested a link between diabetic neuropathy and foot ulcers.

During the mid-nineteenth century, several case reports of neuropathies were either reported or published. In 1828, Auguste François Chomel, in a medical journal, described a remarkable and obscure, epidemic of acute sensorimotor polyneuropathy. It was Robert Graves, an eminent Irish surgeon who established the disease of the peripheral nerves as a cause of paralysis, and years later, in 1854, the first monograph on diphtheritic paralysis was prepared by Maingault M. In 1878, Ernst Julius Remak described the features of polyneuritis due to lead intoxication. Though several case reports were published by Chomel, Maingault, Remak, and others on epidemic polyneuritis, diphtheria, and lead intoxication, they were uncertain about the pathological site and presumed that the central nervous system as the possible site of pathology. In 1880, François Alexis Albert Gombault, a French neurologist, stated that peripheral nerve lesions were responsible for paralysis following lead or arsenic poisoning, beriberi, diabetes, leprosy, and alcoholism.

In 1859, Jean Baptiste Octave Landry, a French physician, provided a detailed report about the ascending paralysis of a 43-year-old patient but surprisingly Landry did not examine the peripheral nerves of the patient. Later in 1864, Louis Dumenil described a case of ascending paralysis and demonstrated the pathological evidence of the atrophy of peripheral nerve tubes' medullary substance. Almost six decades after Landry published his findings of ascending paralysis, Georges Guillain, Jean Alexandre Barré, and André Strohl described an identical condition, characterized by a better prognosis and increased protein content in the Cerebrospinal fluid (CSF) without increased cell count. In 1890, Eichhorst H. was the first to describe chronic and recurrent polyneuritis and later, in 1914, Hoestermann E. reported a case series that closely resembled Chronic Inflammatory Demyelinating Polyradiculopathy (CIDP).

The earliest description of familial neuropathic foot ulcers can be traced back to a report by Auguste Nélaton, a French physician and surgeon, in 1852. Jean-Martin Charcot, a French neurologist and his assistant Pierre Marie, in 1886, published their original case series of patients who were diagnosed to have progressive muscular atrophy and postulated that the primary disturbance was in the spinal cord. Three months later, Howard Henry Tooth, a British neurologist localized the lesion to the peripheral nerves in his thesis titled "Peroneal Type of Progressive Muscular Atrophy." In 1893, Joseph Jules Dejerine and Jules Sottas, French neurologists, reported two similar cases among siblings and titled the disorder "Progressive Hypertrophic Interstitial Neuropathy of Childhood." Right from the time of description, controversy surrounded the classification of the same and subsequent observers opined that the Dejerine-Sottas disease was merely a variant of the peroneal form of progressive muscular atrophy, described earlier by Charcot, Marie, and Tooth.

In 1943, Herbert Seddon, neuroscientist and surgeon, classified peripheral nerve injuries based on the severity, prognosis, and time for recovery. Later, in 1951, Sydney Sunderland, an eminent physician and anatomist, further expanded Seddon's classification, based on the continuity of the connective tissue covering the nerve fibers. Uremic neuropathy was recognized after the introduction of dialysis and renal transplantation programs in the 1960s. During the second half of the twentieth century, Russell Brain, Henson R.A., Urich H., and McLeod J.G. meticulously studied and reported peripheral neuropathies due to paraneoplastic causes.

The ready availability of electron microscopy, advancements in the clinical electrophysiological tests and molecular genetics, the development of quantitative techniques for the histological assessment of peripheral nerves, and immunologic methods for the histopathological examination of nerves have led to a marked expansion in the understanding of the common forms of peripheral neuropathies. Lascelles R.G. and Peter Kynaston Thomas, in 1967, stated repeated demyelination and remyelination as the cause of hypertrophy of nerve trunks and reported that nerve hypertrophies are not exclusive for the Charcot–Marie–Tooth disorder. In 1968, Peter J. Dyck and Edward Lambert categorized the Charcot–Marie–Tooth disease patients into two groups, one with considerably slow nerve conduction velocities and the other with normal or near-normal conduction velocities. Major advances in molecular genetics during the past few decades have led to the identification of several gene loci and disease-causing genes, further expanding the classification of inherited neuropathies.

11.3 Basic Anatomy of Peripheral Nervous System

The peripheral nervous system is built almost entirely from nerves and can be functionally divided into the autonomic and somatic nervous systems. The autonomic system is further subdivided into the sympathetic and parasympathetic sub-systems and the somatic nervous system is subdivided into the sensory and motor divisions. Each peripheral nerve fiber is an extension of a neuron whose cell body is located either within the grey matter of the central nervous system or within the dorsal root ganglia or cranial nerve ganglia of the peripheral nerve. Those peripheral nerves that carry information toward the brain and the spinal cord are called the afferent or sensory neurons, while the ones transmitting impulses from the brain and the spinal cord to the periphery are known as the efferent or motor neurons.

The afferent neurons transmit a variety of impulses from the sensory receptors/ sense organs which include general sensations like touch, pain, temperature, and joint position sense and special sensations like senses of smell, vision, hearing, and balance. The efferent neurons bring neural information towards the effector organs like skeletal muscles, visceral organs, and glands. Based on the anatomical location of exit from the central nervous system, the peripheral nerves can also be classified as "cranial" or "spinal." The cranial nerves arise intracranially from the brain and the brainstem, whilst the spinal nerves emerge from the spinal cord. There are 12 pairs of cranial nerves and 31 pairs of spinal nerves.

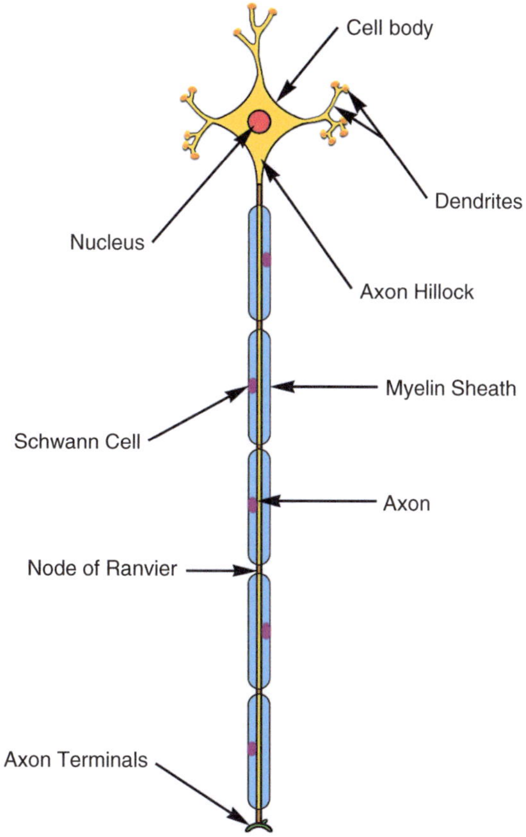

Fig. 11.1 An illustration of basic parts of the neuron

Figure 11.1 illustrates the basic parts of a neuron. Each peripheral nerve contains bundles of nerve fibers, either axons or dendrites, surrounded by connective tissue, i.e., the sensory nerves contain long dendrites of sensory neurons and the motor nerves have long axons of motor neurons. Mixed nerves contain both afferent and efferent nerve fibers. A connective tissue sheath called the epineurial sheath surrounds each nerve. Each bundle of nerve fibers is called a fascicle. Each fascicle is surrounded by a layer of connective tissue called the perineurium. Within the fascicle, each individual nerve fiber, with its myelin and neurilemma, is surrounded by connective tissue called the endoneurium. In addition to the above, the blood vessels are also seen enclosed in its connective tissue wrappings. Figure 11.2 illustrates the epineurial, perineurial, and endoneurial coverings, the fascicles, and the blood vessels of a nerve trunk.

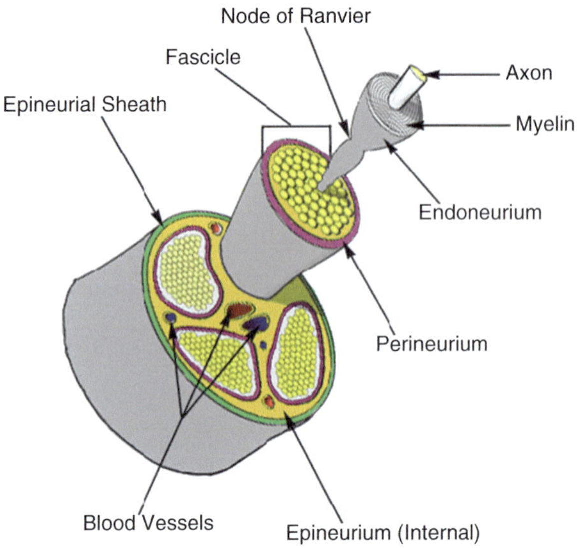

Fig. 11.2 Illustration of the epineurial, perineurial, and endoneurial coverings, the fascicles, and the blood vessels of a nerve trunk

11.4 Pathological Processes and Electrophysiological Changes

The symptoms, the type, and the pattern of distribution of signs, and the electrophysiological study abnormalities provide information about the underlying pathological changes. Though the causes of neuropathy are many, the pathological reactions of peripheral nerves to different types of insults remain limited and are categorized into four: (1) Wallerian degeneration, following axonal severance or injury, (2) Axonal degeneration or axonopathy, (3) Primary neuronal (perikaryal) degeneration or neuronopathy, and (4) Segmental demyelination or myelinopathy. Figure 11.3 illustrates the four categories of peripheral nerve lesions and the descriptions are given below.

Trauma to the nerve due to compression, traction, laceration, and thermal or chemical agents and ischemia causes the disruption of axons and leads to Wallerian degeneration, and immediately results in sensory and motor impairments in the distribution of the injured nerve. In complete nerve injury, needle Electromyography (EMG) reveals a total loss of voluntary activity and for partial lesion, there is a decreased Motor Unit Potential (MUP) recruitment. Until 10 to 11 days post-injury, the axons remain excitable distally, beyond which the distal nerve trunk is typically unexcitable. For motor Nerve Conduction Studies (NCS), the Compound Muscle Action Potential (CMAP) amplitude, evoked by distal stimulation, begins to decline by the second-day post-injury and reaches its nadir by the fifth or sixth day. For sensory NCS, the amplitude of the Sensory Nerve Action Potential (SNAP) remains normal for 5–6 days when evoked distally and then rapidly declines to reach its nadir by the tenth or eleventh-day, post-injury. The temporal sequence of Wallerian

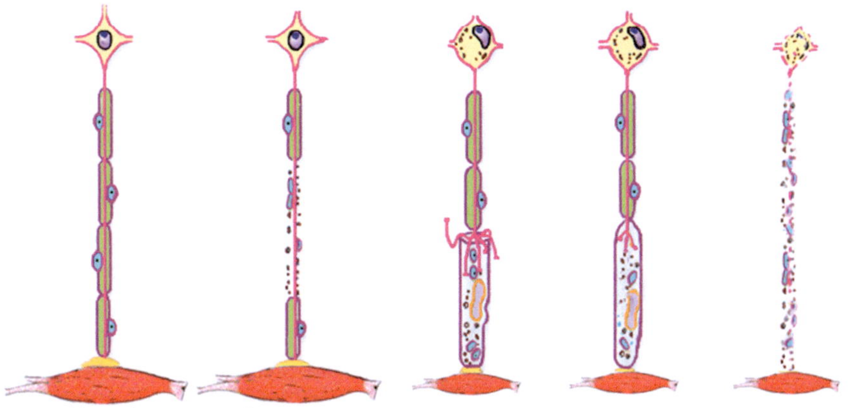

Fig. 11.3 The four categories of peripheral nerve lesions

degeneration is length-dependent and it occurs earlier in the shorter distal nerve stumps compared to the longer stumps. By 10 to 14 days post-injury, mainly in the affected proximal muscles, needle EMG may demonstrate fibrillation potentials and positive sharp waves (denervation potentials), which will be quite noticeable after 3 weeks.

Axonal interruption leads to secondary morphological changes of the nerve cell body, known as chromatolysis, and in addition to the above, there will be a reduction in the proximal axonal caliber. The regeneration process from the proximal stump may begin as early as 24 hours following the nerve transection; however, it may progress slowly at a rate of 1–3 mm/day and is often incomplete. In partial lesions, about one-month post-injury, locally at the muscle level, the needle EMG may demonstrate evidence of axonal sprouting from the intact axons. Typically, the quality of recovery depends on the distance between the site of injury and the nerve cell body, degree of preservation of the Schwann cell, basal lamina tube, nerve sheath, surrounding tissue, and the patient's age.

The distal axonal breakdown seen in axonopathy or axonal degeneration is apparently caused by neuronal metabolic derangement or ischemia due to vascular compromise. Systemic metabolic disorders, exposure to toxins, vasculitis, and certain inherited neuropathies are the usual causes of the same. The axonal degeneration process is accompanied by concomitant myelin sheath breakdown. Typically, the degenerative process starts distally on the nerve fiber and progresses toward the nerve cell body, a term known as "dying-back" or "length-dependent" polyneuropathy. Failure of the perikaryon to synthesize enzymes or structural proteins, changes in axonal transport, or regional disturbances of energy metabolism are certain plausible factors for the selective length-dependent vulnerability of distal axons.

The dying-back polyneuropathy usually presents with symmetrical distal sensory and motor loss in the lower limbs and tends to progress proximally in a graded manner. The stocking-like pattern of sensory loss, distal muscle weakness and atrophy, and loss of distal limb stretch reflexes are the important characteristics of length-dependent polyneuropathy. As the disease ascends, the involvement of hands and distal upper limbs leads to glove-like sensory loss and weakness, and atrophy of the hand muscles. Generally, the axonopathies reveal low-amplitude sensory and motor nerve action potentials; however, the distal latencies and conduction velocities are marginally affected. Needle EMG of distal muscles of the limbs will demonstrate acute and/or chronic denervation changes. Even in axonopathy, the axonal regeneration process proceeds slowly at a rate of 1–3 mm/day and hence the recovery may be delayed and is often incomplete.

In neuronopathy, either the anterior horn cells or the dorsal root ganglion cells may be involved. It is characterized by the loss of nerve cell bodies with the subsequent degeneration of their entire peripheral and central axons. In neuronopathy affecting the anterior horn cells, focal weakness in the absence of sensory loss is the striking feature. Damage to the dorsal root ganglion neurons due to sensory neuronopathy or dorsal polyganglionopathy causes sensory ataxia, sensory loss, and diffuse areflexia. Several toxins like organic mercury compounds and high-dose pyridoxine, deficiency states like vitamin E deficiency, and immune-mediated inflammatory damages seen in paraneoplastic sensory neuronopathy can produce primary sensory neuronal degeneration. Once the pathological processes involving dorsal root ganglion cells are no longer active, the sensory deficits tend to become fixed with minimal or no further recovery.

The term myelinopathy or segmental demyelination denotes the breakdown of myelin with the sparing of axons because of damage to either myelin sheaths or Schwann cells. Mechanical injury by acute or chronic nerve compression or entrapment, immune-mediated demyelinating neuropathies, myelinotoxic agents like diphtheria toxin, and hereditary disorders of Schwann cell or myelin can cause myelinopathy or segmental demyelination. Typically, within weeks the remyelination of the demyelinated segments tends to occur. The freshly formed remyelinated segments have shorter internode distances and thinner-than-normal myelin sheaths, due to which such segments have a slower conduction velocity. The proliferation of several layers of Schwann cells around the axon in response to repeated episodes of demyelination and remyelination, as seen in progressive disorders, can also lead to onion-bulb formation.

Conduction block is the physiological consequence of acquired demyelination. It results in the inability of the action potential to reach the muscle and thereby produces weakness. Since the axon remains intact, obvious muscle atrophy is less likely. Relative sparing of pain and temperature sensations in most of the demyelinating neuropathies explains the preserved function of the thin myelinated and unmyelinated fibers in segmental demyelination. Early generalized loss of deep tendon reflexes, less considerable wasting despite proximal and distal muscle weakness, neuropathic tremor, and palpable and enlarged nerve trunks are all clinical clues suggesting demyelinating polyneuropathy. In demyelination, the sensory and motor Nerve Conduction Velocities (NCVs) are reduced to less than 70% of the lower limits of normal, with relative conservation of CMAP and SNAP amplitudes. Temporal dispersion of CMAPs, marked prolongation of distal motor latencies, and prolongation of F-wave latencies are all features consistent with acquired demyelination. In segmental demyelination, the recovery depends on the magnitude of remyelination and typically the clinical improvement may occur within weeks. In some cases, the coexistence of axonal degeneration with segmental demyelinating neuropathies may demonstrate distal limb atrophy with active denervation and reinnervation changes on needle EMG and in such cases, the clinical improvement will be delayed.

11.5 General Classification of Peripheral Nerve Disorders

Though several patterns of peripheral nerve disorders are known, distinct classifications do exist. The criteria of classification may include motor, sensory, or autonomic, based on the modality of impulses conducted by the nerve fibers, myelinated or unmyelinated based on fiber-size-involved; and distal or proximal based on the topographical distribution. In addition to the above, other criteria of classification are (1) acute, subacute, or chronic, based on the time course; (2) axonal or demyelinating, based on the pathology; (3) inflammatory, hereditary, metabolic, toxic, nutritional, paraneoplastic, and chemotherapy-induced, based on the etiology of the disorder; and (4) mononeuropathy, mononeuropathy multiplex or polyneuropathy, based on the number of nerves affected.

The term "mononeuropathy" means the focal involvement of a single nerve and denotes a local process. Entrapment, physical trauma, compression, vascular injuries, and metastatic or neoplastic infiltration are some of the usual causes of mononeuropathy. Localization of the lesion may be possible by neurological examination; however, in many cases electrodiagnostic tests help provide a more accurate localization and differentiate axonal loss from focal segmental demyelination. The term "multiple mononeuropathies" or "mononeuropathy multiplex" denote the simultaneous or sequential damage to multiple noncontiguous nerves and may give rise to muscle weakness with sensory loss which may simulate the feature of a length-dependent peripheral neuropathy.

Typically, polyneuropathy presents with symmetrical distal motor and/or sensory deficits, which is associated with a graded increase in severity distally and distal reduction of deep tendon reflexes. The sensory deficits usually follow a length-dependent stocking–glove pattern. Though many of the polyneuropathies are fairly symmetrical, some can have an asymmetrical clinical presentation. A small fraction of polyneuropathies, like the neuropathy secondary to acute intermittent porphyria, can have a predominant proximal presentation. The majority of polyneuropathies have a mixed sensorimotor presentation; however, a careful inspection may reveal some degree of autonomic dysfunction in the afflicted patients.

11.6 History and Clinical Examination as Diagnostic Clues

An appropriate and careful history often helps to differentiate acquired neuropathy from hereditary. The symptoms of peripheral nerve disorders arise as a result of motor, sensory, or autonomic disturbances. The clinician should establish the existence of peripheral nervous system disease and be able to differentiate it from the central nervous system, neuromuscular junction, or muscle disorders. It is essential to distinguish the topography of involvement and determine whether the disorder is predominantly motor or sensory or autonomic or of the mixed type. Clinical examination clubbed with appropriate electrophysiological tests also helps to identify whether the myelin sheath, the axon, or the cell body is the target of the disease.

The clinician should inquire and seek both negative and positive symptoms. Weakness, muscle wasting or atrophy, and difficulty walking are a few examples of negative motor symptoms. Whereas, fasciculations, muscle cramps, tremors, and myokymia are instances of positive motor manifestations. Early distal toe and ankle dorsiflexor weakness, which may result in tripping on loose carpets or uneven ground, can be instances of negative motor symptoms of polyneuropathies. The clinician should understand that the sheer complaint of difficulty walking does not discriminate muscle weakness from pyramidal, extrapyramidal, cerebellar, or sensory disturbance. The weakness of fingers, a negative motor symptom of the upper extremity, may cause difficulty to open containers or use a key to lock or unlock.

Tingling, burning, searing, and tight band-like sensations are examples of positive sensory symptoms. Often patients may complain of paresthesia which is an unpleasant sensation arising spontaneously without any obvious stimulus. Spontaneously reported paresthesia is more commonly seen among acquired polyneuropathies compared to inherited neuropathies. Neuropathic pain, a positive sensory symptom, is a cardinal feature of many neuropathies. The neuropathic pain often presents as deep, burning, or stabbing character that may be associated with electric or shooting pains, typically increased at night or during rest periods. Experiencing pain from normally painless stimuli (allodynia) and perceiving enhanced sensitivity to noxious stimuli (hyperalgesia) are also instances of positive sensory symptoms. Negative sensory manifestations may include loss or reduction of pain, temperature, or touch sensation, often quoted by patients as numbness. Loss of kinesthetic and proprioceptive sensations, common negative sensory symptoms of polyneuropathy, can lead to imbalance and gait disturbance. Nevertheless, such negative symptoms can arise as a result of central disorders affecting the dorsal column of the spinal cord due to conditions like vitamin B12 deficiency and malignancies compressing the cord. Symptoms of autonomic dysfunction may help to direct attention toward certain neuropathies that have noticeable autonomic symptoms. It is essential to inquire about symptoms of orthostatic hypotension like dizziness, presyncopal symptoms or syncope, excessive or decreased perspiration, intolerance to heat and bladder, bowel and sexual dysfunctions. Noninvasive autonomic function studies like Quantitative Sudomotor Axon Reflex Test (QSART) and thermoregulatory sweat test can be useful to document the degree of autonomic involvement.

Historical information regarding the duration of illness and the onset and evolution of symptoms provides essential hints to the diagnosis. Information about the time course (acute, subacute, or chronic) and the course of evolution (monophasic, progressive, or relapsing) helps to narrow down the diagnostic possibilities. The Landry–Guillain–Barré–Strohl syndrome, widely known as Guillain–Barré Syndrome (GBS), neuropathy associated with acute porphyria or vasculitis, amyotrophic neuralgia, and certain toxin-induced neuropathies have acute presentations. A relapsing course is often found in neuropathic conditions like CIDP, acute porphyria, Refsum disease, hereditary neuropathy with liability to pressure palsies, hereditary neuralgic amyotrophy, and neuropathy related to repeated exposure to the toxin. A chronic slow progressive course of symptoms over several years and the presence of similar symptoms and bony deformities like high arch foot among immediate blood relatives often suggest an inherited polyneuropathy.

Constitutional symptoms like weight loss, malaise, and lack or loss of appetite can suggest an underlying systemic disorder, a cause of polyneuropathy. It is essential to inquire about coexisting or preceding medical disorders like diabetes mellitus, hypothyroidism, chronic kidney disease, chronic liver disease, malabsorption disorders, malignancy, connective tissue disorders, Human Immunodeficiency Virus (HIV) infection, medications used, personal habits like alcohol consumption, diet, and exposure to solvents, pesticides, or heavy metals.

Determining the anatomical pattern and localization of the disease process is the first step in the examination of neuropathy. In mononeuropathy, the neurological deficit follows the distribution of a single nerve. For instance, in a patient with radial nerve injury near the spiral groove, the neurological examination will reveal painless wrist drop, weakness of extensors of metacarpophalangeal joints, and sensory loss over the dorsum of the hand. In mononeuropathy multiplex, the clinical examination findings will point to simultaneous or sequential damage of two or more noncontiguous peripheral nerves. For instance, confluent mononeuropathy multiplex of median and ulnar nerves may give rise to motor weakness with sensory loss, manifested as impairments in the flexion of the metacarpophalangeal joints of all five digits, extension of the interphalangeal joints of second to fifth digits, and opposition of the thumb, atrophy of the thenar and hypothenar eminence, complete claw-hand and loss or reduced sensation over the volar aspect of the hand and digits. Examination for motor dysfunction should include an assessment of muscle strength, tone, bulk, and tendon reflexes. In case of a suspected root or peripheral nerve injury, individual muscle strength examination is more appropriate than group muscle examination to clinically locate the level of the lesion. Tables 11.1 and 11.2 provide the muscles, roots, actions, and nerve supply of the upper and lower limbs, respectively. The schematic representations of the peripheral nerve cutaneous branches for upper and lower limbs are depicted in Figs. 11.4 and 11.5, respectively. The dermatomes of the trunk and the extremities are depicted in Fig. 11.6.

Table 11.1 Muscles, roots, actions, and nerve supply of the upper limb

Muscle	Root	Action	Nerve
Pectoralis Major-clavicular	C5, C6	Flexion and adduction of arm	Brachial plexus
Pectoralis Major-sternal	C7, C8, T1	Adduction and medial rotation of arm	Brachial plexus
Serratus Anterior	C5, C6, C7	Protracts and stabilizes the scapula	Long thoracic
Supraspinatus	C5, C6	Initial abduction of the arm	Suprascapular
Infraspinatus	C5, C6	External rotation of the flexed arm	Suprascapular
Latissimus Dorsi	C6, C7, C8	Adducts, extends, and internally rotates arm	Thoracodorsal
Rhomboid Major and Minor	C5	Retracts the scapula and rotates glenoid cavity inferiorly	Dorsal scapular
Deltoid	C5, C6	Abduction and elevation of arm up to 90°	Axillary
Biceps Brachii	C5, C6	Flexion of the supinated forearm	Musculocutaneous
Triceps	C6, C7, C8	Extension of the elbow	Radial
Extensor Carpi Radialis Longus	C6	Extension with radial deviation of the wrist	Radial
Brachioradialis	C5, C6	Flexion of the semi-pronated forearm	Radial
Supinator	C6, C7	Supination of the forearm	Posterior interosseous
Extensor Digitorum	C7, C8	Extends metacarpophalangeal joints of medial four digits	Posterior interosseous
Extensor Carpi Ulnaris	C7, C8	Extension with ulnar deviation of the wrist	Posterior interosseous
Extensor Indicis	C7, C8	Extension of proximal phalanx of the index	Posterior interosseous
Abductor Pollicis Longus & Brevis	C7, C8	Abduction of the thumb	Posterior interosseous
Extensor Pollicis Longus & Brevis	C7, C8	Extension of the thumb	Posterior interosseous
Pronator Teres	C6, C7	Pronation of the forearm	Median
Flexor Carpi Radialis	C6, C7	Flexion with radial deviation of the wrist	Median
Flexor Digitorum Superficialis	C7, C8, T1	Proximal interphalangeal joint flexion	Median
Flexor Pollicis Brevis	C8, T1	Flexion of the proximal phalanx of the thumb	Median
Opponens Pollicis	C8, T1	Opposition of the thumb against the fifth finger	Median
First & second Lumbricals	C8, T1	Flexion of the metacarpophalangeal joints with the extension of interphalangeal joints	Median
Flexor Digitorum Profundus	C7, C8	Distal phalanges flexion of the index & middle fingers	Anterior interosseous

Table 11.1 (continued)

Muscle	Root	Action	Nerve
Flexor Pollicis Longus	C8, T1	Flexion of terminal phalanx of the thumb	Anterior interosseous
Flexor Digitorum Profundus	C8, T1	Distal phalanges flexion of ring & little fingers	Ulnar
Abductor Digiti Minimi	C8,T1	Abduction of the little finger at the metacarpophalangeal joint	Ulnar
Opponens Digiti Minimi	C8, T1	Opposition of the little finger against thumb	Ulnar
Third & fourth Lumbricals	C8, T1	Flexion of the metacarpophalangeal joints with the extension of interphalangeal joints	Ulnar
Adductor Pollicis	C8, T1	Adduction of the thumb against the second finger	Ulnar
Flexor Pollicis Brevis	C8, T1	Flexion of the proximal phalanx of thumb	Ulnar
Dorsal and ventral interossei	C8, T1	Abduction and adduction of the fingers	Ulnar

Table 11.2 Muscles, roots, actions, and nerve supply of the lower limb

Muscle	Root	Action	Nerve supply
Iliopsoas	L1, L2, L3	Hip flexion from a semiflexed position	Femoral
Quadriceps Femoris	L2, L3, L4	Knee extension	Femoral
Sartorius	L2, L3	Flexion, weak abduction, and lateral rotation of the hip; flexion of the knee and medial rotation of the leg when knee is in flexion	Femoral
Gracilis	L2, L3, L4	Adduction and flexion of the hip and flexion of the knee with medial rotation of the tibia on the femur	Obturator
Adductor Longus/ Magnus/Brevis	L2, L3, L4	Adduction of the thigh	Obturator
Gluteus Medius	L4, L5, S1	Abduction and internal rotation of the thigh	Superior gluteal
Gluteus Maximus	L5, S1, S2	Extension of the thigh	Inferior gluteal
Biceps Femoris	L5, S1, S2	Knee flexion	Sciatic
Semitendinosus	L5, S1, S2	Knee flexion	Sciatic
Semimembranosus	L5, S1, S2	Knee flexion	Sciatic
Tibialis Anterior	L4, L5	Dorsiflexion of the foot with inversion	Deep peroneal
Extensor Digitorum Longus & Brevis	L5, S1	Dorsiflexion of proximal and distal phalanges of the toes	Deep peroneal
Extensor Hallucis Longus	L5, S1	Dorsiflexion of the great toe	Deep peroneal

(continued)

Table 11.2 (continued)

Muscle	Root	Action	Nerve supply
Peroneus Longus & Brevis	L5, S1	Foot eversion	Superficial peroneal
Gastrocnemius, Soleus	S1, S2	Plantar flexion of the foot	Tibial
Tibialis Posterior	L4, L5	Plantar flexion of the foot with inversion	Tibial
Flexor Digitorum Longus	L5, S1, S2	Flexion of distal phalanges of the toes	Tibial
Flexor Digitorum Brevis	S1, S2	Flexion of proximal phalanges of the toes	Tibial
Flexor Hallucis Longus	L5, S1, S2	Flexion of distal phalanx of the great toe	Tibial
Flexor Hallucis Brevis	S1, S2	Flexion of proximal phalanx of the great toe	Tibial
Perineal muscles	S2, S3, S4	Anal sphincter contraction	Pudendal

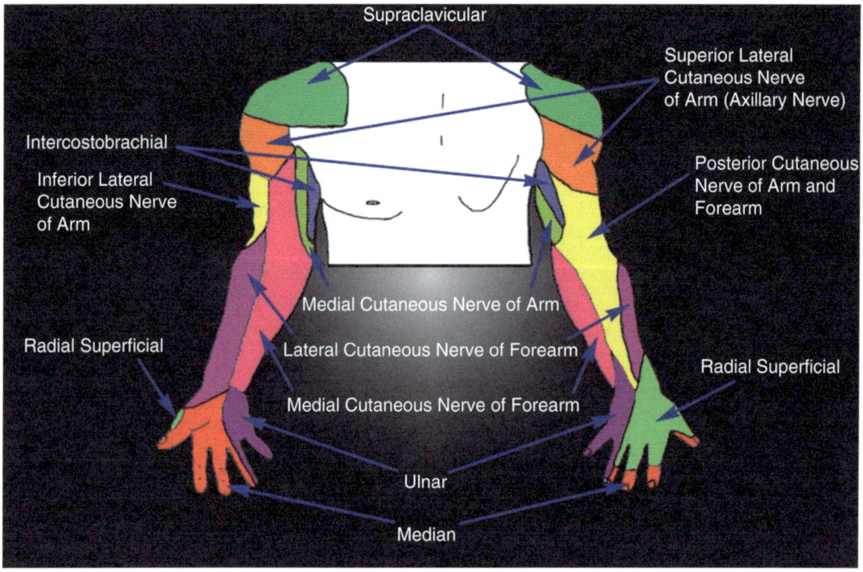

Fig. 11.4 Schematic representations of the cutaneous branches of the upper limbs

Electrophysiological tests can help to ascertain whether the primary pathological process is axonal degeneration or segmental demyelination. About two-thirds of mononeuropathy multiplex patients display a picture of axonal damage. Details regarding the causes of multiple mononeuropathies or mononeuropathy multiplex are provided in Table 11.3.

In polyneuropathy, sensory deficits present in the form of a length-dependent stocking–glove pattern. Once the sensory disturbances of the longest nerves (lower limbs) reach the level of the knees, the patient typically complains of paresthesia at

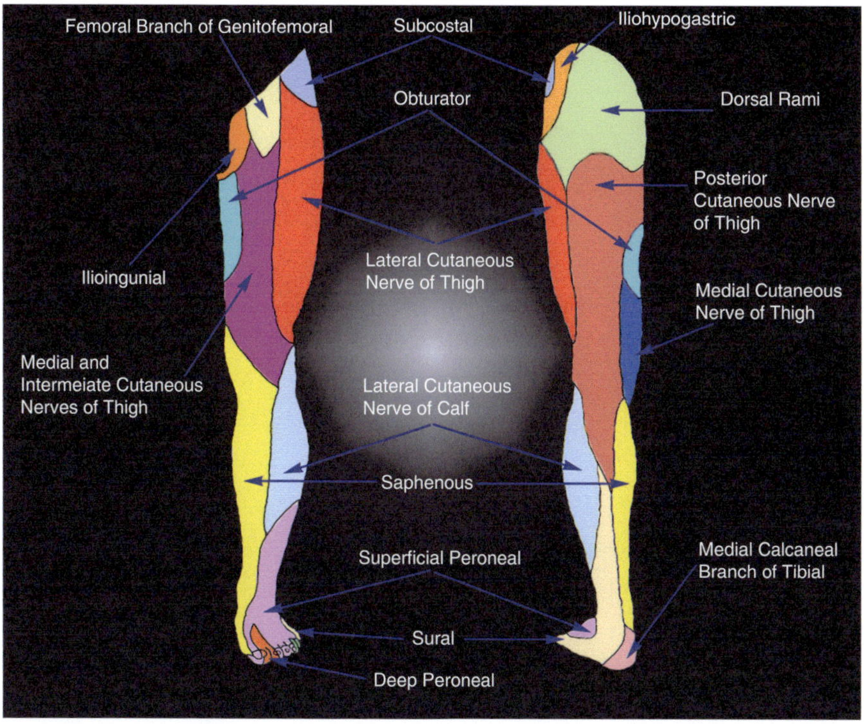

Fig. 11.5 Schematic representations of the cutaneous branches of the lower limbs

the tips of the fingers in the distribution of the second-longest nerves (upper limbs). Further, when the impairment progresses towards the mid-thigh, the involvement of the third-longest nerves (for instance, anterior intercostal and lumbar segmental nerves), produces a tent-shaped area of hypoesthesia on the anterior chest and abdomen. Even the motor weakness follows a dying-back pattern, characterized by considerable weakness of the extensor muscles of the foot compared to the flexors. For instance, heel walking is difficult for patients compared to toe walking in most polyneuropathies.

In general, motor deficits tend to dominate the clinical picture in GBS, CIDPs, hereditary motor and sensory neuropathies, and neuropathies associated with porphyria, lead toxicity, organophosphate intoxication, hypoglycemia, and certain malignancies. The list of polyneuropathies with predominant motor manifestations is presented in Box 11.1. Asymmetrical weakness without sensory loss may suggest multifocal motor neuropathy. The seventh cranial nerve (facial nerve) can be involved in several neuropathies like GBS, Lyme disease, sarcoidosis, HIV infection, and rarely CIDP. Lower limbs are more severely involved compared to the upper limbs in most polyneuropathies, with certain exceptions such as porphyria, multifocal motor neuropathy, Lewis–Sumner variant of CIDP, lead neuropathy, and hereditary motor neuropathy. Polyradiculoneuropathies like GBS and

Fig. 11.6 Schematic representations of the dermatomes of the trunk and the extremities

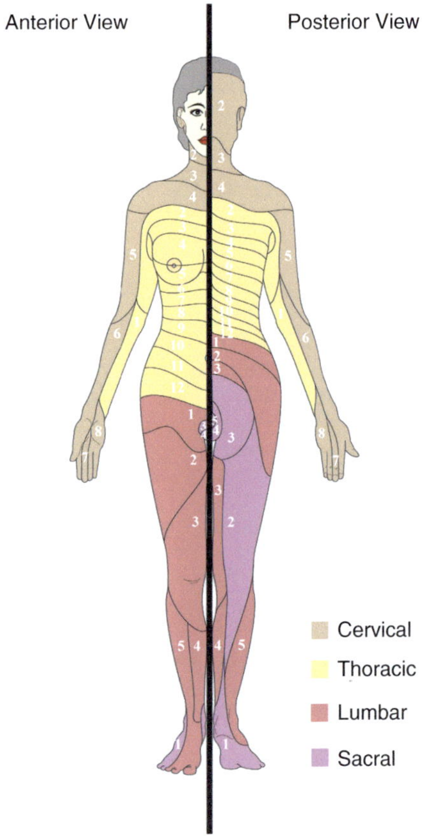

Table 11.3 The causes of mononeuropathy multiplex

Cause	Conditions
Axonal injury	Vasculitis
	Diabetes mellitus
	Sarcoidosis
	Hansen disease (leprosy)
	HIV
Demyelination	Multifocal motor neuropathy
	Multifocal acquired demyelinating sensory and motor neuropathy (Lewis-Sumner syndrome)
	Multiple compression neuropathies (hypothyroidism, diabetes mellitus)
	Hereditary neuropathy with liability to pressure palsies (HNPP)

CIDPs, porphyria, and diabetic lumbar radiculoplexopathy cause both proximal and distal muscle weakness. In case of nerve root involvement, needle EMG may reveal denervation in the paraspinal muscles. Autonomic dysfunction of clinical importance is seen in association with specific acute or chronic sensorimotor polyneuropathies. Sensory involvement can be a principal feature of the polyneuropathies caused by diabetes mellitus, Sjögren syndrome, dysproteinemia, acquired immunodeficiency syndrome, cobalamin deficiency, inherited and idiopathic sensory neuropathies, certain carcinomas, gastro-intestinal disorders, and intoxications.

Box 11.1 List of Polyneuropathies with Predominant Motor Manifestations

Acute motor axonal neuropathy
Multifocal motor neuropathy
Guillain-Barré syndrome
Chronic inflammatory demyelinating polyradiculoneuropathy
Diabetic lumbar radiculoplexopathy
Charcot–Marie-Tooth disease (hereditary motor sensory neuropathies)
Peripheral neuropathy in osteosclerotic myeloma
Lead-induced toxicity

Loss of sensation of all modalities (pain, temperature, touch, joint position sense, and vibration) is a common finding in peripheral neuropathies. However, in many, the impairment can be restricted to selective sensory modalities making it possible to correlate the type of sensory loss with the diameter of the affected nerve fibers. Reduced noxious and thermal sensation along with spontaneous burning pain, painful dysesthesias, and autonomic dysfunction are typical of polyneuropathies especially affecting small fibers. In such conditions, the deep tendon reflexes, balance, and muscle strength are preserved. Loss of tendon reflex, presence of sensory ataxia, and loss of kinesthetic, joint position, and vibration senses are typical of selective large-fiber sensory loss. Those patients with kinesthetic and joint position sensation loss, usually demonstrate involuntary sinuous movements of hands and digits when the upper extremity is outstretched with eyes blindfolded (pseudoathetosis) and/or a Romberg's sign (disproportionate or abrupt loss of balance with eyes closed condition compared to eyes open condition) can be the manifestation. Prominent sensory ataxia, in conjunction with pseudoathetosis or asymmetrical truncal or facial sensory loss, suggests a primary disorder of sensory neurons or polyganglionopathies.

Though palpation of peripheral nerves is an important part of clinical examination, the presence of hypertrophic nerve trunks will not suggest any specific disorder and is found in disorders of the peripheral nerve including leprosy, neurofibromatosis, Charcot–Marie–Tooth disease, acromegaly, Refsum disease, and rarely CIDP. Isolated hypertrophy of a nerve trunk can suggest a neoplastic condition like neurofibroma, schwannoma, or malignant nerve sheath tumor.

11.7 Mononeuropathies-Definition, Classification, and Features

Mononeuropathy, a disorder of a single peripheral nerve may result from compression, traction, laceration, thermal or chemical injury. The injury may affect one or more structural constituents of the nerve and the pathophysiological response to injuries consist of axon loss and/or demyelination. The peripheral nerve injuries due to the aforementioned causes are classified based on the functional status of the nerve and histological findings. Seddon divided peripheral nerve injury into three types: neurapraxia, axonotmesis, and neurotmesis. The classification was based on the severity of tissue injury, prognosis, and time for recovery and remains popular, especially among surgeons, because of its correlation to the outcome. Later, Sunderland revised the classification into five degrees with better prognostic implications. The first degree and second degree of Sunderland's classification corresponds to neurapraxia and axonotmesis, respectively of Seddon's schema. The third, fourth, and fifth degrees of injuries correspond to the neurotmesis of Seddon's classification. In Sunderland's third-degree injury, the axons, Schwann sheaths, and the endoneurium are disrupted within intact nerve fascicles. In the fourth-degree injury, the axon, endoneurium, and perineurium surrounding the fascicles are damaged, and in the fifth-degree injury, the entire nerve trunk is severed, i.e., disruption of the axon, endoneurium, perineurium, and epineurium.

11.7.1 Neurapraxia

Neurapraxia, the first-degree nerve injury, typically results from a brief or mild compression of the nerve that distorts the myelin covering of the axon. It results in segmental demyelination with no anatomical disruption of the axons. The electrophysiological properties of the nerve below the level of the lesion are normal; however, the nerve fails to conduct normally across the lesion due to the conduction block and the findings correlate to neurapraxia. Typically, complete recovery can be anticipated within weeks to a few months, the time required for remyelination, provided the offending cause is eliminated.

11.7.2 Axonotmesis

Axonotmesis is characterized by axonal damage that results in Wallerian degeneration. Distal to the injury, the axons and their encircling myelin sheath degenerate, and the end-organs become denervated. It is a second-degree nerve injury characterized by axonal disruption with intact endoneurium and perineurium and epineurium. Since the nerve regeneration between the site of injury and the target end-organ is well-guided by the intact endoneurial tubes, axonotmesis lesions have a fairly good prognosis.

11.7.3 Neurotmesis

Third-degree nerve injuries have a fair prognosis as the disruption is limited to the axons and the endoneurium, leaving the perineurium and epineurium intact. These lesions may require surgical intervention, particularly when there is a potential for axonal misdirection and neuroma formation. Regarding the fourth-degree nerve injury, other than epineurium, the remaining components (axons, endoneurium, and perineurium) are disrupted. Such lesions have a poor prognosis and often require surgical repair. The fifth-degree injury is the most severe form of nerve injury characterized by the disruption of the nerve and all its supporting structures. The nerve trunk is severed, with a loss of continuity between its proximal and distal stumps. Here in these lesions too, the prognosis is poor and requires surgical repair for a certain amount of functional recovery.

11.7.4 Entrapment Neuropathy

Entrapment neuropathy is a mononeuropathy caused by focal compression or mechanical distortion of a nerve within a fibrous or fibro-osseous tunnel or less frequently by structures like bone, ligament, connective tissues, blood vessels, or space-occupying lesions. In entrapment neuropathy, compression, pressure, sharp angulation, and stretching are key mechanisms that produce nerve injury at certain susceptible anatomical locations. Since compression occurs at specific sites, surgery is often required to release the entrapped nerve, such as the median nerve within the carpal tunnel following moderate to severe Carpal Tunnel Syndrome (CTS). Overuse has been implicated as one of the plausible causes of entrapment neuropathies, particularly in certain professions, like graphic artists, musicians, hairdressers, dental hygienists, cashiers, and stenographers. In chronic entrapment, the mechanical distortion of the nerve fibers can lead to focal demyelination and in more severe cases can lead to Wallerian degeneration. Morphological studies reveal a combination of active demyelination, remyelination, Wallerian degeneration, and axonal regeneration at the site of entrapment. In addition to the abovementioned nerve fiber changes, the condition is also accompanied by endoneurial swelling, collagen proliferation, and perineurial sheath thickening. Though in chronic compression, ischemia is not an important contributing factor for further damage, it plays a critical role in accentuating nerve injury associated with acute compression secondary to space-occupying lesions like hematoma and compartment syndromes.

In mild entrapments, the electrodiagnostic studies may reveal a focal slowing or conduction block across the site of entrapment. However, in severe cases, due to Wallerian degeneration and regeneration, the electrophysiological studies may reveal features of axonal degeneration and are essential for diagnosis and reliable documentation of the site and severity of nerve entrapment. In addition to the electrophysiological studies, plain X-rays, Computed Tomography (CT), ultrasound, and Magnetic Resonance Imaging (MRI) may provide additional information about the structural abnormalities.

11.7.5 Common Mononeuropathies of the Upper Extremities

Table 11.4 provides details of the common mononeuropathies of the upper extremities. Before the elaboration of the common upper extremity mononeuropathies, a brief anatomy of the peripheral nerves is supplemented.

11.7.5.1 Median Nerve

In the axilla, the median nerve (Fig. 11.7) is formed by the fusion of the lateral roots arising from the lateral cord (C6 and C7) and the medial roots arising from the medial cord (C8 and T1) of the brachial plexus. The lateral and medial roots unite either anterior or lateral to the axillary artery and in the arm, the nerve descends

Table 11.4 Details regarding the common mononeuropathies of upper extremities

Nerve	Compression Site	Etiolgical/Predisposing Factors	Clinical features
Suprascapular nerve	Suprascapular notch	Neuralgic amyotrophy, blunt trauma	Supraspinatus muscle and infraspinatus muscle atrophy
Axillary nerve (circumflex nerve)	Axilla	Shoulder anterior dislocation, forced abduction of shoulder joint or surgery	Weakness of arm abduction, sensory loss over lower deltoid ("regimental badge area")
T1 roots (lower trunk of brachial plexus)	Thoracic outlet	Cervical rib, cervical band with enlarged C7 transverse process	Intrinsic hand muscles atrophy (mostly thenar), paresthesias of medial hand and forearm
Musculocutaneous nerve	Axilla/arm (rare)	Shoulder dislocations, neuralgic amyotrophy, stab injury, vigorous exercises like repetitive weight lifting movements or surgery	Weakness of elbow flexors and forearm supination, lateral antebrachial cutaneous territory sensory loss

Table 11.4 (continued)

Nerve	Compression Site	Etiolgical/Predisposing Factors	Clinical features
Median nerve	Wrist (CTS)	Tenosynovitis, wrist joint arthritis, Repetitive wrist maneuvers, repeated use of vibrating hand tools, pregnancy, obesity, diabetes, and thyroid dysfunction	Nocturnal paresthesias, pain, thenar atrophy, thumb weakness, reduced hand dexterity
	Anterior interosseous	Trauma, vigorous exercise, Neuralgic amyotrophy, and compartment syndrome	Weakness of pincer movement of the thumb and index finger, normal sensation, poorly localized pain in the forearm and cubital fossa
	Elbow or proximal forearm (Pronator Teres syndrome)	Prolonged or repetitive elbow motions	Tenderness over pronator teres, presence or absence of muscle weakness innervated by the anterior interosseous nerve, no sensory loss.
Ulnar nerve	Elbow (cubital tunnel syndrome)	Direct pressure on the nerve from leaning on the elbow, trauma, Tardy ulnar palsy (excessive cubitus valgus)	Clawing, Froment's sign, tingling or sensory loss in the palm and fourth and fifth fingers, tenderness in the elbow joint
	Guyon canal	Mechanics, cyclists, handles of canes	Atrophy of the interosseous muscles, intact sensation over the dorsum of the fourth and little fingers
Radial nerve	Axilla	Crutches	Wrist drop, triceps involved, sensory loss extending into the forearm and sometimes arm
	Spiral groove	Abnormal sleep postures, fractures of the humerus	Wrist drop, triceps spared, loss of sensation over of the dorsum of hand
	Posterior interosseous	Synovitis of the elbow joint, trauma, fracture, neuralgic amyotrophy, tumors	Weakness of finger extensors, radial deviation when wrist extended
	Cheiralgia Paresthetica (superficial sensory branch)	Wrist bands, handcuffs	Paresthesias over the dorsum of hand

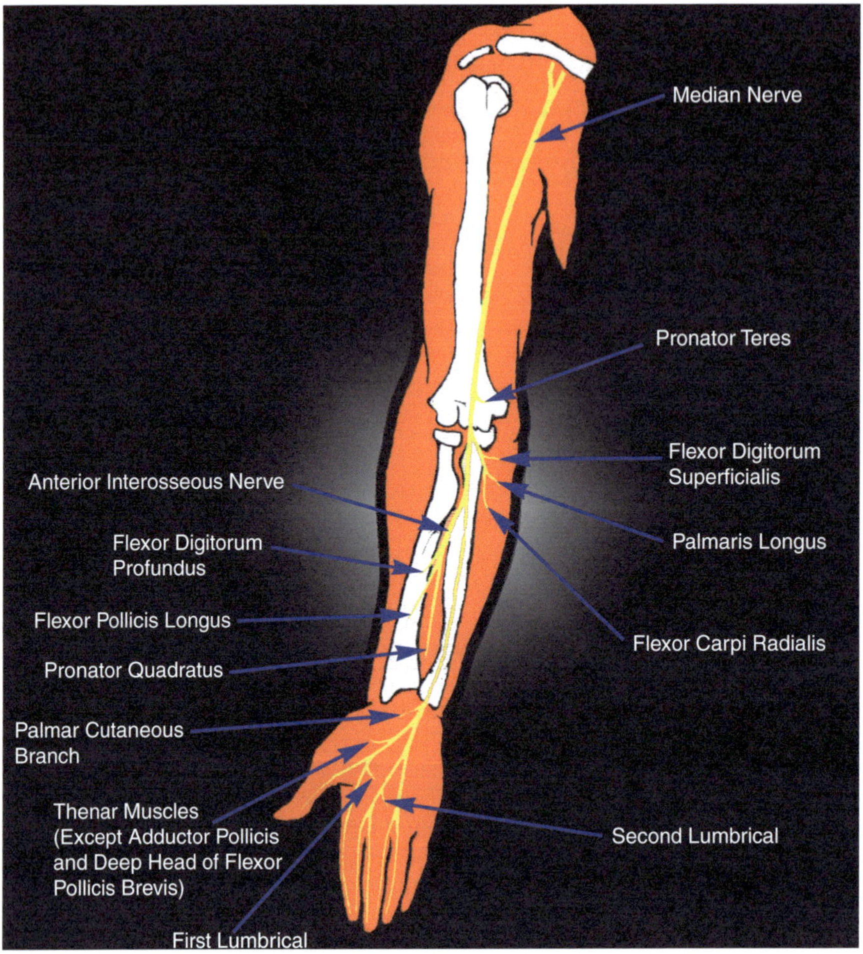

Fig. 11.7 An illustration of the median nerve course and its branches

lateral to the brachial artery. Adjacent to the coracobrachialis muscle insertion, the nerve crosses the brachial artery and then descends medially till the cubital fossa. The nerve enters the anterior compartment of the forearm by passing through the two heads of the pronator teres muscle. In the upper forearm, the nerve gives branches to the pronator teres, flexor carpi radialis, flexor digitorum superficialis and the palmaris longus muscles and a pure motor branch called the anterior interosseous nerve. The anterior interosseous nerve, being the largest branch of the median nerve innervates the flexor pollicis longus, pronator quadratus, and the lateral half of the flexor digitorum profundus. In the forearm, the nerve travels between the flexor digitorum superficialis and flexor digitorum profundus muscles and then enters the wrist through the carpal tunnel, formed by the carpal bones and the flexor retinaculum (transverse carpal ligament). Before entering the tunnel, the palmar cutaneous sensory branch originates and runs subcutaneously to innervate the skin over the thenar eminence. After crossing the tunnel, the nerve divides into the motor and sensory branches. The motor branch supplies the first and second lumbricals and thenar muscles namely the abductor pollicis brevis, opponens pollicis, and superficial head of the flexor pollicis brevis. The sensory division of the median nerve innervates the skin over the thumb, index, middle, and lateral half of the ring fingers.

Carpal Tunnel Syndrome CTS is the commonest entrapment neuropathy and it results from compression of the median nerve in the carpel tunnel through which the nerve and long finger flexor tendons pass. Tenosynovitis of the flexor tendons and arthritis of the carpal bones in the tunnel area often produces pressure on the median nerve. In many patients, the syndrome presents bilaterally but with a greater intensity of symptoms in the dominant hand. The patients may report symptoms like nocturnal pain and paresthesia over the thumb, index, and middle fingers, or the entire hand. They may complain of tingling, numbness, and burning sensations, and the symptoms can often disturb their sleep. Patients may report referred pain radiating towards the forearm or even the shoulder. Activities like driving or holding or carrying weights with the wrist either in flexed or extended posture and excessive use of the hand or wrist may often provoke the symptoms. CTS can be a work-related condition, as repetitive forceful grasping, gripping, or pinching, awkward positions of the hand and wrist, direct pressure over the tunnel, and use of handheld vibrating tools can injure the nerve within the tunnel. The risk for the syndrome is higher among meat packers, butchers, garment workers, musicians, dental hygienists, grocery checkers, and housekeepers. Obesity, pregnancy, diabetes, age, rheumatoid arthritis, hypothyroidism, gout, old wrist fractures, and inflammatory diseases affecting tendons or connective tissues at the wrist level, may predispose the development of CTS.

Clinical examination may reveal objective sensory changes in the sensory distribution of the median nerve. Typically, two-point discrimination, pinprick, and light touch sensations are often impaired, and occasionally hyperesthesia in the thumb, index, and middle fingers, with sparing of the skin over the thenar eminence can be a finding. Weakness and atrophy of the thenar muscles can be a finding in advanced cases of CTS. The Phalen maneuver (passive flexion of the patient's wrist, maintained for one minute) and the reversed Phalen maneuver (passive hyperextension of the wrist) in about 80% of CTS patients may provoke the symptoms. A positive Tinel sign, in which percussion of the nerve at the carpal tunnel causes paresthesia in the distal distribution of the median nerve, is present in about 60% of CTS patients. Both the Phalen maneuver and the Tinel sign are not specific for CTS and may provide false-positive results. Sensory and motor conduction studies of the median nerve will exhibit delayed latencies across the wrist in about 70% of CTS patients. However, in one-third of CTS patients, the abovementioned electrophysiological tests may fail to identify the abnormality, particularly when the symptoms are mild in nature. Recording the median nerve latency of distal segments at short distances and/or comparing this latency with the latency for the ulnar or radial nerve at the same distance may increase the sensitivity of the conduction studies. Detection of the increased cross-sectional area of the median nerve at the carpal tunnel inlet (more than 13 mm girth) using ultrasound suggests thickening of the median nerve. For space-occupying lesions like ganglion, bony deformity, or hemangioma with CTS symptoms, MRI can be useful.

Patients with only mild sensory symptoms may benefit from the usage of splints in a neutral position with nonsteroidal anti-inflammatory drugs and/or intralesional administration of corticosteroids. A comparison of splinting versus surgery suggests that the latter may have a better long-term outcome than the former. Evidence regarding the effectiveness of nonsteroidal anti-inflammatory agents, diuretics, laser, and ultrasound are conflicting, and exercise therapy may not be of any help to alleviate the symptoms of CTS. Incomplete sectioning of the flexor retinaculum, injury to the palmar cutaneous branch of the median nerve while making skin incision, scarring within the carpal tunnel, and incorrect preoperative diagnosis, are some of the reasons for poor surgical results.

Anterior Interosseous Nerve Syndrome It is a pure motor neuropathy due to fascicular lesions of the anterior interosseous nerve. The syndrome is characterized by the isolated palsy of the flexor pollicis longus, pronator quadratus, and the radial half of flexor digitorum profundus. Typically, the syndrome manifests as pain in the forearm or elbow, with weakness of the thumb and index muscles to perform the finger pincer movement. It will be difficult or impossible to bring the distal phalanx of the thumb and index finger to close approximation to form a circle (pinch or O sign). These patients may complain of difficulty or inability to button shirts. When asked to make a fist, these patients will fail to flex the index and middle finger at the distal interphalangeal joints. The most common site of anterior interosseous nerve entrapment occurs near the tendinous edge of the deep head of the pronator teres muscle. The nerve can also be externally compressed or injured following anterior elbow dislocations, complex elbow fractures, penetrating injuries and stab wounds, open reduction and internal fixation of fractures, and fibrous bands attachment to the flexor digitorum superficialis muscle. Electrophysiological tests, together with MRI, may help to identify the possible etiology and assist in the diagnosis of the condition. Median sensory nerve conduction studies will be normal and needle EMG will demonstrate denervation in the muscles supplied by the anterior interosseous nerve.

Prognosis is generally good and spontaneous recovery occurs within 3–12 months in most cases; however, some may require surgical intervention if conservative treatment fails to improve within the initial 3 months. Patients developing symptoms due to torsion injury or inflammation and intraneural adhesions, secondary to the high mobility of the anterior interosseous nerve fascicles during elbow flexion, may benefit from interfascicular neurolysis. For those patients with spontaneous anterior interosseous nerve syndrome, initial conservative treatment may include rest, use of analgesics with anti-inflammatory medications, and physiotherapy. Based on the etiology of the anterior interosseous nerve syndrome, the physiotherapeutic management varies and may consist of stretching exercises, nerve gliding exercises, activity modification, strengthening exercises of hand and wrist muscles, and the Neuromuscular Electrical Stimulation (NMES) of the weak muscles.

Pronator Teres Syndrome The syndrome develops as a result of the compression of the median nerve in the proximal forearm between the humeral and the ulnar heads of the pronator teres muscle, thickened proximal edge (fibrous arcade) of the flexor digitorum superficialis muscle, or thickened biceps aponeurosis (lacertus fibrosus). Compared to CTS, the pronator teres syndrome is rare and women over 40 years of age are more frequently affected. Typically, the patients will present with vague pain in the anterior aspect of the elbow and forearm. Pain tends to begin or worsen during activities involving grasping and/or pronation. They may report an insidious onset of paresthesia and numbness of the palm and the radial three digits and the radial half of the fourth digit, mimicking CTS but without the nocturnal symptoms. These patients may present with vague weakness of the hand and digits and may report poor grip strength or clumsiness. Muscles involved in the complete pronator teres syndrome are the pronator teres, flexor carpi radialis, palmaris longus, flexor digitorum superficialis, and the muscles supplied by the anterior interosseous nerve.

Physical examination may reveal an inability to move the thumb away from the rest of the hand (simian hand deformity) and when asked to make a fist, they may fail to flex the index and middle finger at the distal metacarpophalangeal and interphalangeal joints (sign of Benediction). Certain provocation maneuvers like resistance to pronation or resistance to elbow flexion with the forearm held in supination may produce pain in the proximal forearm. On palpation, the proximal volar aspect of the forearm can be firm and tender and the Tinel sign can be elicited over the median nerve on the pronator teres muscle. Symptoms of pronator teres syndrome and CTS can be hard to differentiate; however, distinction can be made by identifying the numbness of the forearm, absence of nocturnal exacerbation, and provocation maneuvers discussed earlier.

The median sensory and motor NCS are usually normal and will not reveal impaired distal latencies at the wrist that accompany CTS. Needle EMG is generally normal and may not show any definite signs of denervation. In the majority of cases, conservative treatment is found to be effective. Corticosteroid injection(s) into the pronator teres muscle, use of analgesics and non-steroid anti-inflammatory drugs, and immobilization of the arm with the elbow flexed to 90° with the forearm in slight pronation, often relieves the symptoms. The role of surgery is controversial; however, it is indicated in space-occupying lesions and when the conservative treatment fails beyond 3 months.

11.7.5.2 Ulnar Nerve

Within the axilla, through the lower trunk and medial cord of the brachial plexus, the ulnar nerve (Fig. 11.8) derives its fibers from the C8 and T1 roots. After arising from the brachial plexus, the nerve descends in a plane between the axillary artery and axillary vein and further descends the arm medially, lateral to the brachial artery. Near the mid-point of the arm, the nerve penetrates the medial fascial septum to enter the posterior compartment of the arm and then travels through the cubital

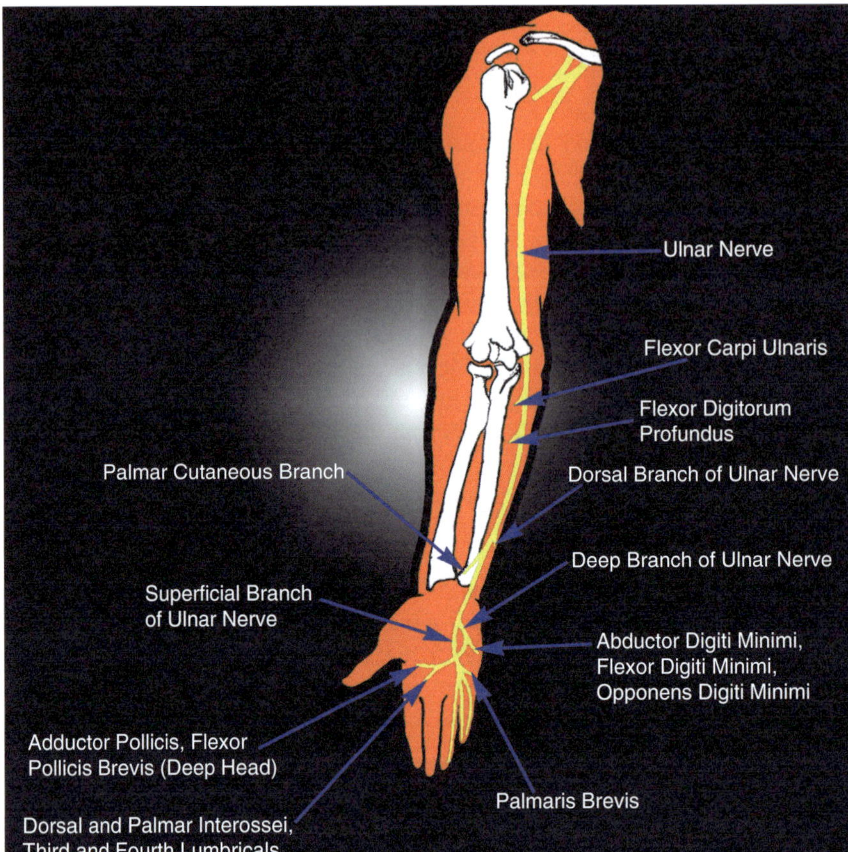

Fig. 11.8 An illustration of the ulnar nerve course and its branches

tunnel, a small space between the olecranon and the medial epicondyle. In the majority, the nerve remains in the ulnar groove but in some with an unusual degree of physiological cubitus valgus, it may sublux anteriorly over the medial epicondyle, particularly when the elbow is flexed. After giving an articular branch to the elbow joint, the nerve enters the forearm. In the proximal region of the forearm, the ulnar nerve gives muscular branches to the flexor carpi ulnaris and the ulnar half of the flexor digitorum profundus. Approximately 5–8 cm proximal to the wrist, the dorsal ulnar cutaneous branch exits to innervate the skin over the dorsal medial hand and the dorsal aspect of the little and ring digits. At the level of the ulnar styloid, the nerve gives off the palmar cutaneous branch which innervates the proximal medial palm. Through the Guyon canal (a canal formed between the hook of the hamate and pisiform bones, located anterior to the flexor retinaculum and posterior to the palmar carpal ligament) the nerve enters the hand. Within the Guyon canal, the nerve divides into terminal superficial and deep palmar branches. The terminal superficial branch supplies the palmaris brevis and the skin on the volar aspect of the little and ring digits. The terminal deep palmar branch is purely motor and innervates the hypothenar muscles, the palmar and dorsal interossei, the third and fourth lumbricals, and the adductor pollicis, and the deep head of the flexor pollicis brevis.

Ulnar Nerve Entrapment at the Elbow Ulnar mononeuropathy is a common compression or entrapment mononeuropathy and its incidence is second common to CTS. The cubital tunnel syndrome is a common cause of ulnar neuropathy and develops as a result of compression of the nerve by the thickened fibrotic flexor carpi ulnaris aponeurosis, near the cubital tunnel entrance. The risk of nerve compression at the elbow is high, following subluxation of the ulnar nerve. Ulnar groove syndrome, another cause of ulnar neuropathy develops due to external pressure to the nerve as a result of prolonged and frequent resting of the flexed elbow on a hard surface like a table or armrest of a wooden chair. The ulnar entrapment can be work-related due to the repeated flexion movement of the elbow. The nerve can undergo direct compression following prolonged unconsciousness, lower end humerus fracture, elbow joint dislocation, cubitus valgus deformity (Tardy ulnar palsy), or a ganglion or lipoma at the elbow.

The lesion of the ulnar nerve at the level of the elbow results in numbness and tingling sensation of the fourth and fifth digits, with variable degrees of hand muscle weakness. In chronic cases, the patients may present with weakness and muscle atrophy with no clear sensory symptoms. Hypothenar muscle atrophy and noticeable atrophy of the first dorsal interosseous and adductor muscle are prominent features. The clinical examination may also reveal the variable weakness of the flexor carpi ulnaris and the ulnar half of the flexor digitorum profundus that moves the distal interphalangeal joints of the ring and little fingers. The weakness of the adductor pollicis, flexor pollicis brevis, and palmar or volar interossei muscles can reduce the grip strength. The weakness of the interossei can result in the loss of ability to forcefully extend the interphalangeal joints, as in the case of finger-flicking movements. Lumbrical muscle weakness leads to clawing of the little and ring fingers, characterized by flexion of the proximal and distal interphalangeal joints and hyperextension of the metacarpophalangeal joints (ulnar clawing). The weakness of the third palmar interosseous muscle causes abduction of the little finger away from the ring finger, which may get caught when the patient tries to put the hand in a pocket (Wartenberg sign). The patient fails to pinch a piece of paper between the thumb and index finger against resistance and when attempting to do so he or she may flex the thumb at the interphalangeal joint due to adductor pollicis muscle weakness (Froment's sign). The presence of Tinel sign at the elbow level, positive provocative tests such as flexion compression test, tenderness of ulnar nerve on palpation and presence of nerve thickening have poor diagnostic values.

Compared to entrapment neuropathies like CTS, diagnostic confirmation of ulnar nerve entrapment at the elbow using electrophysiological tests can be challenging. While performing ulnar motor nerve conduction studies to localize a lesion at the elbow, a 70–90° flexed position of the elbow is preferred to the extended position. Demonstration of focal demyelination across the elbow, as shown by the reduction in motor nerve conduction velocity by more than 10 m/s and/or conduction block as revealed by the localized reduction in CMAP amplitude and area can help to localize the ulnar neuropathy at the elbow level.

For patients with mild to moderate sensory symptoms or symptoms due to work-related causes, conservative treatment should be attempted. Minimizing repetitive elbow flexion and extension movements and avoiding direct pressure on the elbow may alleviate the symptoms. Those patients with a habitual tendency to excessively lean on the elbow may benefit from elbow protectors. Surgery is considered only if conservative treatment fails to ease the symptoms during the initial 3 months. Surgical techniques may include the release of the flexor carpi ulnaris aponeurosis, anterior transposition of the ulnar nerve trunk, and resection of the medial epicondyle.

Ulnar Nerve Entrapment at the Wrist Ulnar compression or entrapment near the distal region or at the Guyon canal or wrist and hand is a less common condition. Apart from direct trauma and laceration, a ganglion cyst is the most common cause of distal nerve compression. Chronic or repetitive external pressures over the nerve by the handle of canes, hand tools, bicycle handlebars, or excessive push-ups are certain causes of distal ulnar neuropathy. Degenerative changes at the wrist joint, rheumatoid arthritis, or vascular anomalies distally at the wrist region can also cause distal ulnar nerve compression. Conservative and surgical management will be similar to those advised for ulnar nerve entrapment for the proximal region.

11.7.5.3 Radial Nerve

The radial nerve (Fig. 11.9) is the largest nerve in the upper extremity and it contains fibers from the nerve roots C5-T1. It is the terminal continuation of the posterior cord that is formed by the union of the posterior division of the upper, middle, and lower trunk of the brachial plexus. In the axilla, the nerve is situated posterior to the axillary artery. After supplying branches to the long and lateral heads of the triceps brachii, the nerve exits the axilla inferiorly. In the upper arm, the radial nerve lies medial to the humerus and innervates all three heads of the triceps muscle and the anconeus muscle. The radial nerve passes obliquely behind the humerus to enter a shallow groove termed the spiral groove (a groove formed deep to the lateral head of the triceps muscle). Prior to the entry to the spiral groove, the nerve gives three sensory branches namely, the posterior cutaneous nerve of the arm, the lower lateral cutaneous nerve of the arm, and the posterior cutaneous nerve of the forearm. After exiting the groove, the nerve lies lateral to the humerus in the anterior compartment of the arm and innervates the brachioradialis and the extensor carpi radialis longus. It then passes anterior to the lateral epicondyle and enters the forearm to innervate the extensor carpi radialis brevis and supinator muscles. In the upper forearm, near the humeroradial joint, the radial nerve divides into its terminal branches, i.e., the

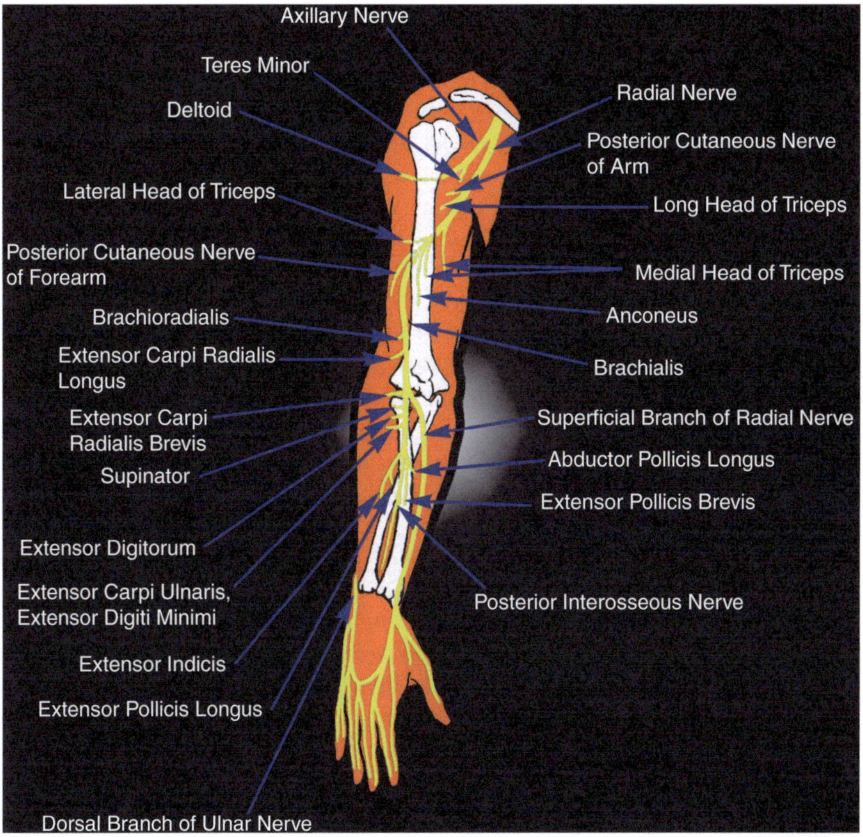

Fig. 11.9 An illustration of the radial nerve course and its branches

superficial radial nerve and the posterior interosseous nerve. The superficial radial nerve is a pure cutaneous terminal branch of the radial nerve and innervates the skin of the proximal two-thirds of the extensor surfaces of the thumb, index, and middle fingers, and the radial half of the ring finger, along with the corresponding dorsum of the hand. The posterior interosseous nerve is a pure motor terminal branch that passes under the proximal edge of the supinator muscle and then travels in the forearm to innervate wrist and finger extensors (except those muscles innervated proximally by the radial nerve).

Radial Nerve Compression in the Arm The compression of the radial nerve at the spiral groove of the humerus often results from prolonged sleep or deep sleep following alcohol intoxication with the arm hanging over the armrest of a chair (Saturday-night palsy). Fractures of the humerus can be another cause of injury to the radial nerve at the arm level. Lesions of radial nerve at the axillary level are less frequent and may result from the improper use of crutches or weight of a partner's head over the arm while sleeping (Honeymoon palsy). Lesion of the radial nerve in the spiral groove or mid-arm causes weakness of the brachioradialis, and wrist and digit extensors. The weakness of the latter two muscle groups leads to wrist drop deformity. However, in such lesions, the motor functions of the triceps are usually preserved as they are innervated superiorly. Sensations are impaired over the dorsal aspect of the hand, thumb, index finger, and middle finger, the cutaneous territory of the radial nerve. Absence of synergistic action of wrist extensors and finger flexors essential to make a strong fist causes a weak or ineffective fist with excessive and unwanted wrist flexion. Radial nerve palsy causes difficulty or inability to extend the metacarpophalangeal joints; however, the extension of the interphalangeal joints is possible owing to intact lumbricals and interossei with their attachments to the dorsal digital expansion. Though lesions of radial nerve at the axillary level are least common, the symptoms will be the same as discussed earlier with additional weakness of the triceps, with sensory loss extending into the extensor surface of the forearm, lateral half of the arm, and over the triceps muscles owing to the involvement of posterior cutaneous nerve of the forearm and lower lateral cutaneous and posterior cutaneous nerves of the arm, respectively.

The electrophysiological tests are essential to confirm the site and extent and estimate the severity and prognosis of the nerve lesion. The absence of SNAP or reduction in SNAP amplitude is common in radial nerve injuries except when the pathology is purely demyelinating at the spiral groove. Conduction block across the spiral groove can also be seen in segmental demyelinating lesions. A mixed lesion (a combination of neurapraxia and axonotmesis) is not uncommon. In case of axonal injury, the needle EMG will show the denervation of extensor carpi radialis longus, extensor carpi radialis brevis, brachioradialis, and wrist and digits extensors. Like other peripheral nerve compression and entrapment lesions, the prognosis depends on the primary pathology. In radial nerve lesions due to demyelination and conduction block, as in most cases of Saturday-night palsy, complete improvement can be anticipated in a few months. In axonal degenerative conditions, a protracted course and often incomplete recovery can be anticipated, thus suggesting a less favorable prognosis.

Posterior Interosseous Neuropathy Injuries to the posterior interosseous nerve, which affects the innervation of the extensor compartment muscles of the forearm, are less common. Generally, it occurs as a result of trauma, fracture, space-occupying lesions like gangliomas and lipomas, or inflammatory conditions like rheumatoid arthritis. The most commonly described sites of compression are the arcade of Frohse (a fibrous band between the two heads of the supinator muscle) and the distal edge of the supinator muscle.

The posterior interosseous neuropathy often presents with insidious onset weakness of the wrist and digital extensors. Clinical examination will reveal a radial deviation of the wrist when the patient is asked to extend the wrist, owing to the preservation of the extensor carpi radialis longus, supplied by the radial nerve. On palpation, there can be abnormal tenderness over the arcade of Frohse or the lateral epicondyle. Resisted supination and pronation of the forearm and resisted extension of the middle finger can produce pain in the aforementioned site. Brachioradialis, extensor carpi radialis longus, and extensor carpi radialis brevis muscles innervated by the radial nerve have normal strength. The condition is generally painless. Patients with isolated posterior interosseous nerve neuropathy will have no sensory deficit, thus delineating this syndrome from radial tunnel syndrome (Box 11.2).

Box 11.2 Features of Radial Tunnel Syndrome

Radial tunnel syndrome
Highly controversial topic consisting of entrapment of the radial nerve or its posterior interosseous branch within the radial tunnel. The nerve may get entrapped in the proximal fibrous band of the extensor carpi radialis, hypertrophied extensor carpi radialis brevis margin, or radial recurrent vessel arch. Throwing, overhead activities, and repetitive pronation and supination movements may cause radial nerve compression and precipitate the symptoms. The patient usually complains of lateral elbow and dorsal forearm pain that may radiate to the wrist and dorsum of the fingers. The condition is more prevalent among women aged 30–50 years. On palpation, there will be abnormal localized tenderness over the radial nerve distal to the lateral epicondyle. Typically, the pain gets aggravated at night and might interfere with sleeping. Traction applied to the nerve by extending the elbow, pronating the forearm, or flexing the wrist can worsen the severity of pain. Electrophysiological tests such as EMG and NCS are typically negative. In many cases, the radio-image studies and ultrasound may not reveal any pathology. Rarely the MRI may show muscle edema or atrophy along the distribution of the radial and posterior interosseous nerves.

Ultrasound, radioimaging, and electrophysiological tests will be useful in confirming the site, extent, and severity of the injury. Conservative treatment consists of rest, activity modification, wrist and/or elbow splinting, physiotherapy, and the use of nonsteroidal anti-inflammatory drugs. The use of therapeutic ultrasound, therapeutic massage, and stretching exercises, strengthening and Range of Motion (ROM) exercises, and neural mobilization techniques are the components of physiotherapy. The prognosis is generally good with conservative measures and surgical decompression is generally indicated if symptoms fail to refract after conservative management for 3 months.

11.7.5.4 Musculocutaneous Nerve

The musculocutaneous nerve (Fig. 11.10), a terminal branch of the lateral cord of the brachial plexus originates from the C5 and C6 roots through the upper trunk. The nerve emerges at the inferior border of the pectoralis minor muscle and then leaves the axilla. It then innervates and penetrates the coracobrachialis muscle and descends anteriorly in the upper arm between the biceps brachii and brachialis muscles. The nerve innervates both the muscles and gives articular branches to the

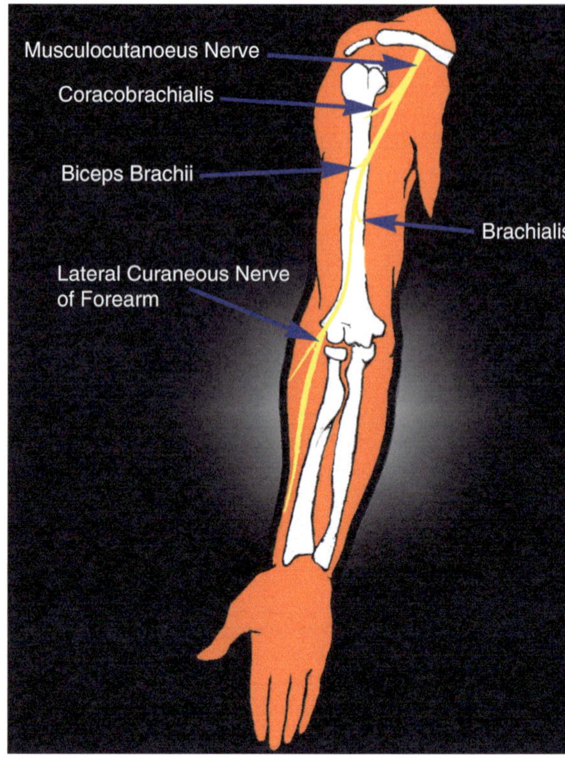

Fig. 11.10 An illustration of the musculocutaneous nerve course and its branches

humerus and the elbow. After piercing the deep fascia, the nerve emerges lateral to the biceps tendon and brachioradialis and continues in the forearm as the lateral cutaneous nerve of the forearm and provides cutaneous innervation to the lateral aspect of the forearm.

Since the nerve is well-protected within the axilla and the arm, injury to the musculocutaneous nerve is relatively uncommon but may get affected in neuralgic amyotrophy. It can also be damaged in shoulder dislocations, stab, and iatrogenic injuries, following general anesthesia or vigorous exercises like repetitive weight lifting movements as seen in carpet carriers and headload carriers. The differential diagnosis of musculocutaneous nerve palsy includes C5 and/or C6 radiculopathy, upper trunk or lateral cord brachial plexopathy, and biceps tendon rupture.

11.7.6 Common Mononeuropathies of the Lower Extremities

Table 11.5 provides details of the common lower extremity mononeuropathies. Brief anatomy of the nerves is supplemented for a better understanding before the elaboration of the lower extremity mononeuropathies.

Table 11.5 Details regarding the common mononeuropathies of lower extremities

Nerve	Compression site	Etiological factors	Clinical features
Sciatic nerve	Sciatic notch	Intramuscular injections, bullet injuries, stab wounds, fractures, dislocations, hematomas	Pain down the thigh, no ankle and/ or foot movements possible, absent ankle jerk
	Hip	Fracture, dislocations, bullet injuries, stab wounds, hematomas	Pain down the thigh, no ankle and/ or foot movements possible, absent ankle jerk
	Piriformis muscle	Fall or blow to the buttock	No muscle weakness, pain while sitting
	Popliteal fossa	Popliteal synovial cysts (Baker's cysts)	Bulge and/or feeling of tightness behind knee, normal hamstrings
Tibial nerve	Medial malleolus (Tarsal tunnel syndrome)	Tenosynovitis, ankle fracture, use of high heel shoes	Sensory loss over the sole of the foot, Tinel's sign at flexor retinaculum
Peroneal nerve	Fibular neck	Weight loss, habitual leg crossing, improper positioning or padding during anesthesia, prolonged bed rest of critically ill, debilitated, or unconscious patients; casts, orthoses, pneumatic compression, antithrombotic stockings, bandages, straps, blunt trauma, proximal fibula fracture, knee dislocation, open injury, and complication of knee surgery	Foot drop, ankle evertor weakness, Sensory loss in the dorsum of the foot
	Anterior compartment	Muscle edema, heavy exercise, trauma, or ischemia	Acute severe lower leg pain, swelling, weakness of foot and toe extensors; foot drop
Femoral nerve	Inguinal ligament	Lithotomy position	Weak knee extensors, absent knee Jerk
	Pelvis	Intraoperative compression	Weak hip flexors, weak knee Extensors, absent knee jerk
Obturator nerve	Obturator canal	Space-occupying lesions, pelvic fracture, surgery	Loss of sensation in the medial thigh region, weak hip adductors

(continued)

Table 11.5 (continued)

Nerve	Compression site	Etiological factors	Clinical features
Lateral femoral cutaneous nerve	Inguinal ligament (Meralgia Paresthetica)	Tight clothing or stockings, obesity, diabetes, pregnancy, and wearing heavy or tight utility belts	Tingling, numbness, and burning pain or loss of sensation in the lateral thigh region

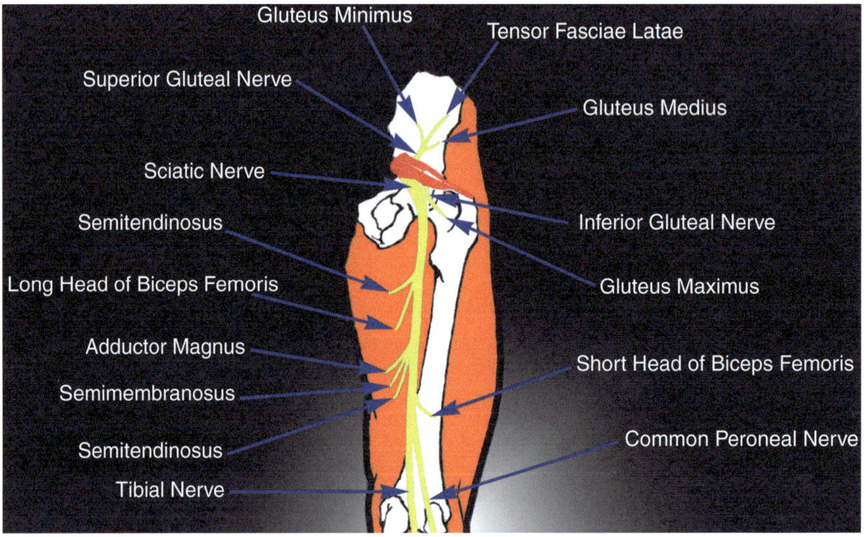

Fig. 11.11 An illustration of the sciatic nerve course and its branches

11.7.6.1 Sciatic Nerve

The sciatic nerve (Fig. 11.11), the largest nerve in the body and a major nerve of the lower extremity, is formed from the nerve roots L4 through S3 of the lumbosacral plexus. It leaves the pelvis and enters the gluteal region through the greater sciatic foramen. The nerve emerges from the pelvis by passing beneath the piriformis muscle and descends in an inferolateral direction. During its course down through the gluteal region, the nerve crosses the posterior surface of the superior and inferior gemelli, obturator internus, and quadratus femoris muscles. Passing deep to the long head of the biceps femoris, the nerve enters the posterior thigh. At the level of the posterior thigh region, the nerve gives branches that innervate the hamstring and adductor magnus muscles. The tibial division of the sciatic nerve innervates all the hamstring muscles except the short head of the biceps femoris which is innervated by the common peroneal division of the sciatic nerve. The sciatic nerve terminates near the superior angle of the popliteal fossa by bifurcating into the tibial and common peroneal nerves.

Entrapment of the Sciatic Nerve at the Sciatic Notch Rarely the sciatic nerve is vulnerable to entrapment as it passes over the sciatic notch. The common causes of sciatic nerve lesions are bullet injuries, stab injuries, fractures, dislocations, hematomas in the posterior thigh compartment, misplaced intramuscular injections at the gluteal region, and complications of replacement surgeries of the hip. Direct compression of the nerve is less common and may result from a prolonged coma, anesthesia, or prolonged sitting on a hard surface. Lesions to the sciatic nerve cause weakness of knee flexors and all the lower limb muscles below the knee level. In addition to the above, the patient will have a sensory loss of the entire foot and leg below the knee, except the medial leg, the region supplied by the saphenous nerve. Pain may be perceived in the proximal thigh, radiating posteriorly and laterally into the leg. The ankle jerk is either depressed or lost on the affected side. Sciatic nerve palsy secondary to injection injuries can begin immediately or hours after the trauma to the nerve.

Proximal lesions of the sciatic nerve typically involve the common peroneal division than the tibial division and for the same reason, it may mimic a more distal common peroneal neuropathy. In such cases, the presence of denervation in the short head of the biceps femoris and tibialis posterior muscles on needle EMG and abnormal sural or medial plantar SNAPs may help to localize the partial proximal sciatic nerve lesions from a distal common peroneal neuropathy. In addition to the above, during the physical examination, close attention must be paid to the muscles that receive non-peroneal innervation, especially the tibialis posterior and flexor digitorum longus (innervated by the tibial nerve) and the hamstrings (innervated by the sciatic nerve). Weakness in any of these muscles with a foot drop suggests a dysfunction beyond the common peroneal nerve distribution.

The isolated sciatic nerve lesion typically spares the sensation over the medial calf and foot (supplied by the saphenous nerve) and posterior thigh (supplied by the posterior cutaneous nerve of the thigh). Sensory involvement of these territories in addition to foot drop suggests proximal widespread lesions affecting other components of the lumbosacral plexus. It is also important to bear in mind that sciatic neuropathy and peroneal neuropathy are not the only two conditions that present with footdrop with sensory disturbance over the lateral calf and dorsum of the foot. Such presentation can often arise as a result of lumbosacral plexopathy, L5 radiculopathy, or a central lesion like a frontal meningioma or anterior cerebral artery infarct.

Symptoms of sciatic nerve neuropathy may not respond well to analgesics and non-steroidal anti-inflammatory drugs. The use of corticosteroids like methylprednisolone can be more promising in the management of neuropathic pain and motor and sensory deficits. Surgery (neurolysis or grafting) can be an option for those patients who fail to respond to conservative treatment during the first 3 months following injection palsy to the sciatic nerve. During the early stage when pain is severe, the use of Transcutaneous Electrical Nerve Stimulation (TENS), gentle stretching, and desensitization techniques may help to alleviate the pain. For those with definite axonal injury, strengthening exercise has to be given with caution. It is advisable to use gradually progressive low-intensity strengthening exercises. Splinting or bracing may be needed to prevent deformity and associated risks and for the same reasons splints like Ankle Foot Orthosis (AFO) can be prescribed to prevent foot drop, skin damage, and falls. Low-intensity cold laser therapy can be used for promoting nerve nutrition and regeneration. The use of NMES may help in the later stages to improve the strength of the muscles and for the re-education of muscles following tendon transplantation.

Piriformis Syndrome Usually, the patient with piriformis syndrome gives a history of buttock trauma and may complain of unilateral hip and/or buttock pain during prolonged sitting, bending at the waist, crossed legs sitting in a figure of four-position or activity that requires hip adduction and internal rotation. The neurological examination and electrophysiological tests are usually normal. Piriformis test (Box 11.3) may reproduce the pain and is considered diagnostic. Radioimaging is usually normal but can less frequently reveal piriformis muscle hypertrophy or abnormal vessels or bands in the region of the muscle. Magnetic resonance neurography may show sciatic nerve hyperintensity at the sciatic notch, a more specific sign of nerve entrapment. Therapeutic exercise mainly focusing on prolonged stretching of the piriformis muscle by placing the hip in flexion, adduction and internal rotation can relieve the symptoms. For those patients with no alleviation with stretching exercise, CT- or MRI-guided corticosteroid injection into the piriformis muscle may help relieve the symptoms.

Box 11.3 Description of Piriformis Test

Piriformis test
The test is preferably performed in the side-lying position on the unaffected side. The symptomatic leg is positioned 60 to 90° flexed at the hip joint and 90° degrees at the knee joint. The examiner should place one hand over the pelvis to stabilize it and the other hand is kept over the lateral aspect of the knee. Following which the examiner should apply gentle hand pressure on the lateral side of the knee in the direction of the examination couch thus performing horizontal adduction of the hip. Such a passive maneuver of hip flexion, adduction, and internal rotation may compress the sciatic nerve and due to which the patient may feel pain or discomfort.

11.7.6.2 Common Peroneal Nerve

The sciatic nerve bifurcates near the apex of the popliteal fossa, to form the common peroneal nerve (also known as the common fibular nerve) and the tibial nerve. The former follows the medial border of the biceps femoris and runs laterally and inferiorly and gives off the lateral cutaneous nerve of the calf, which innervates the skin over the upper third of the lateral aspect of the leg and the fibular communicating nerve which joins the sural nerve. The common peroneal nerve (Fig. 11.12) then winds or wraps around the fibular neck and passes through the origin of the peroneus longus muscle ("fibular tunnel") and here the nerve divides into its terminal branches, the deep and superficial peroneal nerves. The deep peroneal nerve descends initially through the lateral and then the anterior leg compartments and

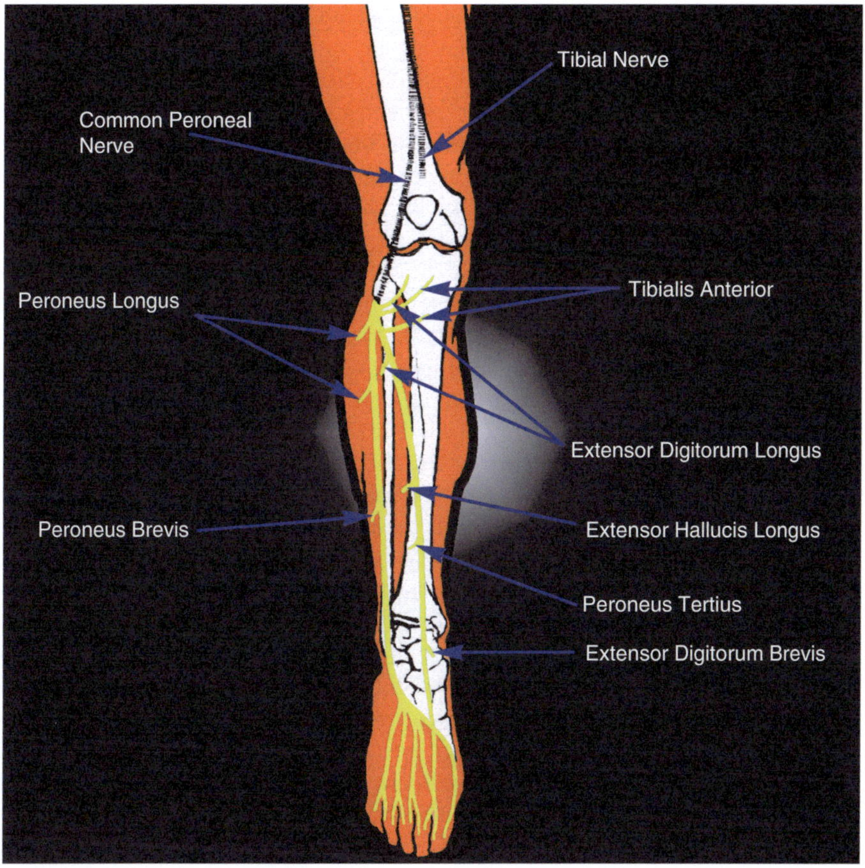

Fig. 11.12 An illustration of the common peroneal nerve course and its branches

innervates the tibialis anterior, extensor hallucis longus, peroneus tertius, and extensor digitorum longus. It then divides close to the ankle joint to innervate the extensor digitorum brevis and the skin over the first webspace. The superficial peroneal nerve innervates the peroneus longus and brevis and the skin of the lower two-thirds of the lateral aspect of the leg and the dorsum of the foot (except the first webspace).

Common Peroneal Neuropathy at the Fibular Neck Compression of the common peroneal nerve is the most frequent compressive neuropathy in the lower limb. The nerve is particularly vulnerable to direct pressure near the fibular neck as it passes through the origin of the peroneus longus muscle. Improper positioning or padding during anesthesia can cause acute common peroneal neuropathy at the fibular neck. Pressure over the nerve due to weight loss, prolonged bed rest of critically ill, debilitated, or unconscious patients, tight casts, orthoses, pneumatic compression, compression stockings, bandages or straps, and habitual leg crossing, are other possible causes of nerve injury. Blunt trauma such as a fracture of the proximal fibula and knee dislocation or an open injury can also account for the same. The nerve can also be injured as a result of a complication of knee surgery, like arthroscopic surgery and lateral meniscus repair.

Damage to the common peroneal nerve leads to weakness of ankle and toe dorsiflexors and foot evertors, with a foot drop and a high stepping gait or a foot slap gait. Cutaneous sensations are impaired over the lateral aspect of the lower two-thirds of the leg and the dorsum of the foot. Compressive mononeuropathy of the peroneal nerve can often be confused with other causes of unilateral foot drop, which may include L5 radiculopathy, sciatic nerve injury with predominant involvement of common peroneal division, and lumbosacral plexopathy.

In case of selective injury to the deep peroneal nerve, foot eversion and foot sensation (except the cutaneous sensations over the first web space of the foot) are typically spared. Anterior tibial compartment syndrome, an acute syndrome characterized by severe lower leg pain, swelling, with weakness of foot and toes, may present like deep peroneal nerve lesion and must be differentiated from the latter. In the abovementioned syndrome, the nerve undergoes compression by muscle swelling secondary to injury, heavy exercise, trauma, or ischemia. This compartment syndrome needs early identification and must undergo immediate fasciotomy to decompress and prevent irreversible nerve and muscle damage.

Tinel sign is a reliable clinical sign to locate the area of nerve compression or entrapment. By tapping along the course of the nerve, particularly around the fibular neck, the examiner may be able to elicit or reproduce the pain, tingling, or paresthesia distal to the point of compression by tapping. Passive straight-leg-raising of the lower limb with the ankle in plantar flexion and the foot in inversion may reproduce symptoms (peroneal nerve tension test). Electrophysiological tests will help to distinguish conduction blocks and axonal degeneration across the fibular head. CMAP records are either taken from the extensor digitorum brevis or the tibialis anterior muscle. The absence of denervation in the short head of the biceps femoris and the presence of denervation in the common peroneal-innervated muscles distally may suggest compression pathology in the fibular neck region. For localization and severity of the lesion, the nerve may be stimulated at the knee and fibular head regions. Radioimaging studies and ultrasound can often help visualize the intraneural ganglia, soft-tissue masses, or tumors. In cases of acute demyelinating lesions, the prognosis is consistently good; however, in axonal lesions and stretch injuries recovery often tends to get delayed.

Foot drop is the main residual symptom following common peroneal nerve compression neuropathy. The use of AFO helps to maintain the foot in a neutral position, improves the gait, and prevents falls. For patients with a higher potential for improvement, graded exercises (active-assisted, active, and progressive resisted exercises) should be given for the ankle dorsiflexors and foot evertors. NMES can be used as an adjunct if the muscles are considerably weak. Muscle strength imbalance and constant plantar flexion attitude of the foot may facilitate tightness and in long-standing cases contractures and deformities and to avoid the same, regular stretching exercises are crucial. Strengthening exercises helps to prevent atrophy and speed up the recovery process and in most cases, the foot drop issue may resolve within 3–12 months. In severe cases, surgery including transplantation of the tibialis posterior may regain active dorsiflexion and/or improve the ankle stability and gait.

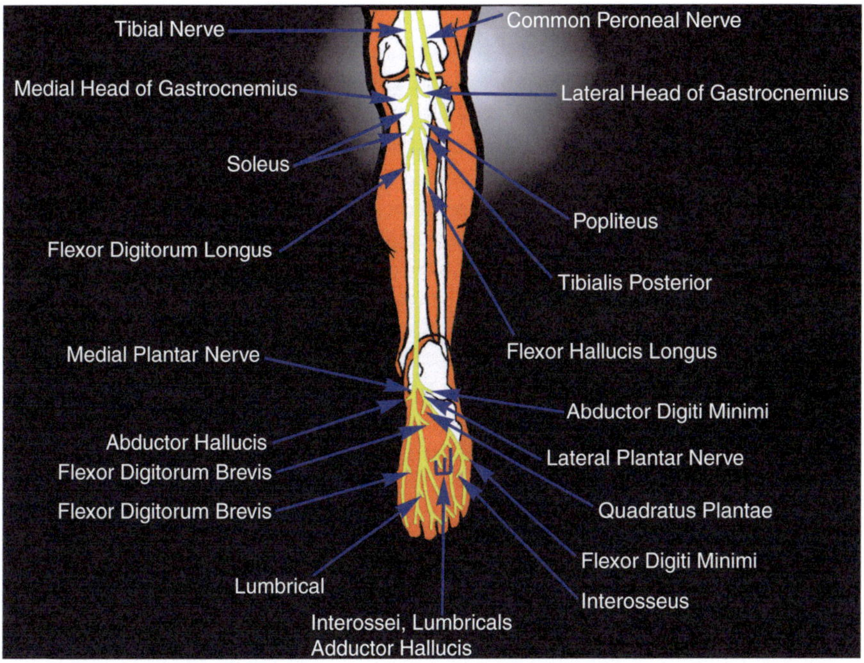

Fig. 11.13 An illustration of the tibial nerve course and its branches

11.7.6.3 Tibial Nerve

The tibial nerve (Fig. 11.13) division of the sciatic nerve innervates all the hamstring muscles except the short head of the biceps femoris. At the apex of the popliteal fossa, the sciatic nerve terminates as the tibial and common peroneal nerve. Near the upper region of the popliteal fossa, the tibial nerve gives off the sural sensory nerve, which is often joined by a branch from the common peroneal nerve, the sural communicating nerve, to supply the skin over the lateral aspect of the lower leg and foot and the little toe. In the proximal region of the leg, the nerve runs beneath the soleus muscle and innervates all the muscles of the posterior compartment (gastrocnemius, soleus, tibialis posterior, flexor digitorum profundus, and flexor hallucis longus). At the ankle, the nerve passes beneath the flexor retinaculum (laciniate ligament) that covers the tarsal tunnel. The contents of the tarsal tunnel are tendons of the tibialis posterior, flexor digitorum longus, and flexor halluces longus muscles, the tibial artery and vein, and the tibial nerve.

Tarsal Tunnel Syndrome Tibial nerve can get entrapped behind or immediately below the medial malleolus. The patients typically experience a burning pain in the toes and the sole of the foot. Involvement of the calcaneal sensory branches can cause pain and numbness over the heel. The clinical examination frequently reveals impaired cutaneous sensation over the plantar area and wasting of the intrinsic foot muscles.

Eversion of the foot may provoke pain and paresthesia. Percussion at the site of nerve compression often elicits pain and paresthesia. The electrophysiological tests may demonstrate a slowing of motor fibers to the abductor hallucis and/or abductor digiti minimi muscles and the involvement of the medial and/or lateral plantar branches of the tibial nerve. Sparing of the sural nerve SNAP, with the aforementioned electrophysiological findings suggests the involvement of the tibial nerve at the tarsal tunnel. Needle EMG will demonstrate the denervation of the abductor hallucis and/or abductor digiti minimi muscles and normal innervation of the gastrocnemius, soleus, biceps femoris, and gluteus maximus muscles. The differential diagnosis includes generalized peripheral neuropathy, S1 radiculopathy, or non-neurological foot painful conditions of foot including plantar fasciitis, stress fracture, and bursitis.

11.7.7 Brachial and Lumbosacral Plexopathies

The brachial plexus of the upper extremity and the lumbosacral plexus of the lower extremity are complex peripheral nervous system structures that contain the nerve fibers originating from the anterior horn cells, sympathetic and parasympathetic ganglia, or dorsal root ganglia. The disorders involving the abovementioned plexi are much less common when compared to other focal peripheral nerve disorders like radiculopathies or mononeuropathies, but the clinical presentations may have similarities. Many conditions such as direct trauma or compression, immune-mediated, inflammatory, or metabolic disorders can cause brachial and lumbosacral plexopathies.

Various diagnostic tools like electroneuromyography, neuroimaging, laboratory testing, and genetic testing, may complement a comprehensive and thorough clinical assessment in identifying the possible etiologies. The extent and the type of nerve injury within the plexus depend on the mechanism and the severity of injury. Typically, in mild or early lesions like nerve compression, recovery may occur rapidly, provided decompression is achieved and the prognosis is overall favorable. In severe injuries to the plexus, the axons and often the supporting structures undergo Wallerian or axonal degeneration leading to prolonged effective recovery and a less favorable prognosis.

11.7.7.1 Brachial Plexopathies

The brachial plexus (Fig. 11.14) is derived from the anterior primary rami of the C5 through T1 roots, whereas the posterior primary ramus innervates the paraspinal muscles. As the brachial plexus courses through the neck, shoulder, and upper arm, it divides into roots, trunks, divisions, cords, and terminal nerves. Table 11.6 provides information about the trunk, cord, and terminal nerves, and the muscles innervated by the brachial plexus.

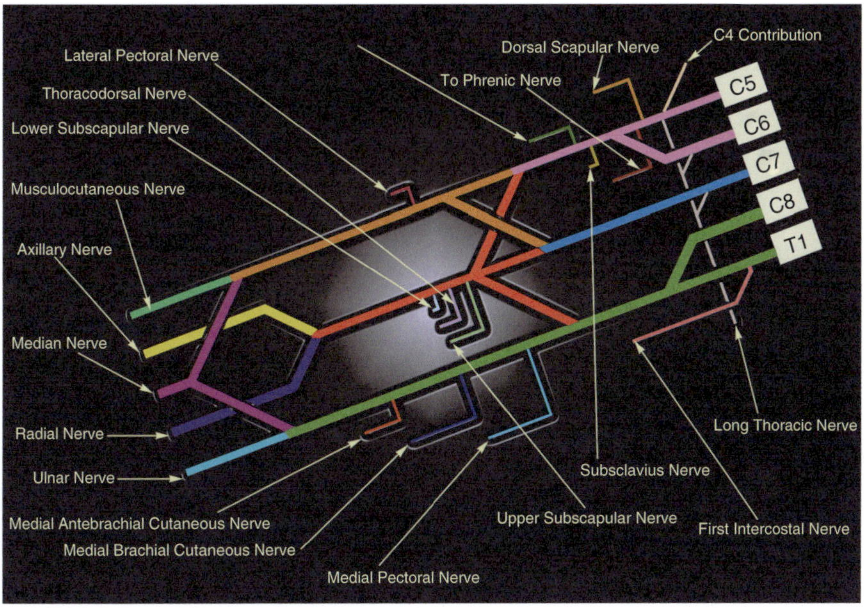

Fig. 11.14 Schematic diagram of brachial plexus

Table 11.6 Details about the trunk, cord, and terminal nerves, and the muscles innervated by the brachial plexus

Direct/Cord	Trunk/Root(s)	Terminal nerve	Muscle(s)
Dorsal scapular nerve	Ventral ramus of C5		Levator Scapulae Rhomboid Major
Long thoracic nerve	Ventral ramus of C5–7		Serratus Anterior
Suprascapular nerve	Upper trunk (C5–6)		Supraspinatus Infraspinatus
Lateral cord	Upper trunk (C5–6)	Musculocutaneous nerve	Biceps Brachialis, Coracobrachialis
		Lateral pectoral nerve	Clavicular fibers of pectoralis major
	Middle trunk (C6–7-8)	Musculocutaneous nerve	Coracobrachialis
		Median nerve	Pronator Teres Flexor Carpi Radialis

(continued)

Table 11.6 (continued)

Direct/Cord	Trunk/Root(s)	Terminal nerve	Muscle(s)
Posterior cord	Upper trunk (C5–6)	Radial nerve	Brachioradialis Supinator Triceps
		Axillary nerve	Deltoid Teres Minor
		Thoracodorsal nerve	Latissimus Dorsi
		Upper and lower subscapular nerve	Subscapularis and Teres Major, respectively
		Musculocutaneous and radial nerves	Brachialis
	Middle trunk (C6–7-8)	Radial nerve	Triceps Anconeus Extensor Carpi Radialis Extensor Digiti Communis Extensor Carpi Ulnaris
		Thoracodorsal nerve	Latissimus Dorsi
	Lower trunk (C8-T1)	Radial nerve	Extensor Indicis Proprius Extensor Carpi Ulnaris Extensor Pollicis Longus Extensor Digitorum Communis Extensor Carpi Ulnaris Triceps
		Thoracodorsal nerve	Latissimus Dorsi
Medial cord	Lower trunk (C8-T1)	Medial nerve	Abductor Pollicis Brevis Opponens Pollicis Flexor Pollicis Longus Pronator Quadratus
		Ulnar nerve	First Dorsal Interosseous Abductor Digiti Minimi Flexor Carpi Ulnaris
		Medial and ulnar nerves	Flexor Digitorum Profundus
		Medial pectoral nerve	Sternal fibers of Pectoralis Major

The roots are generally short, less "interwoven" within the nerve sheath and unprotected by adherent dura or epineurium making them more susceptible to traction injury than the plexus. Long thoracic nerve to the serratus anterior muscle, the nerve to the subclavian muscle, dorsal scapular nerve to the rhomboid muscles and levator scapulae are some of the individual nerves that branch directly from the roots. In the supraclavicular fossa near the lateral border of the anterior and medial scalene muscles, the roots join to form three trunks: upper (C5 and C6 roots), middle (predominantly C7 root) and lower (C8 and T1 roots). The suprascapular nerve, a branches directly off of the upper trunk, innervates the supraspinatus and infraspinatus muscles.

Due to its length and relationship to surrounding structures, the plexus is a vulnerable structure. The apex of the lung, lymph nodes, clavicle, ribs, and blood vessels may all be sites of disease which may extend toward and involve the plexus. In addition to the above, the brachial plexus is susceptible to traction injuries too. The clinical manifestations of brachial plexopathies depend on the site of involvement, the etiology, and the temporal course of the disease. The onset of symptoms can be rapid, subacute (days to weeks), or chronic (months to years). Pain is a common and prominent symptom in many patients and may be experienced as a deep, aching, or burning type. In many, the location of the pain can reflect the portion of the plexus injured, although some patients may not be able to distinctly localize the pain. Typically, the pain tends to get worsened by arm movements and unlike radiculopathies, maneuvers that increase intrathoracic pressure and intracranial pressure (secondary to intracranial venous vasculature congestion) like "Valsalva" do not worsen the symptom. Loss of sensation or presence of positive sensory symptoms like tingling and prickling sensations is common but can be overshadowed by pain.

Muscle weakness is one of the most devastating manifestations of brachial plexopathies. The distribution of weakness can be highly localized or patchy and incomplete and may reflect the site of plexus involvement. For instance, upper trunk injury usually manifests with proximal arm weakness, whereas lower trunk injuries involve the hand. In addition to the neuromuscular features, the brachial plexopathies may present with non-neuromuscular features like Horner's syndrome (involvement of lower trunk plexopathies or T1 root).

Electroneuromyographic studies are valuable and essential component of the evaluation of brachial plexopathies. Electrophysiological tests help to confirm the location of the brachial plexus lesion while excluding conditions like radiculopathies or mononeuropathies. It also helps to identify the segment(s) involved, determine the type (compression, axonal, or demyelinating) and extent of nerve injury and assess any evidence of reinnervation or recovery. Sensory NCS findings help to distinguish a post-ganglionic (brachial plexus) lesion from a preganglionic (cervical root) lesion. Needle EMG is used in conjunction with NCS to further help in localizing the lesion, severity, and degree of axonal loss and reinnervation. Needle EMG abnormalities are seen in muscles supplied by the portion(s) of the plexus involved and the findings include fibrillation potentials/positive sharp waves. In pure brachial plexopathies, paraspinal muscles are spared.

Depending on the clinical scenario, imaging studies are necessary to identify possible compressive structural causes, neoplastic infiltration or abnormalities. Various imaging modalities are used to assess brachial plexopathies, including Computed Tomography (CT), Magnetic Resonance Imaging (MRI), magnetic resonance neurography, ultrasound, and myelography.

Traumatic Brachial Plexopathies Blunt and penetrating injuries are common traumatic causes. Forceful lateral deviation of the head away from a depressed shoulder can induce stretch injuries to the supraclavicular portions of the nerves. In about 50% of traumatic plexopathies, upper trunk and middle trunks are most frequently involved. Burner or Stinger syndrome, a transient stretch injury to the plexus that typically follows sudden forceful trauma to the shoulder (usually during contact sporting activities) and rucksack or backpack palsy, a traction injury to the upper portion of the plexus after wearing a heavy backpack or rucksack, are examples for traumatic brachial plexopathies. Treatment of traumatic plexopathies is challenging and depends on the severity of injury, site of involvement, and degree of reinnervation. A multi-disciplinary team approach is required to manage the pain, weakness, and contractures. Surgical reconstruction may be required to improve functional movements of the upper extremity when spontaneous functional recovery does not occur and the strategies include neurolysis, nerve grafting, or nerve or muscle transfers.

Neuralgic Amyotrophy Neuralgic amyotrophy, also known as Parsonage–Turner syndrome, is an immune-mediated plexopathy of unknown etiology. It has been associated with various conditions, including infections, immunizations, connective tissue diseases, trauma, surgical operations, and pregnancy. In approximately 50% of cases, a precipitating cause has been identified. In those patients who develop symptoms post-immunizations, the onset can be between few days to weeks and may involve either the injected or non-injected limb.

Typically, the patients present with sudden onset of severe shoulder or arm pain. In about 60% of cases, the pain is experienced at night, may involve any region (proximal or distal limb), and limb movements may aggravate the symptom. Usually, the pain may last for hours to weeks before subsiding and in some patients it may persist for a few months. The muscle weakness and atrophy is typically delayed by days to weeks following the onset of pain. The weakness may worsen as the pain subsides. The muscle weakness may involve any distribution of the brachial plexus or a single nerve distribution. Isolated or combined involvement of motor nerves clinically may resemble those of "mononeuropathy multiplex" than a pure "brachial plexopathy."

Electrophysiological tests are important in the evaluation of neuralgic amyotrophy. T2 weighed MRI scan may demonstrate abnormal hyperintensities in the involved segments. MRI and ultrasonography may demonstrate subtle hourglass-like constrictions on the terminal branches of nerves. No specific treatments have been systematically proven to be helpful in reducing the degree of neurologic impairment or improving the prognosis. In the early course of the disease, analgesic medication may be necessary to reduce the pain. Physiotherapy and regular ROM exercises can be advocated to prevent secondary complications due to weakness and immobility. Typically, the recovery begins with improvement in pain. For most patients, the prognosis of neuralgic amyotrophy is favorable. In about 35% a favorable improvement occurs within one year and 75% of cases by the end of the second year.

Neurogenic Thoracic Outlet Syndrome Thoracic Outlet Syndrome (TOS) is a syndrome characterized by pain and numbness in the arm. It is presumably due to transient compression or narrowing of the neuro-vasculature (subclavian vessels and lower trunk of the brachial plexus) as they course through the thoracic outlet. Neurogenic TOS, the most common type, develops when the lower trunk of the brachial plexus gets compressed by a cervical rib (elongated C7 transverse process to the first rib) or a cervical band (a fibrous band extending from a rib). Surgical removal of the cervical rib or resection of the band can resolve the symptoms in most neurogenic TOS patients.

Neoplastic Plexopathies and Radiation-Induced Plexopathies Since the brachial plexus lies in close proximity to the lung, breast, and lymphatic system, neoplastic invasion can damage or cause dysfunction of the brachial plexus. Neoplastic plexopathies are more frequent among breast cancer patients and are also seen in other neoplastic conditions like lung cancer, lymphoma, and sarcoma. Rarely, the metastatic infiltration of the brachial plexus with focal conduction block may mimic CIDP. Imaging studies can confirm the diagnosis of neoplastic invasion of the plexus and MRI with contrast is the primary imaging modality used. Treatment of neoplastic plexopathies consists of localized radiation or chemotherapy. About 50–80% of patients are likely to have partial or complete remission of symptoms following treatment.

Patients who have been treated with radiation near the vicinity of the brachial plexus can also develop a plexopathy known as radiation-induced brachial plexopathy, a rare manifestation of radiation therapy. There is no established treatment to reverse or improve the nerve injury, although surgical interventions such as neurolysis or neurolysis with omental grafting, have been performed in some patients with variable improvement in symptoms.

11.7.7.2 Lumbosacral Plexopathies

The lumbosacral plexus (Fig. 11.15) has less 'merging' of nerve fascicles or formation of trunks or cords compared to the brachial plexus. However, the structure is complex and can be considered as two adjacent plexi: the lumbar and the sacral. The former is derived from the anterior rami of the L1-L4 roots and the latter is formed from the L5-S4 roots. Table 11.7 provides details regarding the nerve branches, muscles innervated, and sensory distribution by the lumbosacral plexus. Since the lumbosacral plexus is a deep structure in the pelvis, it is less susceptible to direct trauma; however, proximity to structures like muscles, vasculature, and intestinal structures predisposes it to injury from disorders involving them.

To assess the lumbar plexus, the femoral motor, lateral femoral cutaneous and saphenous sensory NCSs need to be performed but can be technically challenging. Needle EMG examination of the iliopsoas, adductor longus, and quadriceps muscles may show abnormality. In sacral plexopathies, the fibular and tibial motor, sural, superficial fibular, and plantar sensory NCSs may show abnormality. Needle EMG examination may demonstrate abnormalities in tibialis anterior, peroneus longus, gastrocnemius, and tibialis posterior muscles as well as more proximal sciatic innervated muscles (biceps femoris, semitendinosus, semimembranosus). The presence of needle EMG abnormalities in proximal muscles innervated by nerves derived directly from the plexus like inferior gluteal or superior gluteal nerves helps to distinguish a

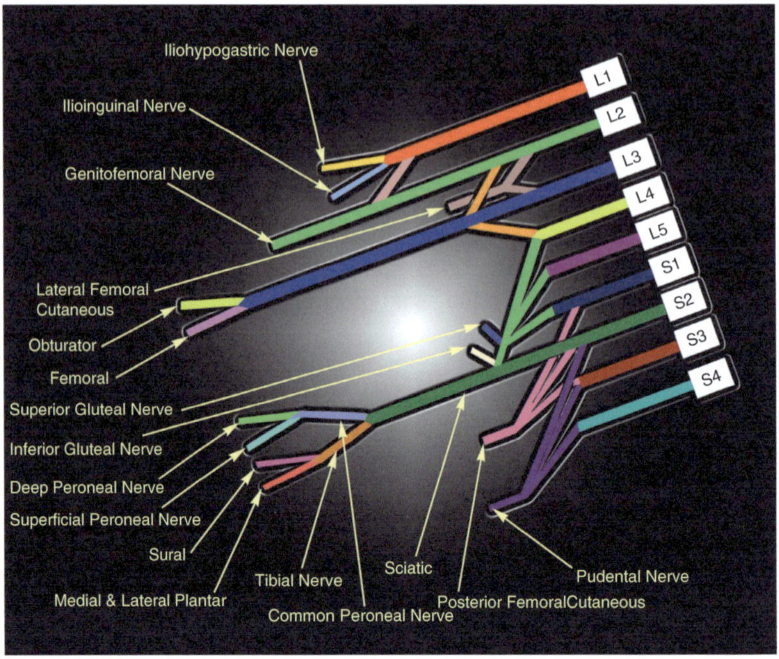

Fig. 11.15 Schematic diagram of lumbosacral plexus

Table 11.7 Details regarding the nerve branches, muscles innervated, and sensory distribution by the lumbosacral plexus

Plexus	Nerve	Root(s)	Muscle(s)	Sensory Distribution
Lumbar plexus	Iliohypogastric	L1–2		Inferior abdominal wall
	Ilioinguinal	L1–2		Superior medial thigh
	Obturator	L2–4	Adductor Magnus, Adductor Longus, Gracilis	
	Femoral	L2,3,4	Rectus Femoris, Vastus Lateralis, Vastus Medialis, & Vastus Intermedius	
	Saphenous	L2,3,4		Medial leg and foot
	Lateral femoral cutaneous	L3–4		Anterolateral thigh
Sacral plexus	Superior gluteal	L4–5	Gluteus Medius, Gluteus Minimus, Tensor Fascia Lata	
	Inferior gluteal	L4-S1	Gluteus Maximus	
	Sciatic	L4-S2	Tibialis Anterior, Peroneus Longus, Gastrocnemius & Soleus, intrinsic muscles of the foot	Lateral leg and foot
	Pudendal	S2,3,4	External anal sphincter	Perineal

sacral plexopathy from a sciatic neuropathy. MRI is the neuroimaging study of choice to evaluate the lumbosacral plexus and the findings include increased T2 signal intensity, focal or diffuse enhancement, and edema of nerve segments.

Though infiltration of the lumbosacral plexus by neoplasms is rare, MRI is more sensitive than CT in detecting subtle infiltrative lesions. Rarely, the plexus can be compressed by large retroperitoneal hematomas, abscesses, or large aneurysms, and may present with pain and weakness, particularly in the thigh region. Trauma and vascular procedures are likely to produce acute and rapid onset of plexopathy symptoms; whereas conditions like abscess or aneurysm formation are likely to produce gradual onset of symptoms. Imaging studies including ultrasound, CT, or MRI are diagnostic.

11.8 Inherited Neuropathies

Inherited or hereditary neuropathies are a complex heterogeneous group of disorders. These disorders generally have insidious onset clinical features and the course is indolent and progresses over years to decades. Due to advancements in molecular genetics, next-generation sequencing, and increased detection of metabolic or molecular defects, a more accurate classification of hereditary neuropathies is now possible. However, for those neuropathies for which the genetic abnormality is yet to be identified, the classification is still based on the clinical phenotype, mode of inheritance, and the class of neurons predominantly affected. Identification of chromosomal loci or causative genes for a given disease phenotype has led to an ever-increasing number of genes coding for a specific gene product necessary for myelin or axonal function.

Inherited neuropathies are common disorders and account for approximately 40% of chronic polyneuropathies. Surprisingly in a large percentage of patients, their inheritance can go unrecognized. Obtaining detailed family histories, a careful screening of skeletal abnormalities like hammer toes, pes cavus, or scoliosis, and more importantly, performing clinical examinations in blood relatives of patients are important in recognizing a previously unsuspected inherited neuropathy. The dearth of positive symptoms often makes patients not volunteer information about their own or relatives' conditions. Even in the presence of a truly negative family history, the likelihood of a hereditary neuropathy cannot be ruled out due to situations like the early death of one or both parents, few blood relatives, or autosomal recessive traits. Hence it is prudent to consider the possibility of a hereditary neuropathy in every patient with chronic 'acquired' neuropathy that remains obscure or uncertain or refractory to medical treatment.

The hereditary forms of peripheral neuropathies include Charcot–Marie–Tooth disease (CMT), also known as Hereditary Motor Sensory Neuropathy (HMSN), hereditary sensory and autonomic neuropathies, hereditary motor neuropathies, and small-fiber neuropathies. The overall prevalence of inherited peripheral neuropathies is 1:2500. Approximately 75% of cases of inherited peripheral neuropathic disorders belong to CMT 1 and other variants of CMT and in this light, the author would like to emphasize and give details of those which are more prevalent.

11.8.1 Charcot–Marie–Tooth Disease

Named after three investigators who described them in the late 1800s (details covered earlier in the historical background), CMT affects approximately one in 2500 individuals and is one of the most common hereditary neurological disorders. Most CMT patients have an autosomal dominant inheritance; in addition to the same, X-linked dominant, autosomal recessive and a few sporadic forms do exist. Roughly two-thirds of the CMT neuropathies are demyelinating and the remaining are primarily the axonal degenerative type. In the majority of the patients having a "typical" CMT phenotype, the disease onset is in childhood or early adulthood and is characterized by distal weakness, sensory loss, absence of reflexes, and presence of foot deformities like pes cavus and hammer toes. Conversely, some patients may develop a severe physical disability in infancy, as seen in Dejerine–Sottas Syndrome (DSS), while others may hardly develop any symptoms.

11.8.1.1 Classification

The classification of CMT is constantly undergoing expansion and revision due to the advancement in molecular genetics. The molecular genetic classification of CMT and related disorders are provided in Table 11.8. Among hereditary neuropathies, more than 30 genes have been identified and so far, more than 50 distinct chromosomal loci also have been identified. Genetic testing for many variants of CMT is now available, which provides an accurate diagnosis and genotypic-phenotypic correlations. Based on electrodiagnostic and neuropathological criteria, Dyck and Lambert have subdivided CMT or HMSNs into dominantly inherited demyelinating HMSN I (CMT 1) and dominantly inherited axonal HMSN II (CMT 2) forms. The remaining HMSN types have also been classified into HMSN III to VII based on the inheritance pattern and accompanying features. CMT 1 is an autosomal dominantly inherited demyelinating neuropathy, CMT 2 is dominantly inherited axonal neuropathy, CMT X is an X-linked neuropathy and CMT 4 is a recessive neuropathy. The CMT 1, 2, and 4 are heterogeneous and several subtypes have been identified by linkage analysis (Table 11.8). DSS (CMT 3) is a severe disabling neuropathy that begins in infancy and has both the autosomal dominant and the autosomal recessive pattern of inheritance.

11.8.1.2 Features

Generally, CMT 1 patients are slow runners in childhood and the foot issues develop when they are in their teenage and often require ankle orthotic support during adult life. Wasting of calf muscles can be profound in certain subtypes of CMT 1 and leads to an inverted champagne bottle appearance of the leg. Variable degrees of hand weakness are likely to occur as the disease progresses and typically lag 10 years behind the development of foot weakness. Even sensory loss is variable and

Table 11.8 Molecular genetic classification of Charcot–Marie–Tooth disease and related disorders

Type	Subtypes	Chromosomal Locus	Gene
CMT 1	CMT 1A	17p11.2	Peripheral myelin protein 22 (PMP22)
	CMT 1B	1q22-q23	Myelin protein zero (MPZ)
	CMT 1C	16p13.1	Lipopolysaccharide-induced tumor necrosis factor-α factor (LITAF)
	CMT 1D	10q21	Early growth response 2 (EGR2)
	CMT 1E	17p11.2	Peripheral myelin Protein-22 (PMP22)
	CMT 1F	8p21	Neuro-filament light chain (NEFL)
CMT 2	CMT 2A	1p35/1p36.2	Microtube motor KIFBß; Mitofusin 2 (MFN2)
	CMT 2B	3q13-q22	RAS associated protein (RAB7)
	CMT 2C	12q24	–
	CMT 2D	7p15	Glycyl tRNA Synthetase (GARS)
	CMT 2E	8p21	Neuro-filament light chain (NEFL)
	CMT 2F	7q11–21	Heat shock protein (HSPB1)
	CMT 2G	12q12-q13.3	–
	CMT 2H	8q13/q21.3	Ganglioside-induced differentiation-associated protein 1 (GDAP1)
	CMT 2 L	12q24	Heat shock protein (HSPB8)
CMT 3	DSS-A	17p11.2	Peripheral myelin Protein-22 (PMP22)
	DSS-B	1q22-q23	Myelin protein zero (MPZ)
	DSS-C	1q21-q22	Early growth response 2 gene (EGR2)
CMT 4	CMT 4A	8q21	Ganglioside-induced differentiation-associated protein 1 (GDAP1)
	CMT 4B	11q22	Myotubularin-related protein 2 (MTMR2)
	CMT 4C	5q23-q33	SH3 domain and tetratricopeptide repeats-containing protein 2 (SH3TC2)
	CMT 4D	8q24	N-myc downstream regulated gene 1 (NDRG1)
	CMT 4E	10q21-q22	Early growth response 2 (EGR2)
	CMT 4F	19q13	Periaxin
	CMT 4G	10q23	–
	CMT 4H	12q11.1-q13.11	FYVE, RhoGEF, and PH domain-containing protein 4 (FGD4)
	CMT 4 J	6q21	Factor-induced gene 4 protein
CMT X	CMT X1	Xq13.1	Connexin-32; (Cx32)
	CMT X2	Xq24	–
HNPP		17p11.2	Peripheral myelin Protein-22 (PMP22)

usually affects both large and small fiber modalities. The clinical examination may reveal enlarged nerve trunks in the arm. Despite the combined weakness of distal muscles of lower limbs and reduced proprioception with concomitant balance problems, many CMT patients remain ambulatory throughout their life. The patients may also report postural tremors and muscle cramps.

Regarding the pathological findings, considerable overlaps do exist among the different subtypes of CMT 1. Segmental demyelination, remyelination, and axonal loss are characteristic features and demyelination is profound in DSS. Onion-bulb formations on nerve trunks are less frequent in children compared to adults. Axonal loss is variable and is seen in both small and large diameter myelinated fibers.

One-third of the autosomal dominant CMT belongs to the CMT 2 type. CMT 2 is heterogeneous and like CMT 1, the disease presents with distal weakness, atrophy, sensory loss, and foot deformities. CMT 2 patients have a wider age range of symptom onset and degree of disability than those with CMT 1. CMT 2 patients are more likely to maintain their deep tendon reflexes. Clinically, it is difficult to accurately distinguish CMT 1 from CMT 2 and electrodiagnostic testing may help to distinguish them. Reduced CMAP and SNAP amplitudes with a normal or mild reduction in conduction velocities are distinguishing features of CMT 2. Needle EMG may reveal evidence of active denervation and partial re-innervation. The sural nerve biopsies may show axonal loss without evidence of demyelination. Even CMT 4, a recessively inherited form, has heterogeneous phenotypes and is usually more severe than the autosomal dominantly inherited disorders. In addition to sensory and motor dysfunctions, CMT 4 patients may have systemic symptoms like cataracts and deafness. CMT 4 subtypes can have both demyelinating and axonal involvement.

11.8.1.3 Management

There are no specific cures for inherited neuropathies. Most patients may benefit from physiotherapy or occupational therapy. Care of the foot and use of orthotics are important and can help patients ambulate independently throughout their lives. Occupational therapy can help those patients who have difficulties with fine movements such as buttoning and zippering. Quite often the patients are uninformed about genetic patterns, the risk of developing the disease, and the reproductive options available, and to address the same, genetic counseling is critical in the management of patients. In addition to the above, genetic counselors can be of invaluable help in obtaining pedigrees from the patients.

In a systematic review on rehabilitation management of the CMT syndrome, 11 studies were finally shortlisted. Of the shortlisted studies for review, five studies were on physiotherapy treatment and six were on orthosis treatment. The physiotherapy trials included a home exercise program, endurance training program, and strength training. The CMT patients revealed a moderate recovery of leg strength and betterment of Activities of Daily Living (ADL) performance with an insignificant improvement for walking speed. Some studies on orthotics verified the hypothesis that AFOs improve posture and walking control and found that AFO prescription is relevant for improving balance and walking performance. In a study on the use of night splint following stretching exercise, the results revealed an improvement in ankle dorsiflexion by 4° after 4 weeks of usage; however, another study on night splint use did not reveal any significant differences in the ROM or muscle strength of inverters or evertors. A study comparing three different AFOs (foot-up splint, push brace for ankle stability and Multifit Achilles drop foot orthosis, a variant of the posterior leaf AFO) versus standard shoes did not show any significant differences in the speed and step length of patients wearing any of the three AFOs and standard shoes. The authors of the systematic review stressed that the majority of the scientific literature available was focusing on the clinical aspects of CMT and neglected the useful and innovative aspects of rehabilitation. According to them, an evidence-based protocol could not be established due to a dearth of literature on rehabilitation and limited comparison trials.

11.9 Immune-Mediated Neuropathies

Immune-mediated neuropathies are a heterogeneous group of disorders caused by immune-mediated damage to peripheral nerves. The disorders can range from a fulminant, life-threatening situation to an asymptomatic, minimally progressive process. GBS and its subtypes including Acute Inflammatory Demyelinating Polyradiculoneuropathy (AIDP) and Acute Motor and Sensory Axonal Neuropathy (AMSAN) encompass the acute onset immune-mediated neuropathies; whereas CIDP, multifocal motor neuropathy, and neuropathies associated with an Immunoglobulin M (IgM) Monoclonal Gammopathy of Undetermined Significance (MGUS) comprise the chronic type. Few of the rare immune-mediated neuropathies identified in the recent past are now considered subtypes of GBS or CIDP. Table 11.9 gives information about the types of immune-mediated neuropathies.

Table 11.9 Types of immune-mediated neuropathies

Type	Subtypes
Guillain–Barré syndrome (GBS) and variants	• Acute inflammatory demyelinating Polyradiculoneuropathy (AIDP) • Acute motor and sensory axonal neuropathy (AMSAN) • Acute motor axonal neuropathy (AMAN) • Miller Fisher syndrome • Miller Fisher/Guillain–Barré overlap syndrome • Acute sensory Neuronopathy • Acute Pandysautonomia • Pharyngeal-cervical-brachial variant
Chronic inflammatory demyelinating Polyradiculoneuropathy and variants	• Chronic inflammatory demyelinating Polyradiculoneuropathy (CIDP) Pure motor Pure sensory • Chronic inflammatory axonal polyneuropathy • Distal acquired demyelinating symmetric neuropathy • Demyelinating neuropathies with MGUS other than IgM • Lewis–Sumner syndrome (multifocal acquired demyelinating sensory and motor neuropathy)
Chronic demyelinating neuropathies different from CIDP	• Multifocal motor neuropathy (MMN) • Demyelinating neuropathies associated with IgM MGUS • Paranodopathies • Chronic ataxic neuropathy with ophthalmoplegia, monoclonal protein, cold agglutinins, and disialosyl antibodies • Chronic ataxic neuropathy with disialosyl antibodies

GBS can often be considered a neurological emergency and patients require hospitalization and intensive care management. Though many may survive, they are often left with long-term disabilities. On the contrary, chronic immune-mediated neuropathies are typically less disabling but may require long-term immunotherapy during the course of the disease. Making the right diagnosis is quite essential to prevent extensive damage to the peripheral nervous system and also aids in identifying the correct treatment option. Overall, diagnosis is based on clinical features, electrophysiological diagnostic tests, and other supportive evidence, including the identification of autoantibodies for some disorders.

11.9.1 Guillain–Barré Syndrome

11.9.1.1 Introduction

Landry–Guillain–Barré–Strohl syndrome (GBS) is an umbrella term to describe several related syndromes (Table 11.9). Among the syndromes, AIDP is the most common form and the disorder was described by Guillain et al., as a syndrome of radiculoneuritis with increased protein content (albumin) in the CSF without increased cell count. The syndrome affects to a variable extent the motor, sensory, and autonomic functions of the peripheral nervous system. Typically in AIDP, the demyelinating features predominate but in other variants, the axonal degeneration predominates. Acute onset, a monophasic course with recovery, increased CSF protein with cytoalbumin disassociation, and overlap of clinical features among the disorders, are the factors unifying the different variants of GBS. Intensive clinical support coupled with immunomodulators like intravenous immunoglobulin and plasma exchange has considerably reduced the length of illness, morbidity, and mortality.

The incidence of AIDP is 1–2/100,000 per year and the disease occurs in all the geographical regions of the world and all age groups with a slightly higher male preponderance. The disorder is generally sporadic in nature; however, occasional epidemiological clusters are observed, following certain vaccination programs.

11.9.1.2 Clinical Features

In approximately 60% of the cases, a history of a mild respiratory or gastrointestinal infection or immunization precedes the neuropathic symptoms by a few weeks. Paresthesia and slight numbness in the toes and fingers are the earliest clinical features seen among typical cases. Weakness is the most prominent symptom of AIDP and it evolves more or less symmetrically over a period of days to a week or two. Weakness classically begins distally in the lower extremities and ascends to the upper extremities and the trunk, with bulbar musculature involvement in some cases. During the early days, the upper limb muscles are usually stronger compared to the lower limb muscles and in some, they are almost completely spared. The disorder may affect both the proximal and the distal limb nerve roots, which include the cranial nerves as well as the phrenic nerve. The severity of weakness can vary from mild gait difficulty to total extremity and respiratory paralysis. In about 5% of cases, within a span of few days, the weakness can progress to total motor and respiratory paralysis and death from respiratory failure. The net effect of the generalized weakness is quadriplegia in one-third of the patients, limited bed mobility in about one-third, and a need for respiratory assistance in another one-third of the patients.

In around 5% of AIDP patients, the disease presents with isolated cranial nerve involvement, followed by descending limb weakness. In severe cases, the ocular motor nerves can be paralyzed and pupils can be non-reactive. AIDP may have dysautonomia, bulbar weakness, and axonal damage as a part of the syndrome, thus overlapping with other subtypes of the syndrome. Oropharyngeal dysfunction is observed in severe cases and in about 50% of the cases, facial diplegia (bilateral facial paralysis) is another important finding. Though facial muscle weakness is typically bilateral, in about 10% of patients, the weakness can be unilateral. Signs of AIDP generally include reduced or absent deep tendon reflexes.

Approximately 50% of AIDP patients complain about muscle pain or aches and are primarily located in the hips, thighs, and back. Such symptoms are often mistaken for lumbar disc disease, low back strain, or other orthopedic conditions. For a few AIDP patients burning sensations in the fingers and toes can be an early symptom and may become a persistent problem. For most patients, the modality and the extent of somatosensory impairment are variable during the early days. By the end of the first week, in many, the deep sensations (kinesthetic and joint position sense and vibration) in the toes and fingers are reduced and are more affected than the superficial sensations (pain and temperature).

Approximately 65% of AIDP patients develop autonomic dysfunction due to the involvement of both sympathetic and parasympathetic systems. Sinus tachycardia is a common feature and may be coupled with bradycardia, orthostatic intolerance, facial flushing, and hypohidrosis and episodic profuse diaphoresis. In a few, the autonomic changes can be fatal, and therefore constant monitoring is crucial in such cases. In approximately 15% of the patients, urinary retention can occur soon after the onset of weakness but urinary catheterization is rarely required for more than a few days as incontinence is generally transient. Development of respiratory insufficiency secondary to diaphragmatic and intercostal muscle weakness is an important complication of AIDP and roughly 10–20% of AIDP patients develop respiratory failure requiring ventilatory support and 2–5% succumb to complications. Typically, the disease has a monophasic course and true relapses are extremely rare.

Severe, progressive paralysis with sensory loss and partial recovery is a classic feature of AMSAN. Griffin et al. coined the name AMSAN and based on the histological examination of specimens, they reported that the variant is characterized by axonal degeneration with periaxonal macrophages and minimal demyelination. A high correlation with antecedent Campylobacter jejuni infection was also noted by them. A primary axonal form of pure motor involvement was reported by McKhann et al. and termed that variant as Acute Motor Axonal Neuropathy (AMAN).

Compared to AMSAN, AMAN is typically associated with acute weakness or paralysis without sensory loss. Like AMSAN, an antecedent Campylobacter infection is common for AMAN. Clinically AMAN is similar to AIDP but characterized pathologically by widespread and severe axonal degeneration. Within a few weeks, the muscle atrophy becomes apparent. The pathology of AMAN suggests an antibody-mediated attack and is characterized by the presence of Immunoglobulin G (IgG) and complement bound to the axolemma of the motor fibers. The muscular weakness in AMAN is more distal and one-fourth of the patients may present with cranial nerve involvement. In AMAN, the sensory examination is normal both clinically and electrophysiologically; however, the reflexes are most often reduced. Many patients with AMAN may recover in a few months; however, those having a rapidly progressive severe weakness may have a poor prognosis. Empirical treatment studies indicate that Intravenous Immunoglobulin G (IVIg) may be effective but not plasma exchange or corticosteroids.

Ophthalmoplegia, ataxia, and areflexia are the triad of the Miller Fisher variant of GBS. Facial muscle weakness, ptosis, and pupillary abnormalities are also frequently seen among these patients. The incidence of the Miller Fisher variant syndrome is probably 0.09/100,000 in the general population. A sensory variant of AIDP termed acute sensory neuronopathy is also reported and is characterized by the acute loss of distal sensory function with the preservation of muscle strength. This condition is characterized by profound sensory ataxia, areflexia, and widespread sensory loss due to the principally large sensory fiber involvement. Slowed or absent sensory conduction and normal muscle and motor nerve conduction are the electrophysiological findings of this variant. Due to the rapid onset, extensive and pure sensory involvement, and poor prognosis, the injury is most likely confined to the dorsal root and Gasserian ganglia and hence the name "sensory neuronopathy."

Acute pandysautonomia, a variant of GBS, first reported by Young et al., is characterized by the acute onset of severe and disabling autonomic failure affecting sympathetic, parasympathetic, and enteric functions. The symptoms include orthostatic hypotension, gastrointestinal issues like nausea, vomiting, severe constipation or diarrhea, reduced sweating, urinary incontinence, and sexual dysfunction. History may reveal an antecedent of acute infection, vaccination, or surgery. The majority of patients have a poor prognosis with a chronic debilitating course and some patients may respond to plasmapheresis with a certain amount of success.

11.9.1.3 Pathology

Segmental demyelination and mononuclear cellular infiltration are the histopathological evidence seen in AIDP. The cellular infiltrations are observed in peripheral nerves, cranial nerves, spinal roots, dorsal root ganglia, and autonomic ganglia. Macrophage infiltration is associated with phagocytosis, segmental demyelination, and retraction of the paranodal sheath. In case of severe demyelinating neuropathy, damage to the adjoining axons may lead to axonal degeneration.

The precise cause for the inflammatory changes is largely unclear. Several viral and bacterial infections have been associated with AIDP, which include cytomegalovirus, Mycoplasma pneumonia, Epstein–Barr virus, HIV, hepatitis A and B viruses, Streptococcus pyogenes, Staphylococcus aureus, and Borrelia burgdorferi. Even vaccines for tetanus, poliomyelitis, influenza, and rabies have been associated with demyelinating neuropathy. Certain cases of vaccine-mediated AIDP are associated with sensitization to myelin basic protein and occur in clusters. For those patients with gastrointestinal symptoms, an increase in Campylobacter jejuni titer is reported and is more likely to cause pure motor involvement with axonal degeneration and poor prognosis.

The immune response in GBS is believed to be secondary to the activation of autoreactive T cells and antibodies due to molecular mimicry. A variety of cytokines, chemokines, and proteolytic enzymes are elevated in the peripheral nerves and CSF of GBS patients. Certain Campylobacter jejuni serotypes are associated with GBS and contain sialylated oligosaccharide structures in their outer core identical to peripheral nerve gangliosides; however, the role of ganglioside antibodies in GBS is still unclear.

Though it has been customary to anatomically classify neuropathies into disorders of axons or myelin, in the recent past, toxic, immune-mediated, and hereditary disorders targeting proteins and ion channels in the nodal regions (nodes of Ranvier), referred to as nodopathies, have been recognized. AMAN, a variant of GBS, is an instance of such nodopathies. Like demyelination, nodopathies are not limited to disrupted conduction and may be associated with subsequent axonal loss. The impairment of the sodium-calcium pump function is hypothesized to lead to intracellular calcium accumulation contributing to eventual axonal degeneration. GBS with autoantibodies associated with the nodal antigen, CIDP, Miller Fisher syndrome, and multifocal motor neuropathy are some other neuropathies in which nodal dysfunction is hypothesized.

11.9.1.4 Laboratory Findings

Albuminocytologic dissociation (an elevated protein level in the CSF without associated pleocytosis) is a typical CSF abnormality found in AIDP. Increased white blood cell count, predominantly the lymphocytes, beyond 50 cells/ml suggests another diagnosis, like HIV or Lyme disease. CSF examination may demonstrate the presence of oligoclonal bands and elevated myelin basic protein. In 10% of AIDP patients, CSF protein may stay normal all through the course of illness.

Electrophysiological studies are crucial and may give insights into the underlying pathology. But, during the first week of symptoms, a repeated evaluation may be required to identify the abnormality. Typically, during the early stage of AIDP, the absence or prolonged H-reflex and F-wave responses may be the only abnormality. In sequential studies, multifocal conduction block and slowing and temporal dispersion may become evident. Due to the multifocal conduction block, the distal latency can be prolonged and the conduction velocity and the CMAP amplitude can be reduced. In suspected cases, the nerve conduction studies should be performed for a minimum of three nerves and preferably the sural, peroneal, and median nerves. Criteria for demyelination are given in Box 11.4 and a minimum of three of the four criteria are suggestive of demyelination. Since many patients may not fulfill these electrodiagnostic criteria during the early stage of illness, other sets of criteria have been suggested. Typically, in AMSAN, unlike AMAN, the stimulation of several nerves reveals no responses or considerable reduction of amplitude in CMAP (below 80%), with sensory nerve involvement.

Needle EMG may be helpful in the initial stage for early diagnosis of AIDP. Many

Box 11.4 Criteria for Identifying Demyelination During Nerve Conduction Studies

Criteria for demyelination

- Conduction velocity <80% of the lower limit of normal (LLN) in two or more nerves, if the amplitude is >80% of LLN "or" to <70% of LLN, if amplitude <80% of LLN
- Partial conduction block or abnormal temporal dispersion in one or more motor nerves.
- Distal latencies prolonged in two or more nerves to >125% of the upper limit of normal (ULN), if the amplitude >80% LLN "or" to >150% of ULN, if the amplitude <80% of LLN.
- Absent F waves or prolonged minimum F-wave latencies in two or more motor nerves to >120% of ULN if the amplitude >80% of LLN "or" to >150% of ULN, if the amplitude <80% of LLN.

patients may demonstrate a reduced motor unit potential recruitment without any configuration change or spontaneous abnormal activities. Needle EMG may reveal myokymic discharges during the initial few weeks. In the case of axonal degeneration, fibrillation potentials and positive waves may appear between 2 and 5 weeks. AMSAN is characterized by an abundance of early diffuse fibrillation potentials by the third week. To exclude other conditions like transverse myelitis or brainstem stroke, radioimaging may be useful, especially when the patient presents with unusual clinical presentations. A serologic test may help in identifying the antibodies to specific gangliosides.

11.9.1.5 Diagnosis and Differential Diagnosis

The diagnosis of AIDP is based on the medical history, clinical examination findings, abnormal CSF findings, and electrophysiological studies. It can often be difficult when CSF studies are normal during the first week with normal or few abnormal electrophysiological findings. In quite a few patients treatment may need to be commenced based on clinical suspicion without confirmatory laboratory testing. In the most common variants of GBS, the AIDP and the AMSAN, weakness is the chief early complaint. In most patients, the legs are initially involved and then the weakness progresses proximally. The National Institute of Neurological Disorders and Stroke (NINDS) had put forward the diagnostic criteria, which require progressive motor weakness of more than one extremity and loss of tendon reflex to make a diagnosis of AIDP. Weakness should be symmetric. However, these criteria generally fail to differentiate AIDP from AMSAN and can only be differentiated based on electrodiagnostic and pathologic findings.

While evaluating an AIDP patient it is essential to consider other causes of rapidly progressive weakness. Atypical presentations like continuous fever, retained reflexes, manifesting a definite spinal cord level, and loss of pupillary reflexes, most likely suggest other diagnoses. Table 11.10 provides details of the differential diagnosis.

Table 11.10 Differential diagnosis for Guillain–Barré syndrome

	CSF findings	ENMG findings	Features
Acute compression of spinal cord	Normal or pleocytosis	Normal findings	Areflexic; flaccidity of muscles, sensory level; progressive appearance of reflexes; absence of bulbar symptoms
Poliomyelitis	Pleocytosis, predominant lymphocyte	Denervation, decreased CMAP amplitude and motor unit recruitment	Flaccidity with absence of sensory symptoms; asymmetric
Transverse myelitis	Pleocytosis may or not be present; increased protein, presence of oligoclonal bands	Normal	Areflexic; flaccidity of muscles, sensory level; progressive appearance of reflexes; absence of bulbar symptoms
Botulism	Normal findings	Post-tetanic amplification; reduced CMAP amplitudes	Pupillary light reflex loss; positive bulbar findings, motor symptoms only; rapid onset of symptoms
Organo-phosphate poisoning	Normal	–	History of exposure, acute weakness with no sensory symptoms
Carcinomatous meningitis	Pleocytosis, abnormal cytology, increased protein	Radicular pattern denervation	Features of multiple radiculopathies; asymmetric sensory and motor findings; suspicious for malignancy
Hypokalemia	Normal	–	Skeletal muscle weakness; cramping; arrhythmia; diarrhea or constipation; low serum potassium level
Uremia	Normal or increased protein	Demyelinating neuropathy	Polyneuritis, restless legs, and cramps; high levels of urea and creatinine in the blood; positive history of renal failure

(continued)

Table 11.10 (continued)

	CSF findings	ENMG findings	Features
Porphyria	Normal	Axonal neuropathy	Muscle pain and weakness, tingling, and numbness; gastrointestinal and psychiatric manifestations, positive urine porphyrin test
Viral hepatitis	Normal or increased protein	Demyelinating neuropathy	Raised liver enzymes; hepatitis
Thallium poisoning	Normal or increased protein	Axonal neuropathy	Alopecia, positive history of exposure
Arsenic poisoning	Normal	Initially demyelinating and then axonal neuropathy	Alopecia, Mees' lines in the nails and rain drop pigmentation of skin, positive history of exposure

11.9.1.6 Treatment

The treatment options currently available are meant to shorten the recovery time and improve prognosis. Plasma exchange was the first treatment found to be favorable. Evidence suggests that the continuous flow plasma exchange technique may be superior to the intermittent flow and albumin may be superior to fresh frozen plasma as the exchange fluid. The exchange is most effective when commenced within a week after disease onset but may continue to have a beneficial impact even when initiated later in the course of the disease. IVIg therapy is equally beneficial when initiated within the first few weeks; however, a combination of plasma exchange and IVIg therapy is not recommended. Unlike CIDP, standalone treatment using corticosteroids is not usually beneficial in AIDP. Both therapies have potential side effects. For plasma exchange, albumin and saline are used as replacement fluids and the administration of both can significantly reduce the concentration of clotting factors, increasing the risk of coagulopathy. To ensure a reasonable flow rate, many patients may require a central venous catheter placement that may considerably increase the risk of developing complications like pneumothorax and septicemia. Hypotension is another complication of plasma exchange.

IVIg has become a common treatment due to the inadequate availability of plasma exchange machines and limited qualified technicians. Though the exact mechanism of action of IVIg is unclear, the actions include neutralization of antibodies, suppression of cytokines, inhibition of complement binding, blockade of Fc receptors, modulation of T cell function, and selective modulation of circulating proinflammatory cytokines. Severe adverse reactions with IVIg are less common and the complications include congestive heart failure, serum sickness, and migraine headache. Typically, neurologists do not treat barely-affected AIDP patients or those who are stable. However, aggressive treatment and closed monitoring will be initiated by the neurologist if the patients are presenting with rapidly progressive weakness with pulmonary function decline.

Respiratory failure, autonomic dysfunction, and secondary medical complications are the causes of mortality among AIDP patients. Intubation should be considered for imminent respiratory failure or to protect the respiratory passage when the bulbar function is compromised. Elective intubation in a controlled environment is preferred to minimize the complications of intubation (aspiration, hypoxia). Factors that can increase the risk of respiratory failure include bulbar weakness and poor clearance of secretions. Respiratory strength should be monitored frequently using bedside spirometry, forced vital capacity, or maximum negative inspiratory force. Those at risk for aspiration should be fed by alternate means and patients who require frequent suctioning to clear the airway from secretions may need future intubation.

Autonomic dysfunction can be difficult to manage and requires intensive monitoring and the dysfunctions include tachycardia, malignant hypertension, significant hypotension, fever, and sweating. Malignant hypertension can predispose to myocardial infarctions. Hypertension can often be hard to treat, especially when antihypertensive agents cause hypotension, and hence rapidly reversible medications in small doses are preferred. Fever and a clear chest X-ray may signal a fever caused by dysautonomia. Deep vein thrombosis and pulmonary embolism are complications of immobility and may require sequential compression devices and heparin or heparinoid agents unless the agents are contraindicated. Nursing care is crucial to avert skin breakdown and therefore frequent change of position and turning and observance for skin breakdown is required. As soon as medically tolerated, ROM exercise should be performed every day to prevent contractures in immobile patients.

Factors that determine poor prognosis in AIDP include age above 60 years, respiratory failure requiring ventilatory support, rapid progression, and mean distal CMAP amplitudes close to 20% of the LLN. Such poor prognostic factors reduce the possibility of independent walking at 6 months, by 80%. The majority of the AIDP patients are likely to regain muscle strength to a certain extent; however, 20% of the cases are left with considerable residual motor weakness even after 1 year. Those patients with an obvious gastrointestinal tract illness preceding their symptoms are more likely to develop AMSAN and have a poor prognosis. Patients with extensive and predominant axonal involvement tend to have a worse prognosis, with the prolonged need for ventilation and higher residual weakness. Patients with AMAN generally have a good prognosis, similar to that of AIDP.

11.9.2 Physiotherapy for Guillain–Barré Syndrome

The management of the sensorimotor deficits due to GBS includes supportive care during the early phase, prevention of long-term medical complications during the acute through subacute phases, and rehabilitation throughout the recovery stage. Since we anticipate a significant return of function within weeks to months for the majority of GBS patients, the focus of therapeutic management should be to help maintain the integrity of functioning systems, attend cramps and pain, encourage compensatory strategies when appropriate, and promote more activity once the disease has reached a plateau. The objectives and means of management will change when the GBS patient moves through the acute phase, the plateau phase once the disease reaches its nadir, and the subacute and chronic phases. For instance, the strategies used for fair muscle strength are certainly different from those used when the strength is poor.

11.9.2.1 Assessment

The comprehensive examination of the patient's sensorimotor and autonomic functions should include components like historical details such as patterns and sequence of onset of the symptom(s), any recent illness, and prior episodes of sensorimotor issues. The scope of a comprehensive clinical examination in any given session depends on the patient's neurological condition and his or her ability to participate. The motor system examination should encompass components like the visual inspection for muscle wasting and deformities, tendon reflexes for the muscle stretch response, Manual Muscle Testing (MMT) to identify the pattern of weakness, functional evaluation of cranial nerves, ROM of the involved extremity joints, static and dynamic balance if appropriate, functional status, and endurance and fatigue.

A battery of screening tests is most appropriate to determine the involvement of the sensorimotor and autonomic systems. Noting the vital signs at baseline, mid and immediately after physical activity, variations in the vitals with respect to posture, and checking the integrity of skin, particularly in bed- or wheelchair-bound patients are essential. For those patients with sensory deficits, a body chart can be used to document the pattern of sensory loss or changes, the modality and location involved, the specific type of sensory change like paresthesia, hypoesthesia, or anesthesia, and the aggravating and relieving factors for the sensory issues.

Muscle strength and ROM evaluation are particularly important to track the course of progression or improvement, identify, predict, and prevent potential contractures and devise an appropriate level of exercise. MMT using modified Medical Research Council (MRC) grading, dynamometry or isokinetic device is helpful to objectively measure the muscle strength and a standard goniometer for ROM evaluation of joints. Since the MMT and ROM evaluation can be time-consuming and extensive during the initial phase post-diagnosis, especially when patients get exhausted quickly due to early fatigue, poor endurance, and considerable pain, the therapist may have to select a few specific muscles and joints each session to complete the assessment. Due to the potential strength difference between the individual muscles, the MMT evaluation should be muscle-specific than testing in groups. Serial examinations of motor and sensory systems, ROM, and functional status help to monitor the changes in the patient's condition. Periodic ROM evaluation of joint(s) may reveal changes in "end feel" due to tightness of the capsular and ligamentous structures. The tightness of the abovementioned structures can add further ROM restriction induced by the shortened muscles and tendon tissues.

The use of numerical rating scales, known for their simplicity, reproducibility, clarity, and sensitivity to minimal changes in pain, may assist in differentiating the difference between muscle weakness, loss of joint ROM, immobility, or pain. Typically, most of the numerical scales have a rating from 0 to 10, with 0 being "no pain" and 10 being "the worst ever imaginable pain." The patient may verbally pick a number or spot a point or draw a circle around the number or line spanning between 0 and 10, which best describes his or her pain intensity. A similar rating scale can be extrapolated for issues such as fatigue and respiratory difficulties. Fatigue can often result from general deconditioning, perception of excessive effort to accomplish activity with weakened muscles, or lack of ability to recruit adequate motor units to sustain muscle contractions for the task. Fatigue can be documented using a questionnaire such as the Fatigue Severity Scale (FSS) or Fatigue Impact Scale (FIS). Standardized scales of independence in ADL, manual dexterity, balance, and gait are certain functional tests that can be included in the comprehensive assessment. Nottingham Health Profile, a generic self-administered questionnaire designed to measure patient's perceived emotional, social, and physical health problems, and Short Form (SF-36) survey questionnaire, a self-reported measure covering eight domains of health including physical, social, emotional, mental, and general health perceptions are few Health-related Quality of Life (HRQOL) measurement tools for Quality of Life (QOL) documentation.

For those GBS patients with possible or definite respiratory and swallowing issues, acute care should include serial evaluation of respiratory or bulbar paralysis. The therapist needs to frequently keep track of the oxygen saturation and Forced Vital Capacity (FVC) levels and any drop below the critical level if noticed, should alert the nursing and/or medical staff to avoid imminent deterioration or life-threatening situation. Several testing measures are available for FVC measurement which includes the use of spirometer, breath-holding test, and single breath count test. An easy bedside test for FVC consists of instructing the patient to take a large breath and count out loud to 25. Failing to cross five counts without pausing for a breath indicates that the patient may be bordering on respiratory failure and a drop of FVC<15 generally warrants elective intubation.

GBS patients who are either intubated or have lower cranial nerve paralysis with oral motor weakness are typically at a high risk of aspiration. A thorough evaluation of oral-motor dysfunction and feeding issues is essential for those patients with severe oral-motor issues and dysphagia. Generally, oral-motor and dysphagia issues are addressed by the multidisciplinary team members consisting of the speech therapist, occupational therapist, and physiotherapist. While feeding, ensure that the patients on the nasogastric tube or percutaneous endoscopic gastrostomy (PEG) tube are positioned in a relatively upright position and maintain the same position for about 30–60 minutes post-feeding to minimize the possibility of aspiration. For those patients with evidence of dysphagia and possible aspiration, comprehensive testing with videofluoroscopy is advisable. Other alternatives for swallowing assessment include techniques like scintigraphy while swallowing a radioactive bolus, fiberoptic endoscopy, ultrasound, and electroglottography.

11.9.2.2 Treatment

The treatment for GBS can be categorized into the acute or the early phase including the plateau, the subacute phase or the recovery phase, and the chronic phase. Though the aforementioned categorization oversimplifies the concept, the reader should realize that the course and progression of the disease and the recovery process once the disease reaches its nadir varies considerably based on the variants of GBS, the extent of involvement, the complications of the disease, and the time delay for medical management. For instance, in many GBS patients, the absence of respiratory muscle involvement with minimal skeletal musculature involvement may speed up the recovery process. On the contrary, those patients with severe respiratory muscle involvement may need weeks of ventilator support in the intensive care unit followed by intense rehabilitation, which may prolong the stay and affect the extent of recovery. Even, variants like AMSAN are known for slow, gradual, and incomplete recovery due to the predominance of axonal degeneration.

Early Phase Management During the initial days post-admission, due to the possibility of respiratory muscle weakness and respiratory failure, GBS patients are typically monitored and treated in the intensive care. Since the acute phase is usually characterized by the progression of the symptoms, the goals of physiotherapy include assisting the pulmonary function, curtailing the acute signs and symptoms, preventing secondary complications like decubitus ulcer, deformities, and contractures, and managing pain. For those patients with weak cough reflex, retention of secretions, and difficulty to expectorate secretions, the therapist needs to provide postural drainage with chest manipulations to maintain the integrity of the airway. For patients with weak respiratory muscles and those on ventilator support based on their potential and participation level, segmental breathing exercises, lung volume expansion exercise techniques, resistive inspiratory training, and gentle stretching of the chest wall have to be provided. The goals of such treatment strategies are to increase the ventilation or oxygenation capacity, reduce oxygen consumption, improve airway clearance of secretions, and enhance exercise tolerance. While training the respiratory muscles, care has to be taken to avoid fatiguing the respiratory muscles. Communication can be difficult and tiresome for many ventilator-dependent patients and in such cases, it is advisable to execute alternative means of communication which may include the use of a communication board, specialized talking tracheostomy tubes with an inflated cuff, infrared eye-blink detector, and touch-sensitive buttons or screen.

To minimize the potential of deformities and contractures, passive gentle stretching exercises, positioning of the limbs with adequate support of pillows and rolls, and use of night splints are appropriate. A few studies have stated that gentle sustained stretch maintained for at least 20 minutes is more beneficial than brief, strong, stretching exercises. Hence, the positional stretch and the use of splints for prolonged positioning should be encouraged instead of short bursts of intermittent, manually applied passive stretching. Many of the facilities presently encourage the use of moldable plastic splints instead of footboards to minimize the tightness of the posterior tibial muscles. For those patients using prefabricated night splints or moldable plastic splints, care should be taken to avoid compression of the common peroneal nerve near the fibular head, especially when the patient's lower limbs are thin due to loss of muscle mass. Hypertonicity, a feature of upper motor neuron lesion, often predisposes for early and progressive deformation of the wrist and the hand. GBS being a lower motor neuron disorder with the muscle tone either flaccid or sub-normal, causes less potential for early and severe wrist and hand muscle tightness and deformities among patients. For the aforementioned reason, the use of a resting splint for the wrist is generally discouraged; instead, a cloth roll or a soft cone should be adequate enough to maintain the functional position of the wrist and the hands.

For many immobile patients, passive ROM exercise may be the only activity possible once the weakness reaches its nadir. While performing routine passive ROM and gentle stretching exercises, ensure the limb is aligned and supported throughout the range. The ROM exercises should cover both accessory and physiological motions, to increase circulation, provide joint lubrication, and maintain soft tissue extensibility in and around the joint. Passive ROM exercises should be performed twice daily for the paralyzed limb(s) and desirable to increase the frequency of sessions, provided no voluntary active movements are possible. Those GBS patients, who can actively move without pain or fatigue, should be encouraged to perform the ROM exercises by themselves and for patients who cannot complete movement through full range independently, the therapist or well-trained and instructed caregiver should provide assistance to help the patient move to the end of the range. While teaching or training ROM exercises to caregivers, the methods of handling, and the care and precautions to be taken to avoid any potential impingement or overstretching injuries need to be emphasized. Overstretching is strictly not advisable as the muscles are likely to be tender with reduced tone. The use of thermal agents (hot fomentation or heat pads) before stretching is advisable for those patients who experience continuous pain. Typically, such patients have a tendency to hold the limbs in a potentially contracture-prone position. The utmost care has to be taken while using thermal agents in those subjects where sensations are diminished. Application of heat is contraindicated if pain and temperature sensations are completely lost.

Medications may not help alleviate pain in many GBS patients and for the same, gentle passive ROM exercise, Interferential Therapy (IFT), and TENS can be the alternatives or adjuncts. Presently no strong evidence exists regarding the benefits of therapeutic currents on the pain associated with GBS. However, the use of the abovementioned therapeutic currents should be considered as a treatment option to help desensitize pain when passive movement or pain medications are not alleviating the same. Reassurance that the pain will subside with gentle ROM exercises and therapeutic currents generally minimize the anxiety that could compound the pain. For those patients with hypersensitivity to light touch and movement, cradling the limb away from the body or bed-spread may minimize intermittent touch contact or movement of sheets or supportive surfaces. A continuous low pressure provided by snuggly wrapping the limbs with elastic bandages may ward off the light and intermittent stimuli that provoke a hypersensitive reaction.

Acclimation to the upright posture, a continuation of the gentle passive ROM exercise, maintenance and improvement of pulmonary function, and prevention of fatigue and overexertion are the common therapeutic goals when the disease reaches the plateau. While acclimatizing to sitting, proper trunk support and postural alignment are of utmost importance. Prevention of decubitus ulcers starts within the first few days of hospitalization, particularly when paralysis of the trunk and limbs is complete or nearly complete. Frequent changes of positions on a regular basis (two hourly) will provide pressure relief. The use of an airbed mattress with a pneumatic pump will help to constantly change the pressure within the mattress to minimize any sustained pressure over any pressure-prone areas. Thin-built patients or those with significant weight loss due to muscle atrophy, with prominent bony areas, may benefit from the use of soft cushions, foams, or pads for pressure relief. Paralysis of the distal musculature, gravity, and persistent recumbent position generally predisposes to edema formation in the lower extremities. An easy and effective strategy to relieve pedal edema and enhance venous drainage is to elevate the distal segment of the extremity to approximately 35–40°. The use of compressive stockings or intermittent compression by a pneumatic sleeve or stocking is recommended, while lying and sitting, to avoid the possibilities of deep vein thrombosis.

For patients with oromotor dysfunction, proper postural support and oromotor training exercises may facilitate the return of oromotor functions. In the majority of the care facilities, speech pathologists or occupational therapists are primarily responsible for initiating a dysphagia treatment program. The therapeutic goal of dysphagia management is to prevent choking and aspiration and encourage effective swallowing of solid and liquid food. Deglutition being a complex process, requires the coordinated reflexive and conscious control of many muscles. Dysphagia management should focus on facilitating head and neck control to maintain an upright posture with a slight forward tilt and oral-motor coordination. During the initial phase of training, thick liquids are used for the conscious swallowing technique and then progressed to thinner liquids as the patient's oral-motor coordination response improves. The training program consists of encouraging a good lip closure, sipping thick fluids through short-length straws, and progressing from thick fluids to semi-soft, moist foods, and eventually solid food. Any sticky or crumbly or hard-to-chew foods are avoidable. The patients are strictly instructed not to talk while attempting to chew or swallow the food. To minimize fatigue during the training, frequent short-session feeding should be encouraged. All the team members should be prepared to use the Heimlich maneuver if the patient shows signs and symptoms of choking like gagging, coughing, wheezing, clutching at the throat, face turning red and lips turning blue, and inability to speak, breathe or swallow. A wall-mounted or portable suction unit should be available to suction the contents and the concerned speech or occupational therapist should immediately alert the nursing and medical staff for further management.

Subacute Phase Management The additional therapeutic goals during this phase include the introduction of a graduated active exercise program while monitoring overuse and fatigue, maximizing functional abilities, resumption of psychosocial roles, and improvement of QOL. Since the acute, subacute, and chronic phases are not water-tight categories, the therapeutic goals may tend to overlap between two consecutive phases. For instance, post-decannulation of tracheostomy and weaning of ventilator support, if the GBS patient's respiratory muscles are still weak, prophylactic breathing exercise, chest physiotherapy, respiratory muscle-strengthening exercise, and encouraging coughing to minimize the retention of secretions, should be continued until those respiratory issues are resolved. Similarly, muscle pain and the tendency for joint stiffness can be issues even during the subacute phase and in such situations, strategies like pain management and joint ROM exercise should be continued until those issues are relieved.

To date, no promising evidence is available associating active exercise with the rate of progression of the disease, remyelination, or axonal regeneration. However, evidence suggests that active exercise can improve function through increased strength and aerobic capacity following the reinnervation of the muscles. The common goal of therapeutic management throughout the rehabilitation phase should be to optimize the ready state of the patient's musculoskeletal system, avoid overwork, enhance cardiovascular and respiratory endurance within the limits of active movement, and encourage and maximize functional recovery as reinnervation occurs.

During the subacute phase, post plateau when the muscle strength begins to return, prescribe limited amounts of low-resistance activities and avoid antigravity strain on the musculature until strength reaches a "fair" grade (3/5) on MMT. In most GBS cases, the strength progressively recovers within a period of 4–6 weeks once the muscle weakness reaches its nadir. The order of recovery of muscles is generally the reverse (descending pattern- proximal towards distal) of progression of weakness (ascending pattern- distal to proximal involvement) that is seen from onset till the weakness reaches the nadir. For example, the girdle, neck, and trunk muscles tend to recover first, before the distal limb muscles show any signs of recovery. Similarly, the recovery of the upper limb tends to progress faster compared to those of the lower limb muscles.

The early recovery of the spinal musculature encourages the GBS patients to use a tilt table for continued acclimation to an erect position and weight-bearing through the lower extremities. During the tilt-table training, to avoid peripheral venous pooling, the use of an abdominal binder and/or compression stockings for the lower extremities is advisable. Due to the potential chance of orthostatic hypotension, therapists must ensure that the patient is well-hydrated prior to the commencement of standing or tilt-table training. In those patients, with a fairly fast recovery, tilt-table management can be circumvented and the direct edge-of-bed sitting and assisted standing can be encouraged, provided the overall muscle strength is fair in the lower extremities. The patient, the caregivers, and the nursing staff should be reminded about the possible taxing of axial musculature when prolonged sitting in bed or wheelchair with or without support is attempted. For those with poor trunk control, an acclimation program of gradual sitting should be instituted until unsupported independent sitting with functional equilibrium reactions is achieved. Foot splints may be needed for the lower extremities to stabilize the ankles while standing. Strengthening the respiratory muscles also minimizes intolerance to upright posture secondary to the weakness of the abovementioned.

Performing strengthening exercises and participating in physical activities for many GBS patients can be challenging, particularly when the functional motor units are limited due to considerable nerve involvement. Active exercise can be commenced as soon as the medical condition is stabilized and the progression of the weakness has reached a plateau. The primary goal in the recovery phase is to direct the strengthening exercises at improving function and not merely muscle strength, i.e., gain in muscle strength should translate into functional activities. The selection of functional activities should be based on the grade of the muscle or group of muscles. In the recovery phase, short periods of non-fatiguing exercises, matched to the patient's strength can be attempted. The difficulty level of the exercise or activity can be gradually increased, provided the patient shows improvement with no deterioration in functional status or strength. In case, a reduction in strength or function is noticed, the continuation of the exercise should be avoided and the patient should be allowed to rest until his status improves. During the initial days of the subacute phase, overuse and fatigue are common issues but if the weakness persists, the details regarding the same should be updated to the consultant neurologist or physician.

In those patients with partial denervation of muscles or extensive segmental demyelination, overworking can produce a profound reduction in muscles' ability to exhibit strength and endurance. In such patients, signs of overuse weakness may produce delayed onset of muscle soreness, typically worsened 1–5 days following the commencement of active or strengthening exercises, associated with a decrease in the maximum amount of muscle force generation. When the strength increases, additional resistance may be applied to those muscles showing good recovery while avoiding strain on muscles that have not yet reached the same level (typically, the distal-most muscles). Researchers have recommended strength training in a gravity-eliminated plane when the muscles are having a grade of 2/5 (poor) on the MRC grading system. Once the patient can move the limb against a resistance equal to the mass of the limb or has a fair grade (3/5) on the MRC grading, he or she can be encouraged to perform free exercises in the antigravity plane. In general, for this population, the exercise should progress like a pyramid with passive ROM at the bottom, active-assisted and free exercise over it, functional specific resistive exercise and progressive resistance strengthening exercise above it, and coordinated functional activities at the top.

A GBS patient with muscle strength below "poor" grade (2−/5) should be encouraged to perform gravity-eliminated exercises like suspension therapy using the axial suspension (Fig. 11.16), a powdered re-education board, or exercise in a hydrotherapy pool. Pendular type of suspension therapy with the point of suspension adjusted further medially/laterally and anteriorly/posteriorly based on the group for which strengthening is devised is the choice of exercise when the strength is 2/5 (poor). To progress muscle strength from 2+/5 towards 3/5 (fair), pendular suspension with additional manual resistance (Fig. 11.17) can be encouraged.

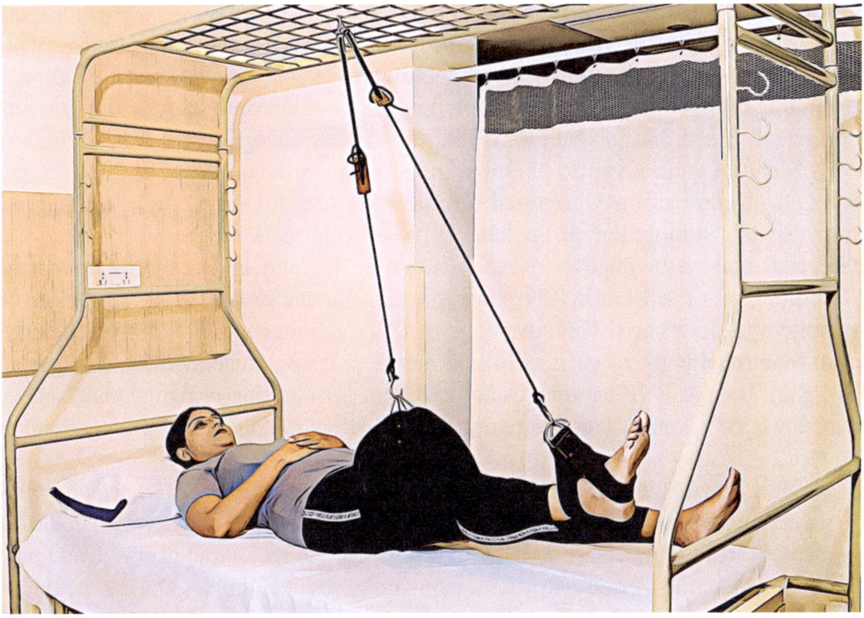

Animated photograph of model with permission

Fig. 11.16 An illustration of axial suspension therapy for the hip abductors

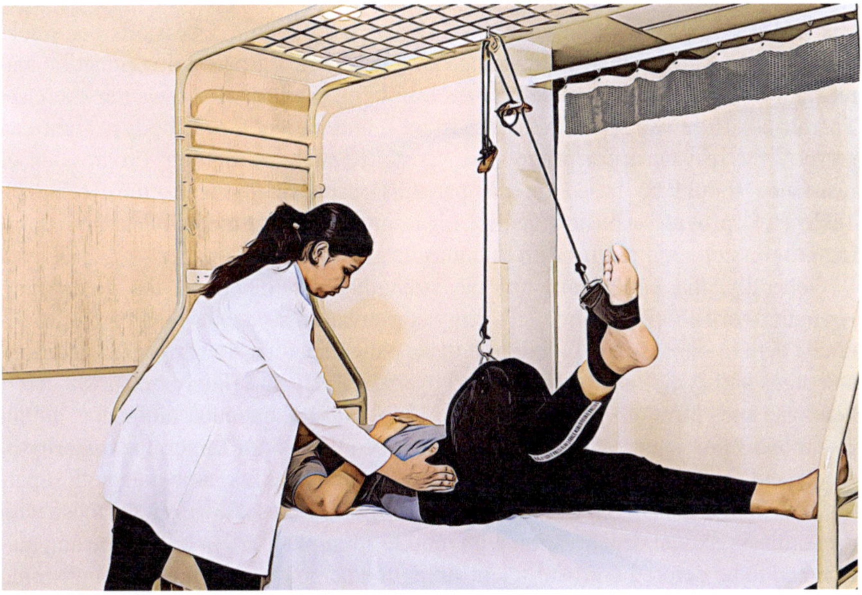

Animated photograph of model and therapist with permission

Fig. 11.17 An illustration of pendular suspension for the hip abductors with manual resistance. Note: The point of suspension is kept medial to the hip joint axis for pendular suspension

Carefully selected, gradual, and progressive supervised suspension therapy will help to strengthen the muscles from poor to fair. For very weak muscles, with strength closer to "trace" (1/5) grade, facilitation techniques like muscle tapping, brushing, and vibration can be attempted in isolation or combination with gravity-eliminated exercises. In addition to the abovementioned facilitatory techniques, NMES can also be given to augment the strength of weak muscles. Even if the strength of contraction is minimal to moderate, Faradic stimulation is preferred over galvanic stimulation, as the latter is more likely to cause chemical burns. The therapist can also choose exercise training techniques like Proprioceptive Neuromuscular Facilitation (PNF) to encourage the maximal desired contraction of specific muscle groups. The advantage of PNF techniques is that they can merge with functional patterns such as rolling, prone progression sequences, transitions, and gait. Though PNF techniques are excellent for eliciting maximal contraction, care must be taken not to overwork the weaker components of the movement pattern.

Studies have shown that non-fatiguing exercises are effective in improving mild to moderate weakness without causing overuse and fatigue. The exercise should be performed strictly under supervision to ensure correct technique of usage and prevent the worsening of muscle or joint pain and excessive fatigue. During the ongoing phase of the subacute phase, the number of repetitions should be kept minimal with frequent and adequate rest periods to avoid overuse weakness and fatigue. The intensity, the number of repetitions, and the sets of exercises should be matched to the patient's muscular strength. For instance, the resistance used for non-fatiguing strengthening exercises should be approximately 60% of 1RM. The other parameters include 5–10 repetitions per set, 1–2 sets per session, 5 sessions per week, concentric movement instead of isometric contraction, 1:2 ratio for contraction and relaxation time, and avoidance of breath-holding while performing the exercise. Therabands, free weights, isokinetic devices, and patient's own body weight can serve as the resistance for progressive resistance strength training. Progression of resistance should be based on the patient's response to the exercise regimen. Table 11.11 provides a basic guideline regarding the type or model of strength training progression required for different muscle grades.

Generally, the upper limb muscle strength recovers before the lower limb strength. Work simplification and energy conservation strategies may be of considerable help to improve the function of those patients with a slower pace of recovery and signs and symptoms of fatigue and overuse. Early and timely diagnosis, intervention, and physiotherapy can make most GBS patients ambulant within 2–6 months post-onset of symptoms. We noticed that in our center, the majority of the patients with fair (3/5) or above fair muscle strength could ambulate with supervision within a span of 4–6 weeks post-onset of symptoms. However, for those with predominant axonal degeneration with muscle strength of 2/5 or below, the ambulation had to be defered until sufficient strength emerged from intense strengthening program. For patients with potential for ambulation, limit the use of wheelchair to minimize the detrimental effects due to its prolonged use.

Table 11.11 The type or mode of strength training progression required for different muscle grades

Muscle strength grade*	Type or model of progression of strength training
1	Facilitation techniques: Muscle tapping, quick icing/stretching, brushing, and vibration; NMES; PNF facilitation techniques
2-	Suspension therapy: Axial suspension without additional resistance, with or without facilitation techniques; hydrotherapy; re-education board in a gravity eliminated plane, active assisted movements in a gravity eliminated plane, PNF techniques; NMES as an adjunct
2 and 2+	Axial suspension therapy with additional manual resistance; pendular suspension with progressive application of manual resistance or adjusting the point of suspension or incorporating isometric or eccentric contraction along with pendular suspension with or without manual resistance; hydrotherapy; re-education board in an incline position; active assisted movements in an antigravity plane; PNF techniques; NMES as an adjunct
3-	Free exercise in an antigravity plane; isometric holds against gravity (no additional resistance); PNF techniques; NMES as an adjunct
3	Graded manual resistance exercise (concentric/isometric/eccentric); PNF techniques
3+,4-	Graded manual resistance exercise (concentric/isometric/eccentric); therabands, functional weight-bearing exercises; PNF techniques
4, 4+ and 5-	Progressive resistance strength training using: Free weights, therabands, isokinetic units, and springs; functional weight-bearing exercises; isokinetic units; closed kinematic exercises

*Based on Kendall and McCreary and Daniel and Worthingham's muscle strength grading system

The selection and prescription of orthotic and adaptive devices should be sensible and judicious. Assistive devices or walking aids like a walker, forearm crutches, or a cane may be prescribed for a limited period until the ambulation can be safely performed, provided the return of muscle strength is considerable. Generally, such devices or equipment are meant to protect the weakened structures from overstretching and overuse and facilitate functional activities within the patient's capacity. Lightweight orthotic aids like a posterior leaf splint or a flexible foot-drop-stop splint worn inside lightweight sports shoes or floaters with Velcro straps can partly enhance the ankle joint stability when the strength of the leg and foot muscles is unsatisfactory. The introduction and discontinuation of the orthotic and adaptive devices should be based on the serial examination findings of strength, ROM, and functional level.

For those subjects with predominant sensory deficits, like in the Miller Fisher variant, sensory reintegration exercises and repeated task practice may assist to redevelop motor engrams based on the altered sensory perception. Frenkel's exercise progressively performed from supine-to-sitting and then standing postures can improvise the coordination secondary to sensory ataxia caused by impaired or lost proprioception. GBS patients with facial muscle weakness require strengthening of the facial muscles which may include traditional facial exercises with feedback (use of a mirror for instance), NMES (Faradic) stimulation of facial nerve and its branches, and PNF techniques. However, care has to be taken to avoid the over-stretching of the already weakened facial muscles. The use of functional weight-bearing exercises, prone progressions, and core strengthening exercises are advisable to improve the axial, girdle, and proximal musculature strength. Balance training exercises like traditional balance exercises, stable and unstable platform exercises, Nintendo Wii and other virtual reality balance training consoles can facilitate and/or improve balance when it is involved.

Chronic Phase Management The general prognosis for GBS is usually good to excellent. In most patients, muscle weakness reverses as the peripheral nervous system recovers. However, some of the GBS patients may show considerable involvement, a slower pace of recovery, and a lengthier phase of immobilization. The therapeutic goals for chronic phase patients with residual weakness are simply an elaboration of the therapeutic goals set for the subacute phase treatment. The goals set for the chronic phase must be realistic and individually tailored to the patient's extent of residual capacity. Setting unrealistic goals can demotivate and discourage the patients and often prolong the disability. The therapist should advise the patient to avoid weight gain as it may place stress on an already compromised motor system. The home environment needs to be customized, to minimize the potential of falls. Loose rugs or carpets, slippery floors, and floor mats, poorly lit rooms, and obstacles on the floor can all predispose for falls.

Typically, the residual weakness is visible in the distal muscles of the extremities like the intrinsic muscles of the feet, toe extensors, ankle dorsiflexors, intrinsic muscles of the hand, finger extensors, and wrist and hand extensors. The muscles of the intermediate joints of the limbs (knee and elbow) may also demonstrate residual weakness. Poor endurance can often be a major roadblock to returning to work even if the patient's muscle strength is satisfactory. Both anecdotal and pragmatic evidence reveals that GBS patients can continue to show deficits during strenuous activities that necessitate maximal endurance. For such patients, incorporating endurance exercises into home exercise programs can discourage them from continuing the minimally active lifestyle despite adequate strength. Endurance training also enhances cardiovascular fitness which is often compromised due to altered muscle function and a sedentary lifestyle. The GBS patients can be encouraged to participate in activities like swimming to improve the upper and lower extremity endurance. However, such activities should strictly be done with adequate safety precautions and supervision. Evidence suggests that strength and endurance improvement can continue for months to years after GBS. Nonetheless, a few prospective studies have revealed that the average strength of major muscle groups failed to match the healthy controls, despite the continuing improvement of muscle strength.

11.9.3 Chronic Inflammatory Demyelinating Polyneuropathy

11.9.3.1 Introduction

CIDP is an acquired demyelinating disorder affecting the nerve roots and the peripheral nerves. The nerve involvement pattern is symmetrical in most cases. CIDP may follow monophasic, relapsing, or progressive courses, which distinguish it from the immune-mediated AIDP. In the past, a variety of names have been used, including chronic relapsing polyneuropathy, chronic relapsing Guillain–Barré polyneuritis, steroid-responsive recurrent polyneuropathy, and chronic inflammatory polyradiculoneuropathy. This neurological condition was formally defined in 1975 and the formal research criteria were laid down in 1991.

The disorder affects more adults than children. The presence of a sensorimotor polyneuropathy with multifocal demyelination on electrophysiological examination is the key to diagnosis. Intravenous immunoglobulin therapy is considered to be the first choice of treatment, though other immunotherapies including plasma exchange and corticosteroids are also effective. CIDP may be associated with MGUS. When associated with IgM, the deficits are principally sensory and response to treatment is unpredictable and has a poor prognosis. On the contrary, when associated with IgG or IgA, deficits and treatment are similar to those with idiopathic CIDP.

11.9.3.2 Clinical Features

The disorder may affect subjects of any age and the mean age in adults ranges between 30 and 50 years. CIDP is slightly more frequent in men than in women. Compared to AIDP, the antecedent events are lower and may include upper respiratory infections, gastroenteritis, other infections, vaccinations, surgery, and trauma. Typically, the disorder develops very slowly and the mean time of onset of symptoms to the initial presentation varies from 1 to 2 years. The clinical criteria for CIDP consider a minimum period of 2 months of symptom progression; however, the range is broad, as some patients may present with acute onset, resembling AIDP, before going on to develop a relapsing course which is typical of CIDP. A relapsing or a progressive course that responds to immunotherapy is the most important feature that distinguishes CIDP from AIDP.

The disease is strikingly heterogeneous and most patients present with sensorimotor symptoms and some may present with predominant motor or sensory symptoms. In approximately one-fifth of the patients, the weakness can be severely disabling, making them unable to carry out the ADL. Paresthesia and numbness in the feet and hands at onset are the somatosensory symptoms commonly reported. Unlike AIDP, aches and muscle pain are less common features of CIDP. Only in a minority of CIDP patients, cranial nerve symptoms, including dysarthria, dysphagia, facial numbness, facial muscle weakness, diplopia, and ptosis are reported. Similarly, symptoms of dysautonomia are uncommon; however, the patients may report urinary dysfunction and impotence. Typically, clinical examination will reveal both proximal and distal weakness, which is generally symmetric in nature, with distal weakness more common and severe than proximal weakness. Tendon reflexes are either diminished or lost and in the majority of cases, the Achilles reflex is not elicitable. Sensory deficits are present in the majority of CIDP patients, with the deep sensations more impaired than the superficial sensations.

11.9.3.3 Pathology

The fact that immunosuppressive therapy causes clinical improvement supports the autoimmune nature of this disorder. Both humoral and cell-mediated responses against a variety of myelin-derived antigens have been detected in these patients. However, there is no definite clarity whether CIDP is due to a single pathogenic mechanism or it represents a syndrome with more than one pathogenic cause. No one specific antigen has been identified as the common target in all the CIDP patients; however, in the serum of some patients, antibodies to P0 glycoprotein have been established and these antibodies could produce conduction block and neuronal demyelination when introduced into the sciatic nerve of rodents. The mechanism of CIDP is possibly mediated by both T and B cells and it is most likely that more than one type of autoantibody is involved.

11.9.3.4 Laboratory Findings

Developing the diagnostic criteria for CIDP has always been challenging for neurologists. A set of research criteria have been proposed that categorize patients as definite, probable, or possible CIDP based upon clinical, electrophysiological, pathologic, and CSF findings. The findings are further divided into mandatory, supportive, and exclusion and the highlight of the set criteria for CIDP diagnosis is listed in Table 11.12. The abnormalities seen in motor NCS are similar to those seen in AIDP and include multifocal demyelination characterized by prolonged distal latencies, slowed conduction velocities, prolonged F-wave latencies, and evidence of partial conduction block or abnormal temporal dispersion. Those patients, who present clinically with only sensory issues with normal strength, may show evidence of demyelination in motor nerves during the electrophysiological tests. Unlike the uniform slowing of conduction seen in CMT I, the motor nerve conduction in CIDP is characterized by abnormal temporal dispersion and conduction block on proximal stimulation. Needle EMG may reveal features of axonal degeneration (presence of fibrillation potentials and positive sharp waves) secondary to acute and chronic partial denervation. The motor unit potentials are usually of larger amplitude, longer duration, and are polyphasic in nature.

In suspected cases of CIDP, serum laboratory studies are informative and essential to rule out other possible medical conditions. Examples: (1) Erythrocyte sedimentation rate is generally normal in CIDP, (2) Anti-DNA titer and other tests for

Table 11.12 Highlight of the set criteria for chronic inflammatory demyelinating polyneuropathy diagnosis

	Mandatory	Supportive	Exclusion
Clinical	• Progressive or relapsing motor and/or sensory dysfunction of two or more limbs, developing over 2 months. • Symmetrical proximal and distal weakness. • Hyporeflexia or areflexia.	• Large-fiber sensory loss more than the small-fiber loss.	• Deformities of hands/feet. • History of drug or toxic exposure, family history of an inherited peripheral neuropathy. • Retinitis pigmentosa. • Sensory level. • Bladder dysfunction.
Electrodiagnostic	• Predominant demyelination process. • Conduction velocity reduction in two or more motor nerves. • Partial conduction blocks or abnormal temporal dispersion. • Prolonged distal latencies. • Absence of F-waves or prolonged minimum F-wave latencies.	• Reduction in sensory conduction velocities. • H-reflex absent.	
Pathologic	Nerve biopsy reveals clear evidence of demyelination and remyelination.	• Onion-bulb formation. • Mononuclear cell infiltration. • Subperineurial or endoneurial edema.	• Vasculitis. • Neurofilamentous swollen axons. • Amyloid deposits. • Intracytoplasmic inclusions in Schwann cells or macrophages.
CSF	Less cell count	• Elevated CSF protein.	

collagen vascular disease are usually normal and (3) The two-hour glucose tolerance test can be useful to distinguish diabetic neuropathy from CIDP. In those patients suspected to be at risk for HIV infection, HIV titers should be drawn. In case of inconclusive nerve conduction findings, tests for other causes of polyneuropathy, including thyroid function tests, vitamin B12 level, Venereal Disease Research Laboratory (VDRL) test, and Lyme titer (based on the geographic location), may be helpful. Serum and urine immunofixation or immunoelectrophoresis should be performed to identify monoclonal gammopathy (a non-cancerous condition with abnormal proteins in the blood). A skeletal X-ray examination should be performed in patients with monoclonal gammopathy. In those patients with elevated IgM levels, antibody titers to myelin-associated glycoprotein should be assessed. Cervical and lumbosacral MRI scans in CIDP patients may show hypertrophy of roots caused by demyelination and remyelination with onion-bulb formation.

11.9.3.5 Differential Diagnosis

The differential diagnosis consists of inherited and acquired neuropathies, neuromuscular transmission disorders, and primary diseases of the muscle. CIDP patients with an insidious onset and very slow progression of symptoms may resemble CMT, whereas those with an acute onset may initially resemble AIDP. Though axonal neuropathies can resemble CIDP, clinically the former has a distal preponderance of sensory loss and weakness and the electrophysiological tests may reveal a lack of conduction block, temporal dispersion, and severely slow conduction velocities. In addition to the above, axonal neuropathies are often painful, compared to CIDP. In case CIDP presents with a predominant proximal weakness, the disease may resemble the neuromuscular junction transmission disorders; however, conditions affecting the neuromuscular junction, inflammatory myopathies, toxic myopathies, and muscular dystrophies, lack of sensory loss and nerve conduction abnormalities, which are typical of CIDP. The CIDP may coexist with other conditions like diabetic neuropathy, HIV infection, demyelination disorders of the central nervous system, and systemic lupus erythematosus and make diagnosis and treatment more challenging.

11.9.3.6 Management

Evidence from a large number of clinical trials, suggests that a variety of immunosuppressive agents including corticosteroids, plasma exchange, IVIg, azathioprine, and cyclophosphamide are effective. The overall response rate is good, with the majority of patients responding to these therapies, either as standalone or in combination. The primary goal is to induce and maintain remission by a minimum dosage of immunosuppressive agents with nominal side effects. Attempts are periodically made to taper the immunosuppressive medication. For some patients, the treatment can be challenging when they do not respond appropriately to initial medications. Corticosteroids represent the earliest effective treatment for CIDP. The mean time for initial improvement for steroids is approximately 2 months and may reach a clinical plateau by around 6 months. Studies related to a combination of azathioprine with corticosteroid have not shown any superiority of combination therapy over standalone therapy.

Compared to steroids, the time taken to show visible improvement is generally short (a few days to weeks) following plasma exchange. The exchange is usually performed two or three times per week until a distinct improvement is noticeable, before tapering the frequency to a lower level. The duration of efficacy varies and some may have a monophasic course with no relapse after the plasma exchange treatment. IVIg appears to demonstrate efficacy in some CIDP patients. Those CIDP patients who are unresponsive or poorly responsive to standard therapy may benefit from interferon-α therapy.

Spontaneous improvement may occur in patients with nominal sensory deficits with no apparent weakness and for the same reason, the neurologists may prefer to observe than intervene for days to weeks. The choice of initial therapy is based on factors including the speed of onset action, side effect profile, concurrent diseases, availability, convenience, and cost. Due to the sustained improvement with treatment and to avoid the adverse effects of long-term corticosteroid treatment, IVIg or plasma exchange as the initial treatment is also preferred. Calcium and vitamin D supplementation and electrolyte and glucose monitoring are prudent. For those patients on long-term corticosteroid therapy, bone density studies should be included and aggressive treatment of osteoporosis may be essential. Immunosuppressive therapy can be discontinued in some; however, most patients require a long-term maintenance therapy.

The prognosis of CIDP has improved with regard to the morbidity and mortality associated with the disorder. Earlier literature states that around 10% of patients died from CIDP, 30% were nonambulatory and only 65% were able to work. Subsequent reports suggest a satisfactory outcome in over three-quarters of patients, with a reduced death rate (3–6%). In 60–70% of patients, a continued long-term immunosuppressive treatment is essential to treat or avert relapses. A major factor influencing the long-term outcome of the disease is the degree of axon loss. Evidence suggests that severe axon loss is linked to a poorer prognosis.

Distal muscle weakness and distal reduced sensory functions are the prominent features of CIDP. In CIDP, the largest afferent nerve fibers which carry the vibration sensations are often involved. The considerable distal weakness of the muscle can affect functional abilities like ambulation, obstacle crossing, negotiating curbs, climbing, and finger dexterity. The involvement of large fiber sensory afferents may affect the balance. Due to the abovementioned, a large section of CIDP patients may quote restricted mobility and imbalance as their main concerns. Fatigue is another common symptom and the patients may perceive it even after rest. The fatigue should be taken into consideration while assessing the impact of treatment in the long-term course of the disease. Research on functional mobility and functional capacity among CIDP patients is scarce. Few studies on home exercise programs and aerobic exercise have shown improvement in muscle strength and betterment in the physical sub-scales of QOL and fatigue. Further research is required to verify whether the disabilities associated with CIDP can affect the ability to perform activities and participation.

11.10 Metabolic Neuropathies

The metabolic neuropathies include a wide range of peripheral nerve disorders associated with systemic diseases of metabolic origin. These disorders are either inherited or acquired. The systemic diseases that are associated with metabolic neuropathies include diabetes mellitus, uremia, hypothyroidism, liver diseases, polycythemia, porphyria, disorders of lipid or glycolipid metabolism, nutritional and vitamin deficiencies, and mitochondrial disorders. The alteration(s) in the structure or function of myelin and/or axons due to metabolic pathway dysregulation is the fundamental basis for peripheral nerve involvement and is the common hallmark of all these diseases.

Diabetes mellitus is the most common cause of metabolic neuropathy, followed by uremia. Some of the systemic diseases which involve peripheral nerves can also affect muscles (metabolic myopathies). The common symptoms of metabolic neuropathies include symmetric polyneuropathy, insidious onset of initial symptoms, and prominent distal involvement (lower extremities more frequent than upper extremities). Mostly, the sensory disturbances exhibit a classic "length-related pattern," with the initial involvement of the toes that advances proximally to the feet and legs in a "stocking-like pattern." The upper limbs are less often affected and in case they are affected, the symptoms advance in the same pattern, with the involvement of the digits that progresses to the hands and forearms in a "glove-like pattern." In the later stages, the sensory symptoms can progress towards the anterior part of the abdomen and trunk ("trunk neuropathy"), leading sometimes to the erroneous diagnosis of myelopathy. Regarding the pathophysiology, little is known about the underlying mechanisms of metabolic neuropathies. As stated earlier, metabolic impairment or dysregulation could be the cause of demyelination or axonal degeneration. The author would like to state that, considering the wide range of peripheral nerve disorders in this section, only a few of the disorders are elaborated below.

11.10.1 Diabetic Neuropathy

11.10.1.1 Introduction

Diabetic neuropathy, a neurodegenerative disease that affects the peripheral nervous system, is the most common disorder of the peripheral nerves worldwide and is a leading cause of disability. According to the World Health Organization (WHO), in 2014, the global prevalence of diabetes among adults was 8.5%. In the United States, an estimated 17 million people have diabetes which is a rough estimate as the disease is undiagnosed in approximately 50% of the population. Its prevalence has been rapidly rising in low- and middle-income countries than in high-income countries. Despite the greatest proportion of people with undiagnosed diabetes mellitus, the African continent has an estimate of 16 million adults living with diabetes with a regional prevalence of 3.1%. In India, 73 million cases of diabetes among adults are estimated and the prevalence in urban areas ranges between 11% and 14%, and in rural India, between 3 and 8% among adults.

Diabetes is a major cause of blindness, renal disease, cardiac arrests, cerebrovascular accidents, and amputations. In almost half of the diabetic patients and among those with prediabetes (people with impaired fasting glucose or glucose tolerance), a distal peripheral neuropathy eventually develops. The role of hyperglycemia is not fully understood but involves a variety of metabolic consequences leading to direct injury to peripheral nerve(s) and microvascular endothelium resulting in nerve ischemia. Both Impaired Fasting Glucose (IFG) and Impaired Glucose Tolerance (IGT) represent intermediate levels of glucose dysregulation between the normal state and frank diabetes. IFG and IGT are associated with other features of insulin resistance including obesity, hyperlipidemia and hypertension, and large vessel atherosclerotic disease. Evidence suggests that both IFG and IGT are also associated with microvascular injury and peripheral neuropathy, traditionally associated with frank diabetes.

A healthy diet, regular physical exercise, maintaining normal body weight, and abstaining from tobacco use are ways to prevent or delay the onset of type 2 diabetes and its complications, which is the only approach for effective glycemic control. According to the popular traditional view, neuropathy develops only after years of sustained hyperglycemia; however, there is convincing evidence that neuropathy may occur early in the course of glucose dysregulation, including prediabetes. Neuropathy, a common cause of disability among diabetic patients causes pain and is a leading risk factor for foot ulcers and lower limb amputation.

Diabetes mellitus is a group of disorders characterized by hyperglycemia related to either a defect in the production of insulin and/or resistance to the action of insulin. Both hyperglycemia and hyperinsulinemia can cause damage to the large and the small blood vessels, located within the retina, kidney, and peripheral nerves. In Type 1 diabetes, the destruction of pancreatic β-cells, often due to an autoimmune mechanism, results in insulin deficiency and hyperglycemia, and those with Type 1 diabetes require insulin therapy for survival. Whereas in Type 2 diabetes, the most common form of diabetes, the increased peripheral resistance to the action of insulin results in hyperinsulinemia and hyperglycemia. Type 2 diabetes can be typically managed with a healthy diet, physical activity and oral hypoglycemic agents but eventually may require insulin therapy. Type 3 diabetes is due to less common mechanisms like genetic defects in β-cell function, drug toxicity, or other endocrinopathies, and Type 4 is classified as gestational diabetes.

11.10.1.2 Spectrum of Disease

There are several varieties of neuropathies classified under the term "diabetic neuropathy," some clearly related to hyperglycemia and subsequent metabolic and ischemic changes, others with compressive etiologies, and still others linked to inflammatory or immune processes. The disorder can affect many types of nerves, including large fiber sensory, small-fiber sensory, autonomic, and motor and the clinical findings may or may not be symmetric. In addition to the above, the disorder is characterized by the involvement of distal nerves as well as large nerve trunks, nerve roots, and cranial nerves. The spectrum of variants under diabetic neuropathy can be classified into chronic and acute (Table 11.13). This wide neuropathic spectrum represents the complex interplay of the involvement of different tissues (large and small vessels) and different fiber types (large and small nerve fibers). Figure 11.18 depicts the spectrum of variants of diabetic neuropathy.

Table 11.13 Subtypes or variants under the spectrum of diabetic neuropathies

Type	Subtype
Chronic neuropathies	Distal symmetric polyneuropathy (DSP) Autonomic neuropathy
Acute neuropathies	Diabetic lumbosacral Radiculoplexus neuropathy (DLRPN) Cranial neuropathy Truncal radiculopathy Insulin neuritis Diabetic neuropathic cachexia (DNC)
Compression Mononeuropathies	Carpal tunnel syndrome (CTS) Peroneal mononeuropathy at the fibular head Ulnar mononeuropathy at the elbow. Meralgia paresthetica

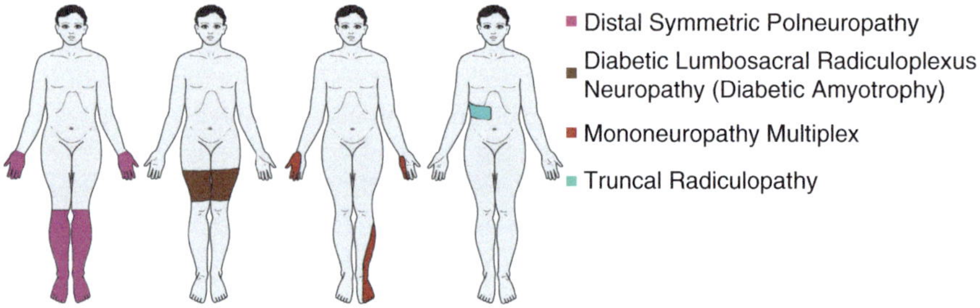

Fig. 11.18 The pictorial depiction of the spectrum of variants of diabetic neuropathy

The Distal Symmetric Polyneuropathy (DSP), a chronic sensorimotor polyneuropathy, is the most common of these syndromes, which typically produces mild distal sensory abnormalities as well as distal weakness. However, some of the rarer neuropathic conditions, like Diabetic Lumbosacral Radiculoplexus Neuropathy (DLRPN), are important to recognize, as they can produce severe pain and weakness with considerable morbidity. Prognoses for these syndromes also vary depending on the underlying pathology, with some being progressive disorders and others being monophasic in nature.

11.10.1.3 Chronic Neuropathies

Distal Symmetric Polyneuropathy Sensory or combined sensorimotor DSP is the most frequent form of diabetic neuropathy, whereas autonomic neuropathy is probably the least common form. The diagnosis of DSP requires the presence of symptoms of sensory loss, paresthesia or neuropathic pain, clinical examination evidence of distal sensory loss, and electrophysiological tests confirmation of distal axonal injury. There is evidence that the abovementioned can be a presenting symptom even for prediabetes.

Typically, the dominant DSP symptoms are sensory in nature. The disorder is characterized by insidious onset distal symmetric sensory loss, often affecting both large and small fibers. The patients may not recognize the sensory loss until postural instability, falls, or foot injuries occur. The majority of the patients may provide positive symptoms like tingling, lancinating, or burning paresthesias or pain in their feet or legs, often worse as they try to fall asleep at night. Acute worsening of pain is known among patients with established diabetes due to poor glycemic control or conversely with rapid pharmacologic re-establishment of euglycemic state. Such findings suggest prominent early involvement of small unmyelinated nerve fibers. Animal model and human studies indicate that transient hyperglycemia increases spontaneous discharge from C fibers (small diameter nociceptive afferents) and is clinically associated with increased neuropathic pain. In DSP, cold sensation mediated by the small unmyelinated nerve fibers is frequently involved before the vibration sense carried by the large myelinated sensory fibers is involved.

Unlike the distal sensory symptoms, the distal lower extremity weakness is rare as an initial complaint and the muscle strength is typically normal or minimally affected, even after several years of sensory symptoms. Often the clinical examination may reveal a loss of bulk for the intrinsic muscles of the foot and electrophysiological tests may demonstrate abnormalities in the absence of obvious motor signs and symptoms. Ankle reflexes are usually diminished or absent. The presence of brisk reflexes in the face of diabetic neuropathy should suggest the possibility of superimposed myelopathy. Approximately 20–50% of diabetic patients may report erectile dysfunction due to autonomic nerve involvement.

Conduction studies usually show low amplitude SNAP with prolonged distal latency. CMAP amplitudes are initially normal but tend to decline as the disease progresses. Early in the disease course, the motor conduction velocities are minimally affected and F-wave responses are prolonged suggesting a mild degree of demyelination. Small fiber injury leading to prominent neuropathic pain and autonomic dysfunction, is often the key manifestation of neuropathy, especially in prediabetes. Quantitative Sensory Testing (QST) for cold detection threshold is abnormal in the majority of DSP patients with prominent small fiber neuropathy. Distal muscles may reveal abnormal spontaneous activities, reduced motor unit recruitment, and a larger motor unit amplitude suggesting axonal loss with compensatory partial re-innervation.

Typically, the pathological findings seen in moderate diabetic neuropathy reflect a combination of axonal injury and segmental demyelination. Sural nerve biopsy reveals atrophy of nerve fascicles due to loss of myelinated and small unmyelinated axons. Axonal damage leads to Wallerian degeneration with the formation of digestion chambers. The presence of monocytes and macrophages suggests remodeling in the absence of primary inflammatory or autoimmune injury. A relative sparing of individual fascicles may indicate a microvascular ischemic injury, a primary pathological mechanism, reinforced by findings like endothelial cell hyperplasia, capillary closure, endothelial degeneration, thrombosis, and basement membrane thickening. The presence of inappropriately thin myelinated axons associated with onion-bulb formations suggests repeated demyelination and remyelination.

Autonomic Neuropathy Though the prevalence of diabetic autonomic neuropathy is considerably low compared to the DSP, autonomic nerve involvement is not uncommon in diabetes. Autonomic neuropathy is seen in approximately 5% of diabetic patients with long-standing disease and is clinically characterized by overt autonomic failures, such as anhidrosis, gustatory sweating (sweating on the forehead, face, scalp, and neck, immediately after ingesting food), orthostatic hypotension, supine blood pressure instability, constipation or diarrhea, bladder paresis, and cardiac dysrhythmias. Erectile dysfunction is a prevalent complication of mild to moderate diabetes among men and has also been associated with a higher risk for myocardial infarction and peripheral vascular disease.

Gastrointestinal manifestations of autonomic neuropathy include lower esophageal sphincter dysfunction, disordered peristalsis leading to constipation, diarrhea, and fecal incontinence. Even vagal nerve dysfunction can lead to esophageal atony and gastroparesis, manifested as early satiety, nausea, vomiting, and epigastric bloating. Poor motility and bacterial overgrowth are the consequences of bowel stasis and lead to diarrhea. Lack of anal sphincter tone also contributes to diarrhea or frank bowel incontinence.

11.10.1.4 Pathogenesis of Diabetic Neuropathy

Though the exact pathogenesis is not fully elucidated, acknowledgment of essential mechanisms facilitates an understanding of potential therapies. The pathogenesis of DSP involves an interaction between the metabolic and microvascular injury to the nerve. In some cases, nerve injury may be accelerated by the reactive autoimmune attack and exacerbated by the failure of the injured axons to undergo efficient regeneration. Vascular injury, an essential feature of all diabetic complications, plays a crucial role in the development and progression of diabetic neuropathy. The severity of neuropathy is associated with the degree of endoneurial capillary occlusion and hypoxia within the nerve. The vascular injury affects the vasoregulatory signaling at the endothelial lining of small and large blood vessels. Initially, the nerve ischemia occurs due to reduced capacity for vasodilation and with time, structural injury occurs to the microarterial vessels ultimately leading to permanent neurovascular ischemia.

The vasoregulation is mainly controlled through nitric oxide, which is synthesized by nitric oxide synthase in the endothelial cells in response to insulin and other stimuli. It plays a pivotal link between metabolic and ischemic nerve injury. Normally, nitric oxide facilitates vasodilatation and also serves to detoxify the Reactive Oxygen Species (ROS) that contribute to tissue injury. In situations of excess ROS, less nitric oxide is available for vasodilation. Several mechanisms are responsible for the depletion of nitric oxide and the production of ROS during hyperglycemia. For instance, hyperglycemia leads to aldose reductase (enzyme) activation that converts excess glucose to sorbitol and fructose (the polyol pathway). The polyol reaction depletes Nicotinamide Adenine Dinucleotide Phosphate

(NADPH), an important cofactor for the synthesis of glutathione and nitric oxide, thus resulting in lowered nitric oxide. An excessive amount of sorbitol and fructose can nonenzymatically react with proteins, lipids, and nucleic acids to produce Advanced Glycation Endproducts (AGEs), which induce ROS generation. Even the ischemia itself can accelerate the formation of ROS due to the reduction of mitochondrial efficiency and reperfusion injury. Hence, the final result is local endothelial damage by the ROS and impaired vasodilation from the depleted nitric oxide. Small arteries are more vulnerable to the abovementioned pathological process, eventually leading to tissue ischemia.

Inhibition of vasodilatory signaling pathways, increased production of prostanoid endothelium-derived constricting factors and angiotensin II promoted vasoconstriction, inappropriate arteriovenous shunting in epineurial arterioles, and a direct occlusion of vessels due to the thickening of the endothelial basal lamina, and fibrin and platelet deposition are the mechanisms underlying endoneurial ischemia. Direct toxic and metabolic damage also has a role in neuronal damage. Long axons may be particularly susceptible to direct metabolic consequences of hyperglycemia. ROS damage proteins and inhibit mitochondrial electron transport. Damage and dysfunction of mitochondria may serve as a common final pathway for the neuronal damage mediated by the polyol, AGE, and hexosamine pathways. Depletion of Nerve Growth Factor (NGF) and Insulin-like Growth Factors (IGF-I) reduces the growth of the small unmyelinated nerve fibers, inhibits axonal sprouting and neuronal function, results in apoptotic death of dorsal root ganglia neurons, and limits Schwann cell repair following injury. In some forms of diabetic neuropathy, immune activation may act as a primary or accelerating factor for nerve injury. For instance, in type 1 diabetes, the atrophy of cervical sympathetic ganglia, with infiltration by lymphocytes, plasma cells, and macrophages, indicates a strong immune response. An autoimmune cause has also been linked to DLRPN. People with diabetes are also more likely to develop CIDP, indicating another autoimmune association.

11.10.1.5 Acute Neuropathies

Diabetic Lumbosacral Radiculoplexus Neuropathy The acute form of diabetic neuropathy may present with asymmetric lower extremity pain and weakness. Though a variety of descriptive terms and eponyms were given in the past for the aforementioned presentation, the advancements in identifying and understanding the precise location of the causative lesion and the pathophysiology have elucidated many uncertainties and supported the use of DLRPN instead. Patients are often initially misdiagnosed as having either lumbar canal stenosis or less common compressive neuropathies like femoral and obturator mononeuropathies. Early and accurate diagnosis of DLRPN can prevent avoidable surgical procedures and facilitate appropriate therapy.

DLRPN occurs most frequently in elderly type 2 diabetic patients and is less often seen in type 1 diabetes. It can present as the first sign of neuropathy in many patients with long-established diabetes. Unlike DSP patients, DLRPN patients have better glycemic control, minimal complications due to other systems involvement, and a lower body mass index. Typically, the DLRPN presents with a sudden onset of asymmetric lower extremity pain involving the thigh more than the lower leg. The neuropathic pain can be deep and aching or sharp and lancinating type. Soon after the pain presentation, the condition progresses toward progressive weakness and atrophy of the involved muscles, often associated with significant weight loss. In the majority of the patients, the symptoms will progress to the contralateral limb and trunk. In about half of the DLRPN patients, autonomic symptoms like sexual, bowel, and bladder dysfunctions and orthostatic hypotension are also seen. Less often, the upper extremities can be involved and may present with distal numbness and weakness of upper limbs, with or without proximal involvement. Though pain is frequent, it is usually not a dominant symptom and the onset of upper limb involvement generally develops weeks or months after the involvement of the lower extremities.

On an average, the symptoms of DLRPN progress over several months and reach their nadir within 6 months. Many patients are disabled due to severe pain and weakness. The severe pain necessitates opiate therapy for pain management and the significant muscle weakness can make patients wheelchair-dependent for ambulation. Several months after the nadir, the symptoms may improve spontaneously and gradually over a period of 1 year. Though many patients recover substantially, most have significant residual symptoms and may require long-term assistance with ambulation.

The condition is usually associated with CSF albuminocytological dissociation and the presence of oligoclonal bands in around one-third of the patients. In about 75% of DLRPN patients, the electrophysiological studies usually reveal a distal diabetic peripheral polyneuropathy, with reduced sensory and motor responses, slowed conduction velocity, and prolonged F-wave latencies in the legs. Compared to CIDP or other acquired demyelinating neuropathies, the demyelinating features are not of sufficient severity. The EMG studies reveal fibrillations and positive sharp waves, reduced recruitment pattern, and increased motor unit potential amplitude in weak limb and paraspinal muscles, indicating a widespread presentation compared to the clinical symptoms. Such a constellation of findings is consistent with the involvement of both spinal root and plexus in addition to peripheral neuropathy, which forms the basis of the term DLRPN.

Nerve biopsy typically reveals a multifocal fiber loss with perivascular inflammation in the tissue. In milder forms, the metabolic factors are most likely involved, whereas, in the more severe forms, the metabolic factors superimposed by an inflammatory process with nerve ischemia are involved. Evidence of inflammatory features compared to metabolic-induced microvascular injury, better glycemic control, and spontaneous improvement in nearly all patients which are unexpected in a metabolic or degenerative process distinguish DLRPN from the rest. The presence of oligoclonal bands also supports an inflammatory process and the positive response to immunomodulatory therapy in some patients is also consistent with an autoimmune etiology.

The differential diagnosis of DLRPN is restricted to a few and may include conditions like CIDP and lumbosacral disc disease. Though CIDP presents with proximal and distal weakness with lower extremity areflexia and CSF albuminocytological dissociation, the asymmetry and severe pain seen among DLRPN are not typical for CIDP. Empirical treatment using corticosteroids or intravenous immunoglobulin may be appropriate, especially given evidence that DLRPN may respond to treatments typically used for CIDP.

Truncal Radiculopathy Typically, the patients present with the acute onset of pain and numbness involving the trunk. Abdominal muscles may become weak, mimicking a hernia. The distribution of sensory loss is variable, ranging from a distinct area within one dermatome to several dermatomes spanning bilaterally. The low thoracic dermatomes are more commonly affected compared to the high thoracic dermatomes which may include the axilla. Both truncal radiculopathy and DLRPN share many features and can often coexist. Older type 2 diabetic patients are most commonly affected and usually, the syndrome resolves spontaneously over a period of a few months.

Due to lesions involving the dorsal root ganglion or its distal process, the skin biopsy reveals epidermal denervation. Clinical improvement is characterized by signs of re-innervation on repeated skin biopsy after several months. EMG studies reveal denervation of the abdominal muscles and the paraspinal muscles at the affected level. Like the DLRPN, the acute nature of onset, spontaneous recovery, and its close association with the abovementioned, supports an inflammatory etiology for truncal radiculopathy. Treatment of isolated truncal radiculopathy is unwarranted given that the disorder is self-limited and causes no disability beyond pain. The focus of treatment is to control the pain and the same strategies advised for the management of pain associated with DSP and DLRPN can be utilized. For the focal discrete painful areas, the use of topical lidocaine via a patch is a therapeutic option.

Cranial Neuropathy Cranial mononeuropathies can develop as a complication of diabetes mellitus. Among the cranial nerves, the oculomotor nerve which supplies the extraocular muscles is the one most commonly involved. The third nerve palsy typically presents with acute onset double vision and ptosis, often associated with periorbital pain. In most cases, the pupils are spared, with associated minor degrees of internal ophthalmoplegia. Most patients recover within several months and for the same reason, the prognosis is excellent. Occasionally patients can have a recurrent episode. Compared to the oculomotor nerve, the trochlear and abducens nerves are much less involved. Seventh cranial nerve palsy (Bell's palsy) can be another instance of cranial neuropathy secondary to diabetes mellitus.

Diabetic Neuropathic Cachexia (DNC) DNC is a less common acute severe painful condition involving the distal limbs and trunk, accompanied by profound weight loss. Depression and impotence are additional features of DNC. The disorder is more common among type 2 diabetes elderly male patients. Without any specific therapy, the condition is known to spontaneously recover over months. The cause of DNC is still unclear. No inflammatory changes have been observed and the nerve biopsy typically demonstrates an axonal degeneration of both the large myelinated and the small unmyelinated fibers with evidence of axonal regeneration.

Insulin Neuritis Treatment-induced diabetic neuropathy, previously known as insulin neuritis, is a reversible painful neuropathic condition. The disorder presents with acute severe distal limb pain and autonomic dysfunction due to peripheral nerve fiber damage. Typically, the pain is of the burning type, often with allodynia. Although the pain is more often reported in the distal limbs, it can affect other areas like the trunk and abdomen. Patients with high pre-treatment glycosylated hemoglobin undergoing rapid re-establishment of glycemic control are likely to develop this condition. The disorder is seen in both type 1 and type 2 diabetes, undergoing insulin and/or oral hypoglycemic treatment with a history of poor glycemic control. Pathogenesis of the condition and the pain accompanying it is poorly understood and the proposed mechanisms include endoneurial ischemia, hypoglycemic microvascular neuronal damage, and regenerating nerve firing. Usually, the disorder has a monophasic course but can recur.

11.10.1.6 Compression Mononeuropathies

Compressive neuropathies are more frequent among diabetic patients with peripheral neuropathy. Among the compressive neuropathies, CTS is the most frequent one. The pathophysiologic relationship between CTS and diabetes is unclear. The risk of CTS correlates with the duration of diabetes, obesity, and body mass index. The obesity associated with diabetes may predispose to fat accumulation in the carpal tunnel. The diabetic nerve may be more sensitive to compressive injury caused by metabolic injury, microvascular dysfunction, and impaired axonal transport. Though treatment options for compressive neuropathies are the same for patients with or without diabetes, the response to the treatment is not always satisfactory in diabetic patients when compared to others. Peroneal mononeuropathy at the fibular head, ulnar mononeuropathy at the elbow, and meralgia paresthetica are some of the compressive mononeuropathies common among diabetics.

11.10.1.7 Pharmacological Treatment

To date, no pharmacological agents or treatment strategies have shown any definite reversal of neuropathy; however, in type I diabetes, pancreas transplantation helps to stabilize the neuropathy with minimal improvement over many years. The medical management for diabetic neuropathy involves three principal aspects: (1) A disease-modifying therapy with glucose control, (2) Surveillance and adaptive measures to prevent secondary tissue injury, and (3) Treatment of neuropathic pain.

Neuropathy being one of the complications of diabetes mellitus, a prompt screening for other micro- and macrovascular complications affecting other systems like renal, cardiac and visual are essential. Aggressive control of blood glucose is central to the treatment of diabetes and remains the only treatment to delay the onset and slow the progression of neuropathy. Evidence suggests that maintenance of normoglycemia by intensive insulin therapy and glucose monitoring, reduces the likelihood of neuropathy by two-thirds, compared to patients with poor glycemic control.

Since strict glucose control alone cannot prevent the progression of neuropathy, there is a need for therapies that address the neuropathy. Agents tested in preclinical and human studies have been based on altering pathways leading to nitric oxide inhibition. Experimental studies on animals using agents like aldose reductase inhibitors that reduce the shunting of glucose into the polyol pathway, medications that decrease ROS, lipid-lowering agents, and transition metal chelators have shown reduction or reversal of nerve injury. Unfortunately, none of the abovementioned agents have been effective in human trials.

Approximately 15% of diabetic patients develop foot ulcers and these ulcers are responsible for 20% of the hospitalizations of the patients. Diabetic peripheral neuropathy causes reduction or loss of protective sensation. Inability to detect a pressure of a 10 g Semmes–Weinstein monofilament, the sensory degree necessary to detect pressure, has been closely correlated with a painless foot injury. The weakness and atrophy of the intrinsic muscles of the foot due to moderate or advanced neuropathy can alter the anatomy of the foot and potentiate friction injury. The autonomic injury associated with neuropathy can reduce foot sweating and increase foot edema, increasing friction. Reduced joint mobility and pes cavus associated with weakness of intrinsic muscles of the foot can increase the force transmission to the heel and the ball of the foot. Stenosis of the small and large blood vessels and loss of vasodilatory regulation can also result in distal limb ischemia, promoting tissue fragility and impairing wound healing.

Regular and periodic evaluation by health professionals significantly reduces the incidence of foot ulcers among diabetic neuropathy patients. In addition to the above, daily surveillance (visual inspection) for foot skin breakdown is essential. Physical activities and exercises are also important for diabetes management. Many patients have difficulty exercising due to exacerbation of foot pain and often worry that physical stress on the feet may worsen their neuropathy. Such patients require reassurance, motivation, proper footwear, and visual inspection of foot to aggressively manage glycemic control, reduce neuropathic pain and facilitate weight loss. Properly fitting footwear can reduce pain in the foot, prevent traumatic injury to the foot and encourage exercise. Often patients require footwear modifications, including high-top shoes for improved ankle fixation, a custom orthotic insole, a rocker at the ball of the foot, and a well-cushioned heel. For those patients with diabetic foot ulcers, apart from the routine surgical wound debridement, cleaning, and dressing, use of footwear and other removable off-loading devices can facilitate faster wound healing. Patients with significant weakness with or without gait ataxia due to loss of kinesthetic and joint position sense should be encouraged to participate in overground and aquatic exercises (details regarding the same discussed later). For those patients with neuropathic pain, glycemic control alone may not manage the same and may require the use of anticonvulsant agents, tricyclic antidepressants, and opiates as an adjunct to alleviate the pain.

11.10.1.8 Physiotherapy for Diabetic Neuropathy

Typically, the light touch sensation is assessed using a 10 g Semmes–Weinstein monofilament. The filament is applied to the skin perpendicularly for a brief while with even pressure until it bends. The metatarsal heads, the heel region, the pad of the great toe, and any bony prominent area on the plantar surface of the foot are the common high-risk locations for skin breakdown. The light touch sensory assessment should cover all those high-risk locations. Apart from the abovementioned sites, additional sites can be tested if the clinician has any concerns. An inability to consistently sense the 10 g filament on even one weight-bearing site, suggests the loss of protective sensation. The large sensory nerve fibers can be assessed by noting the ankle tendon reflex response and the use of a 128 Hz tuning fork. Clinical examination of the alignment and function of the foot and ankle is essential to identify the risk of skin breakdown. The presence of metatarsophalangeal hyperextension and midfoot deformities can lead to bony prominences and potential sites for high pressure and friction, thus increasing the risk of foot ulcers. Midfoot deformity can predispose to joint instability and potentiate extensive multijoint deterioration.

The impairments in tactile, kinaesthetic, joint position and vibration sensations in diabetic peripheral neuropathy contribute to poor balance, gait deviations, and an increased risk of falls. The accelerated arterial disease associated with sensory impairments further increases the susceptibility of lower limbs to injuries and infections and predisposes to diabetic foot or leg amputations. Till recently weight-bearing exercises and physical activities like walking were considered to be a contraindication for diabetic neuropathy due to the concern that such activities might injure the patient's insensitive feet. But current research demonstrates that those activities neither pose any threat nor increase the risk of foot ulcers in diabetic neuropathic patients provided they have no severe foot deformities.

Before 2009, the recommendation by the Standards of Medical Care in Diabetes, published by the American Diabetes Association (ADA), stated that non-weight-bearing activities such as swimming, bicycling or arm exercises should be encouraged in severe peripheral neuropathy to minimize the risk of skin breakdown, infection and joint damage. However, a randomized controlled trial by LeMaster et al. was instrumental in leading to a substantial change in these guidelines. The study concluded that assigning to a walking exercise did not increase the rate of foot ulcers. A subsequent body of research has found positive adaptations to exercise and physical activity among diabetic neuropathic patients, corroborating the findings of LeMaster et al.'s preliminary work. Currently, ADA guidelines do not preclude weight-bearing exercises in diabetic neuropathic patients but provide a cautionary statement that all neuropathic patients should wear proper footwear and undergo a daily examination of their feet to detect any early lesions. Subjects with open or injured foot should be restricted to non-weight-bearing activities. Those therapists handling diabetic foot and diabetic neuropathic patients should incorporate the ADA guidelines in their regular clinical practice settings.

Evidence indicates that diabetic neuropathic patients who are less physically active are more at risk for skin breakdown compared to the more active patients. However, in those diabetic neuropathic patients with insensitive feet, excessive physical stress has been clearly associated with neuropathic skin breakdown and foot ulceration. The existing body of evidence indicates that high levels of localized stress while walking, cause skin breakdown, primarily under the metatarsal heads. Earlier guidelines on insensitive foot mainly focused on protecting it from physical stress. The recommendation consisted of unloading and use of protective footwear and avoiding weight-bearing exercises to promote wound healing and prevent skin breakdown of insensitive feet. According to the Physical Stress Theory (PST) unloading the tissues can clearly promote healing of the wound; however, the absence of physical stress post healing may reduce the tolerance of the tissues for stress and may even reduce the threshold for injury. An increase in the diameter accompanied by the structural organization of the collagen fibrils has been reported as an adaptation post-exposure to compressive and shear stresses. Therefore, contrary to traditional and most current clinical approaches, the PST hypothesizes that diabetic patients with an insensitive foot without any skin lesions or post healing of wound or skin breakdown may benefit from overload stress to become more tolerant to subsequent stress. Hence, the new perspective should be to encourage and progressively increase the weight-bearing activities than avoiding them. With adequate monitoring, weight-bearing exercises can be safe and feasible for diabetic neuropathic patients with an insensitive foot and may lead to positive outcomes.

A study on weight-bearing exercises and non-weight-bearing exercises showed a definite improvement in walking speed and daily step count for the former group compared to the latter. The ADA guidelines were followed and the weight-bearing exercises were performed in standing or walking which included body weight for resistance exercises and treadmill walking. Each exercise session began with 20 minutes of group-specific flexibility and stretching exercises followed by strengthening exercises and aerobic exercises. The study participants of the weight-bearing group were regularly instructed to visually inspect their feet and footwear and recorded the foot skin temperature using a handheld infrared thermometer prior to and following the exercise session. They were also instructed to use their own athletic or walking shoes post-screening for excessive wear, accommodation of bony deformities, and areas of high pressure.

Evidence from animal model studies has shown that exercise is effective in minimizing oxidative stress, re-establishing neurotrophin levels, and reducing inflammation. Animal model studies also prove that exercise can preclude myelin damage and reduce calcium channel dysfunction to improve electrophysiological function. Results from a large body of evidence hence conclude that exercise has a distinct advantage over the single-factor approach by influencing multiple pathways to restore peripheral nerve milieus and enhance nerve regeneration.

It is a well-established fact that diabetes mellitus and associated complications can be prevented by the tight regulation of blood glucose through diet, physical activity, exercise, and medication. Results of recent randomized controlled trials have proved that aerobic exercise enhances physical fitness, glycemic control, and insulin sensitivity among diabetic patients. The clinical trials used cycle ergometers, treadmills, recumbent steppers, and elliptical trainers for aerobic training. Current evidence suggests that moderate-intensity aerobic exercise minimizes cardiovascular risk due to improved vascular health as noted by a faster and enhanced arterial dilatation. A randomized controlled trial by Dixit et al. also found that aerobic exercise had a positive effect on nerve conduction velocity. Few clinical trials have also demonstrated a reduction in pain and betterment in general and physical fatigue following aerobic exercise intervention. Based on the abovementioned established facts, exercise has been recommended for diabetic patients to improve glycemic control and minimize diabetic complications. However, the need and the merits of the exercise are still not instilled in many patients who have been historically advised not to perform exercise due to the concern that it may potentiate further injury.

Regardless of the advancements in the identification of the several molecular pathways involved in pathogenesis, most of the clinical trials targeting specific molecules have revealed only modest benefits in slowing the disease progression and alleviating the pain accompanying the disease. Such clinical outcomes suggest that pharmacological management as a standalone treatment may not be adequate to reverse or halt or slow the progression of diabetic neuropathy. Exercise or physical activity is known to improve the multiple metabolic factors that may affect the health and microvascular function of the nerve, which may indirectly protect the nerve against further damage. A prospective study on diabetic patients without neuropathic features had demonstrated a lower frequency of motor or sensory neuropathy following supervised brisk walking for 4 hours per week for 4 years, compared to the control group. The effect of exercise on the prevention of neuropathy was further supported by subsequent studies on pre-diabetes, those with impaired glucose tolerance, and patients with clinical signs of neuropathy. Lifestyle modifications like diet control and exercise in a cohort study also revealed partial cutaneous reinnervation with a reduction in the neuropathic pain severity.

Musculoskeletal impairments can often go unnoticed in diabetic neuropathy because of the composite interactions of fundamental metabolic and neurological dysfunctions. Diabetes patients may experience premature and progressive sarcopenia due to the excessive intermuscular adipose tissue accumulation in the skeletal muscles. Studies have shown that lower extremity skeletal muscles of diabetic neuropathic patients tend to accumulate excessive volumes of intermuscular adipose tissue, with a concomitant reduction in the muscle volume in the intrinsic and extrinsic muscles of the foot. Loss of muscle volume is accompanied by a reduction in muscle strength. In addition to the above, diabetic neuropathy is also a major determinant for the premature decline in functional abilities leading to the early onset of physical frailty.

Patients with severe, symptomatic neuropathy are more likely to have leg muscle weakness compared to those with less severe diabetic neuropathy. Based on the evaluation findings, the strength training program can be designed to target a specific group or individual muscle. Poor leg and foot muscle strength and reduced rate of force production can predispose to ineffective balance reactions, as normal recovery from perturbation consists of the rapid production of adequate muscle force to restore the body's center of mass over its base of support. Progressive resistance strength training protocols are the most appropriate for enhancing muscle strength and are known to increase muscle mass, reduce pain and disability. Previous studies on foot and ankle exercises have shown a reduced plantar pressure during gait in diabetic neuropathic patients. A protocol consisting of strength training exercises including closed kinetic chain type and flexibility exercises for the foot and ankle can improve lower extremity function, minimize foot complications and enhance the levels of daily living activities. The findings of the randomized controlled trials support both strength and balance training, as they are known to significantly improve balance, strength, and walking speed and reduce the fear of falling for those participating in strength, balance, and functional training programs.

Limited joint mobility of the large and small joints of the spine and extremities is another prevalent musculoskeletal impairment in diabetic patients. It is readily noticeable in the distal joints (hand and foot) and when combined with other neuropathic impairments may contribute to forefoot, midfoot and hindfoot deformities, high vertical and shear pressures, and foot ulceration. The joint deformities and the foot muscle weakness can also reduce the patient's willingness to participate in daily physical activities and may provide a fertile ground for foot ulceration. Footwear should maximize force distribution, minimize pressure and friction, and fit the length of the foot. The toe box must be wide and deep enough to accommodate any forefoot deformities if present. Shoes are preferred over open sandals or floaters and shoes with laces can prevent slipping without being overly tight. Similarly, covered footwear offers greater protection from foreign objects compared to floaters or sandals.

Amputations, foot ulcerations, and skin breakdown are strong predictors of future skin breakdowns. Callus, blisters, or redness manifestations are indicators of friction and/or high pressure, a common precursor to skin breakdown. The presence of dry skin and hair loss in the distal leg and foot region can suggest autonomic neuropathy. Pale skin and reduced skin temperature on palpation can indicate vascular compromise, whereas, red skin and raised skin temperature on palpation may indicate a potential infection or skin irritation. Overgrown and thickened nails can lead to self-inflicted injuries during routine nail care or get injured while performing an exercise or physical activity. To prevent inadvertent injury while performing self-care, professional help can be sorted particularly when sensory neuropathy is associated with callus, nails, blisters, and wounds.

Generally, the distal lower limb musculature weakness happens during the natural progression of diabetic neuropathy. Progressive involvement of large nerve fiber leads to joint position and kinaesthetic sensory loss and together with the weakness of the leg and the foot leads to balance and gait impairments. Both static and dynamic components of balance are involved. Routine or traditional balance exercises have shown beneficial effects for this population. A large body of evidence has already established that static balance can be improved by balance exercise components like single-leg stance, near tandem and tandem stance, and the dynamic by the walk over a beam, reach-out activities, partial squats, and transition movements. Enhanced neuromuscular control and foot sensations are the plausible factors attributing to improved balance following balance training. Studies on several balance training strategies including Tai Chi have also shown improvement in gait parameters, QOL, and neuropathic symptoms. The existing body of evidence suggests that balance exercises are feasible and safe, and have the potential to improve balance and gait among diabetic neuropathic patients. A systematic review that compared balance training interventions including lower limb strengthening exercises, monochromatic infrared energy therapy, vibrating insoles, and the use of a cane for diabetic neuropathy, following a thorough analysis of outcome measures found only strengthening exercises to support the evidence to treat balance dysfunction among diabetic neuropathic patients.

Early diabetic neuropathy is characterized by loss of unmyelinated axons, which may result in pain and numbness. In addition to the above, it is also characterized by a progressive decline in intraepidermal nerve fiber density. Comparison of structured supervised weekly exercise with lifestyle counseling for a period of 1 year, among diabetic patients without neuropathy revealed an increased cutaneous intraepidermal nerve fiber density on proximal thigh biopsy among the exercise group. The results suggested that preclinical injury to unmyelinated axons is potentially reversible and that intraepidermal nerve fiber density can be a useful biomarker for future clinical trials.

Several rehabilitation strategies or treatments are used for the treatment of diabetic neuropathies which include pharmacological drugs, manual therapy, electrotherapy, and exercise therapy. Most of the trials in a systematic review comparing the effectiveness of exercise therapy on gait function among diabetic neuropathic patients, confirmed the effectiveness of exercise therapy on gait function with high methodological quality. The review could not determine the dose needed to achieve the desired improvements in gait function due to variations in the intensity or amount of exercise for the study participants. Even the time taken for therapy varied among the participants and it ranged between 30 minutes to 1 hour per session, two times to five times a week. The review had a study period ranging from a few months to a few years. The finding of this systematic review suggested that multi-component exercise therapy enhanced the gait function in diabetic peripheral neuropathy patients compared to the control groups. The multi-component exercise therapy generally consists of strength training, ROM exercise, balance training, flexibility, and stretching exercises, circuit training, and gait training.

Pharmaceutical clinical trials have demonstrated only modest benefits in alleviating pain, suggesting that pharmacological agents as standalone may not be sufficient to manage the pain. Physical therapy can improve muscle strength, joint mobility, balance, coordination, physical function, and overall QOL. Therapeutic exercises performed regularly, may reduce neuropathic pain and can help control blood sugar levels. A wide variety of electrotherapy agents which are inexpensive, safe, and non-pharmacological can be used for managing the neuropathic pain associated with diabetes. The electrotherapy agents include TENS, low-level laser therapy, IFT, and monochromatic infrared photo energy.

A review evaluated the efficacy of electrotherapeutic modalities for diabetic neuropathic pain. The review article reported that the effects of TENS were consistent and that the beneficial effects were more in those who used it for a prolonged time. Of the 15 articles reviewed, one study used a frequency-modulated electromagnetic neural stimulation (frequency rhythmic electrical modulation system), consisting of modulated electrical stimuli that vary automatically in terms of pulse frequency, duration, and voltage amplitude and reported that frequency-modulated electromagnetic neural stimulation was safe and that it significantly reduced the neuropathic pain among diabetic patients. According to the authors, the articles studied were diverse with several shortcomings and they recommended blinded, randomized controlled trials comprising larger sample sizes, longer duration of treatment, with follow-up assessments.

Exercise, medication, and diet are considered to be the three most significant interventions in diabetic patients. The merits of aerobic and strengthening exercises on glucose control, cardiovascular risk factors, and lipid metabolism are well established; however, there is poor consensus among researchers regarding the best exercise practice. Safe and active training needs to be applied and motivational strategies need to be incorporated to ensure that the patients are interested and motivated to take part in extended training programs with less chance of attrition. Irrespective of the physical exercise type, the training should begin with a brief period of warm-up consisting of stretching and flexibility exercises, lasting for 5–10 minutes and the session should end with a cool-down exercise. Flexibility exercises should address joint ROM limitations which are more likely to occur in the ankle, hip, and shoulder regions. Flexibility exercises should include gentle stretching programs like hamstring stretch, calf stretch, knee to chest stretch, and toe curls. Both the ADA and American College of Sports Medicine (ACSM) recommend 2 or 3 days of large-muscle-group resistance training per week. The training should include a minimum of one set of 5–10 resistance exercises.

Low-level laser therapy can be used for foot ulcers. The potential benefits of low-level laser can be attributed to its anti-inflammatory effect, pain inhibiting activity, and increased microcirculation. Studies have reported that laser therapy is effective for nerve regeneration and reduces neuropathic pain plausibly due to increased microcirculation, a mechanism characterized by the release of cytokines and growth factors aiding in vasodilation and neovascularization.

11.10.2 Uremic Neuropathy

Uremic neuropathy is a common neurological manifestation of chronic kidney disease. It is a distal sensorimotor polyneuropathy caused by uremic toxins. The extent of neuropathy correlates strongly with the severity of renal dysfunction. Although the precise mechanism of uremic neuropathy remains unclear, it is considered to be a dying-back neuropathy with secondary demyelination, like other toxin-induced neuropathies. Typically, the uremic neuropathy associated with kidney disease occurs at a later stage when the glomerular filtration level falls below 12 ml per minute. The clinical presentation and the progression of neuropathy are usually insidious and are reported by a majority of dialysis patients. Uremic neuropathy can be considered an accurate predictor of insufficient dialysis care.

The neuropathic symptoms include tingling and prickling sensation and pain sensation in the lower limbs. Paresthesia is one of the commonest and earliest symptoms. The weakness of lower limbs and muscular atrophy follow the sensory symptoms. During the progression of the disease, the symptoms may accent proximally and involve even the upper limbs. Muscle cramps and restless legs syndrome are frequently reported by the patients. Partial dysfunction of the peripheral nerves and amyloid deposits on the connective tissues and tendons can make the nerves susceptible to local compression, due to which compressive mononeuropathies can occur in the median nerve at the carpal tunnel, ulnar nerve at the elbow, or peroneal nerve at the fibular head. Approximately 50% of the patients may develop autonomic dysfunctions with orthostatic hypotension, impaired sweating, diarrhea, constipation, and impotence. Currently, therapy for uremic neuropathy includes renal dialysis and vitamin supplementation, and erythropoietin administration. In early-stage uremic neuropathy, renal transplantation has shown a positive outcome.

11.11 Nutritional Neuropathies

Prolonged caloric restriction or chronic alcoholism has been historically recognized among patients suffering from nutritional deficiency-related neuropathies. Chronic alcohol intoxication, malnutrition, fad diets, malabsorption syndromes, long-standing parenteral nutrition, certain genetic disorders, parasitic infection, and gastric resection are known conditions or disorders that can result in neuropathies secondary to specific nutritional deficits. Recognizing the clues from the clinical history and physical manifestations associated with neuropathy can be challenging and demanding for initial appreciation.

Even though nutritional neuropathies can assume diverse patterns, most are length-dependent, sensory axonopathies. Neuropathy associated with specific nutritional deficit commonly presents as a constellation of neurological and systemic symptoms and signs that help the clinician to differentiate individual nutritional neuropathies from other etiologies. A careful history taking and detailed clinical examination can establish the presence of nutritional neuropathy. However, serological, electrodiagnostic, or histologic evaluation is essential to identify the underlying cause. Neuropathies associated with nutritional deficiencies are preventable and generally respond best to therapy when diagnosed and treated in the initial stages. Supplementation for specific deficiencies and abstinence from toxic compounds are the two major treatment components involved in the management of nutritional neuropathies. Response to therapy for individual deficiencies may vary from the complete resolution of signs and symptoms to the arrest of disease progression.

11.11.1 Alcoholic Neuropathy

The neuropathy associated with alcohol intoxication is frequently associated with evidence of chronic alcoholism. Though the association is clear, controversy exists regarding the etiology of neuropathy. The disorder is well-established among chronic alcoholics with poor nutritional status. The estimates of neuropathy among patients with alcoholism are wide (between 9 and 67%), as the alcoholic population does not often present for medical evaluation. Current evidence suggests that females are more commonly and severely affected by the neuropathy associated with alcoholism than males. Because the caloric content of alcohol can replace food, the majority who abuse alcohol can develop various nutritional deficiencies. Alcohol is also known to inhibit the absorption and metabolism of thiamine and for the same reason, the neuropathy associated with alcohol intoxication has several similarities to thiamine deficiency neuropathy.

11.11.1.1 Features

The early presenting symptoms of neuropathy associated with alcoholism include painful, burning dysesthesias and lancinating leg pains historically described as "pseudotabes." The pain is typically accompanied by a subjective numbness that progresses in a stocking–glove pattern. Neurological examination reveals distal, symmetric loss of fine touch, noxious, thermal, and vibration sensations. The deep tendon reflexes are reduced or absent at the ankles. As the neuropathy progresses, the weakness of the distal lower extremities develops and can be severe. Usually, the progression of alcoholic neuropathy is gradual but can present subacutely over months. Autonomic abnormalities are prominent and consist of hyperhidrosis, hair loss, and thinning of the skin along with the distal distribution of sensory loss. One-fourth of chronic alcoholic patients may demonstrate sympathetic-mediated orthostasis, parasympathetic-mediated cardiac irregularities, and esophageal dysmotility. Myopathy is also known to commonly accompany neuropathy associated with alcoholism and presents chronically as diffuse muscle cramps, local pain, and a selective loss of the proximal type II muscle fibers. Other neurological findings that may accompany neuropathy include truncal and gait ataxia due to anterior superior cerebellar vermis degeneration and the Wernicke–Korsakoff syndrome. The systemic abnormalities found with chronic alcohol abuse include evidence of hepatic damage that can progress to liver cirrhosis, anemia, and hypoglycemia with ketoacidosis.

Serum alcohol level evaluation can help to establish acute intoxication. Studies on patients diagnosed with alcoholism had demonstrated a good association between the lifetime dose of alcohol and the presence of alcoholic neuropathy. Chronic excessive alcohol use can result in elevated transaminases and gamma-glutamyl transferase with a relative red cell macrocytosis. If hepatic toxicity progresses to liver cirrhosis, transaminases can be low or normal and the serum albumin and clotting factors can be reduced. Electrophysiological tests reveal low amplitude SNAPs and normal motor responses, especially in the lower limbs early in the course of neuropathy. However, the amplitude of the motor responses can be considerably low, particularly when recorded from the weak distal lower limb muscles. The conduction velocities recorded from the upper and lower limbs show only a nominal slowing with prolonged distal latency without conduction block. Reduced myelinated fiber density, especially in the small myelinated and unmyelinated fibers is an important pathological finding. Patients with excessive alcohol consumption but with normal serum thiamine levels have also demonstrated a substantial loss of large myelinated fibers. Evidence of axonal degeneration is seen in unmyelinated fibers too. In long-standing cases of alcoholic neuropathy, abundant axonal sprouting can also be noted.

11.11.1.2 Management

Abstinence from alcohol is the key treatment for alcoholic neuropathy. To produce a noticeable improvement in symptoms like dysesthesias and loss of sensations, months to years of abstinence from alcohol is required and often the improvement can be incomplete. Since most clinical presentations of alcoholic neuropathy are associated with malnutrition, vitamin supplementation with B complex vitamins is also essential.

11.11.2 Thiamine Deficiency Neuropathy

In the 1600s, Western medicine termed the neuropathy associated with thiamine (Vitamin B1) deficiency as beriberi. Extreme weakness was reported as a prominent feature of thiamine deficiency during the repeated epidemics of neuropathy during the next 300 years. It was determined that a diet of white rice stripped of the thiamine-containing pericarp resulted in a well-described thiamine deficiency disorder both in humans and in domestic fowls. Research in the early twentieth century defined thiamine as the essential agent in beriberi and introduced the concept of vitamin and vitamin supplementation in the medical field. Although pure thiamine deficiency is relatively rare, some countries continue to have endemic beriberi resulting from a diet high in polished rice. Fad diets, long-standing dialysis, and gastric surgeries have resulted in patients with signs of both acute and chronic beriberi, although some of the abovementioned may be related to other nutritional deficiencies. Peritoneal dialysis or hemodialysis patients usually have a poor nutritional status with reduced thiamine levels and may develop the clinical characteristics of thiamine deficiency.

11.11.2.1 Features

Sensory loss, extremity weakness, and edema are the features of acute thiamine deficiency. Sensory loss is more profound distally and clinical examination reveals a reduced sensation for all sensory modalities. Even weakness is considerable in distal musculature and physical examination reveals a reduced strength of foot dorsiflexors and wrist extensors, associated with diminished or lost knee and ankle deep tendon reflexes. The patients may often complain of aching pain and muscle cramps in the lower legs. The weakness and sensory loss, with or without peripheral edema, can appear within a few months to a year following a thiamine-restricted diet or may develop insidiously over years without any signs of cardiac abnormalities. Historically, earlier literature has reported weakness of laryngeal, facial, and tongue muscles; however, recent studies have not reported any obvious cranial nerve involvement. Thiamine deficiency induced Wernicke encephalopathy causes confusion, apathy, gait ataxia, and ophthalmoparesis with nystagmus.

Blood and urine assays are less reliable for the diagnosis of thiamine deficiency. For thiamine level evaluation, high-performance liquid chromatography (HPLC), a method for the measurement of thiamine, is preferred. Electrophysiological tests are likely to reveal axonal sensorimotor polyneuropathy in the lower extremities and nerve biopsies demonstrate axonal degeneration (may reveal a preferential degeneration of large myelinated axons). A considerable reduction of lower extremity sensory and motor responses, with a nominal slowing of velocities and distal latencies, are features of NCS. Needle EMG evaluation of distal extremities may reflect evidence of denervation and reinnervation and some neurogenic changes.

11.11.2.2 Management

In suspected cases of acute thiamine deficiency, prompt parenteral administration of thiamine is indicated until all the symptoms resolve. For those chronically malnourished patients with an acute presentation of Wernicke's encephalopathy, the rapid administration of thiamine intravenously generally results in the improvement of sensorium and the reduction of ataxia and extraocular movement abnormalities within a few days. But if left untreated, these abnormalities can become irreversible. Oral administration of thiamine is beneficial for chronic thiamine deficiency and should probably be continued indefinitely. Non-oral means of supplementation are appropriate for those patients who have evidence of malabsorption or are undergoing dialysis.

11.11.3 Vitamin B12 Deficiency Neuropathy

Vitamin B12 (cobalamin) is synthesized by micro-organisms and is present in both animal and dairy products. Cobalamin deficiency has been observed in 5–20% of older adults. Cobalamin is an integral component of two biochemical reactions in humans and is essential for the formation of the myelin sheath. Cobalamin deficiency can be due to malabsorption, pernicious anemia, gastrointestinal surgeries, weight reduction surgery, and strict vegan diets without vitamin B12 supplementation. Chronic exposure to nitrous oxide can also induce vitamin B12 deficiency. B12 deficiency is associated with hematologic, neurologic, and psychiatric manifestations. The neurological consequences include subacute combined degeneration, neuropsychiatric issues, optic neuropathy, and peripheral neuropathy.

11.11.3.1 Features

The disorder presents with acute onset symptoms, concomitantly involving the upper and lower extremities. In many patients, the B12 deficiency may begin with sensory symptoms in the feet and then progress to the upper limbs. "Numb hand syndrome" is the term used when they present with initial sensory symptoms and signs in the upper limbs. When associated with cord involvement (myeloneuropathy), findings often include a considerable loss of proprioception and vibration sense and signs symptoms of corticospinal tract dysfunction (increased tone, brisk tendon jerks, Hoffman's signs in the fingers, and Babinski sign). Histopathological studies reveal breakdown and vacuolization of central nervous system myelin. In contrast to the demyelinating features seen in the spinal cord, features of axonal neuropathy are seen in nerve biopsies and nerve conduction studies in cobalamin deficiency neuropathy.

11.11.3.2 Management

Early diagnosis is crucial since patients with advanced neuropathy may be left with major residual disabilities. The common treatment includes the intramuscular administration of cobalamin daily for a week and then tapering it to a monthly once administration. The residual neurological damage following cobalamin therapy is dependent on the initial severity, the duration of symptoms, and the initial hemoglobin measurements. For those patients with pernicious anemia, oral cobalamin therapy can also be an option. Since vitamin E and vitamin B6 are abundantly available in the diet, both in animal and vegetable food products, the dietary deficiency of vitamin E and B6 is rare. To know more about their clinical features, electrophysiological tests, pathological findings, and management, comprehensive textbooks and articles are available.

11.12 Toxic Neuropathies

Toxic-induced neuropathies are a rare but important cause of acquired peripheral nerve disorder. Ingestion or abuse of therapeutic drugs chemicals or solvents and exposure to workplace chemicals and environmental pollutants can trigger peripheral neuropathy. A large majority of drug or chemical agents can produce distal axonal degeneration in long peripheral nerves and for the same reason, distal axonopathy, a dying-back axonal degeneration, is the most frequent form of toxic neuropathy. Exposure to certain drugs and chemicals can also cause direct nerve cell damage or induce primary demyelination instead of distal axonal degeneration.

The majority of patients present with symptoms of pain, tingling, or numbness in their feet, consistent with dysfunction involving the longest and largest peripheral nerve fibers. Symptoms and signs initially appear in the distal regions and progress proximally as the duration of illness or level of toxic exposure intensifies. The stocking–glove pattern of sensory loss typically precedes the development of a similar distribution of motor weakness. Other than the neurological manifestations of neuropathy, the patients may present with symptoms like hypohidrosis or hyperhidrosis, diarrhea or constipation, urinary incontinence, blurred vision, facial flushes, orthostatic hypotension, sexual dysfunction, muscle cramps, increased heart rate, and hypertension.

Clinical examination may reveal the early loss of the symmetrical ankle jerk. In severe cases, motor dysfunctions like abnormal gait and foot drop will be apparent. Usually, the clinical signs appear weeks after exposure to the toxic substances and may continue to evolve even after exposure to the chemical substances ceases. In most cases, the neurological recovery may span a period of months or years. In addition to the historical and detailed clinical examination, careful laboratory assessment is needed to identify the toxic agents responsible for human neuropathy. Though the treatment options for peripheral neuropathy are limited, the new therapeutic approach for neuropathy induced by heavy metal intoxication is promising.

11.13 Chemotherapy-Induced Neuropathies

Chemotherapy-induced neuropathy is one of the most common side effects of antineoplastic agents, with a prevalence ranging from 19 to 85%. Being highly prevalent among cancer patients, the disorder constitutes a major problem for both cancer patients and survivors. It is also a matter of concern for health care professionals, especially because there is no single effective method of preventing or treating the disorder. Six of the main chemotherapy substances that cause damage to the peripheral sensory, motor, and autonomic neurons are platinum-based antineoplastic agents, vinca alkaloids, epothilones, taxanes, proteasome inhibitors, and immunomodulatory drugs (thalidomide). Among the aforementioned substances, platinum-based agents, taxanes, epothilones, and thalidomide are the most neurotoxic. Chemotherapy substances like proteasome inhibitors and vinca alkaloids are less neurotoxic.

11.13.1 Features

Chemotherapeutic agents can damage the neural system structures and depending on the distinct compound, a variety of large and small fiber, sensory and/or motor, demyelinating and axonal, cranial, and autonomic neuropathies can develop. Clinically, chemotherapy-induced neuropathy is frequently a sensory neuropathy that may be accompanied by motor and autonomic changes of variable intensity and duration. Typically, the sensory symptoms are the first to develop. The symptoms tend to develop distally in the feet and hands with a typical stocking–glove pattern. The symptoms include numbness, tingling, altered tactile sensation, impaired vibration, paresthesias, and dysesthesias. Often the painful sensations are reported as spontaneous burning, shooting, or electric pain. The patients may even report mechanical or thermal allodynia or hyperalgesia and in severe cases, these symptoms can progress to a loss of sensory perception. Motor symptoms are less frequent and may present as distal weakness, impaired movements, and gait and balance disturbances. The motor symptoms can have a marked impact on the QOL and safety and patients with sensorimotor deficits are three times more likely to fall. The autonomic symptoms are also less frequent compared to the sensory deficits and may include orthostatic hypotension, constipation, and altered sexual or urinary function.

The symptoms of chemotherapy-induced peripheral neuropathy typically emerge weeks or months following the cessation of chemotherapy. The severity of symptoms is usually proportional to the cumulative dose of the chemotherapeutic agent. Certain patients may experience paradoxical worsening and/or intensification of symptoms following the completion of treatment, a phenomenon termed as "coasting," where the existing neuropathy either worsens or new symptoms arise. In most cases, post-chemotherapy, the pain and sensory abnormalities may continue for months to years, indicating that many may suffer debilitating neuropathy even when they are cancer-free.

Based on the electrophysiological study findings, chemotherapy-induced neuropathy is usually an axonal sensorimotor neuropathy. The pathomechanism by which these agents damage the nervous system components and cause neuropathy is multifactorial and consists of microtubule disruption, oxidative stress and mitochondrial damage, alteration in ion channel activity, myelin sheath damage, DNA damage, and immunological and inflammatory processes.

11.13.2 Management

The course of chemotherapy-induced neuropathy can be unpredictable and discontinuation or reduction in the chemotherapy dose may alleviate the symptoms but can continue for years with long-term effects on QOL. For instance, breast cancer patients following cessation of taxane had neuropathy symptoms for up to 2 years and colon cancer patients post-cessation of oxaliplatin, a drug containing platinum, had numbness and tingling sensations in hands and feet for about 6 years from the commencement of the treatment. Poor clarity regarding the precise pathophysiology is one of the most important challenges faced in the management and prevention of chemotherapy-induced neuropathy. Though several pharmacological agents have been studied for prevention and treatment, out of the 18 agents investigated, only duloxetine revealed an efficacy for treatment. None of the pharmacological agents investigated had shown consistent, clinically meaningful benefits for the prevention of chemotherapy-induced neuropathy. Oral administration of vitamin E and intravenous administration of calcium or magnesium for oxaliplatin-induced neuropathy did not show any benefit in the prevention of chemotherapy-induced neuropathy. On the contrary, two agents, namely acetyl-L-carnitine and nimodipine worsened the neuropathic pain. Clinical trials investigating the alternative therapies found Scrambler therapy (a noninvasive cutaneous electrical stimulation) effective in reducing chemotherapy-induced neuropathic pain with no toxicity. Evidence also suggests that the use of low-level laser therapy also has a moderate benefit in alleviating neuropathic pain.

Chemotherapy-induced neuropathic symptoms may include chronic pain, severe dysesthesia, sleep disturbances, and sensorimotor dysfunctions with coordination deficits of the extremities. General weakness, incoordination, and risk and fear can discourage patients from performing physical activities. Such risk-avoidance behavior can further precipitate neuropathic issues and can thereby lead to a further reduction in physical performance and can contribute to depression, fear, anxiety, and a reduced QOL.

Current evidence suggests that physical exercise to improve endurance capacity and/or muscular strength and sensorimotor functions has several beneficial effects for normal healthy subjects, elderly subjects, and even cancer patients. Though there is a dearth of literature on physiotherapy management for chemotherapy-induced neuropathy, some researchers have opined that the exercise program can be advocated for such patients after a thorough and frequent assessment for any potential contraindications. The assessment should cover the pulmonary, cardiac, and psychosocial components which include echocardiography, pulmonary function test, musculoskeletal strength, and flexibility tests. If the assessment reveals an increased risk of harm, the patients need to be excluded from the exercise program. Any form of exercise intervention should be performed under strict and regular medical supervision to avoid or minimize side effects and potentially life-threatening complications.

11.14 Infectious Neuropathies

Infectious neuropathy is a heterogeneous group of neuropathies with multiple causes, evidenced by the direct involvement of nerves by the infective agent or an immune-mediated reaction or a toxic effect secondary to the drugs used against a causal infectious agent. Infections by Mycobacterium leprae, HIV, herpes zoster, hepatitis-C virus, and Borrelia burgdorferi (causes Lyme disease) can cause peripheral neuropathies and the features and management of a few are addressed below.

11.14.1 *Leprosy-Related Neuropathies*

Leprosy (Hansen's disease), a chronic granulomatous disease, principally affects the skin and peripheral nerves. Following the identification of the infectious agent Mycobacterium leprae, much has been learned about the natural history of the disease, the clinicopathologic presentations, and the treatment of leprosy. The disorder still affects a large number of people worldwide; however, it is often treatable and/or preventable. In tropical countries like India, Brazil, and central Africa, leprosy is one of the primary causes of infective neuropathy. Early detection, effective treatment, and introduction of leprosy control projects worldwide aimed at prevention and correction of deformities including reconstructive surgery programs and post-surgical physiotherapy to address the backlog of deformities, during the past five decades have overall resulted in a considerable and steady decline in the incidence of the disease and the disabilities associated with it. A comparison of the newly detected cases of leprosy between 2003 (approximately 515,000 cases) and 2009 (about 245,000 cases) reveals a 50% drop in the new cases worldwide.

Leprosy can be classified into three major clinical subtypes based on the extent of host immune response and are (1) Lepromatous or multibacillary type, due to predominant humoral response; (2) Tuberculoid or paucibacillary type, due to predominant cell-mediated immunity; and (3) A borderline or in-between type. Immunological response of the host to the bacilli produces major damage to the nerves intraneurally which attributes to considerable and rapid damage of the nerves in tuberculoid leprosy compared to lepromatous leprosy. A cooler temperature for the bacilli to grow has been suggested as the plausible cause of predilection for the superficial nerve. Since most of the superficial nerves are mixed nerves, the damage leads to paralysis of motor, sensory and autonomic fibers. Nasal mucosa and oral mucosa are the preferential sites of transmission and infection of the disease.

11.14.1.1 Features

Leprosy-related neuropathies are one of the most common treatable neuropathies in the world. The disorder can affect the sensory, motor, and autonomic peripheral nerve fibers. The disorder causes predominantly axonal neuropathy, which is more severe in lower limbs. Loss of sensation can be the earliest and most frequent manifestation. Among the peripheral nerves involved, the tibial nerve is the most common one, causing anesthesia on the soles of the feet, followed by the ulnar, median, and common peroneal. Though the clinical evidence of cranial nerve involvement is below 20%, the facial and trigeminal nerves are the most commonly involved cranial nerves. The granulomatous inflammation of the nerve causes a palpable enlargement of the nerve trunk, which can be often painful. About 4–10% of leprosy patients may present with a pure neuritic form without any skin lesions. Mononeuritis or mononeuritis multiplex are the most common presentations of the pure neuritic type.

Based on the pathogenesis, deformities in leprosy can be categorized into primary or secondary. Primary deformities are attributed directly to the disease process initiated by the infectious organism and secondary deformities are the consequence of primary damage to nerves. Primary deformities are usually encountered in the face, hands, and feet and follow a definite and predictable anatomical pattern. The transient, cosmetic deformities during the active phase of the disease such as raised or hypopigmented patches and nodular infiltration mainly of the face and ears are examples of primary deformities. Typically, these deformities subside with prompt and appropriate medical treatment. The more permanent cosmetic deformities like loss of eyebrows, deformities of the nose, and excess skin on the face and ears, generally require plastic or cosmetic surgery for better outcomes.

Though bilateral facial paralysis is rare, selective unilateral paralysis of the zygomatic branch of the facial nerve is less uncommon and leads to lagophthalmos. In the upper extremity, the ulnar nerve is frequently involved above the medial epicondyle, the median nerve near the flexor retinaculum, and the radial nerve in the spiral groove. Among the aforementioned nerves of the upper extremity, the ulnar nerve is always the first nerve to be involved. Deformities due to an isolated median or radial nerve involvement are uncommon unless they are in conjunction with ulnar paralysis. In the lower extremity, the common peroneal nerve is involved near the proximal region of the fibular neck and the tibial nerve near the tarsal tunnel.

Secondary deformities in leprosy are primarily due to the contractures arising from disuse. Clawed fingers and hands, clawed toes, and eversion of the lower eyelid in longstanding lagophthalmos are instances of secondary deformities following motor paralysis. The biggest and most intractable problem in Hansen's disease is the loss of sensations. Loss of pain sensation and inadequate rest of the affected region causes injuries to be unacknowledged leading to extensive ulceration and destruction of tissue without any subjective discomforts. Deformities due to motor paralysis can lead to the abnormal posturing of the affected body parts and predispose for excessive-high pressure. Pressure necrosis due to the sustained pressure of sharp or irregular fingernails within the clawed hands and prominent metatarsal heads of the feet with clawed toes causes blister and hematoma formations predisposing to trophic ulceration. The absence of sweating and dry skin and the use of inappropriate footwear or barefoot walking can also contribute to trophic ulceration.

The involvement of small dermal nerves can lead to anhydrosis and a glove–stocking pattern of sensory loss. Both the sensory and motor fiber involvements favor trauma and secondary infections. The isolated involvement of sensory fibers may present as a distal symmetric neuropathy with temperature and pain loss with no associated muscle weakness. In such cases, the deep tendon reflexes may be retained and needle EMG may appear normal. A skin lesion overlying a major nerve trunk is associated with a significant increase in the risk of impairment in that nerve.

11.14.1.2 Medical and Surgical Management

The presence of phenolic glycolipid-1 antibodies is essential for the diagnosis of the pure neuritic type. Instead of nerve biopsy, an invasive procedure that may lead to neural deficit, fine needle aspiration cytology of an affected nerve can also be a valuable and less invasive tool in confirming the disorder. The Polymerase Chain Reaction (PCR) technique has been utilized to demonstrate the presence of the bacilli in nerve biopsies, where the diagnosis is inconclusive. Paucibacillary leprosy is treated with dapsone and rifampicin, whereas multibacillary leprosy is treated with the abovementioned plus a third antibiotic, clofazimine. During or following the multidrug treatment, a so-called "reversal reaction" characterized by acute, painful, and disabling neuritis may occur. To minimize or avoid the same, oral corticosteroids like prednisone is usually recommended along with multidrug therapy.

Physical disabilities from leprosy include muscle paralysis, foot drop, claw hand and lagophthalmos, paresthesias, ulcers, and amputations. Evidence supports that post bacterial colonization, the superficial peripheral nerves are more likely to suffer ischemia from inflammation, mechanical stress, and/or trauma, near-certain anatomical vulnerable sites contributing to the development of neuropathy. The nerve compression resulting from pressure in and around selective anatomic sites causes the ischemia of the nerve and paresthesias. In such cases, surgical nerve decompression may reverse the nerve damage and improve patient sensation and strength. Surgical interventions also include nerve stripping and/or transposition in certain cases of neuritis.

11.14.1.3 Physiotherapy for Leprosy-Related Neuropathies

Bacterial invasion causes nerve enlargement (neuritis) and may occur during bouts of acute exacerbation. During the acute or sub-acute neuritis phase effort should be directed toward the care of the inflamed nerve. Restoration of function in the paretic muscles should be the concern during the re-educative phase. During the acute phase, rest and warmth are essential for the nerve. Typically, the limb will be immobilized in a well-padded splint or cast in a relaxed position for 4–6 weeks. Local applications of heat in the form of hot packs and ultrasound may aid to resolve the inflammation and pain during the subacute phase. Once the inflammation subsides and the re-educative phase begins, the treatment should be directed towards preventing tightness and contracture of the muscles, joint stiffness, and muscle atrophy. Active exercises should be instituted as soon as the voluntary movements return and should advance towards resisted exercises to build the strength and endurance of the muscles. Once the strength and endurance improve, the patient should be encouraged to use the affected muscles for functional activities.

Though the primary determinants of prognosis are the type of leprosy and the duration of paralysis, the extent and potentials of recovery are better in low-resistant lepromatous leprosy, less in the borderline type, and least in tuberculoid leprosy.

Irrespective of the leprosy types, a shorter duration of paralysis has a greater chance of recovery compared to a longer duration of paralysis.

In case of permanent paralysis, the primary goal of physiotherapy is to prevent, minimize and correct the possible damage that may occur due to secondary complications and deformities. The rehabilitation of these patients requires considerable co-operation of the patient and it is essential to educate the patient about the deformities and insensitive body parts including prevention and care of the deformities and ulcerations. Dry and hard skin that tends to crack, changes and reduction of weight-bearing surfaces, poor muscle padding, and absence of sensations, predispose to skin breakdown and ulcerations in the hands and feet.

Care for the anesthetic skin should include frequent inspections for injuries because of lack of pain, identify the cause of the injury, and prevent subsequent injuries by protection, insulation, or redistribution of pressure. In case of injury, the part has to be immobilized in a functional position using a splint or cast until healing is satisfactory. Once the swelling of the wound subsides along with clean granulation tissue formation, appropriate offloading footwear can be recommended to encourage walking but with adequate caution and care. Typically, a podiatric surgeon is the primary consultant who manages the wound.

Footwear for leprosy patients should be designed to eliminate areas of high pressure and aid in the redistribution of weight over the entire plantar surface of the skin. The use of microcellular rubber sandals or footwear is most appropriate for those patients with insensitive feet. Patients should be taught to soak their limbs in lukewarm water, trim callosities, and apply a moisturizing cream to keep the skin supple and smooth. Where contractures have occurred, following wax bath or hot fomentation, gentle stretching of the musculature, passive ROM exercise, and assisted active exercises should be encouraged to maintain joint mobility. For those patients with foot drop, AFO can be prescribed to maintain the foot in a neutral position while walking.

Since the postoperative joint ROM following a tendon transfer or reconstruction for correction of a claw hand or toe(s) is proportional to the pre-operative range, necessary measures to maximize pre-operative range prior to surgery is crucial. Preoperatively, it is also essential to teach the patient the action of the muscle so that attempts to contract the same muscle post-tendon transfer will produce the new function designed for it by virtue of its new route and insertion. Typically, post-tendon transfer, the hand or the foot is immobilized for a period of 3–6 weeks, and to prevent edema, the limb will be kept in elevation. To promote faster healing the tendons are usually kept in a slack position. Passive or active exercises for the limb are contra-indicated during the immobilization phase.

Following the removal of plaster, initially, gentle active ROM exercises that produce isolated action of the muscle should be taught and encouraged for short bouts, several times a day. During the early postoperative period, care has to be taken to avoid the overstretching of the transferred muscle. The use of splints may be continued during the postoperative phase to maintain the joint functional position and should be removed during the exercise session and later discontinued once the transferred tendon function is in its full ROM. Strengthening exercises should gradually take over the free or active exercises to promote the smooth gliding of the tendons in their new pathways. Progressive resistance strength training exercises are usually introduced after few months postoperatively. Once the isolated action of the transferred muscle is possible, the patient should be encouraged to integrate it into the functional activities. The period and rate of progression of exercises post-tendon transfer vary from patient to patient, which will depend on the age, interest, and motivation level of the patient and the type of surgery and complications if any.

In leprosy, compression and entrapment of the swollen nerve by the neighboring anatomic structures result in the characteristic features of compression syndromes consisting of impaired mobility, loss of muscle strength, paresthesias, and pain. Neural mobilization technique (slider technique) can be used as a treatment option to reduce the disability and pain and promote QOL among leprosy patients. Analysis of the electromyography function, muscle strength, degree of disability, and pain among leprosy patients following the slider neural mobilization of the lumbosacral roots and peroneal component of sciatic nerve innervating the anterior tibial musculature revealed a significant increase in outcomes compared to the control group. In this study, the slider mobilization techniques were performed in three sets of 30 oscillating movements per minute. Few studies on upper limb neural mobilization techniques also had similar improvements in outcome measures like grip and pinch strength. The plausible beneficial effects of neural mobilization are reduction of edema, normalization of axoplasmic flow, promotion of appropriate nerve mobility, increasing glial and neuronal activity, and reduction in nerve mechanosensitivity.

11.14.2 Herpes Zoster (Shingles)-Related Neuropathies

Varicella which causes chickenpox and herpes zoster which causes shingles are two distinct clinical entities caused by a single member of the herpes-virus family (human alphaherpesvirus 3), varicella-zoster virus. Several decades after the initial chickenpox infection caused by varicella-zoster virus, the dormant virus located in the cranial nerve or dorsal root ganglia can reactivate anywhere on the body to produce an acute, cutaneous viral infection known as shingles. The lifetime risk of shingles is approximately 10–20%. Herpes zoster or shingles is characterized by unilateral radicular pain and a cutaneous vesicular rash that is generally limited to one to three contiguous dermatomes. Post reactivation, the virus can cause additional neurological complications like postherpetic neuralgia, cranial neuropathies, motor radiculopathies of the limbs, and bladder and bowel dysfunction. Among the shingles-related complications, postherpetic neuralgia is the most frequent one. Postherpetic neuralgia, a neuropathic pain syndrome, can persist for three or more months beyond the healing of the cutaneous rashes.

The most frequent site of shingles is the chest, followed by the cutaneous sensory distribution of the trigeminal nerve. The involvement of the latter can lead to zoster keratitis and the involvement of the third, fourth, or sixth cranial nerves can lead to ophthalmoplegia. Herpes zoster in the cervical or lumbar cutaneous distribution may be followed by lower motor neuron-type weakness in the respective dermatomes. Ramsay Hunt syndrome, an acute peripheral facial neuropathy, associated with erythematous vesicular lesions of the skin of the ear canal, tympanic membrane, auricle, and/or the anterior two-thirds of the tongue or hard palate, occurs when shingles affect the seventh cranial nerve. The syndrome is characterized by facial muscle weakness and when compared with idiopathic facial palsy (Bell's palsy), the syndrome is often more severe and is likely to have incomplete recovery. The herpes zoster has no cure but prompt prescription and treatment with antiviral drugs such as acyclovir, famciclovir, and valacyclovir can speed healing and reduce the risk of shingles-related complications. Medications like gabapentin, tricyclic antidepressants like amitriptyline, topical agents such as capsaicin and lidocaine, and the use of corticosteroids can alleviate the severe pain associated with the condition.

11.14.3 HIV-Related Neuropathies

In HIV infection, among the nervous system structures, the peripheral nervous system is the most commonly involved part. The main forms of HIV-related peripheral neuropathies include multiple mononeuropathy, acute or chronic inflammatory neuropathy, polyradiculopathy, and distal symmetric neuropathy. Among the abovementioned, the latter is the commonest form. The signs and symptoms of distal symmetric neuropathy include dysesthesias, paresthesias and numbness, and reduced pain, temperature, and vibration sensations, initially in the feet and later in the hands. The tendon reflex of the ankle in most cases is lost. Electrophysiological tests reveal distal and symmetrical axonal degeneration of sensory and motor fibers. Some factors known to be associated with the development of distal symmetric neuropathy include the patient's age, race, the severity of HIV infection, antiretroviral medications, diabetes, and alcohol abuse. The use of a few antiretroviral drugs like stavudine and didanosine itself can cause features of distal symmetric neuropathy. The axonal degeneration seen in nerve biopsies of HIV patients with distal symmetric neuropathy appears to be due to the induction of proinflammatory cytokines. Studies have also revealed the involvement of the autonomic nervous system. Considerable overlap does exist between the symptoms of HIV and hepatitis C virus neuropathies.

Further Reading

Asbury AK, Cornblath DR. Assessment of current diagnostic criteria for Guillain-Barré syndrome. *Ann Neurol.* 1990;27:S21–4.

Bansal V, Kalita J, Misra UK. Diabetic neuropathy. *Postgrad Med J.* 2006 Feb; 82(964):95–100.

Berger AR, Schaumberg HH. Rehabilitation of focal nerve injuries. J Neurol Rehab. 1988;2:65–91.

Berger AR, Schaumberg HH. Rehabilitation of peripheral neuropathies. *J Neurol Rehab.* 1988;2:25–36.
Boullerne AI. The history of myelin. *Exp Neurol.* 2016 Sep;283(Pt B):431–45.
Bromberg MB, Smith AG, editors. Handbook of peripheral neuropathy. Boca Raton, Florida: Taylor & Francis Group; 2005.
Burns TM. Chronic inflammatory demyelinating polyradiculoneuropathy. *Arch Neurol.* 2004 Jun;61(6):973–5.
Bussmann JB, Garssen MP, van Doorn PA, Stam HJ. Analysing the favourable effects of physical exercise: relationships between physical fitness, fatigue and functioning in Guillain-Barré syndrome and chronic inflammatory demyelinating polyneuropathy. *J Rehabil Med.* 2007 Mar;39(2):121–5.
Chang AT, Boots R, Hodges PW, Paratz J. Standing with assistance of a tilt table in intensive care: a survey of Australian physiotherapy practice. *Aust J Physiother.* 2004;50(1):51–4.
Corrado B, Ciardi G, Bargigli C. Rehabilitation Management of the Charcot-Marie-Tooth Syndrome: A Systematic Review of the Literature. *Medicine (Baltimore).* 2016 Apr;95(17):e3278. Available from: https://doi.org/10.1097/MD.0000000000003278.
Crevenna R, Ashbury FD. Physical interventions for patients suffering from chemotherapy-induced polyneuropathy. *Support Care Cancer.* 2018 Apr;26(4):1017–8.
Dabholkar AT, Dabholkar A, Yardi S. Effect of nervous tissue mobilization on hand function in leprosy. *Indian J Physiother Occup Ther.* 2005;9(1):103.
Dana CL. The symptoms and diagnosis of multiple neuritis; with reports of cases. Based on a clinical lecture delivered at the New York Post Graduate Medical School, January 18, 1887. *JAMA.* 1887;8(11):281–5.
Daroff RB, Jankovic J, Mazziotta JC, Pomeroy SL, editors. Bradley's Neurology in clinical practice. seventh ed. London: Elsevier; 2016.
Davidson I, Wilson C, Walton T, Brissenden S. Physiotherapy and Guillain-Barré syndrome: results of a national survey. *Physiotherapy.* 2009 Sep;95(3):157–63.
de Freitas MR. Infectious neuropathy. *Curr Opin Neurol.* 2007 Oct; 20(5):548–52.
Dennis D, Mullins R. Guillain-Barré syndrome patient's satisfaction with physiotherapy: A two-part observational study. *Physiother Theory Pract.* 2013 May;29(4):301–8.
Donofrio PD, editor. Textbook of peripheral neuropathy. New York: Demos Medical Publishing, LLC; 2012.
Dyck PJ, Thomas PK, editors. Peripheral neuropathy. fourth ed. Philadelphia, Pennsylvania: Elsevier Saunders;2005.
Dyck PJ. Detection, characterization, and staging of polyneuropathy: assessed in diabetics. *Muscle Nerve.* 1988 Jan;11(1):21–32.
Ellis RF, Hing WA. Neural mobilization: a systematic review of randomized controlled trials with an analysis of therapeutic efficacy. *J Man Manip Ther.* 2008;16(1):8–22.
Fearnhead L, Fritz VU. Guillain-Barré syndrome- Rationale for physiotherapy management of the acute severe patient. *S Afr J Physiother.* 1996 Nov; 52(4):85–87.
Ferrante, MA. Brachial plexopathies: classification, causes, and consequences. *Muscle Nerve.* 2004 Nov;30(5):547–68.
Furness MA. Physiotherapy in leprosy. *Aust J Physiother.* 1982 Dec;28(6):9–12.
Gable KL, Attarian H, Allen JA. Fatigue in chronic inflammatory demyelinating polyneuropathy. *Muscle Nerve.* 2020;62:673–80.
Gibbons C, Dyck PJ, Stewart JD, Dyck PJB. Diabetic neuropathy. AANEM 59th Annual Meeting, Orlando, Florida, 2012.
Hammond N, Wang Y, Dimachkie MM, Barohn RJ. Nutritional neuropathies. *Neurol Clin.* 2013 May;31(2):477–89.
Hernberg S. Lead poisoning in a historical perspective. *Am J Ind Med.* 2000 Sep;38(3):244–54.
Herskovitz S, Scelsa SN, Schaumburg HH, editors. Peripheral neuropathies in clinical practice. New York: Oxford University Press, Inc.; 2010.

Hillier T. Diphtheria in London. *Br Med J.* 1864 Sep 24; 2 (195): 347–54.

Hou S, Huh B, Kim HK, Kim KH, Abdi S. Treatment of chemotherapy-induced peripheral neuropathy: Systematic review and recommendations. *Pain Physician.* 2018 Nov;21(6):571–92.

Husain S. An attempt towards prevention and management of disabilities and deformities in leprosy. *Indian J Lepr.* 2011 Jan-Mar;83(1):9–14.

Ites KI, Anderson EJ, Cahill ML, Kearney JA, Post EC, Gilchrist LS. Balance interventions for diabetic peripheral neuropathy: A systematic review. *J Geriatr Phys Ther.* 2011 Jul-Sep;34(3):109–16.

Jahantigh Akbari N, Hosseinifar M, Naimi SS, Mikaili S, Rahbar S. The efficacy of physiotherapy interventions in mitigating the symptoms and complications of diabetic peripheral neuropathy: *A systematic review. J Diabetes Metab Disord.* 2020 Oct 12;19(2):1995–2004.

Jesson T, Runge N, Schmid AB. Physiotherapy for people with painful peripheral neuropathies: a narrative review of its efficacy and safety. Pain Rep. 2020 Sep 23;5(5):e834. Available from: https://doi.org/10.1097/PR9.0000000000000834.

Kieseier BC, Mathey EK, Sommer C, Hartung HP. Immune-mediated neuropathies. *Nat Rev Dis Primers.* 2018 Oct 11;4(1):31.

Kluding PM, Bareiss SK, Hastings M, Marcus RL, Sinacore DR, Mueller MJ. Physical training and activity in people with diabetic peripheral neuropathy: Paradigm shift. *Phys Ther.* 2017 Jan 1;97(1):31–43.

Krishnan AV, Kiernan MC. Uremic neuropathy: clinical features and new pathophysiological insights. *Muscle Nerve.* 2007 Mar;35(3):273–90.

Lezak B, Massel DH, Varacallo M. Peroneal Nerve Injury. 2020 May 18. In: StatPearls [Internet]. Treasure Island (FL): StatPearls Publishing; 2020 Jan.

Llewelyn JG. The diabetic neuropathies: types, diagnosis and management. *J Neurol Neurosurg Psychiatry.* 2003 Jun;74 Suppl 2(Suppl 2):ii15-ii19.

Martin S, Kessler M, editors. Neurologic interventions for physical therapy. second ed. St. Louis, Missouri: Saunders Elsevier; 2007.

McDonald CM. Peripheral neuropathies of childhood. *Phys Med Rehabil Clin N Am.* 2001 May;12(2):473–90.

Melese H, Alamer A, Hailu Temesgen M, Kahsay G. Effectiveness of Exercise Therapy on Gait Function in Diabetic Peripheral Neuropathy Patients: A Systematic Review of Randomized Controlled Trials. *Diabetes Metab Syndr Obes.* 2020 Aug 5;13:2753–64.

Moghekar AR, Moghekar AR, Karli N, Chaudhry V. Brachial plexopathies: etiology, frequency, and electrodiagnostic localization. *J Clin Neuromuscul Dis.* 2007 Sep;9(1):243–7.

Mueller MJ, Tuttle LJ, Lemaster JW, Strube MJ, McGill JB, Hastings MK, Sinacore DR. Weight-bearing versus nonweight-bearing exercise for persons with diabetes and peripheral neuropathy: a randomized controlled trial. *Arch Phys Med Rehabil.* 2013 May;94(5):829–38.

Orsini M, de Freitas MRG, Presto B, Mello MP, Reis CHM, Silveira V, et al. Guideline for neuro-muscular rehabilitation in Guillain-Barré Syndrome: what can we do? *Rev Neurocienc.* 2010;18(4):572–80.

Pearce JMS. Robert Graves and multiple neuritis. *J Neurol Neurosurg Psychiatry.* 1990 Feb;53(2):113.

Pieber K, Herceg M, Paternostro-Sluga T. Electrotherapy for the treatment of painful diabetic peripheral neuropathy: a review. *J Rehabil Med.* 2010 Apr;42(4):289–95.

Ramdharry G. Peripheral nerve disease. *Handb Clin Neurol.* 2018;159:403–15.

Ropper AH, Samuels MA, Klein JP, Prasad S, editors. Adams and Victor's Principles of neurology. tenth ed. New York: McGraw Hill Education; 2014.

Rubin DI. Brachial and lumbosacral plexopathies: A review. *Clin Neurophysiol Pract.* 2020 Aug 13;5:173–93.

Ruhland JL, Shields RK. The effects of a home exercise program on impairment and health-related quality of life in persons with chronic peripheral neuropathies. *Phys Ther.* 1997 Oct;77(10):1026–39.

Santana HHS, Fernandes de Oliveira IAV, Medrado AP, Nunes SK. Neurodynamic mobilization and peripheral nerve regeneration: A narrative review. *Int J Neurorehabilitation*. 2015;2(2):1–7. Available from: https://doi.org/10.4172/2376-0281.1000163

Serrano-Coll H, Vélez JD, Trochez D, Beltrán JC, Suanca D, Monsalve F, et al. Effectiveness of an individual physical rehabilitation programme in a group of patients with Hansen's disease. *Lepr Rev*. 2016;87(3):355–67.

Sindic CJ. Infectious neuropathies. *Curr Opin Neurol.* 2013 Oct;26(5):510–5.

Singleton JR, Marcus RL, Jackson JE, K Lessard M, Graham TE, Smith AG. Exercise increases cutaneous nerve density in diabetic patients without neuropathy. *Ann Clin Transl Neurol.* 2014 Oct;1(10):844–9.

Singleton JR, Smith AG, Marcus RL. Exercise as Therapy for Diabetic and Prediabetic neuropathy. *Curr Diab Rep.* 2015 Dec;15(12):120.

Sommer C, Geber C, Young P, Forst R, Birklein F, Schoser B. Polyneuropathies. *Dtsch Arztebl Int.* 2018 Feb 9;115(6):83–90.

Song CH, Petrofsky JS, Lee SW, Lee KJ, Yim JE. Effects of an exercise program on balance and trunk proprioception in older adults with diabetic neuropathies. *Diabetes Technol Ther.* 2011 Aug;13(8):803–11.

Stewart, T. G. Multiple Neuritis. *Br Med J.* 1925 Sep 12;2:461.

Streckmann F, Zopf EM, Lehmann HC, May K, Rizza J, Zimmer P, et al. Exercise intervention studies in patients with peripheral neuropathy: a systematic review. Sports Med. 2014 Sep;44(9):1289–304.

Sujjay J, Green SR. Historical perspectives of peripheral neuropathy. *Int J Recent Sci Res.* 2019 Mar; 10(3):31377–9.

Thomas PK. Metabolic neuropathy. *J R Coll Physicians Lond.* 1973;7(2):154–60.

Tofthagen C, Visovsky C, Berry DL. Strength and balance training for adults with peripheral neuropathy and high risk of fall: current evidence and implications for future research. Oncol Nurs Forum. 2012 Sep;39(5):E416–24.

Tracy JA, Dyck PJ. The spectrum of diabetic neuropathies. *Phys Med Rehabil Clin N Am.* 2008 Feb;19(1):1–26.

Trivedi MS, Hershman DL, Crew KD. Management of chemotherapy-induced peripheral neuropathy. *Am J Hematol Oncol.* 2015;11:6

Umphred DA, Burton GU, Lazaro RT, Roller ML, editors. Umphred's neurological rehabilitation. sixth ed. St. Louis: Elsevier Mosby; 2013.

Uncini A, Kuwabara S. Nodopathies of the peripheral nerve: an emerging concept. *J Neurol Neurosurg Psychiatry.* 2015 Nov;86(11):1186–95.

Uncini A, Susuki K, Yuki N. Nodo-paranodopathy: beyond the demyelinating and axonal classification in anti-ganglioside antibody-mediated neuropathies. *Clin Neurophysiol.* 2013 Oct;124(10):1928–34.

van Alfen N. Clinical and pathophysiological concepts of neuralgic amyotrophy. *Nat Rev Neurol.* 2011 May 10;7(6):315–22.

Véras LS, Vale RG, Mello DB, Castro JA, Lima V, Silva KN, et al. Degree of disability, pain levels, muscle strength, and electromyographic function in patients with Hansen's disease with common peroneal nerve damage. *Rev Soc Bras Med Trop.* 2012 Jun;45(3):375–9.

Véras LS, Vale RG, Mello DB, Castro JA, Lima V, Trott A, et al. Electromyography function, disability degree, and pain in leprosy patients undergoing neural mobilization treatment. *Rev Soc Bras Med Trop.* 2012 Feb;45(1):83–8.

Wan EL, Rivadeneira AF, Jouvin RM, Dellon AL. Treatment of Peripheral Neuropathy in Leprosy: The Case for Nerve Decompression. Plast Reconstr Surg Glob Open. 2016 Mar 17;4(3):e637. Available from: https://doi.org/10.1097/GOX.0000000000000641.

Westblad ME, Forsberg A, Press R. Disability and health status in patients with chronic inflammatory demyelinating polyneuropathy. *Disabil Rehabil.* 2009;31(9):720–5.

White CM, Pritchard J, Turner-Stokes L. Exercise for people with peripheral neuropathy. *Cochrane Database Syst Rev.* 2004, Issue 4. Art. No.: CD003904. https://doi.org/10.1002/14651858.CD003904.pub2.

Yagihashi S, Mizukami H, Sugimoto K. Mechanism of diabetic neuropathy: Where are we now and where to go? *J Diabetes Investig.* 2011 Jan 24;2(1):18–32.

Yoo M, D'Silva LJ, Martin K, Sharma NK, Pasnoor M, LeMaster JW, Kluding PM. Pilot study of exercise therapy on painful diabetic peripheral neuropathy. *Pain Med.* 2015 Aug;16(8):1482–9.

Zajączkowska R, Kocot-Kępska M, Leppert W, Wrzosek A, Mika J, Wordliczek J. Mechanisms of chemotherapy-induced peripheral neuropathy. *Int J Mol Sci.* 2019 Mar 22;20(6):1451. https://doi.org/10.3390/ijms20061451.

Chapter 12
Muscular Dystrophies

Abraham M. Joshua, Karishma H. S. Keswani, and Zulkifli Misri

12.1 Introduction

The term "myopathies" means a group of disorders of muscles characterized by a primary structural or functional impairment of the skeletal muscles. This word is coined from two Greek words, i.e., "myo" meaning "muscle" and "patheia" meaning "suffering." Myopathies usually affect the skeletal muscles with no involvement of the nervous system. Depending on the type of myopathy, the symptoms may include muscle weakness, undue fatigue, exercise intolerance myalgia, muscle cramps and contractures, myotonia, and myoglobinuria. Muscular weakness is the most common and noticeable clinical indicator of this group of disorders. Weakness is more profound in the proximal muscles, though in some forms of myopathies, the weakness may be more distal in nature, as seen in conditions like myotonic dystrophy and inclusion body myositis. The patients may experience difficulty performing routine activities that require the use of the proximal muscles, such as climbing stairs, getting up from a chair, and lifting objects.

This group of disorders can be broadly classified into hereditary and acquired types. The hereditary conditions include muscular dystrophies, congenital

A. M. Joshua (✉)
Department of Physiotherapy, Kasturba Medical College of Mangalore, Manipal Academy of Higher Education, Manipal, India
e-mail: abraham.joshua@manipal.edu

K. H. S. Keswani
The Rise Network, Mumbai, India
e-mail: karishma@therisenetwork.in

Z. Misri
Department of Neurology, Kasturba Medical College of Mangalore, Manipal Academy of Higher Education, Manipal, India
e-mail: misri.zk@manipal.edu

A. M. Joshua (ed.), *Physiotherapy for Adult Neurological Conditions*,
https://doi.org/10.1007/978-981-19-0209-3_12

myopathies, myotonias and channelopathies, metabolic myopathies, and mitochondrial myopathies. In contrast, the acquired conditions include inflammatory myopathies, endocrine myopathies, drug or toxic induced myopathies, and myopathies associated with systemic illness.

Muscular dystrophies are a group of primary muscle disorders that have a hereditary basis. These are progressive degenerative diseases of the skeletal muscles and the word dystrophy means degenerative disease. The innervation of the affected muscles is intact. The disorder occurs insidiously at all ages. The degree of severity varies with highly selective muscle involvement in the early stages. The common characteristic features of this group of diseases are the symmetrical distribution of muscular weakness and atrophy, intact sensibility, preservation of cutaneous reflexes, and the liability to heredofamilial incidence. They are clinically manifested as muscle weakness, reduced tone, muscle pain, easy fatigability, and muscle cramps and may have a dark tea-colored discoloration of urine due to excretion of myoglobin (a by-product of muscle breakdown). Benign myopathic conditions like nemaline, central core, mitochondrial, and centronuclear diseases are primarily diseases of muscle and often heredofamilial in nature. However, due to the distinctive histochemical and ultrastructural features and slow progression, such disorders are not classified under muscular dystrophies.

12.2 Historical Background

During the second half of the nineteenth century, several physicians and neurologists such as Guillaume Benjamin Amand Duchenne, William Richard Gowers, and Ernst Viktor von Leyden had prepared literature on the clinical features of dystrophies. They even described the autopsy findings of the muscles of the affected individuals. The first possible historical account of muscular dystrophy appeared in 1830, when Charles Bell, a Scottish surgeon, anatomist, physiologist, and neurologist, wrote an essay about an illness that caused progressive weakness in male children. In 1836, Gaetano Conte and L. Gioja, Italian physicians, described two brothers with progressive muscular weakness who were severely disabled by the age of 8–10 years and died at the age of 18 years. The diseased subjects had widespread muscle atrophy, progressive muscle wasting and weakness, enlarged calves and deltoids, normal sensory function and mentation, and contractures of the knees and hips. They reported that the older brother died of cardiac failure. Unfortunately, other than the clinical features, there were no reports on muscle pathology. These are regarded as the earliest case descriptions of what is now known as Duchenne muscular dystrophy. In 1847, W. J. Little, an English physician, also reported about two affected brothers aged 12 and 14, who were unable to walk from the age of 11 and died in the adolescent age. Post-death examination of the muscles of the lower limb revealed the replacement of muscle tissue by fat.

In 1852, an English physician, Edward Meryon, published a detailed article describing the disorder as a degenerative disease of the muscle without changes in the anterior horn cells or the motor roots. Meryon even suggested that a sarcolemmal defect was the root of the disorder. In 1868, Duchenne, the French neurologist, in his seminal work *De la paralysie musculaire pseudo-hypertrophique ou paralysie myo-sclérosique*, gave its first ever detailed clinical description and muscle histology. Due to the high esteem for his work in Faradism and contributions to understanding muscle diseases, subsequently, the disorder was known as Duchenne muscular dystrophy (DMD). However, some of the scientific investigators believe that the first systematic and detailed study of the disorder must be attributed to Meryon.

Gowers, a British neurologist, was the first to produce a comprehensive account of the disorder in English, and it was Wilhelm Heinrich Erb, a German neurologist, in 1891 who crystallized the clinical and histological concept of a group of diseases due to the primary degeneration of muscle, which he called "muscular dystrophies". Soon after DMD was identified, it became evident that more than one form of muscular dystrophy existed and that these diseases affected people of either gender of all ages.

The first description of facioscapulohumeral muscular dystrophy in medical literature appeared in an autopsy report by Jean Cruveilhier, a French surgeon and anatomist, in 1852. Later, in 1855, Duchenne photographed and recorded cases of the same disorder. In 1874, about a decade later, Louis Théophile Joseph Landouzy and Joseph Jules Dejerine, French neurologists, gave the classical description of facioscapulohumeral muscular dystrophy, including the familial nature of the disorder. William Bateson, an English biologist, in the year 1909, was the first who used this disorder to illustrate the relevant Mendelian principles. He noticed the unusual pattern of affected males with unaffected female carriers that fit the pattern for sex-limited segregation. Seitz D., in 1957, distinguished a form of scapuloperoneal dystrophy with cardiomyopathy from the large group of scapuloperoneal muscular dystrophy. In 1961, Fritz E. Dreifuss and Gwendolyn R. Hogan indicated a slow progressive muscular dystrophy that was non-disabling in nature, and Dreifuss, along with Alan Eglin H. Emery, in 1966, provided detailed characteristics of the same unusual type of benign sex-linked humeroperoneal dystrophy, which is now known as Emery-Dreifuss muscular dystrophy. In 1979, R. H. Lindenbaum and fellow researchers identified the abnormal gene of DMD on X-chromosome. During the later half of the twentieth century, many muscular dystrophy variants were identified and advancements in molecular genetics and biochemistry also helped classify the same based on their genes and gene products.

12.3 Muscle Anatomy and Physiology

The normal contractile proteins actin and myosin are arranged with other muscle proteins such as troponin, to form the familiar thick and thin filaments. In the sliding filament model, the thick and thin filaments form an array, which slides back and forth. The reaction between actin and myosin is responsible for the realignment between the two molecules. These contractile proteins are connected to the "outside"" of the cell by means of a protein complex that ultimately links up with the basal lamina. Dystrophin, dystrophin-associated glycoproteins, and dystrophin-associated proteins (discussed later in this chapter) belong to this complex protein chain and are essential for the integrity of the sarcolemmal membrane. Specific defects in the abovementioned chain of proteins account for the occurrence of some but not all muscular dystrophies.

The muscular dystrophies, particularly the Duchenne and Becker type, are caused by abnormalities in the genes responsible for encoding a protein called dystrophin. Dystrophin is a cytoplasmic protein situated close to the sarcolemmal membrane in the myocytes and forms a strong mechanical link to cytoplasmic actin. Thus, it forms an interface between the intracellular contractile tissue and extracellular connective tissue. This protein complex enables the transference of force of contraction to the connective tissue. The absence or abnormality of dystrophin or other related proteins leads to the structural weakness of the cell membrane, making it vulnerable to rupture under mechanical stress, thus explaining the preferential involvement of muscles that are subjected to considerable mechanical stress.

12.4 Classifications of Muscular Dystrophy

Traditionally these disorders are classified on clinical grounds and are recognized by their distinct clinical appearances. The age of onset, the severity of symptoms, the topography of muscles involved, the mode of inheritance, and the pace of progression formed the basis of the traditional classification. Recent advancement in molecular genetics and biochemistry has helped in the discovery of certain genes and their gene products such as the dystrophin gene and dystrophin protein. The discovery of dystrophin deficiency in DMD has presaged a change in both the definition of muscular dystrophies and their classification. Increasing information about the molecular abnormalities in these illnesses has also clarified a number of uncertainties as to their clinical presentations and has necessitated a revision of the traditional form of classification. The newer classifications separate the muscular dystrophies according to the patterns of inheritance and the presence and locus of the abnormal gene and the defective gene product. Table 12.1 provides an outline of muscular dystrophies based on the patterns of inheritance, chromosome locus, and gene product.

In this chapter, the author prefers to describe those muscular dystrophies that are seen in the adult population with the onset of the symptoms during childhood, adolescence, or adult life.

12.5 Becker Muscular Dystrophy

Becker muscular dystrophy (BMD) shares all the characteristics of the DMD but has a milder course. Usually, the illness begins in the first decade, although the parents often notice the first sign of weakness later because of the less severe

Table 12.1 An outline of muscular dystrophies based on patterns of the inheritance, chromosome locus, and abnormal gene product

Disorder	Chromosomal locus	Altered gene product
X-linked recessive		
1. Duchenne or Becker	Xp21	Dystrophin
2. Emery-Dreifuss	Xq28	Emerin
3. Scapuloperoneal.	Xq26	FHL1
Autosomal dominant		
1. Myotonic dystrophy (type 1 myotonic dystrophy)	19q13.2 to 19q13.3	Myotonin protein kinase
2. Proximal myotonic myopathy (type 2 myotonic dystrophy)	3q21	Zinc finger protein 9 (ZNF9)
3. Facioscapulohumeral	4q35	–
4. Oculopharyngeal dystrophy	14q11.2 to 14q13	Poly(A) binding protein 2
Autosomal dominant or autosomal recessive		
1. Limb-girdle muscular dystrophy: Nearly thirty types identified (includes formerly categorized congenital muscular dystrophy like "Fukuyama")	Each one has a distinct chromosomal involvement	Calpain 3, desmin, dysferlin, titin, caveolin, myotilin, telethonin, γ-sarcoglycan, α-sarcoglycan, δ-sarcoglycan, β-sarcoglycan, merosin, fukutin, fukutin-related protein, selenoprotein-N, integrin α-7
2. Distal muscular dystrophies: Includes Gower-Laing distal myopathy, Nonaka myopathy with rimmed vacuoles, Miyoshi myopathy, Welander distal dystrophy, Desmin myopathy	Each one has a distinct chromosomal involvement	GNE kinase-epimerase, desmin, dysferlin, titin

symptoms. The female gender is very rarely affected, either due to the presence of only one X chromosome (Turner's syndrome) or chromosomal translocation. The muscle hypertrophy, contracture, and pattern of weakness are similar to those seen in DMD. When compared to DMD, BMD is approximately one-tenth as common. The prevalence rate is approximately 1 to 2 per 100,000 general population. BMD has a strong familial liability and is an X-linked recessive disorder, and therefore, the disease is practically limited to males and is transmitted by females. Even though the inheritance is X-linked, almost one-third of the cases are sporadic and are presumably due to spontaneous mutation.

12.5.1 Clinical Features

The BMD is delineated clinically as a milder phenotype of DMD. Symptoms may begin as early as the first year or as late as 45 years of age, but in the great majority, the onset is between the ages of 5 and 15 years. Walking may be delayed in infancy. The child is physically less active and more placid in the early years and cannot keep up with play activities when compared to his peers. Toe walking is a common early feature and is often disregarded for many years. The other initial symptoms are difficulty in running or climbing the stairs and later in doing heavy works using the arms. Most patients find it difficult to run after the second decade. In recent years, it has been customary to use the ability to walk unaided beyond the age of 16 years as the point of clinical separation from DMD. An elevated creatine kinase (CK) may be the first clue about the disease. The swayback posture and waddling gait (Box 12.1) become more obvious as these patients age.

Box 12.1 Features of the Dystrophic Gait

Dystrophic gait
It is seen in various myopathies, including DMD and BMD, in which there is weakness of the hip girdle muscles. The patient stands and walks with a pronounced lordosis. While walking, there is marked waddling because of the difficulty in fixing the pelvis in the frontal plane. The patient walks with a broad base and shows an exaggerated rotation of the pelvis, rolling or throwing his hip from side to side with every step to shift the weight of the body. This compensatory lateral movement of the pelvis is due to the weakness of the gluteal muscles. During the early stage of the disease, these patients have mild to moderate lordosis, waddling and protuberant abdomen, absence of heel strike, and mild hyperextension of the knee during the stance phase of the gait. In the advanced stage, toe walking becomes prominent, the knee gets hyperextended during the mid-stance phase, and lordosis, waddling, and protuberant abdomen become more pronounced.

The selective muscle involvement is virtually identical to that in DMD. Iliopsoas, quadriceps, and gluteal muscles are the first to be affected, followed by the pre-tibial muscles. Two distinct patterns of progression are seen in BMD. In the first type, the age of onset is around eight years and most patients have difficulty climbing stairs by age 20. In the second type, the more common and milder form, the age of onset is 12 years or beyond and there is no problem climbing stairs at age 20. As the lower limb weakness progresses, activities such as rising from a chair become increasingly difficult. The extensor muscles are generally weaker than the flexor muscles.

Muscles of the pectoral girdle and the upper limbs are affected after the pelvicrural muscles. Though the weakness and hypertrophy of the muscles are the same as seen in DMD, the symptoms of upper limb weakness may not begin until they are approximately 25 to 30 years of age. The muscles involved in the upper limbs are serratus anterior, lower parts of the pectoralis major, latissimus dorsi, biceps, and brachioradialis (the involvement and progression are more or less in the same order as mentioned above). Since the disease is progressive in nature, the weakness and atrophy will spread to the other muscles of the legs, arms, and trunk. The muscles that are selectively affected include the neck flexors, triceps, anterior tibial, and peroneal muscles. Deltoid, flexors, wrist and finger extensors, hand muscles, and the calves are relatively preserved. The ocular, facial, and bulbar muscles are usually spared even in the late stages.

In the very early stage, true hypertrophy may result in the enlargement of certain muscles such as calf muscles, followed by pseudohypertrophy in the later stage. The muscle hypertrophy can be progressive in the early stages of the disease, and eventually, the size of the muscle may reduce. Certain muscles like the gastrocnemii and vastus lateralis and the deltoids may be consistently large. On palpation, the pseudohypertrophic muscles are firm and resilient (rubbery feel) and, as a rule, are less strong and more hypotonic than the healthy ones. Rarely, all the muscles are at first large and strong (true hypertrophy) .

Muscles of the pelvic girdle, lumbosacral spine, shoulders, arms, and legs become weak and wasted, which accounts for the following clinical peculiarities:

1. The weakness of the abdominal and the paravertebral muscles leads to lordotic posture and protuberant abdomen when standing and rounded back posture while sitting. The lordotic posture disappears when the BMD patient lies down.
2. A bilateral weakness of the extensors of the knees and hips will cause balance impairment, stair-climbing difficulty, difficulty in rising from a chair, and stooped posture. While standing, to stabilize the hips, the BMD patient may keep the hands behind the hips with scapular retraction or hold the hands in the middle guard position.
3. The weakness of the pelvicrural muscles can give rise to wide base standing.
4. The weakness of scapula fixing muscles can cause winging of the scapula.
5. The weakness of the pectoral and the upper arm muscles may promote mild shoulder external rotation and elbow flexion, use of support from the other limb, or compensation by neck and trunk muscles during reach out or elevation of the arm.
6. Space between the lower ribs and the iliac crests diminishes with the affection of the abdominal muscles.
7. The feet assume an equinovarus position due to the shortening of the posterior calf muscles.
8. The unequal weakening of the paravertebral muscles may cause scoliosis. However, BMD patients are less prone to develop scoliosis or kyphoscoliosis, perhaps because they lack wheelchair confinement until after the spine has become fully mature.
9. Hip flexor contractures occur as a result of weak hip extensors and abdominals.
10. The weakness of the quadriceps muscles leads to a progressive and permanent shortening of the hamstrings.

The classical feature of DMD, the patient "straddles as he stands and waddles as he walks," as stated by Samuel Alexander Kinnier Wilson, the British neurologist, can be noticeable in those BMD patients with an early onset of symptoms. As the disease progresses, to rise from a sitting position, the BMD patient may flex his trunk at the hips, puts his hands on his knees, and pushes the trunk upward by working the hands up the thighs. The Gowers's sign (Box 12.2), another classical feature of DMD, can be seen if the BMD patient is instructed to stand up from a seated position on the ground.

Box 12.2 Description of the Gowers's sign

Gowers's sign
While rising from the ground, the patient first assumes a prone position and then a four-point position by extending the arms and legs to the fullest possible extent. Following this, to stabilize the knee, which may buckle due to quadriceps weakness, and to come to an erect position, the patient works each hand alternately up the corresponding thigh and finally extends the spine, i.e., in short, the patient "climbs on his own thighs to stand up" (Fig.12.1)

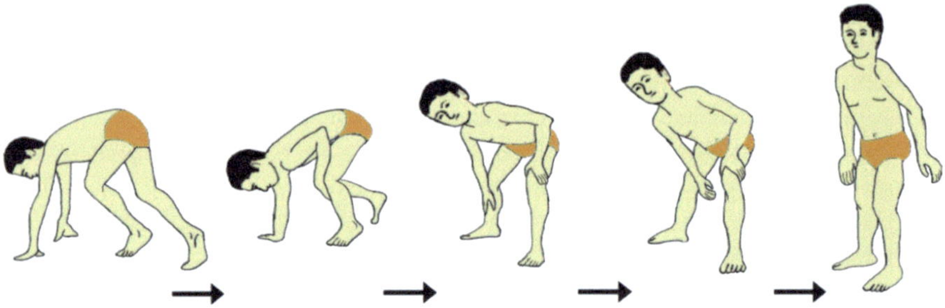

Fig. 12.1 An illustration of the Gowers's sign

Frequently, BMD patients may complain about leg cramps and muscle pains. These symptoms are often associated with exercise and are more severe in BMD than in DMD. The exercise-induced myalgia and cramps, sometimes with episodes of rhabdomyolysis, during the early stage, may suggest a metabolic myopathy. The symptoms of rhabdomyolysis are muscle pain, muscle weakness, dark red or brown urine (myoglobinuria), or decreased urination. Genetic and muscle biopsy studies and the use of electromyography (EMG) can distinguish muscular dystrophies from metabolic myopathies.

The progression of muscular degeneration leads to an increase in the level of physical disability. Fibrous contractures appear as a result of the limbs remaining in one position secondary to the imbalance between the agonists and the antagonists. These contractures are not a significant early functional problem in BMD and account for the habitual posture and loss of ambulation. Many times, the contractures become a problematic issue once the BMD patients are wheelchair dependent and typically occur in the hip flexors, knee flexors, and ankle plantar flexors.

Unlike DMD, the BMD patient's mentation is normal. Smooth muscles are spared, but the cardiac musculature is often affected. Just like the skeletal muscles, the cardiac muscles are also lost and replaced by fibrous tissue. Cardiac arrhythmia may appear in the later stages and a considerable proportion of BMD patients can have cardiomyopathy that can be more disabling than the weakness. Electrocardiogram (ECG) abnormalities can be detected in about 75% of BMD patients. BMD patients with early onset of symptoms are more likely to have a higher rate of electrocardiogram (ECG) abnormalities than those patients with the late onset of symptoms (onset of symptoms those beyond 12 years of age). ECG demonstrates left ventricular dilatation in 37% of BMD patients, and 63% have a subnormal systolic function due to global cardiac hypokinesia. ECG shows prominent "R" waves in the right precordial leads and deep "Q" waves in the left precordial and limb leads. A routine cardiac checkup is required for these patients as the degree of cardiac compromise may not be reflected by clinical symptoms. Significant pulmonary dysfunction is not a hallmark of BMD as compared to DMD. Typically, until the third to fifth decade of life, the forced vital capacity (FVC) does not fall below the predicted level.

Atrophy and weakness of the muscles make the bones thin and demineralized and may give rise to prominent bones, which are vulnerable to fractures, given the poor postural control and balance due to muscle weakness. As the weakness progresses and the contracture becomes severe, the changes associated with it eventually lead to loss of ambulation. Affected limbs are usually loose and flaccid. The deep tendon reflexes are diminished or lost. The average age at which the patient becomes unable to walk is 30–40 years; death may occur by the fifth decade of life, as a result of cardiac decompensation or pulmonary infections and respiratory failure.

12.5.2 Pathology

In the early stages of DMD or BMD, the most distinctive features are the prominent segmental degeneration and the phagocytosis of the single fiber or a group of fibers and evidence of regenerative activity. Necrosis of the muscle fibers initiates a regenerative or restorative process that explains the forking of fibers and clustering of small fibers with prominent nuclei. The mononuclear cells remove the necrotic sarcoplasms and sarcolemma. The histologic changes in muscle biopsy (obtained mostly from rectus abdominis rather than gastrocnemius or deltoid due to worsening caused by immobilization) associated with these conditions are:

- Loss of muscle fibers
- Residual muscle fibers have abnormal variations in size (either unusually large or smaller in size)
- Muscle fibers are haphazardly arranged
- An increased amount of lipocytes and fibrosis
- Presence of central nuclei
- Splitting of muscle fiber

True hypertrophy, usually seen in the early stage, can be the result of the work-induced enlargement of the sound muscle fibers. On the contrary, pseudohypertrophy is due to the lipocytic replacement of degenerated muscle fibers. As the disease advances, true hypertrophy will give way to pseudohypertrophy. Eventually, the muscle fibers degenerate and disappear owing to the exhaustion of regenerative capacity after repeated injuries or due to extensive necrosis. Only a few scattered muscle fibers remain, almost lost in a sea of fat cells in the late stages.

12.5.3 Etiology of BMD

The most important development in the understanding of DMD and BMD has been the discovery of the abnormal gene shared by these disorders and the gene product. Dystrophin (the muscle protein) is the name given to the protein encoded by the dystrophin gene. The biochemical assay of the dystrophin protein and the histochemical demonstration of the same protein near the sarcolemma have made possible the accurate diagnosis of Duchenne and Becker phenotypes (in the Duchenne phenotype, dystrophin is absent, whereas in Becker it is abnormal).

Dystrophin is present in the normal skeletal and cardiac muscles. They are localized to the cytoplasmic face of the sarcolemma, where it interacts with the "F" actin of the cytoskeleton (the filamentous reinforcing structure of the muscle cell). Dystrophin is also tightly bound to a complex of sarcolemmal proteins known as dystrophin-associated proteins. This protein complex (Fig. 12.2) comprises several subcomplexes, namely dystroglycan, sarcoglycan, and syntrophin, and functions as a trans-sarcolemmal link between the subsarcolemmal cytoskeleton and the extracellular matrix. It collectively maintains the muscle fiber integrity during the repeated cycles of contraction and relaxation and while conducting the signals. The loss or abnormality of dystrophin leads to the disruption of the protein complex rendering the muscle plasma membrane susceptible to breaks or tears during muscle

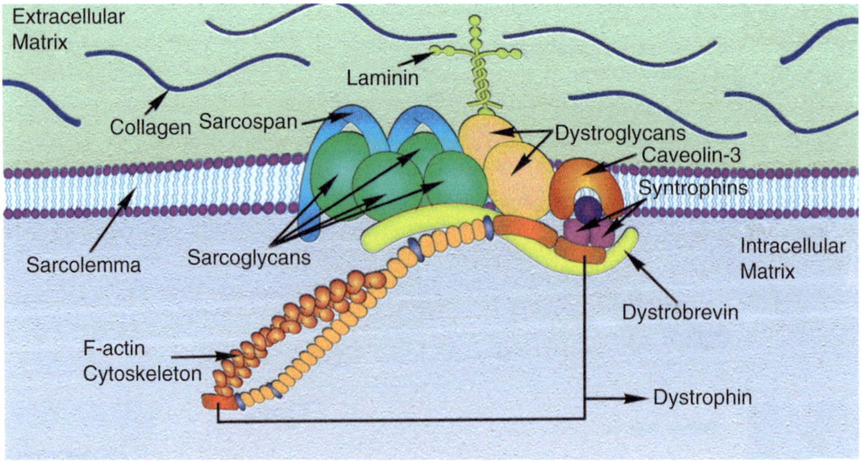

Fig. 12.2 A pictorial illustration of dystrophin and dystrophin protein complexes functioning as a trans-sarcolemmal link between the subsarcolemmal cytoskeleton and the extracellular matrix, collectively to maintain muscle fiber integrity

contraction. The abovementioned presumably means that molecules, such as calcium, would have unlimited access to the muscle fibers and would initiate a whole chain of destructive processes, producing necrosis of the muscle fiber. In dystrophinopathies, the mutation of the dystrophin gene results in the absence or abnormality of the dystrophin muscle protein rendering the muscle membrane susceptible to damage. Though in BMD, the gene product is abnormal, the dystrophin protein preserves enough function to slow down the progress of the illness, as compared to the rate of progression of DMD.

12.5.4 Diagnosis

The breaking down of muscle tissue due to the pathological process has the corollary effect of allowing CK or creatine phosphokinase, a muscle enzyme, from spilling out into the bloodstream. The serum CK value for normal adults is usually within 20–220 IU/L range. In BMD, the serum CK levels are elevated and are approximately 25–200 times to that of normal. Though the serum CK is not typically as high as seen in DMD, the degree of the CK elevation does not discriminate between the two diseases.

The disease is characterized by abnormal EMG and muscle biopsy findings. The EMG demonstrates myopathic features, characterized by a pattern of short duration, low amplitude, polyphasic motor unit potentials during voluntary contraction and some fibrillation potentials, and positive sharp waves at rest. In the later stages, the number of motor units activated diminishes and the tissue may even become inexcitable. Like in DMD, the high-frequency discharges are occasionally seen at rest.

The EMG abnormalities are nonspecific and diagnosis requires the demonstration of a mutation in the dystrophin gene or a reduced quantity or size of dystrophin on muscle biopsy. Genetic testing is the diagnostic procedure of choice and involves analyzing the DNA of any cell (typically the blood cells) to see whether there is a mutation in the dystrophin gene. Dystrophin, the gene product of DMD and BMD, is a diagnostic marker, and without dystrophin analysis, the clinical discrimination between DMD and BMD can be difficult. A complete lack of dystrophin is pathognomonic for DMD, while variable molecular changes are features of BMD. Immunofluorescence detections on cryosections and Western blotting are two methods available for dystrophin assessment. For one-third of the BMD patients, the gene abnormality may not be readily demonstrable, and in such cases for confirmation, the Western blot technique or staining can be carried out on tissues derived from a muscle biopsy that may reveal a severe reduction in the quantity of dystrophin or reduction in molecular weight of dystrophin. If the reduction is considerable and below 5% of the normal quantity of dystrophin, it is suggestive of DMD. If the value is 5% to 20% of the normal, then it is an intermediate form of dystrophinopathy, and if it is more than 20% of the normal, then it is suggestive of BMD. However, the percentage of normal dystrophin noted during staining will not help predict the clinical course of BMD.

The majority of the female carriers (mothers of the BMD patients) remain asymptomatic. Less than 10% of the carriers are symptomatic and manifest features to variable degrees. The symptoms may range from mild muscle weakness to severe gait abnormalities with a history of frequent falls and difficulty in rising from the floor. Female carriers may have enlargement of the calves, elevated CK values, and abnormalities of the EMG and muscle biopsy. ECG may demonstrate features of cardiomyopathy among the BMD female carriers. Occasionally, BMD carriers may present with cardiomyopathy, with no presence of skeletal muscle involvement. It is advisable that all carriers should undergo a regular cardiac assessment to monitor for any cardiomyopathy. Like BMD patients, the diagnosis of BMD carriers is based on clinical symptoms, family history, biochemical markers, echocardiography, and pathological and genetic studies.

12.5.5 Medical and Surgical Management

To date, there is no cure or specific drug treatment for BMD. The administration of prednisone appears to retard the tempo of progression of the disease; however, weight gains and gastrointestinal disorders, excessive hair growth, osteoporosis, and behavioral problems are some of the side effects of prednisone therapy. The management relies on the supportive treatment of the complications caused by progressive weakness. Fasciotomy and tendon lengthening surgeries are indicated if contractures are formed, which interferes with the function or mobility of the patient. In some patients, the insidious onset of respiratory failure can be evident as sleep apnea, requiring ventilatory assistance. In a later stage, patients may require symptomatic treatment for pulmonary infections and cardiac problems. For BMD patients with cardiomyopathy, optimal management remains controversial. The cardiologist, based on the evaluation, may make use of diuretics, β-blockers, and angiotensin-converting enzyme inhibitors. Less frequently, cardiac transplantation may be advised and has proven successful in some patients with this form of the illness.

Prevention The genetic analysis of the dystrophin gene among BMD carriers is typically required for genetic consulting and prenatal diagnosis. Fetal abnormalities can be detected by chorionic villus sampling, a prenatal diagnostic method to determine chromosomal or genetic disorders, or amniocentesis, a procedure to diagnose fetal defects in the early second trimester of pregnancy. The DNA analysis is more accurate and serves as an essential tool for prenatal diagnosis and for identifying female carriers.

Genetic Counseling A known carrier has a 50% chance of giving birth to a male child with BMD or a female carrier. About one-third of patients do not have family histories of the disorder. Such patients have a new mutation in the dystrophin gene. Prenatal testing of a mother with an affected son in subsequent pregnancies must be done as it cannot be ascertained whether the affected child underwent a new mutation or had a genetic transmission.

Alternate Treatment Strategies One of the therapeutic approaches involves the transfer of myoblasts into the affected muscles, to enable the fusion of the normal myoblasts with the dystrophin-deficient muscle fibers. The encoded dystrophin from the normal myoblasts would then rescue the dystrophic fibers. However, the effectiveness of this as a standalone treatment has been disappointing, but when combined with cyclosporine, an immunosuppressant drug has shown some positive effects. The second possibility is to enable a gene transfer through a viral vector (retroviral or adenoviral vectors), but this needs careful examination for safety before human trials can begin. Other approaches include elucidating the function of the dystrophin-glycoprotein complex, replacing dystrophin, and upregulating utrophin and sarcoglycans.

12.5.6 Physiotherapy Management of BMD

Physiotherapy has an important role in helping to maintain the patient's maximum potential function. Its main aims are:

1. To provide a physical assessment and to use that information for decision-making about future care.
2. To minimize the development of contractures: by passive stretching.
3. To maintain muscle strength: by exercise.
4. To prolong mobility and function: by means of splints, calipers, spinal jackets, and wheelchairs.
5. To preserve and/or improvise airway hygiene and respiratory muscle strength.

Prompt therapeutic intervention helps to prolong the functional ability of the patient, and it should begin once the diagnosis is established. An active program of physiotherapy and timely application of braces forms the important components of management.

12.5.6.1 Evaluation

Evaluation helps to gather information about the condition and the present functional status of the patient, which helps in setting the goals and means of management. Unlike DMD patients, these patients do not require frequent periodic evaluation.

Manual Muscle Testing Manual muscle testing remains a valid approach to assess the progression of the disease. It is reliable and sensitive to changes in "muscle strength." It must be a routine part of physiotherapy evaluation. Reduction of muscle strength is insidious in nature and continues to progress several years during the early and advancing stages of the disease. The patient may demonstrate normal performance of various tasks during serial evaluation despite a continuing decline in strength. The speed of progression of the symptoms of the disease varies considerably among BMD patients. Manual muscle testing and the evaluation of the performance of functional tasks help determine when bracing or a wheelchair is needed. Box 12.3 provides the list of muscles that require routine muscle strength examination.

Box 12.3 List of Muscles that Require Routine Muscle Strength Examination

Muscles for routine strength examination	
Upper trapezius	Iliopsoas
Lower trapezius	Quadriceps
Deltoid	Gluteus maximus
Serratus anterior	Gluteus medius and minimus
Pectoralis	Anterior tibial muscles
Latissimus dorsi	Abdominals
Triceps	Erector spinae
Rhomboids	

Evaluating the strength of the serratus anterior is difficult during the advancing stages of the disease. Asking the patient to keep hands at shoulder level with arms horizontally abducted and elbows extended and then observing the scapular winging is the way to assess the muscle strength of the serratus anterior. Similarly, care has to be taken while assessing the power of iliopsoas, since most of the hip flexion strength is generated by the rectus femoris or sartorius. In the later stage of the disease, due to poor strength of the neck flexors, grading the abdominal strength becomes more difficult, particularly when the muscle power is poor or fair.

Range of Motion Evaluation Standard assessment of the range of motion (ROM) of joints with goniometry is essential and it should be done periodically. Early loss of ambulation is more frequently caused by loss of ROM and contractures rather than due to weakness. Measurement of ROM of ankle dorsiflexion, knee extension, and hip extension is imperative.

Functional Evaluation Compared with normal subjects, the BMD patients perform functional activities more slowly. Until around 20–30 years, they can climb stairs, walk, and stand up unassisted. By the age of 30, these patients lose their ability to walk unassisted, and by 40 years, they lose total ambulation even with long leg braces. The patient may develop scoliosis by the age of 40 years due to back muscle weakness. Once they are wheelchair dependent, the respiratory muscles are prone to weakness and may lead to ineffective coughing, which can cause

Table 12.2 The Vignos scale grading for lower extremity functional mobility

Functional ability grades for lower extremity	
Grade	Description
I	Walks and climbs stairs without assistance
II	Walks and climbs stair with the aid of railing
III	Walks and climbs stairs slowly with the aid of railing (over 25 seconds for eight standard steps)
IV	Walks unassisted and rises from chair but cannot climb stairs
V	Walks unassisted but cannot rise from chair or climb stairs
VI	Walks only with assistance or walks independently with long leg braces
VII	Walks in long leg braces but requires assistance for balance
VIII	Stands in long leg braces but unable to walk even with assistance
IX	Is in a wheelchair
X	Is confined to a bed

aspiration of oral and gastric fluids and pulmonary infections. However, the possibilities of developing scoliosis and respiratory complications are lesser than in DMD. The functional ability of the patient's lower extremity can be graded using the Vignos scale (Table 12.2). The muscular dystrophy functional rating scale (MDFRS) is a disease-specific scale developed by Lue et al. It has four domains to rate, namely mobility, basic activities of daily living (ADL), arm function, and impairment. The mobility domain measures the ability of stair climbing, outdoor and indoor mobility, transfers, wheelchair manipulation, and rolling and changing body position in bed. The basic ADL domain measures the ability of feeding, combing hair, personal hygiene, and dressing. The arm function domain measures the ability to perform overhead tasks, carrying objects, clearing a table, writing, turning the pages of a book, and picking up and manipulation of small objects. The impairment domain measures contractures, scoliosis, muscle strength, and respiratory function.

Respiratory and Bulbar Functions An in-depth evaluation of the respiratory and bulbar functions may not be essential for BMD patients compared to DMD. Typically, the respiratory failure and its associated symptoms occur only in the advanced stage, long after ambulation has been lost. Therefore, it is advisable to periodically evaluate the bulbar function, cough effectiveness, and FVC when the patients are wheelchair dependent. The use of handheld incentive spirometer units serves as a device for monitoring the respiratory muscle strength and means for strengthening the same.

12.5.6.2 Physiotherapy Intervention

Although there is no cure for BMD, the disorder's impact can be changed significantly by keeping the body as flexible, upright, and mobile as possible. It can be accomplished with a combination of therapeutic exercises, bracing, and the use of a wheelchair. We must realize that physiotherapy manages the patient's problem and not the disease. The primary problems BMD patients have are:

1. Weakness
2. Reducing range of motion
3. Loss of ambulation
4. Decreasing functional ability
5. Reducing pulmonary function
6. Progressive scoliosis
7. Emotional trauma

Evaluation of the patient helps in identifying and prioritizing the patient's problems, and based on it, the general goals can be developed. During the early and advancing stages, the important goals are:

1. Strengthening exercise of weak muscles
2. Prevention of deformities/contractures
3. Promoting function and mobility
4. Pain management
5. Home program
6. Facilitating the development and assistance of family support

Exercise Program Exercise can promote skeletal muscle strength, facilitate muscle hypertrophy, keep the cardiovascular system healthy, and contribute to feeling better. Studies in muscular dystrophy patients suggest the positive effects of aerobic and strength training. Investigators have used bicycle ergometers, conventional strength training programs, antigravity treadmill training, and partial body weight support system training as a means of treatment. Such studies have proven that low-to-moderate intensity aerobic exercise and strengthening exercise improve parameters like walking distance, dynamic postural balance, and functional mobility and are safe and feasible in BMD and limb-girdle muscular dystrophies. The studies also revealed that serum CK remained unchanged between pre- and post-training sessions.

The regenerative or repair process does not occur well with those having dystrophic muscles particularly when the stabilizing proteins are either abnormal or absent to prevent muscle membrane damage secondary to repeated contractions. For the aforementioned reason, extensive and heavy resisted exercise may not be advisable for most muscular dystrophies and may actually accelerate the progression of the disease. Hence, the exercise must be approached with caution. It is a well-known fact that high-intensity training is a more efficient way than low-intensity training to improve oxidative capacity and reduce the risk of cardiovascular disease among normal subjects. Based on that assumption, a study conducted high-intensity exercise training for BMD, limb-girdle muscular dystrophy, and facioscapulohumeral muscular dystrophy patients and concluded that such exercises were generally well tolerated by the latter two types of muscular dystrophy patients. Contrary to the above, the BMD patients showed elevated plasma CK levels beyond 24 hours post-exercise, suggestive of muscle damage, thus substantiating the deleterious effect of high-intensity exercise.

A common notion has been that any form of exercise given for dystrophinopathy patients, particularly strength and endurance training exercises, could accelerate muscle degeneration. Due to which the BMD patients are often advised to avoid physical exertion, thus encouraging a sedentary lifestyle. Such notions were further supported by earlier studies, which suggested exercise training had a deleterious effect on the dystrophinopathy animal models. In contrast, other dystrophinopathy animal model studies have shown beneficial effects of aerobic exercise with low mechanical force. Twelve weeks of endurance training for BMD patients, for 30 minutes per session, at 65% of their maximal oxygen uptake, revealed no increase in CK levels with training and no histopathological change in muscle biopsies. The study also revealed that the improvements and safety markers were maintained after one year of training, indicating that endurance training is a safe method to increase exercise performance and daily function in BMD patients.

Current molecular and pathophysiological knowledge of dystrophinopathies suggests that regular exercise training in BMD patients may be deleterious if training is eccentric or at maximal exertion levels. Depending on the degree of weakness in a particular muscle, low-to-moderate intensity strengthening exercises may be safe and feasible. Supervised strength or resistance training is advisable and may improve muscle strength and endurance in patients with BMD and limb-girdle muscular dystrophy. Though studies on low-to-moderate resistance training among BMD patients have shown improvement in muscle strength, a meta-analysis by Gianola et al. revealed no significant effect on physical functions, including the time to stand up from a seated position, ascend stairs, and descend stairs. The training program must be designed as per the patient's functional requirements and must include warm-up exercises and adequate rest between the exercise sets. The use of functional tasks is preferred over conventional strengthening exercises using dead weights and sandbags. Functional tasks such as sit-to-stand activities from a chair, climbing up and descending a designated number of steps, stair step-up exercise onto a box of various and progressive heights, squatting or assisted squatting, and lunges can be encouraged.

However, for those muscles that have lost a significant degree of strength, resistive exercise may be contraindicated. In such situations, exercises to maintain or improve flexibility are more appropriate. If the initial muscle strength level is less than 10% of normal, the BMD patients may gain certain training effects by using an antigravity treadmill or treadmill with a partial body weight support system that reduces loads on lower limbs while preserving gait pattern and free ROM.

The muscles that are particularly vulnerable in BMD are those that control the hips, knees, shoulders, and the trunk. It is not advisable to compel the patient to perform all the components of exercises as certain components can be too demanding or challenging for the patient, particularly during the later stages. Pacing the exercise and spreading the components across two or three sessions may minimize overburdening the patient.

Flexibility and resisted exercises can be performed in supine, prone, side-lying, sitting, and standing positions. The amount of resistance given during resisted exercise should be such that the patient just manages to complete the movement. The resisted exercise can be made harder by asking the patient to hold the limb in place (isometric) or by providing external weights or manual resistance. However, it is important not to give resisted exercise if the patient is stressed or fatigued. In the case of exercise-related fatigue, provide adequate rest prior to the next session's exercise. Proprioceptive neuromuscular facilitation (PNF) exercises that promote trunk and extremities rotation can be incorporated. As the condition progresses, the exercises must be made lighter. Eccentric muscle contractions may cause further muscle damage and such exercises are avoidable.

In addition to specific exercises, physical activities and "adaptive" sports can help maintain as much strength as possible and give self-confidence. Swimming and cycling are ideal sports and, depending on muscle strength, such exercises can be encouraged. Swimming and water exercises (aquatic exercises) keep the muscles as toned as possible without causing undue stress. The buoyancy of the water provides assistance and/or resistance to the muscles that are being worked. Flotation aids like a collar and swim disks can help the patient to feel secure enough to exercise in the water. However, it is essential to rule out any cardiac abnormalities before undertaking any "aquatic" exercise program and not leave the patient unattended while swimming or exercising in the water. Recreational activities and other activities, including table tennis, gym ball exercises, and obstacle courses, may be incorporated.

Prevention of Deformities Contractures are a complication of muscular dystrophy as they interfere with the normal functioning of the affected body part. In BMD, flexion contractures of the knee, hip, and ankle joints are common. The elbows may also end up flexed. Gentle stretching exercises, together with ROM exercises, on a regular schedule, can help to keep tendons from shortening prematurely. Thus, the contractures can usually be postponed or slowed down, if not entirely prevented.

The contracture of plantar flexors is one of the earliest problems. Except for the sensation of pulling of the muscles during the stretch, the correct technique of passive stretching is unlikely to cause pain. A daily stretching program with sustained stretch to the heel cord for about 30 seconds, 20 times each, at least twice daily is appropriate. The patient must relax completely (a good idea is to do it after a warm bath when the patient feels relaxed) and should be encouraged not to make any active movement or resist the stretch. As the disease advances, the stretching program has to be expanded due to the possibility of new muscles developing tightness or contractures. Daily stretching of the iliotibial band, hip flexors, and foot invertors is essential during the ambulation stage, and if unchecked, it can lead to loss of ambulation.

To prolong the ambulation, the treatment program must combine passive stretching with lower extremity braces. Rarely surgery is done to relieve contractures and in case of surgical intervention (for Achilles tendon contracture), physiotherapy (post-surgical) and orthotic interventions are necessary to improve and prolong ambulation. Night splints used in conjunction with passive stretching may effectively retard the development of ankle joint contracture. However, splints are not a substitute for passive stretching and should be used in combination with stretching once the contracture is apparent. The night splints prescribed for the patient should be lightweight and comfortable and should fit properly. For those patients who are wheelchair-bound, corsets, jackets, and back braces may retard or prevent the development and progression of scoliosis.

Prolonging Function and Mobility Though active exercise at home will not change the course of the disease, the therapist should encourage patients to perform routine ADL. Performing ADL during the advancing stage of the disease can often be perceived as a strong "exercise" program. Accomplishing tasks such as washing clothes, combing one's hair, dressing, bathing, or performing housework will require more energy from weakened muscles because of the weight of the limbs in space. However, care has to be taken not to encourage any activities that may cause undue fatigue.

As the condition advances, the stride length reduces, the base of support widens, waddling increases, and the lordotic curve exaggerates (the center of gravity moves posterior to the fulcrum of the hip joint by pulling arms back and by exaggerating the lordosis when the patient walks). During this stage, even a mild knee flexion contracture makes ambulation difficult. Braces and standing frames can help to prolong walking and standing in muscular dystrophy. Braces used in BMD include the dynamic ankle-foot orthosis (AFO) and the knee-ankle-foot orthosis (KAFO). Early in the disease, a dynamic AFO can help in walking and can slow the development of contractures. AFOs are sometimes prescribed for nightwear to keep the foot from pointing downward while the patient is sleeping.

Further, during the advancing stage, to extend the ambulation, long leg braces may be prescribed. Walking with long leg braces can often be unsafe and may cause falls and fractures, particularly when it is effortful and needs precarious and cumbersome maneuver. Since the patients tend to walk at a slower pace, many prefer the wheelchair over the long leg brace for ambulation.

Pain Management Pain in the extremities or the back is an extremely common problem seen in BMD. This can trouble the patients even in the night, disturbing their sleep. Use of a firm bed, air or pneumatic mattress, or regular hospital bed with adjustable height for the head- and leg-ends may alleviate the pain. Frequent position changes and the use of sufficient number of pillows to support the limb while sitting or lying may also help. Hot fomentation, pain-relieving modalities, or gentle stretching programs for the limbs can also be attempted.

Respiratory Care The BMD patient should be taught simple breathing exercises, with an emphasis on diaphragmatic and segmental breathing and thoracic chest expansion. Adequate airway clearance is critical for patients with BMD to prevent atelectasis and pneumonia. Careful attention must be paid to bronchial hygiene using mechanically assisted or manual postural drainage techniques followed by good coughing. Expiratory inefficiency may lead to retention of secretions; such patients can be taught breath stacking techniques to increase the effectiveness of cough. Ineffective airway clearance can hasten the onset of respiratory failure, which can be prevent by using mechanical assistive devices such as positive expiratory pressure (PEP) device, flutter, Acapella®, inspiratory muscle training, and incentive spirometry.

Home Program Since most of the time, the patient is at home, prescription of ROM exercises, stretching, splinting, and appropriate postures or positions are essential to ensure functional independence. An effective program of care at home by the family is essential, consisting of simple instructions and positive reinforcement. The patient may need specific orthotic or adaptive equipment during the advancement of the disease. The therapist should be notified about any deterioration in the function or development of a new symptom. Thus, a review can be availed of by the patient regarding his functional and mobility status and modification of exercise, stretching program, or the brace.

Facilitating Development and Assistance of Family Support Being a therapist, it is essential to know how the patient and patient's family are getting accustomed to the disease. When a family member has BMD, all members of the family are affected by caregiving demands and emotional reactions. Many people find help and support from religious sources, families with similar experiences, self-help books, or professional counseling. The professional experts usually suggest the following:

For the BMD Patient

- Answer patients' questions about the disease and symptoms as and when they arise, with honesty and a language they understand.
- Always view the patient as an individual with the disease as one aspect of his life.
- Emphasize what the patient can do and let him find ways to do things he wants. Allow him to find creative ways to participate in sports and other hobbies.
- Treat him like any other member of the family, providing responsibility, hope, and love. Do not overprotect him and do help him to become independent.
- Encourage family members to undertake normal family activities, including vacations and recreations. With imagination and patience, the patient may find ways to do many activities.

For the Caregivers

- Respect each other's emotions and stress levels; be kind and patient.
- Schedule regular breaks from caregiving responsibilities.
- Deal with the disease one day at a time and one crisis at a time.
- Do not focus on future complications and brood about any future crisis.
- Give oneself credit for the effort expended and the difficulty faced while performing the task.
- Build a support team and ask for help whenever the need arises.

Wheelchair-Bound Stage In general, patients with BMD have much greater phenotypic variability; patients may become wheelchair-bound as early as age 20 years or as late as age 70 years. Once wheelchair-bound, patients with dystrophy become much more susceptible to the scourges of the sedentary life, which include scoliosis, contractures, and impaired pulmonary function. A constant sitting position hastens the development of contractures and other complications of immobility, such as fat accumulation and loss of bone mass. For these reasons, it should be delayed until all other means of ambulation have been exhausted. However, the possibilities of BMD patients developing scoliosis and kyphoscoliosis are rare when compared to DMD.

During the wheelchair-bound stage, standing using a standing device for a few hours each day, even with minimal weight-bearing, should be promoted. Upright posture is good for the body in general, and it promotes circulation, improves the health of bones, prevents early pulmonary complications, and straightens the spine. The wheelchair should have the following features:

- Solid seat (to maintain pelvis horizontally aligned and to postpone the development of scoliosis)
- Lateral trunk support (to postpone scoliosis)
- Appropriate width (for comfort)
- Reclining back (to extend the spine and enable pressure relief)
- Seat cushion (for comfort)
- Headrest (to support head due to weakness of neck muscles)
- Seat belt (for safety)
- Adjustable desk arm
- Swing away, elevated, and detachable leg rest

When sitting in a wheelchair, the feet should be at an angle of 90°. The seat of the chair should be firm and ideally not too wide. The back of the chair should be firm and either upright or just slightly slanting backward. The depth of the seat should be the same length as the thigh so that he is encouraged to use the back of the chair and not slump. The armrests should be at a height so that the patient can support his elbow without hunching his shoulders up or leaning sideways. When sitting, the weight should be equally distributed on each buttock. The selection of the correct wheelchair is absolutely essential for the welfare of the patient. Early attention to these factors will help to prevent contractures and scoliosis.

In the wheelchair-bound stage, the patient may be encouraged to stand using a tilt table or edge of the bed with utmost care for a few hours in a day. Also, encouraging the patient to adopt prone lying for an hour a day helps prevent the development of contractures at the hips and knees and scoliosis. Since the patients are prone to weight gain during this stage, which can interfere with transfers and reduce mobility and social activity, adequate active exercises with diet control can be advised.

Table 12.3 Common problems and therapeutic interventions in the late stage

Common problems	Therapeutic interventions
Respiratory problems	Breathing exercise; lung hygiene; chest physiotherapy
Pressure sores	Regularly turning and/or pressure area inspection
Abnormal posture; deformities	ROM exercise; stretching; positioning
Difficulty in speaking and swallowing	Prevent abnormal head posture

Late Stage (Bed-Bound Stage) This stage is characterized by more or less complete dependence of patients for the ADL. Careful monitoring of respiratory function, clearing of secretions with assisted coughing and other methods, and prompt treatment of respiratory infections can help the patient to avoid the worst scenario, i.e., the patient being shifted to an emergency room for medical interventions and ventilator support. The common problems the therapists face and the general means to manage the patients in the late stages are shown in Table 12.3.

12.6 Facioscapulohumeral Muscular Dystrophy (Landouzy–Dejerine Muscular Dystrophy)

The term "facioscapulohumeral" can be ramified, with each division having a separate meaning: *facio* - facial, *scapulo* - shoulder blade, and *humeral* - upper arm. As the term implies, the disorder primarily affects the musculature of the face and shoulder; however, the legs can also be involved. Facioscapulohumeral muscular dystrophy (FSHD) is the second most common inherited muscular dystrophy seen among adults. Its prevalence is around five per 100,000 population and is less common than DMD. FSHD is a slow progressive adult form of muscular dystrophy affecting males and females equally. The disorder is characterized by weakness and atrophy of involved muscles. Long periods of nearly complete arrest can be seen in patients with FSHD.

12.6.1 Clinical Features

The onset of symptoms can be anywhere between 6 and 20 years of age. At first, the condition is manifested by difficulty in raising the arms above the head. While raising or pushing with the arm, the scapula juts backward with the inferomedial corner pointing backward (winging of the scapula). The winging of the scapula is due to the weakness of the scapular muscles (mainly the serratus anterior). The shoulders appear sloping, owing to the weakness of the trapezii and rhomboids, and the patient will be unable to raise the arms overhead due to weakness of deltoids, serratus anterior, and trapezius. Facial muscle weakness may attract medical attention in the early stage. The smile is often flattened and transverse and the patient cannot whistle, bury the eyelashes while closing eyes screw tight, purse the lips, or drink through a straw. The involvement of orbicularis oculi, orbicularis oris, and zygomaticus is the major feature of FSHD. Speech is characteristically indistinct.

Muscles of mastication are spared (muscles innervated by the trigeminal nerve). Similarly, extraocular, pharyngeal, and respiratory muscles are also spared. Pseudohypertrophy of muscles is rare and true hypertrophy of deltoid is common. The lower part of the trapezius and sternal part of the pectoralis are almost invariably affected. As the disease advances, sternomastoid, serrati, rhomboids, erector spinae, latissimus, and eventually, deltoid muscles are involved. "Angel-wing" appearance of scapula (Fig.12.3) associated with prominent clavicles is a common finding and such a scapular appearance is due to the weakness of latissimus dorsi, rhomboids, the lower portion of the trapezius, and spinatus muscles. Atrophy and pseudohypertrophy of shoulder muscles may further constitute to such an appearance. Anterior axillary folds slope down and out as a result of the wasting of the pectoral muscles. Wasting of biceps is less than triceps. The upper arm appears thinner than the forearm ("Popeye" effect of upper extremity). Figure 12.4 illustrates the Popeye effect seen in FSHD. Selective involvement and sparing of specific muscles (hypertrophied infraspinatus muscle, atrophied trapezius muscle, partially preserved deltoid and biceps brachii, and preserved forearm muscles) all along the shoulder girdle and arm leads to poly-hill sign (Fig. 12.5).

Asymmetry of weakness is almost a rule and is common in the early years and may continue throughout the course. The disease can get virtually arrested at any point during its course. The disorder varies in intensity, even within the same family. While some patients have mild facial weakness that can go unnoticed throughout life, others may have total paralysis of the face and severe weakness of many other muscles in the body. Typically, the severe form of FSHD occurs in infancy. Such patients may start using wheelchairs by the time they are 10–15 years old.

Fig. 12.3 An illustration of "angel-wing" with winging of the scapula

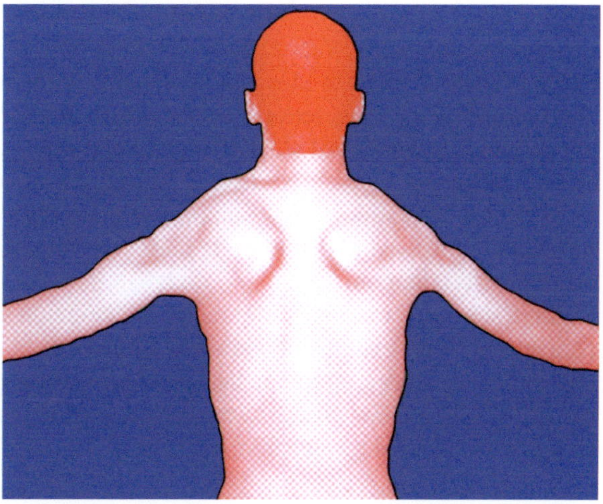

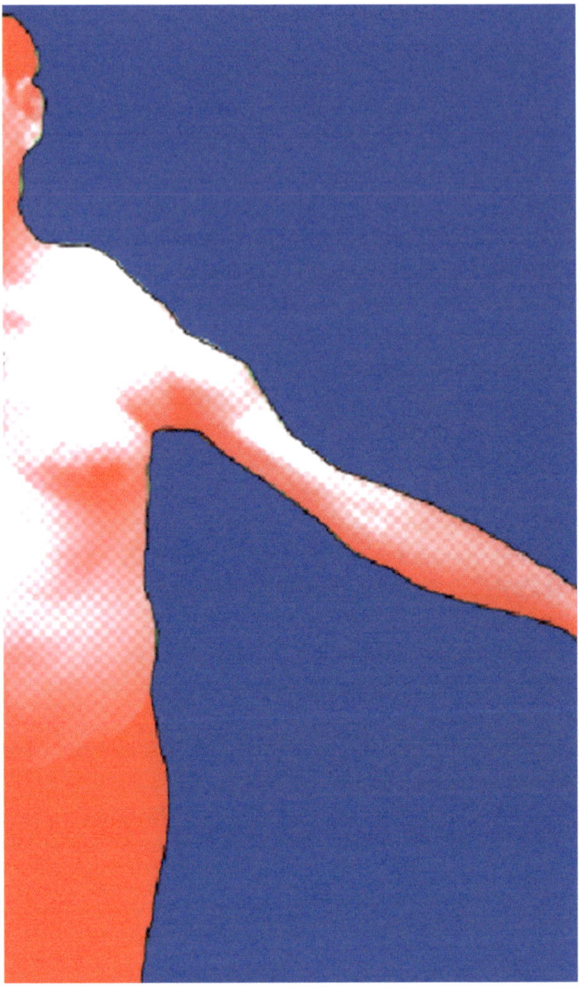

Fig. 12.4 Illustrates the "Popeye effect" seen in FSHD

As in scapuloperoneal muscular dystrophy, the weakness usually begins in the shoulder girdle muscles, serratus anterior, rhomboids, triceps, and biceps. The pelvic and thigh muscles are affected much later; however, early involvement of the anterior tibial and peroneal muscles is typical. Pre-tibial muscle weakness can lead to foot drop. The strength of plantar flexors is preserved till a later stage. Pelvic muscles are involved to a mild degree, and weakness of the hip flexor muscles and quadriceps is common.

Slight lordosis and pelvic instability are common in the later stages, but some patients may have extreme lumbosacral lordosis when standing or walking (a compensatory mechanism used for balance). Approximately 15–20% of the patients may develop scoliosis. Waddling may also be seen in the later stages. Beevor's sign (a sign characterized by upward movement of the umbilicus on active flexion of the

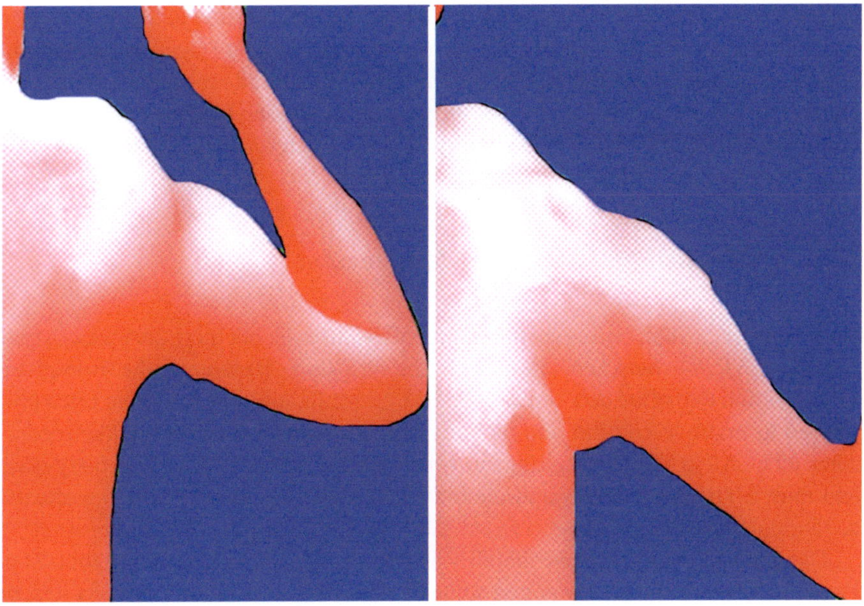

Fig. 12.5 Poly-hill sign in FSHD (posterior and anterior views)

neck in supine position and indicates a selective weakness of the lower abdominal muscles compared to upper abdominals; typically seen in T-10 spinal cord injuries) may be present. Less frequently, FSHD patients may present with congenital absence of one or more muscles or a part of a muscle like pectoralis, brachioradialis, or biceps femoris. Only about a fifth of the cases, over the age of 40 years, will be unable to walk.

Sensory hearing loss is frequent in these patients and loss can often be subclinical. Cardiac involvement is rare and is found in approximately only 5% of FSHD patients with a predilection for atrial tachyarrhythmia. Typically, the mentation is intact; however, few patients may demonstrate central nervous system abnormalities, including epilepsy, a lower volume of cerebral gray matter, and mental retardation. Patients can also present with symptoms of retinal vascular disease due to abnormalities of the peripheral retinal capillary. Similar to other dystrophic disorders, 50% of FSHD patients may present with mild restrictive lung disease with the expiratory muscles affected more than the inspiratory muscles; however, the mild respiratory involvement has no deleterious effect on the life expectancy of the patients.

Molecular genetic testing is available to confirm the diagnosis of FSHD. In FSHD, the CK level is elevated 2–4 times normal in 50% of the patients. Muscle biopsy may show general dystrophic features. Microscopic findings include the presence of degenerating and regenerating muscle fibers, abnormal variation in muscle fiber size, disruption of muscle architecture, and replacement by fat and fibrous tissue. The EMG is typically myopathic with fibrillation and positive sharp waves and repetitive discharges are occasionally seen. In later stages, the motor unit potentials will be small and polyphasic, and show increased recruitment relative to effort. The life expectancy in most of the patients is normal, but in the more rapidly progressive cases, death from respiratory infections can occur in middle life.

12.6.2 Medical and Surgical Management

No treatment can halt or reverse the effects of the disease, but there are means, strategies, or devices to help alleviate many of the symptoms. Because of the widespread inflammation that often accompanies FSHD, anti-inflammatory drugs are often prescribed to improve comfort and mobility. Studies using certain drugs like prednisone, albuterol (a β2 adrenergic agent, which is believed to improve lean body mass and muscle strength), and oxandrolone (which may have muscle-building effects) are currently underway.

Dynamic AFO can improve mobility and prevent falls in patients with foot drop. Physicians or therapists may advise some devices such as back supports, corsets, or girdle support to help to compensate for the weakening muscles in the upper and lower back and scapula region. Surgical fixation of the scapula to the chest wall often improves ROM of the arms, although this gain can be short-lived in patients with rapidly progressive disease. Similarly, surgical transposition of the posterior tibial tendon to the dorsum of the foot is particularly useful in patients who have a marked intorsion of the foot (severe inversion deformity with callus formation on the outer border of the foot making it impossible to walk or use AFO for ambulation), to improvise their gait.

12.6.3 Physiotherapy Intervention

Unlike DMD or BMD, where the involvement is generalized, a thorough and extensive physiotherapeutic intervention is not necessary. In the same vein, a periodic assessment is not as much asked for as in DMD or limb-girdle muscular dystrophy children. In general, evaluation of these patients typically shows facial weakness along with weakness of the shoulder girdle, arms, and feet, all the while helping to formulate the various principles of management and the treatment regime.

In the past, physical exercise and strengthening exercise were not recommended for muscular dystrophy patients because of the hypothetical risk that overuse would induce rhabdomyolysis and precipitate weakness. Recent studies have indicated that moderate aerobic training programs are safe and effective in improving aerobic and muscular capacities to maintain patient's independence. A randomized, controlled study noted that regular endurance training improves fitness, walking speed, and self-assessed health in FSHD patients without causing muscle damage. The study also additionally noted that protein-carbohydrate supplementation did not add any further improvement to training effects. Carefully supervised strengthening program with a graded increase in activity, which does not cause damage to the muscle or increase the weakness, can be attempted. Dynamic AFO may be particularly beneficial for patients with foot drop resulting from weakness of pre-tibial muscles. Neuromuscular electrical stimulation (NMES) was extensively used in the past for the weak muscles among muscular dystrophy patients, including FSHD patients. During the 1990s, the clinical trials showed that NMES training was safe and well tolerated. The serum CK did not change significantly and the pain and fatigue scores reduced during the training. Studies also noted improvement in maximal voluntary strength and function in stimulated muscles. However, due to difficulty in setting up long-term protocols, such programs were progressively abandoned.

In case of scapular instability (secondary to weakness of the scapular stabilizing muscles) with deltoid and supraspinatus muscle sparing, the patient may be able to abduct the arm provided the scapula is stabilized. In assessing the potential effectiveness of surgical fixation of the scapula to the thorax, a preoperative assessment can include manually holding the scapula stable bilaterally (performed by the examiner), while the patient attempts to raise the arms above the head. If the patient can do so, it ascertains the success of the stabilizing operation. If weakness continues to preclude this maneuver, then the surgery is unlikely to improve the patient's function. Postoperative management after scapular fixation requires the shoulder to be managed in a shoulder spica, with the arm abducted in the "salute position" for about 2–3 months. Isometric deltoid exercises should be encouraged with the spica cast in place, and later on, pendular exercises and strengthening exercises of the shoulder muscles are incorporated. Preoperative strength has been noted to return within six months of surgery and shoulder abduction can be improved by approximately 60°. This often leads to an improvement in the strength to lift and carry objects. There may, however, be a decrease in vital capacity secondary to immobilization of the ribs.

Despite the name given to this condition, it is essential not to overlook trunk and lower extremity involvement. Pelvic girdle weakness and specifically hip extensor weakness may give rise to compensatory lordosis. Sometimes the weakness can be so severe that the sacrum may assume a horizontal plane leading to back pain and spondylosis that may jeopardize the patient's ambulatory status. Targeting strengthening exercises to hip and trunk extensors should be considered before such a stage arises. The therapist should be vigilant for the development of hip flexor contractures, which would exacerbate the problem. Flexible spinal support may relieve the back symptoms without obliterating the necessary functional lordosis.

During the later stage, patients may come to rely on a wheelchair for indoor mobility and a suitable orthosis can be tried if the patient is seen to be developing spinal deformity. Despite adequate bracing, if the deformity progresses, surgery could be recommended.

12.7 Scapuloperoneal Muscular Dystrophy

This is a decidedly uncommon muscular dystrophy, which first becomes apparent in early adulthood and predominantly affects the upper limb girdle. It is a progressive degenerative disorder of muscles characterized by weakness and wasting. The muscles typically involved are of the neck, shoulder, upper arm, anterior tibial, and peroneal region. The nature of this disorder has been a matter of controversy; some scientific investigators are claiming it to be progressive muscular dystrophy and others claiming it to be muscular atrophy of spinal or neuropathic origin. Few investigators have established the existence of a pure myopathic form with the mutation in the FHL-1 gene on the X-chromosome.

12.7.1 Clinical Features

The onset of this disorder is usually during early adult life (the mean age of onset is between 15 and 25 years). The condition progresses very slowly but continuous in nature and the rate of progression varies from case to case. Both sexes are equally affected and the condition can also occur in combination with other disorders.

The weakness is at first most prominent in the biceps, triceps, trapezius, rhomboids, and serrati muscles. Scapular winging is a major feature. The deltoid is relatively preserved. The weakness of anterior tibial and peroneal muscles causes foot drop. About 10–20 years is taken for the weakness to spread from upper to lower limbs. In some cases, the lower leg involvement may precede the upper limb-girdle involvement. The proximal muscles of the lower limbs are affected later, especially the hip flexors and quadriceps, with relative preservation of hamstrings and calf muscles. The weakness of the pre-tibial muscles causes difficulty in walking and stair climbing and the patient tends to develop a high stepping gait (Box 12.4).

Box 12.4 Description of High Stepping Gait

High stepping gait
This gait appears in association with foot drop and the patient either drags his foot while walking or, typically, in attempting to compensate for the foot drop and to permit the toes from scraping the floor during the stride phase and avoid tripping, lifts the leg as high as possible (high stepping). Such an exaggerated hip and knee flexion with the foot and toes flopped down will make a characteristic "double-tap" sound before either the heel or ball of the foot strikes the floor. The phase of support on the affected side is shortened. The patient is unable to stand on the heel, and when he stands with his foot projecting over the edge of a step, the forefoot drops.

The facial, forearm, and hand muscles are preserved until the late stages. Rarely, mild weakness may be noticeable in the facial muscles. Hypertrophy of the muscles is absent. Contractures develop in those joints where active ROM is limited but are not an early feature. The intellect is unharmed and the cardiac abnormalities are typically absent. The serum CK level is usually raised to around 20 times of normal value. The EMG findings are similar to those of FSHD.

12.7.2 Management

Just like other muscular dystrophies, this condition also does not have any therapy to cure, halt, or reverse the disease. Genetic counseling may benefit affected patients and their caregivers. The common complications are managed symptomatically. As in FSHD, dynamic AFO can improve mobility and prevent falls in patients with foot drop. Corsets and back support may help keep the alignment of the trunk and function of the pectoral girdle. Similarly, surgical fixation of the scapula and transposition of posterior tibial muscles may be required. Physiotherapy management is comparable to that of FSHD.

12.8 Limb-Girdle Muscular Dystrophy (Erb Dystrophy)

Those patients with muscular dystrophy, who do not fit into the Duchenne or Becker, facioscapulohumeral, or scapuloperoneal categories, may fall in limb-girdle muscular dystrophy (LGMD) group. Advancement in molecular genetic techniques has helped in identifying nearly thirty limb-girdle syndromes (most with a distinct chromosomal locus, gene, and gene products). Less than 10% of cases are inherited as autosomal dominant traits (type 1), and the rest are inherited as autosomal recessive traits (type 2), which are often more severe and resemble DMD. Table 12.4 provides an outline of limb-girdle muscular dystrophies based on the patterns of inheritance, gene product, age of onset, and features.

12.8.1 Pathology

The sarcolemmal membrane of skeletal muscle comprises large complexes of muscle proteins, which are important structural links between the actin cytoskeleton and the extracellular matrix. Actin cytoskeleton dynamics play a central role in regulating multiple aspects of cellular behavior, including the control of cell morphology

Table 12.4 An outline of limb-girdle muscular dystrophies based on the patterns of inheritance, gene product, age of onset, and features

Limb-girdle muscular dystrophies				
Autosomal dominant				
Type	Gene product	Age of onset	Associated/special features	Serum CK
LGMD 1A	Myotilin	18–35 years	Dysarthria; Achilles tendon contractures	Normal to moderately elevated
LGMD 1B	Lamin A/C	Adolescence to adulthood	Cardiac conduction defects, hypertrophy of calf; Achilles tendon contractures	Normal to moderately elevated
LGMD 1C	Caveolin-3	Early childhood	Cramps and muscle pain after exercise; hypertrophy of calf	Moderate to very high
LGMD 1D	Unknown	Early adulthood to sixth decade	Conduction abnormalities and dilated cardiomyopathy	Normal to moderately elevated
LGMD 1E	Unknown	Childhood and later	Weakness progressive from proximal muscles of the upper and lower limbs; dysphagia; contractures; cardiac abnormalities	Normal to moderately elevated
LGMD 1F	Unknown	Variable	–	Normal to high
LGMD 1G	Unknown	Adulthood	–	Normal to very high
LGMD 1H	Unknown	Variable	–	Normal to very high
Autosomal recessive				
LGMD 2A	Calpain	Usually in childhood, but can extend up to 40 years of age	Atrophy of muscles prominent; tendency to walk on tiptoes; waddling gait; hip adductors muscles spared; Achilles contracture; respiratory problems often reported	High to very high
LGMD 2B	Dysferlin	Late teens/early twenties	Muscle weakness progresses very slowly; painful swelling of the calf; early development of contractures may occur	High to very high
LGMD 2C	γ-Sarcoglycan	Childhood or adult	Severe form is often seen in childhood-onset and mild form seen in adult-onset; calf hypertrophy, early involvement of scapular muscles, deltoid, and pelvic girdle; contracture; scoliosis; cardiomyopathy, respiratory abnormalities	High to very high
LGMD 2D	α-Sarcoglycan			
LGMD 2E	β-Sarcoglycan			
LGMD 2F	δ-sarcoglycan			

Table 12.4 (continued)

Limb-girdle muscular dystrophies				
LGMD 2G	Telethonin	Late childhood to adulthood	Distal lower limb weakness; calf hypertrophy	Moderately high to high
LGMD 2H	Tripartite motif-containing protein 32	Childhood	Limb-girdle and facial weakness; myalgia and fatigue; calf hypertrophy	Normal to very high
LGMD 2I	Fukutin-related protein	Childhood- or adult-onset	High clinical variability; may range from mild to severe; associated with cardiomyopathy and respiratory abnormalities; hypertrophy of calf and tongue	High to very high
LGMD 2 J	Titin	10–30 years	Anterior tibial muscle atrophy	High to very high
LGMD 2 K	Protein-O-mannosyltransfer-ase 1	Infancy or early childhood	Cognitive impairment; microcephaly; developmental delay; no structural brain abnormalities	Very high
LGMD 2 L	Anoctamin 5	Adulthood	Asymmetry; muscle hypertrophy common	Very high
LGMD 2 M	Fukutin	Early childhood	Possible facial weakness; calf hypertrophy	Very high
LGMD 2 N	Protein-O-mannosyltransfer-ase 2	Early childhood	Congenital impairment (learning disability) in most cases; no structural brain abnormalities; calf hypertrophy	Very high
LGMD 2O	Protein-O linked	Early childhood	–	–
LGMD 2P	Dystroglycan	Early childhood	–	–
LGMD 2Q	Plectin	Early childhood	–	–

and motility. Abnormality or deficiency of these complexes disrupts the structural links between the actin cytoskeleton and the extracellular matrix resulting in pelvic and pectoral girdle weakness, the features of LGMD. The microscopic structure of the affected muscle is similar to the ones discussed earlier, except for the relative absence of fatty infiltration of the affected muscle.

12.8.2 Clinical Features

In general, both adult and childhood forms are present. Children of both sexes are affected. Though the age of onset, severity, and progression of clinical symptoms may vary greatly from case to case, even among the same family members, there are some common themes recognizable about the main types of LGMD. Those cases of adulthood onset tend to develop mild symptoms and progress slowly. On the contrary, the childhood-onset LGMD tends to progress more rapidly than adolescent or adult-onset cases and develop early severe disability such as difficulty climbing stairs and walking and may become wheelchair dependent.

Progressive wasting and proximal muscle weakness of the hip and shoulder areas are the major symptoms of LGMD. Weakness may spread from the proximal muscles to the distal muscles. Either the shoulder girdle or pelvic girdle muscles may be first affected. Weakness and atrophy may become evident either during late childhood or early adult life and spread from shoulders to hips or vice versa. The weakness of the proximal hip muscles may cause a distinctive waddling gait. The involvement of pectoral and upper arm muscles may cause difficulty in raising the arms over the head or carrying heavy objects. Muscle soreness and joint pain may accompany muscle weakness. Like DMD and BMD, weakness of paravertebral muscles and trunk muscles may cause scoliosis and increased lordosis. Though significant contractures and deformities are unusual, they may cause joint ROM restrictions and may limit functional ability. Hypertrophy of the calf muscles is common.

In some types of LGMD, cardiomyopathy can occur and may cause arrhythmias and cardiac output impairment and predispose to conduction block and cardiac failure. The involvement of the respiratory muscles in some LGMD cases may predispose to breathing difficulties, which may become progressively worse, with difficulty swallowing and speaking. Facial muscles are spared. In most of the LGMDs, mentation is normal. Apart from the muscular involvement, the patient may have sensorineural deafness and retinal telangiectases. This heterogeneous group of disorders has almost invariably slow progress with death ensuing about 30 years or more after the onset of disease (particularly when symptoms arise late).

12.8.3 Investigations

EMG studies will be suggestive of myopathic condition and serum CK range from normal to very high level. Various techniques such as immunostaining, immunofluorescence, or Western blot can be used on muscle biopsy samples to determine the presence and levels of specific muscle proteins within the cells. These tests involve the use of certain antibodies that react to certain muscle proteins, thus helping to demonstrate deficiency of any particular protein, like the caveolin, dysferlin, calpain, or sarcoglycan, based on the LGMD type. In a few subtypes of LGMD, ECG may reveal cardiomyopathy and, if significant, warrants a close follow-up with cardiology consultation and regular ECG, echocardiography, and Holter monitoring.

12.8.4 Medical and Surgical Management

Regarding LGMDs, it is important to reach a precise diagnosis so that the patient and the caregivers are given correct genetic advice, as well as appropriate guidance for the management of complications, which can vary from one type to another. Like BMD, this condition is also symptomatically treated, and there is no specific treatment for this disorder. The physician may advise a weight reduction or weight stabilization diet for some people with LGMD since marked overweight can put greater stress on already weakened muscles. Fasciotomy, tendon lengthening surgeries, and spinal surgeries are indicated if deformities interfere with the function or mobility of the patient. Ventilatory assistance, including intermittent negative or positive pressure ventilation, may be needed when respiratory failure arises.

12.8.5 Physiotherapy Intervention

Several research reports have suggested that supervised aerobic exercise training is harmless and well tolerated and may be considered effective in improving oxidative capacity and muscle function in LGMD patients. Strictly supervised low-to-moderate intensity resistance training and aerobic training resistance training in slowly progressive subtypes of LGMD might maximize muscle and cardiorespiratory function and prevent additional disuse atrophy. However, it is advisable to refrain from strenuous exercise, strength training with protocols of lengthening (eccentric) contractions, and overuse.

Increasing difficulty with walking is the major feature, but some patients remain ambulant for many years after disease onset and continue walking with the aid of calipers and walking aids until they begin to suffer more frequent falls or perhaps a limb fracture. Most patients have great difficulty in accepting that the use of a wheelchair would be both safer and less exhausting. Once they accept a wheelchair, they are rarely able to propel themselves for more than a few meters (owing to the weakness of the pectoral girdle and upper arm muscles) and become exhausted rapidly. Such patients should be offered electrically powered wheelchairs for safety and to preserve energy. Acceptance of such a wheelchair will improve function and morale, obviate the need for a number of aids and appliances, and reduce the possibilities of spinal deformities (due to the muscle weakness and/or a tendency to lean to one side when they propel the wheelchair) occurring.

The physiotherapy principles and management for LGMD are comparable with that of BMD. A monitored strengthening exercise program has been shown to have no adverse effects and, in some cases, show an increase in strength, at least for the short term. A shift from one job to another may be essential if the patient cannot cope up with the physical demands of the current job and those no longer ambulant are usually totally dependent on others for ADL.

12.9 Myotonic Dystrophy

Myotonic dystrophy is the commonest adult form of muscular dystrophy. The disorder was previously considered as a variant of myotonia congenita (Thomsen disease). Two distinct forms of myotonic dystrophy have been identified. Type 1 myotonic dystrophy is an inherited autosomal dominant disorder with an incidence of approximately one per 8000 live births. It is a distal type of myopathy affecting a variety of tissues in addition to skeletal muscles. Type 2 myotonic dystrophy is a proximal form of myotonic dystrophy and an inherited autosomal dominant disorder, with prevalence less than that of type 1 myotonic dystrophy.

Myotonia is defined as an involuntary, temporary stiffness of muscle, which may follow a voluntary contraction or contraction induced by electrical or mechanical stimulation. Clinically, myotonia is seen as slowness in relaxation of grip or dimpling of muscle after percussion of its belly. It is relieved by repeated use of a muscle to an extent but often returns. It is aggravated by cold, emotion, and other factors.

12.9.1 Clinical Features

Though the clinical features of type 1 and type 2 myotonic dystrophies are somewhat similar, the author believes that it is appropriate to discuss the two distinct forms separately than combined. However, the core features, such as the presence of myotonia, muscle weakness and atrophy, and dystrophic changes in the non-muscular tissues, are common for both types.

Features of Type 1 Myotonic Dystrophy Usually, for type 1 myotonic dystrophy, the illness begins in the early teenage life, starting with noticeable weakness of the hands and often with a foot drop. Muscle weakness is the commonest symptom causing referral to the physician with early involvement of cranial and distal limb muscles. The severity of myotonic dystrophy ranges from mild weakness in some adults to profound mental retardation and severe weakness in children. Overall, the disease is a slowly progressive one, gradually spreading from the distal to the proximal muscles of the limbs and muscles of the trunk. The disease tends to increase in severity and appear at a younger age in succeeding generations, a phenomenon known as anticipation.

The condition is characterized by a unique topographical involvement of muscles with muscular atrophy. Certain muscles, namely levator palpebrae, facial and masseter, sternomastoid, forearm, hand, and pre-tibial muscles, are consistently involved. Facial weakness and ptosis that cause thinness and slackness of the facial muscles and wrinkled forehead, respectively, are the earliest problems.

The small muscles of the hands, along with the extensor muscles of the forearm, are first to become atrophied. Due to pre-tibial muscle involvement, foot drop develops, and often, this is an early sign in some cases. Atrophy of masseter causes narrowing of the lower half of the face (mandible appears slender), and malpositioning of teeth is common. Recurrent dislocation of the jaw may be seen, particularly when the patient attempts to open the mouth wide. Sternomastoids are almost invariably thin and weak (this is associated with exaggerated forward curvature of the neck called "swan neck posture"). In the fully developed disease, a characteristic "haggard" appearance is seen with eyes that are hooded, frontal baldness, and the mouth that slacks and is often tented with marked atrophy of temporalis and masseter muscles. Ptosis and occasionally diplopia may be present due to extraocular muscle involvement.

Myotonia is an important finding along with the weakness. It is due to increased excitability of the muscle membrane caused by the dysfunction of muscle ion channels. It can be demonstrated either by sharp percussion of the muscles of the tongue or thenar eminence with a reflex hammer or after firm voluntary contraction, for example, gripping during a handshake. Either maneuver elicits a sustained, involuntary contraction of the muscles, which fades slowly over a matter of seconds. For instance, percussion over the thenar eminence produces a sharp abduction of the thumb with gradual relaxation, allowing the thumb to return to its normal position. Myotonia may be elicited in weak as well as strong muscles. However, myotonia is not a frequent complaint and is not as widespread as in myotonia congenita.

The occurrence of dystrophic changes in non-muscular tissues (lens of the eye, testicle, endocrine system, skin, esophagus, heart, and cerebrum) is another important feature. Those dystrophic changes seen in non-muscular tissues result in lenticular opacities, mild to moderate degree of mental retardation, progressive frontal alopecia (beginning at an early age), testicular atrophy with androgenic deficiency, decreased libido, impotence, sterility, gynecomastia, and ovarian deficiency in females interfering with fertility and menstruation. Cataracts, hyperostosis of the skull, and diabetes mellitus may also occur. Serum CK is normal or mildly elevated. The condition is also characterized by other features such as somnolence (uncontrolled urge to sleep) and uterine muscle weakness interfering with normal parturition and esophageal dilatation. Smooth muscle involvement accounts for a number of problems, which include dysphagia, peristalsis cholecystitis, constipation, and urinary tract symptoms.

As the disease advances, the muscular weakness progresses to shoulder, hip, and leg muscles. Repeated falls are common in the middle ages. The involvement of pharyngeal or pharyngeal muscles leads to monotonous nasal voice, and facial muscle weakness makes pronunciation of words difficult. As time goes by, the weakness of individual muscle becomes severe and the myotonia may be lost.

Contractures are rarely seen. By about 15–20 years after the onset of symptoms, the patient becomes dependent on a wheelchair. Diaphragmatic weakness may lead to alveolar hyperventilation, chronic bronchitis, and bronchiectasis. Conduction abnormalities of the heart are common. Mitral valve prolapse and cardiomyopathy are less frequent abnormalities. Death occurs between the fourth and sixth decade of life from pulmonary infection or heart block or heart failure.

The deep tendon reflexes are reduced or lost. Common EMG finding is the presence of myotonic potentials, a spontaneous high-frequency discharge, which waxes and wanes in both amplitude and frequency, creating a "dive bomber" sound through the loudspeaker. These electrical discharges can be initiated or reproduced by muscle tapping or needle insertion and may arise spontaneously or following active contraction. The individual waveform of this high-frequency discharge may resemble a positive sharp wave or fibrillation potential and the amplitude is typically less fluctuating, whereas the firing rate (frequency of discharge) is more variable, ranging between 20 and 80 Hz. The presence of fibrillation potentials and positive sharp waves at rest, together with reduced size of the motor unit potentials and polyphasic potentials, indicates the dystrophic changes.

Features of Type 2 Myotonic Dystrophy (Proximal Myotonic Myopathy) The onset of symptoms is typical during adult life. Early symptoms consist of the mild weakness of hip extensors, knee flexors, and finger flexors. The disease is characterized by the weakness of the proximal and axial muscles, particularly the neck flexors. Type 2 myotonic dystrophy patients often complain of muscle pain that can fluctuate in severity and be debilitating. The patients may report extensive muscular pain that may include even the abdominals and can be exercise related. Fatigue is common. Facial weakness and ankle dorsiflexor weakness are less common as compared to type 1 myotonic dystrophy.

The myotonia and cardiac issues are often less apparent in type 2 myotonic dystrophy compared to type 1 myotonic dystrophy. The cognitive impairment seen in type 2 is similar to but less severe than that of type 1 myotonic dystrophy. Tremor and calf hypertrophy can be early features of type 2 myotonic dystrophy compared to the other. The features including early cataract, hypersomnia, insulin resistance, testicular atrophy, frontal baldness, and hypogammaglobulinemia are common among both forms of myotonic dystrophy.

12.9.2 Etiology and Pathology

Genetic mutation at two distinct locations on the chromosome makes two distinct forms of myotonic dystrophy. In type 1 myotonic dystrophy, there is an expansion of a trinucleotide (CTG) in a myotonic dystrophy protein kinase gene located on the long arm of chromosome 19. The gene product is myotonin protein kinase and is believed to be involved in sodium- and chloride-channel function. Type 2 myotonic dystrophy, also known as proximal myotonic dystrophy, is the consequence of a tetranucleotide (CCTG) repeat expansion in the zinc finger protein 9 (ZNF9) gene on chromosome 3.

The histological features of skeletal muscle biopsy in type 1 and type 2 myotonic dystrophies are very similar, and in both diseases, affected muscles show a high number of central nuclei and a markedly increased variation in fiber diameter (some unusually large and others considerably small in diameter). In addition to the above, there are several unusual myopathologic features such as excess of intrafusal fibers in muscle spindles, peripherally placed sarcoplasmic masses, basophilic regenerating fibers, splitting fibers, fibrosis, and adipose deposition to a variable degree depending on the extent of muscle involvement.

12.9.3 Medical and Surgical Management

No treatment can cure this disease. The condition is symptomatically treated, like other muscular dystrophies. Although drugs like quinine can alleviate myotonia, it does not affect the muscle strength or progress of the muscular atrophy. Some patients may find the side effects of the drug more distressing than myotonia. A pacemaker may be needed if cardiac conduction abnormalities are present.

In some cases, testosterone therapy may compensate for gonadal insufficiency. Constipation and other digestive tract complaints are symptomatically treated. Surgery for cataracts and either surgery or special eye crutches for drooping eyelids can markedly improve vision. Medications that reduce excessive sleepiness can make life more enjoyable for a person with myotonic dystrophy.

12.9.4 Physiotherapy Intervention

Evaluation of these patients typically shows facial weakness as described above, as well as weakness of the hands and feet. Myotonia may be present and a slow release of a handshake may be another indication of it. Because the patients do not have much trouble in the first few years, aggressive physiotherapy, surgical reconstruction, and night splints are less often needed. Treatment for patients with myotonic dystrophy includes orthoses for distal weakness, including dynamic AFOs or wrist hand orthoses. Wrist splints are less useful. Theoretically, it should give added function to the hand, but most patients prefer not to use them.

Aerobic training on a cycle ergometer for 12 weeks has shown an increase in the maximal oxygen uptake, maximal workload, and muscle fiber diameter associated with the absence of elevation of serum CK among myotonic dystrophic patients. The study also reported improvement in the muscle strength of the participants without any observable negative side effects. Another study on progressive high-resistance strength training, performed three times a week for 12 weeks (training was performed with free weights on the iron shoe, with a target load of 80% of 1RM, three sets of 8 repetitions, consisting of 3 seconds of concentric, isometric, and eccentric phases), among limited study participants, reported an improved muscle strength and a tendency toward an increase in cross-sectional area of type I muscle fibers. Few of the study participants had severe back pain, and due to the limited number of subjects, the researchers commented that it is difficult to draw a definite conclusion from the study.

Prevention of contracture due to muscle weakness is managed as in other muscle disorders. Breathing exercise and postural drainage can be helpful in severe forms to ward off frequent respiratory infections. The myotonia, which is less of a problem than weakness, can be overcome by utilizing the "warm-up" effect, i.e., prior to an activity, patient contracts the muscles several times, so that relaxation in between the contractions becomes faster and the contractions themselves become stronger. This enables the activity to be carried out more effectively. Examples include strategies that the patients may have worked out for themselves, such as chewing gum before an interview, so the speech is fluent, arm exercises prior to playing games, or opening and closing the fist before a handshake. Appropriate strategies could be worked out for different functional activities.

Further Reading

Andersen G, Prahm KP, Dahlqvist JR, Citirak G, Vissing J. Aerobic training and postexercise protein in facioscapulohumeral muscular dystrophy: RCT study. *Neurology.* 2015;85:396–403.

Andersen SP, Sveen ML, Hansen RS, Madsen KL, Hansen JB, Madsen M, et al. Creatine kinase response to high-intensity aerobic exercise in adult-onset muscular dystrophy. *Muscle Nerve.* 2013;48(6):897–901.

Arahata K, Beggs AH, Honda H, Ito S, Ishiura S, Tsukahara T, et al. Preservation of the C-terminus of dystrophin molecule in the skeletal muscle from Becker muscular dystrophy. *J Neurol Sci.* 1991;101:148–56.

Awerbuch GI, Nigro MA, Wishnow R. Beevor's sign and facioscapulohumeral dystrophy. *Arch Neurol.* 1990;47:1208–9.

Berthelsen MP, Husu E, Christensen SB, Prahm KP, Vissing J, Jensen BR. Anti-gravity training improves walking capacity and postural balance in patients with muscular dystrophy. *Neuromuscul Disord.* 2014;24(6):492–8.

Bostock EL, O'Dowd DN, Payton CJ, Smith D, Orme P, Edwards BT, et al. The effects of resistance exercise training on strength and functional tasks in adults with limb-girdle, Becker, and facioscapulohumeral dystrophies. *Front Neurol.* 2019 Nov;10:1216. Available from: https://doi.org/10.3389/fneur.2019.01216.

Colson SS, Benchortane M, Tanant V, Faghan JP, Fournier-Mehouas M, Benaïm C, et al. Neuromuscular electrical stimulation training: a safe and effective treatment for facioscapulohumeral muscular dystrophy patients. *Arch Phys Med Rehabil.* 2010;91(5):697–702.

Daroff RB, Jankovic J, Mazziotta JC, Pomeroy SL, editors. Bradley's neurology in clinical practice. 7th ed. London: Elsevier; 2016.

Denny AP, Heather AK. Are Antioxidants a Potential Therapy for FSHD? A Review of the Literature. *Oxid Med Cell Longev.* 2017 Jan;2017:7020295. Available from: https://doi.org/10.1155/2017/7020295.

Donaghy M, editor. Brain's Diseases of the nervous system. 12th ed. New York: Oxford University Press; 2009.

Emery AE, Emery ML. Edward Meryon (1809–1880) and muscular dystrophy. *J Med Genet:* 1993;30:506–11.

Gianola S, Pecoraro V, Lambiase S, Gatti R, Banfi G, Moja L. Efficacy of muscle exercise in patients with muscular dystrophy: a systematic review showing a missed opportunity to improve outcomes. *PLoS One.* 2013;8(6):e65414. Available from: https://doi.org/10.1371/journal.pone.0065414.

Jensen BR, Berthelsen MP, Husu E, Christensen SB, Prahm KP, Vissing J. Body weight-supported training in Becker and limb girdle 2I muscular dystrophy. *Muscle Nerve.* 2016;54(2):239–43.

Lue YJ, Su CY, Yang RC, Su WL, Lu YM, Lin RF, Chen SS. Development and validation of a muscular dystrophy-specific functional rating scale. *Clin Rehabil.* 2006 Sep;20(9):804–17.

Meola G, Cardani R. Myotonic dystrophy type 2: An update on clinical aspects, genetic and pathomolecular mechanism. *J Neuromuscul Dis.* 2015;2(2):59–71.

Ørngreen MC, Olsen DB, Vissing J. Aerobic training in patients with myotonic dystrophy type 1. *Ann Neurol.* 2005;57:754–7.

O'Sullivan SB, Schmitz TJ, Fulk GD, editors. 6th ed. Physical rehabilitation. Philadelphia: FA Davis Co; 2014.

Pradhan S. Poly-Hill sign in facioscapulohumeral dystrophy. *Muscle Nerve.* 2002 May;25(5):754–5.

Ropper AH, Samuels MA, Klein JP, Prasad S, editors. Adams and Victor's Principles of neurology 10th ed. New York: McGraw Hill Education; 2014.

Roussel MP, Morin M, Gagnon C, Duchesne E. What is known about the effects of exercise or training to reduce skeletal muscle impairments of patients with myotonic dystrophy type 1? A scoping review. *BMC Musculoskelet Disord.* 2019 May;24;20(1):244. Available from: https://doi.org/10.1186/s12891-019-2458-7.

Siciliano G, Simoncini C, Giannotti S, Zampa V, Angelini C, Ricci G. Muscle exercise in limb girdle muscular dystrophies: Pitfall and advantages. *Acta Myol.* 2015 May;34(1):3–8.

Suhas M, Patil S, Nayak S. Climb 6-hills in a coat with Popeye, you shall find facioscapulohumeral dystrophy - Going from phenotype to genotype. *J Med Sci Res.* 2019; 7(4):120–2.

Sveen ML, Andersen SP, Ingelsrud LH, Blichter S, Olsen NE, Jønck S, et al. Resistance training in patients with limb-girdle and Becker muscular dystrophies. *Muscle Nerve.* 2013;47(2):163–9.

Sveen ML, Jeppesen TD, Hauerslev S, Køber L, Krag TO, Vissing J. Endurance training improves fitness and strength in patients with Becker muscular dystrophy. *Brain.* 2008;131(11):2824–31.

Tollbäck A, Eriksson S, Wredenberg A, Jenner G, Vargas R, Borg K, et al. Effects of high resistance training in patients with myotonic dystrophy. *Scand J Rehabil Med.* 1999;31(1):9–16.

Vignos Jr PJ, Spencer Jr GE, Archibald KC. Management of progressive muscular dystrophy in childhood. *JAMA.* 1963;184(2):89–96.

Voit T, Stuettgen P, Cremer M, Goebel HH. Dystrophin as a diagnostic marker in Duchenne and Becker muscular dystrophy. Correlation of immunofluorescence and western blot. *Neuropediatrics.* 1991;22(3):152–62.

Chapter 13
Myasthenia Gravis

Abraham M. Joshua and Rohit Pai

13.1 Introduction

Myasthenia gravis (MG) is the most common and widely known primary neuromuscular junction transmission disorder. The group of diseases affecting neuromuscular junction has certain common striking features, namely fluctuating weakness and easy fatigability of the muscles. Based on the anatomical location of transmission dysfunction, the neuromuscular junction (NMJ) disorders can be presynaptic, synaptic, or postsynaptic. Diseases like Lambert-Eaton myasthenic syndrome and botulism are presynaptic, insecticide poisoning is synaptic, and diseases like MG are postsynaptic. The NMJ disorders are classified into immune-mediated, toxic, or metabolic and congenital myasthenic syndromes. The details regarding the abovementioned are given in Table 13.1. MG is characterized by notable symptoms of weakness and easy fatigability of the ocular muscles, isolated or associated with skeletal muscles in a generalized form.

The term myasthenia can be ramified into "myo" and "asthenia" (Greek words), meaning "muscle" and "weakness," respectively, and "gravis" (Latin word), meaning "serious." Though it was considered a condition of grave prognosis in the past, with the advancement in medical and surgical treatment, MG is no more a serious disorder, rendering the word "gravis" a misnomer. Advancements in intensive care, including ventilators and bronchial hygiene, and the advent of medications like

A. M. Joshua (✉)
Department of Physiotherapy, Kasturba Medical College of Mangalore, Manipal Academy of Higher Education, Manipal, India
e-mail: abraham.joshua@manipal.edu

R. Pai
Department of Neurology, Kasturba Medical College, Mangalore, Manipal Academy of Higher Education, Manipal, India
e-mail: pai.rohit@manipal.edu

A. M. Joshua (ed.), *Physiotherapy for Adult Neurological Conditions*,
https://doi.org/10.1007/978-981-19-0209-3_13

Table 13.1 Classification of neuromuscular junction disorders

Type	Name of the disease/disorder	Anatomical location of dysfunction
Immune-mediated disorders	Myasthenia gravis	Postsynaptic
	Lambert-Eaton myasthenic syndrome	Presynaptic
Toxic/metabolic disorders	Botulism	Presynaptic
	Tetanus	Presynaptic
	Organophosphate and insecticide poisoning	Pre/synaptic/ postsynaptic
	Snake venom poisoning	Pre/postsynaptic
	Arthropod venom poisoning including the black widow spider	Pre/postsynaptic
	Hypermagnesemia	Presynaptic
Congenital myasthenic syndromes	Defective synthesis or packaging of acetylcholine	Presynaptic
	Genes for choline acetyltransferase	Presynaptic
	Paucity of synaptic vesicles	Presynaptic
	Lambert-Eaton-like congenital myasthenic syndromes	Presynaptic
	Deficiency of collagenic tail of acetylcholinesterase	Synaptic
	Quantitative deficiency of acetylcholine receptors	Postsynaptic
	Genetic mutations of Rapsyn	Postsynaptic
	Kinetic abnormalities of acetylcholine receptors	Postsynaptic
	Slow channel syndrome	Postsynaptic
	Fast channel syndrome	Postsynaptic
	Anomaly of muscle Na+ channel	Postsynaptic
	Myasthenic syndrome with plectin deficiency	Postsynaptic

cholinesterase inhibitor, corticosteroids, azathioprine, plasmapheresis, and intravenous immunoglobulin have reduced the mortality from 80% in early the twentieth century to less than 5% in the twenty-first century.

The most striking feature of MG is the fluctuating weakness of certain voluntary muscles, particularly those innervated by the motor nuclei of the brainstem (i.e., ocular, masticator, facial, deglutition, and lingual). The weakness and fatigability of the bulbar muscles in isolation or associated with skeletal muscles may increase with physical effort and rest will restore the muscle power. Administration of anticholinesterase drugs like neostigmine also dramatically improves muscle power. MG has three clinical forms: neonatal, congenital, and acquired; here, the author would like to inform the readers that the former two types will not be elaborated and are advised to refer to appropriate books.

13.2 Historical Background

In 1672, Thomas Willis, an English physician, gave an account of a disease that could be none other than MG. However, many credit Samuel Wilks (1877), an English physician and biographer, for the first description of MG. In the description, Wilks mentioned that the brainstem was free of disease, contrary to the other types of bulbar paralysis. Wilhelm Heinrich Erb (1878), a German neurologist, and Samuel Wulfowicz Goldflam (1893), a Polish neurologist, were among the first who gave a reasonable and complete account of the disorder. It was Goldflam who described the clinical features of three of his patients and emphasized the fatigable quality and the fluctuating nature of the weakness and the spontaneous improvement or remission of the disease. For many years, the disorder was referred to as the Erb-Goldflam syndrome, a disorder characterized by bulbar palsy in the absence of anatomical lesions. In 1886, Hermann Oppenheim, a German neurologist, reported a case of fluctuating weakness initially involving the upper and lower limbs and then spreading to involve the bulbar muscles. Eisenlohr C., in 1886, reported a case of fluctuating ocular weakness, and Shaw L., in 1890, described a case of bulbar paralysis, with more weakness evident as the day advanced and without any structural changes in the medulla. In 1895, Friedrich Jolly, a German neurologist and psychiatrist, coined the name "myasthenia gravis" for the disorder. He originally demonstrated that repeated faradic current stimulation of the relevant motor nerve could induce myasthenic weakness and also found that the "fatigued" muscle could still respond to galvanic current stimulation (Jolly test). Interestingly, Jolly proposed the use of physostigmine as a form of treatment, but it took several years to demonstrate the therapeutic value of the same.

In 1899, Oppenheim reported a patient with MG and a coexisting tumor of the mediastinum, and a few years later, Leopold Lacquer and Carl Weigert noted the relationship between MG and tumors of the thymus gland. In 1905, Edward Farquhar Buzzard, a British physician, published a detailed clinicopathological analysis of the disease, commenting on both the thymic abnormalities and the infiltrations of the lymphocytes in the muscle. It was Lazar Remen (1932), a German physician, and Mary Broadfoot Walker (1934), a Scottish physician, who demonstrated the therapeutic value of the drug physostigmine. In 1949, Benjamin Castleman and Norris E. H., American pathologists, gave a detailed report about the pathologic changes in the thymus gland in MG patients. A few decades later, John A. Simpson and William L. Nastuk and coworkers theorized the role of an autoimmune mechanism in MG. Strauss A. J. L., in 1960, demonstrated the presence of antibodies to muscle striations in the serum of patients suffering from MG, implicating autoimmunity as the pathophysiological process. Within a few decades, several researchers, including Patrick J., Lindstrom J., Lennon V. A., Engel A. G., and Lambert E. H., established the autoimmune origin of the disease based on a series of investigations.

13.3 Epidemiology

The disease can begin at any age, but onset within the first 10 years of life is relatively rare. Epidemiological studies have shown that MG affects all the races worldwide with a wide variability in incidence and prevalence. Though racial differences exist, the biological basis is unclear. In most countries, the prevalence rate is approximately 15–20 cases per 100,000 population. The disease has a biphasic age distribution. The peak age of onset is between the third and fourth decade in women and the sixth and seventh decade in men. Under the age of 40, females are affected 2–3 times as often as males, whereas, in later life, the incidence in males predominates.

MG is frequently found to coexist with other autoimmune disorders such as Graves's disease, rheumatoid arthritis, polymyositis, Sjögren syndrome, psoriasis, and systemic lupus erythematosus. Around 75% of MG patients have thymus gland abnormalities, of which 85% have thymus gland hyperplasia and 15% have thymoma. For those patients with thymoma, the majority are older, and here again, males predominate. Approximately 15% of MG patients have symptoms of hyperthyroidism.

In the modern era, patients with MG have a near-normal life expectancy; however, morbidity resulting from intermittent impairment of muscle strength may cause aspiration and pneumonia, respiratory failure, and falls, reducing the life expectancy. Recent advances in critically ill patients' treatment and care have resulted in a marked decrease in the mortality rate, which was around 30–40% several decades ago. The mortality rate is approximately 3–4%, with principal risk factors being an age older than 40 years, a short history of severe disease, and the associated thymoma. The majority of the patients with appropriate therapy are able to lead a productive life.

13.4 Clinical Course and Manifestations

Patients with MG can present with a wide range of signs and symptoms, depending on the severity of the disease. Even the progression can vary, but in many of these patients, the onset of the symptoms tends to be insidious and in some can be of reasonably rapid progression. Once the disease has begun, a slow progression follows with a general tendency to progress from mild to more severe, over weeks to months.

In MG, certain muscles are more commonly affected, and during its progression, the weakness can spread to other muscles. The initial manifestation of the disease in about half of the cases consists of weakness of the levator palpebrae and the extraocular muscles leading to ptosis and diplopia, respectively. In due course of the illness, 90% of MG patients will develop weakness of the same. Rarely, even patients with severe, generalized weakness may have any associated ocular muscle weakness.

The usual initial complaint is a specific muscle weakness rather than a generalized muscle weakness. Weakness can be present in a variety of different muscles and is usually proximal and symmetric. The weakness tends to spread from the ocular to the facial to the bulbar muscles and then to the truncal and the limb muscles. In about 80% of the cases, the muscles of facial expression, mastication, swallowing (dysphagia), and speech (dysarthria) are involved. The flexors and extensors of the neck, the muscles of the shoulder girdle, and the flexors of the hips are the others that can often be involved. Infrequently, the initial weakness can be limited to single muscle groups like the neck or finger extensors, the hip flexors, or the ankle dorsiflexors.

Clinically, MG is most accurately considered as a fluctuating and fatigable oculo-facio-bulbar palsy. The majority of patients may frequently complain of fluctuant unequally drooping eyelids (ptosis) and intermittent diplopia. Attempts to overcome the ptosis by the nonstop contraction of the frontalis muscle may impart horizontal wrinkles on the forehead and produce a "worried or surprised" look. In the early stage of the disease, the ocular symptoms can be intermittent, typically worsened in the evening or while reading, watching television, or driving in bright sunlight. The presence of unilateral painless ptosis in the absence of ophthalmoplegia is a cardinal feature of MG. Repeated activities like gazing at a revolving drum painted with black and white strips (weakness induced by an optokinetic stimulus) and activities like gazing persistently upward are found to produce progressive paresis of the extraocular muscles.

Similarly, reduced facial mobility and altered facial expressions are the other common complaints. The bilateral facial muscle weakness may produce a mask-like face, which is in addition to the ptosis discussed earlier, a snarl (horizontal smile) instead of a natural smile, an inability to close the eyes firmly, and a difficulty to purse or roll the lips. With moderate weakness of the orbicularis oculi muscle, the eyelashes are not "buried" during the tight closure of eyes. The MG patients will have difficulty or inability to whistle, suck through a straw, or inflate balloons. The corners of the mouth often droop downward at rest, giving a depressed look, and such appearance does not necessarily reflect their feelings. The weakness of the tensor tympani may lead to the muffling of low tones, whereas the weakness of stapedius may produce hyperacusis.

The weakness of the bulbar muscle may cause the jaw to hang open and difficulty to chew tough and fibrous solid food with a sensation of fatigue and discomfort and often require excessive time to consume food. Difficulty to eat after talking and the tendency for the voice to fade and become nasal after a sustained conversation are additional features. The patient may typically start reading a paragraph with normal enunciation, but by the end of the paragraph, the articulation will become soft and broken by labored respirations. The weakness of the palatal muscles can result in a nasal twang to the voice and nasal regurgitation of food, especially liquids. Swallowing may become difficult and aspiration may occur with fluids, giving rise to coughing or choking while drinking. Vocal cord muscle weakness can lead to stridor or "crowing" during attempted deep inspirations and be a sign of impending medical emergency requiring endotracheal intubation. The weakness of the neck muscles is common with difficulty in holding up the head, leading to head lolling. Usually, the neck flexors are more severely affected than the neck extensors.

The weakness of limb musculature may impart a "myopathic" picture, i.e., the involvement of proximal muscles greater than the distal ones, with a general tendency for the upper limb to be involved than the lower limb muscles. In about 20% of MG patients, the weakness begins in the limb or the axial muscles. The weakness of the proximal muscles of the shoulder girdles makes overhead activities difficult and women often complain of inability to tie their hair. In the upper extremities, deltoids, triceps, and extensors of the wrist and fingers are frequently affected. The triceps are more likely to be affected than the biceps. In the lower extremities, the commonly involved muscles include the hip flexors, the quadriceps, and the hamstrings. Typically, the foot dorsiflexors or plantar flexors are less frequently involved than the proximal muscles. The weakness of the erector spinae can be a frequent problem in the advancing stage and patients may complain about early tiring while sustaining an erect posture against gravity. The common sequence of the spread of disease in many of the patients is as follows: ocular, facial, deglutition, masticatory, lingual, neck, shoulders, arms, hands, hips, thigh, lower legs, trunk, and respiration muscles. Though the sequence is highly variable, the muscles innervated by the cranial nerves tend to be severely affected compared to the weakness of the extremities or the trunk.

Typically, the severity of the weakness seen in MG fluctuates over hours being least severe during the morning hours and worse as the day progresses. The disease severity also varies over weeks or months, with exacerbations and remissions. Though remissions and relapses are common, spontaneous remissions are rare. Long or permanent remissions are even less common, and in most patients, the remissions with treatment occur during the first 3 years of the disease. In untreated, long-standing MG, the disease may resemble a chronic myopathy due to the fixed nature of weakness and atrophy of the severely involved muscles.

Vaccination, menstruation, infection, emotional upset, exposure to extreme temperature, pregnancy, puerperium, and certain medications are some of the precipitating factors that may exacerbate, initiate, or hasten weakness. Such weakness can lead to a myasthenic crisis and rapid respiratory compromise, requiring emergency care. In the most advanced cases, all the muscles are weakened, including the diaphragm, abdominals, intercostal muscles, and even the external sphincters of the bladder and bowel. The weakness of the respiratory muscles, namely the diaphragm and the intercostal muscles, may result in carbon dioxide retention secondary to hypoventilation and may produce acute respiratory failure, which requires immediate medical attention. The weakness of pharyngeal muscles may even collapse the upper airway requiring critical or emergency care. Patients with coexisting autoimmune disorders such as rheumatoid arthritis and systemic lupus erythematosus will have skin and joint involvement in addition to the abovementioned clinical features. In patients with coexisting thymic disorders, symptoms of hyperthyroidism will be seen. Many implicate the thymic abnormalities for the pathogenesis of the disease and believe thymectomy as an essential line of treatment for at least some of the patients.

In the majority of the patients with generalized involvement, the interval from onset to maximal weakness is found to be less than 3 years. Although the disorder is progressive, in some with a mild presentation, the symptoms remain restricted to the extraocular and the eyelid muscles. Often, in such cases, the subtle findings, such as ptosis and diplopia, may not be apparent unless the muscle weakness is provoked by repetitive or sustained use of the muscles involved. Usually, one-third of MG patients develop signs and symptoms of generalized myasthenia within a month after onset, about 50% develop generalized myasthenia within 6 months, and around 80% develop generalized disease within 2 years post-onset.

Patients may exhibit a kind of irregular tremor while maintaining posture or performing a sustained contraction, which is similar to that of the normal muscle nearing the point of exhaustion. Attempts to open the jaw against resistance will demonstrate jaw muscle weakness, which is typically difficult when the strength is normal. Presence of "peek sign", a sign characterized by the slight involuntary opening of the eyes when the patient attempts to keep the eyes tightly closed, demonstrates the fatigability of the orbicularis oculi muscles. The clinical examination may reveal Bell's phenomenon (upward rotation of the eyeballs with the inability to close the eye during attempted eyelid closure), suggesting the weakness of orbicularis oculi. Mild to moderate amount of wasting may occur in the severely affected skeletal muscles, which is not uncommon with the weakness. In MG, the smooth and cardiac muscles are not involved and the difference in the antigenicity of cholinergic receptors of the smooth and cardiac muscle may be the key factor sparing them. The patients with MG also have a normal pupillary response to light and accommodation. Table 13.2 gives details about the clinical characteristics of NMJ disorders. The deep tendon reflexes are seldom altered. Typically, sensory examination reveals no abnormality. However, some patients may complain of tingling or numbness sensations in the face or limbs and the physiological basis for the same is unclear. The weakness of the supporting muscles may cause pain and soreness of joints or tendons around the neck, back, or extremities, and the ocular imbalance may cause mild headache or eye strain in some patients.

13.5 Classification of Myasthenia Gravis

Many researchers and experts had classified MG, and among them, the most commonly accepted classification, the Osserman's classification (Table 13.3), is based on severity, clinical course, and drug response, which help in the staging and prognosis of the disorder. Other classifications are based on the presence or absence of a thymoma, antibody level against acetylcholine receptors (AChRs), and association with the human leukocyte antigen (HLA) haplotypes, which will not be dealt further. Currently, the Osserman's classification has been replaced by a scheme suggested by a task force of the Myasthenia Gravis Foundation (MGF). Though Osserman's classification remains useful, the suggested scheme by MGF, based on the clinical presentation, consists of class 1 to class 5 with subtypes (except for class

Table 13.2 Details regarding the clinical characteristics of neuromuscular junction disorders

Name of the disorder	Onset	Ocular symptoms	Bulbar involvement	Reflexes	Autonomic involvement	Sensory symptoms	Gastrointestinal involvement
Myasthenia gravis	Subacute	+	+	Normal	–	–	–
Lambert-Eaton myasthenic syndrome	Subacute	±	±	Reduced	±	±	–
Congenital myasthenic syndrome	Congenital	+	±	Normal/reduced	–	–	–
Botulism	Acute	+	+	Normal/reduced	+	–	+

Table 13.3 Osserman's classification of myasthenia gravis

Type	Severity	Features	Occurrence (%)
Type I	Ocular only	Ocular myasthenia and myasthenia restricted to ocular muscles	15–20
Type II A	Mild generalized	Mild generalized myasthenia with slow progression; no crises; and good response to drug therapy	30
Type II B	Moderate generalized	Moderately severe generalized myasthenia; severe skeletal and bulbar involvement but no crises; drug response less than satisfactory	25
Type III	Acute severe	Acute fulminating myasthenia; symptoms are severe with progressive weakness causing respiratory crises; poor drug response; high incidence of thymoma and high mortality	15
Type IV	Chronic severe	Late stage myasthenia; same as III but progression over 2 years from class I to class II	10

1 and class 5). For further details, readers may refer to an article titled "Myasthenia gravis: recommendations for clinical research standards" by Jaretzki A et al.

Even though the prognosis and response to treatment vary with the pattern of muscle involvement and severity, it is difficult to predict the outcome in an individual case. An increase in the duration of purely ocular myasthenia is associated with a decreasing risk of late generalization of weakness. Late-onset MG has a poor prognosis; however, the long-term outlook for myasthenic children is generally good.

13.6 Pathological Findings

Brain, Spinal Cord, Peripheral Nerves, and Muscles The brain, spinal cord, and peripheral nerves are essentially normal. The muscles are generally intact; however, in fatal cases with extensive paralysis, the respiratory, eye, and esophageal muscles may display segmental necrosis with variable regeneration.

Motor End Plate The number and size of the presynaptic vesicles and the amount of acetylcholine (ACh) molecules in each vesicle are within the normal range. Figures 13.1 and 13.2 illustrate the normal neuromuscular junction and the neuromuscular junction in MG, respectively. The ultrastructural changes at the neuromuscular junction reveal the following:

1. Reduction in the area of the nerve terminal
2. Widening of the primary synaptic cleft
3. Simplification of the postsynaptic region, i.e., sparse, shallow, and abnormally wide or absent secondary synaptic clefts
4. Ultrastructural evidence of an active process of degeneration and repair in the NMJ

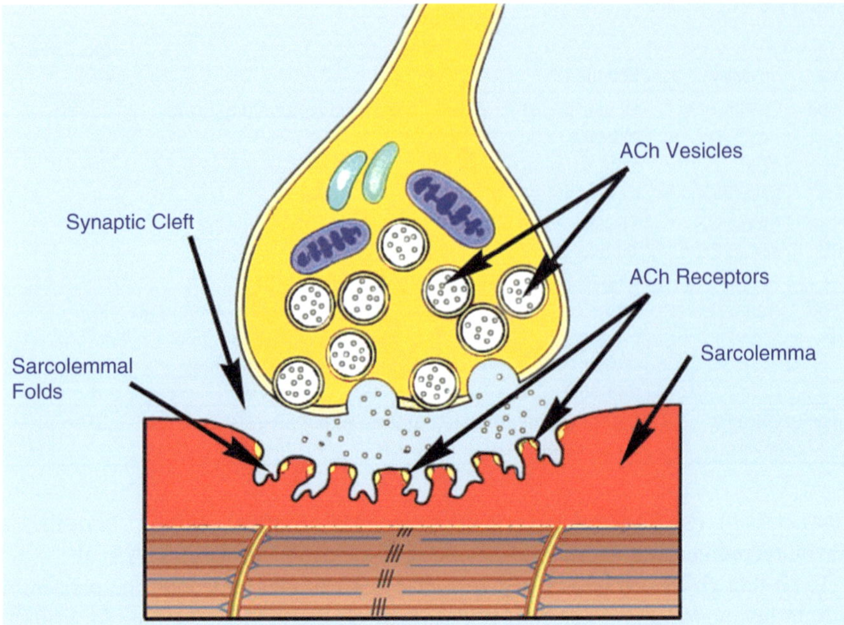

Fig. 13.1 A schematic illustration of the normal neuromuscular junction

13.6.1 Etiopathogenesis

The main cause behind the development of MG remains speculative; however, the end result is a derangement in the immune system regulation (an autoimmune disease). Sensitization to a foreign antigen that has cross-reactivity with the α1 subunit of the nicotinic postsynaptic AChRs at the NMJ, with the destruction and reduction in numbers of receptors has been proposed as the cause of MG. However, the triggering antigen has not yet been identified.

The presence of associated autoimmune disorders such as rheumatoid arthritis and systemic lupus erythematosus, the presence of transient myasthenia-like syndrome for infants born to myasthenic mothers, and the positive therapeutic response to various immunomodulating treatments including plasmapheresis, corticosteroids, intravenous immunoglobulin therapy, and thymectomy are the clinical observations that support the immunogenic mechanisms. In addition to the above, the induction of a myasthenia-like syndrome in mice by injecting a large quantity of immunoglobulin G (IgG) from MG patients is an experimental observation that supports the immunogenic mechanism of MG.

Many consider MG as a B cell-mediated disease (antibodies against acetylcholine receptors are produced by the B cells). Although the B cell is the effector cell producing antibodies, experimental evidence has shown that the autoreactive T cells are necessary for the disease to occur. The immunologic function of the thymus (it is the central organ for T cell-mediated immunity), the close association of MG with

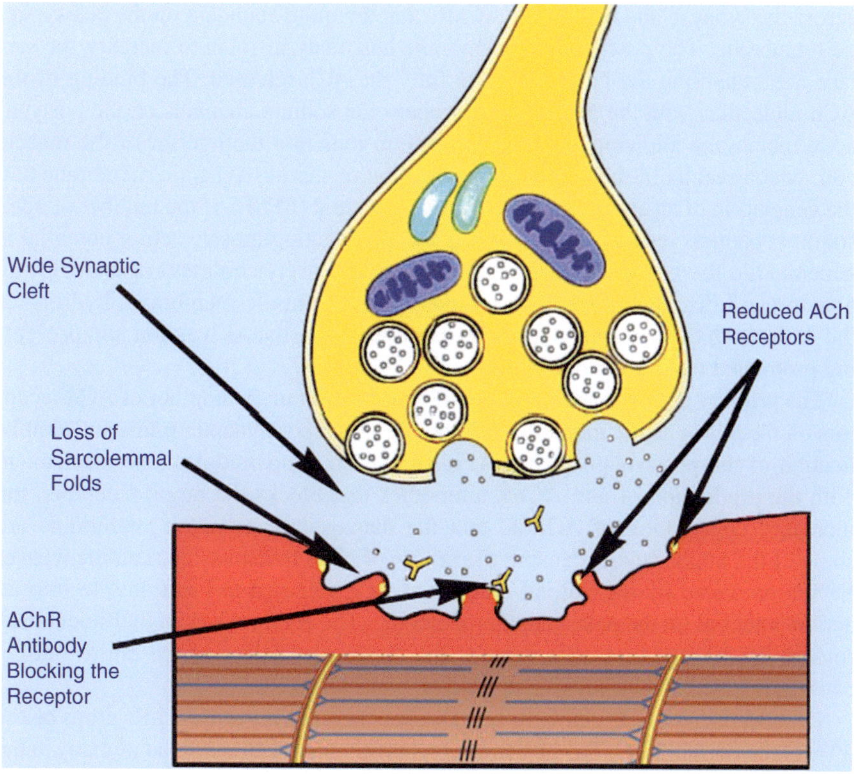

Fig. 13.2 A schematic illustration of the neuromuscular junction in myasthenia gravis

thymic abnormalities such as thymic hyperplasia and thymoma, and the improvement in the clinical condition following thymectomy are well recognized in myasthenic patients suspecting the importance of thymus as the site of autoantibody formation. Epithelial myoid cells normally present in the thymus resemble the skeletal muscle cells and possess AChRs on their surface membrane, which may become antigenic to unleash an autoimmune attack on the motor endplate AChRs, a process which is known as molecular mimicry. However, the stimulus that initiates the autoimmune process has not yet been identified.

To understand MG, a familiarity with normal and abnormal anatomy and physiology of NMJ is necessary. The nerve terminal of the motor nerve enlarges at its end (terminal bulb) and lies within a groove or indentation along the muscle fiber. The nerve membrane of the terminal bulb forms the presynaptic membrane and the muscle membrane forms the postsynaptic membrane. The presynaptic membrane, postsynaptic membrane, and synaptic cleft (space between the two membranes) together constitute the NMJ.

The presynaptic terminal contains vesicles filled with ACh molecules. On the arrival of a nerve action potential, the contents of these vesicles are released into the synaptic cleft in a calcium-dependent manner. The released ACh molecules diffuse across the synapse and bind to the AChRs that are quite abundant on the postsynaptic membrane. The postsynaptic membrane has many in-folds to increase the surface area, enabling the NMJ to utilize fully the ACh released. The binding of the ACh molecules with the AChRs briefly opens the sodium channels on the postsynaptic membrane allowing the entry of sodium ions into the interior of the muscle cell, which results in the partial depolarization of the postsynaptic membrane and the generation of an excitatory postsynaptic potential (EPSP). If the number of open sodium channels reaches a threshold, a self-propagating muscle action potential is generated in the postsynaptic membrane. Soon the acetylcholinesterase (AChE), a glycoprotein enzyme attached to the postsynaptic muscle membrane, hydrolyzes the ACh molecules to acetate and choline terminating the ACh action and preventing prolonged muscle depolarization.

The primary defect in MG is the marked reduction in the number of ACh receptors on the postsynaptic membrane, flattening of the postsynaptic folds, and simplification of the postsynaptic region (Fig. 13.2). The impaired immune mechanism with the production of anti-AChR antibodies that blocks the target receptors, the increased degradation of AChRs, and the damaged postsynaptic membrane are some of the major factors causing the failure of NMJ transmission; i.e., the weaker EPSPs produced by the normal amount of ACh released fail to translate into an action potential on the postsynaptic membrane. The end result is an inefficient neuromuscular transmission that reduces the contractile power of the entire muscle causing myasthenic weakness and fatigue.

Usually, the onset of symptoms begins when the population of AChRs drops below 30% of the normal level. The presence of relatively few AChRs instead of many in the postsynaptic membrane may account for the early involvement of the bulbar muscles. Approximately 80–90% of patients with generalized MG are found to have an anti-AChR antibody. However, the anti-AChR antibody titer levels may not correlate with the severity of the myasthenia, which suggests the existence of more than one subset of antibodies in the generation of MG. Several cellular components, including ion channels and proteins at the NMJ, may be the targets of the autoantibodies leading to abnormal NMJ transmission. Approximately 50% of the generalized MG patients who are negative for anti-AChR antibodies are detected to have antibodies to the muscle-specific tyrosine kinase (MuSK) and the levels of anti-MuSK antibodies are neither correlated with disease severity nor the treatment response.

13.7 Investigations

13.7.1 Neurophysiological Tests

Currently, the two distinct electrophysiologic studies widely used to demonstrate the defect in the NMJ transmission are the repetitive nerve stimulation (RNS) test and the single-fiber electromyography (SFEMG) studies. In the former, a peripheral motor nerve is stimulated repetitively at a rate of 3 pulses per second, to observe any decremental response, which is a classical feature of MG. In the latter test, the variability in the interpotential interval (jitter) and muscle fiber densities are studied.

Conventional EMG needle electrodes like concentric or monopolar electrodes have a larger recording area and record the single motor unit potentials instead of the discriminate individual muscle fibers within the motor unit. On the contrary, the SFEMG needle has only a small recording surface making it possible to record the individual muscle fibers. The presence of a variable interpotential interval between two or more muscle fibers of a motor unit and the normal muscle fiber density are the SFEMG findings, which suggest the NMJ transmission defect. Examination of a weak muscle by SFEMG is more useful than the RNS in demonstrating abnormal neuromuscular transmission. SFEMG findings are abnormal in almost 100% of MG patients, while RNS findings are abnormal in only 44–65%. Jitter analysis is highly sensitive but not specific for MG, as it can be abnormal in a variety of neuromuscular disorders, including motor neuron disease, myopathies, and peripheral neuropathies. Therefore, the diagnostic value of jitter must correlate with the patient's clinical manifestations and other electrodiagnostic findings.

Though SFEMG is more sensitive than RNS in MG, the former is technically more difficult and greatly dependent on the experience and skill of the examiner, making the latter the most frequently performed neurophysiological test for detecting NMJ transmission abnormality. The decremental response induced by RNS test reverses with the administration of pharmacological agents like neostigmine or edrophonium; however, such agents do not normalize the SFEMG findings.

The application of five supramaximal stimuli to a mixed or motor nerve at a rate of 2–3 Hz is the technique to demonstrate the decremental response (slow or low rate RNS). In a postsynaptic disorder like MG, rapid RNS causes no change of compound muscle action potentials (CMAPs), as the depleted stores are compensated by calcium influx. In the slow RNS test, usually, the first five consecutive CMAPs are studied and any decrement over 10% between the largest (typically the first or second CMAP) and smallest amplitude CMAP (typically the fourth or fifth CMAP) is considered abnormal. Patients with MG rarely have a decremental response in a clinically normal muscle. Thus, testing a proximal weak muscle gives a better yield than testing a technically easier unaffected distal muscle; i.e., testing a facial muscle is more useful since most of the patients suffer from eyelid weakness or ptosis. Evidence also suggests that the diagnostic sensitivity is noticeably higher for RNS recording in proximal muscles than in distal muscles.

Patients with MG may report clinically significant improvement in cold temperatures. Typically they report worsening of ptosis in bright sunlight or on a warm day. Therefore, maintaining a constant and perhaps higher-than-ambient temperature during RNS testing is important to bring out the abnormalities of the NMJ function. The temperature of the skin overlying the tested muscle should be at least 34 °C. Tests performed on warm extremities decrease the false-negative results, as cooling improves neuromuscular transmission and may mask the decrement. Administration of AChE inhibitors before the RNS testing and post-tetanic potentiation are the other two essential factors that can affect the RNS results. To avoid the possible masking of the abnormality, it is advisable to temporarily discontinue the usage of AChE for at least a day prior to the test. Promoting post-tetanic exhaustion by voluntarily working the muscle at maximum force for 10 min before the RNS test minimizes the possible errors. Table 13.4 gives the details of the electrophysiological features of neuromuscular junction disorders.

13.7.2 Laboratory Test

The presence of anti-AChR antibody in serum is a reliable test for diagnosing autoimmune MG. The result of the test for the anti-AChR antibody is positive in about 80% of patients with generalized myasthenia and 50% of those with pure ocular

Table 13.4 Details regarding the electrophysiological features of neuromuscular junction disorders

Disorder name	RNS: Decremental response	RNS: Incremental response	Compound muscle action potential amplitude at rest	Single fiber EMG	Fibrillation potentials and/or positive sharp waves	Motor unit potentials
Myasthenia gravis	+	–	Normal	Increased jitter/ blocking	–	Normal or short duration, small amplitude polyphasic
Lambert-Eaton syndrome	+	+	Decreased	Increased jitter/ blocking	–	Normal or short duration, small amplitude polyphasic
Congenital myasthenia	±	–	Normal	Increased jitter/ blocking	–	Normal or short duration, small amplitude polyphasic
Botulism	+	+	Decreased	Increased jitter/ blocking	+	Normal or short duration, small amplitude polyphasic

myasthenia. Usually, the antibody titer tends to be higher in more severe disease, although titer is not predictive of severity in an individual patient. Change in antibody titer correlates with long-term improvement induced by prednisone or azathioprine; however, the same changes may not be observed consistently in patients who undergo thymectomy, thus making serial antibody titer in isolation not reliable to judge a patient's response. In addition to the above, histological studies may reveal lymphofollicular hyperplasia of the thymic medulla in approximately 65% of MG patients and thymoma, in 15%. Currently, no lab tests are available in a time frame that is useful to confirm the emergency diagnosis of MG.

13.7.3 Pharmacological Tests

The edrophonium and neostigmine tests are the two pharmacological tests used for the diagnosis of MG. These pharmacological agents given intravenously tend to cause a transient improvement in the muscle strength of MG patients. Both edrophonium and neostigmine, used for the tests mentioned above, have the AChE inhibitor property. Pharmacological inhibition of AChE increases the ACh concentration at the NMJ, improving the chance for interactions between ACh and its receptor. Though both the drugs are proven to improve muscle weakness, edrophonium (Tensilon) is a short-acting AChE inhibitor, and its effects last only for 4–5 min. On the contrary, neostigmine has a longer duration effect and allows a more careful and sustained evaluation of muscle strength following its administration (objective and subjective improvement occurs within 15–20 min and lasts for more than a few hours).

Before and after the administration of these pharmacological agents, the weakness of certain muscles like levator palpebrae and extraocular muscles and facial muscles has to be evaluated. Subjective improvement alone is not dependable. Sinus bradycardia due to excessive cholinergic stimulation of the heart is a serious complication that has to be countered by necessitating the immediate administration of atropine.

While interpreting these tests, it is important to remember that these drugs can improve weakness in diseases other than MG, such as amyotrophic lateral sclerosis, poliomyelitis, and some peripheral neuropathies. The tensilon test is useful in distinguishing myasthenic crisis from the cholinergic crisis. If the muscle strength fails to improve following the maximum dose of edrophonium, the patient is having a cholinergic crisis or has another cause of weakness that is unrelated to MG. The overdose of anticholinergic causes cholinergic crisis and the acute exacerbation of MG leads to myasthenic crisis. Table 13.5 depicts the differences between myasthenic crisis and cholinergic crisis.

Table 13.5 Comparison between myasthenic crisis and cholinergic crisis

	Myasthenic crisis	Cholinergic crisis
Muscle	Weak	Weak and fasciculation
Heart rate	Tachycardia	Bradycardia
Blood pressure	Increased	Decreased
Pupil	Normal or dilated	Constricted
Skin	Faint and cold	Warm and flushing
Secretion	Normal	Increase
Cause/reason	Under medications of acetylcholinesterase inhibitors; respiratory infection; emotional stress; surgery	Overmedication of AChE inhibitors; chemicals like nerve gas and organophosphate compounds used in pesticides and insecticides
Autonomic	Normal	Parasympathomimetic effect
Other findings	Reduced urine output, absence of cough, and swallowing reflex	Abdominal cramps, blurred vision, pallor
Neostigmine test	MG symptoms improve	MG symptoms worsen
Treatment	Ventilator support and neostigmine	Ventilator support and atropine

13.7.4 Ice Pack Test

Cooling is believed to improve the NMJ transmission. This simple test consists of the light application or placement of ice kept in a surgical glove or wrapped in a towel over the eyelid for about 2 min. A positive test means a clear resolution of the ptosis following its application. The test is positive in approximately 80% of patients with ocular myasthenia. Though the precise mechanism of cooling enhancing the muscle function is unclear, the fall in temperature is believed to affect the NMJ both by decreasing cholinesterase activity and by prompting the efficacy of ACh at eliciting depolarizations at the end plate.

13.7.5 Radiodiagnostic Procedures

It is mandatory to screen every patient with MG for thymoma. Plain anteroposterior and lateral chest x-ray views may identify a thymoma within the mediastinum; however, a negative chest x-ray does not rule out a smaller thymoma, which can be more accurately detected by a CT scan or MRI of the chest. A chest x-ray may also help in determining the presence of aspiration or other types of pneumonia, which commonly occur in patients with MG.

13.8 Diagnosis

The diagnosis of MG is primarily based on the clinical picture, pharmacological and neurophysiological tests, and anti-AChR antibody titer. Fluctuating diplopia, unequally drooping eyelids, a smile looking like a smirk or a snarl, jaw hanging open, difficulty to speak and swallow, nasal voice while speaking, inconsistent weakness of the limbs, rest improving the strength of the weakened muscles, a positive response to pharmacological tests such as edrophonium or neostigmine tests, SFEMG studies revealing variable interpotential interval, RNS test showing a decremental response, and blood serum revealing a high anti-AChR titer will seldom miss the diagnosis of MG. However, in long-standing cases, the symptoms of MG may not vary much and can make the diagnosis difficult.

13.9 Medical and Surgical Management

The outlook for myasthenic patients has improved dramatically in the past few decades due to the advancements in understanding the pathogenesis of MG and developments in immunological treatment strategies. The medical and surgical treatment goal for MG is to help regain the patient's functional status and to improve the condition without recurrence or development of the adverse effects of the treatments. Several factors, including severity, distribution, and rapidity of disease progression, are to be considered before initiating or changing therapy. Agents that inhibit AChE and immunomodulating therapies are the mainstays of treatment. For the milder forms of MG, AChE inhibitors are given initially, and for most of the patients with generalized MG, additional immunomodulating therapy is required. Plasmapheresis and thymectomy are important management techniques for treating MG. Plasmapheresis (plasma exchange) is an effective treatment for managing an exacerbation and preparing for surgery. Thymectomy is an important treatment option for myasthenia when thymoma is present and in patients aged 10–55 years without thymoma but with a generalized myasthenia.

For marked severe weakness and anxious patients who are leading a very active life, agents that may produce rapid, short-term improvement include AChE inhibitor drugs and the short-term immunotherapeutic methods of plasmapheresis or infusion of intravenous immunoglobulin can be the treatment of choice. Such patients generally require hospitalization, and the overall medical status must be thoroughly evaluated. Various immunosuppressive agents tend to produce improvement in weeks to months and peaking within months to a year. Such agents can be considered as the intermediate-term treatment choices and generally include drugs like adrenal corticosteroids and calcineurin inhibitors (cyclosporine A and tacrolimus). Long-term treatment modalities take several months or even a few years to provide effective beneficial effects and such medications include immunosuppressive agents including azathioprine and mycophenolate mofetil and surgical thymectomy. For many of the patients with generalized myasthenia, the neurologists prefer to give a combination of short, intermediate, and long-term treatments. The treatment options for MG fall into six categories:

1. Enhancement of cholinergic transmission
2. Short-term immunotherapy
3. Immunosuppression
4. Thymectomy
5. Treatment of refractory MG
6. Management of associated conditions

Enhancement of Cholinergic Transmission Anticholinesterase agents inhibit the rapid elimination of ACh by the enzyme AChE, thereby prolonging the action of ACh at the postsynaptic membrane and enhancing neuromuscular transmission. There are no substantial differences in efficacy among the various anticholinesterase drugs, including pyridostigmine, rivastigmine, galantamine, and neostigmine. Pyridostigmine is the most widely used medication and its beneficial action begins within 30 min and lasts for 3–6 h. The dosage of this medication is adjusted to meet the needs of the patient, i.e., timed to maximize strength prior to anticipated activities, such as 30 min before meals. Timespan, a long-acting pyridostigmine preparation, is advised at bedtime, for patients who are symptomatic at night or in the early morning. The side effects of anticholinesterase drugs include gastrointestinal hyperactivity with abdominal cramping or diarrhea, increased oral and upper respiratory secretions, exacerbation of asthma, and rarely, bradycardia. The use of anticholinergic medications such as atropine or loperamide can help to overcome these muscarinic side effects without diminishing the nicotinic benefit.

Short-Term Immunotherapy Both plasmapheresis and immunoglobulin therapy help in the removal of antibodies. During plasmapheresis, the plasma, which contains the pathogenic antibodies, is mechanically separated from the blood cells and returned to the patient. Rapid reduction in autoantibodies from the blood following plasmapheresis is known to produce clinical improvement in both AChR antibody and MuSK antibody-positive patients. A typical course of plasmapheresis consists of 5 or 6 exchanges of 2–3 L each, every other day. The indications for intravenous immunoglobulin therapy are generally similar to those for plasmapheresis. Immunoglobulin treatment has the advantage of not requiring special equipment or the difficulty of obtaining adequate venous access and is typically administered over 5 days. Adverse reactions may include headache, fluid overload, and rarely, aseptic meningitis. Many clinicians, based on the clinical impression, consider plasmapheresis to be more effective and more rapid in the treatment of myasthenic crisis.

Immunosuppressive Agents The use of an optimum level of immunosuppressive agents can restore full activity in most MG patients. Currently, there are several immunosuppressive agents, including glucocorticoids, azathioprine, cyclosporine, and tacrolimus. These agents generally produce clinical improvement within a period of 1–3 months. Adrenal corticosteroids, when used judiciously and safely, are known to produce a clinical improvement in many of the patients but have a long list of potential side effects that must be safeguarded. Steroids exert a wide variety of immunosuppressive and anti-inflammatory actions that may contribute to their therapeutic benefit in MG, which includes the inhibition of production of cytokines and interleukins, alteration of trafficking of lymphocytes, and a reduction of antibodies. Approximately one-third of MG patients treated initially with a high dose of steroid can develop a further worsening of weakness during the early stages of treatment. For the above situation, a combination of other immunosuppressive agents with corticosteroids can reduce the steroid dose while maintaining the therapeutic effect on MG. The side effects of steroid therapy include hyperglycemia, hypertension, fluid retention, weight gain, insomnia, gastrointestinal irritation and ulcers, cataract formation, osteoporosis, aseptic necrosis of hip bones, risk of infection, impaired wound healing, and suppression of adrenocorticotropic hormone secreted by the pituitary gland. Because of these side effects, a consistent follow-up is essential.

Because of its long track record and relative safety in most patients, azathioprine, a nonsteroidal immunosuppressive agent, has been widely used for the management of MG. Azathioprine's therapeutic effect supplements that of the corticosteroids and usually allows the steroid dose to be reduced. Though side effects are less compared to steroids, a few of them include flu-like symptoms of fever and malaise, bone marrow depression, or abnormalities of liver function. Cyclosporine A, a cyclic polypeptide produced by a fungus, is another extensively used immunosuppression agent and is approximately as effective as azathioprine in the clinical management of MG. Side effects of cyclosporine A include hypertension, nephrotoxicity, hirsutism, gingival hyperplasia, and gastrointestinal dysfunction.

Thymectomy Since the thymus plays a crucial role in the origin and maintenance of the autoimmune response in MG, thymectomy is indicated either to surgically remove the thymoma or to perform thymectomy as a treatment for MG. Almost all thymomas are histologically benign and do not metastasize but are invasive and may spread locally. Therefore, the surgical removal of the tumor is followed by postoperative radiation therapy and chemotherapy. Generally, thymectomy is known to provide a long-term beneficial effect, although the benefits are seen after an interval of years. For maximum benefit, it is advisable to perform thymectomy early in the disease course. Evidence suggests that the MuSK antibody-positive patients do not benefit from thymectomy. The potential complications from thymectomy include the general risks of anesthesia, impaired wound healing, sternal instability, pleural effusion, atelectasis, pneumonia, pulmonary embolism, paresis of phrenic or recurrent laryngeal nerves, and precipitation of myasthenic crisis.

Treatment of Refractory MG Though most patients respond to immunosuppressive agents, occasionally, a few may fail to respond to appropriate doses of these agents or tolerate their adverse side effects. The use of high-dose cyclophosphamide treatment without bone marrow transplantation can successfully eliminate or significantly reduce the mature immune system, resulting in durable remissions. Administration of high-dose cyclophosphamide treatment helps to "reboot" the immune system so that the newly developed system may recognize the autoantigen as "self" and thereby induce tolerance to it. Following the administration of high-dose cyclophosphamide treatment, granulocyte colony-stimulating factor is administered to enhance the proliferation of hematopoietic cells and reconstitute the immune system.

Management of Associated Conditions The management of MG patients with complex comorbidities can be difficult. Infections of any sort may exacerbate MG and are the most common causes of the myasthenic crisis. As soon as the infection is recognized, it should be treated empirically. For those myasthenic patients with obesity, corticosteroid treatment is a relative contraindication as its use may increase appetite and obesity. The use of corticosteroids can worsen hyperglycemia and may result in fluctuations in serum glucose levels. Corticosteroids and calcineurin inhibitors can increase blood pressure and may need proper monitoring of blood pressure. The use of calcineurin inhibitors, which are nephrotoxic, is a relative contraindication for renal disease. Periodic bone density measurements and administration of bisphosphonates or high-dose vitamin D can minimize the possibility of osteopenia while using the corticosteroids.

13.10 Physiotherapy Management

13.10.1 Evaluation

The therapist plays an important role in evaluating muscular weakness, fatigability, and functional mobility with MG patients. A thorough evaluation helps us to link the impairments and disabilities to appropriate effective treatment and functional prognostication. Each patient is unique and the therapist must utilize a symptom-specific, objective approach with every evaluation. Muscle weakness and fatigue are two components of the physiotherapy evaluation, which require special attention. A close interdisciplinary approach with the primary consultant is imperative to ensure the patient's safety and best functional outcome.

Though the presentation of muscle weakness in generalized MG can be diverse from patient to patient, it usually follows a specific pattern of proximal greater than distal involvement. Those MG patients with significant bulbar involvement can present with neck flexor weakness with difficulty to maintain a normal cervical postural alignment. The patients with the pectoral girdle and proximal upper limb weakness will have problems with daily activities above the shoulder level. Difficulty in ambulating long distances and climbing the stairs are the problems faced when weakness is more in the proximal muscles of the lower limbs. The severity of the weakness can vary, where some patients may have significant generalized weakness and cannot lift their extremities against gravity for more than a few repetitions while others may only be affected by activities that are repetitive.

Traditional manual muscle testing can be used for grading the strength of the affected muscles and can be used in diagnosis and treatment evaluation when the baseline weakness severity and post-treatment changes are significant. The muscle strength testing should include the evaluation of neck flexors, the shoulder flexors, the elbow extensors and flexors, the wrist extensors, the hip extensors, flexors, and abductors, the knee extensors and flexors, and the ankle dorsiflexors. The hand-held dynamometer is another alternative, a quantitative and reliable tool for measuring muscle strength. Hand-held dynamometry also helps in measuring fatigue. A series of 10 repeated strength testing performed using a hand-held dynamometer, with 1-min rest between each contraction and calculating the difference between the best value and the worst value, can identify significant fatigue, provided the difference is more than 50% of the best value.

In addition to the above, the Myasthenia Gravis Fatigue Scale (MGFS) and the Chalder Fatigue Scale (CFS) are a few scales appropriate to measure the fatigue severity in MG patients. Most MG patients with no recent past history of myasthenic crisis can undergo the 3-min walk test for functional evaluation. The distance covered over time can be compared to the normal data, which will provide the individual patient's ambulatory baseline. The Myasthenia Gravis-Activities of Daily Living scale (MG-ADL), a short and easy to use disease-specific patient-reported questionnaire, the Myasthenia Gravis Quality of Life 15 (MG-QOL15), a disease-specific quality of life (QOL) questionnaire, and Myasthenia Gravis Disability Scale (MG-DIS), a tool to quantify the disability based on the International Classification of Functioning, Disability and Health (ICF) framework, are a few of the scales explicitly used for MG. Regarding the bulbar and respiratory function assessment, and for a better understanding, the author recommends the readers to refer to Chap. 8, "Motor Neuron Disease."

13.10.2 Treatment

Stable MG patients are generally treated by the interdisciplinary team consisting of the neurologist, occupational and physical therapists, speech pathologist, nurse, and dietician. The exercise programs given by healthcare professionals should be well-coordinated and be in tune with the patient's total daily activity tolerance. The therapist should motivate the MG patients to learn how to monitor their symptoms at rest, with activity and exercise. Those patients who have an exacerbation of symptoms and are unable to monitor their own symptoms are at high risk of making their disease worse and are not suitable candidates for the active exercise program. It is advised to avoid any prescription of the exercise program until MG is medically stable. Once stabilized, carefully designed consistent exercise will elevate the patient's baseline functional capacity and diminish the effect of the myasthenic exacerbations.

For ocular myasthenic patients, medical management is the mainstay, and physiotherapy intervention is not required unless the disease progresses to a generalized form. For generalized MG patients, the primary role of the therapist is to encourage them to improve and/maintain muscle strength and build up activity tolerance. The therapeutic exercise program should provide gentle, non-resistive activities that are intellectually and psychologically stimulating, which do not induce fatigue in the patient. The practitioner should regularly monitor the patient's muscle strength and take account of the factors contributing to fatigue. If the therapist detects any worsening of the condition, including progression in drooping of the eyes, further weakening of the facial muscles, and alterations of breathing or swallowing, he should report the details to the concerned neurologist or physician.

It is crucial to avoid overexertion and the treatment plan should include work simplification, energy conservation, and the use of adaptive and assistive devices to reduce effort during daily activities. Besides, the therapist may guide and advise regarding architectural barriers, bathroom adaptations, and furniture rearrangements. The therapist may advise regarding the use of arm supports and splints to protect the weakened musculature from overstretching and to maintain functional position. It is also essential to educate the patient regarding the disease and how overexertion, fatigue, emotional stress, and excessive heat or cold can exacerbate the symptoms of the disease. The goals of treatment are as follows:

1. To improve the muscle strength
2. To avoid exhaustion and minimize fatigability
3. To provide energy conservation by work simplification and job modification
4. To promote basic activities of daily living (ADL) and improve quality of life (QOL)
5. To prevent secondary complications
6. To maintain the general condition in the later stages

13.10.2.1 Improvement of Muscle Strength

Due to the lack of adequate evidence on exercise rehabilitation for MG patients and the fear of exacerbating the condition, many healthcare practitioners believe that exercise can be counterproductive for MG patients. On the contrary, a carefully designed exercise program can reduce a medically stable MG patient's fatigue and encourage the patient to return to previous activities, as lack of exercise can diminish the physiological capabilities and increase the fatigue, limiting the patient's ability to engage in meaningful occupations. Low-to-moderate intensity exercises can increase the number and density of the mitochondria and increase skeletal mass. Therefore, improvement or maintenance of muscle strength to minimize the muscle atrophy within the limitations imposed by the disease process is the fundamental exercise prescription principle in the management of neuromuscular disorders. Strength training can be performed using the patient's own body weight and dumbbells may be used as well if tolerated. While undergoing the exercise program, a perceived exertion rate of 12–14 (moderate level intensity) on the Borg rating of perceived exertion (RPE) scale is advised for generalized MG patients. It is also preferable to monitor the oxygen saturation, and throughout the program, a minimum saturation of 90% is to be maintained. It is mandatory to stop the exercise immediately if the oxygen saturation drops below 88% while breathing the room air or supplemental oxygen.

Advise the MG patients to use energy conservation for all daily activities. The patients should not exhaust themselves or drain their energy in the morning; instead ask them to identify the best time of the day to do the exercise. Encourage the patient to perform the exercise when the medication is at its peak level. The strength training should focus on the larger proximal muscle groups and the intensity of exercise should not go beyond moderate intensity. Provide adequate rest or discontinue the exercise if patients' pulse rate rises 30 beats per minute above the baseline value, exhibits shortness of breath at the peak of exercise, or exhibits a worsening of symptoms during exercise. For those patients with a history of severe residual muscle soreness, secondary to the previous day(s) exercise session, the session can be temporarily skipped until the symptoms subside.

Before the commencement of the strengthening and low-to-moderate intensity aerobic exercise programs, the patient should get clearance from the neurologist or the treating physician. The following are a few strengthening and low-to-moderate intensity aerobic exercise programs that could be advised for the patients.

- Weight training program: Machines or weight training equipment with safety mechanisms or light free weights, including elastic bands and sandbags. The number of repetitions in each set, preferably, should be within 10–12 repetitions and the number of sets should not exceed three sets per exercise.
- Walking: Encourage the patient for level ground walking at a comfortable pace determined by himself or herself, in an environment with minimal crowd and obstacles with no temperature extremes.
- Stationary ergometer: MG patients can use either upright or recumbent bicycles and the exertion can be measured and controlled.
- Treadmill: The speed of the treadmill should be adjusted to the patient's walking pace, and if not, it may exhaust and fatigue.
- Swimming: Advise the patients to perform swimming in the shallow end of the pool and not to move toward the deeper end as overexertion, exhaustion, and fatigue can make them weak and be dangerous. Always swim with a partner and never swim in extreme temperatures of water.

13.10.2.2 Avoidance of Exhaustion and Fatigability

Measures to avoid exhaustion and counter fatigue include structured exercise prescriptions, rehabilitation to improve body mechanics to minimize the potential for injury, and energy conservation methods. Self-awareness about one's physical limits, engagement in relaxing occupations, and rest can minimize the possibility of fatigue. Fatigue may worsen from inadequate sleep, poor nutrition, and stress. Fatigue and sleep disturbances often coexist, creating a downward spiral of sleep deprivation and increasing fatigue. Patients with low endurance and fatigue should find methods of improving the amount of sleep. The patient should be encouraged to develop a bedtime routine and schedule, be instructed not to consume alcohol, tobacco, and caffeine within 4 hours of bedtime, and be discouraged to use computers or television just before bedtime. These are some of the strategies advised for sleep-deprived patients. If the exercise program is prepared arbitrarily without taking the patient's exercise tolerance and capacity into consideration, fatigue can set in during the training session. Overdoing strength training exercises can also predispose to fatigue.

As discussed in the earlier section, the use of carefully designed low-to-moderate intensity exercises can prevent the possibility of exhaustion. Few other strategies which can prevent exhaustion and fatigue include the use of lightweight objects, utensils, and tools, the advantage of gravity to assist the movement, the use of adaptive equipment and powered tools, and the utilization of the principles of biomechanics for routine tasks like lever and application of force.

13.10.2.3 Promotion of Energy Conservation

Strategies that reduce the overall energy requirements of the task and level of fatigue are the fundamental principles of energy conservation. The strategies can include modifying the task and/or modifying the environment to ensure the successful completion of daily activities. For MG patients with low endurance or fatigue, energy conservation is a primary intervention approach. These patients are taught to take adequate rest throughout the day and use energy conserving techniques for all possible daily activities. The MG patients may have to change their routine habits, set new goals, monitor progress, and develop problem-solving strategies to promote goal attainment.

Adaptive methods, including keeping frequently used items within easy reach and the use of adaptive equipment like a long-handled sponge to eliminate bending in the shower, can conserve energy. The following are the principles of energy conservation and work simplification.

- Limit the amount of work: Avoid overdoing by assigning heavy-duty jobs like dusting, mopping, and vacuuming to family members or a housekeeper.
- Reduce the homemaking demands and expectations: Use pre-cooked, canned, or frozen foods to decrease the time and energy utilized for meal preparation. Order out or ask for community resources that deliver prepared meals.

- Plan ahead before executing the task: Organize ahead of time to minimize the last-minute rushing or hurrying. Schedule the routine tasks to distribute energy demands and also prioritize the tasks and complete the important tasks before fatigue sets in. Try to alternate the tasks that require more energy with tasks that require less.
- Use efficient methods: Organize the work areas and store the frequently used supplies in a convenient location. Avoid standing when it is possible and slide objects across a counter or table rather than lift them. Identify needed items at the beginning of a task to avoid extra trips. Eliminate steps or tasks that are not essential by combining chores. Use a utility cart to move things from one room to another. Use good body mechanics and work with gravity assisting and not resisting. Avoid heavy lifting such as children, groceries, and laundry.
- Use the correct equipment and techniques: Use assistive equipment to minimize bending and stooping. Use nonskid mats to secure items instead of stabilizing with the weak upper limbs. Adjust the work height and use the tools most appropriate for the job. Use electrical appliances and lightweight utensils and tools to conserve personal energy.
- Use of assistive technologies: Some of the assistance devices that could be used are walkers, handrails, and grab bars and the use of book holder and pencil holder, "U"-shaped cuff, or handle to hold the brush, cutlery, and other lightweight objects.
- Balance activities with adequate rest: Advise them to take rest before getting tired, even while in the middle of a task. Patients are advised to perform energy-demanding tasks early in the day when they are more fresh and energetic. Self-pacing of tasks has to be encouraged to avoid fatigue. Regular 5–10 min rest periods can considerably improve functional endurance during homemaking and other routine tasks.

13.10.2.4 Promotion of Basic ADL

The tips to preserve energy and to promote basic ADL and improve the QOL are listed in Box 13.1. Basic ADL itself can often serve as an aerobic exercise or a strength training exercise. Encourage the patients to perform his or her routine activities to minimize the deconditioning effect of the disease, muscle wasting, and atrophy, and strength and endurance. Overhead reaching activities can be avoided by keeping things at a lower level and within reach of the patient. For those patients with tremors, weight cuff can be applied to them to prevent tremor while maintaining posture. Promoting and challenging the daily activities by progressively reducing the level of assistance and increasing the task difficulty helps the patients to perform the task at ease and with confidence.

Box 13.1 Tips for Preserving Energy

Tips to preserve energy

Tips for housework and routine work

- Use long-handled dusters, mops, and dustpans
- Hire help
- Use an automatic washer and dryer if possible
- Use a lightweight iron and cookware
- Plan the work, around the best times of the day
- Create shortcuts
- Take rest breaks

Tips for meal preparation

- Assemble all ingredients before starting
- Use mixes or pre-packaged foods
- Use cookware to serve from
- Buy utensils that fit comfortably in the hand
- Store frequently used items at chest level to minimize bending or to reach
- Use the dishwasher if possible
- Let dishes soak rather than scrubbing them and air-dry

Dressing tips

- Loose-fitting clothes are more comfortable to wear and remove
- Rest the foot over the knee while sitting to wear the socks and shoes
- Use slip-on shoes or shoes that have Velcro or elastic shoelaces
- For women, fasten the bra in front and turn it to back
- Wear clothes that have press buttons or zippers in front

Personal care and hygiene tips

- Wash hair in a shower rather than over a sink
- Use a terrycloth robe instead of towels to dry off
- Use a chair in the shower or tub
- Use a long-handled sponge, brush, comb to reach the back and overhead
- Use liquid soap or soap on a rope
- Use a raised Western-style toilet seat

Shopping tips

- Make a list and organize before shopping
- Use store assistance
- Identify shops with home deliveries
- Online purchasing of groceries and other items

Parenting tips

- Take advantage of daycare programs
- Teach children to make a game of household chores
- Plan activities or outings at a place that will allow the patient to sit or lie down

Tips for leisure activities

- Plan activities that allow the patient to sit or lie down
- Plan social events at the patient's peak energy time

Activities like staircase climbing have to be done with the utmost caution, and if required, the vitals need to be monitored. Frequent rests and adequate safety are essential to retrain stair climbing. Mild MG patients can also be encouraged to practice sit-to-stand transitions with progressively lower surfaces and decreasing levels of external support. Persistent muscular weakness, fatigue, and respiratory impairments, along with the psychological impact of the disease, can affect the patient's health-related QOL. Encouraging daily activities along with a carefully planned exercise program minimizes the possibility of depression and helps to boost the patient's QOL psychologically.

For those MG patients who are medically stable, the primary treatment philosophy is to progressively challenge the cardiorespiratory, musculoskeletal, and balance systems in a way that emphasizes improving functional mobility. The therapy should emphasize incorporating functional tasks with decreasing levels of assistance and increasing complexity, frequency, and repetition. High-intensity exercises (including running, sprinting, heavy-weight training, and prolonged exercises) are strictly not recommended for MG patients, and bear in mind that the wrong tempo, high-intensity exercise, and high temperature and air humidity can exacerbate the symptoms of the disease.

13.10.2.5 Prevention of Secondary Complications

Balance training is a component of the rehabilitation process for myasthenic patients. Since many of the MG patients tend to lead a sedentary lifestyle compared to healthy people, the consequences include reduced muscle strength, early fatigue, and poor bone density leading to falls and fractures. Balance training strategies should consist of strengthening exercises and endurance training exercises tailored individually to the patient's physical ability as determined by an initial assessment. Examples of such activities or exercises include heel-toe walking, sit-to-stand, and ball catching and throwing. The number of repetitions, altering the speed, introducing dual tasks, or changing the base of support are certain methods of progressing the exercise. The therapist should also encourage static and dynamic sitting exercises at the edge of the bed, including maintaining balance while reaching/leaning in all directions. While standing at the edge of the bed, promote balance training like standing with feet together, reach-out activities, standing on one leg, and partial squatting. The abovementioned exercises can be progressed with eyes open and eyes closed and by decreasing the levels of external support.

In addition to the above, weight-bearing exercise is encouraged as a preventative measure to counter steroid-induced osteoporosis and myopathy. Splints can be used for the upper and lower limbs to maintain the normal position of the joints, particularly when the muscular imbalance causes deformities at the joint level. A lightweight foot-drop-stop splint worn inside a lightweight shoe can preserve the mobility of the patient and, to a certain extent, maintain the flexibility of the ankle joint. The treatment should include positioning, skincare, and passive ROM, and gentle stretching for maintenance of joint mobility and soft tissue extensibility, for those patients who are bed-bound or wheelchair mobile. Educate the patients on the importance of performing the ankle pump exercise regularly to minimize the risk of deep vein thrombosis, provided the ankle musculatures are not considerably weak. For those patients with severe weakness of ankle musculature, pneumatic compression devices or pumps can be recommended to prevent the possibility of deep vein thrombosis.

13.10.2.6 Maintenance of the General Condition in the Later Stages

The Class III and Class IV MG patients (classification based on the task force for Myasthenia Gravis Foundation) are likely to develop exacerbation of the condition and may lead to myasthenic crisis. These patients have decreased respiratory strength and endurance and generally tend to have reduced maximal voluntary ventilation at rest. Those patients with respiratory muscle weakness based on the severity will either require ventilatory support or oxygen therapy. The patients who are dependent on the ventilator for a few weeks might undergo tracheostomy. For such patients, chest care and bronchial hygiene are the main goals of the physiotherapy treatment: chest manipulations and aseptic suctioning need to be performed regularly.

Once the patient is weaned off the ventilator, relaxed deep breathing exercise, pursed-lip breathing exercise, and the use of an inspiratory muscle training device can be encouraged. Pursed-lip breathing can mitigate shortness of breath during physical activities. Some studies have reported that inspiratory muscle training resulted in a significant increase in the respiratory muscle strength and endurance and alleviated dyspnea. Discourage repeated coughing, especially when the patient has considerable respiratory muscle weakness. Repeated coughing can weaken and fatigue the respiratory muscles predisposing the patient back to ventilatory support breathing.

Those patients in Class II of MG are generally afraid of exacerbating their symptoms by engaging in strenuous exercise. Encourage these patients to participate in low-intensity sports, such as golf, bowling, and cricket, as these activities are safe and can improve the maximal muscle force and endurance for patients with mild MG. In a recent case study, resistance exercises in the form of bench press, latissimus pull-down, shoulder press, leg extension, and leg curl, performed three times per week for 15 weeks combined with creatine supplementation, were found to be safe and improved peak leg strength, fat-free mass, and load volume for a mild MG patient. The therapist, taking the patient into confidence, should develop a specific home exercise program. The home program may include lower extremity strengthening exercises consisting of supine bridging exercise, heel slides, straight leg raises, ankle-toe movements, mini squats, and sit-ups at the edge of the bed and upper extremity consisting of biceps curl, latissimus pull-down, triceps push-down, and shoulder press. For those patients with no possibilities or facilities for machines or equipment, resisted exercises using elastic band or exercises utilizing the patient's own body weight with or without added resistance can be practiced. The number of sets should be preferably two and the number of repetitions per set can be ten and the training program can be performed thrice a week, provided the patient can tolerate the program.

Further Reading

Barnett C, Herbelin L, Dimachkie MM, Barohn RJ. Measuring clinical treatment response in myasthenia gravis. *Neurol Clin.* 2018;36:339–53.

Buzzard EF. The clinical history and post-mortem examination of five cases of myasthenia gravis. *Brain.* 1905;28:438–83.

Castleman, B.: The pathology of the thymus gland in myasthenia gravis. *Ann N Y Acad Sci.* 1966;135:496–503.

Daroff RB, Jankovic J, Mazziotta JC, Pomeroy SL, editors. Bradley's Neurology in clinical practice. 7th ed. London: Elsevier; 2016.

Donaghy M, editor. Brain's Diseases of the nervous system. 12th ed. New York: Oxford University Press; 2009.

Engel AG, editor. Myasthenia gravis and myasthenic disorders. 2nd ed. New York: Oxford University Press; 2012.

Hohlfeld R, Wekerle H. The role of the thymus in myasthenia gravis. *Adv Neuroimmunol.* 1994;4:373–86.

Jaretzki A, Barohn RJ, Ernstoff RM, Kaminski HJ, Keesey JC, Penn AS, et al. Myasthenia gravis: recommendations for clinical research standards. Task Force of the Medical Scientific Advisory Board of the Myasthenia Gravis Foundation of America. *Ann Thorac Surg.* 2000 Jul;70(1):327–34.

Kwiatkowska K, Lamtych M, Kubiak K, Badiuk N. Physiotherapy in myasthenia gravis. *J Educ Health Sport.* 2018;8(12):1027–38. Available from: https://doi.org/10.5281/zenodo.3242665, http://ojs.ukw.edu.pl/index.php/johs/article/view/7008

Naumes J, Hafer-Macko C, Foidel S. Exercise and myasthenia gravis: a review of the literature to promote safety, engagement, and functioning. *Int J Neurorehabilitation.* 2016;3(3):218. Available from: https://doi.org/10.4172/2376-0281.1000218

O'Sullivan SB, Schmitz TJ, Fulk GD, editors. 6th ed. Physical rehabilitation. Philadelphia: FA Davis Co; 2014.

Pascuzzi RM. The history of myasthenia gravis. *Neurol Clin.* 1994;12:231–42

Rassler B, Marx G, Hallebach S, Kalischewski P, Baumann I. Long-term respiratory muscle endurance training in patients with myasthenia gravis: first results after four months of training. *Autoimmune Dis.* 2011; Article ID 808607, 7 pages, 2011. https://doi.org/10.4061/2011/808607. Available from: http://www.hindawi.com/journals/ad/2011/808607/cta/ or http://www.ncbi.nlm.nih.gov/pubmed/21869926.

Ropper AH, Samuels MA, Klein JP, Prasad S, editors. Adams and Victor's Principles of neurology 10th ed. New York: McGraw Hill Education; 2014.

Trombly CA, Scott AD. Occupational therapy for physical dysfunction. 7th ed. Baltimore: Lippincott Williams & Wilkins; 2014.

Westerberg E, Molin CJ, Lindblad I, Emtner M, Punga AR. Physical exercise in myasthenia gravis is safe and improves neuromuscular parameters and physical performance-based measures: a pilot study. *Muscle Nerve.* 2017;56:207–14.

Westerberg E, Molin CJ, Sporndly Nees S, Widenfalk J, Punga AR. The impact of physical exercise on neuromuscular function in myasthenia gravis patients: a single-subject design study. *Medicine.* 2018;97:e11510. Available from: https://doi.org/10.1097/MD.0000000000011510

Wong SH, Nitz JC, Williams K, Brauer SG. Effects of balance strategy training in myasthenia gravis: a case series. *Muscle Nerve.* 2014;49(5):654–60.

Chapter 14
Perceptual Disorders

Gargi Kundaliya, Abraham M. Joshua, and Shivananda Pai

14.1 Introduction

Perception means the organization, identification, and interpretation of sensory information to represent and understand the environment. The word "perception" is derived from the Latin word "perceptio," meaning gathering or receiving. It is defined as the integration of sensory impressions into information that is psychologically meaningful. By elaborating and interpreting the neural signals picked up as sensory information from the environment, perception helps us become aware of external stimulation. Since perception makes sense of what we see, hear, and touch, it is a prerequisite for both learning and rehabilitation. The key components of the perception process are a selective focus on the stimulus, persistent attention, elimination of irrelevant stimuli, and identification of stimulus by associating it with personal remembered experience.

A reduction in the number of perceived stimuli in each unit of time, failure to synthesize the stimuli properly, and inability to relate them to the ongoing activities of the mind are some of the common forms of perceptual disturbances. Other forms of disturbances include attentional deficits and sensory distortions that lead to a misinterpretation of stimuli termed as illusions. Failure to correlate the previously

G. Kundaliya (✉)
Aster RV Hospital, Bangalore, India

A. M. Joshua
Department of Physiotherapy, Kasturba Medical College of Mangalore, Manipal Academy of Higher Education, Manipal, India
e-mail: abraham.joshua@manipal.edu

S. Pai
Department of Neurology, Kasturba Medical College of Mangalore, Manipal Academy of Higher Education, Manipal, India
e-mail: dshivanand.pai@manipal.edu

A. M. Joshua (ed.), *Physiotherapy for Adult Neurological Conditions*,
https://doi.org/10.1007/978-981-19-0209-3_14

learned action to a new one can be due to the inability to perceive all the components of a large complex of stimuli, which can be a characteristic feature of delirium or a confused state of mind. The patients with perceptual deficits may present with the inability to perform simple tasks independently or safely, difficulty in switching from one task to the next, and a diminished capacity to locate objects visually that are necessary for task completion. These characteristics can make activities of daily living (ADL) difficult and can often hinder the therapy. In such instances, the therapist should not dismiss the patient as uncooperative, intellectually inferior, or confused.

14.2 Classification and Types of Perceptual Disorder

Perceptual dysfunction(s) can be seen in a variety of neurological disorders and are particularly common among stroke. They can occur in association with neoplasms, cerebral trauma, senile dementia, or Alzheimer's disease. Patients with mental health disorders such as schizophrenia or chronic depression and neurological conditions like Parkinson's disease and multiple sclerosis may also experience perceptual dysfunctions.

The perceptual disorders can be broadly categorized as body scheme, visual discrimination skills, agnosia, and apraxia. Each of these categories is further subdivided for better understanding (Table 14.1).

14.2.1 Body Scheme Disorders

Developing meaningful goal-directed activities requires the integration of proprioceptive, tactile, and other forms of sensory inputs. These inputs provide information about the actual position and configuration of one's body parts, without which the basic feelings of where one's body parts are and their relationship to the environment cannot be understood. This basic awareness of the limits and the spatial layout

Table 14.1 Classification of perceptual disorders

Body scheme/Body image	Visual discrimination skills	Agnosia	Apraxia
• Unilateral neglect • Anosognosia • Asomatognosia • Right-left discrimination • Finger agnosia	• Figure-ground discrimination • Form discrimination • Spatial relations • Position in space • Depth and distance perception • Topographical disorientation • Vertical disorientation	• Visual agnosia • Visual object agnosia. • Simultagnosia • Prosopagnosia • Color agnosia • Tactile agnosia • Auditory agnosia	• Ideomotor apraxia • Ideational apraxia • Oral apraxia • Constructional apraxia • Dressing apraxia

of one's body and its relation with the environment are conceptualized as "body schema." The body schema reflects and determines the posture and the general stance taken to perform effective motor patterns and the lack of the same results in body scheme disorders. The common body scheme disorders are unilateral neglect, asomatognosia, right-left discrimination, finger agnosia, and anosognosia.

14.2.1.1 Unilateral Neglect or Visual-Spatial Inattention

Visual-spatial inattention can be described as the failure to report, respond, or orient to meaningful stimuli given to the side opposite to the brain lesion and cannot be attributed to either motor or sensory defects. It is the most common perceptual disorder and is commonly seen in the right hemispheric lesions, particularly those involving the inferior-posterior regions of the parietal lobe, with or without visual field defects. The lesions in the dorsolateral frontal lobe, cingulate gyrus, thalamus, or putamen may also result in unilateral neglect. Visual-spatial inattention may occur along with sensorimotor deficits such as decreased tactile and proprioceptive sensation, paresis, or paralysis of the arm or leg. Defective interactions between impaired sensation and mental deterioration or deficit in the internal representation of space are a few theoretical bases for visual-spatial inattention. Typically, most of the patients with unilateral spatial inattention have their heads, eyes, and trunk rotated toward the lesion side. Visual-spatial inattention can be graded from mild to severe. The presence of inattention can affect ADL, such as feeding, reading, and writing. The patient might eat only half of his meal, shave half of his face, locate things only in the attended space, or walk toward the attended space and not in a straight line due to which he might collide with the doors or walls in the unattended space.

14.2.1.2 Anosognosia

Anosognosia is a severe form of neglect where the patient does not appreciate the presence or the severity of his paralysis, due to which he may present with simple apathy for his paralysis or may completely deny his paralysis. This observation was first made by the German neuropsychiatrist Gabriel Anton and later termed as "anosognosia" by the French neurologist Joseph Babinski. The lesion area is still unclear, but the supramarginal gyrus is anticipated. If the patient is confronted and asked to move his paralyzed side, he may reply that the "arm is always lazy" or "it is tired" or may move the unaffected side instead. Some patients may feel and behave as though the affected side is nonexistent, and if made aware of the same, they may deny the affected part as theirs. Loss of stored body schemes can be the possible cause or reason for this. Anosognosia can be associated with intellectual deficits, and it is also theorized that both the intellectual deficit and the proprioceptive deficit may play a role in the development of anosognosia. Patients with anosognosia cannot develop a precise picture of their paralysis and will be in a state of denial in the beginning. These patients may look uninterested, distracted, and dull, but days to weeks later, they may become more restless and irritable when the denial decreases.

14.2.1.3 Asomatognosia

Asomatognosia also known as somatoagnosia is the inability to recognize one's own body parts and relationship of body parts to oneself or to others. Typically, the dominant parietal lobe lesion is responsible for it. However, left hemiplegic patients may also present with asomatognosia. Patients with asomatognosia may confuse between the left and right sides of their body, have difficulty using their contralesional limbs, and may fail to distinguish between their body parts and those of the examiner. The deficit causes difficulty in transfers and ADL. The lack of proprioception accompanying this disorder may cause the patient to feel that the affected limb is heavy. These patients may often perceive their body part(s) as abnormally small or large. Patients who suffer from somatoparaphrenia, a specific form of asomatognosia, ignore or deny ownership of body part(s) on the contralesional side.

14.2.1.4 Right-Left Discrimination

Right-left discrimination is the failure to selectively apply the right-left distinction to the symmetrical body parts. It is a specific disorder of spatial dysfunction and involves the parietal lobe. Verbal, sensory, and conceptual skills are the prerequisites to execute this function. These patients have difficulty understanding and executing the commands that consist of "right" and "left" terminologies. The patient fails to distinguish the left from the right side or cannot tell which of his arms is right or left. The patient also fails to follow instructions like "lift your right hand and place it on the left elbow" or directions like "take the second left from the main road."

14.2.1.5 Finger Agnosia

Finger agnosia comprises doubts and ambiguity regarding fingers. Typically, the patient is in a dilemma to name his fingers on command. The disorder is usually found bilaterally, with more involvement of the middle three fingers. Patients with finger agnosia show clumsiness when performing tasks like copying meaningful gestures. Lesions are usually found in either parietal lobes and are more common in the angular gyrus of the left hemisphere. Deficits in finger agnosia are found to correlate heavily with poor dexterity functions. Finger agnosia has been linked with Gerstmann's syndrome, a syndrome characterized by finger agnosia, agraphia, acalculia, and impaired right-left discrimination. Some scientific investigators believe that finger agnosia is a spatial disorder, while others assume it as a secondary impairment associated with aphasia and cognitive deficits.

14.2.2 Visual Discrimination Skills

Vision can be considered as one of the most amazing miracles of life. Approximately 70% of the sensory input of the brain is concerned with vision and more than 50% of the brain is directly or indirectly involved in visual processing. About 32 visual areas have been located in the cerebral cortex. Vision is a journey of light rays being transformed into visual cognition which is required to continuously evaluate the spatial maps, to create an internal representation of the external world, and to allow us to integrate and navigate through our environment. It informs us where our bodies are in space, what is in space, and the relationship of objects to each other in space. Visual perception results when the visual system transforms the small, distorted, upside-down 2D patterns of light received at the retina into a coherent representation of a 3D world. Once processed, the visual information is integrated with other sensory modalities, such as touch and sound, and is involved in the prediction, preparation, and control of the motor movements needed for ADL. The subjects with visual discrimination skills disorders have difficulty perceiving the relationship between themselves and two or more objects. Figure-ground discrimination, form discrimination, spatial relations, and depth and distance perceptions are the visual discrimination skills. The disorders affecting visual discrimination skills typically involve the non-dominant parietal lobe.

14.2.2.1 Figure-Ground Discrimination

Figure-ground perception involves the ability to differentiate between the figure from the background in which it is embedded. Figure 14.1 illustrates an example of a figure-ground perception test. Figure-ground discrimination is required to

Fig. 14.1 An illustration of a figure-ground perception test

organize the layout of the environment and the impairment can cause self-care disability. Lesion involving the parieto-occipital region of the right hemisphere results in figure-ground discrimination issues. The "figure" or foreground is a part of the field of perception that is the center of an individual's attention at any given time and the rest of the field forms the background. Based on differences in depth, luminance, color, alignment, temporal information, texture, or movement provided by the visual system, a normal subject typically separates figures from the background. Some of the researchers hypothesize that memory plays an important role in figure-ground perception. Some recent research indicates that recognition of other nearby objects may occur before the completion of figure-ground organization, a term known as the pre-figural recognition process.

14.2.2.2 Form Discrimination

Form discrimination is a type of perception required to visually locate the objects in the surroundings and for directions to reach a destination. Impaired perception in form discrimination is the failure to differentiate between different forms or shapes, which are important to have an effective interaction with the environment. For instance, a patient may mistake a button for a nickel or a water pitcher for a urinal. The perception of spatial properties is accomplished immediately and effortlessly. Color, mobility, edge, and orientation are all utilized for form discrimination. Damage to the parietal-temporal-occipital region of the non-dominant lobe results in form discrimination disorder. Abstract visual form system and specific visual form system are the two systems required to perceive form discrimination. The former system is hypothesized to perform abstract thinking to distinguish the types of forms and helps to store nonspecific visual information of forms. The latter system identifies the distinct differences between similar objects by producing specific output representations. For instance, when an individual is looking for writing material, he does not specifically look for a paper but may generalize his visual scan for a diary, book, or paper. Whereas if the subject wants a particular book, the specific visual form system processes the item's information which is unique for it, like its specific design on the cover page, color, font, and page texture.

14.2.2.3 Spatial Relations

Spatial relations are important for individuals to orient themselves with the environment and recognize objects, scenes, and gestures. Spatial disorientation is the inability to understand the relationship of one object in space to another object. For instance, a patient might find difficulty arranging the dining set on the table or may have difficulty positioning the hands, legs, and trunk to transfer from chair to bed. Typically, the lesion occurs in the inferior parietal lobe or the parietal-occipital-temporal junction of the right hemisphere. Spatial relations involve the coordination of the perceptual and the conceptual components of space. The perceptual component represents the sets of objects and surfaces. The conceptual component consists of words like above and below to describe the relationship of objects in the environment. Another area of spatial relation explored is the spatial relations types, i.e., the categorical and the coordinate spatial relations. Some researchers conceptualize that categorical spatial relation relates to above-below, left-right, and on-off, whereas the coordinate spatial relation relates to specific locations that can precisely guide the movements.

14.2.2.4 Position in Space

Position in space is the ability to perceive and interpret spatial parameters like up and down, front and behind, and in and out. Patients with the position in space deficit fail to follow instructions such as "keep your foot on the step" or "lift your arm above your shoulder" and often appear puzzled when instructed. Impairment in position in space is also due to the involvement of the non-dominant parietal lobe.

14.2.2.5 Depth and Distance Perception

Depth perception is defined as the ability to visualize the third dimension of depth. It allows the person to locate objects in the visual environment and safely navigate through stairs or while driving. The damage to visual cortices due to cerebrovascular accidents or head injury may impair the depth perception, in the absence of strabismus. The inability or inaccuracy to judge and perceive the distance is related to disorder with distance perception. The lesion area commonly affected is the posterior right hemisphere.

An important cue for depth perception is binocular disparity. The difference in image location picked up by the left and right eyes separately (binocular disparity) is utilized by the brain to extract depth information from the two-dimensional retinal images. Depth perception of an object is dependent on the factors such as texture, the apparent size of the object, and linearity. The patient with depth and distance perception issues may find it difficult to fill a bucket of water, climb a stair, or perform transfers.

14.2.2.6 Topographical Disorientation

Topographical disorientation is the difficulty to comprehend and recollect relationships of places to one another. These patients can have difficulty finding their routes and identifying visual landmarks even after being repeatedly guided. When a patient is asked to describe his room or home layout, he is unable to do so. The involvement of the right retrosplenial cortex is the possible site for topographical disorientation. The condition is also seen among patients with a bilateral parietal lobe or left parietal lesion. Normally, a person requires visual processes and spatial working memory to hold information about where he is and to plan future movements. Adequate attentional processes and stored previous memories of the landmarks are also required. According to Margaret A. Christensen and Raschko B., topographical orientation requires cognitive mapping, which is a mental representation of one's surroundings.

14.2.2.7 Vertical Disorientation

Vertical disorientation is the failure to perceive verticality. The perception of verticality requires vestibular, visual, and somatosensory information. It can affect the ADL, the gait, and the posture of a person. The lesion is in the non-dominant parietal lobe; central vestibular pathways on either side or sensory pathways can result in vertical disorientation. A person with vertical disorientation sees the world tilted away from normal. Assessment of vertical perception, early after stroke, is crucial to plan the posture, balance, and gait components of rehabilitation. For further understanding of the verticality issues among stroke, the readers may refer the Pusher's syndrome in Chap. 3 "Stroke."

14.2.3 Agnosia

Agnosia is a perceptual impairment involving a lack of recognition of common objects perceived by the senses. The word agnosia is taken from a Greek word αγνωσία (agnosia) meaning "non-knowledge." It frequently occurs in many disorders, including cerebrovascular accidents, traumatic brain injury, brain tumors, central nervous system infections, dementia, hypoxic damage of the brain, developmental disorders, and other neurological diseases. The patient may not be able to identify a cup by sight, although he may be able to tell its color or identify it by touch (by the feel of shape and texture). Damage to pathways that connect the primary sensory processing areas (visual, tactile, proprioceptive, and auditory) might attribute to this deficit. The system responsible for processing sensory information is composed of three areas. The primary projection area is responsible for elementary level processing; i.e., it processes the basic sensory information and is responsible for the senses such as sight and hearing. Lesions in the primary areas of the sensory unit lead to sensory loss. The secondary area is known as the projection association cortex, which interprets the incoming sensory information to attach a meaning to the information. For instance, a sound may be perceived as a cat moaning or a baby crying. It is the tertiary area that enables the interpretation of the information from all the sources to produce an appropriate interaction with the environment. Lesions in the secondary area of the sensory unit tend to produce agnosia. Classically, there are two types of agnosia, namely apperceptive and associative. Failure in identification due to perceptual processing deficits at an early stage causes apperceptive agnosia, whereas failure in identification with no deficit in perception causes associative agnosia. Though unable to recognize common objects, the associative agnosia patients can typically draw, match, or copy objects while the apperceptive agnosia patients can neither recognize nor copy.

14.2.3.1 Visual Agnosia

Visual agnosia is defined as a failure to recognize familiar visual stimuli although visual-sensory processing, language, and the general intellectual functions are preserved. The patient cannot name, copy, or recognize the visually presented objects but can readily identify objects by tactile or auditory cues. The patient perceives information but does not add meaning to the perceived knowledge. Visual agnosia is rarely seen in its pure form. Visual agnosia can be either apperceptive or associative. Apperceptive agnosia is due to the distortion of the stimulus, leading to the inability to recognize visually presented objects. In this type, the patient cannot name or copy objects but can identify objects with tactile or auditory cues. The disordered association leads to associative agnosia, and patients with associative agnosia will be unable to recognize familiar objects regardless of having good perception but can copy figures or written material and name objects from verbal definitions. The following section outlines the various visual agnosia types like visual object agnosia, prosopagnosia, simultanagnosia, and color agnosia.

Visual Object Agnosia Visual object agnosia, a rare condition, was first described by Heinrich Lissauer, a German neurologist and psychiatrist, in 1890. It consists of a failure to name and indicate the use of a seen object by spoken or written word or gesture, although the visual acuity skills are intact. The lesion in the occipital-temporal-parietal areas of either hemisphere can lead to visual object agnosia.

Simultagnosia Simultagnosia is the inability to comprehend a whole visual stimulus. These patients can perceive only one thing at a time and the time required to differentiate between two perceptual acts is too long. They can describe specific components of a compound stimulus but cannot combine these elements to formulate the whole picture. For instance, out of the word given, the patient comprehends only a few letters. The lesion is in the dominant occipital lobe.

Prosopagnosia Prosopagnosia is the inability to identify a familiar person by his or her face in the absence of sensory, cognitive, visual acuity, and visual field deficits. The word prosopagnosia is taken from two Greek words: "prósōpon," meaning "face" and "αγνωσία," meaning "non-knowledge." It is also called face blindness, and the condition is characterized by the failure to recognize familiar faces, including one's own face. Here, the subject can appreciate the face as a body part but cannot acknowledge to whom it belongs. The lesion is in the right posterior hemisphere. Apperceptive prosopagnosia and associative prosopagnosia are the two forms of acquired prosopagnosia. The right occipito-temporal region plays a critical role in apperceptive prosopagnosia. Patients with apperceptive prosopagnosia cannot make any sense of faces and cannot recognize or distinguish both familiar and unfamiliar faces. These patients fail to read facial emotions and may be able to recognize people based on non-face clues such as their voices, style of walking, clothing, or hairstyle. Associative prosopagnosia is due to the impaired links between early face perception processes and the semantic information about people in our memories. The right anterior temporal region is likely to play a critical role in associative prosopagnosia. The patients may typically tell whether the photographs of faces are the same or different and derive the age and gender from a face but are unable to identify or provide any information about the person's name, occupation, or address. The impaired functioning of the parahippocampal gyrus is attributed to associative prosopagnosia.

Color Agnosia Color agnosia is a condition where the patient is unable to name or distinguish colors on command despite the color perception, the semantic memory for color information, and color naming being intact. They fail to understand the meaning and its association with any color. If asked to match or associate the colors like yellow, green, and blue to the sky, grass, and moon, the patient fails to associate green with the grass, yellow to the moon, and blue to the sky. Color agnosia is a result of dominant hemisphere lesions, specifically in the left occipito-temporal region of the brain.

14.2.3.2 Tactile Agnosia

Tactile agnosia, often called astereognosis, is a modality-specific disorder characterized by the inability to recognize familiar objects by touch but being able to recognize them by sight. Failure to integrate accurately the acquired somesthetic features into a haptic mental image is the plausible cause for this high-level perceptual disorder. The primary sensations like touch, temperature, and proprioception are intact. The lesion is usually in the parietal-temporal-occipital region of the cerebral cortex.

14.2.3.3 Auditory Agnosia

In auditory agnosia, sound cannot be distinguished from one another and the patient fails to distinguish and differentiate common auditory stimuli such as water running, a siren, a car engine starting, and tire screeching. The patient might typically identify the sounds by using unrelated terms such as motorbike engine as the roar of a lion. Auditory system evaluation may reveal a similar disturbance in the audiogram while testing to discriminate various sounds. The lesion in the dominant temporal lobe results in auditory agnosia. The following are a few of the sub-types of auditory agnosia.

- Phonagnosia: It is the inability to recognize familiar voices. They can recognize words spoken by others. Phonagnosia is caused by damage to certain parts of the sound association region.
- Verbal auditory agnosia or pure word deafness: The inability to comprehend spoken words; however, the patient retains the ability to read, write, and speak in a relatively normal manner.
- Nonverbal auditory agnosia: The inability to comprehend nonverbal sounds and noises, with the sparing of speech comprehension, is the characteristic feature of nonverbal auditory agnosia.
- Amusia: It is the inability to recognize the music. The patient will be unable to comprehend that certain types of sounds represent music and cannot distinguish music from other sounds.

14.2.4 Apraxia

Apraxia is an inability to perform highly complex, goal-directed skilled motor movements and gestures. The term apraxia comes from the Greek word "praxis," meaning the performance of an action or skilled movement. In 1908, Hugo Karl Liepmann, a German neurologist and psychiatrist, first defined apraxia as the inability to perform voluntary acts despite preserved muscle strength. Later, it was defined as the inability to perform purposeful movements in the absence of paresis, ataxia, sensory loss, incoordination, attentional deficits, cognitive impairments, disorders of movement, or disturbance of the muscle tone. Apraxia is mainly due to lesions in the left dominant hemisphere. Both frontal and posterior parietal lesions can result in apraxia due to the disruption of pathways between the two lobes. Certain lesions in the corpus callosum or right hemisphere may also result in apraxia.

Though many theories exist, according to Eric A. Roy and Paula A. Square, two systems, namely conceptualization and production, help us to perform purposeful movements. The conceptual system consists of three categories of knowledge for motor planning, and they are the knowledge of tools and objects in terms of action and their functioning, knowledge of actions in which these tools may be included, and knowledge associated with single actions being seriated into a sequence. The production system is hypothesized to operate various parallel systems. Higher-level processes need attention and keep the action seriated toward the targeted goal. The lower-level processes require minimal attention and are more autonomous. The neurological control can shift from higher centers to lower centers based on the ongoing action. Apraxia can occur independently or coexist with aphasia. Apraxia may take several forms, such as ideomotor, ideational, oral, constructional, and dressing apraxia. The deficit can be associated with spatiotemporal and memory-related issues.

14.2.4.1 Ideomotor Apraxia

Ideomotor apraxia is the failure to imitate gestures or perform purposeful movements or tasks on command, although the patient comprehends the idea of the task. The disruption in the connections between the region of the brain containing ideas and execution of movements causes ideomotor apraxia. It is a common condition having spatial and temporal errors. Spatial errors may involve posture, spatial movements, or spatial orientation. Though unable to follow command(s), these patients can automatically carry out many old motor tasks as they retain the kinesthetic memory patterns. These patients may have a tendency to perseverate (repeat an activity or a segment of a task over and over, even when it is no longer needed or appropriate). Ideomotor apraxia is due to an insult to the left hemisphere involving the parietal region. Subcortical tracts that disconnect the left parietal from the left frontal cortex can lead to right limb apraxia. The interruption of callosal connections between the left and right premotor association cortices leads to left limb apraxia. Based on the location of the lesion, ideomotor apraxia can take two forms. The first results from a lesion in the area storing the "visuokinesthetic engrams," the left parietal lobe, the area representing spatiotemporal gesture. The second results from a disconnection of the intact movement representations from the motor output region.

Ideomotor apraxia is a multidimensional disorder involving different engrams related to the body parts utilized, different conceptual depictions, and different forms for interaction between the body parts and the objects. The following dimensions should be considered while evaluating ideomotor apraxia. The disturbances appear in the movements of appendages like fingers or hands. Axial and trunk movements are often spared. The patient with ideomotor apraxia fails to "blow" on command but may blow the candle when presented casually or the patient fails to button up his shirt on command but wears and buttons it properly when performed as a routine task. While holding the scissors, these patients might laterally orient it to the sagittal plane and sometimes may use fingers as blades to describe the scissor's actions.

14.2.4.2 Ideational Apraxia

Ideational apraxia is the inability to conceptualize, plan, and execute the complex sequences of motor actions using tools or objects. It is the loss of information of tool function and the patient fails to retrieve any memory related to task performance. The patient loses the perception of the object's purpose and can neither spontaneously nor on command formulate an action to use it and the examiner may misdiagnose it as a dilemma issue. The errors in ideational apraxia are in planning complex events, and generally, the simple isolated tasks are unhampered. The lesion typically involves the connections from the sensory areas 5 and 7 in the dominant parietal lobe and the supplementary and premotor cortices of both cerebral hemispheres. Ideational apraxia affects function bilaterally. Though the ideational apraxia patients know the name or use of the tool or object, due to the flaws in the conceptual information, they cannot verbally describe the procedure for using the tool or how the object functions. For instance, if presented and then asked to light a cigarette using a matchbox, the patient may put the matchstick in his mouth instead of the cigarette or try to light a cigarette without lighting the matchstick.

14.2.4.3 Oral Apraxia or Buccofacial Apraxia

Oral apraxia is the difficulty in forming and organizing comprehensible words and performing purposeful movements with the lips, the tongue, the cheeks, the larynx, and the pharynx on command, although the muscles involved remain intact. It is the most common type of apraxia. It is different from dysarthria, as the latter is due to muscle weakness. Patients with oral apraxia will not have difficulty swallowing or chewing activities but when asked to stick their tongue out or pretend to whistle or cough, they will be unable to carry out the activity. However, they may imitate such activities when asked to follow the examiner. The lesion is typically seen in the anterior insula, a small area of the first temporal gyrus, and the frontal and central opercula. It is hypothesized that the problem is either with the production system or the wrong selection of movement at the higher-level processes for motor tasks.

14.2.4.4 Constructional Apraxia

Constructional apraxia is the incapacity to produce two- or three-dimensional designs while drawing, copying, or constructing, whether upon commands or spontaneously. It results from lesions in either hemisphere, but it is hypothesized that patients with right-sided brain damage are likely to show greater incidents of the deficit. However, others hypothesize an equal distribution of the symptoms for both the left and right-side brain-damaged patients. Patients with right-sided apractic lesions are characterized by a visual-spatial disability, such that they cannot exactly locate figures in space, lack perception, and analyze the relationship of the parts with each other. These patients are certain in their drawings. Careful inspection of these drawings may reveal that they are complicated, often unidentifiable with many pieces scattered with appropriate spatial relationships to each other. Their drawings depict the problem of perception and construction. These patients are unlikely to get any help from the presence of a model. On the other hand, patients with left-sided lesions are reluctant to draw, and when insisted, they tend to produce simple drawings. These patients have a problem with execution and generalized comprehension, due to which they cannot draw angles and have poorly defined borders. Usually, models help in improvising their drawings and end up making simple drawings.

14.2.4.5 Dressing Apraxia

Dressing apraxia is a lack of ability to dress oneself. It is due to a disorder in body schemes and spatial relations. The patient puts the clothes backward, upside down, or inside out. After completing the task, the patient may appear puzzled to find the right sleeve worn over the left arm or wonder why the shirt's backside faces the front. Patients with dressing apraxia and neglect will have difficulty dressing on the left side of the body or they might put both legs in the same pant leg.

14.3 Evaluation

A thorough evaluation is essential to understand the way a perceptual deficit impacts task performance which is fundamentally an ADL. Information provided by evaluation also helps to plan the treatment strategy and understand the prognosis of the deficit. At this juncture, the author would like to disclose that the evaluation and treatment of auditory agnosia and oral apraxia are not within this book's scope.

Some of the generalized tests for perceptual disorders are Rivermead Perceptual Assessment Battery (RPAB), the Loewenstein Occupational Therapy Cognitive Assessment (LOTCA), and the Chessington Occupational Therapy Neurological Assessment Battery (COTNAB). The RPAB was designed to examine visual perceptual disorder in patients following a head injury or stroke. It consists of 16 performance tests that examine form discrimination, color constancy, body image, inattention, and spatial awareness, figure-ground discrimination, object completion, and sequencing.

LOTCA is another test that was developed for stroke and traumatic brain injury patients. It takes 35–45 min to administer and is composed of two subtests that examine orientation, visual and spatial perception, motor praxis, visuomotor organization, and thinking operations. The orientation subcomponent is related to orientation to place and time. The visual and spatial perception assesses the individual's ability to identify pictures of everyday objects, objects photographed from unusual angles, distinguish between overlapping figures, and recognize the spatial relations between objects. The motor praxis subcomponent assesses the individual's ability to imitate motor actions and use of objects and perform symbolic actions. Whereas the visuomotor organization subcomponent assesses the individual's ability to copy geometric figures, reproduce a 2D model, copy block design, and reproduce a puzzle. The individual's ability to complete tasks, sort, categorize, and sequence is assessed by the thinking operation component. COTNAB is a battery designed to examine the cognitive and perceptual deficits in patients 16 years of age and above following stroke or head injury. The battery consists of 12 tests divided into four parts examining visual perception, constructional ability, sensory-motor ability, and the ability to follow instructions.

14.3.1 Unilateral Neglect

The Behavioral Inattention Test (BIT) provides the therapist the information on how neglect affects the patient's ability to perform the ADL. It consists of 15 subtests, of which nine are activity-based and six are pen- and paper-based. To name a few, the paper-based subtests include line crossing, star cancellation, and line bisection tests. The activity-based subtests include menu reading, coin-sorting, and article reading. In each subtest, the number of omissions is recorded, and based on the omissions, the examiner decides about the patient's neglect. The star cancellation test is a test where a patient is presented with a sheet of paper containing 56 small stars, 52 large stars, 13 letters, and 10 short words. He is asked to strike out all the small stars. The standardized scoring method is detailed in the BIT manual.

The modified Albert's Test (MAT) is a screening tool used to detect the presence of unilateral neglect in patients with stroke. The test items in the modified version of Albert's test consist of a series of 40 black lines, each about 2 cm long, randomly oriented on a white paper sheet of 11 × 8.6-inch dimension paper in 6 rows (Fig. 14.2). The test sheet is centered to the patient's midline, and then the examiner instructs the patient to cross out all of the lines after demonstrating the crossing out of the five central lines. The patient is encouraged to cross out all the lines as much as he can. If several lines are left uncrossed and more than 70% of uncrossed lines are on the one side, unilateral neglect is indicated. The neglect may be quantified in terms of the percentage of lines left uncrossed. The figure provides an example of an illustration of the modified Albert's test.

Fig. 14.2 A sketch to show the series of 40 black lines used in the modified Albert's test

The Clock Drawing Test (CDT) is used to assess visuospatial and praxis abilities and may determine the presence of both attention and executive dysfunctions. Here, the individual is given a fully drawn clock with a certain time pre-marked and is asked to replicate the drawing as closely as possible (Fig. 14.3). The clock reading test is a modified version of the copy command CDT, which asks the patient to read aloud the indicated time on a clock drawn by the examiner. Both tests are good for assessing parietal lobe lesions, such as those that may result in hemineglect.

The line bisection tests require patients to estimate and indicate the midpoint of a horizontal line presented on a piece of paper placed in front of the patient (Fig. 14.4). The sheet of paper with the lines is presented to the patient's midline, and he is instructed to bisect the line right at the center by a pencil using the preferred or unaffected hand. It takes less than 2 min to administer the test. The test is typically scored by measuring the deviation of the bisection line from the true center of the line. A deviation toward the side of the lesion (unaffected side) is usually indicative of neglect, although the magnitude of deviation can vary.

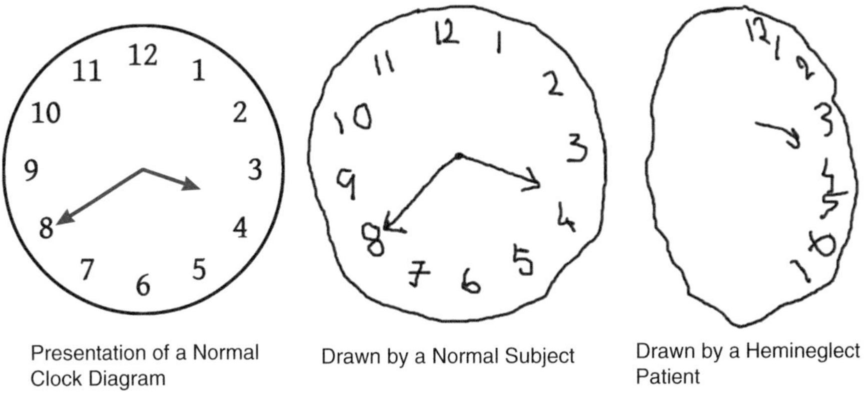

Fig. 14.3 An illustration of the Clock Drawing Test

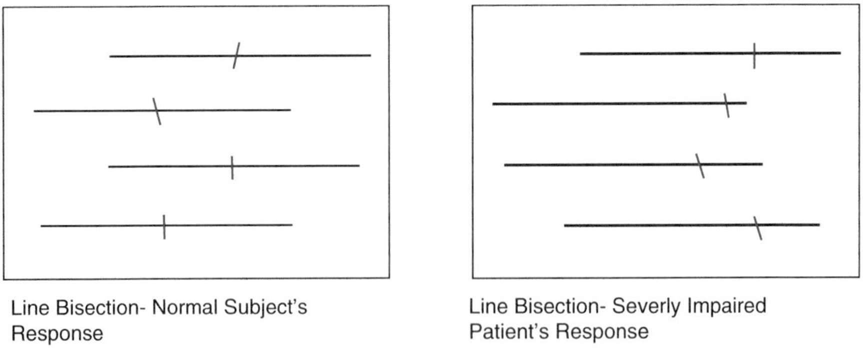

Fig. 14.4 An illustration of line bisection test

The Bell's test is a cancellation test that allows for a quantitative and qualitative assessment of visual neglect in the near extra-personal space. For this test, the patient is asked to circle with a pencil all the 35 bells embedded within 280 distracters on an 11 × 8.5-inch page. Before the commencement of the test, the examiner should place the paper at the patient's midline with the black dot on the bottom of the page aligned to the patient's mid-sagittal plane. If the patient omits to circle the bells in the last column on the left or right, we can estimate that their neglect is mild. Omissions in the columns near the center can suggest greater neglect. The total number of circled bells and the time taken to complete the test are recorded. The test has a maximum score of 35, and an omission of 6 or more bells on the right or left half of the page indicates unilateral neglect. Typically, the test can be administered in less than 5 min. Figure 14.5 illustrates the bells and the distractors of the Bell's test.

The Draw-A-Man Test, a test developed by Florence Goodenough in 1926, has been widely used to identify the presence of unilateral spatial neglect in adult patients post-stroke. Patients are given a blank piece of paper (8.5 × 11) entitled "Draw an Entire Man" and a pencil and are asked to draw an entire man from memory. A maximum score of 10 can be given based on the examiner's ability to distinguish homogenous unilateral body parts from homogenous bilateral body parts. An illustration of a Draw-A-Man Test is depicted in Fig. 14.6.

The functional test consists of observing the functional activities performed by the patient. While observing the task, the examiner should ask oneself whether the activities are restricted only to one side of the patient's body, such as combing only half the hair or shaving only one side of the face. If the patient shows no signs of

Fig. 14.5 An illustration of the bells and distractors of the Bell's Test

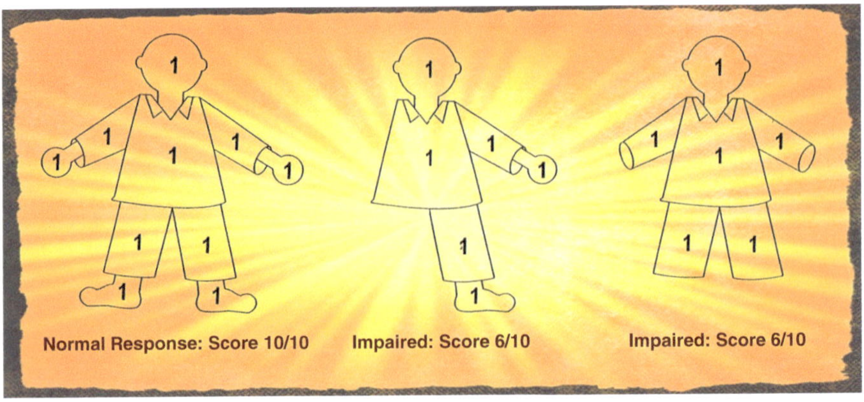

Fig. 14.6 A depiction of a Draw-A-Man Test

neglect during self-care activities, an intact score is provided. If the patient demonstrates the signs of unilateral neglect in some activities, it is considered impaired, and if all self-care activities show signs of unilateral neglect, it is considered severely impaired.

14.3.2 Anosognosia

The bedside test for anosognosia consists of asking the patient to find what is wrong with his affected limb and the reason why he cannot move the limb. If the patient gives inappropriate reasons for the limb immobility, he is likely to have anosognosia. For example, if the patient justifies by saying "the left hand is sleeping" or "it does not want to move on its own."

14.3.3 Asomatognosia

The examiner names the body parts or shows a human body chart and asks the patient to point the same body parts on him or the body chart presented. For instance, commands like "touch your left hand" or "touch your right leg" will be given. Or else, the examiner may perform some movements and the patient is asked to imitate the same movement. In this case, mirror-image movements are considered normal. A face puzzle can also be used to check if the patient has asomatognosia. In this test, the patient is asked to put together the puzzle pieces to make it a complete face. Scoring is non-standardized for the abovementioned tests and the tests are considered intact if the patient answers or performs it correctly in a fixed time.

14.3.4 Right-Left Discrimination

The patient is asked to point in response to the various commands given by the examiner. In A. Jean Ayers's right-left discrimination test, the examiner gives a series of commands sitting right opposite to the patient, such as "show me your left hand" and "show me your left foot." Ayers's test is given for the pediatric population and the scoring is based on the time limits. Two points if correct within 3 s and one point if correct within 4–10 s. If a patient first gives a wrong answer and then corrects himself, the score is based on the time of the correct response. If the command is repeated, the score is no more than one. Another test is "point to the body parts on command." For this test, the therapist asks to point or indicate the body parts on him or a human figure doll. The scoring is non-standardized and is intact if the patient correctly indicates all parts named within a limited time. The Bergen left-right discrimination test is yet another test that is used for right-left discrimination.

14.3.5 Finger Agnosia

The finger localization or identification is a test for finger agnosia. For this test, the patient must place his hands down on the table. A picture of two hands is placed in front of the patient with the fingers in the picture aligned in the same direction as the patient's fingers. The examiner touches the patient's fingers one at a time and asks him to name or point to the finger on the picture. This test can be performed with eyes closed too. To compare the results, a combination of five attempts with vision and five without vision is useful. The finger localization is a non-standardized test and is considered to be intact if the patient points all the fingers correctly within a certain time. The other testing methods are to name the fingers, ask the patient to touch or point to the named finger, or imitate the finger movements performed by the examiner.

14.3.6 Figure-Ground Discrimination

A. Jean Ayres's Southern California Figure-Ground Visual Perception Test is a test where the patient must identify three pictures of objects from a figure having six pictures. The time limit to name or point is 1 min for all three pictures and the test has a standardized scoring method. If the patients commit five or more errors, the test is discontinued. Another test is a functional test where the patient is asked to find a specific item from a cluster of similar things, for example, finding a spoon from a cutlery set.

14.3.7 Form Discrimination

In the form board test, the patient is presented with ten forms and a test plate. He is given one form at a time and asked to match each shape on the board. For aphasic patients, a demonstration is given by the therapist. The forms matched are taken out of eyesight, so that he does not use the elimination process. The therapist should look for spatial neglect, poor planning, or conceptual skills, and a general inability to deal with objects. The form discrimination will be considered severely impaired if the subject is unable to match less than five forms, impaired if he matches five to nine forms, and intact if he matches all correctly. The functional test for form discrimination consists of differentiating objects from one another from a group of objects with similar forms. For example, a cluster including pencil, pen, toothbrush, and a wooden stick of the same size is placed and the patient is asked to identify all. The objects must be kept in different angulations while identifying. Prior to the test, the visual object agnosia must be ruled out by asking the patient to name some familiar objects. The scoring is subjective and is considered intact if the patient displays no difficulty in distinguishing forms during occupational performance tasks.

14.3.8 Spatial Relations

In the "position blocks" test, the examiner places two small cube blocks on a table in front of the patient and instructs him to place each block in a different position with respect to the other. For this test, commands like above, below, front, and behind will be given by the therapist. The test is a non-standardized one. If the patient correctly places the cubes as per the command, the spatial relations will be considered intact. The RPAB and Arnadottir Occupational Therapy Activities of Daily Living Neurobehavioral Evaluation (A-ONE) are some of the standardized tests that can assess spatial relations.

14.3.9 Position in Space

For the position in space test, two objects are kept in a relationship with each other, and the patient is asked to describe their relationship with each other. For example, if a patient is presented with the pen placed over a book and he can narrate where the pen is placed, it is certain that the patient's position in space is intact. The other way of testing is by asking the patient to copy the examiner's way of manipulating two objects. For instance, the examiner places a toothbrush over the toothpaste tube and then instructs the patient to repeat the same. If the patient succeeds in positioning and aligning the set in the same way the examiner did, the patient's position in space sense can be considered to be intact.

14.3.10 Depth and Distance Perception

For distance perception, the patient is asked to reach out for an object held in front of him. The patient is said to have impaired distance perception if he overshoots or undershoots, but the movements are smooth and not jerky, inferring the absence of incoordination. For depth perception, the patient is asked to step onto a stair and see if he steps correctly without tripping.

14.3.11 Topographical Disorientation

The functional test for topographical disorientation consists of asking the patient to find the way back to his home or the ward after being shown the way several times. Another alternative is to ask the patient to draw the layout of his house or direction to reach his place, and if he fails to do so, it might suggest topographical disorientation. Before evaluating the topographical disorientation, the examiner must rule out the possibilities of memory and cognitive issues.

14.3.12 Vertical Disorientation

Vertical disorientation can be identified with a simple bedside test. Here, the patient is given a stick and asked to rotate it 360° and position it back to its original position. If the patient has vertical disorientation, he fails to judge the verticality of the stick, and typically the stick will be placed at an angle that is as per the patient's vertical perception.

14.3.13 Visual Agnosia

The standardized tests for visual agnosia are the Structured Observational Test of Function (SOTOF) and the A-ONE. For visual object agnosia, "object recognition" is the test used. Here, the patient is presented with a few common objects like a toothbrush, pen, pencil, coins, and watch and is asked to identify them or describe their usage. Scoring is based on the patient's potential to recognize the object or pick up the object named. The Visual Identification of Objects, a subtest of the LOTCA, can be an alternative to test visual object agnosia. For this test, the patient is shown eight cards of regular everyday objects such as a watch, key, glasses, scissors, shoe, chair, teapot, and bicycle and is asked to recognize and name every object. For patients with dysarthria or aphasia, the examiner may ask them to spot the object when asked. If the patient has deficits in the reception of information, then the therapist shows him four similar objects presented on two boards. Then the patient is asked to match the objects presented in front with the objects on the boards. A score of 1 is allotted if the patient identifies less than four objects, 2 points if the patient identifies five to eight objects by exact matching, 3 points if the patient identifies at least four objects by naming, understanding, or matching, and 4 points if he identifies all the objects by naming, understanding, or matching.

For simultagnosia, the patient is presented with pictures containing several objects or people and then asked to identify what they see. These patients will have difficulty identifying multiple things at a time and fail to identify the whole picture or the entire group of people. For prosopagnosia, the functional test consists of the patient identifying family members or close friends from photographs or the ones within his visual field.

Tactile agnosia can be assessed by testing the stereognosis, where the patient is given different familiar objects such as a spoon, key, pen, button, and scissors in hand and asked to identify them with touch with eyes blindfolded. Alternatively, the patient can also be asked to identify the object by touch and/or manipulation of different familiar objects kept in a tray, with the eyes blindfolded. Test for ahylognosis (difficulty or inability to distinguish various textures by touch), a component of tactile agnosia, can be tested by placing different textured materials like silk, plastic, and cotton in the hand of the patient. Here again, the patient will be asked to describe the texture of the material (soft, hard, or smooth) by palpation with the eyes blindfolded.

14.3.14 Ideomotor Apraxia and Ideational Apraxia

The testing should be always done for both arms and hands. The testing objects should be the same for all the subtests. The Goodglass and Kaplan test consists of standardized movements such as brushing, shaving, and hammering. It has a hierarchy of difficult tasks. Initially, the examiner observes the patient's routine actions such as shaving, washing, and using culinary or gestures like waving goodbye and blowing a kiss. If not, the examiner may ask the patient to perform the same actions on command or else imitate him. If he can do these simple tasks, he is given a little more difficult task such as hammering a nail or opening a bottle with an opener. The patient can pick up the tools and demonstrate how to put them to use. The tools for testing ideational apraxia are similar to that of ideomotor apraxia. The only difference between these two forms is that in ideomotor apraxia the patient can perform activity spontaneously but not in ideational apraxia.

The other test for ideational apraxia is the Praxis Test-Santa Clara Valley Medical Center. It is a 10-item test consisting of buccal-facial, unilateral and bilateral limb, and total body tasks. The test items are performed on command which is either an imitation or the use of a real object. The Praxis Subtest of the LOTCA is another test for assessing ideational apraxia. It contains three parts: motor imitation, utilization of objects, and symbolic actions. A score of 1 point is assigned if the patient is unable to perform any task, 2 points if he can imitate movements, 3 points if he can imitate movements, and manipulate the objects, and 4 points if he performs all tasks.

14.3.15 Constructional Apraxia

"Copying two- or three-dimensional designs" is one of the tests used for constructional apraxia. Here, the examiner gives the patient a paper and a pencil and asks him to copy the design (drawing a house, flower, clock, or geometric design) as shown on the stimulus card. For every stimulus card copied, a new paper sheet is provided to the patient to attempt for the next item. The scoring of this test is non-standardized. Each drawing is scored on a scale of 1–3. A score of 1 is given when the drawing is correct in terms of spatial arrangement, with no extra lines. A score of 2 is obtained when the drawing is partially defective, but the figure can be recognized. If the drawing is unrecognizable, a score of 3 is given.

The Bender Visual-Motor Gestalt Test, also known as the Bender-Gestalt test, is a test where the patient is given a blank piece of paper and a pencil. He is shown nine cards, one at a time, and asked to draw the design as depicted in the card. The patient is asked to draw all nine designs on one piece of paper. Bender-Gestalt test is standardized and the scoring method is illustrated in the manual provided for the same. Copying the matchstick designs is another test, and for this test, the examiner makes designs using two to nine wooden kitchen matches with varying complexity and asks the patient to copy them. The matchstick design test is also a non-standardized test. An example of a matchstick design test is depicted in Fig. 14.7.

14.3.16 Dressing Apraxia

This test is strictly a functional one. The examiner may observe the patient wearing a shirt or an overcoat. Does he find trouble identifying which is the right or left side of the shirt or inside or outside of the shirt? Does he fail to align the buttons while wearing the dress? If either of these happens, then the patient has dressing apraxia provided that the examiner has ruled out the possibilities of weakness or paralysis of the limbs.

14.4 Treatment

Early recognition and timely management may encourage functional independence and minimize safety issues among patients with perceptual deficits. Two types of approaches are used for managing the various perceptual disorders, namely, the remedial approach and the adaptive approach.

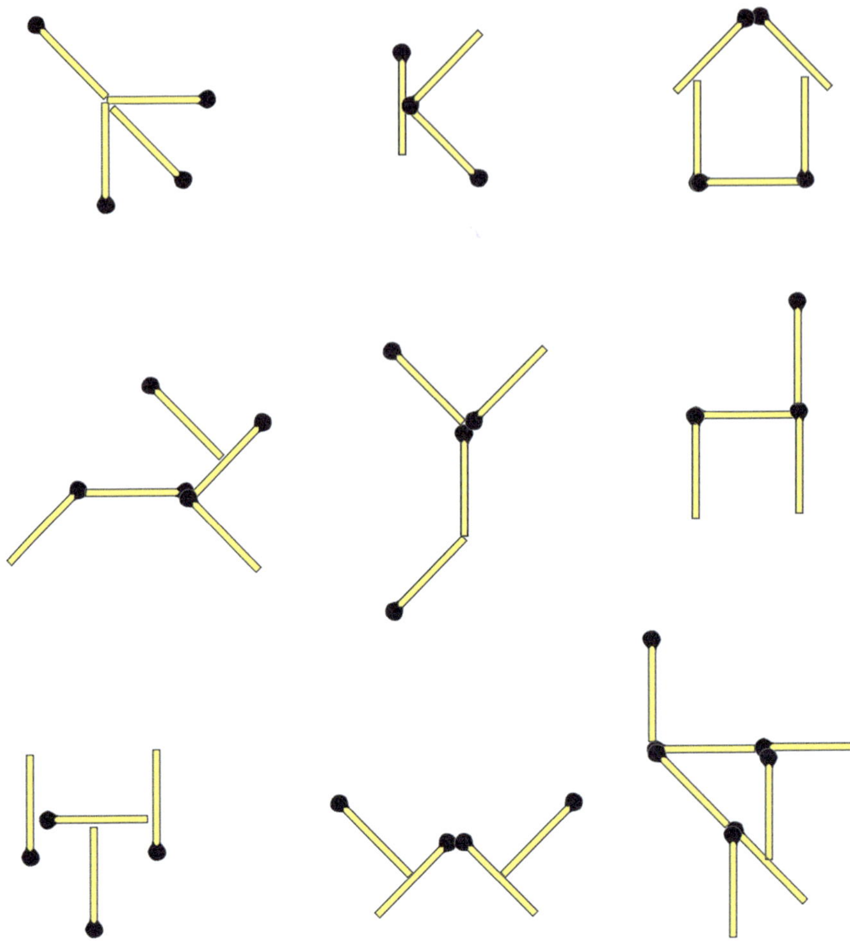

Fig. 14.7 An illustration of the matchstick design test items

14.4.1 Remedial Approach

The remedial approach, also known as the restorative approach, aims at restoring the deficits causing disability and improving the specific perceptual components that are missing. This approach is called a bottom-up approach, which assumes that the patient can generalize the learned core functions to the other daily activities that consist of the same core functions. This approach assumes that learning the perceptual requirements of one task will enhance other tasks' performance with similar perceptual requirements. The approach uses repetition of tasks and exercises which focus on promoting neural plasticity and functional recovery. Some of the remedial approaches are perceptual-motor training, sensory integration, Affolter, and neuro-development treatment. Perceptual-motor training focuses specifically on perceptual functioning. The sensory integration techniques address the integration of the basic sensorimotor functions (tactile, proprioceptive, and vestibular), which proceed in a developmental sequence. The underlying assumption is that, by providing opportunities for controlled sensory input, the therapy can affect the CNS processing of sensory information and elicit specific desired motor responses. The treatment techniques include rubbing or icing to provide sensory input, resistance and weight-bearing to impart proprioceptive input, and spinning for vestibular input. Neurodevelopmental treatment deals with proprioceptive and kinesthetic perceptions as they relate to functional movement patterns. These approaches provide training in the perceptual processing components of functional behavior with perceptual drills or specific sequences of sensorimotor exercises.

14.4.2 Adaptive Approach

The adaptive approach is a top-down approach where the therapist focuses on targeted functional outcomes rather than working on the basic skills which are lost. This approach encourages the adaptation of the environment to make the most of the patient's abilities and is used when it is impossible to restore the missing perceptual functional components required to carry out ADL. This approach promotes compensation such as assistance from caregivers, substitution or use of other intact sensory systems of the body, or assistive devices to improvise the functions of the patient. The patient must have some awareness of his deficits so that he can adapt to new strategies. The new strategies must be learned in different environments and should be practiced to become automatic. For example, if the patient has a visual field deficit, the therapist should educate the patient about the visual problem. Following this, the patient should be shown how to turn the head to compensate for the deficit. The drawback of this approach is that methods learned in one task may not help in generalizing the concept to perform another task. The functional approach, the occupational performance approach, and the dynamic interactional approach are some of the adaptive approaches regularly used.

14.4.2.1 Visual-Spatial Inattention/Unilateral Neglect

Conventional treatment methods include the use of prismatic glasses (Fresnel prisms) and a variety of optical aids to help the patients compensate their visual difficulties by training visual exploration toward the involved side. The aids include a 28°–30° wide-angle lens, mirrors attached to the spectacle frame, and closed-circuit television monitor systems. By causing an optical deviation of the visual field to either the left or the right side, the prismatic glasses or equivalent visual aids affect spatial representation. Over time, the visuomotor system will adapt to this new deviation, understanding that the true location of the objects is not as they appear in the goggles. This forced deviation can help individuals with neglect attend better to the contralesional visual hemisphere. The participants who are adapted to the visual deviation tend to show a bias to one side for the visuomotor behavior, even when the glasses are removed.

Other strategies include the use of vibratory stimulation to the right side of the neck, encouraging activities using blocks and shapes for the affected side, and providing verbal and auditory (bell) cues to encourage the patient to look toward the affected side. Techniques like eye patching also increase eye movements toward the contralateral space and encourage the development of the intentional control of attention in the short term and the development of the automatic shifts of attention over a longer time. Mirror therapy can be another strategy to reduce the neglect, and for the same, a mirror box is used. The patient's affected limb is kept hidden behind the mirror and the unaffected limb is placed in front of the mirror. The patient is instructed to perform voluntary movements of the nonparetic wrist and fingers while looking into the mirror. Such kind of placement of limbs will make the patient see the reflection of the unaffected hand as the movement of the affected hand in the mirror. While moving the nonparetic limb, the patients are encouraged to do the same movements in the paretic hand.

The compensatory approach is advisable only when no remedial treatment is effective in reducing the severity of neglect. This approach consists of placing all necessary belongings on the less affected side. Meanwhile, there is a need to increase the patient's and caregivers' knowledge and awareness about the deficit. The patient should be taught about compensation with head and eye movements and encourage activities that lead to eye movements from the unaffected side to the affected side. The placement of a red ribbon on the left margin of the reading material may encourage the patient to start reading from the line. Addressing and providing demonstrations from the unaffected side may draw the patient's attention to scan the vision toward the affected side.

14.4.2.2 Anosognosia

Most cases of anosognosia tend to disappear over time, while some cases can last indefinitely. The remedial approach consists of providing instant, objective, and specific feedback during the task. The use of cues may help to guide the patient when issues are anticipated while performing the task. Videotaping can help to increase the level of awareness. The compensatory approach includes educating the caregivers about the deficits and making them aware of the safety measures to be taken while performing kitchen tasks or driving. Educating about safety is of paramount importance because the patients typically do not acknowledge their disability and will therefore refuse to be careful. The patient should be taught to recognize problems using feedback from the surroundings when he is performing the action. It will help the patient develop anticipatory awareness and formulate a compensation strategy to minimize his problems. For example, if the patient has difficulty in memorizing a task, he should keep a diary to note down all the related important information.

14.4.2.3 Asomatognosia

A remedial approach that aims to associate sensory input with an adaptive motor response may alleviate this perceptual deficit. This approach facilitates body awareness through the sensory stimulation of the body part affected. For instance, the therapist can rub a rough cloth over a body part, and meanwhile, the patient can be encouraged to name or point the part where the stimulus is applied. Progression is made when the patient verbally identifies the body part(s) or points to the picture of them as the therapist touches them. The patient can be asked to play face and body parts puzzles and quizzes related to body parts. The neurodevelopmental concepts can also be used by including bilateral activities in daily treatment sessions and the practice of functional activities such as transfers or dressing in a variety of ways. The compensatory approach consists of educating the family members about the condition and how they can assist the patient to perform his daily tasks. If the patient can localize some body parts, the therapist can use these as reference points to localize the adjacent parts.

14.4.2.4 Right-Left Discrimination

The remedial approach for tackling the right-left discrimination issues consists of the application of tactile inputs on the dominant side like weighted cuffs. When the compensatory approach is used to overcome the right-left discrimination issues, the use of words like "right" and "left" has to be discouraged and instead cues or pointing to the parts has to be encouraged. For instance, to distinguish the left arm, the caregiver or therapist can use words like "the arm with the watch" to address the same. The use of markings on one side may also help to overcome the right-left discrimination issues.

14.4.2.5 Finger Agnosia

The remedial approach includes using the patient's tactile sensory system. A rough cloth or a thin Velcro strap can be rubbed on the affected areas with the eyes open and then with the eyes closed. The patient is then asked to identify the finger. Repeated sensory inputs can be provided to areas such as thumbpads or fingertips which are used commonly for daily tasks. The patient is made to practice tasks that are relevant to his daily life, which may include keyboard typing. Alternatively, the adaptive approach comprises highlighting the areas which require safety, educating family members about the deficit, and providing environmental adaptations such as adaptive knobs.

14.4.2.6 Figure-Ground Discrimination

The remedial approach includes arranging three dissimilar objects and then visually identifying the objects when asked by the therapist. The complexity of the task can be increased by increasing the number of objects in the set presented to the patient. The compensatory approach includes the reaching and tactile feeling of the objects. For instance, if the patient is locating a door handle or lock on the orthosis, he should be encouraged to touch and locate with verbal cues instead of using vision. To make the tasks easier, red/radium tapes can be used over the locks, loops, and stair ends. Systematically arranging or placing the objects within a home environment minimizes the figure-ground discrimination issues. The patient should be educated about his deficit and should learn to organize his work. If he is searching for an item, he should carefully examine the entire area in detail.

14.4.2.7 Form Discrimination

The patient should be assisted in understanding different objects. He should regularly practice identifying and describing similar objects using tactile cues. In the compensatory approach, the patient is first made aware of his deficit and then encouraged to identify the objects after labeling or assigning a permanent and specific location. The intact sensory systems such as vision and touch can be used for identifying objects.

14.4.2.8 Spatial Relations

By providing external cues and instructions, the remedial approach helps the patient to improve his ability to position himself to other objects such as a wheelchair or bed. They are given verbal cues with spatial contexts such as "sit opposite to me." The patient can be taught spatial relations by arranging the blocks the same way the therapist has arranged them or by solving puzzles related to spatial relations. Alternatively, the therapist can place objects in different places in front of the patient and then ask him to describe the spatial relation of him to the objects. It can also be taught by computer software specifically designed to improvise spatial relations. The adaptive approach consists of placing the objects in a fixed place so that it is easier to locate the objects.

14.4.2.9 Position in Space

In the remedial approach, two or three identical objects are placed in the same orientation and another object is placed next to them in another orientation. The patient must identify the odd one out and place it back in the orientation of the other aligned objects.

14.4.2.10 Depth and Distance Perception

The remedial approach is where the patient is made to practice placing the feet on different blocks repeatedly with height differences of 2–8 inches. The Affolter approach (Box 14.1) can be used which emphasizes the guidance of tasks with tactile-kinesthetic cues, for instance, asking the patient to touch and understand the depth of the chair before initiating a bed to chair transfer. Computer software programs designed to teach depth and distance perception can be another remediation approach. For patients who are not benefited by the remedial strategies, taping the edges of the stairs and use of verbal cues may overcome the depth and distance perceptual issues.

Box 14.1 Details Regarding Affolter Approach

Affolter approach
• Developed by Felicie Affolter, a speech therapist and psychotherapist
• Approach is based on the premise that learning takes place only when patient challenged
• Nonverbal guidance to facilitate perceptual-cognitive interaction
• Approach emphasizes on appropriate input than successful output
• Development of sensory-motor skills to augment complex cognitive and perceptual skills
• Mistakes/errors are allowed during the learning process and should encourage learning from mistakes
• Learning occurs by tactile-kinesthetic input than verbal or visual input
• Tactile input the primary input for interaction
• Guidance progressed from maximal toward minimal assistance

14.4.2.11 Topographical Disorientation

Practicing repeated navigation from one spot to another location with verbal cues may overcome topographical disorientation. In the beginning, the patient can be trained to practice short distances and simple routes and later progress to difficult and long routes. The use of dots for marking routes and signboards or landmarks, which are realistic and practical, are the compensatory strategies.

14.4.2.12 Vertical Disorientation

The patient is made to compensate for the deficit by using other sensations such as touch so that he can carry out the transition and mobility activities, which require vertical perception. The use of mirror feedback may help overcome the issues pertinent to vertical disorientation and may improve the patient's posture and mobility.

14.4.2.13 Visual Agnosia

For prosopagnosia, the patient can be shown different photographs to differentiate between faces. The therapist can initially help the patient to identify the faces by giving visual cues specific to that photograph, whereas the compensatory approach involves using the other sensory inputs such as auditory and tactile cues.

14.4.2.14 Visual Object Agnosia

The remedial approach consists of practicing the recognition of the familiar objects required for functional independence. Tactile and kinesthetic cues during the task may augment the learning process. After completing the task, the patient can be encouraged to name the tools used for the task. If the remedial strategies fail to improve the agnosia, the patient should be encouraged to use the intact alternative sensory systems to manipulate the object. The labels or tags placed on the objects required for daily usage can be considered a compensatory approach to overcome visual object agnosia.

14.4.2.15 Prosopagnosia

Treatment strategies like face-matching tasks can be initiated to overcome prosopagnosia. If the patient cannot carry out face-matching tasks, then pictures with names can be presented. These pictures can be presented in different environments, with different people, and the patient can be assisted in identifying the faces with specific features such as the color of hair, voice, way of walking, height, and build.

14.4.2.16 Tactile Agnosia

Since the brain has the potential for functional recovery through reorganization, sensory retraining can be attempted. To appreciate sensory input from the receptors, stroking of the hook side of the Velcro can be attempted. Once the patient can appreciate the moving stimulus, progress the training toward appreciating the stationary stimulus necessary for holding objects. Activities to recognize objects' shapes with eyes closed by manipulation and matching with the items kept in front should also be attempted. When the above techniques fail, the compensatory approaches that are used include improvising the awareness of the problem, utilizing other senses such as visual cues, and making the patient aware of the other specific characteristics of that object.

14.4.2.17 Ideomotor Apraxia and Ideational Apraxia

The remedial approach to treat apraxia includes many strategies, such as the use of short commands and one command and wait for the response. Every new task should be broken into subtasks. Each subtask should be guided manually, precisely, and repeatedly before the whole task is attempted. The patient is advised to practice the same technique at home and the family members are taught the same and advised to guide and encourage the patient. The use of pictures in a sequence to carry out ADL can be a compensatory approach for training tasks. Compensatory strategies and visual and verbal mediation strategies, combined with appropriate tactile and kinesthetic inputs, may provide multimodal stimulation to overcome apraxia. It involves cueing the patients to look at what they are doing and encouraging them to verbalize what they want to do and what they have done.

14.4.2.18 Constructional Apraxia

The remedial approach consists of the Affolter approach that uses tactile and kinesthetic cues to understand a 3D model with the patient's hands before constructing it. For example, if the patient is working on work-related nuts and bolts assembly tasks, the therapist places his hand over the patient's and guides him through the necessary sequence. As the patient takes over the task, the guidance is reduced. The dynamic interactional model of cognitive rehabilitation can also be used to identify the task characteristics that can be altered to improve the execution. For example, the patient may assemble objects, provided the objects are arranged in a defined manner or may complete the task which is partially completed.

14.4.2.19 Dressing Apraxia

Dressing apraxia can be remediated using the Affolter approach where the patient is asked to dress up on his own with visual and verbal cues. If the patient does not initiate, give him a verbal cue "get dressed please." Then wait and look for a response; if not, then give a nonverbal tactile cue during the dressing task. Reduce the guidance if the patient can perform independently. The use of weight-bearing, weight shifting, and handling techniques of the neurodevelopmental approach may augment the training process. The compensatory strategy consists of using the dynamic interactional approach to either alter the number of apparels placed in front of the patient (from many to one at a time) or dress in different ways. The functional adaptive approach consists of labeling the back of the shirt to distinguish it from the front or asking the patient to feel the buttonhole by passing a finger through it. Educate the family members about the disability and train them on the compensatory approaches.

Further Reading

Arbuse DI. The Gerstmann syndrome; case report and review of the literature. *J Nerv Ment Dis.* 1947 Apr;105(4):359–71.

Bechinger D, Tallis R. Perceptual disorders in neurological disease part 1. *Br J Occup Ther.* 1986;49(9):282–4.

Bradley WG, Daroff RB, Fenichel GH, Jankovic J. Neurology in clinical practice. Principles of diagnosis and management. 4th ed. Elsevier Health Sciences; 2004.

Coslett HB. Apraxia, neglect and agnosia. *Continuum (Minneap Minn).* 2018;24(3):768–82.

Crosson B, Barco PP, Velozo CA, Bolesta MM, Cooper PV, Wertz D et al. Awareness and compensation in postacute head injury rehabilitation. *J Head Trauma Rehabil.* 1989;4(3):46–54.

Finney GR. Perceptual-Motor Dysfunction. *Continuum (Minneap Minn)* 2015;21(3):678–89.

Gainotti G. History of Anosognosia. *Front Neurol Neurosci.* 2019;44:75–82.

Greene JD. Apraxia, agnosias, and higher visual function abnormalities. *J Neurol Neurosurg Psychiatry.* 2005 Dec;76(5):25–34.

Gurd J, Kischka U, Marshall J. Assessment of perceptual disorders The Handbook of Clinical Neuropsychology. 2nd ed.; 2010.

Heilman KM, Watson RT, Valenstein E. Neglect and related disorders. In: Heilman KM, Valenstein E, editors. Clinical neuropsychology. 3rd ed. New York: Oxford University Press; 1993. p.279–336.

John Spillane. Bickerstaff's Neurological Examination in Clinical Practice. 6th ed. New Delhi: Blackwell science; 2005.

Marsh NV, Kersel DA. Screening tests for visual neglect following stroke. *Neuropsychol Rehabil.* 1993;3(3): 245–57.

Meador KJ, Loring DW, Feinberg TE, Lee GP, Nichols ME. Anosognosia and asomatognosia during intracarotid amobarbital inactivation. *Neurology.* 2000 Sep 26;55(6):816–20.

Meerwaldt JD. Spatial disorientation in right-hemisphere infarction: a study of the speed of recovery. *J Neurol Neurosurg Psychiatry.* 1983 May;46(5):426–9.

Piscielli C, Perennou D. Visual verticality perception after stroke. A systematic review of methodological approaches and suggestions for standardization. *Ann Rehabil Med.* 2017; 60(3): 208–16.

Ropper AH, Samuels MA, Klein JP, Prasad S, editors. Adams and Victor's Principles of neurology. 10th ed. New York: McGraw Hill Education; 2014.

Rowland LP, editor. Merritt's Neurology. 10th ed. Washington: Lippincott Williams & Wilkins Publishers; 2000.

Umphred DA, Burton GU, Lazaro RT, Roller ML, editors. Umphred's neurological rehabilitation. 6th ed. St. Louis: Elsevier Mosby; 2013.

West C, Bowen A, Hesketh A, Vail A. Interventions for motor apraxia following stroke. *Cochrane Database Syst Rev.* 2008 Jan 23;2008(1):CD004132. https://doi.org/10.1002/14651858.CD004132.pub2. PMID: 18254038; PMCID: PMC6464830.

Yelnik AP, Lebreton FO, Bonan IV, Colle FM, Meurin FA, Guichard JP, et al. Perception of verticality after recent cerebral hemispheric stroke. *Stroke.* 2002; 33(9):2247–53.

Zoltan B. Vision, Perception and Cognition. A manual for the evaluation and treatment of the neurologically impaired adult. 3rd ed. New Jersey: Slack Incorporated; 1996.

Chapter 15
Neurobehavioral Issues in Adult Neurological Conditions

S. Malarmathi, A. T. Safeekh, and Abraham M. Joshua

15.1 Introduction

Apart from the regular neurological manifestations, cognitive and behavioral symptoms are an integral part of most neurological conditions due to the involvement of the higher cortical centers. The early identification and management of those neurobehavioral symptoms play a very important role in the neurophysiotherapy as those symptoms interfere with compliance and prognosis. The neurobehavioral manifestations of the neurological conditions share certain common presentations irrespective of the socio-cultural background and the personality traits due to the specific pathological involvement of the areas of the brain involved. The onset, course, and prognosis of neurobehavioral symptoms in neurological disorders widely vary, wherein certain neurological disorders, behavioral and psychiatric symptoms could be the presenting complaints. In many others, they develop during the disorder, and in some as a consequence of the pharmacotherapy, radiotherapy, or surgical management of the primary neurological condition.

Syndromes of the localized brain lesions can present with apraxia, agnosia, visuospatial impairment, cognitive and attention deficits, amnesia, disorders of mood and emotions, abnormal social behavior, and abnormal beliefs and experiences. Many times, the patients' insight into their deficits will be limited. The primary step in managing neurobehavioral issues should include the assessment of

S. Malarmathi (✉) · A. T. Safeekh
Department of Psychiatry, Father Muller Medical College, Mangalore, India
e-mail: malarsannasi@fathermuller.in; safeekh@fathermuller.in

A. M. Joshua
Department of Physiotherapy, Kasturba Medical College of Mangalore, Manipal Academy of Higher Education, Manipal, India
e-mail: abraham.joshua@manipal.edu

A. M. Joshua (ed.), *Physiotherapy for Adult Neurological Conditions*,
https://doi.org/10.1007/978-981-19-0209-3_15

neurological deficits, cognitive functions, insight, psychiatric symptoms, psychosocial stressors, biological functions, disability, and compliance. The following section of the chapter would like to address the neurobehavioral issues seen in common neurological conditions. The definitions or meanings of certain terminologies used in the current chapter are provided in Table 15.1.

Table 15.1 Definitions or meanings of certain terminologies

Terminology	Definition or meaning
Anxiety	An unpleasurable emotional state, associated with psychophysiological changes in response to an intrapsychic conflict
Behavior therapy	A collection of techniques based on learning principles to modify maladaptive behavior
Bipolar/affective/mood disorder	Behavioral disorder characterized by recurrent episodes of mania and depression
Cognitive behavioral therapy	A form of psychological intervention which helps the patient to alter maladaptive thoughts and behaviors by correcting the cognitive distortions
Cognitive remediation training	A collection of procedures designed to improve the neurocognitive functions using various techniques which can be delivered through various modalities including computerized programs
Compulsion	Repetitive, purposeful physical mental actions that the individual feels compelled to do
Delusions	Firm but false belief contrary to reality and held in spite of the evidence and common sense
Depression	A psychological state marked by lowered activity, gloomy thoughts, anxiety, feeling of worthlessness, and an inability to deal effectively with life
Family therapy	Group psychotherapy involving two or more members of a family rather than any single member
Hallucinations	Perception without an external stimulus, that is, hearing of voices in a completely silent room
Mania	A disorder characterized by elated/irritable mood with great excitement, activity, energy levels, increased self-esteem, and decreased need for sleep
Memory training	A collection of techniques used to restore memory functioning. Intervention includes the teaching of the use of mnemonics, the skilled use of prompts, reinforcement or a combination of these techniques
Obsession	Persistent recurrence of an unwanted idea, thought, image, or impulse
Obsessive-compulsive disorder	A disorder marked by persistent obsession, compulsion or both
Panic attack	An acute intense, overwhelming episode of anxiety, often associated with a feeling of impending doom and autonomic symptoms
Phobia	An irrational fear of an object, situation or activity
Premack's principle	The technique is a positively reinforcing contingency in which the instrumental or to be rewarded behavior occurs with a lower momentary probability than does the contingently available or rewarding behavior
Psychosis/psychotic disorder	Is characterized by hallucinations, delusions, and/extreme psychomotor agitation/retardation

Table 15.1 (continued)

Terminology	Definition or meaning
Psychotherapy	Treatment of mental disorders and behavioral problems by psychological methods
Reminiscence therapy	A treatment method used in dementia to improve memory by using all the special senses
Schizophrenia	A disorder characterized by psychotic symptoms in which there are fundamental disturbances of thought and perception
Snoezelen therapy	Also known as controlled multi-sensory environment therapy is a therapy used for patients with dementia by exposing them to high levels of stimulus using various modalities like taped music, aroma, bubble tubes, fiber optic spray and moving shapes projected across the walls
Supportive therapy	A psychological treatment designed to remove symptoms by reinforcing the existing personality defenses

15.2 Traumatic Brain Injury

Traumatic brain injury (TBI) is a non-degenerative, non-congenital insult to the brain from an external mechanical force, possibly leading to permanent or temporary impairment of cognitive, physical, and psychosocial functions, with an associated diminished or altered state of consciousness. Traumatic injury of the brain can range in severity from concussion to coma and can be categorized from mild, moderate to severe.

Apart from the mortality, the physical manifestations and psychiatric complications of head injury could also be quite devastating, which could be due to the physical damage caused by the brain injury or at times a psychological sequel that develops due to the physical disability. Structural and functional brain disturbances of TBI can play a causal, predisposing, or precipitating factor in many psychiatric disorders. Complications of TBI are some of the most common conditions that require neurorehabilitation. TBI can be fatal in its severe type and can have persistent neurological and psychological problems in the survivors.

During the acute phase of the TBI, behavioral symptoms will be predominated by cognitive deficits presenting as impaired attention, memory, and consciousness, post-traumatic delirium, post-traumatic amnesia, confabulations, and even coma depending upon the extent of the injury. The long-term neurobehavioral consequences of TBI include a wide range of neuropsychiatric symptoms from persisting cognitive deficits in attention, memory, and executive functioning, dementia, psychosis, mood disorders, anxiety disorders to personality changes and neurological deficits like motor deficits, which require physiotherapy. The cognitive deficits and personality changes often make physical rehabilitation a challenging task.

15.2.1 Neuropsychiatric Manifestations

Cognitive Impairments Head injury is associated with cognitive impairments as it can affect the lobe functions and the deficits will depend on the area of the lesion. The deficits are observed in the majority of cognitive functions, including simple and complex, and most primarily the areas of memory and executive functioning. On neuropsychological assessment, the impairments will be evident in areas like attention, memory, and learning, which include retrograde amnesia, post-traumatic amnesia, and difficulty in new learning and memory, mental speed, language, reasoning, self-awareness, and perception. The cognitive impairment has long-term consequences and it affects a person in such a way that he/she finds it difficult to carry out the Activities of Daily Living (ADL). Most people with mild deficits can carry out day-to-day functioning, but patients with moderate to severe brain injuries are likely to have more difficulties with cognitive skills and professional skills. Though the cognitive functions are likely to improve during the initial few years, research suggests that minimal amounts of recovery can continue throughout life.

Psychosis Though the incidence of psychosis is supposed to be rare in TBI, some studies have even reported rates as high as 10%. Few studies reported TBI as a risk factor for psychosis, where higher incidence rates were found in persons with schizophrenia. These studies suggest that persons who develop psychotic disorder due to TBI have a premorbid vulnerability, including a genetic predisposition for schizophrenia or other neurological or psychiatric conditions. Psychosis can occur in both close and open head injury with any kind of severity, from mild to severe. Auditory hallucinations and delusion of persecution are the most common symptoms reported.

Depression Depression is the common sequel of TBI and the prevalence ranges from 3 to 95% in various studies. The symptoms of depression can be seen immediately following a head injury, which could be due to the lesion, whereas, for later-onset depression, the psychosocial factors play an important role as a risk factor. Depression can range from mild to severe in patients with TBI and manifest with symptoms like fatigue, frustration, and reduced concentration. The majority of persons recover after a few months and the risk of suicide is found to be only 1% in the first 15 years following TBI. Depression can often worsen the Quality of Life (QOL) and even predispose to higher rates of morbidity and mortality.

Anxiety Disorders Generalized Anxiety Disorder (GAD) can co-exist with depression and the patient often reports a persistent state of anxiety and tension about the trauma followed by TBI. Phobia is the most common sequela, followed by TBI. The patient typically develops fear towards the object, vehicle, or situation that led to the accident and develops avoidance towards the mode of transportation that led to TBI. They often experience fearfulness and tremors while trying to drive and tend to be extra cautious while driving. Post-TBI, some meet the criteria for Post-traumatic Stress Disorder (PTSD), which is characterized by re-living the trauma, where patients have nightmares, flashbacks of the traumatic event, following which they develop avoidance behavior and report heightened arousal. There are conflicting findings regarding the correlation between PTSD and the severity of TBI. Obsessive-Compulsive Disorder (OCD) is rarely seen in TBI with a prevalence rate of 1.6% and case studies have demonstrated that the OCD symptoms are not reported often because of the significant cognitive deficits and lack of insight. Various conversion and dissociative symptoms are reported in many patients following TBI, including fugue, amnesia, Ganser state, motor paralysis, anesthesia and speech, sight, and hearing disturbances. The diagnosis of conversion and dissociative disorders in TBI often becomes challenging as these symptoms can occur as a direct consequence of the TBI.

Personality Changes Premorbid personality traits may or may not be preserved post-TBI. The personality changes following TBI depend highly on the severity and the localization of the pathology. TBI patients can present with the symptoms of motivation, decreased interest in plcasurable activities, avoidance behavior, emotional changes like apathy, blunted affect, euphoria, anger, agitation, irritability and disinhibited behavior, sexual disturbances, and regressive behavior. Apart from all the behavioral changes, anger is the most distressing, which can be provoked or unprovoked, resulting in immense distress to the caregiver and the patient if the insight is preserved. The personality changes often remain permanent and it generates dependency on the caregivers and the patient gradually isolates himself/herself from the social surroundings. Short-lasting explosive episodes of sudden rage following minor provocation described as "organic aggression syndrome or episodic dyscontrol syndrome" have been reported in a few cases following severe TBI. The case report of Phineas Gage (1823–1860), an American railroad worker who had a complete personality change after an iron rod went through his head damaging a large part of the left frontal lobe, who miraculously survived and lived for another 12 years is considered the prototype for post-head injury personality, where his friends described that he was "no more Gage" after the head injury.

15.2.2 Non-pharmacological Management

Once the patient is medically stable, the psychosocial intervention has to be planned along with neurorehabilitation, considering the nature and intensity of the deficits. The treatment aims to assess the current functioning of the patient and design the rehabilitation plan accordingly. The majority of mild head injury patients do well without any specific intervention. Rehabilitation should be meticulously planned from the early stage for the moderate to severe TBI patients, focusing on the baseline assessment, psychometric assessment followed by behavioral and psychological management, which will improve the compliance towards physiotherapy. Usually, the interdisciplinary team consisting of the neurologist, clinical psychologist, physiotherapist, occupational therapist, rehabilitation nurse, speech and language pathologist, social worker, and vocational counselor are an integral part of the team helping the patient in the recovery.

Management for cognitive impairments includes neuropsychological assessment and memory assessments, which can give a record of functional and dysfunctional brain areas, and the treatment is tailored according to the patient's needs. Cognitive rehabilitation helps reduce the disability and improve the functions of the patient, and contingency management can be beneficial for those with severe cognitive impairment or poor motivation. Behavior therapy is recommended for behavioral and emotional problems like amotivation, irritability, and aggressiveness, whereas cognitive behavioral therapy aims at rectifying maladaptive thinking. Lifestyle modification helps patients engage in activities, keep their morale high and improve the QOL. Often the family is involved in the intervention to make caregivers understand the principles employed in the intervention to reduce the risk of unwanted behaviors. Generally, it involves the psychoeducation of the family members about the potential effects of expressed emotion and the importance of providing a non-confrontational supportive environment. Many outreach programs in the community aided by the government and non-governmental organizations, health workers, and run by the local authorities and volunteers are found to be effective in reducing disability where patients are retrained in acquiring living skills, vocational guidance, and employment opportunities.

15.3 Stroke

Stroke is defined as a sudden loss of blood supply to the brain leading to permanent tissue damage. The interruption of blood flow could be due to a blood clot, a ruptured artery or a blood vessel. The stroke could be ischemic (due to the absence of blood flow) or hemorrhagic (due to bleeding in or around the brain). While around one-third of the patients succumb, the remaining reports persisting multiple neurological problems. Emotional, behavioral, and cognitive impairments are the common neuropsychological complications of a stroke.

15.3.1 Neuropsychiatric Manifestations

Delirium Delirium is an acute disorder of consciousness and cognition, which is transient with fluctuating intensity. Delirium is frequently seen in many patients following a stroke, which resolves soon. It can also develop much later in stroke patients due to reasons like metabolic abnormalities, electrolyte imbalance, and infections. Patients with delirium can have disturbances in consciousness manifested by a reduced clarity of awareness about the surroundings and impaired attention. In addition to the above, the patients will also have memory deficits, disorientation, and language disturbances. Many can develop perceptual disturbances like illusions and hallucinations. Change in the normal circadian rhythm producing reversal of the sleep-wake cycle is also quite common in delirium.

Cognitive Impairments Cognitive impairments affect about three-fourths of stroke patients and often lead to permanent disability. Assessment and management of cognitive impairments are vital in the rehabilitation of stroke as they interfere with compliance and progress. Impairments in the cognitive domains include impairment in attention, executive functioning, memory, language, visuospatial abilities, aphasia, and anosognosia. A progressive decline in cognitive abilities is common among patients who develop dementia following a stroke. Vascular dementia is the second most common cause of dementia after Alzheimer's disease. About 20 to 30% develop dementia within 3 months following a stroke and its prevalence increases with age. The course may vary as some show partial improvement, while others show a stepwise deterioration.

Psychosis and Mania Post-stroke psychosis is rare and has been considered a complication in cerebral infarction occurring at a minimal rate. No clear relation between the lesion sites and the types of delusion are found in the psychosis cases reported following stroke. Unlike depression, mania after a stroke is rare and the prevalence rate is around 1%. Studies suggest that a lesion in the right fronto-temporal lobe is frequently associated with manic symptoms. Post-stroke manic patients exhibit the classic symptoms of pressured speech, sleep disturbances, motor agitation, and social and sexual disinhibition. Individuals with genetic vulnerability are more prone to develop post-stroke mania.

Post-stroke Depression The most distressing element of stroke is depression. Studies suggest a positive correlation between depression and the left frontal lobe lesions. It occurs in approximately one-fourth of the patients. It is advisable to look for symptoms in the acute stage of stroke as it often goes undetected as certain symptoms of depression could be the clinical features of the stroke itself. Depression has an impact on individual functioning, as the daily routine gets changed and the patient remains dependent on others for their ADL like eating, toileting, and ambulation that leads to irritability and negative automatic thoughts (loss of the previous role) and worries regarding executing the daily skills, make it more disabling. Grief is also commonly seen post-stroke, but the symptoms typically last for a short duration and are less pervasive (less than 6 weeks). Cognitive deficits or dementia and depression frequently co-exist.

Anxiety Anxiety affects almost one-fourth of the post-stroke patients and the most common conditions seen are phobia and GAD. Anxiety symptoms often increase the disability as the patient becomes less ambulant and does not make an effort to walk following the fear of falling. It also contributes to poor physical functioning and leads to decreased social interaction as they perceive that others may judge them. The anxious thoughts predominantly consist of anticipatory anxiety and the worries about the loss of premorbid level of functioning, uncertainty regarding the future, and financial burdens that compound misery.

Post-stroke Emotional Incontinence It includes catastrophic reaction and pseudobulbar syndrome. A catastrophic reaction is characterized by excessive anger, despair, frustration, denial, and tearfulness, which is often associated with post-stroke depression, where the patient goes into a paroxysm of anger and sobbing with accompanying gestures. The patient may refuse to carry on with any verbal explanation for this behavior. This condition is more often seen in patients who have a personal and family history of depression. The sobbing is usually short-lasting, whereas the negativism can persist for a considerably longer time. Patients with catastrophic reactions also show excessive anxiety, and many have shown an inability to control anger or aggression. Pseudobulbar palsy also sometimes produces emotional outbursts and it is observed in approximately 15% of the post-stroke patients. This presents with episodes of laughing and crying, often unprovoked. The symptoms often decline on their own over a period without any treatment.

Impairment of Self: Anosognosia and Aprosodia Anosognosia is a lack of ability to perceive the realities of one's condition. Anosognosia could be "perceptive," where the person will have a decreased awareness of impairment or "behavioral," where the person will not be aware of the behavior. It is important to identify this symptom as it interferes with the treatment, compliance, and rehabilitation processes. It is uncommon for anosognosia to persist beyond 3 months after stroke. The frequency of anosognosia is reported to vary from 20 to 40%. Aprosodia is a lack of emotional inflection that could be receptive or expressive (motor). The receptive aprosodia is an inability to recognize affective tone or speech often seen with right parietal lesions. Whereas expressive or motor aprosodia patients appear inexpressive, which could result in an impairment in communication.

Apathy Almost 11% of the post-stroke patients reports of apathy without depression. Apathy can be described as a lack of interest, feeling, and motivation, and about one-fourth of the post-stroke patients report apathy. Patients will have cognitive or emotional apathy or often both. Apathy is considered as one of the most disabling sequelae as it interferes with the physical and psychological rehabilitation process, leading to less improvement in physical ability and functional independence. Apathy is classically associated with frontal lesions, particularly in the dorsolateral prefrontal convexity, as well as in the basal ganglia, thalamus, and related networks. It can also be associated with old age and cognitive impairments. The extreme state of apathy like abulia and akinetic mutism is the one that is most difficult to manage by the caregiver.

Impulsiveness, Irritability, and Aggressiveness Aggressiveness is considered a defect of emotional regulation and induces predetermined acts of violence towards others. As with many behaviors, there are many factors involved in the development of impulsivity and aggression (social factors, prior exposure to violence, presence of mood disorders, medication, and history of alcohol use disorder). Along with many other factors, anger also acts as a risk factor for stroke, which might worsen after stroke. Post-stroke irritability is more common in young adults with left-sided strokes and also in patients with aphasia, and this condition might continue up to even 1 year after stroke.

Hyposexuality It is not uncommon for stroke patients to have a change in sexual behavior and the most common complaint is the lack of sexual desire. Apart from the lack of libido, erectile dysfunction and premature ejaculation are the other frequent sexual disorders seen following stroke and the etiology could be organic or psychosocial and at times, the side effects of the medications. Hypersexuality is a less frequent abnormal sexual behavior reported among stroke patients than hyposexuality. Overall, the change in sexual behavior leads to poor QOL, strained interpersonal relationships with spouse, low self-esteem, anxiety, and depression.

Post-stroke Fatigue Fatigue can be defined as a reversible decrease or loss of abilities associated with a heightened sensation of physical or mental strain, even without conspicuous effort. Due to the feeling of extreme exhaustion, the patient will find carrying on with routine ADL a struggle and is often a long-standing symptom. "Fatigue" could be objective (an observable decrease in the physical or mental tasks undertaken) as well as subjective (a feeling of exhaustion and a reluctance to put on an effort). At times post-stroke fatigue could be the only prominent sequela of stroke and almost 30–70% of patients complain of fatigue.

Hypersomnia According to some studies, the frequency of sleep disturbances ranges from 1.1 to 27% among post-stroke patients. The severity of sleep disturbances can vary from mild to severe, and the nature could be transient or permanent, which depends on the determinants like the location and extent of the damage. Studies indicate that the size of the lesion is a major determinant of hypersomnia.

15.3.2 Non-pharmacological Management

Psychological and behavioral techniques have only a limited role in the management of behavioral symptoms following stroke. Cognitive behavioral therapy has shown certain benefits as a mode of treatment for stroke patients to treat depression and anxiety. Treatment modalities like back massage and relaxation techniques can be helpful. Motivational interviewing and problem-solving therapy will act as a protective measure for depression and anxiety. Other psychological interventions, such as distress management, group support, and music therapy, have shown limited beneficial effects. Cognitive rehabilitation is recommended for cognitive deficits that improve certain domains of cognitive functioning, but the improvement cannot be generalized and applied in the other areas of cognitive functioning too.

15.4 Huntington's Disease

Huntington's disease (HD) is a hereditary, degenerative, disabling disorder caused by autosomal dominant mutation. The adult Huntington disorder patient becomes symptomatic in their 30s and 40s, but in many cases, symptoms begin during adulthood and progressively worsen until the affected person cannot live and function independently. If the symptoms find their expression before the age of 20, then it is called juvenile HD. The common symptoms of HD are uncontrollable movements, abnormality in balance while walking, slurred speech, difficulty in swallowing, thinking difficulties, dysregulation of emotions, and personality changes. HD also impairs the cognitive process and produces many psychiatric symptoms.

15.4.1 Neuropsychiatric Manifestations

Cognitive Impairments Typically patients will report impairment in divided attention and difficulty in learning and memory and slowness in processing speed. The neuropsychological profile of HD patients indicates that they can learn new information, but the rate of learning can be relatively slow. They find difficulty in retrieving information, and recall is found to be impaired. Psychiatric conditions can also contribute to cognitive changes. Sustaining attention and impulsivity, which manifest as behavioral symptoms like anger, outbursts, and sexual promiscuity, can also have problems in shifting their focus from one task to another, learning and memory, word-finding difficulty, and preservation. Many can lack insight for self-awareness, known as organic denial or anosognosia (inability to recognize their disabilities). Due to the lack of insight, many HD patients will neither seek medical help nor be willing for medication.

Affective and Psychotic Disturbances The prevalence rate of psychosis in HD is considerably low, as at times the psychotic symptoms get missed or masked because of their transient nature and antipsychotics which are prescribed for some other indications. Schizophrenic symptoms are reported very rarely, though delusions of persecution are common along with religiosity and grandiosity. Ideas or delusions of reference seem to become more pronounced as the involuntary movements become noticeable and attract attention.

Depression Depressive disorders are one of the common psychiatric conditions that occur in HD. The patients might experience depression during the prodromal stage or the course of their illness and the reasons could be either situational factors or organic variables and often both together. HD and depression share many common symptoms; hence it becomes challenging for the clinician to determine whether the symptoms are of depression, HD, or a combination of both. In the prodromal stage, the patient presents himself to the clinician with symptoms of depression, but in the advanced stages of HD, patients will be brought by the caregiver as the patient might not be aware of his symptoms. The rate of suicide is high in HD ranging from 0.6 to 10.1%. Most of the time, impulsivity and anger act as risk factors for suicide.

Anxiety Disorders Like depressive disorders, anxiety disorders can also present at any clinical stage of HD. The patient can have symptoms of panic disorder, social anxiety, or PTSD. Studies indicate that anxiety co-exists with depression, and the prevalence rate is found to be more among women. The anxiety episodes often get triggered by cognitive factors, physical factors, behavioral disturbances, and environmental factors and worsen as the disease progresses.

Apathy It is the most widely recognized symptom that influences the patient to the maximum and leads to an expanded rate of disability during the progression of the disease. Apathy can manifest alone or can present along with depression. The patients will lack an emotional investment in activities and surroundings; they might appear to be disengaged with difficulty in initiating behaviors or activities, characterized by "hard to get started" and sluggishness. Apathy can be hard to distinguish from depression, and apathetic patients need not be essentially depressed.

Sleep Disorder The circadian rhythm often gets disturbed in the early stage of HD. The commonly seen symptoms are difficulty in falling asleep, disturbed sleep, increase in the daytime nap, and early morning awakening. As the disease progresses, the sleep cycle gets worsened, where anxiety and depression play a precipitating role in sleep disturbances. Poor sleep impacts the patients' socio-occupational life and they report fatigue, drowsiness, irritability, and apathy. Excessive worries about sleep disturbances can further compromise the overall QOL.

Changes in Sexual Behavior Many patients report sexual disturbances followed by HD, where depression or the medication prescribed for depression can also play a precipitating and maintaining role for sexual disturbances apart from the disease itself. The predominant sexual disturbances are hyposexuality and inhibited orgasm in both men and women. This condition often remains untreated because of the taboo and the patients become reluctant to discuss the issues. Some patients also report sexual assault, promiscuity, and disinhibition, such as inappropriate sexual remarks and exhibitionism, which are uncommon.

Anger, Agitation, and Irritability Anger, agitation, and irritability are the commonly manifested behaviors among HD patients and worsen as the disease progresses and become challenging for the individual to carry on with social interaction. The behavior gets worsened as the disease alters the person's thoughts and feelings. Studies show that irritability is linked to aggression and impulsivity. Psychiatric conditions and comorbid medical conditions along with psychological and social factors play a prominent role in the exacerbation of anger episodes. Aggressive behaviors can be particularly distressing and troublesome to the primary caregivers because such behavior actuates fear and irritability in the caregiver, influencing the management and worsening the burden of care.

15.4.2 Non-pharmacological Management

The cognitive and behavioral changes in HD are most disabling and distressing for both the patients and the primary caregivers. Non-pharmacological treatments like cognitive behavioral therapy, support groups, coping strategies, peer support, and lifestyle modification adjunct to pharmacological treatment are found to be successful in treating depression, anger, anxiety, insomnia, and behavioral problems. All the abovementioned modalities can greatly improve a person's QOL. It is crucial that the treatment also focuses on the management of the caregiver's issues.

15.5 Parkinson's Disease

It is a neurodegenerative disorder with movement abnormalities, such as resting tremors, slowness, rigidity, loss of balance, dystonia, anosmia, and difficulty with speech and writing. The clinical symptoms of Parkinson's disease (PD) often get complicated by comorbid psychiatric syndromes and by the behavioral side effects of antiparkinson medications.

15.5.1 Neuropsychiatric Manifestations

Psychosis Psychotic symptoms are extremely rare in untreated PD, whereas psychosis is a common sequel seen in Parkinson's following the treatment with levodopa, dopamine agonists, and anticholinergic drugs. Hallucinations ranging from mild to severe in various modalities, visual being the most common, are reported. Delusions, mostly persecutory, always accompany hallucinations. Patients with psychotic symptoms warrant clinical attention to reduce the morbidity and will require pharmacotherapy.

Depression Depression is the commonest psychiatric presentation of Parkinson's, occurring in about 40–50% of the patients at one time or another throughout the disease. Depression could be either a part of the disease process itself or can develop as a response to the psychosocial and physical burden of the chronic debilitating disease. The diagnosis of depression becomes challenging in PD as some symptoms overlap with the Parkinson's motor symptoms like bradykinesia and masked faces (hypomimia). Studies report that PD patients often complain of dysphoria, pessimism, somatic symptoms, anhedonia (reduced interest in activities an individual used to enjoy before with decreased ability to feel pleasure), reduced reactivity to external stimuli, emptiness, and hopelessness. However, symptoms like mood swings and guilt are rare and the prevalence of suicide is low.

Anxiety Anxiety disorders are less studied in PD compared to depression. Studies suggest that approximately 40% of PD patients experience anxiety during illness, which frequently co-occurs with depression. Patients may report GAD, panic disorder, and social phobia, with a prevalence rate of about 30%. PD patients may experience anxiety and anticipatory anxiety due to their physical disabilities (movement abnormalities that can lead to falls) and the fear of negative evaluation by others due to the motor impairments that impact and restrict their social lives. These factors also influence the patient's overall function and induce an immense burden on the family members due to the patient's dependency on the caregivers.

Sleep Disturbance Sleep disturbances are common in PD and approximately three-fourths of the patients report the same. Unlike depression, sleep disturbances are not much researched. Disturbance in the maintenance of sleep is the most common complaint reported. Delay in sleep initiation, daytime sleepiness, sleep apnea, restless leg syndrome, and rapid eye movement behavior disorders are common issues. In PD patients, sleep disorders can also be due to old age or comorbid medical and psychiatric conditions. As the disease advances, sleep disturbances can impair the overall socio-occupational life of the patient. Antiparkinson drugs also can cause insomnia; hence they should be avoided at bedtime. Disturbed sleep or poor quality of sleep is known to impair cognitive functions like attention and judgment.

Fatigue Patients with PD commonly report fatigue. Physical and mental fatigue interferes with movement and ambulation and can also impair the cognitive process. Apart from PD, psychiatric comorbidities like depression and anxiety can further worsen fatigue. Due to the physical disabilities caused by the movement abnormalities, patients might find even completing simple motor tasks or daily activities tiresome, worsening the QOL.

Personality Changes Personality changes are not so evident in PD and the changes observed often depend on contributing factors like stress, psychological vulnerability, low self-esteem, and the environment. The personality changes seen include irritability, suspiciousness, egocentricity, obsessive traits, and hypochondriasis. The suspiciousness seen as a part of the personality changes has to be differentiated from delusions and can affect physical rehabilitation treatment as the patient might not trust the treatment team. Obsessive traits might get exaggerated if the patient had a premorbid personality with obsessive traits. Some also report impulsivity or compulsive behaviors like hoarding, gambling, and obsessive sexual thoughts.

Cognitive Impairment The cognitive functioning of patients with PD varies as some remain intellectually intact despite gross physical disablement. PD patients can have mild to severe cognitive impairment and even dementia. Prominent impairment is seen in the areas of divided attention, which is also accompanied by distraction where the patient finds difficulty paying attention to two tasks at the same time, like having an inability to focus attention in a group conversation. The impairment in memory makes it difficult for the patient to remember the information and in planning and executing tasks. They struggle to make decisions and feel overwhelmed due to anxiety. The cognitive changes in behavior can interfere with the patient's day-to-day functioning and hamper the QOL of both the patient and his family members. The brain changes could also contribute towards the motor symptoms that could manifest as slowness in memory and thinking. Medications like anticholinergics might also further worsen these deficits. Although 50% of the PD patients report cognitive disturbances, not all patients diagnosed with PD will end up with dementia.

15.5.2 Non-pharmacological Management

A clinical interview and a reliable history from the caregivers about the patient are mandatory. Baseline assessment includes the use of Mini-mental state examination (MMSE) or the Montreal cognitive assessment (MoCA) for screening. A review of the medication is an absolute necessity to rule out the possibility of iatrogenic etiology of the symptoms and consider the change in pharmacotherapy. Non-pharmacological treatment might help take the patient off the dopamine agonists that are given primarily to treat motor symptoms, which are likely to cause compulsive and impulsive behavior. Cognitive remediation therapy emphasizes teaching alternative ways to compensate for cognitive deficits and is found to be useful in patients with PD.

15.6 Epilepsy

Seizures are characterized by a sudden, excessive, disorderly electrical discharge of brain tissues producing physical, cognitive, and behavioral symptoms, while epilepsy is characterized by recurrent unprovoked seizures. The International League Against Epilepsy (ILAE) 2017 classifies epilepsy broadly into focal onset, generalized onset, and unknown onset (either motor or non-motor). Focal onset epilepsy could be either with or without impaired awareness. One-third of the epileptic patients undergoing neurosurgery report psychiatric symptoms. The behavioral disorders associated with epilepsy can express any time during the seizures (pre-ictal, ictal, post-ictal, and inter-ictal stages).

15.6.1 Neuropsychiatric Manifestations

Psychosis Schizophrenia, like psychosis during the inter-ictal phase, has been reported in patients who have chronic epilepsy. Earlier studies have reported a higher prevalence of schizophrenia-like psychosis in epilepsy, but Qin P et al. in 2005 found that the relative risk of schizophrenia in persons with epilepsy was 2.48 (95% CI 2.20–2.80) whereas the relative risk for non-affective psychosis was 2.93

(95% CI 2.69–3.20). Patients with chronic epilepsy are two to three times more likely to develop schizophrenia than people without the disorder.

Depression Depression is the most frequently reported condition in epilepsy and almost 30–40% of patients with epilepsy report clinical depression. The prevalence of depression is found to be much higher in those patients who require neurosurgery for epilepsy. Depression is often correlated to the types of seizures. Usually, depression remains untreated and unrecognized in patients with epilepsy and depression often leads to a poor QOL. The suicide rate is higher in epilepsy and the probability is much higher in temporal lobe epilepsy.

Sexual Disturbances Disturbances in sexual functioning are often reported among patients with epilepsy, where the men report erectile dysfunction and women report arousal and orgasmic dysfunction. Though hyposexuality is the most common, hypersexuality, deviation in sexual interest such as fetishism and transvestism are also reported. Psychosocial factors play a vital role as causal and maintaining factors, apart from the neurophysiological changes caused by epilepsy and antiepileptic drugs. Sexual dysfunction can also precipitate strained interpersonal relationships with the spouse, poor self-esteem, low self-confidence, and the fear of having an epileptic attack during sex.

Personality Changes Chronic uncontrolled epilepsy has a lasting effect on the behavior and the personality of the patients and the most common features are irresponsibility, impulsivity, emotional instability, poor frustration tolerance, and self-centeredness along with mild cognitive impairments. Wendy S. Matthews and Gabor Barabas found that a poor sense of mastery of skills in adults with epilepsy was associated with anxiety, poor self-esteem, feelings of helplessness, and an increased risk of suicide.

Cognitive Changes The cognitive changes that commonly occur in epilepsy are attention impairment, memory deficits, and mental slowness. The cognitive profile varies according to the seizure focus. Patients with temporal lobe epilepsy may have deficits in the domains of episodic memory, whereas patients with frontal lobe epilepsy may demonstrate impairment in psychomotor speed, attention, working memory, response inhibition, and planning.

15.6.2 Non-pharmacological Management

The first and foremost principle in the management of behavioral disorders in epilepsy is to control the seizures using antiepileptics. Non-pharmacological management includes the EEG biofeedback technique, which reduces the frequency of seizures, whereas progressive muscle relaxation may reduce anxiety. Cognitive behavioral therapy is a choice of treatment to address "seizure phobia," a term used to describe the fear of having further episodes of seizures. Negative automatic thoughts and schemas about seizures are challenged, and the patients are encouraged not to be engaged in risky and avoidable behaviors. Being diagnosed with epilepsy itself can act as a stress factor and the negative cognitions related to it have to be addressed. Psychotherapy can play a major role in addressing the real and perceived stigma, thereby facilitating an emphatic relationship with the patients. The helplessness and frustration of a restricted lifestyle like avoiding driving, swimming, or standing near the balcony or near high altitude areas have to be managed. Psychosocial stressors (parental and societal rejection, negative attitudes, stigma prevailing in the family, overprotected parents, coping with the limitation) are addressed in the counseling sessions. Patients often benefit from learning coping skills and problem-solving abilities. Psychoeducation and counseling will be adequate to address the distress of some families, whereas others may need more structured and focused therapies.

15.7 Dementia

Dementia is a clinical syndrome characterized by intellectual decline with age. It is often gradual, persistent, progressive, and irreversible. Dementia is caused by the damage of brain cells due to the multiple reasons and differences that exist in its clinical presentation, course, and outcome depending upon the etiology. There is a danger that it can go undiagnosed in the initial stages. The neuroanatomical correlates behind intellectual deficits in dementias are yet to be clearly understood. The "subcortical dementia" is characterized by memory loss, inability to utilize acquired knowledge (i.e., calculating and abstracting ability), personality changes like apathy and inertia, slowness in thinking processes, and severe motor abnormalities. Though the cortical dementias have many features of subcortical dementias, they are differentiated from the latter due to the presence of specific cognitive deficits like aphasia, apraxia, and agnosia. However, reports have shown that cortical lesions may be important in the pathogenesis of subcortical dementia and subcortical lesions may be significant in cortical dementia.

Dementia can be seen in various neurological disorders, including Alzheimer's disease, fronto-temporal dementia, Lewy body disease, vascular dementia and dementia due to substance abuse and medication use, CNS involvement due to HIV infection, PD, prion disease, and HD. The age of onset, clinical features, course, and prognosis of dementia depends upon the etiology.

The hallmark of dementia is intellectual decline, which includes difficulty in decision making, loss of recent memory (complaining that they have not taken food and requesting to have it again), forgetfulness, and impairment in goal-directed behavior. Deterioration in memory and affective changes are prominent. Neurological symptoms like aphasia, apraxia, visuospatial dysfunction, and executive cognitive dysfunction are also seen in dementia.

15.7.1 Neuropsychiatric Manifestations

Depression Depression is the most common psychiatric disorder reported in dementia (9–68%). Both the conditions share some common features like cognitive deficits, sleep disturbances, and poor socio-occupational functioning. Patients with dementia find it difficult to verbalize the depressive symptoms and it is often underreported as it gets masked by dementia. The prevalence of depression increases as the disease advances and the cognitive deficits worsen. Depression can also act as a risk factor for dementia.

Psychosis Hallucinations and delusions usually occur in the advanced stages of the condition where the nature of the delusions is less complex and organized, and the content would be more of suspiciousness, abandonment, and misidentification (like patients complaining of someone coming to harm him, calling a known person as an imposter, someone is hiding in the home, suspecting the spouse on fidelity). When they are associated with depression, patients can have guilt and suicidal ideas. Though hallucinations in all modalities are reported in dementia, they are not as frequent as delusions except in Lewy body dementia, where visual hallucinations are common.

Sleep Patients with dementia often report sleep disturbances and it worsens as the disease progresses. Patients complain of insomnia, disturbances in the sleep-wake cycle, fragmented sleep, and rapid eye movement sleep behavior disorder. Many complain of daytime sleepiness and nighttime awakening. The contributing factors could be a daytime nap, increased caffeine and water intake, lack of physical activity, urinating at night, and medications.

Behavioral Problems As the disease progresses, the behavioral and psychological symptoms start worsening with further deterioration in cognitive impairments and also with psychiatric comorbidities. Complaints like agitation, disinhibition (sexual), apathy, wandering behavior, restlessness, irritability, perseverative behavior, and stereotypic or compulsive/ritualistic behavior, are also seen in certain dementias. Such behaviors often pose a risk to a patient's life and burden and anxiety to the family members. Patients will need continuous supervision and become entirely dependent on the caregiver.

15.7.2 Non-pharmacological Management

The patient's mental state examination can be assessed by the MMSE and General Practitioner Assessment of Cognition (GPCOG) scale. The presence of psychotic symptoms, depression, or anxiety should also be elicited by the practitioners by taking an in-depth clinical history. The patient's level of insight should also be explored, as a lack of insight can interfere with any kind of intervention for dementia and comorbid conditions.

In the early stage, the patients can be managed at home with the support of caregivers and community support groups. When the condition worsens, these patients might require residential care or a day-care center facility, and sometimes a day hospital might also be needed to provide respite and support. Psychosocial intervention is warranted as pharmacotherapy has only a limited effect on improving the ADL. Psychological treatment modalities like reminiscence therapy, reality orientation, multi-sensory stimulation, Snoezelen therapy, memory training, music therapy, aromatherapy, massage, light exercise, activity programs, distraction techniques, pet and doll therapy can be of some help.

Family members managing the ADL of the affected member should be psychoeducated regarding communication, dressing, bathing, eating, physical exercise, incontinence, disturbance in sleep, wandering behavior, agitation, falls, and sensory impairment. Supportive therapy addresses stress, anxiety, depression, stigma, and perceived burden among caregivers. Online forums like dementia helplines and many non-governmental organizations can provide a round-the-clock helpline for Alzheimer's patients and family members. These organizations are dedicated to providing care, support, and research in dementia across the countries.

15.8 Brain Tumors

The intracranial tumor is a mass of tissue due to abnormal cell growth, which could be either benign or malignant. The presence and types of physical or behavioral symptoms due to tumors will depend on the stage, progression, and location of the tumors. Many also develop behavioral symptoms along with the neurological and cognitive deficits as sequel of chemotherapy, radiotherapy, or surgical management of the tumors.

Cognitive deficits and behavioral disturbances can be one of the complications of neurosurgery for tumors, and the extent of deficits depends on the location and area involved. Radiation or chemotherapy and some medications to prevent seizures (anticonvulsants) can also induce cognitive and behavioral side effects. In many cases, the earlier cognitive functions get worsened, or patients develop deficits in attention, memory, executive functioning, and difficulty in information processing. Patients may also experience depression and agitation.

15.8.1 Neuropsychiatric Manifestations

Cognitive Deficits The prevalence of cognitive deficits among patients diagnosed with intracranial tumors is reported to be between 60 and 90%. The cognitive deficits include memory deficits, confusion, impaired executive functioning, decreased attention span or concentration, organizational inability, poor arithmetic or language skills, motor incoordination, and difficulty in new learning and information processing.

Psychosis Psychosis is often a common feature of brain tumors and is often characterized by hallucinations. About 22% of the patients with intracranial tumors have reported psychotic symptoms. The presence of tumors in the cortical, pituitary, pineal, and posterior locations is more associated with psychotic symptoms, whereas a few have reported psychotic manifestations with temporal lobe tumors.

Mood Disorder, Depression, and Apathy About 2.5–15.4% of the patients with brain tumors are found to have depression and are most commonly seen with left frontal tumors. According to Mainio A et al., patients with 44% of primary and metastatic tumors reported depressive symptoms which were more common in frontal lobe tumors. The patients also reported cognitive dysfunction, functional impairment, and a poor QOL, resulting in reduced survival. In addition to depression, the patients with right frontal tumors can even present with mania.

Personality Changes and Apathy Some tumors can lead to personality changes like sexual disturbances (hypersexuality), disinhibited behavior and aggression. Apathy reported in tumors independent of depression could also lead to a dependency on others for the ADL. Patients with apathy will have chronic fatigue, amotivation, a lack of interest in learning new tasks, a lack of concern concerning oneself, and a lack of emotional responsivity.

15.8.2 Neurobehavioral Changes Following Treatment

The most common personality disturbances following intracranial surgery for tumors include irritability, impulsivity, moodiness, inflexibility, and being easily overwhelmed. It is difficult to comment on post-surgery psychological complications as most of the time, pre-surgery cognitive testing is not being conducted. Hence it becomes challenging to comment on the post-surgical cognitive changes.

Studies indicate that deficits often get masked by motor and speech deficits. Many patients with glioma reported cognitive impairments following neurosurgery but most of them were transient. Though the cognitive functioning showed a significant improvement following surgery, many did not reach the premorbid level. More than half of the patients who received radiation therapy have reported transient cognitive deficits. Chemotherapy is also found to be associated with cognitive impairments, irritability, depression, anxiety and in rare cases, even psychosis and mania.

15.8.3 Non-pharmacological Management

Patients with tumors or who underwent surgical treatment often get benefited from supportive therapy and cognitive behavioral therapy as an adjunct to pharmacotherapy. Behavior interventions like relaxation programs and systematic desensitization help in decreasing the anxiety. Neuropsychological assessment and mental status examination can help to chart the deficits and retaining can be planned accordingly. Computer-based cognitive rehabilitation therapy can also address the cognitive retaining aspects.

15.9 Psychopharmacology in Neurological Conditions

Though the use of psychotropics in the management of neurobehavioral manifestations of neurological conditions is often a necessity, a judicious selection of the drug and regular monitoring has to be performed due to the underlying brain pathology and altered pharmacokinetics and pharmacodynamics. These patients are more prone to adverse drug effects, idiosyncratic drug reactions, and a higher incidence of drug interactions with the medications they will be receiving for the primary neurological conditions and other medical comorbidities. The dictum of "start low, go slow" for the elderly in pharmacotherapy will also have to be followed in conditions like stroke and should be regularly monitored for side effects.

Atypical antipsychotics like olanzapine and quetiapine are preferred over typical antipsychotics like haloperidol and trifluoperazine, as they have less propensity to produce extrapyramidal side effects, movement disorders, and anticholinergic side effects for the management of psychotic symptoms. But the higher incidences of metabolic syndrome and cardiac complications, which are more common with atypical antipsychotics, have to be taken into consideration while choosing atypical antipsychotics. Increased risk of falling is high among patients on antipsychotics and higher among patients on sedatives and hypnotics. This could be either due to sedation or postural hypotension, which is a common side effect of many antipsychotics. Though a wide range of anti-depressants is available with various modes and mechanisms of action, escitalopram and sertraline are preferred as they have fewer adverse effects and drug interactions. Hyponatremia is often reported with sertraline when used in medically ill patients.

While using mood stabilizers (lithium, valproate, oxcarbazepine, and carbamazepine), regular monitoring has to be done for blood dyscrasias and other renal, hepatic and endocranial abnormalities depending upon the drug. Valproate is known to produce hyperammonemia, which can cause cognitive impairment and delirium. The long-term use of benzodiazepines (alprazolam, clonazepam, lorazepam, oxazepam, and diazepam) and sedatives like zolpidem and zopiclone are not advisable as they are known to produce cognitive deficits and dependence. Anticholinergic (trihexyphenidyl, procyclidine, benztropine, and diphenhydramine) drugs have to be preferably avoided as they can impair cognitive functions and have been known to induce delirium when administered in higher doses.

Further Reading

Anderson KE, van Duijn E, Craufurd D, Drazinic C, Edmondson M, Goodman N, et al. Clinical management of neuropsychiatric symptoms of Huntington disease: expert-based consensus guidelines on agitation, anxiety, apathy, psychosis and sleep disorders. *J Huntingtons Dis.* 2018;7(3):355–66.

Barrowclough B. Suicide and epilepsy. In Reynolds FH, Trimble MR, editors. Epilepsy and psychiatry. Edinburgh: Churchill-Livingstone; 1981. p72–6.

Blumer D. Postictal depression: significance for the neurobehavioral disorder of epilepsy. *J Epilepsy.* 1992;5:214–9.

Carota A, Rossetti AO, Karapanayiotides T, Bogousslavsky J. Catastrophic reaction in acute stroke: a reflex behavior in aphasic patients. *Neurology.* 2001 Nov;57(10):1902–5.

Chen JJ, Marsh L. Anxiety in Parkinson's disease: identification and management. *Ther Adv Neurol Disord.* 2014;7(1):52–9.

Chun HY, Whiteley WN, Dennis MS, Mead GE, Carson AJ. Anxiety after stroke: the importance of subtyping. *Stroke.* 2018;49(3):556–64.

Cummings JL. Frontal-subcortical circuits and human behavior.*Arch Neurol* 1993;50:873–80.

David AS, Kopelman MD, Lovestone S, Mellers JDC, editors. Lishman's Organic Psychiatry: A textbook of neuropsychiatry. 4th ed.West Sussex, UK: Blackwell publishing Ltd; 2009.

Deb S, Lyons I, Koutzoukis C, Ali I, McCarthy G. Rate of psychiatric illness 1 year after traumatic brain injury. *Am J Psychiatry.* 1999 Mar;156(3):374–8.

Duong S, Patel T, Chang F. Dementia: What pharmacists need to know. *Can Pharm J (Ott).* 2017;150(2):118–129.

Exner C, Boucsein K, Lange C, Winter H, Weniger G, Steinhoff BJ, et al. Neuropsychological performance in frontal lobe epilepsy. *Seizure.* 2002 Jan;11(1):20–32.

Fiedorowicz JG, Mills JA, Ruggle A, Langbehn D, Paulsen JS. Suicidal behavior in prodromal Huntington disease. *Neurodegener Dis.* 2011;8(6):483–90.

Filley CM, Kleinschmidt-DeMasters BK. Neurobehavioral presentations of brain neoplasms. *West J Med.*1995;163:19–25.

Gehrke AK, Baisley MC, Sonck AL, Wronski SL, Feuerstein M. Neurocognitive deficits following primary brain tumor treatment: Systematic review of a decade of comparative studies. *J Neuro Oncol.* 2013;115:135–42.

Godefroy O, Bogousslavsky J, editors. The behavioral and cognitive neurology of stroke. first south Asian ed. New Delhi, India: Cambridge University Press; 2008.

Helmstaedter C, Kemper B, Elger CE. Neuropsychological aspects of frontal lobe epilepsy. *Neuropsychologia.* 1996 May;34(5):399–406.

Hermann BP, Seidenberg M, Schoenfeld J, Davies K. Neuropsychological characteristics of the syndrome of mesial temporal lobe epilepsy. *Arch Neurol.* 1997 Apr;54(4):369–76.

Huffman J, Stern TA. Acute psychiatric manifestations of stroke: a clinical case conference. *Psychosomatics.* 2003 Jan–Feb;44(1):65–75.

Jung J, Kam S, Choi S, Jae S, Lee S, Hyun J. Sexual dysfunction in male stroke patients: correlation between brain lesions and sexual function. *Urology.* 2008;71(1):99–103.

Kanner AM. Depression in epilepsy: prevalence, clinical semiology, pathogenic mechanisms, and treatment. *Biol Psychiatry.* 2003;54:388–98.

LyketsosCG, KozauerN, Rabins PV. Psychiatric manifestations of neurologic disease: where are we headed? *Dialogues Clin Neurosci.* 2007 Jun;9(2):111–24.

Madhusoodanan S, Opler MG, Moise D, Gordon J, Danan DM, Sinha A, et al. Brain tumor location and psychiatric symptoms: is there any association? a meta-analysis of published case studies. *Expert Rev Neurother.* 2010;10:1529–36.

Mainio A, Hakko H, Timonen M, Niemelä A, Koivukangas J, Räsänen P. Depression in relation to survival among neurosurgical patients with a primary brain tumor: a 5-year follow-up study. *Neurosurgery.* 2005;56: 1234–41.

Matthews WS, Barabas G. Suicide and epilepsy: a review of the literature. Psychosomatics. 1981 Jun;22(6):515–24.

Menza M, Dobkin RD, Marin H, Bienfait K. Sleep disturbances in Parkinson's disease. *Mov Disord.* 2010;25(Suppl 1):117–22.

Nuti A, Ceravolo R, Piccinni A, Dell'Agnello G, Bellini G, Gambaccini G, et al. Psychiatric comorbidity in a population of Parkinson's disease patients. *Eur J Neurol.* 2004 May;11(5):315–20.

Qin P, Xu H, Laursen TM, Vestergaard M, Mortensen PB. Risk for schizophrenia and schizophrenia-like psychosis among patients with epilepsy: population based cohort study. *BMJ.* 2005 Jul 2;331(7507):23.

Rongve A, Boeve BF, Aarsland D. Frequency and correlates of caregiver-reported sleep disturbances in a sample of persons with early dementia. *J Am Geriatr Soc.* 2010 Mar;58(3):480–6.

Rooney AG, Carson A, Grant R. Depression in cerebral glioma patients: a systematic review of observational studies. *J Natl Cancer Inst.* 2011;103: 61–76.

Schwarzbold M, Diaz A, Martins ET, Rufino A, Amante LN, Thais ME, et al. Psychiatric disorders and traumatic brain injury. *Neuropsychiatr Dis Treat.* 2008 Aug; 4(4): 797–816.

Starkstein SE, Fedoroff JP, Price TR, Leiguarda R, Robinson RG. Apathy following cerebrovascular lesions. *Stroke.* 1993 Nov;24(11):1625–30.

Stein MB, Heuser IJ, Juncos JL, Uhde TW. Anxiety disorders in patients with Parkinson's disease. *Am J Psychiatry.* 1990 Feb;147(2):217–20.

Surma-aho O, Niemelä M, Vilkki J, Kouri M, Brander A, Salonen O, et al. Adverse long-term effects of brain radiotherapy in adult low-grade glioma patients. *Neurology.* 2001;56:1285–90.

Upton D, Thompson PJ. General neuropsychological characteristics of frontal lobe epilepsy. *Epilepsy Research.*1996;23:169–77.

Vázquez A, Jiménez-Jiménez FJ, García-Ruiz P, García-Urra D. "Panic attacks" in Parkinson's disease. A long-term complication of levodopa therapy. *ActaNeurol Scand.* 1993 Jan;87(1):14–8.

Warden DL, Labbate LA, Salazar AM, Nelson R, Sheley E, Staudenmeier J, et al. Posttraumatic stress disorder in patients with traumatic brain injury and amnesia for the event? *J Neuropsychiatry Clin Neurosci.* 1997 Winter;9(1):18–22.

Weitzner MA, Kanfer S, Booth-Jones M. Apathy, and pituitary disease: it has nothing to do with depression. *J Neuropsychiatry Clin Neurosci* 2005;17:159–66.

Wellisch DK, Kaleita TA, Freeman D, Cloughesy T, Goldman J. Predicting major depression in brain tumor patients. *Psychooncology.*2002;11:230–8.

Williams WH, Evans JJ, Wilson BA, Needham P. Brief report: prevalence of post-traumatic stress disorder symptoms after severe traumatic brain injury in a representative community sample. *Brain Inj.* 2002 Aug;16(8):673–9.

Chapter 16
Gadgets and Technologies in Adult Neurological Physiotherapy

Amreen Mahmood, B. Unnikrishnan, Abraham M. Joshua, and Sanjay Parmar

16.1 Introduction

Adult neurological physical rehabilitation has been a fast-evolving field with recent advances and changing evidence. There has been an upsurge in the use of technology for motor rehabilitation and the rehabilitation of people with neurological conditions has witnessed a paradigm shift in the last two decades. There have been panel discussions and interviews with experts asking if technology is going to replace physical and occupational therapists. This chapter explores the core concepts of technology in rehabilitation, the distinct types of technology used in neurological rehabilitation, and their applications.

Neurological conditions lead to persistent disabilities and require an intensive exercise program. In the current situation, both the high-income and the low- and middle-income countries are facing a challenge to provide intensive rehabilitation to people with neurological disabilities. The lower-income countries do not have adequate infrastructure and access to structured facilities. The rehabilitation

A. Mahmood (✉) · A. M. Joshua
Department of Physiotherapy, Kasturba Medical College of Mangalore, Manipal Academy of Higher Education, Manipal, India
e-mail: amreen.mahmood@manipal.edu; abraham.joshua@manipal.edu

B. Unnikrishnan
Department of Community Medicine, Kasturba Medical College of Mangalore, Manipal Academy of Higher Education, Mangalore, India
e-mail: unnikrishnan.b@manipal.edu

S. Parmar
SDM College of Physiotherapy, SDM University, Dharwad, India

A. M. Joshua (ed.), *Physiotherapy for Adult Neurological Conditions*,
https://doi.org/10.1007/978-981-19-0209-3_16

pathway for patients from lower-income countries consists of inpatient exercises, followed by a prescription of a home exercise program. In some cases, the patients opt for supervised exercises delivered by therapists in their homes. Home-based rehabilitation is also discontinued due to a lack of finances or lack of further improvement in patients' outcomes. Similarly, even in high-income countries, very few regions have access to focused rehabilitation centers, adequate workforce, or planning to deliver intensive programs. A systematic review reported that people with stroke receive an average of 22 min of exercise per day. Moreover, the duration of exercise sessions is inversely proportional to the time since stroke, that is, people with stroke exercise for a lesser duration in their chronic stages compared to the acute and subacute stages.

Both animal and human studies have proved that exercise induces plastic changes in the brain and thereby improves the cognitive and motor functions. Aerobic exercises increase the neurotrophic and growth factors that support structural and functional changes in the gray and white matter of the central nervous system after a neurological insult. It has also been found that higher-intensity exercises improve brain-derived neurotrophic factor levels when compared to lower-intensity exercises. The high-intensity exercises of more than 6 h daily for 5 days a week have been found to enhance motor recovery compared to the low-intensity exercises delivered for 1 h per day for 3–5 days a week, in all stages of stroke recovery. Evidence also supports higher repetition for promoting true recovery and motor functions in stroke and Spinal Cord Injury (SCI). Likewise, rigorous rehabilitation is associated with better recovery in other neurological disabilities such as Traumatic Brain Injury (TBI) and Parkinson's Disease (PD). Despite the proven benefits of intensive exercises in promoting motor recovery, it is practically difficult to deliver such a rehabilitation program by using only conventional rehabilitation methods.

The use of technologies has received a lot of attention in the recent past for rehabilitating patients with chronic neurological illnesses. Technology has the potential to offer intensive rehabilitation, which is not otherwise practical in clinical scenarios. The use of technology in rehabilitation can deliver high-intensity repetitive movements for weak arms or legs after any neurological illness. It can provide rigorous task practice without much human effort. The use of technology for rehabilitation can also assist in the accurate and objective evaluation of patients' conditions and deliver exercises in a controlled and graded manner. Technology involving gaming and simulation of real-life scenarios can make rehabilitation sessions more fun, engaging, meaningful, and challenging. However, for the success of any technology-based rehabilitation, it must be affordable, feasible, and appropriate and should show a measurable change in the patient's condition. Due to the apparent benefits of technology, there has been extensive research on technology-based rehabilitation in neurological conditions. Following are some of the technologies and gadgets that have been used for improving motor outcomes in neurological physical therapy.

16.2 Virtual Reality

The history of virtual reality dates to 1957, when Sensorama the first virtual reality system was invented and patented in 1962. It was developed for delivering the multi-sensory theater and was considered to be the future of cinemas. In 1968, the first head-mounted device was developed by Professor Ivan Sutherland, an American computer scientist, along with his student. It was in the 1980s that the term 'virtual reality' gained popularity. Virtual Reality (VR) was used in military training in the USA and for simulations in the National Aeronautics and Space Administration. The mass production of VR devices started in the 1990s. Currently, several tech companies produce diverse types of VR systems such as personal computer-tethered, mobile-tethered, and non-tethered systems for both gaming and serious games.

VR refers to a simulated environment generated by advanced technology in which the participant interacts with the virtual objects and environment. The different types of sensory feedback that are used in VR systems include:

- Visual
- Auditory
- Optokinetic
- Haptic or touch

The core concept of VR is the feeling of the presence of oneself in the virtual environment and the extent of interaction between oneself and the simulated environment. The attributes of presence include–a sense of being present, a focus on a virtual environment compared to a real environment, and the memory of a person visiting the simulated location. There has been enough evidence stating that the greater presence of a person in a virtual scenario leads to a better performance of virtual activities and more clinical effectiveness. Multiple factors influence the performance of an individual in VR systems. The VR system-related factors include the amount and quality of sensory feedback, the use of 2D or 3D technology, and the number of wearable sensors that restrict movement. The individual-related factors include age, gender, previous experience with VR and disability. The activity-related factors include the relevance or meaningfulness of the task and whether it is realistic and credible.

Another core concept of VR is the sense of immersion in the virtual environment. Immersion is the degree to which a person focuses attention on the virtual environment over the real environment. Immersion of any VR device depends on the quality, resolution, and speed of the system. Earlier, VR systems were classified as immersive or non-immersive. Current literature supports the concept of immersion as a continuum that is the VR systems vary from being less immersive to more immersive. However, the more immersive systems do not guarantee better performance or more user engagement.

A common complication encountered during VR sessions is cybersickness. Many people have experienced nausea, vomiting, headache, dizziness, and blurred vision during the VR sessions which can last up to 12 h after the session. In some cases, people have experienced imbalance, eye–hand incoordination, and sleep disturbances. Cybersickness occurs due to the complex and discordant multi-sensory information received by the person. Such side effects are often more reported for head-mounted devices such as Oculus Rift and HTC Vive units compared to other devices.

16.2.1 Instrumentation of Virtual Reality Devices

The design and the instrumentation of VR systems depend on the goals of rehabilitation and the needs of the user. However, all VR devices consist of input and output tools, as shown in Fig. 16.1. The input tools transmit information from the user to the system, while the output tools deliver information from the system to the user. VR systems also have an integrated software program along with the input and output tools.

The output can be delivered through visual, auditory, haptic, vestibular, and olfactory feedback. The visuals are usually delivered using head-mounted devices that are more immersive and give a feeling of proximity to the virtual scene. Another way to deliver visual feedback is through projection screens, in which the virtual scene is projected on one or multiple screens to give a sense of immersion. Another cost-effective and less immersive tool includes computer monitors or desktops. Auditory feedback facilitates the feeling of presence and aids in interacting with a virtual environment. It can be provided using headsets or inbuilt speakers. The more expensive devices use haptic feedback that helps the user manipulate objects in a

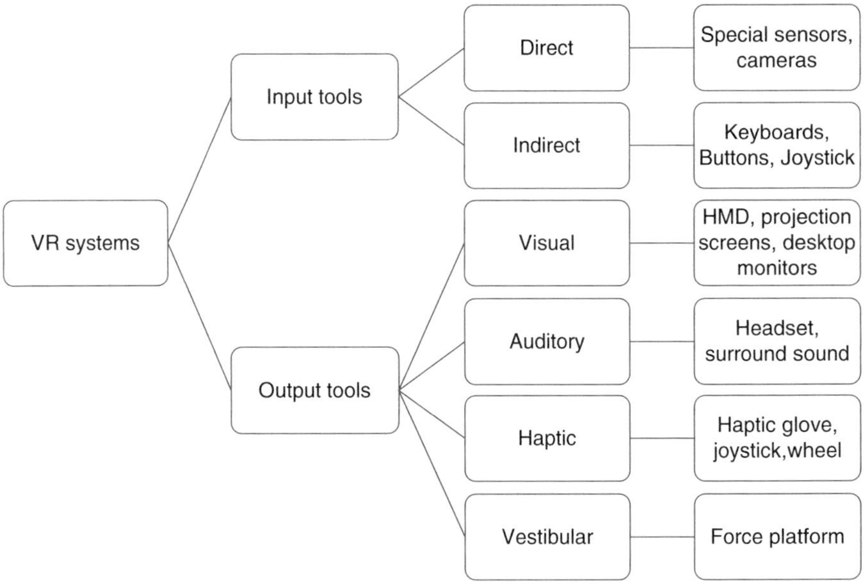

Fig. 16.1 Instrumentation in virtual reality systems

virtual world and give a more realistic experience. Haptic feedback gives a sense of pressure and force while interacting with virtual objects. Haptic gloves, joysticks, and haptic wheels help in grading the pressure applied by the user. Some systems provide vestibular feedback through pressure plates and standing platforms. Though a few systems use olfactory feedback, they are still less prevalent and under-researched in rehabilitation.

Input information is equally important in VR devices for real-time interaction of users with the virtual world. Input tools can be categorized into direct and indirect. In the direct method, the user performs free and natural movements that are captured using specialized motion sensors and cameras. In the indirect method, the user manipulates objects using an interface such as buttons, mouse, keyboard, joystick, or hand probe. These methods can help in assessing the amount of pressure used by the person and modify the task accordingly.

Currently, commercial VR systems are available for clinical use. However, designing a successful VR device requires an understanding of the patients' age, their context and condition, and performance abilities. It also depends on the cost, space, and training required for a particular VR system.

16.2.2 Model for Rehabilitation in Virtual Reality

The purpose of using VR is to promote independence in Activities of Daily Living (ADL) and develop confidence in performing activities in real life for people with neurological dysfunctions. It simulates real-life scenarios so that the person can perform those tasks with ease and safety. The model of VR has been divided into three realms that is an inner interaction zone, a middle transfer space, and an outer real-world space (Fig. 16.2). Therefore, VR is used to gradually train a person's

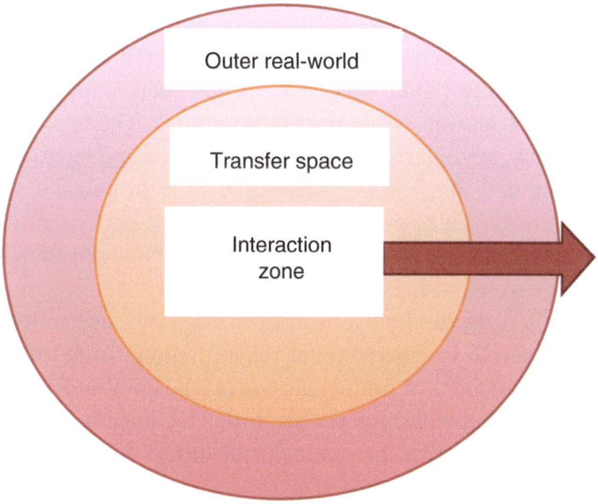

Fig. 16.2 Model of rehabilitation in virtual reality

skills from the inner interaction zone to the outer real-world space with the help of a therapist. The characteristics of the VR system and the individual interacting with it determine the quality and success of interaction and transfer to the outer world.

16.2.3 Advantages of Virtual Reality in Rehabilitation

VR became a popular rehabilitation tool because of its innumerable advantages over conventional treatment. As stated earlier, it can provide intensive rehabilitation with higher repetition and frequency of exercises. It can be tailored to each person's capabilities and the task can be gradually progressed in complexity and duration. VR allows practicing real-life tasks in a safer environment. Therefore, it offers ecological validity and safety. The multi-sensory feedback provided during task performance assists in motor learning and provides knowledge of performance and knowledge of results. The sensory feedback can be controlled and adapted based on the specific task and environment. Dual tasks and cognitive training can be easily delivered using VR-based platforms. The specialized input system helps in the precise and standardized evaluation of the patients' outcomes. Objective and real-time monitoring are also possible through VR devices. Another important benefit of VR is the provision of fun, motivating, and engaging tasks that encourage patients to perform regular exercises. VR can be used for gamification of daily activities which further enhances adherence to rehabilitation.

16.2.4 Applications of Virtual Reality in Neurological Physical Rehabilitation

VR has been extensively used in the rehabilitation of cognitive, executive, and motor functions. In neurological conditions, VR has been implemented for improving reaching and grasp, hand strength, wrist movements, speed, and motor coordination using haptic feedback. It has been employed in training for strength and range of motion of the upper extremities. The use of indirect input tools such as a joystick, keyboard, and mouse can only be applied to the distal upper extremities. In stroke and PD, VR has been used to train balance and gait. It provides multiple options for video games to perform dual tasks by training cognitive and motor skills simultaneously. In people with SCI, VR has been found to improve functional independence as well as balance.

VR is especially useful for training ADL due to its ecological validity, which is not feasible in conventional rehabilitation. Some of the VR applications, such as virtual kitchens or finding a route in a hospital/university or vending machines, have been implemented for people with TBI since it offers a safer environment to train real-life activities. In vestibular disorders, simulation of shopping stores has been effective for reducing vertigo, dizziness, and imbalance. Another important ADL that has gained popularity in VR-based rehabilitation is street-crossing for people with TBI or spatial disorientation. Simulation of car driving has been tried for people with TBI. Driving skills can gradually be progressed to complex tasks by introducing traffic, highways, and narrow paths. VR has also found applications for hemineglect and post-stroke attention training.

16.3 Augmented Reality

Augmented reality (AR) is an extension of VR where users can interact with computer-generated images that are superimposed over the real environment. Like VR, AR allows the practice of daily activities in a safer environment by simulating the real-life world. The tasks can be adjusted to the users' abilities and limitations. Moreover, by using AR in hospital and rehabilitation centers, it can provide an opportunity to practice tasks in a patient's home and context-specific environment.

There have been confusions regarding the definition of AR and many researchers use VR and AR interchangeably. However, the basic difference between them is that VR involves complete simulation of the real world, whereas AR enhances the real environment by introducing objects and images that are virtually embedded within the real world. AR offers more flexibility and feedback on daily activities. According to Azuma et al., AR should meet the following criteria: combine real and virtual objects in a real scenario, should be in real-time, and should align virtual objects with real objects. In AR, users do not have to wear any input devices such as a glove or joystick, instead, the proprioceptive sensations of the real objects lead to a more natural operation.

16.3.1 Applications of Augmented Reality in Neurological Physical Rehabilitation

AR has been used in stroke rehabilitation for upper and lower extremity training. Daily activities and fine motor skills have been used for training upper limbs, while for the lower extremities, AR has been used to train gait, balance, and posture. Mirror therapy has been incorporated with AR to train the affected arm by exercising the unaffected arm. This is known as augmented reflection technology. The principles and techniques are the same as conventional mirror therapy with the addition of augmented feedback and gaming activities. Few pilot studies have tested the effectiveness of AR to train cognitive and motor functions in children with cerebral palsy, autism, and cognitive decline. Like VR, AR also has the potential to provide engaging exercise sessions. However, it has more therapeutic applications since it incorporates real-life objects that can engender more motivation, skill acquisition, and better transferability to the real-life context compared to VR in which the practice of task is in a completely virtual world. To sum up, AR provides more control and sense of reality than VR because the user interacts with a real-world object in a natural environment.

16.4 Exergames

Exergames in plain language mean the use of video games for exercising. Other names for exergames include activity promoting video games, active video games, interactive video games, and dance simulation. According to the Exergame Network, exergames combine video game technology and exercises to promote an active and healthy lifestyle. Having a healthy lifestyle requires health-related as well as skill-related fitness. Health-related fitness can be achieved by improving endurance, muscle strength, cardiovascular fitness, bone density, and muscle mass. Skill-related fitness includes agility, balance, coordination, reaction time, speed, and power training. Traditionally, video games incorporated eye–hand coordination and reaction time but did not include physical activity and strength training. Therefore, video games that allow players to be more physically active and enhance motor functions are included in the category of exergames.

Exercise-oriented games emerged with the advent of VR in the 1980s. PowerPad and Family Fun Fitness are a few of the earliest exergames developed by Bandai Company and Nintendo, respectively. Dance Dance Revolution was developed in the 1990s that requires players to move in a specific fashion depending on the music played. In 2006, Nintendo introduced the Wii Fit, the Wii Balance Board, and the Wii Remote exergames. Microsoft Kinect devices such as Xbox came after 2010 and had a multitude of gaming options. With the advancement of mobile-based technology and the addition of AR, more options for exergames such as Pokémon GO and Yourself! Fitness were marketed.

Exergames research is trending due to its multiple benefits, including the ease of use at home, low-cost, target-oriented activities, instant feedback, and its enjoyment value for patients. It aids in tracking movements of the body and assists in physical mobility using digital solutions. The gamification of exergames promotes physical, cognitive, and psychosocial health. Gamification is important for creating interest, motivation, efficiency, and behavior change. It comprises three ingredients: dynamics, mechanics, and components.

Dynamics in gamification: Dynamics includes (1) storytelling for enhancing motivation, (2) determining the relationship between players such as competition, social interaction, or friendship, (3) setting up rules such as time limitation and (4) explaining goals at the beginning such as catching ten fishes or collecting three diamonds. Figure 16.3 shows the various aspects of dynamics.

Mechanics in gamification: Mechanics provide positive reinforcement by using factors such as luck, reward, or challenge. Figure 16.4 demonstrates the several types of mechanics used in games.

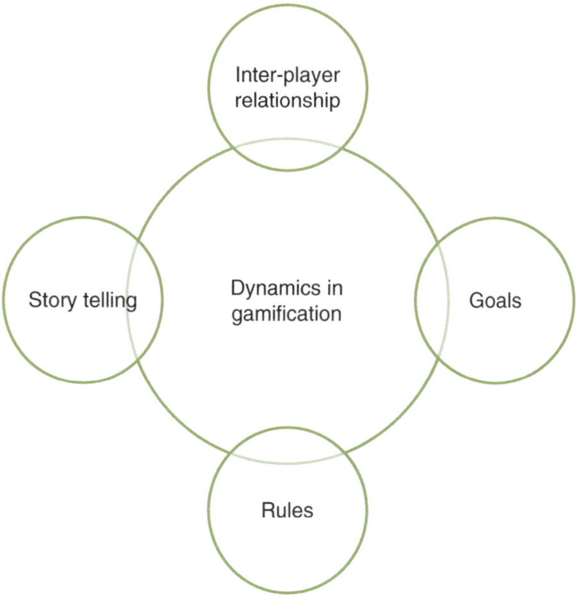

Fig. 16.3 Dynamics in gamification

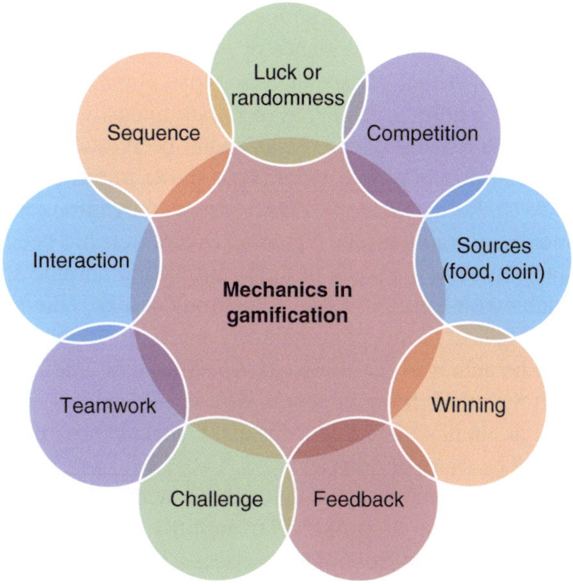

Fig. 16.4 Mechanics used in gamification

Components of gamification: The end-users directly interact with components of gamification. Components include Avatars which are the virtual profile of players, rewards and scores, badges, leadership for visibility in a social environment, levels of progress, or tasks.

Therefore, potent video game systems should develop a user-specific exercise protocol that includes elements of games or gamification, expert-supervised selection of games, accurate measurement and evaluation of movement, immediate and long-term feedback of performance and can be used in a home setting.

16.4.1 Applications of Exergames in Neurological Physical Rehabilitation

There are a plethora of games available for neurological rehabilitation. Exergames have been used for training the upper limb and lower limb motor functions, static and dynamic balance, heart rate, energy expenditure, and the perceived exertion in various neurological conditions. A common method is recreating therapeutic tasks and incorporating feedback, rewards, and scoring for performance. Exergames such as Gate Crossing, Bike Rehab, BeTheBall, and PicsWalk have been used for the lower limb rehabilitation of stroke patients. Similarly, VRheab games have been used for upper limb training in various neurological conditions. Mystic Isle and Priest in the Air are a few games for stroke rehabilitation. For vestibular rehabilitation, the Cardboard Break and the Follow Me have been tested, while the Imaginator has been tried for sensory processing disorders.

Even though the evidence of exergames on SCI is limited, preliminary findings have shown that games such as Move Boxing, Move Gladiator Duel, and Move Kayaking help in achieving moderate-intensity exercises in wheelchair users with SCI. The feasibility of exergames has been established in PD. Use of the Nintendo Wii Platform in people with PD has shown improvement in their Unified Parkinson's Disease Rating Scale (UPDRS) scores, cognition, and static balance even after 2 months of training. The Nintendo GameCycle has been shown to achieve moderate-intensity arm workouts in Spina Bifida and chronic spinal cord injury patients. In adults with Cerebral Palsy and chronic stroke, Nintendo Wii, Wii Tennis, and Wii Boxing lead to moderate-intensity exercises. Light- to moderate-intensity exercises can be achieved by Nintendo Wii, Sony PlayStation, EyeToy, Wii Canoeing, and Wii Sword Fighting in people with stroke. Games involving dynamics, mechanics, and aesthetics yield better results. Cybersickness is a common side effect experienced by users of exergames.

A systematic review explored the evidence of exergames on the health-related quality of life in stroke, PD, multiple sclerosis, spastic hemiparesis, cognitive impairment, and Huntington's disease. Wii platform, Microsoft Kinect console, and Dance Dance Revolution were delivered in healthcare centers and the results showed a small significant effect on the health-related quality of life outcomes. It is hypothesized that exergames delivered under therapists' supervision in healthcare facilities ensure controlled training of exercise intensity, better attention and motivation, reinforcement, and compliance.

16.5 Robotics

Human beings have invariably used external devices to reduce effort and enhance efficiency while performing any strenuous work. Similarly, robotics has found a lot of attention and demand after disability when the efficiency of movements reduces and the need for efforts increases. The use of robotics started in the 1970s when "prosthetic man" was developed. The pivotal moment in rehabilitation robotics was the development of hand and limb prostheses with the use of myoelectric control in the early 1970s.

Robotics have been extensively used in healthcare for imaging, surgical procedures, and therapeutic purposes. Robotics has been used for exercise training, as a prosthesis or orthosis, and for assisting in physical training programs in rehabilitation. Robots are useful in delivering intensive rehabilitation and higher repetitions which is instrumental for inducing neuroplasticity. As defined by Hillman M.R., "Rehabilitation robotics is the application of robotic technology to the rehabilitation needs of people with disabilities as well as the growing elderly population."

16.5.1 Classification of Rehabilitation Robots

The rehabilitation robots have been used for multiple purposes and have been classified based on their functions, such as assistance in manipulating objects, mobility, therapeutic assistance and caregiving.

For manipulating objects, wheelchair-mounted robots have been used for picking up objects and placing them. MANUS is a popular robot arm for wheelchair users. Other robots in this category include Helping Hand, KAIST Rehabilitation Engineering System (KARES), PLAYBOT for children, Manipulation arm, and the Chamaeleon. Even though the robots aid in exploring the environment, they still cannot perform complex tasks and depend on the user for intricate maneuvers.

Another function for which robots are utilized is in the form of workstations where they are placed on a desk or fixed location. Workstation robots serve specific functions and do not require adapting to different tasks, and therefore are easier to operate. They can be used for tasks such as food preparation, computer-related activities, telephoning, etc. Some examples of workstation robots are Desktop Vocational Assistant Robot (DeVAR), Professional Vocational Assistant Robot (ProVAR), MASTER, AFMASTER, and Robotic Automation & Intelligence Discovery (RAID).

Robotic feeders have also been used for people with disabilities that allow them to eat at their own pace with the help of the robot. HANDY 1 is one of the popular robots in this category. Other feeding robots are Intelligent Soft-Arm Control (ISAC) and My Spoon. Robots can be used as upper limb orthoses that can suppress tremors and train upper limb activities. However, due to their complicated design, they are not yet widely employed in rehabilitation.

Enhanced mobility can be achieved with robotic wheelchairs that are either fully or semi-automatic. The fully automatic wheelchairs can be used for uneven terrain or climbing stairs. The semi-automatic wheelchair is less complex and gives feedback to the users for obstacle clearance. There are a few robots that guide walking in the visually impaired by sensing obstacles and helping in navigating around them. Examples of robotic walkers are the Personal Adaptive Mobility Aid (PAM-AID), the HITOMI, the GuideCane, and the Robotic Cane. A powered walking assistance system aids elderly people in sitting down, getting up, walking, and preventing falls.

Robots are also used in delivering therapeutic exercises both in physical and occupational rehabilitation. The purpose of using robots in therapy is to provide an intensive treatment regimen with multiple sets and repetitions that are not feasible by therapists' efforts. Most commonly the upper limb exercises are trained by using robots. Nowadays, robots for lower limb training are also available. MIT-Manus, the ARM Guide and Rehabilitation Robot (REHABOT) are proposed robots for therapy. Some of the commercially available robotic devices for upper limb rehabilitation are Armeo® Spring (Hokoma), Amadeo (Tyromotion), Hand of Hope, ARM/WRIST, InMotion Arm®, and ReoGo™.

Another key area where robotic technology is being explored is in the shifting or transferring the individuals with severe disabilities in nursing procedures or whenever the entire body weight must be supported. Such robots reduce the burden on caregivers during demanding activities.

16.5.2 Attributes of Rehabilitation Robots

- Safety while using robots for rehabilitation is a primary concern. Robots are fully mechanized and the amount of force or range could vary depending on the complexity of the system. Accidents are likely to happen when the robotic device exceeds the limited range or pain threshold of patients. Therefore, safety while maneuvering is a key factor for designing rehabilitation robotics.
- A human–machine interface is another crucial domain while designing a robotic device since it is used by people with physical and mental disabilities. The users can control the device by voice communication, head, shoulder, or eye movements, electromyography, and cortical mapping. A user-friendly and simple interface is easily accepted among this population.
- Cost and ease of use are a few other attributes of rehabilitation robotics that are evaluated by end-users. A robot that serves multiple functions is typically difficult to operate but expensive. A simple robot that serves a specific function is more precise and affordable. The wear-and-tear of robots in the long term should also be considered when using them for rehabilitation.

16.5.3 Applications of Robotics Neurological Physical Rehabilitation

Robotic devices have been used in motor rehabilitation for passive range of movements, assisting a weaker muscle, or for resistance training. There can be single-joint or multi-joint training depending on the type of robotics device. The treatment can be progressed by decreasing assistance, increasing resistance, and amplitude of movement. Some devices also assist in bimanual exercises. Gamification and VR have also been integrated into robotic training to facilitate participant engagement and health outcomes.

Electromechanical devices and robot-assisted arm therapy have shown some evidence for improving arm function, strength, and ADL in people with stroke, therefore, it can be used as an adjunct with conventional motor rehabilitation. Active Learning Program for Stroke incorporates salience and transference components of motor learning along with repetition, intensity, and specificity for promoting upper limb function in a home- and community-living people with stroke and is found feasible. Evidence suggests that electromechanical-assisted gait training delivered along with standard exercises had higher odds of achieving independent walking, post-stroke.

In high SCI, there is a need to strengthen the upper extremities. The robotic devices for training upper limbs in SCI are either exoskeletons or end-effector robotic devices targeting shoulder-elbow or wrist-hand, respectively. Powered exoskeletons are useful for promoting weight-bearing and providing repetitive and intensive locomotor training in lower SCI. Robotic-assisted gait training has been found to improve mobility in people with SCI. Indego™ and ReWalk™ are the two lower limb robotics that is available for community mobility in thoracic SCI. Exoskeleton-assisted walking has shown improvement in cardiovascular health, bone mineral density, bowel function, and quality of life in people with SCI.

Robot-assisted intensive gait training has shown significant improvements in gait speed, endurance, and freezing episodes in people with PD. Evidence on improving balance and postural instability using robot-assisted gait training in PD is high. There is a moderate level of evidence on the effectiveness of robot-assisted training on other motor outcomes in PD, however, more research is needed to determine its effect on specific outcomes in PD.

16.6 Body Weight Support Treadmill Training

Walking impairments are common after a neurological insult and require intensive rehabilitation to retrain walking. Therefore, the use of technology in the form of Body Weight Support Treadmill Training (BWSTT) has gained popularity for gait training in neurological conditions. BWSTT includes a treadmill and an overhead harness that supports the body weight of patients. The system can unload body weight off the patient during the early phases when patients are unable to bear weight and can gradually increase the load, once the patients can attain adequate lower limb strength and balance. Walking on a treadmill allows a greater number of steps in a lesser time duration compared to overground walking, thus making it a repetitive and task-oriented practice. The BWSTT also leads to improved strength in the lower limbs and balance ability.

This treatment technique emerged in 1987 when Barbeau H. and Rossignol S. used treadmill training in T13 level spinalized cats and observed an improvement in their locomotion. After this experiment, multiple studies published on locomotory training in spinalized cats reported improvement in the gait pattern. The mechanism of action lies in the activation of circuitries in the spinal cord and central pattern generators that improve hindlimb locomotion with repetitive and alternate stepping in SCI. The experience-dependent sensory feedback also facilitates motor learning by inducing neuroplasticity.

16.6.1 Body Weight Support System

The body weight support system primarily consists of three parts—a harness that is worn by the patient, pulleys, and ropes from overhead suspension, and a counterpoise system. Based on the types of counterpoise systems, they can be divided into static systems, passive counterweight systems, passive elastic systems, and active dynamic systems.

- A static system is connected to a winch using ropes and pulleys that unload body weight.
- A passive counterweight system uses the inertia and gravitational force of counterweights to unload body weight.
- A passive elastic system uses a spiral metal spring or cord. The tension in the elastic components supports body weight.
- An active dynamic system uses active force-generating units such as pneumatic, hydraulic, or electromagnetic units that support body weight during walking.

Some of the commercially available BWSTT systems are Ergo Trainer, Safe Gait 360, Lite Gait, and Andago. The advantages of BWSTT are: (1) to provide an opportunity for task-specific and repetitive practice, (2) to ensure patients' safety and stability while ambulating, (3) the speed of walking can be varied on a treadmill depending on the patients' motor control, (4) the amount of therapists' assistance required for gait training can be reduced and (5) the amount of load on lower extremities can be graded according to patients' abilities. The disadvantages of BWSTT are: (1) expensive to purchase and maintain, (2) requires human support to operate the system, (3) not suitable for home and community setting, and (4) not portable and has limited accessibility in healthcare centers.

16.6.2 Applications of BWSTT in Neurological Physical Rehabilitation

BWSTT has been extensively researched for gait training in stroke. A Cochrane review including 3105 participants reported that BWSTT improves walking speed and endurance but does not improve independent walking when compared to conventional physiotherapy treatment. However, in non-ambulatory early subacute stroke, BWSTT leads to independent walking within 6 months of stroke compared to assisted ground walking. BWSTT is found to be associated with lesser and more symmetrical activation of the medial sensory-motor cortex, which means that ambulation is controlled by lower spinal centers and movement gets more automatic augmenting gait performance.

BWSTT has beneficial effects on walking distance, strength, and functional mobility but is not found to improve walking speed in SCI. Improvement in gait is possible even in chronic SCI when measured using electromyography. However, BWSTT was not found to have any superior effects on gait in acute incomplete SCI compared to training. BWSTT improves femoral artery compliance and has some effect on heart rate variability and blood pressure variability in people with complete SCI and could be used for promoting their cardiovascular health.

In people with PD, BWSTT was found to significantly improve walking distance, balance, pain, fatigue, motor functions, and ADL compared to physical therapy. Partial weight-supported treadmill training has also been found to improve the UPDRS scores, balance, and limits of stability in PD compared to overground gait training. BWSTT has also been seen to have long-term effects on gait patterns in PD.

16.7 Telerehabilitation

Telerehabilitation is a new branch of telemedicine and comprises delivering rehabilitation services at a distance using telecommunication and computer software. Telerehabilitation pertains to clinical services and includes diagnosis, assessment, and treatment. The three aspects of telerehabilitation include (a) training and counseling for providing education, training, or counseling to patients and caregivers, (b) monitoring and assessment for regular monitoring and progress of health outcomes, and (c) therapeutic interventions for delivering exercise sessions and therapy.

The first paper on telerehabilitation was published in 1998, with an upswing seen in early 2000 and then after 2008 due to the rapidly evolving technology and developments. Telerehabilitation has been attempted in cardiac, neurological, and cognitive conditions that require long-term management. Telerehabilitation has also been used for monitoring the health conditions of patients remotely, such as cardiovascular health, blood pressure, oxygen saturation, etc. The COVID-19 pandemic has provided unprecedented opportunities for clinicians and researchers to explore telerehabilitation.

Telerehabilitation is useful in neurological conditions that need treatment to be delivered early after an injury, require intensive rehabilitation to facilitate neuroplasticity, and must be continued for a prolonged period to maximize recovery. Delivering rehabilitation services in inaccessible and remote areas to improve health outcomes is the primary focus of telerehabilitation. People with severe disabilities can be monitored with telemonitoring and it allows for recording the patients' activities, pain, and mood, thus promoting self-management. Barriers to telerehabilitation include licensing, data confidentiality, financial issues, and lack of infrastructure to support telerehabilitation, especially in low-income countries.

16.7.1 Design of Telerehabilitation

Deploying telerehabilitation at home requires devices and systems that can capture patients' data and transmit it to a central office. Designing telerehabilitation equipment requires the knowledge of sensorimotor function, mental and psychological aspects of patients. Therefore, simple and easy-to-use interfaces are key features in telerehabilitation systems. Even perceptual, motor, and cognitive impairments need to be considered while designing telerehabilitation system. Services can be delivered from hospitals, clinics, or service providers' offices to a home setting. Real-time telerehabilitation can be provided by using store-and-forward modalities. Figure 16.5 shows the important aspects of designing telerehabilitation systems.

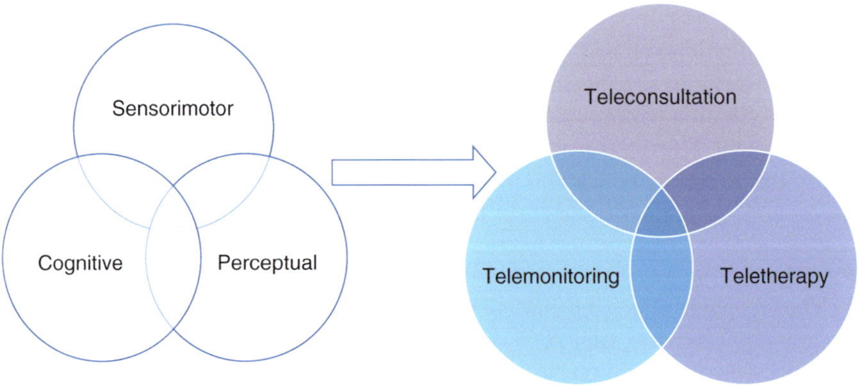

Fig. 16.5 Considerations of human aspects for designing different domains of telerehabilitation

16.7.2 Conceptual Models of Telerehabilitation

1. The first model is a traditional hub-and-spoke that connects community outreach centers and patients to rehabilitation experts in comprehensive care facilities (Fig. 16.6)
2. The second model involves two interfaces, one for patients and one for home nurses that assist in caregiving. The home nurse needs to update charts, records, and vital signs which are transmitted to experts in the main center for tele-counseling and training (Fig. 16.7)

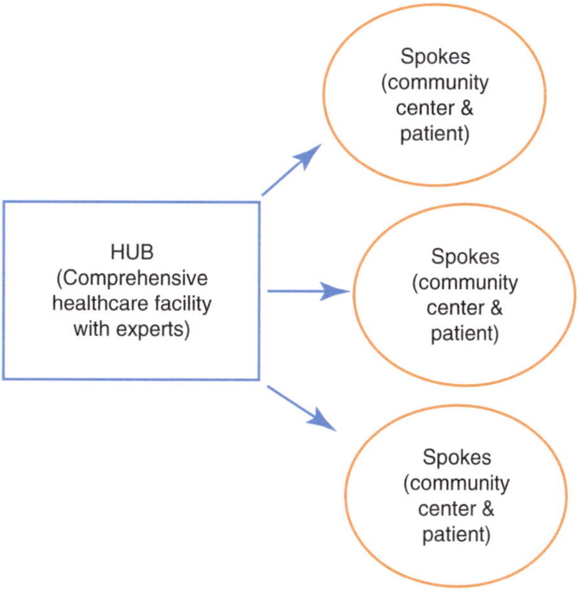

Fig. 16.6 Traditional hub-and-spoke model

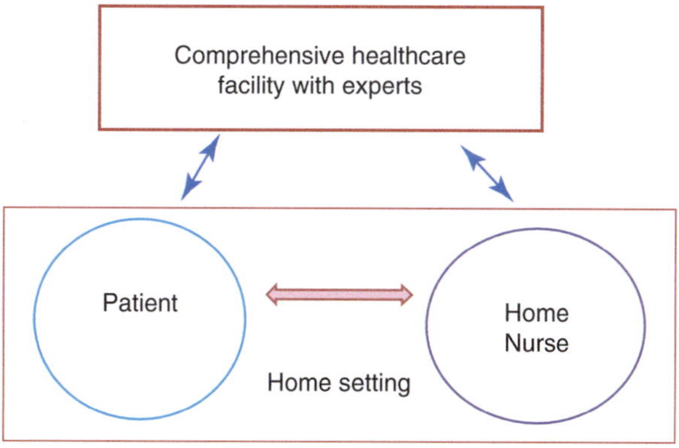

Fig. 16.7 Model for training and counseling

3. The third model is for telemonitoring and assessment and provides an option for real-time teleconferencing between patient and healthcare provider as well as asynchronous sessions using exergames or unsupervised exercise programs (Fig. 16.8)
4. The fourth model is for teletherapy where therapists deliver exercise sessions to patients in their homes using equipment, gaming, or VR systems. It may include both synchronous and asynchronous sessions. (Fig. 16.9)

16.7.3 Considerations for Telerehabilitation

It is imperative that certain factors are considered before delivering telerehabilitation. It is crucial to know what is going to be delivered in telerehabilitation session and how, the person who is going to deliver the session and target participants. Inclusion of acceptable and appropriate treatment options should be considered prior to implementing telerehabilitation. The evidence and effectiveness of a particular tele-intervention, outcome assessment, and plan for progression are some of the factors that have to be well-thought-out before giving the intervention. Moreover, a

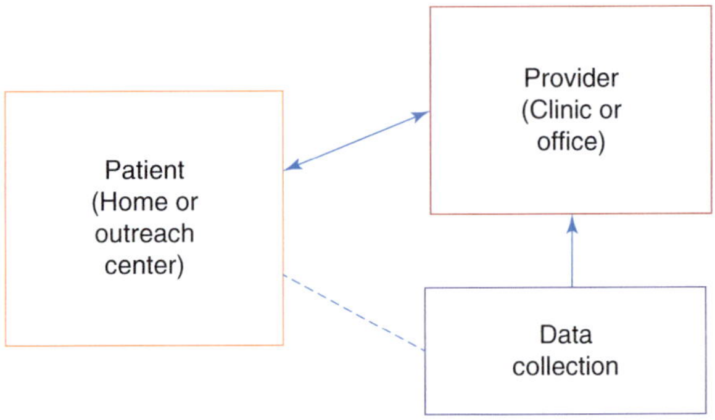

Fig. 16.8 Telemonitoring model using real-time conferencing

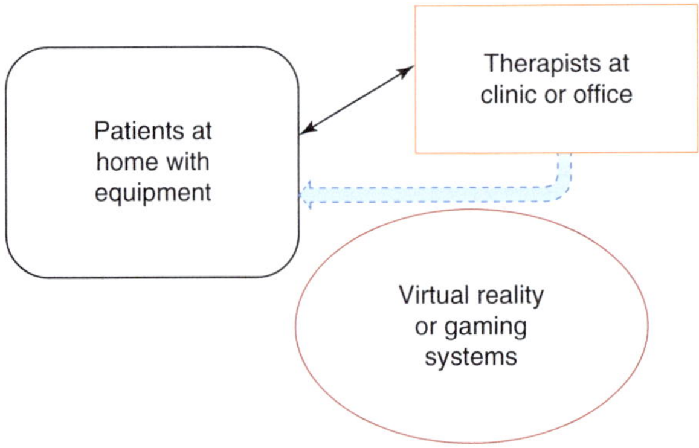

Fig. 16.9 Teletherapy model

comprehensive knowledge of the patients' barriers, the facilitators of participation in telerehabilitation, their experiences, and outcomes should be incorporated while planning the treatment session. Healthcare providers' barriers and facilitators, their knowledge and experience in telerehabilitation, training, and support required for telerehabilitation also influence the implementation of treatment program. The cost of delivering telerehabilitation, reimbursements, and scope of practice are some of the considerations for accomplishing successful telerehabilitation. Recently, the University of Central Lancashire and East Lancashire Hospitals NHS Trust has provided the following guidelines for delivering real-time telerehabilitation in remote areas.

1. Set up—It includes the availability of hardware devices and software for teleconsultation such as laptops, iPad, headphones, etc. Preparatory contact with patients and family members to understand their context, impairments, and available equipment. Assisting patients in finding resources such as standardized assessment tools, session plans, and exercise charts. Setting up a conducive environment for teleconsultation that includes private and quiet locations, proper positioning, and internet connectivity.
2. Connect—After calling patients during teleconsultation, check for internet connection for clear audio and video call, confirm the identity of the patient including date of birth and postcode and check the environment for safety and presence of family members for patients who require assistance.
3. Get started—Check in with patients about their general health and well-being and clarify the purpose for the teleconsultation. Be patient and actively listen to their concerns. Identify any red flags or symptoms requiring immediate attention. Ask about their problem areas and priorities.
4. Assessment—Initial assessment must include functional performance evaluation and tests that can be remotely done without risk of falls or injuries. The use of questionnaires is also recommended. Assessment should include tests for balance, mobility, upper limb, cognition, general well-being, and activities for daily living.
5. During the consultation—Exercises should be demonstrated and made to practice with verbal cues and feedback. Clear instructions should be given and identify what works best for patients. Building the self-efficacy of patients is primarily important in telerehabilitation where the patients must learn self-management.
6. Next step—After the consultation and exercise session, review options should be decided with the patients. Time and form of review should be discussed, and methods to escalate future issues should be identified. Be realistic in setting up treatment goals.
7. Closing the calls—Closing the session is equally important. Summarize the key points of the session, highlight the follow-up actions to be undertaken by patients, ask for their feedback to improve anything for the next sessions.
8. After a session—It is good practice to send out emails or reminders to patients after the session. Document the details of the session, including patients' feedback and their performance. Mention points that went well and those that did not. Highlight any follow-up issues and time for the next session.

16.7.4 Applications of Telerehabilitation in Neurological Conditions

Telerehabilitation has been delivered in stroke management, but the actual intervention varies across different studies. The dosage and parameters of interventions also vary in different studies. Most commonly, audio and video conferencing have been used in telerehabilitation and partial supervision could be provided. A few studies have used 3D motion capture and biofeedback systems to evaluate patients' movements. Telerehabilitation has been used for improving motor functions, ADL, independence, quality of life, and self-efficacy of stroke survivors and has shown equal effectiveness as usual care. However, due to variation in the time since stroke, post-stroke complications, and different technology, the findings cannot be generalized.

In people with multiple sclerosis, attending clinics gets increasingly difficult due to disabilities, therefore, telerehabilitation has to be used in this population to deliver clinical services daily. A telerehabilitation program delivered along with usual care is found effective for improving motor functions, cognitive functions, and quality of life in people with PD. The adherence rate and patients' satisfaction are also higher with telerehabilitation.

Many tele-based healthcare models have been developed for people with SCI due to difficulty in accessibility and transport. The telerehabilitation models deliver holistic treatment from an acute phase of spinal injury till discharge to home. The Queensland Spinal Cord Injuries Service, the SCI Hospital in Home model, and the Spinal Cord Injury Disease Management Protocol are some of the comprehensive interdisciplinary telehealth programs developed for SCI. In wheelchair users with spinal cord injury, an intensive upper limb telerehabilitation program is effective for improving upper limb strength, shoulder pain, and function. Wound care and management of pressure ulcers in SCI patients could be successfully delivered using telerehabilitation.

The internet-based programs are feasible and have a positive impact on TBI outcomes. A systematic review reported the efficacy of telerehabilitation in TBI with improvements in depression, post-traumatic symptoms, sleep quality, and global functions. Telebehavioral interventions for caregivers of TBI patients have shown positive effects on caregiver depression, stress, anxiety, burden, problem-solving, and self-efficacy.

Thus, telerehabilitation has a mounting evidence of effectiveness in different conditions. It has higher attendance, compliance, and user satisfaction; however, the evidence of cost-effectiveness is still lacking.

16.8 Take-Home Message

In recent times, many other assistive technologies and gadgets have been deployed for treatment purposes that are not discussed in this chapter. These include transcranial direct current stimulation, transcranial magnetic stimulation, wearable devices such as accelerometers and pedometers, pressure sensing footplates, and real-time movement capture devices. However, the evidence of such recent advancements is still developing.

Technologies have changed the style of neurological rehabilitation. Technology-based rehabilitation has provided excellent scope for intensive and repetitive practice, can be used in homes and communities, and is fun and engaging compared to the conventional rehabilitation techniques. It has also helped in reducing the load on therapists. However, all technology-based rehabilitation equipment requires cost, training, and safety precautions. Even though the use of technology in rehabilitation has enhanced the treatment outcomes and promoted recovery, it can only be used as an adjunct to therapist-supervised conventional rehabilitation. There is absolutely no doubt that the clinical assessment, goal setting, and treatment plan based on the therapists' skills and knowledge cannot be superseded by technology.

Further Reading

Ada L, Dean CM, Morris ME, Simpson JM, Katrak P. Randomized trial of treadmill walking with body weight support to establish walking in subacute stroke: the MOBILISE trial. *Stroke*. 2010 Jun;41(6):1237–42.

Agostini M, Moja L, Banzi R, Pistotti V, Tonin P, Venneri A, et al. Telerehabilitation and recovery of motor function: a systematic review and meta-analysis. *J Telemed Telecare*. 2015 Jun;21(4):202–13.

Appleby E, Gill ST, Hayes LK, Walker TL, Walsh M, Kumar S. Effectiveness of telerehabilitation in the management of adults with stroke: A systematic review. *PLoS One*. 2019 Nov 12;14(11):e0225150.

Atan T, Özyemisci Taşkıran Ö, Bora Tokçaer A, Kaymak Karataş G, Karakuş Çalışkan A, Karaoğlan B. Effects of different percentages of body weight-supported treadmill training in Parkinson's disease: a double-blind randomized controlled trial. *Turk J Med SCI*. 2019 Aug 8;49(4):999–1007.

Ballester BR, Ward NS, Brander F, Maier M, Kelly K, Verschure PFMJ. Relationship between intensity and recovery in post-stroke rehabilitation: a retrospective analysis. *J Neurol Neurosurg Psychiatry*. 2021 Jun 24:jnnp-2021-326948.

Barry G, Galna B, Rochester L. The role of exergaming in Parkinson's disease rehabilitation: a systematic review of the evidence. *J Neuroeng Rehabil*. 2014 Mar 7;11:33.

Beare B, Doogan CE, Douglass-Kirk P, Leff AP, Ward N. Neuro-Rehabilitation OnLine (N-ROL): description and evaluation of a group-based telerehabilitation programme for acquired brain injury. *J Neurol Neurosurg Psychiatry*. 2021 Jun 23:jnnp-2021-326809.

Chew E, Turner DA. Can a Robot Bring Your Life Back? A Systematic Review for Robotics in Rehabilitation. *Adv Exp Med Biol*. 2019;1170:1–35.

Corrêa AGD, de Assis GA, do Nascimento M, de Deus Lopes R. Perceptions of clinical utility of an Augmented Reality musical software among health care professionals. *Disabil Rehabil Assist Technol*. 2017 Apr;12(3):205–216.

Cugusi L, Prosperini L, Mura G. Exergaming for Quality of Life in Persons Living with Chronic Diseases: A Systematic Review and Meta-analysis. *PM & R*. 2021 Jul;13(7):756–780.

Ditor DS, Macdonald MJ, Kamath MV, Bugaresti J, Adams M, McCartney N, et al. The effects of body-weight supported treadmill training on cardiovascular regulation in individuals with motor-complete SCI. *Spinal Cord*. 2005 Nov;43(11):664–73.

Dobkin BH. Motor rehabilitation after stroke, traumatic brain, and spinal cord injury: common denominators within recent clinical trials. *Curr Opin Neurol.* 2009 Dec;22(6):563–9.

Duncan PW, Sullivan KJ, Behrman AL, Azen SP, Wu SS, Nadeau SE, et al. Body-weight-supported treadmill rehabilitation after stroke. *N Engl J Med.* 2011 May 26;364(21):2026–36.

El-Sayes J, Harasym D, Turco CV, Locke MB, Nelson AJ. Exercise-Induced Neuroplasticity: A Mechanistic Model and Prospects for Promoting Plasticity. *The Neuroscientist.* 2019 Feb;25(1):65–85.

Fasoli SE, Adans-Dester CP. A Paradigm Shift: Rehabilitation Robotics, Cognitive Skills Training, and Function After Stroke. *Front Neurol.* 2019 Oct 15;10:1088.

Ferreira B, Menezes P. Gamifying motor rehabilitation therapies: challenges and opportunities of immersive technologies. *Information.* 2020 Feb; *11*(2), 88.

Frey M, Colombo G, Vaglio M, Bucher R, Jörg M, Riener R. A novel mechatronic body weight support system. *IEEE Trans Neural Syst Rehabilitation Eng.* 2006 Sep;14(3):311–21.

Gorman C, Gustafsson L. The use of augmented reality for rehabilitation after stroke: a narrative review. *Disabil Rehabil Assist Technol.* 2020 Jul 14:1–9.

Hailey D, Roine R, Ohinmaa A, Dennett L. The status of telerehabilitation in neurological applications. *J Telemed Telecare.* 2013 Sep;19(6):307–10.

Isernia S, Di Tella S, Pagliari C, Jonsdottir J, Castiglioni C, Gindri P, et al. Effects of an Innovative Telerehabilitation Intervention for People With Parkinson's Disease on Quality of Life, Motor, and Non-motor Abilities. *Front Neurol.* 2020 Aug 13;11: 846.

Kairy, D., Lehoux, P., Vincent, C., Visintin, M. A systematic review of clinical outcomes, clinical process, healthcare utilization and costs associated with telerehabilitation. *Disabil Rehabil.* 2009; *31*(6): 427–447.

Khademi M, Hondori HM, Dodakian L, Cramer S, Lopes CV. Comparing "pick and place" task in spatial Augmented Reality versus non-immersive Virtual Reality for rehabilitation setting. *Annual International Conference of the IEEE Engineering in Medicine and Biology Society.* 2013;2013:4613–6.

Lathan CE, Kinsella A, Rosen MJ, Winters J, Trepagnier C. Aspects of human factors engineering in home telemedicine and telerehabilitation systems. *Telemed J.* 1999;5(2):169–75.

Martinez M, Delivet-Mongrain H, Leblond H, Rossignol S. Effect of locomotor training in completely spinalized cats previously submitted to a spinal hemisection. *J Neurosci.* 2012 Aug 8;32(32):10961–70.

Martinez-Martin E, Cazorla M. Rehabilitation Technology: Assistance from Hospital to Home. *Comput Intell Neurosci.* 2019 Jun 2;2019: 1431509.

Massetti T, da Silva TD, Crocetta TB, Guarnieri R, de Freitas BL, Bianchi Lopes P, et al. The Clinical Utility of Virtual Reality in Neurorehabilitation: A Systematic Review. *J Cent Nerv Syst Dis.* 2018 Nov 27;10:1179573518813541.

Mat Rosly M, Mat Rosly H, Davis Oam GM, Husain R, Hasnan N. Exergaming for individuals with neurological disability: a systematic review. *Disabil Rehabil.* 2017 Apr;39(8):727–735.

Mehrholz J, Kugler J, Pohl M. Locomotor training for walking after spinal cord injury. *Cochrane Database of Systematic Reviews.* 2008 Apr 16;(2):CD006676.

Mehrholz J, Thomas S, Elsner B. Treadmill training and body weight support for walking after stroke. *Cochrane Database of Systematic Reviews.* 2017 Aug 17;8(8):CD002840.

Mehrholz J, Harvey LA, Thomas S, Elsner B. Is body-weight-supported treadmill training or robotic-assisted gait training superior to overground gait training and other forms of physiotherapy in people with spinal cord injury? A systematic review. *Spinal Cord.* 2017 Aug;55(8):722–729.

Mehrholz J, Pohl M, Platz T, Kugler J, Elsner B. Electromechanical and robot-assisted arm training for improving activities of daily living, arm function, and arm muscle strength after stroke. *Cochrane Database of Systematic Reviews.* 2015 Nov 7;2015(11):CD006876.

Mehrholz J, Thomas S, Kugler J, Pohl M, Elsner B. Electromechanical-assisted training for walking after stroke. *Cochrane database of systematic reviews.* 2020 Oct 22;10(10):CD006185.

Mekki M, Delgado AD, Fry A, Putrino D, Huang V. Robotic Rehabilitation and Spinal Cord Injury: a Narrative Review. *Neurotherapeutics*. 2018 Jul;15(3):604–617.

Moore JL, Roth EJ, Killian C, Hornby TG. Locomotor training improves daily stepping activity and gait efficiency in individuals poststroke who have reached a "plateau" in recovery. *Stroke*. 2010 Jan;41(1):129–35.

Mubin O, Alnajjar F, Jishtu N, Alsinglawi B, Al Mahmud A. Exoskeletons With Virtual Reality, Augmented Reality, and Gamification for Stroke Patients' Rehabilitation: Systematic Review. *JMIR Rehabilitation and Assistive Technologies*. 2019 Sep 8;6(2):e12010.

Musselman KE, Shah M, Zariffa J. Rehabilitation technologies and interventions for individuals with spinal cord injury: translational potential of current trends. *J Neuroeng Rehabil*. 2018 May 16;15(1):40.

Pamukcu BS, Çakir H. Design Considerations for 3-D Exergames. *Journal of Learning and Teaching in Digital Age*. 2021; *6*(1): 58–71.

Pollock A, Farmer SE, Brady MC, Langhorne P, Mead GE, Mehrholz J, et al. Interventions for improving upper limb function after stroke. *Cochrane Database of Systematic Reviews*. 2014 Nov 12;2014(11):CD010820.

Peretti A, Amenta F, Tayebati SK, Nittari G, Mahdi SS. Telerehabilitation: Review of the State-of-the-Art and Areas of Application. *JMIR Rehabilitation and Assistive Technologies*. 2017 Jul 21;4(2):e7.

Rose DK, Nadeau SE, Wu SS, Tilson JK, Dobkin BH, Pei Q, et al. Locomotor Training and Strength and Balance Exercises for Walking Recovery After Stroke: Response to Number of Training Sessions. *Phys Ther*. 2017 Nov 1;97(11):1066–1074.

Schiza E, Matsangidou M, Neokleous K, Pattichis CS. Virtual Reality Applications for Neurological Disease: A Review. Front Robot AI. 2019 Oct 16; 6:100.

Shaw MT, Best P, Frontario A, Charvet LE. Telerehabilitation benefits patients with multiple sclerosis in an urban setting. J Telemed Telecare. 2021 Jan;27(1): 39–45.

Sulzer J, Karfeld-Sulzer LS. Our child's TBI: a rehabilitation engineer's personal experience, technological approach, and lessons learned. *Journal of Neuroengineering and Rehabilitation*. 2021 Apr 7;18(1):59.

Tao G, Garrett B, Taverner T, Cordingley E, Sun C. Immersive virtual reality health games: a narrative review of game design. *J Neuroeng Rehabil*. 2021 Feb 11;18(1):31.

Tchero H, Tabue Teguo M, Lannuzel A, Rusch E. Telerehabilitation for Stroke Survivors: Systematic Review and Meta-Analysis. *J Med Internet Res*. 2018 Oct 26;20(10):e10867.

Tejima, N. Rehabilitation robotics: a review. *Adv Robot. 2001; 14*(7): 551–64.

Vesmarovich S, Walker T, Hauber RP, Temkin A, Burns R. Use of telerehabilitation to manage pressure ulcers in persons with spinal cord injuries. *Adv Wound Care*. 1999 Jun;12(5):264–9.

Ward NS, Brander F, Kelly K. Intensive upper limb neurorehabilitation in chronic stroke: outcomes from the Queen Square programme. *J Neurol Neurosurg Psychiatry*. 2019 May;90(5):498–506.

Weiss PL, Kizony R, Feintuch U, Katz N. Virtual reality in neurorehabilitation. In: Selzer ME, Clarke S, Cohen LG, Duncan P, Gage F. editors. Textbook of Neural Repair and Rehabilitation-Medical Rehabilitation. Cambridge: Cambridge University Press. Cambridge. 2006. pp. 182–197.

Index

A. M. Joshua (ed.), *Physiotherapy for Adult Neurological Conditions*, https://doi.org/10.1007/978-981-19-0209-3

D

E

H

J

K

Q

R

MIX
Papier aus verantwortungsvollen Quellen
Paper from responsible sources
FSC® C105338

If you have any concerns about our products,
you can contact us on
ProductSafety@springernature.com

In case Publisher is established outside the EU,
the EU authorized representative is:
Springer Nature Customer Service Center GmbH
Europaplatz 3, 69115 Heidelberg, Germany

Printed by Libri Plureos GmbH
in Hamburg, Germany